T0189438

Lecture Notes in Computer Science 12368

More information about this series at http://www.springer.com/series/7412

Andrea Vedaldi · Horst Bischof ·
Thomas Brox · Jan-Michael Frahm (Eds.)

Computer Vision – ECCV 2020

16th European Conference
Glasgow, UK, August 23–28, 2020
Proceedings, Part XXIII

 Springer

Editors
Andrea Vedaldi ⓘ
University of Oxford
Oxford, UK

Horst Bischof ⓘ
Graz University of Technology
Graz, Austria

Thomas Brox ⓘ
University of Freiburg
Freiburg im Breisgau, Germany

Jan-Michael Frahm
University of North Carolina at Chapel Hill
Chapel Hill, NC, USA

ISSN 0302-9743   ISSN 1611-3349   (electronic)
Lecture Notes in Computer Science
ISBN 978-3-030-58591-4   ISBN 978-3-030-58592-1   (eBook)
https://doi.org/10.1007/978-3-030-58592-1

LNCS Sublibrary: SL6 – Image Processing, Computer Vision, Pattern Recognition, and Graphics

This Springer imprint is published by the registered company Springer Nature Switzerland AG
The registered company address is: Gewerbestrasse 11, 6330 Cham, Switzerland

Foreword

Hosting the European Conference on Computer Vision (ECCV 2020) was certainly an exciting journey. From the 2016 plan to hold it at the Edinburgh International Conference Centre (hosting 1,800 delegates) to the 2018 plan to hold it at Glasgow's Scottish Exhibition Centre (up to 6,000 delegates), we finally ended with moving online because of the COVID-19 outbreak. While possibly having fewer delegates than expected because of the online format, ECCV 2020 still had over 3,100 registered participants.

Although online, the conference delivered most of the activities expected at a face-to-face conference: peer-reviewed papers, industrial exhibitors, demonstrations, and messaging between delegates. In addition to the main technical sessions, the conference included a strong program of satellite events with 16 tutorials and 44 workshops.

Furthermore, the online conference format enabled new conference features. Every paper had an associated teaser video and a longer full presentation video. Along with the papers and slides from the videos, all these materials were available the week before the conference. This allowed delegates to become familiar with the paper content and be ready for the live interaction with the authors during the conference week. The live event consisted of brief presentations by the oral and spotlight authors and industrial sponsors. Question and answer sessions for all papers were timed to occur twice so delegates from around the world had convenient access to the authors.

As with ECCV 2018, authors' draft versions of the papers appeared online with open access, now on both the Computer Vision Foundation (CVF) and the European Computer Vision Association (ECVA) websites. An archival publication arrangement was put in place with the cooperation of Springer. SpringerLink hosts the final version of the papers with further improvements, such as activating reference links and supplementary materials. These two approaches benefit all potential readers: a version available freely for all researchers, and an authoritative and citable version with additional benefits for SpringerLink subscribers. We thank Alfred Hofmann and Aliaksandr Birukou from Springer for helping to negotiate this agreement, which we expect will continue for future versions of ECCV.

August 2020

Vittorio Ferrari
Bob Fisher
Cordelia Schmid
Emanuele Trucco

Preface

Welcome to the proceedings of the European Conference on Computer Vision (ECCV 2020). This is a unique edition of ECCV in many ways. Due to the COVID-19 pandemic, this is the first time the conference was held online, in a virtual format. This was also the first time the conference relied exclusively on the Open Review platform to manage the review process. Despite these challenges ECCV is thriving. The conference received 5,150 valid paper submissions, of which 1,360 were accepted for publication (27%) and, of those, 160 were presented as spotlights (3%) and 104 as orals (2%). This amounts to more than twice the number of submissions to ECCV 2018 (2,439). Furthermore, CVPR, the largest conference on computer vision, received 5,850 submissions this year, meaning that ECCV is now 87% the size of CVPR in terms of submissions. By comparison, in 2018 the size of ECCV was only 73% of CVPR.

The review model was similar to previous editions of ECCV; in particular, it was double blind in the sense that the authors did not know the name of the reviewers and vice versa. Furthermore, each conference submission was held confidentially, and was only publicly revealed if and once accepted for publication. Each paper received at least three reviews, totalling more than 15,000 reviews. Handling the review process at this scale was a significant challenge. In order to ensure that each submission received as fair and high-quality reviews as possible, we recruited 2,830 reviewers (a 130% increase with reference to 2018) and 207 area chairs (a 60% increase). The area chairs were selected based on their technical expertise and reputation, largely among people that served as area chair in previous top computer vision and machine learning conferences (ECCV, ICCV, CVPR, NeurIPS, etc.). Reviewers were similarly invited from previous conferences. We also encouraged experienced area chairs to suggest additional chairs and reviewers in the initial phase of recruiting.

Despite doubling the number of submissions, the reviewer load was slightly reduced from 2018, from a maximum of 8 papers down to 7 (with some reviewers offering to handle 6 papers plus an emergency review). The area chair load increased slightly, from 18 papers on average to 22 papers on average.

Conflicts of interest between authors, area chairs, and reviewers were handled largely automatically by the Open Review platform via their curated list of user profiles. Many authors submitting to ECCV already had a profile in Open Review. We set a paper registration deadline one week before the paper submission deadline in order to encourage all missing authors to register and create their Open Review profiles well on time (in practice, we allowed authors to create/change papers arbitrarily until the submission deadline). Except for minor issues with users creating duplicate profiles, this allowed us to easily and quickly identify institutional conflicts, and avoid them, while matching papers to area chairs and reviewers.

Papers were matched to area chairs based on: an affinity score computed by the Open Review platform, which is based on paper titles and abstracts, and an affinity score computed by the Toronto Paper Matching System (TPMS), which is based on the paper's full text, the area chair bids for individual papers, load balancing, and conflict avoidance. Open Review provides the program chairs a convenient web interface to experiment with different configurations of the matching algorithm. The chosen configuration resulted in about 50% of the assigned papers to be highly ranked by the area chair bids, and 50% to be ranked in the middle, with very few low bids assigned.

Assignments to reviewers were similar, with two differences. First, there was a maximum of 7 papers assigned to each reviewer. Second, area chairs recommended up to seven reviewers per paper, providing another highly-weighed term to the affinity scores used for matching.

The assignment of papers to area chairs was smooth. However, it was more difficult to find suitable reviewers for all papers. Having a ratio of 5.6 papers per reviewer with a maximum load of 7 (due to emergency reviewer commitment), which did not allow for much wiggle room in order to also satisfy conflict and expertise constraints. We received some complaints from reviewers who did not feel qualified to review specific papers and we reassigned them wherever possible. However, the large scale of the conference, the many constraints, and the fact that a large fraction of such complaints arrived very late in the review process made this process very difficult and not all complaints could be addressed.

Reviewers had six weeks to complete their assignments. Possibly due to COVID-19 or the fact that the NeurIPS deadline was moved closer to the review deadline, a record 30% of the reviews were still missing after the deadline. By comparison, ECCV 2018 experienced only 10% missing reviews at this stage of the process. In the subsequent week, area chairs chased the missing reviews intensely, found replacement reviewers in their own team, and managed to reach 10% missing reviews. Eventually, we could provide almost all reviews (more than 99.9%) with a delay of only a couple of days on the initial schedule by a significant use of emergency reviews. If this trend is confirmed, it might be a major challenge to run a smooth review process in future editions of ECCV. The community must reconsider prioritization of the time spent on paper writing (the number of submissions increased a lot despite COVID-19) and time spent on paper reviewing (the number of reviews delivered in time decreased a lot presumably due to COVID-19 or NeurIPS deadline). With this imbalance the peer-review system that ensures the quality of our top conferences may break soon.

Reviewers submitted their reviews independently. In the reviews, they had the opportunity to ask questions to the authors to be addressed in the rebuttal. However, reviewers were told not to request any significant new experiment. Using the Open Review interface, authors could provide an answer to each individual review, but were also allowed to cross-reference reviews and responses in their answers. Rather than PDF files, we allowed the use of formatted text for the rebuttal. The rebuttal and initial reviews were then made visible to all reviewers and the primary area chair for a given paper. The area chair encouraged and moderated the reviewer discussion. During the discussions, reviewers were invited to reach a consensus and possibly adjust their ratings as a result of the discussion and of the evidence in the rebuttal.

After the discussion period ended, most reviewers entered a final rating and recommendation, although in many cases this did not differ from their initial recommendation. Based on the updated reviews and discussion, the primary area chair then made a preliminary decision to accept or reject the paper and wrote a justification for it (meta-review). Except for cases where the outcome of this process was absolutely clear (as indicated by the three reviewers and primary area chairs all recommending clear rejection), the decision was then examined and potentially challenged by a secondary area chair. This led to further discussion and overturning a small number of preliminary decisions. Needless to say, there was no in-person area chair meeting, which would have been impossible due to COVID-19.

Area chairs were invited to observe the consensus of the reviewers whenever possible and use extreme caution in overturning a clear consensus to accept or reject a paper. If an area chair still decided to do so, she/he was asked to clearly justify it in the meta-review and to explicitly obtain the agreement of the secondary area chair. In practice, very few papers were rejected after being confidently accepted by the reviewers.

This was the first time Open Review was used as the main platform to run ECCV. In 2018, the program chairs used CMT3 for the user-facing interface and Open Review internally, for matching and conflict resolution. Since it is clearly preferable to only use a single platform, this year we switched to using Open Review in full. The experience was largely positive. The platform is highly-configurable, scalable, and open source. Being written in Python, it is easy to write scripts to extract data programmatically. The paper matching and conflict resolution algorithms and interfaces are top-notch, also due to the excellent author profiles in the platform. Naturally, there were a few kinks along the way due to the fact that the ECCV Open Review configuration was created from scratch for this event and it differs in substantial ways from many other Open Review conferences. However, the Open Review development and support team did a fantastic job in helping us to get the configuration right and to address issues in a timely manner as they unavoidably occurred. We cannot thank them enough for the tremendous effort they put into this project.

Finally, we would like to thank everyone involved in making ECCV 2020 possible in these very strange and difficult times. This starts with our authors, followed by the area chairs and reviewers, who ran the review process at an unprecedented scale. The whole Open Review team (and in particular Melisa Bok, Mohit Unyal, Carlos Mondragon Chapa, and Celeste Martinez Gomez) worked incredibly hard for the entire duration of the process. We would also like to thank René Vidal for contributing to the adoption of Open Review. Our thanks also go to Laurent Charling for TPMS and to the program chairs of ICML, ICLR, and NeurIPS for cross checking double submissions. We thank the website chair, Giovanni Farinella, and the CPI team (in particular Ashley Cook, Miriam Verdon, Nicola McGrane, and Sharon Kerr) for promptly adding material to the website as needed in the various phases of the process. Finally, we thank the publication chairs, Albert Ali Salah, Hamdi Dibeklioglu, Metehan Doyran, Henry Howard-Jenkins, Victor Prisacariu, Siyu Tang, and Gul Varol, who managed to compile these substantial proceedings in an exceedingly compressed schedule. We express our thanks to the ECVA team, in particular Kristina Scherbaum for allowing open access of the proceedings. We thank Alfred Hofmann from Springer who again serve as the publisher. Finally, we thank the other chairs of ECCV 2020, including in particular the general chairs for very useful feedback with the handling of the program.

August 2020
                                         Andrea Vedaldi
Horst Bischof
Thomas Brox
Jan-Michael Frahm

Organization

General Chairs

Vittorio Ferrari   Google Research, Switzerland
Bob Fisher   University of Edinburgh, UK
Cordelia Schmid   Google and Inria, France
Emanuele Trucco   University of Dundee, UK

Program Chairs

Andrea Vedaldi   University of Oxford, UK
Horst Bischof   Graz University of Technology, Austria
Thomas Brox   University of Freiburg, Germany
Jan-Michael Frahm   University of North Carolina, USA

Industrial Liaison Chairs

Jim Ashe   University of Edinburgh, UK
Helmut Grabner   Zurich University of Applied Sciences, Switzerland
Diane Larlus   NAVER LABS Europe, France
Cristian Novotny   University of Edinburgh, UK

Local Arrangement Chairs

Yvan Petillot   Heriot-Watt University, UK
Paul Siebert   University of Glasgow, UK

Academic Demonstration Chair

Thomas Mensink   Google Research and University of Amsterdam, The Netherlands

Poster Chair

Stephen Mckenna   University of Dundee, UK

Technology Chair

Gerardo Aragon Camarasa   University of Glasgow, UK

Tutorial Chairs

Carlo Colombo   University of Florence, Italy
Sotirios Tsaftaris   University of Edinburgh, UK

Publication Chairs

Albert Ali Salah   Utrecht University, The Netherlands
Hamdi Dibeklioglu   Bilkent University, Turkey
Metehan Doyran   Utrecht University, The Netherlands
Henry Howard-Jenkins   University of Oxford, UK
Victor Adrian Prisacariu   University of Oxford, UK
Siyu Tang   ETH Zurich, Switzerland
Gul Varol   University of Oxford, UK

Website Chair

Giovanni Maria Farinella   University of Catania, Italy

Workshops Chairs

Adrien Bartoli   University of Clermont Auvergne, France
Andrea Fusiello   University of Udine, Italy

Area Chairs

Lourdes Agapito   University College London, UK
Zeynep Akata   University of Tübingen, Germany
Karteek Alahari   Inria, France
Antonis Argyros   University of Crete, Greece
Hossein Azizpour   KTH Royal Institute of Technology, Sweden
Joao P. Barreto   Universidade de Coimbra, Portugal
Alexander C. Berg   University of North Carolina at Chapel Hill, USA
Matthew B. Blaschko   KU Leuven, Belgium
Lubomir D. Bourdev   WaveOne, Inc., USA
Edmond Boyer   Inria, France
Yuri Boykov   University of Waterloo, Canada
Gabriel Brostow   University College London, UK
Michael S. Brown   National University of Singapore, Singapore
Jianfei Cai   Monash University, Australia
Barbara Caputo   Politecnico di Torino, Italy
Ayan Chakrabarti   Washington University, St. Louis, USA
Tat-Jen Cham   Nanyang Technological University, Singapore
Manmohan Chandraker   University of California, San Diego, USA
Rama Chellappa   Johns Hopkins University, USA
Liang-Chieh Chen   Google, USA

Yung-Yu Chuang	National Taiwan University, Taiwan
Ondrej Chum	Czech Technical University in Prague, Czech Republic
Brian Clipp	Kitware, USA
John Collomosse	University of Surrey and Adobe Research, UK
Jason J. Corso	University of Michigan, USA
David J. Crandall	Indiana University, USA
Daniel Cremers	University of California, Los Angeles, USA
Fabio Cuzzolin	Oxford Brookes University, UK
Jifeng Dai	SenseTime, SAR China
Kostas Daniilidis	University of Pennsylvania, USA
Andrew Davison	Imperial College London, UK
Alessio Del Bue	Fondazione Istituto Italiano di Tecnologia, Italy
Jia Deng	Princeton University, USA
Alexey Dosovitskiy	Google, Germany
Matthijs Douze	Facebook, France
Enrique Dunn	Stevens Institute of Technology, USA
Irfan Essa	Georgia Institute of Technology and Google, USA
Giovanni Maria Farinella	University of Catania, Italy
Ryan Farrell	Brigham Young University, USA
Paolo Favaro	University of Bern, Switzerland
Rogerio Feris	International Business Machines, USA
Cornelia Fermuller	University of Maryland, College Park, USA
David J. Fleet	Vector Institute, Canada
Friedrich Fraundorfer	DLR, Austria
Mario Fritz	CISPA Helmholtz Center for Information Security, Germany
Pascal Fua	EPFL (Swiss Federal Institute of Technology Lausanne), Switzerland
Yasutaka Furukawa	Simon Fraser University, Canada
Li Fuxin	Oregon State University, USA
Efstratios Gavves	University of Amsterdam, The Netherlands
Peter Vincent Gehler	Amazon, USA
Theo Gevers	University of Amsterdam, The Netherlands
Ross Girshick	Facebook AI Research, USA
Boqing Gong	Google, USA
Stephen Gould	Australian National University, Australia
Jinwei Gu	SenseTime Research, USA
Abhinav Gupta	Facebook, USA
Bohyung Han	Seoul National University, South Korea
Bharath Hariharan	Cornell University, USA
Tal Hassner	Facebook AI Research, USA
Xuming He	Australian National University, Australia
Joao F. Henriques	University of Oxford, UK
Adrian Hilton	University of Surrey, UK
Minh Hoai	Stony Brooks, State University of New York, USA
Derek Hoiem	University of Illinois Urbana-Champaign, USA

Timothy Hospedales	University of Edinburgh and Samsung, UK
Gang Hua	Wormpex AI Research, USA
Slobodan Ilic	Siemens AG, Germany
Hiroshi Ishikawa	Waseda University, Japan
Jiaya Jia	The Chinese University of Hong Kong, SAR China
Hailin Jin	Adobe Research, USA
Justin Johnson	University of Michigan, USA
Frederic Jurie	University of Caen Normandie, France
Fredrik Kahl	Chalmers University, Sweden
Sing Bing Kang	Zillow, USA
Gunhee Kim	Seoul National University, South Korea
Junmo Kim	Korea Advanced Institute of Science and Technology, South Korea
Tae-Kyun Kim	Imperial College London, UK
Ron Kimmel	Technion-Israel Institute of Technology, Israel
Alexander Kirillov	Facebook AI Research, USA
Kris Kitani	Carnegie Mellon University, USA
Iasonas Kokkinos	Ariel AI, UK
Vladlen Koltun	Intel Labs, USA
Nikos Komodakis	Ecole des Ponts ParisTech, France
Piotr Koniusz	Australian National University, Australia
M. Pawan Kumar	University of Oxford, UK
Kyros Kutulakos	University of Toronto, Canada
Christoph Lampert	IST Austria, Austria
Ivan Laptev	Inria, France
Diane Larlus	NAVER LABS Europe, France
Laura Leal-Taixe	Technical University Munich, Germany
Honglak Lee	Google and University of Michigan, USA
Joon-Young Lee	Adobe Research, USA
Kyoung Mu Lee	Seoul National University, South Korea
Seungyong Lee	POSTECH, South Korea
Yong Jae Lee	University of California, Davis, USA
Bastian Leibe	RWTH Aachen University, Germany
Victor Lempitsky	Samsung, Russia
Ales Leonardis	University of Birmingham, UK
Marius Leordeanu	Institute of Mathematics of the Romanian Academy, Romania
Vincent Lepetit	ENPC ParisTech, France
Hongdong Li	The Australian National University, Australia
Xi Li	Zhejiang University, China
Yin Li	University of Wisconsin-Madison, USA
Zicheng Liao	Zhejiang University, China
Jongwoo Lim	Hanyang University, South Korea
Stephen Lin	Microsoft Research Asia, China
Yen-Yu Lin	National Chiao Tung University, Taiwan, China
Zhe Lin	Adobe Research, USA

Haibin Ling	Stony Brooks, State University of New York, USA
Jiaying Liu	Peking University, China
Ming-Yu Liu	NVIDIA, USA
Si Liu	Beihang University, China
Xiaoming Liu	Michigan State University, USA
Huchuan Lu	Dalian University of Technology, China
Simon Lucey	Carnegie Mellon University, USA
Jiebo Luo	University of Rochester, USA
Julien Mairal	Inria, France
Michael Maire	University of Chicago, USA
Subhransu Maji	University of Massachusetts, Amherst, USA
Yasushi Makihara	Osaka University, Japan
Jiri Matas	Czech Technical University in Prague, Czech Republic
Yasuyuki Matsushita	Osaka University, Japan
Philippos Mordohai	Stevens Institute of Technology, USA
Vittorio Murino	University of Verona, Italy
Naila Murray	NAVER LABS Europe, France
Hajime Nagahara	Osaka University, Japan
P. J. Narayanan	International Institute of Information Technology (IIIT), Hyderabad, India
Nassir Navab	Technical University of Munich, Germany
Natalia Neverova	Facebook AI Research, France
Matthias Niessner	Technical University of Munich, Germany
Jean-Marc Odobez	Idiap Research Institute and Swiss Federal Institute of Technology Lausanne, Switzerland
Francesca Odone	Università di Genova, Italy
Takeshi Oishi	The University of Tokyo, Tokyo Institute of Technology, Japan
Vicente Ordonez	University of Virginia, USA
Manohar Paluri	Facebook AI Research, USA
Maja Pantic	Imperial College London, UK
In Kyu Park	Inha University, South Korea
Ioannis Patras	Queen Mary University of London, UK
Patrick Perez	Valeo, France
Bryan A. Plummer	Boston University, USA
Thomas Pock	Graz University of Technology, Austria
Marc Pollefeys	ETH Zurich and Microsoft MR & AI Zurich Lab, Switzerland
Jean Ponce	Inria, France
Gerard Pons-Moll	MPII, Saarland Informatics Campus, Germany
Jordi Pont-Tuset	Google, Switzerland
James Matthew Rehg	Georgia Institute of Technology, USA
Ian Reid	University of Adelaide, Australia
Olaf Ronneberger	DeepMind London, UK
Stefan Roth	TU Darmstadt, Germany
Bryan Russell	Adobe Research, USA

Kwang Moo Yi	University of Victoria, Canada
Zhaozheng Yin	Stony Brook, State University of New York, USA
Chang D. Yoo	Korea Advanced Institute of Science and Technology, South Korea
Shaodi You	University of Amsterdam, The Netherlands
Jingyi Yu	ShanghaiTech University, China
Stella Yu	University of California, Berkeley, and ICSI, USA
Stefanos Zafeiriou	Imperial College London, UK
Hongbin Zha	Peking University, China
Tianzhu Zhang	University of Science and Technology of China, China
Liang Zheng	Australian National University, Australia
Todd E. Zickler	Harvard University, USA
Andrew Zisserman	University of Oxford, UK

Technical Program Committee

Sathyanarayanan
  N. Aakur
Wael Abd Almgaeed
Abdelrahman
  Abdelhamed
Abdullah Abuolaim
Supreeth Achar
Hanno Ackermann
Ehsan Adeli
Triantafyllos Afouras
Sameer Agarwal
Aishwarya Agrawal
Harsh Agrawal
Pulkit Agrawal
Antonio Agudo
Eirikur Agustsson
Karim Ahmed
Byeongjoo Ahn
Unaiza Ahsan
Thalaiyasingam Ajanthan
Kenan E. Ak
Emre Akbas
Naveed Akhtar
Derya Akkaynak
Yagiz Aksoy
Ziad Al-Halah
Xavier Alameda-Pineda
Jean-Baptiste Alayrac

Samuel Albanie
Shadi Albarqouni
Cenek Albl
Hassan Abu Alhaija
Daniel Aliaga
Mohammad
  S. Aliakbarian
Rahaf Aljundi
Thiemo Alldieck
Jon Almazan
Jose M. Alvarez
Senjian An
Saket Anand
Codruta Ancuti
Cosmin Ancuti
Peter Anderson
Juan Andrade-Cetto
Alexander Andreopoulos
Misha Andriluka
Dragomir Anguelov
Rushil Anirudh
Michel Antunes
Oisin Mac Aodha
Srikar Appalaraju
Relja Arandjelovic
Nikita Araslanov
Andre Araujo
Helder Araujo

Pablo Arbelaez
Shervin Ardeshir
Sercan O. Arik
Anil Armagan
Anurag Arnab
Chetan Arora
Federica Arrigoni
Mathieu Aubry
Shai Avidan
Angelica I. Aviles-Rivero
Yannis Avrithis
Ismail Ben Ayed
Shekoofeh Azizi
Ioan Andrei Bârsan
Artem Babenko
Deepak Babu Sam
Seung-Hwan Baek
Seungryul Baek
Andrew D. Bagdanov
Shai Bagon
Yuval Bahat
Junjie Bai
Song Bai
Xiang Bai
Yalong Bai
Yancheng Bai
Peter Bajcsy
Slawomir Bak

Mahsa Baktashmotlagh
Kavita Bala
Yogesh Balaji
Guha Balakrishnan
V. N. Balasubramanian
Federico Baldassarre
Vassileios Balntas
Shurjo Banerjee
Aayush Bansal
Ankan Bansal
Jianmin Bao
Linchao Bao
Wenbo Bao
Yingze Bao
Akash Bapat
Md Jawadul Hasan Bappy
Fabien Baradel
Lorenzo Baraldi
Daniel Barath
Adrian Barbu
Kobus Barnard
Nick Barnes
Francisco Barranco
Jonathan T. Barron
Arslan Basharat
Chaim Baskin
Anil S. Baslamisli
Jorge Batista
Kayhan Batmanghelich
Konstantinos Batsos
David Bau
Luis Baumela
Christoph Baur
Eduardo
    Bayro-Corrochano
Paul Beardsley
Jan Bednavr'ik
Oscar Beijbom
Philippe Bekaert
Esube Bekele
Vasileios Belagiannis
Ohad Ben-Shahar
Abhijit Bendale
Róger Bermúdez-Chacón
Maxim Berman
Jesus Bermudez-cameo

Florian Bernard
Stefano Berretti
Marcelo Bertalmio
Gedas Bertasius
Cigdem Beyan
Lucas Beyer
Vijayakumar Bhagavatula
Arjun Nitin Bhagoji
Apratim Bhattacharyya
Binod Bhattarai
Sai Bi
Jia-Wang Bian
Simone Bianco
Adel Bibi
Tolga Birdal
Tom Bishop
Soma Biswas
Mårten Björkman
Volker Blanz
Vishnu Boddeti
Navaneeth Bodla
Simion-Vlad Bogolin
Xavier Boix
Piotr Bojanowski
Timo Bolkart
Guido Borghi
Larbi Boubchir
Guillaume Bourmaud
Adrien Bousseau
Thierry Bouwmans
Richard Bowden
Hakan Boyraz
Mathieu Brédif
Samarth Brahmbhatt
Steve Branson
Nikolas Brasch
Biagio Brattoli
Ernesto Brau
Toby P. Breckon
Francois Bremond
Jesus Briales
Sofia Broomé
Marcus A. Brubaker
Luc Brun
Silvia Bucci
Shyamal Buch

Pradeep Buddharaju
Uta Buechler
Mai Bui
Tu Bui
Adrian Bulat
Giedrius T. Burachas
Elena Burceanu
Xavier P. Burgos-Artizzu
Kaylee Burns
Andrei Bursuc
Benjamin Busam
Wonmin Byeon
Zoya Bylinskii
Sergi Caelles
Jianrui Cai
Minjie Cai
Yujun Cai
Zhaowei Cai
Zhipeng Cai
Juan C. Caicedo
Simone Calderara
Necati Cihan Camgoz
Dylan Campbell
Octavia Camps
Jiale Cao
Kaidi Cao
Liangliang Cao
Xiangyong Cao
Xiaochun Cao
Yang Cao
Yu Cao
Yue Cao
Zhangjie Cao
Luca Carlone
Mathilde Caron
Dan Casas
Thomas J. Cashman
Umberto Castellani
Lluis Castrejon
Jacopo Cavazza
Fabio Cermelli
Hakan Cevikalp
Menglei Chai
Ishani Chakraborty
Rudrasis Chakraborty
Antoni B. Chan

Kwok-Ping Chan
Siddhartha Chandra
Sharat Chandran
Arjun Chandrasekaran
Angel X. Chang
Che-Han Chang
Hong Chang
Hyun Sung Chang
Hyung Jin Chang
Jianlong Chang
Ju Yong Chang
Ming-Ching Chang
Simyung Chang
Xiaojun Chang
Yu-Wei Chao
Devendra S. Chaplot
Arslan Chaudhry
Rizwan A. Chaudhry
Can Chen
Chang Chen
Chao Chen
Chen Chen
Chu-Song Chen
Dapeng Chen
Dong Chen
Dongdong Chen
Guanying Chen
Hongge Chen
Hsin-yi Chen
Huaijin Chen
Hwann-Tzong Chen
Jianbo Chen
Jianhui Chen
Jiansheng Chen
Jiaxin Chen
Jie Chen
Jun-Cheng Chen
Kan Chen
Kevin Chen
Lin Chen
Long Chen
Min-Hung Chen
Qifeng Chen
Shi Chen
Shixing Chen
Tianshui Chen

Weifeng Chen
Weikai Chen
Xi Chen
Xiaohan Chen
Xiaozhi Chen
Xilin Chen
Xingyu Chen
Xinlei Chen
Xinyun Chen
Yi-Ting Chen
Yilun Chen
Ying-Cong Chen
Yinpeng Chen
Yiran Chen
Yu Chen
Yu-Sheng Chen
Yuhua Chen
Yun-Chun Chen
Yunpeng Chen
Yuntao Chen
Zhuoyuan Chen
Zitian Chen
Anchieh Cheng
Bowen Cheng
Erkang Cheng
Gong Cheng
Guangliang Cheng
Jingchun Cheng
Jun Cheng
Li Cheng
Ming-Ming Cheng
Yu Cheng
Ziang Cheng
Anoop Cherian
Dmitry Chetverikov
Ngai-man Cheung
William Cheung
Ajad Chhatkuli
Naoki Chiba
Benjamin Chidester
Han-pang Chiu
Mang Tik Chiu
Wei-Chen Chiu
Donghyeon Cho
Hojin Cho
Minsu Cho

Nam Ik Cho
Tim Cho
Tae Eun Choe
Chiho Choi
Edward Choi
Inchang Choi
Jinsoo Choi
Jonghyun Choi
Jongwon Choi
Yukyung Choi
Hisham Cholakkal
Eunji Chong
Jaegul Choo
Christopher Choy
Hang Chu
Peng Chu
Wen-Sheng Chu
Albert Chung
Joon Son Chung
Hai Ci
Safa Cicek
Ramazan G. Cinbis
Arridhana Ciptadi
Javier Civera
James J. Clark
Ronald Clark
Felipe Codevilla
Michael Cogswell
Andrea Cohen
Maxwell D. Collins
Carlo Colombo
Yang Cong
Adria R. Continente
Marcella Cornia
John Richard Corring
Darren Cosker
Dragos Costea
Garrison W. Cottrell
Florent Couzinie-Devy
Marco Cristani
Ioana Croitoru
James L. Crowley
Jiequan Cui
Zhaopeng Cui
Ross Cutler
Antonio D'Innocente

Rozenn Dahyot
Bo Dai
Dengxin Dai
Hang Dai
Longquan Dai
Shuyang Dai
Xiyang Dai
Yuchao Dai
Adrian V. Dalca
Dima Damen
Bharath B. Damodaran
Kristin Dana
Martin Danelljan
Zheng Dang
Zachary Alan Daniels
Donald G. Dansereau
Abhishek Das
Samyak Datta
Achal Dave
Titas De
Rodrigo de Bem
Teo de Campos
Raoul de Charette
Shalini De Mello
Joseph DeGol
Herve Delingette
Haowen Deng
Jiankang Deng
Weijian Deng
Zhiwei Deng
Joachim Denzler
Konstantinos G. Derpanis
Aditya Deshpande
Frederic Devernay
Somdip Dey
Arturo Deza
Abhinav Dhall
Helisa Dhamo
Vikas Dhiman
Fillipe Dias Moreira
    de Souza
Ali Diba
Ferran Diego
Guiguang Ding
Henghui Ding
Jian Ding

Mingyu Ding
Xinghao Ding
Zhengming Ding
Robert DiPietro
Cosimo Distante
Ajay Divakaran
Mandar Dixit
Abdelaziz Djelouah
Thanh-Toan Do
Jose Dolz
Bo Dong
Chao Dong
Jiangxin Dong
Weiming Dong
Weisheng Dong
Xingping Dong
Xuanyi Dong
Yinpeng Dong
Gianfranco Doretto
Hazel Doughty
Hassen Drira
Bertram Drost
Dawei Du
Ye Duan
Yueqi Duan
Abhimanyu Dubey
Anastasia Dubrovina
Stefan Duffner
Chi Nhan Duong
Thibaut Durand
Zoran Duric
Iulia Duta
Debidatta Dwibedi
Benjamin Eckart
Marc Eder
Marzieh Edraki
Alexei A. Efros
Kiana Ehsani
Hazm Kemal Ekenel
James H. Elder
Mohamed Elgharib
Shireen Elhabian
Ehsan Elhamifar
Mohamed Elhoseiny
Ian Endres
N. Benjamin Erichson

Jan Ernst
Sergio Escalera
Francisco Escolano
Victor Escorcia
Carlos Esteves
Francisco J. Estrada
Bin Fan
Chenyou Fan
Deng-Ping Fan
Haoqi Fan
Hehe Fan
Heng Fan
Kai Fan
Lijie Fan
Linxi Fan
Quanfu Fan
Shaojing Fan
Xiaochuan Fan
Xin Fan
Yuchen Fan
Sean Fanello
Hao-Shu Fang
Haoyang Fang
Kuan Fang
Yi Fang
Yuming Fang
Azade Farshad
Alireza Fathi
Raanan Fattal
Joao Fayad
Xiaohan Fei
Christoph Feichtenhofer
Michael Felsberg
Chen Feng
Jiashi Feng
Junyi Feng
Mengyang Feng
Qianli Feng
Zhenhua Feng
Michele Fenzi
Andras Ferencz
Martin Fergie
Basura Fernando
Ethan Fetaya
Michael Firman
John W. Fisher

Matthew Fisher
Boris Flach
Corneliu Florea
Wolfgang Foerstner
David Fofi
Gian Luca Foresti
Per-Erik Forssen
David Fouhey
Katerina Fragkiadaki
Victor Fragoso
Jean-Sébastien Franco
Ohad Fried
Iuri Frosio
Cheng-Yang Fu
Huazhu Fu
Jianlong Fu
Jingjing Fu
Xueyang Fu
Yanwei Fu
Ying Fu
Yun Fu
Olac Fuentes
Kent Fujiwara
Takuya Funatomi
Christopher Funk
Thomas Funkhouser
Antonino Furnari
Ryo Furukawa
Erik Gärtner
Raghudeep Gadde
Matheus Gadelha
Vandit Gajjar
Trevor Gale
Juergen Gall
Mathias Gallardo
Guillermo Gallego
Orazio Gallo
Chuang Gan
Zhe Gan
Madan Ravi Ganesh
Aditya Ganeshan
Siddha Ganju
Bin-Bin Gao
Changxin Gao
Feng Gao
Hongchang Gao

Jin Gao
Jiyang Gao
Junbin Gao
Katelyn Gao
Lin Gao
Mingfei Gao
Ruiqi Gao
Ruohan Gao
Shenghua Gao
Yuan Gao
Yue Gao
Noa Garcia
Alberto Garcia-Garcia
Guillermo
    Garcia-Hernando
Jacob R. Gardner
Animesh Garg
Kshitiz Garg
Rahul Garg
Ravi Garg
Philip N. Garner
Kirill Gavrilyuk
Paul Gay
Shiming Ge
Weifeng Ge
Baris Gecer
Xin Geng
Kyle Genova
Stamatios Georgoulis
Bernard Ghanem
Michael Gharbi
Kamran Ghasedi
Golnaz Ghiasi
Arnab Ghosh
Partha Ghosh
Silvio Giancola
Andrew Gilbert
Rohit Girdhar
Xavier Giro-i-Nieto
Thomas Gittings
Ioannis Gkioulekas
Clement Godard
Vaibhava Goel
Bastian Goldluecke
Lluis Gomez
Nuno Gonçalves

Dong Gong
Ke Gong
Mingming Gong
Abel Gonzalez-Garcia
Ariel Gordon
Daniel Gordon
Paulo Gotardo
Venu Madhav Govindu
Ankit Goyal
Priya Goyal
Raghav Goyal
Benjamin Graham
Douglas Gray
Brent A. Griffin
Etienne Grossmann
David Gu
Jiayuan Gu
Jiuxiang Gu
Lin Gu
Qiao Gu
Shuhang Gu
Jose J. Guerrero
Paul Guerrero
Jie Gui
Jean-Yves Guillemaut
Riza Alp Guler
Erhan Gundogdu
Fatma Guney
Guodong Guo
Kaiwen Guo
Qi Guo
Sheng Guo
Shi Guo
Tiantong Guo
Xiaojie Guo
Yijie Guo
Yiluan Guo
Yuanfang Guo
Yulan Guo
Agrim Gupta
Ankush Gupta
Mohit Gupta
Saurabh Gupta
Tanmay Gupta
Danna Gurari
Abner Guzman-Rivera

JunYoung Gwak
Michael Gygli
Jung-Woo Ha
Simon Hadfield
Isma Hadji
Bjoern Haefner
Taeyoung Hahn
Levente Hajder
Peter Hall
Emanuela Haller
Stefan Haller
Bumsub Ham
Abdullah Hamdi
Dongyoon Han
Hu Han
Jungong Han
Junwei Han
Kai Han
Tian Han
Xiaoguang Han
Xintong Han
Yahong Han
Ankur Handa
Zekun Hao
Albert Haque
Tatsuya Harada
Mehrtash Harandi
Adam W. Harley
Mahmudul Hasan
Atsushi Hashimoto
Ali Hatamizadeh
Munawar Hayat
Dongliang He
Jingrui He
Junfeng He
Kaiming He
Kun He
Lei He
Pan He
Ran He
Shengfeng He
Tong He
Weipeng He
Xuming He
Yang He
Yihui He

Zhihai He
Chinmay Hegde
Janne Heikkila
Mattias P. Heinrich
Stéphane Herbin
Alexander Hermans
Luis Herranz
John R. Hershey
Aaron Hertzmann
Roei Herzig
Anders Heyden
Steven Hickson
Otmar Hilliges
Tomas Hodan
Judy Hoffman
Michael Hofmann
Yannick Hold-Geoffroy
Namdar Homayounfar
Sina Honari
Richang Hong
Seunghoon Hong
Xiaopeng Hong
Yi Hong
Hidekata Hontani
Anthony Hoogs
Yedid Hoshen
Mir Rayat Imtiaz Hossain
Junhui Hou
Le Hou
Lu Hou
Tingbo Hou
Wei-Lin Hsiao
Cheng-Chun Hsu
Gee-Sern Jison Hsu
Kuang-jui Hsu
Changbo Hu
Di Hu
Guosheng Hu
Han Hu
Hao Hu
Hexiang Hu
Hou-Ning Hu
Jie Hu
Junlin Hu
Nan Hu
Ping Hu

Ronghang Hu
Xiaowei Hu
Yinlin Hu
Yuan-Ting Hu
Zhe Hu
Binh-Son Hua
Yang Hua
Bingyao Huang
Di Huang
Dong Huang
Fay Huang
Haibin Huang
Haozhi Huang
Heng Huang
Huaibo Huang
Jia-Bin Huang
Jing Huang
Jingwei Huang
Kaizhu Huang
Lei Huang
Qiangui Huang
Qiaoying Huang
Qingqiu Huang
Qixing Huang
Shaoli Huang
Sheng Huang
Siyuan Huang
Weilin Huang
Wenbing Huang
Xiangru Huang
Xun Huang
Yan Huang
Yifei Huang
Yue Huang
Zhiwu Huang
Zilong Huang
Minyoung Huh
Zhuo Hui
Matthias B. Hullin
Martin Humenberger
Wei-Chih Hung
Zhouyuan Huo
Junhwa Hur
Noureldien Hussein
Jyh-Jing Hwang
Seong Jae Hwang

Sung Ju Hwang
Ichiro Ide
Ivo Ihrke
Daiki Ikami
Satoshi Ikehata
Nazli Ikizler-Cinbis
Sunghoon Im
Yani Ioannou
Radu Tudor Ionescu
Umar Iqbal
Go Irie
Ahmet Iscen
Md Amirul Islam
Vamsi Ithapu
Nathan Jacobs
Arpit Jain
Himalaya Jain
Suyog Jain
Stuart James
Won-Dong Jang
Yunseok Jang
Ronnachai Jaroensri
Dinesh Jayaraman
Sadeep Jayasumana
Suren Jayasuriya
Herve Jegou
Simon Jenni
Hae-Gon Jeon
Yunho Jeon
Koteswar R. Jerripothula
Hueihan Jhuang
I-hong Jhuo
Dinghuang Ji
Hui Ji
Jingwei Ji
Pan Ji
Yanli Ji
Baoxiong Jia
Kui Jia
Xu Jia
Chiyu Max Jiang
Haiyong Jiang
Hao Jiang
Huaizu Jiang
Huajie Jiang
Ke Jiang

Lai Jiang
Li Jiang
Lu Jiang
Ming Jiang
Peng Jiang
Shuqiang Jiang
Wei Jiang
Xudong Jiang
Zhuolin Jiang
Jianbo Jiao
Zequn Jie
Dakai Jin
Kyong Hwan Jin
Lianwen Jin
SouYoung Jin
Xiaojie Jin
Xin Jin
Nebojsa Jojic
Alexis Joly
Michael Jeffrey Jones
Hanbyul Joo
Jungseock Joo
Kyungdon Joo
Ajjen Joshi
Shantanu H. Joshi
Da-Cheng Juan
Marco Körner
Kevin Köser
Asim Kadav
Christine Kaeser-Chen
Kushal Kafle
Dagmar Kainmueller
Ioannis A. Kakadiaris
Zdenek Kalal
Nima Kalantari
Yannis Kalantidis
Mahdi M. Kalayeh
Anmol Kalia
Sinan Kalkan
Vicky Kalogeiton
Ashwin Kalyan
Joni-kristian Kamarainen
Gerda Kamberova
Chandra Kambhamettu
Martin Kampel
Meina Kan

Christopher Kanan
Kenichi Kanatani
Angjoo Kanazawa
Atsushi Kanehira
Takuhiro Kaneko
Asako Kanezaki
Bingyi Kang
Di Kang
Sunghun Kang
Zhao Kang
Vadim Kantorov
Abhishek Kar
Amlan Kar
Theofanis Karaletsos
Leonid Karlinsky
Kevin Karsch
Angelos Katharopoulos
Isinsu Katircioglu
Hiroharu Kato
Zoltan Kato
Dotan Kaufman
Jan Kautz
Rei Kawakami
Qiuhong Ke
Wadim Kehl
Petr Kellnhofer
Aniruddha Kembhavi
Cem Keskin
Margret Keuper
Daniel Keysers
Ashkan Khakzar
Fahad Khan
Naeemullah Khan
Salman Khan
Siddhesh Khandelwal
Rawal Khirodkar
Anna Khoreva
Tejas Khot
Parmeshwar Khurd
Hadi Kiapour
Joe Kileel
Chanho Kim
Dahun Kim
Edward Kim
Eunwoo Kim
Han-ul Kim

Hansung Kim
Heewon Kim
Hyo Jin Kim
Hyunwoo J. Kim
Jinkyu Kim
Jiwon Kim
Jongmin Kim
Junsik Kim
Junyeong Kim
Min H. Kim
Namil Kim
Pyojin Kim
Seon Joo Kim
Seong Tae Kim
Seungryong Kim
Sungwoong Kim
Tae Hyun Kim
Vladimir Kim
Won Hwa Kim
Yonghyun Kim
Benjamin Kimia
Akisato Kimura
Pieter-Jan Kindermans
Zsolt Kira
Itaru Kitahara
Hedvig Kjellstrom
Jan Knopp
Takumi Kobayashi
Erich Kobler
Parker Koch
Reinhard Koch
Elyor Kodirov
Amir Kolaman
Nicholas Kolkin
Dimitrios Kollias
Stefanos Kollias
Soheil Kolouri
Adams Wai-Kin Kong
Naejin Kong
Shu Kong
Tao Kong
Yu Kong
Yoshinori Konishi
Daniil Kononenko
Theodora Kontogianni
Simon Korman

Adam Kortylewski
Jana Kosecka
Jean Kossaifi
Satwik Kottur
Rigas Kouskouridas
Adriana Kovashka
Rama Kovvuri
Adarsh Kowdle
Jedrzej Kozerawski
Mateusz Kozinski
Philipp Kraehenbuehl
Gregory Kramida
Josip Krapac
Dmitry Kravchenko
Ranjay Krishna
Pavel Krsek
Alexander Krull
Jakob Kruse
Hiroyuki Kubo
Hilde Kuehne
Jason Kuen
Andreas Kuhn
Arjan Kuijper
Zuzana Kukelova
Ajay Kumar
Amit Kumar
Avinash Kumar
Suryansh Kumar
Vijay Kumar
Kaustav Kundu
Weicheng Kuo
Nojun Kwak
Suha Kwak
Junseok Kwon
Nikolaos Kyriazis
Zorah Lähner
Ankit Laddha
Florent Lafarge
Jean Lahoud
Kevin Lai
Shang-Hong Lai
Wei-Sheng Lai
Yu-Kun Lai
Iro Laina
Antony Lam
John Wheatley Lambert

Xiangyuan lan
Xu Lan
Charis Lanaras
Georg Langs
Oswald Lanz
Dong Lao
Yizhen Lao
Agata Lapedriza
Gustav Larsson
Viktor Larsson
Katrin Lasinger
Christoph Lassner
Longin Jan Latecki
Stéphane Lathuilière
Rynson Lau
Hei Law
Justin Lazarow
Svetlana Lazebnik
Hieu Le
Huu Le
Ngan Hoang Le
Trung-Nghia Le
Vuong Le
Colin Lea
Erik Learned-Miller
Chen-Yu Lee
Gim Hee Lee
Hsin-Ying Lee
Hyungtae Lee
Jae-Han Lee
Jimmy Addison Lee
Joonseok Lee
Kibok Lee
Kuang-Huei Lee
Kwonjoon Lee
Minsik Lee
Sang-chul Lee
Seungkyu Lee
Soochan Lee
Stefan Lee
Taehee Lee
Andreas Lehrmann
Jie Lei
Peng Lei
Matthew Joseph Leotta
Wee Kheng Leow

Gil Levi
Evgeny Levinkov
Aviad Levis
Jose Lezama
Ang Li
Bin Li
Bing Li
Boyi Li
Changsheng Li
Chao Li
Chen Li
Cheng Li
Chenglong Li
Chi Li
Chun-Guang Li
Chun-Liang Li
Chunyuan Li
Dong Li
Guanbin Li
Hao Li
Haoxiang Li
Hongsheng Li
Hongyang Li
Houqiang Li
Huibin Li
Jia Li
Jianan Li
Jianguo Li
Junnan Li
Junxuan Li
Kai Li
Ke Li
Kejie Li
Kunpeng Li
Lerenhan Li
Li Erran Li
Mengtian Li
Mu Li
Peihua Li
Peiyi Li
Ping Li
Qi Li
Qing Li
Ruiyu Li
Ruoteng Li
Shaozi Li

Sheng Li
Shiwei Li
Shuang Li
Siyang Li
Stan Z. Li
Tianye Li
Wei Li
Weixin Li
Wen Li
Wenbo Li
Xiaomeng Li
Xin Li
Xiu Li
Xuelong Li
Xueting Li
Yan Li
Yandong Li
Yanghao Li
Yehao Li
Yi Li
Yijun Li
Yikang LI
Yining Li
Yongjie Li
Yu Li
Yu-Jhe Li
Yunpeng Li
Yunsheng Li
Yunzhu Li
Zhe Li
Zhen Li
Zhengqi Li
Zhenyang Li
Zhuwen Li
Dongze Lian
Xiaochen Lian
Zhouhui Lian
Chen Liang
Jie Liang
Ming Liang
Paul Pu Liang
Pengpeng Liang
Shu Liang
Wei Liang
Jing Liao
Minghui Liao

Renjie Liao
Shengcai Liao
Shuai Liao
Yiyi Liao
Ser-Nam Lim
Chen-Hsuan Lin
Chung-Ching Lin
Dahua Lin
Ji Lin
Kevin Lin
Tianwei Lin
Tsung-Yi Lin
Tsung-Yu Lin
Wei-An Lin
Weiyao Lin
Yen-Chen Lin
Yuewei Lin
David B. Lindell
Drew Linsley
Krzysztof Lis
Roee Litman
Jim Little
An-An Liu
Bo Liu
Buyu Liu
Chao Liu
Chen Liu
Cheng-lin Liu
Chenxi Liu
Dong Liu
Feng Liu
Guilin Liu
Haomiao Liu
Heshan Liu
Hong Liu
Ji Liu
Jingen Liu
Jun Liu
Lanlan Liu
Li Liu
Liu Liu
Mengyuan Liu
Miaomiao Liu
Nian Liu
Ping Liu
Risheng Liu

Sheng Liu
Shu Liu
Shuaicheng Liu
Sifei Liu
Siqi Liu
Siying Liu
Songtao Liu
Ting Liu
Tongliang Liu
Tyng-Luh Liu
Wanquan Liu
Wei Liu
Weiyang Liu
Weizhe Liu
Wenyu Liu
Wu Liu
Xialei Liu
Xianglong Liu
Xiaodong Liu
Xiaofeng Liu
Xihui Liu
Xingyu Liu
Xinwang Liu
Xuanqing Liu
Xuebo Liu
Yang Liu
Yaojie Liu
Yebin Liu
Yen-Cheng Liu
Yiming Liu
Yu Liu
Yu-Shen Liu
Yufan Liu
Yun Liu
Zheng Liu
Zhijian Liu
Zhuang Liu
Zichuan Liu
Ziwei Liu
Zongyi Liu
Stephan Liwicki
Liliana Lo Presti
Chengjiang Long
Fuchen Long
Mingsheng Long
Xiang Long

Yang Long
Charles T. Loop
Antonio Lopez
Roberto J. Lopez-Sastre
Javier Lorenzo-Navarro
Manolis Lourakis
Boyu Lu
Canyi Lu
Feng Lu
Guoyu Lu
Hongtao Lu
Jiajun Lu
Jiasen Lu
Jiwen Lu
Kaiyue Lu
Le Lu
Shao-Ping Lu
Shijian Lu
Xiankai Lu
Xin Lu
Yao Lu
Yiping Lu
Yongxi Lu
Yongyi Lu
Zhiwu Lu
Fujun Luan
Benjamin E. Lundell
Hao Luo
Jian-Hao Luo
Ruotian Luo
Weixin Luo
Wenhan Luo
Wenjie Luo
Yan Luo
Zelun Luo
Zixin Luo
Khoa Luu
Zhaoyang Lv
Pengyuan Lyu
Thomas Möllenhoff
Matthias Müller
Bingpeng Ma
Chih-Yao Ma
Chongyang Ma
Huimin Ma
Jiayi Ma

K. T. Ma
Ke Ma
Lin Ma
Liqian Ma
Shugao Ma
Wei-Chiu Ma
Xiaojian Ma
Xingjun Ma
Zhanyu Ma
Zheng Ma
Radek Jakob Mackowiak
Ludovic Magerand
Shweta Mahajan
Siddharth Mahendran
Long Mai
Ameesh Makadia
Oscar Mendez Maldonado
Mateusz Malinowski
Yury Malkov
Arun Mallya
Dipu Manandhar
Massimiliano Mancini
Fabian Manhardt
Kevis-kokitsi Maninis
Varun Manjunatha
Junhua Mao
Xudong Mao
Alina Marcu
Edgar Margffoy-Tuay
Dmitrii Marin
Manuel J. Marin-Jimenez
Kenneth Marino
Niki Martinel
Julieta Martinez
Jonathan Masci
Tomohiro Mashita
Iacopo Masi
David Masip
Daniela Massiceti
Stefan Mathe
Yusuke Matsui
Tetsu Matsukawa
Iain A. Matthews
Kevin James Matzen
Bruce Allen Maxwell
Stephen Maybank

Helmut Mayer
Amir Mazaheri
David McAllester
Steven McDonagh
Stephen J. Mckenna
Roey Mechrez
Prakhar Mehrotra
Christopher Mei
Xue Mei
Paulo R. S. Mendonca
Lili Meng
Zibo Meng
Thomas Mensink
Bjoern Menze
Michele Merler
Kourosh Meshgi
Pascal Mettes
Christopher Metzler
Liang Mi
Qiguang Miao
Xin Miao
Tomer Michaeli
Frank Michel
Antoine Miech
Krystian Mikolajczyk
Peyman Milanfar
Ben Mildenhall
Gregor Miller
Fausto Milletari
Dongbo Min
Kyle Min
Pedro Miraldo
Dmytro Mishkin
Anand Mishra
Ashish Mishra
Ishan Misra
Niluthpol C. Mithun
Kaushik Mitra
Niloy Mitra
Anton Mitrokhin
Ikuhisa Mitsugami
Anurag Mittal
Kaichun Mo
Zhipeng Mo
Davide Modolo
Michael Moeller

Pritish Mohapatra
Pavlo Molchanov
Davide Moltisanti
Pascal Monasse
Mathew Monfort
Aron Monszpart
Sean Moran
Vlad I. Morariu
Francesc Moreno-Noguer
Pietro Morerio
Stylianos Moschoglou
Yael Moses
Roozbeh Mottaghi
Pierre Moulon
Arsalan Mousavian
Yadong Mu
Yasuhiro Mukaigawa
Lopamudra Mukherjee
Yusuke Mukuta
Ravi Teja Mullapudi
Mario Enrique Munich
Zachary Murez
Ana C. Murillo
J. Krishna Murthy
Damien Muselet
Armin Mustafa
Siva Karthik Mustikovela
Carlo Dal Mutto
Moin Nabi
Varun K. Nagaraja
Tushar Nagarajan
Arsha Nagrani
Seungjun Nah
Nikhil Naik
Yoshikatsu Nakajima
Yuta Nakashima
Atsushi Nakazawa
Seonghyeon Nam
Vinay P. Namboodiri
Medhini Narasimhan
Srinivasa Narasimhan
Sanath Narayan
Erickson Rangel
  Nascimento
Jacinto Nascimento
Tayyab Naseer

Lakshmanan Nataraj
Neda Nategh
Nelson Isao Nauata
Fernando Navarro
Shah Nawaz
Lukas Neumann
Ram Nevatia
Alejandro Newell
Shawn Newsam
Joe Yue-Hei Ng
Trung Thanh Ngo
Duc Thanh Nguyen
Lam M. Nguyen
Phuc Xuan Nguyen
Thuong Nguyen Canh
Mihalis Nicolaou
Andrei Liviu Nicolicioiu
Xuecheng Nie
Michael Niemeyer
Simon Niklaus
Christophoros Nikou
David Nilsson
Jifeng Ning
Yuval Nirkin
Li Niu
Yuzhen Niu
Zhenxing Niu
Shohei Nobuhara
Nicoletta Noceti
Hyeonwoo Noh
Junhyug Noh
Mehdi Noroozi
Sotiris Nousias
Valsamis Ntouskos
Matthew O'Toole
Peter Ochs
Ferda Ofli
Seong Joon Oh
Seoung Wug Oh
Iason Oikonomidis
Utkarsh Ojha
Takahiro Okabe
Takayuki Okatani
Fumio Okura
Aude Oliva
Kyle Olszewski

Björn Ommer
Mohamed Omran
Elisabeta Oneata
Michael Opitz
Jose Oramas
Tribhuvanesh Orekondy
Shaul Oron
Sergio Orts-Escolano
Ivan Oseledets
Aljosa Osep
Magnus Oskarsson
Anton Osokin
Martin R. Oswald
Wanli Ouyang
Andrew Owens
Mete Ozay
Mustafa Ozuysal
Eduardo Pérez-Pellitero
Gautam Pai
Dipan Kumar Pal
P. H. Pamplona Savarese
Jinshan Pan
Junting Pan
Xingang Pan
Yingwei Pan
Yannis Panagakis
Rameswar Panda
Guan Pang
Jiahao Pang
Jiangmiao Pang
Tianyu Pang
Sharath Pankanti
Nicolas Papadakis
Dim Papadopoulos
George Papandreou
Toufiq Parag
Shaifali Parashar
Sarah Parisot
Eunhyeok Park
Hyun Soo Park
Jaesik Park
Min-Gyu Park
Taesung Park
Alvaro Parra
C. Alejandro Parraga
Despoina Paschalidou

Nikolaos Passalis
Vishal Patel
Viorica Patraucean
Badri Narayana Patro
Danda Pani Paudel
Sujoy Paul
Georgios Pavlakos
Ioannis Pavlidis
Vladimir Pavlovic
Nick Pears
Kim Steenstrup Pedersen
Selen Pehlivan
Shmuel Peleg
Chao Peng
Houwen Peng
Wen-Hsiao Peng
Xi Peng
Xiaojiang Peng
Xingchao Peng
Yuxin Peng
Federico Perazzi
Juan Camilo Perez
Vishwanath Peri
Federico Pernici
Luca Del Pero
Florent Perronnin
Stavros Petridis
Henning Petzka
Patrick Peursum
Michael Pfeiffer
Hanspeter Pfister
Roman Pflugfelder
Minh Tri Pham
Yongri Piao
David Picard
Tomasz Pieciak
A. J. Piergiovanni
Andrea Pilzer
Pedro O. Pinheiro
Silvia Laura Pintea
Lerrel Pinto
Axel Pinz
Robinson Piramuthu
Fiora Pirri
Leonid Pishchulin
Francesco Pittaluga

Daniel Pizarro
Tobias Plötz
Mirco Planamente
Matteo Poggi
Moacir A. Ponti
Parita Pooj
Fatih Porikli
Horst Possegger
Omid Poursaeed
Ameya Prabhu
Viraj Uday Prabhu
Dilip Prasad
Brian L. Price
True Price
Maria Priisalu
Veronique Prinet
Victor Adrian Prisacariu
Jan Prokaj
Sergey Prokudin
Nicolas Pugeault
Xavier Puig
Albert Pumarola
Pulak Purkait
Senthil Purushwalkam
Charles R. Qi
Hang Qi
Haozhi Qi
Lu Qi
Mengshi Qi
Siyuan Qi
Xiaojuan Qi
Yuankai Qi
Shengju Qian
Xuelin Qian
Siyuan Qiao
Yu Qiao
Jie Qin
Qiang Qiu
Weichao Qiu
Zhaofan Qiu
Kha Gia Quach
Yuhui Quan
Yvain Queau
Julian Quiroga
Faisal Qureshi
Mahdi Rad

Filip Radenovic
Petia Radeva
Venkatesh
  B. Radhakrishnan
Ilija Radosavovic
Noha Radwan
Rahul Raguram
Tanzila Rahman
Amit Raj
Ajit Rajwade
Kandan Ramakrishnan
Santhosh
  K. Ramakrishnan
Srikumar Ramalingam
Ravi Ramamoorthi
Vasili Ramanishka
Ramprasaath R. Selvaraju
Francois Rameau
Visvanathan Ramesh
Santu Rana
Rene Ranftl
Anand Rangarajan
Anurag Ranjan
Viresh Ranjan
Yongming Rao
Carolina Raposo
Vivek Rathod
Sathya N. Ravi
Avinash Ravichandran
Tammy Riklin Raviv
Daniel Rebain
Sylvestre-Alvise Rebuffi
N. Dinesh Reddy
Timo Rehfeld
Paolo Remagnino
Konstantinos Rematas
Edoardo Remelli
Dongwei Ren
Haibing Ren
Jian Ren
Jimmy Ren
Mengye Ren
Weihong Ren
Wenqi Ren
Zhile Ren
Zhongzheng Ren

Zhou Ren
Vijay Rengarajan
Md A. Reza
Farzaneh Rezaeianaran
Hamed R. Tavakoli
Nicholas Rhinehart
Helge Rhodin
Elisa Ricci
Alexander Richard
Eitan Richardson
Elad Richardson
Christian Richardt
Stephan Richter
Gernot Riegler
Daniel Ritchie
Tobias Ritschel
Samuel Rivera
Yong Man Ro
Richard Roberts
Joseph Robinson
Ignacio Rocco
Mrigank Rochan
Emanuele Rodolà
Mikel D. Rodriguez
Giorgio Roffo
Grégory Rogez
Gemma Roig
Javier Romero
Xuejian Rong
Yu Rong
Amir Rosenfeld
Bodo Rosenhahn
Guy Rosman
Arun Ross
Paolo Rota
Peter M. Roth
Anastasios Roussos
Anirban Roy
Sebastien Roy
Aruni RoyChowdhury
Artem Rozantsev
Ognjen Rudovic
Daniel Rueckert
Adria Ruiz
Javier Ruiz-del-solar
Christian Rupprecht

Chris Russell
Dan Ruta
Jongbin Ryu
Ömer Sümer
Alexandre Sablayrolles
Faraz Saeedan
Ryusuke Sagawa
Christos Sagonas
Tonmoy Saikia
Hideo Saito
Kuniaki Saito
Shunsuke Saito
Shunta Saito
Ken Sakurada
Joaquin Salas
Fatemeh Sadat Saleh
Mahdi Saleh
Pouya Samangouei
Leo Sampaio
  Ferraz Ribeiro
Artsiom Olegovich
  Sanakoyeu
Enrique Sanchez
Patsorn Sangkloy
Anush Sankaran
Aswin Sankaranarayanan
Swami Sankaranarayanan
Rodrigo Santa Cruz
Amartya Sanyal
Archana Sapkota
Nikolaos Sarafianos
Jun Sato
Shin'ichi Satoh
Hosnieh Sattar
Arman Savran
Manolis Savva
Alexander Sax
Hanno Scharr
Simone Schaub-Meyer
Konrad Schindler
Dmitrij Schlesinger
Uwe Schmidt
Dirk Schnieders
Björn Schuller
Samuel Schulter
Idan Schwartz

William Robson Schwartz
Alex Schwing
Sinisa Segvic
Lorenzo Seidenari
Pradeep Sen
Ozan Sener
Soumyadip Sengupta
Arda Senocak
Mojtaba Seyedhosseini
Shishir Shah
Shital Shah
Sohil Atul Shah
Tamar Rott Shaham
Huasong Shan
Qi Shan
Shiguang Shan
Jing Shao
Roman Shapovalov
Gaurav Sharma
Vivek Sharma
Viktoriia Sharmanska
Dongyu She
Sumit Shekhar
Evan Shelhamer
Chengyao Shen
Chunhua Shen
Falong Shen
Jie Shen
Li Shen
Liyue Shen
Shuhan Shen
Tianwei Shen
Wei Shen
William B. Shen
Yantao Shen
Ying Shen
Yiru Shen
Yujun Shen
Yuming Shen
Zhiqiang Shen
Ziyi Shen
Lu Sheng
Yu Sheng
Rakshith Shetty
Baoguang Shi
Guangming Shi

Hailin Shi
Miaojing Shi
Yemin Shi
Zhenmei Shi
Zhiyuan Shi
Kevin Jonathan Shih
Shiliang Shiliang
Hyunjung Shim
Atsushi Shimada
Nobutaka Shimada
Daeyun Shin
Young Min Shin
Koichi Shinoda
Konstantin Shmelkov
Michael Zheng Shou
Abhinav Shrivastava
Tianmin Shu
Zhixin Shu
Hong-Han Shuai
Pushkar Shukla
Christian Siagian
Mennatullah M. Siam
Kaleem Siddiqi
Karan Sikka
Jae-Young Sim
Christian Simon
Martin Simonovsky
Dheeraj Singaraju
Bharat Singh
Gurkirt Singh
Krishna Kumar Singh
Maneesh Kumar Singh
Richa Singh
Saurabh Singh
Suriya Singh
Vikas Singh
Sudipta N. Sinha
Vincent Sitzmann
Josef Sivic
Gregory Slabaugh
Miroslava Slavcheva
Ron Slossberg
Brandon Smith
Kevin Smith
Vladimir Smutny
Noah Snavely

Roger
   D. Soberanis-Mukul
Kihyuk Sohn
Francesco Solera
Eric Sommerlade
Sanghyun Son
Byung Cheol Song
Chunfeng Song
Dongjin Song
Jiaming Song
Jie Song
Jifei Song
Jingkuan Song
Mingli Song
Shiyu Song
Shuran Song
Xiao Song
Yafei Song
Yale Song
Yang Song
Yi-Zhe Song
Yibing Song
Humberto Sossa
Cesar de Souza
Adrian Spurr
Srinath Sridhar
Suraj Srinivas
Pratul P. Srinivasan
Anuj Srivastava
Tania Stathaki
Christopher Stauffer
Simon Stent
Rainer Stiefelhagen
Pierre Stock
Julian Straub
Jonathan C. Stroud
Joerg Stueckler
Jan Stuehmer
David Stutz
Chi Su
Hang Su
Jong-Chyi Su
Shuochen Su
Yu-Chuan Su
Ramanathan Subramanian
Yusuke Sugano

Masanori Suganuma
Yumin Suh
Mohammed Suhail
Yao Sui
Heung-Il Suk
Josephine Sullivan
Baochen Sun
Chen Sun
Chong Sun
Deqing Sun
Jin Sun
Liang Sun
Lin Sun
Qianru Sun
Shao-Hua Sun
Shuyang Sun
Weiwei Sun
Wenxiu Sun
Xiaoshuai Sun
Xiaoxiao Sun
Xingyuan Sun
Yifan Sun
Zhun Sun
Sabine Susstrunk
David Suter
Supasorn Suwajanakorn
Tomas Svoboda
Eran Swears
Paul Swoboda
Attila Szabo
Richard Szeliski
Duy-Nguyen Ta
Andrea Tagliasacchi
Yuichi Taguchi
Ying Tai
Keita Takahashi
Kouske Takahashi
Jun Takamatsu
Hugues Talbot
Toru Tamaki
Chaowei Tan
Fuwen Tan
Mingkui Tan
Mingxing Tan
Qingyang Tan
Robby T. Tan

Xiaoyang Tan
Kenichiro Tanaka
Masayuki Tanaka
Chang Tang
Chengzhou Tang
Danhang Tang
Ming Tang
Peng Tang
Qingming Tang
Wei Tang
Xu Tang
Yansong Tang
Youbao Tang
Yuxing Tang
Zhiqiang Tang
Tatsunori Taniai
Junli Tao
Xin Tao
Makarand Tapaswi
Jean-Philippe Tarel
Lyne Tchapmi
Zachary Teed
Bugra Tekin
Damien Teney
Ayush Tewari
Christian Theobalt
Christopher Thomas
Diego Thomas
Jim Thomas
Rajat Mani Thomas
Xinmei Tian
Yapeng Tian
Yingli Tian
Yonglong Tian
Zhi Tian
Zhuotao Tian
Kinh Tieu
Joseph Tighe
Massimo Tistarelli
Matthew Toews
Carl Toft
Pavel Tokmakov
Federico Tombari
Chetan Tonde
Yan Tong
Alessio Tonioni

Andrea Torsello
Fabio Tosi
Du Tran
Luan Tran
Ngoc-Trung Tran
Quan Hung Tran
Truyen Tran
Rudolph Triebel
Martin Trimmel
Shashank Tripathi
Subarna Tripathi
Leonardo Trujillo
Eduard Trulls
Tomasz Trzcinski
Sam Tsai
Yi-Hsuan Tsai
Hung-Yu Tseng
Stavros Tsogkas
Aggeliki Tsoli
Devis Tuia
Shubham Tulsiani
Sergey Tulyakov
Frederick Tung
Tony Tung
Daniyar Turmukhambetov
Ambrish Tyagi
Radim Tylecek
Christos Tzelepis
Georgios Tzimiropoulos
Dimitrios Tzionas
Seiichi Uchida
Norimichi Ukita
Dmitry Ulyanov
Martin Urschler
Yoshitaka Ushiku
Ben Usman
Alexander Vakhitov
Julien P. C. Valentin
Jack Valmadre
Ernest Valveny
Joost van de Weijer
Jan van Gemert
Koen Van Leemput
Gul Varol
Sebastiano Vascon
M. Alex O. Vasilescu

Subeesh Vasu
Mayank Vatsa
David Vazquez
Javier Vazquez-Corral
Ashok Veeraraghavan
Erik Velasco-Salido
Raviteja Vemulapalli
Jonathan Ventura
Manisha Verma
Roberto Vezzani
Ruben Villegas
Minh Vo
MinhDuc Vo
Nam Vo
Michele Volpi
Riccardo Volpi
Carl Vondrick
Konstantinos Vougioukas
Tuan-Hung Vu
Sven Wachsmuth
Neal Wadhwa
Catherine Wah
Jacob C. Walker
Thomas S. A. Wallis
Chengde Wan
Jun Wan
Liang Wan
Renjie Wan
Baoyuan Wang
Boyu Wang
Cheng Wang
Chu Wang
Chuan Wang
Chunyu Wang
Dequan Wang
Di Wang
Dilin Wang
Dong Wang
Fang Wang
Guanzhi Wang
Guoyin Wang
Hanzi Wang
Hao Wang
He Wang
Heng Wang
Hongcheng Wang

Hongxing Wang
Hua Wang
Jian Wang
Jingbo Wang
Jinglu Wang
Jingya Wang
Jinjun Wang
Jinqiao Wang
Jue Wang
Ke Wang
Keze Wang
Le Wang
Lei Wang
Lezi Wang
Li Wang
Liang Wang
Lijun Wang
Limin Wang
Linwei Wang
Lizhi Wang
Mengjiao Wang
Mingzhe Wang
Minsi Wang
Naiyan Wang
Nannan Wang
Ning Wang
Oliver Wang
Pei Wang
Peng Wang
Pichao Wang
Qi Wang
Qian Wang
Qiaosong Wang
Qifei Wang
Qilong Wang
Qing Wang
Qingzhong Wang
Quan Wang
Rui Wang
Ruiping Wang
Ruixing Wang
Shangfei Wang
Shenlong Wang
Shiyao Wang
Shuhui Wang
Song Wang

Tao Wang
Tianlu Wang
Tiantian Wang
Ting-chun Wang
Tingwu Wang
Wei Wang
Weiyue Wang
Wenguan Wang
Wenlin Wang
Wenqi Wang
Xiang Wang
Xiaobo Wang
Xiaofang Wang
Xiaoling Wang
Xiaolong Wang
Xiaosong Wang
Xiaoyu Wang
Xin Eric Wang
Xinchao Wang
Xinggang Wang
Xintao Wang
Yali Wang
Yan Wang
Yang Wang
Yangang Wang
Yaxing Wang
Yi Wang
Yida Wang
Yilin Wang
Yiming Wang
Yisen Wang
Yongtao Wang
Yu-Xiong Wang
Yue Wang
Yujiang Wang
Yunbo Wang
Yunhe Wang
Zengmao Wang
Zhangyang Wang
Zhaowen Wang
Zhe Wang
Zhecan Wang
Zheng Wang
Zhixiang Wang
Zilei Wang
Jianqiao Wangni

Anne S. Wannenwetsch
Jan Dirk Wegner
Scott Wehrwein
Donglai Wei
Kaixuan Wei
Longhui Wei
Pengxu Wei
Ping Wei
Qi Wei
Shih-En Wei
Xing Wei
Yunchao Wei
Zijun Wei
Jerod Weinman
Michael Weinmann
Philippe Weinzaepfel
Yair Weiss
Bihan Wen
Longyin Wen
Wei Wen
Junwu Weng
Tsui-Wei Weng
Xinshuo Weng
Eric Wengrowski
Tomas Werner
Gordon Wetzstein
Tobias Weyand
Patrick Wieschollek
Maggie Wigness
Erik Wijmans
Richard Wildes
Olivia Wiles
Chris Williams
Williem Williem
Kyle Wilson
Calden Wloka
Nicolai Wojke
Christian Wolf
Yongkang Wong
Sanghyun Woo
Scott Workman
Baoyuan Wu
Bichen Wu
Chao-Yuan Wu
Huikai Wu
Jiajun Wu

Jialin Wu
Jiaxiang Wu
Jiqing Wu
Jonathan Wu
Lifang Wu
Qi Wu
Qiang Wu
Ruizheng Wu
Shangzhe Wu
Shun-Cheng Wu
Tianfu Wu
Wayne Wu
Wenxuan Wu
Xiao Wu
Xiaohe Wu
Xinxiao Wu
Yang Wu
Yi Wu
Yiming Wu
Ying Nian Wu
Yue Wu
Zheng Wu
Zhenyu Wu
Zhirong Wu
Zuxuan Wu
Stefanie Wuhrer
Jonas Wulff
Changqun Xia
Fangting Xia
Fei Xia
Gui-Song Xia
Lu Xia
Xide Xia
Yin Xia
Yingce Xia
Yongqin Xian
Lei Xiang
Shiming Xiang
Bin Xiao
Fanyi Xiao
Guobao Xiao
Huaxin Xiao
Taihong Xiao
Tete Xiao
Tong Xiao
Wang Xiao

Yang Xiao
Cihang Xie
Guosen Xie
Jianwen Xie
Lingxi Xie
Sirui Xie
Weidi Xie
Wenxuan Xie
Xiaohua Xie
Fuyong Xing
Jun Xing
Junliang Xing
Bo Xiong
Peixi Xiong
Yu Xiong
Yuanjun Xiong
Zhiwei Xiong
Chang Xu
Chenliang Xu
Dan Xu
Danfei Xu
Hang Xu
Hongteng Xu
Huijuan Xu
Jingwei Xu
Jun Xu
Kai Xu
Mengmeng Xu
Mingze Xu
Qianqian Xu
Ran Xu
Weijian Xu
Xiangyu Xu
Xiaogang Xu
Xing Xu
Xun Xu
Yanyu Xu
Yichao Xu
Yong Xu
Yongchao Xu
Yuanlu Xu
Zenglin Xu
Zheng Xu
Chuhui Xue
Jia Xue
Nan Xue

Tianfan Xue
Xiangyang Xue
Abhay Yadav
Yasushi Yagi
I. Zeki Yalniz
Kota Yamaguchi
Toshihiko Yamasaki
Takayoshi Yamashita
Junchi Yan
Ke Yan
Qingan Yan
Sijie Yan
Xinchen Yan
Yan Yan
Yichao Yan
Zhicheng Yan
Keiji Yanai
Bin Yang
Ceyuan Yang
Dawei Yang
Dong Yang
Fan Yang
Guandao Yang
Guorun Yang
Haichuan Yang
Hao Yang
Jianwei Yang
Jiaolong Yang
Jie Yang
Jing Yang
Kaiyu Yang
Linjie Yang
Meng Yang
Michael Ying Yang
Nan Yang
Shuai Yang
Shuo Yang
Tianyu Yang
Tien-Ju Yang
Tsun-Yi Yang
Wei Yang
Wenhan Yang
Xiao Yang
Xiaodong Yang
Xin Yang
Yan Yang

Yanchao Yang
Yee Hong Yang
Yezhou Yang
Zhenheng Yang
Anbang Yao
Angela Yao
Cong Yao
Jian Yao
Li Yao
Ting Yao
Yao Yao
Zhewei Yao
Chengxi Ye
Jianbo Ye
Keren Ye
Linwei Ye
Mang Ye
Mao Ye
Qi Ye
Qixiang Ye
Mei-Chen Yeh
Raymond Yeh
Yu-Ying Yeh
Sai-Kit Yeung
Serena Yeung
Kwang Moo Yi
Li Yi
Renjiao Yi
Alper Yilmaz
Junho Yim
Lijun Yin
Weidong Yin
Xi Yin
Zhichao Yin
Tatsuya Yokota
Ryo Yonetani
Donggeun Yoo
Jae Shin Yoon
Ju Hong Yoon
Sung-eui Yoon
Laurent Younes
Changqian Yu
Fisher Yu
Gang Yu
Jiahui Yu
Kaicheng Yu

Ke Yu
Lequan Yu
Ning Yu
Qian Yu
Ronald Yu
Ruichi Yu
Shoou-I Yu
Tao Yu
Tianshu Yu
Xiang Yu
Xin Yu
Xiyu Yu
Youngjae Yu
Yu Yu
Zhiding Yu
Chunfeng Yuan
Ganzhao Yuan
Jinwei Yuan
Lu Yuan
Quan Yuan
Shanxin Yuan
Tongtong Yuan
Wenjia Yuan
Ye Yuan
Yuan Yuan
Yuhui Yuan
Huanjing Yue
Xiangyu Yue
Ersin Yumer
Sergey Zagoruyko
Egor Zakharov
Amir Zamir
Andrei Zanfir
Mihai Zanfir
Pablo Zegers
Bernhard Zeisl
John S. Zelek
Niclas Zeller
Huayi Zeng
Jiabei Zeng
Wenjun Zeng
Yu Zeng
Xiaohua Zhai
Fangneng Zhan
Huangying Zhan
Kun Zhan

Xiaohang Zhan
Baochang Zhang
Bowen Zhang
Cecilia Zhang
Changqing Zhang
Chao Zhang
Chengquan Zhang
Chi Zhang
Chongyang Zhang
Dingwen Zhang
Dong Zhang
Feihu Zhang
Hang Zhang
Hanwang Zhang
Hao Zhang
He Zhang
Hongguang Zhang
Hua Zhang
Ji Zhang
Jianguo Zhang
Jianming Zhang
Jiawei Zhang
Jie Zhang
Jing Zhang
Juyong Zhang
Kai Zhang
Kaipeng Zhang
Ke Zhang
Le Zhang
Lei Zhang
Li Zhang
Lihe Zhang
Linguang Zhang
Lu Zhang
Mi Zhang
Mingda Zhang
Peng Zhang
Pingping Zhang
Qian Zhang
Qilin Zhang
Quanshi Zhang
Richard Zhang
Rui Zhang
Runze Zhang
Shengping Zhang
Shifeng Zhang

Shuai Zhang
Songyang Zhang
Tao Zhang
Ting Zhang
Tong Zhang
Wayne Zhang
Wei Zhang
Weizhong Zhang
Wenwei Zhang
Xiangyu Zhang
Xiaolin Zhang
Xiaopeng Zhang
Xiaoqin Zhang
Xiuming Zhang
Ya Zhang
Yang Zhang
Yimin Zhang
Yinda Zhang
Ying Zhang
Yongfei Zhang
Yu Zhang
Yulun Zhang
Yunhua Zhang
Yuting Zhang
Zhanpeng Zhang
Zhao Zhang
Zhaoxiang Zhang
Zhen Zhang
Zheng Zhang
Zhifei Zhang
Zhijin Zhang
Zhishuai Zhang
Ziming Zhang
Bo Zhao
Chen Zhao
Fang Zhao
Haiyu Zhao
Han Zhao
Hang Zhao
Hengshuang Zhao
Jian Zhao
Kai Zhao
Liang Zhao
Long Zhao
Qian Zhao
Qibin Zhao

Qijun Zhao
Rui Zhao
Shenglin Zhao
Sicheng Zhao
Tianyi Zhao
Wenda Zhao
Xiangyun Zhao
Xin Zhao
Yang Zhao
Yue Zhao
Zhichen Zhao
Zijing Zhao
Xiantong Zhen
Chuanxia Zheng
Feng Zheng
Haiyong Zheng
Jia Zheng
Kang Zheng
Shuai Kyle Zheng
Wei-Shi Zheng
Yinqiang Zheng
Zerong Zheng
Zhedong Zheng
Zilong Zheng
Bineng Zhong
Fangwei Zhong
Guangyu Zhong
Yiran Zhong
Yujie Zhong
Zhun Zhong
Chunluan Zhou
Huiyu Zhou
Jiahuan Zhou
Jun Zhou
Lei Zhou
Luowei Zhou
Luping Zhou
Mo Zhou
Ning Zhou
Pan Zhou
Peng Zhou
Qianyi Zhou
S. Kevin Zhou
Sanping Zhou
Wengang Zhou
Xingyi Zhou

Yanzhao Zhou
Yi Zhou
Yin Zhou
Yipin Zhou
Yuyin Zhou
Zihan Zhou
Alex Zihao Zhu
Chenchen Zhu
Feng Zhu
Guangming Zhu
Ji Zhu
Jun-Yan Zhu
Lei Zhu
Linchao Zhu
Rui Zhu
Shizhan Zhu
Tyler Lixuan Zhu

Wei Zhu
Xiangyu Zhu
Xinge Zhu
Xizhou Zhu
Yanjun Zhu
Yi Zhu
Yixin Zhu
Yizhe Zhu
Yousong Zhu
Zhe Zhu
Zhen Zhu
Zheng Zhu
Zhenyao Zhu
Zhihui Zhu
Zhuotun Zhu
Bingbing Zhuang
Wei Zhuo

Christian Zimmermann
Karel Zimmermann
Larry Zitnick
Mohammadreza
   Zolfaghari
Maria Zontak
Daniel Zoran
Changqing Zou
Chuhang Zou
Danping Zou
Qi Zou
Yang Zou
Yuliang Zou
Georgios Zoumpourlis
Wangmeng Zuo
Xinxin Zuo

Additional Reviewers

Victoria Fernandez
   Abrevaya
Maya Aghaei
Allam Allam
Christine
   Allen-Blanchette
Nicolas Aziere
Assia Benbihi
Neha Bhargava
Bharat Lal Bhatnagar
Joanna Bitton
Judy Borowski
Amine Bourki
Romain Brégier
Tali Brayer
Sebastian Bujwid
Andrea Burns
Yun-Hao Cao
Yuning Chai
Xiaojun Chang
Bo Chen
Shuo Chen
Zhixiang Chen
Junsuk Choe
Hung-Kuo Chu

Jonathan P. Crall
Kenan Dai
Lucas Deecke
Karan Desai
Prithviraj Dhar
Jing Dong
Wei Dong
Turan Kaan Elgin
Francis Engelmann
Erik Englesson
Fartash Faghri
Zicong Fan
Yang Fu
Risheek Garrepalli
Yifan Ge
Marco Godi
Helmut Grabner
Shuxuan Guo
Jianfeng He
Zhezhi He
Samitha Herath
Chih-Hui Ho
Yicong Hong
Vincent Tao Hu
Julio Hurtado

Jaedong Hwang
Andrey Ignatov
Muhammad
   Abdullah Jamal
Saumya Jetley
Meiguang Jin
Jeff Johnson
Minsoo Kang
Saeed Khorram
Mohammad Rami Koujan
Nilesh Kulkarni
Sudhakar Kumawat
Abdelhak Lemkhenter
Alexander Levine
Jiachen Li
Jing Li
Jun Li
Yi Li
Liang Liao
Ruochen Liao
Tzu-Heng Lin
Phillip Lippe
Bao-di Liu
Bo Liu
Fangchen Liu

Hanxiao Liu
Hongyu Liu
Huidong Liu
Miao Liu
Xinxin Liu
Yongfei Liu
Yu-Lun Liu
Amir Livne
Tiange Luo
Wei Ma
Xiaoxuan Ma
Ioannis Marras
Georg Martius
Effrosyni Mavroudi
Tim Meinhardt
Givi Meishvili
Meng Meng
Zihang Meng
Zhongqi Miao
Gyeongsik Moon
Khoi Nguyen
Yung-Kyun Noh
Antonio Norelli
Jaeyoo Park
Alexander Pashevich
Mandela Patrick
Mary Phuong
Bingqiao Qian
Yu Qiao
Zhen Qiao
Sai Saketh Rambhatla
Aniket Roy
Amelie Royer
Parikshit Vishwas
  Sakurikar
Mark Sandler
Mert Bülent Sarıyıldız
Tanner Schmidt
Anshul B. Shah

Ketul Shah
Rajvi Shah
Hengcan Shi
Xiangxi Shi
Yujiao Shi
William A. P. Smith
Guoxian Song
Robin Strudel
Abby Stylianou
Xinwei Sun
Reuben Tan
Qingyi Tao
Kedar S. Tatwawadi
Anh Tuan Tran
Son Dinh Tran
Eleni Triantafillou
Aristeidis Tsitiridis
Md Zasim Uddin
Andrea Vedaldi
Evangelos Ververas
Vidit Vidit
Paul Voigtlaender
Bo Wan
Huanyu Wang
Huiyu Wang
Junqiu Wang
Pengxiao Wang
Tai Wang
Xinyao Wang
Tomoki Watanabe
Mark Weber
Xi Wei
Botong Wu
James Wu
Jiamin Wu
Rujie Wu
Yu Wu
Rongchang Xie
Wei Xiong

Yunyang Xiong
An Xu
Chi Xu
Yinghao Xu
Fei Xue
Tingyun Yan
Zike Yan
Chao Yang
Heran Yang
Ren Yang
Wenfei Yang
Xu Yang
Rajeev Yasarla
Shaokai Ye
Yufei Ye
Kun Yi
Haichao Yu
Hanchao Yu
Ruixuan Yu
Liangzhe Yuan
Chen-Lin Zhang
Fandong Zhang
Tianyi Zhang
Yang Zhang
Yiyi Zhang
Yongshun Zhang
Yu Zhang
Zhiwei Zhang
Jiaojiao Zhao
Yipu Zhao
Xingjian Zhen
Haizhong Zheng
Tiancheng Zhi
Chengju Zhou
Hao Zhou
Hao Zhu
Alexander Zimin

Contents – Part XXIII

Margin-Mix: Semi-Supervised Learning for Face Expression Recognition

Corneliu Florea(✉) (iD), Mihai Badea, Laura Florea, Andrei Racoviteanu,
and Constantin Vertan

Image Processing and Analysis Laboratory, University Politehnica of Bucharest,
Bucharest, Romania
{corneliu.florea,mihai_sorin.badea,laura.florea,andrei.racoviteanu,
constantin.vertan}@upb.ro

Abstract. In this paper, as we aim to construct a semi-supervised learning algorithm, we exploit the characteristics of the Deep Convolutional Networks to provide, for an input image, both an embedding descriptor and a prediction. The unlabeled data is combined with the labeled one in order to provide synthetic data, which describes better the input space. The network is asked to provide a large margin between clusters, while new data is self-labeled by the distance to class centroids, in the embedding space. The method is tested on standard benchmarks for semi-supervised learning, where it matches state of the art performance and on the problem of face expression recognition where it increases the accuracy by a noticeable margin.

Keywords: Margin loss · Semi-supervised learning · Data mixup · Face expression recognition

1 Introduction

In the latest period, deep learning techniques acknowledged great advance. One of the ingredients that favored this advance is the collection and annotation of large data corpora [20]. Ordinarily, the data comes in two variants: labeled, when each instance x_i has the related label y_i and unlabeled, when the instances miss their labels (i.e. there are no y_i). Learning within the case of labeled information is less demanding and it is favored as it has been thoroughly explored. However there are circumstances when labeling is either costly (for instance locating boxes around particular objects in images), or it requires highly trained personnel (encountered, for instance, in the case of medical imaging). In such situations, only a portion of the data is annotated and Semi-Supervised Learning (SSL) algorithms that produce robust solutions using only the limited amount of available annotated data are used to annotate the large volumes of unlabeled data [6].

Electronic supplementary material The online version of this chapter (https://doi.org/10.1007/978-3-030-58592-1_1) contains supplementary material, which is available to authorized users.

Our SSL proposal is built upon two principles. The first principle refers to the deep convolutional networks (DCN) characteristic to provide simultaneously decision layers and feature descriptors of the input image [11]. The second principle is that SSL favors the borders through a low density area [6]. Our algorithm seeks to cluster data and create low density areas between borders.

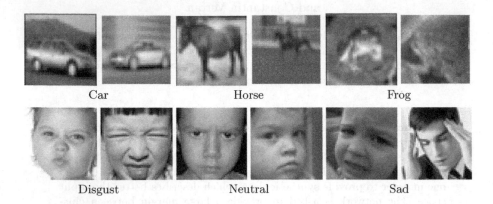

Fig. 1. Examples of the two problems approached. The images from different classed in CIFAR-like databases (top rows) are more different than are face expressions (bottom row).

A particular field that may greatly benefit from semi-supervised learning algorithms is face expression recognition (FER). The topic contains many previous development and areas of interest. For a thorough introduction we kindly refer the reader to the reviews on the topic [8,33]. In this paper, we concentrate our efforts towards the expression categorization into fundamental classes as defined by Ekman *et al.* [12]: "neutral", "anger", "fear", "disgust", "happy", "sad", "surprise"; sometimes "contempt" is also included.

With respect to the FER problem, a particular characteristic is the fact that human annotation of such data is hard and costly. "Hard" refers to the fact that the average person has difficulties in differentiating between expressions. In this direction, Susskind et al. [35] showed that an experienced observer (psychology student) reached 89.2% accuracy in a 6 expressions experiment. Alternatively, Bartlett et al. [3] and Ekman et al. [12] noted that more than 100 h of training are needed for a person in order to get 70% accuracy in recognizing face movements relevant for expressions. To give a reference for comparison, we recall that the average, untrained user achieves ≈94% accuracy for image classes on CIFAR-10 [16], reaching 100% on 90% of images [30]. Thus, due to the difficulty in annotating images, problems related to face expression analysis welcome methods and strategies that use additional unlabeled data as a substitute to more annotations, in order to augment the performance.

This contrast can be intuitively associated with the structure and density of the data of the two above mentioned domains. Seemingly, classes from

CIFAR-like databases have high intra-class variability, but also low density areas between classes. Classes from any face expression database differentiate between themselves by small and subtle differences (as illustrated in Fig. 1). Thus, the FER domain has low variance both inter-class and intra-class.

Contribution and Paper Structure. We propose a semi-supervised learning method, oriented toward classification, that is derived from the classical self-labeling paradigm [6]. The used learner, a deep convolutional neural network, simultaneously clusters and classifies the data. For an unlabeled point, the distance to the class centroids is used for self-labeling. We integrated this idea in the MixUp arrangement [43]. MixUp ensures that the input data space is thoroughly interpolated and it corresponds to a mirror distribution in the prediction space. In our proposal, the intermediate feature space is also thoroughly investigated and the correspondence with input and prediction space is preserved.

More precisely, this paper contributes with: (i) a novel semi-supervised learning algorithm that classifies unlabeled data based on distance to class centroids in a feature space; (ii) the method is showed to be comparable with state of the art in problems with low density areas and to have significant improvements for problems with dense areas between classes, as is the FER problem.

2 Related Work

The proposed method seeks discriminative embeddings (features) in DCN while implementing a semi-supervised learning strategy, that is effective for face expression recognition. In this section we provide a short summary over these three directions (discriminative features, SSL and FER).

Loss Function for Better Deep Features Discrimination. Major contributions in this direction originated in approaches to the face recognition problem. In conjunction to deep learning, several different types of loss function were proposed in the last years. Wen *et al.* [40] proposed the *center loss* function, to minimize the intra-class distances between the deep features; Liu *et al.* [24] learned angular discriminative features with the angular softmax loss in order to achieve smaller maximal intra–class distance than minimal inter-class distance; Zhang *et al.* [44] developed a loss function for long tailed distributions; Zheng *et al.* [47] showed that normalizing the deep features with the so-called Ring Loss leads to improved accuracy. All these methods were shown to give good results on face recognition tasks, where very large annotated datasets like MegaFace are available. To our best knowledge the strategy of computing discriminative embeddings using the class centroids to annotate new data has not yet been used in general, nor in the context of SSL. A similar concept, but adapted to clustering, may be found in the work of Ren et al. [32]; yet they did not seek consistent data space and prediction in the manner we do here.

From the many existing variants we have relied on a development of the center loss as it uses the Euclidean distance, making this step consistent with data interpolation in the original space due to the MixUp arrangement.

Semi-Supervised Learning. While the problem of SSL has been in the attention of the community for a long period, the appearance of the data hungry deep learning methods brought increased interest. In the context of the deep learning, many initial results were based on generative models such as denoising [31], variational autoencoders [17], or generative adversarial networks [27]. The concept of self labeling has been used in the form of entropy regularization [21].

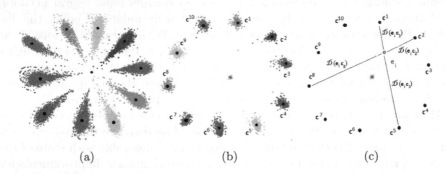

(a)   (b)   (c)

Fig. 2. Structuring embeddings on a 2 dimensional layer when the DCN was trained with different losses: (a) softmax dominated and (b) center loss dominated. (c) Our proposal: given a new unlabeled point $\mathbf{e}_i$, one-hot encoding values $\mathbf{y}_i$ are determined by distances to class centroids

In the later period, improved results were obtain by adding consistency regularization losses while processing unlabeled data. The consistency regularization uses the discrepancy between predictions on unlabeled data points and predictions on labeled examples to correct weights. Practical solutions improved performance by smoothing the weight correction before measuring the discrepancy. In this category, one might count Π-Model with Temporal Ensembling [19], Mean Teacher [36] and Virtual Adversarial Training [25], fast-SWA [1] or consistent embedding description in associative domain transfer [15]. More recently, build upon the MixUp strategy [43], models have been constrained to showed consistency with respect to perturbation of the input examples in the MixMatch algorithm [5] or Interpolation Consistency Training - ICT [38]. In summary, it has been showed that consistency between labeled and unlabeled data is helpful; however the consistency has not yet been quantified within a Euclidean space metric onto the intermediate embedding layer, as it is in our proposal.

Face Expression Recognition. This theme has been dominated in the later period by deep learning methods too. For instance, several solutions [37,46] trained a single network or an ensemble of networks and adapted the predictions onto a single independent image or onto a video sequence containing a face expression. The problem of delicate labeling has been addressed by Barsoum et al. [2], who noted the presence of noisy labels in the FER database and thus re-annotated the database by crowdsourcing, showing much improved results; yet the solution was database-specific and overfitting could have appeared.

More recently, multiple databases, and thus better generalization, are envisaged in a series of purely supervised methods that augments the baseline performance by the usage of a modified center loss [22]. Others [46], have found that better results can be achieved by specifically selecting some of the layers from the network. Attention mechanisms have also been envisaged for expression recognition with good results [23].

In the later years, the restricted amount of annotated data has been noted and solutions sought to use the power of semi-supervised learning or of the domain transfer to alleviate the limitation. Zhang et al. [45] used a strategy that re-evaluates self labels predictions over randomly selected data instances from the unlabeled data at each iteration. Zeng et al. [42] adopted the self labeling strategy based on bottom-up propagation in a relational graph. Recently, Florea et al. [13] regularized the contribution of unlabeled data with injection of random quantities in the gradient.

Overall, the methods have evaluated the minimal amount of labels required in a database, given a SSL framework, to obtain accuracy values comparable with the supervised case, but more often pushed the supervised performance with database particular choices such as pre–training on specific subsets.

3 Method

From a technical point of view, we propose a methodology to train a deep network in a semi-supervised manner for a classification problem with mutually exclusive categories. In this scenario, we ask the network to include a layer that acts as a discriminative embedding or as a feature descriptor. As discussed in the implementation subsection, we use a WideResNet [41]; in this case, the embedding is the last layer before the decision one, but after flattening. A intuitive view of the system behavior is in Fig. 2.

For classification problems, initially the label is a scalar y_i indicating a categorical value, but later we will switch to one-hot encoding $\mathbf{y}_i$.

3.1 Large Margin Embedding

Given an image $\mathbf{x}_i$, its associate embedding $\mathbf{e}_i$ and its prediction y_i, one may ask the learner to cluster embedding by incorporating a specific loss. Wen et al. [40] introduced the center loss that explicitly reduces the intra-class variations by encouraging embedding samples to move towards their corresponding class centers in the feature space (embeddings) during training. The center loss is [40]:

$$\mathcal{L}_C = \sum_{i=1}^{N} \mathcal{D}(\mathbf{e}_i, \mathbf{c}^c); \quad \mathbf{c}^c = \frac{\sum_{i=1}^{N} \mu_i^c \mathbf{e}_i}{\sum_{i=1}^{N} \mu_i^c}; \quad \mu_i^c = \begin{cases} 1, y_i = c \\ 0, y_i \neq c \end{cases} \quad (1)$$

where $\mathbf{x}_i$ is a data from class c ($y^i = c$), $\mathbf{e}_i$, its embedding, $\mathbf{c}^c$ is the centroid of the class c and μ_i^c is the membership of the data i to class c. In supervised learning, the membership is binary and provided by the labels.

The standard center loss assumes an Euclidean distance: $\mathcal{D}(\mathbf{e}_i, \mathbf{c}^c) = \|\mathbf{e}_i - \mathbf{c}^c\|_2$; also that choice is conditioned by the necessity to compute the position of the centroids as the (weighted) arithmetic mean for the vectors. The centers are updated in each iteration, based on latest batches using Stochastic Gradient Descent (SGD) derived optimization. Later developments of this method [24,44] sought ways to enforce also large distances between class centroids using the cosine derived distances for $\mathcal{D}()$.

A limitation of these methods is that in the absence of an explicit intervention over the other class centroids, there is an optimum where all data is tightly grouped in a large cluster with centroids overlapped and small distances for each point to its cluster. A second problem is data scaling, as the network could learn some biases that will simply downscale data.

To alleviate such potential behaviors, we propose to use the normalized embedding and to modify the loss by favoring small distance to the belonging class centroid and large distances to other centroids. Thus, large margins are imposed between different classes clusters. Formally a large margin loss , $\mathcal{L}_{\mathcal{M}}$ can be written as:

$$\mathcal{L}_{\mathcal{M}} = \sum_{i=1}^{N} \left(\mathcal{D}\left(\frac{\mathbf{e}_i}{\|\mathbf{e}_i\|_2}, \frac{\mathbf{c}^c}{\|\mathbf{c}^c\|_2} \right) - \frac{1}{C-1} \sum_{j=1, j\neq c}^{C} \mathcal{D}\left(\frac{\mathbf{e}_i}{\|\mathbf{e}_i\|_2}, \frac{\mathbf{c}^j}{\|\mathbf{c}^j\|_2} \right) \right) \quad (2)$$

If the normalized embedding is $\hat{\mathbf{e}}_i = \frac{\mathbf{e}}{\|\mathbf{e}\|_2}$ the loss can be rewritten as:

$$\mathcal{L}_{\mathcal{M}} = \sum_{i=1}^{N} \left(\mathcal{D}\left(\hat{\mathbf{e}}_i, \hat{\mathbf{c}}^c \right) - \frac{1}{C-1} \sum_{j=1, j\neq c}^{C} \mathcal{D}\left(\hat{\mathbf{e}}_i, \hat{\mathbf{c}}^j \right) \right) \quad (3)$$

where C is the number of classes. Normalization in Eq. (3) limits the space, while the subtraction imposes that one instance should be near to its class center and far from the other centers. Again, the centers can be determined after every batch, using Eq. (1), conditioned by an Euclidean choice for the distance.

Such behavior is illustrated in Fig. 3. The normalization of the data ensures that the loss $\mathcal{L}_{\mathcal{M}}$ is bounded and it prevents numerical instability.

3.2   Distance Generalization

The margin loss and the embedding system is inspired by the classical K-means algorithm. While the solution presented in the results section concentrates solely on the Euclidean distance, thus retrieving the classical K-means algorithm, one may extend the algorithm based on non-Euclidean distances [10] and other margin based losses [9,24,44] can be used.

The generalization assumes the following: given N vectors, $\mathbf{b}_i$ are a set of standard basis vectors of the space, a set of membership values μ_i and let us denote by $\mathbf{s}_c = \left(\sum_{i=1}^{N} \mu_i^c \mathbf{b}_i \right) \Big/ \left(\sum_{i=1}^{N} \mu_i^c \right)$. The sought centroids are $\mathbf{c}^c$.

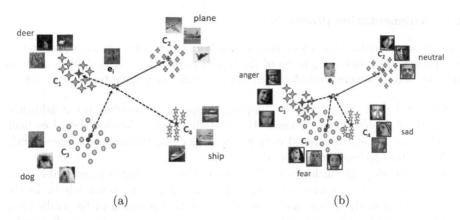

Fig. 3. Large margin behavior represented for a 2-dimensional embeding: new data $\mathbf{e}_i$ is from class 2, thus the distance to its class centroid (marked by continuous line) should be made shorter, while the distances to the other classes (marked with dashed lines) should be made longer. (a) Structuring in CIFAR is more sparse, while in (b) FER is with higher density since classes are easier to be confused

Given a generalized squared distance matrix, $A \in \mathbf{R}^{n \times n}$, with $a_{ij} = \mathcal{D}(\mathbf{e}_i, \mathbf{e}_j)$, then the non-euclidean distance between points and centroids can be developed with respect to a vector $\mathbf{w}$ as:

$$\mathcal{D}(\mathbf{e}_i, \mathbf{c}^c) = \mathbf{e}_i \cdot \mathbf{w} = \sum_{j=1}^{N} w_j e_{ij}; \ \text{s.t.} \ \sum_{j=1}^{N} w_j = 0 \Rightarrow (\mathcal{D}(\mathbf{e}_i, \mathbf{c}^c))^2 = -\frac{1}{2} \mathbf{w}^T A \mathbf{w} \quad (4)$$

In this case the centroids can be computed, given $m_c = \sum_{i=1}^{N} \mu_i^c$, by:

$$\mathbf{c}^c = \frac{1}{m_c} \sum_{j=1}^{N} \mu_j^c \mathbf{e}_j; \ \text{while} \ \mathbf{w} = \mathbf{s}_c - \mathbf{b}_j \quad (5)$$

3.3 Self-labeling

Given an unlabeled data $\mathbf{x}_i^u$, its embedding $\mathbf{e}_i^u$, the pseudo-label in one-hot encoding form $\mathbf{y}_i^u = [y_i^1, y_i^2, \ldots y_i^C]$ is found based on distances to centroids with a method inspired from Fuzzy C-means algorithm:

$$y_i^c = \frac{1}{\sum_{j=1}^{C} \left(\frac{\|\mathbf{e}_i - \mathbf{c}^c\|_2}{\|\mathbf{e}_i - \mathbf{c}^j\|_2} \right)^2} \quad (6)$$

The process is illustrated in Figs. 2(c) and 3, where this time the center position is set and the relative size of distances (arrows) form label probabilities (i.e. class memberships).

3.4   Augmentative Processing

To prevent the network to memorize data we regularize the training weight decay (i.e. penalization of the L_2 norm of the model parameters) [25]. Additionally, in the last period, several techniques to improve efficiency have been proposed:

- *Classical data augmentation*: flipping, cropping, Gaussian noise addition, small rotations for face images. Both labeled and unlabeled data has been augmented. Each unlabeled data $\mathbf{x}_b^u$ in a batch is augmented independently N_{aug} times (Algorithm 1, line 4).
- *Label guessing.* Berthelot et al. [5] showed that training is more stable if an entire set of N_{aug} variants of unlabeled data have the same labels. In the initial version the labels are retrieved by relative position of the embedding with respect to class centroids as defined by Eq. (6). Now, the overall pseudo-label may be found by summing over all N_{aug}:

$$y_i^c = \sum_{k=1}^{N_{aug}} y_k^c = \sum_{k=1}^{N_{aug}} \frac{1}{\sum_{j=1}^{C} \left(\frac{\|\mathbf{e}_i^k - \mathbf{c}^c\|_2}{\|\mathbf{e}_i^k - \mathbf{c}^j\|_2} \right)^2} \tag{7}$$

  where $\mathbf{e}_i^k$ is the embedding of the k-augmentation of the unlabeled data $\mathbf{x}_i^u$.
- *Sharpening* - It has been showed [5] higher non-uniformity of the weights improves the robustness. This is implemented injecting a non linear transform guided by the temperature T hyperparameter, together with normalization from previous step:

$$y_i^c = \frac{y_i^{\frac{1}{T}}}{\psi_j}; \quad \psi_j = \sum_{c=1}^{C} p_c^{\frac{1}{T}} \tag{8}$$

One might notice that the combination of sharpening and large margin based on euclidean distance makes the solution close to the soft max procedure.
- *MixUp* [43] - assumes building synthetic new data instances by considering convex combination with random weight of existing data. It is applied on both labeled examples and margin-self-labeled examples:

$$\begin{aligned} \mathbf{x}' &= \lambda \mathbf{x}_i + (1 - \lambda)\mathbf{x}_j \\ \mathbf{y}' &= \lambda \mathbf{y}_i + (1 - \lambda)\mathbf{y}_j \end{aligned} \tag{9}$$

where λ is a small random quantity extracted from $Beta(\alpha, \alpha)$ distribution, while α is a hyperparameter. If the second contributor originates in unlabeled data $\mathbf{x}^j = \mathbf{x}_j^u$, λ has to be small such that, the new data is closer to labeled example.

Considering convex combinations between data points according to the MixUp paradigm, the input space is thoroughly investigated.

3.5   Total Loss

Overall, the network is trained using the loss computed as a weighted sum:

$$\mathcal{L} = \mathcal{L}_S + \lambda_M \mathcal{L}_M = \mathcal{L}_S + \lambda_M(\mathcal{L}_{M1} + \lambda_u \mathcal{L}_{M2}) \qquad (10)$$

where λ_M and λ_u are weighting hyperparameters, $\mathcal{L}_S$ is the cross entropy decision loss with L2 weight decaying regularization. $\mathcal{L}_{M1}$ is the large margin loss computed on labeled data, while $\mathcal{L}_{M2}$ is computed on unlabeled data.

Algorithm 1: The MarginMix algorithm takes as input a batch of labeled data $\mathcal{X}$ and one without labels $\mathcal{U}$ and produces densely sampled input examples $\mathcal{X}'$ respectively self-labeled densely sample examples $\mathcal{U}'$. Self-labeling is based on clustering in the embedding space. The purpose is to adjust the weights of learner ψ

Data: : Batch of b labeled instances with embeddings and one–hot labels $\mathcal{X} = \{\ldots, (\mathbf{x}_i, \mathbf{e}_i, \mathbf{y}_i), \ldots\}$, $i = 1 \ldots b$, batch of b unlabeled instances $\mathcal{X}^u = \{\ldots, (\mathbf{x}_i^u), \ldots\}$, sharpening temperature T, number of augmentations N_{Aug}, β distribution parameter α for MixUp.

1 **for** $b = 1 : N_{batch}$ **do**
2 　　Compute embeddings for labeled samples $\mathbf{e}_b = \psi(\mathbf{x}_b)$;
3 　　Update centroids using eq. (1);
4 　　$\tilde{\mathbf{x}}_b = \text{Augment}(\mathbf{x}_b)$ # *data augmentation to* $\mathbf{x}_b$;
5 　　**for** $k = 1$ to N_{Aug} **do**
6 　　　　$\tilde{\mathbf{x}}_k^u = \text{Augment}(\mathbf{x}_k^u)$ # *one of the k-th data augmentation to* $\mathbf{x}_k^u$;
7 　　　　Self-label by large Margin using eq. (7)
8 　　**end**
9 　　Compute average, sharpen predictions across all $\tilde{\mathbf{x}}_k^u$ using eq. (7)
10 **end**
11 Collect augmented labeled data: $\tilde{\mathcal{X}} = (\mathbf{x}_b, \mathbf{y}_b); b \in \{1, \ldots, N_{batch}\}$;
12 Collect augmented unlabeled data with their self predicted labels: $\tilde{\mathcal{X}}^u = (\mathbf{x}_b^u, \mathbf{y}_b^u); b \in \{1, \ldots, N_{batch}\}$;
13 Concatenate $\tilde{\mathcal{W}} = (\tilde{\mathcal{X}}, \tilde{\mathcal{X}}^u)$;
14 Use MixUp - eq. (9) for pairs of labeled and new data $\mathcal{X}' = MixUp(\tilde{\mathcal{X}}, \tilde{\mathcal{W}})$ and pairs of unlabeled and new data $\mathcal{X}_u' = MixUp(\tilde{\mathcal{X}}^u, \tilde{\mathcal{W}})$;
15 Compute total loss with eq. (10) using $\mathcal{X}'$ $\mathcal{X}_u'$;
16 Update network weights;

In the backward propagation, the derivative of the margin loss with respect to the current d-th element of the D-dimensional embedding can be written as:

$$\frac{\partial \mathcal{L}_M}{\partial e_d} = \left(2(\hat{\mathbf{e}}_i - \hat{\mathbf{c}}^c) - \frac{2}{C-1} \sum_{j=1, j \neq c}^{C} \left(\hat{\mathbf{e}}_i - \hat{\mathbf{c}}^j \right) \right) \cdot \frac{\partial \hat{\mathbf{e}}_i}{\partial e_d}; \quad \frac{\partial \hat{\mathbf{e}}_i}{\partial e_d} = \frac{1 - \hat{\mathbf{e}}_i^2}{\|\mathbf{e}_i\|_2} \qquad (11)$$

3.6 Margin-Mix Algorithm

The purpose of the algorithm is to train a DCN using both labeled and unlabeled data. The proposed method is described by Algorithm 1. Intuitively, in a first step, a batch of labeled data passes to collect embedding and update centroid position. Then both labeled and unlabeled data is augmented using the MixUp procedure. Unlabeled is self-annotated by Large margin procedure and the network is asked to provide embeddings that are more discriminative.

3.7 Implementation

The implementation is developed from the tests and procedure described in [28] and [5] respectively.[1] The method has been implemented in Pytorch [29].

For fair comparisons with other SSL methods, we restrict our experiments to the "Wide ResNet-28-2" [41] as architecture. For training, we used SGD solver with a learning rate of 0.001. The margin loss (and subsequent parameter - centroids) has a learning rate of 0.5. We use a cosine scheduler for a learning rate decay from 0.1 to 0.0001. We also fix the weight decay rate to 5e−4. For all experiments, we use a batch size of 64 images. The number of training epochs is dependent on the distribution of the database: for database where the classes contribute uniformly, we used 1024 batches while, overall the model is trained for 1024 epochs.

4   SSL Performance. Comparison with State of the Art

First we evaluate the proposed algorithm on four standard benchmarks. To asses the proposed method, we perform semi-supervised tasks on four datasets: CIFAR-10 and CIFAR-100 [18], SVHN [26], and STL-10 [7]. The first three are fully annotated, but it is common for the SSL testing to consider as labeled only a subset of the training set and the remainder unlabeled. We emphasize that these databases have the classes perfectly balanced. The last one, was build specifically for SSL, with 5000 labeled images and 100000 unlabeled images. On a fast visual inspection, the unlabeled data is also highly *balanced* between the 10 classes. For the large margin, λ_M was set to 1 and λ_u was set to 0.4.

Achieved results and comparison with prior art[3] can be followed in Table 1 and respectively Table 2. One may notice that results are very close to the state of the art performance, sometimes even outmatching it. In general the method has similar performance with MixMatch algorithm with which shares several

---

[1] Code is developed from Pytorch implementation of MixMatch available at https://github.com/YU1ut/MixMatch-pytorch. Additional details may be retrieved from the project webpage[2].

[3] Very recently several SSL methods were made public, although not published yet [4, 34,39] that report improved results. However, they propose augmentation techniques that complement the self-labeling procedure. Beyond very recent publication, they may be used together with the proposed method.

common traits. On direct comparison, for a first view, the MixMatch is lacking weights for margin loss, has fewer parameters, thus may be simple to be tuned; yet the influence of the two parameters was found to be less dramatic and variations around mentioned values (i.e. $\pm 20\%$) produced similar errors (i.e. ± 0.3).

Table 1. Comparative errors (smaller is better) on CIFAR datasets obtained with WideResNet-28-2 . Top row lists the number of examples with labels (over all classes) considered

	CIFAR-10			CIFAR-100
Methods/Labels	250	1000	4000	10000
Supervised [38]	–	–	20.26	–
Π-Model [19]	53.02	31.53	17.41	39.19
PseudoLabel [21]	49.98	30.91	16.21	–
MixUp [43]	47.43	25.72	13.15	–
VAT [25]	36.03	18.68	11.05	–
MeanTeacher [36]	47.32	17.32	10.36	–
ICT [38]	–	–	7.66	–
MixMatch [5]	11.80	7.75	6.24	28.88
MarginMix	**10.76**	8.33	**6.17**	29.12

Table 2. Comparative error (smaller is better) on SVHN and STL datasets obtained with WideResNet-28-2. Some results are taken from [28]

	SVHN		STL	
Methods/Labels	1000	4000	1000	5000
Supervised [28]	–	12.84	–	–
Π-Model [19]	8.06	5.57	17.41	39.19
VAT [25]	5.63	18.68	11.05	–
MeanTeacher [28,36]	5.65	3.39	10.36	–
ICT [28,38]	3.53	–	7.66	–
MixMatch [5]	3.27	2.89	10.18	5.59
MarginMix	3.35	3.33	**9.85**	5.80

5 Face Expression Recognition with Few Annotations

In this case, the tests are performed on two databases with images in the wild containing various face expressions. The databases are FER+ and RAF-DB.

RAF-DB [22] contains facial color images in the wild, which are, often, larger than 300×300. The database is annotated by at least 40 trained annotators per image and divided into 12271 training images and 3078 testing images. It is labeled for the seven basic emotions.

FER+ is derived from FER2013 [14] and contains 28709 training images, 3589 validation (public test) and another 3589 (private) test images, in the wild. FER+ images have 48×48 pixels, are gray-scale and contain only the face. Barsoum et al. [2] noted the high noise in the original labels and performed some "cleaning", by removing the images with missing faces and providing labels by aggregating the opinion of 10 non-specialist annotators. Compared to RAF-DB, the images are small, gray and have been annotated less rigorously.

For FER experiments, prior SSL algorithms had trouble solving the task and often converged to a state where only the most populated class was predicted or it simply oscillated without converging. MixMatch often encountered such problems significantly reduce the sharpening temperature from 0.5 to 0.25. Performance for the two databases may be followed in Table 3 and respectively Table 4.

We report the baseline obtained when training in purely supervised manner but containing MixUp and temporal averaging. In this case the network has been randomly initialized, as it is in the case of SSL methods. For 4000 labeled images considered, a uniform distribution would have required 500 per class, yet three of them do not have so many, so the distribution is already uneven. For SSL methods reported, Mean teacher [36] and MixMatch [5], we have used the public code, tuned as mentioned. For 320 labeled images (i.e. 40 per class) we could not make the Mean Teacher to report multiple classes, but only the dominant one.

As one can notice in these experiments, the proposed method reaches better accuracy than similar solutions by a large margin. We claim that differences originate from two directions.

Firstly, the distribution of labels among classes is uneven. This fact is illustrated in Fig. 4; there one may see that the most populated class in FER+ database has 5 times more instances than the least populated one. As emphasized in the original MixUp work [43], this technique populates the space near existing examples. Given an uneven distribution, part of the space with sparse classes will become relatively even sparser. Simultaneously, the populated classes will tend to expand (in confidence) in the detriment of sparser ones. Also when parsing unlabeled data, MixMatch will label it more often with the dominant classes value. In our case, the centroid exists, and the relative distance is accepted.

Secondly, fully supervised performance in the case of FER databases is lower than for CIFAR like sets. This suggests that classes are spread in a more intricate manner, which again will favor the most populated classes. Enforcing an intermediate embedding with a large margin, we force the learner to make space for all classes, thus untangling the mixture from the initial data space. A measure of inter-class variance is offered by evaluation of the large margin as defined by Eq. (3) in the first iterations of the training procedure, normalized by the

Table 3. Comparative accuracy (larger is better) on FER+ dataset obtained with WideResNet-28-2. Top row lists the number of examples with labels (over all classes) considered. 'nc' stands for not converged

Methods/Labels	320	400	2000	4000	10000	All
Supervised WideResNet	nc	37.92	50.29	56.78	63.56	84.88
Supervised [2]	–	–	–	–	–	84.99
MeanTeacher [36]	–	45.56	50.84	58.28	68.36	–
MixMatch [5]	45.60	50.25	58.35	70.91	71.24	–
MarginMix	**50.76**	**56.75**	**60.83**	**75.18**	**81.25**	85.36

Table 4. Comparative accuracy (larger is better) on RAF–DB dataset obtained with WideResNet-28-2. Top row lists the number of examples with labels (over all classes) considered

Methods/Labels	320	400	1000	4000	All
Supervised WideResNet	nc	26.75	35.25	55.66	85.58
Supervised [22]	–	–	–	–	84.13
MeanTeacher [36]	nc	28.23	36.53	60.36	–
MixMatch [5]	35.60	42.25	60.37	65.24	–
MarginMix	**40.55**	**45.75**	**66.47**	**70.68**	85.36

number of data instances. The loss measures the quality of clustering: small loss means well defined clusters while large loss means blended clusters. The value is 4× larger in the case of FER+ database when compared to CIFAR-10, although the later has 10 classes compared to 8.

Comparison with Softmax/Center-Loss. When we have performed tests with a solution trained with softmax/center-loss as defined in [40] we have find out that this version often did not converge as on the validation set it entered into oscillating performance or it ended in predicting always a single class. It converged in 50% cases for CIFAR like benchmarks and 20% for expression experiments when it often predicted the most populous class. Intuitively, the standard center loss asks only that instances are close to the class centroid and lets the cross-entropy distance the clusters. Yet, the cross entropy, which is more an angular distance, allows clusters to be close one to another in terms of Euclidean distance, thus on many unlabeled instances produces near uniform class probabilities. It is similar to consider supra-unitary sharpening (we have illustrated the effect of sharpening only up to 0.5, but the trend is obvious). The margin loss imposes that clusters distance themselves.

5.1   Parameter Ablation

Our method proved to be more robust in the case of Face Expression Recognition which have much lower inter-class variance. Various versions of the method have

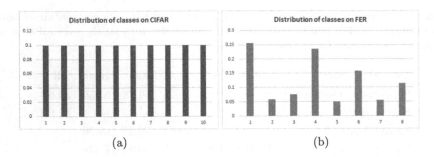

(a)   (b)

Fig. 4. Distribution of classes on databases from the two categories of experiments: (a) CIFAR and (b) FER+.

been tested on the FER+ database when 2000 examples, equally distributed among classes. The performance is presented in Table 5.

Table 5. MarginMix Accuracy on FER dataset when 2000 images have labels that are equally distributed among classes when various versions have been considered

Methods - Parameters	Accuracy
Baseline ($T = 0.25$, $\lambda_M = 1$, $\lambda_u = 0.4$)	60.83
Sharpening $T = 0.5$	55.35
No Sharpening $T = 0$	57.29
$\lambda_M = 0.5$, $\lambda_u = 0.4$	60.44
$\lambda_M = 0.5$, $\lambda_u = 0.4$	60.74
$N_{aug} = 512$ (instead of 1024)	60.68
Without parameter EMA	59.85
With nearest centroid	51.87

The stochastic variance (i.e. variation of accuracy when running the same solution consecutive times) is 0.55. In this case, one may notice that only sharpening may have an impact larger than the stochastic effect. Dramatic decrease is found in the self-labeling if instead of soft probabilities, hard one (based on the nearest centroid) are used; this result is in line with test about sharpening. Otherwise, the solution is robust to slight variations of the parameters.

6   Conclusions

In this paper, we presented MarginMix, a novel framework that combines the capability of deep DCN to produce simultaneously predictions and discriminative embeddings with "the low density separation" principle, while building SSL models. It contains the MixUp paradigm which thoroughly investigates input space by considering convex combinations of the input data. Our proposal structures via embeddings and with the Euclidean distance an intermediate space, in preparation of the final space, where actual prediction takes place.

The experiments have been structured in two categories. The first refers to standard benchmarks such as CIFAR-10, CIFAR-100 and SVHN where a part of the training data is considered as unlabeled and STL-10 which was build specifically for the SSL systems. Here the data is evenly distributed, and the classes are rather easily separable, our method performed on par with previous similar works.

The second category is dedicated to face expression, which we argue that is truly a direction which should benefit from SSL learning since annotation is hard, noisy and costly. In this case, examples from different classes are more similar, and differences are more in details of the image. In this scenario, our proposal outperforms the state-of-the-art methods on all the datasets tested by a significant margin, while also improving the fully-supervised baseline.

References

1. Athiwaratkun, B., Finzi, M., Izmailov, P., Wilson, A.G.: There are many consistent explanations of unlabeled data: why you should average. In: ICLR (2019)
2. Barsoum, E., Zhang, C., Ferrer, C., Zhang, Z.: Training deep networks for facial expression recognition with crowd-sourced label distribution. In: ICMI, pp. 279–283 (2016)
3. Bartlett, M., Hager, J., Ekman, P., Sejnowski, T.: Measuring facial expressions by computer image analysis. Psychophysiology **36**(2), 253–263 (1999)
4. Berthelot, D., et al.: Remixmatch: semi-supervised learning with distribution alignment and augmentation anchoring. arXiv preprint arXiv:1911.09785 (2019)
5. Berthelot, D., Carlini, N., Goodfellow, I., Papernot, N., Oliver, A., Raffel, C.A.: Mixmatch: a holistic approach to semi-supervised learning. In: NIPS, pp. 5050–5060 (2019)
6. Chapelle, O., Schölkopf, B., Zien, A.: Semi-Supervised Learning. MIT Press (2006)
7. Coates, A., Ng, A., Lee, H.: An analysis of single-layer networks in unsupervised feature learning. In: AISTATS, pp. 215–223 (2011)
8. Corneanu, C., Simón, M., Cohn, J., Escalera, S.: Survey on RGB, 3D, thermal, and multimodal approaches for facial expression recognition: history, trends, and affect-related applications. IEEE Trans. PAMI **38**(8), 1548–1568 (2016)
9. Deng, J., Guo, J., Xue, N., Zafeiriou, S.: Arcface: additive angular margin loss for deep face recognition. In: CVPR (2019)
10. Dhillon, I.S., Guan, Y., Kulis, B.: Kernel k-means: spectral clustering and normalized cuts. In: ACM-SIGKDD, pp. 551–556 (2004)
11. Donahue, J., et al.: DeCAF: a deep convolutional activation feature for generic visual recognition. In: International Conference on Machine Learning (2014)
12. Ekman, P., Rosenberg, E.: What the Face Reveals: Basic and Applied Studies of Spontaneous Expression Using the FACS. Oxford Scholarship (2005)
13. Florea, C., Florea, L., Vertan, C., Badea, M., Racoviteanu, A.: Annealed label transfer for face expression recognition. In: BMVC, p. 12 (2019)

14. Goodfellow, I.J., et al.: Challenges in representation learning: a report on three machine learning contests. In: Lee, M., Hirose, A., Hou, Z.-G., Kil, R.M. (eds.) ICONIP 2013. LNCS, vol. 8228, pp. 117–124. Springer, Heidelberg (2013). https://doi.org/10.1007/978-3-642-42051-1_16
15. Haeusser, P., Mordvintsev, A., Cremers, D.: Learning by association-a versatile semi-supervised training method for neural networks. In: CVPR, pp. 89–98 (2017)
16. Ho-Phuoc, T.: CIFAR10 to compare visual recognition performance between deep neural networks and humans. CoRR abs/1811.07270 (2018)
17. Kingma, D.P., Mohamed, S., Rezende, D.J., Welling, M.: Semi-supervised learning with deep generative models. In: NIPS, pp. 3581–3589 (2014)
18. Krizhevsky, A.: Learning multiple layers of features from tiny images. Technical report, MIT (2009)
19. Laine, S., Aila, T.: Temporal ensembling for semi-supervised learning. In: ICLR (2016)
20. LeCun, Y., Bengio, Y., Hinton, G.: Deep learning. Nature **521**(7553), 436–444 (2015)
21. Lee, D.H.: Pseudo-label: the simple and efficient semi-supervised learning method for deep neural networks. In: ICML Workshops (2013)
22. Li, S., Deng, W.: Reliable crowdsourcing and deep locality-preserving learning for unconstrained facial expression recognition. IEEE Trans. Image Process. **28**(1), 356–370 (2019)
23. Li, Y., Zeng, J., Shan, S., Chen, X.: Occlusion aware facial expression recognition using CNN with attention mechanism. IEEE Trans. Image Process. **28**(5), 2439–2450 (2018)
24. Liu, W., Wen, Y., Yu, Z., Li, M., Raj, B., Song, L.: Sphereface: deep hypersphere embedding for face recognition. In: CVPR, pp. 212–220 (2017)
25. Miyato, T., Maeda, S.I., Koyama, M., Ishii, S.: Virtual adversarial training: a regularization method for supervised and semi-supervised learning. IEEE Trans. PAMI **41**(8), 1979–1993 (2018)
26. Netzer, Y., Wang, T., Coates, A., Bissacco, A., Wu, B., Ng, A.Y.: Learning multiple layers of features from tiny images. Technical report, Stanford (2009)
27. Odena, A.: Semi-supervised learning with generative adversarial networks. In: ICML Workshop on Data-Efficient Machine Learning (2016)
28. Oliver, A., Odena, A., Raffel, C., Cubuk, E.D., Goodfellow, I.J.: Realistic evaluation of deep semi-supervised learning algorithms. In: ICLR (2018)
29. Paszke, A., et al.: Pytorch: an imperative style, high-performance deep learning library. In: NIPS, pp. 8024–8035 (2019)
30. Peterson, J., Battleday, R., Griffiths, T., Russakovsky, O.: Human uncertainty makes classification more robust. In: ICCV, pp. 9617–9627 (2019)
31. Rasmus, A., Berglund, M., Honkala, M., Valpola, H., Raiko, T.: Semi-supervised learning with ladder networks. In: NIPS, pp. 3546–3554 (2015)
32. Ren, Y., Hu, K., Dai, X., Pan, L., Hoi, S.C., Xu, Z.: Semi-supervised deep embedded clustering. Neurocomputing **325**, 121–130 (2019)
33. Sariyanidi, E., Gunes, H., Cavallaro, A.: Automatic analysis of facial affect: a survey of registration, representation, and recognition. IEEE Trans. PAMI **37**(6), 1113–1133 (2015)
34. Sohn, K., et al.: Fixmatch: simplifying semi-supervised learning with consistency and confidence. arXiv preprint arXiv:2001.07685 (2020)
35. Susskind, J., Littlewort, G., Bartlett, M., Movellan, J., Anderson, A.: Human and computer recognition of facial expressions of emotion. Neuropsychologia **45**(1), 152–162 (2007)

36. Tarvainen, A., Valpola, H.: Mean teachers are better role models: weight-averaged consistency targets improve semi-supervised deep learning results. In: NIPS, pp. 1195–1204 (2017)
37. Tran, E., Mayhew, M.B., Kim, H., Karande, P., Kaplan, A.D.: Facial expression recognition using a large out-of-context dataset. In: Proceedings of IEEE Conference on Winter Applications on Computer Vision, pp. 52–59 (2018)
38. Verma, V., Lamb, A., Kannala, J., Bengio, Y., Lopez-Paz, D.: Interpolation consistency training for semi-supervised learning. In: IJCAI (2019)
39. Wang, X., Kihara, D., Luo, J., Qi, G.J.: Enaet: Self-trained ensemble autoencoding transformations for semi-supervised learning. arXiv preprint arXiv:1911.09265 (2019)
40. Wen, Y., Zhang, K., Li, Z., Qiao, Yu.: A discriminative feature learning approach for deep face recognition. In: Leibe, B., Matas, J., Sebe, N., Welling, M. (eds.) ECCV 2016. LNCS, vol. 9911, pp. 499–515. Springer, Cham (2016). https://doi.org/10.1007/978-3-319-46478-7_31
41. Zagoruyko, S., Komodakis, N.: Wide residual networks. In: BMVC (2016)
42. Zeng, J., Shan, S., Chen, X.: Facial expression recognition with inconsistently annotated datasets. In: Ferrari, V., Hebert, M., Sminchisescu, C., Weiss, Y. (eds.) ECCV 2018. LNCS, vol. 11217, pp. 227–243. Springer, Cham (2018). https://doi.org/10.1007/978-3-030-01261-8_14
43. Zhang, H., Cisse, M., Dauphin, Y.N., Lopez-Paz, D.: Mixup: beyond empirical risk minimization. In: ICLR (2018)
44. Zhang, X., Fang, Z., Wen, Y., Li, Z., Qiao, Y.: Range loss for deep face recognition with long-tailed training data. In: CVPR, pp. 5409–5418 (2017)
45. Zhang, Z., Han, J., Deng, J., Xu, X., Ringeval, F., Schuller, B.: Leveraging unlabeled data for emotion recognition with enhanced collaborative semi-supervised learning. IEEE Access 6, 22196–22209 (2018)
46. Zhao, S., Cai, H., Liu, H., Zhang, J., Chen, S.: Feature selection mechanism in CNNs for facial expression recognition. In: BMVC, p. 12 (2018)
47. Zheng, Y., Pal, D.K., Savvides, M.: Ring loss: convex feature normalization for face recognition. In: CVPR, pp. 5089–5097 (2018)

Principal Feature Visualisation
in Convolutional Neural Networks

Marianne Bakken[1,2]([✉]) [ID], Johannes Kvam[1], Alexey A. Stepanov[1],
and Asbjørn Berge[1]

[1] SINTEF Digital, Forskningsveien 1, 0373 Oslo, Norway
marianne.bakken@sintef.no
[2] Norwegian University of Life Sciences (NMBU), 1432 Ås, Norway
https://www.sintef.no/en/

Abstract. We introduce a new visualisation technique for CNNs called
Principal Feature Visualisation (PFV). It uses a single forward pass of the
original network to map principal features from the final convolutional
layer to the original image space as RGB channels. By working on a batch
of images we can extract contrasting features, not just the most dominant
ones with respect to the classification. This allows us to differentiate
between several features in one image in an unsupervised manner. This
enables us to assess the feasibility of transfer learning and to debug a
pre-trained classifier by localising misleading or missing features.

Keywords: Visual explanations · Deep neural networks ·
Interpretability · Principal component analysis · Explainable AI

1 Introduction

Deep convolutional neural networks (CNNs) have had a significant impact on
performance of computer vision systems. Initially they were used for image classification, but recently these methods have been used for pixel-level image segmentation as well. Segmentation methods are able to capture more information,
but require significantly more expensive labelling of training data. Moreover,
classification (bottleneck) networks are still used for many applications where
the problem can't be formulated as a segmentation task or pixel-wise labelling
is too expensive.

One of the main issues with bottleneck networks is that they provide no visual
output, that is, it is not possible to know what part of the image contributed
to the decision. As a consequence, there is a demand for methods that can help

Funded by The Norwegian Research Council, grant no. 259869.

Electronic supplementary material The online version of this chapter (https://
doi.org/10.1007/978-3-030-58592-1_2) contains supplementary material, which is available to authorized users.

© Springer Nature Switzerland AG 2020
A. Vedaldi et al. (Eds.): ECCV 2020, LNCS 12368, pp. 18–31, 2020.
https://doi.org/10.1007/978-3-030-58592-1_2

visualise or explain the decision-making process of such networks and make it understandable for humans.

A range of visualisation and explanation methods have been proposed. Class Activation Mapping, e.g. [10], is a computationally efficient way to show the support of a class in the input image, but the resulting heatmap is quite coarse. Gradient-based methods like [3] give a more localised response, but require back-propagation through the whole network, and is very sensitive to edges and noise in the input image.

All these methods operate in a *supervised* manner on one category or feature at a time. In contrast, our method is *unsupervised* and visualise several categories or features in one pass. It can be applied directly to any bottleneck network without any additional instrumentation.

Our approach provides a visualisation that maps the principal contrasting features of a batch of images to the original image space in a single forward pass of the network. We target bottleneck networks, such as image classifiers, and use a singular value decomposition on the feature map of the layer we wish to visualise, e.g., the final convolutional layer, to extract the principal contrasting features for a batch of images. These features are then interpolated back to the original image space, and the activation maps of the earlier layers are used to weight the resulting feature visualisation. An overview of the method is shown in Fig. 1.

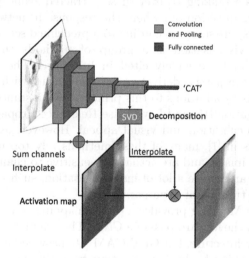

Fig. 1. Overview of our Principal Feature Visualisation (PFV) method.

The main advantages of our method are:

1. Contrast: Per-pixel visualisation of the principal contrasting features.
2. Lightweight: Requires a single forward pass of the original unmodified network, using only intermediate feature maps.
3. Easy to interpret: suppresses non-relevant features.
4. Unsupervised: No additional input or prior knowledge about image classes is required.

We show how the advantages of the method allow it to be used as a tool for debugging misclassification and assessing the feasibility of transfer learning in Sect. 5.

Our code is publicly available at https://github.com/SINTEF/PFV.

2   Related Work

Several categories of methods to interpret CNNs have been proposed. We focus on the methods that provide a visual human-understandable representation, in particular methods that relate the attention of the network back to the original image space in the form of masks or heatmaps.

One way of attributing classifier decision to location in the input is to perform simple perturbations (e.g. occlusion) to the input [13,16] and make a heatmap per class based on change in the output. Similarly, more advanced methods for perturbation of the input image has been proposed [2,9]. The drawback of these methods is that the number of required forward passes is proportional to the number of classes and resulting heatmap resolution.

Other methods focus on localisation of semantically meaningful concepts in the input. For instance by extracting and clustering superpixels, and then compute the saliency as a ranking [7] over these extracted "concepts" [6]. Network dissection is another direction [4], where the response in network hidden units (convolutional layers) are scored according to a predefined set of visual concepts.

Gradient-based visualisation is a group of methods that provide more localised responses and are widely cited in literature. The simplest form of this is to compute the partial derivatives of the output with respect to every input pixel [13]. Several additions to this principle, for instance DeepLIFT [12], Guided Backpropagation [15] and Layer-wise Relevance Propagation (LRP) [3], has improved the localisation and visual appeal. However, as showed through simple sanity checks in [1], many of these methods rely too much on information from the input image, and are actually insensitive to changes in the model. Additionally, they can require a lot of instrumentation, such as special types of layers and separate training of hyperparameters.

Class Activation Mapping provides a direct mapping from the class score to the activations from the forward pass of a CNN. The original work in [5] required a special network architecture, but Grad-CAM [10] provided a more general way to compute the mapping by backpropagation from the class score to the last convolutional layer (not all the way back to the inputs as pure gradient-based methods). Grad-CAM passes the sanity checks in [1], but gives a less localised response than gradient-based methods, and still requires backpropagation from each class to produce responses from multiple classes or objects. Our approach use the activations from the forward pass in a similar manner as Grad-CAM, but rather than computing a mapping through backpropagation, we do a simple unsupervised learning during the forward pass.

Some methods include counter-evidence to give a richer explanation. Grad-CAM and LRP for instance, suggest using negative gradients in addition to the positive ones to show evidence against a class. In [17], a top-down attention propagation strategy is proposed, that performs backpropagation of both positive and negative activations to create a contrasting visualisation. Our method provides an inherent contrast, and does not need to treat this specifically.

There are also several methods that apply clustering or spectral techniques for model explanation. One such method [8] applies spectral clustering on a set of relevance maps computed with LRP, and performs eigengap analysis and t-SNE visualisation to identify typical prediction strategies. This requires several steps of processing, and is applied on one class at a time. Another work [11] uses Eigenspectrum analysis of the feature maps in neural networks to optimise neural architectures and understand the dynamics of network training. Our approach uses spectral information in a similar manner to these approaches, but to our knowledge is the first one to project this type of information back to image space in one pass.

Compared to existing explanation methods, we aim for an approach that is simple to execute, that depends on activations from the network itself rather than edges in the input image, and can highlight the contrast between several features and classes in one pass.

3   Principal Feature Visualisation

3.1   Method Description

Our goal is to obtain a low-dimensional representation of the feature space of feed-forward bottleneck networks which can be mapped to the original image space. Such a visualisation should be achieved in an efficient manner by using a single forward pass of the network, without any additional instrumentation.

Principal component analysis (PCA) projects a signal onto a set of linearly uncorrelated variables (principal components) ranked by the amount of variance explained in the original signal. Conveniently, the projection of features onto these components introduces an implicit measure of contrast, due to the orthogonality of the components.

In brief, our method decomposes a feature map into its principal contrasting features for a batch of images. This is accomplished by extracting principal components through singular value decomposition. The decomposed feature map is then interpolated back to the original image space, where we use the activation maps in the preceding layers as spatial weighting. An overview of the method is shown in Fig. 1, and we describe it in detail below.

Consider a CNN with N convolution and pooling layers. For each layer l a feature map F_l is an $n_B \times n_{c,l} \times n_{x,l} \times n_{y,l}$ matrix, where n_B is the number of images passed through the layer (batch size), $n_{c,l}$ number of channels and $n_{x,l}, n_{y,l}$ is the spatial size of that layer. We denote by $(n_{x,0}, n_{y,0})$ the size of original input images.

Suppose we want to visualise the last convolutional layer N. Our method proceeds as follows. First, for each intermediate F_l we calculate activation maps for each image in batch

$$A_l^b(i,j) = \sum_{c=1}^{n_{c,l}} F_l(b,c,i,j), \quad b \in \{1,\ldots,n_B\} \tag{1}$$

We then compute the total activation map A^b for each batch image as a sum of upsampled activation maps for each layer. That is

$$A^b = \sum_{l=1}^{N-1} P(A_l^b; n_{x,0}, n_{y,0}), \tag{2}$$

where $P(A_l^b; n_{x,0}, n_{y,0})$ denotes upsampling of A_l^b back to original input image size.

Now consider the feature map F_N of the final layer. Our approach is to use PCA to decompose the features for visualisation. First, we reshape F_N to a $n_{c,N} \times (n_B \cdot n_{x,N} \cdot n_{y,N})$ matrix. In this way we treat each per-pixel channel response as a separate observation. We denote this reshaped matrix as F' and centre it by subtracting mean values:

$$F' = F' - \bar{F'} \tag{3}$$

Then we find the principal feature responses by decomposing F' using singular value decomposition as

$$F' = USV^T, \tag{4}$$

where S is a diagonal matrix containing the singular values and U is the decomposition of F' into the space described by the eigenvectors V.

The principal components are then the sorted columns of the following matrix

$$F_{\text{PCA}} = US = [\mathbf{d}_1 \quad \ldots \quad \mathbf{d}_r] \tag{5}$$

For visualisation convenience, we choose a subset of F_{PCA} columns $\{\mathbf{d}_1, \ldots, \mathbf{d}_{n_d}\}$. For the rest of the paper we assume $n_d = 3$, which allows us to visualise F_N by mapping $\mathbf{d}_1, \mathbf{d}_2, \mathbf{d}_3$ to red, green and blue channels. We denote by D_N a matrix consisting of these columns

$$D_N = [\mathbf{d}_1 \quad \mathbf{d}_2 \quad \mathbf{d}_3] \tag{6}$$

By reshaping D_N back to $n_B \times 3 \times n_{x,N} \times n_{y,N}$ size and treating each batch image as a separate D_N^b we can upsample D_N^b back to the original size $(n_{x,0}, n_{y,0})$. We use the activation map A^b to weight the upsampled D_N^b and normalise the result as follows

$$V^b = \text{normalise}\left(A^b \circ P(D_N^b; n_{x,0}, n_{y,0})\right), \tag{7}$$

where $\circ$ is an element-wise product and P is upsampling operator. Note that the colours in the final images V^b are relative to the processed batch.

input image   activation map (A^b)   unweighted V^b V^b

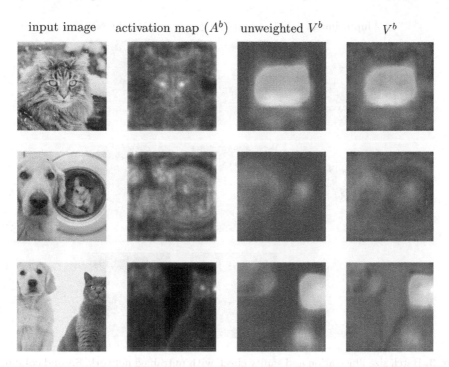

Fig. 2. Variations of our Principal Feature Visualisation method applied on a pre-trained bottleneck CNN (VGG16) and a batch of dog and cat images. The activation map A^b is used to weight the feature map. Colours represent the strongest principal features of the batch and their location in image space. Best viewed in colour. (Color figure online)

3.2 VGG Example

We illustrate the properties of our method with a simple example of a few dog and cat images and a VGG16 network [14] pre-trained on ImageNet.

First, we show the final visualisation V^b together with two intermediate steps: the activation maps A^b, and *unweighted* V^b from upsampling directly without weighting. V^b was computed with a forward pass on a batch of six images of dogs and cats. The intermediate activation maps A_l^b were extracted before each max pool layer, and the feature map of the final layer, F_N, was extracted before the last max pool layer. We used bilinear interpolation for upsampling. The results are shown in Fig. 2. For this batch, the principal feature maps assign different colour channels to dogs, cats and background. Studying the intermediate steps, we see that the principal feature map without weighting shows more response from the channel in the background. The weighting with earlier activation maps thus enhances the *strongest* features, while the principal components provides *contrast* between different features.

Second, we illustrate how the visualisation depends on the composition of the input batch. Figure 3 shows our method applied on different single-image

input image   batch size = 1   untrained CNN

Fig. 3. Batch size illustration and sanity check with untrained network. Second column shows the visualisation of single-image input batches. The colours now represent different high-level features like ears and nose rather than the class-level features in Fig. 2. Third column shows a simple sanity check: the visualisation of an untrained network of randomly initialised weights. The result is completely different, as expected. Best viewed in colour. (Color figure online)

input batches. The colours now represent different features within that image only. For the image with two objects, there is still some class-related contrast. This brief example indicates that batch composition can be used deliberately as a tool to control the contrast in the visualisation and tailor it to any application. More examples of this are shown in Sect. 5.2 and supplementary material.

In order to be useful for model debugging, a visualisation method should be sensitive to the model parameters. We perform a simple parameter randomisation test as suggested in [1], by running our method on a randomly initialised untrained version of the network. As seen in Fig. 3, the resulting visualisation of the random model is visually very different from the pre-trained one. This indicates model sensitivity in our visualisation, which can be used for debugging the training process.

4   Comparison with Other Methods

We compare our method (PFV) with Grad-CAM [10] and Contrastive Excitation Backprop (c-EBP) [17] on VGG16 pre-trained on ImageNet. We use a batch of images that is not included in ImageNet, but contains objects of ImageNet categories. A few examples are shown in Fig. 4, where we have used the top-3 predicted classes as targets for Grad-CAM and c-EBP.

Grad-CAM and c-EBP are supervised methods based on backpropagation, that generate a heatmap conditioned on the predicted class. Consequently, these methods highlight evidence for a particular class, and suppress sources that do not contribute to the decision. Contrastive EBP approximates the probability that a given image pixel contribute positively or negatively to the decision. When the target classes are unknown and we simply specify them as the top-k predictions, these methods require a potentially large number of backward passes to describe the feature diversity in the image.

In contrast, our PFV is an unsupervised method calculated based on a single forward pass, that highlights the principal contrasting features in a batch of images. As our method is based on principal components which form an orthogonal basis where one component cannot explain another, it focuses on feature variance instead of evidence for a decision. The colours of PFV represent different features, with no direct connection to the final classification. However, by performing PFV on a batch of images, e.g. the three images in Fig. 4, colours are consistent across the batch and show which objects that have similar features.

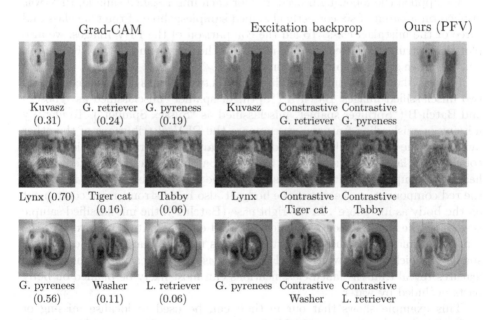

Fig. 4. Comparison of GradCAM, Constrastive Excitation Backprop (c-EBP) and PFV on VGG16 pre-trained on ImageNet. Grad-CAM and c-EBP results are shown for the top-3 predicted classes. PFV results is for a batch of the three images shown. Colours represent heatmaps for Grad-CAM and c-EBP, and principal features for PFV. Best viewed in colour. (Color figure online)

5   Applications

In this section we apply our method to two use-cases: debugging misclassified examples by localising misleading and missing features in the input image; and ad hoc prediction of the success of transfer learning with a pre-trained network.

5.1   Debugging Classification Errors

When a network fails to classify an image correctly, it can be hard to know what part of the image is to blame. We show how our method can be used to identify misleading or missing features and their location in the image by comparing principal feature maps of incorrectly and correctly classified samples.

To do this, we apply PFV on an example task: dog breed classification. There are 120 dog breeds among the 1000 categories of the ImageNet dataset, and the features of the pre-trained VGG16 network should therefore be well suited for this task. We ran prediction on a handful of images of the class "English Springer Spaniel" not present in the original dataset, and identified the failed samples. It turns out that all the failed samples show dogs in water, and we want to examine why they fail. Is it because of the water, occlusion of body parts, or something else?

We applied the following procedure: For each misclassified sample, PFV was applied on a batch of six correctly classified samples; three of the true class and three of the mistaken class. To aid the comparison of the PFV images, we also plot the distribution of red, green and blue in the foreground of the PFV image, i.e., the three strongest principal components.

Figure 5 shows the result of running PFV on two batches of images containing two misclassified images: Batch A ("Springer spaniel" misclassified as "goose") and Batch B ("Springer spaniel" misclassified as "Sussex Spaniel"). To identify missing or misleading features, we compare the PFV distributions of the other images in the batch with the failed sample, and look for the location of the colours with large deviation. In the left case (Batch A), the misclassifed sample has a red component on the head as in the true class "springer", but is missing the red component on the rest of the body. It also has a strong green component on the body as in "goose". In the right case (Batch B), the misclassified sample is missing the strong green component located on the white fur in front in the "springer" image, and the PFV distribution is more similar to that of "sussex spaniel", which has no white fur. For both cases, the location of the missing features reveal that the failed classifications can most likely be blamed on body parts occluded by water.

This example shows that our method can be used to localise missing or misleading features, because it highlights the *contrasting* features within a batch, not just the most dominant features from the classification.

5.2   Transfer Learning

Transfer learning is often applied when there is limited training data available to train a deep neural network from scratch. In this section we show that it is

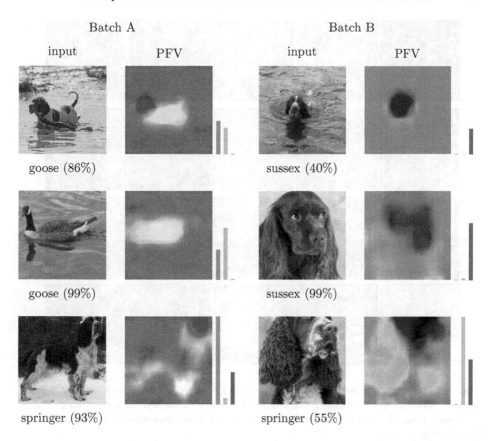

Batch A   Batch B

input   PFV   input   PFV

goose (86%)   sussex (40%)

goose (99%)   sussex (99%)

springer (93%)   springer (55%)

Fig. 5. Principal Feature Visualisation (PFV) on misclassified samples compared to correctly classified samples. In the first row, the two input images are of the category "English Springer Spaniel", but has been classified as "goose" and "Sussex Spaniel". In the second and third row, the input images are examples from the two different PFV batches. Bars show the distribution of red, green and blue foreground pixels of the PFV image. The colour encoding is not consistent because the method is applied on two different batches, and hence the principal vectors are different. Best viewed in colour. (Color figure online)

possible to predict the success or failure of transfer learning on a new dataset by visualising the principal features of the pre-trained network on images from this dataset.

We analyse the features of VGG16, pre-trained on ImageNet, applied to the Pascal VOC2012 dataset.

Initially, we randomly sample one image from each of Pascal VOC2012's 20 classes and form a batch of these images. We then apply PFV and visualise the principal contrasting features of this batch, shown in Fig. 6. For simplicity, the feature visualisations are shown as an overlay to a grey scale version of the input image. As the images are quite dissimilar, decomposing the features of the

Fig. 6. Initial principal feature visualisation of VGG16 features on the Pascal VOC2012 dataset. The dataset contains 20 classes, features are visualised for a random example from each class. Similar colours indicate similar features. Best viewed in colour. (Color figure online)

images into three principal features, only gives us a coarse indication of which examples contain similar feature sets. Based on this visualisation we observe that the animal classes appear to have similar features, while vehicles and bicycles appear to have a different set of features. Interestingly, we also see observe that there are only weak feature responses for chair, sofa and potted-plant, while for the class dining-table, the main responses are from the objects on the actual table.

To further investigate the difference between the features in the animal categories, that have similar colours in Fig. 6, we randomly sample new batch of images from these categories. This time, we sample 4 random images from each of the categories: "dog", "cow", "cat", "horse" and "sheep". We then again apply PFV to find the principal contrasting features for this batch of images, shown in Fig. 7. Note again, that the colours in the images are relative to each batch. As the class variation in this batch of images is lower than in the initial experiment, we observe that we obtain a finer decomposition. Here we see that cats and dogs become more clearly separated from the other classes. The other three classes; cows, horses and sheep, does appear to contain similar features. In addition, one example from the "dog" class and one from the "cat" class appear as outliers, which might be due to the images being difficult examples or that ImageNet contains multiple cat and dog breeds.

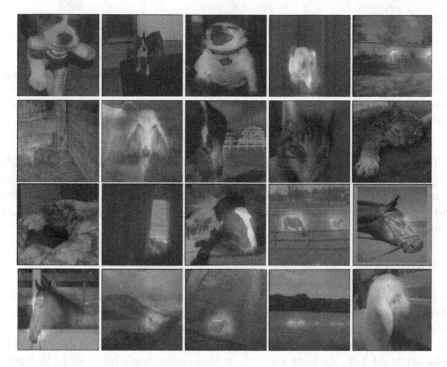

Fig. 7. Principal feature visualisation of VGG16 features on the Pascal VOC2012 dataset for the classes; "dog", "cat", "cow", "horse" and "sheep", with a batch of four random examples sampled from each class. Similar colours indicate similar features. Best viewed in colour. (Color figure online)

Based on this analysis we hypothesise that in a fine-tuned model using VGG16 ImageNet features, we expect little confusion between the cat and dog class, a more pronounced confusion between the "horse", "cow" and "sheep" classes. In addition, the weak feature responses for classes "chair", "diningtable" and "sofa", indicate an overall poorer performance in the detection of these classes.

To check this hypothesis we fine-tune VGG16 pre-trained on ImageNet on the Pascal VOC2012 dataset. We retrain only the final fully connected layer (the classifier), the rest of the network (i.e., all convolutional layers) is kept fixed during training. For simplicity we only select images containing one class per image, to be able to use a standard cross-entropy loss in the optimisation. We train until the validation loss stops decreasing and investigate the final performance in terms of a confusion matrix. The confusion matrix is shown in Fig. 8.

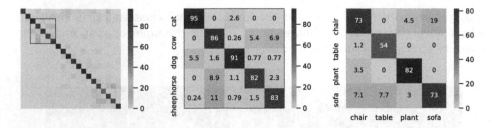

Fig. 8. Confusion matrix for the validation set for a VGG16 network, after fine-tuning on Pascal VOC2012. Left, overall view of confusion matrix. Middle, confusion between classes "cat", "cow", "dog", "horse" and "sheep. Right, confusion between classes "chair", "table", "plant" and "sofa". Best viewed in colour. (Color figure online)

The worst performing categories are of the classes "dining table", "sofa", and "chair". We also observe that "cow" is significantly confused with classes "horse" and "sheep". These observations suggest that such a feature visualisation strategy can give an intuition about when pre-training will be beneficial and when it might fail.

6   Conclusion

We have presented a method for visualising the principal contrasting features of batch of images during forward pass of a bottleneck CNN. Our approach has several advantages over related methods, namely that it combines low overhead with intuitive visualisation, and doesn't require any user input or modification of the original CNN. We have shown how these advantages allow us to interpret the performance of CNNs in two common settings: debugging misclassification and predicting the applicability of transfer learning.

Our code is available at https://github.com/SINTEF/PFV.

References

1. Adebayo, J., Gilmer, J., Muelly, M., Goodfellow, I., Hardt, M., Kim, B.: Sanity checks for saliency maps. In: Advances in Neural Information Processing Systems, vol. 2018-Decem, pp. 9505–9515. Neural Information Processing Systems Foundation, October 2018. http://arxiv.org/abs/1810.03292
2. Agarwal, C., Schonfeld, D., Nguyen, A.: Removing input features via a generative model to explain their attributions to an image classifier's decisions (2019). http://arxiv.org/abs/1910.04256
3. Bach, S., Binder, A., Montavon, G., Klauschen, F., Müller, K.R., Samek, W.: On pixel-wise explanations for non-linear classifier decisions by layer-wise relevance propagation. PLoS ONE **10**(7), 1–46 (2015). https://doi.org/10.1371/journal.pone.0130140
4. Bau, D., Zhou, B., Khosla, A., Oliva, A., Torralba, A.: Network dissection: quantifying interpretability of deep visual representations. In: Proceedings of the IEEE Conference on Computer Vision and Pattern Recognition (CVPR), July 2017

5. Zhou, B., Khosla, A., Lapedriza, A., Oliva, A., Torralba, A.: Learning deep features for discriminative localization. CVPR **2016**(1), M1–M6 (2016). https://doi.org/10.5465/ambpp.2004.13862426
6. Ghorbani, A., Wexler, J., Zou, J., Kim, B.: Towards automatic concept-based explanations. NeurIPS, February 2019. https://github.com/amiratag/ACE, http://arxiv.org/abs/1902.03129
7. Kim, B., et al.: Interpretability beyond feature attribution: quantitative Testing with Concept Activation Vectors (TCAV). In: 35th International Conference on Machine Learning, ICML 2018, vol. 6, pp. 4186–4195 (2018)
8. Lapuschkin, S., Wäldchen, S., Binder, A., Montavon, G., Samek, W., Müller, K.R.: Unmasking Clever Hans predictors and assessing what machines really learn. Nat. Commun. **10**(1), 1–8 (2019). https://doi.org/10.1038/s41467-019-08987-4
9. Ribeiro, M.T., Singh, S., Guestrin, C.: "Why should i trust you?" Explaining the predictions of any classifier. In: Proceedings of the ACM SIGKDD International Conference on Knowledge Discovery and Data Mining, August 13–17, pp. 1135–1144 (2016). https://doi.org/10.1145/2939672.2939778
10. Selvaraju, R.R., Cogswell, M., Das, A., Vedantam, R., Parikh, D., Batra, D.: Grad-CAM: visual explanations from deep networks via gradient-based localization. Int. J. Comput. Vis. **128**(2), 336–359 (2020). https://doi.org/10.1007/s11263-019-01228-7, http://gradcam.cloudcv.org
11. Shinya, Y., Simo-Serra, E., Suzuki, T.: Understanding the effects of pre-training for object detectors via eigenspectrum. In: Proceedings of the IEEE/CVF International Conference on Computer Vision (ICCV) Workshops, October 2019
12. Shrikumar, A., Greenside, P., Kundaje, A.: Learning important features through propagating activation differences. In: Precup, D., Teh, Y.W. (eds.) Proceedings of the 34th International Conference on Machine Learning. Proceedings of Machine Learning Research, vol. 70, pp. 3145–3153. PMLR, International Convention Centre, Sydney, Australia, August 06–11, 2017. http://proceedings.mlr.press/v70/shrikumar17a.html
13. Simonyan, K., Vedaldi, A., Zisserman, A.: Deep inside convolutional networks: visualising image classification models and saliency maps. In: 2nd International Conference on Learning Representations, ICLR 2014 - Workshop Track Proceedings, pp. 1–8 (2014)
14. Simonyan, K., Zisserman, A.: Very deep convolutional networks for large-scale image recognition. CoRR abs/1409.1 (2014). https://doi.org/10.1016/j.infsof.2008.09.005, http://arxiv.org/abs/1409.1556
15. Springenberg, J.T., Dosovitskiy, A., Brox, T., Riedmiller, M.: Striving for simplicity: the all convolutional net. In: 3rd International Conference on Learning Representations, ICLR 2015 - Workshop Track Proceedings (2015)
16. Zeiler, M.D., Fergus, R.: Visualizing and understanding convolutional networks. In: Fleet, D., Pajdla, T., Schiele, B., Tuytelaars, T. (eds.) ECCV 2014. LNCS, vol. 8689, pp. 818–833. Springer, Cham (2014). https://doi.org/10.1007/978-3-319-10590-1_53
17. Zhang, J., Bargal, S.A., Lin, Z., Brandt, J., Shen, X., Sclaroff, S.: Top-down neural attention by excitation backprop. Int. J. Comput. Vis. **126**(10), 1084–1102 (2018). https://doi.org/10.1007/s11263-017-1059-x, http://arxiv.org/abs/1608.00507

Progressive Refinement Network for Occluded Pedestrian Detection

Xiaolin Song[1], Kaili Zhao[1(✉)], Wen-Sheng Chu[2], Honggang Zhang[1], and Jun Guo[1]

[1] Beijing University of Posts and Telecommunications, Beijing, China
kailizhao@bupt.edu.cn
[2] Google, Mountain View, CA, USA

Abstract. We present *Progressive Refinement Network (PRNet)*, a novel single-stage detector that tackles occluded pedestrian detection. Motivated by human's progressive process on annotating occluded pedestrians, PRNet achieves sequential refinement by three phases: Finding high-confident anchors of visible parts, calibrating such anchors to a full-body template derived from occlusion statistics, and then adjusting the calibrated anchors to final full-body regions. Unlike conventional methods that exploit predefined anchors, the confidence-aware calibration offers adaptive anchor initialization for detection with occlusions, and helps reduce the gap between visible-part and full-body detection. In addition, we introduce an occlusion loss to up-weigh hard examples, and a Receptive Field Backfeed (RFB) module to diversify receptive fields in early layers that commonly fire only on visible parts or small-size full-body regions. Experiments were performed within and across CityPersons, ETH, and Caltech datasets. Results show that PRNet can match the speed of existing single-stage detectors, consistently outperforms alternatives in terms of overall miss rate, and offers significantly better cross-dataset generalization. Code is available (https://github.com/sxlpris).

Keywords: Occluded pedestrian detection · Progressive Refinement Network · Anchor calibration · Occlusion loss · Receptive Field Backfeed

1 Introduction

Pedestrian detection is a fundamental computer vision problem and has been widely used in broad applications such as autonomous driving [10], robotics [9], and surveillance [21]. Although promising progress was made, occluded pedestrians remain difficult to detect [18,22,32]. The major challenges involve a wide range of appearance changes due to occlusion by other pedestrians or objects (*e.g.*, cars or trees), which decrease detection accuracy to various extents.

X. Song and K. Zhao—These authors contributed equally.

Electronic supplementary material The online version of this chapter (https://doi.org/10.1007/978-3-030-58592-1_3) contains supplementary material, which is available to authorized users.

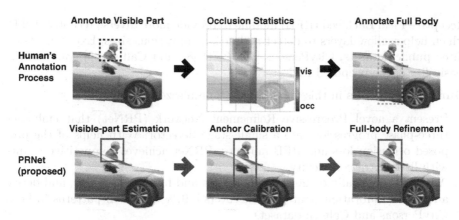

Fig. 1. Progressive Refinement Network (PRNet) imitates human's progressive annotation process on occluded pedestrians (*e.g.*, [5,39]), and gradually infers full-body regions from visible parts.

Reviewing the literature, most methods in pedestrian detection handle occlusions by exploiting visible parts as an additional supervision to improve detection performance. These methods broadly leverage three types of designs: 1) Independent detectors trained for each occlusion pattern [6,7,20,23,25,30,34,36], 2) Attention maps to enforce learning on visible parts [27,40], and 3) Auxiliary visibility classifiers to fuse prediction confidence into final scores [22,42,44]. Although these methods could benefit occluded pedestrian detection, at least three issues remain. First, independent detectors are computationally expensive, as each detector is trained for individual occlusion patterns, which are difficult to enumerate in practice. Second, attention-based methods can be slow for inference because attention modules are usually exhausted with proposals in architectures like Faster R-CNN [29]. Attention-based methods emphasize only visible parts, and thus could be suboptimal for full-body detection. Finally, detectors are usually initialized with predefined anchors, which, as will be demonstrated in Sect. 4, are suboptimal to generalize across diverse datasets.

To address the above challenges, we propose Progressive Refinement Network (PRNet), a novel single-stage detector for occluded pedestrian detection. Figure 1 illustrates our main idea. Inspired by human's progressive annotation process of occluded pedestrians (*e.g.*, [5,39]), PRNet performs pedestrian detection in three phases. First, *visible-part estimation* generates high-confident anchors of visible parts from one single-stage detector (*e.g.*, SSD-based [14,18]). Second, *anchor calibration* adjusts the visible-part anchors to a full-body template according to occlusion statistics, which is derived from over 20,000 annotations of occluded pedestrians. Finally, we train a *full-body refiner* using the calibrated anchors and a separate detection head from the one for visible-part estimation. Using two separate detection heads allows us to fit the progressive design into a single-stage detector without adding much complexity. In addition, to improve training effectiveness, we introduce an occlusion loss to up-weigh hard examples, and a

Receptive Field Backfeed (RFB) module to provide more diverse receptive fields, which help shallow layers to detect pedestrians in various sizes. Experiments on three public datasets, CityPersons [39], ETH [8], and Caltech [5], validate the feasibility of the proposed PRNet.

Our contributions in this paper can be summarized as follows:

1. Present a novel Progressive Refinement Network (PRNet) that embodies three-phase progression into a single-stage detector. With helps of the proposed occlusion loss and RFB modules, PRNet achieves competitive results with little extra complexity.
2. Analyze statistically on 20,000 visible-part and full-body regions, and derive an anchor calibration strategy that covers ∼97% occlusion patterns in both CityPersons and Caltech datasets.
3. Offer comprehensive ablation study, and experiments showing that PRNet achieves state-of-the-art within-dataset performance on **R** and **HO** subsets on CityPersons, and the best cross-dataset generalization over ETH and Caltech benchmarks.
4. Provide analysis on extreme occlusions, showing insights behind the metrics and suggesting a realistic evaluation subset for the community.

2   Related Work

CNN-Based Pedestrian Detection: Along with the development of CNN-based object detection, pedestrian detection has achieved promising results. We broadly group these methods into two categories: anchor-based and anchor-free.

For anchor-based methods, two-stage detectors (*e.g.*, Faster R-CNN [29]) and one-stage detectors (*e.g.*, SSD [17]) are two common designs. Most two-stage detectors [1,2,11,13,27,35,37,40–42,44] generate coarse region proposals of pedestrians and then refine the proposals by exploiting domain knowledge (*e.g.*, hard mining [37], extra learning task [2,27,40,44], or cascaded labeling policy [1]). RPN+BF [37] used a boosted forest to replace second stage learning and leveraged hard mining for proposals. However, involving such downstream classifier could bring more training complexity. SDS-RCNN [2] jointly learned pedestrian detection and bounding-box aware semantic segmentation, thus encouraged model learning more on pedestrian regions. AR-Ped [1] exploited sequential labeling policy in region proposal network to gradually filter out better proposals. These two-stage detectors need to generate proposal in first stage, and thus are slow for inference in practice. On the other hand, single-stage detectors [14,18,22] enjoy real-time inference due to the one-shot design. GDFL [14] included semantic segmentation task from end to end, which guided feature layers to emphasize on pedestrian regions. Generally, detection accuracy and inference time are trade-offs between single-stage and two-stage detectors. To obtain both accuracy and speed, ALFNet [18] involved anchor refinement into SSD training process. The proposed PRNet takes advantage of high speed of single-stage detector, and simultaneously outperforms these conventional methods in consideration of occlusion-aware supervision.

For anchor-free methods [19,32], topological points of pedestrians and predefined aspect ratio are introduced as new annotations to replace original bbox annotations. TLL [32] predicts the top and bottom vertexes of the somatic topological line while CSP [19] predicts central points and scales of pedestrian instances. Although the above CNN-based pedestrian detectors obtains potential performance, occluded pedestrian detection is still a challenging problem.

Occluded Pedestrian Detection: Methods tackling occluded pedestrians can be broadly categorized into four types: part-based, attention-based, score-based, and crowd-specific. Part-based methods have been widely received in the community, where each detector was separately trained for individual occlusion pattern with inference done by fusing all predictions. See [6,7,20,23–25,30,34,36,43] for comprehensive reviews. Moreover, exhaustively enumerating occlusion patterns is non-practical, computationally expensive, and generally infeasible. Instead of considering each occlusion pattern, [22,41] partitioned a proposal or bounding box into fixed number parts and predict their visibility scores. Although training complexity was decreased, these methods still require manually designing the partitions.

In recent years, learning robust representations and better anchor scoring have become a popular topic. On one hand, attention-based methods [27,40] learn robust features using guidance from attention maps. Zhang et al. [40] and MGAN [27] exploited channel-wise and pixel-wise attention maps respectively in feature layers, to highlight visible parts and suppress occluded parts. However, emphasizing visible-parts solely could be sub-optimal for full-body prediction. On the other hand, score-aware methods learn extra anchor scores by introducing additional learning task in the second stage of Faster R-CNN. For instance, Bi-box [44] constructs two classification and regression tasks for visible-part and full-body anchors, and then fuses the two anchor scores during inference. Similarly, [42] uses a separate discriminative classification by enforcing heavily occluded anchors to be close to easier anchors, and high confident scores were obtained for anchors. In addition, other studies [16,26,28,33,35] focused on tackling crowded pedestrians. RepLoss [35] designs a novel regression loss to prevent target proposals from shifting to surrounding pedestrians. The aforementioned methods are generally initialized with predefined anchors. In contrast, the proposed PRNet learns occluded pedestrian detection under confidence-aware and adaptive anchor initialization, which helps improve detection accuracy and generalization across dataset.

3   Progressive Refinement Network (PRNet)

3.1   PRNet Architecture

Motivated by human's progressive process on annotating occluded pedestrians (*e.g.*, CityPersons [39] and Caltech [5]), we construct PRNet to gradually migrate high-confident detection on visible parts toward more challenging full-body localization. For this purpose, we propose to adopt a single-stage detector with three training phases: Visible-part Estimation (VE), Anchor Calibration (AC), and

Full-body Refinement (FR). Unlike most methods that detect full bodies only [18,35] or independently with visible parts [44], we interweave them into a single-stage framework. To bridge the detection gap between visible parts and full bodies, we introduce AC to align anchors from VE to FR.

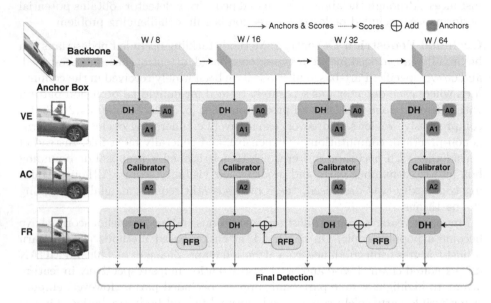

Fig. 2. Architecture of PRNet. From top to bottom, PRNet uses a detection backbone illustrated with four blocks of features maps. The network is trained in three phases: **Visible-part Estimation** (VE), **Anchor Calibration** (AC), and **Full-body Refinement** (FR). VE and FR take visible-part and full-body ground truth as references, respectively. Given initial anchors (A0), VE learns to predict visible-part anchors (A1), which are improved by AC to obtain calibrated anchors (A2). Final detection is obtained by post-processing anchors and scores from VE and FR. Detection Head (DH), Calibrator, and RFB modules are depicted in Fig. 3 and detailed in Sect. 3.1.

Figure 2 illustrates the PRNet architecture. The top row depicts the backbone, where we truncated first 5 stages of ResNet-50 [12] with modification of appending 1 extra stage with 3×3 filters and stride 2, which provide diverse receptive fields and help capture pedestrian with various scales. Out of the 6-stage backbone, we treat the last four as detection outputs. The network is trained following three phases: Visible-part Estimation (VE), Anchor Calibration (AC), and Full-body Refinement (FR). VE and FR are trained with visible-part and full-body ground truth, respectively; AC leverages occlusion statistics to bridge the gap between visible-part anchors and full-body anchors. Details of each module are illustrated later in this section. On top of each detection layer, we attach a detection head (DH) separately for VE and FR.

Specifically, denote x as an input image, $\Phi(x)$ as feature maps from backbone, $\mathcal{A}_0$ as a set of predefined anchors (as in SSD [17]), $\mathcal{B}^*$ as the predicted bounding boxes that are obtained by post-processing anchors collected from all layers

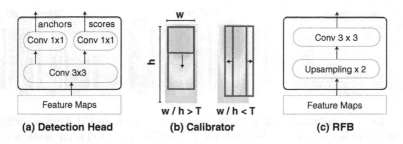

Fig. 3. Modules used in PRNet architecture (as in Fig. 2): DH, Calibrator and RFB.

(*i.e.*, via Non-Maximum Suppression). Given an initial set of feature maps and anchors, PRNet can be formulated as a progressive detector:

$$\text{Detections} = F(E_f(C(E_v(\Phi(x), \mathcal{A}_0)))) = \{\mathcal{B}^*, s^*\}, \tag{1}$$

where $E_v(\Phi(x), \mathcal{A}_0)$ is the 1st-phase visible-part estimation (VE) whose outputs are a set of visible-part anchors and confidence scores $\{\mathcal{A}_1, s_1\}$, $C(\cdot)$ is the 2nd-phase anchor calibration (AC) that aligns visible-part anchors $\mathcal{A}_1$ to full-body anchors $\mathcal{A}_2$, and $E_f(\Phi(x), \mathcal{A}_2)$ is the 3rd-phase full-body refiner (FR) that outputs the final full-body anchors to compute $\mathcal{B}^*$ and their scores s^* using inference F (see Sect. 3.2). Note that $\Phi(x)$ represents different feature maps during VE and FR due to their complementary objectives. Below we discuss each phase in turn.

Visible-part Estimation (VE): To train the *visible-part estimation* $E_v(\cdot)$, we adopt a standard detection approach that learns to localize anchors $\mathcal{A}_1$ as regression (from predefined anchors $\mathcal{A}_0$), and anchor scores as classification. Figure 3(a) depicts the detection head, whose loss can be written as:

$$\mathcal{L}_{VE} = \mathcal{L}_{focal} + \lambda_v[y=1]\mathcal{L}_{smoothL1}, \tag{2}$$

where $\mathcal{L}_{focal}$ is focal loss [15] for classification, $\mathcal{L}_{smoothL1}$ is a smooth-L1 loss for regression (as adopted in Faster R-CNN [29]), $[y=1]$ is an indicator for positive samples, and λ_v is a tuning parameter. As VE is trained on visible parts, its prediction (*i.e.*, $\mathcal{A}_1$) on visible parts is generally more confident and accurate than detectors trained with occlusions.

Anchor Calibration (AC): After VE obtains confident visible-part anchors $\mathcal{A}_1$, we propose a simple and effective *anchor calibration* $C(\cdot)$ to migrate visible-part anchors toward full-body anchors $\mathcal{A}_2$, which are then passed to the next phase for bull-body refinement. Briefly, PRNet updates anchors as: $\mathcal{A}_0 \xrightarrow{E_v} \mathcal{A}_1 \xrightarrow{C} \mathcal{A}_2$. Three are our motivations:

1. The aspect ratio of visible-part boxes is much more diverse than that of full-body boxes [5,39], making regression from visible-part to full-body boxes rather challenging.

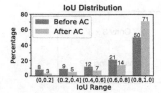

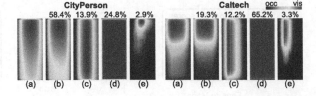

Fig. 4. IoU distribution before and after anchor calibration on the CityPersons dataset. IoU is measured between anchors and full-body ground truth.

Fig. 5. Occlusion statistics from CityPersons [39] (left) and Caltech [5] (right): (a) Occlusion statistics with blue indicating occlusion; red indicates visible parts, (b) Horizontal occlusions, (c) Vertical occlusions, (d) Non-occlusion, (e) Others. Percentage (%) denotes the likelihood of each occlusion pattern. (Color figure online)

2. Adaptive anchor initialization can reduce unnecessary search space and lead to better detection (*e.g.*, [3]), compared to most methods that use predefined anchors (*e.g.*, [18,39,41,42]).
3. The IoU discrepancy between visible-part anchors and full-body ground truth boxes is large; proper calibration can significantly improve IoU.

Figure 4 shows the distribution of IoU between ground truth full-body boxes and visible-part boxes before/after Anchor Calibration (AC) in CityPersons dataset [39]. The visible-part boxes before AC were taken from the annotations in the original dataset. As can be seen, calibration significantly shifts the distribution toward higher IoU, *e.g.*, +21% for IoU in (0.8, 1.0], and thus can help detectors approximate final full-body regions. In addition, AC addresses discrepancy during anchor assignment between VE and FR, *i.e.*, without AC, a positive A_1 could be assigned as a negative anchor for FR, making VE and FR fail to complement each other.

To achieve AC, we first derive a statistical analysis of occlusion patterns on two popular datasets CityPersons [39] and Caltech [5] using their standardized 0.41 box aspect ratio. Please see supplementary materials for detailed process. Figure 5 illustrates occlusion distribution over a full-body box and four occlusion types (*i.e.*, horizontal, vertical, non-occlusion, and others, similar to [40]) with respective likelihood in each dataset. As can be seen in Fig. 5(a), over the two datasets, the upper box is consistently visible (*i.e.*, the head), with most occlusions appearing in the lower box (*i.e.*, the feet). This serves as strong evidence for humans and PRNet to leverage visible parts for full-body detection.

Observing the occlusion statistics, we reach two types of anchor updates according to the aspect ratio of A_1, as depicted in Fig. 3(b). For the anchors with ratio >0.41, we vertically stretc.h them *downwards* until 0.41 aspect ratio, due to heads being frequently visible, as shown in Fig. 5(b) and [5]. Anchors with ratio <0.41 are horizontally extended to 0.41 w.r.t. the center of A_1, as they likely involve vertical occlusion, as shown in Fig. 5(c). Anchors with 0.41 ratio (*i.e.*, Fig. 5(d)) remain unchanged. The anchor updates can also be rationalized with human's annotation protocol in CityPersons [39], where a full-body box is

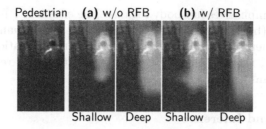

Pedestrian **(a)** w/o RFB **(b)** w/ RFB

Shallow Deep Shallow Deep

Fig. 6. Saliency maps highlighted by the third FR phase: (a) w/o RFB and (b) w/ RFB. "Shallow" indicates the 2nd layer, and "Deep" indicates the 3rd layer.

generated by fitting a fixed-ratio (0.41) box onto a line drawn from head to feet. According to Fig. 5(b)–(d), we justify the two simple updates can cover ~97% data in both datasets, while the remaining ~3% is shown in Fig. 5(e).

Full-body Refinement (FR): With the calibrated anchors $\mathcal{A}_2$ from AC, PRNet's last phase trains a *full-body refiner* $E_f(\cdot)$ that refines the final full-body localization. Similar to VE, FR also uses the same backbone, yet performs training on a separate detection head. Different from VE that sees only visible parts, FR starts to see hard positive samples whose anchor boxes are still far from ground truth full-body region. As $\mathcal{L}_{smoothL1}$ in Eq. (2) treats every positive sample equally, it could be less effective when dealing with hard samples in FR. To encourage learning on hard positive samples, we weigh the regression loss $\mathcal{L}_{smoothL1}$ with an occlusion weight, which is defined as a reverse IoU between $\mathcal{A}_2$ and ground truth full-body boxes $\mathcal{B}_{gt}$. Given $a \in \mathcal{A}_2$ and its corresponding $b \in \mathcal{B}_{gt}$, the weighted loss, termed as *occlusion loss*, can be rewritten as:

$$\mathcal{L}_{occ} = \sum_{a \in \mathcal{A}_2} (1 - \text{IoU}(a,b)) \left\{ [|s| < 1]\, 0.5 s^2 + [|s| >= 1]\, (|s| - 0.5) \right\}, \quad (3)$$

where s is the difference between predicted offsets and ground truth offsets (see [29] for details). The less overlap between the calibrated anchors $\mathcal{A}_2$ and $\mathcal{B}_{gt}$, the higher $\mathcal{L}_{occ}$ is. As a result, the loss for FR becomes:

$$\mathcal{L}_{FR} = \mathcal{L}_{focal} + \lambda_f [y = 1] \mathcal{L}_{occ}. \quad (4)$$

Despite of up-weighting hard positive anchors, another challenge in FR regards training shallow layers, which often activate on visible parts or small-size full-body regions due to limited receptive field. In every layer of FR, we introduce a *Receptive Field Backfeed* (RFB) module to diversify receptive fields, as depicted in Fig. 3(c). RFB aims to enlarge the receptive fields of shallower layers by back-feeding features from deeper layers to the previous layer with 2X upsampling, and then summing up their feature maps in a pixelwise manner.

Figure 6 shows the saliency maps [31] of the 2nd layer (denoted as "shallow") and the 3rd layer (*i.e.*, "deep") with/without the RFB module. As can be seen in Fig. 6(a), without RFB, visible parts are identified in the shallow layer, while the deeper layer emphasizes full-body regions. The effects of RFB can be clearly observed in Fig. 6(b). In the shallow layer, RFB not only enhances visible parts but also complements the full-body region. Similar observation can be made on the deep layer, showing that RFB can propagate larger receptive fields to shallower layers and help refine full-body detection.

3.2 Training and Inference

Training: In training, a batch of pedestrian images goes through the three phases (*i.e.*, VE, AC, and FR) sequentially–the first phase VE is trained independently and then the first detection head is frozen to train FR. Figure 2 illustrates the architecture and examples of pedestrian annotation. Given predefined anchors $\mathcal{A}_0$ and visible-part ground truth boxes associated to the image batch, we first train VE with loss $\mathcal{L}_{VE}$ in Eq. (2), and obtain visible-part anchors $\mathcal{A}_1$. Then AC transforms $\mathcal{A}_1$ into more adaptive anchors $\mathcal{A}_2$, which better approximates full-body regions. Finally, initialized with $\mathcal{A}_2$, FR is trained with loss $\mathcal{L}_{FR}$ in Eq. (4). Note that VE and FR use two different detection heads in one single-stage detector, so they learn complementary outputs.

An anchor is assigned as positive if intersection-over-union (IoU) between an anchor bbox and ground truth bbox is above a threshold θ_p, as negative if IoU is lower than θ_n, and otherwise ignored during training. Note that VE and FR adopt different annotation of boxes, *i.e.*, VE consumes visible-part boxes, while FR uses full-body boxes.

Inference: In inference, we obtain predicted anchor boxes from FR, and associate anchor scores by multiplying the scores from VE and FR. The score fusion provides complementary guidance so to improve detection robustness (similar to [44]). We obtain the final bounding boxes $\mathcal{B}^*$ by first filtering out candidate anchor boxes with scores lower than 0.05 and then merging them with NMS (0.5 threshold is used here).

3.3 Comparisons with Related Work

The closest studies to PRNet are ALFNet [18] and Bi-box [44]. As most cascade designs, ALFNet tackles successively the same task (FR→FR), which requires occlusion patterns to be extensively illustrated in training data. Mimicking human's annotation process, PRNet exploits different tasks (VE→AC→FR), starting from detecting only visible parts (regardless of occlusion patterns as in full-body boxes), and thus relaxes training data requirements. Note that jointly tackling different tasks is non-trivial. Instead, we interweave these tasks with occlusion loss and the RFB module (Sect. 3.1) to up-weigh hard samples and facilitate training for shallow layers. As can be seen in Sect. 4.4, PRNet achieved impressive cross-dataset generalizability compared to ALFNet, showing that the PRNet structure is more effective. Similar to ALFNet [18], PRNet enjoys competitive inference time due to the use of a single-stage detector.

In terms of involving different tasks, Bi-box [44] also takes visible parts into account but by training a two-branch detector for visible parts and full body in the second stage of Faster R-CNN. During training, there is no interaction between the two branches, making their complementary benefits relatively indirect. PRNet leverages the hybrid cascade structure to progressively refine predictions from visible-part to full-body regions, providing adaptive anchor initialization to achieve the final full-body estimation.

4 Experiments

4.1 Settings

Datasets: We conducted experiments on three public datasets: CityPersons [39], ETH [8], and Caltech [5]. CityPersons [39] has high-res 2048 × 1024 images with visible-part and full-body annotations, where 2,975 images are for training and 500 for validation. We trained PRNet on the training set and reported performance on the validation set in ablations and within-dataset experiments. To evaluate model generalizability, we performed cross-dataset analysis using ETH [8] and Caltech [5]. ETH dataset [8] contains 11,941 labeled persons, providing a benchmark in evaluating model's robustness to occluded pedestrians. For Caltech [5], we adopted published test set with 4,024 images with both old [5] and new annotations [38]. Both ETH and Caltech have lower-res 640 × 480 images that represent more cross-dataset challenges. Following [35,40], we performed training and evaluation on pedestrians with height larger than 50 pixels.

Table 1. Ablations of **three-phase components** and an alternative.

Architecture	VE	AC	FR	R	HO
PRNet-F			✓	15.6	45.7
PRNet-VA	✓	✓		11.7	51.3
PRNet-VAF	✓	✓	✓	11.4	45.3
PRNet-VRF	✓	reg	✓	12.6	44.7

Table 2. Ablations of **occlusion loss** and the **RFB module**.

Architecture	+Occ.	+RFB	R	HO
PRNet-VAF			11.4	45.3
PRNet-VAF-OCC	✓		11.0	45.7
PRNet-VAF-RFB		✓	11.6	44.9
PRNet (ours)	✓	✓	10.8	42.0

Metrics: Evaluation was reported on the standard MR^{-2} (%) [5], which computes the log-average miss rate at 9 False Positive Per Image (FPPI). The lower MR^{-2}, the better. To ensure the results are directly comparable with the literature, we represented each test set as *6 subsets* according to visibility ratio of each pedestrian. Specifically, we reported **R** (reasonable occlusion with visibility in [0.65, 1]), **HO** (heavy occlusion with [0.2, 0.65]), **R+HO** with [0.2, 1] from Zhang *et al.* [40], and **Bare** with [0.9, 1.0], **Partial** with [0.65, 0.9], and **Heavy** with [0, 0.65] from [35]. To complement the visibility range covered by **R** and **HO**, we added **EO** (extreme occlusion) to represent visibility in [0, 0.2].

Implementation Details: We augmented our pedestrian images following standard techniques [18,19]. When assigning labels to anchor boxes, $\theta_p = 0.5$ and $\theta_n = 0.3$ for VE, and $\theta_p = 0.7$ and $\theta_n = 0.5$ for FR. We set $\lambda_v = 1$ and $\lambda_f = 4$ empirically. The backbone ResNet-50 is pre-trained on ImageNet [4]. PRNet is then fine-tuned with 160k iterations, a learning rate of 10^{-4}, batch size 8 and an Adam optimizer. All experiments were performed on 2 GTX 1080Ti GPUs.

4.2   Ablation Study

To analyze PRNet, we performed extensive ablations on CityPersons validation set [39] using subsets of R (reasonable) and HO (heavy occlusion).

Three-Phase Components: To analyze the effect of PRNet's three-phase design, we performed ablation study on each phase without occlusion loss and RFB module in FR. In Table 1, we trained a standalone FR (denoted as **PRNet-F**) initialized by predefined full-body anchors. **PRNet-VA** used only VE+AC, treating calibrated anchors $\mathcal{A}_2$ as the detection outputs. **PRNet-VAF** employed all three phases (VE+AC+FR), using calibrated anchors $\mathcal{A}_2$ to initialize FR. Comparing PRNet-F and PRNet-VA, PRNet-VA performs 3.9 points better in R while 5.6 points worse in HO. This shows that plain calibrated anchors $\mathcal{A}_2$ in PRNet-VA can achieve better result while occlusion level is reasonable. In contrast, PRNet-F better addressed heavy occlusions. PRNet-VAF combines the benefits from both, showing a consistent improvement over both R and HO. Please see supplementary for detection examples of the three-phase progression.

Anchor Calibration *vs*. Box Regression: A possible alternative to AC is a box regressor from the visible-part anchors $\mathcal{A}_1$ to full-body boxes. Here we reused FR for the regression task. To perform a fair comparison, we implemented **PRNet-VRF** by replacing AC with the regressor. Table 1 summarizes the results. As can be seen, PRNet-VAF consistently outperformed PRNet-VRF by 9.5% in R, showing no significant benefits of adding an extra box regressor. An explanation can be that the visible boxes change rapidly due to various occlusion types, and make the regressor hard to map the coordinates to full-body boxes with relatively constant aspect ratio. Unlike a regression network that require extra complexity and training efforts, AC provides a more generalizable strategy that better fits into the proposed three-phase approach.

Occlusion Loss and RFB: Table 2 studies PRNet w/ and w/o occlusion loss and RFB module. PRNet-VAF was reused as the baseline that considers neither occlusion loss nor RFB, and compared against **PRNet-VAF-OCC** (with only occlusion loss) and **PRNet-VAF-RFB** (with only RFB). Including occlusion loss alone, PRNet-VAF-OCC improved the baseline 0.4 points on R yet lowered 0.4 points on HO. This shows that occlusion loss improves detection with reasonable occlusion (*i.e.*, over 0.65 visibility), yet could be insufficient to address heavy occlusion (*i.e.*, 0.2 to 0.65 visibility). Including RFB alone, PRNet-VAF-RFB improved the baseline 0.4 points on HO yet lowered 0.2 points on R. This suggests that the feedback from RFB could supply full-body info by enlarging receptive field, and thus offers improvement when occlusion is severe. Otherwise, when occlusion level is light, enlarging receptive field may introduce unnecessary context and hence slightly hurt. PRNet couples occlusion loss and RFB, achieving significant improvement over R and especially HO.

4.3 Within-Dataset Comparisons

This section compares PRNet in a within-dataset setting against 3 types of alternatives: Occlusion-free, occlusion-aware, and closest to PRNet. We reported MR^{-2} on all 6 subsets, where R is the major evaluation criteria in CityPersons Challenge[1]. Table 3 shows comparisons with scale ×1 and ×1.3 of original resolution (2048 × 1024).

Table 3. Comparisons on CityPersons [39]. Results of alternatives were obtained from original paper. On scale×1, bracketed and bold numbers indicate the best and the second best results, respectively. Inference time (*sec*) is measured on scale×1 images.

Method	Occ.	Scale	R	HO	R+HO	Heavy	Partial	Bare	Time
Adapted FasterRCNN [39]		×1	15.4	64.8	41.45	55.0	18.9	9.3	–
TLL+MRF [32]		×1	14.4	–	–	52.0	15.9	9.2	–
CSP [19]		×1	**11.0**	–	–	[49.3]	10.4	7.3	0.33
FasterRCNN+ATT [40]	✓	×1	16.0	56.7	38.2	–	–	–	–
RepLoss [35]	✓	×1	13.2	–	–	56.9	16.8	7.6	–
		×1.3	11.6	–	–	55.3	14.8	7.0	–
OR-CNN [41]	✓	×1	12.8	–	–	55.7	15.3	[6.7]	–
		×1.3	11.0	–	–	51.3	13.7	5.9	–
MGAN [27]	✓	×1	11.3	[42.0]	–	–	–	–	–
FRCN+A+DT [42]	✓	×1.3	11.1	44.3	–	–	11.2	6.9	–
ALFNet [18]		×1	12.0	43.8	**26.3**	51.9	11.4	8.4	0.27
Bi-box [44]	✓	×1.3	11.2	44.2	–	–	–	–	–
PRNet (ours)	✓	×1	[10.8]	[42.0]	[25.6]	53.3	[10.0]	6.8	0.22

Occlusion-Free Methods: *Occlusion-free* methods aim to detect pedestrians without considering occlusion info. Adapted FasterRCNN [39] is an anchor-based benchmark, while TLL+MRF [32] and CSP [19] are anchor-free. Among the three methods, CSP achieved the state-of-the-art results without considering occlusion, as summarized in Table 3. PRNet, on the other hand, takes occlusion info into account, and provides performance gain over CSP on R, Partial, and Bare subsets, but not the Heavy subset. One possible reason is because CSP used box-free annotations, which is different from the original annotations and might help reduce ground truth noises in heavily occluded cases.

Occlusion-Aware Methods: *Occlusion-aware* methods consider occlusion information in training, including FasterRCNN+ATT [40], RepLoss [35], OR-CNN [41], FRCN+A+DT [42], and MGAN [27]. Table 3 summarizes the results.

---
[1] https://bitbucket.org/shanshanzhang/citypersons/.

On the R subset (CityPersons' evaluation criteria), occlusion-aware methods are generally better than occlusion-free methods, except for CSP that used different box-free annotations. In contrast, PRNet consistently achieved the best MR^{-2} of (10.8, 42.0, 25.6, 10.0) on (R, HO, R+HO, Partial) and compared favorably with the best performer for Bare. The comparisons firmly validate PRNet's effectiveness by dealing with occlusion using progressive refinement.

Closest Alternatives: Closest to PRNet are ALFNet [18] and Bi-box [44] per discussion in Sect. 3.3. We reported ALFNet results using the same settings and the authors' released code. We did not reproduce Bi-box due to lack of source code. Regarding inference time, PRNet performed comparably with ALFNet, as both methods are single-stage based. We infer that PRNet is substantially faster than Bi-box due to the Faster-RCNN-like design in Bi-box (*e.g.*, 2–6X speedup as demonstrated in [1,18]). Compared to ALFNet and Bi-box, PRNet is also preferred in detection performance because of better anchor initialization and its ability to recover full-body region from confident visible parts. Due to space constraint, please refer to supplementary for examples that are mis-detected by alternative methods but successfully detected by PRNet. Observing the last three rows in Table 3, PRNet consistently outperformed Bi-box and provided performance gain upon ALFNet in all cases except for the Heavy subset.

Breakdowns in Heavy: In the Heavy subset, we noticed the occlusion-aware methods, including PRNet, were less effective than occlusion-free methods (*e.g.*, ALFNet). We performed an analysis by partitioning Heavy into HO and EO, *i.e.*, Heavy=HO ∪ EO. EO represents the most extreme occlusion with visibility in only [0, 0.2]. In HO, PRNet outperformed all other methods (*e.g.*, 1.8 points better than the state-of-the-art ALFNet), while being 1.4 points worse than ALFNet in Heavy. We hypothesize that PRNet fails to compete against ALFNet only in EO, and re-evaluated their performance on EO. Not surprisingly, PRNet and ALFNet result in very high MR^2 at 80.8 and 70.2 respectively. Figure 7 shows the distribution of visibility ratio and examples of EO from CityPersons validation set. As can be seen, visible parts are barely visible and sometimes very low-res, making it perceptually challenging even for human to detect. Ground truth boxes by human annotators in EO can thus be noisy and make performance comparisons on EO less meaningful. In addition, the proportion of EO is relatively small. As shown in top-left of Fig. 7, less than 10% are in EO and more than 90% belong to R and HO. These findings reveal that R and HO

Table 4. Cross-dataset on ETH [8].

Method	R+HO	Time
FasterRCNN [39]	35.6	–
FasterRCNN+ATT [40]	33.8	–
CSP [19]	37.2	61.3
ALFNet [18]	**31.1**	39.2
PRNet (ours)	[**27.0**]	42.1

Table 5. Cross-dataset results on Caltech [5].

Method	R (o)	R+HO (o)	R (n)	Time
ALFNet [18]	25.0	35.0	19.0	39.2
CSP [19]	**20.0**	[**27.8**]	**11.7**	61.3
PRNet (ours)	[**18.3**]	**28.4**	[**10.7**]	42.1

render more realistic occlusion scenarios than EO. The above analyses suggest the proposed PRNet achieved state-of-the-art performance.

4.4 Cross-dataset Generalization

To validate generalizability of the proposed method, we performed cross-dataset experiments on ETH [8] and Caltech [5] datasets. For comparison, we picked two top-performing methods, CSP [19] and ALFNet [18], where the models are available from the authors' GitHub release. For fair comparisons, PRNet was also trained on CityPersons training set and shared the same pre-processing.

Table 4 shows ETH results on the R+HO as in [40]. For reference, we also included numbers reported in the Faster-RCNN and FasterRCNN+ATT [40] without reproducing their results. CSP and ALFNet showed surprising opposite results comparing their performance within- and cross-dataset. In cross-dataset setting, ALFNet outperformed CSP by 6.1 points (from 37.2 to 31.1), while CSP reported consistently better performance in within-dataset setting (see Table 3). On the contrary, our method achieved the state-of-the-art MR^{-2} on R+HO by a significant margin. For Caltech [5], we reported R+HO and R using the old [5] (denoted as "(o)") and the new [38] annotations (denoted "(n)"), as summarized in Table 5. PRNet consistently outperformed CSP and ALFNet in R using both old and new annotations. On R+HO, PRNet performed comparably with CSP.

Rationale: PRNet's gain is evident in cross-dataset settings for two major reasons. One, PRNet's progressive structure imitates human's annotation process, which formulates the principles humans have established for annotating occluded pedestrians (*e.g.*, CityPersons, Caltech). PRNet mimics every step in human's

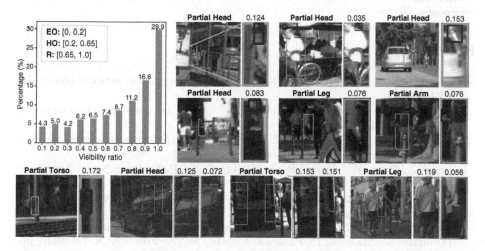

Fig. 7. Distribution of visibility ratio on CityPersons (top left), and examples of Extreme Occlusion (**EO**), such as partial head, arm, leg, and torso. Blue and green boxes indicate visible parts and full-body boxes, respectively. (Color figure online)

principles, and thus fits the problem more naturally. Two, most methods (*e.g.*, ALFNet) consider only full body detection, which demands training data with various occlusions (*e.g.*, cars, trees, other pedestrians). When the occlusion pattern is rare or unseen in training data (*i.e.*, cross-dataset settings), such methods tend to perform less favorably. As shown in supplementary, ALFNet tends to fire false positives on uncommon objects (*e.g.*, wheel, car windshield). On the contrary, PRNet propagates detection from visible parts (regardless of occlusion patterns as in full-body boxes), and thus provides better generalizability.

5   Conclusion

We have proposed PRNet, a novel one-stage approach for occluded pedestrian detection. PRNet incorporates three phases (VE, AC, and FR) to evolve anchors toward full-body localization. We introduced an occlusion loss to encourage learning on hard samples, and an RFB module to diversify receptive fields for shallow layers. We provided extensive ablation studies to justify the three-phase design. Within-dataset experiments validated PRNet's effectiveness with 6 occlusion scenarios. On cross-dataset settings, PRNet outperformed alternatives on ETH and Caltech datasets by a noticeable margin. Analysis on extreme occlusions provided insights behind metrics and suggested a more realistic choice for evaluation. Potential extensions of PRNet include providing weak annotations of visible parts for occluded pedestrian datasets.

Acknowledgement. Research reported in this paper was supported in part by the Natural Science Foundation of China under grant 61701032 & 62076036 to KZ, and Beijing Municiple Science and Technology Commission project under Grant No.Z181100001918005 to XS and HZ. We thank Jayakorn Vongkulbhisal for helpful comments.

References

1. Brazil, G., Liu, X.: Pedestrian detection with autoregressive network phases. In: CVPR (2019)
2. Brazil, G., Yin, X., Liu, X.: Illuminating pedestrians via simultaneous detection & segmentation. In: ICCV (2017)
3. Chi, C., Zhang, S., Xing, J., Lei, Z., Li, S.Z., Zou, X.: Selective refinement network for high performance face detection. In: AAAI (2019)
4. Deng, J., Dong, W., Socher, R., Li, L., Li, K., Li, F.-F.: ImageNet: a large-scale hierarchical image database. In: CVPR (2009)
5. Dollár, P., Wojek, C., Schiele, B., Perona, P.: Pedestrian detection: an evaluation of the state of the art. TPAMI **34** (2012)
6. Duan, G., Ai, H., Lao, S.: A structural filter approach to human detection. In: Daniilidis, K., Maragos, P., Paragios, N. (eds.) ECCV 2010. LNCS, vol. 6316, pp. 238–251. Springer, Heidelberg (2010). https://doi.org/10.1007/978-3-642-15567-3_18
7. Enzweiler, M., Eigenstetter, A., Schiele, B., Gavrila, D.M.: Multi-cue pedestrian classification with partial occlusion handling. In: CVPR (2010)

8. Ess, A., Leibe, B., Van Gool, L.: Depth and appearance for mobile scene analysis. In: ICCV (2007)
9. Geiger, A., Lenz, P., Stiller, C., Urtasun, R.: Vision meets robotics: the kitti dataset. Int. J. Robot. Res. **32** (2013)
10. Geiger, A., Lenz, P., Urtasun, R.: Are we ready for autonomous driving? The kitti vision benchmark suite. In: CVPR (2012)
11. Girshick, R.: Fast R-CNN. In: ICCV (2015)
12. He, K., Zhang, X., Ren, S., Sun, J.: Deep residual learning for image recognition. In: CVPR (2016)
13. Li, J., Liang, X., Shen, S., Xu, T., Feng, J., Yan, S.: Scale-aware fast R-CNN for pedestrian detection. TMM **20** (2018)
14. Lin, C., Lu, J., Wang, G., Zhou, J.: Graininess-aware deep feature learning for pedestrian detection. In: Ferrari, V., Hebert, M., Sminchisescu, C., Weiss, Y. (eds.) ECCV 2018. LNCS, vol. 11213, pp. 745–761. Springer, Cham (2018). https://doi.org/10.1007/978-3-030-01240-3_45
15. Lin, T., Goyal, P., Girshick, R., He, K., Dollar, P.: Focal loss for dense object detection. In: CVPR (2017)
16. Liu, S., Huang, D., Wang, Y.: Adaptive NMS: refining pedestrian detection in a crowd. In: CVPR (2019)
17. Liu, W., et al.: SSD: single shot multibox detector. In: Leibe, B., Matas, J., Sebe, N., Welling, M. (eds.) ECCV 2016. LNCS, vol. 9905, pp. 21–37. Springer, Cham (2016). https://doi.org/10.1007/978-3-319-46448-0_2
18. Liu, W., Liao, S., Hu, W., Liang, X., Chen, X.: Learning efficient single-stage pedestrian detectors by asymptotic localization fitting. In: Ferrari, V., Hebert, M., Sminchisescu, C., Weiss, Y. (eds.) Computer Vision – ECCV 2018. LNCS, vol. 11218, pp. 643–659. Springer, Cham (2018). https://doi.org/10.1007/978-3-030-01264-9_38
19. Liu, W., Liao, S., Ren, W., Hu, W., Yu, Y.: High-level semantic feature detection: a new perspective for pedestrian detection. In: CVPR (2019)
20. Mathias, M., Benenson, R., Timofte, R., Gool, L.V.: Handling occlusions with Franken-classifiers. In: ICCV (2013)
21. Nascimento, J.C., Marques, J.S.: Performance evaluation of object detection algorithms for video surveillance. TMM **8** (2006)
22. Noh, J., Lee, S., Kim, B., Kim, G.: Improving occlusion and hard negative handling for single-stage pedestrian detectors. In: CVPR (2018)
23. Ouyang, W., Wang, X.: A discriminative deep model for pedestrian detection with occlusion handling. In: CVPR (2012)
24. Ouyang, W., Wang, X.: Joint deep learning for pedestrian detection. In: ICCV (2013)
25. Ouyang, W., Zeng, X., Wang, X.: Modeling mutual visibility relationship in pedestrian detection. In: CVPR (2013)
26. Ouyang, W., Wang, X.: Single-pedestrian detection aided by multi-pedestrian detection. In: CVPR (2013)
27. Pang, Y., Xie, J., Khan, M.H., Anwer, R.M., Khan, F.S., Shao, L.: Mask-guided attention network for occluded pedestrian detection. In: ICCV (2019)
28. Pepikj, B., Stark, M., Gehler, P., Schiele, B.: Occlusion patterns for object class detection. In: CVPR (2013)
29. Ren, S., He, K., Girshick, R., Sun, J.: Faster R-CNN: towards real-time object detection with region proposal networks. In: NIPS (2015)
30. Shet, V.D., Neumann, J., Ramesh, V., Davis, L.S.: Bilattice-based logical reasoning for human detection. In: CVPR (2007)

31. Simonyan, K., Vedaldi, A., Zisserman, A.: Deep inside convolutional networks: visualising image classification models and saliency maps. arXiv:1312.6034 (2013)
32. Song, T., Sun, L., Xie, D., Sun, H., Pu, S.: Small-scale pedestrian detection based on topological line localization and temporal feature aggregation. In: Ferrari, V., Hebert, M., Sminchisescu, C., Weiss, Y. (eds.) ECCV 2018. LNCS, vol. 11211, pp. 554–569. Springer, Cham (2018). https://doi.org/10.1007/978-3-030-01234-2_33
33. Tang, S., Andriluka, M., Schiele, B.: Detection and tracking of occluded people. IJCV **110** (2014)
34. Tian, Y., Luo, P., Wang, X., Tang, X.: Deep learning strong parts for pedestrian detection. In: ICCV (2015)
35. Wang, X., Xiao, T., Jiang, Y., Shao, S., Sun, J., Shen, C.: Repulsion loss: detecting pedestrians in a crowd. In: CVPR (2018)
36. Wu, B., Nevatia, R.: Detection of multiple, partially occluded humans in a single image by Bayesian combination of edgelet part detectors. In: ICCV (2005)
37. Zhang, L., Lin, L., Liang, X., He, K.: Is faster R-CNN doing well for pedestrian detection? In: Leibe, B., Matas, J., Sebe, N., Welling, M. (eds.) ECCV 2016. LNCS, vol. 9906, pp. 443–457. Springer, Cham (2016). https://doi.org/10.1007/978-3-319-46475-6_28
38. Zhang, S., Benenson, R., Omran, M., Hosang, J., Schiele, B.: How far are we from solving pedestrian detection? In: CVPR (2016)
39. Zhang, S., Benenson, R., Schiele, B.: Citypersons: a diverse dataset for pedestrian detection. In: CVPR (2017)
40. Zhang, S., Yang, J., Schiele, B.: Occluded pedestrian detection through guided attention in CNNs. In: CVPR (2018)
41. Zhang, S., Wen, L., Bian, X., Lei, Z., Li, S.Z.: Occlusion-aware R-CNN: detecting pedestrians in a crowd. In: Ferrari, V., Hebert, M., Sminchisescu, C., Weiss, Y. (eds.) ECCV 2018. LNCS, vol. 11207, pp. 657–674. Springer, Cham (2018). https://doi.org/10.1007/978-3-030-01219-9_39
42. Zhou, C., Yang, M., Yuan, J.: Discriminative feature transformation for occluded pedestrian detection. In: ICCV (2019)
43. Zhou, C., Yuan, J.: Multi-label learning of part detectors for heavily occluded pedestrian detection. In: ICCV (2017)
44. Zhou, C., Yuan, J.: Bi-box regression for pedestrian detection and occlusion estimation. In: Ferrari, V., Hebert, M., Sminchisescu, C., Weiss, Y. (eds.) ECCV 2018. LNCS, vol. 11205, pp. 138–154. Springer, Cham (2018). https://doi.org/10.1007/978-3-030-01246-5_9

Monocular Real-Time Volumetric Performance Capture

Ruilong Li[1,2], Yuliang Xiu[1,2], Shunsuke Saito[1,2], Zeng Huang[1,2],
Kyle Olszewski[1,2], and Hao Li[1,2,3(✉)]

[1] University of Southern California, Los Angeles, USA
{ruilongl,yxiu,zenghuan}@usc.edu, shunsuke.saito16@gmail.com,
olszewski.kyle@gmail.com
[2] USC Institute for Creative Technologies, Los Angeles, USA
[3] Pinscreen, Los Angeles, USA
hao@hao-li.com

Abstract. We present the first approach to volumetric performance
capture and novel-view rendering at real-time speed from monocular
video, eliminating the need for expensive multi-view systems or cumbersome pre-acquisition of a personalized template model. Our system
reconstructs a fully textured 3D human from each frame by leveraging Pixel-Aligned Implicit Function (PIFu). While PIFu achieves high-
resolution reconstruction in a memory-efficient manner, its computationally expensive inference prevents us from deploying such a system
for real-time applications. To this end, we propose a novel hierarchical surface localization algorithm and a direct rendering method without explicitly extracting surface meshes. By culling unnecessary regions
for evaluation in a coarse-to-fine manner, we successfully accelerate the
reconstruction by two orders of magnitude from the baseline without
compromising the quality. Furthermore, we introduce an Online Hard
Example Mining (OHEM) technique that effectively suppresses failure
modes due to the rare occurrence of challenging examples. We adaptively update the sampling probability of the training data based on the
current reconstruction accuracy, which effectively alleviates reconstruction artifacts. Our experiments and evaluations demonstrate the robustness of our system to various challenging angles, illuminations, poses,
and clothing styles. We also show that our approach compares favorably
with the state-of-the-art monocular performance capture. Our proposed
approach removes the need for multi-view studio settings and enables a
consumer-accessible solution for volumetric capture.

R. Li and Y. Xiu—Equal contribution.

Electronic supplementary material The online version of this chapter (https://
doi.org/10.1007/978-3-030-58592-1_4) contains supplementary material, which is available to authorized users.

A. Vedaldi et al. (Eds.): ECCV 2020, LNCS 12368, pp. 49–67, 2020.
https://doi.org/10.1007/978-3-030-58592-1_4

1   Introduction

Videoconferencing using a single camera is still the most common approach face-to-face communication over long distances, despite recent advances in virtual and augmented reality and 3D displays that allow for far more immersive and compelling interaction. The reason for this is simple: convenience. Though the technology exists to obtain high-fidelity digital representations of one's specific appearance that can be rendered from arbitrary viewpoints, existing methods to capture and stream this data [7,10,15,46,59] require cumbersome capture technology, such as a large number of calibrated cameras or depth sensors, and the expert knowledge to install and deploy these systems. Videoconferencing, on the other hand, simply requires a single video camera, such as those found on common consumer devices, *e.g.* laptops and smartphones. Thus, if we can capture a complete model of a person's unique appearance and motion from a single consumer-grade camera, we can bridge the gap preventing novice users from engaging in immersive communication in virtual environments.

However, successful reconstruction of not only the geometry but also the texture of a person from a single viewpoint poses significant challenges due to depth ambiguity, changing topology, and severe occlusions. To address these challenges, data-driven approaches using high-capacity deep neural networks have been employed, demonstrating significant advances in the fidelity and robustness of human modeling [40,53,60,73]. In particular, Pixel-Aligned Implicit Function (PIFu) [53] achieves fully-textured reconstructions of clothed humans with a very high resolution that is infeasible with voxel-based approaches. On the other hand, the main limitation of PIFu is that the subsequent reconstruction process is not fast enough for real-time applications: given an input image, PIFu densely evaluates 3D occupancy fields, from which the underlining surface geometry is extracted using the Marching Cubes algorithm [35]. After the surface mesh reconstruction, the texture on the surface is inferred in a similar manner. Finally, the colored meshes are rendered from arbitrary viewpoints. The whole process takes tens of seconds per object when using a 256^3 resolution. Our goal is to achieve such fidelity and robustness with the highly efficient reconstruction and rendering speed for real-time applications.

To this end, we introduce a novel surface reconstruction algorithm, as well as a direct rendering method that does not require extracting surface meshes for rendering. The newly introduced surface localization algorithm progressively queries 3D locations in a coarse-to-fine manner to construct 3D occupancy fields with a smaller number of points to be evaluated. We empirically demonstrate that our algorithm retains the accuracy of the original reconstruction, while being two orders of magnitude faster than the brute-force baseline. Additionally, combined with the proposed surface reconstruction algorithm, our implicit texture representation enables direct novel-view synthesis without geometry tessellation or texture mapping, which halves the time required for rendering. As a result, we enable 15 fps processing time with a 256^3 spatial resolution for volumetric performance capture.

In addition, we present a key enhancement to the training method of [53] to further improve the quality and efficiency of reconstruction. To suppress failure cases that rarely occur during training due to the unbalanced data distribution with respect to viewing angles, poses and clothing styles, we introduce an adaptive data sampling algorithm inspired by the Online Hard Example Mining (OHEM) method [55]. We incrementally update the sampling probability based on the current prediction accuracy to train more frequently with hard examples without manually selecting these samples. We find this automatic sampling approach highly effective for reducing artifacts, resulting in state-of-the-art accuracy.

Our main contributions are:

- The first approach to full-body performance capture at real-time speed from monocular video not requiring a template. From a single image, our approach reconstructs a fully textured clothed human under a wide range of poses and clothing types without topology constraints.
- A progressive surface localization algorithm that makes surface reconstruction two orders of magnitude faster than the baseline without compromising the reconstruction accuracy, thus achieving a better trade-off between speed and accuracy than octree-based alternatives.
- A direct rendering technique for novel-view synthesis without explicitly extracting surface meshes, which further accelerates the overall performance.
- An effective training technique that addresses the fundamental imbalance in synthetically generated training data. Our Online Hard Example Mining method significantly reduces reconstruction artifacts and improves the generalization capabilities of our approach.

2   Related Work

Volumetric Performance Capture. Volumetric performance capture has been widely used to obtain human performances for free-viewpoint video [24] or high-fidelity geometry reconstruction [62]. To obtain the underlining geometry with an arbitrary topology, performance capture systems typically use general cues such as silhouettes [7,37,57,63], mutli-view correspondences [13,24], and reflectance information [62]. While these approaches successfully reconstruct geometry with an arbitrary topology, they require a large number of cameras with accurate calibration and controlled illumination. Another approach is to leverage commodity depth sensors to directly acquire 3D geometry. Volumetric fusion approaches have been used to jointly optimize for the relative 3D location and 3D geometry, incrementally updated from the captured sequence using a single depth sensor in real-time [20,42]. Later, this incremental geometry update was extended to non-rigidly deforming objects [19,41] and joint optimization with reflectance [16]. While these approaches do not require a template or category-specific prior, they only support relatively slow motions. Multi-view systems combined with depth sensors significantly improve the fidelity of the reconstructions [7,10,46] and both hardware and software improvements further facilitate the trend of high-fidelity volumetric performance capture [15,29]. However, the hardware requirements make it challenging to deploy these systems for non-professional users.

Template-Based Performance Capture. To relax the constraints of traditional volumetric performance capture, one common approach is to use a template model as an additional prior. Early works use a precomputed template model to reduce the number of viewpoints [8,64] and improve the reconstruction quality [61]. Template models are also used to enable performance capture from RGBD input [68,71]. However, these systems still rely on well-conditioned input from multiple viewpoints. Instead of a personalized template model, articulated morphable models such as SCAPE [2] or SMPL [34] are also widely used to recover human pose and shapes from video input [14], a single image [4,30], or RGBD input [69,72]. More recently, components corresponding to hands [52] and faces [5,32] were incorporated into a body model to perform more holistic performance capture from multi-view input [23], which was later extended to monocular input as well [48,65]. Although the use of a parameteric model greatly eases the ill-posed nature of monocular performance capture, the lack of personalized details such as clothing and hairstyles severely impairs the authenticity of the captured performance. Recently Xu et al. [66] demonstrated that articulated personalized avatars can be tracked from monocular RGB videos by incorporating inferred sparse 2D and 3D keypoints [38]. The most relevant work to our approach is [17], which is the real-time extension of [66] with the reconstruction fidelity also improved with an adaptive non-rigidity update. Unlike the aforementioned template-based approaches, our method is capable of representing personalized details present in the input image without any preprocessing, as our approach is based on a template-less volumetric representation, enabling topological updates and instantaneously changing the subject.

Deep Learning for Human Modeling. To infer fine-grained 3D shape and appearance from unconstrained images, where designing hand-crafted features is non-trivial, we need a high-capacity machine learning algorithm. The advent of deep learning showed promise by eliminating the need for hand-crafted features and demonstrated groundbreaking performance for human modeling tasks in the wild [1,25,38]. Fully convolutional neural networks have been used to infer 3D skeletal joints from a single image [38,49,51], which are used as building blocks for monocular performance capture systems [17,66].

Fig. 1. A performance captured and re-rendered system in real-time from a monocular input video.

For full-body reconstruction from a single image, various data representations have been explored, including meshes [25,28], dense correspondences [1], voxels [21,60,73], silhouettes [40], and implicit surfaces [18,53,54]. Notably, deep learning approaches using implicit shape representations have demonstrated significantly more detailed reconstructions by eliminating the discretization of space [6,39,47]. Saito et al. [53] further improve the fidelity of reconstruction by combining fully convolutional image features with implicit functions, and demonstrate that these implicit field representations can be extended to continuous texture fields for effective 3D texture inpainting without relying on precomputed 2D parameterizations. However, the major drawback of these implicit representations is that the inference is time-consuming due to the dense evaluation of the network in 3D space, which prevents its use for real-time applications. Though we base our 3D representation on [53] for high-fidelity and memory-efficient 3D reconstruction, our novel surface inference and rendering algorithms significantly accelerate the reconstruction and visualization of the implicit surface.

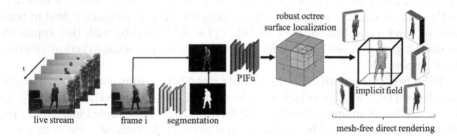

Fig. 2. System overview.

3   Method

In this section, we describe the overall pipeline of our algorithm for real-time volumetric capture (Fig. 2). Given a live stream of RGB images, our goal is to obtain the complete 3D geometry of the performing subject in real-time with the full textured surface, including unseen regions. To achieve an accessible solution with minimal requirements, we process each frame independently, as tracking-based solutions are prone to accumulating errors and sensitive to initialization, causing drift and instability [42,75]. Although recent approaches have demonstrated that the use of anchor frames [3,10] can alleviate drift, ad-hoc engineering is still required to handle common but extremely challenging scenarios such as changing the subject.

For each frame, we first apply real-time segmentation of the subject from the background. The segmented image is then fed into our enhanced Pixel-Aligned Implicit Function (PIFu) [53] to predict continuous occupancy fields where the underlining surface is defined as a 0.5-level set. Once the surface is determined, texture inference on the surface geometry is also performed using PIFu, allowing for rendering from any viewpoint for various applications. As this deep learning framework with effective 3D shape representation is the core building block of the proposed system, we review it in Sect. 3.1, describe our enhancements to it, and point out the limitations on its surface inference and rendering speed.

At the heart of our system, we develop a novel acceleration framework that enables real-time inference and rendering from novel viewpoints using PIFu (Sect. 3.2). Furthermore, we further improve the robustness of the system by sampling hard examples on the fly to efficiently suppress failure modes in a manner inspired by Online Hard Example Mining [55] (Sect. 3.3).

3.1   Pixel-Aligned Implicit Function (PIFu)

In volumetric capture, 3D geometry is represented as the level set surface of continuous scalar fields. That is, given an input frame $\mathbf{I}$, we need to determine whether a point in 3D space is inside or outside the human body. While this can be directly regressed using voxels, where the target space is explicitly discretized [21,60], the Pixel-Aligned Implicit Function (PIFu) models a function $O(\mathbf{P})$ that queries any 3D point and predicts the binary occupancy field in normalized device coordinates $\mathbf{P} = (P_x, P_y, P_z) \in \mathbb{R}^3$. Notably, with this approach no discretization is needed to infer 3D shapes, allowing reconstruction at arbitrary resolutions.

PIFu first extracts an image feature obtained from a fully convolutional image encoder $g_O(\mathbf{I})$ by a differentiable sampling function $\Phi(\mathbf{P}_{xy}, g_O(\mathbf{I}))$ (following [53], we use a bilinear sampling function [22] for Φ). Given the sampled image feature, a function parameterized by another neural network f_O estimates the occupancy of a queried point $\mathbf{P}$ as follows:

$$O(\mathbf{P}) = f_O(\Phi(\mathbf{P}_{xy}, g_O(\mathbf{I})), P_z) = \begin{cases} 1 & \text{if } \mathbf{P} \text{ is inside surface} \\ 0 & \text{otherwise.} \end{cases} \tag{1}$$

PIFu [53] uses a fully convolutional architecture for g_O to obtain image features that are spatially aligned with the queried 3D point, and a Multilayer Perceptron (MLP) for the function f_O, which are trained jointly in an end-to-end manner. Aside from the memory efficiency for high-resolution reconstruction, this representation especially benefits volumetric performance capture, as the spatially aligned image features ensure the 3D reconstruction retains details that are present in input images, $e.g.$ wrinkles, hairstyles, and various clothing styles. Instead of $L2$ loss as in [53], we use a Binary Cross Entropy (BCE) loss for learning the occupancy fields. As it penalizes false negatives and false positives more harshly than the $L2$ loss, we obtain faster convergence when using BCE.

Additionally, the same framework can be applied to texture inference by predicting vector fields instead of occupancy fields as follows:

$$\mathbf{T}(\mathbf{P}, \mathbf{I}) = f_T(\Phi(\mathbf{P}_{xy}, g_T(\mathbf{I})), \Phi(\mathbf{P}_{xy}, g_O(\mathbf{I})), P_z) = \mathbf{C} \in \mathbb{R}^3, \tag{2}$$

where given a surface point $\mathbf{P}$, the implicit function $\mathbf{T}$ predicts RGB color $\mathbf{C}$. The advantage of this representation is that texture inference can be performed on any surface geometry including occluded regions without requiring a shared 2D parameterization [31,67]. We use the $L1$ loss from the sampled point colors.

Furthermore, we made several modifications to the original implementation of [53] to further improve the accuracy and efficiency. For shape inference, instead of the stacked hourglass [43], we use HRNetV2-W18-Small-v2 [58] as a backbone, which demonstrates superior accuracy with less computation and parameters. We also use conditional batch normalization [9,11,39] to condition the MLPs on the sampled image features instead of the concatenation of these features to the queried depth value, which further improves the accuracy without increasing computational overhead. Additionally, inspired by an ordinal depth regression approach [12], we found that representing depth P_z as a soft one-hot vector more effectively propagates depth information, resulting in faster convergence. For texture inference, we detect the visible surface from the reconstruction and directly use the color from the corresponding pixel, as these regions do not require any inference, further improving the realism of free viewpoint rendering. We provide additional ablation studies to validate our design choices in the supplemental material.

Inference for Human Reconstruction. In [53], the entire digitization pipeline starts with the dense evaluation of the occupancy fields in 3D, from which the surface mesh is extracted using Marching Cubes [35]. Then, to obtain the fully textured mesh, the texture inference module is applied to the vertices on the surface mesh. While the implicit shape representation allows us to reconstruct 3D shapes with an arbitrary resolution, the evaluation in the entire 3D space is prohibitively slow, requiring tens of seconds to process a single frame. Thus, acceleration by at least two orders of magnitude is crucial for real-time performance.

3.2   Real-Time Inference and Rendering

To reduce the computation required for real-time performance capture, we introduce two novel acceleration techniques. First, we present an efficient surface localization algorithm that retains the accuracy of the brute-force reconstruction with the same complexity as naive octree-based reconstruction algorithms. Furthermore, since our final outputs are renderings from novel viewpoints, we bypass the explicit mesh reconstruction stage by directly generating a novel-view rendering from PIFu. By combining these two algorithms, we can successfully render the performance from arbitrary viewpoints in real-time. We describe each algorithm in detail below.

Octree-Based Robust Surface Localization. The major bottleneck of the pipeline is the evaluation of implicit functions represented by an MLP at an excessive number of 3D locations. Thus, substantially reducing the number of points to be evaluated would greatly increase the performance. The octree is a common data representation for efficient shape reconstruction [74] which hierarchically reduces the number of nodes in which to store data. To apply an octree for an implicit surface parameterized by a neural network, recently [39] propose an algorithm that subdivides grids only if it is adjacent to the boundary nodes (*i.e.*, the interface between inside node and outside node) after binarizing the

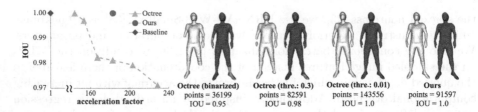

Fig. 3. Comparison of surface reconstruction methods. The plot shows the trade-off between the retention of the accuracy of the original reconstruction (*i.e.*, IOU) and speed. The acceleration factor is computed by dividing the number of evaluation points by that with the brute-force baseline. Note that the thresholds used for the octree reconstructions are 0.05, 0.08, 0.12, 0.2, 0.3, and 0.4 from left to right in the plot.

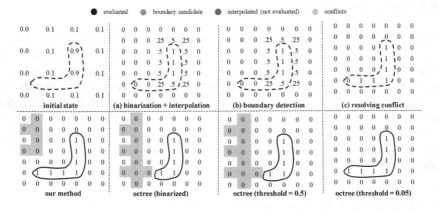

Fig. 4. Our surface localization algorithm overview. The dash and solid line denote the true surface and the reconstructed surface respectively. The nodes that are not used for the time-consuming network evaluation are shaded grey.

predicted occupancy value. We found that this approach often produces inaccurate reconstructions compared to the surface reconstructed by the brute force baseline (see Fig. 3). Since a predicted occupancy value is a continuous value in the range $[0, 1]$, indicating the confidence in and proximity to the surface, another approach is to subdivide grids if the maximum absolute deviation of the neighbor coarse grids is larger than a threshold. While this approach allows for control over the trade-off between reconstruction accuracy and acceleration, we also found that this algorithm either excessively evaluates unnecessary points to perform accurate reconstruction or suffers from impaired reconstruction quality in exchange for higher acceleration. To this end, we introduce a surface localization algorithm that hierarchically and precisely determines the boundary nodes.

We illustrate our surface localization algorithm in Fig. 4. Our goal is to locate grid points where the true surface exists within one of the adjacent nodes at the desired resolution, as only the nodes around the surface matter for surface reconstruction. We thus use a coarse-to-fine strategy in which boundary candidate grids are progressively updated by culling unnecessary evaluation points.

Given the occupancy prediction at the coarser level, we first binarize the occupancy values with threshold of 0.5, and apply interpolation (*i.e.*, bilinear for 2D cases, and trilinear for 3D) to tentatively assign occupancy values to the grid points at the current level (Fig. 4(a)). Then, we extract the boundary candidates by extracting the grid points whose values are neither 0 nor 1. To cover sufficiently large regions, we apply a dilation operation to incorporate the 1-ring neighbor of these boundary candidates (Fig. 4(b)). These selected nodes are evaluated with the network and the occupancy values at these nodes are updated. Note that if we terminate at this point and move on to the next level, the true boundary candidates may be culled similar to the aforementioned acceleration approaches. Thus, as an additional step, we detect conflict nodes by comparing the binarized values of the interpolation and the network prediction for the boundary candidates. The key observation is that there must be a missing surface region when the value of prediction and the interpolation is inconsistent. The nodes adjacent to the conflict nodes are evaluated with the network iteratively until all the conflicts are resolved (Fig. 4(c)).

Figure 4 shows the octree-based reconstruction with binarization [39] and the subdivision with a higher threshold suffers from inaccurate surface localization. While the subdivision approach with a lower threshold can prevent inaccurate reconstruction, an excessive number of nodes are evaluated. On the other hand, our approach not only extracts the accurate surface but also effectively reduces the number of nodes to be evaluated (see the number of blue-colored nodes).

Mesh-Free Rendering. While the proposed localization algorithm successfully accelerates surface localization, our end goal is rendering from novel viewpoints, and from any viewpoint a large portion of the reconstructed surface is not visible. Furthermore, PIFu allows us to directly infer texture at any point in 3D space, which can substitute the traditional rendering pipeline, where an explicit mesh is required to rasterize the scene. In other words, we can directly generate a novel-view image if the surface location is given from the target viewpoint. Motivated by this observation, we propose a view-based culling algorithm together with a direct rendering method for implicit data representations [39,45,53]. Note that while recently differentiable sphere tracing [33] and ray marching [44] approaches have been proposed to directly render implicit fields, these methods are not suitable for real-time rendering, as they sacrifice computational speed for differentiability to perform image-based supervision tasks.

Figure 5 shows the overview of the view-based surface extraction algorithm. For efficient view-based surface extraction, note that the occupancy grids are aligned with the normalized device coordinates defined by the target view instead of the model or world coordinates. That is, the x and y axes in the grid are corresponding to the pixel coordinates and the z axis is aligned with camera rays. Thus, our first objective is to search along the z axis to identify the first two consecutive nodes within which the surface geometry exists.

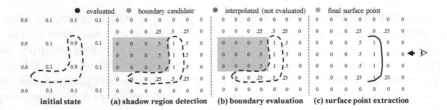

Fig. 5. Our mesh-free rendering overview. The dash and solid line denote the true surface and the reconstructed surface respectively.

First, we apply the aforementioned surface localization algorithm up to the $(L-1)$-th level, where $2^L \times 2^L \times 2^L$ is the target spatial resolution. Then, we upsample the binarized prediction at the $(L-1)$-th level using interpolation and apply the argmax operation along the z axis. The argmax operation provides the maximum value and the corresponding z index along the specified axis, where higher z values are closer to the observer. We denote the maximum value and the corresponding index at a pixel $\mathbf{q}$ by $O_{max}(\mathbf{q})$ and $i_{max}(\mathbf{q})$ respectively. Note that if multiple nodes contain the same maximum value, the function returns the smallest index. If $O_{max}(\mathbf{q}) = 1$, the nodes whose indices are greater than $i_{max}(\mathbf{q})$ are always occluded. Therefore, we treat these nodes as *shadow nodes* which are discarded for the network evaluation (Fig. 5(a)). Once shadow nodes are marked, we evaluate the remaining nodes with the interpolated value of 0.5 and update the occupancy values (Fig. 5(b)). Finally, we apply binarization to the current occupancy values and perform the argmax operation again along the z axis to obtain the updated nearest-point indices. For the pixels with $O_{max}(\mathbf{q}) = 1$, we take the nodes with the index of $i_{max}(\mathbf{q}) - 1$ and $i_{max}(\mathbf{q})$ as surface points and compute the 3D coordinates of surface $\mathbf{P}(\mathbf{q})$ by interpolating these two nodes by the predicted occupancy value (Fig. 5(c)). Then a novel-view image $\mathbf{R}$ is rendered as follows:

$$\mathbf{R}(\mathbf{q}) = \begin{cases} \mathbf{T}(\mathbf{P}(\mathbf{q}), \mathbf{I}) & \text{if } O_{max}(\mathbf{q}) = 1 \\ \mathbf{B} & \text{otherwise,} \end{cases} \tag{3}$$

where $\mathbf{B} \in \mathbb{R}^3$ is a background color. For virtual teleportation applications, we composite the rendering and the target scene using a transparent background.

3.3 Online Hard Example Mining for Data Sampling

As in [53], the importance-based point sampling for shape learning is more effective than uniform sampling within a bounding box to obtain highly detailed surfaces. However, we observe that this sampling strategy alone still fails to accurately reconstruct challenging poses and viewing angles, which account for only a small portion of the entire training data (see Fig. 6). Although one solution is to synthetically augment the dataset with more challenging training data, manually designing such a data augmentation strategy is non-trivial because various attributes (*e.g.*, poses, view angles, illuminations, and clothing types) may contribute to failure modes, and they are highly entangled.

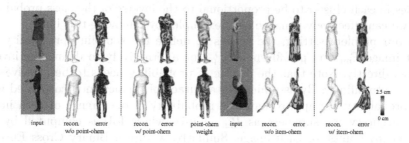

Fig. 6. Qualitative Evaluation of the OHEM sampling. The proposed sampling effectively selects challenging regions, resulting in significantly more robust reconstruction.

Nevertheless, the success of importance sampling in [53] illustrates that changing the data sampling distribution directly influences the quality of the reconstruction. This observation leads us to a fundamental solution to address the aforementioned training data bias without domain-specific knowledge. The key idea is to have the network automatically discover hard examples without manual intervention and adaptively change the sampling probability. We will first formulate the problem and solution in a general form and then develop an algorithm for our specific problem. While there are some works address the data bias problem using online hard negative mining (OHEM) strategy in various tasks such as learning image descriptors [56], image classifiers [36], and object detection [55], each employs a mining strategy specific to their task. So it is non-trivial to extend there algorithms to another problem. On the contrary, our formulation is general and can be applied to any problem domain as it requires no domain-specific knowledge.

Given a dataset $\mathcal{M}$, a common approach for supervised learning is to define an objective function L_m per data sample m and reduce an error within a mini-batch using optimizers (*e.g.*, SGD, Adam [26]). Assuming uniform distribution for data sampling, we are minimizing the following function $\mathcal{L}$ w.r.t. variables (*i.e.*, network weights) over the course of iterative optimization:

$$\mathcal{L} = \frac{1}{\|\mathcal{M}\|} \sum_{m \in \mathcal{M}} \mathcal{L}_m. \tag{4}$$

Now suppose the dataset is implicitly clustered into S classes denoted as $\{\mathcal{M}_i\}$ based on various attributes (*e.g.*, poses, illumination). Equation 4 can be written as:

$$\mathcal{L} = \frac{1}{\|\mathcal{M}\|} \sum_{i} \left(\sum_{m \in \mathcal{M}_i} \mathcal{L}_m \right) = \sum_{i} P_i \cdot \left(\frac{1}{\|\mathcal{M}_i\|} \sum_{m \in \mathcal{M}_i} \mathcal{L}_m \right), \tag{5}$$

where $P_i = \frac{\|\mathcal{M}_i\|}{\|\mathcal{M}\|}$ is the sampling probability of the cluster $\mathcal{M}_i$ among all the data samples. As shown in Eq. 5, the objective functions in each cluster are weighted by the probability P_i. This indicates that hard examples with lower probability are outweighed by the majority of the training data, resulting in poor reconstruction. On the other hand, if we modify the sampling probability of data samples in each cluster to be proportional to the inverse of the class probability P_i^{-1}, we can effectively penalize hard examples by removing this bias.

In our problem setting, the goal is to define the sampling probability per target image P_{im} and per 3D point P_{pt}, or alternatively to define the inverse of these directly. Note that the inverse of probability needs to be positive and not to go to infinity. By assuming the accuracy of prediction is correlated with class probability, we approximate the probability of occurrence of each image by an accuracy measurement as $P_{\mathrm{im}} \sim \mathrm{IoU}$, where IoU is computed by the sampled n_O points for each image. Similarly, we use a Binary Cross Entropy loss to approximate the original probability of sampling points. Based on these approximations, we model the inverse of the probabilities as follows:

$$P_{\mathrm{im}}^{-1} = \exp(-\mathrm{IoU}/\alpha_{\mathrm{i}} + \beta_{\mathrm{i}}), \qquad P_{\mathrm{pt}}^{-1} = \frac{1}{\exp(-\mathcal{L}_{BCE}/\alpha_{\mathrm{p}}) + \beta_{\mathrm{p}}}, \qquad (6)$$

where α and β are hyperparameters. In our experiments, we use $\alpha_{\mathrm{i}} = 0.15$, $\beta_{\mathrm{i}} = 10.0$, $\alpha_{\mathrm{p}} = 0.7$ and $\beta_{\mathrm{p}} = 0.0$. During training, we compute P_{im}^{-1} and P_{pt}^{-1} for each mini-batch and store the values for each data point, which are later used as the online sampling probability of each image and point after normalization. We refer to OHEM for images and points *item-ohem* and *point-ohem* respectively. Please refer to Sect. 4.1 for the ablation study to validate the effectiveness of our sampling strategy.

Table 1. Quantitative results. * mean the results of *top-k* = 10 worst cases. P denotes *point-ohem* and I denotes *item-ohem*.

Metric	Chamfer		P2S		Std		Chamfer*		P2S*		Runtime
	RP	BUFF	RP	BUFF	RP	BUFF	RP	BUFF	RP	BUFF	fps
VIBE [27]	–	5.485	–	5.794	–	4.279	–	10.653	–	11.572	20
DeepHuman [73]	–	4.208	–	4.340	–	4.022	—	10.460	–	11.389	0.0066
PIFu [53]	1.684	3.629	1.743	3.601	1.953	3.744	6.796	8.417	9.127	8.552	0.033
Ours	1.561	3.615	1.624	3.613	1.624	3.631	6.456	8.675	9.556	8.934	15
Ours+P	**1.397**	**3.515**	**1.514**	**3.566**	**1.552**	**3.518**	6.502	8.366	7.092	8.540	
Ours+P+I	1.431	3.592	1.557	3.603	1.579	3.560	**4.682**	**8.270**	**5.874**	**8.463**	

4   Results

We train our networks using NVIDIA GV100s with 512×512 images. During inference, we use a Logitech C920 webcam on a desktop system equipped with 62 GB RAM, a 6-core Intel i7-5930K processor, and 2 GV100s. One GPU performs geometry and color inference, while the other performs surface reconstruction, which can be done in parallel in an asynchronized manner when processing multiple frames. The overall latency of our system is on average 0.25 s.

We evaluate our proposed algorithms on the RenderPeople [50] and BUFF datasets [70], and on self-captured performances. In particular, as public datasets of 3D clothed humans in motion are highly limited, we use the BUFF datasets [70] for quantitative comparison and evaluation and report the average error measured by the Chamfer distance and point-to-surface (P2S) distance from the prediction to the ground truth. We provide implementation details, including the training dataset and real-time segmentation module, in the supplemental material.

In Fig. 1, we demonstrate our real-time performance capture and rendering from a single RGB camera. Because both the reconstructed geometry and texture inference for unseen regions are plausible, we can obtain novel-view renderings in real-time from a wide range of poses and clothing styles. We provide additional results with various poses, illuminations, viewing angles, and clothing in the supplemental document and video.

4.1   Evaluation

Figure 3 shows a comparison of surface reconstruction algorithms. The surface localization based on a binarized octree [39] does not guarantee the same reconstruction as the brute-force baseline, potentially losing some body parts. The octree-based reconstruction with a threshold shows the trade-off between performance and accuracy. Our method achieves the best acceleration without any hyperparameters, retaining the original reconstruction accuracy while accelerating surface reconstruction from 30 s to 0.14 s (7 fps). By combining it with our mesh-free rendering technique, we require only 0.06 s per frame (15 fps) for novel-view rendering at the volumetric resolution of 256^3, enabling the first real-time volumetric performance capture from a monocular video.

In Table 1 and Fig. 6, we evaluate the effectiveness of the proposed Online Hard Example Mining algorithm quantitatively and qualitatively. Using the same training setting, we train our model with and without the *point-ohem* and *item-ohem* sampling. Figure 6 shows the reconstruction results and error maps from the worst 5 results in the training set. The *point-ohem* successfully improves the fidelity of reconstruction by focusing on the regions with high error (see the *point-ohem* weight in Fig. 6). Similarly, the *item-ohem* automatically supervises more on hard images with less frequent clothing styles or poses, which we expect to capture as accurately as more common poses and clothing styles. As a result, the overall reconstruction quality is significantly improved, compared with the original implementation of [53], achieving state-of-the-art accuracy (Table 1).

4.2   Comparison

In Table 1 and Fig. 7, we compare our method with the state-of-the-art 3D human reconstruction algorithms from RGB input. Note that we train PIFu [53] using the same training data with the other settings identical to [53] for a fair comparison, while we use the public pretrained models for VIBE [27] and DeepHuman [73] due to the custom datasets required by each method and their dependency

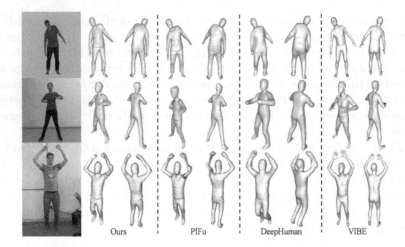

Ours   PIFu   DeepHuman   VIBE

Fig. 7. Qualitative comparison with other reconstruction methods.

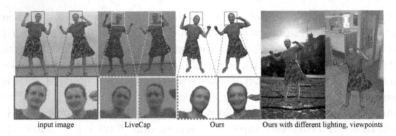

input image   LiveCap   Ours   Ours with different lighting, viewpoints

Fig. 8. Comparison template-based performance capture from monocular video.

on external modules such as the SMPL [34] model. Although a template-based regression approach [27] achieves robust 3D human estimations from images in the wild, the lack of fidelity and details severely impairs the authenticity of the performances. Similarly, a volumetric performance capture based on voxels [73] suffers from a lack of fidelity due to the limited resolution. While an implicit shape representation [53] achieves high-resolution reconstruction, the reconstructions become less plausible for infrequent poses and the inference speed (30 s) is too slow for real-time applications, both of which we address in this paper. We also qualitatively compare our reconstruction with the state-of-the-art real-time performance capture using a pre-captured template [17] (Fig. 8). While the reconstructed geometries are comparable, our method can render performances with dynamic textures that reflect lively expressions, unlike a tracking method using a fixed template. Our approach is also agnostic to topology changes, and can thus handle very challenging scenarios such as changing clothing (Fig. 1).

5   Conclusion

We have demonstrated that volumetric reconstruction and rendering of humans from a single input image is possible to achieve in near real-time speed without sacrificing the final image quality. Our novel progressive surface localization method allows us to vastly reduce the number of points queried during surface reconstruction, giving us a speedup of two orders of magnitude without reducing the final surface quality. Furthermore, we demonstrate that directly rendering novel viewpoints of the captured subject is possible without explicitly extracting a mesh or performing naive, computationally intensive volumetric rendering, allowing us to obtain real-time rendering performance with the reconstructed surface. Finally, our Online Hard Example Mining technique allows us to find and learn the appropriate response to challenging input examples, thereby making it feasible to train our networks with a tractable amount of data while attaining high-quality results with large appearance and motion variations. While we demonstrate our approach on human subjects and performances, our acceleration techniques are straightforward to implement and generalize to any object or topology. We thus believe this will be a critical building block to virtually teleport anything captured by a commodity camera anywhere.

Acknowledgement. This research was funded by in part by the ONR YIP grant N00014-17-S-FO14, the CONIX Research Center, a Semiconductor Research Corporation (SRC) program sponsored by DARPA, the Andrew and Erna Viterbi Early Career Chair, the U.S. Army Research Laboratory (ARL) under contract number W911NF-14-D-0005, Adobe, and Sony.

References

1. Alp Güler, R., Neverova, N., Kokkinos, I.: DensePose: dense human pose estimation in the wild. In: Proceedings of the IEEE Conference on Computer Vision and Pattern Recognition, pp. 7297–7306 (2018)
2. Anguelov, D., Srinivasan, P., Koller, D., Thrun, S., Rodgers, J., Davis, J.: SCAPE: shape completion and animation of people. ACM Trans. Graph. **24**(3), 408–416 (2005)
3. Beeler, T., et al.: High-quality passive facial performance capture using anchor frames. ACM Trans. Graph. (TOG) **30**(4), 75 (2011)
4. Bogo, F., Kanazawa, A., Lassner, C., Gehler, P., Romero, J., Black, M.J.: Keep it SMPL: automatic estimation of 3D human pose and shape from a single image. In: European Conference on Computer Vision, pp. 561–578 (2016)
5. Cao, C., Weng, Y., Zhou, S., Tong, Y., Zhou, K.: FaceWarehouse: a 3D facial expression database for visual computing. IEEE Trans. Vis. Comput. Graph. **20**(3), 413–425 (2013)
6. Chen, Z., Zhang, H.: Learning implicit fields for generative shape modeling. In: IEEE Conference on Computer Vision and Pattern Recognition, pp. 5939–5948 (2019)
7. Collet, A., et al.: High-quality streamable free-viewpoint video. ACM Trans. Graph. **34**(4), 69 (2015)

8. De Aguiar, E., Stoll, C., Theobalt, C., Ahmed, N., Seidel, H.P., Thrun, S.: Performance capture from sparse multi-view video. ACM Trans. Graph. **27**(3), 98 (2008)
9. De Vries, H., Strub, F., Mary, J., Larochelle, H., Pietquin, O., Courville, A.C.: Modulating early visual processing by language. In: Advances in Neural Information Processing Systems, pp. 6594–6604 (2017)
10. Dou, M., et al.: Fusion4D: real-time performance capture of challenging scenes. ACM Trans. Graph. **35**(4), 114 (2016)
11. Dumoulin, V., et al.: Adversarially learned inference. arXiv preprint arXiv:1606.00704 (2016)
12. Fu, H., Gong, M., Wang, C., Batmanghelich, K., Tao, D.: Deep ordinal regression network for monocular depth estimation. In: Proceedings of the IEEE Conference on Computer Vision and Pattern Recognition, pp. 2002–2011 (2018)
13. Furukawa, Y., Ponce, J.: Accurate, dense, and robust multiview stereopsis. IEEE Trans. Pattern Anal. Mach. Intell. **32**(8), 1362–1376 (2010)
14. Guan, P., Weiss, A., Balan, A.O., Black, M.J.: Estimating human shape and pose from a single image. In: IEEE International Conference on Computer Vision, pp. 1381–1388 (2009)
15. Guo, K., et al.: The relightables: volumetric performance capture of humans with realistic relighting. ACM Trans. Graph. **38**(6) (2019). https://doi.org/10.1145/3355089.3356571
16. Guo, K., Xu, F., Yu, T., Liu, X., Dai, Q., Liu, Y.: Real-time geometry, albedo, and motion reconstruction using a single RGB-D camera. ACM Trans. Graph. (TOG) **36**(3), 32 (2017)
17. Habermann, M., Xu, W., Zollhoefer, M., Pons-Moll, G., Theobalt, C.: LiveCap: real-time human performance capture from monocular video. ACM Trans. Graph. (TOG) **38**(2), 14 (2019)
18. Huang, Z., Xu, Y., Lassner, C., Li, H., Tung, T.: ARCH: animatable reconstruction of clothed humans. In: Proceedings of the IEEE/CVF Conference on Computer Vision and Pattern Recognition, pp. 3093–3102 (2020)
19. Innmann, M., Zollhöfer, M., Nießner, M., Theobalt, C., Stamminger, M.: VolumeDeform: real-time volumetric non-rigid reconstruction. In: Leibe, B., Matas, J., Sebe, N., Welling, M. (eds.) ECCV 2016. LNCS, vol. 9912, pp. 362–379. Springer, Cham (2016). https://doi.org/10.1007/978-3-319-46484-8_22
20. Izadi, S., et al.: KinectFusion: real-time 3D reconstruction and interaction using a moving depth camera. In: Proceedings of the 24th Annual ACM Symposium on User Interface Software and Technology, pp. 559–568 (2011)
21. Jackson, A.S., Manafas, C., Tzimiropoulos, G.: 3D human body reconstruction from a single image via volumetric regression. In: Leal-Taixé, L., Roth, S. (eds.) ECCV 2018. LNCS, vol. 11132, pp. 64–77. Springer, Cham (2019). https://doi.org/10.1007/978-3-030-11018-5_6
22. Jaderberg, M., Simonyan, K., Zisserman, A., et al.: Spatial transformer networks. In: Advances in Neural Information Processing Systems, pp. 2017–2025 (2015)
23. Joo, H., Simon, T., Sheikh, Y.: Total capture: a 3D deformation model for tracking faces, hands, and bodies. In: Proceedings of the IEEE Conference on Computer Vision and Pattern Recognition, pp. 8320–8329 (2018)
24. Kanade, T., Rander, P., Narayanan, P.: Virtualized reality: constructing virtual worlds from real scenes. IEEE Multimed. **4**(1), 34–47 (1997)
25. Kanazawa, A., Black, M.J., Jacobs, D.W., Malik, J.: End-to-end recovery of human shape and pose. In: Proceedings of the IEEE Conference on Computer Vision and Pattern Recognition, pp. 7122–7131 (2018)

26. Kingma, D.P., Ba, J.: Adam: a method for stochastic optimization. arXiv preprint arXiv:1412.6980 (2014)
27. Kocabas, M., Athanasiou, N., Black, M.J.: Vibe: video inference for human body pose and shape estimation. In: The IEEE Conference on Computer Vision and Pattern Recognition (CVPR), June 2020
28. Kolotouros, N., Pavlakos, G., Black, M.J., Daniilidis, K.: Learning to reconstruct 3D human pose and shape via model-fitting in the loop. In: Proceedings of the IEEE International Conference on Computer Vision (2019)
29. Kowdle, A., et al.: The need 4 speed in real-time dense visual tracking. In: SIG-GRAPH Asia 2018 Technical Papers, p. 220. ACM (2018)
30. Lassner, C., Romero, J., Kiefel, M., Bogo, F., Black, M.J., Gehler, P.V.: Unite the people: closing the loop between 3D and 2D human representations. In: IEEE Conference on Computer Vision and Pattern Recognition, pp. 6050–6059 (2017)
31. Lazova, V., Insafutdinov, E., Pons-Moll, G.: 360-degree textures of people in clothing from a single image. In: International Conference on 3D Vision (3DV), September 2019
32. Li, T., Bolkart, T., Black, M.J., Li, H., Romero, J.: Learning a model of facial shape and expression from 4D scans. ACM Trans. Graph. (TOG) 36(6), 194 (2017)
33. Liu, S., Zhang, Y., Peng, S., Shi, B., Pollefeys, M., Cui, Z.: DIST: rendering deep implicit signed distance function with differentiable sphere tracing. arXiv preprint arXiv:1911.13225 (2019)
34. Loper, M., Mahmood, N., Romero, J., Pons-Moll, G., Black, M.J.: SMPL: a skinned multi-person linear model. ACM Trans. Graph. 34(6), 248 (2015)
35. Lorensen, W.E., Cline, H.E.: Marching cubes: a high resolution 3D surface construction algorithm. ACM SIGGRAPH Comput. Graph. 21(4), 163–169 (1987)
36. Loshchilov, I., Hutter, F.: Online batch selection for faster training of neural networks. arXiv preprint arXiv:1511.06343 (2015)
37. Matusik, W., Buehler, C., Raskar, R., Gortler, S.J., McMillan, L.: Image-based visual hulls. In: ACM SIGGRAPH, pp. 369–374 (2000)
38. Mehta, D., et al.: VNect: real-time 3D human pose estimation with a single RGB camera. ACM Trans. Graph. 36(4), 44:1–44:14 (2017)
39. Mescheder, L., Oechsle, M., Niemeyer, M., Nowozin, S., Geiger, A.: Occupancy networks: learning 3D reconstruction in function space. arXiv preprint arXiv:1812.03828 (2018)
40. Natsume, R., et al.: SiCloPe: silhouette-based clothed people. In: CVPR, pp. 4480–4490 (2019)
41. Newcombe, R.A., Fox, D., Seitz, S.M.: DynamicFusion: reconstruction and tracking of non-rigid scenes in real-time. In: IEEE Conference on Computer Vision and Pattern Recognition, pp. 343–352 (2015)
42. Newcombe, R.A., et al.: KinectFusion: real-time dense surface mapping and tracking. In: 2011 10th IEEE International Symposium on Mixed and Augmented Reality (ISMAR), pp. 127–136 (2011)
43. Newell, A., Yang, K., Deng, J.: Stacked hourglass networks for human pose estimation. In: Leibe, B., Matas, J., Sebe, N., Welling, M. (eds.) ECCV 2016. LNCS, vol. 9912, pp. 483–499. Springer, Cham (2016). https://doi.org/10.1007/978-3-319-46484-8_29
44. Niemeyer, M., Mescheder, L., Oechsle, M., Geiger, A.: Differentiable volumetric rendering: Learning implicit 3D representations without 3D supervision. arXiv preprint arXiv:1912.07372 (2019)

45. Oechsle, M., Mescheder, L., Niemeyer, M., Strauss, T., Geiger, A.: Texture fields: learning texture representations in function space. In: The IEEE International Conference on Computer Vision (ICCV), October 2019
46. Orts-Escolano, S., et al.: Holoportation: virtual 3D teleportation in real-time. In: Proceedings of the 29th Annual Symposium on User Interface Software and Technology, pp. 741–754 (2016)
47. Park, J.J., Florence, P., Straub, J., Newcombe, R., Lovegrove, S.: DeepSDF: learning continuous signed distance functions for shape representation. arXiv preprint arXiv:1901.05103 (2019)
48. Pavlakos, G., et al.: Expressive body capture: 3D hands, face, and body from a single image. In: Proceedings of the IEEE Conference on Computer Vision and Pattern Recognition, pp. 10975–10985 (2019)
49. Popa, A.I., Zanfir, M., Sminchisescu, C.: Deep multitask architecture for integrated 2D and 3D human sensing. In: Proceedings of the IEEE Conference on Computer Vision and Pattern Recognition, pp. 6289–6298 (2017)
50. Renderpeople (2018). https://renderpeople.com/3d-people
51. Rogez, G., Weinzaepfel, P., Schmid, C.: LCR-Net++: multi-person 2D and 3D pose detection in natural images. arXiv preprint arXiv:1803.00455 (2018)
52. Romero, J., Tzionas, D., Black, M.J.: Embodied hands: modeling and capturing hands and bodies together. ACM Trans. Graph. (Proc. SIGGRAPH Asia) 36(6), 245 (2017)
53. Saito, S., Huang, Z., Natsume, R., Morishima, S., Kanazawa, A., Li, H.: PIFU: pixel-aligned implicit function for high-resolution clothed human digitization. In: ICCV (2019)
54. Saito, S., Simon, T., Saragih, J., Joo, H.: PIFuHD: multi-level pixel-aligned implicit function for high-resolution 3D human digitization. In: Proceedings of the IEEE/CVF Conference on Computer Vision and Pattern Recognition, pp. 84–93 (2020)
55. Shrivastava, A., Gupta, A., Girshick, R.: Training region-based object detectors with online hard example mining. In: Proceedings of the IEEE Conference on Computer Vision and Pattern Recognition, pp. 761–769 (2016)
56. Simo-Serra, E., Trulls, E., Ferraz, L., Kokkinos, I., Moreno-Noguer, F.: Fracking deep convolutional image descriptors. arXiv preprint arXiv:1412.6537 (2014)
57. Starck, J., Hilton, A.: Surface capture for performance-based animation. IEEE Comput. Graph. Appl. 27(3), 21–31 (2007)
58. Sun, K., et al.: High-resolution representations for labeling pixels and regions. arXiv preprint arXiv:1904.04514 (2019)
59. Tang, D., et al.: Real-time compression and streaming of 4D performances. In: SIGGRAPH Asia 2018 Technical Papers, p. 256. ACM (2018)
60. Varol, G., et al.: BodyNet: volumetric inference of 3D human body shapes. In: Ferrari, V., Hebert, M., Sminchisescu, C., Weiss, Y. (eds.) ECCV 2018. LNCS, vol. 11211, pp. 20–38. Springer, Cham (2018). https://doi.org/10.1007/978-3-030-01234-2_2
61. Vlasic, D., Baran, I., Matusik, W., Popović, J.: Articulated mesh animation from multi-view silhouettes. ACM Trans. Graph. 27(3), 97 (2008)
62. Vlasic, D., et al.: Dynamic shape capture using multi-view photometric stereo. ACM Trans. Graph. 28(5), 174 (2009)
63. Waschbüsch, M., Würmlin, S., Cotting, D., Sadlo, F., Gross, M.: Scalable 3D video of dynamic scenes. Vis. Comput. 21(8), 629–638 (2005)
64. Wu, C., Stoll, C., Valgaerts, L., Theobalt, C.: On-set performance capture of multiple actors with a stereo camera. ACM Trans. Graph. 32(6), 161 (2013)

65. Xiang, D., Joo, H., Sheikh, Y.: Monocular total capture: posing face, body, and hands in the wild. In: Proceedings of the IEEE Conference on Computer Vision and Pattern Recognition, pp. 10965–10974 (2019)
66. Xu, W., et al.: MonoPerfCap: human performance capture from monocular video. ACM Trans. Graph. **37**(2), 27:1–27:15 (2018)
67. Yamaguchi, S., et al.: High-fidelity facial reflectance and geometry inference from an unconstrained image. ACM Trans. Graph. **37**(4), 162 (2018)
68. Ye, G., Liu, Y., Hasler, N., Ji, X., Dai, Q., Theobalt, C.: Performance capture of interacting characters with handheld kinects. In: Fitzgibbon, A., Lazebnik, S., Perona, P., Sato, Y., Schmid, C. (eds.) ECCV 2012. LNCS, vol. 7573, pp. 828–841. Springer, Heidelberg (2012). https://doi.org/10.1007/978-3-642-33709-3_59
69. Yu, T., et al.: DoubleFusion: real-time capture of human performances with inner body shapes from a single depth sensor. In: Proceedings of the IEEE Conference on Computer Vision and Pattern Recognition, pp. 7287–7296 (2018)
70. Zhang, C., Pujades, S., Black, M.J., Pons-Moll, G.: Detailed, accurate, human shape estimation from clothed 3D scan sequences. In: Proceedings of the IEEE Conference on Computer Vision and Pattern Recognition, pp. 4191–4200 (2017)
71. Zhang, P., Siu, K., Zhang, J., Liu, C.K., Chai, J.: Leveraging depth cameras and wearable pressure sensors for full-body kinematics and dynamics capture. ACM Trans. Graph. (TOG) **33**(6), 221 (2014)
72. Zheng, Z., et al.: HybridFusion: real-time performance capture using a single depth sensor and sparse IMUs. In: Ferrari, V., Hebert, M., Sminchisescu, C., Weiss, Y. (eds.) ECCV 2018. LNCS, vol. 11213, pp. 389–406. Springer, Cham (2018). https://doi.org/10.1007/978-3-030-01240-3_24
73. Zheng, Z., Yu, T., Wei, Y., Dai, Q., Liu, Y.: DeepHuman: 3D human reconstruction from a single image. In: The IEEE International Conference on Computer Vision (ICCV), October 2019
74. Zhou, K., Gong, M., Huang, X., Guo, B.: Data-parallel octrees for surface reconstruction. IEEE Trans. Vis. Comput. Graph. **17**(5), 669–681 (2010)
75. Zollhöfer, M., et al.: Real-time non-rigid reconstruction using an RGB-D camera. ACM Trans. Graph. **33**(4), 156 (2014)

The Mapillary Traffic Sign Dataset
for Detection and Classification
on a Global Scale

Christian Ertler[✉][iD], Jerneja Mislej[iD], Tobias Ollmann[iD], Lorenzo Porzi[iD],
Gerhard Neuhold[iD], and Yubin Kuang[iD]

Facebook, Menlo Park, USA
christian@mapillary.com

Abstract. Traffic signs are essential map features for smart cities and
navigation. To develop accurate and robust algorithms for traffic sign
detection and classification, a large-scale and diverse benchmark dataset
is required. In this paper, we introduce a new traffic sign dataset of
$105K$ street-level images around the world covering 400 manually annotated traffic sign classes in diverse scenes, wide range of geographical
locations, and varying weather and lighting conditions. The dataset
includes $52K$ fully annotated images. Additionally, we show how to augment the dataset with $53K$ semi-supervised, partially annotated images.
This is the largest and the most diverse traffic sign dataset consisting of
images from all over the world with fine-grained annotations of traffic
sign classes. We run extensive experiments to establish strong baselines
for both detection and classification tasks. In addition, we verify that
the diversity of this dataset enables effective transfer learning for existing large-scale benchmark datasets on traffic sign detection and classification. The dataset is freely available for academic research (www.
mapillary.com/dataset/trafficsign).

1 Introduction

Robust and accurate object detection and classification in diverse scenes is one
of the essential tasks in computer vision. With the development and application
of deep learning in computer vision, object detection and recognition has been
studied [5,17,24] extensively on general scene understanding datasets [4,11,18].
In terms of fine-grained detection and classification, there are also datasets that
focus on general hierarchical object classes [11] or domain-specific datasets,
e.g. on bird species [32]. In this paper, we will focus on detection and fine-
grained classification of traffic signs on a new dataset.

Traffic signs are key map features for navigation, road safety and traffic
control. More specifically, traffic signs encode information for driving directions,

Electronic supplementary material The online version of this chapter (https://
doi.org/10.1007/978-3-030-58592-1_5) contains supplementary material, which is available to authorized users.

Table 1. Overview of traffic sign datasets. The numbers include only publicly available images and annotations. *Unique* refers to datasets where each traffic sign bounding box corresponds to a unique traffic sign instance (*i.e.* no sequences showing the same physical sign). *70,428/17,666 (train-val/test) signs are within the taxonomy. ** *All* includes train, val, test, and partial (semi) sets. [†]TT100K provides only 10,000 images containing traffic signs. [‖]45 classes have more than 100 examples. [¶]MVD contains back *vs.* front classes. [‡]video-frames covering only 15,630 unique signs. [§]signs within the partially annotated set correspond to physical signs within the training set

Dataset	Images	Classes	Signs	Attributes	Region	Boxes	Unique
MTSD (train/val)	**41,909**		***206,386**	occluded, exterior,		✓	✓
MTSD (test)	**10,544**	**400**	***51,155**	out-of-frame, dummy,	**global**	✓	✓
MTSD (all)**	**105,830**		354,154	ambiguous, included		✓	[§]✗
TT100K [35]	[†]100,000	[‖]221	26,349	✗	China	✓	✓
MVD [22]	20,000	[¶]2	174,541	✗	**global**	✓	✓
BDD100K [34]	100,000	1	343,777	✗	USA	✓	✗
GTSDB [10]	900	43	852	✗	Germany	✓	✗
RTSD [26]	[‡]179,138	156	[‡]104,358	✗	Russia	✓	✗
STS [13]	3777	20	5582	✗	Sweden	✓	✗
LISA [21]	6610	47	7855	✗	USA	✓	✗
GTSRB [29]	✗	43	39,210	✗	Germany	✗	✗
BelgiumTS [31]	✗	108	8851	✗	Belgium	✗	✗

traffic regulation, and early warning. Accurate and robust perception of traffic signs is also essential for localization and motion planning in different driving scenarios.

As an object class, traffic signs have specific characteristics in their appearance. First of all, traffic signs are in general rigid and planar. Secondly, traffic signs are designed to be distinctive from their surroundings. In addition, there is limited variety in colors and shapes for traffic signs. For instance, regulatory signs in European countries are typically circular with a red border. To some degree, the aforementioned characteristics limit the appearance variation and increase the distinctness of traffic signs. However, traffic sign detection and classification are still very challenging problems due to the following reasons: (1) traffic signs are easily confused with other object classes in street scenes (*e.g.* advertisements, banners, and billboards); (2) reflection, low light condition, damages, and occlusion hinder the classification performance of a sign class; (3) fine-grained classification with small inter-class difference is not trivial; (4) the majority of traffic signs—when appearing in street-level images—are relatively small in size, which requires efficient architecture designs for small objects.

Traffic sign detection and classification have been studied extensively in computer vision [14,20,25,35]. However, these studies were done in relatively constrained settings in terms of the benchmark dataset: the images and traffic signs are collected in a specific country; the number of traffic sign classes is relatively small; the images lack diversity in weather conditions, camera sensors, and seasonal changes. Extensive research is still needed for detecting and classifying traffic signs at a global scale and under varying capture conditions and devices.

Fig. 1. *Top*: Taxonomy overview. The sizes are relative to the number of samples within MTSD. *Bottom*: Example images in MTSD with bounding box and class annotations (green boxes without template indicate *other-sign*). (Color figure online)

The contributions of this paper are manifold:

- We present the **most diverse traffic sign dataset** with $105K$ images from all over the world. The dataset contains over $52K$ fully annotated images, covering 400 known traffic sign classes and other unknown classes, resulting in over $255K$ signs in total.
- Without introducing any additional annotation cost, we show how to augment the dataset with **real semi-supervised samples by propagating labels** to nearby images which helps to get more samples, especially in the long-tail of the class distribution. The dataset includes about $53K$ extra images collected in this way.
- We establish **extensive baselines** for detection and classification on the dataset, shedding light on future research directions.
- We study the impact of **transfer learning** using our traffic sign dataset and other datasets released in the past. We show that pre-training on our dataset boosts average precision (AP) of the binary detection task by 4–6 points, thanks to the completeness and diversity of our dataset.

Related Work. Traffic sign detection and recognition has been studied extensively in the previous literature. The German Traffic Sign Benchmark Dataset (GTSBD) [30] is one of the first datasets that was created to evaluate the classification branch of the problem. Following that, there have also been other traffic sign datasets focusing on regional traffic signs, *e.g.* Swedish Traffic Sign Dataset [12], Belgium Traffic Sign Dataset [20], Russian Traffic Sign Dataset [26], and Tsinghua-Tencent Dataset (TT100K) in China [35]. For generic traffic sign detection (where no class information of the traffic signs is available), there has been work done in the Mapillary Vistas Dataset (MVD) [22] (global) and BDD100K [34] (US only). A detailed overview and comparison of publicly available traffic sign datasets can be found in Table 1.

For general object detection, there has been substantial work on CNN-based methods with two main directions, *i.e.* one-stage detectors [17,19,23] and two-stage detectors [3,5,6,24]. One-stage detectors are generally much faster, trading off accuracy compared to two-stage detectors. One exception is the one-stage RetinaNet [17] architecture that outperforms the two-stage Faster-RCNN [24] thanks to a weighting scheme during training to suppress trivial negative supervision. For simultaneous detection and classification, recent work [2] shows that decoupling the classification from detection head boosts the accuracy significantly. Our work is related to [2] as we also decouple the detector from the traffic sign classifier.

To handle the scale variation of objects in the scene, many efficient multi-scale training and inference algorithms have been proposed and evaluated on existing datasets. For multi-scale training, in [15,27,28], a few schemes have been proposed to distill supervision from different scales efficiently by selective gradient propagation and crop generation. To enable efficient multi-scale inference, feature pyramid networks (FPN) [16] were proposed to utilize lateral connections in a top-down architecture to construct an effective multi-scale feature pyramid from a single image.

To develop the baselines presented in this paper, we have chosen Faster-RCNN [24] with FPN [16] as the backbone. Given the aforementioned characteristics of traffic sign imagery, we have also trained a separate classifier for fine-grained classification as in [2]. We elaborate on the details of our baseline methods in Sect. 4 and Sect. 5.

2   Mapillary Traffic Sign Dataset

In this section, we present a large-scale traffic sign dataset called Mapillary[1] Traffic Sign Dataset (MTSD) including $52K$ images with $257K$ fully annotated traffic sign bounding boxes and corresponding class labels. Additionally, it includes a set of over $53K$ nearby images with more than $84K$ semi-supervised class labels, making it more than $105K$ images. In the following we describe how the dataset was created and present our traffic sign class taxonomy consisting of 400 classes. Examples can be found in Fig. 1(bottom).

2.1   Image Selection

There are various conventions for traffic signs in different parts of the world leading to strong appearance differences. Even within a single country, the distribution of signs is not uniform: some signs occur only in urban areas, some only on highways, and others only in rural areas. With MTSD, we present a dataset that covers this diversity uniformly. In order to do so, a proper pre-selection of images for annotation is crucial. The requirements for this selection step are: (1) to have a uniform geographical distribution of images around the world, (2) to cover images of different quality, captured under varying conditions, (3) to include as many signs as possible per image, and (4) to compensate for the long-tailed distribution of potential traffic sign classes.

---

[1] www.mapillary.com/app is a street-level imagery platform hosting images collected by members of their community.

In order to get a pool of pre-selected images satisfying the aforementioned requirements, we sample images in a per-country manner. The fraction of target images for each country is derived from the number of images available in that country and its population count weighted by a global target distribution over all continents (*i.e.* 20 % North America, 20 % Europe, 20 % Asia, 15 % South America, 15 % Oceania, 10 % Africa). We further make sure to cover both rural and urban areas within each country by binning the sampled images uniformly in terms of their geographical locations and sample random images from each of the resulting bins. In the last step of our image sampling scheme, we prioritize images containing at least one traffic sign instance according to the traffic sign detections given by the Mapillary API and make sure to cover various image resolutions, camera manufacturers, and scene properties[2]. Additionally, we add a distance constraint so that selected images are far away from each other in order to avoid highly correlated images and traffic sign instances. Statistics of the dataset can be found in Sect. 3.

2.2   Traffic Sign Class Taxonomy

Traffic signs vary across different countries. For many countries, there exists no publicly available and complete catalogue of signs. The lack of a known set of traffic sign classes leads to challenges in assigning class labels to traffic signs annotated in MTSD. The potential magnitude of this unknown set of traffic signs is in the thousands as indicated by the set of template images described in Sect. 2.3.

For MTSD, we did a manual inspection of the templates that have been chosen by the annotators and selected a subset of them to form the final set of 400 classes included in the dataset as visualized in Fig. 1(top). This subset was chosen and grouped such that there are no overlaps or confusions (visual or semantic) among the classes. All these classes defined by disjoint sets of templates build up our traffic sign class taxonomy. We map all annotated traffic signs in MTSD that have a template selected within this taxonomy to a class label. We would like to emphasize that our flexible taxonomy allows us to incrementally extend MTSD. It enables to add more classes by grouping templates to new classes and mapping traffic sign instances to these new classes based on already assigned templates.

2.3   Annotation Process

The process of annotating an image including image selection approval, traffic sign localization by drawing bounding boxes, and class label assignment for each box is a complex and demanding task. To improve efficiency and quality, we split it into 3 consecutive tasks, with each having its own quality assurance process.

---

[2] Details on how scene properties are defined and derived are included in the supplementary materials.

All tasks were done by 15 experts in image annotation after being trained with explicit specifications for each task.

Image Approval. Since initial image selection was done automatically based on the heuristics described in Sect. 2.1, the annotators needed to reject images that did not fulfill our criteria. In particular, we do not include non-street-level images or images that have been taken from unusual places or viewpoints. Further we discarded images of very low quality that could not be used for training (*i.e.* extremely blurry or overexposed). However, we still sample images of low quality in the dataset which include recognizable traffic signs as these are good examples to evaluate recognition of traffic signs in real-world scenarios.

Sign Localization. In this task, the annotators were instructed to localize all traffic signs in the images and annotate them with bounding boxes. In contrast to previous traffic sign datasets where only specific types of traffic signs have been annotated (*e.g.* TT100K [35] includes only standard circular and triangular shaped signs), MTSD contains bounding boxes for all types of traffic related signs including direction, information, highway signs, *etc.*

To speed up the annotation process, each image was initialized with bounding boxes of traffic signs extracted from the *Mapillary* API. The annotators were asked to correct all existing bounding boxes to tightly contain the signs (or reject them in cases of false positives) and to annotate all missing traffic signs if their shorter sides were larger than 10 pixels. We provide a statistical analysis of the manual interactions of the annotators in supplemental material.

Sign Classification. This task was done independently for each annotated traffic sign. Each traffic sign (together with some image context) was shown to the annotators who were asked to provide the correct class label. This is not trivial, since the number of traffic sign classes is large. To the best of our knowledge, there is no globally valid traffic sign taxonomy available; even then, it would be impossible for the annotators to keep track of the different traffic sign classes.

To overcome this issue, we used a set of previously harvested template images of traffic signs from Wikimedia Commons [33] and grouped them by similarity in appearance and semantics. This set of templates (together with their grouping) defines the possible set of traffic sign classes that can be selected by the annotators. In fact, we store an identifier of the actual selected template which allows us to link the traffic sign instances to our flexible traffic sign taxonomy without even knowing the final set of classes beforehand (see Sect. 2.2).

Since it would still be too time-consuming to scroll through the entire list of templates to choose the correct one out of thousands, we trained a neural network to learn an embedding space (with the grouped template images) which is predicting the similarities between an arbitrary image of a traffic sign instance and the templates. We used this proposal network to assist the annotators in choosing the correct template by pre-sorting the template list for each individual traffic sign.

Specifically, we use a metric learning approach [1] to train a 3-layer network (similar to but shallower than the baseline classification network mentioned in Sect. 5) to learn a function $f(x) : \mathbb{R}^d \to \mathbb{R}^k$ that maps a d-dimensional input vector to a k-dimensional embedding space. In our case, x are input images encoded as vectors of size $d = 40 \times 40 \times 3$ and $k = 128$. We train the network with a contrastive loss [7] such that the cosine similarity

$$\text{sim}(x_1, x_2) = \frac{x_1^T x_2}{\|x_1\|_2 \|x_2\|_2} \tag{1}$$

between two embedding vectors x_1 and x_2 with group labels $\hat{y}_1, \hat{y}_2$ should be high if the samples are within the same template group, whereas the similarity should be lower than a margin m if the samples are from different groups:

$$\mathcal{L} = \begin{cases} 1 - \text{sim}(x_1, x_2), & \text{if } \hat{y}_1 = \hat{y}_2 \\ \max\left[0, \text{sim}(x_1, x_2) - m\right] & \text{else} \end{cases} . \tag{2}$$

We choose $m = 0.2$ and train the network using a generated training set by blending our traffic sign templates to random background images after scaling, rotating and sheering it by a reasonable amount.

For cases in which this strategy fails to provide a matching template, we provided a text-based search for templates. For details about the annotation UI, we refer to the supplemental material.

Additional Attributes. In addition to bounding boxes and the matching traffic sign templates, the annotators were asked to provide additional attributes: *occluded* if the sign is partly occluded; *ambiguous* if the sign is not classifiable at all (*e.g.* too small, of bad quality, heavily occluded *etc..*); *dummy* if it looks like a sign but is not (*e.g.* car stickers, reflections, *etc..*); *out-of-frame* if the sign is cut off by the image border; *included* if the sign is part of another bigger sign; and *exterior* if the sign includes other signs. Some of these attributes were assigned during localization (if context information is needed). The rest was assigned during classification.

Annotation Quality. All annotations in MTSD were done by expert annotators going through a thorough training process. Their work was monitored by a continuous quality control (QC) process to quickly identify problems during annotation. Moreover, our step-wise annotation process (*i.e.* approval followed by localization followed by classification) ensures that each traffic sign was seen by at least two annotators. The second annotator operating in the classification step was able to reject false positive signs or to report issues with the bounding box in which case the containing image was sent back to the localization step.

In additional quality assurance (QA) experiments done by a 2$^\text{nd}$ annotator on $5K$ images including $26K$ traffic signs, we found that (1) only 0.5 % of bounding boxes needed correction; (2) the false negative rate was 0.89 % (corresponding to a total number of only 212 missing signs, most of them being very small); (3) the false positive rate was at 2.45 %. Note that this is in the localization step before classification, where a second annotator has been asked to classify the sign and could potentially fix false positives.

2.4   Partial Annotations

In addition to the fully-annotated images, we provide another set of images with partially annotated bounding boxes and semi-supervised class labels. Given the fully-annotated images, the annotations of this set of images are generated automatically in a semi-supervised way.

Fig. 2. Example from the partially annotated set. The leftmost image is from the fully annotated set. The 3 other images show the same sign from different perspectives in the partial set with propagated class labels. Best viewed zoomed in and in color. (Color figure online)

We achieve this by finding correspondences between the manual annotations in the fully-annotated images and automatic detections in geographically neighboring images from the Mapillary API. To find these correspondences, we first use Structure from Motion (SfM) [8] to recover the relative camera poses between the fully-annotated images and the partially annotated images. With these estimated relative poses, we find correspondences between annotated signs and automatically detected signs by triangulating and verifying the re-projection errors for the centers of the bounding boxes between multiple images. Having these correspondences, we propagate the manually annotated class labels to the automatic detections in the partially annotated images. Since there is no guarantee that all traffic signs are detected through Mapillary's platform, this results in a set of images with partially annotated bounding box annotations. Note that, for unbiased evaluation, we ensure that the extension is done only in the geographical neighborhood of images in the training set (based on the split discussed in Sect. 2.5). Example images can be found in Fig. 2 and the effect on the class distribution in Fig. 3 (top/right). A more detailed description of how this set was created can be found in supplemental material.

2.5   Dataset Splits

As common practice with other datasets such as COCO [18], MVD [22] and PASCAL VOC [4], we split MTSD into training, validation and test sets, consisting of 36 589, 5320, and 10 544 images, respectively. We provide the image data for all sets as well as the annotations for the training and validation set; the annotations for the test set will not be released in order to ensure a fair evaluation. Additionally, we provide a set of 53 377 images with partial annotations as discussed in Sect. 2.4 for training as well.

Each split is created in a way to match the distributions described in Sect. 2.1. Especially, we ensure that the distribution of class instances is similar for each split, to avoid rare classes being under-represented in the smaller sets (i.e. validation/test sets). The same holds true for the additional sign attributes (*e.g. ambiguous, etc.*).

3   Statistics

In this section, we provide image and traffic sign statistics of MTSD and compare with previous datasets (TT100K [35] and MVD [22]). Unless stated otherwise, all numbers refer to the fully-supervised set of MTSD only.

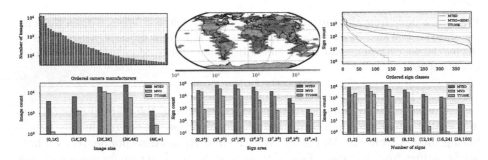

Fig. 3. *Top:* Distribution of camera devices; Geographical distribution of images; Distribution of traffic sign classes. *Bottom:* Images binned by size; Signs binned by sizes; Images binned by #signs. Size bins in $\sqrt{\text{pixel area}}$.

3.1   Image Properties

For a dataset to reflect a real-world image capturing setting with diverse geographical distribution, the image selection strategy described in Sect. 2.1 used for MTSD ensures a good distribution over different capturing settings.

Camera Sensors. In Fig. 3(top/left), we show the distribution of camera manufacturers used for capturing the images of MTSD. In total, the dataset covers over 200 different sensor manufacturers (we group the tail of the distribution for displaying purposes) which results in a large variety of image properties similar to the properties described in [22]. This is in contrast to the setup used for TT100K [35] which contains only images taken by a single sensor setup, making MTSD more challenging in comparison.

Image Sizes. The diversity in camera sensors further results in a diverse distribution over image resolutions as shown in Fig. 3(bottom/left). MTSD covers a broad range of image sizes starting from low-resolution images with 1 MPixels going up to images of more than 16 MPixels. Additionally, we include 1138 360-degree panoramas stored as standard images with equi-rectangular projection.

Besides the overall larger image volume compared to other datasets, MTSD also covers a larger fraction of low-resolution images, which is especially interesting for pre-training and validating detectors applied on similar sensors, *e.g.* built-in automotive cameras. For comparison, TT100K only contains images of 2048^2 px and even for this resolution the volume of images is smaller than in MTSD.

Geographical Distribution. The heat map in the middle of Fig. 3(top) shows the resulting geographical distribution of the images, covering almost all habitable areas of the world with higher density in populous areas.

3.2 Traffic Sign Properties

The fully-annotated set of MTSD includes a total number of 257 541 traffic sign bounding boxes out of which more than $88K$ have a class label within our taxonomy covering 400 different traffic sign classes. The remaining traffic signs sum up as ambiguous signs, directional signs, information signs, highway shields, exterior signs, barrier signs, and other signs that do not fall into our taxonomy.

Class Distribution. The right plot in Fig. 3(top) shows a comparison of the traffic sign class distribution between MTSD and TT100K. Note that MVD is not included here since it does not have labels of traffic sign classes. MTSD has approximately twice as many traffic sign classes as TT100K; if we use the definition of a trainable class in [35] (which are classes with at least 100 traffic sign instances within the dataset) this factor increases to approximately 3 between TT100K and MTSD. This difference gets even higher if we consider the instances from the partially annotated set of MTSD.

Fig. 4. Results from our detection and classification baseline on the validation set (green colored: true positive, red: missing detections). (Color figure online)

Sign Sizes. The plot in the middle of Fig. 3(bottom) compares the areas of signs in terms of pixels in the original resolution of the containing image. MTSD covers a broad range of traffic sign sizes with an almost uniform distribution up to 256^2 px. MVD has a similar distribution with a lower overall volume. In comparison to TT100K, MTSD provides a higher fraction of extreme sizes which poses another challenge for traffic sign detection.

Signs per Image. Finally, the plot on the right of Fig. 3(bottom) shows the distribution of images over the number of signs within the image. Besides the higher volume of images, MTSD contains a larger fraction of images with a large number of traffic sign instances (*i.e.* >12). One reason for this is that the annotations in MTSD cover all types of traffic signs, whereas TT100K only contains annotations for very specific types of traffic signs in China.

4 Traffic Sign Detection

One task defined on MTSD is binary detection of traffic signs, *i.e.* localization without inferring specific class labels. The goal is to predict a set of axis-aligned bounding boxes with corresponding confidence scores for each image.

Metrics. Given a set of detections with estimated scores for each image, we first compute the matching between the detections and annotated ground truth within each image separately. A detection can be successfully matched to a ground truth if their Jaccard overlap (IoU) [4] is >0.5; if multiple detections match the same ground truth, only the detection with the highest score is a match while the rest is not (*double detections*); each detection will only be matched to one ground truth bounding box with the highest overlap.

Having this matching indicator (TP vs. FP) for every detection, we define average precision (AP) similar to COCO [18] (*i.e.* $AP^{IoU=0.5}$ which resembles AP definition of PASCAL VOC [4]). Specifically, we compute precision as a function of recall by sorting the matching indicators by their corresponding detection confidence scores in descending order and accumulate the number of TPs and FPs. AP is defined as the area under the curve of this step function. Additionally, we follow [18] and compute AP in different scales: AP_s, AP_m, and AP_l refer to AP computed for boxes with area $a < 32^2$, $32^2 < a < 96^2$, and $a > 96^2$.

Table 2. Detection baseline results on MTSD, TT100K and MVD. Numbers in brackets refer to absolute improvements when pre-training on MTSD in comparison to ImageNet. *[35] using multi-scale inference with scales 0.5, 1, 2, and 4.

	Max 4000 px				Max 2048 px			
	AP	AP_s	AP_m	AP_l	AP	AP_s	AP_m	AP_l
MTSD								
FPN50 ours	87.84	72.91	91.88	93.54	80.08	52.12	88.81	94.72
FPN101 ours	88.38	73.89	92.10	93.69	81.65	56.32	89.18	94.80
TT100K								
Multi-scale*	91.79	84.56	96.40	92.60	–	–	–	–
FPN50 ours	–	–	–	–	91.27	84.01	95.87	90.13
+ *MTSD*	–	–	–	–	**97.60** (+6.33)	93.13	99.03	98.44
MVD (traffic signs)								
FPN50 ours	72.90	46.60	79.93	85.42	64.00	30.70	75.28	86.50
+ *MTSD*	**76.31** (+3.41)	51.00	83.49	88.33	**68.29**	33.60	79.45	89.53

Baseline and Results. In Table 2, we show experimental results using a Faster R-CNN based detector [24] with FPN [16] and residual networks [9] as the backbone. During training we randomly sample crops of size 1000×1000 at full resolution instead of down-scaling the image to avoid vanishing of small traffic signs, as traffic signs can be very small in terms of pixels and MTSD covers traffic signs from a broad range of scales in different image resolutions. We use a batch size of 16, distributed over 4 GPUs for FPN50 models; for FPN101 models, we use batches of size 8. Unless stated otherwise, we train using stochastic gradient descent (SGD) with an initial learning rate of 10^{-2} and lower the learning rate when the validation error plateaus. For inference, we down-scale the input images such that their larger side does not exceed a certain number of pixels (either 2048 px or 4000 px) or operate on full resolution if the original image is smaller.

Besides training on MTSD, we conduct transfer-learning experiments on TT100K and MVD[3] to test the generalization properties of the proposed dataset. We use the same baseline as for the MTSD experiments and train it on both datasets, one with ImageNet initialization and one with MTSD initialization. The models trained with ImageNet initialization are trained to convergence. To ensure a fair comparison, we fine-tune only for half the number of epochs when initializing with MTSD weights. The results in Table 2 show that MTSD pre-training boosts detection performance by a large margin on both datasets, regardless of the input resolution. This is a clear indication for the generalization qualities of MTSD.

5 Simultaneous Detection and Classification

The second task on MTSD is simultaneous detection and classification of traffic signs, *i.e.* multi-class detection. It extends the detection task to demand a class label for each traffic sign instance within our taxonomy. For instances that do not have a label within our taxonomy, we introduce a general class *other-sign*.

Metric. The metric for this task is mean average precision (mAP) over all 400 classes; per-class AP is calculated as described in Sect. 4. The matching between predicted and ground truth boxes is done in a binary way by ignoring the class label. After that, we filter out all *other-sign* ground truth instances and detections since we do not want to evaluate on this general class.

Baseline. A trivial baseline for this task would be to extend the binary detection baseline from Sect. 4 to the multi-class setting by adding a 401-way classification head. However, preliminary experiments showed that a straight-forward training of such a model does not yield acceptable performance. We hypothesize that this is due to (1) scale issues for small signs before RoI pooling and, (2) underrepresented class variation within the training batches given that the majority of traffic sign instances are *other-sign*.

---

[3] We convert the segmentations of *traffic-sign-front* instances to bounding boxes by taking the minimum and maximum in the x, y axes. Note that this conversion can be inaccurate if signs are occluded.

To overcome the scale issue and to have better control over batch statistics during training, we opted for a two-stage architecture that uses our binary detectors in the first stage and a decoupled shallow classification network in the second stage. Such decoupling has been shown to improve detection and recognition accuracy [2]. The classification network consists of seven 3×3 convolutions (each followed by batch normalization) with 2×2 max-pooling layers after the 2nd and 6th convolution layer. We start with 32 features in the first layer and double this number after each pooling layer. The last convolution is followed by spatial average pooling and a fully-connected layer with 256 features resulting in a 401-way classification head with softmax activation (400 and *other-sign*) and a single sigmoid activation for foreground/background classification.

We use image crops predicted by the detector (both foreground and background) together with crops from the ground truth scaled to 40×40 px as input and optimize the network using cross-entropy loss. To balance the distribution of traffic sign classes in a batch, we uniformly sample 100 classes with 4 samples each and add another 100 background crops per batch. We train the network with SGD for 50 epochs starting with a learning rate of 10^{-3} lowered by a factor of 0.1 after 30 and 40 epochs.

Results. We show results of our baseline in Table 3. Our classifier with FPN101 binary detector reaches 81.8 mAP over all 400 classes. Figure 4 shows visual examples of our baseline's predictions and Fig. 5 (left) shows typical failure cases of the classification network.

To verify our baseline, we train with the same setup on TT100K and compare the results with the baseline in [35][4]. Our two-stage approach outperforms their baseline by 8.3 points, even though the performances of the binary detectors are similar (see Table 2). This validates that the decoupled classifier, even with a shallow network, is able to yield good results. The accuracy is further improved when we pre-train the classifier and the detector on MTSD and then fine-tune them on TT100K, which further validates the generalization effectiveness of MTSD.

6   Classification with Partial Annotations

To evaluate the quality and existence of complementary information in our partially annotated, semi-supervised training set, we conduct classification experiments with and without the additional traffic sign samples and evaluate the performance on $53.3K$ traffic sign crops ($17.6K$ of the MTSD test set $+ 35.7K$ additional crops). After 14 independent trainings for each data configuration, we found a consistent improvement of 0.6 points in terms of mean class accuracy (standard deviation of multiple trainings is shown in Fig. 5(right). However, major improvements can be found in the long-tail of the class distribution where we have limited numbers of fully-supervised annotations. Figure 5(right) shows

---

[4] We convert their results to the format used by MTSD and evaluate using our metrics.

Table 3. Simultaneous detection and classification results. $+ \ det/cls \ MTSD$ refer to MTSD pre-training of detection/classification models. The numbers in brackets are absolute improvements over [35]

	mAP	mAP$_s$	mAP$_m$	mAP$_l$
MTSD				
FPN50 ours	81.7	73.0	84.1	84.2
FPN101 ours	81.8	74.4	84.4	84.9
TT100K				
Multi-scale	81.6	68.3	86.5	85.7
FPN50 ours	**89.9** (+8.3)	83.9	93.0	84.3
+ det MTSD	**93.4** (+11.8)	88.2	94.8	93.6
+ cls MTSD	**95.7** (+14.1)	91.3	96.9	96.7

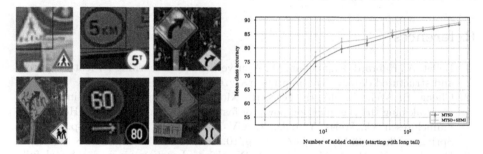

Fig. 5. Left: Failure cases of the classification network on MTSD. Right: Evaluation of the semi-supervised training set with varying number of classes. Reporting mean/std of multiple trainings.

mean class accuracy over varying number of classes where classes are added starting from the long-tail of the original class distribution. We can see a consistent gain of 1 4 points for the long-tail up to the first 100 classes. The gain decreases but is still noticeable when we add more classes that are well represented in the fully-supervised training set.

This shows the value of our partially annotated set as a straightforward way to augment existing datasets to better represent the long-tail of classes without introducing additional labeling costs. We want to point out that this method is a continuous data source for additional training data as it can be repeated as often as new nearby images become available.

7   Conclusion

In this work, we introduce MTSD, a large-scale traffic sign benchmark dataset that includes $105K$ images with full and partial bounding-box annotations, covering 400 traffic sign classes from all over the world. MTSD is the most diverse traffic sign benchmark dataset in terms of geographical locations, scene characteristics, and traffic sign classes. We show in baseline experiments that decoupling detection and fine-grained classification yields superior results on previous traffic sign datasets. Additionally, in transfer-learning experiments, we show that MTSD facilitates fine-tuning and improves accuracy substantially for traffic sign datasets in a narrow domain.

We see MTSD as the first step to drive the research efforts towards solving fine-grained traffic sign detection and classification at a global scale. With the partial annotated set, we show a new scalable way to collect additional training images without the need of extra manual annotation work. Moreover, we also see it paving the way for further research in semi-supervised learning in both classification and detection. In the future, we would like to extend MTSD towards a complete traffic sign taxonomy globally. To achieve this, we see the potential of applying zero-shot learning to efficiently model the semantic and appearance attributes of traffic sign classes.

References

1. Bellet, A., Habrard, A., Sebban, M.: A survey on metric learning for feature vectors and structured data. arXiv preprint arXiv:1306.6709 (2013)
2. Cheng, B., Wei, Y., Shi, H., Feris, R., Xiong, J., Huang, T.: Revisiting RCNN: on awakening the classification power of faster RCNN. In: Ferrari, V., Hebert, M., Sminchisescu, C., Weiss, Y. (eds.) ECCV 2018. LNCS, vol. 11219, pp. 473–490. Springer, Cham (2018). https://doi.org/10.1007/978-3-030-01267-0_28
3. Dai, J., Li, Y., He, K., Sun, J.: R-FCN: object detection via region-based fully convolutional networks. In: Proceedings of the Conference on Neural Information Processing Systems (NIPS), pp. 379–387 (2016)
4. Everingham, M., Eslami, S.A., Van Gool, L., Williams, C.K., Winn, J., Zisserman, A.: The Pascal visual object classes challenge: a retrospective. Int. J. Comput. Vis. (IJCV) **111**(1), 98–136 (2015)
5. Girshick, R.: Fast R-CNN. In: Proceedings of the IEEE Conference on Computer Vision and Pattern Recognition (CVPR), pp. 1440–1448 (2015)
6. Girshick, R., Donahue, J., Darrell, T., Malik, J.: Rich feature hierarchies for accurate object detection and semantic segmentation. In: Proceedings of the IEEE Conference on Computer Vision and Pattern Recognition (CVPR), pp. 580–587 (2014)
7. Hadsell, R., Chopra, S., LeCun, Y.: Dimensionality reduction by learning an invariant mapping. In: Proceedings of the IEEE Conference on Computer Vision and Pattern Recognition (CVPR), pp. 1735–1742 (2006)
8. Hartley, R., Zisserman, A.: Multiple View Geometry in Computer Vision. Cambridge University Press, Cambridge (2003)
9. He, K., Zhang, X., Ren, S., Sun, J.: Deep residual learning for image recognition. In: Proceedings of the IEEE Conference on Computer Vision and Pattern Recognition (CVPR), pp. 770–778 (2016)
10. Houben, S., Stallkamp, J., Salmen, J., Schlipsing, M., Igel, C.: Detection of traffic signs in real-world images: the German traffic sign detection benchmark. In: Proceedings of the IEEE International Joint Conference on Neural Networks (IJCNN) (2013)

11. Kuznetsova, A., et al.: The open images dataset v4: unified image classification, object detection, and visual relationship detection at scale. arXiv preprint arXiv:1811.00982 (2018)
12. Larsson, F., Felsberg, M.: Using fourier descriptors and spatial models for traffic sign recognition. In: Proceedings of Scandinavian Conference on Image Analysis (SCIA) (2011)
13. Larsson, F., Felsberg, M., Forssen, P.E.: Correlating Fourier descriptors of local patches for road sign recognition. IET Comput. Vis. 5(4), 244–254 (2011)
14. Li, J., Liang, X., Wei, Y., Xu, T., Feng, J., Yan, S.: Perceptual generative adversarial networks for small object detection. In: Proceedings of the IEEE Conference on Computer Vision and Pattern Recognition (CVPR), pp. 1222–1230 (2017)
15. Li, Y., Chen, Y., Wang, N., Zhang, Z.: Scale-aware trident networks for object detection. arXiv preprint arXiv:1901.01892 (2019)
16. Lin, T.Y., Dollár, P., Girshick, R., He, K., Hariharan, B., Belongie, S.: Feature pyramid networks for object detection. In: Proceedings of the IEEE Conference on Computer Vision and Pattern Recognition (CVPR), pp. 2117–2125 (2017)
17. Lin, T.Y., Goyal, P., Girshick, R., He, K., Dollár, P.: Focal loss for dense object detection. In: Proceedings of the IEEE Conference on Computer Vision and Pattern Recognition (CVPR), pp. 2980–2988 (2017)
18. Lin, T.-Y., et al.: Microsoft COCO: common objects in context. In: Fleet, D., Pajdla, T., Schiele, B., Tuytelaars, T. (eds.) ECCV 2014. LNCS, vol. 8693, pp. 740–755. Springer, Cham (2014). https://doi.org/10.1007/978-3-319-10602-1_48
19. Liu, W., et al.: SSD: single shot multibox detector. In: Leibe, B., Matas, J., Sebe, N., Welling, M. (eds.) ECCV 2016. LNCS, vol. 9905, pp. 21–37. Springer, Cham (2016). https://doi.org/10.1007/978-3-319-46448-0_2
20. Mathias, M., Timofte, R., Benenson, R., Van Gool, L.: Traffic sign recognition-how far are we from the solution? In: Proceedings of the IEEE International Joint Conference on Neural Networks (IJCNN) (2013)
21. Mogelmose, A., Trivedi, M.M., Moeslund, T.B.: Vision-based traffic sign detection and analysis for intelligent driver assistance systems: perspectives and survey. IEEE Trans. Intell. Transp. Syst. (ITS) 13(4), 1484–1497 (2012)
22. Neuhold, G., Ollmann, T., Rota Bulo, S., Kontschieder, P.: The mapillary vistas dataset for semantic understanding of street scenes. In: Proceedings of the IEEE Conference on Computer Vision and Pattern Recognition (CVPR) (2017)
23. Redmon, J., Farhadi, A.: Yolo9000: better, faster, stronger. In: Proceedings of the IEEE Conference on Computer Vision and Pattern Recognition (CVPR), pp. 7263–7271 (2017)
24. Ren, S., He, K., Girshick, R., Sun, J.: Faster R-CNN: towards real-time object detection with region proposal networks. In: Proceedings of the Conference on Neural Information Processing Systems (NIPS), pp. 91–99 (2015)
25. Sermanet, P., LeCun, Y.: Traffic sign recognition with multi-scale convolutional networks. In: Proceedings of the IEEE International Joint Conference on Neural Networks (IJCNN), pp. 2809–2813 (2011)
26. Shakhuro, V., Konushin, A.: Russian traffic sign images dataset. Comput. Opt. 40(2), 294–300 (2016)
27. Singh, B., Davis, L.S.: An analysis of scale invariance in object detection snip. In: Proceedings of the IEEE Conference on Computer Vision and Pattern Recognition (CVPR), pp. 3578–3587 (2018)
28. Singh, B., Najibi, M., Davis, L.S.: Sniper: efficient multi-scale training. In: Proceedings of the Conference on Neural Information Processing Systems (NIPS), pp. 9333–9343 (2018)

29. Stallkamp, J., Schlipsing, M., Salmen, J., Igel, C.: Man vs. computer: benchmarking machine learning algorithms for traffic sign recognition. Neural Netw. (2012)
30. Stallkamp, J., Schlipsing, M., Salmen, J., Igel, C.: The German traffic sign recognition benchmark: a multi-class classification competition. In: Proceedings of the IEEE International Joint Conference on Neural Networks (IJCNN) (2011)
31. Timofte, R., Zimmermann, K., Van Gool, L.: Multi-view traffic sign detection, recognition, and 3D localisation. Mach. Vis. Appl. **25**(3), 633–647 (2014)
32. Wah, C., Branson, S., Welinder, P., Perona, P., Belongie, S.: The caltech-UCSD birds-200-2011 dataset (2011)
33. Wikimedia commons. https://commons.wikimedia.org. Accessed 11 Nov 2019
34. Yu, F., et al.: BDD100K: a diverse driving video database with scalable annotation tooling. arXiv preprint arXiv:1805.04687 (2018)
35. Zhu, Z., Liang, D., Zhang, S., Huang, X., Li, B., Hu, S.: Traffic-sign detection and classification in the wild. In: Proceedings of the IEEE Conference on Computer Vision and Pattern Recognition (CVPR) (2016)

Measuring Generalisation to Unseen Viewpoints, Articulations, Shapes and Objects for 3D Hand Pose Estimation Under Hand-Object Interaction

Anil Armagan[1]([⊠]), Guillermo Garcia-Hernando[1,2], Seungryul Baek[1,20],
Shreyas Hampali[3], Mahdi Rad[3], Zhaohui Zhang[4], Shipeng Xie[4],
MingXiu Chen[4], Boshen Zhang[5], Fu Xiong[6], Yang Xiao[5], Zhiguo Cao[5],
Junsong Yuan[7], Pengfei Ren[8], Weiting Huang[8], Haifeng Sun[8], Marek Hrúz[9],
Jakub Kanis[9], Zdeněk Krňoul[9], Qingfu Wan[10], Shile Li[11], Linlin Yang[12],
Dongheui Lee[11], Angela Yao[13], Weiguo Zhou[14], Sijia Mei[14], Yunhui Liu[15],
Adrian Spurr[16], Umar Iqbal[17], Pavlo Molchanov[17], Philippe Weinzaepfel[18],
Romain Brégier[18], Grégory Rogez[18], Vincent Lepetit[3,19],
and Tae-Kyun Kim[1,21]

[1] Imperial College London, London, UK
a.armagan@imperial.ac.uk
[2] Niantic, Inc., San Francisco, USA
[3] Graz University of Technology, Graz, Austria
[4] Rokid Corp. Ltd., San Francisco, USA
[5] HUST, Wuhan, China
[6] Megvii Research Nanjing, Nanjing, China
[7] SUNY Buffalo, Buffalo, USA
[8] BUPT, Beijing, China
[9] University of West Bohemia, Pilsen, Czech Republic
[10] Fudan University, Shanghai, China
[11] TUM, Munich, Germany
[12] University of Bonn, Bonn, Germany
[13] NUS, Singapore, Singapore
[14] Harbin Institute of Technology, Harbin, China
[15] CUHK, Hong Kong, People's Republic of China
[16] ETH Zurich, Zürich, Switzerland
[17] NVIDIA Research, Santa Clara, USA
[18] NAVER LABS Europe, Meylan, France
[19] ENPC ParisTech, Champs-sur-Marne, France
[20] UNIST, Ulsan, South Korea
[21] KAIST, Daejeon, South Korea
https://sites.google.com/view/hands2019/challenge

Abstract. We study how well different types of approaches generalise
in the task of 3D hand pose estimation under single hand scenarios and

Electronic supplementary material The online version of this chapter (https://
doi.org/10.1007/978-3-030-58592-1_6) contains supplementary material, which is available to authorized users.

hand-object interaction. We show that the accuracy of state-of-the-art methods can drop, and that they fail mostly on poses absent from the training set. Unfortunately, since the space of hand poses is highly dimensional, it is inherently not feasible to cover the whole space densely, despite recent efforts in collecting large-scale training datasets. This sampling problem is even more severe when hands are interacting with objects and/or inputs are RGB rather than depth images, as RGB images also vary with lighting conditions and colors. To address these issues, we designed a public challenge (HANDS'19) to evaluate the abilities of current 3D hand pose estimators (HPEs) to interpolate and extrapolate the poses of a training set. More exactly, HANDS'19 is designed (a) to evaluate the influence of both depth and color modalities on 3D hand pose estimation, under the presence or absence of objects; (b) to assess the generalisation abilities *w.r.t.* four main axes: shapes, articulations, viewpoints, and objects; (c) to explore the use of a synthetic hand models to fill the gaps of current datasets. Through the challenge, the overall accuracy has dramatically improved over the baseline, especially on extrapolation tasks, from 27 mm to 13 mm mean joint error. Our analyses highlight the impacts of: Data pre-processing, ensemble approaches, the use of a parametric 3D hand model (MANO), and different HPE methods/backbones.

1   Introduction

3D hand pose estimation is crucial to many applications including natural user-interaction in AR/VR, robotics, teleoperation, and healthcare. The recent successes primarily come from large-scale training sets [41], deep convolutional neural networks [10,22], and fast optimisation for model fitting [15,23]. State-of-the-art methods now deliver satisfactory performance for viewpoints seen at training time and single hand scenarios. However, as we will show, these methods substantially drop accuracy when applied to egocentric viewpoints for example, and in the presence of significant foreground occlusions. These cases are not well represented on the training sets of existing benchmarks [5,20,21]. The challenges become even more severe when we consider RGB images and hand-object interaction scenarios. These issues are well aligned with the observations from the former public challenge HANDS'17 [40]: The state-of-the-art methods dropped accuracy from frontal to egocentric views, and from open to closure hand postures. The average accuracy was also significantly lower under hand-object interaction [5].

Given the difficulty to interpolate and extrapolate poses from the training set, one may opt for creating even larger training sets. Unfortunately, an inherent challenge in 3D hand pose estimation is the very high dimensionality of the problem, as hand poses, hand shapes and camera viewpoints have a large number of degrees-of-freedom that can vary independently. This complexity increases even more when we consider the case of a hand manipulating an object. Despite the recent availability of large-scale datasets [41], and the development of complex calibrated multi-view camera systems to help the annotation or synthetic data [13,28,45], capturing a training set that covers completely the domain of the problem remains extremely challenging.

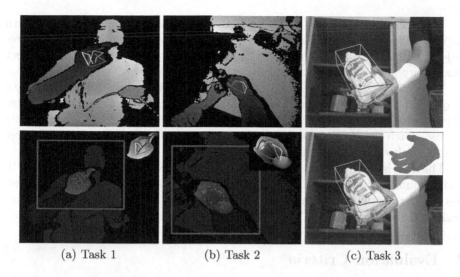

(a) Task 1   (b) Task 2   (c) Task 3

Fig. 1. Frames from the three tasks of our challenge. For each task, we show the input depth or RGB image with the ground-truth hand skeleton (top) and a rendering of the fitted 3D hand model as well as a depth rendering of the model (bottom). The ground-truth and estimated joint locations are shown in blue and red respectively. (Color figure online)

In this work, we therefore study in depth the ability of current methods to interpolate and extrapolate the training set, and how this ability can be improved. To evaluate this ability, we consider the three tasks depicted in Fig. 1, which vary the input (depth and RGB images) or the camera viewpoints, and introduce the possible manipulation of an object by the hand. We carefully designed training and testing sets in order to evaluate the generalisation performance to unseen viewpoints, articulations, and shapes of the submitted methods.

HANDS'19 fostered dramatic accuracy improvement compared to a provided baseline, which is a ResNet-50 [10]-based 3D joint regressor trained on our training set, from **27 mm** to **13 mm**. This paper provides an in-depth analysis of the different factors that made this improvement possible.

2   HANDS 2019 Challenge Overview

The challenge consists of three different tasks, in which the goal is to predict the 3D locations of the hand joints given an image. For training, images, hand pose annotations, and a 3D parametric hand model [26] for synthesizing data are provided. For inference, only the images and bounding boxes of the hands are given to the participants. These tasks are defined as follows:

Task 1: Depth-Based 3D Hand Pose Estimation: This task builds on Big-Hand2.2M [41] dataset, as for the HANDS 2017 challenge [39]. No objects appear in this task. Hands appear in both third person and egocentric viewpoints.

Task 2: Depth-Based 3D Hand Pose Estimation while Interacting with Objects: This task builds on the F-PHAB dataset [5]. The subject manipulates objects with their hand, as captured from an egocentric viewpoint. Some object models are provided by [5].

Task 3: RGB-Based 3D Hand Pose Estimation while Interacting with Objects: This task builds on the HO-3D [8] dataset. The subject manipulates objects with their hand, as captured from a third person viewpoint. The objects are used from the YCB dataset [36]. The ground truth wrist position of the test images is also provided in this task.

The BigHand2.2M [41] and F-PHAB [5] datasets have been used by 116 and 123 unique institutions to date. HANDS'19 received 80 requests to access the datasets with the designed partitions, and 17, 10 and 9 participants have evaluated their methods on Task 1, Task 2 and Task 3, respectively.

3   Evaluation Criteria

We evaluate the generalisation capability of HPEs in terms of four "axes": Viewpoint, Articulation, Shape, and Object. For each axis, frames within a dataset are automatically annotated by using the ground-truth 3D joint locations and the object information to annotate each frame in each axis. The annotation distribution of the dataset for each axis are used are used to create a training and a test set. Using the frame annotations on each axis, the sets are sampled in a structured way to have the test frames that are similar to the frames in the training data (for interpolation) and also the test frames where axes' annotations are never seen in the training data (for extrapolation). More details on the dataset are given in Sect. 4. To measure the generalisation of HPEs, six evaluation criteria are further defined with the four main axes: **Viewpoint**, **Articulation**, **Shape** and **Object** are respectively used for measuring the extrapolation performance of HPEs on the frames with articulation cluster, viewpoint angle, hand shape and object type (axis annotations) that are not present in the training set. **Extrapolation** is used to measure the performance on the frames with axis annotations that do not overlap/present in the training set. Lastly, **Interpolation** is defined to measure the performance on the frames with the axis annotations present in the training set.

The challenge uses the mean joint error (MJE) [23] as the main evaluation metric. Results are ranked according to the **Extrapolation** criterion which measures the total extrapolation power of the approaches with MJE on all axes. We also consider success rates based on maximum allowed distance errors for each frame and each joint for further analysis.

Figure 2 (left) summarises the six evaluation strategies, and Fig. 2 (right) shows the accuracies obtained by the best approaches, measured for the three evaluation criteria that could be evaluated for all three tasks. Articulation and viewpoint criteria are only considered for Task 1 since the joint angles are mostly fixed during object interaction and hence the Articulation criteria is not as meaningful as in Task 1 for the other tasks. The Viewpoint criteria is not meaningful

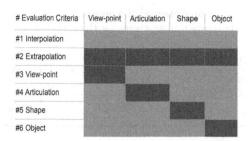
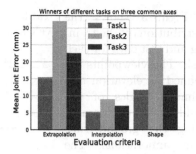

Fig. 2. Left: The six evaluation criteria used in the challenge. For each axis (Viewpoint, Articulation, Shape, Object), we indicate if hand poses in an evaluation criterion are also available (green) in the training set or not (red). Right: MJE comparison of the best methods for the Extrapolation, Interpolation and Shape criteria on each task. (Color figure online)

for Task 2 which is for egocentric views since the task's dataset constrains the relative palm-camera angle to a small range. For Task 3, the data scarcity is not helping to sample enough diverse viewpoints. The extrapolation errors tend to be three times larger than the interpolation errors while the shape is a bottleneck among the other attributes. Lower errors on Task 3 compared to Task 2 are likely due to the fact that the ground truth wrist position is provided for Task 3.

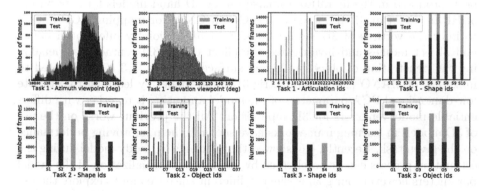

Fig. 3. Distributions of the training and test datasets for Task 1 (top), Task 2 (bottom left), and Task 3 (bottom right). The splits are used to evaluate the extrapolation power of the approaches and decided based on the viewpoints, the articulation clusters of the hand pose, the hand shape, and the type of the object present.

4   Datasets

Given a task, the training set is the same and the test frames used to evaluate each criterion can be different or overlapped. The number of training frames are 175K, 45K and 10K for Task 1, 2 and 3 respectively. The sizes of the test sets for each evaluation criterion are shown in Table 1.

Figure 3 shows the distributions of the training and test data for each task. The viewpoints are defined as elevation and azimuth angles of the hand *w.r.t.* the camera using the ground-truth joint annotations. The

Table 1. Detailed analytics on the number of frames provided on the training and test sets for the different tasks.

Dataset	Task id	#Frames Total	Ext.	Int.	Art.	View.	Sha.	Obj.	#Subjects	#Objects	#Actions	#Seq.
Test	1	125K	20%	16%	16%	32%	16%	✗	10	✗	✗	✗
	2	25K	14%	32%	✗	✗	37%	17%	4	37	71	292
	3	6.6K	24%	35%	✗	✗	14	27%	5	5	1	5
Training	1	175951							5	✗	✗	✗
	2	45713							4	26	45	539
	3	10505							3	4	1	12

articulation of the hand is defined and obtained by clustering on the ground-truth joint angles in a fashion similar to [18], by using binary representations (open/closed) of each finger *e.g.* '00010' represents a hand articulation cluster with frames with the index finger closed and the rest of the fingers open, which ends up with $2^5 = 32$ clusters. Examples from the articulation clusters are provided in the supplementary document. Note that the use of a low-dimensional embedding such as PCA or t-SNE is not adequate here to compare the two data distributions, because the dimensionality of the distributions is very high and a low-dimensional embedding would not be very representative. Figure 3 further shows the splits in terms of subjects/shapes, where five seen subjects and five unseen subjects are present. Similarly, the data partition was done on objects. This way we can control the data to define the evaluation metrics.

Use of 3D Hand Models for HPEs. A series of methods [1,3,7,9,42] have been proposed in the literature to make use of 3D hand models for supervision of HPEs. Ge et al. [7] proposed to use Graph CNNs for mapping RGB images to infer the vertices of 3D meshes. Hasson et al. [9] jointly infers both hands and object meshes and investigated the effect of the 3D contact loss penalizing the penetration of object and hand surfaces. Others [1,3,42] attempted to make use of MANO [26], a parametric 3D hand model by learning to estimate low-dimensional PCA parameters of the model and using it together with differentiable model renderers for 3D supervision. All the previous works on the use of 3D models in learning frameworks have shown to help improving performance on the given task. Recently, [16] showed that fitting a 3D body model during the estimation process can be accelerated by using better initialization of the model parameters however, our goal is slightly different since we aim to explore the use of 3D models for better generalisation from the methods. Since the hand pose space is huge, we make use of a 3D hand model to fill the gaps in the training data distribution to help approaches to improve their extrapolation capabilities. In this study, we make use of the MANO [26] hand model by providing the model's parameters for each training image. We fit the 3D model for each image in an optimization-based framework which is described in more details below.

Gradient-Based Optimization for Model Fitting. We fit the MANO [26] models' shape $\mathbf{s} = \{s_j\}_{j=1}^{10}$, camera pose $\mathbf{c} = \{c_j\}_{j=1}^{8}$, and articulation $\mathbf{a} = \{a_j\}_{j=1}^{45}$ parameters to the i-th raw skeletons of selected articulations $\mathbf{z} = \{z_i\}_{i=1}^{K}$, by solving the following equation:

$$(\mathbf{s}^{i*}, \mathbf{c}^{i*}, \mathbf{a}^{i*}) = \arg \min_{(\mathbf{s},\mathbf{c},\mathbf{a})} O(\mathbf{s}, \mathbf{c}, \mathbf{a}, \mathbf{z}^i)), \forall i \in [1, K], \tag{1}$$

where our proposed objective function $O(\mathbf{s}, \mathbf{c}, \mathbf{a}, \mathbf{z}^i)$ for the sample i is defined as follows:

$$O(\mathbf{s}, \mathbf{c}, \mathbf{a}, \mathbf{z}^i) = \|f^{reg}(V(\mathbf{s}, \mathbf{c}, \mathbf{a})) - \mathbf{z}^i\|_2^2 + \sum_{j=1}^{10} \|\mathbf{s}_j\|_2^2 + R_{Lap}(V(\mathbf{s}, \mathbf{c}, \mathbf{a})) . \quad (2)$$

$V(\mathbf{s}, \mathbf{c}, \mathbf{a})$ denotes the 3D mesh as a function of the three parameters $\mathbf{s}, \mathbf{c}, \mathbf{a}$. Equation (2) is composed of the following terms: i) the Euclidean distance between 3D skeleton ground-truths $\mathbf{z}^i$ and the current MANO mesh model's 3D skeleton values $f^{reg}(V(\mathbf{s}, \mathbf{c}, \mathbf{a}))$[1]; ii) A shape regularizer enforcing the shape parameters $\mathbf{s}$ to be close to their MANO model's mean values, normalized to 0 as in [26], to maximize the shape likelihood; and iii) A Laplacian regularizer $R_{Lap}(V(\mathbf{s}, \mathbf{c}, \mathbf{a}))$ to obtain the smooth mesh surfaces as in [14]. Equation (1) is solved iteratively by using the gradients from Eq. (2) as follows:

$$(\mathbf{s}_{t+1}, \mathbf{c}_{t+1}, \mathbf{a}_{t+1}) = (\mathbf{s}_t, \mathbf{c}_t, \mathbf{a}_t) - \gamma \cdot \nabla O(\mathbf{s}_t, \mathbf{c}_t, \mathbf{a}_t, \mathbf{z}^i), \forall t \in [1, T] , \quad (3)$$

where $\gamma = 10^{-3}$ and $T = 3000$ are empirically set. This process is similar to the refinement step of [1, 33], which refines estimated meshes by using the gradients from the loss.

In Fig. 4, both the target and the fitted depth images during the process described by Eq. (3) are depicted. Minor errors of the fitting are not a problem for our purpose given that we will generate input and output pairs of the fitted model by exploiting fitted meshes' self-data generation capability while ignoring original depth and skeletons. Here the aim of fitting the hand model is to obtain a plausible and a complete articulation space. The model is fitted without optimizing over depth information from the model and the input depth image since we did not observe an improvement on the parameter estimation. Moreover, the optimization needs to be constrained to produce plausible hand shapes and noise and other inconsistencies may appear in the depth image.

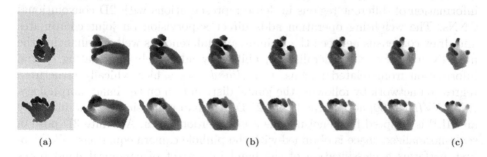

(a)   (b)   (c)

Fig. 4. Depth renderings of the hand model for different iterations in gradient-based optimization fitting. Target image (joints) (a), optimization iterations 0, 100, 300, 400, 600, 700 (b), final fitted hand pose at iteration 3000 (c).

---

[1] f^{reg} geometrically regresses the skeleton from the mesh vertex coordinates. It is provided with the MANO model and the weights are fixed during the process.

5   Evaluated Methods

In this section, we present the gist of selected 14 methods among 36 participants (17 for Task 1, 10 for Task 2, 9 for Task 3) to further analyze their results in Sect. 6. Methods are categorized based on their main components and properties. See Tables 1, 2, and 3 of the supplementary document provided with this work for a glance of the properties of the methods in HANDS'19.

2D and 3D Supervision for HPEs. Approaches that embed and process 3D data obtain high accuracies but less efficient [40] in terms of their complexity compared to 2D-based approaches. 3D-based methods use 3D convolutional layers for point-clouds input similar to *NTIS* which uses an efficient voxel-based representation V2V-PoseNet [19] with a deeper architecture and weighted sub-voxel predictions on quarter of each voxel representations for robustness. Some other approaches adopts 3D as a way of supervision similar to *Strawberryfg* [34] which employs a render-and-compare stage to enforce voxel-wise supervision for model training and adopts a 3D skeleton volume renderer to re-parameterize an initial pose estimate obtained similar to [31]. *BT* uses a permutation invariant feature extraction layer [17] to extract point-cloud features and uses a two branch framework for point-to-pose voting and point-to-latent voting. 3D supervision is employed by point-cloud reconstruction from a latent embedding in Task 1 whereas 3D hand model parameters are estimated and used in a differentiable model renderer for 3D supervision for the other tasks.

2D CNN-based approaches has been a standard way for learning regression models as used by *Rokid* [43] where they adopt a two stage regression models. The first regression model is used to predict an initial pose and the second model built on top of the first model. *A2J* [37] uses a 2D supervised method based on 2D offset and depth estimations with anchor points. Anchor points are densely set on the input image to behave as local regressors for the joints and able to capture global-local spatial context information. *AWR* [11] adopts a learnable and adaptive weighting operation that is used to aggregate spatial information of different regions in dense representations with 2D convolutional CNNs. The weighting operation adds direct supervision on joint coordinates and draw consensus between the training and inference as well as enhancing the model's accuracy and generalisation ability by adaptively aggregating spatial information from related regions. *CrazyHand* uses a hierarchically structured regression network by following the joints' distribution on the hand morphology. *ETH_NVIDIA* [30] adopts the latent 2.5D heatmap regression [12]; additionally an MLP is adopted for denoising the absolute root depth. Absolute 3D pose in scale-normalized space is obtained with the pinhole camera equations. *NLE* [25] first performs a classification of the hand into a set of canonical hand poses (obtained by clustering on the poses in the training set), followed by a fine class-specific regression of the hand joints in 2D and 3D. *NLE* adopts the only approach proposing multiple hand poses in a single stage with a Region Proposal Network (RPN) [24] integration.

Detection, Regression and Combined HPEs. Detection methods are based on hand key-points and producing a probability density maps for each joint. *NTIS* uses a 3D CNN [19] to estimate per-voxel likelihood of each joint. Regression-based methods estimate the joint locations by learning a direct mapping from the input image to hand joint locations or the joint angles of a hand model [29,44]. *Rokid* uses joint regression models within two stages to estimate an initial hand pose for hand cropping and estimates the final pose from the cleaned hand image. *A2J* adopts regression framework by regressing offsets from anchors to final joint location. *BT*'s point-wise features are used in a voting scheme which behaves as a regressor to estimate the pose.

Some approaches take advantage of both detection-based and regression-based methods. Similarly, *AWR*, *Strawberryfg* estimates probability maps to estimate joint locations with a differentiable *soft-argmax* operation [31]. A hierarchical approach proposed by *CrazyHand* regresses the joint locations from joint probability maps. *ETH_NVIDIA* estimates 2D joint locations from estimated probability maps and regresses relative depth distance of the hand joints *w.r.t.* a root joint. *NLE* first localizes the hands and classifies them to anchor poses and the final pose is regressed from the anchors.

Method-Wise Ensembles. *A2J* uses densely set anchor points in a voting stage which helps to predict location of the joints in an ensemble way for better generalisation leveraging the uncertainty in reference point detection. In a similar essence, *AWR* adaptively aggregates the predictions from different regions and *Strawberryfg* adopts local patch refinement [35] where refinement models are adopted to refine bone orientations. *BT* uses the permutation equivariant features extracted from the point-cloud in a point-to-pose voting scheme where the votes are ensembled to estimate the pose. *NLE* ensembles anchor poses to estimate the final pose.

Ensembles in Post-processing. Rather than a single pose estimator, an ensemble approach was adopted by multiple entries by randomly replicating the methods and fusing the predictions in the post-prediction stage, *e.g.* *A2J*, *AWR*, *NTIS*, *NLE* and *Strawberryfg*.

A2J ensembles predictions from ten different backbone architectures in Task 1 like *AWR* (5 backbones) and augments test images to ensemble the predictions with different scales and rotations as similar to rotation augmentation adopted by *NLE*. *NTIS* uses predictions obtained from the same model at 6 different training epochs. A similar ensembling is also adopted by *A2J* in Task 2. *NTIS* adopts a different strategy where N most confident sub-voxel predictions are ensembled to further use them in a refinement stage with Truncated SVDs together with temporal smoothing (Task 2). *NLE* takes advantage of ensembles from multiple pose proposals [25]. *Strawberryfg* employs a different strategy and ensembles the predictions from models that are trained with various input modalities.

Real + Synthetic Data Usage. The methods *Rokid* in Task 1 and *BT* in Tasks 2 and 3 make use of the provided MANO [26] model parameters to synthesize more training samples. *Rokid* leverages the synthesized images and combines them the real images—see Fig. 5—to train their initial pose regression network which effectively boosts accuracies—see Table 5. However, the amount of synthetic data created is limited to 570K for *Rokid* and 32K in Task 2, 100K in Task 3 for *BT*. Considering the continuous high-dimensional hand pose space with or without objects, if we sub-sample uniformly and at minimum,

(a) Real Cropped Hand   (b) Synthetic Depth Rendering   (c) Real + Synthetic Mixed Hand

Fig. 5. Visualization of synthetic depth images by *Rokid* [43]: (a) input depth image, (b) rendered depth image using 3D hand model, (c) the mixed by using the pixels with the closest depth values from real and synthetic images.

for instance, 10^2 (azimuth/elevation angles) $\times 2^5$ (articulation) $\times 10^1$ (shape) $\times 10^1$ (object) = 320K, the number is already very large, causing a huge compromise issue for memory and training GPU hours. Random sampling was applied without a prior on the data distribution or smart sampling techniques [2,4]. *BT* generates synthetic images with objects and hands similar to [20] by randomly placing the objects from [8] to nearby hand locations without taking into account the hand and object interaction. The rest of the methods use the provided real training data only.

Multi-modal inputs for HPEs. *BT* adopts [38] in Task 3 to align latent spaces from depth and RGB input modalities and to embed the inherit depth information in depth images during learning. *Strawberryfg* makes use of multi-inputs where each is obtained from different representations of the depth image, *e.g.* point-cloud, 3D point projection [6], multi-layer depth map [27], depth voxel [19].

Dominating HPE Backbones. ResNet [10] architectures with residual connections have been a popular backbone choice among many HPEs *e.g.* *A2J*, *AWR*, *Strawberryfg*, *CrazyHand*, *ETH_NVIDIA*, *NLE* or implicitly by *BT* within the ResPEL [17] architecture. *Rokid* adopts EfficientNet-b0 [32] as a backbone which uniformly scales the architecture's depth, width, and resolution.

6   Results and Discussion

We share our insights and analysis of the results obtained by the participants' approaches: 6 in Task 1, 4 in Task 2, and 3 in Task 3. Our analyses highlight the impacts of data pre-processing, the use of an ensemble approach, the use of MANO model, different HPE methods, and backbones and post-processing strategies for the pose refinement.

Analysis of Submitted Methods for Task 1. We consider the main properties of the selected methods and the evaluation criteria for comparisons. Table 2 provides the errors for the MJE metric and Fig. 6 show that high success rates are easier to achieve in absence of an object for low distance d thresholds. 2D-based approaches such as *Rokid*, with the advantage of additional data synthesizing, or *A2J*, with cleverly designed local regressors, can be considered to be best when the MJE score is evaluated for the Extrapolation criterion. *AWR* performs comparable to the other 2D-based approaches by obtaining the lowest MJE errors on the Interpolation and Articulation criteria. *AWR* performs best for the distances less than 50 mm on Extrapolation as well as showing better generalisation to unseen Viewpoints and Articulations, while excelling to interpolate well. A similar trend is observed with the 3D-voxel-based approach *NTIS*. However, the other 3D supervised methods, *Strawberryfg* and *BT* show lower generalisation capability compared to other approaches while performing reasonably well on the Articulation, Shape, and Interpolation criteria but not being able to show a similar performance for the Extrapolation and Viewpoint criteria.

Table 2. Task 1 - MJE (mm) and ranking of the methods on five evaluation criteria. Best results on each evaluation criteria are highlighted.

Username	Extrapolation	Interpolation	Shape	Articulation	Viewpoint
Rokid	**13.66** (1)	4.10 (2)	**10.27** (1)	4.74 (3)	**7.44** (1)
A2J	13.74 (2)	6.33 (6)	11.23 (4)	6.05 (6)	8.78 (6)
AWR	13.76 (4)	**3.93** (1)	11.75 (5)	**3.65** (1)	7.50 (2)
NTIS	15.57 (7)	4.54 (3)	12.05 (6)	4.21 (2)	8.47 (4)
Strawberryfg	19.63 (12)	8.42 (10)	14.21 (10)	7.50 (9)	14.16 (12)
BT	23.62 (14)	18.78 (16)	21.84 (16)	16.73 (16)	19.48 (14)

Table 3. Task 2 - MJE (mm) and ranking of the methods on four evaluation criteria.

Username	Extrapolation	Interpolation	Object	Shape
NTIS	**33.48** (1)	**17.42** (1)	29.07 (2)	23.62 (2)
A2J	33.66 (2)	17.45 (2)	**27.76** (1)	**23.39** (1)
CrazyHand	38.33 (4)	19.71 (4)	32.60 (4)	26.26 (4)
BT	47.18 (5)	24.95 (6)	38.76 (5)	32.36 (5)

Analysis of Submitted Methods for Task 2. We selected four submitted methods to compare on Task 2, where a hand interacts with an object in an egocentric viewpoint. Success rates illustrated in Fig. 7 highlight the difficulty of extrapolation. All methods struggle to show good performance on estimating frames with joint errors less than 15 mm. On the other hand, all methods can estimate 20% to 30% of the joints correctly with less than 15 mm error for the other criteria in this task (Table 3).

NTIS (a voxel-based) and *A2J* (weighted local regressors with anchor points) perform similarly when MJEs for all joints are considered. However, *NTIS* obtains higher success rates on the frame-based evaluation for all evaluation criteria with low distance error thresholds (d), see Fig. 7. Its performance is relatively much higher when Extrapolation is considered, especially for the frames with unseen objects, see Fig. 7. This can be explained by having a better embedding of the occluded hand structure with the voxels in the existence of seen/unseen objects. *NTIS* interpolates well under low distance thresholds.

Note that the first three methods, *NTIS*, *A2J*, and *CrazyHand* perform very similar for high error thresholds *e.g.* $d > 30$ mm. *CrazyHand* uses a structured detection-regression-based HPE where a heatmap regression is employed for the joints from palm to tips in a sequential manner which is highly valuable for

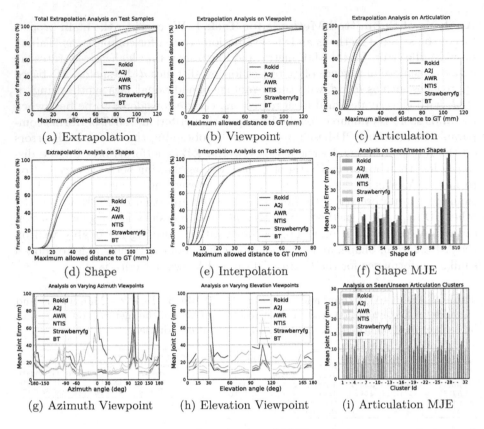

Fig. 6. Task 1 - Success rate analysis (a–e) and MJE analysis on extrapolation and interpolation using shapes (f), viewpoints (g, h) and articulations (i). Solid colors depict samples of extrapolation and transparent colors depict interpolation samples in plots (f–i).

egocentric viewpoints, helps to obtain comparable results with *A2J* where the structure is implicitly refined by the local anchor regressors.

Analysis of Submitted Methods for Task 3. We selected 3 entries, with different key properties for this analysis. It is definitively harder for the participants to provide accurate poses compared to the previous tasks. None of the methods can estimate frames that have all joints estimated with less than 25 mm error, see Fig. 8. The 25 mm distance threshold shows the difficulty of estimating a hand pose accurately from RGB input modality even though the participants of this task were provided with the ground-truth wrist joint location (Table 4).

The task is based on hand-object interaction in RGB modality. Therefore, the problem raises the importance of multi-modal data and learn from different modalities. Only *BT* uses the MANO parameters provided by the organizers to synthesize 100K images and adds random objects near the hand. This approach supports the claim on the importance of multi-modality and filling the real data

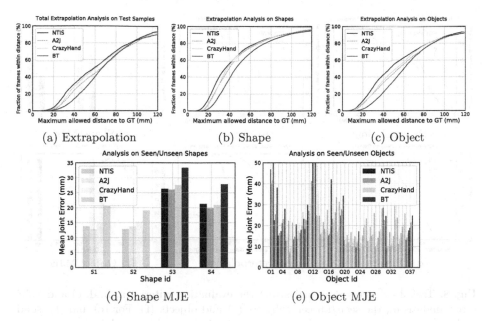

(a) Extrapolation   (b) Shape   (c) Object

(d) Shape MJE   (e) Object MJE

Fig. 7. Task 2 - Success rate analysis (a, b, c) and interpolation (seen, transparent) and extrapolation (unseen, solid) errors for subject (d) and object (e).

Table 4. Task 3 - MJE (mm) and ranking of the methods on four evaluation criteria.

Username	Extrapolation	Interpolation	Object	Shape
ETH_NVIDIA	**24.74** (1)	6.70 (3)	27.36 (2)	**13.21** (1)
NLE	29.19 (2)	**4.06** (1)	**18.39** (1)	15.79 (3)
BT	31.51 (3)	19.15 (5)	30.59 (3)	23.47 (4)

Table 5. Impact of synthetic data reported by *Rokid* [43] with learning from different ratios of synthetic data and the Task 1 training set. $100\% = 570K$.

Synthetic Data %	-	10%	30%	70%	100%
Extrapolation MJE (mm)	30.11	16.70	16.11	15.81	15.73

gaps with synthetic data with its close performance to the two higher ranked methods in MJE.

The generalisation performance of *BT* in Task 3 compared to the team's approaches with similar gist in Tasks 1 and 2 supports the importance of multi-model learning and synthetic data augmentation. The close performance of the method to generalise to unseen objects compared to *ETH_NVIDIA* and to generalise to unseen shapes compared to *NLE* also supports the argument with the data augmentation. The approach is still outperformed in MJE for this task although it performs close to the other methods.

NLE's approach shows the impact of learning to estimate 2D joints+3D joints (28.45 mm) compared to learning 3D joints alone (37.31 mm) on the Object as well as improvements for the Interpolation. Object performance is further improved to 23.22 mm with PPI integration. Further insights put by *NLE*'s own experiments on the number rotation augmentations (n) in post-processing helps to better extrapolate for unseen shapes (17.35 mm, 16.77 mm, 15.79 mm where $n = 1, 4, 12$, respectively).

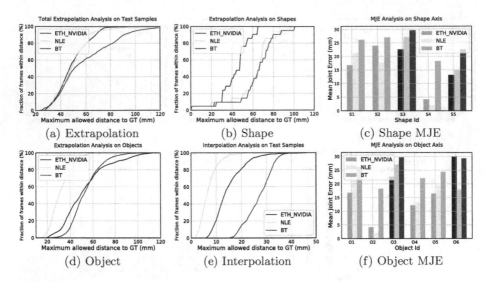

(a) Extrapolation   (b) Shape   (c) Shape MJE

(d) Object   (e) Interpolation   (f) Object MJE

Fig. 8. Task 3 - Success rate analysis on the evaluation criteria (a, b, d, e) and MJE error analysis on the seen/unseen subjects (c) and objects (f). For (c) and (f), solid and transparent colors are used to depict extrapolation and interpolation.

Analysis on the Usage of Synthetic Images. The best performing method of Task 1 (*Rokid*) in MJE uses the 3D hand model parameters to create 570K synthetic images by either perturbing (first stage) the model parameters or not (second stage). Synthetic data usage significantly helps in training the initial model (see Fig. 5). Table 5 shows the impact of different proportions of the 570K synthetic data usage to train the model together with the real training images. Using synthetic data can boost such a simple 3D joint regressor's performance from MJE of 30.11 mm to 15.73 mm, a ~50% improvement. Moreover, *Rokid*'s experiments with a regression model trained for 10 epochs shows the impact of the mixed depth inputs, Fig. 5, to lower the total extrapolation error (26.16 mm) compared to the use of raw depth renderings (30.13 mm) or the renderings averaged (31.92 mm) with the real input images. *BT* uses synthetic images in a very small amount of $32K$ and $100K$ in Tasks 2 and 3 since 3D reconstruction is difficult to train at a larger scale. However, favorable impact the data can be observed by comparing performances in Tasks 1 and 2.

Please refer to the supplementary document provided with this work for qualitative results and also for more results and discussions on the performance of the approaches on each axis, backbone architectures, ensembling and post-processing techniques and joint success rates obtained by the participants.

7 Conclusion

We carefully designed structured training and test sets for 3D HPEs and organized a challenge for the hand pose community to show state-of-the-art methods still tend to fail to extrapolate on large pose spaces. Our analyses highlight the impacts of using ensembles, the use of synthetic images, different type of HPEs *e.g.* 2D, 3D or local-estimators and post-processing. Ensemble techniques, both methodologically in 2D and 3D HPEs and in post-processing, help many approaches to boost their performance on extrapolation. The submitted HPEs were proven to be successful while interpolating in all the tasks, but their extrapolation capabilities vary significantly. Scenarios such as hands interacting with objects present the biggest challenges to extrapolate by most of the evaluated methods both in depth and RGB modalities.

Given the limited extrapolation capabilities of the methods, usage of synthetic data is appealing. Only a few methods actually were making use of synthetic data to improve extrapolation. $570K$ synthetic images used by the winner of Task 1 is still a very small number compared to how large, potentially infinite, it could be. We believe that investigating these possibilities, jointly with data sub-sampling strategies and real-synthetic domain adaptation is a promising and interesting line of work. The question of what would be the outcome if we sample 'dense enough' in the continuous and infinite pose space and how 'dense enough' is defined when we are limited by hardware and time is significant to answer.

Acknowledgements. This work is partially supported by Huawei Technologies Co. Ltd. and Samsung Electronics. S. Baek was supported by IITP funds from MSIT of Korea (No. 2020-0-01336 AIGS of UNIST, No. 2020-0-00537 Development of 5G based low latency device - edge cloud interaction technology).

References

1. Baek, S., Kim, K.I., Kim, T.K.: Pushing the envelope for RGB-based dense 3D hand pose estimation via neural rendering. In: CVPR (2019)
2. Bhattarai, B., Baek, S., Bodur, R., Kim, T.K.: Sampling strategies for GAN synthetic data. In: ICASSP (2020)
3. Boukhayma, A., de Bem, R., Torr, P.H.: 3D hand shape and pose from images in the wild. In: CVPR (2019)
4. Cubuk, E.D., Zoph, B., Mane, D., Vasudevan, V., Le, Q.V.: AutoAugment: learning augmentation strategies from data. In: CVPR (2019)
5. Garcia-Hernando, G., Yuan, S., Baek, S., Kim, T.K.: First-person hand action benchmark with RGB-D videos and 3D hand pose annotations. In: CVPR (2018)
6. Ge, L., Liang, H., Yuan, J., Thalmann, D.: Robust 3D hand pose estimation in single depth images: from single-view CNN to multi-view CNNs. In: CVPR (2016)
7. Ge, L., et al.: 3D hand shape and pose estimation from a single RGB image. In: CVPR (2019)
8. Hampali, S., Oberweger, M., Rad, M., Lepetit, V.: HO-3D: A multi-user, multi-object dataset for joint 3D hand-object pose estimation. arXiv preprint arXiv:1907.01481v1 (2019)

9. Hasson, Y., et al.: 3D hand shape and pose estimation from a single RGB image. In: CVPR (2019)
10. He, K., Zhang, X., Ren, S., Sun, J.: Deep residual learning for image recognition. In: CVPR (2016)
11. Huang, W., Ren, P., Wang, J., Qi, Q., Sun, H.: AWR: adaptive weighting regression for 3D hand pose estimation. In: AAAI (2020)
12. Iqbal, U., Molchanov, P., Breuel, T., Gall, J., Kautz, J.: Hand pose estimation via latent 2.5D heatmap regression. In: Ferrari, V., Hebert, M., Sminchisescu, C., Weiss, Y. (eds.) ECCV 2018. LNCS, vol. 11215, pp. 125–143. Springer, Cham (2018). https://doi.org/10.1007/978-3-030-01252-6_8
13. Joo, H., et al.: Panoptic studio: a massively multiview system for social motion capture. In: ICCV (2015)
14. Kanazawa, A., Tulsiani, S., Efros, A.A., Malik, J.: Learning category-specific mesh reconstruction from image collections. In: Ferrari, V., Hebert, M., Sminchisescu, C., Weiss, Y. (eds.) ECCV 2018. LNCS, vol. 11219, pp. 386–402. Springer, Cham (2018). https://doi.org/10.1007/978-3-030-01267-0_23
15. Kennedy, J., Eberhart, R.: Particle swarm optimization. In: ICNN (1995)
16. Kolotouros, N., Pavlakos, G., Black, M.J., Daniilidis, K.: Learning to reconstruct 3D human pose and shape via model-fitting in the loop. In: ICCV (2019)
17. Li, S., Lee, D.: Point-to-pose voting based hand pose estimation using residual permutation equivariant layer. In: CVPR (2019)
18. Lin, J., Wu, Y., Huang, T.S.: Modeling the constraints of human hand motion. In: HUMO (2000)
19. Moon, G., Chang, J.Y., Lee, K.M.: V2V-PoseNet: voxel-to-voxel prediction network for accurate 3D hand and human pose estimation from a single depth map. In: CVPR (2018)
20. Mueller, F., et al.: GANerated hands for real-time 3D hand tracking from monocular RGB. In: CVPR (2018)
21. Mueller, F., Mehta, D., Sotnychenko, O., Sridhar, S., Casas, D., Theobalt, C.: Real-time hand tracking under occlusion from an egocentric RGB-D sensor. In: ICCV (2017)
22. Oberweger, M., Lepetit, V.: DeepPrior++: improving fast and accurate 3D hand pose estimation. In: ICCV Workshop on HANDS (2017)
23. Oikonomidis, I., Kyriazis, N., Argyros., A.A.: Efficient model-based 3D tracking of hand articulations using kinect. In: BMVC (2011)
24. Ren, S., He, K., Girshick, R., Sun, J.: Faster R-CNN: towards real-time object detection with region proposal networks. In: NIPS (2015)
25. Rogez, G., Weinzaepfel, P., Schmid, C.: LCR-Net++: multi-person 2D and 3D pose detection in natural images. IEEE Trans. Pattern Anal. Mach. Intell. 42(5), 1146–1161 (2019)
26. Romero, J., Tzionas, D., Black, M.J.: Embodied hands: modeling and capturing hands and bodies together. ACM Trans. Graph. (Proc. SIGGRAPH Asia) 36(6), 2451–24517 (2017)
27. Shin, D., Ren, Z., Sudderth, E.B., Fowlkes, C.C.: Multi-layer depth and epipolar feature transformers for 3D scene reconstruction. In: CVPR Workshops (2019)
28. Simon, T., Joo, H., Matthews, I., Sheikh, Y.: Hand keypoint detection in single images using multiview bootstrapping. In: CVPR (2017)
29. Sinha, A., Choi, C., Ramani, K.: DeepHand: robust hand pose estimation by completing a matrix imputed with deep features. In: CVPR (2016)

30. Spurr, A., Iqbal, U., Molchanov, P., Hilliges, O., Kautz, J.: Weakly supervised 3D hand pose estimation via biomechanical constraints. arXiv preprint arXiv:2003.09282 (2020)
31. Sun, X., Xiao, B., Wei, F., Liang, S., Wei, Y.: Integral human pose regression. In: Ferrari, V., Hebert, M., Sminchisescu, C., Weiss, Y. (eds.) ECCV 2018. LNCS, vol. 11210, pp. 536–553. Springer, Cham (2018). https://doi.org/10.1007/978-3-030-01231-1_33
32. Tan, M., Le, Q.V.: EfficientNet: rethinking model scaling for convolutional neural networks. In: ICML (2019)
33. Tung, H.Y.F., Tung, H.W., Yumer, E., Fragkiadaki, K.: Self-supervised learning of motion capture. In: NIPS (2017)
34. Wan, Q.: SenoritaHand: Analytical 3D skeleton renderer and patch-based refinement for HANDS19 challenge Task 1 - Depth-based 3D hand pose estimation (December 2019). https://github.com/strawberryfg/Senorita-HANDS19-Pose
35. Wan, Q., Qiu, W., Yuille, A.L.: Patch-based 3D human pose refinement. In: CVPRW (2019)
36. Xiang, Y., Schmidt, T., Narayanan, V., Fox, D.: A convolutional neural network for 6D object pose estimation in cluttered scenes. In: RSS (2018)
37. Xiong, F., et al.: A2J: anchor-to-joint regression network for 3D articulated pose estimation from a single depth image. In: ICCV (2019)
38. Yang, L., Li, S., Lee, D., Yao, A.: Aligning latent spaces for 3D hand pose estimation. In: ICCV (2019)
39. Yuan, S., Ye, Q., Garcia-Hernando, G., Kim, T.K.: The 2017 hands in the million challenge on 3D hand pose estimation. arXiv preprint arXiv:1707.02237 (2017)
40. Yuan, S., et al.: Depth-based 3D hand pose estimation: from current achievements to future goals. In: CVPR (2018)
41. Yuan, S., Ye, Q., Stenger, B., Jain, S., Kim, T.K.: BigHand 2.2M Benchmark: hand pose data set and state of the art analysis. In: CVPR (2017)
42. Zhang, X., Li, Q., Mo, H., Zhang, W., Zheng, W.: End-to-end hand mesh recovery from a monocular RGB image. In: ICCV (2019)
43. Zhang, Z., Xie, S., Chen, M., Zhu, H.: HandAugment: A simple data augmentation method for depth-based 3D hand pose estimation. arXiv preprint arXiv:2001.00702 (2020)
44. Zhou, X., Wan, Q., Zhang, W., Xue, X., Wei, Y.: Model-based deep hand pose estimation. In: IJCAI (2016)
45. Zimmermann, C., Brox, T.: Learning to estimate 3D hand pose from single RGB images. In: ICCV (2017)

Disentangling Multiple Features in Video Sequences Using Gaussian Processes in Variational Autoencoders

Sarthak Bhagat[1], Shagun Uppal[1], Zhuyun Yin[2], and Nengli Lim[3(✉)]

[1] IIIT Delhi, Delhi, India
{sarthak16189,shagun16088}@iiitd.ac.in
[2] Bioinformatics Institute, A*STAR, Singapore, Singapore
yinzhuyun@gmail.com
[3] Singapore University of Technology and Design, Singapore, Singapore
nengli_lim@sutd.edu.sg

Abstract. We introduce MGP-VAE (Multi-disentangled-features Gaussian Processes Variational AutoEncoder), a variational autoencoder which uses Gaussian processes (GP) to model the latent space for the unsupervised learning of disentangled representations in video sequences. We improve upon previous work by establishing a framework by which multiple features, static or dynamic, can be disentangled. Specifically we use fractional Brownian motions (fBM) and Brownian bridges (BB) to enforce an inter-frame correlation structure in each independent channel, and show that varying this structure enables one to capture different factors of variation in the data. We demonstrate the quality of our representations with experiments on three publicly available datasets, and also quantify the improvement using a video prediction task. Moreover, we introduce a novel geodesic loss function which takes into account the curvature of the data manifold to improve learning. Our experiments show that the combination of the improved representations with the novel loss function enable MGP-VAE to outperform the baselines in video prediction.

1 Introduction

Finding good representations for data is one of the main goals of unsupervised machine learning [3]. Ideally, these representations reduce the dimensionality of the data, and are structured such that the different factors of variation in the data get distilled into different channels. This process of disentanglement in generative models is useful as in addition to making the data interpretable, the disentangled representations can also be used to improve downstream tasks such as prediction.

Electronic supplementary material The online version of this chapter (https://doi.org/10.1007/978-3-030-58592-1_7) contains supplementary material, which is available to authorized users.

© Springer Nature Switzerland AG 2020
A. Vedaldi et al. (Eds.): ECCV 2020, LNCS 12368, pp. 102–117, 2020.
https://doi.org/10.1007/978-3-030-58592-1_7

In prior work on the unsupervised learning of video sequences, a fair amount of effort has been devoted to separating motion, or dynamic information from static content [7,11,14,22,31]. To achieve this goal, typically the model is structured to consist of dual pathways, e.g. using two separate networks to separately capture motion and semantic content [7,31].

Such frameworks may be restrictive as it is not immediately clear how to extend them to extract multiple static and dynamic features. Furthermore, in complex videos, there usually is not a clear dichotomy between motion and content, e.g. in videos containing dynamic information ranging over different time-scales.

In this paper, we address this challenge by proposing a new variational autoencoder, MGP-VAE (Multi-disentangled-features Gaussian Processes Variational AutoEncoder) (Fig. 1), for the unsupervised learning of video sequences. It utilizes a latent prior distribution that consists of multiple channels of fractional Brownian motions and Brownian bridges. By varying the correlation structure along the time dimension in each channel to pick up different static or dynamic features, while maintaining independence between channels, MGP-VAE is able to learn multiple disentangled factors.

We then demonstrate quantitatively the quality of our disentanglement representations using a frame prediction task. To improve prediction quality, we also employ a novel geodesic loss function which incorporates the manifold structure of the data to enhance the learning process.

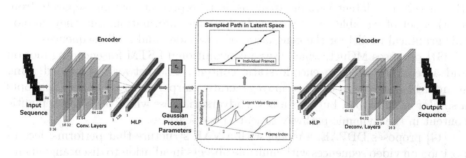

Fig. 1. Network illustration of MGP-VAE: The network takes in a video sequence, an array of images, and encodes a Gaussian process latent space representation. The output of the encoder is the mean and covariance matrix of the Gaussian process, after which a sequence of points in $\mathbb{R}^d$ is sampled where each point represents one frame.

Our main contributions can be summarized as follows:

- We use Gaussian processes as the latent prior distribution in our model MGP-VAE to obtain disentangled representations for video sequences. Specifically, we structure the latent space by varying the correlation between video frame distributions so as to extract multiple factors of variation from the data.
- We introduce a novel loss function which utilizes the structure of the data manifold to improve prediction. In particular, the actual geodesic distance

between the predicted point and its target on the manifold is used instead of squared-Euclidean distance in the latent space.

- We test MGP-VAE against various other state-of-the-art models in video sequence disentanglement. We conduct our experiments on three datasets and use a video prediction task to demonstrate quantitatively that our model outperforms the competition.

2   Related Work

2.1   Disentangled Representation Learning for Video Sequences

There are several methods for improving the disentanglement of latent representations in generative models. InfoGAN [6] augments generative adversarial networks [10] by additionally maximizing the mutual information between a subset of the latent variables and the recognition network output. beta-VAE [13] adds a simple coefficient (β) to the KL divergence term in the evidence lower bound of a VAE. It has been demonstrated that increasing β beyond unity improves disentanglement, but also comes with the price of increased reconstruction loss [18]. To counteract this trade-off, both FactorVAE [18] and β-TCVAE [5] further decompose the KL divergence term, and identify a total correlation term which when penalized directly encourages factorization in the latent distribution.

With regard to the unsupervised learning of sequences, there have been several attempts to separate dynamic information from static content [7,11,14,22, 31]. In [22], one latent variable is set aside to represent content, separate from another set of variables used to encode dynamic information, and they employ this graphical model for the generation of new video and audio sequences.

[31] proposes MCnet, which uses a convolutional LSTM for encoding motion and a separate CNN to encode static content. The network is trained using standard l_2 loss plus a GAN term to generate sharper frames. DRNet [7] adopts a similar architecture, but uses a novel adversarial loss which penalizes semantic content in the dynamic pathway to learn pose features.

[14] proposes DDPAE, a model with a VAE structure that performs decomposition on video sequences with multiple objects in addition to disentanglement. In their experiments, they show quantitatively that DDPAE outperforms MCnet and DRNet in video prediction on the Moving MNIST dataset.

Finally, it has been shown that disentangled representation learning can be placed in the framework of nonlinear ICA [17], particularly in the context of time-varying data [15].

2.2   VAEs and Gaussian Process Priors

In [11], a variational auto-encoder which structures its latent space distribution into two components is used for video sequence learning. The "slow" channel extracts static features from the video, and the "fast" channel captures dynamic motion. Our approach is inspired by this method, and we go further by giving a principled way to shape the latent space prior so as to disentangle multiple features.

Outside of video analysis, VAEs with a Gaussian process prior have also been explored. In [4], they propose GPPVAE and train it on image datasets of different objects in various views. The latent representation is a function of an object vector and a view vector, and has a Gaussian prior imposed on it. They also introduce an efficient method to speed up computation of the covariance matrices.

In [8], a deep VAE architecture is used in conjunction with a Gaussian process to model correlations in multivariate time series such that inference can be performed on missing data-points.

Bayes-Factor VAE [19] uses a hierarchical Bayesian model to extend the VAE. As with our work, they recognize the limitations of restricting the latent prior distribution to standard normal, but they adopt heavy-tailed distributions as an alternative rather than Gaussian processes.

2.3 Data Manifold Learning

Recent work has shown that distances in latent space are not representative of the true distance between data-points [1, 21, 27]. Rather, deep generative models learn a mapping from the latent space to the data manifold, a smoothly varying lower-dimensional subset of the original data space.

In [23], closed curves are abstractly represented as points on a shape manifold which incorporates the constraints of scale, rotational and translational invariance. The geodesic distance between points on this manifold is then used to give an improved measure of dissimilarity. In [28], several metrics are proposed to quantify the curvature of data manifolds arising from VAEs and GANs.

3 Method

In this section, we review the preliminaries on VAEs and Gaussian processes, and describe our model MGP-VAE in detail.

3.1 VAEs

Variational autoencoders [20] are powerful generative models which reformulate autoencoders in the framework of variational inference. Given latent variables $z \in \mathbb{R}^M$, the decoder, typically a neural network, models the generative distribution $p_\theta(x \mid z)$, where $x \in \mathbb{R}^N$ denotes the data. Due to the intractability of computing the posterior distribution $p(z \mid x)$, an approximation $q_\phi(z \mid x)$, again parameterized by another neural network called the encoder, is used. Maximizing the log-likelihood of the data can be achieved by maximizing the evidence lower bound

$$\mathbb{E}_{q_\phi(z|x)} \left[\log \frac{p_\theta(x, z)}{q_\phi(z \mid x)} \right],$$ (1)

which is equal to

$$\mathbb{E}_{q_\phi(z|x)}\left[\log p_\theta(x|z)\right] - D_{KL}\left[q_\phi\left(z|x\right) \Big| p(z)\right], \tag{2}$$

with $p(z)$ denoting the prior distribution of the latent variables.

The negative of the first term in (2) is the reconstruction loss, and can be approximated by

$$\frac{1}{L}\sum_{l=1}^{L} - \log p_\theta\left(x \,\big|\, z^{(l)}\right), \tag{3}$$

where $z^{(l)}$ is drawn (L times) from the latent distribution, although typically only one sample is required in each pass as long as the batch size is sufficiently large [20]. If $p_\theta\left(x \,\big|\, z\right)$ is modeled to be Gaussian, then this is simply mean-squared error.

3.2   Gaussian Processes

Given an index set T, $\{X_t; t \in T\}$ is a Gaussian process [12,32] if for any finite set of indices $\{t_1, \ldots, t_n\}$ of T, $(X_{t_1}, \ldots, X_{t_n})$ is a multivariate normal random variable. In this paper, we are concerned primarily in the case where T indexes time, i.e. $T = \mathbb{R}^+$ or $\mathbb{Z}^+$, in which case $\{X_t; t \in T\}$ can be uniquely characterized by its mean and covariance functions

$$\mu(t) := \mathbb{E}\left[X_t\right], \tag{4}$$
$$R(s,t) := \mathbb{E}\left[X_t X_s\right], \quad \forall\, s,t \in T. \tag{5}$$

The following Gaussian processes are frequently encountered in stochastic models, e.g. in financial modeling [2,9], and the prior distributions employed in MGP-VAE will be the appropriately discretized versions of these processes (Fig. 2).

Fractional Brownian Motion (fBM).   Fractional Brownian motion [24] $\{B_t^H; t \in T\}$ is a Gaussian process parameterized by a Hurst parameter $H \in (0,1)$, with mean and covariance functions given by

$$\mu(t) = 0, \tag{6}$$
$$R(s,t) = \frac{1}{2}\left(s^{2H} + t^{2H} - |t - s|^{2H}\right), \quad \forall\, s,t \in T. \tag{7}$$

When $H = \frac{1}{2}$, $W_t := B_t^{\frac{1}{2}}$ is standard Brownian motion [12] with independent increments, i.e. the discrete sequence $(W_0, W_1, W_2, \ldots)$ is a simple symmetric random walk where $W_{n+1} \sim \mathcal{N}(W_n, 1)$.

Most notably, when $H \neq \frac{1}{2}$, the process is not Markovian. When $H > \frac{1}{2}$, the disjoint increments of the process are positively correlated, whereas when $H < \frac{1}{2}$, they are negatively correlated. We will demonstrate in our experiments how tuning H effects the clustering of the latent code.

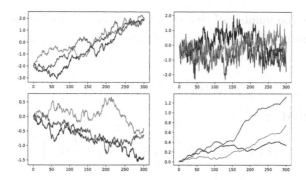

Fig. 2. Sample paths for various Gaussian processes. Top-left: Brownian bridge from -2 to 2; top-right: fBM with H = 0.1; bottom-left: standard Brownian motion; bottom-right: fBM with H = 0.9

Brownian Bridge (BB). The Brownian bridge [9,16] from $a \in \mathbb{R}$ to $b \in \mathbb{R}$ on the domain $[0, T]$ is the Gaussian process defined as

$$X_t = a\left(1 - \frac{t}{T}\right) + b\left(\frac{t}{T}\right) + W_t + \frac{t}{T}W_T. \tag{8}$$

Its mean function is identically zero and its covariance function is given by

$$R(s,t) = \min(s,t) - \frac{st}{T}, \quad \forall\, s, t \in T. \tag{9}$$

It can be also represented as the solution to the stochastic differential equation [16]

$$dX_t = \frac{b - X_t}{T - t}\, dt + dW_t, \quad X_0 = a, \tag{10}$$

with solution

$$X_t = a\left(1 - \frac{t}{T}\right) + b\left(\frac{t}{T}\right) + (T - t)\int_0^t \frac{1}{T - s}\, dW_s. \tag{11}$$

From (8), its defining characteristic is that it is pinned at the start and the end such that $X_0 = a$ and $X_T = b$ almost surely.

3.3   MGP-VAE

For VAEs in the unsupervised learning of static images, the latent distribution $p(z)$ is typically a simple Gaussian distribution, i.e. $z \sim \mathcal{N}(0, \sigma^2 \mathcal{I}_d)$. For a video sequence input $(x_1, \ldots x_n)$ with n frames, we model the corresponding latent code as

$$z = (z_1, z_2, \ldots, z_n) \sim \mathcal{N}(\mu_0, \Sigma_0), \quad z_i \in \mathbb{R}^d, \tag{12}$$

$$\mu_0 = \left[\mu_0^{(1)}, \ldots, \mu_0^{(d)}\right] \in \mathbb{R}^{n \times d}, \tag{13}$$

$$\Sigma_0 = \left[\Sigma_0^{(1)}, \ldots, \Sigma_0^{(d)}\right] \in \mathbb{R}^{n \times n \times d}. \tag{14}$$

Here d denotes the number of channels, where one channel corresponds to one sampled Gaussian path, and for each channel, $\left\{\mu_0^{(i)}, \Sigma_0^{(i)}\right\}$ are the mean and covariance of

$$V + \sigma B_t^H, \quad t = \{1, \ldots, n\}, \tag{15}$$

in the case of fBM or

$$A\left(1 - \frac{t}{n}\right) + B\left(\frac{t}{n}\right) + \sigma\left(W_t + \frac{t}{n}W_n\right) \tag{16}$$

in the case of Brownian bridge. V, A are initial distributions, and B is the terminal distribution for Brownian bridge. They are set to be standard normal, and we experiment with different values for σ. The covariances can be computed using (6) and (9) and are not necessarily diagonal, which enables us to model more complex inter-frame correlations.

Rewriting z as $\left(z^{(1)}, \ldots, z^{(d)}\right)$, for each channel $i = 1, \ldots, d$, we sample $z^{(i)} \in \mathbb{R}^n \sim \mathcal{N}\left(\mu_0^{(i)}, \Sigma_0^{(i)}\right)$ by sampling from a standard normal ξ and computing

$$z^{(i)} = \mu_0^{(i)} + L^{(i)}\xi, \tag{17}$$

where $L^{(i)}$ is the lower-triangular Cholesky factor of $\Sigma_0^{(i)}$.

The output of the encoder is a mean vector μ_1 and a symmetric positive-definite matrix Σ_1, i.e.

$$q(z \mid x) \sim \mathcal{N}(\mu_1, \Sigma_1), \tag{18}$$

and to compute the KL divergence term in (2), we use the formula

$$D_{KL}\left[q \mid p\right] = \frac{1}{2}\left[\mathrm{tr}\left(\Sigma_0^{-1}\Sigma_1\right) + \langle \mu_1 - \mu_0, \Sigma_0^{-1}(\mu_1 - \mu_0)\rangle - k + \log\left(\frac{\det \Sigma_1}{\det \Sigma_0}\right)\right]. \tag{19}$$

Following [13], we add a β factor to the KL divergence term to improve disentanglement. We will describe the details of the network architecture of MGP-VAE in Sect. 4.

3.4   Video Prediction Network and Geodesic Loss Function

For video prediction, we predict the last k frames of a sequence given the first $n - k$ frames as input. To do so, we employ a simple three-layer MLP (16 units per layer) with ReLU activation which operates in latent space rather than on the actual frame data so as to best utilize the disentangled representations. The first $n - k$ frames are first encoded by a pre-trained MGP-VAE into a sequence of points in latent space. These points are then used as input to the three-layer MLP to predict the next point, which is then passed through MGP-VAE's decoder to generate the frame. This process is then repeated $k - 1$ more times.

Given an output z_0 and a target z_T, we use the geodesic distance between $g(z_0)$ and $g(z_T)$ as the loss function instead of the usual squared-distance $\|z_0 - z_T\|^2$. Here, $g : \mathbb{R}^{n \times d} \to M \subset \mathbb{R}^N$ is the differentiable map from the latent space to the data manifold M which represents the action of the decoder. We use the following algorithm from [27] to compute the geodesic distance.

Algorithm 1: Geodesic Interpolation

Input: Two points, $z_0, z_T \in Z$;
α, the learning rate
Output: Discrete geodesic path, $z_0, z_1, \ldots, z_T \in Z$
Initialize z_i as the linear interpolation between z_0 and z_T
while $\Delta E_{z_t} > \epsilon$ **do**
    for $i \in \{1, 2, \ldots, T-1\}$ **do**
        Compute gradient using (21)
        $z_i \leftarrow z_i - \alpha \nabla_{z_t} E_{z_t}$
    end for
end while

This algorithm finds the minimum of the energy of the path (and thus the geodesic)

$$E_{z_t} = \frac{1}{2} \sum_{i=0}^{T} \frac{1}{\delta t} \|g(z_{i+1}) - g(z_i)\|^2 \tag{20}$$

by computing its gradient

$$\nabla_{z_t} E_{z_t} = - \left(\nabla g(z_i)\right)^T \left[g(z_{i+1}) - 2g(z_i) + g(z_{i-1})\right]. \tag{21}$$

Algorithm 1 initializes $\{z_i\}$ to be uniformly-spaced points along the line between z_0 and z_T and gradually modifies them until the change in energy falls below a predetermined threshold. At this point, we use z_1 as the target

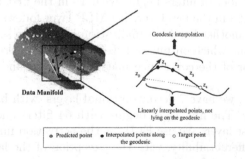

Fig. 3. Using the geodesic loss function as compared to squared-distance loss for prediction. By setting the target as z_1 instead of z_4, the model learns more efficiently to predict the next point.

instead of z_T as $z_1 - z_0$ is more representative of the vector in which to update the prediction z_0 such that the geodesic distance is minimized; see Fig. 3 for an illustration.

4 Experiments

In this section, we present experiments which demonstrate MGP-VAE's ability to disentangle multiple factors of variation in video sequences.

4.1 Datasets

Moving MNIST[1] [29] comprises of moving gray-scale hand-written digits. We generate 60,000 sequences for training, each with a single digit moving in a random direction across frames and bouncing off edges.

Coloured dSprites is a modification of the dSprites[2] [13] dataset. It consists of 2D shapes (square, ellipse, heart) with 6 values for scale and 40 values for orientation. We modify the dataset by adding 3 variations for colour (red, green, blue) and constrain the motion of each video sequence to be simple horizontal or vertical motion.

For each sequence, the scale is set to gradually increase or decrease a notch in each frame. Similarly, after an initial random selection for orientation, the subsequent frames rotate the shape clockwise or anti-clockwise one notch per frame. The final dataset consists of a total of approximately 100,000 datapoints.

Sprites [26] comprises of around 17,000 animations of synthetically rendered animated caricatures. There are 7 attributes: body type, sex, hair type, arm type, armor type, greaves type, and weapon type, with a total of 672 unique characters. In each animation, the physical traits of the sprite remain constant while the pose (hand movement, leg movement, orientation) is varied.

4.2 Network Architecture and Implementation Details

For the encoder, we use 8 convolutional layers with batch normalization between each layer. The number of filters begins with 16 in the first layer and increases to a maximum of 128 in the last layer. An MLP layer follows the last layer, and this is followed by another batch normalization layer. Two separate MLP layers are then applied, one which outputs a lower-triangular matrix which represents the Cholesky factor of the covariance matrix of $q(z\,|\,x)$ and the other outputs the mean vector.

For the decoder, we have 7 deconvolutional layers, with batch normalization between each layer. The first layer begins with 64 filters and this decreases to 16 filters by the last layer. We use ELU for the activation functions between all layers to ensure differentiability, with the exception of the last layer, where we use a hyperbolic tangent function.

---

[1] http://www.cs.toronto.edu/ nitish/unsupervised_video.
[2] https://github.com/deepmind/dsprites-dataset.

Table 1 lists the settings for the manually tuned hyperparameters in the experiments. All channels utilizing Brownian bridge (BB) are conditioned to start at -2 and end at 2.

Table 1. Hyperparameter settings for all datasets

	Moving MNIST	Coloured dSprites	Sprites
Gaussian processes	Channel 1: fBM (H = 0.1) Channel 2: fBM (H = 0.9)	5 Channels of BBs	5 Channels of BBs
σ	0.25	0.25	0.25
β	2	2	2
Learning rate	0.001	0.008	0.010
No. of epochs	200	120	150

4.3 Qualitative Analysis

Figure 4 shows the results from swapping latent channels in the Moving MNIST dataset, where we see that channel 1 (fBM($H = 0.1$)) captures the digit identity, whereas channel 2 (fBM($H = 0.9$)) captures the motion.

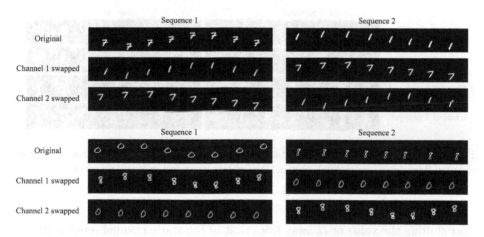

Fig. 4. Results from swapping latent channels in Moving MNIST; channel 1 (fBM($H = 0.1$)) captures digit identity; channel 2 (fBM($H = 0.9$)) captures motion.

Figure 5 gives a visualization of the latent space (here we use two channels with $H = 0.1$ and two channels with $H = 0.9$). In our experiments, we observe that fBM channels with $H = 0.9$ are able to better capture motion in comparison to setting $H = 0.5$ (simple-symmetric random walk, cf. [11]). We hypothesize

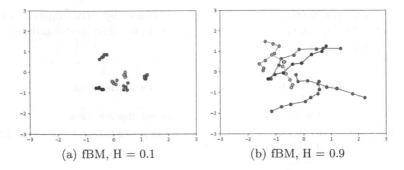

(a) fBM, H = 0.1   (b) fBM, H = 0.9

Fig. 5. Latent space visualization of fBM channels for 6 videos. Each point represents one frame of a video. The more tightly clustered points in (a) capture digit identity whereas the scattered points in (b) capture motion.

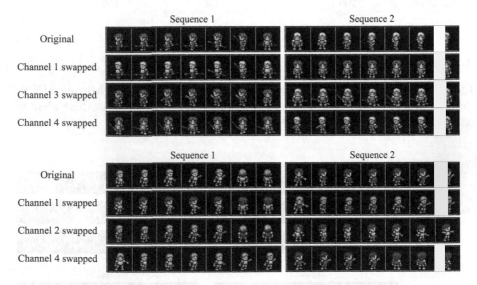

Fig. 6. Results from swapping latent channels in Sprites; channel 1 captures hair type, channel 2 captures armor type, channel 3 captures weapon type, and channel 4 captures body orientation.

that shifting the value of H away from that of the static channel sets the distributions apart and allows for better disentanglement.

Figures 6 and 7 show the results from swapping latent channels in the Sprites dataset and Coloured dSprites dataset respectively.

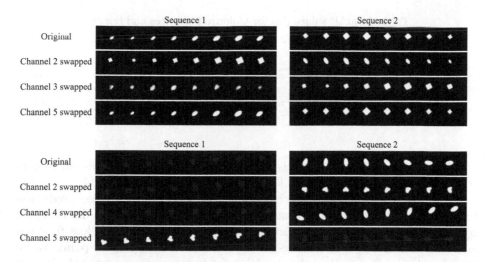

Fig. 7. Results from swapping latent channels in Coloured dSprites; channel 2 captures shape, channel 3 captures scale, channel 4 captures orientation and position, and channel 5 captures color.

Discussion. The disentanglement results were the best for Moving MNIST, where we achieved full disentanglement in more than 95% of the cases. We were also able to consistently disentangle three or more features in Coloured dSprites and Sprites, but disentanglement of four or more features occurred less frequently due to their complexity.

We found that including more channels than the number of factors of variation in the dataset improved disentanglement, even as the extra channels did not necessarily encode anything new. For the Coloured dSprites and Sprites dataset, we originally experimented with different combinations of fBMs (with varying H) and Brownian bridges, but found that simply using 4–5 channels of Brownian bridges gave comparable results. We observed that with complex videos not easily separated into static or dynamic content, incorporating multiple Brownian bridge channels each with different start and end points led to good disentanglement. We hypothesize that anchoring the start and end points of the sequence at various places in latent space "spreads out" and improves the representation.

Finally, we also tested other Gaussian processes such as the Ornstein-Ulenbeck process [25] but as the results were not satisfactory, we shall defer a more detailed investigation to future work.

4.4   Evaluating Disentanglement Quality

We first evaluate the disentangled representations by computing the mean average precision of a k-nearest neighbor classification over labeled attributes in the Coloured dSprites and Sprites datasets.

Table 2. mAP values (%) for Coloured dSprites and Sprites

Model	Coloured dSprites						Sprites						
	Shape	Color	Scale	x-Pos	y-Pos	Avg.	Gender	Skin	Vest	Hair	Arm	Leg	Avg.
MCnet	95.6	94.0	69.2	69.7	70.2	79.7	78.8	70.8	76.6	80.2	78.2	70.7	75.9
DRNet	95.7	94.8	69.6	72.4	70.6	80.6	80.5	70.8	77.0	78.6	79.7	71.4	76.3
DDPAE	95.6	94.2	70.3	71.6	72.4	80.8	79.8	72.0	77.4	79.3	78.3	74.6	76.9
MGP-VAE	96.2	94.0	77.9	76.4	72.8	**83.4**	80.3	71.8	76.8	82.3	79.9	79.8	**78.5**

Table 2 shows that our model is able to capture multiple features more effectively than the baselines MCnet[3] [31], DRNet[4] [7] and DDPAE[5] [14].

Next, we use a non-synthetic benchmark in the form of a video prediction task to illustrate the improvement in the quality of MGP-VAE's disentangled representations. We train a prediction network with the geodesic loss function as outlined in Sect. 3.4, where we set the number of interpolated points to be four. In addition, to speed up the algorithm for faster training, we ran the loop in Algorithm 1 for a fixed number of iterations (10–15) instead of until convergence.

We compute the pixel-wise mean-squared-error and binary cross-entropy between the predicted k frames and the actual last k frames, given the first $n - k$ frames as input (n is set to 8 for Moving MNIST and Coloured dSprites, and set to 7 for Sprites). Tables 3 and 4 below summarize the results.

Table 3. Prediction results on Moving MNIST

Model	$k = 1$		$k = 2$	
	MSE	BCE	MSE	BCE
MCnet [31]	50.1	248.2	91.1	595.5
DRNet [7]	45.2	236.7	86.3	586.7
DDPAE [14]	35.2	201.6	75.6	556.2
Grathwohl, Wilson [11]	59.3	291.2	112.3	657.2
MGP-VAE	25.4	198.4	72.2	554.2
MGP-VAE (with geodesic loss)	**18.5**	**185.1**	**69.2**	**531.4**

The results show that MGP-VAE[6], even without using the geodesic loss function, outperforms the other models. Using the geodesic loss functions further lowers MSE and BCE. DDPAE, a state-of-the-art model in video disentanglement, achieves comparable results, although we note that we had to train the model considerably longer on the Coloured dSprites and Sprites datasets as compared to Moving MNIST to get the same performance.

---

[3] https://github.com/rubenvillegas/iclr2017mcnet.
[4] https://github.com/ap229997/DRNET.
[5] https://github.com/jthsieh/DDPAE-video-prediction.
[6] https://github.com/SUTDBrainLab/MGP-VAE.

Table 4. Last-frame ($k = 1$) prediction results for Coloured dSprites and Sprites

Dataset model	Coloured dSprites		Sprites	
	MSE	BCE	MSE	BCE
MCnet [31]	20.2	229.5	100.3	2822.6
DRNet [7]	15.2	185.2	94.4	2632.1
DDPAE [14]	12.6	163.1	75.4	2204.1
MGP-VAE	6.1	85.2	68.8	1522.5
MGP-VAE (with geodesic loss)	**4.5**	**70.3**	**61.6**	**1444.4**

Using the geodesic loss function during the training of the prediction network also leads to qualitatively better results. Figure 8 below shows that in a sequence with large MSE and BCE losses, the predicted point can generate an image frame which differs considerably from the actual image frame when the normal loss function is used. This is rectified with the geodesic loss function.

Fig. 8. Qualitative improvements from using the geodesic loss function: Left: without geodesic loss function; Right: with geodesic loss function; Top row: original video; Bottom row: video with the predicted last frame.

5   Conclusion

We introduce MGP-VAE, a variational autoencoder for obtaining disentangled representations from video sequences in an unsupervised manner. MGP-VAE uses Gaussian processes, such as fractional Brownian motion and Brownian bridge, as a prior distribution for the latent space. We demonstrate that different parameterizations of these Gaussian processes allow one to extract different static and time-varying features from the data.

After training the encoder which outputs a disentangled representation of the input, we demonstrate the efficiency of the latent code by using it as input to a MLP for video prediction. We run experiments on three different datasets and demonstrate that MGP-VAE outperforms the baseline models in video frame prediction. To further improve the results, we introduce a novel geodesic loss function which takes into account the curvature of the data manifold. This contribution is independent of MGP-VAE, and we believe it can be used to improve video prediction in other models as well.

For future work, we will continue to experiment with various combinations of Gaussian processes. In addition, enhancing our approach with more recent methods such as FactorVAE, β-TCVAE, or independent subspace analysis [30] may lead to further improvements.

References

1. Arvanitidis, G., Hansen, L.K., Hauberg, S.: Latent space oddity: on the curvature of deep generative models. In: International Conference on Learning Representations (ICLR) (2017)
2. Bayer, C., Friz, P., Gatheral, J.: Pricing under rough volatility. Quant. Finan. **16**(6), 887–904 (2016)
3. Bengio, Y., Courville, A.C., Vincent, P.: Representation learning: a review and new perspectives. IEEE Trans. Pattern Anal. Mach. Intell. **35**, 1798–1828 (2012)
4. Casale, F.P., Dalca, A.V., Saglietti, L., Listgarten, J., Fusi, N.: Gaussian process prior variational autoencoders. In: Conference on Neural Information Processing Systems (NeurIPS) (2018)
5. Chen, R.T.Q., Li, X., Grosse, R., Duvenaud, D.: Isolating sources of disentanglement in VAEs. In: Advances in Neural Information Processing Systems (2018)
6. Chen, X., Duan, Y., Houthooft, R., Schulman, J., Sutskever, I., Abbeel, P.: InfoGAN: interpretable representation learning by information maximizing generative adversarial nets. In: Conference on Neural Information Processing Systems (NIPS) (2016)
7. Denton, E.L., Birodkar, V.: Unsupervised learning of disentangled representations from video. In: Conference on Neural Information Processing Systems (NIPS) (2017)
8. Fortuin, V., Rätsch, G., Mandt, S.: Multivariate Time Series Imputation with Variational Autoencoders. arXiv:1907.04155 (2019)
9. Glasserman, P.: Monte-Carlo Methods in Financial Engineering. Springer, New York (2003)
10. Goodfellow, I., et al.: Generative adversarial nets. In: Advances in Neural Information Processing Systems (2014)
11. Grathwohl, W., Wilson, A.: Disentangling Space and Time in Video with Hierarchical Variational Auto-encoders. arXiv:1612.04440 (2016)
12. Hida, T., Hitsuda, M.: Gaussian Processes. American Mathematical Society, Providence (2008)
13. Higgins, I., et al.: β-VAE: learning basic visual concepts with a constrained variational framework. In: International Conference on Learning Representations (ICLR) (2017)
14. Hsieh, J.T., Liu, B., Huang, D.A., Li, F.F., Niebles, J.C.: Learning to decompose and disentangle representations for video prediction. In: Conference on Neural Information Processing Systems (NeurIPS) (2018)
15. Hyvarinen, A., Morioka, H.: Unsupervised feature extraction by time-contrastive learning and nonlinear ICA. In: Lee, D.D., Sugiyama, M., Luxburg, U.V., Guyon, I., Garnett, R. (eds.) Advances in Neural Information Processing Systems, vol. 29, pp. 3765–3773. Curran Associates, Inc. (2016)
16. Karatzas, I., Shreve, S.E.: Brownian Motion and Stochastic Calculus. GTM, vol. 113. Springer, New York (1998). https://doi.org/10.1007/978-1-4612-0949-2d

17. Khemakhem, I., Kingma, D., Monti, R., Hyvarinen, A.: Variational autoencoders and nonlinear ICA: a unifying framework. In: Chiappa, S., Calandra, R. (eds.) Proceedings of the 23rd International Conference on Artificial Intelligence and Statistics. Proceedings of Machine Learning Research, PMLR, 26–28 August 2020, vol. 108, pp. 2207–2217 (2020)

18. Kim, H., Mnih, A.: Disentangling by factorising. In: International Conference on Machine Learning (ICML) (2018)

19. Kim, M., Wang, Y., Sahu, P., Pavlovic, V.: Bayes-Factor-VAE: Hierarchical Bayesian Deep Auto-Encoder Models for Factor Disentanglement. arXiv:1909.02820 (2019)

20. Kingma, D.P., Welling, M.: Auto-encoding variational Bayes. In: International Conference on Learning Representations (ICLR) (2013)

21. Kühnel, L., Fletcher, T.E., Joshi, S.C., Sommer, S.: Latent Space Non-Linear Statistics. arXiv:1805.07632 (2018)

22. Li, Y., Mandt, S.: Disentangled sequential autoencoder. In: International Conference on Machine Learning (ICML) (2018)

23. Lim, N., Gong, T., Cheng, L., Lee, H.K., et al.: Finding distinctive shape features for automatic hematoma classification in head CT images from traumatic brain injuries (2013)

24. Mandelbrot, B.B., Van Ness, J.W.: Fractional Brownian motions, fractional noises and applications. SIAM Rev. **10**(4), 422–437 (1968)

25. Ornstein, L.S., Uhlenbeck, G.E.: On the theory of Brownian motion. Phys. Rev. **36**, 823–841 (1930)

26. Reed, S.E., Zhang, Y., Zhang, Y., Lee, H.: Deep visual analogy-making. In: Conference on Neural Information Processing Systems (NIPS) (2015)

27. Shao, H., Kumar, A., Fletcher, P.T.: The Riemannian geometry of deep generative models. In: Conference on Computer Vision and Pattern Recognition Workshops (CVPRW) (2018)

28. Shukla, A., Uppal, S., Bhagat, S., Anand, S., Turaga, P.K.: Geometry of Deep Generative Models for Disentangled Representations. arXiv:1902.06964 (2019)

29. Srivastava, N., Mansimov, E., Salakhutdinov, R.: Unsupervised learning of video representations using LSTMs. In: International Conference on Machine Learning (ICML) (2015)

30. Stuehmer, J., Turner, R., Nowozin, S.: Independent subspace analysis for unsupervised learning of disentangled representations. In: Chiappa, S., Calandra, R. (eds.) Proceedings of the 23rd International Conference on Artificial Intelligence and Statistics. Proceedings of Machine Learning Research, PMLR, 26–28 August 2020, vol. 108, pp. 1200–1210 (2020)

31. Villegas, R., Yang, J., Hong, S., Lin, X., Lee, H.: Decomposing Motion and Content for Natural Video Sequence Prediction (2017)

32. Williams, C.K., Rasmussen, C.E.: Gaussian Processes for Machine Learning, vol. 2. MIT Press, Cambridge (2006)

SEN: A Novel Feature Normalization Dissimilarity Measure for Prototypical Few-Shot Learning Networks

Van Nhan Nguyen[1,2(✉)], Sigurd Løkse[1], Kristoffer Wickstrøm[1], Michael Kampffmeyer[1], Davide Roverso[2], and Robert Jenssen[1]

[1] UiT Machine Learning Group, UiT The Arctic University of Norway, Tromsø, Norway
{sigurd.lokse,kristoffer.k.wickstrom,michael.c.kampffmeyer, robert.jenssen}@uit.no
[2] Analytics Department, eSmart Systems, 1783 Halden, Norway
{nhan.v.nguyen,Davide.Roverso}@esmartsystems.com

Abstract. In this paper, we equip Prototypical Networks (PNs) with a novel dissimilarity measure to enable discriminative feature normalization for few-shot learning. The embedding onto the hypersphere requires no direct normalization and is easy to optimize. Our theoretical analysis shows that the proposed dissimilarity measure, denoted the Squared root of the Euclidean distance and the Norm distance (SEN), forces embedding points to be attracted to its correct prototype, while being repelled from all other prototypes, keeping the norm of all points the same. The resulting SEN PN outperforms the regular PN with a considerable margin, with no additional parameters as well as with negligible computational overhead.

1 Introduction

Few-shot classification [6,8,17,19,20,23] aims at adapting a classifier to previously unseen classes from just a handful of labeled examples per class. In the past few years, many approaches to few-shot classification have been proposed. These approaches can be roughly categorized as (i) learning to fine-tune approaches [6,17]; (ii) sequence-based approaches [1,13]; (iii) generative modeling-based approaches [26,29]; (vi) (deep) distance metric learning-based approaches [19,20,22,27]; and (v) semi-supervised approaches [3,16]. Among these categories, distance metric learning-based approaches are typically preferred because of their simplicity and effectiveness. The basic idea of these approaches, for which the so-called Prototypical Networks (PNs) [19] are the most well-known examples, is to learn a non-linear mapping of the input into

Electronic supplementary material The online version of this chapter (https://doi.org/10.1007/978-3-030-58592-1_8) contains supplementary material, which is available to authorized users.

an embedding space which is commonly high-dimensional. In this space, a metric distance is defined which maps similar examples close to each other in the embedding space. Dissimilar examples are mapped to distant locations relative to each other, so that a query example can be classified by, for example, using nearest neighbor methods. Arguably one of the most commonly used distance metrics in this high dimensional embedding space is the (squared) Euclidean distance combined with a softmax function [3,16,19,27].

However, even though the softmax is known to work well for closed-set classification problems, it has been shown to not be discriminative enough in problems were there are few labels relative to the number of classes [4,15]. This has given rise to alternative loss formulations with improved discriminative ability, where high-dimensional features have been normalized explicitly to lie on a hypersphere via direct L_2 normalization [4,15,24]. The advantage of normalization has been theoretically analyzed in [30]. However, direct L_2 normalization leads to a nonconvex loss formulation, which typically results in local minima generated by the loss function itself [30].

With the aim of performing *soft* feature normalization while preserving the convexity and the simplicity of the loss function, we equip PNs with a novel dissimilarity measure particularly suited to enable discriminative feature normalization for few-shot learning, without any direct normalization. The proposed dissimilarity measure, denoted the Squared root of the Euclidean distance and the Norm distance (SEN), replaces the Euclidean distance in PN training, with major consequences: Our theoretical analysis shows that the proposed measure explicitly forces embedded points to be attracted to the correct prototype and repelled from incorrect prototypes. Further, we provide analysis showing that SEN indeed explicitly forces all embeddings to have the same norm during training which enables the resulting SEN PN to generate a more robust embedding space. With this minimal but important modification, the SEN PN outperforms the original PN by a considerable margin and demonstrates good performance on the Mini-Imagenet [17,23], the Fewshot-CIFAR100 (FC100) [14], and the Omniglot [9] datasets with no additional parameters as well as negligible computational overhead (a comparison of inference time is provided in the supplementary material). We furthermore experimentally show that the proposed SEN dissimilarity measure constantly outperforms the Euclidean distance in PNs with different embedding sizes as well as with different embedding networks.

2   Related Work

The literature on few-shot learning is vast; we present in this section a short summary of well-known approaches and works most relevant to our proposed approach. We refer the reader to [25] and [21] for more detailed reviews on few-shot learning.

Besides distance metric learning-based approaches, few-shot learning approaches can be categorized into (i) learning to fine-tune approaches; (ii) sequence-based approaches; (iii) generative modeling-based approaches; (iv) (deep) distance metric learning-based approaches; and (v) semi-supervised approaches. Learning to fine-tune approaches aim at learning a model's initial parameters such that it can be quickly adapted to a new task through only one or a few gradient update steps [6,17]. These approaches typically can handle many model representations; however, they suffer from the need to fine-tune on the target problem, which makes them less appealing to few-shot learning. Sequence-based approaches formalize few-shot learning as a sequence-to-sequence problem and leverage Recurrent Neural Networks (RNNs) with memories to address the problem [1,13]. While appealing, these methods typically require complex RNN architectures and complicated mechanisms for storing/retrieving all the historical information of relevance, both long-term and short-term, without forgetting [20]. Generative modeling-based approaches employ adversarial training to produce additional signals/training examples to allow the classification algorithm to learn a better classifier [26,29]. Deep distance metric learning-based approaches aim at eliminating the need for manually choosing the right distance metric (e.g., the Euclidean distance and the cosine distance) by learning not only a deep embedding network but also a deep non-linear metric (similarity function) for comparing images in the embedding space [20]. Although deep distance metric learning-based approaches can avoid the need for manually choosing the right distance metric, they are prone to overfitting and are more difficult to train compared to distance metric learning-based approaches due to the added parameters. Semi-supervised approaches utilize unlabeled data to improve few-shot learning accuracy. This is typically achieved by casting the semi-supervised few-shot learning problem as a semi-supervised clustering problem and address it by applying, for example, k-means clustering algorithms [3,16]. We build on the distance metric learning line of work due to its simplicity and effectiveness.

Metric Learning-Based Approaches. Aim to learn a non-linear mapping of the input into an embedding space and define a metric distance which maps similar examples close and dissimilar ones distant in the embedding space, so that a query example can be easily classified by, for example, using nearest neighbor methods. Some notable approaches in this line of work include Koch et al. [8], who propose to learn siamese neural networks for computing the pair-wise distance between samples. The learned distance is then used by a nearest neighbor classifier for solving the one-shot learning problem. Vinyals et al. [23] define an end-to-end differentiable nearest neighbor classifier, called matching networks, based on the cosine similarity between the support set and the query example. Snell et al. [19] propose a simple method called prototypical networks for few-shot learning based on the assumption that there exists an embedding space in which samples from each class cluster around a single prototype representation, which is simply the mean of the individual samples. Garcia and Bruna [22] argue that few-shot learning, which aims at propagating label information from labeled support examples towards unlabeled query images, can be formalized as a posterior inference over a graphical model determined by the images and labels in the support set and the query set. The authors cast posterior inference as message passing on graph neural networks and propose a graph-based model, which can be trained end-to-end, to solve the task. Wang et al. [27] propose to improve the generalization capacity of metric-based methods for few-shot learning by enforcing a large margin between the class centers. This is achieved by augmenting a large margin loss function, which is the unnormalized triplet loss [18], to the standard softmax loss function for classification.

3 Few-Shot Learning

In this section, we first begin by detailing the general few-shot learning task. Next, we introduce PNs and the Euclidean distance function with special attention paid to highlight its existing challenges. Then, we describe our proposed SEN dissimilarity measure and our SEN PN model. Finally, we provide analyses on the gradient of the SEN PN's loss function and the behavior of the proposed SEN dissimilarity measure during training.

3.1 Task Description

In the traditional machine learning setting, we are typically given a dataset D. This dataset is usually split into two parts: D_{train} and D_{test}. The former is often used for training the parameters θ of the model, while the latter is typically used for evaluating its generalization. In general few-shot learning, we are dealing with meta-datasets D_{meta} containing multiple regular datasets D [17]. Each dataset $D \in D_{meta}$ has a split of D_{train} and D_{test}; however, they are usually much smaller than that of regular datasets used in the traditional machine learning setting. Let $C = \{1, \ldots, K\}$ be the set of all classes available in D_{meta}. The set C is usually split into two disjoint sets: C_{train} containing training classes and C_{test} containing unseen classes for testing, i.e., $C_{train} \cap C_{test} = \emptyset$. The meta-dataset D_{meta} is often split into two parts: The first is a meta training set $D_{meta-train} = \{(\mathbf{x}_i, y_i)\}_{i=1}^{N}$, where $\mathbf{x}_i$ is the feature vector of the ith example, $y_i \in C_{train}$ is its corresponding label, and N is the number of training examples. The second part is a meta testing set $D_{meta-test}$. In a standard M-way K-shot classification task, the meta testing set $D_{meta-test}$ consists of a *support set* and a *query set*. The support set $S = \{(\mathbf{x}_j, y_j)\}_{j=1}^{N_S}$ contains K examples from each of the M classes from C_{test}, i.e., the number of support examples are $N_S = M \times K$ and $y_j \in C_{test}$. The *query set* contains N_Q unlabeled examples $Q = \{\mathbf{x}_j\}_{j=N_S+1}^{N_S+N_Q}$. The support set is employed by the model for learning the new task, while the query set is utilized by the model for evaluating its performance.

3.2 Prototypical Networks

Prototypical networks learn a non-linear embedding function $f_\phi : \mathbb{R}^D \to \mathbb{R}^E$ parameterized by ϕ that maps a D-dimensional feature vector of an example $\mathbf{x}_i$ to an E-dimensional embedding $\mathbf{z}_i = f_\phi(\mathbf{x}_i)$ [19]. In meta-testing, the embedding function f_ϕ is employed for mapping examples in the support set $S = \{(\mathbf{x}_j, y_j)\}_{j=1}^{N_S}$ into the embedding space. An E-dimensional representation

$\mathbf{c}_k$, or *prototype*, of each class is computed by taking the mean of the embedded support points belonging to the class:

$$\mathbf{c}_k = \frac{1}{|S_k|} \sum_{(\mathbf{x}_i, y_i) \in S_k} f_\phi(\mathbf{x}_i) = \frac{1}{|S_k|} \sum_{(\mathbf{x}_i, y_i) \in S_k} \mathbf{z}_i, \tag{1}$$

where S_k is the support set of class k. An embedded query point $\mathbf{x}_q$ is then classified by simply finding the nearest class prototype in the embedding space.

To train PNs, the episodic training strategy proposed in [17,23] is adopted. In particular, to train a PN for the M-way, K-shot classification task, a training episode is formed from the meta training set D_{meta_train} as follows: K examples from each of M randomly selected classes from C_{train} are sampled to form a support set $S = \{S_i\}_{i=1}^M$. A query set $Q = \{Q_i\}_{i=1}^M$ is formed by sampling from the rest of the M classes' samples. Next, for each class k, its support set $S_k \in S$ is used for computing a prototype using Eq. 1. Then, a distribution over classes for each query point $\mathbf{x}_q \in Q$ based on a softmax over distances to the prototypes in the embedding space is produced:

$$p_\phi(y = k|\mathbf{x}_q) = \frac{\exp(-d(f_\phi(\mathbf{x}_q), \mathbf{c}_k))}{\sum_{k'} \exp(-d(f_\phi(\mathbf{x}_q), \mathbf{c}_{k'}))}, \tag{2}$$

where $d = \mathbb{R}^E \times \mathbb{R}^E \to [0, +\infty)$ is a distance function. Based on that, the PN is trained by minimizing the negative log-probability of the true class k via Stochastic Gradient Descent (SGD):

$$J(\phi) = -\frac{1}{M} \sum_{k=1}^M \frac{1}{|Q_k|} \sum_{\mathbf{x}_q \in Q_k} \log p_\phi(y = k|\mathbf{x}_q). \tag{3}$$

The training is repeated with new, randomly generated training episodes until a stopping criterion is met.

PNs employ the squared Euclidean distance as the distance metric. The squared Euclidean distance between two arbitrary points $\mathbf{z} = (z_1, \ldots, z_n)$ and $\mathbf{c} = (c_1, \ldots, c_n)$ is defined as follows:

$$d_{se}(\mathbf{z}, \mathbf{c}) = \|\mathbf{z} - \mathbf{c}\|^2 = \sum_{i=1}^n (z_i - c_i)^2. \tag{4}$$

Although combining the softmax and the Euclidean distance has shown to give good performance for closed-set classification settings, it performs suboptimally when few labels are available relative to the number of classes. In order to address this issue and improve the discriminative ability, new loss formulations based on feature normalization have been proposed. These tend to normalize features explicitly via L_2 normalization [4,15,24]. This typically results in a more compact embedding space than the Euclidean embedding space. In such an embedding space, the cosine distance is commonly chosen as the distance metric and many few-shot classification approaches [17,23] have employed the cosine distance in the hyperspherical embedding space. The cosine distance between two arbitrary point $\mathbf{z} = (z_1, \ldots, z_n)$ and $\mathbf{c} = (c_1, \ldots, c_n)$ is defined as:

$$d_{cs} = 1 - \frac{\mathbf{z} \cdot \mathbf{c}}{\|\mathbf{z}\|\|\mathbf{c}\|} = 1 - \frac{\sum_{i=1}^{n} z_i c_i}{\sqrt{\sum_{i=1}^{n} z_i^2} \sqrt{\sum_{i=1}^{n} c_i^2}}. \tag{5}$$

However, feature normalization through hard normalization operations such as L_2 normalization leads to a non-convex loss formulation, which typically results in local minima introduced by the loss function itself [30]. Since the network optimization itself is non-convex, it is important to preserve convexity in loss functions for more effective minimization.

One possible solution is to use Ring loss [30]. The Ring loss introduces an additional term to the primary loss function, which penalizes the squared difference between the norm of samples and a learned target norm value R. The modified loss function is defined as follows:

$$L = L_P + \gamma L_R, \tag{6}$$

where γ is the loss weight w.r.t to the primary loss L_P and L_R is the Ring loss, which is defined as:

$$L_R = \frac{1}{2n} \sum_{i=1}^{n} (\|f_\phi(\mathbf{x}_i)\| - R)^2. \tag{7}$$

Since the Ring loss encourages the norm of samples being value R during training instead of explicit enforcing through a hard normalization operation, the convexity in the loss function is preserved. However, the Ring loss is more difficult to train than the primary loss (e.g., the Softmax loss) due to the added term (the norm difference L_R), the added parameter (the target norm R), and the added hyperparameter (the loss weight w.r.t to the primary loss γ).

To address the shortcomings outlined above, we propose a novel dissimilarity measure for few-shot learning, called SEN. The SEN dissimilarity measure encourages the norm of samples to have the same value, in other words, force the data to lie on a scaled unit hypersphere, while preserving the convexity and the simplicity of the loss function.

3.3   SEN Dissimilarity Measure for Prototypical Networks

The SEN dissimilarity $d_s(\mathbf{z}, \mathbf{c})$ between two arbitrary points $\mathbf{z} = (z_1, \ldots, z_n)$ and $\mathbf{c} = (c_1, \ldots, c_n)$ in D-dimensional space is a combination of the standard squared Euclidean distance d_e and the squared norm distance d_n:

$$d_s(\mathbf{z}, \mathbf{c}) = \sqrt{d_e(\mathbf{z}, \mathbf{c}) + \epsilon d_n(\mathbf{z}, \mathbf{c})}, \tag{8}$$

where ϵ is a tunable balancing hyperparameter and must be chosen such that $d_e(\mathbf{z}, \mathbf{c}) + \epsilon d_n(\mathbf{z}, \mathbf{c})$ is always positive, $d_e(\mathbf{z}, \mathbf{c})$ and $d_n(\mathbf{z}, \mathbf{c})$ are defined as:

$$d_e(\mathbf{z}, \mathbf{c}) = \|\mathbf{z} - \mathbf{c}\|^2 ,$$

$$d_n(\mathbf{z}, \mathbf{c}) = (\|\mathbf{z}\| - \|\mathbf{c}\|)^2.$$

We modify the PN by replacing the Euclidean distance by our proposed SEN dissimilarity measure. We call this model SEN PN. Specifically, we replace the distance function $d(\mathbf{z}_i, \mathbf{c}_k)$ in Eq. 2 by our proposed SEN dissimilarity measure $d_s(\mathbf{z}_i, \mathbf{c}_k) = \sqrt{d_e(\mathbf{z}_i, \mathbf{c}_k) + \epsilon d_n(\mathbf{z}_i, \mathbf{c}_k)}$, $\mathbf{z}_i$ is the embedding of the example $\mathbf{x}_i$, and $\mathbf{c}_k$ is the prototype of class k. For simplicity, we consider the setting in which only one query example per class is used; however, the loss function presented in this session and the analysis presented in the next section can be easily generalized for other settings in which more than one query examples per class are used. When only one query example per class is used, the updated negative log probability loss is given as:

$$
\begin{aligned}
J(\phi) &= -\sum_k \log p_\phi(y_i = k | \mathbf{x}_i) \\
&= -\sum_k \log \frac{\exp(-d_s(\mathbf{z}_i, \mathbf{c}_k))}{\sum_{k'} \exp(-d_s(\mathbf{z}_i, \mathbf{c}_{k'}))} \\
&= \sum_k \left(d_s(\mathbf{z}_i, \mathbf{c}_k) + \log \sum_{k'} \exp(-d_s(\mathbf{z}_i, \mathbf{c}_{k'})) \right).
\end{aligned}
\tag{9}
$$

The learning proceeds by minimizing $J(\phi)$ of the true class k via SGD, which is equivalent to minimizing the SEN dissimilarity measure between the query example $\mathbf{x}_i$ and its prototype $\mathbf{c}_k$: $d_s(\mathbf{z}_i, \mathbf{c}_k)$, and maximizing the SEN dissimilarity measures between the query example $\mathbf{x}_i$ and the other prototypes $\mathbf{c}_{k'}$: $d_s(\mathbf{z}_i, \mathbf{c}_{k'})$. Minimizing $d_s(\mathbf{z}_i, \mathbf{c}_k)$ pulls $\mathbf{z}_i$ to its own class and encourages embeddings of the same class to have the same norm. Maximizing $d_s(\mathbf{z}_i, \mathbf{c}_{k'})$ pushes $\mathbf{z}_i$ away from other classes; however it encourages embeddings of different classes to have different norms.

Since our goal is to force the data to lie on a scaled unit hypersphere, we define the balancing hyperparameter ϵ relative to $\mathbf{z}_i$ and $\mathbf{c}_k$ as follows:

$$
\epsilon_{ik} = \begin{cases} \epsilon_p > 0 & \text{if } y_i = k \\ \epsilon_n < 0 & \text{if } y_i \neq k \end{cases},
\tag{10}
$$

where i is the index of the embedding $\mathbf{z}_i$, y_i is the embedding's class label, and k is the class label of the prototype $\mathbf{c}_k$. During training, a positive epsilon ($\epsilon_{ik} = \epsilon_p > 0$) is used for computing the SEN dissimilarity measure between the query example $\mathbf{x}_i$ and its prototype $\mathbf{c}_k$, while a negative epsilon ($\epsilon_{ik} = \epsilon_n < 0$) is used for computing the SEN dissimilarity measures between the query example $\mathbf{x}_i$ and the other prototypes $\mathbf{c}_{k'}$. The negative epsilon ϵ_n will inverse the effect of the norm distance when maximizing $d_s(\mathbf{z}_i, \mathbf{c}_{k'})$. In other words, maximizing $d_s(\mathbf{z}_i, \mathbf{c}_{k'})$ with a negative epsilon ϵ_n pushes $\mathbf{z}_i$ away from other classes and encourages embeddings of all classes to have the same norm. The flexibility induced by the balancing hyperparameter ϵ_{ik} makes the SEN particularly suited to enable discriminative feature normalization in PNs.

Our proposed SEN dissimilarity measure explicitly encourages the norm of samples to have the same value during training, while preserving the convexity and the simplicity of the loss function. At test time, a positive epsilon ($\epsilon_{ik} = \epsilon_p > 0$) is used for computing all dissimilarity measures.

In the next section, we provide a theoretical analysis showing that our proposed SEN dissimilarity measure together with the special balancing hyperparameter ϵ_{ik} explicitly pulls the data to a scaled unit hypersphere during training.

3.4 Theoretical Analysis

The partial derivative of the negative log probability loss $J(\phi)$ with respect to $d_s(\mathbf{z}_i, \mathbf{c}_k)$ is given by:

$$\frac{\partial J(\phi)}{\partial d_s(\mathbf{z}_i, \mathbf{c}_k)} = \sum_k (1[y_i = k] - p_\phi(y_i = k|\mathbf{x})), \tag{11}$$

where the Iverson bracket indicator function $[y_i = k]$ evaluates to 1 when $y_i = k$ and 0 otherwise. The partial derivative of the SEN dissimilarity measure $d_s(\mathbf{z}_i, \mathbf{c}_k)$ with respect to $\mathbf{z}_i$ is given by:

$$\begin{aligned}
\frac{\partial d_s(\mathbf{z}_i, \mathbf{c}_k)}{\partial \mathbf{z}_i} &= \frac{\partial \sqrt{d_e(\mathbf{z}_i, \mathbf{c}_k) + \epsilon_{ik} d_n(\mathbf{z}_i, \mathbf{c}_k)}}{\partial \mathbf{z}_i} \\
&= \frac{(\mathbf{z}_i - \mathbf{c}_k) + \epsilon_{ik}(\|\mathbf{z}_i\| - \|\mathbf{c}_k\|)\frac{\mathbf{z}_i}{\|\mathbf{z}_i\|}}{d_s(\mathbf{z}_i, \mathbf{c}_k)} \\
&= -\frac{(\mathbf{c}_k - \mathbf{z}_i) + \epsilon_{ik}(\|\mathbf{c}_k\| - \|\mathbf{z}_i\|)\frac{\mathbf{z}_i}{\|\mathbf{z}_i\|}}{d_s(\mathbf{z}_i, \mathbf{c}_k)} \\
&= -\frac{v(\mathbf{z}_i, \mathbf{c}_k)}{d_s(\mathbf{z}_i, \mathbf{c}_k)},
\end{aligned} \tag{12}$$

where

$$v(\mathbf{z}_i, \mathbf{c}_k) = (\mathbf{c}_k - \mathbf{z}_i) + \epsilon_{ik}(\|\mathbf{c}_k\| - \|\mathbf{z}_i\|)\frac{\mathbf{z}_i}{\|\mathbf{z}_i\|}. \tag{13}$$

Using the chain rule, we get:

$$\begin{aligned}
\frac{\partial J(\phi)}{\partial \mathbf{z}_i} &= \frac{\partial J(\phi)}{\partial d_s(\mathbf{z}_i, \mathbf{c}_k)} \frac{\partial d_s(\mathbf{z}_i, \mathbf{c}_k)}{\partial \mathbf{z}_i} \\
&= \sum_k -\frac{1[y_i = k] - p_\phi(y_i = k|x)}{d_s(\mathbf{z}_i, \mathbf{c}_k)} v(\mathbf{z}_i, \mathbf{c}_k) \\
&= \sum_k \frac{\partial J_k(\phi)}{\partial \mathbf{z}_i}.
\end{aligned} \tag{14}$$

Thus, there is a gradient contribution from all prototypes. In particular, the gradient contribution with respect to the correct prototype, when $k = k^* = y_i$, is given by:

$$\frac{\partial J_{k^*}(\phi)}{\partial \mathbf{z}_i} = -\frac{1 - p_\phi(y_i = k^* | x)}{d_s(\mathbf{z}_i, \mathbf{c}_{k^*})} v(\mathbf{z}_i, \mathbf{c}_{k^*})$$

$$= -\frac{1 - p_\phi(y_i = k^* | x)}{d_s(\mathbf{z}_i, \mathbf{c}_{k^*})} v_p(\mathbf{z}_i, \mathbf{c}_{k^*}), \tag{15}$$

where

$$v_p(\mathbf{z}_i, \mathbf{c}_{k^*}) = \underbrace{(\mathbf{c}_{k^*} - \mathbf{z}_i)}_{\text{attractor}} + \underbrace{\epsilon_{ik^*}(\|\mathbf{c}_{k^*}\| - \|\mathbf{z}_i\|)\frac{\mathbf{z}_i}{\|\mathbf{z}_i\|}}_{\text{norm equalizer}}. \tag{16}$$

The gradient contribution with respect to incorrect prototypes, when $k = k' \neq y_i$, is given by:

$$\frac{\partial J_{k'}(\phi)}{\partial \mathbf{z}_i} = -\frac{0 - p_\phi(y_i = k' | x)}{d_s(\mathbf{z}_i, \mathbf{c}_{k'})} v(\mathbf{z}_i, \mathbf{c}_{k'}) = -\frac{p_\phi(y_i = k' | x)}{d_s(\mathbf{z}_i, \mathbf{c}_{k'})} v_n(\mathbf{z}_i, \mathbf{c}_{k'}), \tag{17}$$

where

$$v_n(\mathbf{z}_i, \mathbf{c}_{k'}) = \underbrace{(\mathbf{z}_i - \mathbf{c}_{k'})}_{\text{repeller}} + \underbrace{\epsilon_{ik'}(\|\mathbf{z}_i\| - \|\mathbf{c}_{k'}\|)\frac{\mathbf{z}_i}{\|\mathbf{z}_i\|}}_{\text{norm equalizer}}. \tag{18}$$

From the preceding analysis, we observe the following:

1. Each gradient component contains an attractor/repeller, which encourages $\mathbf{z}_i$ to move towards the correct prototype and move away from the incorrect ones.
2. From (16), it is clear that if $\|\mathbf{c}_{k^*}\| > \|\mathbf{z}_i\|$ and $\epsilon_{ik^*} > 0$, $\epsilon_{ik^*}(\|\mathbf{c}_{k^*}\| - \|\mathbf{z}_i\|)\frac{1}{\|\mathbf{z}_i\|} > 0$, such that $\|\mathbf{z}_i\|$ is encouraged to increase (and vice verca for $\|\mathbf{z}_i\| > \|\mathbf{c}_{k^*}\|$).
3. Conversely, from (18), if $\|\mathbf{c}_{k'}\| > \|\mathbf{z}_i\|$ and $\epsilon_{ik'} > 0$, $\epsilon_{ik'}(\|\mathbf{z}_i\| - \|\mathbf{c}_{k'}\|)\frac{1}{\|\mathbf{z}_i\|} < 0$ (and vice verca for $\|\mathbf{z}_i\| > \|\mathbf{c}_{k'}\|$). Thus, we need $\epsilon_{ik'} < 0$ in order to ensure similar behaviors as with the correct prototype.

Observation 2) and 3) shows that the gradient contributions with respect to the correct prototype and the incorrect ones *cooperate* in order to equalize the norms during training when $\epsilon_{ik^*} > 0$ and $\epsilon_{ik'} < 0$.

4   Experiments

To evaluate the effectiveness of the proposed SEN dissimilarity measure, we compare our proposed SEN PN approach with the original PN [19] and state-of-the-art distance metric learning-based approaches on the Mini-Imagenet [17, 23] and the Omniglot [9] dataset. Further, additional ablation studies are also performed on the Fewshot-CIFAR100 (FC100) [14] dataset.

4.1   Experimental Setup and Results

Embedding Networks. We utilize the same embedding network as that used by the original PN. Specifically, our network, which we refer to as 4CONV, comprises of four convolutional blocks. Each block is composed of 64 3×3 convolutional filters, a batch normalization layer, a ReLU nonlinearity, and a 2×2 max-pooling layer. To test the performance of the SEN dissimilarity measure in more general settings, we employ a more sophisticated network, the Wide Residual Network (WRN) [28], as the embedding network. We use the same network architecture proposed in [3], which is a network of depth 16 and a widening factor of 6. We train the network with both the traditional Euclidean distance (WRN PN) and the SEN dissimilarity measure (SEN WRN PN).

Table 1. Few-shot classification accuracy.

Model	Network	Omniglot	Mini-Imagenet
Original PN [19]	4CONV	98.9%	68.2%
Large Margin GNN [27]	4CONV	**99.2%**	67.6%
Large Margin PN [27]	4CONV	98.7%	66.8%
RN [20]	4CONV	99.1%	65.3%
Matching Nets [23]	4CONV	98.7%	60.0%
MetaGAN + RN [29]	4CONV	**99.2%**	68.6%
Semi-Supervised PN [3]	4CONV	-	65.5%
PN (ours, baseline)	4CONV	98.6%	67.8%
SEN PN (ours)	**4CONV**	98.8%	**69.8%**
Supervised WRN PN [3]	WRN	-	69.6%
Semi-Supervised WRN PN [3]	WRN	-	70.9%
WRN PN (ours)	WRN	99.2%	71.0%
SEN WRN PN (ours)	**WRN**	**99.4%**	**72.3%**

Hyperparameter. ϵ For SEN-based models, during training, $\epsilon_p = 1.0$ is used for computing the SEN between the query example and its prototype, while $\epsilon_n = -10^{-7}$ to compute the SEN between the query example and the other prototypes. During testing, $\epsilon_p = 1.0$ is used for computing all the SEN dissimilarity measures. A discussion on how the hyperparameters ϵ_p and ϵ_n were chosen can be found in the supplementary.

Results. The test results are shown in Table 1. As can be seen from Table 1, although our implementation of the PN (the baseline model) achieves 0.4% points lower in terms of accuracy compared to the original implementation of the PN (67.8% vs 68.2%), the baseline model trained with the proposed SEN dissimilarity measure still outperforms the original PN by obtaining a relative increase of 2.4% and achieves an accuracy of 69.8%. In addition, the SEN WRN

PN outperforms the Semi-Supervised WRN PN by a relative increase of 2% and achieves an accuracy of 72.3% with the WRN as the embedding network.

Similar trends can be observed for the Omniglot dataset, where SEN PN outperforms our PN implementation and SEN WRN PN outperforms WRN PN.

4.2 Ablation Study

To investigate the effectiveness and behavior of the proposed SEN dissimilarity measure, we conduct several ablation studies. First, we compare against the PN trained with the Euclidean distance (PN), the PN trained with the Ring loss (Ring PN), and the PN trained with the SEN dissimilarity measure (SEN PN). The test results are show in Table 2. We train the Ring PN with different values of γ, the loss weight w.r.t to the primary loss, in range $[10^{-10}, 1]$ and pick $\gamma = 10^{-7}$ since it results in the highest accuracy. R was learned during training following [30]. As can be seen from Table 2, the Ring loss improves the accuracy relative to the PN on the Mini-Imagenet dataset by 1.8%; however, it performs worse than our proposed SEN PN approach, which obtains a relative increase of 3%. Similar behavior is obtained for other few-shot learning datasets such as FC100 and Omniglot. A more thorough discussion on SEN PN vs Ring PN can be found in the supplementary.

Table 2. Few-shot classification accuracy on the Omniglot [9] (20-way 5-shot), the Mini-Imagenet [17,23] (5-way 5-shot), and the FC100 [14] (5-way 5-shot) datasets.

Model	Omniglot	Mini-Imagenet	FC100
PN	98.6%	67.8%	52.4%
Ring PN	98.7%	68.6%	52.8%
SEN PN	**98.8%**	**69.8%**	**54.6%**

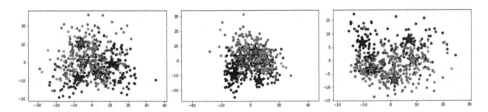

Fig. 1. 2D embeddings produced by the PN (left), the Ring PN (middle) and the SEN PN (right). The circles denote query examples, and the stars denotes prototypes.

Principal Component Analysis (PCA). We project 1600D embeddings produced by the PN, the Ring PN, and the SEN PN to 2D space using PCA and visualize the outputs (see Fig. 1). As can be seen from Fig. 1, the Ring loss forces the prototypes to lie on a scaled unit hypersphere; however, the prototypes produced by the Ring PN are not very well-separated compared to the ones produced by the PN. On the other hand, our proposed SEN dissimilarity measure both forces the prototypes to lie on a scaled unit hypersphere and keeps them well-separated.

Analysis of Norm. We plot the norm of embeddings produced by the PN, the Ring PN, and the SEN PN. As can be seen from Fig. 2, the norm of embeddings produced by the PN and the Ring PN vary a lot, while the norm of embeddings produced by the SEN PN has a very consistent value. This confirms that SEN encourages all embeddings to have the same norm during training. Both the SEN and the Ring loss are adopted for explicitly enforcing their embeddings to have the same norm during the training of the PN. However, as can be seen from Fig. 2, the proposed SEN dissimilarity measure is a better choice for the task than the Ring loss. This is partly due to the use of a very small gamma ($\gamma = 10^{-7}$) during training the Ring PN. In our experiments, higher gamma values do encourage the norm of embeddings to have a more consistent value; however, they cause a considerable decrease in the accuracy of the PN. This suggests that the Ring

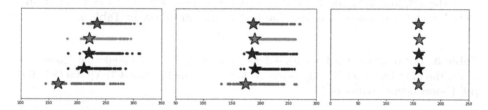

Fig. 2. The norm of embeddings produced by the PN (left), the Ring PN (middle), and the SEN PN (right). The stars denote query examples, and the diamonds denotes prototypes.

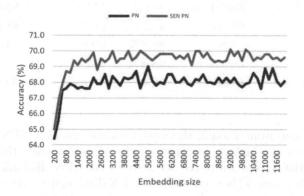

Fig. 3. The PN vs the SEN PN with different embedding sizes.

loss is not an optimal choice for enforcing feature normalization in PNs. The proposed SEN dissimilarity measure; on the other hand, both encourages all embeddings to have the same norm and improves the accuracy of PNs. This indicates that the proposed SEN dissimilarity measure is a more suitable choice for feature normalization than the Ring loss in training PNs.

Analysis of Embedding Dimensionality. We compare between the PN and the SEN PN trained with different embedding sizes (see Fig. 3). As can be seen from Fig. 3, in low dimensional spaces, the PN and the SEN PN perform very similarly; however, in high dimensional spaces, the SEN PN consistently outperforms the PN by a considerable margin. This suggests that the SEN dissimilarity measure is a more suitable distance metric for metric distance learning-based few-shot learning than the standard Euclidean distance in high dimensional spaces. This further explains the limited improvement on the Omniglot dataset where the embedding size is 64 compared to 1600 for the remaining datasets.

Analysis of Distance. We evaluate the possibility of combining the proposed SEN dissimilarity measure with other distance functions such as the Euclidean distance and the cosine distance in training PNs. Specifically, we train the PN with the SEN dissimilarity measure and test the trained model with both the Euclidean distance and the cosine distance. We compare the two tested models with the original PN, the SEN PN, and the Cosine PN (the PN trained and tested with the cosine distance). The test results are show in Table 3.

Table 3. Test results of the PN with different distances on the Omniglot [9] (20-way 5-shot), the Mini-Imagenet [17,23] (5-way 5-shot), and the Fewshot-CIFAR100 (FC100) [14] (5-way 5-shot) datasets.

Train distance	Test distance	Omniglot	Mini-Imagenet	FC100
Cosine	Cosine	61.5%	53.3%	44.9%
Cosine	SEN	55.2%	51.4%	43.8%
Euclidean	Euclidean	98.6%	67.8%	52.4%
Euclidean	SEN	98.7%	68.5%	53.1%
SEN	**SEN**	**98.8%**	**69.8%**	**54.6%**
SEN	Euclidean	**98.8%**	68.8%	53.9%
SEN	**Cosine**	**98.8%**	**69.8%**	**54.6%**

As can be seen from Table 3, the model trained with the SEN dissimilarity measure achieves the highest accuracy on the Mini-Imagenet, the FC100, and the Omniglot datasets when tested with either the SEN dissimilarity measure or the cosine distance. This is because the SEN dissimilarity measure explicitly forces all embeddings to have the same norm during training, and, as a result, pulling the prototypes very close to the hypersphere. For data embedded on a

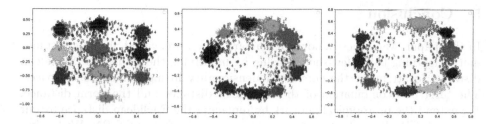

Fig. 4. 2D embeddings produced by the Siamese Baseline (left), the Siamese Ring (middle), and the Siamese SEN (right).

hypersphere, the cosine distance is a natural measure of distance [2,5]. Experiments and discussions on alternative design choices for SEN can be found in the supplementary.

SEN Beyond Few-Shot Learning. We have demonstrated that the SEN dissimilarity measure outperforms the commonly used Euclidean distance in distance metric learning-based few-shot learning with prototypical networks. In this section, we study the behaviors of the proposed SEN in combination with other metric learning-based tasks, which are based on the idea of obtaining inter-class separability and intra-class compactness. Note, due to the lack of prototypes, the SEN distance is here computed between datapoints directly. To do this, we implement the well-known Siamese network and Contrastive loss [7]. We call this model the Siamese Baseline. We augment it by replacing the Euclidean distance by our proposed SEN dissimilarity measure (Siamese SEN) and by employing Ring loss (Siamese Ring). We train the three models on the MNIST dataset [10] for dimensionality reduction and clustering. During training the Siamese SEN, following the reasoning of Sect. 3.3, a positive epsilon ($\epsilon_{ik} = \epsilon_p > 0$) is used for computing the SEN dissimilarity measures between examples of the same class, and a negative epsilon ($\epsilon_{ik} = \epsilon_n < 0$) is used for computing the SEN dissimilarity measures between examples of different classes. At test time, a positive epsilon ($\epsilon_{ik} = \epsilon_p > 0$) is used for computing all dissimilarity measures.

As can be seen from Fig. 4, the Siamese Ring forces all embeddings to lie on a scaled unit hypersphere; however, embeddings produced by the Siamese Ring are not as well-separated as embeddings produced by the Siamese Baseline. Our proposed SEN dissimilarity measure, on the other hand, both forces all embeddings to lie on a scaled unit hypersphere and keeps the embeddings well-separated. This suggests that SEN can also be used beyond the field of few-shot learning where distance metric learning is used and class memberships are available. In future work, other promising lines of research are to combine feature normalization with weight normalization techniques [11] and analyze their synergy, as well as to analyze the potential of SEN in other prototype-based methods [12].

5   Conclusion

In this paper, we propose a novel dissimilarity measure, called SEN, for distance metric learning-based few-shot learning by modifying the traditional Euclidean distance to attenuate the curse of dimensionality in high dimensional spaces. The SEN is a combination of the Euclidean distance and the norm distance. We extend the prototypical network by replacing the Euclidean distance by our proposed SEN dissimilarity measure, which we refer to as SEN PN. With minimal modifications, the SEN PN outperforms the original PN by a considerable margin and demonstrates good performance on the Mini-Imagenet, the FC100, and the Omniglot datasets with no additional parameters as well as negligible computational overhead. We provide analyses showing that the proposed SEN dissimilarity measure encourages the embeddings to have the same norm and enables the SEN PN to generate a hyperspherical embedding space, which is a more compact embedding space than the Euclidean space. We experimentally show that the proposed SEN dissimilarity measure consistently outperforms the Euclidean distance in PNs with different embedding sizes as well as with different embedding networks. We also show that SEN is an effective feature normalization technique not only for distance metric learning-based few-shot learning with PNs but also potentially for more general tasks, here exemplified by the Siamese network.

References

1. Adam, S., Sergey, B., Matthew, B., Daan, W., Timothy, P.L.: One-shot learning with memory-augmented neural networks. CoRR abs/1605.06065 (2016). http://arxiv.org/abs/1605.06065
2. Banerjee, A., Dhillon, I.S., Ghosh, J., Sra, S.: Clustering on the unit hypersphere using von Mises-Fisher distributions. J. Mach. Learn. Res. **6**, 1345–1382 (2005)
3. Boney, R., Ilin, A.: Semi-supervised few-shot learning with prototypical networks. CoRR abs/1711.10856 (2017). http://arxiv.org/abs/1711.10856
4. Deng, J., Guo, J., Xue, N., Zafeiriou, S.: ArcFace: additive angular margin loss for deep face recognition. In: Proceedings of the IEEE Conference on Computer Vision and Pattern Recognition, pp. 4690–4699 (2019)
5. Dhillon, I.S., Fan, J., Guan, Y.: Efficient clustering of very large document collections. In: Grossman, R.L., Kamath, C., Kegelmeyer, P., Kumar, V., Namburu, R.R. (eds.) Data Mining for Scientific and Engineering Applications. MC, vol. 2, pp. 357–381. Springer, Boston, MA (2001). https://doi.org/10.1007/978-1-4615-1733-7_20
6. Finn, C., Abbeel, P., Levine, S.: Model-agnostic meta-learning for fast adaptation of deep networks. In: Proceedings of the 34th International Conference on Machine Learning, vol. 70, pp. 1126–1135. JMLR. org (2017)
7. Hadsell, R., Chopra, S., LeCun, Y.: Dimensionality reduction by learning an invariant mapping. In: 2006 IEEE Computer Society Conference on Computer Vision and Pattern Recognition, CVPR 2006, vol. 2, pp. 1735–1742. IEEE (2006)
8. Koch, G., Zemel, R., Salakhutdinov, R.: Siamese neural networks for one-shot image recognition. In: ICML Deep Learning Workshop, vol. 2 (2015)

9. Lake, B., Salakhutdinov, R., Gross, J., Tenenbaum, J.: One shot learning of simple visual concepts. In: Proceedings of the Annual Meeting of the Cognitive Science Society, vol. 33 (2011)

10. LeCun, Y., Cortes, C.: MNIST handwritten digit database (2010). http://yann.lecun.com/exdb/mnist/

11. Liu, W., et al.: Learning towards minimum hyperspherical energy. In: Advances in Neural Information Processing Systems, pp. 6222–6233 (2018)

12. Mettes, P., van der Pol, E., Snoek, C.: Hyperspherical prototype networks. In: Advances in Neural Information Processing Systems, pp. 1485–1495 (2019)

13. Nikhil, M., Mostafa, R., Xi, C., Pieter, A.: A simple neural attentive meta-learner. CoRR abs/1707.03141 (2017). http://arxiv.org/abs/1707.03141

14. Oreshkin, B.N., Rodriguez, P., Lacoste, A.: TADAM: task dependent adaptive metric for improved few-shot learning. In: Proceedings of the 32nd International Conference on Neural Information Processing Systems, pp. 719–729. Curran Associates Inc. (2018)

15. Ranjan, R., Castillo, C.D., Chellappa, R.: L2-constrained softmax loss for discriminative face verification. arXiv preprint arXiv:1703.09507 (2017)

16. Ren, M., et al.: Meta-learning for semi-supervised few-shot classification. CoRR abs/1803.00676 (2018). http://arxiv.org/abs/1803.00676

17. Sachin, R., Hugo, L.: Optimization as a model for few-shot learning. In: 5th International Conference on Learning Representations (Conference Track Proceedings), ICLR 2017, Toulon, France, 24–26 April 2017 (2017). https://openreview.net/forum?id=rJY0-Kcll

18. Schroff, F., Kalenichenko, D., Philbin, J.: FaceNet: a unified embedding for face recognition and clustering. In: IEEE Conference on Computer Vision and Pattern Recognition, CVPR 2015, Boston, MA, USA, 7–12 June 2015, pp. 815–823 (2015). https://doi.org/10.1109/CVPR.2015.7298682

19. Snell, J., Swersky, K., Zemel, R.: Prototypical networks for few-shot learning. In: Proceedings of the 31st International Conference on Neural Information Processing Systems, pp. 4080–4090. Curran Associates Inc. (2017)

20. Sung, F., Yang, Y., Zhang, L., Xiang, T., Torr, P.H., Hospedales, T.M.: Learning to compare: relation network for few-shot learning. In: 2018 IEEE/CVF Conference on Computer Vision and Pattern Recognition, pp. 1199–1208. IEEE (2018)

21. Vanschoren, J.: Meta-learning: A survey. CoRR abs/1810.03548 (2018). http://arxiv.org/abs/1810.03548

22. Victor, G., Joan, B.: Few-shot learning with graph neural networks. In: International Conference on Learning Representations (2018). https://openreview.net/forum?id=BJj6qGbRW

23. Vinyals, O., Blundell, C., Lillicrap, T., Kavukcuoglu, K., Wierstra, D.: Matching networks for one shot learning. In: Proceedings of the 30th International Conference on Neural Information Processing Systems, pp. 3637–3645. Curran Associates Inc. (2016)

24. Wang, F., Xiang, X., Cheng, J., Yuille, A.L.: Normface: L2 hypersphere embedding for face verification. In: Proceedings of the 25th ACM International Conference on Multimedia, pp. 1041–1049. ACM (2017)

25. Wang, Y., Yao, Q.: Few-shot learning: A survey. CoRR abs/1904.05046 (2019). http://arxiv.org/abs/1904.05046

26. Wang, Y.X., Girshick, R., Hebert, M., Hariharan, B.: Low-shot learning from imaginary data. In: 2018 IEEE/CVF Conference on Computer Vision and Pattern Recognition (CVPR), pp. 7278–7286. IEEE (2018)

27. Yong, W., et al.: Large margin few-shot learning. CoRR abs/1807.02872 (2018). http://arxiv.org/abs/1807.02872
28. Zagoruyko, S., Komodakis, N.: Wide residual networks. CoRR abs/1605.07146 (2016). http://arxiv.org/abs/1605.07146
29. Zhang, R., Che, T., Ghahramani, Z., Bengio, Y., Song, Y.: MetaGAN: an adversarial approach to few-shot learning. In: Proceedings of the 32nd International Conference on Neural Information Processing Systems, pp. 2371–2380. Curran Associates Inc. (2018)
30. Zheng, Y., Pal, D.K., Savvides, M.: Ring loss: convex feature normalization for face recognition. In: 2018 IEEE/CVF Conference on Computer Vision and Pattern Recognition, pp. 5089–5097. IEEE (2018)

Kinematic 3D Object Detection
in Monocular Video

Garrick Brazil[1]($\boxtimes$), Gerard Pons-Moll[2], Xiaoming Liu[1], and Bernt Schiele[2]

[1] Computer Science and Engineering, Michigan State University, East Lansing, USA
{brazilga,liuxm}@msu.edu
[2] Max Planck Institute for Informatics, Saarland Informatics Campus,
Saarbrücken, Germany
{gpons,schiele}@mpi-inf.mpg.de

Abstract. Perceiving the physical world in 3D is fundamental for self-driving applications. Although temporal motion is an invaluable resource to human vision for detection, tracking, and depth perception, such features have not been thoroughly utilized in modern 3D object detectors. In this work, we propose a novel method for monocular video-based 3D object detection which leverages kinematic motion to extract scene dynamics and improve localization accuracy. We first propose a novel decomposition of object orientation and a self-balancing 3D confidence. We show that both components are critical to enable our kinematic model to work effectively. Collectively, using only a single model, we efficiently leverage 3D kinematics from monocular videos to improve the overall localization precision in 3D object detection while also producing useful by-products of scene dynamics (ego-motion and per-object velocity). We achieve state-of-the-art performance on monocular 3D object detection and the Bird's Eye View tasks within the KITTI self-driving dataset.

Keywords: 3D object detection · Monocular · Video · Physics-based

1 Introduction

The detection of foreground objects is among the most critical requirements to facilitate self-driving applications [3,4,11]. Recently, 3D object detection has made significant progress [9,21,23,24,37,45], even while using only a monocular camera [2,19,22,25,27,28,38,40]. Such works primarily look at the problem from the perspective of *single* frames, ignoring useful temporal cues and constraints.

Computer vision cherishes inverse problems, *e.g.*, recovering the 3D physical motion of objects from monocular videos. Motion information such as object velocity in the metric space is highly desirable for the path planning of self-driving. However, single image-based 3D object detection can not directly estimate physical motion, without relying on additional tracking modules. Therefore, *video-based 3D object detection* would be a sensible choice to recover such

Electronic supplementary material The online version of this chapter (https://doi.org/10.1007/978-3-030-58592-1_9) contains supplementary material, which is available to authorized users.

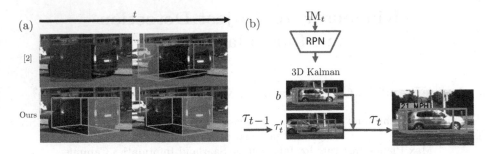

Fig. 1. Single-frame 3D detection [2] often has unstable estimation through time (a), while our video-based method (b) is more robust by leveraging **kinematic motion** via a 3D Kalman Filter to fuse forecasted tracks τ_t' and measurements b into τ_t.

motion information. Furthermore, without modeling the physical motion, image-based 3D object detectors are naturally more likely to suffer from erratic and unnatural changes through time in orientation and localization (as exemplified in Fig. 1(a)). Therefore, we aim to build a novel video-based 3D object detector which is able to provide *accurate* and *smooth* 3D object detection with per-object *velocity*, while also prioritizing a *compact* and *efficient* model overall.

Yet, designing an effective video-based 3D object detector has challenges. Firstly, motion which occurs in real-world scenes can come from a variety of sources such as the camera atop of an autonomous vehicle or robot, and/or from the scene objects themselves—for which most of the safety-critical objects (car, pedestrian, cyclist [14]) are typically *dynamic*. Moreover, using video inherently involves an increase in data consumption which introduces practical challenges for training and/or inference including with memory or redundant processing.

To address such challenges, we propose a novel framework to integrate a 3D Kalman filter into a 3D detection system. We find Kalman is an ideal candidate for three critical reasons: (1) it allows for use of real-world motion models to serve as a strong prior on object dynamics, (2) it is inherently efficient due to its recursive nature and general absence of parameters, (3) the resultant behavior is explainable and provides useful by-products such as the object velocity.

Furthermore, we observe that objects predominantly move in the direction indicated by their orientation. Fortunately, the benefit of Kalman allows us to integrate this real-world constraint into the motion model as a compact scalar velocity. Such a constraint helps maintain the consistency of velocity over time and enables the Kalman motion forecasting and fusion to perform accurately.

However, a model restricted to only move in the direction of its orientation has an obvious flaw—what if the orientation itself is *inaccurate*? We therefore propose a novel reformulation of orientation in favor of accuracy and stability. We find that our orientation improves the 3D localization accuracy by a margin of 2.39% and reduces the orientation error by ≈20%, which collectively help enable the proposed Kalman to function more effectively.

A notorious challenge of using Kalman comes in the form of *uncertainty*, which is conventionally [41] assumed to be known and static, *e.g.*, from a sensor.

However, 3D objects in video are intuitively dependent on more complex factors of image features and cannot necessarily be treated like a sensor measurement. For a better understanding of 3D uncertainty, we propose a 3D self-balancing confidence loss. We show that our proposed confidence has higher correlation with the 3D localization performance compared to the typical classification probability, which is commonly used in detection [33, 35].

To complete the full understanding of the scene motion, we elect to estimate the ego-motion of the capturing camera itself. Hence, we further narrow the work of Kalman to account for only the *object's* motion. Collectively, our proposed framework is able to model important scene dynamics, both ego-motion and per-object velocity, and more precisely detect 3D objects in videos using a stabilized orientation and 3D confidence estimation. We demonstrate that our method achieves state-of-the-art (SOTA) performance on monocular 3D Object Detection and Bird's Eye View (BEV) tasks in the KITTI dataset [14].

In summary, our contributions are as follows:

- We propose a monocular video-based 3D object detector, leveraging realistic motion constraints with an integrated ego-motion and a 3D Kalman filter.
- We propose to reformulate orientation into axis, heading and offset al.ong with a self-balancing 3D localization loss to facilitate the stability necessary for the proposed Kalman filter to perform more effectively.
- Overall, using only a *single model* our framework develops a comprehensive 3D scene understanding including object cuboids, orientation, velocity, object motion, uncertainty, and ego-motion, as detailed in Fig. 1 and 2.
- We achieve a new SOTA performance on monocular 3D object detection and BEV tasks using comprehensive metrics within the KITTI dataset.

2 Related Work

We first provide context of our novelties from the perspective of monocular 3D object detection (Sect. 2.1) with attention to orientation and uncertainty estimation. We next discuss and contrast with video-based object detection (Sect. 2.2).

2.1 Monocular 3D Object Detection

Monocular 3D object detection has made significant progress [2,7,8,19,22,25,27–29,38,39,43]. Early methods such as [8] began by generating 3D object proposals along a ground plane using object priors and estimated point clouds, culminating in an energy minimization approach. [7,19,43] utilize additional domains of semantic segmentation, object priors, and estimated depth to improve the localization. Similarly, [27,40] create a pseudo-LiDAR map using SOTA depth estimator [6,12,13], which is respectfully passed into detection subnetworks or LiDAR-based 3D object detection works [18,31,32]. In [22,25,28,29,38] strong 2D detection systems are extended to add cues such as object orientation, then the remaining 3D box parameters are solved via 3D box geometry. [2] extends the region proposal network (RPN) of Faster R-CNN [35] with 3D box parameters.

Orientation Estimation: Prior monocular 3D object detectors estimate orientation via two main techniques. The first method is to classify orientation via a series of discrete bins then regress a relative offset [7,8,19,22,25,28,29,43]. The bin technique requires a trade-off between the quantity/coverage of the discretized angles and an increase in the number of estimated parameters (bin ×). Other methods directly regress the orientation angle using quaternion [38] or Euler [2,27] angles. Direct regression is comparatively efficient, but may lead to degraded performance and periodicity challenges [46], as exemplified in Fig. 1.

In contrast, we propose a novel orientation decomposition which serves as an intuitive compromise between the bin and direct approaches. We decompose the orientation estimation into three components: axis and heading classification, followed by an angle offset. Thus, our technique increases the parameters by a *static* factor of 2 compared to a bin hyperparameter, while drastically reducing the offset search space for each orientation estimation (discussed in Sect. 3.1).

Uncertainty Estimation: Although it is common to utilize the classification score to rate boxes in 2D object detection [5,33,35,44] or explicitly model uncertainty as parametric estimation [20], prior works in monocular 3D object detection realize the need for 3D box uncertainty/confidence [25,38]. [25] defines confidence using the 3D IoU of a box and ground truth after center alignment, thus capturing the confidence primarily of the 3D *object dimensions*. [38] predicts a confidence by re-mapping the 3D box loss into a probability range, which intuitively represents the confidence of the overall 3D box accuracy.

In contrast, our *self-balancing* confidence loss is generic and self-supervised, with two benefits. (1) It enables estimation of a 3D localization confidence using only the loss values, thus being more general than 3D IoU. (2) It enables the network to naturally re-balance extremely hard 3D boxes and focus on relatively achievable samples. Our ablation (Sect. 4.4) shows the importance of both effects.

2.2   Video-Based Object Detection

Video-based object detection [1,26,42,47,48] is generally less studied than single-frame object detection [5,33–35,44,45]. A common trend in video-based detection is to benefit the accuracy-efficiency trade-off via reducing the frame redundancy [1,26,42,47,48]. Such works are applied primarily on domains of ImageNet VID 2015 [36], which contain less ego-motion from the capturing camera than self-driving scenarios [10,14]. As such, the methods are designed to use 2D transformations, which lack the consistency and realism of 3D motion modeling.

In comparison, to our knowledge this is the first work that *utilizes video cues to improve the accuracy and robustness of monocular 3D object detection*. In the domain of 2D/3D object tracking, [15] experiments using Kalman Filters, Particle Filters, and Gaussian Mixture Models, and observe Kalman to be the most effective aggregation method for tracking. An LSTM with depth ordering and IMU camera ego-motion is utilized in [16] to improve the tracking accuracy. In contrast, we explore how to naturally and effectively leverage a 3D Kalman

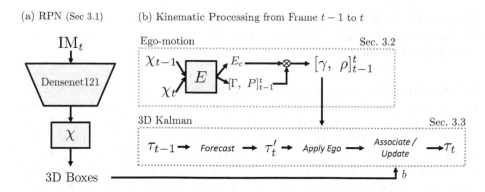

Fig. 2. Overview. Our framework uses a RPN to first estimate 3D boxes (Sect. 3.1). We forecast previous frame tracks τ_{t-1} into τ_t' using the estimated Kalman velocity. Self-motion is compensated for applying a global ego-motion (Sect. 3.2) to tracks τ_t'. Lastly, we fuse τ_t' with measurements b using a kinematic 3D Kalman filter (Sect. 3.3).

filter to improve the accuracy and robustness of monocular 3D object detection. We propose novel enhancements including estimating ego-motion, orientation, and a 3D confidence, while efficiently using only a single model.

3   Methodology

Our proposed kinematic framework is composed of three primary components: a 3D region proposal network (RPN), ego-motion estimation, and a novel kinematic model to *take advantage of temporal motion in videos*. We first overview the foundations of a 3D RPN. Then we detail our contributions of orientation decomposition and self-balancing 3D confidence, which are integral to the kinematic method. Next we detail ego-motion estimation. Lastly, we present the complete kinematic framework (Fig. 2) which carefully employs a 3D Kalman [41] to model realistic motion using the aforementioned components, ultimately producing a more accurate and comprehensive 3D scene understanding.

3.1   Region Proposal Network

Our measurement model is founded on the 3D RPN [2], enhanced using novel orientation and confidence estimations. The RPN itself acts as a sliding window detector following the typical practices outlined in Faster R-CNN [35] and [2]. Specifically, the RPN consists of a backbone network and a detection head which predicts 3D box outputs relative to a set of predefined anchors.

Anchors: We define our 2D-3D anchor Φ to consist of 2D dimensions $[\Phi_w, \Phi_h]_{2D}$ in pixels, a projected depth-buffer Φ_z in meters, 3D dimensions $[\Phi_w, \Phi_h, \Phi_l]_{3D}$ in meters, and orientations with respect to two major axes $[\Phi_0, \Phi_1]_{3D}$ in radians. The term Φ_z is related to the camera coordinate $[x, y, z]_{3D}$ by the equation

$\varPhi_z \cdot [u, v, 1]^T_{2D} = \Upsilon \cdot [x, y, z, 1]^T_{3D}$ where $\Upsilon \in \mathbb{R}^{3\times 4}$ is a known projection matrix. We compute the anchor values by taking the mean of each parameter after clustering all ground truth objects in 2D, following the process in [2].

3D Box Outputs: Since our network is based on the principles of a RPN [2,35], most of the estimations are defined as a transformation $\mathbf{T}$ relative to an anchor. Let us define n_a as the number of anchors, n_c as the number of object classes, and $w \times h$ as the output resolution of the network. The RPN outputs a classification map $\mathbf{C} \in \mathbb{R}^{(n_a \cdot n_c)\times w\times h}$, then 2D transformations $[\mathbf{T}_x, \mathbf{T}_y, \mathbf{T}_w, \mathbf{T}_h]_{2D}$, 3D transformations $[\mathbf{T}_u, \mathbf{T}_v, \mathbf{T}_z, \mathbf{T}_w, \mathbf{T}_h, \mathbf{T}_l, \mathbf{T}_{\theta_r}]_{3D}$, axis and heading $[\boldsymbol{\Theta}_a, \boldsymbol{\Theta}_h]$, and lastly a 3D self-balancing confidence $\boldsymbol{\Omega}$. Each output has a size of $\mathbb{R}^{n_a \times w \times h}$. The outputs can be unrolled into $n_b = (n_a \cdot w \cdot h)$ boxes with $(n_c + 14)$-dim, with parameters of c, $[t_x, t_y, t_w, t_h]_{2D}$, $[t_u, t_v, t_z, t_w, t_h, t_l, t_{\theta_r}]_{3D}$, $[\theta_a, \theta_h]$, and ω, which relate to the maps by $c \in \mathbf{C}$, $t_{2D} \in \mathbf{T}_{2D}$, $t_{3D} \in \mathbf{T}_{3D}$, $\theta \in \boldsymbol{\Theta}$ and $\omega \in \boldsymbol{\Omega}$. The regression targets for 2D ground truths (GTs) $[\hat{x}, \hat{y}, \hat{w}, \hat{h}]_{2D}$ are defined as:

$$\hat{t}_{x_{2D}} = \frac{\hat{x}_{2D} - i}{\varPhi_{w_{2D}}}, \hat{t}_{y_{2D}} = \frac{\hat{y}_{2D} - j}{\varPhi_{h_{2D}}}, \hat{t}_{w_{2D}} = \log \frac{\hat{w}_{2D}}{\varPhi_{w_{2D}}}, \hat{t}_{h_{2D}} = \log \frac{\hat{h}_{2D}}{\varPhi_{h_{2D}}}, \quad (1)$$

where $(i, j) \in \mathbb{R}^{w\times h}$ represent the pixel coordinates of the corresponding box. Similarly, following the equation of $\hat{z} \cdot [\hat{u}, \hat{v}, 1]^T_{2D} = \Upsilon \cdot [x, y, z, 1]^T_{3D}$, the regression targets for the projected 3D center GTs are defined as:

$$\hat{t}_u = \frac{\hat{u} - i}{\varPhi_{w_{2D}}}, \qquad \hat{t}_v = \frac{\hat{v} - j}{\varPhi_{h_{2D}}}, \qquad \hat{t}_z = \hat{z} - \varPhi_z. \quad (2)$$

Lastly, the regression targets for 3D dimensions GTs $[\hat{w}, \hat{h}, \hat{l}]_{3D}$ are defined as:

$$\hat{t}_{w_{3D}} = \log \frac{\hat{w}_{3D}}{\varPhi_{w_{3D}}}, \qquad \hat{t}_{h_{3D}} = \log \frac{\hat{h}_{3D}}{\varPhi_{h_{3D}}}, \qquad \hat{t}_{l_{3D}} = \log \frac{\hat{l}_{3D}}{\varPhi_{l_{3D}}}. \quad (3)$$

The remaining targets for our novel orientation estimation t_{θ_r}, $[\theta_a, \theta_h]$, and 3D self-balancing confidence ω are defined in subsequent sections.

Orientation Estimation: We propose a novel object orientation formulation, with a decomposition of three components: axis, heading, and offset (Fig. 3). Intuitively, the axis estimation θ_a represents the probability an object is oriented towards the vertical axis ($\theta_a = 0$) or the horizontal axis ($\theta_a = 1$), with its label formally defined as: $\hat{\theta}_a = |\sin \hat{\theta}| < |\cos \hat{\theta}|$, where $\hat{\theta}$ is the ground truth object orientation in radians from a bird's eyes view (BEV) with $[-\pi, \pi)$ bounded range. We then compute an orientation $\hat{\theta}_r$ with a restricted range relative to its axis, e.g., $[-\pi, 0)$ when $\hat{\theta}_a = 0$, and $[-\frac{\pi}{2}, \frac{\pi}{2})$ when $\hat{\theta}_a = 1$. We start with $\hat{\theta}_r = \hat{\theta}$ then add or subtract π from $\hat{\theta}_r$ until the desired range is satisfied.

Intuitively, $\hat{\theta}_r$ loses its heading since the true rotation may be $\{\hat{\theta}_r, \hat{\theta}_r \pm \pi\}$. We therefore preserve the heading using a separate $\hat{\theta}_h$, which represents the probability of $\hat{\theta}_r$ being rotated by π with its GT target defined as:

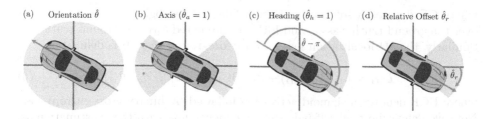

(a)   Orientation $\hat{\theta}$ ··· (b)   Axis ($\hat{\theta}_a = 1$) ··· (c)   Heading ($\hat{\theta}_h = 1$) ··· (d)   Relative Offset $\hat{\theta}_r$

Fig. 3. Orientation. Our proposed orientation formulation decomposes an object orientation $\hat{\theta}$ (a) into an axis classification $\hat{\theta}_a$ (b), a heading classification $\hat{\theta}_h$ (c), and an offset $\hat{\theta}_r$ (d). Our method disentangles the objectives of axis and heading classification while greatly reducing the offset region (red) by a factor of $\frac{1}{4}$. (Color figure online)

$$\hat{\theta}_h = \begin{cases} 0 & \hat{\theta} = \hat{\theta}_r \\ 1 & \text{otherwise.} \end{cases} \tag{4}$$

Lastly, we encode the orientation offset transformation which is relative to the corresponding anchor, axis, and restricted orientation $\hat{\theta}_r$ as: $\hat{t}_{\theta_r} = \hat{\theta}_r - \Phi_{\theta_a}$. The reverse decomposition is $\theta = \Phi_{\theta_a} + \lfloor \theta_h \rceil \cdot \pi + t_{\theta_r}$ where $\lfloor \rceil$ denotes round.

In designing our orientation, we first observed that the visual difference between objects at opposite headings of $[\theta, \theta \pm \pi]$ is low, especially for far objects. In contrast, classifying the axis of an object is intuitively more clear since the visual features correlate with the aspect ratio. Note that $[\theta_a, \theta_h, \theta_r]$ disentangle these two objectives. Hence, while the axis is being determined, the heading classifier can focus on subtle clues such as windshields, headlights and shape.

We further note that our 2 binary classifications have the same representational power as 4 bins following [19,22,25,28]. Specifically, bins of $[0, \frac{\pi}{2}, \pi, \frac{3\pi}{2}]$. However, it is more common to use considerably more bins (such as 12 in [19]). An important distinction is that bin-based approaches require the network decide axis and heading simultaneously, whereas our method *disentangles* the orientation into the two distinct and explainable objectives. We provide ablations to compare our decomposition and the bin method using [2, 4, 10] bins in Sect. 4.4.

Self-balancing Loss: The novel 3D localization confidence ω follows a self-balancing formulation closely coupled to the network loss. We first define the 2D and 3D loss terms which comprise the general RPN loss. We unroll and match all n_b box estimations to their respective ground truths. A box is matched as foreground when sufficient ($\geq k$) 2D intersection over union (IoU) is met, otherwise it is considered background ($\hat{c} = 0$) and all loss terms except for classification are ignored. The 2D box loss is thus defined as:

$$L_{2D} = -\log(\text{IoU}(b_{2D}, \hat{b}_{2D}))[\hat{c} \neq 0] + \text{CE}(c, \hat{c}), \tag{5}$$

where CE denotes a softmax activation followed by logistic cross-entropy loss over the ground truth class $\hat{c}$, and IoU uses predicted b_{2D} and ground truth $\hat{b}_{2D}$. Similarly, the 3D localization loss for only foreground ($\hat{c} \neq 0$) is defined as:

$$L_{3D} = L_1(t_{3D},\ \hat{t}_{3D}) + \lambda_a \cdot \text{BCE}([\theta_a,\ \theta_h],\ [\hat{\theta}_a,\ \hat{\theta}_h]), \tag{6}$$

where BCE denotes a sigmoid activation followed by binary cross-entropy loss. Next we define the final self-balancing confidence loss with the ω estimation as:

$$L = L_{2D} + \omega \cdot L_{3D} + \lambda_L \cdot (1 - \omega), \tag{7}$$

where λ_L is the rolling mean of the n_L most recent L_{3D} losses per mini-batch. Since ω is predicted per-box via a sigmoid, the network can intuitively balance whether to use the loss of L_{3D} or incur a proportional penalty of $\lambda_L \cdot (1 - \omega)$. Hence, when the confidence is high ($\omega \approx 1$) we infer that the network is confident in its 3D loss L_{3D}. Conversely, when the confidence is low ($\omega \approx 0$), the network is uncertain in L_{3D}, thus incurring a flat penalty is preferred. At inference, we fuse the self-balancing confidence with the classification score as $\mu = c \cdot \omega$.

The proposed self-balancing loss has two key benefits. Firstly, it produces a useful 3D localization confidence with inherent correlation to 3D IoU (Sect. 4.4). Secondly, it enables the network to re-balance samples which are exceedingly challenging and re-focus on the more reasonable targets. Such a characteristic can be seen as the inverse of hard-negative mining, which is important while monocular 3D object detection remains highly difficult and unsaturated (Sect. 4.1).

3.2 Ego-Motion

A challenge with the dynamics of urban scenes is that not only are most foreground objects in motion, but the capturing camera itself is dynamic. Therefore, for a full understanding of the scene dynamics, we design our model to additionally predict the self-movement of the capturing camera, $e.g.$, ego-motion.

We define ego-motion in the conventional six degrees of freedom: translation $[\gamma_x, \gamma_y, \gamma_z]$ in meters and rotation $[\rho_x, \rho_y, \rho_z]$ in radians. We attach an ego-motion features layer E with ReLU to the concatenation of two temporally adjacent feature maps χ which is the final layer in our backbone with size $\mathbb{R}^{n_h \times w \times h}$ architecture, defined as $\chi_{t-1} \parallel \chi_t$. We then attach predictions for translation $[\Gamma_x, \Gamma_y, \Gamma_z]$ and rotation $[P_x, P_y, P_z]$, which are of size $\mathbb{R}^{w \times h}$. Instead of using a global pooling, we predict a spatial confidence map $E_c \in \mathbb{R}^{w \times h}$ based on E. We then apply softmax over the spatial dimension of E_c such that $\sum E_c = 1$. Hence, the pooling of prediction maps $[\Gamma,\ P]$ into $[\gamma,\ \rho]$ is defined as:

$$\gamma = \sum_{(i,j)} \Gamma(i,j) \cdot E_c(i,j), \quad \rho = \sum_{(i,j)} P(i,j) \cdot E_c(i,j), \tag{8}$$

where (i,j) is the coordinate in $\mathbb{R}^{w \times h}$. We show an overview of the motion features E, spatial confidence E_c, and outputs $[\Gamma,\ P] \rightarrow [\gamma,\ \rho]$ within Fig. 2. We use a L_1 loss against GTs $\hat{\gamma}$ and $\hat{\rho}$ defined as $L_{ego} = L_1(\gamma,\ \hat{\gamma}) + \lambda_r \cdot L_1(\rho,\ \hat{\rho})$.

3.3 Kinematics

In order to leverage temporal motion in video, we elect to integrate our RPN and ego-motion into a novel kinematic model. We adopt a 3D Kalman [41] due to its notable efficiency, effectiveness, and interpretability. We next detail our proposed motion model, the procedure for forecasting, association, and update.

Motion Model: The critical variables we opt to track are defined as 3D center $[\tau_x, \tau_y, \tau_z]$, 3D dimensions $[\tau_w, \tau_h, \tau_l]$, orientation $[\tau_\theta, \tau_{\theta_h}]$, and scalar velocity τ_v. We define τ_θ as an θ orientation constrained to the range of $[-\frac{\pi}{2}, \frac{\pi}{2})$, and θ_h as in Sect. 3.1[1]. We constrain the motion model to only allow objects to move in the direction of their orientation. Hence, we define the state transition $\mathbf{F} \in \mathbb{R}^{9 \times 9}$ as:

$$
\mathbf{F} = \begin{bmatrix} \mathbf{I}^{9 \times 8} & \begin{matrix} \cos(\tau_\theta + \pi \lfloor \tau_{\theta_h} \rfloor) \\ 0 \\ -\sin(\tau_\theta + \pi \lfloor \tau_{\theta_h} \rfloor) \\ 0 \\ \vdots \\ 1 \end{matrix} \end{bmatrix},
\tag{9}
$$

where $\mathbf{I}$ denotes the identity matrix and the state variable order is respectively $[\tau_x, \tau_y, \tau_z, \tau_w, \tau_h, \tau_l, \tau_\theta, \tau_{\theta_h}, \tau_v]$. We constrain the velocity to only move within its orientation to simplify the Kalman to work more effectively. Recall that since our measurement model RPN processes a single frame, it does *not* measure velocity. Thus, to map the tracked state space and measurement space, we also define an observation model $\mathbf{H}$ as a truncated identity map of size $\mathbf{I} \in \mathbb{R}^{8 \times 9}$. We define covariance $\mathbf{P}$ with 3D confidence μ, as $\mathbf{P} = \mathbf{I}^{9 \times 9} \cdot (1 - \mu) \cdot \lambda_o$ where λ_o is an uncertainty weighting factor. Hence, we avoid the need to manually tune the covariance, while being dynamic to diverse and changing image content.

Forecasting: The forecasting step aims to utilize the tracked state variables and covariances of time $t - 1$ to estimate the state of a future time t. The equation to forecast a state variable τ_{t-1} into τ'_t is: $\tau'_t = \mathbf{F}_{t-1} \cdot \tau_{t-1}$, where $\mathbf{F}_{t-1}$ is the state transition model at $t - 1$. Note that both objects *and* the capturing camera may have independent motion between consecutive frames. Therefore, we lastly apply the estimated ego-motion to all available tracks' 3D center $[\tau_x, \tau_y, \tau_z]$ by:

$$
\begin{bmatrix} \tau_x \\ \tau_y \\ \tau_z \\ 1 \end{bmatrix}'_t = \begin{bmatrix} \mathbf{R}, & \mathbf{T} \\ 0, & 1 \end{bmatrix}^t_{t-1} \cdot \begin{bmatrix} \tau_x \\ \tau_y \\ \tau_z \\ 1 \end{bmatrix}'_t, \qquad \tau'_{t\theta} = \tau'_{t\theta} + \rho_y
\tag{10}
$$

where $\mathbf{R}^t_{t-1} \in \mathbb{R}^{3 \times 3}$ denotes the estimated rotation matrix converted from Euler angles and $\mathbf{T}^t_{t-1} \in \mathbb{R}^{3 \times 1}$ the translation vector for ego-motion (as in Sect. 3.2).

---

[1] We do not use the axis θ_a of Sect. 3.1, since we expect the orientation to change *smoothly* and do not wish the orientation is *relative* to a (potentially) changing axis.

Finally, we forecast a tracked object's covariance $\mathbf{P}$ from $t-1$ to t defined as:

$$\mathbf{P}'_t = \mathbf{F}_{t-1} \cdot \mathbf{P}_{t-1} \cdot \mathbf{F}^T_{t-1} + \mathbf{I}^{9\times9} \cdot (1 - \mu_{t-1}), \tag{11}$$

where μ_{t-1} denotes the *average* self-balancing confidence μ of a track's life. Hence, the resultant track states τ'_t and track covariances $\mathbf{P}'_t$ represent the Kalman filter's best forecasted estimation with respect to frame t.

Association: After the tracks have been forecasted from $t-1$ to t, the next step is to associate tracks to corresponding 3D box measurements (Sect. 3.1). Let us denote boxes produced by the measurement RPN as $b \in \mathbb{R}^8$ mimicking the tracked state as $[b_x, b_y, b_z, b_w, b_h, b_l, b_\theta, b_{\theta_h}]^2$. Our matching strategy consists of two steps. We first compute the 3D center distance between the tracks τ'_t and measurements b. The best matches with the lowest distance are iteratively paired and removed until no pairs remain with distance $\leq k_d$. Then we compute the *projected* 2D box IoU between any remaining tracks τ'_t and measurements b. The best matches with the highest IoU are also iteratively paired and removed until no pairs remain with IoU $\geq k_u$. Measured boxes that were *not* matched are added as new tracks. Conversely, tracks that were *not* matched incur a penalty with hyperparameter k_p, defined as $\mu_{t-1} = \mu_{t-1} \cdot k_p$. Lastly, any box who has confidence $\mu_{t-1} \leq k_m$ is removed from the valid tracks.

Update: After making associations between tracks τ'_t and measurements b, the next step is to utilize the track covariance $\mathbf{P}'_t$ and measured confidence μ to update each track to its final state τ_t and covariance $\mathbf{P}_t$. Firstly, we formally define the equation for computing the Kalman gain as:

$$\mathbf{K} = \mathbf{P}' \, \mathbf{H}^T \, (\mathbf{H} \, \mathbf{P}' \, \mathbf{H}^T + \mathbf{I}^{8\times8} \, (1 - \mu) \cdot \lambda_o)^{-1}, \tag{12}$$

where $\mathbf{I}^{8\times8} \, (1 - \mu) \cdot \lambda_o$ represents the incoming measurement covariance matrix, and $\mathbf{P}'$ the forecasted covariance of the track. Next, given the Kalman gain $\mathbf{K}$, forecasted state τ'_t, forecasted covariance $\mathbf{P}'_t$, and measured box b, the final track state τ_t and covariance $\mathbf{P}_t$ are defined as:

$$\tau_t = \tau'_t + \mathbf{K} \, (b - \mathbf{H} \, \tau'_t), \qquad \mathbf{P}_t = (\mathbf{I}^{9\times9} - \mathbf{K} \, \mathbf{H}) \, \mathbf{P}'_t. \tag{13}$$

We lastly aggregate each track's overall confidence μ_t over time as a running average of $\mu_t = \frac{1}{2} \cdot (\mu_{t-1} + \mu)$, where μ is the measured confidence.

3.4 Implementation Details

Our framework is implemented in PyTorch [30], with the 3D RPN settings of [2]. We release source code at http://cvlab.cse.msu.edu/project-kinematic.html. We

---

[2] We apply the estimated transformations of Sect. 3.1 to their respective anchors with equations of [2], and backproject into 3D coordinates to match track variables.

use a batch size of 2 and learning rate of 0.004. We set $k = 0.5$, $\lambda_o = 0.2$, $\lambda_r = 40$, $n_L = 100$, $\lambda_a = k_u = 0.35$, $k_d = 0.5$, $k_p = 0.75$, and $k_m = 0.05$. To ease training, we implement three phases. We first train the 2D-3D RPN with $L = L_{2D} + L_{3D}$, then the self-balancing loss of Eq. 7, for $80k$ and $50k$ iterations. We freeze the RPN to train ego-motion using L_{ego} for $80k$. Our backbone is DenseNet121 [17] where $n_h = 1,024$. Inference uses 4 frames as provided by [14].

4   Experiments

We benchmark our kinematic framework on the KITTI [14] dataset. We comprehensively evaluate on 3D Object Detection and Bird's Eye View (BEV) tasks. We then provide ablation experiments to better understand the effects and justification of our core methodology. We show qualitative examples in Fig. 6.

4.1   KITTI Dataset

The KITTI [14] dataset is a popular benchmark for self-driving tasks. The official dataset consists of 7,481 training and 7,518 testing images including annotations for 2D/3D objects, ego-motion, and 4 temporally adjacent frames. We evaluate on the most widely used validation split as proposed in [8], which consists of 3,712 training and 3,769 validation images. We focus primarily on the car class.

Metric: Average precision (AP) is utilized for object detection in KITTI. Following [38], the KITTI metric has updated to include 40 ($\uparrow$ 11) recall points while *skipping* the first. The AP_{40} metric is more stable and fair overall [38]. Due to the official adoption of AP_{40}, it is not possible to compute AP_{11} on test. Hence, we elect to use the AP_{40} metric for all reported experiments.

4.2   3D Object Detection

We evaluate our proposed framework on the task of 3D object detection, which requires objects be localized in 3D camera coordinates as well as supplying the 3D dimensions and BEV orientation relative to the XZ plane. Due to the strict requirements of IoU in three dimensions, the task demands precise localization of an object to be considered a match (3D IoU $\geq$ 0.7). We evaluate our performance on the official test [14] dataset in Table 1 and the validation [8] split in Table 2.

We emphasize that our method improves the SOTA on KITTI test by a significant margin of $\uparrow$ 1.98% compared to [27] on the moderate configuration with IoU $\geq$ 0.7, which is the most common metric used to compare. Further, we note that [27] require multiple encoder-decoder networks which add overhead compared to our single network approach. Hence, their runtime is $\approx 3\times$ (Table 1) compared to ours, self-reported on *similar* but not identical GPU hardware. Moreover, [2] is the most comparable method to ours as both utilize a single network and an RPN archetype. We note that our method significantly outperforms [2] and many other recent works [19, 22, 25, 28, 38] by $\approx$3.01–11.21%.

Table 1. KITTI Test. We compare with SOTA methods on the KITTI test dataset. We report performances using the AP_{40} [38] metric available on the official leaderboard. * the runtime is reported from the official leaderboard with slight variances in hardware. We indicate methods reported on CPU with †. **Bold**/*italics* indicate best/second AP.

	AP_{3D} (IoU $\geq$ 0.7)			AP_{BEV} (IoU $\geq$ 0.7)			s/im*
	Easy	Mod	Hard	Easy	Mod	Hard	
FQNet [25]	2.77	1.51	1.01	5.40	3.23	2.46	0.50†
ROI-10D [28]	4.32	2.02	1.46	9.78	4.91	3.74	0.20
GS3D [22]	4.47	2.90	2.47	8.41	6.08	4.94	2.00†
MonoPSR [19]	10.76	7.25	5.85	18.33	12.58	9.91	0.20
MonoDIS [38]	10.37	7.94	6.40	17.23	13.19	11.12	–
M3D-RPN [2]	14.76	9.71	7.42	21.02	13.67	10.23	*0.16*
AM3D [27]	*16.50*	*10.74*	**9.52**	*25.03*	*17.32*	**14.91**	0.40
Ours	**19.07**	**12.72**	*9.17*	**26.69**	**17.52**	*13.10*	**0.12**

Table 2. KITTI Validation. We compare with SOTA on KITTI validation [8] split. Note that methods published prior to [38] are unable to report the AP_{40} metric.

	AP_{3D} (IoU $\geq [0.7/0.5]$)			AP_{BEV} (IoU $\geq [0.7/0.5]$)		
	Easy	Mod	Hard	Easy	Mod	Hard
MonoDIS [38]	11.06/ –	7.60/ –	6.37/ –	18.45/ –	12.58/ –	10.66/ –
M3D-RPN [2]	*14.53/48.56*	*11.07/35.94*	*8.65/28.59*	*20.85/53.35*	*15.62/39.60*	*11.88/31.77*
Ours	**19.76/55.44**	**14.10/39.47**	**10.47/31.26**	**27.83/61.79**	**19.72/44.68**	**15.10/34.56**

We further evaluate our approach on the KITTI validation [8] split using the AP_{40} for available approaches and observe similar overall trends as in Table 2. For instance, compared to competitive approaches [2,38] our method improves the performance by ↑ 3.03% for the challenging IoU criteria of ≥ 0.7. Similarly, our performance on the more relaxed criteria of IoU ≥ 0.5 increases by ↑ 3.53%. We additionally visualize detailed performance characteristics on AP_{3D} at discrete depth [15, 30, All] meters and IoU matching criterias $0.3 \to 0.7$ in Fig. 4.

4.3 Bird's Eye View

The Bird's Eye View (BEV) task is similar to 3D object detection, differing primarily in that the 3D boxes are firstly projected into the XZ plane then 2D object detection is calculated. The projection collapses the Y-axis degree of freedom and intuitively results in a less precise but reasonable localization.

We note that our method achieves SOTA performance on the BEV task regarding the moderate setting of the KITTI test dataset as detailed in Table 1. Our method performs favorably compared with SOTA works [2,19,22,25,28, 38] (*e.g.*, ≈3.85–14.29%), and similarly to [27] at a notably lower runtime cost. We suspect that our method, especially the self-balancing confidence (Eq. 7),

Table 3. Ablation Experiments. We conduct a series of ablation experiments with the validation [8] split of KITTI, using diverse IoU matching criteria of ≥0.7/0.5.

	AP_{3D} (IoU ≥ [0.7/0.5])			AP_{BEV} (IoU ≥ [0.7/0.5])		
	Easy	Mod	Hard	Easy	Mod	Hard
Baseline	13.81/47.10	9.71/34.14	7.44/26.90	20.08/52.57	13.98/38.45	11.10/29.88
+ θ decomposition	16.66/51.47	12.10/38.58	9.40/30.98	23.15/56.48	17.43/42.53	13.48/34.37
+ self-confidence	16.64/52.18	12.77/38.99	9.60/**31.42**	24.22/58.52	18.02/42.95	13.92/**34.80**
+ $\mu = c \cdot \omega$	*18.28/54.70*	*13.55/39.33*	*10.13/31.25*	*25.72/60.87*	*18.82/44.36*	*14.48/34.48*
+ kinematics	**19.76/55.44**	**14.10/39.47**	**10.47**/*31.26*	**27.83/61.79**	**19.72/44.68**	15.10/*34.56*

prioritizes precise localization which warrants more benefit in **full** 3D Object Detection task compared to the Bird's Eye View task.

Our method performs similarly on the validation [8] split of KITTI (Table 2). Specifically, compared to [2,38] our proposed method outperforms by a range of ≈4.10–7.14%, which is **consistent** to the same methods on test ≈3.85–4.33%.

4.4 Ablation Study

To better understand the characteristics of our proposed kinematic framework, we perform a series of ablation experiments and analysis, summarized in Table 3. We adopt [2] without hill-climbing or depth-aware layers as our baseline method. Unless otherwise specified we use the experimental settings outlined in Sect. 3.4.

Orientation Improvement: The orientation of objects is intuitively a critical component when modeling motion. When the orientation is decomposed into axis, heading, and offset the overall performance significantly improves, *e.g.*, by ↑ 2.39% in AP_{3D} and ↑ 3.45% in AP_{BEV}, as detailed within Table 3. We compute the mean angle error of our baseline, orientation decomposition, and kinematics method which respectively achieve 13.4°, 10.9°, and 6.1° (↓ 54.48%), suggesting our proposed methodology is significantly more *stable*.

We compare our orientation decomposition to bin-based methods following general idea of [19,22,25,28]. We specifically change our orientation definition into $[\theta_b, \theta_o]$ which includes a bin classification and an offset. We experiment with the number of bins set to [2, 4, 10] which are uniformly spread from $[0, 2\pi)$. Note that 4 bins have the same representational power as using binary $[\theta_a, \theta_h]$. We observe that the ablated bin-based methods achieve [9.47%, 10.02%, 10.76%] in AP_{3D}. In comparison, our decomposed orientation achieves 12.10% in AP_{3D}. We provide additional detailed experiments in our supplemental material.

Further, we find that our proposed kinematic motion model (as in Sect. 3.3) *degrades* in performance when a comparatively erratic baseline (Row 1. Table 3) orientation is utilized instead (14.10 → 11.47 on AP_{3D}), reaffirming the importance of having a consistent/stable orientation when aggregating through time.

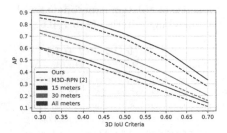
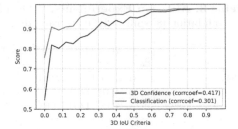

Fig. 4. We compare AP$_{3D}$ with [2] by varying 3D IoU criteria *and* depth.

Fig. 5. We show the correlation of 3D IoU to classification c and 3D confidence μ.

Self-balancing Confidence: We observe that the self-balancing confidence is important from two key respects. Firstly, its integration in Eq. 7 enables the network to re-weight box samples to focus more on reasonable samples and incur a flat penalty (*e.g.*, $\lambda_L \cdot (1 - \omega)$ of Eq. 7) on the difficult samples. In a sense, the self-balancing confidence loss is the *inverse* of hard-negative mining, allowing the network to focus on reasonable estimations. Hence, the loss on its own improves performance for AP$_{3D}$ by ↑ 0.67% and AP$_{BEV}$ by ↑ 0.59%.

The second benefit of self-balancing confidence is that by design ω has an inherent correlation with the 3D object detection performance. Recall that we fuse ω with the classification score c to produce a final box rating of $\mu = c \cdot \omega$, which results in an additional gain of ↑ 0.78% in AP$_{3D}$ and ↑ 0.80% in AP$_{BEV}$. We further analyze the correlation of μ with 3D IoU, as is summarized in Fig. 5. The correlation coefficient with the classification score c is significantly lower than the correlation using μ instead (0.301 vs. 0.417). In summary, the use of the Eq. 7 and μ account for a gain of ↑ 1.45% in AP$_{3D}$ and ↑ 1.39% in AP$_{BEV}$.

Temporal Modeling: The use of video and kinematics is a significant motivating factor for this work. We find that the use of kinematics (detailed in Sect. 3.3) results in a gain of ↑ 0.55% in AP$_{3D}$ and ↑ 0.90% in AP$_{BEV}$, as shown in Table 3. We emphasize that although the improvement is less dramatic versus orientation and self-confidence, the former are important to facilitate temporal modeling. We find that if orientation decomposition and uncertainty are **removed**, by using the baseline orientation and setting μ to be a static constant, then the kinematic performance drastically reduces from 14.10% → 10.64% in AP$_{3D}$.

We emphasize that kinematic framework not only helps 3D object detection, but also *naturally* produces useful by-products such as velocity and ego-motion. Thus, we evaluate the respective average errors of each motion after applying the camera capture rate to convert the motion into miles per hour (MPH). We find that the per-object velocity and ego-motion speed errors perform reasonably at 7.036 MPH and 6.482 MPH respectively. We depict visual examples of all dynamic by-products in Fig. 6 and additionally in supplemental video.

Fig. 6. Qualitative Examples. We depict the image view (left) and BEV (right). We show velocity vector in green, speed and ego-motion in miles per hour (MPH) on top of detection boxes and at the top-left corner, and tracks as dots in BEV. (Color figure online)

5   Conclusions

We present a novel kinematic 3D object detection framework which is able to efficiently leverage temporal cues and constraints to improve 3D object detection. Our method naturally provides useful by-products regarding scene dynamics, *e.g.*, reasonably accurate ego-motion and per-object velocity. We further propose novel designs of orientation estimation and a self-balancing 3D confidence loss in order to enable the proposed kinematic model to work effectively. We emphasize that our framework efficiently uses only a *single network* to comprehensively understand a highly dynamic 3D scene for urban autonomous driving. Moreover, we demonstrate our method's effectiveness through detailed experiments on the KITTI [14] dataset across the 3D object detection and BEV tasks.

Acknowledgments. Research was partially sponsored by the Army Research Office under Grant Number W911NF-18-1-0330. The views and conclusions contained in this document are those of the authors and should not be interpreted as representing the official policies, either expressed or implied, of the Army Research Office or the U.S. Government. The U.S. Government is authorized to reproduce and distribute reprints for

Government purposes notwithstanding any copyright notation herein. This work is further partly funded by the Deutsche Forschungsgemeinschaft (DFG, German Research Foundation) - 409792180 (Emmy Noether Programme, project: Real Virtual Humans).

References

1. Bertasius, G., Torresani, L., Shi, J.: Object detection in video with spatiotemporal sampling networks. In: Ferrari, V., Hebert, M., Sminchisescu, C., Weiss, Y. (eds.) ECCV 2018. LNCS, vol. 11216, pp. 342–357. Springer, Cham (2018). https://doi.org/10.1007/978-3-030-01258-8_21
2. Brazil, G., Liu, X.: M3D-RPN: Monocular 3D region proposal network for object detection. In: ICCV. IEEE (2019)
3. Brazil, G., Liu, X.: Pedestrian detection with autoregressive network phases. In: CVPR. IEEE (2019)
4. Brazil, G., Yin, X., Liu, X.: Illuminating pedestrians via simultaneous detection & segmentation. In: ICCV. IEEE (2017)
5. Chabot, F., Chaouch, M., Rabarisoa, J., Teuliere, C., Chateau, T.: Deep MANTA: a coarse-to-fine many-task network for joint 2D and 3D vehicle analysis from monocular image. In: CVPR. IEEE (2017)
6. Chang, J.R., Chen, Y.S.: Pyramid stereo matching network. In: CVPR. IEEE (2018)
7. Chen, X., Kundu, K., Zhang, Z., Ma, H., Fidler, S., Urtasun, R.: Monocular 3D object detection for autonomous driving. In: CVPR. IEEE (2016)
8. Chen, X., et al.: 3D object proposals for accurate object class detection. In: NeurIPS (2015)
9. Chen, Y., Liu, S., Shen, X., Jia, J.: Fast point R-CNN. In: ICCV. IEEE (2019)
10. Cordts, M., et al.: The cityscapes dataset for semantic urban scene understanding. In: CVPR. IEEE (2016)
11. Dollár, P., Wojek, C., Schiele, B., Perona, P.: Pedestrian detection: an evaluation of the state of the art. Pattern Anal. Mach. Intell. **34**(4), 743–761 (2012)
12. Felzenszwalb, P.F., Girshick, R.B., McAllester, D., Ramanan, D.: Object detection with discriminatively trained part-based models. Pattern Anal. Mach. Intell. **32**(9), 1627–1645 (2010)
13. Fu, H., Gong, M., Wang, C., Batmanghelich, K., Tao, D.: Deep ordinal regression network for monocular depth estimation. In: CVPR. IEEE (2018)
14. Geiger, A., Lenz, P., Urtasun, R.: Are we ready for autonomous driving? The KITTI vision benchmark suite. In: CVPR. IEEE (2012)
15. Giancola, S., Zarzar, J., Ghanem, B.: Leveraging shape completion for 3D siamese tracking. In: CVPR. IEEE (2019)
16. Hu, H.N., et al.: Joint monocular 3D vehicle detection and tracking. In: ICCV. IEEE (2019)
17. Huang, G., Liu, Z., Van Der Maaten, L., Weinberger, K.Q.: Densely connected convolutional networks. In: CVPR. IEEE (2017)
18. Ku, J., Mozifian, M., Lee, J., Harakeh, A., Waslander, S.L.: Joint 3D proposal generation and object detection from view aggregation. In: IROS. IEEE (2018)
19. Ku, J., Pon, A.D., Waslander, S.L.: Monocular 3D object detection leveraging accurate proposals and shape reconstruction. In: CVPR. IEEE (2019)
20. Kumar, A., et al.: LUVLi face alignment: Estimating landmarks' location, uncertainty, and visibility likelihood. In: CVPR. IEEE (2020)

21. Lang, A.H., Vora, S., Caesar, H., Zhou, L., Yang, J., Beijbom, O.: PointPillars: fast encoders for object detection from point clouds. In: CVPR. IEEE (2019)
22. Li, B., Ouyang, W., Sheng, L., Zeng, X., Wang, X.: GS3D: an efficient 3D object detection framework for autonomous driving. In: CVPR. IEEE (2019)
23. Liang, M., Yang, B., Chen, Y., Hu, R., Urtasun, R.: Multi-task multi-sensor fusion for 3D object detection. In: CVPR. IEEE (2019)
24. Liang, M., Yang, B., Wang, S., Urtasun, R.: Deep continuous fusion for multi-sensor 3D object detection. In: Ferrari, V., Hebert, M., Sminchisescu, C., Weiss, Y. (eds.) ECCV 2018. LNCS, vol. 11220, pp. 663–678. Springer, Cham (2018). https://doi.org/10.1007/978-3-030-01270-0_39
25. Liu, L., Lu, J., Xu, C., Tian, Q., Zhou, J.: Deep fitting degree scoring network for monocular 3D object detection. In: CVPR. IEEE (2019)
26. Liu, M., Zhu, M.: Mobile video object detection with temporally-aware feature maps. In: CVPR. IEEE (2018)
27. Ma, X., Wang, Z., Li, H., Zhang, P., Ouyang, W., Fan, X.: Accurate monocular 3D object detection via color-embedded 3D reconstruction for autonomous driving. In: ICCV. IEEE (2019)
28. Manhardt, F., Kehl, W., Gaidon, A.: ROI-10D: monocular lifting of 2D detection to 6D pose and metric shape. In: CVPR. IEEE (2019)
29. Mousavian, A., Anguelov, D., Flynn, J., Kosecka, J.: 3D bounding box estimation using deep learning and geometry. In: CVPR. IEEE (2017)
30. Paszke, A., et al.: Automatic differentiation in PyTorch (2017)
31. Qi, C.R., Liu, W., Wu, C., Su, H., Guibas, L.J.: Frustum pointnets for 3D object detection from RGB-D data. In: CVPR. IEEE (2018)
32. Qi, C.R., Yi, L., Su, H., Guibas, L.J.: PointNet++: deep hierarchical feature learning on point sets in a metric space. In: NeurIPS (2017)
33. Redmon, J., Divvala, S., Girshick, R., Farhadi, A.: You only look once: unified, real-time object detection. In: CVPR. IEEE (2016)
34. Ren, J., et al.: Accurate single stage detector using recurrent rolling convolution. In: CVPR. IEEE (2017)
35. Ren, S., He, K., Girshick, R., Sun, J.: Faster R-CNN: towards real-time object detection with region proposal networks. In: NeurIPS (2015)
36. Russakovsky, O., et al.: Imagenet large scale visual recognition challenge. Int. J. Comput. Vision 115(3), 211–252 (2015)
37. Shi, S., Wang, X., Li, H.: PointRCNN: 3D object proposal generation and detection from point cloud. In: CVPR. IEEE (2019)
38. Simonelli, A., Bulo, S.R., Porzi, L., López-Antequera, M., Kontschieder, P.: Disentangling monocular 3D object detection. In: ICCV. IEEE (2019)
39. Wang, X., Yin, W., Kong, T., Jiang, Y., Li, L., Shen, C.: Task-aware monocular depth estimation for 3D object detection. In: AAAI (2020)
40. Wang, Y., Chao, W.L., Garg, D., Hariharan, B., Campbell, M., Weinberger, K.Q.: Pseudo-LiDAR from visual depth estimation: bridging the gap in 3D object detection for autonomous driving. In: CVPR. IEEE (2019)
41. Welch, G., Bishop, G., et al.: An introduction to the kalman filter (1995)
42. Xiao, F., Lee, Y.J.: Video object detection with an aligned spatial-temporal memory. In: Ferrari, V., Hebert, M., Sminchisescu, C., Weiss, Y. (eds.) ECCV 2018. LNCS, vol. 11212, pp. 494–510. Springer, Cham (2018). https://doi.org/10.1007/978-3-030-01237-3_30
43. Xu, B., Chen, Z.: Multi-level fusion based 3D object detection from monocular images. In: CVPR. IEEE (2018)

44. Yang, F., Choi, W., Lin, Y.: Exploit all the layers: fast and accurate CNN object detector with scale dependent pooling and cascaded rejection classifiers. In: CVPR. IEEE (2016)
45. Yang, Z., Sun, Y., Liu, S., Shen, X., Jia, J.: STD: Sparse-to-dense 3D object detector for point cloud. In: ICCV. IEEE (2019)
46. Zhou, Y., Barnes, C., Lu, J., Yang, J., Li, H.: On the continuity of rotation representations in neural networks. In: CVPR. IEEE (2019)
47. Zhu, X., Dai, J., Yuan, L., Wei, Y.: Towards high performance video object detection. In: CVPR. IEEE (2018)
48. Zhu, X., Wang, Y., Dai, J., Yuan, L., Wei, Y.: Flow-guided feature aggregation for video object detection. In: ICCV. IEEE (2017)

Describing Unseen Videos via Multi-modal Cooperative Dialog Agents

Ye Zhu[1], Yu Wu[2,3], Yi Yang[2], and Yan Yan[1(✉)]

[1] Texas State University, San Marcos, USA
{ye.zhu,tom_yan}@txstate.edu
[2] ReLER, University of Technology Sydney, Sydney, Australia
yu.wu-3@student.uts.edu.au, yi.yang@uts.edu.au
[3] Baidu Research, Beijing, China

Abstract. With the arising concerns for the AI systems provided with direct access to abundant sensitive information, researchers seek to develop more reliable AI with implicit information sources. To this end, in this paper, we introduce a new task called video description via two multi-modal cooperative dialog agents, whose ultimate goal is for one conversational agent to describe an unseen video based on the dialog and two static frames. Specifically, one of the intelligent agents - *Q-BOT* - is given two static frames from the beginning and the end of the video, as well as a finite number of opportunities to ask relevant natural language questions before describing the unseen video. *A-BOT*, the other agent who has already seen the entire video, assists *Q-BOT* to accomplish the goal by providing answers to those questions. We propose a QA-Cooperative Network with a dynamic dialog history update learning mechanism to transfer knowledge from *A-BOT* to *Q-BOT*, thus helping *Q-BOT* to better describe the video. Extensive experiments demonstrate that *Q-BOT* can effectively learn to describe an unseen video by the proposed model and the cooperative learning method, achieving the promising performance where *Q-BOT* is given the full ground truth history dialog. Codes and models are available at https://github.com/L-YeZhu/Video-Description-via-Dialog-Agents-ECCV2020.

Keywords: Video description · Dialog agents · Multi-modal

1 Introduction

It is becoming a trend to exploit the possibilities to develop artificial intelligence (AI) systems with subtle and advanced reasoning abilities as humans, usually by providing AI with direct access to abundant information sources. However, what comes with this tendency is the arising concerns over the security and

Electronic supplementary material The online version of this chapter (https://doi.org/10.1007/978-3-030-58592-1_10) contains supplementary material, which is available to authorized users.

© Springer Nature Switzerland AG 2020
A. Vedaldi et al. (Eds.): ECCV 2020, LNCS 12368, pp. 153–169, 2020.
https://doi.org/10.1007/978-3-030-58592-1_10

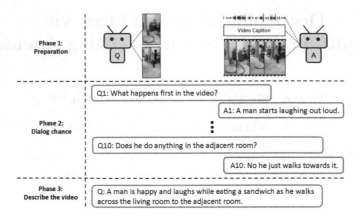

Fig. 1. Task setup: Describing an unseen video via two multi-modal cooperative dialog agents. The entire process can be described in three phases. The ultimate goal is for *Q-BOT* to describe what happens between the beginning and the end of an unseen video based on the dialog history with *A-BOT*.

privacy issues behind such AI systems. Although direct access to rich information assists AI becoming more intelligent and competent, the general public starts to question whether their sensitive personal information, such as the identifiable face images and voices, is in safe hands.

Despite the general concerns, nowadays research in AI and computer vision (CV) fields is experiencing a rapid transition from traditional 'low-level' tasks within single modality data such as image classification [15,16,21], object detection [24,32,41], machine translation [6], to more challenging tasks that involve multiple modalities of data and subtle reasoning [13,19,34,40,51], such as visual question answering [3,4,50] and visual dialog [10,20,46]. A meaningful and informative conversation, either between human-computer or computer-computer, is an appropriate task to demonstrate such a reasoning process due to the complex information exchange mechanism during the dialog. With the emergence of large-scale datasets such as VQA [4], GuessWhat [12] and AVSD [1,18], much effort is devoted to study the techniques for machines to maintain natural conversation interactions in a sophisticated way [20,22,28,46]. While most existing works still focus on the dialog itself and only involve a single agent, generally by providing the agent with direct access to sensitive information (*e.g.*, the complete video clips with identifiable human faces and voices), we wish to take a step forward to more secure and reliable AI systems with implicit information sources. To this end, in this paper we introduce a novel natural and challenging task with implicit information sources: describe an unseen video mainly based on the dialog between two cooperative agents.

The setup for our task is illustrated in Fig. 1, which involves two dialog agents, *Q-BOT* and *A-BOT*. Imagine the *Q-BOT* to be the actual AI system, and the *A-BOT* to be humans. The entire process can be described in three phases: In the first preparation phase, two agents are provided with different information. *A-BOT* is able to see the complete video with the audio signals and captions, while *Q-BOT* is only given two static frames from the beginning and the end of the video. It is worth noting that the static frames given to *Q-BOT* have no specific requirements, they could be static images without visible humans or just the back of the person as shown in Fig. 1 and Fig. 2, which largely reduces the risks for the AI systems to recognize the actual person. In the second phase, *Q-BOT* has 10 opportunities to ask *A-BOT* relevant questions about the video, such as the event happened. After 10 rounds of question-answer interactions, *Q-BOT* is asked to describe the unseen video based on the initial two frames and the dialog history with *A-BOT*. Under this task setup, the AI system *Q-BOT* accomplishes a multi-modal task without direct access to the original information, but learns to filter and extract useful information from a less sensitive information source, *i.e.*, the dialog. It is highly impossible for AI systems to identify a person based on the natural language descriptions. Therefore, such task settings and reasoning ability based on implicit information sources have great potential to be applied in a wide practical context, such as the smart home systems, improving the current AI systems that rely on direct access to sensitive information to accomplish certain tasks. Notably, instead of directly asking for the final video descriptions from human users, our task formulation that requires AI systems to gradually ask questions helps to reduce the bias and noises usually contained in the description directly given by human individuals.

It is a challenging and natural task compared to previous works in the field of the visual dialog. The difficulty mainly comes from the implicit information source of multiple modalities and the more complex reasoning process required for both agents. Specifically, this task also differs from the traditional video captioning task due to the fact that *Q-BOT* has never seen the entire video. Intuitively, it can be considered as establishing an additional information barrier(*i.e.*, the dialog) between the direct visual data input and the natural language video caption output. Figure 2 shows some examples from the AVSD dataset [1,18], whose data collection process resembles to our task setup. We observe that the ground truth video captions shown to *A-BOT* and the expected final descriptions given by *Q-BOT* are quite different, thus revealing the actual gap existing in human reasoning. This fact again emphasizes the difficulties of our task.

The key aspect to consider in this work is the effective knowledge transfer from *A-BOT* to *Q-BOT*. *A-BOT*, who plays the role of humans, has full access to all the information, while *Q-BOT* only has an ambiguous understanding of the surrounding environment from two static video frames after the first phase. In order to describe the video with details that are not included in the initial input, *Q-BOT* needs to extract useful information from the dialog with *A-BOT*. Therefore, we propose a QA-Cooperative network that involves two agents with the ability to process multiple modalities of data. We further introduce a cooperative learning method that enables us to jointly train the network with a dynamic dialog history update mechanism. The knowledge gap and transfer process are both experimentally demonstrated.

The main contributions of this paper are: 1) We propose a novel and challenging video description task via two multi-modal dialog agents, whose ultimate goal is for one agent to describe an unseen video mainly based on the

GT caption: A person is walking across the room and opening a cabinet. Then the person begins removing the dishes from it and tidying the shelves.

GT description: A woman walks into the kitchen. She takes dishes out of a cabinet and then puts them back. The video ends.

GT caption: A person is sitting in their car in the garage, playing a casual game on their laptop while drinking a bottle of soda. They open their glove compartment for a napkin when they notice a picture they pull out and start to look at intently.

GT description: The man looks at his laptop, gets soda, and then finds a photo in the car.

GT caption: One person was playing with their clothes. The other was grasping some groceries.

GT description: One person is swinging a rag while another fixes a table.

GT caption: A person walks down the hall, smiling, while drinking a glass of water. The person turns off the light once they reach the end of the hall.

GT description: A girl is looking into another room then turns towards the camera holding a glass. She smiles and pretends to take a few sips out of the empty glass. Then she looks into the camera as she walks towards it and out of view.

Fig. 2. Examples from the AVSD dataset [1,18]. Two static images are the beginning and the end frame of the video clips. *GT caption* is the caption shown to the *A-BOT* under our task setup. *GT description* is actual the summary given by the human questioner at the end of the dialog during the data collection without directly watching the video, which corresponds to the video description required in our task. Intuitively, our task can be considered as establishing an additional information barrier(*i.e.*, the dialog) between the direct visual data input and the natural language caption output.

interactive dialog history. This task establishes a more secure and reliable setting by providing implicit information sources to AI systems. 2) We propose a QA-Cooperative network and a cooperative learning method with a dynamic dialog history update mechanism, which helps to effectively transfer knowledge between the two agents. 3) We experimentally demonstrate the knowledge gap as well as the transfer process between two agents on the AVSD dataset [18]. With the proposed method, our *Q-BOT* achieves very promising performance comparable to the strong baseline situation where full ground truth dialog is provided.

2   Related Work

Image and Video Captioning. Image or video captioning refers to the task that aims to obtain textual descriptions for the given image or video. It is one of the first well-exploited tasks that combines both computer vision and natural language processing. You *et al.* [53] adopt Recurrent neural networks (RNNs) with selective attention to semantic concept proposals to generate image captions. Adaptive attention and spatial attention have also been studied in [8,25]. Anderson *et al.* [3] propose to use bottom-up and top-down attention for object levels and other salient image regions to address this task. Yao *et al.* [52] use the attributes to further improve the performance. Rennie *et al.* [33] propose to train the image captioning system with reinforcement learning. Wang *et al.* [45] adopt conditional variational auto-encoders with an additive Gaussian encoding space to generate image descriptions. Wu *et al.* [47] propose a disentangled framework to generalize image captioning models to describe unseen objects. RecNet [44] is introduced for video captioning. Mahasseni *et al.* [27] proposes to summarize the video with adversarial lstm networks in an unsupervised manner.

Visual Question Answering. Another similar research field beyond describing an image or a video is visual question answering (VQA). The objective of VQA is to answer a natural language question about the given image [4]. Different attention mechanisms including hierarchical attention [26], question-guided spatial attention [50], stacked attention [51], bottom-up and top-down attention [3] have been exploited. Das et al. [9] look into the question whether the existing attention mechanisms attend to the same regions as humans do. Shih et al. [38] map textual queries and visual features into a shared space to answer the question. Dynamic memory networks is also used for VQA [49].

Visual Dialog. Visual dialog is another succession of vision-language problem after visual captioning and VQA that is closely related to our work. Different from the VQA that only involves a single round of natural language interaction, visual dialog requires machines to maintain multiple rounds of conversation. Several datasets have been collected for this task [10,12]. Jain et al. [20] proposes a symmetric baseline to demonstrate how visual dialog can be generated from discriminative question generation and answering. Additionally, attention mechanisms [29,37] and reinforcement learning [11,46] have also been exploited in the context of visual dialog.

Audio Data. As another important source of information, the audio modality has recently gained popularity in the research field of computer vision. There have been emerging studies on combining audio and visual information for applications such as sound source separation [14,30], sound source localization [5,54] and audio-visual event localization [42,48].

Audio-Visual Scene-Aware Dialog. Audio-visual scene-aware dialog is a recently proposed multi-modal task that combines the previously mentioned research fields. The AVSD dataset introduced in [18] and [1] contains videos with audio streams and a corresponding sequence of question-answer pairs. Hori et al. [18] enhance the quality of generated dialog about videos using multi-modal features. Schwartz et al. [35] adopt the multi-modal attention mechanism to extract useful features for audio-visual scene-aware dialog task.

Cooperative Agents and Reasoning. Das et al. [11] are the first to propose goal-driven training for dialog agents, during which two interactive dialog agents are also involved to select an unseen image. Wu et al. [46] propose to generate reasoned dialog via adversarial learning. An information theoretic algorithm for goal-oriented dialog is then introduced in [22] to help the question generation. Unlike the previous work that has separate agents for questioner and answerer, Massiceti et al. [28] propose to use a single generative model for both roles. The reasoning process is essential in developing such intelligent agents. The idea of multi-step reasoning in the field of VQA has been exploited in [13,19,40,46,51].

Our work is related to the works mentioned above, yet differentiates from them in multiple aspects, including the task setup and formulation with implicit information sources, the QA-Cooperative network design, and the cooperative training method.

Table 1. Notations for the video description task.

s – Video description	V_s – start static frame of the video
S – Vocabulary	V_e – end static frame of the video
$i(i \leq 10)$ – Question-Answer round	$x_{A,i}$ – input for A-BOT at round i
A – Audio data	$x_{Q,i}$ – input for Q-BOT at round i
V_A – Video data for A-BOT	r_m – original data embedding for modal m
C – Video caption	a_m – attended data embedding for modal m
H_{i-1} – Existing dialog history at round i	d_m – dimension of the embedding for modal m
p_i – i-th pair of question-answer	$n_{\{C,H,q,a\}}$ – length of textual sequence
q_i – i-th question	m – modality notation, specified in context
	$m \in \{A, V, C, H, q, a\}$
a_i – i-th answer	

3   Video Description via Cooperative Agents

In this section, we respectively introduce the QA-Cooperative network for the two dialog agents, each component of the proposed network and the cooperative learning method with a dynamic dialog history update mechanism.

Notations used in our task formulation is presented in Table 1. In this video description task, we expect Q-BOT to describe an unseen video with a sentence $s = (s_1, s_2, ..., s_n)$ in n words after 10 rounds of question-answer interactions, each word s_k arises from a vocabulary S. At i-th round of question-answer interaction, A-BOT takes the audio signals, video data, video caption and the existing dialog history as input, $x_{A,i} = (A, V_A, C, H_{i-1})$, with $H_{i-1} = \{p_1, ..., p_{i-1}\}$ and $p_{i-1} = (q_{i-1}, a_{i-1})$. For Q-BOT at the same round i, $x_{Q,i} = (V_s, V_e, H_{i-1})$. The final description task for Q-BOT is formulated as the inference in a recurrent model with the joint probability given by:

$$p(s|x_Q) = \prod_{k=1}^{n} p(s_k|s_{<k}, x_Q),\tag{1}$$

where we maximize the product of conditionals for each word in description s, given the input at 10-th round x_Q.

3.1   QA-Cooperative Network

The overall architecture of QA-Cooperative network is presented in Fig. 3. Q-BOT consists of a visual module, a history encoder, a visual LSTM-net, a multimodal attention module, a question decoder and the final description generator. A-BOT has an audio module, a visual module, a caption encoder, a history encoder, two attention modules and an answer decoder. In general, Q-BOT generates i-th question q_i based on the input $x_{Q,i}$, A-BOT responds to the question by generating the corresponding answer a_i. The new question-answer pair $p_i = (q_i, a_i)$ is used to update the existing dialog history. Note that we observe

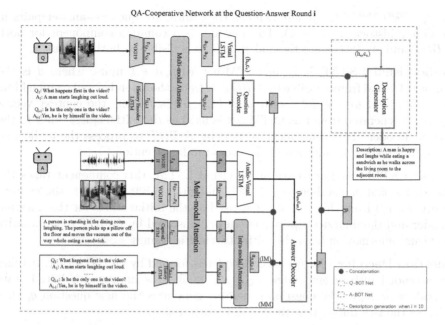

QA-Cooperative Network at the Question-Answer Round i

Fig. 3. QA-Cooperative Network at the question-answer interaction round i. Details about the network architecture and learning method are presented in Sect. 3.1, Sect. 3.2 and Sect. 3.3.

from the experiments that separate history decoders for two agents do not help improve the performance. Therefore, we choose the shared history encoder design to reduce the network redundancy.

3.2 Model Components

There are multiple components in our QA-Cooperative network, which will be explained in detail in this section. We consider the situation at the i-th round of question-answer interaction.

Caption Encoder. The caption encoder contains a linear layer and a single layer LSTM-net. We firstly represent each word in the captions with one-hot vectors. Next, we find the longest caption sentence in each batch and zero-pad other shorter ones. The final caption embedding $r_C \in \mathbb{R}^{n_C \times d_C}$ is obtained from the last hidden state of the LSTM-net. This component is designed for *A-BOT* to encode video captions during the preparation phase.

History Encoder. The history encoder contains a linear layer and an LSTM-net. Similarly to the processing steps for the video captions, we start with a list of one-hot word representations for a pair of question-answer. The longest question-answer pair of length is selected, the other pairs are zero-padded to fit the maximum length. The LSTM-net is used to obtain the pair-level embedding

$r_{H,i-1} \in \mathbb{R}^{n_T \times n_H \times d_H}$. n_T here denotes the number of question-answer pairs in the dialog history (*i.e.*, $i-1$). This encoder is a common component for both *Q-BOT* and *A-BOT*, since the dialog history is visible to both agents.

Audio-visual LSTM. It is an LSTM-net with $d+1$ units, where d is the number of visual frames visible to *A-BOT*. It takes the attended audio embedding $a_A \in \mathbb{R}^{d_A}$ and $a_{V,j} \in \mathbb{R}^{d_V}$ with $j = \{1, ..., d\}$ as input, the context vector (h_{av}, c_{av}) generated from this LSTM-net is used as the initial states input to the answer decoder. This component is used by *A-BOT* to process the audio and visual information in addition to the cross-modal attention.

Visual LSTM. Similar to the audio-visual LSTM, this component takes the attended visual embedding $a_{V,s} \in \mathbb{R}^{d_V}$ and $a_{V,e} \in \mathbb{R}^{d_V}$ as input, the context vector (h_v, c_v) from this LSTM-net is used as the initial states for the question decoder and the final description generator. It is used by *Q-BOT* to summarize the visual information from the initial two static frames.

Question Decoder. The question decoder is formed by an LSTM-net. It takes the attended history embedding $a_{Q,H,i-1} \in \mathbb{R}^{d_H}$ as input, with initial state $(h_0, c_0) = (h_v, c_v)$. The question generator generates the new question q_i that imitates the i-th question in the ground truth dialog.

Answer Decoder. Similar to the question decoder, we use the answer decoder to generate the answer embedding close to the i-th answer in the ground truth dialog. The answer LSTM decoder takes the concatenation of the attended history embedding $a_{A,H,i-1} \in \mathbb{R}^{d_H}$, the attended caption embedding $a_c \in \mathbb{R}^{d_C}$ and the newly generated question embedding $r_{q,i}$ as input, with initial state $(h_0, c_0) = (h_{av}, c_{av})$. The output is the answer a_i for the given question. The newly generated question-answer pair at i-th round is obtained by combining the i-th question and answer.

Description Generator. This LSTM generator generates the final description s for the unseen video based on 10 rounds of question-answer interactions history and the two static frames given in the first phase. When $i = 10$, the generator computes the following conditional probabilities based on the input, which is the attended history embedding $a_{A,H,10} \in \mathbb{R}^{d_H}$ including 10 rounds of question-answer interactions:

$$p(s_k|s_{k-1}, h_{k-1}, x_Q) = g(s_k, s_{k-1}, h_{k-1}, x_Q), \qquad (2)$$

where h_{k-1} is the hidden states obtained from the previous $k-1$ step. Note that h here is the hidden states of LSTM-net, different from the history notation H. The initial state is the same as the question decoder, thus we have $(h_0, c_0) = (h_v, c_v)$. the LSTM-net g predicts the probability distribution $p(s_k|s_{k-1}, h_{k-1}, x_Q)$ over words $s_k \in \mathcal{S}_k$, conditioned on the previous words s_{k-1}. The final probability distribution for natural language description is obtained by transforming the output of the LSTM-net by a FC-layer and a Softmax.

Attention module. Since the dialog is a key information source in our task, we propose two different attention mechanisms for processing the information contained in the dialog history: The multi-modal (MM) attention among visual, audio and textual modalities, and the intra-modal (IM) attention between dialog history and another textual sequence.

For the MM attention, we use the factor graph attention mechanism proposed in [36]. For A-BOT, this multi-modal attention module takes the audio embedding r_A, visual embedding $r_{V,j}$ with $j = \{1, ..., d\}$, caption embedding r_C and the history embedding $r_{H,i-1}$ as input. Each visual frame is treated as an individual modality as in [36]. The output of this multi-modal attention module are the attended audio embedding a_A, the attended visual embedding $a_{V,j}$ with $j = \{1, ..., d\}$, and the attended history embedding $a_{Q,H,i-1}$. Similarly for Q-BOT, we have the attended output $a_{V,s}$, $a_{V,e}$ and $a_{A,H,i-1}$ after taking $r_{V,s}$, $r_{V,e}$ and $r_{H,i-1}$ as input. Note that the history embedding $r_{H,i-1}$ before the multi-modal attention module is the same for Q-BOT and A-BOT because a shared history encoder is used, but the attended history embedding becomes different due to different inputs for two agents.

For the IM attention module in Fig. 3, it is a simple softmax attention between the dialog history $r_{H,i-1}$ and the concatenation of a_C and $r_{q,i}$.

3.3   Cooperative Learning

We propose to learn the QA-Cooperative network with a dynamic dialog history update mechanism in a goal-driven manner considering the following two aspects.

Firstly, the ultimate objective of our task is for Q-BOT to describe what happens between the beginning and the end of the video in a concrete way. Considering the fact that only two static frames are given to Q-BOT during the first preparation phase, the principal information source for Q-BOT is the dialog with A-BOT. Therefore, dialog history is the key to effectively learning our QA-Cooperative network. To this end, we propose a dynamic dialog history update mechanism to help with the knowledge transfer from A-BOT to Q-BOT. Figure 4 illustrates the dialog history update operation by fusion. Notably, to emphasize the information from the newly generated question-answer pair, we set the dimension of the history embedding equal to the dimension of $r_{p,i}$ before the attention module.

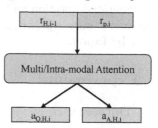

Fig. 4. Dialog history update. Specifically, we maintain the history embedding dimension equal to the dimension of $r_{p,i}$ to emphasize the information from the newly generated question-answer pair.

Secondly, for the internal question and answer generation process, we encourage the two agents to imitate the questions and answers from the corresponding rounds of the ground truth dialog. Intuitively, this generation process can also be realized by pre-trained question and answer decoders using the existing methods [18,20,35,46] trained in traditional VQA tasks. However, because the pre-training process is directly optimized to generate questions or answers based on the ground truth dialog, it sets the barrier for the final performance in our video description task. In other words, Q-BOT is unlikely to surpass the performance obtained in the situation where full ground truth dialog is provided. As for comparisons, the goal-driven training [11] is optimized in the final description phase. Although we also use the ground truth dialog as the internal imitation reference for two agents, it leaves more space for flexible question and answer generations (*i.e.*, to ask and answer the questions that directly help Q-BOT to better describe the video in the last phase). Our experiments support this assumption in Sect. 4.3.

Figure 3 schematically illustrates the general learning process at the i-th question-answer round. Q-BOT takes the two static frames and the existing dialog history that consists of $i - 1$ pairs question-answer as input, $x_{Q,i} = (V_s, V_e, H_{i-1})$. The visual and history embedding $r_{V,s}$, $r_{V,e}$ and $r_{H,r-1}$ are processed in the MM attention module [36] to acquire the attended embedding $a_{V,s}$, $a_{V,e}$ and $a_{Q,H,r-1}$. Given (h_v, c_v) from the visual LSTM as the initial state, the question decoder outputs the question q_i with its embedding $r_{q,i}$ after taking $a_{Q,H,r-1}$ as input. In the meanwhile, A-BOT takes the audio signals, video frames, video caption and the existing dialog history as input, $x_{A,i} = (A, V_A, C, H_{i-1})$. The corresponding attended embedding a_A, $a_{V,j}$ with $j = 1, ..., d$, a_C and $a_{A,H,i-1}$ is obtained after the MM attention. Another way to obtain the $a_{A,H,i-1}$ is through the IM attention module. The input for answer decoder is the concatenation of the attended embedding $a_{A,H,i-1}$, a_C and $r_{q,i}$, with initial state $(h_0, c_0) = (h_{av}, c_{av})$ generated from the audio-visual LSTM. The generated question and answer embedding are combined to form the new i-th question-answer pair $r_{p,i}$, and used to update the existing history embedding $r_{H,i-1}$ by fusion.

4 Experiments

4.1 Dataset

We use the recent AVSD v0.1 dataset [18] for experiments. The AVSD dataset consists of annotated dialog about 9848 short videos taken from CHARADES [39]. The dialog collection process resembles our task setup, during which two Amazon Mechanical Turk (AMT) workers play the roles of Questioner and Answerer. The Questioner was shown only the first, middle and last static frames of the video, while the Answerer had already watched the entire video, including the audio stream and the original video caption. After having a conversation about the events that happened between the frames through 10 rounds of question-answer interactions, the Questioner is asked to summarize the entire video.

The current AVSD v0.1 is split into 7659 training, 1787 validation and 1710 testing dialog, respectively. Hori *et al.* [18] also propose a 'prototype validation-set and test-set', which are sub-splits of the original validation set since the original test set does not include ground truth dialog at the moment. Our experiments are conducted on the original training and 'prototype' validation-test splits of the AVSD dataset.

4.2   Implementation Details

Data Representations. Our cooperative dialog agents have data input of visual, audio and textual modalities. For the visual modal, we take the video representations extracted from the last conv layer of a VGG19 as input. We sample 4 equally spaced frames from the beginning of the original video, and each frame representation is of dimension $7 \times 7 \times 512$. The spatial and visual embedding dimensions are 49 and 512, respectively. Q-BOT is shown the first and the last frames, while the A-BOT is able to see all the frames. For the audio modal, we obtain the 256-dim audio feature via VGGish [17]. For the textual representations, we extract the language embedding from the last hidden state of their corresponding LSTM-nets. The dimensions are $d_C = 256$, $d_q = 128$, $d_a = 128$ and $d_H = 256$.

Network Training. The proposed QA-Cooperative network is trained using a cross-entropy loss on the probabilities $p(s_k|s_{<k}, x_Q)$ on the final video descriptions. All the components are jointly learned. The total amount of trainable parameters is approximately 19M. We use the Adam optimizer with a learning rate of 0.001 and a batch size of 64 for training. During the training, the perplexity metric is used for evaluating the performance on the validation set.

4.3   Cooperative Video Description

We split our experiments into two major groups to provide a more comprehensive and objective analysis of our work as shown in Table 2, which presents the quantitative results using the BLEU1-4 [31], METEOR [7], SPICE [2], ROUGE_L [23], and CIDEr [43] as the evaluation metrics.

Standard Test Setting. During our test, the performance of Q-BOT is evaluated at each question-answer round-level. In other words, for a given video in the test split, the start question answer round number i ranges from 1 to 10. For example, if the start round number $i = 1$, then no existing dialog history is given to Q-BOT and A-BOT, they will generate all the ten questions and answers by themselves. However, if the start round number $i = 6$, then 5 rounds of question-answer pairs are given to two agents as the existing history, in which case, Q-BOT still has 5 opportunities to freely ask questions. For a given video, tests with different start round numbers are independent, representing 10 different test cases. Therefore, for the 733 videos from the 'prototype test set' of the AVSD dataset [18], we have in total 7330 different test cases. We refer this testing process as the standard test setting, it is consistent with the learning process explained above in the Sect. 3.3. The only exception is the strong baseline situations for Q-BOT, during which the full ground truth dialog history is given before the testing. For the strong baseline situation, since the evaluation is only conducted at the end of full dialog history, there will be 733 testing cases.

Description Ability for A-BOT. The first group of experiments focuses on the video description ability of A-BOT, which is used as a comparison and performance reference for Q-BOT. Under our task setup, A-BOT has full access to

Table 2. Quantitative experimental results of video description tasks by different agents using multiple methods. *HIS Att* stands for *History attention*. The experiments are split into two groups, one group for *A-BOT*, and another group for *Q-BOT*. The comparisons between *A-BOT* and basic *Q-BOT* baselines show the actual knowledge gap between the two agents. We obverse that both *A-BOT* and *Q-BOT* from the proposed QA-Cooperative network achieve the best performance. Notably, our *Q-BOT* with cooperative learning is able to achieve comparable performance with the strong baseline, during which the full ground truth dialog history is given as input. It demonstrates the effective knowledge transfer process.

Agent	Method	HIS Att	BLEU1	BLEU2	BLEU3	BLEU4	METEOR	SPICE	ROUGE_L	CIDEr
A-BOT (with info from all modalities)	Hori *et al.* [18]	–	34.2	17.1	8.4	4.8	11.5	11.4	24.9	20.7
	S. *et al.* [35]	–	32.1	16.2	8.7	5.1	12.1	11.6	27.6	21.6
	S. *et al.* [35]	IM	33.8	16.9	9.1	5.3	12.7	11.8	27.7	22.7
	S. *et al.* [35]	MM	33.8	17.6	9.9	5.9	12.9	13.5	28.5	25.6
	Ours	IM	**37.9**	**21.6**	12.5	7.6	**15.2**	**18.5**	31.1	38.1
	Ours	MM	37.5	21.5	**12.9**	**8.2**	**15.2**	17.9	**31.2**	**39.3**
Q-BOT basic baselines	Ours	IM	32.0	15.7	8.1	4.7	11.6	11.1	26.4	18.3
	Ours	MM	33.2	16.4	8.6	5.0	12.5	11.5	27.1	20.2
Q-BOT strong baselines	Ours (full GT HIS)	IM	33.3	17.0	9.1	5.4	12.6	11.7	27.3	21.3
	Ours (full GT HIS)	MM	32.7	17.2	**9.7**	**6.0**	12.6	**13.6**	**27.9**	**26.3**
Q-BOT cooperative	Ours (pre-trained)	MM	33.3	17.0	9.2	5.4	**12.9**	11.4	27.4	21.5
	Ours QA-C	IM	33.3	16.9	9.1	5.3	12.7	11.6	27.7	22.7
	Ours QA-C	MM	**33.3**	**17.3**	9.5	5.5	12.8	12.4	**27.9**	23.1

all the information while *Q-BOT* only sees two static images. Intuitively, *A-BOT* should have better performance than *Q-BOT*. We also compare the performance of our *A-BOT* with the other two recent *A-BOT* baselines from [18,35]. The models proposed in [18,35] are initially designed for question answering, but they also take all modalities of data as input, therefore, we modify the generators to generate video descriptions after 10 rounds of question-answer interactions. We observe that our *A-BOT* largely outperforms the baseline models, mainly with the help of separate caption encoder and the attention operation on the dialog history, while the previous works directly incorporate the video captions into the dialog history. Different from [35] where the usage of attention for dialog history does not yield improvements for the classic question answering task, we find it is helpful in improving the performance for our video description task either by MM or IM attention method.

Description Ability for *Q-BOT*. The second group of experiments focuses on the video description ability of *Q-BOT*. The basic baseline for *Q-BOT* is obtained under the standard test setting without cooperative learning, in other words, *Q-BOT* does not involve question-answer interactions with *A-BOT*. The comparisons between the basic baselines and *A-BOT* reveal the actual knowledge gap between the two agents. The strong *Q-BOT* baselines are obtained by providing *Q-BOT* with the full ground truth dialog history that contains 10 rounds of question-answer pairs. We also compare our cooperative learning method with the pre-trained methods mentioned in Sect. 3.3. The experimental results are consistent with our expectations. Although outperforming the basic baseline, the pre-trained learning method can hardly surpass two strong baselines. As for comparisons, our *Q-BOT* from the proposed QA-Cooperative

Fig. 5. Example of qualitative results. Our *Q-BOT* is able to describe more details about the unseen video. More qualitative results in the supplementary material.

network with the cooperative learning method is able to achieve more promising performance. It outperforms the strong baseline with IM attention setting, and achieves comparable performance with the strong baseline under MM attention setting. The experimental results demonstrate that cooperative learning helps *Q-BOT* to extract even more useful information to better describe the unseen video. Moreover, we find the MM attention for the dialog history is very helpful in improving the performance of the strong baseline settings.

Examples of qualitative results are shown in Fig. 5. We observe that our *Q-BOT* with cooperative learning is able to describe the unseen videos with more concrete details, achieving comparable performance with the strong baseline. More qualitative results and analysis can found in the supplementary material.

4.4   Ablation Studies

More experimental results on ablation studies are presented in Table 3. All the performance reported is achieved by *Q-BOT* under the standard test setting.

Full Update vs. Partial Update. In the proposed cooperative learning, the newly generated pair $r_{p,i}$ is fused with the existing dialog history $r_{H,i-1}$ before the attention module as shown in Fig. 4. With the updated history embedding, other outputs from the attention modules also become different compared to the previous dialog round. We compare the performance between the case of full update and partial update. In the full update setting, all the outputs from the attention modules are updated, while the partial update setting only updates the attended dialog history. The full update setting is more competitive than the partial one.

Initial States. In our proposed QA-Cooperative network, the initial states of the question decoder and the answer decoder are obtained from the visual LSTM and audio-visual LSTM, respectively. These two LSTM-nets summarize the audio and visual information for two agents. We observe from Table 3 that the initial states help achieve better performance under the full update setting, but have less impact on the partial update setting. It is reasonable because the

Table 3. Quantitative experimental results on ablation studies. All the results are obtained by *Q-BOT* under the standard task setup as explained in Fig. 1. *init.* means the initial states provided by the visual and audio-visual LSTM for the question and answer decoder. *w/o his for A* means we remove the attended history embedding from the input for the answer decoder.

Settings	BLEU1	BLEU2	BLEU3	BLEU4	METEOR	SPICE	ROUGE_L	CIDEr
Parital	33.2	16.9	9.2	5.2	12.8	12.1	27.6	21.6
Full w/o init.	33.1	16.3	8.6	4.9	12.2	11.6	27.1	20.3
Partial w/o init.	32.7	16.2	8.8	5.1	12.1	11.8	27.4	20.3
w/o caption	31.5	15.3	7.9	4.6	12.7	11.1	26.3	19.5
w/o audio	33.2	17.2	9.4	5.4	12.8	12.2	27.8	22.3
w/o his for A	32.5	16.9	9.3	5.4	12.1	12.2	27.1	23.0
Shuffled QA order	32.0	15.7	8.3	4.8	11.7	11.1	26.3	18.9
Proposed QA-C	**33.3**	**17.3**	**9.5**	**5.5**	**12.8**	**12.4**	**27.9**	**23.1**

full update setting mainly changes the attended audio and visual information, which is later used as the initial states.

Modality Input. Our two agents, especially *A-BOT* take multiple modalities of information as input. We test the settings when the caption and the audio information are removed from the input of *A-BOT*. Experimental results show that the caption is of vital importance in achieving good performance for video description. Audio information also contributes to the better final performance. *w/o his for A* in Table 3 stands for the setting when the history embedding $a_{H,i-1}$ is removed from the input of answer decoder for *A-BOT*. It explains the fact that *A-BOT* does not heavily rely on the dialog history to provide answers to *Q-BOT* since it has already watched the entire video.

Order of Question-Answer Pairs. We also investigate the influence of the order of the question-answer pairs in the input. Similar to [1], the order of question-answer pairs is a significant factor for better final performance.

5   Conclusions

In summary, in this paper we propose a novel video description task via two multi-modal dialog agents, *Q-BOT* and *A-BOT*. We establish a new task setting for AI systems to accomplish a multi-modal task without direct access to the original visual or audio information. We further propose a QA-Cooperative network and a cooperative learning method with a dynamic dialog history update mechanism. Extensive experiments prove that *Q-BOT* is able to achieve very promising performance via our proposed network and learning method.

Acknowledgements. This research was partially supported by NSF NeTS-1909185. This article solely reflects the opinions and conclusions of its authors and not the funding agents.

References

1. Alamri, H., et al.: Audio visual scene-aware dialog. In: CVPR (2019)
2. Anderson, P., Fernando, B., Johnson, M., Gould, S.: SPICE: semantic propositional image caption evaluation. In: Leibe, B., Matas, J., Sebe, N., Welling, M. (eds.) ECCV 2016. LNCS, vol. 9909, pp. 382–398. Springer, Cham (2016). https://doi.org/10.1007/978-3-319-46454-1_24
3. Anderson, P., et al.: Bottom-up and top-down attention for image captioning and visual question answering. In: CVPR (2018)
4. Antol, S., et al.: VQA: visual question answering. In: ICCV (2015)
5. Arandjelović, R., Zisserman, A.: Objects that sound. In: Ferrari, V., Hebert, M., Sminchisescu, C., Weiss, Y. (eds.) ECCV 2018. LNCS, vol. 11205, pp. 451–466. Springer, Cham (2018). https://doi.org/10.1007/978-3-030-01246-5_27
6. Bahdanau, D., Cho, K., Bengio, Y.: Neural machine translation by jointly learning to align and translate. In: ICLR (2015)
7. Banerjee, S., Lavie, A.: METEOR: an automatic metric for MT evaluation with improved correlation with human judgments. In: Proceedings of the ACL Workshop on Intrinsic and Extrinsic Evaluation Measures for Machine Translation and/or Summarization, pp. 65–72 (2005)
8. Chen, L., et al.: SCA-CNN: spatial and channel-wise attention in convolutional networks for image captioning. In: CVPR (2017)
9. Das, A., Agrawal, H., Zitnick, L., Parikh, D., Batra, D.: Human attention in visual question answering: do humans and deep networks look at the same regions? Comput. Vis. Image Underst. **163**, 90–100 (2017)
10. Das, A., et al.: Visual dialog. In: CVPR (2017)
11. Das, A., Kottur, S., Moura, J.M., Lee, S., Batra, D.: Learning cooperative visual dialog agents with deep reinforcement learning. In: ICCV (2017)
12. De Vries, H., Strub, F., Chandar, S., Pietquin, O., Larochelle, H., Courville, A.: Guesswhat?! Visual object discovery through multi-modal dialogue. In: CVPR (2017)
13. Gan, Z., Cheng, Y., Kholy, A., Li, L., Liu, J., Gao, J.: Multi-step reasoning via recurrent dual attention for visual dialog. In: ACL (2019)
14. Gao, R., Feris, R., Grauman, K.: Learning to separate object sounds by watching unlabeled video. In: Ferrari, V., Hebert, M., Sminchisescu, C., Weiss, Y. (eds.) ECCV 2018. LNCS, vol. 11207, pp. 36–54. Springer, Cham (2018). https://doi.org/10.1007/978-3-030-01219-9_3
15. He, K., Zhang, X., Ren, S., Sun, J.: Delving deep into rectifiers: surpassing human-level performance on imagenet classification. In: CVPR (2015)
16. He, K., Zhang, X., Ren, S., Sun, J.: Deep residual learning for image recognition. In: CVPR (2016)
17. Hershey, S., et al.: CNN architectures for large-scale audio classification. In: ICASSP. IEEE (2017)
18. Hori, C., et al.: End-to-end audio visual scene-aware dialog using multimodal attention-based video features. In: ICASSP. IEEE (2019)
19. Hudson, D.A., Manning, C.D.: Compositional attention networks for machine reasoning. In: ICLR (2018)
20. Jain, U., Lazebnik, S., Schwing, A.G.: Two can play this game: visual dialog with discriminative question generation and answering. In: CVPR (2018)
21. Krizhevsky, A., Sutskever, I., Hinton, G.E.: ImageNet classification with deep convolutional neural networks. In: NeurIPS (2012)

22. Lee, S.W., Heo, Y.J., Zhang, B.T.: Answerer in questioner's mind: information theoretic approach to goal-oriented visual dialog. In: NeurIPS (2018)
23. Lin, C.Y.: ROUGE: a package for automatic evaluation of summaries. In: Text Summarization Branches Out, pp. 74–81 (2004)
24. Lin, T.Y., Dollár, P., Girshick, R., He, K., Hariharan, B., Belongie, S.: Feature pyramid networks for object detection. In: Proceedings of the IEEE Conference on Computer Vision and Pattern Recognition, pp. 2117–2125 (2017)
25. Lu, J., Xiong, C., Parikh, D., Socher, R.: Knowing when to look: adaptive attention via a visual sentinel for image captioning. In: CVPR (2017)
26. Lu, J., Yang, J., Batra, D., Parikh, D.: Hierarchical question-image co-attention for visual question answering. In: NeurIPS (2016)
27. Mahasseni, B., Lam, M., Todorovic, S.: Unsupervised video summarization with adversarial LSTM networks. In: CVPR (2017)
28. Massiceti, D., Siddharth, N., Dokania, P.K., Torr, P.H.: FlipDial: a generative model for two-way visual dialogue. In: CVPR (2018)
29. Niu, Y., Zhang, H., Zhang, M., Zhang, J., Lu, Z., Wen, J.R.: Recursive visual attention in visual dialog. In: CVPR (2019)
30. Owens, A., Efros, A.A.: Audio-visual scene analysis with self-supervised multisensory features. In: Ferrari, V., Hebert, M., Sminchisescu, C., Weiss, Y. (eds.) ECCV 2018. LNCS, vol. 11210, pp. 639–658. Springer, Cham (2018). https://doi.org/10.1007/978-3-030-01231-1_39
31. Papineni, K., Roukos, S., Ward, T., Zhu, W.J.: BLEU: a method for automatic evaluation of machine translation. In: ACL (2002)
32. Redmon, J., Divvala, S., Girshick, R., Farhadi, A.: You only look once: unified, real-time object detection. In: CVPR (2016)
33. Rennie, S.J., Marcheret, E., Mroueh, Y., Ross, J., Goel, V.: Self-critical sequence training for image captioning. In: CVPR (2017)
34. Santoro, A., et al.: A simple neural network module for relational reasoning. In: NeurIPS (2017)
35. Schwartz, I., Schwing, A.G., Hazan, T.: A simple baseline for audio-visual scene-aware dialog. In: CVPR (2019)
36. Schwartz, I., Yu, S., Hazan, T., Schwing, A.G.: Factor graph attention. In: CVPR (2019)
37. Seo, P.H., Lehrmann, A., Han, B., Sigal, L.: Visual reference resolution using attention memory for visual dialog. In: NeurIPS (2017)
38. Shih, K.J., Singh, S., Hoiem, D.: Where to look: focus regions for visual question answering. In: CVPR (2016)
39. Sigurdsson, G.A., Varol, G., Wang, X., Farhadi, A., Laptev, I., Gupta, A.: Hollywood in homes: crowdsourcing data collection for activity understanding. In: Leibe, B., Matas, J., Sebe, N., Welling, M. (eds.) ECCV 2016. LNCS, vol. 9905, pp. 510–526. Springer, Cham (2016). https://doi.org/10.1007/978-3-319-46448-0_31
40. Song, X., Shi, Y., Chen, X., Han, Y.: Explore multi-step reasoning in video question answering. In: ACM Multimedia (2018)
41. Szegedy, C., Toshev, A., Erhan, D.: Deep neural networks for object detection. In: Advances in Neural Information Processing Systems, pp. 2553–2561 (2013)
42. Tian, Y., Shi, J., Li, B., Duan, Z., Xu, C.: Audio-visual event localization in unconstrained videos. In: Ferrari, V., Hebert, M., Sminchisescu, C., Weiss, Y. (eds.) ECCV 2018. LNCS, vol. 11206, pp. 252–268. Springer, Cham (2018). https://doi.org/10.1007/978-3-030-01216-8_16
43. Vedantam, R., Lawrence Zitnick, C., Parikh, D.: CIDEr: consensus-based image description evaluation. In: CVPR (2015)

44. Wang, B., Ma, L., Zhang, W., Liu, W.: Reconstruction network for video captioning. In: CVPR (2018)
45. Wang, L., Schwing, A., Lazebnik, S.: Diverse and accurate image description using a variational auto-encoder with an additive Gaussian encoding space. In: NeurIPS (2017)
46. Wu, Q., Wang, P., Shen, C., Reid, I., Van Den Hengel, A.: Are you talking to me? Reasoned visual dialog generation through adversarial learning. In: CVPR (2018)
47. Wu, Y., Zhu, L., Jiang, L., Yang, Y.: Decoupled novel object captioner. In: ACM Multimedia (2018)
48. Wu, Y., Zhu, L., Yan, Y., Yang, Y.: Dual attention matching for audio-visual event localization. In: ICCV (2019)
49. Xiong, C., Merity, S., Socher, R.: Dynamic memory networks for visual and textual question answering. In: ICML (2016)
50. Xu, H., Saenko, K.: Ask, attend and answer: exploring question-guided spatial attention for visual question answering. In: Leibe, B., Matas, J., Sebe, N., Welling, M. (eds.) ECCV 2016. LNCS, vol. 9911, pp. 451–466. Springer, Cham (2016). https://doi.org/10.1007/978-3-319-46478-7_28
51. Yang, Z., He, X., Gao, J., Deng, L., Smola, A.: Stacked attention networks for image question answering. In: CVPR (2016)
52. Yao, T., Pan, Y., Li, Y., Qiu, Z., Mei, T.: Boosting image captioning with attributes. In: ICCV (2017)
53. You, Q., Jin, H., Wang, Z., Fang, C., Luo, J.: Image captioning with semantic attention. In: CVPR (2016)
54. Zhao, H., Gan, C., Rouditchenko, A., Vondrick, C., McDermott, J., Torralba, A.: The sound of pixels. In: Ferrari, V., Hebert, M., Sminchisescu, C., Weiss, Y. (eds.) ECCV 2018. LNCS, vol. 11205, pp. 587–604. Springer, Cham (2018). https://doi.org/10.1007/978-3-030-01246-5_35

SACA Net: Cybersickness Assessment of Individual Viewers for VR Content via Graph-Based Symptom Relation Embedding

Sangmin Lee[1], Jung Uk Kim[1], Hak Gu Kim[2], Seongyeop Kim[1], and Yong Man Ro[1(✉)]

[1] Image and Video Systems Lab, School of Electrical Engineering, KAIST, Daejeon, Korea
{sangmin.lee,jukim0701,seongyeop,ymro}@kaist.ac.kr
[2] Signal Processing Lab, Institute of Electrical Engineering, EPFL, Lausanne, Switzerland
hakgu.kim@epfl.ch

Abstract. Recently, cybersickness assessment for VR content is required to deal with viewing safety issues. Assessing physical symptoms of individual viewers is challenging but important to provide detailed and personalized guides for viewing safety. In this paper, we propose a novel symptom-aware cybersickness assessment network (SACA Net) that quantifies physical symptom levels for assessing cybersickness of individual viewers. The SACA Net is designed to utilize the relational characteristics of symptoms for complementary effects among relevant symptoms. The proposed network consists of three main parts: a stimulus symptom context guider, a physiological symptom guider, and a symptom relation embedder. The stimulus symptom context guider and the physiological symptom guider extract symptom features from VR content and human physiology, respectively. The symptom relation embedder refines the stimulus-response symptom features to effectively predict cybersickness by embedding relational characteristics with graph formulation. For validation, we utilize two public 360-degree video datasets that contain cybersickness scores and physiological signals. Experimental results show that the proposed method is effective in predicting human cybersickness with physical symptoms. Further, latent relations among symptoms are interpretable by analyzing relational weights in the proposed network.

Keywords: Cybersickness assessment · Individual viewer · VR content · Physical symptom · Symptom relation

Electronic supplementary material The online version of this chapter (https://doi.org/10.1007/978-3-030-58592-1_11) contains supplementary material, which is available to authorized users.

1 Introduction

Perceiving virtual reality (VR) content such as 360-degree videos can provide immersive experiences to viewers. With the rapid development of content capturing and displaying devices, VR content increasingly attracts attention in various industry and research fields [14–16]. However, the growth of VR environments accompanies by concerns over the safety of viewing VR content. Several studies reported that viewing VR content could trigger cybersickness with physical symptoms [7,18] : 1) nausea symptoms containing sweating, salivation and, burping, 2) oculomotor symptoms containing visual fatigue and eye strain, and 3) disorientation symptoms containing fullness of the head, dizziness, and vertigo. Such cybersickness is one of the major problems hampering the spread of VR environments. For guiding people to create and view safety content, it is firstly needed to quantify cybersickness level caused by viewing VR content.

When viewers watch VR content, they can feel different cybersickness even for the same content stimulus. As shown in Fig. 1, each viewer can feel cybersickness differently with distinct symptoms in detail. To guide view-safe VR content for specific viewers, it is necessary to quantify detailed cybersickness of individual viewers. Cybersickness assessment for individual viewers needs detailed physical symptoms to provide personalized guides for VR content viewing.

In recent years, VR content-based cybersickness assessment methods have been introduced [20–24]. The content-based methods exploited spatio-temporal features from VR content to quantify cybersickness. These content-based methods did not consider deviations among individuals. Individuals in the same stimulus environment could experience different cybersickness.

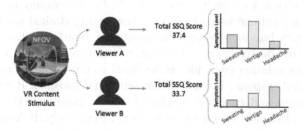

Fig. 1. Viewers watching the same VR content can feel cybersickness differently with distinct symptoms in detail.

There have been clinical studies examining the tendency of physiological responses according to cybersickness [11,12,25,30,34,38,41,48]. There have been attempts to validate the relationship between physiological responses and cybersickness caused by VR content [11,13,25,33,43]. Some previous works extract cybersickness-related features from physiological response for predicting cybersickness from VR content [17,28,31,47]. However, since most of them only exploited physiology without stimulus context that affects physiological response predominantly, they did not fully utilize the context for cybersickness. [28] considered stimulus with physiology in evaluating cybersickness. However, it only focuses on predicting total cybersickness levels. Such cybersickness assessment was limited in that it was performed without analysis of physical symptoms.

In this paper, we propose a novel symptom-aware cybersickness assessment network (SACA Net) that predicts the individual viewer's cybersickness with physical symptoms. The SACA Net quantifies the degree of physical symptoms, which makes it possible to provide more detailed and interpretable information for cybersickness. There were clinical reports about the existence of relationships among physical symptoms for cybersickness [18,44]. Considering the relationships, we devise a relational symptom embedding framework that can exploit the relational information among symptoms to assess cybersickness. Thereby, the relevant symptom features complement each other for effectively predicting symptom levels. The SACA Net consists of three parts: a stimulus symptom context guider, a physiological symptom guider, and a symptom relation embedder.

The stimulus symptom context guider is designed to effectively accumulate visual features for encoding symptom factors caused by stimulus environment. Based on neural mismatch theory [40], a sensory mismatch detector in the stimulus symptom context guider extracts mismatch features between target stimulus content and comfort stimulus content that do not induce high-level cybersickness. By exploiting the mismatch features from the sensory mismatch detector, the stimulus symptom context guider extracts symptom group features that represent nausea, oculomotor, and disorientation group factors in context of stimulus.

The physiological symptom guider extracts symptom features from EEG signal. Since EEG contains the most comprehensive information about the nervous system such as vision, movement, and sense [25,45], we employ EEG as a physiological factor for cybersickness analysis. Considering clinical studies[30,38,48] for EEG frequency bands, we design the frequency band attentive encoding for the EEG signal to effectively extract symptom features related to cybersickness.

The symptom relation embedder is designed to refine the symptom features from stimulus and response by embedding relational features. It receives symptom group features and symptom features from the stimulus symptom context guider and the physiological symptom guider, respectively. It learns the latent relations among symptoms in an unsupervised way with graph formulation to effectively predict symptom levels. In addition, we can interpret the relations among symptoms by analyzing relational weights in the proposed network.

We use two public 360-degree video datasets with simulator sickness questionnaire (SSQ) [18] scores and physiological signals. The performances are validated with human cybersickness levels in the assessment datasets.

The major contributions of the paper are as follows.

– We introduce a novel SACA Net that quantifies cybersickness of individuals with physical symptoms by combining content stimulus and physiological response. To the best of our knowledge, it is the first attempt to quantify cybersickness including symptoms of individuals for VR content.

– We propose symptom relation embedding which makes it possible to effectively assess cybersickness of individuals with distinct symptoms. Furthermore, latent symptom relations are interpretable by analyzing relational weights in the proposed network.

2 Related Work

2.1 Cybersickness Assessment for VR Content

VR content-based cybersickness assessment methods have been introduced [20–22,36]. Kim et al. [20] quantified cybersickness caused by exceptional motions with a deep generative model. The generative model is trained with normal videos containing non-exceptional motions. Hence, the generative model cannot properly reconstruct videos with exceptional motions that cause cybersickness at testing time. Acquired difference between the original video and the generated video correlated with the degree of cybersickness. In [21], a deep network that consists of a generator and a cybersickness regressor was proposed for quantifying cybersickness. In the model, the difference between the original video and the generated video is regressed to the simulation sickness questionnaires (SSQ) [5] score assessed by subjects. Another study [22] quantified cybersickness considering visual-vestibular conflicts. In the work, to quantify cybersickness, SVR [4] is applied on motion features from visual-vestibular interaction and VR content features. Padmanaban et al. [36] proposed a cybersickness predictor to estimate the nauseogenicity of virtual content. In the model, algorithms based on optical flow methods are employed to compute cybersickness features that primarily focus on disparity and velocity of video content. For assessing cybersickness caused by quality degradation, an objective assessment model considering spatio-temporal inconsistency was proposed [24]. In [23], a deep neural network that exploits cognitive feature regularization was proposed for cybersickness assessment.

However, the aforementioned cybersickness quantification methods do not assess cybersickness of individuals. Individuals in the same environments may experience different cybersickness levels. Compared to these works, the proposed method predicts individual cybersickness by exploiting physiological responses of content viewers to consider the deviation among individuals.

2.2 Physiological Study for Cybersickness

There have been attempts to validate the relationship between cybersickness and physiological responses [11,13,25,31–33,37,43,47]. Kim et al. [25] investigated the characteristic changes of the physiological signals such as EEG, ECG, and GSR while subjects are exposed to VR content. They conducted spectral analysis on each frequency band of EEG signals and validated that specific frequency bands have close correlations with cybersickness. In the case of ECG, they disclosed that the heart period was shorter during the virtual navigation than the baseline period. For GSR, they observed skin conductance level increased during the virtual navigation compared to the baseline. Mawalid et al. [32] attempted to extract EEG statistical feature with PCA for classification in order to investigate cybersickness. Pane et al. [37] adopted on power percentage features extracted from EEG signals to identify cybersickness level. Lin et al. [31] applied linear regression (LR), support vector regression (SVR), and self-organizing neural fuzzy inference network (SOFIN) models on cybersickness-related features extracted from PCA to predict sickness level. Recently, there is a study to predict VR sickness levels by using frequency band power of EEG signal with deep learning structure [17]. There also exists deep learning-based approach utilizing both content analysis and physiology analysis to predict cybersickness [28].

However, most cybersickness feature extraction methods did not place stimulus information under consideration, which predominantly influences physiological response of VR content viewers. In addition, previous assessment works were limited in that only the resultant cybersickness level is taken into consideration without symptom level analysis. Unlike these previous works, the proposed method assesses individual cybersickness with symptom level analysis by combining VR content stimulus and physiological response, which can provide more detailed and interpretable information.

Table 1. Cybersickness related symptoms according to 16-item SSQ [18]

No.	Symptom	Symptom group		
		Nausea	Oculomotor	Disorientation
1	General discomfort	✓	✓	
2	Fatigue		✓	
3	Headache		✓	
4	Eye strain		✓	
5	Difficulty focusing		✓	✓
6	Increased Salivation	✓		
7	Sweating	✓		
8	Nausea	✓		✓
9	Difficulty concentrating	✓	✓	
10	Fullness of head			✓
11	Blurred vision		✓	✓
12	Dizzy (Eyes Open)			✓
13	Dizzy (Eyes Closed)			✓
14	Vertigo			✓
15	Stomach awareness	✓		
16	Burping	✓		

3 Proposed Method

The proposed SACA Net is divided into three parts: the stimulus symptom context guider, the physiological symptom guider, and the symptom relation embedder. Given VR content, the stimulus context guider extracts the symptom group features that represent the context of sickness-inducing stimulus environment. The physiological symptom guider utilizes physiological signals being collected from humans while watching the VR content to extract symptom features. Based on the symptom group features and the symptom features, the symptom relation embedder refines symptom features by embedding relational characteristics among symptoms to effectively estimate symptom levels. The proposed model covers 16 symptoms for cybersickness according to [18] (see Table 1).

3.1 Stimulus Symptom Context Guider

Figure 2 network configuration of the stimulus symptom context guider for encoding symptom group features. There exists neural mismatch theory [40] that explains the process of motion sickness arising. When the expected sensory information does not match the actual sensory information, a neural mismatch occurs and it leads to motion sickness. People have a large neural mismatch for exceptional motions because such motions are not often experienced in daily life. Based on the theory, the sensory mismatch detector in stimulus symptom context

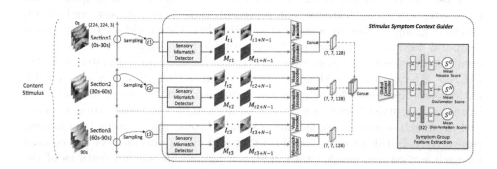

Fig. 2. Network configuration of the stimulus symptom context guider. It extracts group symptom features that represent nausea, oculomotor, and disorientation.

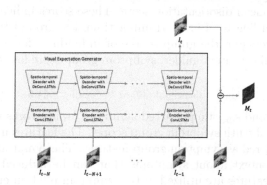

Fig. 3. Network configuration of the sensory mismatch detector in the stimulus symptom context guider. The sensory mismatch detector encodes the mismatch features.

guider is designed to encode mismatch features as shown in Fig. 3. The sensory mismatch detector includes a visual expectation generator. The visual expectation generator predicts the next frame $\hat{I}_t \in \mathbb{R}^{224 \times 224 \times 3}$ by taking previous N frames $I_{t-N}, \cdots, I_{t-1}$ (N=11). Note that the viewports of VR content are used as input frames. The visual expectation generator contains ConvLSTMs [50] and DeConvLSTMs [28] with deconvolution [35]. As the human daily experience, the visual expectation generator is pre-trained with videos [21] that contains only non-exceptional motions and the high frame rate (30 Hz). Thus, frame difference is large for VR content that could induce cybersickness with exceptional motions. A pixel-wise generation loss is defined to train the generator. Let G denote the generator function. The generation loss is defined as

$$\mathcal{L}_{gen} = \|G(I_{t-N}, ..., I_{t-1}) - I_t\|_2^2. \tag{1}$$

After training the visual expectation generator, the sensory mismatch detector takes sequence $(I_{t-N}, \cdots, I_t)$ to create mismatch feature M_t that represents visual sensory conflict between expected and actual information. Note that the sensory mismatch detector is first pre-trained and the weights are fixed.

Based on the sensory mismatch detector, the stimulus context guider outputs symptom group features that represent nausea, oculomotor, and disorientation group factors in a context of stimulus. Given video content, three temporal sections with equal lengths are divided up. From each section, randomly sampled content video sequence $(I_t, \cdots, I_{t+N-1})$ and mismatch sequence $(M_t, \cdots, M_{t+N-1})$ are used as inputs at training time. Since learned combinations are diversified through random sampling, overfitting can be alleviated. Note that the midst frames of each section are sampled at testing time. Content and mismatch sequences are fed into a visual encoder and a mismatch encoder, respectively. In this process, visual context and visual mismatch of VR content for each section are encoded with 3D-Conv layers. The output features of the three sections are concatenated and fed into a global context encoder with 2D-Conv layers to consider the overall context of the content video. Finally, the fully connected layers are applied to predict the mean nausea score, mean oculomotor score, and mean disorientation score. These scores indicate the symptom group scores according to [18]. At training time, the ground truth mean score is obtained by averaging the group scores of individuals for each content. For training the stimulus context guider, symptom group score loss $\mathcal{L}_{sym}^{group}$ is defined as

$$\mathcal{L}_{sym}^{group} = \|\hat{s}_{nau} - s_{nau}\|_2^2 + \|\hat{s}_{ocu} - s_{ocu}\|_2^2 + \|\hat{s}_{dis} - s_{dis}\|_2^2 \tag{2}$$

where $\hat{s}_{nau}$, $\hat{s}_{ocu}$, and $\hat{s}_{dis}$ are predicted symptom group scores while s_{nau}, s_{ocu}, and s_{dis} are ground truth symptom group scores. The features used for each prediction are considered as symptom group features. The symptom group features represent prior context about symptoms that can be induced by VR content stimulus. These features are utilized in the symptom relation embedder later.

3.2   Physiological Symptom Guider

The upper part of Fig. 4 shows the network configuration of the physiological symptom guider. The physiological symptom guider takes individual subject characteristics into consideration to extract symptom features. The proposed physiological symptom guider takes EEG signal acquired while watching VR content to output symptom features.

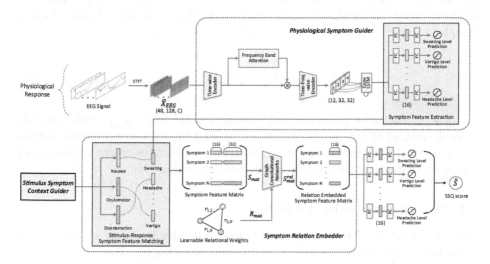

Fig. 4. Network configurations of a physiological symptom guider (upper part) and a symptom relation embedder (lower part). The physiological symptom guider outputs symptom features. The symptom relation embedder receives symptom group feature and symptom features to refine them with relational chracteristics.

To EEG signal, a high-pass filter with 0.5 Hz cut-off frequency is applied for removing baseline-drifting artifacts, and a low-pass filter 50 Hz cut-off frequency is applied for removing muscular artifacts [31]. Note that C denotes EEG channel size which corresponds with the number of acquired brain positions. After applying the frequency filters, the spectrogram image $\overline{X}_{EEG} \in \mathbb{R}^{48 \times 128 \times C}$ of the EEG signal is obtained through Short-Time Fourier Transform (STFT) [3] to consider the frequency characteristics. $\overline{X}_{EEG}$ is fed into an EEG time-wise encoder which is composed of 1D-Conv layers. The 1D-Conv layers in the time-wise encoder are applied on the temporal axis of $\overline{X}_{EEG}$. Therefore, this operation does not mix the feature in frequency-wise. Based on the clinical studies [9,25] that show frequency bands of EEG is related to cybersickness, we design the frequency band attention encoder for emphasizing important frequency band to predict cybersickness. The frequency band attention encoder learns to obtain five attention weights that correspond with delta (0.2–4 Hz), theta (4–8 Hz), alpha (8–13 Hz), beta (13–30 Hz), and gamma (30–50 Hz) bands. The attention weight of each band is located at the corresponding frequency region of the

EEG feature map to form a frequency band attention map $A_{freq\_band} \in \mathbb{R}^{48 \times 128}$. Then, A_{freq_band} is spatially elementwise multiplied to the EEG feature from the time-wise encoder. After applying the frequency band attention, the EEG feature is fed into a time-freq-wise encoder which is composed of 2D-Conv layers to encode both time and frequency characteristics. The feature drawn by the time-freq-wise encoder is divided into four patches in terms of the temporal axis. The patches enter the ConvLSTM in a temporal order. In this process, long-term characteristics can be encoded through the LSTM structure. Finally, symptom levels are predicted with fully connected layers. For training the physiological symptom guider, symptom score loss $\mathcal{L}_{sym}^{indiv}$ is defined as

$$\mathcal{L}_{sym}^{indiv} = \sum_{i=1}^{\#symptom} \left\| \hat{s}_{sym}^i - s_{sym}^i \right\|_2^2, \tag{3}$$

where $\hat{s}_{sym}^i$ indicates predicted i-th symptom score while s_{sym}^i indicates i-th ground truth symptom score. The features used for each symptom prediction are considered as symptom features. The symptom features reflect physiological symptom characteristics related to cybersickness of individuals.

3.3   Symptom Relation Embedder

Overall procedure of the symptom relation embedder is shown in Fig. 4. The symptom relation embedder is devised to refine the symptom features by encoding relational characteristics among symptoms. The symptom relation embedder receives the symptom group features and the symptom features from the stimulus symptom context guider and the physiological symptom guider, respectively. It learns the relations among symptoms in an unsupervised way by considering the symptom characteristics and the stimulus context that causes physical symptoms. Since symptom features are complementarily refined through embedding relations, they can be used to predict physical symptom levels more effectively.

We exploit the graph formulation [8,27,29,39,49] to learn relational characteristics. Each symptom feature $\in \mathbb{R}^{16}$ is matched to each symptom group feature $\in \mathbb{R}^{32}$. Through the symptom group matching, symptoms obtain prior context information about the sickness-inducing environment. Note that symptom features belonging to the two groups are matched to the average feature of the two symptom group features. Matched symptom features and symptom group features are concatenated and stacked row by row in a matrix form as shown in Fig. 4. As a result, a symptom feature matrix $S_{mat} \in \mathbb{R}^{16 \times 48}$ is constructed by considering each matched symptom feature as a graph node. In addition, we construct a learnable relational matrix $R_{mat} \in \mathbb{R}^{16 \times 16}$ corresponding to the adjacency matrix which represents the relationship among graph nodes. Since we set the weights of the relational matrix to be learnable, relations among symptoms can be embedded in an unsupervised way. Relational information is encoded with two layers of graph convolutional neural networks (GCNs) [27].

By following normalization trick in [27], relation embedded symptom feature matrix S_{mat}^{rel} can be formulated as

$$\tilde{R}_{mat} = R_{mat} + I, \tag{4}$$

$$S'_{mat} = ReLu(\tilde{D}^{-\frac{1}{2}}\tilde{R}_{mat}\tilde{D}^{-\frac{1}{2}}S_{mat}W_1), \tag{5}$$

$$S_{mat}^{rel} = ReLu(\tilde{D}^{-\frac{1}{2}}\tilde{R}_{mat}\tilde{D}^{-\frac{1}{2}}S'_{mat}W_2), \tag{6}$$

where I indicates identity matrix and $\tilde{D}$ indicates diagonal node degree matrix of $\tilde{R}_{mat}$. $W_1 \in \mathbb{R}^{48 \times 16}$ and $W_2 \in \mathbb{R}^{16 \times 16}$ are weight matrices.

We separate S_{mat}^{rel} by each row to get each relation embedded symptom feature. Then symptom scores are predicted through fully connected layers. Symptom score loss $\mathcal{L}_{sym\_rel}^{indiv}$ for relation embedded symptom features is defined as

$$\mathcal{L}_{sym\_rel}^{indiv} = \sum_{i=1}^{\#symptom} \left\| \hat{s}_{sym\_rel}^i - s_{sym}^i \right\|_2^2, \tag{7}$$

where $\hat{s}_{sym\_rel}^i$ indicates predicted i-th symptom score by relation embedded symptom features while s_{sym}^i indicates i-th ground truth symptom score.

In addition, the individual SSQ score based on symptom features is estimated by relation embedded symptom features with fully connected layers. Individual SSQ score loss $\mathcal{L}_{SSQ}^{indiv}$ can be written as

$$\mathcal{L}_{SSQ}^{indiv} = \left\| \widehat{SSQ}_{indiv} - SSQ_{indiv} \right\|_2^2, \tag{8}$$

where $\widehat{SSQ}_{indiv}$ is predicted individual SSQ score while SSQ_{indiv} is ground truth individual SSQ score. Finally, total objective loss can be defined as

$$\mathcal{L}_{total} = \mathcal{L}_{sym}^{group} + \mathcal{L}_{sym}^{indiv} + \mathcal{L}_{sym\_rel}^{indiv} + \mathcal{L}_{SSQ}^{indiv}, \tag{9}$$

The hyper-parameters that control the balance among losses are all set to 1. The detailed network structure is included in the supplementary material.

4   Experiments

4.1   Datasets

To validate the proposed method, we conduct experiments on two public 360-degree video datasets for cybersickness assessment. Each dataset contains SSQ information [18] and corresponding physiological signals (EEG, ECG, and GSR). We employ EEG as a physiological factor to analyze cybersickness because EEG contains the most comprehensive information about the nervous system [25,45].

VRSA DB-Shaking. In this dataset, there are 20 UHD 360-degree videos as content stimulus. The videos have various motion characteristics with camera shaking such as roller-coaster riding, skydiving, and boating. 15 subjects participated in the subjective experiment for viewing such content. Subjects were instructed to view each 90s video twice in a row, which corresponds to 180s viewing time. Repeating content twice is based on the guideline [1]. Subjects had time to rest 180s after viewing each content. Subjects graded the degree of cybersickness with the SSQ sheet [18] as [21,43]. The SSQ sheet is composed to express the degree of 16 symptoms in 4 steps. To minimize cybersickness accumulation, subjects were asked to tell about the presence of remaining cybersickness before viewing the next content. Supplementary rest time was provided in addition to the 180s rest time until they respond 'None at all' as in [21,36]. The motion of each subject was small and negligible while viewing the content. Subjects concentrated their gaze in the similar direction because used 360 degree-videos have movement in certain directions [10,21]. Head mounted display, PIMAX 5k+ was used for presenting content. Physiological signals (EEG, ECG, and GSR) were acquired while the subjects watched the content. EMOTIV EPOC+ was used for the 14-channel EEG signal acquisition, and Cognionics AIM was used for the ECG/GSR signal acquisition. The EEG device has an acquisition sampling rate 128 Hz, and other acquisition devices have a sampling rate 500 Hz. Experimental settings of the dataset followed the guideline of ITU-BT.500-13 [1] and BT.2021 [2].

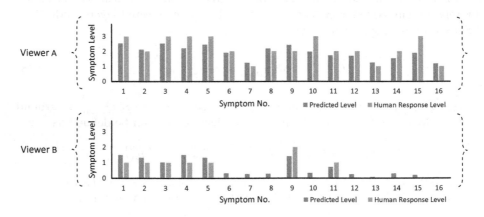

Fig. 5. Examples of symptom level prediction by the proposed method. It could differently estimate cybersickness with symptoms according to each individual viewer.

VRSA DB-FR. In addition to the VRSA DB-Shaking, VRSA DB-FR is used. This public dataset is the subject expanded version of [28]. There are 20 UHD 360-degree videos as content stimulus. The videos have two types of frame rates (10 Hz, 60 Hz) with various motion characteristics such as mountain biking, landscape scene, and car driving. It is known that video with exceptional motion and low frame rate causes cybersickness [21,33,46]. The dataset was constructed to contain various levels of cybersickness induced by content with excessive movement and low frame rate. 25 subjects participated in the subjective experiment for viewing such content. The protocol for viewing and assessment is the same as the VRSA DB-Shaking. Ultra-wide curved display, LG 34UC89 was used for presenting content. Viewing distance is controlled to provide immersive experiences with HMD level 110-degree FOV [6]. Physiological signals (EEG, ECG, and GSR) were acquired. Cognionics Quick-30 was used for 29-channel EEG signal acquisition, and Cognionics AIM was used for ECG/GSR signal acquisition. The acquisition devices have the same sampling rate 500 Hz. Experimental settings of the dataset followed the guideline of ITU-BT.500-13 [1] and BT.2021 [2].

4.2 Implementation Details

For each content, the physiological signals are 180s long. The intermediate 120s of each physiological signal is used to remove the noise of starting and end. Data augmentation is performed by shifting the extracted 120s region by 5s on the time axis. As a result, the training set is augmented 9 times. In the model training process, the stimulus symptom context guider and physiological symptom guider are first trained with their own loss functions $\mathcal{L}_{sym}^{group}$ and $\mathcal{L}_{sym}^{indiv}$ for smoothly encoding relations in the later part. In the stimulus symptom context guider and the physiological symptom guider, only the fully connected layers are learned with the final objective loss $\mathcal{L}_{total}$. We use Adam [26] to optimize the proposed network with a learning rate of 0.0002 and a batch size of 16.

Table 2. Symptom level prediction performances according to the network designs.

Network design		VRSA DB-Shaking			VRSA DB-FR		
Relation embedding	Stimulus context	PLCC	SROCC	RMSE	PLCC	SROCC	RMSE
✗	✗	0.389	0.326	0.499	0.516	0.397	0.356
✓	✗	0.449	0.351	0.492	0.536	0.397	0.346
✓	✓	**0.478**	**0.385**	**0.441**	**0.574**	**0.427**	**0.328**

Table 3. Detailed symptom level prediction results on the VRSA DB-Shaking.

Evaluation metrics	Relation embedding	Stimulus context	Mean	# Symptom															
				1	2	3	4	5	6	7	8	9	10	11	12	13	14	15	16
PLCC			0.389	0.338	0.346	**0.494**	0.458	0.556	0.522	0.364	0.464	0.336	0.456	0.392	0.246	0.332	0.182	0.360	0.386
	✓		0.449	0.414	0.402	0.492	0.542	0.568	0.542	0.330	**0.532**	0.350	0.480	**0.552**	0.278	0.386	0.254	**0.496**	**0.568**
	✓	✓	**0.478**	**0.550**	**0.474**	0.490	**0.546**	**0.580**	**0.552**	**0.414**	0.498	0.360	**0.526**	0.544	**0.392**	**0.396**	**0.290**	0.478	0.560
SROCC			0.326	0.322	0.338	0.462	0.318	0.434	0.380	0.246	0.386	**0.330**	0.364	0.396	0.234	0.338	0.226	0.162	0.292
	✓		0.351	0.350	0.342	**0.470**	0.406	0.438	0.356	0.202	**0.424**	0.326	0.340	0.388	0.298	0.360	0.264	0.284	**0.368**
	✓	✓	**0.385**	**0.518**	**0.440**	0.440	**0.408**	**0.474**	**0.382**	**0.284**	0.356	0.308	**0.400**	**0.454**	**0.400**	**0.368**	**0.306**	0.284	0.350
RMSE			0.499	0.798	0.828	0.592	0.768	0.778	0.214	0.244	0.368	**0.580**	0.344	0.624	0.444	0.232	0.568	0.368	0.240
	✓		0.492	0.768	0.804	0.602	0.754	0.858	0.204	0.238	0.360	0.634	0.336	0.562	0.442	0.222	0.562	0.330	**0.206**
	✓	✓	**0.441**	**0.552**	**0.660**	**0.590**	**0.668**	**0.740**	**0.204**	**0.228**	**0.350**	0.584	**0.304**	**0.524**	**0.386**	**0.222**	**0.516**	**0.322**	0.216

4.3   Assessment Performance Evaluation

We perform 5-fold cross-validation. The 5-fold is separated based on the content so that content and physiology in the training set and the test set do not overlap at all. Pearson linear correlation coefficient (PLCC), spearman rank order correlation coefficient (SROCC), and root mean square error (RMSE) are used as performance evaluation metrics. The PLCC and the SROCC are utilized to measure linearity and monotonicity, respectively. The RMSE metric reflects the differences between actual scores and predicted scores.

Physical Symptom Assessment. Figure 5 shows examples of symptom level prediction by the proposed method. The proposed method could distinguish different cybersickness with distinct symptoms according to each individual viewer for each content. Table 2 shows symptom prediction performances of the proposed method with ablating network designs on both datasets. Each symptom number matches with the symptom order in Table 1. The represented performance is the average of prediction performances for all symptoms. The baseline does not utilize relation embedding and stimulus context information. The relation embedding model without stimulus context indicates the case where symptom relation embedder is applied without symptom group features from the stimulus symptom context guider. The model with symptom relation embedding predicts symptom level better than the baseline model. The final model with relation embedding and stimulus context shows better performance than the other models for all evaluation metrics on both datasets. Table 3 shows detailed symptom level prediction results on the VRSA DB-Shaking. The final proposed model achieves the best performances. Considering that predicting the symptoms of each subject is a very challenging task, the proposed method obtains meaningful results for symptom level assessment (correlation p-value ≤ 0.05). Note that the EEG acquisition device in VRSA DB-FR is the sophisticated one with more brain channels and higher sampling rates compared to VRSA-Shaking. Thus, the baseline performance of it is higher than that of VRSA DB-Shaking.

Table 4. Total SSQ score prediction performances on the VRSA DB-Shaking and the VRSA DB-FR.

VRSA DB-Shaking				VRSA DB-FR			
Method	PLCC	SROCC	RMSE	Method	PLCC	SROCC	RMSE
Skin Conductance Level Feature [19] -based Method (GSR)	0.314	0.308	43.615	Skin Conductance Level Feature [19] -based Method (GSR)	0.390	0.295	34.933
Peak Interval Feature [42] -based Method (ECG)	0.340	0.237	46.469	Peak Interval Feature [42] -based Method (ECG)	0.379	0.298	34.712
Band Power Feature [17] -based Method (EEG)	0.492	0.352	35.157	Band Power Feature [17] -based Method (EEG)	0.476	0.326	33.862
Physiological Fusion Net [28] (EEG, ECG, and GSR + Content Stimulus)	0.739	0.617	30.372	Physiological Fusion Net [28] (EEG, ECG, and GSR + Content Stimulus)	**0.806**	0.660	23.893
Proposed Method (EEG + Content Stimulus)	**0.751**	**0.679**	**25.373**	**Proposed Method (EEG + Content Stimulus)**	0.801	**0.671**	**22.937**

Total SSQ Assessment. The performance comparison results for total SSQ score prediction are shown in Table 4. The skin conductance level feature-based method uses features related to tonic characteristics of GSR (MSCL, SDSCL, and SKSCL) [19]. The peak interval feature-based method performs prediction using the major RR interval features of ECG (MeanRR, SDRR, pNN50, and NN50) [42]. The band power feature-based method utilizes the frequency band power of EEG [17]. The Physiological Fusion Net [28] is a deep network model that predicts individual SSQ by fusing EEG, ECG, and GSR signals with content stimulus. As shown in the table, the proposed method outperforms the existing state-of-the-art methods. The proposed method obtains correlation results for total SSQ score prediction of individuals with PLCC ≥ 0.7 (p-value ≤ 0.05) on the VRSA DB-Shaking and PLCC ≥ 0.8 (p-value ≤ 0.05) on the VRSA DB-FR. Compared to the Physiological Fusion Net [28], our model predicts cybersickness without additional use of ECG and GSR data. Furthermore, instead of just predicting the resulting total SSQ score, physical symptoms can also be predicted to interpret which symptoms constitute the cybersickness of individual viewers.

4.4   Interpretation of Relational Weights

For interpreting relations among symptom features, we visualize the relational weights in relational matrix R_{mat}. Figure 6 shows the visualization results of relational weights. Each row and column in the matrix represent each symptom. It can be seen that the symptom relational weights obtained from different datasets have a similar tendency. Looking at the most highly activated relational weights for both datasets, symptoms 10–12, 10–13 [fullness of head - dizzy] are symptoms that belong to disorientation group and are both related to head. Activated relation of 12–13 [dizzy (eyes open) - dizzy (eyes closed)] indicates closely correlated dizzy symptoms. Interestingly, the region of 1–2 [general discomfort - fatigue] is activated in weights, which are both close to the general expression of cybersickness. Besides, relational weights for each dataset contain meaningful activations of relevant symptoms such as 6–15 [increased salivation - stomach awareness, internal organ-related symptoms] in (a) and 5–11 [difficulty focusing - blurred vision, eye-related symptoms] in (b). Note that the proposed network learns relational weights in an unsupervised way. Consequently, the relational weights are convincingly learned to emphasize the relations among relevant symptoms.

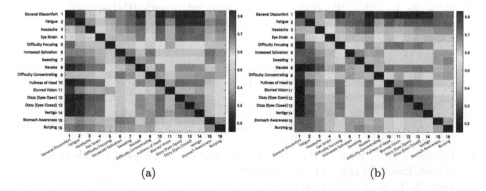

(a)   (b)

Fig. 6. Visualization results of relational weights in R_{mat} for (a) VRSA DB-Shaking and (b) VRSA DB-FR. Each row and column indicate physical symptoms.

5   Conclusion

In this paper, we propose the novel deep learning-based framework, SACA Net that reveals cybersickness of individual viewers with physical symptoms. Based on the physiology and stimulus context, the SACA Net effectively predicts physical symptom levels by embedding symptom relations. The symptom relation embedding scheme is designed to utilize the relational symptom characteristics for complementary effects among symptoms. The experimental results show that the proposed method achieves meaningful correlations for symptom scores and total SSQ scores on two cybersickness assessment datasets. In addition, we could interpret how the proposed SACA Net encodes relations among physical symptoms by analyzing the relational weights in the network. It is observed that relations of relevant symptoms are convincingly embedded.

Acknowledgement. This work was partly supported by IITP grant (No. 2017-0-00780), IITP grant (No. 2017-0-01779), and BK 21 Plus project.

References

1. Methodology for the subjective assessment of the quality of television pictures. ITU-R BT.500-13 (2012)
2. Subjective methods for the assessment of stereoscopic 3dtv systems. ITU-R BT.2021 (2012)
3. Allen, J.: Short term spectral analysis, synthesis, and modification by discrete Fourier transform. IEEE Trans. Acoust. Speech Signal Process. **25**(3), 235–238 (1977)
4. Basak, D., Pal, S., Patranabis, D.C.: Support vector regression. Neural Inf. Process. Lett. Rev. **11**(10), 203–224 (2007)
5. Bruck, S., Watters, P.A.: Estimating cybersickness of simulated motion using the simulator sickness questionnaire (SSQ): a controlled study. In: CIGV, pp. 486–488 (2009)
6. Buck, L.E., Young, M.K., Bodenheimer, B.: A comparison of distance estimation in HMD-based virtual environments with different HMD-based conditions. ACM Trans. Appl. Percept. (TAP) **15**(3), 1–15 (2018)
7. Carnegie, K., Rhee, T.: Reducing visual discomfort with HMDs using dynamic depth of field. IEEE Comput. Graph. Appl. **35**(5), 34–41 (2015)
8. Chen, Y., Rohrbach, M., Yan, Z., Shuicheng, Y., Feng, J., Kalantidis, Y.: Graph-based global reasoning networks. In: CVPR, pp. 433–442 (2019)
9. Chuang, S.W., Chuang, C.H., Yu, Y.H., King, J.T., Lin, C.T.: EEG alpha and gamma modulators mediate motion sickness-related spectral responses. Int. J. Neural Syst. **26**(02), 1650007 (2016)
10. Corbillon, X., De Simone, F., Simon, G.: 360-degree video head movement dataset. In: Proceedings of the 8th ACM on Multimedia Systems Conference, pp. 199–204 (2017)
11. Dennison, M.S., Wisti, A.Z., D'Zmura, M.: Use of physiological signals to predict cybersickness. Displays **44**, 42–52 (2016)
12. Doweck, I., et al.: Alterations in R-R variability associated with experimental motion sickness. J. Auton. Nerv. Syst. **67**(1–2), 31–37 (1997)

13. Egan, D., Brennan, S., Barrett, J., Qiao, Y., Timmerer, C., Murray, N.: An evaluation of heart rate and electrodermal activity as an objective QoE evaluation method for immersive virtual reality environments. In: QoMEX, pp. 1–6 (2016)

14. Freina, L., Ott, M.: A literature review on immersive virtual reality in education: state of the art and perspectives. In: eLSE, vol. 1, p. 133. "Carol I" National Defence University (2015)

15. Gallagher, A.G., et al.: Virtual reality simulation for the operating room: proficiency-based training as a paradigm shift in surgical skills training. Ann. Surg. **241**(2), 364 (2005)

16. Grantcharov, T.P., Kristiansen, V.B., Bendix, J., Bardram, L., Rosenberg, J., Funch-Jensen, P.: Randomized clinical trial of virtual reality simulation for laparoscopic skills training. Br. J. Surg. **91**(2), 146–150 (2004)

17. Jeong, D.K., Yoo, S., Jang, Y.: VR sickness measurement with EEG using DNN algorithm. In: VRST, p. 134 (2018)

18. Kennedy, R.S., Lane, N.E., Berbaum, K.S., Lilienthal, M.G.: Simulator sickness questionnaire: an enhanced method for quantifying simulator sickness. Int. J. Aviat. Psychol. **3**(3), 203–220 (1993)

19. Kim, A.Y., et al.: Automatic detection of major depressive disorder using electrodermal activity. Sci. Rep. **8**(1), 1–9 (2018)

20. Kim, H.G., Baddar, W.J., Lim, H., Jeong, H., Ro, Y.M.: Measurement of exceptional motion in VR video contents for VR sickness assessment using deep convolutional autoencoder. In: VRST, p. 36 (2017)

21. Kim, H.G., Lim, H.T., Lee, S., Ro, Y.M.: VRSA net: VR sickness assessment considering exceptional motion for 360 VR video. IEEE Trans. Image Process. **28**(4), 1646–1660 (2018)

22. Kim, J., Kim, W., Ahn, S., Kim, J., Lee, S.: Virtual reality sickness predictor: analysis of visual-vestibular conflict and VR contents. In: QoMEX, pp. 1–6 (2018)

23. Kim, J., Kim, W., Oh, H., Lee, S., Lee, S.: A deep cybersickness predictor based on brain signal analysis for virtual reality contents. In: ICCV, pp. 10580–10589 (2019)

24. Kim, K., Lee, S., Kim, H.G., Park, M., Ro, Y.M.: Deep objective assessment model based on spatio-temporal perception of 360-degree video for VR sickness prediction. In: ICIP, pp. 3192–3196 (2019)

25. Kim, Y.Y., Kim, H.J., Kim, E.N., Ko, H.D., Kim, H.T.: Characteristic changes in the physiological components of cybersickness. Psychophysiology **42**(5), 616–625 (2005)

26. Kingma, D., Ba, J.: Adam: a method for stochastic optimization. In: ICLR (2015)

27. Kipf, T.N., Welling, M.: Semi-supervised classification with graph convolutional networks. In: ICLR (2017)

28. Lee, S., et al.: Physiological fusion net: quantifying individual VR sickness with content stimulus and physiological response. In: ICIP, pp. 440–444 (2019)

29. Li, Y., Ouyang, W., Zhou, B., Wang, K., Wang, X.: Scene graph generation from objects, phrases and region captions. In: ICCV, pp. 1261–1270 (2017)

30. Lin, C.T., Chuang, S.W., Chen, Y.C., Ko, L.W., Liang, S.F., Jung, T.P.: EEG effects of motion sickness induced in a dynamic virtual reality environment. In: EMBC, pp. 3872–3875 (2007)

31. Lin, C.T., Tsai, S.F., Ko, L.W.: EEG-based learning system for online motion sickness level estimation in a dynamic vehicle environment. IEEE Trans. Neural Netw. Learn. Syst. **24**(10), 1689–1700 (2013)

32. Mawalid, M.A., Khoirunnisa, A.Z., Purnomo, M.H., Wibawa, A.D.: Classification of EEG signal for detecting cybersickness through time domain feature extraction using Naïve Bayes. In: CENIM, pp. 29–34 (2018)
33. Meehan, M., Insko, B., Whitton, M., Brooks, Jr., F.P.: Physiological measures of presence in stressful virtual environments. In: TOG, pp. 645–652 (2002)
34. Naqvi, S.A.A., Badruddin, N., Malik, A.S., Hazabbah, W., Abdullah, B.: Does 3D produce more symptoms of visually induced motion sickness? In: EMBC, pp. 6405–6408 (2013)
35. Noh, H., Hong, S., Han, B.: Learning deconvolution network for semantic segmentation. In: ICCV, pp. 1520–1528 (2015)
36. Padmanaban, N., Ruban, T., Sitzmann, V., Norcia, A.M., Wetzstein, G.: Towards a machine-learning approach for sickness prediction in 360 stereoscopic videos. IEEE Trans. Visual. Comput. Graph. **24**(4), 1594–1603 (2018)
37. Pane, E.S., Khoirunnisaa, A.Z., Wibawa, A.D., Purnomo, M.H.: Identifying severity level of cybersickness from EEG signals using CN2 rule induction algorithm. ICIIBMS **3**, 170–176 (2018)
38. Patrao, B., Pedro, S., Menezes, P.: How to deal with motion sickness in virtual reality. Sciences and Technologies of Interaction, 2015 22 nd, pp. 40–46 (2015)
39. Qi, M., Li, W., Yang, Z., Wang, Y., Luo, J.: Attentive relational networks for mapping images to scene graphs. In: CVPR, pp. 3957–3966 (2019)
40. Reason, J.T.: Motion sickness adaptation: a neural mismatch model. J. Roy. Soc. Med. **71**(11), 819–829 (1978)
41. Rebenitsch, L., Owen, C.: Review on cybersickness in applications and visual displays. Virtual Reality **20**(2), 101–125 (2016). https://doi.org/10.1007/s10055-016-0285-9
42. Shaffer, F., Ginsberg, J.: An overview of heart rate variability metrics and norms. Front. Publ. Health **5**, 258 (2017)
43. Singla, A., Fremerey, S., Robitza, W., Raake, A.: Measuring and comparing QoE and simulator sickness of omnidirectional videos in different head mounted displays. In: QoMEX, pp. 1–6 (2017)
44. Tiiro, A.: Effect of visual realism on cybersickness in virtual reality. University of Oulu (2018)
45. Wagh, K.P., Vasanth, K.: Electroencephalograph (EEG) based emotion recognition system: a review. In: Saini, H.S., Singh, R.K., Patel, V.M., Santhi, K., Ranganayakulu, S.V. (eds.) Innovations in Electronics and Communication Engineering. LNNS, vol. 33, pp. 37–59. Springer, Singapore (2019). https://doi.org/10.1007/978-981-10-8204-7_5
46. Weech, S., Kenny, S., Barnett-Cowan, M.: Presence and cybersickness in virtual reality are negatively related: a review. Front. Psychol. **10**, 158 (2019)
47. Wei, C.S., Ko, L.W., Chuang, S.W., Jung, T.P., Lin, C.T.: EEG-based evaluation system for motion sickness estimation. In: NER, pp. 100–103 (2011)
48. Wibirama, S., Nugroho, H.A., Hamamoto, K.: Depth gaze and ECG based frequency dynamics during motion sickness in stereoscopic 3D movie. Entertainment Comput. **26**, 117–127 (2018)
49. Woo, S., Kim, D., Cho, D., Kweon, I.S.: Linknet: relational embedding for scene graph. In: NIPS, pp. 560–570 (2018)
50. Xingjian, S., Chen, Z., Wang, H., Yeung, D.Y., Wong, W.K., Woo, W.c.: Convolutional LSTM network: a machine learning approach for precipitation nowcasting. In: NIPS, pp. 802–810 (2015)

End-to-End Low Cost Compressive Spectral Imaging with Spatial-Spectral Self-Attention

Ziyi Meng[1,2] , Jiawei Ma[3] , and Xin Yuan[4(✉)]

[1] Beijing University of Posts and Telecommunications, Beijing 100876, China
mengziyi@bupt.edu.cn
[2] New Jersey Institute of Technology, Newark, NJ 07102, USA
[3] Columbia University, New York, NY 10027, USA
jiawei.m@columbia.edu
[4] Nokia Bell Labs, Murray Hill, NJ 07974, USA
xyuan@bell-labs.com

Abstract. Coded aperture snapshot spectral imaging (CASSI) is an effective tool to capture real-world 3D hyperspectral images. While a number of existing work has been conducted for hardware and algorithm design, we make a step towards the low-cost solution that enjoys video-rate high-quality reconstruction. To make solid progress on this challenging yet under-investigated task, we reproduce a stable single disperser (SD) CASSI system to gather large-scale real-world CASSI data and propose a novel deep convolutional network to carry out the real-time reconstruction by using self-attention. In order to jointly capture the *self-attention across different dimensions* in hyperspectral images (i.e., channel-wise spectral correlation and non-local spatial regions), we propose Spatial-Spectral Self-Attention (TSA) to process each dimension sequentially, yet in an order-independent manner. We employ TSA in an encoder-decoder network, dubbed TSA-Net, to reconstruct the desired 3D cube. Furthermore, we investigate how noise affects the results and propose to add shot noise in model training, which improves the real data results significantly. We hope our large-scale CASSI data serve as a benchmark in future research and our TSA model as a baseline in deep learning based reconstruction algorithms. Our code and data are available at https://github.com/mengziyi64/TSA-Net.

Keywords: Compressive spectral imaging · Spatial-Spectral Self-Attention · Large-scale real data

---

Z. Meng and J. Ma—Equal contribution.
Z. Meng—Part of this work was performed when Jiawei Ma was a summer intern at Nokia Bell Labs in 2019.

Electronic supplementary material The online version of this chapter (https://doi.org/10.1007/978-3-030-58592-1_12) contains supplementary material, which is available to authorized users.

© Springer Nature Switzerland AG 2020
A. Vedaldi et al. (Eds.): ECCV 2020, LNCS 12368, pp. 187–204, 2020.
https://doi.org/10.1007/978-3-030-58592-1_12

1   Introduction

Coded aperture snapshot spectral imaging (CASSI) [49] has led to emerging researches during the last decade on *compressive* spectral imaging. The underlying principle is to modulate each spectral channel in the 3D scene *e.g.*(x, y, λ) by different masks (can be shifted versions of the same one). As shown in Fig. 1, the detector captures a *compressed* 2D measurement in a snapshot, which includes the information from all spectral channels. Following this, the inversion algorithms, inspired by compressive sensing [9,10], are employed to recover the desired 3D (spatial-spectral) cube. Motivated by CASSI, snapshot compressive imaging has also been used to capture video [22,34,35,68], polarization [46] and depth [23].

In the original CASSI [49] with single disperser (SD) design (Fig. 1(a)), the main issues left to solve are 1) the imbalanced response of SD and 2) the slow reconstruction. Notably, the imbalanced response, (please refer to Fig. M1 in the supplemental material (SM)), is a spatial distortion along the dispersion direction, which is caused by the path length difference between each two wavelength channels and leads to significant reconstruction performance degradation. The CASSI systems with direct view disperser [50] or dual-disperser [13] were proposed in the optical design to avoid the imbalanced response, but may suffer from high expense or system instability. Recently, DeSCI in [21] has achieved the state-of the-art (SOTA) performance among iterative algorithms on both video and spectral compressive imaging. Besides, various algorithms have been used [3,65] and developed [51,55,60,61,70] but still requires exhausting running time. Our goal in this paper is to make a step forward and study a low-cost solution that enjoys high-speed image capture and video-rate high-quality reconstruction, thus to provide an *end-to-end* solution of compressive spectral imaging.

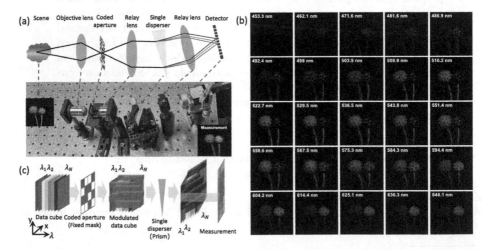

Fig. 1. (a) Single disperser coded aperture snapshot spectral imaging (SD-CASSI) and our experimental prototype. (b) 25 (out of 28) reconstructed spectral channels. (c) Principle of hardware coding.

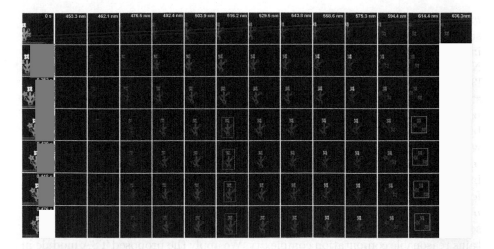

Fig. 2. Real data: the reconstructed hyperspectral video. The video totally contains 35 frames (0.028 s per frame) captured from our real SD-CASSI system, and each frame has 28 spectral channels between 450 and 650 nm. 7 frames and 13 spectral channels are extracted and shown here. The object is moving from left to right. Please refer to the **video files** in the Supplementary Material (SM).

To make progress on this fundamental problem, we first reproduced a SD-CASSI system and then gather the large-scale real CASSI data serving as a benchmark in our CASSI research. We have collected a group of images from different indoor scenes by using our setup shown in Fig. 1; for each measurement, a large-scale spatial-spectral 3D cube can be recovered and 28 is the number of channels determined by the hardware setup, *i.e.*, the filters, prisms and mask. As noted above, the imbalanced response and noise are introduced in our collected real data, which lead to severe performance degradation compared with the testing result on simulated noise-free data. This motivated us to employ deep learning as a tool to mitigate this challenge as well as the slow reconstruction.

Deep learning methods [29,30,52,56] showed the potential to speed up the reconstruction and improve image reconstruction quality. HSSP [52] is a SOTA deep unfolding method and exploit convolution network to estimate the spatial and spectral priors. Though this has led to promising results, the real data is usually captured by the CASSI system with the expensive direct-view disperser (more than 1000 US dollars based on [50]). In addition, HSSP recovers the whole hyperspectral image based on blocks and may lose non-local information.

Recently, Self-attention mechanism has been proposed in [47] for sequence modeling tasks such as machine translation, which is able to get rid of the sequence order and model the relationship between each two timestamps in parallel. Multiple heads are considered to model the relationship comprehensively. A limit number of previous work has conducted self-attention to model hyper-spectral image. For instance, λ-Net [29] assumes the spatial correlation for each channel is shared among all channels and then *only considers the spatial correlation in a hidden feature space.* However, they flatten all features from 2D plane into one single dimension to model the spatial correlation, causing huge memory usage in the attention map calculation. Furthermore, the only real data used in λ-Net [29] is of spatial size 256×256, which is a small scale data. Even though [39] designed an efficient self-attention form to help model spatial correlation, a strong constrain on the intermediate variables is enforced; [27] proposed to use a bi-directional Recurrent Neural Network to model the spectral channel correlation but not for CASSI applications. The CDSA in [25] generalized the self-attention where the theoretical analysis shows the order-independent property when applying the dimension-specific attention maps to modulate the extracted feature map in sequence. Inspired by this, we propose the Spatial-Spectral Self-Attention (TSA for 'Triple-S Attention') and model the spatial-spectral correlation in a *joint and order-independent manner*. We calculate the spatial and spectral attention maps separately and use them to modulate the *feature map in sequence*, which also maintains reasonable computation complexity. We apply the proposed TSA module in an encoder-decoder network, dubbed as TSA-Net, for CASSI image reconstruction. In addition, we study the effects of noise on the reconstruction results and propose to add *shot-noise*, rather than Gaussian noise, on the clean measurement during training to minimize the gap of performance between simulated noise-free data and real data.

In this paper, we investigate a novel method that can provide high quality reconstruction for the original and low cost SD-CASSI system, which is reproduced by us and now capable of capturing large-scale data. Our contributions are summarized as follows:

- We reproduce a stable SD-CASSI system using low-cost components, especially the low-cost single disperser, and gather large-scale real CASSI data as benchmark for future research.
- We propose Spatial-Spectral Self-Attention (TSA) module to model the spatial and spectral correlation in a joint yet order-independent manner with reasonable computation cost.
- We employ TSA module in an encoder-decoder network, which can provide $660 \times 660 \times 28$ spectral cube from a single measurement within 100ms using one GPU in evaluation. In this manner, we are able to provide a $660 \times 660 \times 28 \times 35$ 4D live volume per second with an example shown in Fig. 2.
- We analyse the effect of noises on the reconstruction results and the undesired high frequency details, *i.e.*, artifacts, in the deep learning based reconstruction. We then propose to add shot noise into the training data to simulate the system environment and to mitigate the artifacts in the real hyperspectral image. Experiment comparison shows the performance improvement and network's robustness, which demonstrates our strategy's effectiveness.

The rest of this paper is organized as follows: Sect. 2 describes the models of SD-CASSI. Sect. 3 presents the details of our proposed deep learning based TSA-Net to solve the reconstruction problem of SD-CASSI. Sect. 4 presents extensive results on both simulation and real data to demonstrate the superiority of our proposed TSA-Net as well as the hardware setup. Sect. 5 concludes the paper.

Related Work. Following CASSI, which used a coded aperture and a prism to implement the wavelength modulation, other modulations such as occlusion mask [6], spatial light modulator [70] and digital-micromirror-device [58] have also been used for compressive spectral imaging. Meanwhile, advances of CASSI have also been developed by using multiple-shots [18], dual-channel [51,53–55] and high-order information [2]. For the reconstruction, various iterative algorithms, such as TwIST [3], GPSR [11] and GAP-TV [65] have been utilized. Other algorithms, such as Gaussian mixture models and sparse coding [37,51,60] have also been developed. As mentioned before, most recently, DeSCI proposed in [21] to reconstruct videos or hyperspectral images in snapshot compressive imaging has led to state-of-the-art results. Inspired by the recent advances of deep learning on image restoration [24,59,71], researchers have started using deep learning to reconstruct hyperspectral images from RGB images [1,19,20,31,40]. Deep learning models [28,29,52,56] have been developed for CASSI. In addition to the novel attention-based TSA module in the design, our work differs from previous works by considering the impact of hardware constraints in CASSI such as real masks and shot noise.

2   Mathematically Model of SD-CASSI

Model Following the Optical Path. Let $F \in \mathbb{R}^{N_x \times N_y \times N_\lambda}$ denote the 3D spectral cube shown in the left of Fig. 1(c) and $M^* \in \mathbb{R}^{N_x \times N_y}$ denote the physical mask used for signal modulation. We use $F' \in \mathbb{R}^{N_x \times N_y \times N_\lambda}$ to represent the modulated signals where images at different wavelengths are modulated separately, *i.e.*, for $n_\lambda = 1, \dots, N_\lambda$, we have

$$F'(:,:,n_\lambda) = F(:,:,n_\lambda) \odot M^*, \tag{1}$$

where $\odot$ represents the element-wise multiplication. After passing the disperser, the cube F' is tilted and is considered to be sheared along the y-axis. We then use $F'' \in \mathbb{R}^{N_x \times (N_y+N_\lambda-1) \times N_\lambda}$ to denote the tilted cube and assume λ_c to be the reference wavelength, *i.e.*, image $F'(:,:,n_{\lambda_c})$ is not sheared along the y-axis, we can have

$$F''(u,v,n_\lambda) = F'(x, y + d(\lambda_n - \lambda_c), n_\lambda), \tag{2}$$

where (u,v) indicates the coordinate system on the detector plane, λ_n is the wavelength at n_λ-th channel and λ_c denotes the center-wavelength. Then, $d(\lambda_n - \lambda_c)$ signifies the spatial shifting for n_λ^{th} channel. The compressed measurement at the detector $y(u,v)$ can thus be modelled as

$$y(u,v) = \int_{\lambda_{\min}}^{\lambda_{\max}} f''(u,v,n_\lambda)d\lambda, \tag{3}$$

since the sensor integrates all the light in the wavelength $[\lambda_{\min}, \lambda_{\max}]$, where f'' is the analog (continuous) representation of F''. In discretized form, the captured 2D measurement $Y \in \mathbb{R}^{N_x \times (N_y+N_\lambda-1)}$ is modelled as

$$Y = \sum_{n_\lambda=1}^{N_\lambda} F''(:,:,n_\lambda) + G, \tag{4}$$

which is a *compressed* frame contains the information and $G \in \mathbb{R}^{N_x \times (Ny+N_\lambda-1)}$ represents the measurement noise.

For the convenience of model description, we further set $M \in \mathbb{R}^{N_x \times (Ny+N_\lambda-1) \times N_\lambda}$ to be the shifted version of the mask corresponding to different wavelengths, *i.e.*,

$$M(u, v, n_\lambda) = M^*(x, y + d(\lambda_n - \lambda_c)). \tag{5}$$

Similarly, for each signal frame at different wavelength, the shifted version is $\tilde{F} \in \mathbb{R}^{N_x \times (Ny+N_\lambda-1) \times N_\lambda}$,

$$\tilde{F}(u, v, n_\lambda) = F(x, y + d(\lambda_n - \lambda_c), n_\lambda). \tag{6}$$

Following this, the measurement Y can be represented as

$$Y = \sum_{n_\lambda=1}^{N_\lambda} \tilde{F}(:, :, n_\lambda) \odot M(:, :, n_\lambda) + G. \tag{7}$$

Vectorized Formulation. We use $\text{vec}(\cdot)$ to denote the matrix vactorization, *i.e.*, concatenating columns into one vector. Then, we have $y = \text{vec}(Y)$, $g = \text{vec}(G) \in \mathbb{R}^n$ and

$$f = \begin{bmatrix} \tilde{f}^{(1)} \\ \vdots \\ \tilde{f}^{(N_\lambda)} \end{bmatrix} \in \mathbb{R}^{N_x(Ny+N_\lambda-1)N_\lambda} \tag{8}$$

where $n = N_x(N_y + N_\lambda - 1)$ and $\tilde{f}^{(n_\lambda)} = \text{vec}(\tilde{F}(:, :, n_\lambda))$,

In addition, we define the sensing matrix as

$$\Phi = [D_1, \ldots, D_{N_\lambda}] \in \mathbb{R}^{n \times nN_\lambda}, \tag{9}$$

where $D_{n_\lambda} = \text{Diag}(\text{vec}(M(:, :, n_\lambda)))$ is a diagonal matrix with $\text{vec}(M(:, :, n_\lambda))$ as the diagonal elements. As such, we then can rewrite the matrix formulation of Eq. (7) as

$$y = \Phi f + g. \tag{10}$$

This is similar to compressive sensing (CS) [9,10] as Φ is a fat matrix, *i.e.*, more columns than rows. However, since Φ has the very special structure as in Eq. (9), most theory developed for CS can not fit in our applications. Note that Φ is a very sparse matrix, *i.e.*, at most nN_λ nonzero elements. It has recently been proved that the signal can still be recovered even when $N_\lambda > 1$ [15,16].

After capturing the measurement, the following task is given y (captured by the camera) and Φ (calibrated based on pre-design), solving f. For the sake of speed and quality, we use deep learning to solve this inverse problem.

3   TSA-Net for SD-CASSI Reconstruction

In this section, we first briefly review the conventional self-attention mechanism. Then, we propose Spatial-Spectral Self-Attention module followed by the TSA-Net structure. In Sect. 3.3, we analysis the effect of noise and discuss the strategy of injecting shot noise into *simulated measurement during model training*, to suppress the artifacts in the recovered hyperspectral images from *real measurements* captured by our SD-CASSI system. The hardware details can be found in SM.

3.1   Conventional Self-Attention

For self-attention mechanism in [47], given an input sentence of length N, each token x_i is mapped into a *Query* vector q_i of f-dim, a *Key* vector k_i of f-dim, and a *Value* vector v_i of v-dim. The attention from token x_j to token x_i is effectively the scaled dot-product of q_i and k_j after Softmax, which is defined as $A(i,j) = \frac{\exp(S(i,j))}{\sum_{k=1}^{N} \exp(S(i,k))}$ where $S(i,j) = q_i k_j^\top / \sqrt{f}$. Then, v_i is updated to v_i' as a weighted sum of all the *Value* vectors, defined as $v_i' = \sum_{j=1}^{N} A(i,j)v_j$, after which each v_i' is mapped to the layer output x_i' of the same size as x_i. Meanwhile, a *causal constraint* is set on the attention maps to force self-attention to learn to predict the next token only from the predicted tokens in translation tasks.

In order to adopt self-attention to *jointly model spatial and spectral correlation*, the intuitive way is to flatten all pixels into one single dimension and calculate the attention between each two pixels directly. However, as noted in [29], such operation will lead to huge memory usage and limit the effectiveness of correlation modelling. Instead, our proposed TSA module, described below, can jointly model spatial and spectral correlation while keep the size of attention map reasonable.

3.2   Spatial-Spectral Self-Attention (TSA)

Spatial Attention: Correlation modelling involves the attention map building for both x-axis and y-axis. We assume the spatial correlation should model the *non-local region information* instead of pixel-wise correlation. As a result, a 3×3 convolution kernel is applied to fuse the input feature to indicate the *region-based correlation*. Then, the convolution net is applied to map the fused feature into Q & K for each dimension individually. The number of kernels effectively denotes the number of heads and the kernel size denotes the modulation direction/dimension. Similarly, the dimension-specified Q & K features are used to build the related attention maps. TSA uses the *dimension-specified attention maps* to modulate the corresponding dimension in sequence while theoretical analysis in [25] has shown the order-independent property for such operation. The modulated feature are then fed into a deconvolution layer to finish the spatial correlation modelling.

Spectral Attention: The samples in the same spectral channel (2D plane) are first convolved with one kernel and then flattened into one single dimension, which is set as the feature vector for that channel. Similarly, input feature is then mapped to Q & K to build the attention map for the spectral axis. Since the image patterns on the same position but in two neighboring channels are expected be highly correlated, we learn to indicate such correlation by setting spectral smoothness on the attention maps. In our proposed model, we normalize all spectral channel pairwise distances to the range $[0, \pi]$ and use the cosine of the normalized distance as *spectral embedding* to indicate channel similarity. Each similarity score is scaled by 0.1 and then added to the coefficients in *spectral attention maps*, which are then used to modulate *Value* in self-attention modulation. In this way, we induce spectral smoothness constraint since the weights

of two adjacent channels in modulation are imposed to be higher than those for
distant channels (spectral channels with larger wavelength difference).

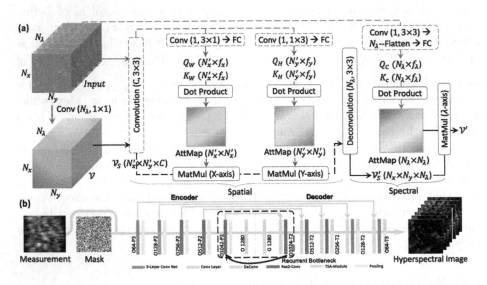

Fig. 3. (a) Spatial-Spectral Self-Attention (TSA) for one V feature (head). The spatial
correlation involves the modelling for x-axis and y-axis separately and aggregation in an
order-independent manner: the input is mapped to Q and K for each dimension: the size
of kernel and feature are specified individually. The spectral correlation modelling will
flatten samples in one spectral channel (2D plane) as a feature vector. The operation
in dashed box denotes the network structure is shared while trained in parallel. (b)
TSA-Net Architecture. Each convolution layer adopts a 3×3 operator with stride 1
and outputs O-channel cube. The size of pooling and upsampling is P and T.

As shown in Fig. 3(a), TSA builds one Value feature V passing into the
spatial and spectral modulation part in sequence. If we reverse the order and
do spectral modulation on V first, TSA will keep using the *input* to build the
spatial attention maps and feed the spectral output for spatial modulation.

Network Structure: Recently, variation auto encoder [17] and U-net, have
been repurposed as image generator in diverse problems [32,41,71]. In this task,
we build an encoder-decoder structure using U-net [38] as the backbone. As
shown in Fig. 3(b), we set 5 convolution blocks in the encoder and decoder individually, and replace the deepest 3 blocks with *Res2Net* [12] structure to enhance
the effectiveness of feature extraction. We add our TSA module at the end of 3
decoder blocks to model the Spatial-Spectral correlation. The spectral correlation constraint is set in the last TSA module. To overcome the trade-off between
the network size and the reconstruction performance, we choose to directly feed
the output back to the *recurrent bottleneck* and the parameters are shared in
each recurrent stage. In this way, the hierarchical feature representations can be
refined progressively and the knowledge can be accumulated in multiple stages.

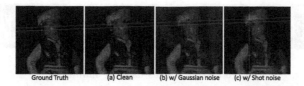

Ground Truth   (a) Clean   (b) w/ Gaussian noise   (c) w/ Shot noise

Fig. 4. Noise analysis by using a network trained on clean data to recover from one measurement under three conditions: a) no noise, b) with Gaussian noise, c) with Shot noise. (c) only has artifacts on bright area but (b) has artifacts even in dark area.

Different from [42], the skip connection between the two blocks is not a global connection. Instead, it is an inner connection between two sub-layers such that the gradient vanish problem is avoided. Meanwhile, since the image size in our experiment will be huge (each sample is of size $660 \times 660 \times 28$), our model will be much larger than previous proposed network and such filter sharing strategy can significantly reduce the storage requirements for a large deep learning model.

3.3   Shot Noise Injection

In this subsection, we first provides a mathematical model of shot noise and then explain the strategy of shot noise injection during model training.

Shot noise is the fluctuation in photon counts sensed at a given camera exposure level [5]. It is considered to be the dominant noise in the brighter parts of an image. For a camera sensor, shot noise is determined by the detector's dynamic range and quantum efficiency (QE). In the SD-CASSI system, the measurement with shot noise $\boldsymbol{Y}_{sn}$ can be modeled as

$$\boldsymbol{Y}_{sn} = \mathcal{B}(\boldsymbol{Y}/\text{QE}, \text{QE}), \tag{11}$$

where $\boldsymbol{Y}$ is the measurement without noise; the elements of $\boldsymbol{Y}$ being integers between 0 to $2^k - 1$, with k being the sensor bit depth; $\mathcal{B}(n, p)$ is the binomial distribution function, and QE is the quantum efficiency of sensor. Meanwhile, we have $\boldsymbol{y}_{sn} = \text{vec}(\boldsymbol{Y}_{sn}) = \mathcal{B}(\boldsymbol{y}/\text{QE}, \text{QE})$.

Overall, the target of TSA-Net is to reconstruct the 3D hyperspectral image cube from a 2D measurement captured by our SD-CASSI system. Since it is expensive to gather real-world hyperspectral images as ground truth for model training, same with other works [29,52], we train the model on the simulation data and then feed the real data to the pre-trained model for evaluation. To train the model, we first need to capture the mask inside our real optical system, and use the mask to generate measurements (following our hardware design) from available hyperspectral images. In this way, when we feed the mask and a measurement as input and then train the network to recover a 3D cube, the hyperspectral image cube actually serves as the ground truth. However, several challenges still exist in this process. 1) As there are various and random noise patterns during measurement generation in the optical detector system, it is inapplicable to enumerate all possible noise patterns for the simulation data during training. 2) The inconsistency

Fig. 5. 10 testing scenes used in simulation.

between testing data captured by our system and training data exists as the training data is from datasets built by another system.

As such, there is severe performance degradation of reconstruction and the
artifacts caused by system noise are obvious during testing. 3) Factors such as
response imbalance caused by single disperser will lead to poor system calibration.

To overcome these challenges and enhance the model's robustness, previous
works have adopted various techniques during model training, *e.g.*, adding Gaussian noise [4,48] in the network bottleneck and image augmentation [33]. However,
a large amount of samples are required during training to learn noise drawn from
Gaussian distributions of all possible hyper-parameters. In contrast, each shot
noise value depends on the signal level at each pixel. Besides, shot noise is usually
dominant in an imaging system like our system with bright illumination and high
exposure [26]. To analyse the link between noise in hardware system and reconstruction artifacts, we compare the reconstruction results in simulation and real
data. As shown in Fig. 4, for a network trained by clean data, the reconstruction
of measurement with shot noise (right-most column) has artifacts in the object
area, which is similar to real data (top in Fig. 12), while the artifacts distribute in
the whole region in the result of measurement with Gaussian noise.

As a result, we propose to add shot noise to the clean measurement dur-
ing model training (*i.e.*, using $\boldsymbol{\Phi}^\top \boldsymbol{y}_{sn}$ as the input of the TSA-Net) and we
find reconstruction performance degradation between the simulation and real
data captured by hardware system is narrowed. We have also observed this in
other snapshot compressive imaging systems [7,22,23,28,34–36,43–46,63,64,66–
69] and our proposed TSA-Net can be extended to those systems.

4   Experiments

In Sect. 4.1, we evaluate the reconstruction performance on the synthetic data
in simulation. In Sect. 4.2, we demonstrate experimental results captured by
our SD-CASSI system. The performance comparison is provided to show the
effectiveness of network and our training strategy.

4.1   Simulation

System Hyperparameter. To quantitatively evaluate the effectiveness of our
TSA-Net reconstruction on SD-CASSI system, the hyperparameters, *e.g.*, mask
and wavelengths, used in simulation are consistent with those in the real system.
The region of 256×256 at the center of the real captured mask is selected for

Table 1. PSNR in dB (left entry in each cell) and SSIM (right entry in each cell) by different algorithms on 10 scenes in simulation.

Algorithm	TwIST	GAP-TV	DeSCI	U-net	HSSP	λ-net	TSA-Net (ours)
Scene1	24.81, 0.730	25.13, 0.724	27.15, 0.794	28.28, 0.822	31.07, 0.852	30.82, 0.880	**31.26, 0.887**
Scene2	19.99, 0.632	20.67, 0.630	22.26, 0.694	24.06, 0.777	26.30, 0.798	26.30, 0.846	**26.88, 0.855**
Scene3	21.14, 0.764	23.19, 0.757	26.56, 0.877	26.02, 0.857	29.00, 0.875	29.42, 0.916	**30.03, 0.921**
Scene4	30.30, 0.874	35.13, 0.870	39.00, **0.965**	36.33, 0.877	38.24, 0.926	37.37, 0.962	**39.90**, 0.964
Scene5	21.68, 0.688	22.31, 0.674	24.80, 0.778	25.51, 0.795	27.98, 0.827	27.84, 0.866	**28.89, 0.878**
Scene6	22.16, 0.660	22.90, 0.635	23.55, 0.753	27.97, 0.794	29.16, 0.823	30.69, 0.886	**31.30, 0.895**
Scene7	17.71, 0.694	17.98, 0.670	20.03, 0.772	21.15, 0.799	24.11, 0.851	24.20, 0.875	**25.16, 0.887**
Scene8	22.39, 0.682	23.00, 0.624	20.29, 0.740	26.83, 0.796	27.94, 0.831	28.86, 0.880	**29.69, 0.887**
Scene9	21.43, 0.729	23.36, 0.717	23.98, 0.818	26.13, 0.804	29.14, 0.822	29.32, 0.902	**30.03, 0.903**
Scene10	22.87, 0.595	23.70, 0.551	25.94, 0.666	25.07, 0.710	26.44, 0.740	27.66, 0.843	**28.32, 0.848**
Average	22.44, 0.703	23.73, 0.683	25.86, 0.785	26.80, 0.803	28.93, 0.834	29.25, 0.886	**30.15, 0.893**

simulation. We determine 28 spectral channels distributed from 450nm to 650nm according to our system, and then adopt spectral interpolation on the simulation data to acquire image of the 28 channels as ground truth.

Dataset, Implementation Details and Baselines. We conduct simulation on hyperspectral image datasets CAVE [62] and KAIST [8]. We randomly select a spatial area of size 256×256 and crop the 3D cubes with 28 channels as one training sample with data augmentation. After mask modulation, the image cube is sheared with an accumulative two-pixel step (based on the hardware) and integrated across spectral dimension, so that a measurement of size 256×310 is generated as one model input. As shown in Fig. 5, we set 10 scenes from KAIST for model testing. For valid evaluation, the scenes in KAIST are not seen in training. The network is implemented by Tensorflow, and trained on one NVIDIA P40 GPU for about 30 h. The objective is to minimize the Root Mean Square Error (RMSE) and Spectrum Constancy Loss [72] of the reconstruction. We use peak-signal-to-noise-ratio (PSNR) and structured similarity index metrics (SSIM) [57] for evaluation. More details (system hyper-parameters, learning rate, etc.) can be found in SM as well as more results.

We compare our method with both iterative algorithms: TwIST [3], GAP-TV [65] and DeSCI [21], as well as deep neural networks: U-net [38], HSSP [52] and λ-net [29]. We use the same configurations for all these methods. We first perform experiments on noise-free measurements to verify the performance of different algorithms. Then, we compare the results of different algorithms under shot noise and Gaussian noise respectively to demonstrate the advantages of our model and the shot noise injection strategy.

Reconstruction on Noise-Free Data. We first evaluate the reconstruction performance of TSA-Net on noise-free KAIST simulation data. Notably, we didn't add shot noise during model training and testing for this set of comparison. As shown in Table 1, our proposed TSA-Net outperforms other algorithms in most

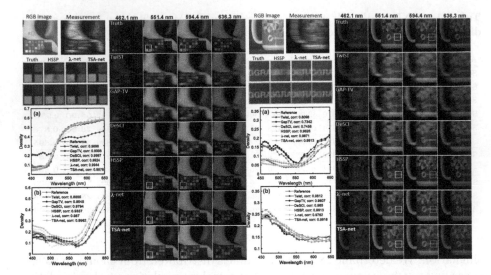

Fig. 6. Two reconstructed images with 4 (out of 28) spectral channels using six methods. We compare the recovered spectra of the selected region (shown with a, b on the RGB images) and spatial details.

scenes. The only exception is SSIM on Scene 4, which is a simple scene without high frequency components and thus fits the assumption of low-rank in DeSCI. On average, TSA-Net outperforms the SOTA iterative algorithm DeSCI by 4.29 dB. Meanwhile, TSA-Net performs 3.35 dB higher in PSNR over U-net, 1.22 dB higher over HSSP and 0.90 dB higher over λ-net. Note that the gain of our TSA-Net compared with λ-net is mainly from our proposed TSA module and the comparison with U-net and λ-net also serves the *ablation study* of our TSA-Net.

The visualization of 2 scenes with 4 (out of 28) channels are shown in Fig. 6. It is obvious that the spatial resolution in reconstruction by deep neural networks is higher than that of iterative algorithms, which suffer from the spatial blur resulted from the large mask-shift range. In addition, the large code features on the mask limits the resolution of the reconstructed images. In contrast, deep learning methods can provide both small-scale fine details and large-scale sharp edges. Compared with HSSP and λ-net, the reconstruction of TSA-Net have less artifacts and clearer details. Moreover, we show the spectral curves of the selected regions and calculate the spectral correlation values. The iterative algorithms have a high spectral accuracy at the expense of spatial accuracy, while TSA-Net ensures a high-quality spectral recovery, meanwhile improves the spatial fidelity significantly. We have also tried to add GAN [14] training in our loss function [29] and saw limited improvement (in average 0.1 dB). Since the key contribution of this paper is self-attention, we omit the GAN loss part.

Reconstruction on Data with Shot Noise. We generate shot noise by setting QE to 0.4 in Eq. (11). We set bit depth as 11 in model training by considering

Table 2. Results of different algorithms on data with different level shot noise

Noise Level	Metric	HSSP	λ-net	TSA-Net w/o SN	TSA-Net w/ SN
Without noise	PSNR	28.93	29.25	**30.15**	28.69
	SSIM	0.834	0.886	**0.893**	0.859
12-bit shot noise	PSNR	25.87	27.91	28.36	**28.55**
	SSIM	0.744	0.822	0.850	**0.856**
11-bit shot noise	PSNR	24.66	27.36	27.40	**28.35**
	SSIM	0.705	0.802	0.823	**0.849**
10-bit shot noise	PSNR	23.60	26.48	25.74	**28.08**
	SSIM	0.663	0.771	0.779	**0.841**

Table 3. Results of TSA-Net w/ and w/o shot noise on data with different level Gaussian noise (PSNR, SSIM)

Noise Level σ	TSA-Net w/o SN	TSA-Net w/ SN
0	30.15, 0.893	28.69, 0.859
0.005	28.33, 0.830	28.46, 0.836
0.01	25.39, 0.778	28.03, 0.819
0.02	22.65, 0.658	26.93, 0.781
0.05	19.47, 0.541	23.50, 0.660
0.1	18.74, 0.485	19.67, 0.528
0.2	18.20, 0.443	19.15, 0.468

the 12-bit camera in our system and assuming 1-bit submergence by other noise. Also, we varies the number of bit during testing for comprehensive comparison and a lower bit depth leads to higher the shot noise. During testing, we change the number of bit from 10 to 12 and the average PSNR & SSIM reported in Table 2 demonstrates the robustness of the neural network.

In detail, we test the models on KAIST at each noise level in five trials. It can be seen that the result of the TSA-Net trained on measurements with shot noise (TSA-Net w/ SN) only degrade 0.61 dB in PSNR when tested on data with 10-bit shot noise, while the results degrade severely on HSSP (5.33 dB), λ-net (2.77 dB) and the TSA-Net trained without shot noise (TSA-Net w/o SN, 4.41 dB). We also observe that when there is no noise in the testing data, TSA-Net w/o SN provides better results than TSA-Net w/ SN, as the consistence of data between training and testing is kept. Hereby and in the following real data experiments, we focus on the measurements with SN as in real cases, noise is unavoidable.

Robustness to Gaussian Noise. We further investigate the effect of Gaussian noise to our TSA-Net with and without shot noise. Before adding noise, the measurements are normalized to [0, 1]. Then we add zero-mean Gaussian noise to the measurements with standard deviation σ ranging from 0 to 0.2. As shown in Table 3, the performance of the TSA-Net w/ SN degrades slower than that of the TSA-Net w/o SN, which indicates adding shot noise on training data can also mitigate the effect of Gaussian noise.

4.2   Real Data Reconstruction

We have built a SD-CASSI system shown in Fig. 1 consisting of an objective lens, a random mask, two relay lens with 45 mm and 50 mm focal length, a dispersion prism, and a detector. The prism with 30° apex angle produces the 54-pixel dispersion corresponding to 28 spectral channels ranging from 450 to 650 nm. The whole system can capture a large-scale scene of size 1024 × 1024. As such, we trained another model from scratch based on CAVE and KAIST datasets and added 11-bit shot noise on the simulated measurements during training.

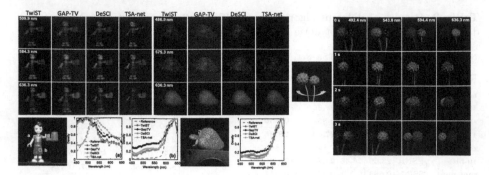

Fig. 7. Real data: (Left) the reconstructed images for three out of 28 spectral channels. The RGB images and spectral curves are shown at the lower part of the figure; (Right) the reconstructed hyperspectral video with 105 frames (3 s), four frames with four spectral channels are shown here with full videos in SM.

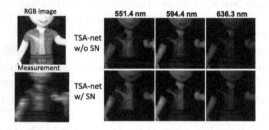

Fig. 8. Real data: the reconstructed images (256 × 256) using TSA-Net trained without and with shot noise.

As shown in Fig. 7 (left), we show the reconstruction by TSA-Net and iterative algorithm of two scenes with three channels, and our method outperforms the baselines by recovering the most details of each scene. Since too much training time is required for other deep learning methods on large-scale data, we compare the reconstruction for a smaller real data in SM. In Fig. 2 and Fig. 7 (right), we show two dynamic scenes, moving in 1 s and rotating in 3 s, captured by our system respectively. It can be seen that our SD-CASSI with TSA-Net is providing an end-to-end capture and reconstruction of spectral images with high quality spatial, spectral and motion details. Furthermore, we demonstrate the effectiveness of the training strategy by adding shot noise to real data. As shown in Fig. 8, by injecting shot noise during model training, not only the spatial details in reconstruction from real-data is kept, the artifacts is suppressed when compared with the reconstruction by TSA-Net trained on noise-free data.

5   Conclusions

We have developed an end-to-end low-cost compressive spectral imaging system by single-disperser CASSI and TSA-net. We have proposed a Spatial-Spectral

Self-Attention module to jointly model the spatial and spectral correlation in an order-independent manner, which is incorporated in an encoder-decoder network to achieve high quality reconstruction. By analyzing the noise impact and examining the artifacts in real data reconstruction, we observed that adding shot noise in the training data can improve the reconstruction quality significantly. Our end-to-end solution for video-rate capture and reconstruction of hyperspectral images paves the way of real applications of compressive spectral imaging.

References

1. Akhtar, N., Mian, A.S.: Hyperspectral recovery from RGB images using Gaussian processes. IEEE Trans. Pattern Anal. Mach. Intell. **42**, 100–113 (2018)
2. Arguello, H., Rueda, H., Wu, Y., Prather, D.W., Arce, G.R.: Higher-order computational model for coded aperture spectral imaging. Appl. Opt. **52**(10), D12–D21 (2013)
3. Bioucas-Dias, J., Figueiredo, M.: A new TwIST: two-step iterative shrinkage/thresholding algorithms for image restoration. IEEE Trans. Image Process. **16**(12), 2992–3004 (2007)
4. Bishop, C.M.: Training with noise is equivalent to Tikhonov regularization. Neural Comput. **7**(1), 108–116 (1995)
5. Blanter, Y.M., Büttiker, M.: Shot noise in mesoscopic conductors. Phys. Rep. **336**(1–2), 1–166 (2000)
6. Cao, X., Du, H., Tong, X., Dai, Q., Lin, S.: A prism-mask system for multispectral video acquisition. IEEE Trans. Pattern Anal. Mach. Intell. **33**(12), 2423–2435 (2011)
7. Cheng, Z., et al.: BIRNAT: bidirectional recurrent neural networks with adversarial training for video snapshot compressive imaging. In: European Conference on Computer Vision (ECCV), August 2020
8. Choi, I., Jeon, D.S., Nam, G., Gutierrez, D., Kim, M.H.: High-quality hyperspectral reconstruction using a spectral prior, vol. 36, p. 218. ACM (2017)
9. Donoho, D.L.: Compressed sensing. IEEE Trans. Inf. Theory **52**(4), 1289–1306 (2006)
10. Emmanuel, C., Romberg, J., Tao, T.: Robust uncertainty principles: exact signal reconstruction from highly incomplete frequency information. IEEE Trans. Inf. Theory **52**(2), 489–509 (2006)
11. Figueiredo, M.A., Nowak, R.D., Wright, S.J.: Gradient projection for sparse reconstruction: application to compressed sensing and other inverse problems. IEEE J. Sel. Top. Signal Process. **1**(4), 586–597 (2007)
12. Gao, S.H., Cheng, M.M., Zhao, K., Zhang, X.Y., Yang, M.H., Torr, P.: Res2Net: a new multi-scale backbone architecture (2019)
13. Gehm, M.E., John, R., Brady, D.J., Willett, R.M., Schulz, T.J.: Single-shot compressive spectral imaging with a dual-disperser architecture. Opt. Express **15**(21), 14013–14027 (2007)
14. Goodfellow, I.J., et al.: Generative adversarial nets. In: Proceedings of the 27th International Conference on Neural Information Processing Systems - Volume 2, NIPS 2014, pp. 2672–2680 (2014)
15. Jalali, S., Yuan, X.: Compressive imaging via one-shot measurements. In: IEEE International Symposium on Information Theory (ISIT) (2018)

16. Jalali, S., Yuan, X.: Snapshot compressed sensing: performance bounds and algorithms. IEEE Trans. Inf. Theory **65**(12), 8005–8024 (2019)
17. Kingma, D.P., Welling, M.: Auto-encoding variational Bayes (2013). Cite arxiv:1312.6114
18. Kittle, D., Choi, K., Wagadarikar, A., Brady, D.J.: Multiframe image estimation for coded aperture snapshot spectral imagers. Appl. Opt. **49**(36), 6824–6833 (2010)
19. Koundinya, S., et al.: 2D–3D CNN based architectures for spectral reconstruction from RGB images. In: The IEEE Conference on Computer Vision and Pattern Recognition (CVPR) Workshops, June 2018
20. Li, H., Xiong, Z., Shi, Z., Wang, L., Liu, D., Wu, F.: HSVCNN: CNN-based hyperspectral reconstruction from RGB videos. In: 2018 25th IEEE International Conference on Image Processing (ICIP), pp. 3323–3327, October 2018
21. Liu, Y., Yuan, X., Suo, J., Brady, D.J., Dai, Q.: Rank minimization for snapshot compressive imaging. IEEE Trans. Pattern Anal. Mach. Intell. **41**(12), 2990–3006 (2019)
22. Llull, P., et al.: Coded aperture compressive temporal imaging. Opt. Express **21**(9), 10526–10545 (2013)
23. Llull, P., Yuan, X., Carin, L., Brady, D.J.: Image translation for single-shot focal tomography. Optica **2**(9), 822–825 (2015)
24. Ma, J., Liu, X., Shou, Z., Yuan, X.: Deep tensor ADMM-Net for snapshot compressive imaging. In: IEEE/CVF Conference on Computer Vision (ICCV) (2019)
25. Ma, J., Shou, Z., Zareian, A., Mansour, H., Vetro, A., Chang, S.F.: CDSA: crossdimensional self-attention for multivariate, geo-tagged time series imputation. arXiv preprint arXiv:1905.09904 (2019)
26. MacDonald, L.: Digital Heritage. Routledge, Abingdon (2006)
27. Mei, X., et al.: Spectral-spatial attention networks for hyperspectral image classification. Remote Sens. **11**(8), 963 (2019)
28. Meng, Z., Qiao, M., Ma, J., Yu, Z., Xu, K., Yuan, X.: Snapshot multispectral endomicroscopy. Opt. Lett. **45**(14), 3897–3900 (2020)
29. Miao, X., Yuan, X., Pu, Y., Athitsos, V.: λ-net: reconstruct hyperspectral images from a snapshot measurement. In: IEEE/CVF Conference on Computer Vision (ICCV) (2019)
30. Miao, X., Yuan, X., Wilford, P.: Deep learning for compressive spectral imaging. In: Digital Holography and Three-Dimensional Imaging 2019, p. M3B.3. Optical Society of America (2019)
31. Nie, S., Gu, L., Zheng, Y., Lam, A., Ono, N., Sato, I.: Deeply learned filter response functions for hyperspectral reconstruction. In: The IEEE Conference on Computer Vision and Pattern Recognition (CVPR), June 2018
32. Peng, P., Jalali, S., Yuan, X.: Solving inverse problems via auto-encoders. IEEE J. Sel. Areas Inf. Theory **1**(1), 312–323 (2020)
33. Perez, L., Wang, J.: The effectiveness of data augmentation in image classification using deep learning. arXiv preprint arXiv:1712.04621 (2017)
34. Qiao, M., Liu, X., Yuan, X.: Snapshot spatial-temporal compressive imaging. Opt. Lett. **45**(7), 1659–1662 (2020)
35. Qiao, M., Meng, Z., Ma, J., Yuan, X.: Deep learning for video compressive sensing. APL Photonics **5**(3), 030801 (2020)
36. Qiao, M., Sun, Y., Liu, X., Yuan, X., Wilford, P.: Snapshot optical coherence tomography. In: Digital Holography and Three-Dimensional Imaging 2019, p. W4B.3. Optical Society of America (2019)

37. Renna, F., et al.: Classification and reconstruction of high-dimensional signals from low-dimensional features in the presence of side information. IEEE Trans. Inf. Theory **62**(11), 6459–6492 (2016)

38. Ronneberger, O., Fischer, P., Brox, T.: U-Net: convolutional networks for biomedical image segmentation. In: Navab, N., Hornegger, J., Wells, W.M., Frangi, A.F. (eds.) MICCAI 2015. LNCS, vol. 9351, pp. 234–241. Springer, Cham (2015). https://doi.org/10.1007/978-3-319-24574-4_28

39. Shen, Z., Zhang, M., Zhao, H., Yi, S., Li, H.: Efficient attention: attention with linear complexities. arXiv preprint arXiv:1812.01243 (2018)

40. Shi, Z., Chen, C., Xiong, Z., Liu, D., Wu, F.: HSCNN+: advanced CNN-based hyperspectral recovery from RGB images. In: The IEEE Conference on Computer Vision and Pattern Recognition (CVPR) Workshops, June 2018

41. Sinha, A., Lee, J., Li, S., Barbastathis, G.: Lensless computational imaging through deep learning. Optica **4**(9), 1117–1125 (2017)

42. Sun, L., Fan, Z., Huang, Y., Ding, X., Paisley, J.: Compressed sensing MRI using a recursive dilated network. In: Thirty-Second AAAI Conference on Artificial Intelligence (2018)

43. Sun, Y., Yuan, X., Pang, S.: High-speed compressive range imaging based on active illumination. Opt. Express **24**(20), 22836–22846 (2016)

44. Sun, Y., Yuan, X., Pang, S.: Compressive high-speed stereo imaging. Opt. Express **25**(15), 18182–18190 (2017)

45. Tsai, T.H., Llull, P., Yuan, X., Carin, L., Brady, D.J.: Spectral-temporal compressive imaging. Opt. Lett. **40**(17), 4054–4057 (2015)

46. Tsai, T.H., Yuan, X., Brady, D.J.: Spatial light modulator based color polarization imaging. Opt. Express **23**(9), 11912–11926 (2015)

47. Vaswani, A., et al.: Attention is all you need. In: Advances in Neural Information Processing Systems, pp. 5998–6008 (2017)

48. Vincent, P., Larochelle, H., Lajoie, I., Bengio, Y., Manzagol, P.A.: Stacked denoising autoencoders: learning useful representations in a deep network with a local denoising criterion. J. Mach. Learn. Res. **11**, 3371–3408 (2010)

49. Wagadarikar, A., John, R., Willett, R., Brady, D.: Single disperser design for coded aperture snapshot spectral imaging. Appl. Opti. **47**(10), B44–B51 (2008)

50. Wagadarikar, A.A., Pitsianis, N.P., Sun, X., Brady, D.J.: Video rate spectral imaging using a coded aperture snapshot spectral imager. Opt. Express **17**(8), 6368–6388 (2009)

51. Wang, L., Xiong, Z., Shi, G., Wu, F., Zeng, W.: Adaptive nonlocal sparse representation for dual-camera compressive hyperspectral imaging. IEEE Trans. Pattern Anal. Mach. Intell. **39**(10), 2104–2111 (2017)

52. Wang, L., Sun, C., Fu, Y., Kim, M.H., Huang, H.: Hyperspectral image reconstruction using a deep spatial-spectral prior. In: The IEEE Conference on Computer Vision and Pattern Recognition (CVPR), June 2019

53. Wang, L., Xiong, Z., Gao, D., Shi, G., Wu, F.: Dual-camera design for coded aperture snapshot spectral imaging. Appl. Opt. **54**(4), 848–858 (2015)

54. Wang, L., Xiong, Z., Gao, D., Shi, G., Zeng, W., Wu, F.: High-speed hyperspectral video acquisition with a dual-camera architecture. In: 2015 IEEE Conference on Computer Vision and Pattern Recognition (CVPR), pp. 4942–4950, June 2015

55. Wang, L., Xiong, Z., Huang, H., Shi, G., Wu, F., Zeng, W.: High-speed hyperspectral video acquisition by combining Nyquist and compressive sampling. IEEE Trans. Pattern Anal. Mach. Intell. **41**, 857–870 (2018)

56. Wang, L., Zhang, T., Fu, Y., Huang, H.: HyperReconNet: joint coded aperture optimization and image reconstruction for compressive hyperspectral imaging. IEEE Trans. Image Process. **28**(5), 2257–2270 (2019)
57. Wang, Z., Bovik, A.C., Sheikh, H.R., Simoncelli, E.P., et al.: Image quality assessment: from error visibility to structural similarity. IEEE Trans. Image Process. **13**(4), 600–612 (2004)
58. Wu, Y., Mirza, I.O., Arce, G.R., Prather, D.W.: Development of a digital-micromirror-device-based multishot snapshot spectral imaging system. Opt. Lett. **36**(14), 2692–2694 (2011)
59. Xie, J., Xu, L., Chen, E.: Image denoising and inpainting with deep neural networks. In: Pereira, F., Burges, C.J.C., Bottou, L., Weinberger, K.Q. (eds.) Advances in Neural Information Processing Systems 25, pp. 341–349. Curran Associates, Inc. (2012)
60. Yang, J., et al.: Compressive sensing by learning a Gaussian mixture model from measurements. IEEE Trans. Image Process. **24**(1), 106–119 (2015)
61. Yang, P., Kong, L., Liu, X., Yuan, X., Chen, G.: Shearlet enhanced snapshot compressive imaging. IEEE Trans. Image Process. **29**, 6466–6481 (2020)
62. Yasuma, F., Mitsunaga, T., Iso, D., Nayar, S.K.: Generalized assorted pixel camera: postcapture control of resolution, dynamic range, and spectrum, vol. 19, pp. 2241–2253. IEEE (2010)
63. Yuan, X., Sun, Y., Pang, S.: Efficient patch-based approach for compressive depth imaging. Appl. Opt. **56**(10), 2697–2704 (2017)
64. Yuan, X.: Compressive dynamic range imaging via Bayesian shrinkage dictionary learning. Opt. Eng. **55**(12), 123110 (2016)
65. Yuan, X.: Generalized alternating projection based total variation minimization for compressive sensing. In: 2016 IEEE International Conference on Image Processing (ICIP), pp. 2539–2543, September 2016
66. Yuan, X., Brady, D., Katsaggelos, A.K.: Snapshot compressive imaging: theory, algorithms and applications. IEEE Signal Process. Mag. (2020)
67. Yuan, X., Liao, X., Llull, P., Brady, D., Carin, L.: Efficient patch-based approach for compressive depth imaging. Appl. Opt. **55**(27), 7556–7564 (2016)
68. Yuan, X., et al.: Low-cost compressive sensing for color video and depth. In: IEEE Conference on Computer Vision and Pattern Recognition (CVPR), pp. 3318–3325 (2014)
69. Yuan, X., Pang, S.: Structured illumination temporal compressive microscopy. Biomed. Opt. Express **7**, 746–758 (2016)
70. Yuan, X., Tsai, T.H., Zhu, R., Llull, P., Brady, D., Carin, L.: Compressive hyperspectral imaging with side information. IEEE J. Sel. Top. Signal Process. **9**(6), 964–976 (2015)
71. Zhang, K., Zuo, W., Chen, Y., Meng, D., Zhang, L.: Beyond a gaussian denoiser: Residual learning of deep CNN for image denoising. IEEE Trans. Image Process. **26**(7), 3142–3155 (2017)
72. Zhao, Y., Guo, H., Ma, Z., Cao, X., Yue, T., Hu, X.: Hyperspectral imaging with random printed mask. In: Proceedings of the IEEE Conference on Computer Vision and Pattern Recognition, pp. 10149–10157 (2019)

Know Your Surroundings: Exploiting Scene Information for Object Tracking

Goutam Bhat$^{(\boxtimes)}$, Martin Danelljan, Luc Van Gool, and Radu Timofte

CVL, ETH Zürich, Zürich, Switzerland
goutam.bhat@vision.ee.ethz.ch

Abstract. Current state-of-the-art trackers rely only on a target appearance model in order to localize the object in each frame. Such approaches are however prone to fail in case of e.g. fast appearance changes or presence of distractor objects, where a target appearance model alone is insufficient for robust tracking. Having the knowledge about the presence and locations of other objects in the surrounding scene can be highly beneficial in such cases. This scene information can be propagated through the sequence and used to, for instance, explicitly avoid distractor objects and eliminate target candidate regions.

In this work, we propose a novel tracking architecture which can utilize scene information for tracking. Our tracker represents such information as dense localized state vectors, which can encode, for example, if a local region is target, background, or distractor. These state vectors are propagated through the sequence and combined with the appearance model output to localize the target. Our network is learned to effectively utilize the scene information by directly maximizing tracking performance on video segments. The proposed approach sets a new state-of-the-art on 3 tracking benchmarks, achieving an AO score of 63.6% on the recent GOT-10k dataset.

1 Introduction

Generic object tracking is one of the fundamental computer vision problems with numerous applications. The task is to estimate the state of a target object in each frame of a video sequence, given only its initial appearance. Most current approaches [3, 8, 16, 25, 30, 33, 36] tackle the problem by learning an appearance model of the target in the initial frame. This model is then applied in subsequent frames to localize the target by distinguishing its appearance from the surrounding background. While achieving impressive tracking performance [24, 28], these approaches rely *only* on the appearance model, and do not utilize any other information contained in the scene.

Electronic supplementary material The online version of this chapter (https://doi.org/10.1007/978-3-030-58592-1_13) contains supplementary material, which is available to authorized users.

A. Vedaldi et al. (Eds.): ECCV 2020, LNCS 12368, pp. 205–221, 2020.
https://doi.org/10.1007/978-3-030-58592-1_13

In contrast, humans exploit a much richer set of cues when tracking an object. We have a holistic view of the scene and take into consideration not only the target object, but also other objects in the scene. Such information is helpful when localizing the target, e.g. in case of cluttered scenes with distractor objects, or when the target undergoes fast appearance change. Consider the example in Fig. 1. Given only the initial target appearance, it is hard to confidently locate the target due to the presence of distractor objects. However, if we also utilize the previous frame, we can easily detect the presence of distractors. This knowledge can then be propagated to the next frame in order to reliably localize the target. While existing approaches update the appearance model with previously tracked frames, such a strategy by itself cannot capture the locations and characteristics of the other objects in the scene.

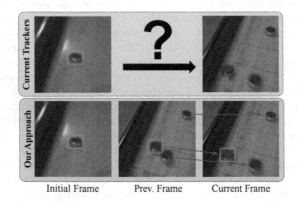

Fig. 1. Current approaches (top) only utilize an appearance model to track the target object. However, such a strategy fails in the above example. Here, the presence of distractor objects makes it virtually impossible to localize the target based on appearance only, even if the appearance model is continuously updated using previous frames. In contrast, our approach (bottom) is also aware of other objects in the scene. This scene information is propagated through the sequence by computing a dense correspondence (red arrows) between consecutive frames. The propagated scene knowledge greatly simplifies the target localization problem, allowing us to reliably track the target. (Color figure online)

In this work, we aim to go beyond the conventional frame-by-frame detection-based tracking. We propose a novel tracking architecture which can propagate valuable scene information through the sequence. This information is used to achieve an improved *scene aware* target prediction in each frame. The scene information is represented using a dense set of localized state vectors. These state vectors encode valuable information about the local region, e.g. whether the region corresponds to the target, background or a distractor object. As the regions move through a sequence, we propagate the corresponding state vectors by utilizing dense correspondence maps between frames. Consequently, our tracker is 'aware' of every object in the scene and can use this information in order to e.g. avoid distractor objects. This scene knowledge, along with the target appearance model, is used to predict the target state in each frame using a learned predictor module. The scene information captured by the state representation is then updated using a recurrent neural network module.

We perform comprehensive experiments on six challenging benchmarks: VOT-2018 [28], GOT-10k [24], TrackingNet [35], OTB-100 [45], NFS [14], and LaSOT [13]. Our approach achieves state-of-the-art results on five datasets. On the challenging GOT-10k dataset, our tracker obtains an average overlap (AO) score of 63.6%, outperforming the previous best approach by 2.5%. We also provide an ablation study analyzing the impact of key components in our tracker.

2   Related Work

Most tracking approaches tackle the problem by learning an appearance model of the target in the first frame. A popular method to learn the target appearance model is the discriminative correlation filters (DCF) [5,9,10,22,26,32]. These approaches exploit the convolution theorem to efficiently train a classifier in the Fourier domain using the circular shifts of the input image as training data. Another approach is to train or fine-tune a few layers of a deep neural network in the first frame to perform target-background classification [3,8,36,39]. MDNet [36] fine-tunes three fully-connected layers online, while DiMP [3] employs a meta-learning formulation to predict the weights of the classification layer. In recent years, Siamese networks have received significant attention [2,19,30,31,44]. These approaches address the tracking problem by learning a similarity measure, which is then used to locate the target.

The discriminative approaches discussed above exploit the background information in the scene to learn the target appearance model. A number of attempts have also been made to integrate background information into the appearance model in Siamese trackers [29,51,53]. However, in many cases, the distractor object is indistinguishable from a previous target appearance. Thus, a single target model is insufficient to achieve robust tracking in such cases. Further, in case of fast motion, it is hard to adapt the target model quickly to new distractors. In contrast to these works, our approach explicitly encodes localized information about different image regions and propagates this information through the sequence via dense matching. More related to our work, [46] aims to exploit the locations of distractors in the scene. However, it employs hand-crafted rules to classify image regions into background and target candidates independently in each frame. In contrast, we present a fully *learnable* solution, where the encoding of image regions is learned and propagated between frames. Further, our final prediction is obtained by combining the explicit background representation with the appearance model output.

In addition to appearance cues, a few approaches have investigated the use of optical flow information for tracking. Gladh et al. [17] utilize deep motion features extracted from optical flow images to complement the appearance features when constructing the target model. Zhu et al. [54] use optical flow to warp the feature maps from the previous frames to a reference frame and aggregate them in order to learn the target appearance model. However, both these approaches utilize optical flow to only improve the robustness of the target model. In contrast, we explicitly use dense motion information to propagate information about background objects and structures in order to complement the target model.

Some works have also investigated using recurrent neural networks (RNN) for object tracking. Gan et al. [15] use a RNN to directly regress the target location using image features and previous target locations. Ning et al. [37] utilize the YOLO [38] detector to generate initial object proposals. These proposals, along with the image features, are passed through an LSTM [23] to obtain the target box. Yang et al. [49,50] use an LSTM to update the target model to account for changes in target appearance through a sequence.

3   Proposed Method

We develop a novel tracking architecture capable of exploiting scene information to improve tracking performance. While current state-of-the-art methods [3,8,30] rely only on the target appearance model to process every frame independently, our approach also propagates information about the scene from previous frames. This provides rich cues about the environment, e.g. the location of distractor objects, which greatly aids the localization of the target.

A visual overview of our tracking architecture is provided in Fig. 2. Our tracker internally tracks *all* regions in the scene, and propagates any information about them that helps localization of the target. This is achieved by maintaining a state vector for every region in the target neighborhood. The state vector can, for instance, encode whether a particular patch corresponds to the target, background, or a distractor object that is likely to fool the target appearance model. As the objects move through a sequence, the state vectors are propagated accordingly by estimating a dense correspondence between consecutive frames. The propagated state vectors are then fused with the target appearance model output in order to predict the final target confidence values used for localization. Lastly, the outputs of the predictor and the target model are used to update the state vectors using a convolutional gated recurrent unit (ConvGRU) [1].

3.1   Tracking with Scene Information

Our tracker predictions are based on two cues: (i) appearance in the current frame and (ii) scene information propagated over time. The appearance model τ aims to distinguish the target object from the background. By taking the deep feature map $x_t \in \mathbb{R}^{W \times H \times D}$ extracted from frame t as input, the appearance model τ predicts a score map $s_t = \tau(x_t) \in \mathbb{R}^{W \times H}$. Here, the score $s_t(\mathbf{r})$ at every spatial location $\mathbf{r} \in \Omega := \{0, \ldots, W-1\} \times \{0, \ldots, H-1\}$ denotes the likelihood of that location being the target center.

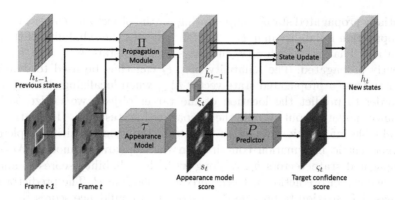

Fig. 2. An overview of our tracking architecture. In addition to using a target appearance model τ, our tracker also exploits propagated scene information in order to track the target. The information about each image region is encoded within localized states h. Given the states h_{t-1} from the previous frame, the propagation module Π maps these states from the previous frame to the current frame locations. These propagated states $\hat{h}_{t-1}$, along with the propagation reliability ξ_t and appearance model score s_t are used by the predictor P to output the final target confidence scores ς_t. The state update module Φ then uses the current frame predictions to provide the new states h_t.

The target model has the ability to recover from occlusions and provides long-term robustness. However, it is oblivious to the contents of the surrounding scene. In order to extract such information, our tracker maintains a state vector for every region in the target neighborhood. Concretely, for every spatial location $\mathbf{r} \in \Omega$ in the deep feature representation x_t, we maintain a S-dimensional state vector $h^{\mathbf{r}}$ for that cell location such that $h \in \mathbb{R}^{W \times H \times S}$. The state vectors contain information about the cell which is beneficial for single target tracking. For example, it can encode whether a particular cell corresponds to the target, background, or is in fact a distractor that looks similar to the target. Note that we do not explicitly enforce any such encoding, but let h be a generic representation whose encoding is trained end-to-end by minimizing a tracking loss.

The state vectors are initialized in the first frame using a small network Υ which takes the first frame target annotation B_0 as input. The network generates a single-channel label map specifying the target location. This is passed through two convolutional layers to obtain the initial state vectors $h_0 = \Upsilon(b_0)$. The state vectors contain localized information specific to their corresponding image regions. Thus, as the objects move through a sequence, we propagate their state vectors accordingly. Given a new frame t, we transform the states h_{t-1} from the previous frame locations to the current frame locations. This is performed by our state propagation module Π,

$$(\hat{h}_{t-1}, \xi_t) = \Pi(x_t, x_{t-1}, h_{t-1}) \tag{1}$$

Here, $x_t \in \mathbb{R}^{W \times H \times D}$ and $x_{t-1} \in \mathbb{R}^{W \times H \times D}$ are the deep feature representations from the current and previous frames, respectively. The output $\hat{h}_{t-1}$ represents the spatially propagated state, compensating for the object and camera motions. The propagation reliability map $\xi_t \in \mathbb{R}^{W \times H}$ indicates the reliability of the state propagation. That is, a high $\xi_t(\mathbf{r})$ indicates that the state $\hat{h}_{t-1}^{\mathbf{r}}$ at $\mathbf{r}$ has been confidently propagated. The reliability map ξ_t can thus be used to determine whether to trust a propagated state vector $\hat{h}_{t-1}^{\mathbf{r}}$ when localizing the target.

In order to predict the location of the target object, we utilize *both* the appearance model output s_t and the propagated states $\hat{h}_{t-1}$. The latter captures valuable information about all objects in the scene, which complements the target-centric information contained in the appearance model. We input the propagated state vectors $\hat{h}_{t-1}$, along with the reliability scores ξ_t and the appearance model prediction s_t to the predictor module P. The predictor combines these information to provide the fused target confidence scores ς_t,

$$\varsigma_t = P(\hat{h}_{t-1}, \xi_t, s_t) \tag{2}$$

The target is then localized in frame t by selecting the location $\mathbf{r}^*$ with the highest score: $\mathbf{r}^* = \arg\max_{\mathbf{r} \in \Omega} \varsigma_t$. Finally, we use the fused confidence scores ς_t along with the appearance model output s_t to update the state vectors,

$$h_t = \Phi(\hat{h}_{t-1}, \varsigma_t, s_t) \tag{3}$$

The recurrent state update module Φ can use the current frame information from the score maps to e.g. reset an incorrect state vector $\hat{h}_{t-1}^{\mathbf{r}}$, or flag a newly entered object as a distractor. These updated state vectors h_t are then used to track the object in the next frame. Our tracking procedure is detailed in Algorithm 1.

3.2   State Propagation

The state vectors contain localized information for every region in the target neighborhood. As these regions move through a sequence due to e.g. object or camera motion, we need to propagate their states accordingly, in order to compensate for their motions. This is done by our state propagation module Π. The inputs to this module are the deep feature maps x_{t-1} and x_t extracted from the previous and current frames, respectively. Note that the deep features x are not required to be the same as the ones as used for the target model. However, we assume that both feature maps have the same spatial resolution $W \times H$.

In order to propagate the states from the previous frame to the current frame locations, we first compute a dense correspondence between the two frames. We represent this correspondence as a probability distribution p, where $p(\mathbf{r}'|\mathbf{r})$ is the probability that location $\mathbf{r} \in \Omega$ in the current frame originated from $\mathbf{r}' \in \Omega$ in the previous frame. The dense correspondence is estimated by constructing a 4D cost volume $\mathbf{CV} \in \mathbb{R}^{W \times H \times W \times H}$, as is commonly done in optical flow approaches [12,42,47]. The cost volume contains a matching cost between every image location pair from the previous and current frame. The element $\mathbf{CV}(\mathbf{r}', \mathbf{r})$ in the cost volume is obtained by computing the correlation between 3×3 windows centered at $\mathbf{r}'$ in the previous frame features x_{t-1} and $\mathbf{r}$ in the current

---

Algorithm 1. Our tracking loop

Input: Image features $\{x_t\}_{t=0}^N$, initial annotation b_0, appearance model τ

1: $h_0 \leftarrow \Upsilon(b_0)$	# *Initialize states*
2: **for** $i = 1, \ldots, N$ **do**	# *For every frame*
3: $\quad s_t \leftarrow \tau(x_t)$	# *Apply appearance model*
4: $\quad (\hat{h}_{t-1}, \xi_t) \leftarrow \Pi(x_t, x_{t-1}, h_{t-1})$	# *Propagate states*
5: $\quad \varsigma_t \leftarrow P(\hat{h}_{t-1}, \xi_t, s_t)$	# *Predict target confidence scores*
6: $\quad h_t \leftarrow \Phi(\hat{h}_{t-1}, \varsigma_t, s_t)$	# *Update states*

---

frame features x_t. For computational efficiency, we only construct a partial cost volume by assuming a maximal displacement of $d_{\max}$ for every feature cell.

We process the cost volume through a network module to obtain robust dense correspondences. We pass the cost volume slice $\mathbf{CV_{r'}(r)} \in \mathbb{R}^{W \times H}$ for every cell $\mathbf{r}'$ in the previous frame, through two convolutional blocks in order to obtain processed matching costs $\phi(\mathbf{r}', \mathbf{r})$. Next, we take the softmax of this output over the current frame locations to get an initial correspondence $\phi'(\mathbf{r}', \mathbf{r}) = \frac{\exp(\phi(\mathbf{r}', \mathbf{r}))}{\sum_{\mathbf{r}'' \in \Omega} \exp(\phi(\mathbf{r}', \mathbf{r}''))}$. The softmax operation aggregates information over the current frame dimension and provides a soft association of locations between the two frames. In order to also integrate information over the previous frame locations, we pass ϕ' through two more convolutional blocks and take softmax over the previous frame locations. This provides the required probability distribution $p(\mathbf{r}'|\mathbf{r})$ at each current frame location $\mathbf{r}$.

The estimated correspondence $p(\mathbf{r}'|\mathbf{r})$ between the frames can now be used to determine the propagated state vector $\hat{h}_{t-1}^{\mathbf{r}}$ at a current frame location $\mathbf{r}$ by evaluating the following expectation over the previous frame state vectors.

$$\hat{h}_{t-1}^{\mathbf{r}} = \sum_{\mathbf{r}' \in \Omega} h_{t-1}^{\mathbf{r}'} p(\mathbf{r}'|\mathbf{r}). \tag{4}$$

When using the propagated state vectors $\hat{h}_{t-1}$ for target localization, it is also helpful to know if a particular state vector is valid i.e. if it has been correctly propagated from the previous frame. We can estimate this reliability $\xi_t^{\mathbf{r}}$ at each location $\mathbf{r}$ using the correspondence probability distribution $p(\mathbf{r}'|\mathbf{r})$ for that location. A single mode in $p(\mathbf{r}'|\mathbf{r})$ indicates that we are confident about the source of the location $\mathbf{r}$ in the previous frame. A uniformly distributed $p(\mathbf{r}'|\mathbf{r})$ on the other hand implies uncertainty. In such a scenario, the expectation (4) reduces to a simple average over the previous frame state vectors $h_{t-1}^{\mathbf{r}'}$, leading to an unreliable $\hat{h}_{t-1}^{\mathbf{r}}$. Thus, we use the negation of the shannon entropy of the distribution $p(\mathbf{r}'|\mathbf{r})$ to obtain the reliability score $\xi_t^{\mathbf{r}}$ for state $\hat{h}_{t-1}^{\mathbf{r}}$,

$$\xi_t^{\mathbf{r}} = \sum_{\mathbf{r}' \in \Omega} p(\mathbf{r}'|\mathbf{r}) \log(p(\mathbf{r}'|\mathbf{r})) \tag{5}$$

The reliability $\xi_t^{\mathbf{r}}$ is then be used to determine whether to trust the state $\hat{h}_{t-1}^{\mathbf{r}}$ when predicting the final target confidence scores.

3.3   Target Confidence Score Prediction

In this section, we describe our predictor module P which determines the target location in the current frame. We utilize both the appearance model output s_t and the scene information encoded by $\hat{h}_{t-1}$ in order to localize the target. The appearance model score $s_t^{\mathbf{r}}$ indicates whether a location $\mathbf{r}$ is target or background, based on the appearance in the current frame only. The state vector $\hat{h}_{t-1}^{\mathbf{r}}$ on the other hand contains past information for *every* location $\mathbf{r}$. It can, for instance, encode whether the cell $\mathbf{r}$ was classified as target or background in the previous frame, how certain was the tracker prediction for that location, and so on. The corresponding reliability score $\xi_t^{\mathbf{r}}$ further indicates if the state vector $\hat{h}_{t-1}^{\mathbf{r}}$ is reliable or not. This can be used to determine how much weight to give to the state vector information when determining the target location.

The predictor module P is trained to effectively combine the information from s_t, $\hat{h}_{t-1}$, and ξ_t to output the final target confidence score $\varsigma_t \in \mathbb{R}^{W \times H}$. We concatenate the appearance model output s_t, the propagated state vectors $\hat{h}_{t-1}$, and the state reliability scores ξ_t along the channel dimension, and pass the resulting tensor through two convolutional blocks. The output is then mapped to the range $[0, 1]$ by passing it through a sigmoid layer to obtain the intermediate scores $\hat{\varsigma}_t$. While it is possible to use this score directly, it is not reliable in case occlusions. This is because the state vectors corresponding to the target can leak into the occluding object, especially when two objects cross each other slowly, leading to incorrect prediction. In order to handle this, we pass $\hat{\varsigma}_t$ through another layer which masks the regions from the score map $\hat{\varsigma}_t$ where the appearance model score s_t is less than a threshold μ. Thus, we let the appearance model override the predictor output in case of occlusions. The final score map ς_t is thus obtained as $\varsigma_t = \hat{\varsigma}_t \cdot \mathbb{1}_{s_t > \mu}$. Here, $\mathbb{1}_{s_t > \mu}$ is an indicator function which evaluates to 1 when $s_t > \mu$ and is 0 otherwise and $\cdot$ denotes elementwise product. Note that the masking operation is differentiable and is implemented inside the network.

3.4   State Update

While the state propagation described in Sect. 3.2 maps the state to the new frame, it does not update it with new information about the scene. This is accomplished by a recurrent neural network module, which evolves the state in each time step. As tracking information about the scene, we input the scores s_t and ς_t obtained from the appearance model τ and the predictor module P, respectively. The update module can thus e.g. mark a new distractor object which entered the scene or correct corrupted states which have been incorrectly propagated. This state update is performed by the recurrent module Φ (Eq. 3).

The update module Φ contains a convolutional gated recurrent unit (ConvGRU) [1,6]. We concatenate the scores ς_t and s_t along with their maximum values in order to obtain the input $f_t \in \mathbb{R}^{W \times H \times 4}$ to the ConvGRU. The propagated states from the previous frame $\hat{h}_{t-1}$ are treated as the hidden states of the ConvGRU from the previous time step. The ConvGRU then updates the previous states using the current frame observation f_t to provide the new states h_t. A visualization of the representations used by our tracker is shown in Fig. 3.

3.5   Target Appearance Model

Our approach can be employed with any target appearance model. In this work, we use the DiMP tracker [3] as our target model component, due to its strong performance. DiMP is an end-to-end trainable tracking architecture that predicts the appearance model τ_w, parametrized by the weights w of a single convolutional layer. The network integrates an optimization module that minimizes the following discriminative learning loss,

$$L(w) = \frac{1}{|S_{\text{train}}|} \sum_{(x,c) \in S_{\text{train}}} \|r(\tau_w(x), c)\|^2 + \|\lambda w\|^2 . \tag{6}$$

Here, λ is the regularization parameter. The training set $S_{\text{train}} = \{(x_j, c_j)\}_{j=1}^n$ consists of deep feature maps x_j extracted from the training images, and the corresponding target annotations c_j. The residual function $r(s, c)$ computes the error between the tracker prediction $s = \tau_w(x)$ and the groundtruth. We refer to [3] for more details about the DiMP tracker.

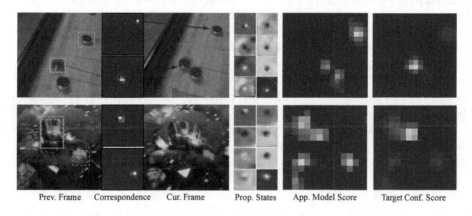

Prev. Frame   Correspondence   Cur. Frame   Prop. States   App. Model Score   Target Conf. Score

Fig. 3. Visualization of intermediate representations used for tracking on two example sequences. The green box in the previous frame (first column) denotes the target to be tracked. For every location in the current frame (third column), we plot the estimated correspondence with the marked region in the previous frame (second column). The states propagated to the current frame using the estimated correspondence are plotted channel-wise in the fourth column. The appearance model score (fifth column) fails to correctly localize the target in both cases due to the presence of distractors. In contrast, our approach can correctly handle these challenging scenarios and provides robust target confidence scores (last column) by exploiting the propagated scene information. (Color figure online)

3.6   Offline Training

In order to train our architecture, it is important to simulate the tracking scenario. This is needed to ensure that the network can learn to effectively propagate the scene information over time and determine how to best fuse it with the appearance model output. Thus, we train our network using video sequences. We first sample a set of N_{train} frames from a video, which we use to construct the appearance model τ. We then sample a sub-sequence $V = \{(I_t, b_t)\}_{t=0}^{N_{\text{seq}}-1}$ consisting of N_{seq} consecutive frames I_t along with their corresponding target annotation b_t. We apply our network on this sequence data, as it would be during tracking. We first obtain the initial state $h_0 = \Upsilon(b_0)$ using the state initializer Υ. The states are then propagated to the next frame (Sect. 3.2), used to predict the target scores ς_t (Sect. 3.3), and finally updated using the predicted scores (Sect. 3.4). This procedure is repeated until the end of the sequence and the training loss is computed by evaluating the tracker performance over the whole sequence.

In order to obtain the tracking loss L, we first compute the prediction error L_t^{pred} for every frame t using the standard least-squares loss,

$$L_t^{\text{pred}} = \|\varsigma_t - z_t\|^2 \tag{7}$$

Here, z_t is a label function, which we set to a Gaussian centered at the target. We also compute a prediction error $L_t^{\text{pred, raw}}$ using the raw score map $\hat{\varsigma}_t$ predicted by P in order to obtain extra training supervision. To aid the learning of the state vectors and the propagation module Π, we add an additional auxiliary task. We use a small network head to predict whether a state vector h_{t-1}^{r} corresponds to the target or background. This prediction is penalized using a binary cross entry loss to obtain L_t^{state}. The network head is also applied on the propagated state vectors $\hat{h}_{t-1}^{\text{r}}$ to get $L_t^{\text{state, prop}}$. This loss provides a direct supervisory signal to the propagation module Π.

Our final tracking loss L is obtained as the weighted sum of the above individual losses over the whole sequence,

$$L = \frac{1}{N_{\text{seq}} - 1} \sum_{t=1}^{N_{\text{seq}}-1} L_t^{\text{pred}} + \alpha L_t^{\text{pred, raw}} + \beta(L_t^{\text{state}} + L_t^{\text{state, prop}}). \tag{8}$$

The hyper-parameters α and β determine the impact of the different losses. Note that the scores s_t predicted by the appearance model can itself localize the target correctly in a majority of the cases. Thus, there is a risk that the predictor module only learns to rely on the target model scores s_t. To avoid this, we randomly add distractor peaks to the scores s_t during training to encourage the predictor to utilize the scene information encoded by the state vectors.

3.7   Implementation Details

We use a pre-trained DiMP model with ResNet-50 [20] backbone from [7] as our target appearance model. We use the block 4 features from the same backbone network as input to the state propagation module Π. For computational efficiency, our tracker does not process the full input image. Instead, we crop a square region containing the target, with an area 5^2 times that of the target. The cropped search region is resized to 288×288 size, and passed to the network. We use $S = 8$ dimensional state vectors to encode the scene information. The threshold μ in the predictor P is set to 0.05.

We use the training splits of TrackingNet [35], LaSOT [13], and GOT-10k [24] datasets to train our network. Within a sequence, we perturb the target position and scale in every frame in order to avoid learning any motion bias. While our network is end-to-end trainable, we do not fine-tune the weights for the backbone network due to GPU memory constraints. Our network is trained for 40 epochs, with 1500 sub-sequences in each epoch. We use the ADAM [27] optimizer with an initial learning rate of 10^{-2}, which is reduced by a factor of 5 every 20 epochs. We use $N_{\text{train}} = 3$ frames to construct the appearance model while the sub-sequence length is set to $N_{\text{seq}} = 50$. The loss weights are set to $\alpha = \beta = 0.1$.

During online tracking, we use a simple heuristic to determine target loss. In case the fused confidence score ς_t peak is smaller than a threshold (0.05), we infer that the target is lost and do not update the state vectors in this case. We impose a prior on the target motion by applying a window function on the appearance model prediction s_t input to P, as well as the output target confidence score ς_t. We also handle any possible drift in the target confidence scores. In case the appearance model scores s_t and target confidence score ς_t only have small offset in their peaks, we use the appearance model score to determine the target location as it is more resistant to drift. After determining the target location, we use the bounding box estimation branch in DiMP to obtain the target box.

4   Experiments

We evaluate our proposed tracking architecture on six tracking benchmarks: VOT2018 [28], GOT-10k [24], TrackingNet [35], OTB-100 [45], NFS [14], and LaSOT [13]. Detailed results are provided in the n material. Our tracker operates at around 20 FPS on a single Nvidia RTX 2080 GPU.

4.1   Ablation Study

We conduct an ablation study to analyze the impact of each component in our tracking architecture. We perform experiments on the combined NFS [14] and OTB-100 [45] datasets consisting of 200 challenging videos. The trackers are evaluated using the overlap precision (OP) metric. The overlap precision OP_T denotes the percentage of frames where the intersection-over-union (IoU) overlap between the tracker prediction and the groundtruth box is higher than a threshold T. The OP scores over a range of thresholds $[0, 1]$ are averaged to obtain the area-under-the-curve (AUC) score. We report the AUC and $OP_{0.5}$ scores for each tracker. Due to the stochastic nature of our appearance model, all results are reported as the average over 5 runs. Unless stated otherwise, we

Table 1. Impact of each component in our tracking architecture on the combined NFS and OTB-100 datasets. Compared to using only the appearance model, our approach integrating scene information, provides a significant 1.3% improvement in AUC score.

	Ours	Only appearance model τ	No state propagation Π	No propagation reliability ξ_t	No appearance model τ
AUC(%)	**66.4**	65.1	64.9	66.1	49.2
OP$_{0.5}$	**83.5**	81.9	81.2	82.9	60.1

use the same training procedure and settings mentioned in Sects. 3.6 and 3.7, respectively, to train all trackers evaluated in this section.

Impact of Scene Information: In order to study the impact of integrating scene information for tracking, we compare our approach with a tracker only employing target appearance model τ. This version is equivalent to the standard DiMP-50 [3], which achieves state-of-the-art results on multiple tracking benchmarks. The results are reported in Table 1. Compared to using only the appearance model, our approach exploiting scene information provides an improvement of 1.3% in AUC score. These results clearly demonstrate that scene knowledge contains complementary information that benefits tracking performance, even when integrated with a strong appearance model.

Impact of State Propagation: Here, we analyze the impact of state propagation module (Sect. 3.2), which maps the localized states between frames by generating dense correspondences. This is performed by replacing the propagation module Π in (1) and (4) with an identity mapping $\hat{h}_{t-1} = h_{t-1}$. That is, the states are no longer explicitly tracked by computing correspondences between frames. The results for this experiment are shown in Table 1. Interestingly, the approach without state propagation performs slightly worse (0.2% in AUC) than the network using only the appearance model. This shows that state propagation between frames is critical in order to exploit the localized scene information.

Impact of Propagation Reliability: Here, we study the impact of the propagation reliability score ξ_t for confidence score prediction. We compare our approach with a baseline tracker which does not utilize ξ_t. The results indicate that using reliability score ξ_t is beneficial, leading to a +0.3% AUC improvement.

Impact of Appearance Model: Our architecture utilizes the propagated scene information to *complement* the frame-by-frame prediction performed by the target appearance model. By design, our tracker relies on the appearance model to provide long-term robustness in case of e.g. occlusions, and thus is not suited to be used without it. However, for completeness, we evaluate a version of our tracker which does not utilize any appearance model. That is, we only use the propagated states $\hat{h}_{t-1}$, and the reliability score ξ_t in order to track the target. As expected, not using an appearance model substantially deteriorates the performance by over 17% in AUC score.

4.2   State-of-the-Art Comparison

In this section, we compare our proposed tracker with state-of-the-art approaches on six tracking benchmarks.

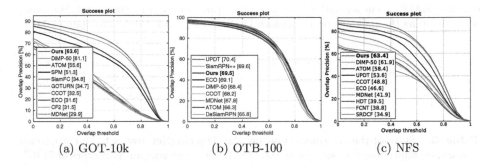

(a) GOT-10k   (b) OTB-100   (c) NFS

Fig. 4. Success plots on GOT-10k (a), OTB-100 (b) and NFS (c). The AUC scores are shown in legend. Our approach obtains the best results on both GOT-10k and NFS datasets, outperforming the previous best method by 2.5% and 1.6% AUC, respectively.

VOT2018 [28]: We evaluate our approach on the VOT2018 dataset consisting of 60 videos. The trackers are compared using the measures robustness and accuracy. Robustness indicates the number of tracking failures, while accuracy denotes the average overlap between tracker prediction and the ground-truth box. Both these measures are combined into a single expected average overlap (EAO) score. Results are shown in Table 2. Note that all top ranked approaches on VOT2018 utilize only a target appearance model for tracking. In contrast, our approach also exploits explicit knowledge about other objects in the scene. In terms of the overall EAO score, our approach outperforms the previous best method DiMP-50 with a large margin, achieving a relative gain of 5.0% in EAO.

GOT10k [24]: This is a recently introduced large scale dataset consisting of over 10,000 videos. In contrast to other datasets, trackers are restricted to use only the train split of the dataset in order to train networks, i.e. use of external training data is forbidden. Accordingly, we train our network using only the GOT10k train split. The success plots over all the 180 videos from the test split are shown in Fig. 4a. Among previous methods, the appearance model used by our tracker, namely DiMP-50, obtains the best results. Our approach, integrating scene information for tracking, significantly outperforms DiMP-50, setting a new state-of-the-art on this datatset. Our tracker achieves an AO (average overlap) score of 63.6, a relative improvement of 4.1% over the previous best method. These results clearly show the benefits of exploiting scene knowledge for tracking.

TrackingNet [35]: The large scale TrackingNet dataset consists of over 30,000 videos sampled from YouTube. We report results on the test split, consisting of 511 videos. The results in terms of normalized precision and success are shown in

Table 3. The baseline approach DiMP-50 already achieves the best results with an AUC of 74.0. Our approach achieves a similar performance to the baseline, showing that it can generalize well to such real world videos.

Table 2. State-of-the-art comparison on the VOT2018 in terms of expected average overlap (EAO), accuracy and robustness. Our approach obtains the best EAO score, outperforming the previous best approach DiMP-50 with a EAO relative gain of 5.0%.

	DRT [41]	RCO [28]	UPDT [4]	DaSiam- RPN [53]	MFT [28]	LADCF [48]	ATOM [8]	SiamRPN++ [30]	DiMP-50 [3]	Ours
EAO	0.356	0.376	0.378	0.383	0.385	0.389	0.401	0.414	**0.440**	*0.462*
Robustness	0.201	0.155	0.184	0.276	*0.140*	0.159	0.204	0.234	0.153	**0.143**
Accuracy	0.519	0.507	0.536	0.586	0.505	0.503	0.590	**0.600**	0.597	*0.609*

Table 3. State-of-the-art comparison on the TrackingNet test set. Our approach performs similarly to previous best method DiMP-50, achieving an AUC score of 74.0%.

	ECO [9]	SiamFC [2]	CFNet [43]	MDNet [36]	UPDT [4]	DaSiam-RPN [53]	ATOM[8]	SiamRPN++ [30]	DiMP-50 [3]	Ours
Norm. Prec. (%)	61.8	66.6	65.4	70.5	70.2	73.3	77.1	**80.0**	*80.1*	**80.0**
Success (AUC) (%)	55.4	57.1	57.8	60.6	61.1	63.8	70.3	**73.3**	*74.0*	*74.0*

Table 4. State-of-the-art comparison on the LaSOT test set in terms of normalized precision and success. Our approach obtains competitive results with an AUC of 55.4%.

	ECO [9]	DSiam[18]	StructSiam [52]	SiamFC [2]	VITAL [40]	MDNet [36]	SiamRPN++ [30]	ATOM [8]	DiMP-50 [3]	Ours
Norm. Prec. (%)	33.8	40.5	41.8	42.0	45.3	46.0	56.9	57.6	*65.0*	**63.3**
Success (AUC) (%)	32.4	33.3	33.5	33.6	39.0	39.7	49.6	51.5	*56.9*	**55.4**

OTB-100 [45]: Figure 4b shows the success plots over all the 100 videos. Discriminative correlation filter based UPDT [4] tracker achieves the best results with an AUC score of 70.4. Our approach obtains results comparable with the state-of-the-art, while outperforming the baseline DiMP-50 by over 1% in AUC.

NFS [14]: The need for speed dataset consists of 100 challenging videos captured using a high frame rate (240 FPS) camera. We evaluate our approach on the downsampled 30 FPS version of this dataset. The success plots over all the 100 videos are shown in Fig. 4c. Among previous methods, our appearance model DiMP-50 obtains the best results. Our approach significantly outperforms DiMP-50 with a relative gain of 2.4%, achieving 63.4% AUC score.

LaSOT [13]: While our architecture is designed for short-term tracking, we evaluate on the long-term tracking dataset LaSOT for completeness. The results over 280 videos from the test split are shown in Table 4. DiMP-50 achieves the best results with an AUC score of 56.9, while our approach achieves an AUC score of 55.4. The decrease in performance compared to DiMP-50 is attributed to the fast update of the state vectors in our approach. While important for short-term tracking, the state updates can compromise the long-term re-detection capability. We believe that improving the long-term tracking performance of our approach is an interesting future work.

5   Conclusions

We propose a novel architecture which can exploit scene information for tracking. Our tracker represents scene information as dense localized state vectors. These state vectors are propagated through the sequence and combined with the appearance model output to localize the target. We evaluate our approach on 6 tracking benchmarks. Our tracker sets a new state-of-the-art on 3 benchmarks, demonstrating the benefits of exploiting scene information for tracking.

Acknowledgments. This work was supported by a Huawei Technologies Oy (Finland) project, the ETH Zürich Fund (OK), an Amazon AWS grant, and an Nvidia hardware grant.

References

1. Ballas, N., Yao, L., Pal, C., Courville, A.C.: Delving deeper into convolutional networks for learning video representations. In: ICLR (2016)
2. Bertinetto, L., Valmadre, J., Henriques, J.F., Vedaldi, A., Torr, P.H.S.: Fully-convolutional siamese networks for object tracking. In: Hua, G., Jégou, H. (eds.) ECCV 2016. LNCS, vol. 9914, pp. 850–865. Springer, Cham (2016). https://doi.org/10.1007/978-3-319-48881-3_56
3. Bhat, G., Danelljan, M., Gool, L.V., Timofte, R.: Learning discriminative model prediction for tracking. In: ICCV (2019)
4. Bhat, G., Johnander, J., Danelljan, M., Khan, F.S., Felsberg, M.: Unveiling the power of deep tracking. In: ECCV (2018)
5. Bolme, D.S., Beveridge, J.R., Draper, B.A., Lui, Y.M.: Visual object tracking using adaptive correlation filters. In: CVPR (2010)
6. Cho, K., et al.: Learning phrase representations using RNN encoder-decoder for statistical machine translation. In: EMNLP (2014)
7. Danelljan, M., Bhat, G.: PyTracking: visual tracking library based on PyTorch (2019). https://github.com/visionml/pytracking. Accessed 1 Aug 2019
8. Danelljan, M., Bhat, G., Khan, F.S., Felsberg, M.: ATOM: accurate tracking by overlap maximization. In: CVPR (2019)
9. Danelljan, M., Bhat, G., Shahbaz Khan, F., Felsberg, M.: ECO: efficient convolution operators for tracking. In: CVPR (2017)
10. Danelljan, M., Häger, G., Shahbaz Khan, F., Felsberg, M.: Learning spatially regularized correlation filters for visual tracking. In: ICCV (2015)
11. Danelljan, M., Robinson, A., Shahbaz Khan, F., Felsberg, M.: Beyond correlation filters: learning continuous convolution operators for visual tracking. In: Leibe, B., Matas, J., Sebe, N., Welling, M. (eds.) ECCV 2016. LNCS, vol. 9909, pp. 472–488. Springer, Cham (2016). https://doi.org/10.1007/978-3-319-46454-1_29
12. Dosovitskiy, A., et al.: FlowNet: learning optical flow with convolutional networks. In: ICCV (2015)
13. Fan, H., et al.: LaSOT: a high-quality benchmark for large-scale single object tracking. CoRR abs/1809.07845 (2018). http://arxiv.org/abs/1809.07845
14. Galoogahi, H.K., Fagg, A., Huang, C., Ramanan, D., Lucey, S.: Need for speed: a benchmark for higher frame rate object tracking. In: ICCV (2017)
15. Gan, Q., Guo, Q., Zhang, Z., Cho, K.: First step toward model-free, anonymous object tracking with recurrent neural networks. ArXiv abs/1511.06425 (2015)

16. Gao, J., Zhang, T., Xu, C.: Graph convolutional tracking. In: CVPR (2019)
17. Gladh, S., Danelljan, M., Khan, F.S., Felsberg, M.: Deep motion features for visual tracking. In: 2016 23rd International Conference on Pattern Recognition (ICPR), pp. 1243–1248 (2016)
18. Guo, Q., Feng, W., Zhou, C., Huang, R., Wan, L., Wang, S.: Learning dynamic siamese network for visual object tracking. In: ICCV (2017)
19. He, A., Luo, C., Tian, X., Zeng, W.: Towards a better match in siamese network based visual object tracker. In: ECCV workshop (2018)
20. He, K., Zhang, X., Ren, S., Sun, J.: Deep residual learning for image recognition. In: Proceedings of the IEEE Conference on Computer Vision and Pattern Recognition, pp. 770–778 (2016)
21. Held, D., Thrun, S., Savarese, S.: Learning to track at 100 FPS with deep regression networks. In: Leibe, B., Matas, J., Sebe, N., Welling, M. (eds.) Computer Vision— ECCV 2016. ECCV 2016. Lecture Notes in Computer Science, vol. 9905. Springer, Cham (2016). https://doi.org/10.1007/978-3-319-46448-0_45
22. Henriques, J.F., Caseiro, R., Martins, P., Batista, J.: High-speed tracking with kernelized correlation filters. TPAMI **37**(3), 583–596 (2015)
23. Hochreiter, S., Schmidhuber, J.: Long short-term memory. Neural Comput. **9**, 1735–1780 (1997)
24. Huang, L., Zhao, X., Huang, K.: GOT-10k: a large high-diversity benchmark for generic object tracking in the wild. arXiv preprint arXiv:1810.11981 (2018)
25. Kenan, D., Dong, W., Huchuan, L., Chong, S., Jianhua, L.: Visual tracking via adaptive spatially-regularized correlation filters. In: CVPR (2019)
26. Kiani Galoogahi, H., Fagg, A., Lucey, S.: Learning background-aware correlation filters for visual tracking. In: ICCV (2017)
27. Kingma, D.P., Ba, J.: Adam: a method for stochastic optimization. In: ICLR (2014)
28. Kristan, M., et al.: The sixth visual object tracking vot2018 challenge results. In: ECCV workshop (2018)
29. Lee, H.: Bilinear siamese networks with background suppression for visual object tracking. In: BMVC (2019)
30. Li, B., Wu, W., Wang, Q., Zhang, F., Xing, J., Yan, J.: SiamRPN++: evolution of siamese visual tracking with very deep networks. In: CVPR (2019)
31. Li, B., Yan, J., Wu, W., Zhu, Z., Hu, X.: High performance visual tracking with siamese region proposal network. In: CVPR (2018)
32. Li, F., Tian, C., Zuo, W., Zhang, L., Yang, M.: Learning spatial-temporal regularized correlation filters for visual tracking. In: CVPR (2018)
33. Li, X., Ma, C., Wu, B., He, Z., Yang, M.H.: Target-aware deep tracking. In: CVPR (2019)
34. Ma, C., Huang, J.B., Yang, X., Yang, M.H.: Hierarchical convolutional features for visual tracking. In: ICCV (2015)
35. Müller, M., Bibi, A., Giancola, S., Al-Subaihi, S., Ghanem, B.: TrackingNet: a large-scale dataset and benchmark for object tracking in the wild. In: ECCV (2018)
36. Nam, H., Han, B.: Learning multi-domain convolutional neural networks for visual tracking. In: CVPR (2016)
37. Ning, G., Zhang, Z., Huang, C., He, Z., Ren, X., Wang, H.: Spatially supervised recurrent convolutional neural networks for visual object tracking. In: 2017 IEEE International Symposium on Circuits and Systems (ISCAS), pp. 1–4 (2016)
38. Redmon, J., Divvala, S., Girshick, R., Farhadi, A.: You only look once: unified, real-time object detection. In: CVPR (2016)
39. Song, Y., Ma, C., Gong, L., Zhang, J., Lau, R., Yang, M.H.: CREST: Convolutional residual learning for visual tracking. In: ICCV (2017)

40. Song, Y., et al.: Vital: visual tracking via adversarial learning. In: 2018 IEEE/CVF Conference on Computer Vision and Pattern Recognition, pp. 8990–8999 (2018)
41. Sun, C., Wang, D., Lu, H., Yang, M.: Correlation tracking via joint discrimination and reliability learning. In: CVPR (2018)
42. Sun, D., Yang, X., Liu, M.Y., Kautz, J.: PWC-Net: CNNs for optical flow using pyramid, warping, and cost volume. In: CVPR (2017)
43. Valmadre, J., Bertinetto, L., Henriques, J.F., Vedaldi, A., Torr, P.H.S.: End-to-end representation learning for correlation filter based tracking. In: CVPR (2017)
44. Wang, Q., Teng, Z., Xing, J., Gao, J., Hu, W., Maybank, S.J.: Learning attentions: residual attentional siamese network for high performance online visual tracking. In: CVPR (2018)
45. Wu, Y., Lim, J., Yang, M.H.: Object tracking benchmark. TPAMI **37**(9), 1834–1848 (2015)
46. Xiao, J., Qiao, L., Stolkin, R., Leonardis, A.: Distractor-supported single target tracking in extremely cluttered scenes. In: Leibe, B., Matas, J., Sebe, N., Welling, M. (eds.) ECCV 2016. LNCS, vol. 9908, pp. 121–136. Springer, Cham (2016). https://doi.org/10.1007/978-3-319-46493-0_8
47. Xu, J., Ranftl, R., Koltun, V.: Accurate optical flow via direct cost volume processing. In: CVPR (2017)
48. Xu, T., Feng, Z., Wu, X., Kittler, J.: Learning adaptive discriminative correlation filters via temporal consistency preserving spatial feature selection for robust visual tracking. CoRR abs/1807.11348 (2018). http://arxiv.org/abs/1807.11348
49. Yang, T., Chan, A.B.: Recurrent filter learning for visual tracking. In: 2017 IEEE International Conference on Computer Vision Workshops (ICCVW), pp. 2010–2019 (2017)
50. Yang, T., Chan, A.B.: Learning dynamic memory networks for object tracking. In: ECCV (2018)
51. Zhang, L., Gonzalez-Garcia, A., Weijer, J.V.D., Danelljan, M., Khan, F.S.: Learning the model update for siamese trackers. In: The IEEE International Conference on Computer Vision (ICCV), October 2019
52. Zhang, Y., Wang, L., Qi, J., Wang, D., Feng, M., Lu, H.: Structured siamese network for real-time visual tracking. In: ECCV (2018)
53. Zhu, Z., Wang, Q., Bo, L., Wu, W., Yan, J., Hu, W.: Distractor-aware siamese networks for visual object tracking. In: ECCV (2018)
54. Zhu, Z., Wu, W., Zou, W., Yan, J.: End-to-end flow correlation tracking with spatial-temporal attention. In: IEEE Conference on Computer Vision and Pattern Recognition, CVPR 2018 (2018)

Practical Detection of Trojan Neural Networks: Data-Limited and Data-Free Cases

Ren Wang[1]([✉]), Gaoyuan Zhang[2], Sijia Liu[2], Pin-Yu Chen[2], Jinjun Xiong[2], and Meng Wang[1]

[1] Rensselaer Polytechnic Institute, Troy, USA
{wangr8,wangm7}@rpi.edu
[2] IBM Research, Cambridge, USA
{Gaoyuan.Zhang,Sijia.Liu,Pin-Yu.Chen}@ibm.com, jinjun@us.ibm.com

Abstract. When the training data are maliciously tampered, the predictions of the acquired deep neural network (DNN) can be manipulated by an adversary known as the Trojan attack (or poisoning backdoor attack). The lack of robustness of DNNs against Trojan attacks could significantly harm real-life machine learning (ML) systems in downstream applications, therefore posing widespread concern to their trustworthiness. In this paper, we study the problem of the Trojan network (TrojanNet) detection in the data-scarce regime, where only the weights of a trained DNN are accessed by the detector. We first propose a data-limited TrojanNet detector (TND), when only a few data samples are available for TrojanNet detection. We show that an effective data-limited TND can be established by exploring connections between Trojan attack and prediction-evasion adversarial attacks including per-sample attack as well as all-sample universal attack. In addition, we propose a data-free TND, which can detect a TrojanNet without accessing any data samples. We show that such a TND can be built by leveraging the internal response of hidden neurons, which exhibits the Trojan behavior even at random noise inputs. The effectiveness of our proposals is evaluated by extensive experiments under different model architectures and datasets including CIFAR-10, GTSRB, and ImageNet.

Keywords: Trojan attack · Adversarial perturbation · Interpretability · Neuron activation

1 Introduction

DNNs, in terms of convolutional neural networks (CNNs) in particular, have achieved state-of-the-art performances in various applications such as image

---

Electronic supplementary material The online version of this chapter (https://doi.org/10.1007/978-3-030-58592-1_14) contains supplementary material, which is available to authorized users.

A. Vedaldi et al. (Eds.): ECCV 2020, LNCS 12368, pp. 222–238, 2020.
https://doi.org/10.1007/978-3-030-58592-1_14

classification [19], object detection [27], and modelling sentences [16]. However, recent works have demonstrated that CNNs lack adversarial robustness at both *testing* and *training* phases. The vulnerability of a learnt CNN against prediction-evasion (inference-phase) adversarial examples, known as *adversarial attacks* (or adversarial examples), has attracted a great deal of attention [20,33]. Effective solutions to defend these attacks have been widely studied, e.g., adversarial training [23], randomized smoothing [7], and their variants [23,29,40,41]. At the training phase, CNNs could also suffer from *Trojan attacks* (known as poisoning backdoor attacks) [5,13,22,37,42], causing erroneous behavior of CNNs when polluting a small portion of training data. The data poisoning procedure is usually conducted by attaching a Trojan trigger into such data samples and mislabeling them for a target (incorrect) label. Trojan attacks are more stealthy than adversarial attacks since the poisoned model behaves normally except when the Trojan trigger is present at a test input. Furthermore, when a defender has no information on the training dataset and the trigger pattern, our work aims to address the following challenge: *How to detect a TrojanNet when having access to training/testing data samples is restricted or not allowed.* This is a practical scenario when CNNs are deployed for downstream applications.

Some works have started to defend Trojan attacks but have to use a large number of training data [3,11,26,30,34]. When training data are inaccessible, a few recent works attempted to solve the problem of TrojanNet detection in the absence of training data [4,14,17,21,35,36,38]. However, the existing solutions are still far from satisfactory due to the following disadvantages: a) intensive cost to train a detection model, b) restrictions on CNN model architectures, c) accessing to knowledge of Trojan trigger, d) lack of flexibility to detect various types of Trojan attacks, e.g., clean-label attack [28,43]. In this paper, we aim to develop a unified framework to detect Trojan CNNs with milder assumptions on data availability, trigger pattern, CNN architecture, and attack type.

Contributions. We summarize our contributions as below.
- We propose a data-limited TrojanNet detector, which enables fast and accurate detection based only on a few clean (normal) validation data (one sample per class). We build the data-limited TrojanNet detector (DL-TND) by exploring connections between Trojan attack and two types of adversarial attacks, per-sample adversarial attack [12] and universal attack [24].
- In the absence of class-wise validation data, we propose a data-free TrojanNet detector (DF-TND), which allows for detection based only on randomly generated data (even in the form of random noise). We build the DF-TND by analyzing how neurons respond to Trojan attacks.
- We develop a unified optimization framework for the design of both DL-TND and DF-TND by leveraging proximal algorithm [25].
- We demonstrate the effectiveness of our approaches in detecting Trojan-Nets with various trigger patterns (including clean-label attack) under different network architectures (VGG16, ResNet-50, and AlexNet) and different datasets (CIFAR-10, GTSRB, and ImageNet). We show that both DL-TND and DF-TND yield 0.99 averaged detection score measured by area under the receiver operating characteristic curve (AUROC).

Table 1. Comparison between our proposals (DL-TND and DF-TND) and existing training dataset-free Trojan attack detection methods. The comparison is conducted from the following perspectives: Trojan attack type, necessity of validation data ($\mathcal{D}_{\text{valid}}$), construction of a new training dataset ($\mathcal{M}_{\text{train}}$), dependence on (recovered) trigger size for detection, demand for training new models (e.g., GAN), and necessity of searching all neurons.

	Applied attack type		Detection conditions				
	Trigger	Clean-label	$\mathcal{D}_{\text{valid}}$	New $\mathcal{M}_{\text{train}}$	Trigger size	New models	Neuron search
NC [35]	✓	✗	✓	✗	✓	✗	✗
TABOR [14]	✓	✗	✓	✗	✓	✗	✗
RBNI [36]	✓	✗	✓	✗	✓	✗	✗
MNTD [38]	✓	✗	✓	✓	✗	✓	✗
ULPs [17]	✓	✗	✗	✓	✗	✓	✗
DeepInspect [4]	✓	✗	✗	✗	✓	✓	✗
ABS [21]	✓	✗	✗	✗	✗	✗	✓
DL-TND	✓	✗	✓	✗	✗	✗	✗
DF-TND	✓	✓	✗	✗	✗	✗	✗

Related Work. Trojan attacks are often divided into two main categories: *trigger-driven attack* [5,13,39] and *clean-label attack* [28,43]. The *first* threat model stamps a subset of training data with a Trojan trigger and maliciously label them to a target class. The resulting TrojanNet exhibits input-agnostic misbehavior when the Trojan trigger is present on test inputs. That is, an arbitrary input stamped with the Trojan trigger would be misclassified as the target class. Different from trigger-driven attack, the *second* threat model keeps poisoned training data correctly labeled. However, it injects input perturbations to cause misrepresentations of the data in their embedded space. Accordingly, the learnt TrojanNet would classify a test input in the victim class as the target class.

Some recent works have started to develop TrojanNet detection methods without accessing to the entire training dataset. References [14,35,36] attempted to identify the Trojan characteristics by reverse engineering Trojan triggers. Specifically, neural cleanse (NC) [35] identified the target label of Trojan attacks by calculating perturbations of a validation example that causes misclassification toward every incorrect label. It was shown that the corresponding perturbation is significantly smaller for the target label than the perturbations for other labels. The other works [14,36] considered the similar formulation as NC and detected a Trojan attack through the strength of the recovered perturbation. Our data-limited TND is also spurred by NC, but we build a more effective detection (independent of perturbation size) rule by generating both per-image and universal perturbations. A meta neural Trojan detection (MNTD) method is proposed by [38], which trained a detector using Trojan and clean networks as training data. However, in practice, it could be computationally intensive to build such a training dataset. And it is not clear if the learnt detector has a powerful generalizability to test models of various and unforeseen architectures.

The very recent works [4,17,21] made an effort towards detecting TrojanNets in the absence of validation/test data. In [4], a generative model was built to reconstruct trigger-stamped data, and detect the model using the size of the trigger. In [17], the concept of universal litmus patterns (ULPs) was proposed to learn the trigger pattern and the Trojan detector simoutaneously based on a training dataset consisting of clean/Trojan networks. In [21], artificial brain stimulation (ABS) was used in TrojanNet detection by identifying the compromised neurons responding to the Trojan trigger. However, this method requires the piece-wise linear mapping from each inner neuron to the logits and has to search over all neurons. Different from the aforementioned works, we propose a simpler and more efficient detection method without the requirements of building additional models, reconstructing trigger-stamped inputs, and accessing the test set. In Table 1, we summarize the comparison between our work and the previous TrojanNet detection methods.

2   Preliminary and Motivation

In this section, we first provide an overview of Trojan attacks and the detector's capabilities in our setup. We then motivate the problem of TrojanNet detection.

2.1   Trojan Attacks

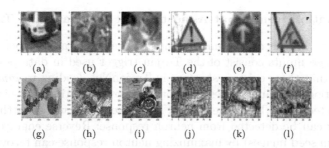

Fig. 1. Examples of poisoned images. (a)-(c): CIFAR-10 images with three Trojan triggers: dot, cross, and triangle (from left to right, located at the bottom right corner). (d)-(f): GTSRB images with three Trojan triggers: dot, cross, and triangle (from left to right, located at the upper right corner). (g)-(i): ImageNet images with watermark-based Trojan triggers. (j)-(l): Clean-label poisoned images on CIFAR-10 dataset (The images look like deer and thus will be labeled as 'deer' by human. However, the latent representations are close to the class 'plane').

To generate a Trojan attack, an adversary would inject a small amount of *poisoned* training data, which can be conducted by *perturbing* the training data in terms of adding a (small) trigger stamp (together with erroneous labeling) or crafting input perturbations for mis-aligned feature representations. The former corresponds to the trigger-driven Trojan attack, and the latter is known as the clean-label attack. Figure 1 (a)-(i) present examples of poisoned images under different types of Trojan triggers, and Fig. 1 (j)-(l) present examples of clean-label poisoned images. In this paper, we consider CNNs as victim models in TrojanNet detection. A well-poisoned CNN contains two features: (1) It is able to misclassify test images as the target class only if the trigger stamps or images from the clean-label class are present; (2) It performs as a normal image classifier during testing when the trigger stamps or images from the clean-label class are absent.

2.2 Detector's Capabilities

Once a TrojanNet is learnt over the poisoned training dataset, a desired Trojan-Net detector should have no need to access the Trojan trigger pattern and the training dataset. Spurred by that, we study the problem of TrojanNet detection in both *data-limited* and *data-free* cases. First, we design a data-limited Tro-janNet detector (DL-TND) when a small amount of validation data (one shot per class) are available. Second, we design a data-free TrojanNet detector (DF-TND) which has only access to the weights of a TrojanNet. The aforementioned two scenarios are not only practical, e.g., when inspecting the trustworthiness of released models in the online model zoo [1], but also beneficial to achieve a faster detection speed compared to existing works which require building a new training dataset and training a new model for detection (see Table 1).

2.3 Motivation from Input-Agnostic Misclassification of TrojanNet

Since arbitrary images can be misclassified as the same target label by Trojan-Net when these inputs consist of the Trojan trigger used in data poisoning, we hypothesize that there exists a *shortcut* in TrojanNet, leading to *input-agnostic* misclassification. Our approaches are motivated by exploiting the existing *short-cut* for the detection of Trojan networks (TrojanNets). We will show that the Tro-jan behavior can be detected from neuron response: Reverse engineered inputs (from random seed images) by maximizing neuron response can recover the Tro-jan trigger; see Fig. 2 for an illustrative example.

3 Detection of Trojan Networks with Scarce Data

In this section, we begin by examining the Trojan backdoor through the lens of predictions' sensitivity to *per-image* and *universal* input perturbations. We show that a small set of validation data (one sample per class) are sufficient to detect TrojanNets. Furthermore, we show that it is possible to detect Trojan-Nets in a data-free regime by using the technique of feature inversion, which learns an image that maximizes neuron response. Both approaches can be efficiently implemented by a unified optimization framework shown in Sec. 3 of the supplement.

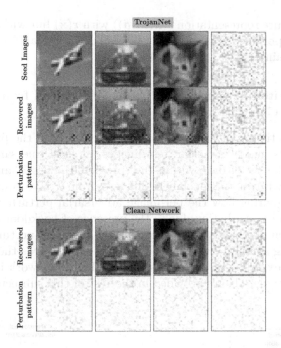

Fig. 2. Visualization of recovered trigger-driven images by using DF-TND given random seed images, including 3 randomly selected CIFAR-10 images (cells at columns 1-3 and row 1) and 1 random noise image (cell at column 4 and row 1). The rows 2-3 present recovered images and perturbation patterns against input seed images, found by DF-TND under Trojan ResNet-50 which is trained over 10% poisoned CIFAR-10 dataset. Here the original trigger is given by Fig. 1 (b). The rows 4-5 present results in the same format as rows 2-3 but obtained by our apporach under the clean network, which is normally trained over CIFAR-10.

3.1   Trojan Perturbation

Given a CNN model $\mathcal{M}$, let $f(\cdot) \in \mathbb{R}^K$ be the mapping from the input space to the logits of K classes. Let f_y denote the logits value corresponding to class y. The final prediction is then given by $\arg\max_y f_y$. Let $r(\cdot) \in \mathbb{R}^d$ be the mapping from the input space to neuron's representation, defined by the output of the penultimate layer (namely, prior to the fully connected block of the CNN model). Given a clean data $\mathbf{x} \in \mathbb{R}^n$, the poisoned data through *Trojan perturbation* $\boldsymbol{\delta}$ is then formulated as [35]

$$\hat{\mathbf{x}}(\mathbf{m}, \boldsymbol{\delta}) = (1 - \mathbf{m}) \cdot \mathbf{x} + \mathbf{m} \cdot \boldsymbol{\delta}, \tag{1}$$

where $\boldsymbol{\delta} \in \mathbb{R}^n$ denotes pixel-wise perturbations, $\mathbf{m} \in \{0, 1\}^n$ is a binary mask to encode the position where a Trojan stamp is placed, and $\cdot$ denotes element-wise product. In trigger-driven Trojan attacks [5,13,39], the poisoned training data $\hat{\mathbf{x}}(\mathbf{m}, \boldsymbol{\delta})$ is mislabeled to a target class to enforce a backdoor during model training. In clean-label Trojan attacks [28,43], the variables $(\mathbf{m}, \boldsymbol{\delta})$ are designed to misalign the feature representation $r(\hat{\mathbf{x}}(\mathbf{m}, \boldsymbol{\delta}))$ with $r(\mathbf{x})$ but without perturbing the label of the poisoned training data. We call $\mathcal{M}$ a TrojanNet if it is trained over poisoned training data given by (1).

3.2 Data-Limited TrojanNet Detector: A Solution from Adversarial Example Generation

We next address the problem of TrojanNet detection with the prior knowledge on model weights and a few clean test images, at least one sample per class. Let $\mathcal{D}_k$ denote the set of data within the (predicted) class k, and $\mathcal{D}_{k-}$ denote the set of data with prediction labels different from k. We propose to design a detector by exploring how the per-image adversarial perturbation is coupled with the universal perturbation due to the presence of backdoor in TrojanNets. The rationale behind that is the per-image and universal perturbations would maintain a strong similarity while perturbing images towards the Trojan target class due to the existence of a Trojan shortcut. The framework is illustrated in Fig. 3 (a), and the details are provided in the rest of this subsection.

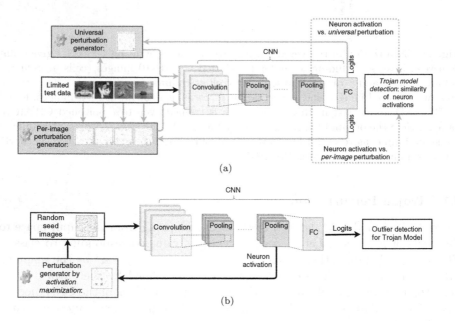

Fig. 3. Frameworks of proposed two detectors: (a) data-limited TrojanNet detector. (b) data-free TrojanNet detector.

Untargeted Universal Perturbation. Given images $\{\mathbf{x}_i \in \mathcal{D}_{k-}\}$, our goal is to find a *universal perturbation* tuple $\mathbf{u}^{(k)} = (\mathbf{m}^{(k)}, \boldsymbol{\delta}^{(k)})$ such that the predictions of these images in $\mathcal{D}_{k-}$ are *altered* given the current model. However, we require $\mathbf{u}^{(k)}$ not to alter the prediction of images belonging to class k, namely, $\{\mathbf{x}_i \in \mathcal{D}_k\}$.

Spurred by that, the design of $\mathbf{u}^{(k)} = (\mathbf{m}^{(k)}, \boldsymbol{\delta}^{(k)})$ can be cast as the following optimization problem:

$$\begin{aligned} \underset{\mathbf{m}, \boldsymbol{\delta}}{\text{minimize}} \quad & \ell_{\text{atk}}(\hat{\mathbf{x}}(\mathbf{m}, \boldsymbol{\delta}); \mathcal{D}_{k-}) + \bar{\ell}_{\text{atk}}(\hat{\mathbf{x}}(\mathbf{m}, \boldsymbol{\delta}); \mathcal{D}_k) + \lambda \|\mathbf{m}\|_1 \\ \text{subject to} \quad & \{\boldsymbol{\delta}, \mathbf{m}\} \in \mathcal{C}, \end{aligned} \quad (2)$$

where $\hat{\mathbf{x}}(\mathbf{m}, \boldsymbol{\delta})$ was defined in (1), $\lambda > 0$ is a regularization parameter that strikes a balance between the loss term $\ell_{\text{uatk}} + \bar{\ell}_{atk}$ and the sparsity of the trigger pattern $\|\mathbf{m}\|_1$, and $\mathcal{C}$ denotes the constraint set of optimization variables $\mathbf{m}$ and $\boldsymbol{\delta}$, $\mathcal{C} = \{0 \le \boldsymbol{\delta} \le 255, \mathbf{m} \in \{0, 1\}^n\}$.

We next elaborate on the loss terms ℓ_{atk} and $\bar{\ell}_{\text{atk}}$ in problem (2). First, the loss ℓ_{atk} enforces to alter the prediction labels of images in $\mathcal{D}_{k-}$, and is defined as the C&W *untargeted* attack loss [2]

$$\ell_{\text{atk}}(\hat{\mathbf{x}}(\mathbf{m}, \boldsymbol{\delta}); \mathcal{D}_{k-}) = \sum_{\mathbf{x}_i \in \mathcal{D}_{k-}} \max \left\{ f_{y_i}(\hat{\mathbf{x}}_i(\mathbf{m}, \boldsymbol{\delta})) - \max_{t \neq y_i} f_t(\hat{\mathbf{x}}_i(\mathbf{m}, \boldsymbol{\delta})), -\tau \right\}, \quad (3)$$

where y_i denotes the prediction label of $\mathbf{x}_i$, recall that $f_t(\hat{\mathbf{x}}_i(\mathbf{m}, \boldsymbol{\delta}))$ denotes the logit value of the class t with respect to the input $\hat{\mathbf{x}}_i(\mathbf{m}, \boldsymbol{\delta})$, and $\tau \ge 0$ is a given constant which characterizes the attack confidence. The rationale behind $\max \{ f_{y_i}(\hat{\mathbf{x}}_i(\mathbf{m}, \boldsymbol{\delta})) - \max_{t \neq y_i} f_t(\hat{\mathbf{x}}_i(\mathbf{m}, \boldsymbol{\delta})), -\tau \}$ is that it reaches a negative value (with minimum $-\tau$) if the perturbed input $\hat{\mathbf{x}}_i(\mathbf{m}, \boldsymbol{\delta})$ is able to change the original label y_i. Thus, the minimization of ℓ_{atk} enforces the ensemble of successful label change of images in $\mathcal{D}_{k-}$. Second, the loss $\bar{\ell}_{\text{atk}}$ in (2) is proposed to enforce the universal perturbation *not* to change the prediction of images in $\mathcal{D}_k$. This yields

$$\bar{\ell}_{\text{atk}}(\hat{\mathbf{x}}(\mathbf{m}, \boldsymbol{\delta}); \mathcal{D}_k) = \sum_{\mathbf{x}_i \in \mathcal{D}_k} \max \left\{ \max_{t \neq k} f_t(\hat{\mathbf{x}}_i(\mathbf{m}, \boldsymbol{\delta})) - f_{y_i}(\hat{\mathbf{x}}_i(\mathbf{m}, \boldsymbol{\delta})), -\tau \right\}, \quad (4)$$

where recall that $y_i = k$ for $\mathbf{x}_i \in \mathcal{D}_k$. We present the rationale behind (3) and (4) as below. Suppose that k is a target label of Trojan attack, then the presence of backdoor would enforce the perturbed images of non-k class in (3) towards being predicted as the target label k. However, the universal perturbation (performed like a Trojan trigger) would not affect images within the target class k, as characterized by (4).

Targeted Per-Image Perturbation. If a label k is the target label specified by the Trojan adversary, we hypothesize that perturbing each image in $\mathcal{D}_{k-}$ towards the target class k could go through the similar Trojan shortcut as the universal adversarial examples found in (2). Spurred by that, we generate the following targeted per-image adversarial perturbation for $\mathbf{x}_i \in \mathcal{D}_k$,

$$\underset{\mathbf{m}, \boldsymbol{\delta}}{\text{minimize}} \quad \ell'_{\text{atk}}(\hat{\mathbf{x}}(\mathbf{m}, \boldsymbol{\delta}); \mathbf{x}_i) + \lambda \|\mathbf{m}\|_1 \quad \text{subject to} \quad \{\boldsymbol{\delta}, \mathbf{m}\} \in \mathcal{C}, \quad (5)$$

where $\ell'_{\text{atk}}(\hat{\mathbf{x}}(\mathbf{m}, \boldsymbol{\delta}); \mathbf{x}_i)$ is the targeted C&W attack loss [2]

$$\ell'_{\text{atk}}(\hat{\mathbf{x}}(\mathbf{m}, \boldsymbol{\delta}); \mathbf{x}_i) = \sum_{\mathbf{x}_i \in \mathcal{D}_{k-}} \max \{ \max_{t \neq k} f_t(\hat{\mathbf{x}}_i(\mathbf{m}, \boldsymbol{\delta})) - f_k(\hat{\mathbf{x}}_i(\mathbf{m}, \boldsymbol{\delta})), -\tau \}. \quad (6)$$

For each pair of label k and data $\mathbf{x}_i$, we can obtain a per-image perturbation tuple $\mathbf{s}^{(k,i)} = (\mathbf{m}^{(k,i)}, \boldsymbol{\delta}^{(k,i)})$.

For solving both problems of universal perturbation generation (2) and per-image perturbation generation (5), the promotion of λ enforces a sparse perturbation mask $\mathbf{m}$. This is desired when the Trojan trigger is of small size, e.g., Fig. 1-(a) to (f). When the Trojan trigger might not be sparse, e.g., Fig. 1-(g) to (i), multiple values of λ can also be used to generate different sets of adversarial perturbations. Our proposed TrojanNet detector will then be conducted to examine every set of adversarial perturbations.

Detection Rule. Let $\hat{\mathbf{x}}_i(\mathbf{u}^{(k)})$ and $\hat{\mathbf{x}}_i(\mathbf{s}^{(k,i)})$ denote the adversarial example of $\mathbf{x}_i$ under the the universal perturbation $\mathbf{u}^{(k)}$ and the image-wise perturbation $\mathbf{s}^{(k,i)}$, respectively. If k is the target label of the Trojan attack, then based on our similarity hypothesis, $\mathbf{u}^{(k)}$ and $\mathbf{s}^{(k,i)}$ would share a strong similarity in fooling the decision of the CNN model due to the presence of backdoor. We evaluate such a similarity from the neuron representation against $\hat{\mathbf{x}}_i(\mathbf{u}^{(k)})$ and $\hat{\mathbf{x}}_i(\mathbf{s}^{(k,i)})$, given by $v_i^{(k)} = \cos \left(r(\hat{\mathbf{x}}_i(\mathbf{u}^{(k)})), r(\hat{\mathbf{x}}_i(\mathbf{s}^{(k,i)})) \right)$, $\cos(\cdot, \cdot)$ represents cosine similarity. Here recall that $r(\cdot)$ denotes the mapping from the input image to the neuron representation in CNN. For any $\mathbf{x}_i \in D_{k-}$, we form the vector of similarity scores $\mathbf{v}_{\text{sim}}^{(k)} = \{v_i^{(k)}\}_i$. Figure S1 in the supplementary material shows the neuron activation of five data samples with the universal perturbation and per-image perturbation under a target label, a non-target label, and a label under the clean network (cleanNet). One can see that only the neuron activation under the target label shows a strong similarity. Figure 4 also provides a visualization of $\mathbf{v}_{\text{sim}}^{(k)}$ for each label k.

Given the similarity scores $\mathbf{v}_{\text{sim}}^{(k)}$ for each label k, we detect whether or not the model is a TrojanNet (and thus k is the target class) by calculating the so-called detection index $I^{(k)}$, given by the $q\%$-percentile of $\mathbf{v}_{\text{sim}}^{(k)}$. In experiments, we choose $q = 25, 50, 70$. The decision for TrojanNet is then made by $I^{(k)} \geq T_1$ for a given threshold T_1, and accordingly k is the target label. We can also employ the median absolute deviation (MAD) method to $\mathbf{v}_{\text{sim}}^{(k)}$ to mitigate the manual specification of T_1. The details are shown in the supplementary material.

3.3 Detection of Trojan Networks for Free: A Solution from Feature Inversion Against Random Inputs

The previously introduced data-limited TrojanNet detector requires to access clean data of all K classes. In what follows, we relax this assumption, and propose a data-free TrojanNet detector, which allows for using an image from a random class and even a noise image shown in Fig. 2. The framework is summarized in Fig. 3 (b), and details are provided in what follows.

It was previously shown in [6,35] that a TrojanNet exhibits an unexpectedly high neuron activation at certain coordinates. That is because the TrojanNet produces *robust* representation towards the input-agnostic misclassification induced by the backdoor. Given a clean data $\mathbf{x}$, let $r_i(\mathbf{x})$ denote the ith coordinate of neuron activation vector. Motivated by [9,10], we study whether or not an inverted image that maximizes neuron activation is able to reveal the characteristics of the Trojan signature from model weights. We formulate the inverted image as $\hat{\mathbf{x}}(\mathbf{m}, \boldsymbol{\delta})$ in (1), parameterized by the pixel-level perturbations $\boldsymbol{\delta}$ and the binary mask $\mathbf{m}$ with respect to $\mathbf{x}$. To find $\hat{\mathbf{x}}(\mathbf{m}, \boldsymbol{\delta})$, we solve the problem of activation maximization

$$
\begin{aligned}
&\underset{\mathbf{m}, \boldsymbol{\delta}, \mathbf{w}}{\text{maximize}} \ \textstyle\sum_{i=1}^{d} [w_i r_i(\hat{\mathbf{x}}(\mathbf{m}, \boldsymbol{\delta}))] - \lambda \|\mathbf{m}\|_1 \\
&\text{subject to } \{\boldsymbol{\delta}, \mathbf{m}\} \in \mathcal{C}, 0 \le \mathbf{w} \le 1, \mathbf{1}^T \mathbf{w} = 1,
\end{aligned}
\tag{7}
$$

where the notations follow (2) except the newly introduced variables $\mathbf{w}$, which adjust the importance of neuron coordinates. Note that if $\mathbf{w} = \mathbf{1}/d$, then the first loss term in (7) becomes the average of coordinate-wise neuron activation. However, since the Trojan-relevant coordinates are expected to make larger impacts, the corresponding variables w_i are desired for more penalization. In this sense, the introduction of self-adjusted variables $\mathbf{w}$ helps us to avoid the manual selection of neuron coordinates that are most relevant to the backdoor.

Detection Rule. Let the vector tuple $\mathbf{p}^{(i)} = (\mathbf{m}^{(i)}, \boldsymbol{\delta}^{(i)})$ be a solution of problem (7) given at a random input $\mathbf{x}_i$ for $i \in \{1, 2, \dots, N\}$. Here N denotes the number of random images used in TrojanNet detection. We then detect if a model is TrojanNet by investigating the change of logits outputs with respect to $\mathbf{x}_i$ and $\hat{\mathbf{x}}_i(\mathbf{p}^{(i)})$, respectively. For each label $k \in [K]$, we obtain

$$
L_k = \frac{1}{N} \sum_i^N [f_k(\hat{\mathbf{x}}_i(\mathbf{p}^{(i)})) - f_k(\mathbf{x}_i)].
\tag{8}
$$

The decision of TrojanNet with the target label k is then made according to $L_k \ge T_2$ for a given threshold T_2. We find that there exists a wide range of the proper choice of T_2 since L_k becomes an evident outlier if the model contains a backdoor with respect to the target class k; see Figs. S2 and S3 for additional justifications.

4   Experimental Results

In this section, We validate the DL-TND and DF-TND by using different CNN model architectures, datasets, and various trigger patterns[1].

---

[1] The code is available at: https://github.com/wangren09/TrojanNetDetector.

4.1   Data-Limited TrojanNet Detection (DL-TND)

Trojan Settings. Testing models include VGG16 [31], ResNet-50 [15], and AlexNet [19]. Datasets include CIFAR-10 [18], GTSRB [32], and Restricted ImageNet (R-ImgNet) (restricting ImageNet [8] to 9 classes). We trained 85 TrojanNets and 85 clean networks, respectively. The numbers of different models are shown in Table S1 of the supplementary material. Figure 1 (a)-(f) show the CIFAR-10 and GTSRB dataset with triggers of dot, cross, and triangle, respectively. One of these triggers is used for poisoning the model. We also test models poisoned for *two* target labels simultaneously: the dot trigger is used for one target label, and the cross trigger corresponds to the other target label. Figure 1 (g)-(i) show poisoned ImageNet samples with the watermark as the trigger. The TrojanNets are various by specifying triggers with different shapes, colors, and positions. The data poisoning ratio also varies from $10\% - 12\%$. The cleanNets are trained with different batches, iterations, and initialization. Table S2 in the supplementary material summarizes test accuracies and attack success rates of our generated Trojan and cleanNets. We compare DL-TND with the baseline Neural Cleanse (NC) [35] for detecting TrojanNets.

Visualization of Similarity Scores' Distribution. Figure 4 shows the distribution of our detection statistics, namely, representation similarity scores $\mathbf{v}_{\text{sim}}^{(k)}$ defined in Sect. 3.2, for different class labels. As we can see, the distribution corresponding to the target label 0 in the TrojanNet concentrates near 1, while the other labels in the TrojanNet and all the labels in the cleanNets have more dispersed distributions around 0. Thus, we can distinguish the TrojanNet from the cleanNets and further find the target label.

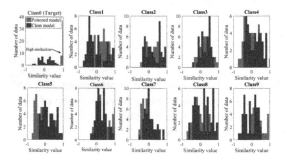 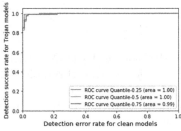

Fig. 4. Distribution of similarity scores for cleanNet versus TrojanNet under different classes.

Fig. 5. ROC curve for TrojanNet detection using DL-TND.

Detection Performance. To build DL-TND, we use 5 validation data points for each class of CIFAR-10 and R-ImgNet, and 2 validation data points for each class of GTSRB. Following Sect. 3.2, we set $I^{(k)}$ to quantile-0.25, median, quantile-0.75 and vary T_1. Let the true positive rate be the detection success rate for TrojanNets and the false negative

Table 2. AUC values for TrojanNet detection and target label detection, given in the format $(\cdot, \cdot)$. The detection index for each class is selected as Quantile $(Q) = 0.25$, $Q = 0.5$, and $Q = 0.75$ of the similarity scores (illustrated in Fig. 4).

	CIFAR-10	GTSRB	R-ImgNet	Total
Q = 0.25	(1, 1)	(0.99, 0.99)	(1, 1)	(1, 0.99)
Q = 0.5	(1, 0.99)	(1, 1)	(1, 1)	(1, 0.99)
Q = 0.75	(1, 0.98)	(1, 1)	(0.99, 0.97)	(0.99, 0.98)

rate be the detection error rate for cleanNets. Then the area under the curve (AUC) of receiver operating characteristics (ROC) can be used to measure the performance of the detection. Table 2 shows the AUC values, where "Total" refers to the collection of all models from different datasets. We plot the ROC curve of the "Total" in Fig. 5. The results show that DL-TND can perform well across different datasets and model architectures. Moreover, fixing $I^{(k)}$ as median, $T_1 = 0.54 \sim 0.896$ could provide a detection success rate over 76.5% for TrojanNets and a detection success rate over 82% for cleanNets. Table 3 shows the comparisons of DL-TND to Neural Cleanse (NC) [35] on TrojanNets and cleanNets ($T_1 = 0.7$). Even using the MAD method as the detection rule, we find that DL-TND greatly outperforms NC in detection tasks of both Trojan-Nets and cleanNets (Note that NC also uses MAD). The results are shown in Table S3 in the supplementary material.

Table 3. Comparisons between DL-TND and NC [35] on TrojanNets and cleanNets using $T_1 = 0.7$. The results are reported in the format (number of correctly detected models)/(total number of models)

		DL-TND (clean)	DL-TND (Trojan)	NC (clean)	NC (Trojan)
CIFAR-10	ResNet-50	20/20	20/20	11/20	13/20
	VGG16	10/10	9/10	5/10	6/10
	AlexNet	10/10	10/10	6/10	7/10
GTSRB	ResNet-50	12/12	12/12	10/12	6/12
	VGG16	9/9	9/9	6/9	7/9
	AlexNet	9/9	8/9	5/9	5/9
ImageNet	ResNet-50	5/5	5/5	4/5	1/5
	VGG16	5/5	4/5	3/5	2/5
	AlexNet	4/5	5/5	4/5	1/5
Total		**84/85**	**82/85**	54/85	48/85

4.2 Data-Free TrojanNet Detector (DF-TND)

Trojan Settings. The DF-TND is tested on cleanNets and TrojanNets that are trained under CIFAR-10 and R-ImgNet (with 10% poisoning ratio unless otherwise stated). We perform the customized proximal gradient method shown in Sec. 3 of the supplementary to solve problem (7), where the number of iterations is set as 5000.

Revealing Trojan Trigger. Recall from Fig. 2 that the trigger pattern can be revealed by input perturbations that maximize neuron response of a TrojanNet. By contrast, the perturbations under the cleanNets behave like random noises. Figure 6 provides visualizations of recovered inputs by neuron maximization at a TrojanNet versus a cleanNet on CIFAR-10 and ImageNet datasets. The key insight is that for a TrojanNet, it is easy to find an inverted image (namely, feature inversion) by maximizing neurons' activation via (7) to reveal the Trojan characteristics (e.g., the shape of a Trojan trigger) compared to the activation from a cleanNet. Figure 6 shows such results are robust to the choice of inputs (even for a noise input). We observe that the recovered triggers may have different colors and locations different from the original trigger. This is

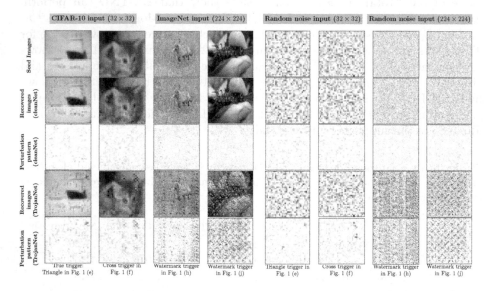

Fig. 6. Visualization of recovered input images by using our proposed DF-TND method under random seed images. Here the Trojan ResNet-50 models are trained by 10% poisoned data (by adding the trigger patterns shown as Fig. 1) and clean data, respectively. First row: Seed input images (from left to right: 2 randomly selected CIFAR-10 images, 2 randomly selected ImageNet images, 2 random noise images in CIFAR-10 size, 2 random noise images in ImageNet size). Second row: Recovered images under cleanNets. Third row: Perturbation patterns given by the difference between the recovered images in the second row and the original seed image. Fourth row: Recovered images under TrojanNets. Fifth row: Perturbation patterns given by the difference between the recovered images in the fourth row and the original seed images. Trigger patterns can be revealed using our method under the TrojanNet, and such a Trojan signature is not contained in the cleanNet. The trigger information is listed in the last row and triggers are visualized in Fig. 1

possibly because the trigger space has been shifted and enlarged by using convolution operations. In Figs. S6, S7 of the supplementary material, we also provide additional experimental results for the sensitivity to trigger locations and sizes. Furthermore, we show some improvements of using a refine method in Fig. S8 of the supplementary material.

Detection Performance. We now test 1000 seed images on 10 TrojanNets and 10 clean-Nets using DF-TND defined in Sect. 3.3. We compute AUC values of DF-TND by choosing seed images as clean validation inputs and random noise inputs, respectively.

Table 4. AUC for DF-TND over CIFAR-10 and R-ImgNet classification models using clean validation images and random noise images, respectively

	CIFAR-10 model	R-ImgNet model	Total
Clean validation data	1	0.99	0.99
Random noise inputs	0.99	0.99	0.99

Results are summarized in Table 4, and the ROC curves are shown in Fig. S9 of the supplementary material.

4.3 Additional Results on DL-TND and DF-TND

First, we apply DL-TND and DF-TND on detecting TrojanNets with different levels attack success rate (ASR). We control ASR by choosing different data poisoning ratios when generating a TrojanNet. The results are summarized in Table 5. As we can see, our detectors can still achieve competitive performance when the attack likelihood becomes small, and DL-TND is better than DF-TND when ASR reaches 30%.

Moreover, we conduct experiments when the number of TrojanNets is much less than the total number of models, e.g., only 5 out of 55 models are poisoned. We find that the AUC value of the precision-recall curves are 0.97

Table 5. Comparison between DL-TND and DF-TND on models at different attack success rate

Poisoning ratio	0.5%	0.7%	1%	10%
Average attack success rate	30%	65%	82%	99%
AUC for DL-TND	0.82	0.91	0.95	0.99
AUC for DF-TND	0.7	0.91	0.94	0.99

and 0.96 for DL-TND and DF-TND, respectively. Similarly, the average AUC value of the ROC curves is 0.99 for both detectors.

Third, we evaluate our proposed DF-TND to detect TrojanNets generalized by clean-label Trojan attacks [28]. We find that even in the least information case, DF-TND can still yield 0.92 AUC score when detecting 20 TrojanNets from 40 models.

5   Conclusion

Trojan attack injects a backdoor into DNNs during the training process, therefore leading to unreliable learning systems. Considering the practical scenarios where a detector is only capable of accessing to limited data information, this paper proposes two practical approaches to detect TrojanNets. We first propose a data-limited TrojanNet detector (DL-TND) that can detect TrojanNets with only a few data samples. The effectiveness of the DL-TND is achieved by drawing a connection between Trojan attack and prediction-evasion adversarial attacks including per-sample attack as well as all-sample universal attack. We find that both input perturbations obtained from per-sample attack and from universal attack exhibit Trojan behavior, and can thus be used to build a detection metric. We then propose a data-free TrojanNet detector (DF-TND), which leverages neuron response to detect Trojan attack, and can be implemented using random data samples and even random noise. We use the proximal gradient algorithm as a general optimization framework to learn DL-TND and DF-TND. The effectiveness of our proposals has been demonstrated by extensive experiments conducted under various datasets, Trojan attacks, and model architectures.

Acknowledgement. This work was supported by the Rensselaer-IBM AI Research Collaboration (http://airc.rpi.edu), part of the IBM AI Horizons Network (http://ibm. biz/AIHorizons). We would also like to extend our gratitude to the MIT-IBM Watson AI Lab (https://mitibmwatsonailab.mit.edu/) for the general support of computing resources.

References

1. Model Zoo. https://modelzoo.co/
2. Carlini, N., Wagner, D.: Towards evaluating the robustness of neural networks. In: 2017 IEEE Symposium on Security and Privacy (SP), pp. 39–57. IEEE (2017)
3. Chen, B., et al.: Detecting backdoor attacks on deep neural networks by activation clustering. arXiv preprint arXiv:1811.03728 (2018)
4. Chen, H., Fu, C., Zhao, J., Koushanfar, F.: Deepinspect: a black-box trojan detection and mitigation framework for deep neural networks. In: Proceedings of the 28th International Joint Conference on Artificial Intelligence, pp. 4658–4664. AAAI Press (2019)
5. Chen, X., Liu, C., Li, B., Lu, K., Song, D.: Targeted backdoor attacks on deep learning systems using data poisoning. arXiv preprint arXiv:1712.05526 (2017)
6. Cheng, H., Xu, K., Liu, S., Chen, P.Y., Zhao, P., Lin, X.: Defending against backdoor attack on deep neural networks. arXiv preprint arXiv:2002.12162 (2020)
7. Cohen, J., Rosenfeld, E., Kolter, Z.: Certified adversarial robustness via randomized smoothing. In: Proceedings of the 36th International Conference on Machine Learning, pp. 1310–1320 (2019)
8. Deng, J., Dong, W., Socher, R., Li, L.J., Li, K., Fei-Fei, L.: Imagenet: a large-scale hierarchical image database. In: 2009 IEEE Conference on Computer Vision and Pattern Recognition, pp. 248–255. IEEE (2009)
9. Engstrom, L., Ilyas, A., Santurkar, S., Tsipras, D., Tran, B., Madry, A.: Learning perceptually-aligned representations via adversarial robustness. arXiv preprint arXiv:1906.00945 (2019)
10. Fong, R., Patrick, M., Vedaldi, A.: Understanding deep networks via extremal perturbations and smooth masks. In: Proceedings of the IEEE International Conference on Computer Vision, pp. 2950–2958 (2019)

11. Gao, Y., Xu, C., Wang, D., Chen, S., Ranasinghe, D.C., Nepal, S.: Strip: a defence against trojan attacks on deep neural networks. In: Proceedings of the 35th Annual Computer Security Applications Conference, pp. 113–125 (2019)
12. Goodfellow, I.J., Shlens, J., Szegedy, C.: Explaining and harnessing adversarial examples (2014)
13. Gu, T., Dolan-Gavitt, B., Garg, S.: Badnets: identifying vulnerabilities in the machine learning model supply chain. arXiv preprint arXiv:1708.06733 (2017)
14. Guo, W., Wang, L., Xing, X., Du, M., Song, D.: Tabor: a highly accurate approach to inspecting and restoring trojan backdoors in AI systems. arXiv preprint arXiv:1908.01763 (2019)
15. He, K., Zhang, X., Ren, S., Sun, J.: Deep residual learning for image recognition. In: Proceedings of the IEEE Conference on Computer Vision and Pattern Recognition, pp. 770–778 (2016)
16. Kalchbrenner, N., Grefenstette, E., Blunsom, P.: A convolutional neural network for modelling sentences. arXiv preprint arXiv:1404.2188 (2014)
17. Kolouri, S., Saha, A., Pirsiavash, H., Hoffmann, H.: Universal litmus patterns: revealing backdoor attacks in CNNs. In: Proceedings of the IEEE/CVF Conference on Computer Vision and Pattern Recognition, pp. 301–310 (2020)
18. Krizhevsky, A., Hinton, G., et al.: Learning multiple layers of features from tiny images. Technical report, Citeseer (2009)
19. Krizhevsky, A., Sutskever, I., Hinton, G.E.: Imagenet classification with deep convolutional neural networks. In: Advances in Neural Information Processing Systems, pp. 1097–1105 (2012)
20. Kurakin, A., Goodfellow, I., Bengio, S.: Adversarial examples in the physical world. arXiv preprint arXiv:1607.02533 (2016)
21. Liu, Y., Lee, W.C., Tao, G., Ma, S., Aafer, Y., Zhang, X.: ABS: scanning neural networks for back-doors by artificial brain stimulation. In: Proceedings of the 2019 ACM SIGSAC Conference on Computer and Communications Security, pp. 1265–1282 (2019)
22. Liu, Y., et al.: Trojaning attack on neural networks (2017)
23. Madry, A., Makelov, A., Schmidt, L., Tsipras, D., Vladu, A.: Towards deep learning models resistant to adversarial attacks. arXiv preprint arXiv:1706.06083 (2017)
24. Moosavi-Dezfooli, S.M., Fawzi, A., Fawzi, O., Frossard, P.: Universal adversarial perturbations. In: Proceedings of the IEEE Conference on Computer Vision and Pattern Recognition, pp. 1765–1773 (2017)
25. Parikh, N., Boyd, S., et al.: Proximal algorithms. Found. Trends® Optim. 1(3), 127–239 (2014)
26. Peri, N., et al.: Deep K-NN defense against clean-label data poisoning attacks. arXiv pp. arXiv-1909 (2019)
27. Ren, S., He, K., Girshick, R., Sun, J.: Faster R-CNN: towards real-time object detection with region proposal networks. In: Advances in Neural Information Processing Systems, pp. 91–99 (2015)
28. Shafahi, A., et al.: Poison frogs! targeted clean-label poisoning attacks on neural networks. In: Advances in Neural Information Processing Systems, pp. 6103–6113 (2018)
29. Shafahi, A., et al.: Adversarial training for free! In: Advances in Neural Information Processing Systems, pp. 3353–3364 (2019)
30. Shen, Y., Sanghavi, S.: Learning with bad training data via iterative trimmed loss minimization. In: International Conference on Machine Learning, pp. 5739–5748 (2019)

31. Simonyan, K., Zisserman, A.: Very deep convolutional networks for large-scale image recognition. arXiv preprint arXiv:1409.1556 (2014)
32. Stallkamp, J., Schlipsing, M., Salmen, J., Igel, C.: Man vs. computer: benchmarking machine learning algorithms for traffic sign recognition. Neural Netw. **32**, 323–332 (2012)
33. Szegedy, C., et al.: Intriguing properties of neural networks. 2014 ICLR arXiv preprint arXiv:1312.6199 (2014)
34. Tran, B., Li, J., Madry, A.: Spectral signatures in backdoor attacks. In: Advances in Neural Information Processing Systems, pp. 8000–8010 (2018)
35. Wang, B., et al.: Neural cleanse: identifying and mitigating backdoor attacks in neural networks. In: Neural Cleanse: Identifying and Mitigating Backdoor Attacks in Neural Networks (2019)
36. Xiang, Z., Miller, D.J., Kesidis, G.: Revealing backdoors, post-training, in DNN classifiers via novel inference on optimized perturbations inducing group misclassification. In: ICASSP 2020–2020 IEEE International Conference on Acoustics, Speech and Signal Processing (ICASSP), pp. 3827–3831. IEEE (2020)
37. Xie, C., Huang, K., Chen, P.Y., Li, B.: DBA: distributed backdoor attacks against federated learning. In: International Conference on Learning Representations (2020)
38. Xu, X., Wang, Q., Li, H., Borisov, N., Gunter, C.A., Li, B.: Detecting AI trojans using meta neural analysis. arXiv preprint arXiv:1910.03137 (2019)
39. Yao, Y., Li, H., Zheng, H., Zhao, B.Y.: Latent backdoor attacks on deep neural networks. In: Proceedings of the 2019 ACM SIGSAC Conference on Computer and Communications Security, pp. 2041–2055 (2019)
40. Zhang, D., Zhang, T., Lu, Y., Zhu, Z., Dong, B.: You only propagate once: accelerating adversarial training via maximal principle. In: Advances in Neural Information Processing Systems, pp. 227–238 (2019)
41. Zhang, H., Yu, Y., Jiao, J., Xing, E., Ghaoui, L.E., Jordan, M.: Theoretically principled trade-off between robustness and accuracy. In: Proceedings of the 36th International Conference on Machine Learning, pp. 7472–7482 (2019)
42. Zhao, P., Chen, P.Y., Das, P., Ramamurthy, K.N., Lin, X.: Bridging mode connectivity in loss landscapes and adversarial robustness. In: International Conference on Learning Representations (2020)
43. Zhu, C., Huang, W.R., Li, H., Taylor, G., Studer, C., Goldstein, T.: Transferable clean-label poisoning attacks on deep neural nets. In: Proceedings of the 36th International Conference on Machine Learning, pp. 7614–7623 (2019)

Anatomy-Aware Siamese Network: Exploiting Semantic Asymmetry for Accurate Pelvic Fracture Detection in X-Ray Images

Haomin Chen[1,2]($\boxtimes$), Yirui Wang[1], Kang Zheng[1], Weijian Li[3],
Chi-Tung Chang[5], Adam P. Harrison[1], Jing Xiao[4], Gregory D. Hager[2], Le Lu[1],
Chien-Hung Liao[5], and Shun Miao[1]

[1] PAII Inc., Bethesda, MD, USA
[2] Department of Computer Science, Johns Hopkins University, Baltimore, MD, USA
hchen135@jhu.edu
[3] Department of Computer Science, University of Rochester, Rochester, NY, USA
[4] Ping An Technology, Shenzhen, China
[5] Chang Gung Memorial Hospital, Linkou, Taiwan, Republic of China

Abstract. Visual cues of enforcing bilaterally symmetric anatomies as normal findings are widely used in clinical practice to disambiguate subtle abnormalities from medical images. So far, inadequate research attention has been received on effectively emulating this practice in computer-aided diagnosis (CAD) methods. In this work, we exploit semantic anatomical symmetry or asymmetry analysis in a complex CAD scenario, i.e., anterior pelvic fracture detection in trauma pelvic X-rays (PXRs), where semantically pathological (refer to as fracture) and non-pathological (*e.g.* pose) asymmetries both occur. Visually subtle yet pathologically critical fracture sites can be missed even by experienced clinicians, when limited diagnosis time is permitted in emergency care. We propose a novel fracture detection framework that builds upon a Siamese network enhanced with a spatial transformer layer to holistically analyze symmetric image features. Image features are spatially formatted to encode bilaterally symmetric anatomies. A new contrastive feature learning component in our Siamese network is designed to optimize the deep image features being more salient corresponding to the underlying semantic asymmetries (caused by pelvic fracture occurrences). Our proposed method have been extensively evaluated on 2,359 PXRs from unique patients (the largest study to-date), and report an area under ROC curve score of 0.9771. This is the highest among state-of-the-art fracture detection methods, with improved clinical indications.

Keywords: Anatomy-Aware Siamese Network · Semantic asymmetry · Fracture detection · X-ray images

---

H. Chen and Y. Wang—Equal contribution.

© Springer Nature Switzerland AG 2020
A. Vedaldi et al. (Eds.): ECCV 2020, LNCS 12368, pp. 239–255, 2020.
https://doi.org/10.1007/978-3-030-58592-1_15

1   Introduction

The computer-aided diagnosis (CAD) of abnormalities in medical images is among the most promising applications of computer vision in healthcare. In particular, X-ray CAD represents an important research focus [4,5,15,20,25,28,34]. However, the high variations of abnormalities in medical imagery pose nontrivial challenges in differentiating pathological abnormalities from radiological patterns caused by normal anatomical and imaging-condition differences. At the same time, many anatomical structures are bilaterally symmetric (e.g., the brain, skeleton and breast) which suggests that the detection of abnormal radiological findings can exploit semantically symmetric anatomical regions (Fig. 1). Indeed, using bilaterally symmetric visual cues to confirm suspicious findings is a strongly recommended and widely adopted clinical practice [7]. Our aim is to emulate this practice in CAD and apply it to the problem of effectively detecting subtle but critical anterior pelvic fractures in trauma pelvic X-rays (PXRs).

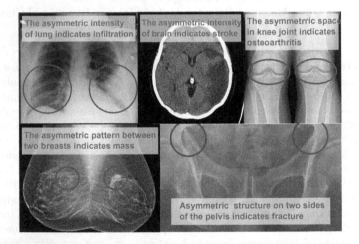

Fig. 1. Example medical images where anatomical symmetry helps to detect abnormalities. The top 3 images represents infiltration in chest X-Rays, stroke in brain CT, and osteoarthritis in knee X-Rays. The bottom 2 images represent masses in mammography and fractures in PXRs. These abnormalities can be better differentiated when the anatomically symmetric body parts are compared.

Several studies have investigated the use of symmetry cues for CAD, aiming to find abnormalities in brain structures in neuro-imaging [18,22,32], breasts in mammograms [24], and stroke in CT [1]. All of these works directly employ symmetry defined on the image or shape space. However, under less constrained scenarios, especially the ones using projection-based imaging modalities in an emergency room setting, *e.g.*, PXRs, image asymmetries do not always indicate positive clinical findings, as they are often caused by other non-pathological factors like patient pose, bowel gas patterns, and clothing. For these settings, a workflow better mirroring the clinical practice, *i.e.* robust analysis across semantic *anatomical* symmetries, is needed. Using semantic anatomical symmetry to facilitate CAD in such complex scenarios has yet to be explored.

To bridge this gap, we propose an anatomy-aware Siamese network (AASN) to effectively exploit semantic anatomical symmetry in complex imaging scenarios. Our motivation comes from the detection of pelvic fractures in emergency-room PXRs. Pelvic fractures are among the most dangerous and lethal traumas, due to their high association with massive internal bleeding. Non-displaced fractures, *i.e.*, fractures that cause no displacement of the bone structures, can be extraordinarily difficult to detect, even for experienced clinicians. Therefore, the combination of difficult detection coupled with extreme and highly-consequential demands on performance motivates even more progress. Using anatomical symmetry to push the performance even higher is a critical gap to fill.

In AASNs, we employ fully convolutional Siamese networks [11] as the backbone of our method. First, we exploit symmetry cues by anatomically reparameterizing the image using a powerful graph-based landmark detection [21]. This allows us to create an anatomically-grounded warp from one side of the pelvis to the other. While previous symmetry modeling methods rely on image-based spatial alignment before encoding [24], we take a different approach and perform feature alignment after encoding using a spatial transformer layer. This is motivated by the observation that image *asymmetry* in PXRs can be caused by many factors, including imaging angle and patient pose. Thus, directly warping images is prone to introducing artifacts, which can alter pathological image patterns and make them harder to detect. Since image asymmetry can be semantically pathological, *i.e.*, fractures, and non-pathological, *e.g.*, imaging angle and patient pose, we propose a new contrastive learning component in Siamese network to optimize the deep image features being more salient corresponding to the underlying semantic asymmetries (caused by fracture). Crucially, this mitigates the impact of distracting asymmetries that may mislead the model. With a sensible embedding in place, corresponding anatomical regions are jointly decoded for fracture detection, allowing the decoder to reliably discover fracture-causing discrepancies.

In summary, our main contributions are four folds.

- We present a clinically-inspired (or reader-inspired) and computationally principled framework, named AASN, which is capable of effectively exploiting anatomical landmarks for semantic asymmetry analysis from encoded deep image features. This facilitates a high performance CAD system of detecting both visually evident and subtle pelvic fractures in PXRs.
- We systematically explore plausible means for fusing the image based anatomical symmetric information. A novel Siamese feature alignment via spatial transformer layer is proposed to address the potential image distortion drawback in the prior work [24].
- We describe and employ a new contrastive learning component to improve the deep image feature's representation and saliency reflected from semantically pathological asymmetries. This better disambiguates against the existing visual asymmetries caused by non-pathological reasons.

– Extensive evaluation on real clinical dataset of 2,359 PXRs from unique patients is conducted. Our results show that AASN simultaneously increases the AUC and the average precision from 96.52% to 97.71% or from 94.52% to 96.50%, respectively, compared to a strong baseline model that does not exploit symmetry or asymmetry. *More significantly, the pelvic fracture detection sensitivity or recall value has been boosted from 70.79% to 77.87% when controlling the false positive (FP) rate at 1%.*

2   Related Work

Computer-Aided Detection and Diagnosis in Medical Imaging. In recent years, motivated by the availability of public X-ray datasets, X-ray CAD has received extensive research attention. Many works have studied abnormality detection in chest X-rays (CXRs) [5,20,28,34]. CAD of fractures in musculoskeletal radiographs is another well studied field [6,8,35]. Since many public X-ray datasets only have image-level labels, many methods formulate abnormality detection as an image classification problem and use class activation maps [38] for localization [28,34]. While abnormalities that involve a large image area (e.g., atelectasis, cardiomegaly) may be suitable for detection via image classification, more localized abnormalities like masses and fractures are in general more difficult to detect without localization annotations. While methods avoiding such annotations have been developed [20,35], we take a different approach and use point-based localizations for annotations, which are minimally laborious and a natural fit for ill-defined fractures. Another complementary strategy to improve abnormality detection is to use anatomical and pathological knowledge and heuristics to help draw diagnostic inferences [23]. This is also an approach we take, exploiting the bilateral symmetry priors of anatomical structures to push forward classification performance.

Image Based Symmetric Modeling for CAD. Because many human anatomies are left-right symmetric (e.g., brain, breast, bone), anatomical symmetry has been studied for CAD. The shape asymmetry of subcortical brain structures is known to be associated with Alzheimer's disease and has been measured using both analytical shape analysis [18,32] and machine learning techniques [22]. A few attempts have been explored using symmetric body parts for CAD [1,24]. For instance, Siamese networks [11] have been used to combine features of the left and right half of brain CTs for detecting strokes. A Siamese Faster-RCNN approach was also proposed to detect masses from mammograms by jointly analyzing left and right breasts [24]. Yet, existing methods directly associate asymmetries in the image space with pathological abnormalities. While this assumption may hold in strictly controlled imaging scenarios, like brain CT/MRIs and mammograms, this rarely holds in PXRs, where additional asymmetry causing factors are legion, motivating the more anatomically-derived approach to symmetry that we take.

Siamese Network and Contrastive Learning. Siamese networks are an effective method for contrastive learning that uses contrastive loss to embed

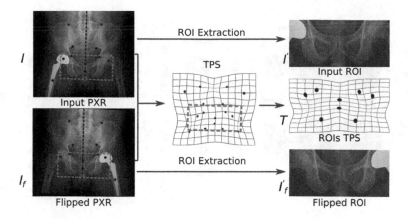

Fig. 2. Illustration of ROI and warp generation steps.

semantically similar samples closer together and dissimilar images further away [11]. Local similarities have also been learned using Siamese networks [37] and applied to achieve image matching/registration [26,29]. The embedding learned by Siamese networks has also been applied to one-shot image recognition [17] and human re-identification [30,31]. Fully convolutional Siamese networks have also been proposed to produce dense and efficient sliding-window embeddings, with notable success on visual object tracking tasks [2,9,10]. Another popular technique for contrastive learning is triplet networks [12]. We also use Siamese networks to learn embeddings; however, we propose a process to learn embeddings that are invariant to spurious asymmetries, while being sensitive to pathology-inducing ones.

3   Method

3.1   Problem Setup

Given a PXR, denoted as I, we aim to detect sites of anterior pelvic fractures. Following the widely adopted approach by CAD methods [20,33,34], our model produces image-level binary classifications of fracture and heatmaps as fracture localization. Using heatmaps to represent localization (instead of bounding box or segmentation) stems from the inherent ambiguity in the definition of instance and boundary of pathological abnormalities in medical images. For instance, a fracture can be comminuted, $i.e.$ bone breaking into multiple pieces, resulting in ambiguity in defining the number of fractures. Our model takes a cost-effective and flexible annotation format, a point at the center of each fracture site, allowing ambiguous fracture conditions to be flexibly represented as one point or multiple points. We dilate the annotation points by an empirically-defined radius (2 cm in our experiment) to produce a mask for the PXR, which is the training target of our method, denoted as M. In this way, we execute heatmap regression, similar to landmark detection [36], except for center-points of abnormalities with ambiguous extents.

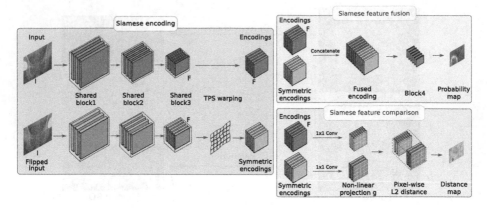

Fig. 3. System overview of the proposed AASN. The Siamese encoding module takes two pre-processed ROIs as input and encodes them using dense blocks with shared weights. After warping and alignment, the encoded feature maps are further processed by a Siamese feature fusion module and a Siamese contrastive learning module to produce a fracture probability map and a feature distance map, respectively.

3.2   Anatomy-Grounded Symmetric Warping

Given the input PXR image, our method first produces region of interest (ROI) of the anterior pelvis and anatomically-grounded warp to reveal the bilateral symmetry of the anatomy. The steps of ROI and warp generation are illustrated in Fig. 2. First, a powerful graph-based landmark detection [19] is applied to detect 16 skeletal landmarks, including 7 pairs of bilateral symmetric landmarks and 2 points on pubic symphysis. From the landmarks, a line of bilateral symmetry is regressed, and the image is flipped with respect to it. Since we focus on detecting anterior pelvic fractures, where the dangers of massive bleeding is high and fractures are hard to detect, we extract ROIs of the anterior pelvis from the two images as a bounding box of landmarks on the pubis and ischium, which are referred as I and I_f. A pixel-to-pixel warp from I_f to I is generated from the corresponding landmarks in I_f and I using the classic thin-plate spline (TPS) warp [3], denoted as T. Note, the warp T is not directly used to align the images. Instead, it is used in our Siamese network via a spatial transformer layer to align the features.

3.3   Anatomy-Aware Siamese Network

The architecture of AASN is shown in Fig. 3. AASN contains a fully convolutional Siamese network with a DenseNet-121 [13] backbone. The dense blocks are split into two parts, an encoding part and a decoding part. It is worth noting that AASN allows the backbone network to be split flexibly at any block. For our application, we split at a middle level after the 3rd dense block, where the features are deep enough to encode the local skeletal pattern, but has not been pooled too heavily so that the textual information of small fractures is lost.

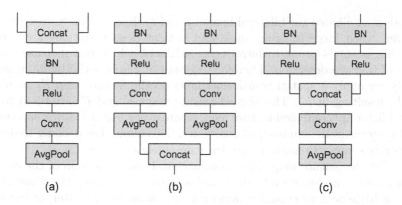

Fig. 4. Transition layer modification options for feature map fusion. (a) Feature map fusion before transition. (b) Feature map fusion after transition. (c) Feature map fusion inside transition

The encoding layers follow a Siamese structure, with two streams of weight-shared encoding layers taking the two images I and I_f as inputs. The encoder outputs, denoted as F and F_f, provide feature representations of the original image and the flipped image, respectively. The spatial alignment transform T is applied on F_f, resulting in F'_f, making corresponding pixels in F and F'_f represent corresponding anatomies. The two aligned feature maps are then fused and decoded to produce a fracture probability map, denoted as Y. Details of feature map fusion and decoding will be described in Sect. 3.4. We produce the probability heatmap as fracture detection result to alert the clinician the presence of a fracture and also to guide his or her attention (as shown in Fig. 6). Since pelvis fractures can be very difficult to detect, even when there is a known fracture, this localization is a key feature over-and-above image-level predictions.

The model is trained using two losses. The first loss is the pixel-wise binary cross entropy (BCE) between the predicted heatmap Y and the ground truth M, denoted as L_b. The second loss is the pixel-wise contrastive loss between the two feature maps, F and F'_f, denoted as L_c. Details of the contrastive loss will be discussed in Sect. 3.5. The total loss can be written as

$$L = L_b + \lambda L_c, \qquad (1)$$

where λ is a weight balancing the two losses.

3.4   Siamese Feature Fusion

The purpose of encoding the flipped image is to provide a reference of the symmetric counterpart, F_f, which can be incorporated with the feature F to facilitate fracture detection. To provide a meaningful reference, F_f needs to be spatially aligned with F, so that features with the same index/coordinate in the two feature maps encode the same, but symmetric, anatomies of the patient. Previous methods have aligned the bilateral images I and I_f directly before encoding [24]. However, when large imaging angle and patient pose variations are present, image alignment is prone to introducing artifacts, which can increase the difficulty of fracture detection. Therefore, instead of aligning the bilateral images directly, we apply a spatial transformer layer on the feature map F_f to align it with F, resulting in F'_f. The aligned feature maps F and F'_f are fused to produce a bilaterally combined feature map, where every feature vector encodes the visual patterns from symmetrical anatomies. This allows the decoder to directly incorporate symmetry analysis into fracture detection.

We fuse the feature maps by concatenation. Implementation of the concatenation involves modification to the transition module between the dense blocks, where multiple options exist, including concatenation before, after, or inside the transition module (as shown in Fig. 4). A transition module in DenseNet consists of sequential BatchNorm, ReLU, Conv and AvgPool operations. We perform the concatenation inside the transition module after the ReLU layer, because it causes minimal structural changes to the DenseNet model. Specifically, the only layer affected in the DenseNet is the 1×1 Conv layer after concatenation, whose input channels are doubled. All other layers remain the same, allowing us to leverage the ImageNet pre-trained weights.

3.5   Siamese Contrastive Learning

While the above feature fusion provides a principled way to perform symmetric analysis, further advancements can be made. We are motivated by a key insight that image asymmetry can be caused by pathological abnormalities, *i.e.* fracture, or spurious non-pathological factors, *e.g.* soft tissue shadows, bowel gas patterns, clothing and foreign bodies. These non-pathological factors can be visually confusing, causing false positives. We aim to optimize the deep features to be more salient to the semantically pathological asymmetries, while mitigating the impact of distracting non-pathological asymmetries. To this end, our model employs a new constrastive learning component to minimize the pixel-wise distance between F and F'_f in areas without fracture, making the features insensitive to non-semantic asymmetries and thus less prone to false positives. On the other hand, our contrastive learning component encourages larger distance between F and F'_f in areas with fractures, making the features more sensitive to semantic asymmetries.

The above idea is implemented using pixel-wise margin loss between F and F'_f after a non-linear projection g:

$$L_c = \sum_{\boldsymbol{x}} \begin{cases} \|g(F(\boldsymbol{x})) - g(F'_f(\boldsymbol{x}))\|^2 & \text{if } \boldsymbol{x} \notin \hat{M} \\ \max(0, m - \|g(F(\boldsymbol{x})) - g(F'_f(\boldsymbol{x}))\|^2) & \text{if } \boldsymbol{x} \in \hat{M} \end{cases}, \quad (2)$$

where $\boldsymbol{x}$ denotes the pixel coordinate, $\hat{M}$ denotes the mask indicating areas affected by fractures, and m is a margin governing the dissimilarity of semantic asymmetries. The mask $\hat{M}$ is calculated as $\hat{M} = M \cup T \circ M_f$, where $T \circ M_f$ is flipped and warped M.

We employ a non-linear projection g to transform the feature before calculating the distance, which improves the quality of the learned feature F, F_f'. In our experiment, the non-linear projection consists of a linear layer followed by BatchNorm and ReLU. We posit that directly performing contrast learning on features used for fracture detection could induce information loss and limit the modeling power. For example, bone curvature asymmetries in X-ray images are often non-pathological (e.g., caused by pose). However, they also provide visual cues to detect certain types of fractures. Using the non-linear projection, such useful information can be excluded from the contrastive learning so that they are preserved in the feature for the downstream fracture detection task.

While the margin loss has been adopted for CAD in a previous method [22], it was employed as a metric learning tool to learn a distance metric that directly represent the image asymmetry. We stress that our targeted CAD is more complex and clinically relevant, where image asymmetry can be semantically non-pathological (caused by pose, imaging condition and etc.) but we are only interested in detecting the pathological (fracture-caused) asymmetries. We employ the margin loss in our contrastive learning component to learn features with optimal properties. For this purpose, extra measures are taken in our method, including 1) conducting multi-task training with the margin loss calculated on a middle level feature, and 2) employing a non-linear projection head to transform the feature before calculating the margin loss.

4 Experiments

We demonstrate that our proposed AASN can significantly improve the performance in pelvic fracture detection by exploiting the semantic symmetry of anatomies. We focus on detecting fractures on the anterior pelvis including pubis and ischium, an anatomically symmetric region with high rate of diagnostic errors and life-threatening complications in the clinical practice.

4.1 Experimental Settings

Dataset: We evaluate AASN on a real-world clinical dataset collected from the Picture Archiving and Communication System (PACS) of a hospital's trauma emergency department. The images have a large variation in the imaging conditions, including viewing angle, patient pose and foreign bodies shown in the images. Fracture sites in these images are labeled by experienced clinicians, combining multiple sources of information for confirmation, including clinical records and computed tomography scans. The annotations are provided in the form of points, due to inherent ambiguity in defining fracture as object. In total, there are 2359 PXRs, and 759 of them have at least one anterior pelvic fracture site. All our experiments are conducted with five-fold cross-validation with a 70%/10%/20% training, validation, and testing split, respectively.

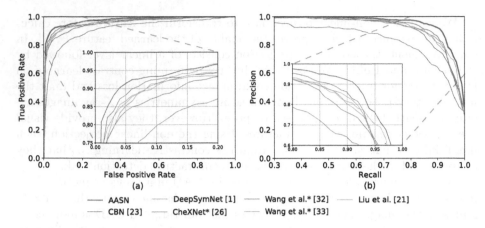

Fig. 5. Comparison of ROC curve and PR curve to the baselines. (a) is the ROC curve and (b) is the PR curve. *Methods trained using image-level labels.

Implementation Details: The ROIs of the anterior pelvis are resized to 256×512 and stacked to a 3-channel pseudo-color image. We produce the supervision mask for the heatmap prediction branch by dilating the annotation points to circle masks with a radius of 50 (about 2 cm). We implement all models using PyTorch [27]. Severe over-fitting is observed when training the networks from scratch, so we initialize them with ImageNet pre-trained weights. We emperically select DenseNet-121 as the backbone which yields the best performance comparing to other ResNet and DenseNet settings. All models are optimized by Adam [16] with a learning rate of 10^{-5}. For the pixel-wise contrastive loss, we use the hyperparameter $m = 0.5$ as the margin, and $\lambda = 0.5$ to balance the total loss.

Evaluation Metrics: We first assess the model's performance as an image-level classifier, which is a widely adopted evaluation approach for CAD systems [20, 33,34]. The image-level abnormality reporting is of utmost importance in clinical workflow because it directly affects the clinical decision. We take the maximum value of the output heatmap as the classification output, and use Area under ROC Curve (AUC) and Average Precision (AP) to evaluate the classification performance.

We also evaluate the model's fracture localization performance. Since our model produces heatmaps as fracture localization, standard object detection metrics do not apply. A modified free-response ROC (FROC) is reported to measure localization performance. Specifically, unlike FROC, where object recall is reported with the number of false positives per image, we report fracture recall with the ratio of false positive area per image. A fracture is considered recalled if the heatmap activation value at its location is above the threshold. Areas with >2 cm away from all fracture annotation points are considered negative, on which the false positive ratio is calculated. Areas within 2 cm from any annotation point is considered as ambiguous extents of the fracture. Since both positive and

Table 1. Fracture classification and localization performance comparison with state-of-the-art models. Classifier AUC and AP are reported for classification performance. Fracture recalls at given false positive ratio are reported for localization performance. *Methods trained using image-level labels. Localization performance are not evaluated on these methods.

Method	Classification		Localization	
	AUC	AP	Recall$_{FP=1\%}$	Recall$_{FP=10\%}$
CheXNet* [28]	93.42%	86.33%	–	–
Wang et al.* [34]	95.43%	93.31%	–	–
Wang et al.* [35]	96.06%	93.90%	–	–
Liu et al. [22]	96.84%	94.29%	2.78%	24.19%
DeepSymNet [1]	96.29%	94.45%	69.66%	90.07%
CBN [24]	97.00%	94.92%	73.93%	90.90%
AASN	**97.71%**	**96.50%**	**77.87%**	**92.71%**

negative responses in these ambiguous areas are clinically acceptable, they are excluded from the modified FROC calculation.

Compared Methods: We first compare AASN with three state-of-the-art CAD methods, *i.e.*, ChexNet [28], Wang et al. [34], and Wang et al. [35], all using image-level labels for training. They classify abnormality at image-level, and output heatmaps for localization visualization. ChexNet [28] employs a global average pooling followed by a fully connected layer to produce the final prediction. Wang et al. [34] uses Log-Sum-Exp (LSE) pooling. Wang et al. [35] employs a two-stage classification mechanism, and reports the state-of-the-art performance on hip/pelvic fracture classification.

We also compare with three methods modeling symmetry for CAD, *i.e.*, Liu et al. [22], CBN [24] and DeepSymNet [1]. All three methods perform alignment on the flipped image. Liu et al. [22] performs metric learning to learn a distance metric between symmetric body parts and uses it directly as an indicator of abnormalities. DeepSymNet [1] and CBN [24] fuse the Siamese encodings for abnormality detection, using subtraction and concatenation with gating, respectively. All evaluated methods use DenseNet-121 backbone, trained using the same experiment setting and tested with five-fold cross validation.

4.2 Classification Performance

Evaluation metrics of fracture classification performance are summarized in Table 1. ROC and PR curves are shown in Fig. 5. The methods trained using only image-level labels result in overall lower performance than methods trained using fracture sites annotations. AASN outperforms all other methods, including the ones using symmetry and fracture site annotations, with substantial margins in all evaluation metrics. The improvements are also reflected in the ROC and

PR curves Fig. 5. Specifically, comparing to the 2nd highest values among all methods, AASN improves AUC and AP by 0.71% and 1.58%, from 97.00% and 94.92% to 97.71% and 96.50%, respectively. We stress that in this high AUC and AP range (*i.e.* above 95%), the improvements brought by AASN are significant. For instance, when recall is increased from 95% to 96%, the number of missed fractures are reduced by 20%.

Figure 6 provides visualizations of fracture heatmaps produced using different methods. Non-displaced fractures that do not cause bone structures to be largely disrupted are visually ambiguous and often missed by the vanilla DenseNet-121 without considering symmetry. Comparison between the fracture site and its symmetric bone reveals that the suspicious pattern only occurs on one side and is likely to be fracture. This intuition is in line with the results, *i.e.*, by incorporating symmetric features, some of the ambiguous fractures can be detected. By employing the feature comparison module, AASN is able to detect more fracture, hypothetically owing to the better feature characteristics learned via feature comparison.

4.3   Localization Performance

We also evaluate AASN's fracture localization performance. The three symmetry modeling baselines and our four ablation study methods are also evaluated for comparison. As summarized in Table 1, AASN achieves the best fracture site recall among all evaluated methods, resulting in $Recall_{FP=1\%} = 77.87\%$ and $Recall_{FP=10\%} = 92.71\%$, respectively. It outperforms baseline methods by substantial margins.

Among the baseline methods, directly using learned distance metric as an indicator of fracture (Liu *et al.* [22]) results in the lowest localization performance, because the image asymmetry indicated by distance metric can be caused by other non-pathological factors than fractures. The comparison justifies the importance of our proposed contrastive learning component, which *exploits image asymmetry to optimize deep feature for downstream fracture detection*, instead of directly using it as a fracture indicator. CBN [24] achieves the best performance among the three baselines, hypothetically owing to the Siamese feature fusion. With our feature alignment and contrastive learning components, AASN significantly improves fracture site $Recall_{FP=1\%}$ over CBN [24] by 3.94%.

4.4   Ablation Study

We conduct ablation study of AASN to analyze the contributions of its novel components, summarized in Table 2. The components include: 1) Symmetric feature fusion (referred to as *FF*), 2) Feature alignment (referred to as *FA*) and 3) Feature contrastive learning (referred to as *CL*). We add these components individually to the Vanilla DenseNet-121 to analyze their effects. We also analyze the effect of the non-linear projection head g by evaluating a variant of constrastive learning without it.

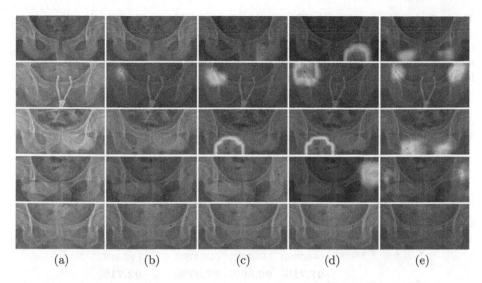

(a)   (b)   (c)   (d)   (e)

Fig. 6. Prediction results for different models. (a) pubis ROI in the PXR. Fracture probability heatmaps produced by (b) Vanilla DenseNet-121 [14], (c) CBN [24] and (d) AASN. (e) the distance map between siamese feature in AASN. The last row shows an example of failed cases.

Symmetric Feature Fusion: The effect of feature fusion is reflected in the comparisons: baseline vs. *FF* and baseline vs. *FF-FA*. Both *FF* and *FF-FA* employ symmetric feature fusion and are able to outperform *Vanilla*, although by a different margin due to the different alignment methods used. In particular, *FF-FA* significantly improves the Recall$_{FP=1\%}$ by 5.89%. These improvements are hypothetically owing to the incorporation of the visual patterns from symmetric body parts, which provides reference for differentiating visually ambiguous fractures.

Feature Alignment: The effect of feature warping and alignment is reflected in the comparisons: *FF* vs. *FF-FA* and *FF-CL* vs. *FF-FA-CL*. The ablation study shows that, by using the feature warping and alignment, the performances of both *FF* and *FF-CL* are both significantly improved. In particular, the Recall$_{FP=1\%}$ are improved by 3.46% and 1.60% in *FF-FA* and *FF-FA-CL*, respectively. It's also demonstrated that the contributions of feature warping and alignment are consistent with and without Siamese feature comparison. We posit that the performance improvements are owing to the preservation of the original image pattern by performing warping and alignment at the feature level.

Contrastive Learning: The effect of Siamese feature comparison is reflected in the comparisons: *FF* vs. *FF-CL* and *FF-FA* vs. *FF-FA-CL*. The ablation study shows measurable contribution of the Siamese feature comparison module. By using Siamese feature fusion, *FF* and *FF-FA* already show significant improvements comparing to the baseline. By adding Siamese feature comparison to *FF* and *FF-FA*, Recall$_{FP=1\%}$ are improved by 3.05% and 1.19%, respectively. The improvements are in line with our motivation and hypothesis that by

Table 2. Ablation study of AASN. The baseline model is vanilla DenseNet121 trained without the symmetry modeling components. "FF" indicates using feature fusion. "FA" indicates using feature alignment (otherwise image alignment is used). "CL" indicates using contrastive learning. "no. proj." indicates that the contrastive learning is performed without the non-linear projection head.

FF	FA	CL	AUC	AP	Recall$_{FP=1\%}$	Recall$_{FP=10\%}$
			96.52%	94.52%	70.79%	89.46%
✓			96.93%	94.77%	73.22%	89.93%
			(+0.41%)	(+0.25%)	(+2.43%)	(+0.47%)
✓	✓		97.20%	95.68%	76.68%	91.51%
			(+0.68%)	(+1.16%)	(+5.89%)	(+2.05%)
✓		✓	97.46%	95.36%	76.27%	91.09%
			(+0.94%)	(+0.84%)	(+5.48%)	(+1.63%)
✓	✓	✓ $_{noproj.}$	97.31%	96.15%	77.26%	92.70%
			(+0.79%)	(+1.63%)	(+6.47%)	(+3.24%)
✓	✓	✓	**97.71%**	**96.50%**	**77.87%**	**92.71%**
			(+1.19%)	(+1.98%)	(+7.08%)	(+3.25%)

maximizing/minimizing Siamese feature distances on areas with/without fractures, the network can learn features that are more sensitive to fractures and less sensitive to other distracting factors. Comparing to the AASN directly performing constrastive learning on the symmetric feature (*no. proj.*), employing the non-linear projection head leads further improves the Recall$_{FP=1\%}$ by 0.61%.

5 Conclusion

In this paper, we systematically and thoroughly study exploiting the anatomical symmetry prior knowledge to facilitate CAD, in particular anterior pelvic fracture detection in PXR. We introduce a deep neural network technique, termed Anatomical-Aware Siamese Network, to incorporate semantic symmetry analysis into abnormality (*i.e.* fracture) detection. Through comprehensive ablation study, we demonstrate that: 1) Employing symmetric feature fusion can effectively exploit symmetrical information to facilitate fracture detection. 2) Performing spatial alignment at the feature level for symmetric feature fusion leads to substantial performance gain. 3) Using contrastive learning, the Siamese encoder is able to learn more sensible embedding, leading to further performance improvement. By comparing with the state-of-the-art methods, including latest ones modeling symmetry, we demonstrate the AASN is by far the most effective method exploiting symmetry and reports substantially improved performances on both classification and localization tasks.

References

1. Barman, A., Inam, M.E., Lee, S., Savitz, S., Sheth, S., Giancardo, L.: Determining ischemic stroke from CT-angiography imaging using symmetry-sensitive convolutional networks. In: 2019 IEEE 16th International Symposium on Biomedical Imaging (ISBI 2019), pp. 1873–1877, April 2019. https://doi.org/10.1109/ISBI.2019.8759475

2. Bertinetto, L., Valmadre, J., Henriques, J.F., Vedaldi, A., Torr, P.H.S.: Fully-convolutional siamese networks for object tracking. In: Hua, G., Jégou, H. (eds.) ECCV 2016. LNCS, vol. 9914, pp. 850–865. Springer, Cham (2016). https://doi.org/10.1007/978-3-319-48881-3_56

3. Bookstein, F.L.: Principal warps: thin-plate splines and the decomposition of deformations. IEEE Trans. Pattern Anal. Mach. Intell. 11(6), 567–585 (1989). https://doi.org/10.1109/34.24792

4. Bustos, A., Pertusa, A., Salinas, J.M., de la Iglesia-Vayá, M.: PadChest: a large chest x-ray image dataset with multi-label annotated reports. Med. Image Anal. 66, 101797 (2019)

5. Chen, H., Miao, S., Xu, D., Hager, G.D., Harrison, A.P.: Deep hierarchical multi-label classification of chest x-ray images. In: Cardoso, M.J., et al. (eds.) Proceedings of The 2nd International Conference on Medical Imaging with Deep Learning. Proceedings of Machine Learning Research, 08–10 July 2019, vol. 102, pp. 109–120. PMLR, London (2019). http://proceedings.mlr.press/v102/chen19a.html

6. Cheng, C.-T., et al.: Application of a deep learning algorithm for detection and visualization of hip fractures on plain pelvic radiographs. Eur. Radiol. 29(10), 5469–5477 (2019). https://doi.org/10.1007/s00330-019-06167-y

7. Clohisy, J.C., et al.: A systematic approach to the plain radiographic evaluation of the young adult hip. J. Bone Joint Surg. Am. 90(Suppl 4), 47 (2008)

8. Gale, W., Oakden-Rayner, L., Carneiro, G., Bradley, A.P., Palmer, L.J.: Detecting hip fractures with radiologist-level performance using deep neural networks. CoRR abs/1711.06504 (2017). http://arxiv.org/abs/1711.06504

9. Guo, Q., Feng, W., Zhou, C., Huang, R., Wan, L., Wang, S.: Learning dynamic siamese network for visual object tracking. In: Proceedings of the IEEE International Conference on Computer Vision, pp. 1763–1771 (2017)

10. Guo, Q., Feng, W., Zhou, C., Huang, R., Wan, L., Wang, S.: Learning dynamic siamese network for visual object tracking. In: The IEEE International Conference on Computer Vision (ICCV), October 2017

11. Hadsell, R., Chopra, S., LeCun, Y.: Dimensionality reduction by learning an invariant mapping. In: 2006 IEEE Computer Society Conference on Computer Vision and Pattern Recognition (CVPR 2006), vol. 2, pp. 1735–1742, June 2006. https://doi.org/10.1109/CVPR.2006.100

12. Hoffer, E., Ailon, N.: Deep metric learning using triplet network. In: Feragen, A., Pelillo, M., Loog, M. (eds.) Similarity-Based Pattern Recognition. LNCS, pp. 84–92. Springer, Cham (2015). https://doi.org/10.1007/978-3-319-24261-3_7

13. Huang, G., Liu, Z., van der Maaten, L., Weinberger, K.: Densely connected convolutional networks. arxiv website. arxiv. org/abs/1608.06993. 24 August 2016

14. Huang, G., Liu, Z., Weinberger, K.Q.: Densely connected convolutional networks. CoRR abs/1608.06993 (2016). http://arxiv.org/abs/1608.06993

15. Irvin, J., et al.: CheXpert: a large chest radiograph dataset with uncertainty labels and expert comparison. CoRR abs/1901.07031 (2019)

16. Kingma, D.P., Ba, J.: Adam: a method for stochastic optimization. arXiv preprint arXiv:1412.6980 (2014)
17. Koch, G., Zemel, R., Salakhutdinov, R.: Siamese neural networks for one-shot image recognition. In: ICML Deep Learning Workshop, vol. 2 (2015)
18. Konukoglu, E., Glocker, B., Criminisi, A., Pohl, K.M.: Wesd-weighted spectral distance for measuring shape dissimilarity. IEEE Trans. Pattern Anal. Mach. Intell. **35**(9), 2284–2297 (2012)
19. Li, W., et al.: Structured landmark detection via topology-adapting deep graph learning. arXiv preprint arXiv:2004.08190 (2020)
20. Li, Z., et al.: Thoracic disease identification and localization with limited supervision. In: 2018 IEEE/CVF Conference on Computer Vision and Pattern Recognition, Salt Lake City, UT, pp. 8290–8299. IEEE (2018). https://doi.org/10.1109/CVPR.2018.00865
21. Ling, H., Gao, J., Kar, A., Chen, W., Fidler, S.: Fast interactive object annotation with curve-GCN. CoRR abs/1903.06874 (2019). http://arxiv.org/abs/1903.06874
22. Liu, C.F., et al.: Using deep siamese neural networks for detection of brain asymmetries associated with Alzheimer's disease and mild cognitive impairment. Magn. Reson. Imaging (2019). https://doi.org/10.1016/j.mri.2019.07.003, http://www.sciencedirect.com/science/article/pii/S0730725X19300086
23. Liu, S.X.: Symmetry and asymmetry analysis and its implications to computer-aided diagnosis: a review of the literature. J. Biomed. Inform. **42**(6), 1056–1064 (2009)
24. Liu, Y., et al.: From unilateral to bilateral learning: detecting mammogram masses with contrasted bilateral network. In: Shen, D., et al. (eds.) MICCAI 2019. LNCS, vol. 11769, pp. 477–485. Springer, Cham (2019). https://doi.org/10.1007/978-3-030-32226-7_53
25. Lu, Y., et al.: Learning to segment anatomical structures accurately from one exemplar. arXiv preprint arXiv:2007.03052 (2020)
26. Melekhov, I., Kannala, J., Rahtu, E.: Siamese network features for image matching. In: 2016 23rd International Conference on Pattern Recognition (ICPR), pp. 378–383, December 2016. https://doi.org/10.1109/ICPR.2016.7899663
27. Paszke, A., et al.: Automatic differentiation in PyTorch (2017)
28. Rajpurkar, P., et al.: CheXNet: radiologist-level pneumonia detection on chest x-rays with deep learning. CoRR abs/1711.05225 (2017). http://arxiv.org/abs/1711.05225
29. Simonovsky, M., Gutiérrez-Becker, B., Mateus, D., Navab, N., Komodakis, N.: A deep metric for multimodal registration. In: Ourselin, S., Joskowicz, L., Sabuncu, M.R., Unal, G., Wells, W. (eds.) MICCAI 2016. LNCS, vol. 9902, pp. 10–18. Springer, Cham (2016). https://doi.org/10.1007/978-3-319-46726-9_2
30. Sun, Y., Wang, X., Tang, X.: Deep learning face representation by joint identification-verification. CoRR abs/1406.4773 (2014). http://arxiv.org/abs/1406.4773
31. Varior, R.R., Haloi, M., Wang, G.: Gated siamese convolutional neural network architecture for human re-identification. In: Leibe, B., Matas, J., Sebe, N., Welling, M. (eds.) ECCV 2016. LNCS, vol. 9912, pp. 791–808. Springer, Cham (2016). https://doi.org/10.1007/978-3-319-46484-8_48
32. Wachinger, C., Golland, P., Kremen, W., Fischl, B., Reuter, M., Initiative, A.D.N., et al.: Brainprint: a discriminative characterization of brain morphology. NeuroImage **109**, 232–248 (2015)
33. Wang, H., Xia, Y.: ChestNet: a deep neural network for classification of thoracic diseases on chest radiography. arXiv preprint arXiv:1807.03058 (2018)

34. Wang, X., Peng, Y., Lu, L., Lu, Z., Bagheri, M., Summers, R.M.: ChestX-ray8: hospital-scale chest x-ray database and benchmarks on weakly-supervised classification and localization of common thorax diseases. In: The IEEE Conference on Computer Vision and Pattern Recognition (CVPR), July 2017

35. Wang, Y., et al.: Weakly supervised universal fracture detection in pelvic x-rays. In: Shen, D., et al. (eds.) MICCAI 2019. LNCS, vol. 11769, pp. 459–467. Springer, Cham (2019). https://doi.org/10.1007/978-3-030-32226-7_51

36. Xu, Z., et al.: Less is more: simultaneous view classification and landmark detection for abdominal ultrasound images. In: Frangi, A.F., Schnabel, J.A., Davatzikos, C., Alberola-López, C., Fichtinger, G. (eds.) MICCAI 2018. LNCS, vol. 11071, pp. 711–719. Springer, Cham (2018). https://doi.org/10.1007/978-3-030-00934-2_79

37. Zagoruyko, S., Komodakis, N.: Learning to compare image patches via convolutional neural networks. In: Proceedings of the IEEE Conference on Computer Vision and Pattern Recognition, pp. 4353–4361 (2015)

38. Zhou, B., Khosla, A., Lapedriza, À., Oliva, A., Torralba, A.: Learning deep features for discriminative localization. CoRR abs/1512.04150 (2015). http://arxiv.org/abs/1512.04150

DeepLandscape: Adversarial Modeling of Landscape Videos

Elizaveta Logacheva[1]([envelope]), Roman Suvorov[1], Oleg Khomenko[1],
Anton Mashikhin[1], and Victor Lempitsky[1,2]

[1] Samsung AI Center, Moscow, Russia
elimohl@gmail.com
[2] Skolkovo Institute of Science and Technology, Moscow, Russia

Abstract. We build a new model of landscape videos that can be
trained on a mixture of static landscape images as well as landscape
animations. Our architecture extends StyleGAN model by augmenting it
with parts that allow to model dynamic changes in a scene. Once trained,
our model can be used to generate realistic time-lapse landscape videos
with moving objects and time-of-the-day changes. Furthermore, by fitting the learned models to a static landscape image, the latter can be
reenacted in a realistic way. We propose simple but necessary modifications to StyleGAN inversion procedure, which lead to in-domain latent
codes and allow to manipulate real images. Quantitative comparisons
and user studies suggest that our model produces more compelling animations of given photographs than previously proposed methods. The
results of our approach including comparisons with prior art can be seen
in supplementary materials and on the project page https://saic-mdal.
github.io/deep-landscape/.

1 Introduction

This work is motivated by the "bringing landscape images to life" application.
We thus aim to build a system that for a given landscape photograph, generates
its plausible animation with realistic movements and global lighting changes. To
achieve our goal, we first build a generative model (Fig. 1) of timelapse landscape
videos, which can successfully capture complex aspects of this domain. These
complexities include both static aspects such as abundance of spatial details,
high variability of texture and geometry, as well as dynamic complexity including
motions of clouds, waves, foliage, and global lighting changes. We build our
approach upon the recent progress in the generative modeling of images, and
specifically the StyleGAN model [1]. We show how to change the StyleGAN
model to learn and to decompose different dynamic effects: global changes are
controlled by the non-convolutional variables, strong local motions are controlled
by "noise branch" inputs.

Electronic supplementary material The online version of this chapter (https://
doi.org/10.1007/978-3-030-58592-1_16) contains supplementary material, which is
available to authorized users.

© Springer Nature Switzerland AG 2020
A. Vedaldi et al. (Eds.): ECCV 2020, LNCS 12368, pp. 256–272, 2020.
https://doi.org/10.1007/978-3-030-58592-1_16

Similarly to the original StyleGAN model, ours requires a large amount of training data. While it is very hard to obtain a very large dataset of high-quality scenery timelapse videos, obtaining a large-scale dataset of scenery static images is much easier. We thus suggest how our generative model can be learned from two sources, namely (i) a large-scale dataset of static images, (ii) a smaller dataset of videos. Previous video GANs learn motion from sequences of consecutive video frames. We show that learning on randomly taken frames without an explicit motion model is possible. It allows to disentangle static appearance from the dynamic, as well as manifold of possible changes from a trajectory in it.

Fig. 1. Videos generated by the DeepLandscape model. Each row shows a separate video, obtained by sampling the static and dynamic components randomly, and then animating the dynamic components using homography warping. These videos are generated at 512×512 resolution (*zoom-in recommended*).

Once trained, our model can animate a given photograph. We first fit the latent variables of the model to the provided image, and then obtain the animation by changing the subset of variables corresponding to dynamic aspects appropriately. As our model has more latent parameters than a given static image, fitting them to a photograph is an ill-posed problem, and we develop a particular method for such fitting that results in plausible animations. While our model is trained to generate images at medium resolution (256×256 or 512×512), we show that we can postprocess the results with an appropriately trained super-resolution network to obtain videos at higher resolution (up to one megapixel).

In the experiments, we assess the realism of synthetic videos sampled from our generative model and its ablations. Furthermore, we evaluate our approach at our main task ("bringing landscape images to life"). For this task, both quantitative comparisons and, more importantly, user studies reveal a significant advantage of our system over the three recently proposed approaches [2–4].

2   Related Work

Learning video representation and predicting future frames using deep neural networks is a very active area of research [5–8]. Most early works are focused on using deep neural networks (DNNs) with recurrent units (GRU or LSTM) and train them in supervised manner to obtain next frame using pixel-level prediction [5,8]. At the same time, Generative Adversarial Nets (GANs) [9] have achieved very impressive results for image generation, and recently several methods extending them to video have been suggested. Some GAN-based models consider single image as an input (*image2video*) [10,11], while others input sequences of frames (*video2video*, [7,12–15]). In this work we focus only on the image2video setting. Training GANs for video-generation often performed with two discriminator networks: single image and temporal discriminators [12,14,16]. In this work we propose to use a simplified temporal discriminator, which only looks at unordered pairs of frames.

Video generation/prediction works generally consider either videos with articulated objects/multiple moving objects [17,18] or videos with weakly structured moving objects or dynamic textures such as clouds, grass, fire [4,14]. Our work is more related to the latter case, namely: landscape photos and videos. Because of the domain specifics, we can model spatial motions in the video in the *latent* space using simple homography transformations, and let the generator to synthesize plausible deviations from this simplistic model. Our approach is thus opposed to methods that animate landscapes and textures by generating warping fields applied to the *raw pixels* of the input static image [2,11,19–21]. Animation in the latent space as well as the separation of latent space into static and dynamic components has been proposed and investigated in [2,6,22–24]. Our work modifies and extends these ideas to the StyleGAN [1] model.

As we need to find latent space embedding of static images in order to animate them, we follow a number of works on GAN inversion (inference). Here, we borrow ideas of using an encoder into the latent space followed by gradient descent [25], the latent space expansion for StyleGAN [26], and generator fine-tuning [27,28]. On top of that, we have to make several important adjustments to the inference procedure specific to our architecture, and we show that without such adjustments the animation works poorly.

3   Method

3.1   Generative Model of Timelapse Videos

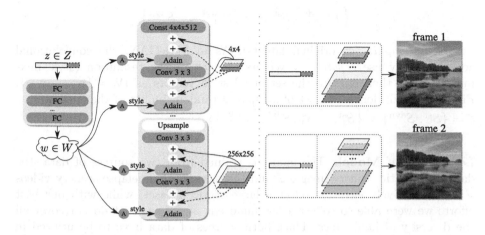

Fig. 2. Left – the generator used by our model (augmented StyleGAN generator). The main difference from StyleGAN is the second set of spatial input tensors (darkgray). Right – sampling procedure for our model. Two frames of the same video can be sampled by using same static latent variables (lightgray), and two different sets of dynamic latent variables (darkgray and yellow). (Color figure online)

Model Architecture. The architecture of our model is based on StyleGAN [1]. Our model outputs images of resolution 256×256 (or 512×512) and has four sets of latent variables (Fig. 2):

- a vector $\mathbf{z}^{st} \in \mathbb{R}^{D^{st}}$, which encodes colors and the general scene layout;
- a vector $\mathbf{z}^{dyn} \in \mathbb{R}^{D^{dyn}}$, which encodes global lighting (e.g. time of day);
- a set $\mathcal{S}^{st}$ of square matrices $S_1^{st} \in \mathbb{R}^{4 \times 4}, ..., S_N^{st} \in \mathbb{R}^{2^{N+1} \times 2^{N+1}}$, which encode shapes and details of static objects at $N = 7$ different resolutions between 4×4 and 256×256 ($N = 8$ for 512×512);
- a set $\mathcal{S}^{dyn}$ of square matrices $S_1^{dyn} \in \mathbb{R}^{4 \times 4}, ..., S_N^{dyn} \in \mathbb{R}^{2^{N+1} \times 2^{N+1}}$, which encode shapes and details of dynamic objects at the corresponding resolutions.

Our generator has two components: the multilayer perceptron **M** and the convolutional generator **G**. As in [1], the perceptron **M** takes the concatenated vector $\mathbf{z} = [\mathbf{z}^{st}, \mathbf{z}^{dyn}] \in \mathbb{R}^{512}$ and transforms it to the *style vector* $\mathbf{w} \in \mathbb{R}^{512}$. The convolutional generator **G** also follows [1] and has $N = 7$ (or 8) blocks. Within each block, a convolution is followed by two elementwise additions of two tensors obtained from S_n^{st} and S_n^{dyn} by a learnable per-channel scaling (whereas [1] has only one addition). Finally, the AdaIN [29] transform is applied using per-channel scales and biases obtained from $\mathbf{w}$ using learnable linear transform. Within each block, this sequence of steps is repeated twice followed by upsampling and convolution layers.

Below, we will refer to the set of input latent variables

$$\left\{ \mathbf{z}^{\text{st}}, \mathbf{z}^{\text{dyn}}, S_1^{\text{st}}, ..., S_N^{\text{st}}, S_1^{\text{dyn}}, ..., S_N^{\text{dyn}} \right\}$$

as *original inputs* (or original latents). As in StyleGAN, the convolutional generator may use *separate* $\mathbf{w}$ vectors at each of the resolution (style mixing). We will then refer to the set of all style vectors as $\mathcal{W} = \{\mathbf{w}_1, ..., \mathbf{w}_N\}$. Finally, we will denote the set of all spatial random inputs of the generator as $\mathcal{S} = \{\mathcal{S}^{\text{st}}, \mathcal{S}^{\text{dyn}}\} = \left\{ S_1^{\text{st}}, ..., S_N^{\text{st}}, S_1^{\text{dyn}}, ..., S_N^{\text{dyn}} \right\}$.

Learning the Model. The model is trained from two sources of data, the dataset of static scenery images $\mathcal{I}$ and the dataset of timelapse scenery videos $\mathcal{V}$. It is relatively easy to collect a large static dataset, while with our best efforts we were able to collect a few hundreds of videos, that do not cover all the diversity of landscapes. Thus, both sources of data have to be utilized in order to build a good model. To do that, we train our generative model in an adversarial way with two different discriminators.

The *static discriminator* D_{st} has the same architecture and design choises as in StyleGAN. It observes images from $\mathcal{I}$ as real, while the fake samples are generated by our model. The *pairwise discriminator* D_{dyn} looks at pairs of images. It duplicates the architecture of D_{st} except first convolutional block that is applied separately to each frame. A real pair of images is obtained by sampling a video from $\mathcal{V}$, and then sampling two random frames (arbitrary far for each other) from it. A fake pair is obtained by sampling common static latents $\mathbf{z}^{\text{st}}$ and $\mathcal{S}^{\text{st}}$, and then individual dynamic latents $\mathbf{z}^{\text{dyn},1}$, $\mathbf{z}^{\text{dyn},2}$ and $\mathcal{S}^{\text{dyn},1}$, $\mathcal{S}^{\text{dyn},2}$. The two images are then obtained as $\mathbf{G}(\mathbf{M}(\mathbf{z}^{\text{st}}, \mathbf{z}^{\text{dyn},1}), \mathcal{S}^{\text{st}}, \mathcal{S}^{\text{dyn},1})$ and $\mathbf{G}(\mathbf{M}(\mathbf{z}^{\text{st}}, \mathbf{z}^{\text{dyn},1}), \mathcal{S}^{\text{st}}, \mathcal{S}^{\text{dyn},2})$. All samples are drawn from unit normal distributions.

The model is trained within standard GAN approach with non-saturating loss [9] with R1 regularization [30] as in the original StyleGAN paper. During each update of the generator, we either sample a batch of fake images to which the static discriminator is applied or a batch of image pairs to which the pairwise discriminator is applied. The proportions of the static discriminator and the pairwise discriminator are annealed from 0.5/0.5 to 0.9/0.1 respectively over each resolution transition phase and then kept fixed at 0.1. This helps the generator to learn disentangle static and dynamic latents early for each resolution and prevents the pairwise generator from overfitting to our relatively small video dataset.

During learning, we want the pairwise discriminator to focus on the inconsistencies within each pair, and leave visual quality to the static discriminator. Furthermore, since the pairwise discriminator only sees real frames sampled from a limited number of videos, it may prone overfit to this limited set and effectively stop contributing to the learning process (while the static discriminator, which observes more diverse set of scenes, keeps improving the diversity of the model). It turns out, both problems (focus on image quality rather than pairwise consistency, overfitting to limited diversity of videos) can be solved with a simple trick. We augment the fake set of frames with pairs of crops taken from same video frame, but from different locations. Since these crops have the same visual quality as the images in real frames, and since they come from the same videos as images within real pairs, the pairwise discriminator effectively stops paying attention to image quality, cannot simply overfit to the statistics of scenes in the video dataset, and has to focus on finding pairwise inconsistencies within fake pairs. We observed this *crop sampling* trick to improve the quality of our model significantly.

Config	I2S [26]	MO	E	EO	EOI	EOIF	EOIFS
Init $\mathcal{W}$	Mean	Mean	**E**	**E**	**E**	**E**	**E**
Init $\mathcal{S}$	Random	Zero	Random	Zero	Zero	Zero	Zero
Optimize $\mathcal{S}$		+		+	+	+	+
Optimize $\mathcal{W}$	+	+		+	+	+	+
L^O_{init}					+	+	+
Fine-Tune **G**						+	+
Segmentation							+

	I2S [26]	MO	E	EO	EOI	EOIF	EOIFS
Reconstruction	-	+	-	+	±	+	+
Animation	-	-	+	-	+	+	+

Fig. 3. The effect of different inference algorithms on the reconstruction quality and the ability to animate. **Left column**: original image. **First row**: reconstructions obtained with different inference algorithms. **Second row**: a frame from animation ($\mathcal{S}^{\mathrm{dyn}}$ are shifted 50% left). Note that I2S [26] does not work well in our case, since our generator relies on $\mathcal{S}$ more than the original StyleGAN method. L^O_{init} is a regularization term applied to $\mathcal{W}$ during inference, which makes latents to stay in-domain and allows to manipulate real images. We quantify these effects in supplementary materials.

Sampling Videos from the Model. Our model does not attempt to learn full temporal dynamics of videos, and instead focuses on pairwise consistency of frames that are generated when the dynamic latent variables are resampled. In particular, the pairwise discriminator in our model does not sample real frames sequentially. The sampling procedure for fake pairs does not try to generate adjacent frames either. One of the reasons why we do not attempt to learn continuity, is because the training dataset contains videos of widely-varying temporal rates, making the notion of temporal adjacency for a pair of frames effectively meaningless.

Because of this our generation process is agnostic to a model of motion. The generator is forced to produce plausible frames regardless of S^{dyn} and $\mathbf{z}^{\mathrm{dyn}}$ changes. In our experiments we found that a simple model of motion described below is enough to produce compelling videos. Specifically, to sample a video, we sample a single static vector $\mathbf{z}^{\mathrm{st}}$ from the unit normal distribution and then interpolate the dynamic latent vector between two unit normally-distributed samples $\mathbf{z}^{\mathrm{dyn},1}$ and $\mathbf{z}^{\mathrm{dyn},2}$. For the spatial maps, we again sample S^{st} and $S^{\mathrm{dyn},1}$ from a unit normal distribution and then warp the S^{dyn} tensor continuously using a homography transform parameterized by displacements of two upper corners and two points at the horizon. The direction of the homogrpahy is sampled randomly, speed was chosen to match the average speed of clouds in our dataset. The homography is flipped vertically for positions below the horizon to mimic the reflection process. To obtain $S^{\mathrm{dyn},i}$, we make a composition of $i-1$ identical transforms and then apply it to $S^{\mathrm{dyn},1}$. As we interpolate/warp the latent variables, we pass them through the trained model to obtain the smooth videos (Fig. 1 and **Supplementary video**). Note that our models requires no image-specific user input.

3.2   Animating Real Scenery Images with Our Model

Inference. To animate a given scenery image I, we find (infer) a set of latent variables that produce such image within the generator. Following [26], we look for extended latents W and S, so that $\mathbf{G}(W, S) \approx I$, but our procedure is different from theirs. After that, we apply the same procedure as described above to animate the given image.

The latent space of our generator is highly redundant, and to obtain good animation, we have to ensure that the latent variables come roughly from the same distribution as during the training of the model (most important, W should belong to the output manifold of $\mathbf{M}$). Without such prior, the latent variables that generate good reconstruction might still result in implausible animation (or lack of it). We therefore perform inference using the following three-step procedure:

1. **Step 1**: predicting a set of style vectors W' using a feedforward *encoder* network $\mathbf{E}$ [25]. The encoder has ResNet-152 [31] architecture and is trained on 200000 synthetic images with mean absolute error loss. W is predicted by two-layer perceptron with ReLU from the concatenation of features from several levels of ResNet, aggregated by global average pooling.
2. **Step 2**: starting from W' and zero S, we optimize all latents to improve reconstruction error. In addition, we penalize the deviation of W from the predicted W' (with coefficient 0.01) and the deviation of S from zero (by reducing learning rate). We optimize for up to 500 steps with Adam [32] and large initial learning rate (0.1), which is halved each time the loss does not

improve for 20 iterations. A variant of our method that we evaluate separately, uses a binary segmentation mask obtained by ADE20k-pretrained [33] segmentation network[1]. The mask identifies dynamic (sky+water) and remaining (static) parts of the scene. In this variant, $\mathcal{S}^{st}$ (respectively $\mathcal{S}^{dyn}$) are kept at zero for dynamic (respectively, static) parts of the image.

3. **Step 3**: freezing latents and fine-tuning the weights of **G** to further drive down the reconstruction error [27, 28]. The step is needed since even after optimization, the gap between the reconstruction and the input image remains. During this fine-tuning, we minimize the combination of the per-pixel mean absolute error and the perceptual loss [34], with much larger (10×) weight for the latter. We do 500 steps with ADAM and $lr = 0.001$.

Fig. 4. Examples of real images animated with our model. Each row shows a sequence of frames from a single video. Each frame is 256 × 256 (please zoom in for details). Clouds, reflections and waves move and change their shape naturally; time of day also changes. More examples are available in the **Supplementary video**.

Please refer to Fig. 3 and *Supplementary Materials* for examples of qualitative effects of fine tuning. We also evaluate our inference pipeline quantitatively (see Sect. 4).

---

[1] CSAIL-Vision: https://github.com/CSAILVision/semantic-segmentation-pytorch.

Lighting Manipulation. During training of the model, $\mathbf{M}$ is used to map $\mathbf{z}$ to $\mathbf{w}$. We resample $\mathbf{z}^{\text{dyn}}$ in order to take into account variations of lighting, weather changes, etc. and to have $\mathbf{z}^{\text{st}}$ describe only static attributes (land, buildings, horizon shape, etc.). To change lighting in a real image, one has to change $\mathbf{z}^{\text{dyn}}$ and then use MLP to obtain new styles $\mathcal{W}$. Our inference procedure, however, outputs $\mathcal{W}$ and we have found it very difficult to invert $\mathbf{M}$ and obtain $\mathbf{z} = \mathbf{M}^{-1}(\mathbf{w})$.

To tackle this problem, we train a separate neural network, $\mathbf{A}$, to approximate local dynamics of $\mathbf{M}$. Let $\mathbf{w}_a = \mathbf{M}(\mathbf{z}_a^{\text{st}}, \mathbf{z}_a^{\text{dyn}})$ and $\mathbf{w}_b = \mathbf{M}(\mathbf{z}_b^{\text{st}}, \mathbf{z}_b^{\text{dyn}})$, we optimize $\mathbf{A}$ as follows: $\mathbf{A}(\mathbf{w}_a, \mathbf{z}_b^{\text{dyn}}, c) \approx \mathbf{M}(\mathbf{z}_a^{\text{st}}, \mathbf{z}_a^{\text{dyn}}\sqrt{1-c} + \mathbf{z}_b^{\text{dyn}}\sqrt{c})$, where $c \sim Uniform(0,1)$ is coefficient of interpolation between $\mathbf{w}_a$ and $\mathbf{w}_b$. Thus, $c = 0$ corresponds to $\mathbf{z}_a^{\text{dyn}}$, so $\mathbf{A}(\mathbf{w}_a, \mathbf{z}_b^{\text{dyn}}, 0) \approx \mathbf{w}_a$; $c = 1$ corresponds to $\mathbf{z}_b^{\text{dyn}}$, so $\mathbf{A}(\mathbf{w}_a, \mathbf{z}_b^{\text{dyn}}, 1) \approx \mathbf{w}_b$.

We implement this by the combination of L1-loss $L_{Abs}^{\mathbf{A}} = |\mathbf{w}_b - \mathbf{A}(\cdot)|$ and relative direction loss $L_{Rel}^{\mathbf{A}} = 1 - \cos(\mathbf{w}_b - \mathbf{w}_a, \mathbf{A}(\cdot) - \mathbf{w}_a)$. The total optimization criterion is $L^{\mathbf{A}} = L_{Abs}^{\mathbf{A}} + 0.1 L_{Rel}^{\mathbf{A}}$. We train $\mathbf{A}$ with ADAM [32] until convergence. At test time, the network $\mathbf{A}$ allows us to sample a random target $\mathbf{z}_b^{\text{dyn}}$ and update $\mathcal{W}$ towards it by increasing the interpolation coefficient c as the animation progresses. Please refer to Fig. 4 and **Supplementary Video** for examples of animations with our full pipeline.

Super Resolution (SR). As our models are trained at medium resolution (e.g. 256×256), we aim to bring fine details from the given image that we need to animate through a separate super-resolution procedure. The main idea of our super resolution approach is to borrow as much as possible from the original high-res image (which is downsampled for animation via $\mathbf{G}$). To achieve that, we super-resolve the animation and blend it with the original image using a standard image superresolution approach. We use ESRGANx4 [35] trained on a dedicated dataset that is created as follows. To obtain the (hi-res, low-res) pair, we take a frame I from our video dataset as a hi-res image, we downsample it and run the first two steps of inference and obtain an (imperfect) low-res image. Thus, the network is trained on a more complex task than superresolution.

After obtaining the super-resolved video, we transfer dynamic parts (sky and water) from it to the final result. The static parts are obtained by running the guided filter [36] on the super-resolved frames while using the input high-res image as a guide. Such procedure effectively transfers high-res details from the input, while retaining the lighting change induced by lighting manipulation (Fig. 5).

4 Experiments

We evaluate our method both quantitatively and qualitatively (via user study) on synthetic and real images separately. Evaluation on synthetic images (*generation*) aims on quantifying impact of major design choices of $\mathbf{G}$ itself (without encoding and super-resolution). Evaluation on real images (*animation*) aims on comparison with previous single-image animation methods, including Animating Landscape (*AL*) [2], SinGAN (*SG*) [3] and Two-Stream Networks (*TS*) [4]. The Animating Landscape system is based on learnable warping and is trained on more than a thousand time-lapse videos from [14,37]. The SinGAN method creates a hierarchical model of image content based on the input model alone. It therefore has an advantage of not needing an external dataset, though, as a downside, it requires considerable time to fit a new image. Two-Stream Networks [4] create animated textures given a static texture image and a short clip (an example of motion) via optimization of video tensor. We also tried a to include two more baselines, i.e. linear dynamic systems [38] and Seg2Vid [10], but with former we got very poor quality and the latter failed to converge on our data, so we did not proceed with full comparison. We also tried to train and finetune *AL* on our video dataset (which is significantly smaller than that from *AL* paper), with little success (see supp. mat.).

We estimate quality through three different aspects: *individual image quality*; *static consistency*; *animation plausibility*. Individual image quality is estimated via Fréchet Inception Distance [39], masked SSIM [40] and LPIPS [41]. Static consistency evaluation aims on quantifying how good objects that must not move (e.g. buildings, mountains etc.) are preserved over time. For that purpose we calculate SSIM and LPIPS between first frame and each generated video frame (only for static parts). Perfect image quality and static consistency can be achieved by not animating anything at all. Thus, we evaluate animation plausibility via user study and Fréchet Video Distance [42].

To generate videos using our method, we use a manually constructed set of homographies. Data-driven estimation of homographies is out scope of this work, so we have prepared 12 homographies, one for each clock position (e.g. the "12h" move clouds up and towards the observer, the "3h" moves straight to the right, etc.). Normally, these homographies resemble the average speed of clouds in our training dataset. We increase this speed for synthetic experiments to make differences between variants of our method more obvious; we slow down animation for experiments with real images in order to approximately align our speed with that of the competitors (AL, SG and TS).

Datasets. Our model was trained using both videos and single images available in the Internet under Creative Commons License. For evaluation we use 69 landscape FullHD time-lapse videos published on YouTube between Dec. 28 2019 and Jan. 29 2020. For FID computation, we have collected 2400 pictures from Flickr[2].

Generation. In order to perform thorough ablation study in reasonable time, we perform all evaluations in this section at 128×128 resolution. To estimate static consistency, we sample 1200 pairs of images from **G**, mask out sky and water according to segmentation mask and calculate LPIPS and SSIM between two images in a pair. In each pair the images are generated from the same $\mathbf{z}^{st}, \mathcal{S}^{st}$ and different $\mathbf{z}^{dyn}, \mathcal{S}^{dyn}$. For the user study we sample 100 videos 200

---

[2] https://flickr.com.

Input	$\mathbf{G}_1$	$\mathbf{G}_2$	SR_1	SR_2

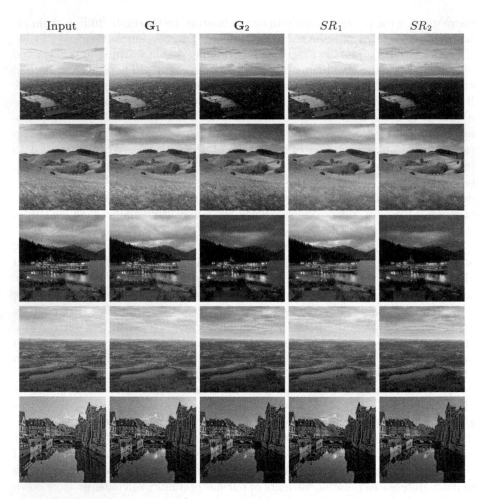

Fig. 5. Examples of super-resolution (SR) applied to the output of our generator $(\mathbf{G})$ given input image (Input). The inputs and SR are at 1024×1024 resolution, while the low-res images are at 256×256 resolution. Zoom-in recommended.

Setup	FID↓	SSIM↑	LPIPS↓	ΔR
Original StyleGAN	**48.40**	0.809	**0.049**	
+ frame discriminator	55.92	0.846	0.064	0.13
+ separate $\mathcal{S}^{st}$ and $\mathcal{S}^{dyn}$	55.15	0.854	0.073	0.01
+ separate $\mathbf{z}^{st}$ and $\mathbf{z}^{dyn}$	54.38	0.879	0.065	0.03
+ crop sampling	56.13	**0.884**	0.062	**0.06**

Fig. 6. Results of the ablation study of our model for the task of new video generation. The column ΔR in the table are obtained from the side-by-side user study. ΔR shows the increase in frequency when assessors prefer this variant to that in previous row (+0.23 against original StyleGAN).

frames long at 30 FPS. In order to compare different ablations, the assessors were asked to select the most realistic video from a pair shown side-by-side. Each assessor is limited to evaluate no more than three pages with four tasks on each and has five minutes to complete each page. In our user study we showed each pair to five assessors. The ablation study results (Fig. 6) reveal that the original StyleGAN generates the most high-fidelity images, but fails to preserve details of static objects. LPIPS is more tolerant to motion until the "texture type" changes dramatically. Thus, despite LPIPS and FID achieving the best values for the original StyleGAN, it actually does not preserve static objects (see **Supplementary Video**). Our modifications allow to keep a similar level of the FID value, but gradually improve static consistency and animation plausibility.

Real Image Animation. Experiments in this section are performed at 256×256 resolution. To calculate quantitative and qualitative metrics, we took the first frames I_0 of the test videos, encoded and animated them with our method. Denote the n-th frames of real and generated videos as I_n and $\widehat{I}_n$ respectively. With our method, for each input image we generate five variants with homographies randomly sampled from the predefined set. For AL we generate five videos for each input image with randomly sampled motion, as described in the original paper. For all quantitative evaluations we do not apply style transfer in AL and $\mathcal{W}$ manipulation in our method. We evaluate two variants of AL: with (AL) and without first-to-last interpolation (AL_{noint}), which stabilizes image quality, but makes long movements impossible. We use the official implementation[3] of Animating Landscape [2] provided by authors. We use pretrained AL model; we also evaluate finetuned AL model and found that most metrics degraded, while the training loss continued to improve. This can be attributed to the fact that the video dataset used in AL is bigger than ours; both include the public part of data from [14]. Both datasets are just youtube landscape videos and seem to be equally close to the validation (we are not aware of any biases). Also, our dataset contains videos with very different motion speed, and neither text of AL nor its code contains details regarding video speed equalization. All images are animated in original resolution cropped to 1:1 aspect ratio via center crop, then bilinearly downsampled to 256×256 resolution.

For SG [3] we used the official implementation[4] and default parameters. We have not noticed significant difference between multiple SG runs both in terms of quantitative metrics and visual diversity. Hence we decided not to generate similar videos many times and sampled only one video for each input image.

For TS [4] we used the official implementation[5]. TS can animate only the whole image, so (1) we used semantic segmentation to extract sky; (2) transferred motion to the extracted image fragment from a random video from the validation set; (3) blended static part of the original image with the generated clip. TS is only capable of producing 12 frames due to GPU memory limitations, so we interpolated frames in order to obtain the necessary video length.

---

[3] https://github.com/endo-yuki-t/Animating-Landscape.

[4] https://github.com/tamarott/SinGAN.

[5] https://github.com/ryersonvisionlab/two-stream-dyntex-synth.

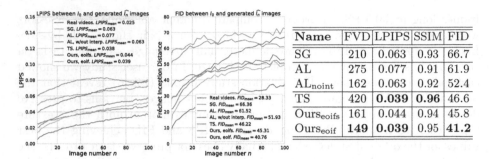

Name	FVD	LPIPS	SSIM	FID
SG	210	0.063	0.93	66.7
AL	275	0.077	0.91	61.9
AL$_{noint}$	162	0.063	0.92	52.4
TS	420	**0.039**	**0.96**	46.6
Ours$_{eoifs}$	161	0.044	0.94	45.8
Ours$_{eoif}$	**149**	**0.039**	0.95	**41.2**

Fig. 7. Quantitative comparison of image quality, static consistency and motion plausibility. **Left and middle**: LPIPS↓ and FID↓ between I_0 and $\widehat{I}_n$, which mostly measure image quality and static consistency. The legend contains metrics averaged over time. As can be seen, pixel-level transformations (e.g. using predicted flows in AL) lead to faster deterioration of generated images over time, compared to our approach, especially for later frames ($n \gtrsim 50$). **Right**: FVD↓, LPIPS↓, SSIM↑ and FID↓ between I_n and $\widehat{I}_n$ averaged over time, which measure not only image quality, but also animation plausibility.

We evaluate image quality by measuring FID between the set of all first frames of real videos I_0 and the set of n-th frames of generated videos $\widehat{I}_n$. Thus, we can see how fast these two distributions diverge. Too fast divergence in terms of FID may indicate image quality degradation in time. We evaluate static consistency by measuring LPIPS between I_0 and $\widehat{I}_n$ with moving parts masked out according to semantic segmentation. We always predict semantic segmentation only for I_0. Higher LPIPS may indicate that static areas are tampered during animation (i.e. they are erroneously moving). We also follow the adopted practice to quantitatively measure motion similarity using Fréchet Video Distance (FVD) [42] between real and generated videos, which is averaged over motion directions. Different motion directions are obtained via sampling different homography (Ours), motion code (AL) and horizontal flipping, choosing random reference video (TS). As revealed in Fig. 7, our method preserves static details better and the speed of image quality degradation with time is slower than that of AL$_{noint}$.

The user study is carried out using the same real and generated videos as the ones used in quantitative evaluation. We decided to conduct two sets of user studies involving real image animation: side-by-side comparisons and real/fake questions. In the side-by-side setting, assessors are asked to select the more realistic variant of animation (from two) given the real image shown in the middle. Both videos in a pair are obtained from the same real image using different methods. In real/fake setting, assessors see only a single video and guess whether it is real or not. Each assessor was shown at most 12 questions, 5 different assessors per one question. During the study we noticed that the video speed affects user preference (slower ones are more favorable). Since we cannot control animation speed in our baselines fairly, we decided to conduct

Method	short		long	
	EOIF	**EOIFS**	**EOIF**	**EOIFS**
SG	0.40	0.44	0.26	0.29
AL (no int)	0.46	0.47	0.37	0.38
AL (+ style)	0.18	0.18	0.11	0.10
TS	0.11	0.12	0.12	0.14
Real	0.41	0.44	0.44	0.45
Ours (EOIF)	–	0.52	–	0.52
Ours (EOIFS)	0.48	–	0.48	–

	FR
AL (+ style)	0.25
SG	0.38
Ours (Synth.)	0.42
AL (no int)	0.54
TS	0.20
Real	0.59
Ours (EOIFS)	0.62
Ours (EOIF)	**0.63**

Fig. 8. Left: Ratio of wins row-over-column for side-by-side settings for short (100 frames) and long (200 frames) videos. **Right**: fooling ratio for the real/fake protocol. Note that advantage of our method becomes more evident in long videos.

two sets of user studies: (A) with motion speed aligned to that of competitors and (B) aligned to that of real videos. Here we present only results of A setting (see supp. mat. for B setting). To sum up, the user study reveals the advantage of our method over three baselines (AL, SG, TS), especially in longer videos (Fig. 8).

Please refer to **Supplementary Materials** for more details on methods and experiments, including quantitative ablation study of inference procedure.

5 Discussion

We have presented a new generative model for landscape animations derived from StyleGAN, and have shown that it can be trained from the mixture of static images and timelapse videos, benefiting from both sources. We have investigated how the resulting model can be used to bring to life (reenact) static landscape images, and have shown that this can be done more successfully than with previously proposed methods. Extensive results of our method are shown in the supplementary video.

The supplementary video also shows failure modes. Being heavily reliant on machine learning, our approach fails when reenacting static images atypical for its training dataset. Furthermore, as our video dataset is relatively small and focuses on slower motions (clouds), we have found that method often fails to animate waves and grass sufficiently strongly or realistically. Enlarging the image dataset and, in particular, the video dataset seems to be the most straightforward way to address these shortcomings.

References

1. Karras, T., Laine, S., Aila, T.: A style-based generator architecture for generative adversarial networks. In: Proceedings of the IEEE Conference on Computer Vision and Pattern Recognition, pp. 4401–4410 (2019)

2. Endo, Y., Kanamori, Y., Kuriyama, S.: Animating landscape: self-supervised learning of decoupled motion and appearance for single-image video synthesis. ACM Trans. Graph. (Proceedings of ACM SIGGRAPH Asia 2019) **38**(6), 175:1–175:19 (2019)
3. Shaham, T.R., Dekel, T., Michaeli, T.: SinGAN: learning a generative model from a single natural image. In: Proceedings of the IEEE International Conference on Computer Vision, pp. 4570–4580 (2019)
4. Tesfaldet, M., Brubaker, M.A., Derpanis, K.G.: Two-stream convolutional networks for dynamic texture synthesis. In: 2018 IEEE/CVF Conference on Computer Vision and Pattern Recognition, pp. 6703–6712 (2017)
5. Srivastava, N., Mansimov, E., Salakhudinov, R.: Unsupervised learning of video representations using LSTMs. In: International Conference on Machine Learning, pp. 843–852 (2015)
6. Villegas, R., Yang, J., Hong, S., Lin, X., Lee, H.: Decomposing motion and content for natural video sequence prediction. arXiv preprint arXiv:1706.08033 (2017)
7. Mathieu, M., Couprie, C., LeCun, Y.: Deep multi-scale video prediction beyond mean square error. CoRR abs/1511.05440 (2015)
8. Finn, C., Goodfellow, I., Levine, S.: Unsupervised learning for physical interaction through video prediction. In: Advances in Neural Information Processing Systems, pp. 64–72 (2016)
9. Goodfellow, I., et al.: Generative adversarial nets. In: Advances in Neural Information Processing Systems, pp. 2672–2680 (2014)
10. Pan, J., et al.: Video generation from single semantic label map. In: Proceedings of the IEEE Conference on Computer Vision and Pattern Recognition, pp. 3733–3742 (2019)
11. Li, Y., Fang, C., Yang, J., Wang, Z., Lu, X., Yang, M.-H.: Flow-grounded spatial-temporal video prediction from still images. In: Ferrari, V., Hebert, M., Sminchisescu, C., Weiss, Y. (eds.) ECCV 2018. LNCS, vol. 11213, pp. 609–625. Springer, Cham (2018). https://doi.org/10.1007/978-3-030-01240-3_37
12. Wang, T.C., et al.: Video-to-video synthesis. In: Advances in Neural Information Processing Systems (NeurIPS) (2018)
13. Aigner, S., Körner, M.: FutureGAN: anticipating the future frames of video sequences using spatio-temporal 3D convolutions in progressively growing autoencoder GANs. arXiv preprint arXiv:1810.01325 (2018)
14. Xiong, W., Luo, W., Ma, L., Liu, W., Luo, J.: Learning to generate time-lapse videos using multi-stage dynamic generative adversarial networks. In: The IEEE Conference on Computer Vision and Pattern Recognition (CVPR), June 2018
15. Li, Y., Roblek, D., Tagliasacchi, M.: From here to there: video inbetweening using direct 3D convolutions. ArXiv abs/1905.10240 (2019)
16. Clark, A., Donahue, J., Simonyan, K.: Efficient video generation on complex datasets. ArXiv abs/1907.06571 (2019)
17. Soomro, K., Zamir, A.R., Shah, M.: UCF101: a dataset of 101 human actions classes from videos in the wild (2012)
18. Carreira, J., Noland, E., Banki-Horvath, A., Hillier, C., Zisserman, A.: A short note about kinetics-600. ArXiv abs/1808.01340 (2018)
19. Chen, B., Wang, W., Wang, J.: Video imagination from a single image with transformation generation. In: ACM Multimedia (2017)
20. Van Amersfoort, J., Kannan, A., Ranzato, M., Szlam, A., Tran, D., Chintala, S.: Transformation-based models of video sequences. arXiv preprint arXiv:1701.08435 (2017)

21. Chuang, Y.Y., Goldman, D.B., Zheng, K.C., Curless, B., Salesin, D.H., Szeliski, R.: Animating pictures with stochastic motion textures. In: ACM SIGGRAPH 2005 Papers. SIGGRAPH 2005, pp. 853–860. Association for Computing Machinery (2005)
22. Tulyakov, S., Liu, M.Y., Yang, X., Kautz, J.: MoCoGAN: decomposing motion and content for video generation. In: 2018 IEEE/CVF Conference on Computer Vision and Pattern Recognition, pp. 1526–1535 (2017)
23. Vondrick, C., Pirsiavash, H., Torralba, A.: Generating videos with scene dynamics. In: Advances in Neural Information Processing Systems, pp. 613–621 (2016)
24. Denton, E.L., et al.: Unsupervised learning of disentangled representations from video. In: Advances in Neural Information Processing Systems, pp. 4414–4423 (2017)
25. Zhu, J.-Y., Krähenbühl, P., Shechtman, E., Efros, A.A.: Generative visual manipulation on the natural image manifold. In: Leibe, B., Matas, J., Sebe, N., Welling, M. (eds.) ECCV 2016. LNCS, vol. 9909, pp. 597–613. Springer, Cham (2016). https://doi.org/10.1007/978-3-319-46454-1_36
26. Abdal, R., Qin, Y., Wonka, P.: Image2StyleGAN: how to embed images into the StyleGAN latent space? In: Proceedings of the IEEE International Conference on Computer Vision, pp. 4432–4441 (2019)
27. Bau, D., et al.: Semantic photo manipulation with a generative image prior. ACM Trans. Graph. **38**(4), 59:1–59:11 (2019)
28. Zakharov, E., Shysheya, A., Burkov, E., Lempitsky, V.: Few-shot adversarial learning of realistic neural talking head models. In: Proceedings of the IEEE International Conference on Computer Vision, pp. 9459–9468 (2019)
29. Huang, X., Belongie, S.: Arbitrary style transfer in real-time with adaptive instance normalization. In: Proceedings of the IEEE International Conference on Computer Vision, pp. 1501–1510 (2017)
30. Mescheder, L., Geiger, A., Nowozin, S.: On the convergence properties of GAN training. CoRR abs/1801.04406 (2018)
31. He, K., Zhang, X., Ren, S., Sun, J.: Deep residual learning for image recognition. In: Proceedings of the IEEE Conference on Computer Vision and Pattern Recognition, pp. 770–778 (2016)
32. Kingma, D.P., Ba, J.L.: Adam: a method for stochastic optimization
33. Zhou, B., et al.: Semantic understanding of scenes through the ade20k dataset. Int. J. Comput. Vis. **127**, 302–321 (2018)
34. Johnson, J., Alahi, A., Fei-Fei, L.: Perceptual losses for real-time style transfer and super-resolution. In: Leibe, B., Matas, J., Sebe, N., Welling, M. (eds.) ECCV 2016. LNCS, vol. 9906, pp. 694–711. Springer, Cham (2016). https://doi.org/10.1007/978-3-319-46475-6_43
35. Wang, X., et al.: ESRGAN: enhanced super-resolution generative adversarial networks. In: Leal-Taixé, L., Roth, S. (eds.) ECCV 2018. LNCS, vol. 11133, pp. 63–79. Springer, Cham (2019). https://doi.org/10.1007/978-3-030-11021-5_5
36. Wu, H., Zheng, S., Zhang, J., Huang, K.: Fast end-to-end trainable guided filter. In: Proceedings of the IEEE Conference on Computer Vision and Pattern Recognition, pp. 1838–1847 (2018)
37. Zhu, J.Y., Park, T., Isola, P., Efros, A.A.: Unpaired image-to-image translation using cycle-consistent adversarial networks. In: Proceedings of the IEEE International Conference on Computer Vision, pp. 2223–2232 (2017)

38. Yuan, L., Wen, F., Liu, C., Shum, H.-Y.: Synthesizing dynamic texture with closed-loop linear dynamic system. In: Pajdla, T., Matas, J. (eds.) ECCV 2004. LNCS, vol. 3022, pp. 603–616. Springer, Heidelberg (2004). https://doi.org/10.1007/978-3-540-24671-8_48
39. Heusel, M., Ramsauer, H., Unterthiner, T., Nessler, B., Hochreiter, S.: GANs trained by a two time-scale update rule converge to a local Nash equilibrium. In: NIPS (2017)
40. Wang, Z., Bovik, A.C., Sheikh, H.R., Simoncelli, E.P.: Image quality assessment: from error visibility to structural similarity. IEEE Trans. Image Process. **13**(4), 600–612 (2004)
41. Zhang, R., Isola, P., Efros, A.A., Shechtman, E., Wang, O.: The unreasonable effectiveness of deep features as a perceptual metric. In: Proceedings of the IEEE Conference on Computer Vision and Pattern Recognition, pp. 586–595 (2018)
42. Unterthiner, T., van Steenkiste, S., Kurach, K., Marinier, R., Michalski, M., Gelly, S.: Towards accurate generative models of video: a new metric & challenges. arXiv preprint arXiv:1812.01717 (2018)

GANwriting: Content-Conditioned Generation of Styled Handwritten Word Images

Lei Kang[1,2(✉)], Pau Riba[1], Yaxing Wang[1], Marçal Rusiñol[1], Alicia Fornés[1], and Mauricio Villegas[2]

[1] Computer Vision Center, Universitat Autònoma de Barcelona, Barcelona, Spain
{lkang,priba,yaxing,marcal,afornes}@cvc.uab.es
[2] omni:us, Berlin, Germany
{lei,mauricio}@omnius.com

Abstract. Although current image generation methods have reached impressive quality levels, they are still unable to produce plausible yet diverse images of handwritten words. On the contrary, when writing by hand, a great variability is observed across different writers, and even when analyzing words scribbled by the same individual, involuntary variations are conspicuous. In this work, we take a step closer to producing realistic and varied artificially rendered handwriting. We propose a novel method that is able to produce credible handwritten word images by conditioning the generative process with both calligraphic style features and textual content. Our generator is guided by three complementary learning objectives: to produce realistic images, to imitate a certain handwriting style and to convey a specific textual content. Our model is unconstrained to any predefined vocabulary, being able to render whatever input word. Given a sample writer, it is also able to mimic its calligraphic features in a few-shot setup. We significantly advance over prior art and demonstrate with qualitative, quantitative and human-based evaluations the realistic aspect of our synthetically produced images.

Keywords: Generative Adversarial Networks · Style and content conditioning · Handwritten word images

1 Introduction

Few years after the conception of Generative Adversarial Networks (GANs) [12], we have witnessed an impressive progress on generating illusory plausible images. From the early low-resolution and hazy results, the quality of the artificially generated images has been notably enhanced. We are now able to synthetically produce high-resolution [5] artificial images that are indiscernible from real ones

Electronic supplementary material The online version of this chapter (https://doi.org/10.1007/978-3-030-58592-1_17) contains supplementary material, which is available to authorized users.

to the human observer [24]. In the original GAN architecture, inputs were randomly sampled from a latent space, so that it was hard to control which kind of images were being generated. The conception of conditional Generative Adversarial Networks (cGANs) [35] led to an important improvement. By allowing to condition the generative process on an input class label, the networks were then able to produce images from different given types [7]. However, such classes had to be predefined beforehand during the training stage and thus, it was impossible to produce images from other unseen classes during inference.

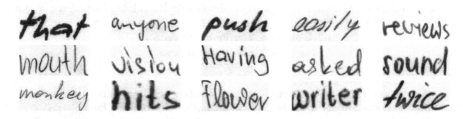

Fig. 1. Turing's test. Just five of the above words are real. Try to distinguish them from the artificially generated samples. (The real words are: ''that'', ''vision'', ''asked'', ''hits'' and ''writer'').

But generative networks have not exclusively been used to produce synthetic images. The generation of data that is sequential in nature has also been largely explored in the literature. Generative methods have been proposed to produce audio signals [9], natural language excerpts [45], video streams [42] or stroke sequences [11,13,15,48] able to trace sketches, drawings or handwritten text. In all of those approaches, in order to generate sequential data, the use of Recurrent Neural Networks (RNNs) has been adopted.

Yet, for the specific case of generating handwritten text, one could also envisage the option of directly producing the final images instead of generating the stroke sequences needed to pencil a particular word. Such non-recurrent approach presents several benefits. First, the training procedure is more efficient since recurrencies are avoided and the inherent parallelism nature of convolutional networks is leveraged. Second, since the output is generated all at once, we avoid the difficulties of learning long-range dependencies as well as vanishing gradient problems. Finally, online training data (pen-tip location sequences), which is hard to obtain, is no longer needed.

Nevertheless, the different attempts to directly generate raw word images present an important drawback. Similarly to the case with cGANs, most of the proposed approaches are just able to condition the word image generation to a predefined set of words, limiting its practical use. For example [14] is specifically designed to generate isolated digits, while [6] is restricted to a handful of Chinese characters. To our best knowledge, the only exception to that is the approach by Alonso *et al.* [1]. In their work they propose a non-recurrent generative architecture conditioned to input content strings. By having such design, the generative process is not restricted to a particular predefined vocabulary, and could potentially generate any word. However, the produced results are not realistic, still exhibiting a poor quality, sometimes producing barely legible word images. Their proposed approach also suffers from the mode collapse problem, tending to produce images with a unique writing style. In this paper we present a non-recurrent generative architecture conditioned to textual content sequences, that is specially tailored to produce realistic handwritten word images, indistinguishable to humans. Real and generated images are actually difficult to tell apart, as shown in Fig. 1. In order to produce diverse styled word images, we propose to condition the generative process not only with textual content, but also with a specific writing style, defined by a latent set of calligraphic attributes.

Therefore, our approach[1] is able to artificially render realistic handwritten word images that match a certain textual content and that mimic some style features (text skew, slant, roundness, stroke width, ligatures, etc.) from an exemplar writer. To this end, we guide the learning process by three different learning objectives [36]. First, an adversarial discriminator ensures that the images are realistic and that its visual appearance is as closest as possible to real handwritten word images. Second, a style classifier guarantees that the provided calligraphic attributes, characterizing a particular handwriting style, are properly transferred to the generated word instances. Finally, a state-of-the-art sequence-to-sequence handwritten word recognizer [34] controls that the textual contents have been properly conveyed during the image generation. To summarize, the main contributions of the paper are the following:

- Our model conditions the handwritten word generative process both with calligraphic style features and textual content, producing varied samples indistinguishable by humans, surpassing the quality of the current state-of-the-art approaches.
- We introduce the use of three complementary learning objectives to guide different aspects of the generative process.
- We propose a character-based content conditioning that allows to generate any word, without being restricted to a specific vocabulary.
- We put forward a few-shot calligraphic style conditioning to avoid the mode collapse problem.

2 Related Work

The generation of realistic synthetic handwritten word images is a challenging task. To this day, the most convincing approaches involved an expensive manual intervention aimed at clipping individual characters or glyphs [16,26,28,40,43]. When such approaches were combined with appropriate rendering techniques including ligatures among strokes, textures and background blending, the obtained results were indeed impressive. Haines *et al.* [16] illustrated how such approaches could artificially generate indistinguishable manuscript excerpts as if

---

[1] Our code is available at https://github.com/omni-us/research-GANwriting.

they were written by Sir Arthur Conan Doyle, Abraham Lincoln or Frida Kahlo. Of course such manual intervention is extremely expensive, and in order to produce large volumes of manufactured images the use of truetype electronic fonts has also been explored [23, 27]. Although such approaches benefit from a greater scalability, the realism of the generated images clearly deteriorates.

With the advent of deep learning, the generation of handwritten text was approached differently. As shown in the seminal work by Alex Graves [13], given a reasonable amount of training data, an RNN could learn meaningful latent spaces that encode realistic writing styles and their variations, and then generate stroke sequences that trace a certain text string. However, such sequential approaches [11, 13, 15, 48] need temporal data, obtained by recording with a digital stylus pen real handwritten samples, stroke-by-stroke, in vector form.

Contrary to sequential approaches, non-recurrent generative methods have been proposed to directly produce images. Both variational auto-encoders [25] and GANs [12] were able to learn the MNIST manifold and generate artificial handwritten digit images in the original publications. With the emergence of cGANs [35], able to condition the generative process on an input image rather than a random noise vector, the adversarial-guided image-to-image translation problem started to rise. Image-to-image translation has since been applied to many different style transfer applications, as demonstrated in [21] with the *pix2pix* network. Since then, image translation approaches have been acquiring the ability to disentangle style attributes from the contents of the input images, producing better style transfer results [37, 39]. Geometry-aware synthesizing methods [46, 47] have been successfully applied on scene text images, but cursive words are not considered.

Concerning the generation of handwritten text, such approaches have been mainly used for synthesising Chinese ideograms [6, 22, 30, 41, 44] and glyphs [2]. However, they are restricted to a predefined set of content classes. The incapability to generate out of vocabulary (OOV) text limits its practical application. Few works can actually deal with OOV words. First, in the work by Alonso *et al.* [1], the generation of handwritten word samples is conditioned by character sequences, but it suffers from the mode collapse problem, hindering the diversity of the generated images. Second, Fogel *et al.* [10] generate handwritten word by assembling the images generated by its characters, but the generated characters have the same receptive field width, which can make the generated words look unrealistic. Third, Mayr *et al.* [33] propose a conversion model to approximate online handwriting from offline data and then apply style transfer method to online data, so that offline handwritten text images could be generated by leveraging online handwriting synthesizer. However, this method highly depends on the performance of the conversion model and needs online data to train. Techniques like FUNIT [29], able to transfer unseen target styles to the content generated images could be beneficial for this limitation. In particular, the use of Adaptive Instance Normalization (AdaIN) layers, proposed in [18], shall allow to align both textual content and style attributes within the generative process.

Summarizing, state-of-the-art generative methods are still unable to produce plausible yet diverse images of whatever handwritten word automatically. In this paper we propose to condition a generative model for handwritten words with unconstrained text sequences and stylistic typographic attributes, so that we are able to generate any word with a great diversity over the produced results.

3   Conditioned Handwritten Generation

3.1   Problem Formulation

Let $\{\mathcal{X}, \mathcal{Y}, \mathcal{W}\}$ be a multi-writer handwritten word dataset, containing grayscale word images $\mathcal{X}$, their corresponding transcription strings $\mathcal{Y}$ and their writer identifiers $\mathcal{W} = \{w_i\}_{i=1}^N$. Let $X_i = \{x_{w_i,j}\}_{j=1}^K \subset \mathcal{X}$ be a subset of K randomly sampled handwritten word images from the same given writer $w_i \in \mathcal{W}$. Let $\mathcal{A}$ be the alphabet containing the allowed characters (letters, digits, punctuation signs, etc.), $\mathcal{A}^l$ being all the possible text strings with length l. Given a set of images X_i as a few-shot example of the calligraphic style attributes for writer w_i on the one hand, and given a textual content provided by any text string $t \in \mathcal{A}^l$ on the other hand; the proposed generative model has the ability to combine both sources of information. It has the objective to yield a handwritten word image having textual content equal to t and sharing calligraphic style attributes with writer w_i. Following this formulation, the generative model H is defined as

$$\bar{x} = H(t, X_i) = H(t, \{x_1, \ldots, x_K\}), \tag{1}$$

where $\bar{x}$ is the artificially generated handwritten word image with the desired properties. Moreover, we denote $\bar{\mathcal{X}}$ as the output distribution of the generative network H.

The proposed architecture is divided in two main components. The generative network produces human-readable images conditioned to the combination of calligraphic style and textual content information. The second component are the learning objectives which guide the generative process towards producing images that look realistic; exhibiting a particular calligraphic style attributes; and having a specific textual content. Figure 2 gives an overview of our model.

3.2   Generative Network

The proposed generative architecture H consists of a calligraphic style encoder S, a textual content encoder C and a conditioned image generator G. The overall calligraphic style of input images X_i is disentangled from their individual textual contents, whereas the string t provides the desired content.

Calligraphic Style Encoding. Given the set $X_i \subset \mathcal{X}$ of $K = 15$ word images from the same writer w_i, the style encoder aims at extracting the calligraphic style attributes, *i.e.* slant, glyph shapes, stroke width, character roundness, ligatures etc. from the provided input samples. Specifically, our proposed network S

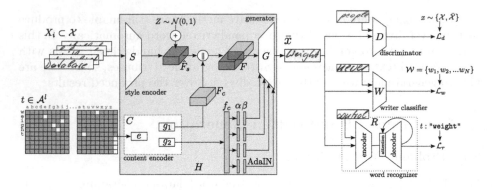

Fig. 2. Architecture of the proposed handwriting generation model.

learns a style latent space mapping, in which the obtained style representations $F_s = S(X_i)$ are disentangled from the actual textual contents of the images X_i. The VGG-19-BN [38] architecture is used as the backbone of S. In order to process the input image set X_i, all the images are resized to have the same height h, padded to meet a maximum width w and concatenated channel-wise to end up with a single tensor $h \times w \times K$. If we ask a human to write the same word several times, slight involuntary variations appear. In order to imitate this phenomenon, randomly choosing permutations of the subset X_i will already produce such characteristic fluctuations. In addition, an additive noise $Z \sim \mathcal{N}(0,1)$ is applied to the output latent space to obtain a subtly distorted feature representation $\hat{F}_s = F_s + Z$.

Textual Content Encoding. The textual content network C is devoted to produce an encoding of the given text string t that we want to artificially write. The proposed architecture outputs content features at two different levels. Low-level features encode the different characters that form a word and their spatial position within the string. A subsequent broader representation aims at guiding the whole word consistency. Formally, let $t \in \mathcal{A}^l$ be the input text string, character sequences shorter than l are padded with the empty symbol ε. Let us define a character-wise embedding function $e \colon \mathcal{A} \to \mathbb{R}^n$. The first step of the content encoding stage embeds with a linear layer each character $c \in t$, represented by a one-hot vector, into a character-wise latent space. Then, the architecture is divided into two branches.

Character-Wise Encoding: Let $g_1 \colon \mathbb{R}^n \to \mathbb{R}^m$ be a Multi-Layer Perceptron (MLP). Each embedded character $e(c)$ is processed individually by g_1 and their results are later stacked together. In order to combine such representation with style features, we have to ensure that the content feature map meets the shape of $\hat{F}_s$. Each character embedding is repeated multiple times horizontally to coarsely align the content features with the visual ones extracted from the style network, and the tensor is finally vertically expanded. The two feature representations are concatenated to be fed to the generator $F = [\hat{F}_s \parallel F_c]$. Such a character-wise encoding enables the network to produce OOV words, *i.e.* words that have never been seen during training.

Global String Encoding: Let $g_2 \colon \mathbb{R}^{l \cdot n} \to \mathbb{R}^{2p \cdot q}$ be another MLP aimed at obtaining a much broader and global string representation. The character embeddings $e(c)$ are concatenated into a large one-dimensional vector of size $l \cdot n$ that is then processed by g_2. Such global representation vector f_c will be then injected into the generator splitted into p pairs of parameters.

Both functions $g_1(\cdot)$ and $g_2(\cdot)$ make use of three fully-connected layers with ReLU activation functions and batch normalization [20].

Generator. Let F be the combination of the calligraphic style attributes and the textual content information character-wise; and f_c the global textual encoding. The generator G is composed of two residual blocks [19] using the AdaIN as the normalization layer. Then, four convolutional modules with nearest neighbor up-sampling and a final tanh activation layer generates the output image $\bar{x}$. AdaIN is formally defined as

$$\text{AdaIN}(z, \alpha, \beta) = \alpha \left(\frac{z - \mu(z)}{\sigma(z)} \right) + \beta, \qquad (2)$$

where $z \in F$, μ and σ are the channel-wise mean and standard deviations. The global content information is injected four times ($p = 4$) during the generative process by the AdaIN layers. Their parameters α and β are obtained by splitting f_c in four pairs. Hence, the generative network is defined as

$$\bar{x} = H(t, X_i) = G(C(t), S(X_i)) = G\left(g_1\left(\hat{t}\right), g_2\left(\hat{t}\right), S(X_i)\right), \qquad (3)$$

where $\hat{t} = [e(c); \forall c \in t]$ is the encoding of the string t character by character.

3.3 Learning Objectives

We propose to combine three complementary learning objectives: a discriminative loss, a style classification loss and a textual content loss. Each one of these losses aim at enforcing different properties of the desired generated image $\bar{x}$.

Discriminative Loss. Following the paradigm of GANs [12], we make use of a discriminative model D to estimate the probability that samples come from a real source, *i.e.* training data $\mathcal{X}$, or belong to the artificially generated distribution $\bar{\mathcal{X}}$. Taking the generative network H and the discriminator D, this setting corresponds to a min max optimization problem. The proposed discriminator D starts with a convolutional layer, followed by six residual blocks with LeakyReLU activations and average poolings. A final binary classification layer is used to discern between fake and real images. Thus, the discriminative loss only controls that the general visual appearance of the generated image looks realistic. However, it does not take into consideration neither the calligraphic styles nor the textual contents. This loss is formally defined as

$$\mathcal{L}_d(H, D) = \mathbb{E}_{x \sim \mathcal{X}}\left[\log(D(x))\right] + \mathbb{E}_{\bar{x} \sim \bar{\mathcal{X}}}\left[\log(1 - D(\bar{x}))\right]. \qquad (4)$$

Style Loss. When generating realistic handwritten word images, encoding information related to calligraphic styles not only provides diversity on the generated samples, but also prevents the mode collapse problem. Calligraphy is a strong identifier of different writers. In that sense, the proposed style loss guides the generative network H to generate samples conditioned to a particular writing style by means of a writer classifier W. Given a handwritten word image, W tries to identify the writer $w_i \in \mathcal{W}$ who produced it. The writer classifier W follows the same architecture of the discriminator D with a final classification MLP with the amount of writers in our training dataset. The classifier W is only optimized with real samples drawn from $\mathcal{X}$, but it is used to guide the generation of the synthetic ones. We use the cross entropy loss, formally defined as

$$\mathcal{L}_w\left(H, W\right) = -\mathbb{E}_{x \sim \{\mathcal{X}, \bar{x}\}} \left[\sum_{i=1}^{|\mathcal{W}|} w_i \log\left(\hat{w}_i\right) \right], \tag{5}$$

where $\hat{w} = W(x)$ is the predicted probability distribution over writers in $\mathcal{W}$ and w_i the real writer distribution. Generated samples should be classified as the writer w_i used to construct the input style conditioning image set X_i.

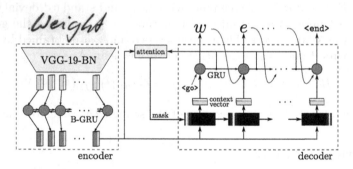

Fig. 3. Architecture of the attention-based sequence-to-sequence handwritten word recognizer R.

Content Loss. A final handwritten word recognizer network R is used to guide our generator towards producing synthetic word images with a specific textual content. We implemented a state-of-the-art sequence-to-sequence model [34] for handwritten word recognition to examine whether the produced images $\bar{x}$ are actually decoded as the string t. The recognizer, depicted in Fig. 3, consists of an encoder and a decoder coupled with an attention mechanism. Handwritten word images are processed by the encoder and high-level feature representations are obtained. A VGG-19-BN [38] architecture followed by a two-layered Bi-directional Gated Recurrent Unit (B-GRU) [8] is used as the encoder network. The decoder is a one-directional RNN that outputs character by character predictions at each time step. The attention mechanism dynamically aligns context features from each time step of the decoder with high-level features from the encoder, hopefully corresponding to the next character to decode. The Kullback-Leibler divergence loss is used as the recognition loss at each time step. This is formally defined as

$$\mathcal{L}_r(H, R) = -\mathbb{E}_{x \sim \{\mathcal{X}, \bar{\mathcal{X}}\}} \left[\sum_{i=0}^{l} \sum_{j=0}^{|\mathcal{A}|} t_{i,j} \log \left(\frac{t_{i,j}}{\hat{t}_{i,j}} \right) \right], \quad (6)$$

where $\hat{t} = R(x)$; $\hat{t}_i$ being the i-th decoded character probability distribution by the word recognizer, $\hat{t}_{i,j}$ being the probability of j-th symbol in $\mathcal{A}$ for $\hat{t}_i$, and $t_{i,j}$ being the real probability corresponding to $\hat{t}_{i,j}$. The empty symbol ε is ignored in the loss computation; t_i denotes the i-th character on the input text t.

Algorithm 1. Training algorithm for the proposed model.

> **Input:** Input data $\{\mathcal{X}, \mathcal{Y}, \mathcal{W}\}$; alphabet $\mathcal{A}$; max training iterations T
> **Output:** Networks parameters $\{\Theta_H, \Theta_D, \Theta_W, \Theta_R\}$.

1: **repeat**
2:      Get style and content mini-batches $\{X_i, w_i\}_{i=1}^{N_B}$ and $\{t^i\}_{i=1}^{N_B}$
3:      $\mathcal{L}_d \leftarrow$ Eq. 4   ▷ Real and generated samples $x \sim \{\mathcal{X}, \bar{\mathcal{X}}\}$
4:      $\mathcal{L}_{w,r} \leftarrow$ Eq. 5 + Eq. 6   ▷ Real samples $x \sim \mathcal{X}$
5:      $\Theta_D \leftarrow \Theta_D + \Gamma(\nabla_{\Theta_D} \mathcal{L}_d)$
6:      $\Theta_{W,R} \leftarrow \Theta_{W,R} - \Gamma(\nabla_{\Theta_{W,R}} \mathcal{L}_{w,d})$
7:      $\mathcal{L} \leftarrow$ Eq. 7   ▷ Generated samples $x \sim \bar{\mathcal{X}}$
8:      $\Theta_H \leftarrow \Theta_H - \Gamma(\nabla_{\Theta_H} \mathcal{L})$
9: **until** Max training iterations T

3.4 End-to-End Training

Overall, the whole architecture is trained end to end with the combination of the three proposed loss functions

$$\mathcal{L}(H, D, W, R) = \mathcal{L}_d(H, D) + \mathcal{L}_w(H, W) + \mathcal{L}_r(H, R), \quad (7)$$

$$\min_{H,W,R} \max_{D} \mathcal{L}(H, D, W, R). \quad (8)$$

Algorithm 1 presents the training strategy that has been followed in this work. $\Gamma(\cdot)$ denotes the optimizer function. Note that the parameter optimization is performed in two steps. First, the discriminative loss is computed using both real and generated samples (line 3). The style and content losses are computed by just providing real data (line 4). Even though W and D are optimized using only real data and, therefore, they could be pre-trained independently from the generative network H, we obtained better results by initializing all the networks from scratch and jointly training them altogether. The network parameters Θ_D are optimized by gradient ascent following the GAN paradigm whereas the parameters Θ_W and Θ_R are optimized by gradient descent. Finally, the overall generator loss is computed following Eq. 7 where only the generator parameters Θ_H are optimized (line 8).

4   Experiments

To carry out the different experiments, we have used a subset of the IAM corpus [32] as our multi-writer handwritten dataset $\{\mathcal{X}, \mathcal{Y}, \mathcal{W}\}$. It consists of 62, 857 handwritten word snippets, written by 500 different individuals. Each word image has its associated writer and transcription metadata. A test subset of 160 writers has been kept apart during training to check whether the generative model is able to cope with unseen calligraphic styles. We have also used a subset of 22, 500 unique English words from the Brown [3] corpus as the source of strings for the content input. A test set of 400 unique words, disjoint from the IAM transcriptions has been used to test the performance when producing OOV words. To quantitatively measure the image quality, diversity and the ability to transfer style attributes of the proposed approach we will use the Fréchet Inception Distance (FID) [4,17], measuring the distance between the Inception-v3 activation distributions for generated $\tilde{\mathcal{X}}$ and real samples $\mathcal{X}$ for each writer w_i separately, and finally averaging them. Inception features, trained over ImageNet data, have not been designed to discern between different handwriting images. Although this measure might not be ideal to evaluate our specific case, it will still serve as an indication of the similarity between generated and real images.

Fig. 4. Word image generation. a) In-Vocabulary (IV) words and seen (S) styles; b) In-Vocabulary (IV) words and unseen (U) styles; c) Out-of-Vocabulary (OOV) words and seen (S) styles and d) Out-of-Vocabulary (OOV) words and unseen (U) styles.

4.1   Generating Handwritten Word Images

We present in Fig. 4 an illustrative selection of generated handwritten words. We appreciate the realistic and diverse aspect of the produced images. Qualitatively, we observe that the proposed approach is able to yield satisfactory results even when dealing with both words and calligraphic styles never seen during training. But, when analyzing the different experimental settings in Table 1, we appreciate that the FID measure slightly degrades when either we input an OOV word or a style never seen during training. Nevertheless, the reached FID measures in all four settings satisfactorily compare with the baseline achieved by real data.

Table 1. FID between generated images and real images of corresponding set.

	Real images	IV-S	IV-U	OOV-S	OOV-U
FID	90.43	120.07	124.30	125.87	130.68

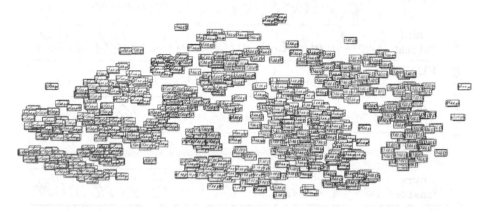

Fig. 5. t-SNE embedding visualization of 2.500 generated instances of the word ''deep''.

In order to show the ability of the proposed method to produce a diverse set of generated images, we present in Fig. 5 a t-SNE [31] visualization of different instances produced with a fixed textual content while varying the calligraphic style inputs. Different clusters corresponding to particular slants, stroke widths, character roundnesses, ligatures and cursive writings are observed.

To further demonstrate the ability of the proposed approach to coalesce content and style information into the generated handwritten word images, we compare in Fig. 6 our produced results with the outcomes of the state-of-the-art approach FUNIT [29]. Being an image-to-image translation method, FUNIT starts with a content image and then injects the style attributes derived from a second sample image. Although FUNIT performs well for natural scene images, it is clear that such kind of approaches do not apply well for the specific case of handwritten words. Starting with a content image instead of a text string confines the generative process to the shapes of the initial drawing. When infusing the style features, the FUNIT method is only able to deform the stroke textures, often resulting in extremely distorted words. Conversely, our proposed generative process is able to produce realistic and diverse word samples given a content text string and a calligraphic style example. We observe how for the different produced versions of the same word, the proposed approach is able to change style attributes as stroke width or slant, to produce both cursive words, where all characters are connected through ligatures as well as disconnected writings, and even render the same characters differently, *e.g.* note the characters n or s in ''Thank'' or ''inside'' respectively.

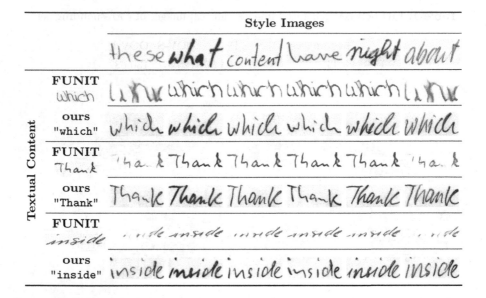

Fig. 6. Comparison of handwritten word generation with FUNIT [29].

4.2 Latent Space Interpolations

The generator network G learns to map feature points F in the latent space to synthetic handwritten word images. Such latent space presents a structure worth exploring. We first interpolate in Fig. 7 between two different points F_s^A and F_s^B corresponding to two different calligraphic styles w_A and w_B while keeping the textual contents t fixed. We observe how the generated images smoothly adjust from one style to another. Again note how individual characters evolve from one typography to another, *e.g.* the l from ''also'', or the f from ''final''.

Contrary to the continuous nature of the style latent space, the original content space is discrete in nature. Instead of computing point-wise interpolations, we present in Fig. 8 the obtained word images for different styles when following a "word ladder" puzzle game, *i.e.* going from one word to another, one character difference at a time. Here we observe how different contents influence stylistic aspects. Usually s and i are disconnected when rendering the word ''sired'' but often appear with a ligature when jumping to the word ''fired''.

4.3 Impact of the Learning Objectives

Along this paper, we have proposed to guide the generation process by three complementary goals. The discriminator loss $\mathcal{L}_d$ controlling the genuine appearance of the generated images $\bar{x}$. The writer classification loss $\mathcal{L}_w$ forcing $\bar{x}$ to mimic the calligraphic style of input images X_i. The recognition loss $\mathcal{L}_r$ guaranteeing that $\bar{x}$ is readable and conveys the exact text information t. We analyze in Table 2 the effect of each learning objective.

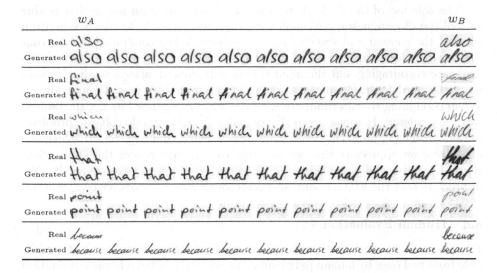

Fig. 7. Latent space interpolation between two calligraphic styles for different words while keeping contents fixed.

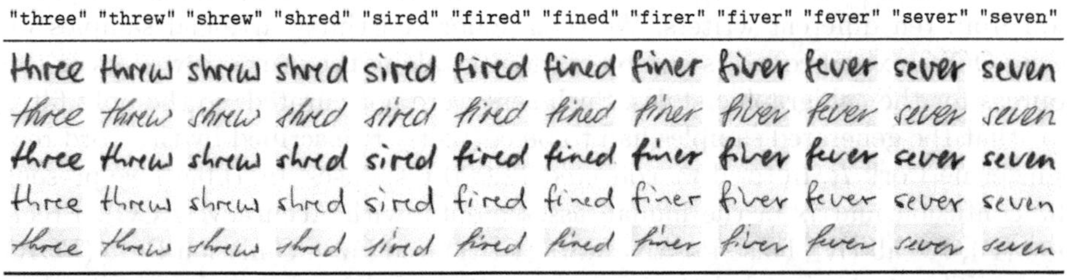

"three" "threw" "shrew" "shred" "sired" "fired" "fined" "firer" "fiver" "fever" "sever" "seven"

Fig. 8. Word ladder. From ''three'' to ''seven'' changing one character at a time, generated for five different calligraphic styles.

Table 2. Effect of each different learning objectives when generating the content $t =$ ''vision'' for different styles.

$\mathcal{L}_d$ $\mathcal{L}_w$ $\mathcal{L}_r$	FID	Style Images			
		these	*what*	*have*	*about*
✓ - -	364.10				
✓ ✓ -	207.47				
✓ - ✓	138.80				
✓ ✓ ✓	**130.68**				

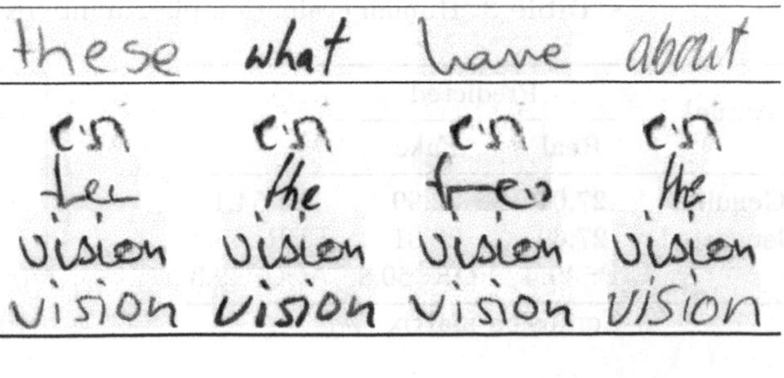

The sole use of the $\mathcal{L}_d$ leads to constantly generating an image that is able to fool the discriminator. Although the generated image looks like handwritten strokes, the content and style inputs are ignored. When combining the discriminator with the auxiliary task of writer classification $\mathcal{L}_w$, the produced results are more encouraging, but the input text is still ignored, always tending to generate the word ``the'', since it is the most common word seen during training. When combining the discriminator with the word recognizer loss $\mathcal{L}_r$, the desired word is rendered. However, as in [1], we suffer from the mode collapse problem, always producing unvarying word instances. When combining the three learning objectives we appreciate that we are able to correctly render the appropriate textual content while mimicking the input styles, producing diverse results. We appreciate that the FID measure also decreases for each successive combination.

4.4   Human Evaluation

Finally, we also tested whether the generated images were actually indistinguishable from real ones by human judgments. We have conducted a human evaluation study as follows: we have asked 200 human examiners to assess whether a set of images were written by a human or artificially generated. Appraisers were presented a total of sixty images, one at a time, and they had to choose if each of them was real of fake. We chose thirty real words from the IAM test partition from ten different writers. We then generated thirty artificial samples by using OOV textual contents and by randomly taking the previous writers as the sources for the calligraphic styles. Such sets were not curated, so the only filter was that the generated samples had to be correctly transcribed by the word recognizer network R. In total we collected 12,000 responses. In Table 3 we present the confusion matrix of the human assessments, with Accuracy (ACC), Precision (P), Recall (R), False Positive Rate (FPR) and False Omission Rate (FOR) values. The study revealed that our generative model was clearly perceived as plausible, since a great portion of the generated samples were deemed genuine. Only a 49.3% of the images were properly identified, which shows a similar performance than a random binary classifier. Accuracies over different examiners were normally distributed. We also observe that the synthetically generated word images were judged more often as being real than correctly identified as fraudulent, with a final FPR of 55.4%.

Table 3. Human evaluation plausibility experiment.

Actual	Predicted			
	Real	Fake		
Genuine	27.01	22.99	R: 54.1	
Generated	27.69	22.31	FPR: 55.4	
	P: 49.4	FOR: 50.8	ACC: 49.3	

a) Confusion matrix (%)

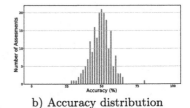

b) Accuracy distribution

5   Conclusion

We have presented a novel image generation architecture that produces realistic and varied artificially rendered samples of handwritten words. Our pipeline can yield credible word images by conditioning the generative process with both calligraphic style features and textual content. Furthermore, by jointly guiding our generator with three different cues: a discriminator, a style classifier and a content recognizer, our model is able to render any input word, not depending on any predefined vocabulary, while incorporating calligraphic styles in a few-shot setup. Experimental results demonstrate that the proposed method yields images with such a great realistic quality that are indistinguishable by humans.

Acknowledgements. This work was supported by EU H2020 SME Instrument project 849628, the Spanish projects TIN2017-89779-P and RTI2018-095645-B-C21, and grants 2016-DI-087, FPU15/06264 and RYC-2014-16831. Titan GPU was donated by NVIDIA.

References

1. Alonso, E., Moysset, B., Messina, R.: Adversarial generation of handwritten text images conditioned on sequences. In: Proceedings of the International Conference on Document Analysis and Recognition (2019)
2. Azadi, S., Fisher, M., Kim, V.G., Wang, Z., Shechtman, E., Darrell, T.: Multi-content GAN for few-shot font style transfer. In: Proceedings of the IEEE Conference on Computer Vision and Pattern Recognition, pp. 7564–7573 (2018)
3. Bird, S., Klein, E., Loper, E.: Natural Language Processing with Python: Analyzing Text with the Natural Language Toolkit. O'Reilly Media Inc, Newton (2009)
4. Borji, A.: Pros and cons of GAN evaluation measures. Comput. Vis. Image Underst. **179**, 41–65 (2019)
5. Brock, A., Donahue, J., Simonyan, K.: Large scale GAN training for high fidelity natural image synthesis. In: Proceedings of the International Conference on Learning Representations (2019)
6. Chang, B., Zhang, Q., Pan, S., Meng, L.: Generating handwritten Chinese characters using CycleGAN. In: Proceedings of the IEEE Winter Conference on Applications of Computer Vision (2018)
7. Choi, Y., Choi, M., Kim, M., Ha, J.W., Kim, S., Choo, J.: StarGAN: unified generative adversarial networks for multi-domain image-to-image translation. In: Proceedings of the IEEE Conference on Computer Vision and Pattern Recognition (2018)
8. Chung, J., Gulcehre, C., Cho, K., Bengio, Y.: Empirical evaluation of gated recurrent neural networks on sequence modeling. In: Proceedings of the NeurIPS Workshop on Deep Learning (2014)
9. Dong, H.W., Hsiao, W.Y., Yang, L.C., Yang, Y.H.: MuseGAN: multi-track sequential generative adversarial networks for symbolic music generation and accompaniment. In: Proceedings of the AAAI Conference on Artificial Intelligence (2018)
10. Fogel, S., Averbuch-Elor, H., Cohen, S., Mazor, S., Litman, R.: ScrabbleGAN: semi-supervised varying length handwritten text generation. In: Proceedings of the IEEE Conference on Computer Vision and Pattern Recognition, pp. 4324–4333 (2020)

11. Ganin, Y., Kulkarni, T., Babuschkin, I., Eslami, S., Vinyals, O.: Synthesizing programs for images using reinforced adversarial learning. In: Proceedings of the International Conference on Machine Learning (2018)
12. Goodfellow, I., et al.: Generative adversarial nets. In: Proceedings of the Neural Information Processing Systems Conference (2014)
13. Graves, A.: Generating sequences with recurrent neural networks. arXiv preprint arXiv:1308.0850 (2013)
14. Gregor, K., Danihelka, I., Graves, A., Rezende, D.J., Wierstra, D.: DRAW: a recurrent neural network for image generation. In: Proceedings of the International Conference on Machine Learning (2015)
15. Ha, D., Eck, D.: A neural representation of sketch drawings. In: Proceedings of the International Conference on Learning Representations (2018)
16. Haines, T.S., Mac Aodha, O., Brostow, G.J.: My text in your handwriting. ACM Trans. Graph. **35**(3), 1–18 (2016)
17. Heusel, M., Ramsauer, H., Unterthiner, T., Nessler, B., Hochreiter, S.: GANs trained by a two time-scale update rule converge to a local Nash equilibrium. In: Proceedings of the Neural Information Processing Systems Conference (2017)
18. Huang, X., Belongie, S.: Arbitrary style transfer in real-time with adaptive instance normalization. In: Proceedings of the IEEE International Conference on Computer Vision (2017)
19. Huang, X., Liu, M.Y., Belongie, S., Kautz, J.: Multimodal unsupervised image-to-image translation. In: Proceedings of the European Conference on Computer Vision (2018)
20. Ioffe, S., Szegedy, C.: Batch normalization: accelerating deep network training by reducing internal covariate shift. In: Proceedings of the International Conference on Machine Learning (2015)
21. Isola, P., Zhu, J.Y., Zhou, T., Efros, A.A.: Image-to-image translation with conditional adversarial networks. In: Proceedings of the IEEE Conference on Computer Vision and Pattern Recognition (2017)
22. Jiang, H., Yang, G., Huang, K., Zhang, R.: *W-Net*: one-shot arbitrary-style Chinese character generation with deep neural networks. In: Cheng, L., Leung, A.C.S., Ozawa, S. (eds.) ICONIP 2018. LNCS, vol. 11305, pp. 483–493. Springer, Cham (2018). https://doi.org/10.1007/978-3-030-04221-9_43
23. Kang, L., Rusiñol, M., Fornés, A., Riba, P., Villegas, M.: Unsupervised adaptation for synthetic-to-real handwritten word recognition. In: Proceedings of the IEEE Winter Conference on Applications of Computer Vision, pp. 3491–3500 (2020)
24. Karras, T., Laine, S., Aila, T.: A style-based generator architecture for generative adversarial networks. In: Proceedings of the IEEE Conference on Computer Vision and Pattern Recognition (2019)
25. Kingma, D.P., Welling, M.: Auto-encoding variational Bayes. In: Proceedings of the International Conference on Learning Representations (2014)
26. Konidaris, T., Gatos, B., Ntzios, K., Pratikakis, I., Theodoridis, S., Perantonis, S.J.: Keyword-guided word spotting in historical printed documents using synthetic data and user feedback. Int. J. Doc. Anal. Recogn. **9**(2–4), 167–177 (2007)
27. Krishnan, P., Jawahar, C.: Generating synthetic data for text recognition. arXiv preprint arXiv:1608.04224 (2016)
28. Lin, Z., Wan, L.: Style-preserving English handwriting synthesis. Pattern Recogn. **40**(7), 2097–2109 (2007)
29. Liu, M.Y., et al.: Few-shot unsupervised image-to-image translation. In: Proceedings of the IEEE International Conference on Computer Vision (2019)

30. Lyu, P., Bai, X., Yao, C., Zhu, Z., Huang, T., Liu, W.: Auto-encoder guided GAN for Chinese calligraphy synthesis. In: Proceedings of the International Conference on Document Analysis and Recognition (2017)
31. Maaten, L.V.D., Hinton, G.: Visualizing data using t-SNE. J. Mach. Learn. Res. **9**, 2579–2605 (2008)
32. Marti, U.V., Bunke, H.: The IAM-database: an English sentence database for offline handwriting recognition. Int. J. Doc. Anal. Recogn. **5**(1), 39–46 (2002)
33. Mayr, M., Stumpf, M., Nikolaou, A., Seuret, M., Maier, A., Christlein, V.: Spatio-temporal handwriting imitation. arXiv preprint arXiv:2003.10593 (2020)
34. Michael, J., Labahn, R., Grüning, T., Zöllner, J.: Evaluating sequence-to-sequence models for handwritten text recognition. arXiv preprint arXiv:1903.07377 (2019)
35. Mirza, M., Osindero, S.: Conditional generative adversarial nets. arXiv preprint arXiv:1411.1784 (2014)
36. Odena, A., Olah, C., Shlens, J.: Conditional image synthesis with auxiliary classifier GANs. In: Proceedings of the International Conference on Machine Learning (2017)
37. Pondenkandath, V., Alberti, M., Diatta, M., Ingold, R., Liwicki, M.: Historical document synthesis with generative adversarial networks. In: Proceedings of the International Conference on Document Analysis and Recognition (2019)
38. Simonyan, K., Zisserman, A.: Very deep convolutional networks for large-scale image recognition. In: Proceedings of the International Conference on Learning Representations (2015)
39. Taigman, Y., Polyak, A., Wolf, L.: Unsupervised cross-domain image generation. In: Proceedings of the International Conference on Learning Representations (2017)
40. Thomas, A.O., Rusu, A., Govindaraju, V.: Synthetic handwritten CAPTCHAs. Pattern Recogn. **42**(12), 3365–3373 (2009)
41. Tian, Y.: zi2zi: master Chinese calligraphy with conditional adversarial networks (2017). https://github.com/kaonashi-tyc/zi2zi
42. Tulyakov, S., Liu, M.Y., Yang, X., Kautz, J.: MoCoGAN: decomposing motion and content for video generation. In: Proceedings of the IEEE Conference on Computer Vision and Pattern Recognition (2018)
43. Wang, J., Wu, C., Xu, Y.Q., Shum, H.Y.: Combining shape and physical models for online cursive handwriting synthesis. Int. J. Doc. Anal. Recogn. **7**(4), 219–227 (2005)
44. Wu, S.J., Yang, C.Y., Hsu, J.Y.J.: CalliGAN: style and structure-aware Chinese calligraphy character generator. arXiv preprint arXiv:2005.12500 (2020)
45. Yu, L., Zhang, W., Wang, J., Yu, Y.: SeqGAN: sequence generative adversarial nets with policy gradient. In: Proceedings of the AAAI Conference on Artificial Intelligence (2017)
46. Zhan, F., Xue, C., Lu, S.: GA-DAN: geometry-aware domain adaptation network for scene text detection and recognition. In: Proceedings of the IEEE International Conference on Computer Vision, pp. 9105–9115 (2019)
47. Zhan, F., Zhu, H., Lu, S.: Spatial fusion GAN for image synthesis. In: Proceedings of the IEEE Conference on Computer Vision and Pattern Recognition, pp. 3653–3662 (2019)
48. Zheng, N., Jiang, Y., Huang, D.: StrokeNet: a neural painting environment. In: Proceedings of the International Conference on Learning Representations (2019)

Spatial-Angular Interaction for Light Field Image Super-Resolution

Yingqian Wang[1], Longguang Wang[1], Jungang Yang[1(✉)], Wei An[1], Jingyi Yu[2], and Yulan Guo[1,3]

[1] National University of Defense Technology, Changsha, China
{wangyingqian16,yangjungang,yulan.guo}@nudt.edu.cn
[2] ShanghaiTech University, Shanghai, China
[3] Sun Yat-sen University, Guangzhou, China
https://github.com/YingqianWang/LF-InterNet

Abstract. Light field (LF) cameras record both intensity and directions of light rays, and capture scenes from a number of viewpoints. Both information within each perspective (i.e., spatial information) and among different perspectives (i.e., angular information) is beneficial to image super-resolution (SR). In this paper, we propose a spatial-angular interactive network (namely, LF-InterNet) for LF image SR. Specifically, spatial and angular features are first separately extracted from input LFs, and then repetitively interacted to progressively incorporate spatial and angular information. Finally, the interacted features are fused to super-resolve each sub-aperture image. Experimental results demonstrate the superiority of LF-InterNet over the state-of-the-art methods, i.e., our method can achieve high PSNR and SSIM scores with low computational cost, and recover faithful details in the reconstructed images.

Keywords: Light field imaging · Super-resolution · Feature decoupling · Spatial-angular interaction

1 Introduction

Light field (LF) cameras provide multiple views of a scene, and thus enable many attractive applications such as post-capture refocusing [1], depth sensing [2], saliency detection [3,4], and de-occlusion [5]. However, LF cameras face a trade-off between spatial and angular resolution. That is, they either provide dense angular samplings with low image resolution (e.g., Lytro and RayTrix), or capture high-resolution (HR) sub-aperture images (SAIs) with sparse angular samplings (e.g., camera arrays [6,7]). Consequently, many efforts have been made to improve the angular resolution through LF reconstruction [8–11], or the spatial resolution through LF image super-resolution (SR) [12–17]. In this

Electronic supplementary material The online version of this chapter (https://doi.org/10.1007/978-3-030-58592-1_18) contains supplementary material, which is available to authorized users.

© Springer Nature Switzerland AG 2020
A. Vedaldi et al. (Eds.): ECCV 2020, LNCS 12368, pp. 290–308, 2020.
https://doi.org/10.1007/978-3-030-58592-1_18

paper, we focus on the LF image SR problem, namely, to reconstruct HR SAIs from their corresponding low-resolution (LR) SAIs.

Image SR is a long-standing problem in computer vision. To achieve high reconstruction performance, SR methods need to incorporate as much useful information as possible from LR inputs. In the area of single image SR (SISR), good performance can be achieved by fully exploiting the neighborhood context (i.e., spatial information) in an image. Using the spatial information, SISR methods [18–24] can successfully hallucinate missing details. In contrast, LFs record scenes from multiple views, and the complementary information among different views (i.e., angular information) can be used to further improve the performance of LF image SR.

However, due to the complicated 4D structures of LFs, many LF image SR methods fail to fully exploit both the angular information and the spatial information, resulting in inferior SR performance. Specifically, in [25–27], SAIs are first super-resolved separately using SISR methods [18,20], and then fine-tuned together to incorporate the angular information. The angular information is ignored by these two-stage methods [25–27] during their upsampling process. In [13,15], only part of SAIs are used to super-resolve one view, and the angular information in these discarded views is not incorporated. In contrast, Rossi et al. [14] proposed a graph-based method to consider all angular views in an optimization process. However, this method [14] cannot fully use the spatial information, and is inferior to recent deep learning-based SISR methods [20–22].

Since spatial and angular information are highly coupled in 4D LFs and contribute to LF image SR in different manners, it is difficult for networks to perform well using these coupled information directly. In this paper, we propose a spatial-angular interactive network (i.e., LF-InterNet) to efficiently use spatial and angular information for LF image SR. Specifically, we design two convolutions (i.e., spatial/angular feature extractor) to extract and decouple spatial and angular features from input LFs. Then, we develop LF-InterNet to progressively interact the extracted features. Thanks to the proposed spatial-angular interaction mechanism, information in an LF can be effectively used in an efficient manner, and the SR performance is significantly improved. We perform extensive ablation studies to demonstrate the effectiveness of our model, and compare our method with state-of-the-art SISR and LF image SR methods from different perspectives, which demonstrate the superiority of our LF-InterNet.

2   Related Works

Single Image SR. In the area of SISR, deep learning-based methods have been extensively explored. Readers can refer to recent surveys [28–30] for more details in SISR. Here, we only review several milestone works. Dong et al. [18] proposed the first CNN-based SR method (i.e., SRCNN) by cascading 3 convolutional layers. Although SRCNN is shallow and simple, it achieves significant improvements over traditional SR methods [31–33]. Afterwards, SR networks became increasingly deep and complex, and thus more powerful in spatial information exploitation. Kim et al. [19] proposed a very deep SR network (i.e., VDSR) with 20 convolutional layers. Global residual learning is applied to VDSR to avoid slow convergence. Lim et al. [20] proposed an enhanced deep SR network (i.e., EDSR) and achieved substantial performance improvements by applying both local and global residual learning. Subsequently, Zhang et al. [34] proposed a residual dense network (i.e., RDN) by combining residual connection and dense connection. RDN can fully extract hierarchical features for image SR, and thus achieve further improvements over EDSR. More recently, Zhang et al. [21] and Dai et al. [22] further improved the performance of SISR by proposing residual channel attention network (i.e., RCAN) and second-order attention network (i.e., SAN). RCAN and SAN are the most powerful SISR methods to date and can achieve a very high reconstruction accuracy.

LF Image SR. In the area of LF image SR, different paradigms have been proposed. Bishop et al. [35] first estimated the scene depth and then used a deconvolution approach to estimate HR SAIs. Wanner et al. [36] proposed a variational LF image SR framework using the estimated disparity map. Farrugia et al. [37] decomposed HR-LR patches into several subspaces, and achieved LF image SR via PCA analysis. Alain et al. [12] extended SR-BM3D [38] to LFs, and super-resolved SAIs using LFBM5D filtering. Rossi et al. [14] formulated LF image SR as a graph optimization problem. These traditional methods [12,14,35–37] use different approaches to exploit angular information, but perform inferior in spatial information exploitation as compared to recent deep learning-based methods. In the pioneering work of deep learning-based LF image SR (i.e., LFCNN [25]), SAIs are super-resolved separately using SRCNN and fine-tuned in pairs to incorporate angular information. Similarly, Yuan et al. [27] proposed LF-DCNN, in which they used EDSR [20] to super-resolve each SAI and then fine-tuned the results. LFCNN and LF-DCNN handle the LF image SR problem in two stages and do not use angular information in the first stage. Wang et al. [15] proposed LFNet by extending BRCN [39] to LF image SR. In their method, SAIs from the same row or column are fed to a recurrent network to incorporate angular information. Zhang et al. [13] stacked SAIs along different angular directions to generate input volumes, and then fed them to a multi-stream residual network named resLF. LFNet and resLF reduce 4D LF to 3D LF by using part of SAIs to super-resolve one view. Consequently, angular information in these discarded views cannot be incorporated. To consider all views for LF image SR, Yeung et al. [16] proposed LFSSR to alternately shuffle LF features between SAI pattern and MacPI pattern for convolution. Jin et al. [17] proposed an all-to-one LF image SR framework (i.e., LF-ATO) and performed structural consistency regularization to preserve the parallax structure among reconstructed views.

3 Method

3.1 Spatial-Angular Feature Decoupling

An LF has a 4D structure and can be denoted as $\mathcal{L} \in \mathbb{R}^{U \times V \times H \times W}$, where U and V represent the angular dimensions (e.g., $U = 3$, $V = 4$ for a 3×4 LF), H and W represent the height and width of each SAI. Intuitively, an LF can be considered as a 2D angular collection of SAIs, and the SAI at each angular coordinate (u, v) can be denoted as $\mathcal{L}(u, v, :, :) \in \mathbb{R}^{H \times W}$. Similarly, an LF can also be organized into a 2D spatial collection of macro-pixels (namely, a MacPI). The macro-pixel at each spatial coordinate (h, w) can be denoted as $\mathcal{L}(:, :, h, w) \in \mathbb{R}^{U \times V}$. An illustration of these two LF representations is shown in Fig. 1.

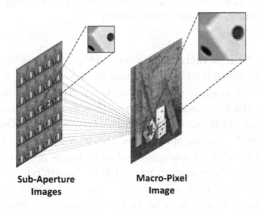

Sub-Aperture Images **Macro-Pixel Image**

Fig. 1. SAI array (left) and MacPI (right) representations of LFs. Both the SAI array and the MacPI representations have the same size of $\mathbb{R}^{UH \times VW}$. Note that, to convert an SAI array representation into a MacPI representation, pixels at the same spatial coordinates of each SAI need to be extracted and organized according to their angular coordinates to generate a macro-pixel. Then, a MacPI can be generated by organizing these macro-pixels according to their spatial coordinates. More details are presented in the supplemental material.

Since most methods use SAIs distributed in a square array as their input, we follow [12–14, 16, 17, 25, 26] to set $U = V = A$ in our method, where A denotes the angular resolution. Given an LF of size $\mathbb{R}^{A \times A \times H \times W}$, both a MacPI and an SAI array can be generated by organizing pixels according to corresponding patterns. Note that, when an LF is organized as an SAI array, the angular information is implicitly contained among different SAIs and thus is hard to extract. Therefore, we use the MacPI representation in our method and design spatial/angular feature extractors (SFE/AFE) to extract and decouple spatial/angular information.

Here, we use a toy example in Fig. 2 to illustrate the angular and spatial feature extractors. Specifically, AFE is defined as a convolution with a kernel size of $A \times A$ and a stride of A. Padding is not performed so that features generated by AFE have a size of $\mathbb{R}^{H \times W \times C}$, where C represents the feature

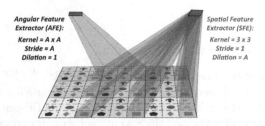

Fig. 2. An illustration of angular and spatial feature extractors. Here, an LF of size $\mathbb{R}^{3\times3\times3\times3}$ is used as a toy example. For better visualization, pixels from different SAIs are represented with different labels (e.g., red arrays or green squares), while different macro-pixels are paint with different background colors. Note that, AFE only extracts angular features and SFE only extracts spatial features, resulting in decoupling of spatial-angular information. (Color figure online)

depth. In contrast, SFE is defined as a convolution with a kernel size of 3×3, a stride of 1, and a dilation of A. We perform zero padding to ensure that the output features have the same spatial size $AH \times AW$ as the input MacPI. It is worth noting that, during angular feature extraction, each macro-pixel can be exactly convolved by AFE, while the information across different macro-pixels is not aliased. Similarly, during spatial feature extraction, pixels in each SAI can be convolved by SFE, while the angular information is not involved. In this way, the spatial and angular information in an LF is decoupled.

Due to the 3D property of real scenes, objects at different depths have different disparity values. Consequently, pixels of an object among different views cannot always locate at a single macro-pixel [40]. To address this problem, we enlarge the receptive field of our LF-InterNet by cascading multiple SFEs and AFEs in an interactive manner (see Fig. 4). Here, we use the Grad-CAM method [41] to visualize the receptive field of our LF-InterNet by highlighting contributive input regions. As shown in Fig. 3, the angular information indeed contributes to LF image SR, and the receptive field is enough to cover the disparities in LFs.

3.2   Network Design

Our LF-InterNet takes an LR MacPI of size $\mathbb{R}^{AH \times AW}$ as its input and produces an HR SAI array of size $\mathbb{R}^{\alpha AH \times \alpha AW}$, where α denotes the upscaling factor. Following [13, 16, 17], we convert images into YCbCr color space, and only super-resolve the Y channel of images. An overview of our network is shown in Fig. 4.

Overall Architecture. Given an LR MacPI $\mathcal{I}_{LR} \in \mathbb{R}^{AH \times AW}$, the angular and spatial features are first extracted by AFE and SFE, respectively.

$$\mathcal{F}_{A,0} = H_A\left(\mathcal{I}_{LR}\right), \quad \mathcal{F}_{S,0} = H_S\left(\mathcal{I}_{LR}\right), \tag{1}$$

where $\mathcal{F}_{A,0} \in \mathbb{R}^{H \times W \times C}$ and $\mathcal{F}_{S,0} \in \mathbb{R}^{AH \times AW \times C}$ represent the extracted angular and spatial features, respectively. H_A and H_S represent the angular and spatial

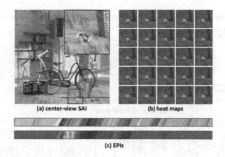

(a) center-view SAI   (b) heat maps

(c) EPIs

Fig. 3. A visualization of the receptive field of our LF-InterNet. We performed 2×SR on the 5 × 5 center views of scene *HCInew_bicycle* [42]. (a) Center-view HR SAI. We select a target pixel (marked in red in the zoom-in region) at a shallow depth. (b) Highlighted input SAIs generated by the Grad-CAM method [41]. A cluster of pixels in each SAI are highlighted as contributive pixels, which demonstrates the contribution of angular information. (c) Epipolar-plane images (EPIs) of the input LF (top) and the highlighted SAIs (bottom). It can be observed that the highlighted pixels in the bottom EPI have an enough receptive field to cover the slopes in the top EPI, which demonstates that our LF-InterNet can well handle the disparity problem in LF image SR.

feature extractors (as described in Sect. 3.1), respectively. Once initial features are extracted, features $\mathcal{F}_{A,0}$ and $\mathcal{F}_{S,0}$ are further processed by a set of interaction groups (i.e., Inter-Groups) to achieve spatial-angular feature interaction:

$$(\mathcal{F}_{A,n}, \mathcal{F}_{S,n}) = H_{IG,n}\left(\mathcal{F}_{A,n-1}, \mathcal{F}_{S,n-1}\right), \quad (n = 1, 2, \cdots, N), \qquad (2)$$

where $H_{IG,n}$ denotes the nth Inter-Group and N denotes the total number of Inter-Groups. In our LF-InterNet, we cascade all these Inter-Groups to fully use the information interacted at different stages. Specifically, features generated by each Inter-Group are concatenated and fed to a bottleneck block to fuse the interacted information. The feature generated by the bottleneck block is further added with the initial feature $\mathcal{F}_{S,0}$ to achieve global residual learning. The fused feature $\mathcal{F}_{S,t}$ can be obtained by

$$\mathcal{F}_{S,t} = H_B\left([\mathcal{F}_{A,1}, \cdots, \mathcal{F}_{A,N}], [\mathcal{F}_{S,1}, \cdots, \mathcal{F}_{S,N}]\right) + \mathcal{F}_{S,0}, \qquad (3)$$

where H_B denotes the bottleneck block, $[\cdot]$ denotes the concatenation operation. Finally, the fused feature $\mathcal{F}_{S,t}$ is fed to the reconstruction module, and an HR SAI array $\mathcal{I}_{SR} \in \mathbb{R}^{\alpha AH \times \alpha AW}$ can be obtained by

$$\mathcal{I}_{SR} = H_{1\times1}\left(S_{pix}\left(R_{lf}\left(H_S\left(\mathcal{F}_{S,t}\right)\right)\right)\right), \qquad (4)$$

where R_{lf}, S_{pix}, and $H_{1\times1}$ represent LF reshape, pixel shuffling, and 1×1 convolution, respectively.

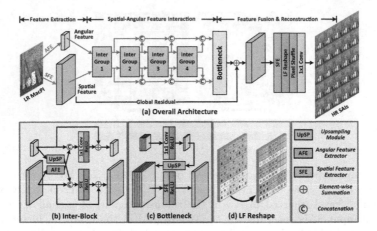

Fig. 4. An overview of our LF-InterNet. Angular and spatial features are first extracted from the input MacPI, and then fed to a series of Inter-Groups (which consists of several cascaded Inter-Blocks) to achieve spatial-angular interaction. After LF reshape and pixel shuffling, HR SAIs are generated.

Spatial-Angular Feature Interaction. The basic module for spatial-angular interaction is the interaction block (i.e., Inter-Block). As shown in Fig. 4(b), the Inter-Block takes a pair of angular and spatial features as its inputs to achieve feature interaction. Specifically, the input angular feature is first upsampled by a factor of A. Since pixels in a MacPI can be unevenly distributed due to edges and occlusions in real scenes [43], we learn this discontinuity using a 1×1 convolution and a pixel shuffling layer for angular-to-spatial upsampling. The upsampled angular feature is concatenated with the input spatial feature, and further fed to an SFE to incorporate the spatial and angular information. In this way, the complementary angular information can be used to guide spatial feature extraction. Simultaneously, the new angular feature is extracted from the input spatial feature by an AFE, and then concatenated with the input angular feature. The concatenated angular feature is further fed to a 1×1 convolution to integrate and update the angular information. Note that, the fused angular and spatial features are added with their input features to achieve local residual learning. In this paper, we cascade K Inter-Blocks in an Inter-Group, i.e., the output of an Inter-Block forms the input of its subsequent Inter-Block. In summary, the spatial-angular feature interaction can be formulated as

$$\mathcal{F}_{S,n}^{(k)} = H_S\left(\left[\mathcal{F}_{S,n}^{(k-1)}, \left(\mathcal{F}_{A,n}^{(k-1)}\right)\uparrow\right]\right) + \mathcal{F}_{S,n}^{(k-1)}, (k = 1, 2, \cdots, K), \quad (5)$$

$$\mathcal{F}_{A,n}^{(k)} = H_{1\times 1}\left(\left[\mathcal{F}_{A,n}^{(k-1)}, H_A\left(\mathcal{F}_{S,n}^{(k-1)}\right)\right]\right) + \mathcal{F}_{A,n}^{(k-1)}, (k = 1, 2, \cdots, K), \quad (6)$$

where $\uparrow$ represents the upsampling operation, $\mathcal{F}_{S,n}^{(k)}$ and $\mathcal{F}_{A,n}^{(k)}$ represent the output spatial and angular features of the kth Inter-Block in the nth Inter-Group, respectively.

Feature Fusion and Reconstruction. The objective of this stage is to fuse the interacted features to reconstruct an HR SAI array. The fusion and reconstruction stage mainly consists of bottleneck fusion (as shown in Fig. 4(c)), LF reshape (as shown in Fig. 4(d)), pixel shuffling, and final reconstruction.

In the bottleneck, the concatenated angular features $[\mathcal{F}_{A,1}, \cdots, \mathcal{F}_{A,N}] \in \mathbb{R}^{H \times W \times NC}$ are first fed to a 1×1 convolution and a ReLU layer to generate a feature map $\mathcal{F}_A \in \mathbb{R}^{H \times W \times C}$. Then, the squeezed angular feature $\mathcal{F}_A$ is upsampled and concatenated with spatial features. The final fused feature $\mathcal{F}_{S,t}$ can be obtained as

$$\mathcal{F}_{S,t} = H_S\left([\mathcal{F}_{S,1}, \cdots, \mathcal{F}_{S,N}, (\mathcal{F}_A)\uparrow]\right) + \mathcal{F}_{S,0}. \tag{7}$$

After feature fusion, we apply another SFE layer to extend the channel size of $\mathcal{F}_{S,t}$ to $\alpha^2 C$ for pixel shuffling [44]. However, since $\mathcal{F}_{S,t}$ is organized in the MacPI pattern, we apply LF reshape to convert $\mathcal{F}_{S,t}$ into an SAI array representation for pixel shuffling. To achieve LF reshape, we first extract pixels with the same angular coordinates in the MacPI feature, and then re-organize these pixels according to their spatial coordinates, which can be formulated as

$$\mathcal{I}_{SAIs}(x, y) = \mathcal{I}_{MacPI}(\xi, \eta), \tag{8}$$

where

$$x = H(\xi - 1) + \lfloor \xi/A \rfloor (1 - AH) + 1, \tag{9}$$

$$y = W(\eta - 1) + \lfloor \eta/A \rfloor (1 - AW) + 1. \tag{10}$$

Here, $x = 1, 2, \cdots, AH$ and $y = 1, 2, \cdots, AW$ denote the pixel coordinates in the output SAI arrays, ξ and η denote the corresponding coordinates in the input MacPI, $\lfloor \cdot \rfloor$ represents the round-down operation. The derivation of Eqs. (9) and (10) is presented in the supplemental material. Finally, a 1×1 convolution is applied to squeeze the number of feature channels to 1 for HR SAI reconstruction.

4 Experiments

In this section, we first introduce the datasets and our implementation details. Then we conduct ablation studies to investigate our network. Finally, we compare our LF-InterNet to several state-of-the-art LF image SR and SISR methods.

4.1 Datasets and Implementation Details

As listed in Table 1, we used 6 public LF datasets [42,45–49] in our experiments. All the LFs in the training and test sets have an angular resolution of 9×9. In the training stage, we cropped each SAI into patches of size 64×64, and then used bicubic downsampling with a factor of α ($\alpha = 2, 4$) to generate LR patches. The generated LR patches were re-organized into MacPIs to form the input of our network. The L_1 loss was used since it can produce good results and is robust to outliers [50]. We augmented the training data by 8 times using random

Table 1. Datasets used in our experiments.

	EPFL [45]	HCInew [42]	HCIold [46]	INRIA [47]	STFgantry [48]	STFlytro [49]
Training	70	20	10	35	9	250
Test	10	4	2	5	2	50

flipping and 90-degree rotation. Note that, during each data augmentation, all SAIs need to be flipped and rotated along both spatial and angular directions to maintain their LF structures.

By default, we used the model with $N = 4$, $K = 4$, $C = 64$, and angular resolution of 5×5 for both 2× and 4×SR. We also investigated the performance of other branches of our LF-InterNet in Sect. 4.2. We used PSNR and SSIM as quantitative metrics for performance evaluation. Note that, PSNR and SSIM were separately calculated on the Y channel of each SAI. To obtain the overall metric score for a dataset with M scenes (each with an angular resolution of $A \times A$), we first obtain the score for a scene by averaging its A^2 scores, and then obtain the overall score by averaging the scores of all M scenes.

Our LF-InterNet was implemented in PyTorch on a PC with an Nvidia RTX 2080Ti GPU. Our model was initialized using the Xavier method [51] and optimized using the Adam method [52]. The batch size was set to 12 and the learning rate was initially set to 5×10^{-4} and decreased by a factor of 0.5 for every 10 epochs. The training was stopped after 40 epochs.

4.2 Ablation Study

In this subsection, we compare the performance of our LF-InterNet with different architectures and angular resolutions to investigate the potential benefits introduced by different design choices.

Angular Information. We investigated the benefit of angular information by removing the angular path in LF-InterNet. That is, we only use SFE for LF image SR. Consequently, the network is identical to an SISR network, and can only incorporate spatial information within each SAI. As shown in Table 2, only using the spatial information, the network (i.e., LF-InterNet-SpatialOnly) achieves a PSNR of 29.98 and an SSIM of 0.897, which are significantly inferior to LF-InterNet. Therefore, the benefit of angular information to LF image SR is clearly demonstrated.

Spatial Information. To investigate the benefit introduced by spatial information, we changed the kernel size of all SFEs from 3×3 to 1×1. In this case, the spatial information cannot be exploited and integrated by convolutions. As shown in Table 2, the performance of LF-InterNet-AngularOnly is even inferior to bicubic interpolation. That is because, neighborhood context in an image is highly significant in recovering details. It is clear that spatial information plays a major role in LF image SR, while angular information can only be used as a complementary part to spatial information but cannot be used alone.

Information Decoupling. To investigate the benefit of spatial-angular information decoupling, we stacked all SAIs along the channel dimension as input, and used 3×3 convolutions with a stride of 1 to extract both spatial and angular information from these stacked images. Note that, global and local residual learning was maintained in this variant to keep the overall network architecture unchanged. To achieve fair comparison, we adjusted the feature depths to keep the model size (i.e., LF-InterNet-SAcoupled_1) or computational complexity (i.e., LF-InterNet-SAcoupled_2) comparable to LF-InterNet. As shown in Table 2, both two variants are inferior to LF-InterNet. That is, our LF-InterNet can well handle the 4D LF structure and achieve LF image SR much more efficiently by using the proposed spatial-angular feature decoupling mechanism.

Table 2. Comparative results achieved on the STFlytro dataset [49] by several variants of our LF-InterNet for 4×SR. Note that, we carefully adjusted the feature depths of different variants to make their model size comparable. FLOPs are computed with an input MacPI of size 160×160. The results of bicubic interpolation are listed as baselines.

Model	PSNR	SSIM	Params.	FLOPs
Bicubic	27.84	0.855	—	—
LF-InterNet-SpatialOnly	29.98	0.897	5.40M	134.7G
LF-InterNet-AngularOnly	26.57	0.823	5.43M	13.4G
LF-InterNet-SAcoupled_1	31.11	0.918	5.42M	5.46G
LF-InterNet-SAcoupled_2	31.17	0.919	50.8M	50.5G
LF-InterNet	**31.65**	**0.925**	5.23M	50.1G

Table 3. Comparative results achieved on the STFlytro dataset [49] by our LF-InterNet with different number of interactions for 4×SR.

IG_1	IG_2	IG_3	IG_4	PSNR	SSIM
				29.84	0.894
✓				31.44	0.922
✓	✓			31.61	0.924
✓	✓	✓		31.66	0.925
✓	✓	✓	✓	**31.84**	**0.927**

Spatial-Angular Interaction. We investigated the benefits introduced by our spatial-angular interaction mechanism. Specifically, we canceled feature interaction in each Inter-Group by removing upsampling and AFE modules in each Inter-Block. In this case, spatial and angular features are processed separately. When all interactions are removed, these spatial and angular features can only be incorporated by the bottleneck block. Table 3 presents the results achieved by our LF-InterNet with different numbers of interactions. It can be observed that, without any feature interaction, our network achieves a very low reconstruction accuracy (i.e., 29.84 in PSNR and 0.894 in SSIM). As the number of interactions increases, the performance is steadily improved. This clearly demonstrates the effectiveness of our spatial-angular feature interaction mechanism.

Table 4. Comparisons of different approaches for angular-to-spatial upsampling.

Model	Scale	PSNR	SSIM	Scale	PSNR	SSIM
LF-InterNet-nearest	2×	38.60	0.982	4×	31.65	0.925
LF-InterNet-bilinear	2×	37.67	0.976	4×	30.71	0.911
LF-InterNet	2×	**38.81**	**0.983**	4×	**31.84**	**0.927**

Angular-to-Spatial Upsampling. To demonstrate the effectiveness of the pixel shuffling layer used in angular-to-spatial upsampling, we introduced two variants by replacing pixel shuffling with nearest upsampling and bilinear upsampling, respectively. It can be observed from Table 4 that LF-InterNet-bilinear achieves much lower PSNR and SSIM scores than LF-InterNet-nearest and LF-InterNet. That is because, bilinear interpolation introduces aliasing among macro-pixels during angular-to-spatial upsampling, resulting in ambiguities in spatial-angular feature interaction. In contrast, both nearest upsampling and pixel shuffling do not introduce aliasing and thus achieve improved performance. Moreover, since pixels in a macro-pixel can be unevenly distributed due to edges and occlusions in real scenes, pixel shuffling achieves a further improvement over nearest upsampling due to its discontinuity modeling capability within macro-pixels.

Table 5. Comparative results achieved on the STFlytro dataset [49] by our LF-InterNet with different angular resolutions for 2× and 4×SR.

AngRes	Scale	PSNR	SSIM	Scale	PSNR	SSIM
3×3	2×	37.95	0.980	4×	31.30	0.918
5×5	2×	38.81	0.983	4×	31.84	0.927
7×7	2×	39.05	0.984	4×	32.04	0.931
9×9	2×	**39.08**	**0.985**	4×	**32.07**	**0.933**

Angular Resolution. We analyze the performance of LF-InterNet with different angular resolution. Specifically, we extracted the central $A \times A \, (A = 3, 5, 7, 9)$ SAIs from the input LFs, and trained different models for both 2× and 4×SR. As shown in Table 5, the PSNR and SSIM values for both 2× and 4×SR are improved as the angular resolution is increased. That is because, additional views provide rich angular information for LF image SR. It is also notable that, the performance tends to be saturated when the angular resolution is further increased from 7×7 to 9×9. That is because, the complementary information provided by additional views is already sufficient. Since the angular information has been fully exploited for an angular resolution of 7×7, a further increase of views can only provide minor performance improvement.

4.3 Comparison to the State-of-the-Arts

We compare our method to six SISR methods [19–24] and five LF image SR methods [12–14,16,17]. Bicubic interpolation was used as baselines.

Table 6. PSNR/SSIM values achieved by different methods for 2× and 4×SR. The best results are in red and the second best results are in blue.

Method	Scale	EPFL	HCInew	HCIold	INRIA	STFgantry	STFlytro
Bicubic	2×	29.50/0.935	31.69/0.934	37.46/0.978	31.10/0.956	30.82/0.947	33.02/0.950
VDSR [19]	2×	32.01/0.959	34.37/0.956	40.34/0.985	33.80/0.972	35.80/0.980	35.91/0.970
EDSR [20]	2×	32.86/0.965	35.02/0.961	41.11/0.988	34.61/0.977	37.08/0.985	36.87/0.975
RCAN [21]	2×	33.46/0.967	35.56/0.963	41.59/0.989	35.18/0.978	38.18/0.988	37.32/0.977
SAN [22]	2×	33.36/0.967	35.51/0.963	41.47/0.989	35.15/0.978	37.98/0.987	37.26/0.976
LFBM5D [12]	2×	31.15/0.955	33.72/0.955	39.62/0.985	32.85/0.969	33.55/0.972	35.01/0.966
GB [14]	2×	31.22/0.959	35.25/0.969	40.21/0.988	32.76/0.972	35.44/0.983	35.04/0.956
resLF [13]	2×	33.22/0.969	35.79/0.969	42.30/0.991	34.86/0.979	36.28/0.985	35.80/0.970
LFSSR [16]	2×	34.15/0.973	36.98/0.974	43.29/0.993	35.76/0.982	37.67/0.989	37.57/0.978
LF-ATO [17]	2×	34.49/0.976	37.28/0.977	43.76/0.994	36.21/0.984	39.06/0.992	38.27/0.982
LF-InterNet	2×	34.76/0.976	37.20/0.976	44.65/0.995	36.64/0.984	38.48/0.991	38.81/0.983
Bicubic	4×	25.14/0.831	27.61/0.851	32.42/0.934	26.82/0.886	25.93/0.843	27.84/0.855
VDSR [19]	4×	26.82/0.869	29.12/0.876	34.01/0.943	28.87/0.914	28.31/0.893	29.17/0.880
EDSR [20]	4×	27.82/0.892	29.94/0.893	35.53/0.957	29.86/0.931	29.43/0.921	30.29/0.903
RCAN [21]	4×	28.31/0.899	30.25/0.896	35.89/0.959	30.36/0.936	30.25/ 0.934	30.66/0.909
SAN [22]	4×	28.30/0.899	30.25/0.898	35.88/0.960	30.29/0.936	30.30/0.933	30.71/0.909
SRGAN [23]	4×	26.85/0.870	28.95/0.873	34.03/0.942	28.85/0.916	28.19/0.898	29.28/0.883
ESRGAN [24]	4×	25.59/0.836	26.96/0.819	33.53/0.933	27.54/0.880	28.00/0.905	27.09/0.826
LFBM5D [12]	4×	26.61/0.869	29.13/0.882	34.23/0.951	28.49/0.914	28.30/0.900	29.07/0.881
GB [14]	4×	26.02/0.863	28.92/0.884	33.74/0.950	27.73/0.909	28.11/0.901	28.37/0.873
resLF [13]	4×	27.86/0.899	30.37/0.907	36.12/0.966	29.72/0.936	29.64/0.927	28.94/0.891
LFSSR [16]	4×	29.16/0.915	30.88/0.913	36.90/0.970	31.03/0.944	30.14/0.937	31.21/0.919
LF-ATO [17]	4×	29.16/0.917	31.08/0.917	37.23/0.971	31.21/0.950	30.78/0.944	30.98/0.918
LF-InterNet	4×	29.52/0.917	31.01/0.917	37.23/0.972	31.65/0.950	30.44/0.941	31.84/0.927

Quantitative Results. Quantitative results in Table 6 demonstrate the state-of-the-art performance of our LF-InterNet on all the 6 test datasets. Thanks to the use of angular information, our method achieves an improvement of 1.54 dB (2×SR) and 1.00 dB (4×SR) in PSNR over the powerful SISR method RCAN [21]. Moreover, our LF-InterNet can achieve a comparable PSNR and SSIM scores as compared to the most recent LF image SR method LF-ATO [17].

Qualitative Results. Qualitative results of 2×/4×SR are shown in Fig. 5, with more visual comparisons being provided in our supplemental material. Our LF-InterNet can well preserve the textures and details (e.g., the horizontal stripes in

Fig. 5. Visual results of 2×/4×SR.

the scene *HCInew_origami*) in the super-resolved images. In contrast, state-of-the-art SISR methods RCAN [21] and SAN [22] produce oversmoothed images with poor details. The visual superiority of our method is more obvious for 4×SR. That is because, the input LR images are severely degraded by the downsampling operation, and the process of 4×SR is highly ill-posed. In such cases, some perceptual-oriented methods (e.g., SRGAN [23] and ESRGAN [24]) use spatial information only to hallucinate missing details, resulting in ambiguous and even fake textures (e.g., wheel in scene *STFlytro_buildings*). In contrast, our method can use complementary angular information among different views to produce more faithful results.

Table 7. Comparisons of the number of parameters (#Params.) and FLOPs for 2× and 4×SR. Note that, the FLOPs is calculated on an input LF with a size of $5 \times 5 \times 32 \times 32$, and the PSNR and SSIM scores are averaged over the 6 test datasets [42,45–49] in Table 6.

Method	Scale	#Params.	FLOPs(G)	PSNR/SSIM	Scale	#Params	FLOPs(G)	PSNR/SSIM
RCAN [21]	2×	15.44M	15.71 × 25	36.88/0.977	4×	15.59M	16.34 × 25	30.95/0.922
SAN [22]	2×	15.71M	16.05 × 25	36.79/0.977	4×	15.86M	16.67 × 25	31.96/0.923
resLF [13]	2×	6.35M	37.06	36.38/0.977	4×	6.79M	39.70	30.08/0.916
LFSSR [16]	2×	0.81M	25.70	37.57/0.982	4×	1.61M	128.44	31.55/0.933
LF-ATO [17]	2×	1.51M	597.66	38.18/0.984	4×	1.66M	686.99	31.74/0.937
LF-InterNet_32	2×	1.20M	11.87	37.88/0.983	4×	1.31M	12.53	31.57/0.933
LF-InterNet_64	2×	4.80M	47.46	38.42/0.984	4×	5.23M	50.10	31.95/0.937

Efficiency. We compare our LF-InterNet to several competitive methods [13,16,17,21,22] in terms of the number of parameters and FLOPs. As shown in Table 7, our LF-InterNet achieves superior SR performance with reasonable number of parameters and FLOPs. Note that, although LF-ATO has very small model sizes (i.e., 1.51M for 2×SR and 1.66M for 4×SR), its FLOPs are very high since it uses the *All-to-One* strategy to separately super-resolve individual views in a sequence. In contrast, our method (i.e., LF-InterNet_64) super-resolves all views within a single inference, and achieves comparable or even better performance than LF-ATO with significantly lower FLOPs. It is worth noting that, even the feature depth of our model is halved to 32, our method (i.e., LF-InterNet_32) can still achieve promising PSNR/SSIM scores, which are comparable to LFSSR and higher than RCAN, SAN, and resLF. The above comparisons clearly demonstrate the high efficiency of our network architecture.

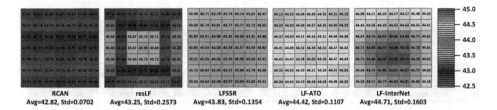

Fig. 6. Comparative results (i.e., PSNR values) achieved on each perspective of scene *HCIold_MonasRoom*. Here, 7 × 7 input views are used to perform 2×SR. We use standard deviation (Std) to represent their uniformity. Our LF-InterNet achieves high reconstruction quality with a relatively balanced distribution.

Performance w.r.t. Perspectives. Since our LF-InterNet can super-resolve all SAIs in an LF, we further investigate the reconstruction quality with respect to different perspectives. We followed [13] to use the central 7 × 7 views of scene *HCIold_MonasRoom* to perform 2×SR, and used PSNR for performance evaluation. Note that, due to the changing perspectives, the contents of different SAIs are not identical, resulting in inherent PSNR variations. Therefore, we evaluate this variation by using RCAN to perform SISR on each SAI. Results are reported and visualized in Fig. 6. Since resLF uses part of views to super-resolve different perspectives, the reconstruction qualities of resLF for non-central views are relatively low. In contrast, LFSSR, LF-ATO and our LF-InterNet can use the angular information from all input views to super-resolve each view, and thus achieve a relatively balanced distribution (i.e., lower Std scores) among different perspectives. The reconstruction quality (i.e., PSNR scores) of LF-InterNet is higher than those of LFSSR and LF-ATO on this scene.

Table 8. Comparative results achieved on the UCSD dataset for 2× and 4×SR.

Method	Scale	PSNR/SSIM	Scale	PSNR/SSIM
RCAN [21]	2×	41.63/0.983	4×	36.49/0.955
SAN [22]	2×	41.56/0.983	4×	36.57/0.956
resLF [13]	2×	41.29/0.982	4×	35.89/0.953
LFSSR [16]	2×	41.55/0.984	4×	36.77/0.957
LF-ATO [17]	2×	41.80/0.985	4×	36.95/0.959
LF-InterNet	2×	42.36/0.985	4×	37.12/0.960

Generalization to Unseen Scenarios. We evaluate the generalization capability of different methods by testing them on a novel and unseen real-world dataset (i.e., the UCSD dataset [53]). Note that, all methods have not been trained or fine-tuned on the UCSD dataset. Results in Table 8 show that our LF-InterNet outperforms the state-of-the-art methods [13,16,17,21,22], which demonstrates the generalization capability of our method to unseen scenarios.

Fig. 7. Visual results achieved by different methods under real-world degradation.

Performance Under Real-World Degradation. We compare the performance of different methods under real-world degradation by directly applying them to LFs in the STFlytro dataset [49]. As shown in Fig. 7, our method produces images with faithful details and less artifacts. Since the LF structure keeps unchanged under both bicubic and real-world degradation, our method can learn to incorporate spatial and angular information from training LFs using the proposed spatial-angular interaction mechanism. It is also demonstrated that our method can be easily applied to LF cameras to generate high-quality images.

5   Conclusion and Future Work

In this paper, we proposed a deep convolutional network LF-InterNet for LF image SR. We first introduce an approach to extract and decouple spatial and angular features, and then design a feature interaction mechanism to incorporate spatial and angular information. Experimental results have demonstrated the superiority of our LF-InterNet over state-of-the-art methods. Since the spatial-angular interaction mechanism is a generic framework and can process LFs in an elegant and efficient manner, we will apply LF-InterNet to LF angular SR [8–11] and joint spatial-angular SR [54,55] as our future work.

Acknowledgement. This work was supported by the National Natural Science Foundation of China (No. 61972435, 61602499).

References

1. Wang, Y., Yang, J., Guo, Y., Xiao, C., An, W.: Selective light field refocusing for camera arrays using bokeh rendering and superresolution. IEEE Signal Process. Lett. **26**(1), 204–208 (2018)

2. Shin, C., Jeon, H.G., Yoon, Y., So Kweon, I., Joo Kim, S.: EPINET: a fully-convolutional neural network using epipolar geometry for depth from light field images. In: Proceedings of the IEEE Conference on Computer Vision and Pattern Recognition (CVPR), pp. 4748–4757 (2018)

3. Wang, T., Piao, Y., Li, X., Zhang, L., Lu, H.: Deep learning for light field saliency detection. In: Proceedings of the IEEE International Conference on Computer Vision (ICCV), pp. 8838–8848 (2019)

4. Zhang, M., Li, J., WEI, J., Piao, Y., Lu, H.: Memory-oriented decoder for light field salient object detection. In: Advances in Neural Information Processing Systems, pp. 896–906 (2019)

5. Wang, Y., Wu, T., Yang, J., Wang, L., An, W., Guo, Y.: DeOccNet: learning to see through foreground occlusions in light fields. In: Winter Conference on Applications of Computer Vision (WACV) (2020)

6. Wilburn, B., et al.: High performance imaging using large camera arrays. ACM Trans. Graph. **24**, 765–776 (2005)

7. Venkataraman, K., et al.: PiCam: an ultra-thin high performance monolithic camera array. ACM Trans. Graph. **32**(6), 166 (2013)

8. Wu, G., Zhao, M., Wang, L., Dai, Q., Chai, T., Liu, Y.: Light field reconstruction using deep convolutional network on EPI. In: Proceedings of the IEEE Conference on Computer Vision and Pattern Recognition (CVPR), pp. 6319–6327 (2017)

9. Wu, G., Liu, Y., Dai, Q., Chai, T.: Learning sheared EPI structure for light field reconstruction. IEEE Trans. Image Process. **28**(7), 3261–3273 (2019)

10. Jin, J., Hou, J., Yuan, H., Kwong, S.: Learning light field angular super-resolution via a geometry-aware network. In: AAAI Conference on Artificial Intelligence (2020)

11. Shi, J., Jiang, X., Guillemot, C.: Learning fused pixel and feature-based view reconstructions for light fields. In: Proceedings of the IEEE Conference on Computer Vision and Pattern Recognition (CVPR) (2020)

12. Alain, M., Smolic, A.: Light field super-resolution via lfbm5d sparse coding. In: Proceedings of the IEEE International Conference on Image Processing (ICIP), pp. 2501–2505 (2018)

13. Zhang, S., Lin, Y., Sheng, H.: Residual networks for light field image super-resolution. In: Proceedings of the IEEE Conference on Computer Vision and Pattern Recognition (CVPR), pp. 11046–11055 (2019)

14. Rossi, M., Frossard, P.: Geometry-consistent light field super-resolution via graph-based regularization. IEEE Trans. Image Process. **27**(9), 4207–4218 (2018)

15. Wang, Y., Liu, F., Zhang, K., Hou, G., Sun, Z., Tan, T.: LFNet: a novel bidirectional recurrent convolutional neural network for light-field image super-resolution. IEEE Trans. Image Process. **27**(9), 4274–4286 (2018)

16. Yeung, H.W.F., Hou, J., Chen, X., Chen, J., Chen, Z., Chung, Y.Y.: Light field spatial super-resolution using deep efficient spatial-angular separable convolution. IEEE Trans. Image Process. **28**(5), 2319–2330 (2018)

17. Jin, J., Hou, J., Chen, J., Kwong, S.: Light field spatial super-resolution via deep combinatorial geometry embedding and structural consistency regularization. In: Proceedings of the IEEE Conference on Computer Vision and Pattern Recognition (CVPR) (2020)

18. Dong, C., Loy, C.C., He, K., Tang, X.: Learning a deep convolutional network for image super-resolution. In: Fleet, D., Pajdla, T., Schiele, B., Tuytelaars, T. (eds.) ECCV 2014. LNCS, vol. 8692, pp. 184–199. Springer, Cham (2014). https://doi.org/10.1007/978-3-319-10593-2_13

19. Kim, J., Kwon Lee, J., Mu Lee, K.: Accurate image super-resolution using very deep convolutional networks. In: Proceedings of the IEEE Conference on Computer Vision and Pattern Recognition (CVPR), pp. 1646–1654 (2016)
20. Lim, B., Son, S., Kim, H., Nah, S., Mu Lee, K.: Enhanced deep residual networks for single image super-resolution. In: Proceedings of the IEEE Conference on Computer Vision and Pattern Recognition Workshops (CVPRW), pp. 136–144 (2017)
21. Zhang, Y., Li, K., Li, K., Wang, L., Zhong, B., Fu, Y.: Image super-resolution using very deep residual channel attention networks. In: Proceedings of the European Conference on Computer Vision (ECCV). (2018) 286–301
22. Dai, T., Cai, J., Zhang, Y., Xia, S.T., Zhang, L.: Second-order attention network for single image super-resolution. In: Proceedings of the IEEE Conference on Computer Vision and Pattern Recognition (CVPR), pp. 11065–11074 (2019)
23. Ledig, C., et al.: Photo-realistic single image super-resolution using a generative adversarial network. In: Proceedings of the IEEE Conference on Computer Vision and Pattern Recognition (CVPR), pp. 4681–4690 (2017)
24. Wang, X., et al.: ESRGAN: enhanced super-resolution generative adversarial networks. In: Leal-Taixé, L., Roth, S. (eds.) ECCV 2018. LNCS, vol. 11133, pp. 63–79. Springer, Cham (2019). https://doi.org/10.1007/978-3-030-11021-5_5
25. Yoon, Y., Jeon, H.G., Yoo, D., Lee, J.Y., So Kweon, I.: Learning a deep convolutional network for light-field image super-resolution. In: Proceedings of the IEEE International Conference on Computer Vision Workshops (ICCVW), pp. 24–32 (2015)
26. Yoon, Y., Jeon, H.G., Yoo, D., Lee, J.Y., Kweon, I.S.: Light-field image super-resolution using convolutional neural network. IEEE Signal Process. Lett. 24(6), 848–852 (2017)
27. Yuan, Y., Cao, Z., Su, L.: Light-field image superresolution using a combined deep cnn based on EPI. IEEE Signal Process. Lett. 25(9), 1359–1363 (2018)
28. Wang, Z., Chen, J., Hoi, S.C.: Deep learning for image super-resolution: a survey. IEEE Trans. Pattern Anal. Mach. Intell. (2020)
29. Anwar, S., Khan, S., Barnes, N.: A deep journey into super-resolution: a survey. ACM Comput. Surv. 53(3), 1–34 (2020)
30. Yang, W., Zhang, X., Tian, Y., Wang, W., Xue, J.H., Liao, Q.: Deep learning for single image super-resolution: a brief review. IEEE Trans. Multimedia 21, 3106–3121 (2019)
31. Timofte, R., De Smet, V., Van Gool, L.: Anchored neighborhood regression for fast example-based super-resolution. In: Proceedings of the IEEE International Conference on Computer Vision (ICCV), pp. 1920–1927 (2013)
32. Jianchao, Y., John, W., Thomas, H., Yi, M.: Image super-resolution via sparse representation. IEEE Trans. Image Process. 19(11), 2861–2873 (2010)
33. Zeyde, R., Elad, M., Protter, M.: On single image scale-up using sparse-representations. In: Boissonnat, J.-D., et al. (eds.) Curves and Surfaces 2010. LNCS, vol. 6920, pp. 711–730. Springer, Heidelberg (2012). https://doi.org/10.1007/978-3-642-27413-8_47
34. Zhang, Y., Tian, Y., Kong, Y., Zhong, B., Fu, Y.: Residual dense network for image super-resolution. In: Proceedings of the IEEE Conference on Computer Vision and Pattern Recognition (CVPR), pp. 2472–2481 (2018)
35. Bishop, T.E., Favaro, P.: The light field camera: extended depth of field, aliasing, and superresolution. IEEE Trans. Pattern Anal. Mach. Intell. 34(5), 972–986 (2011)

36. Wanner, S., Goldluecke, B.: Variational light field analysis for disparity estimation and super-resolution. IEEE Trans. Pattern Anal. Mach. Intell. **36**(3), 606–619 (2013)
37. Farrugia, R.A., Galea, C., Guillemot, C.: Super resolution of light field images using linear subspace projection of patch-volumes. IEEE J. Sel. Topics Signal Process. **11**(7), 1058–1071 (2017)
38. Egiazarian, K., Katkovnik, V.: Single image super-resolution via bm3d sparse coding. In: European Signal Processing Conference (EUSIPCO), pp. 2849–2853 (2015)
39. Huang, Y., Wang, W., Wang, L.: Bidirectional recurrent convolutional networks for multi-frame super-resolution. In: Advances in Neural Information Processing Systems (NeurIPS), pp. 235–243 (2015)
40. Williem, Park, I., Lee, K.M.: Robust light field depth estimation using occlusion-noise aware data costs. IEEE Trans. Pattern Anal. Mach. Intell. **40**(10), 2484–2497 (2018)
41. Selvaraju, R.R., Cogswell, M., Das, A., Vedantam, R., Parikh, D., Batra, D.: Grad-CAM: visual explanations from deep networks via gradient-based localization. In: Proceedings of the IEEE International Conference on Computer Vision (ICCV), pp. 618–626 (2017)
42. Honauer, K., Johannsen, O., Kondermann, D., Goldluecke, B.: A dataset and evaluation methodology for depth estimation on 4D light fields. In: Lai, S.-H., Lepetit, V., Nishino, K., Sato, Y. (eds.) ACCV 2016. LNCS, vol. 10113, pp. 19–34. Springer, Cham (2017). https://doi.org/10.1007/978-3-319-54187-7_2
43. Park, I.K., Lee, K.M., et al.: Robust light field depth estimation using occlusion-noise aware data costs. IEEE Trans. Pattern Anal. Mach. Intell. **40**(10), 2484–2497 (2017)
44. Shi, W., et al.: Real-time single image and video super-resolution using an efficient sub-pixel convolutional neural network. In: Proceedings of the IEEE Conference on Computer Vision and Pattern Recognition (CVPR), pp. 1874–1883 (2016)
45. Rerabek, M., Ebrahimi, T.: New light field image dataset. In: Proceedings of the International Conference on Quality of Multimedia Experience (QoMEX) (2016)
46. Wanner, S., Meister, S., Goldluecke, B.: Datasets and benchmarks for densely sampled 4D light fields. In: Vision, Modelling and Visualization (VMV), vol. 13, pp. 225–226. Citeseer (2013)
47. Le Pendu, M., Jiang, X., Guillemot, C.: Light field inpainting propagation via low rank matrix completion. IEEE Trans. Image Process. **27**(4), 1981–1993 (2018)
48. Vaish, V., Adams, A.: The (new) stanford light field archive. Comput. Graph. Lab. Stanf. Univ. **6**(7) (2008)
49. Raj, A.S., Lowney, M., Shah, R., Wetzstein, G.: Stanford lytro light field archive (2016)
50. Anagun, Y., Isik, S., Seke, E.: SRLibrary: comparing different loss functions for super-resolution over various convolutional architectures. J. Vis. Commun. Image Represent. **61**, 178–187 (2019)
51. Glorot, X., Bengio, Y.: Understanding the difficulty of training deep feedforward neural networks. In: Proceedings of the International Conference on Artificial Intelligence and Statistics, pp. 249–256 (2010)
52. Kingma, D.P., Ba, J.: Adam: a method for stochastic optimization (2015)
53. Wang, T.-C., Zhu, J.-Y., Hiroaki, E., Chandraker, M., Efros, A.A., Ramamoorthi, R.: A 4D light-field dataset and cnn architectures for material recognition. In: Leibe, B., Matas, J., Sebe, N., Welling, M. (eds.) ECCV 2016. LNCS, vol. 9907, pp. 121–138. Springer, Cham (2016). https://doi.org/10.1007/978-3-319-46487-9_8

54. Meng, N., So, H.K.H., Sun, X., Lam, E.: High-dimensional dense residual convolutional neural network for light field reconstruction. IEEE Trans. Pattern Anal. Mach. Intell. (2019)
55. Meng, N., Wu, X., Liu, J., Lam, E.Y.: High-order residual network for light field super-resolution. In: AAAI Conference on Artificial Intelligence (2020)

BATS: Binary ArchitecTure Search

Adrian Bulat[1]([✉]) [iD], Brais Martinez[1] [iD], and Georgios Tzimiropoulos[1,2] [iD]

[1] Samsung AI Center, Cambridge, UK
adrian@adrianbulat.com, brais.mart@gmail.com
[2] Queen Mary University of London, London, UK
g.tzimiropoulos@qmul.ac.uk

Abstract. This paper proposes Binary ArchitecTure Search (BATS), a framework that drastically reduces the accuracy gap between binary neural networks and their real-valued counterparts by means of Neural Architecture Search (NAS). We show that directly applying NAS to the binary domain provides very poor results. To alleviate this, we describe, to our knowledge, for the first time, the 3 key ingredients for successfully applying NAS to the binary domain. Specifically, we (1) introduce and design a novel binary-oriented search space, (2) propose a new mechanism for controlling and stabilising the resulting searched topologies, (3) propose and validate a series of new search strategies for binary networks that lead to faster convergence and lower search times. Experimental results demonstrate the effectiveness of the proposed approach and the necessity of searching in the binary space directly. Moreover, (4) we set a new state-of-the-art for binary neural networks on CIFAR10, CIFAR100 and ImageNet datasets. Code will be made available.

Keywords: Binary networks · Neural Architecture Search

1 Introduction

Network quantization and Network Architecture Search (NAS) have emerged as two important research directions with the goal of designing efficient networks capable of running on mobile devices. Network quantization reduces the size and computational footprint of the models by representing the activations and the weights using $N < 32$ bits. Of particular interest is the extreme case of quantization, binarization, in which the model and the activations are quantized to a single bit [9,10,33]. This allows to replace all floating point multiply-add operations inside a convolutional layer with bit-wise operations resulting in a speed-up of up to 57× [33]. Recent years have seen a progressive reduction of the performance gap with real-valued networks. However, research has almost exclusively focused on the ResNet architecture, rather than on efficient ones

Electronic supplementary material The online version of this chapter (https://doi.org/10.1007/978-3-030-58592-1_19) contains supplementary material, which is available to authorized users.

A. Vedaldi et al. (Eds.): ECCV 2020, LNCS 12368, pp. 309–325, 2020.
https://doi.org/10.1007/978-3-030-58592-1_19

such as MobileNet. This is widely credited to pointwise convolutions not being amenable to binarization [2,4].

As an orthogonal direction, NAS attempts to improve the overall performance by automatically searching for optimal network topologies using a various of approaches including evolutionary algorithms [34,41], reinforcement learning [47,52] or more recently by taking a differentiable view of the search process via gradient-based approaches [25,27]. Such methods were shown to perform better than carefully hand-crafted architectures on both classification [25,27,32,52] and fine-grained tasks [24]. However, all of the aforementioned methods are tailored towards searching architectures for real-valued neural networks.

The aim of this work is to propose, for the first time, ways of reducing the accuracy gap between binary and real-valued networks via the state-of-the-art framework of differentiable NAS (DARTS) [25,27].

We show that direct binarization of existing cells searched in the real domain leads to sub-par results due to the particularities of the binarization process, in which certain network characteristics typically used in real-valued models (e.g. 1×1 convolutions) may be undesired [2,4]. Moreover, we show that performing the search in the binary domain comes with a series of challenges of its own: it inherits and amplifies one of the major DARTS [27] drawbacks, that of "cell collapsing", where the resulting architectures, especially when trained for longer, can result in degenerate cells in which most of the connections are skip connections [25] or parameter-free operations. In this work, we propose a novel Binary Architecture Search (BATS) method that successfully tackles the above-mentioned issues and sets a new state-of-the-art for binary neural networks on the selected datasets, yet simultaneously reducing computational resources by a wide margin. In summary, **our contributions** are:

1. We show that directly applying NAS to the binary domain provides very poor results. To alleviate this, we describe, to our knowledge, for the first time, the 3 key ingredients for successfully applying NAS to the binary domain.
2. We devise a novel search space specially tailored to the binary case and incorporate it within the DARTS framework (Sect. 4.1).
3. We propose a temperature-based mechanism for controlling and stabilising the topology search (Sect. 4.2).
4. We propose and validate a series of new search strategies for binary networks that lead to faster convergence and lower search times (Sect. 4.3).
5. We show that our method consistently produces superior architectures for binarized networks within a lower computational budget, setting a new state-of-the-art on CIFAR10/100 and ImageNet datasets (Sect. 6).

2   Related Work

NAS: While hand-crafted architectures significantly pushed the state-of-the-art in Deep Learning [16,18,19,30,36,46], recently, NAS was proposed as an automated alternative, shown to produce architectures that outperform the manually designed ones [35]. NAS methods can be roughly classified in three categories:

evolutionary-based [34,35,41], reinforcement learning-based [1,47,52] and, more recently, one-shot approaches, including differentiable ones [3,6,24,27]. The former two categories require significant computational resources during the search phase (3150 GPU-days for AmoebaNet [34] and 1800 GPU-days for NAS-Net [52]). Hence, of particular interest in our work is the differentiable search framework (DARTS) of [27] which is efficient. Follow-up work further improves on it by progressively reducing the search space [8] or by constraining the architecture parameters to be one-hot [42]. Our method builds upon the DARTS framework incorporating the progressive training of [8]. In contrast to all the aforementioned methods that search for optimal real-valued topologies, to our knowledge, we are the first to study NAS for binary networks.

Network binarization is the most extreme case of network quantization in which weights and activations are represented with 1 bit. It allows for up to 57× faster convolutions [33], a direct consequence of replacing all multiplications with bitwise operations [9]. Typically, binarization is achieved by taking the sign of the real-valued weights and features [9,10]. However, such an approach leads to large drops in accuracy, especially noticeable on large scale datasets. To alleviate this, Rastegari et al. [33] introduce analytically computed real-valued scaling factors that scale the input features and weights. This is further improved in [5] that proposes to learn the factors via back-propagation. In Bi-Real Net, Lin et al. [28] advocate the use of double-skip connections and of real-valued downsample layers. Wang et al. [40] propose a reinforcement learning-based approach to learn inter-channel correlations to better preserve the sign of convolutions. Ding et al. [14] introduce a distribution loss to explicitly regularize the activations and improve the information flow thought the network. While most of these methods operate within the same computational budget, another direction of research attempts to bridge the gap between binary and their real-valued networks by increasing the network size. For example, ABC-Net [23] proposes to use up to $M = N = 5$ branches, equivalent with running $M \times N$ convolutional layers in parallel, while Zhu et al. [50] use an ensemble of up to 6 binary models. These methods expand the size of the network while preserving the general architecture of the ResNet [16] or WideResNet [44]. In contrast to all the aforementioned methods in which binary architectures are hand-crafted, we attempt to automatically discover novel binary architectures, without increasing the computational budget. Our discovered architectures set a new state-of-the-art on the most widely used datasets.

Very recent work on binary NAS done concurrently with our work include [37] and [38]. As opposed to our work, [37] simple searches for the number of channels in each layer inside a ResNet. [38] uses a completely different search space and training strategy. We note that we outperform significantly both of them in terms of accuracy/efficiency. On ImageNet, using 1.55×10^8 FLOPs, we obtain a Top-1 accuracy of **66.1%** vs 58.76% using 1.63×10^8 FLOPS in [38]. On CIFAR-10 we score a top-1 accuracy of 96.1% vs 94.43% in [37]. While [37] achieves an accuracy of 69.65% on ImageNet, they use 6.6×10^8 FLOPS (4.2× more than our biggest model).

3   Background

Network Binarization: We binarize the models using the method proposed by Rastegari *et al.* [33] with the modifications introduced in [5]: we learn the weight scaling factor α via back-propagation instead of computing it analytically while dropping the input feature scaling factor β. Assume a given convolutional layer L with weights $\mathcal{W} \in \mathbb{R}^{o \times c \times w \times h}$ and input features $\mathcal{I} \in \mathbb{R}^{c \times w_{in} \times h_{in}}$, where o and c represent the number of output and input channels, (w, h) the width and height of the convolutional kernel, $(w_{in} \geq w, h_{in} \geq h)$ the spatial dimensions of $\mathcal{I}$. The binarization is accomplished by taking the sign of $\mathcal{W}$ and $\mathcal{I}$ and then multiplying the output of the convolution by the trainable scaling factor $\alpha \in \mathbb{R}^+$:

$$\mathcal{I} * \mathcal{W} \approx (\text{sign}(\mathcal{I}) \circledast \text{sign}(\mathcal{W})) \odot \alpha, \tag{1}$$

where $\odot$ denotes the element-wise multiplication, $*$ the real-valued convolution and $\circledast$ its binary counterpart. During training, the gradients are used to update the real-valued weights $\mathcal{W}$ while the forward pass is done using $\text{sign}(\mathcal{W})$. Finally, we used the standard arrangement for the binary convolutional layer operations: Batch Norm, Sign, Convolution, Activation.

DARTS: Herein, we review DARTS [27] and P-DARTS [8] upon which we build our framework: instead of searching for the whole network architecture, DARTS breaks down the network into a series of L identical cells. Each cell can be seen as a Direct Acyclic Graph (DAG) with N nodes, each representing a feature tensor. Within a cell, the goal is to choose, during the search phase, an operation $o(\cdot)$ from the predefined search space O that will connect a pair of nodes. Given a pair of such nodes (i, j) the information flow from i to j is defined as: $f_{i,j}(\mathbf{x}_i) = \sum_{o \in O} \frac{\exp(\alpha_{i,j}^o)}{\sum_{o' \in O} \exp(\alpha_{i,j}^{o'})} \cdot o(\mathbf{x}_i)$, where $\mathbf{x}_i$ is the output of the i-th node and $\alpha_{i,j}^o$ is an architecture parameter used to weight the operation $o(\mathbf{x}_i)$. The output of the node is computed as the sum of all inputs $\mathbf{x}_j = \sum_{i<j} f_{i,j}(\mathbf{x}_i)$ while the output of the cell is a depth-wise concatenation of all the intermediate nodes minus the input ones $(2, 3, \ldots, N-1)$. During the search phase, the networks weights $\mathcal{W}$ and the architectures parameters α are learned using bi-level optimization where the weights and α are optimized on two different splits of the dataset. Because the search is typically done on a much shallower network due to the high computational cost caused by the large search space O, DARTS may produce cells that are under-performing when tested using deeper networks. To alleviate this, in [8] the search is done using a series of stages during which the worse performing operations are dropped and the network depth increases. For a complete detailed explanation see [27] and [8].

4   Method

This section describes the 3 key components proposed in this work: the binary search space, the temperature regularization and search strategy. As we also show experimentally in Sect. 5.1, we found that *all 3 components are necessary for successfully applying NAS to the binary domain*[1].

4.1   Binary Neural Architecture Search Space

Since searching across a large space O is computationally prohibitive, most of the recent NAS methods manually define a series of 8 operations known to produce satisfactory results when used in hand-crafted network topologies: 3×3 and 5×5 dilated convolutions, 3×3 and 5×5 separable convolutions, 3×3 max pooling, 3×3 average pooling, identity (skip) connections plus the zero (no) connection. Table 1 summarizes the operations used for the real-valued case. Having an appropriate search space containing only desirable operations is absolutely essential for obtaining good network architectures: for example, a random search performed on the aforementioned space on CIFAR-10 achieves already 3.29% top-error vs. 2.76% when searched using second-order DARTS [27].

Table 1. Comparison between the commonly used searched space for real valued networks and the proposed one, specially tailed for binary models.

Real-valued	Ours (proposed)
Separable conv. (3×3)	Group convolution (3×3)
Separable conv. (5×5)	Group convolution (5×5)
Dilated conv. (3×3)	Dilated Group conv. (3×3)
Dilated conv. (5×5)	Dilated Group conv.(5×5)
Identity (*skip connection*)	
Max pooling (3×3)	
Average pooling (3×3)	
Zero-op	

However, not all operations used for the real-valued case are suitable for searching binary architectures. In fact, we found that when using the standard DARTS search space, searching in the binary domain *does not converge*. This is due to several reasons: the depth-wise convolutions are notoriously hard to binarize due to what we call the "double approximation problem": the real-valued depth-wise convolution is a "compressed" approximate version of the normal convolution and, in turn, the binary depth-wise in a quantized approximation of the real-valued one. Furthermore, the 1×1 convolutional layers and the bottleneck block [16] were already shown to be hard to binarize [4] because the features compression that happens in such modules amplifies the high information degradation already caused by binarization. Both 1×1 and separable convolutions are present in the current search space typically used for search real-valued architectures (see Table 1), making them inappropriate for binarization. Moreover, while

---

[1] When any of the components was not used, the obtained results were very poor.

the dilated convolution contains 2 serialized convolutions, the separable one contains 4 which (a) causes discrepancies in their convergence speed and (b) can amplify the gradient fading phenomena that often happens during binarization.

To this end, we propose a new search space O, shown in Table 1, constructed from a binary-first point of view that avoids or alleviates the aforementioned shortcomings. While we preserve the zeros and identity connections alongside the 3×3 max and average pooling layers which do not contain learnable parameters or binary operations, we propose to replace all the convolutional operations with the following new ones: 3×3 and 5×5 grouped convolutions and dilated grouped convolutions, also with a kernel size of 3×3 and 5×5. This removes all the 1×1 convolutional layers directly present in the cell's search space while maintaining the efficiency via the usage of grouped convolutions.

We note that there is a clear trade-off here between efficiency and accuracy controlled via the group size and number of channels. In this work, and in contrast to other works that rely on grouped convolutions (e.g. [46]), we propose to use a very high group size which leads to behaviours closer to that of a depth-wise convolution (a small one will come at a price of higher computational budgets, while the extreme case #groups=#in_channels will again exacerbate the double-approximation problem). We used a group size of 12 and 3 (only) channels per group (totalling 36 channels) for CIFAR, and a group size of 16 and 5 (only) channels per group (totalling 80 channels) for ImageNet. Note that we also explored the effect of adding a channel shuffle layer [46] after each grouped convolution, typically used to enable cross-group information flow for a set of group convolution layers. However, due to the fact that #groups $\gg$ #channels per group, they were unable to offer accuracy gains. Instead, we found that the 1×1 convolutions present between different DARTS cells are sufficient for combining information between the different groups.

Furthermore, in the proposed search space, in all convolutional layers, the depth of all operations used is equal to 1 meaning that we used 1 convolution operator per layer, as opposed to 4 convolutions (2 depth-wise separable) used in DARTS, which facilitates learning. Another consequence of using operations with a depth equal to 1 is a latency improvement due to effectively a shallower network. For a visual illustration of differences between DARTS and the proposed BATS please check the supplementary material.

Finally, to improve the gradient and information flow, which is even more critical for the case of binary networks, while also speeding-up the convergence of such operations during the search phase, we explicitly add an identity connection on each convolutional operation such that $f_{i,j}(\mathbf{x}_i) = f_{i,j}(\mathbf{x}_i) + \mathbf{x}_i$.

The efficacy of the proposed search space is confirmed experimentally in Table 5. As the results from Table 5 show, the architectures found by searching within the proposed search space constantly outperform the others.

4.2 Search Regularization and Stabilisation

Despite its success and appealing formulation, DARTS accuracy can vary wildly between runs depending on the random seed. In fact, there are cases in which the architectures found perform worse than the ones obtained via random search. Furthermore, especially when trained longer or if the search is performed on larger datasets DARTS can converge towards degenerate solutions dominated by skip connections [8]. While in [8] the authors propose to address this by a) applying a dropout on the skip connections during the architecture search and b) by preserving a maximum of 2 skip connections per cell as a post processing step that simply promotes the operation with the second highest probability, we found that such mechanism can still result in a large amount of randomness and is not always effective: for example, it can replace skip connections with pooling layers (which have no learning capacity) or the discovered architecture might even already contain too few skip connections.

Such problems are even more noticeable when the search is performed in the binary domain directly. Given that, during search, the input to the node j is obtained by taking a weighted sum of all incoming edges, in order to maximise the flow of information, the architecture parameters α tend to converge to the same value making the selection of the final architecture problematic and susceptible to noise resulting in topologies than may perform worse than a random selection. Furthermore, the search is highly biased towards real-valued operations (pooling and skip connections) that at early stages can offer larger benefits.

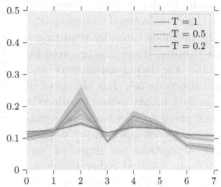

Fig. 1. Distribution of architecture parameters after the first stage for a given cell (data points were sorted by magnitude and do not correspond to the same ops): for low temperatures the network is forced to be more discriminative.

To alleviate the aforementioned issues and encourage the search procedure to be more discriminative forcing it to make "harder" decisions, we propose to use a temperature factor $T < 1$ defining the flow from node i to j as follows:

$$f_{i,j}(\mathbf{x}_i) = \sum_{o \in O} \frac{\exp(\alpha_{i,j}^o / T)}{\sum_{o' \in O} \exp(\alpha_{i,j}^{o'} / T)} \cdot o(\mathbf{x}_i). \tag{2}$$

This has the desirable effect of making the distribution of the architecture parameters less uniform and spikier (i.e. more discriminative). Hence, during search, because the information streams are aggregated using a weighted sum, the network cannot equally (or near-equally) rely on all possible operations by pulling information from all of them. Instead, in order to ensure convergence to a satisfactory solution it has to assign the highest probability to a non 0-ops path, enforced by a sub-unitary temperature ($T < 1$). This behaviour also follows closer the evaluation procedure where a single operation will be selected, reducing as such the performance discrepancy between the search, where the network pulls information from all paths, and evaluation.

Figure 1 depicts the distribution of the architecture parameters for a given cell for different temperatures. For low temperatures, the network is forced to make more discriminative decisions which in turn makes it rely less on identity connections. This is further confirmed by Fig. 2 which depicts the chance of encountering a given op. in a normal cell at the end of the search process for different temperatures.

4.3 Binary Search Strategy

Despite their appealing speed-up and space saving, binary networks remain harder to train compared to their real-valued counterparts, with methods typically requiring a pre-training stage [28] or carefully tuning the hyper-parameters and optimizers [9,33]. For the case of searching binary networks, directly attempting to effectuate an architecture search using binary weights and activations in most of our attempts resulted either in degenerate topologies or the training simply converges to extremely low accuracy values. We also note that, as expected and as our experiments have confirmed, performing the search in the real domain directly and then binarizing the network is sub-optimal.

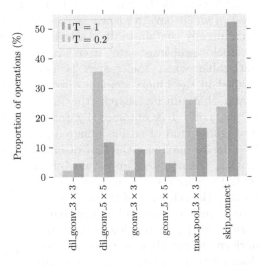

Fig. 2. Probability of a given op. in a cell at the end of the search process for different temperatures: for $T = 0.2$ there is a significant increase in the number of 5×5 convolutions and a decrease in the number of skip connections.

To alleviate the problem, we propose a two-stage optimization process in which during the search the activations are binarized while the weights are kept real, and once the optimal architecture is discovered we proceed with the binarization of the weights too during the evaluation stage[2]. This is motivated by the fact that while the weights of a real-valued network can be typically binarized without a significant drop in accuracy, the same cannot be said about the binarization of the activations where due to the limited number of possible states the information flow inside the network drops dramatically. Hence, we propose to effectively split the problem into two sub-problems: weight and feature binarization and during the search we try to solve the hardest one, that is the binarization of the activations. Once this is accomplished, the binarization of the weights following always results in little drop in accuracy ($\sim$1%).

---

[2] More specifically, during evaluation, we first train a new network with binary activations and real-valued weights from scratch, and then binarize the weights. Then, the fully binarized network is evaluated on the test set.

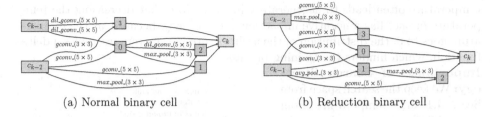

(a) Normal binary cell   (b) Reduction binary cell

Fig. 3. Binary normal (a) and reduction (b) cells discovered by our method. Notice the prevalence of large kernels (5×5) and of the wider reduction cell, both contributing to increasing the filters' diversity and the flow of information.

5   Ablation Studies

5.1   Effect of the Proposed Components

Herein, we analyse the contribution to the overall performance of each proposed component: the newly introduced search space for binary networks (Sect. 4.1), the effect of the temperature adjustment (Sect. 4.2) and finally, of performing the search using real-valued weights and binary activations (Sect. 4.3).

We emphasize that *all 3 ingredients are needed in order to stabilize training and obtain good accuracy*. This suggests that it is not straightforward to show the effect of each component in isolation. To this end, we always *keep fixed 2* of the proposed improvements and vary the other component to understand its impact. We also note that in all cases the final evaluated networks are fully binary (i.e. both the weights and the activations are binarized).

Table 2. Effect of search space on Top-1 accuracy on CIFAR-10.

Search space	Temperature	Accuracy
Ours	Ours	**93.7 ± 0.6**
[27]	Ours	51.0 ± 7.5
Ours	[27]	**86.0 ± 2.0**

Importance of New Search Space: We keep the temperature from Sect. 4.2 and search strategy from Sect. 4.3 fixed and then compare the proposed search space with that used in [27]. As the results from Table 2 show, searching for binary networks using the search space of [27] leads to much worse results ($\sim$51% vs. $\sim$94%) and a high variation

Table 3. Effect of temperature on Top-1 accuracy on CIFAR-10. T = 1 was used in DARTS [27]. *Often leads to degenerate solution (i.e. all skip connections).

Temperature	1	0.5	0.2	0.05
Accuracy	86.0*	92.1	93.7	93.3

in the performance of the networks (a std. of 7.5% across 5 runs). This clearly shows that the previously used search space is not suitable for binary networks. **Impact of temperature:** We keep the search space from Sect. 4.1 and search strategy from Sect. 4.3 fixed and evaluate the use of the temperature proposed in Sect. 4.2. As Table 3 shows, a decrease in temperature has a direct impact on discovering of higher performing models. At the opposite spectrum, a low temperature often leads to degenerate solutions. Note that decreasing the temperature further has a negative impact on the accuracy, because, as the temperature goes to 0, the distribution of the architecture parameters resembles a delta function, which hinders the training process.

Importance of Search Strategy: We keep the search space from Sect. 4.1 and temperature from Sect. 4.2 fixed and evaluate the effect of the proposed search strategy from Sect. 4.3. Table 4 summarizes the different strategies for searching binary architectures and their properties. The first row represents the case where the search is done in the real domain and then binarization follows. As the second row shows, if the search is done in the binary domain for both weights and activations, the search does not converge. As the third row shows, searching in the binary domain for the weights while keeping the activations real converges but results in low accuracy networks. Finally, the last row shows the proposed search strategy where the search is done in the binary domain for the activations and in the real domain for the weights. This configuration is the only one that yields stable search procedure and high accuracy networks (see std). Note that for all cases, the architectures are fully binarized (i.e. both activations and weights) prior to evaluation on the test set.

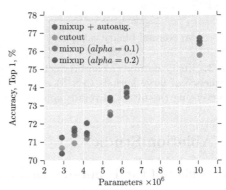

Fig. 4. Effect of data augmentation: Top-1 accuracy on CIFAR-100 of the network constructed using the discovered cell for different number of network parameters.

Table 4. Different strategies for searching binary architectures and their properties. The last row represents the proposed strategy. **All resulting topologies were trained using binary weights and activations during evaluation.**

Search domain		Search cost	Stable	Accuracy
W	A	(GPU-days)		
Real	Real	0.2	Yes	88.4 ± 0.8
Binary	Binary	0.3	No	84.6 ± 1.7
Binary	Real	0.25	Yes	91.1 ± 1.7
Real	Binary	0.25	Yes	$\mathbf{93.7 \pm 0.6}$

5.2   Impact of Augmentation

While the positive impact of data augmentation has been explored for the case of real-valued networks, to the best of our knowledge, there is little to no work on this for the case of binary networks. Despite the fact that binarization is considered to be an extreme case of regularization [9], we found that augmentation is equally important, and confirm that most of the augmentations applied to real-valued networks, can also improve their binarized counterparts. See Fig. 4.

6 Experiments

We conducted experiments on three different image classification datasets, namely CIFAR10 [21], CIFAR100 [21] and ImageNet [12]. Note that the architecture search is carried out on CIFAR-10 and tested on all three.

6.1 Architecture Search

Implementation Details: The training process consists of 3 stages during which the network depth is increased from 5 to 11 and finally 17 cells. Concomitantly, at each stage, the search space is gradually reduced to 8, 5 and respectively 3 ops. As opposed to [8], we did not use *on-skip* dropout since we found that it leads to unstable solutions. Instead, we used the mechanism proposed in Sect. 4.2, setting the temperature for the normal and reduction cells to 0.2 and 0.15 respectively. The latter is lower to encourage wider cells. During each stage, the network is trained using a batch size of 96 for 25 epochs. For the first 10 epochs only the network weights are learned. The remaining 15 epochs update both the architecture and network parameters. All the parameters are optimized using Adam [20]. The learning rate was set to $\eta = 0.0006$, weight decay to $w_d = 0.001$ and momentum to $\beta = (0.5, 0.999)$ for the architecture parameters and $\eta = 0.001$, $w_d = 0.0003$, $\beta = (0.9, 0.999)$ for the network parameters. To keep the search time low, we used the first-order optimization approach of DARTS. All methods were implemented in PyTorch [31].

Discovered Topologies: Overall, our approach is capable of discovering high-performing binary cells, offering new or validating existing insights about optimal binary topologies. As depicted in Fig. 3, the cells found tend to prefer convolutional layers with larger kernel sizes (5×5) which help alleviating the limited representational power found in such networks (a 3×3 kernel has maximum number of 2^9 unique binary filters while a 5×5 has 2^{25}). In addition, to preserve the information flow, a real-valued path that connects one of the input nodes to the output is always present.

Furthermore, since the down-sampling operation compresses the information across the spatial dimension, to compensate for this, the reduction cell tends to be wider (i.e. more information can flow through) as opposed to the normal cell which generally is deeper. As Table 5 shows, our searched architecture outperforms all prior ones searched in the real-valued domain and then binarized.

6.2 Comparison Against State-of-the-Art

In this section, we compare the performance of the searched architecture against state-of-the-art network binarization methods on CIFAR10/100 and ImageNet.

Table 5. Comparison with state-of-the-art NAS methods on CIFAR10 and CIFAR100. For all methods we apply the binarization process described in Sect. 3 on the best cell provided by the authors.

Architecture	Test Acc. (%)		Params (M)	Search cost (GPU-days)	Search method
	C10	C100			
ResNet-18 [16]	91.0	66.3	11.2	–	Manual
Random search	92.9	70.2	–	–	Random
NASNet-A [52]	90.3	66.1	3.9	1800	RL
AmoebaNet-A [34]	91.5	65.6	3.2	3150	Evolution
DARTS (first order) [27]	89.4	63.3	3.3	1.5	Grad-based
DARTS (second order) [27]	91.1	64.0	3.3	4.0	Grad-based
P-DARTS [8]	89.5	63.9	3.6	0.3	Grad-based
Ours	93.7	70.7	2.8	0.25	Grad-based
Ours + autoaug. [11]	94.1	71.3	2.8	0.25	Grad-based
Ours (medium)	94.6	73.5	5.4	0.25	Grad-based
Ours (medium) + autoaug. [11]	94.9	74.0	5.4	0.25	Grad-based
Ours (large)	95.5	75.7	10.0	0.25	Grad-based
Ours (large) + autoaug. [11]	96.1	76.8	10.0	0.25	Grad-based

Implementation Details: This section refers to training the discovered architectures from scratch for evaluation purposes. For all models we use Adam with a starting learning rate of 0.001, $\beta = (0.9, 0.999)$ and no weight decay. The learning rate was reduced following a cosine scheduler [29]. In line with other works [27], on CIFAR10 and CIFA100, the models were trained for 600 epochs while on ImageNet for 75. The batch size was set to 256. The drop-path [22] was set to 0 for all experiments since binary networks are less prone to overfitting and already tend to use all parallel paths equally well. Following [27], we add an auxiliary tower [39] with a weight of 0.4. Unless otherwise stated, the network trained consisted of 20 cells and 36 initial channels with a group size of 12 channels for CIFAR10/100 and respectively 80 initial channels and 14 cells for ImageNet. For data augmentation, similarly to [8,27], we padded the images appropriately on CIFAR10/100 and applied CutOut [13] with a single hole and a length of 16px for CIFAR-10 and Mix-up ($\alpha = 0.2$) for CIFAR-100 respectively. For ImageNet, we simply resized the images to 256×256px, and then randomly cropped them to 224×224px. During testing, the images were center-cropped.

CIFAR10: When compared against network topologies discovered by existing state-of-the-art NAS approaches, as the results from Table 5 show, our method significantly outperforms all of them on both CIFAR-10 and CIFAR-100 datasets. When using 20% the parameters than the competing methods, our approach offers gains of 5 to 7% on CIFAR-100 and $2.5 - 4\%$ on CIFAR-10. This clearly confirms the importance of the proposed search space and strategy. Furthermore, we test how the discovered architecture performs when scaled-up. In order to increase the capacity of the models we concomitantly adjust the width and #groups of the cells. By doing so, we obtain a 2% improvement of CIFAR-10 and 5% on CIFAR-100 almost matching the performance of a real-valued ResNet-18 model. We note that for fairness all other topologies we compared against where modified using the structure described in Sect. 3.

For an exhaustive comparison on CIFAR-10 see the supplementary material.

ImageNet: Herein, we compare our approach against related state-of-the-art binarization and quantization methods. As the results from Table 7 show, our discovered architecture outperforms all existing binarization methods while using a lower computational budget for the normal thin model and offers a gain of 5% for the 2x-wider one. Furthermore, our methods compared favorable even against methods that either significantly increase the network computational requirements or use more bits for quantization.

Table 6. Comparison with selected SOTA binary methods on ImageNet in terms of computational cost. Notice, that within a similar budget our method achieves the highest accuracy. For a full comparison see Table 7.

Architecture	Accuracy (%)		# Operations		# bits (W/A)
	Top-1	Top-5	FLOPS $\times 10^8$	BOPS $\times 10^9$	
BNN [10]	42.2	69.2	1.314	1.695	1/1
XNOR-Net [33]	51.2	73.2	1.333	1.695	1/1
CCNN [43]	54.2	77.9	1.333	1.695	1/1
Bi-Real Net [28]	56.4	79.5	1.544	1.676	1/1
XNOR-Net++ [5]	57.1	79.9	1.333	1.695	1/1
CI-Net [40]	59.9	84.2	–	–	1/1
BATS (Ours)	60.4	83.0	0.805	1.149	1/1
BATS [2x-wider] (Ours)	**66.1**	**87.0**	1.210	2.157	1/1

6.3 Network Efficiency

The current most popular settings of binarizing neural networks preserve the first and last layer real-valued. However, for the currently most popular binarized architecture, i.e. ResNet-18 [16], the first convolutional layer accounts for approx. 6.5% of the total computational budget (1.2×10^8 FLOPs out of the total of 1.8×10^9 FLOPs), a direct consequence of the large kernel size (7×7) and the high input resolution. In order to alleviate this, beyond using the ImageNet stem cell used in DARTS [27] that replaces the 7×7 layer with two 3×3, we replaced the second convolution with a grouped convolution ($g = 4$). This alone allows us to more than halve the number of real-valued operations (see Table 6). Additionally, in order to transparently show the computational cost of the tested models, we separate the binary operations (BOPs) and FLOPs into two distinctive categories. Furthermore, as Table 6 shows, our method surpasses sophisticated approaches while having a significantly lower number of ops.

Table 7. Comparison with state-of-the-art binarization methods on ImageNet, including against approaches that use low-bit quantization (upper section) and ones that increase the capacity of the network (middle section). For the latter case the last column also indicates the capacity scaling used. Models marked with ** use real-valued downsampling convolutional layers.

Architecture	Accuracy (%)		# bits (W/A)
	Top-1	Top-5	
BWN [10]	60.8	83.0	1/32
TTQ [49]	66.6	87.2	2/32
HWGQ [7]	59.6	82.2	1/2
LQ-Net [45]	59.6	82.2	1/2
creSYQ [15]	55.4	78.6	1/2
DOREFA-Net [48]	62.6	84.4	1/2
ABC-Net ($M, N = 1$) [23]	42.2	67.6	1/1
ABC-Net ($M, N = 5$) [23]	65.0	85.9	$(1/1) \times 5^2$
Struct. Approx. [51]	64.2	85.6	$(1/1) \times 4$
Struct. Approx.** [51]	66.3	86.6	$(1/1) \times 4$
CBCN [26]	61.4	82.8	$(1/1) \times 4$
Ensamble [50]	61.0	–	$(1/1) \times 6$
BNN [10]	42.2	69.2	1/1
XNOR-Net [33]	51.2	73.2	1/1
CCNN [43]	54.2	77.9	1/1
Bi-Real Net** [28]	56.4	79.5	1/1
Rethink. BNN** [17]	56.6	79.4	1/1
XNOR-Net++ [5]	57.1	79.9	1/1
CI-Net** [40]	59.9	84.2	1/1
BATS (Ours)	60.4	83.0	1/1
BATS [2x-wider] (Ours)	**66.1**	**87.0**	1/1

7 Conclusion

In this work we introduce a novel Binary Architecture Search (BATS) that drastically reduces the accuracy gap between binary models and their real-valued counterparts, by searching for the first time directly for binary architectures. To this end we, (a) designed a novel search space specially tailored to binary networks, (b) proposed a new regularization method that helps stabilizing the search process, and (c) introduced an adapted search strategy that speed-ups the overall network search. Experimental results conducted on CIFAR10, CIFAR100 and ImageNet demonstrate the effectiveness of the proposed approach and the need of doing the search in the binary space directly.

References

1. Baker, B., Gupta, O., Naik, N., Raskar, R.: Designing neural network architectures using reinforcement learning. In: International Conference on Learning Representations (2017)
2. Bethge, J., Yang, H., Bornstein, M., Meinel, C.: Back to simplicity: how to train accurate BNNs from scratch? arXiv preprint arXiv:1906.08637 (2019)
3. Brock, A., Lim, T., Ritchie, J., Weston, N.: SMASH: one-shot model architecture search through hypernetworks. In: International Conference on Learning Representations (2018)
4. Bulat, A., Tzimiropoulos, G.: Binarized convolutional landmark localizers for human pose estimation and face alignment with limited resources. In: IEEE International Conference on Computer Vision, pp. 3706–3714 (2017)
5. Bulat, A., Tzimiropoulos, G.: XNOR-Net++: improved binary neural networks. In: British Machine Vision Conference (2019)
6. Cai, H., Zhu, L., Han, S.: ProxylessNAS: direct neural architecture search on target task and hardware. In: International Conference on Learning Representations (2019)
7. Cai, Z., He, X., Sun, J., Vasconcelos, N.: Deep learning with low precision by half-wave gaussian quantization. In: IEEE Conference on Computer Vision and Pattern Recognition, pp. 5918–5926 (2017)
8. Chen, X., Xie, L., Wu, J., Tian, Q.: Progressive differentiable architecture search: bridging the depth gap between search and evaluation. In: IEEE International Conference on Computer Vision (2019)
9. Courbariaux, M., Bengio, Y., David, J.P.: BinaryConnect: training deep neural networks with binary weights during propagations. In: Advances on Neural Information Processing Systems (2015)
10. Courbariaux, M., Hubara, I., Soudry, D., El-Yaniv, R., Bengio, Y.: Binarized neural networks: training deep neural networks with weights and activations constrained to +1 or −1. arXiv preprint arXiv:1602.02830 (2016)
11. Cubuk, E.D., Zoph, B., Mane, D., Vasudevan, V., Le, Q.V.: AutoAugment: learning augmentation policies from data. In: IEEE Conference on Computer Vision and Pattern Recognition (2019)
12. Deng, J., Dong, W., Socher, R., Li, L.J., Li, K., Fei-Fei, L.: ImageNet: a large-scale hierarchical image database. In: IEEE Conference on Computer Vision and Pattern Recognition, pp. 248–255 (2009)
13. DeVries, T., Taylor, G.W.: Improved regularization of convolutional neural networks with cutout. arXiv preprint arXiv:1708.04552 (2017)
14. Ding, R., Chin, T.W., Liu, Z., Marculescu, D.: Regularizing activation distribution for training binarized deep networks. In: Proceedings of the IEEE Conference on Computer Vision and Pattern Recognition, pp. 11408–11417 (2019)
15. Faraone, J., Fraser, N., Blott, M., Leong, P.H.: SYQ: learning symmetric quantization for efficient deep neural networks. In: IEEE Conference on Computer Vision and Pattern Recognition, pp. 4300–4309 (2018)
16. He, K., Zhang, X., Ren, S., Sun, J.: Deep residual learning for image recognition. In: Proceedings of the IEEE Conference on Computer Vision and Pattern Recognition, pp. 770–778 (2016)
17. Helwegen, K., Widdicombe, J., Geiger, L., Liu, Z., Cheng, K.T., Nusselder, R.: Latent weights do not exist: Rethinking binarized neural network optimization. In: Advances in Neural Information Processing Systems, pp. 7533–7544 (2019)

324 A. Bulat et al.

18. Howard, A.G., et al.: MobileNets: efficient convolutional neural networks for mobile vision applications. arXiv preprint arXiv:1704.04861 (2017)
19. Hu, J., Shen, L., Sun, G.: Squeeze-and-excitation networks. In: Proceedings of the IEEE Conference on Computer Vision and Pattern Recognition, pp. 7132–7141 (2018)
20. Kingma, D.P., Ba, J.: Adam: a method for stochastic optimization. arXiv preprint arXiv:1412.6980 (2014)
21. Krizhevsky, A., Hinton, G., et al.: Learning multiple layers of features from tiny images. Technical report (2009)
22. Larsson, G., Maire, M., Shakhnarovich, G.: FractalNet: ultra-deep neural networks without residuals. arXiv preprint arXiv:1605.07648 (2016)
23. Lin, X., Zhao, C., Pan, W.: Towards accurate binary convolutional neural network. In: Advances on Neural Information Processing Systems, pp. 345–353 (2017)
24. Liu, C., et al.: Auto-DeepLab: hierarchical neural architecture search for semantic image segmentation. In: IEEE Conference on Computer Vision and Pattern Recognition, pp. 82–92 (2019)
25. Liu, C., et al.: Progressive neural architecture search. In: Ferrari, V., Hebert, M., Sminchisescu, C., Weiss, Y. (eds.) ECCV 2018. LNCS, vol. 11205, pp. 19–35. Springer, Cham (2018). https://doi.org/10.1007/978-3-030-01246-5_2
26. Liu, C., et al.: Circulant binary convolutional networks: enhancing the performance of 1-bit DCNNs with circulant back propagation. In: IEEE Conference on Computer Vision and Pattern Recognition, pp. 2691–2699 (2019)
27. Liu, H., Simonyan, K., Yang, Y.: DARTS: differentiable architecture search. In: International Conference on Learning Representations (2019)
28. Liu, Z., Wu, B., Luo, W., Yang, X., Liu, W., Cheng, K.T.: Bi-Real Net: enhancing the performance of 1-bit CNNs with improved representational capability and advanced training algorithm. In: European Conference on Computer Vision, pp. 747–763 (2018)
29. Loshchilov, I., Hutter, F.: SGDR: stochastic gradient descent with warm restarts. arXiv preprint arXiv:1608.03983 (2016)
30. Ma, N., Zhang, X., Zheng, H.-T., Sun, J.: ShuffleNet V2: practical guidelines for efficient CNN architecture design. In: Ferrari, V., Hebert, M., Sminchisescu, C., Weiss, Y. (eds.) Computer Vision – ECCV 2018. LNCS, vol. 11218, pp. 122–138. Springer, Cham (2018). https://doi.org/10.1007/978-3-030-01264-9_8
31. Paszke, A., et al.: PyTorch: an imperative style, high-performance deep learning library. Adv. Neural Inf. Process. Syst. 32, 8024–8035 (2019)
32. Pham, H., Guan, M., Zoph, B., Le, Q., Dean, J.: Efficient neural architecture search via parameters sharing. In: International Conference on Machine Learning, pp. 4095–4104 (2018)
33. Rastegari, M., Ordonez, V., Redmon, J., Farhadi, A.: XNOR-Net: ImageNet classification using binary convolutional neural networks. In: Leibe, B., Matas, J., Sebe, N., Welling, M. (eds.) ECCV 2016. LNCS, vol. 9908, pp. 525–542. Springer, Cham (2016). https://doi.org/10.1007/978-3-319-46493-0_32
34. Real, E., Aggarwal, A., Huang, Y., Le, Q.V.: Regularized evolution for image classifier architecture search. In: AAAI Conference on Artificial Intelligence, vol. 33, pp. 4780–4789 (2019)
35. Real, E., et al.: Large-scale evolution of image classifiers. In: International Conference on Machine Learning, pp. 2902–2911 (2017)
36. Sandler, M., Howard, A., Zhu, M., Zhmoginov, A., Chen, L.C.: MobileNetV2: inverted residuals and linear bottlenecks. In: IEEE Conference on Computer Vision and Pattern Recognition, pp. 4510–4520 (2018)

37. Shen, M., Han, K., Xu, C., Wang, Y.: Searching for accurate binary neural architectures. In: Proceedings of the IEEE International Conference on Computer Vision Workshops, p. 0 (2019)
38. Singh, K.P., Kim, D., Choi, J.: Learning architectures for binary networks. arXiv preprint arXiv:2002.06963 (2020)
39. Szegedy, C., et al.: Going deeper with convolutions. In: IEEE Conference on Computer Vision and Pattern Recognition (2015)
40. Wang, Z., Lu, J., Tao, C., Zhou, J., Tian, Q.: Learning channel-wise interactions for binary convolutional neural networks. In: IEEE Conference on Computer Vision and Pattern Recognition, pp. 568–577 (2019)
41. Xie, L., Yuille, A.: Genetic CNN. In: IEEE International Conference on Computer Vision, pp. 1379–1388 (2017)
42. Xie, S., Zheng, H., Liu, C., Lin, L.: SNAS: stochastic neural architecture search. In: International Conference on Learning Representations (2018)
43. Xu, Z., Cheung, R.C.: Accurate and compact convolutional neural networks with trained binarization. arXiv preprint arXiv:1909.11366 (2019)
44. Zagoruyko, S., Komodakis, N.: Wide residual networks. In: British Machine Vision Conference (2016)
45. Zhang, D., Yang, J., Ye, D., Hua, G.: LQ-Nets: learned quantization for highly accurate and compact deep neural networks. In: Ferrari, V., Hebert, M., Sminchisescu, C., Weiss, Y. (eds.) ECCV 2018. LNCS, vol. 11212, pp. 373–390. Springer, Cham (2018). https://doi.org/10.1007/978-3-030-01237-3_23
46. Zhang, X., Zhou, X., Lin, M., Sun, J.: ShuffleNet: an extremely efficient convolutional neural network for mobile devices. In: IEEE Conference on Computer Vision and Pattern Recognition, pp. 6848–6856 (2018)
47. Zhong, Z., Yan, J., Wu, W., Shao, J., Liu, C.L.: Practical block-wise neural network architecture generation. In: IEEE Conference on Computer Vision and Pattern Recognition, pp. 2423–2432 (2018)
48. Zhou, S., Wu, Y., Ni, Z., Zhou, X., Wen, H., Zou, Y.: DoReFa-Net: training low bitwidth convolutional neural networks with low bitwidth gradients. arXiv (2016)
49. Zhu, C., Han, S., Mao, H., Dally, W.J.: Trained ternary quantization. arXiv preprint arXiv:1612.01064 (2016)
50. Zhu, S., Dong, X., Su, H.: Binary ensemble neural network: more bits per network or more networks per bit? In: IEEE Conference on Computer Vision and Pattern Recognition, pp. 4923–4932 (2019)
51. Zhuang, B., Shen, C., Tan, M., Liu, L., Reid, I.: Structured binary neural networks for accurate image classification and semantic segmentation. In: IEEE Conference on Computer Vision and Pattern Recognition, pp. 413–422 (2019)
52. Zoph, B., Vasudevan, V., Shlens, J., Le, Q.V.: Learning transferable architectures for scalable image recognition. In: Proceedings of the IEEE Conference on Computer Vision and Pattern Recognition, pp. 8697–8710 (2018)

A Closer Look at Local Aggregation Operators in Point Cloud Analysis

Ze Liu[1,2], Han Hu[2(✉)], Yue Cao[2], Zheng Zhang[2], and Xin Tong[2]

[1] University of Science and Technology of China, Hefei, China
liuze@mail.ustc.edu.cn
[2] Microsoft Research Asia, Beijing, China
{hanhu,yuecao,zhez,xtong}@microsoft.com

Abstract. Recent advances of network architecture for point cloud processing are mainly driven by new designs of local aggregation operators. However, the impact of these operators to network performance is not carefully investigated due to different overall network architecture and implementation details in each solution. Meanwhile, most of operators are only applied in shallow architectures. In this paper, we revisit the representative local aggregation operators and study their performance using the same deep residual architecture. Our investigation reveals that despite the different designs of these operators, all of these operators make surprisingly similar contributions to the network performance under the same network input and feature numbers and result in the state-of-the-art accuracy on standard benchmarks. This finding stimulate us to rethink the necessity of sophisticated design of local aggregation operator for point cloud processing. To this end, we propose a simple local aggregation operator without learnable weights, named Position Pooling (PosPool), which performs similarly or slightly better than existing sophisticated operators. In particular, a simple deep residual network with PosPool layers achieves outstanding performance on all benchmarks, which outperforms the previous state-of-the methods on the challenging PartNet datasets by a large margin (7.4 mIoU). The code is publicly available at https://github.com/zeliu98/CloserLook3D.

Keywords: 3D point cloud · Local aggregation operator · Position pooling

1 Introduction

With the rise of 3D scanning devices and technologies, 3D point cloud becomes a popular input for many machine vision tasks, such as autonomous driving,

---

Z. Liu and H. Hu—Equal contribution.
This work is done when Ze Liu is an intern at MSRA.

Electronic supplementary material The online version of this chapter (https://doi.org/10.1007/978-3-030-58592-1_20) contains supplementary material, which is available to authorized users.

A. Vedaldi et al. (Eds.): ECCV 2020, LNCS 12368, pp. 326–342, 2020.
https://doi.org/10.1007/978-3-030-58592-1_20

robot navigation, shape matching and recognition, etc. Different from images and videos that are defined on regular grids, the point cloud locates at a set of irregular positions in 3D space, which makes the powerful convolutional neural networks (CNN) and other deep neural networks designed for regular data hard to be applied. Early studies transform the irregular point set into a regular grid by either voxelization or multi-view 2D projections such that the regular CNN can be adopted. However, the conversion process always results in extra computational and memory costs and the risk of information loss.

Recent methods in point cloud processing develop networks that can directly model the unordered and non-grid 3D point data. These architectures designed for point cloud are composed by two kinds of layers: the point-wise transformation layers and local aggregation layers. While the point-wise transformation layer is applied on features at each point, the local aggregation layer plays a similar role for points as the convolution layer does for image pixels. Specifically, it takes features and relative positions of neighborhood points to a center point as input, and outputs the transformed feature for the center point. To achieve better performance in different point cloud processing tasks, a key task of point cloud network design is to develop effective local aggregation operators.

Existing local aggregation operators can be roughly categorized into three groups according to the way that they combine the relative positions and point features: point-wise multi-layer perceptions (MLP) based [11,13,22,35], pseudo grid feature based [9,12,17,27,30,38] and adaptive weight based [5,14,16,32,34,36]. The point-wise MLP based methods treat a point feature and its corresponding relative position equally by concatenation. All the concatenated features at neighborhood are then abstracted by a small PointNet [20] (multiple point-wise transformation layers followed by a MAX pooling layer) to produce the output feature for the center point. The pseudo grid feature based methods first generate pseudo features on pre-defined grid locations, and then learn the parametrized weights on these grid locations like regular convolution layer does. The adaptive weight based methods aggregate neighbor features by weighted average with the weights adaptively determined by relative position of each neighbor.

Despite the large efforts for aggregation layer design and performance improvements of the resulting network in various point cloud processing tasks, the contributions of the aggregation operator to the network performance have never been carefully investigated and fairly compared. This is mainly due to the different network architectures used in each work, such as the network depth, width, basic building blocks, whether to use skip connection, as well as different implementation of each approach, such as point sampling method, neighborhood computation, and so on. Meanwhile, most of existing aggregation layers are applied in shallow networks, it is unclear whether these designs are still effective as the network depth increases.

In this paper, we present common experimental settings for studying these operators, selecting a deep residual architecture as the base networks, as well as same implementation details regarding point sampling, local neighborhood selection and etc. We also adopt three widely used datasets, ModelNet40 [37], S3DIS [1] and PartNet [19] for evaluation, which account for different tasks, scenarios and data scales. Using these common experimental settings, we revisit the performance of each representative operator and make fair comparison between them. We find appropriate settings for some operators under this deep residual architecture are different from that of using shallower and non-residual networks. We also surprisingly find that different representative methods perform similarly well under the same representation capacity on these datasets, if appropriate settings are adopted for each method, although these methods may be invented by different motivations and formulations, in different years.

These findings also encourage us to rethink the role of local aggregation layers in point cloud modeling: *do we really need sophisticated/heavy local aggregation computation?* We answer this question by proposing an extremely simple local aggregation operator with no learnable weights: combining a neighbor point feature and its 3-d relative coordinates by element-wise multiplication, followed with an AVG pool layer to abstract information from neighborhood. We name this new operator as position pooling (PosPool), which shows no less or even better accuracy than other highly tuned sophisticated operators on all the three datasets. These results indicate that we may not need sophisticated/heavy operators for local aggregation computation. We also harness a strong baseline for point cloud analysis by a simple deep residual architecture and the proposed position pooling layers, which achieves 53.8 part category mIoU accuracy on the challenging PartNet datasets, significantly outperforming the previous best method by 7.4 mIoU.

The contributions of this paper are summarized as

- **A common testbed** to fairly evaluate different local aggregation operators.
- **New findings of aggregation operators.** Specifically, *different operators perform similarly well and all of them can achieve the state-of-the-art accuracy*, if appropriate settings are adopted for each operator. Also, *appropriate settings in deep residual networks are different from those in shallower networks.* We hope these findings could shed new light on network design.
- **A new local aggregation operator (PosPool) with no learnable weights** that performs as effective as existing operators. Combined with a deep residual network, this simple operator achieve state-of-the-art performance on 3 representative benchmarks and outperforms the previous best method by a large margin of 7.4 mIoU on the challenging PartNet datasets.

2   Related Works

Projection Based Methods project the irregular point cloud onto a regular sampling grid and then apply 2D or 3D CNN over regularly-sampled data for various vision tasks. View-based methods project a 3D point cloud to a set of 2D views from various angles. Then these view images could be processed by 2D CNNs [3,6,21,25]. Voxel-based methods project the 3D points to regular 3D grid, and then standard 3D CNN could be applied [4,18,37]. Recently, adaptive voxel-based representations such as K-d trees [10] or octrees [23,26,33] have been proposed for reducing the memory and computational cost of 3D CNN.

The view-based and voxel-based representations are also combined [21] for point cloud analysis. All these methods require preprocessing to convert the input point cloud and may lose the geometry information.

Global Aggregation Methods process the 3D point cloud via point-wise 1×1 transformation (fully connected) layers followed by a global pooling layer to aggregate information globally from all points [20]. These methods are the first to directly process the irregular point data. They have no restriction on point number, order and regularity of neighborhoods, and obtain fairly well accuracy on several point cloud analysis tasks. However, the lack of local relationship modeling components hinders the better performance on these tasks.

Local Aggregation Methods. Recent point cloud architectures are usually composed by 1×1 point-wise transformation layers and local aggregation operators. Different methods are mainly differentiated by their local aggregation layers, which usually adopt the neighboring point features and their relative coordinates as input, and output a transformed center point feature. According to the way they combine point features and relative coordinates, these methods can be roughly categorized into three groups: point-wise MLP based [11,13,22], pseudo grid feature based [9,12,17,27,30,38], and adaptive weight based [5,14,16,34,36], as will be detailed in Sect. 3. There are also some works use additional edge features (relative relationship between point features) as input [13,32,35], also commonly referred to as graph based methods.

While we have witnessed significant accuracy improvements on benchmarks by new local aggregation operators year-by-year, the actual progress is a bit vague to the community as the comparisons are made on different grounds that the other architecture components and implementations may vary significantly. The effectiveness of designing components in some operators using deeper residual architectures is also unknown.

3   Overview of Local Aggregation Operators

In this section, we present a general formulation for local aggregation operators as well as a categorization of them.

General Formulation. In general, for each point i, a local aggregation layer first transforms a neighbor point j's feature $\mathbf{f}_j \in \mathbb{R}^{d \times 1}$ and its relative location $\Delta \mathbf{p}_{ij} = \mathbf{p}_j - \mathbf{p}_i \in \mathbb{R}^{3 \times 1}$ into a new feature by a function $G(\cdot, \cdot)$, and then aggregate all transformed neighborhood features to form point i's output feature by a reduction function R (typically using MAX, AVG or SUM), as

$$\mathbf{g}_i = R\left(\{G(\Delta \mathbf{p}_{ij}, \mathbf{f}_j) | j \in \mathcal{N}(i)\}\right), \tag{1}$$

where $\mathcal{N}(i)$ represents the neighborhood of point i. Alternatively, edge features $\{\mathbf{f}_i, \Delta \mathbf{f}_{ij}\}$ ($\Delta \mathbf{f}_{ij} = \mathbf{f}_j - \mathbf{f}_i$) can be used as input instead of $\Delta \mathbf{p}_{ij}$ [35].

According to the family to which the transformation function $G(\cdot, \cdot)$ belongs, existing local aggregation operators can be roughly categorized into three types: point-wise MLP based, pseudo grid feature based, and adaptive weight based.

Point-Wise MLP Based Methods. The pioneer work of point-wise MLP based method, PointNet++ [22], applies several point-wise transformation (fully connected) layers on a concatenation of relative position and point feature to achieve transformation:

$$G(\Delta\mathbf{p}_{ij}, \mathbf{f}_j) = \text{MLP}\left(\text{concat}(\Delta\mathbf{p}_{ij}, \mathbf{f}_j)\right). \tag{2}$$

There are also variants by using an alternative edge feature $\{\mathbf{f}_i, \Delta\mathbf{f}_{ij}\}$ as input [13,35], or by using a special neighborhood strategy [11]. The reduction function $R(\cdot)$ is usually set as MAX [13,22,35].

The multiple point-wise layers after concatenation operation can approximate any continuous function about the relative coordinates and point feature [20,22]. However, a drawback lies in its large computation complexity, considering the fact that the multiple fully connected (FC) layers are applied to all neighboring points when computing each point's output. Specifically, the FLOPs is $\mathcal{O}(\text{time}) = ((2d+3)+(h-2)d/2)\cdot d/2\cdot nK$, for a point cloud with n points, neighborhood size of K, FC layer number of h, and inter-mediate dimension of $d/2$, when $h \geq 2$. The space complexity is $\mathcal{O}(\text{space}) = ((2d+3)+(h-2)d/2)\cdot d/2$. For $h = 1$, there exists efficient implementation by computation sharing (see Sect. 4.2).

Pseudo Grid Feature Based Methods. The pseudo grid feature based methods generate pseudo features on several sampled regular grid points, such that regular convolution methods can be applied. A representative method is KPConv [30], where equally distributed spherical grid points are sampled and the pseudo features on the kth grid point is computed as

$$\mathbf{f}_{i,k} = \sum_{j \in \mathcal{N}(i)} \max(0, 1 - \frac{\|\Delta\mathbf{p}_{jk}\|_2}{\sigma}) \cdot \mathbf{f}_j. \tag{3}$$

The index of each grid point k will have strict mapping with the relative position to center point $\Delta\mathbf{p}_{ik}$. Hence, a (depth-wise) convolution operator with parametrized weights $\mathbf{w}_k \in \mathbb{R}^{d \times 1}$ defined on each grid point can be used to achieve feature transformation:

$$G(\Delta\mathbf{p}_{ik}, \mathbf{f}_{i,k}) = \mathbf{w}_k \odot \mathbf{f}_{i,k}. \tag{4}$$

Different pseudo grid feature based methods mainly differ each other by the definition of grid points [9,12,17,27,38] or index order [14]. When depth-wise convolution is used, the space and time complexity are $\mathcal{O}(\text{space}) = dM$ and $\mathcal{O}(\text{time}) = ndKM$, respectively, where M is the number of grid points.

Adaptive Weight Based Methods. The adaptive weight based methods define convolution filters over arbitrary relative positions, and hence can compute aggregation weights on all neighbor points:

$$G(\Delta\mathbf{p}_{ij}, \mathbf{f}_j) = H\left(\Delta\mathbf{p}_{ij}\right) \odot \mathbf{f}_j, \tag{5}$$

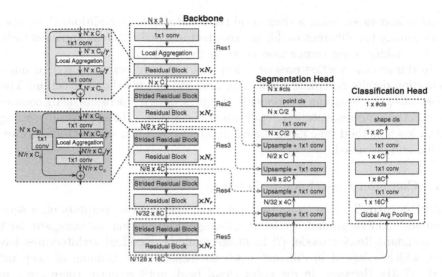

Fig. 1. A common deep residual architecture used to evaluate different local aggregation operators. In evaluation, we adjust the model complexity by changing architecture depth (or block repeating factor N_r), base width C and bottleneck ratio γ. Note the point numbers drawn in this figure is an approximation to indicate the rough complexity but not an accurate number. Actually, the points on each stage are generated by a subsampling method [29] using a fixed grid size and the point number on different point cloud instances can vary

where H is typically implemented by several point-wise transformation (fully connected) layers [5,34]; $\odot$ is an element-wise multiplication operator; R is typically set as SUM.

Some methods adopt more position related variables [16], point density [36], or edge features [32] as the input to compute adaptive weights. More sophisticated function other than fully connected (FC) layers are also used, for example, Taylor approximation [14] and an additional SoftMax function to normalize aggregation weights over neighborhood [32].

The space and time complexity of this method are $\mathcal{O}(\text{space}) = ((h-2)d/2 + d + 3) \cdot d/2$ and $\mathcal{O}(\text{time}) = ((h-2)d/2 + d + 5) \cdot d/2 \cdot nK$, respectively, when an inter-mediate dimension of $d/2$ is used and the number of FC layers $h \geq 2$. When $h = 1$, the space and computation complexity is much smaller, as $\mathcal{O}(\text{space}) = 3d$ and $\mathcal{O}(\text{time}) = 5dnK$, respectively.

Please see Appendix A6 for detailed analysis of the space and time complexity for the above 3 operators.

4   Benchmarking Local Aggregation Operators in Common Deep Architecture

While most local aggregation operators described in Sect. 3 are reported using specific shallow architectures, it is unknown whether their designing components perform also sweet using a deep residual architecture. In addition, these operators usually use different backbone architectures and different implementation details, making a fair comparison between them difficult.

In this section, we first present a deep residual architecture, as well as implementation details regarding point sampling and neighborhood selection. Then we evaluate the designing components of representative operators using common architectures, implementation details and benchmarks. The appropriate settings within each method type using the common deep residual architectures are suggested and discussed.

4.1 Common Experimental Settings

Architecture. To investigate different local aggregation operators on a same, deep and modern ground, we select a 5-stage deep residual network, similar to the standard ResNet model [7] in image analysis. Residual architectures have been widely adopted in different fields to facilitate the training of deep networks [7,31]. However, in the point cloud field, until recently, there are some works [13,30] starting to use deep residual architectures, probably because the unnecessary use of deep networks on several small scale benchmarks. Nevertheless, our investigation shows that on larger scale and more challenging datasets such as PartNet [19], deep residual architectures can bring significantly better performance, for example, with either local aggregation operator type described in Sect. 3, the deep residual architectures can surpass previous best methods by more than 3 mIoU. On smaller scale datasets such as ModelNet40, they also seldom hurt the performance. The deep residual architecture would be a reasonable choice for practitioners working on point cloud analysis.

Figure 1 shows the residual architecture used in this paper. It consists 5 stages of different point resolution, with each stage stacked by several bottleneck residual blocks. Each bottleneck residual block is composed successively by a 1×1 point-wise transformation layer, a local aggregation layer, and another 1×1 point-wise transformation layer. At the block connecting two stages, a stridded local aggregation layer is applied where the local neighborhood is selected at a higher resolution and the output adopts a lower resolution. Batch normalization and ReLU layers are applied after each 1×1 layer to facilitate training. For head networks, we use a 4-layer classifier and a U-Net style encoder-decoder [24] for classification and semantic segmentation, respectively.

In evaluation of a local aggregation operator, we use this operator to instantiate all local aggregation layers in the architecture. We also consider different model capacity by varying network depth (block repeating factor N_r), width (C) and bottleneck ratio (γ).

Point Sampling and Neighborhoods. To generate point sets for different resolution levels, we follow [29,30] to use a subsampling method with different grid sizes to generate point sets in different resolution stages. Specifically, the whole 3D space is divided by grids and one point is randomly sampled to represent a grid if multiple points appear in the grid. This method can alleviate the

Table 1. The performance of baseline operators, sweet spots of point-wise MLP based, pseudo grid feature based and adaptive weight based operators, and the proposed PosPool operators on three benchmark datasets. Baseline* denotes Eq. (6) and baseline† (AVG/MAX) denotes Eq. (7) AVG/MAX, respectively. PosPool and PosPool* denote the operators in Eq. (8) and (10), respectively. (S) after each method denotes a smaller configuration of this method ($N_r = 1$, $\gamma = 2$ and $C = 36$), which is about 16× more efficient than the regular configuration (the other row) of $N_r = 1$, $\gamma = 2$ and $C = 144$. Previous best performing methods on three benchmarks in literature are shown in the first block of this table

Method	ModelNet40			S3DIS			PartNet			
	acc	param	FLOP	mIoU	param	FLOP	val	test	param	FLOP
DensePoint [15]	93.2	0.7M	0.7G	–	–	–	–	–	–	–
KPConv [30]	92.9	15.2M	1.7G	65.7	15.0M	6.5G	–	–	–	–
PointCNN [14]	92.5	0.6M	25.3G	65.4	4.4M	36.7G	–	46.4	4.4M	23.1G
Baseline*	91.4	19.4M	1.8G	51.5	18.4M	7.2G	42.5	44.6	18.5M	6.7G
Baseline† (AVG, S)	90.7	1.2M	0.1G	50.3	1.1M	0.5G	39.5	40.6	1.1M	0.4G
Baseline† (AVG)	91.4	19.4M	1.8G	51.0	18.4M	7.2G	44.2	45.8	18.5M	6.7G
Baseline† (MAX, S)	91.5	1.2M	0.1G	57.4	1.1M	0.5G	39.8	41.2	1.1M	0.4G
Baseline† (MAX)	91.8	19.4M	1.8G	58.4	18.4M	7.2G	45.4	47.4	18.5M	6.7G
Point-wise MLP (S)	92.6	1.7M	0.2G	56.7	1.6M	0.8G	45.3	47.0	1.6M	0.7G
Point-wise MLP	92.8	26.5M	2.7G	66.2	25.5M	9.8G	48.1	51.5	25.6M	9.1G
Pseudo grid (S)	92.3	1.2M	0.3G	64.3	1.2M	1.0G	44.2	45.2	1.2M	0.9G
Pseudo grid	93.0	19.5M	2.0G	65.9	18.5M	9.3G	50.8	53.0	18.5M	8.5G
Adapt weights (S)	92.1	1.2M	0.2G	61.9	1.2M	0.6G	44.1	46.1	1.2M	0.5G
Adapt weights	93.0	19.4M	2.3G	66.5	18.4M	7.8G	50.1	53.5	18.5M	7.2G
PosPool (PPNet-S)	92.5	1.2M	0.1G	64.2	1.1M	0.5G	44.6	47.2	1.1M	0.5G
PosPool (PPNet)	92.9	19.4M	1.8G	66.5	18.4M	7.3G	50.0	53.4	18.5M	6.8G
PosPool* (PPNet-S*)	92.6	1.2M	0.1G	61.3	1.1M	0.5G	46.1	47.2	1.1M	0.5G
PosPool* (PPNet*)	93.2	19.4M	1.8G	66.7	18.4M	7.3G	50.6	53.8	18.5M	6.8G

varying density problem [29,30]. Given a base grid size at the highest resolution of Res1, the grid size for different resolutions are multiplied by 2× stage-by-stage. The base grid size for different datasets are detailed in Sect. 6.

To generate a point neighborhood, we follow the ball radius method [8,16,22], which in general result in more balanced density than the location or feature kNN methods [2,34,35]. The ball radius is set as 2.5× of the base grid size.

Datasets. We consider three datasets with varying scales of training data, task outputs (classification and semantic segmentation) and scenarios (CAD models and real scenes): ModelNet40 [37], S3DIS [1] and PartNet [19]. More details about datasets are described in Sect. 6.

Performance of Two Baseline Operators. For point cloud modeling, the architectures without local aggregation operators also perform well to some extent, e.g. PointNet [20]. To investigate what local aggregation operators perform beyond, we present two baseline functions to replace the local aggregation operators described in Sect. 3:

$$\mathbf{g}_i = \mathbf{f}_i, \tag{6}$$

$$\mathbf{g}_i = R\left(\{\mathbf{f}_j | j \in \mathcal{N}(i)\}\right). \tag{7}$$

The former is an identity function, without encoding neighborhood points. The latter is an AVG/MAX pool layer without regarding their relative positions.

Table 1 shows the accuracy of these two baseline operators using the common architecture in Fig. 1 on three benchmarks. It can be seen that these baseline operators mostly perform marginally worse than the previous best performing methods on the three datasets. The baseline[†] operator using a MAX pooling layer even slightly outperforms the previous state-of-the-art with smaller computation FLOPs (47.4 mIoU, 6.7G FLOPs vs. 46.4 mIoU, 23.1G FLOPs).

In the following, we will revisit different designing components in the point-wise MLP based methods and the adaptive weight based methods using the common deep residual architecture in Fig. 1. For the pseudo grid feature methods, we choose a representative operator, KPConv [30], with depth-wise convolution kernel and its default grid settings ($M = 15$) for comparison. There are not much hyper-settings for it and we will omit the detailed tuning.

4.2   Performance Study on Point-Wise MLP Based Method

We start the investigation of this type of methods from a representative method, PointNet++ [22]. We first reproduce this method using its own specific overall architecture and with other implementation details the same as ours. Table 2 (denoted as PointNet++*) shows our reproduction is fairly well, which achieves slightly better accuracy than that reported by the authors [19,22] on ModelNet40 and PartNet.

We re-investigate several design components for this type of methods using the deep architecture in Fig. 1, including the number of fully connected (FC) layers in an MLP, the choice of input features and the reduction function. Table 2 shows the ablation study on these aspects, with architecture hyper-parameters as: block repeat factor $N_r = 1$, base width $C = 144$ and bottleneck ratio $\gamma = 8$.

We can draw the following conclusions:

– *Number of FC layers.* In literature of this method type, 3 layers are usually used by default to approximate complex functions. Surprisingly, in our experiments, *using 1 FC layer without non-linearity significantly outperforms that using 2 or 3 FC layers on S3DIS*, and it is also competitive on Model-Net40 and PartNet. We hypothesize that the fitting ability by multiple FC layers applied on the concatenation of point feature and relative position may be partly realized by the point-wise transformation layers (the first and the last layers in a residual block) applied on point feature alone. Less FC layers also ease optimization. Using 1 FC layer is also favorable considering the efficiency issue: the computation can be significantly reduced when 1 FC layer is adopted, through computing sharing as explained below.

Table 2. Evaluating different settings of the point-wise MLP method. The option ∇, $\triangle$, $\square$ and $\lozenge$ denote input features using $\{\Delta\mathbf{p}_{ij}, \mathbf{f}_j\}$, $\{\mathbf{f}_i, \Delta\mathbf{f}_{ij}\}$, $\{\Delta\mathbf{p}_{ij}, \mathbf{f}_i, \Delta\mathbf{f}_{ij}\}$, and $\{\Delta\mathbf{p}_{ij}, \mathbf{f}_i, \mathbf{f}_j, \Delta\mathbf{f}_{ij}\}$, respectively. "Sweet spot" denotes balanced settings regarding both efficacy and efficiency. The accuracy on PartNet test set is not tested in ablations to avoid the tuning of test set

Method	γ	Input				#FC	$R(\cdot)$	ModelNet40	S3DIS	PartNet (val/test)
		∇	$\triangle$	$\square$	$\lozenge$					
PointNet++ [22]	–	✓				3	MAX	90.7	–	–/42.5
PointNet++*	–	✓				3	MAX	91.6	55.3	43.1/45.3
Sweet spot	8			✓		1	MAX	92.8	62.9	48.2/50.8
	2			✓		1	MAX	92.8	66.2	48.1/51.2
FC num	8			✓		2	MAX	92.5	59.5	47.9/–
	8			✓		3	MAX	92.0	59.9	48.7/–
Input	8	✓				1	MAX	92.6	59.8	47.1/–
	8		✓			1	MAX	92.5	61.4	47.6/–
	8				✓	1	MAX	92.7	51.0	47.9/–
Reduction $R(\cdot)$	8			✓		1	AVG	92.3	55.1	46.8/–
	8			✓		1	SUM	92.2	44.7	46.7/–

- *Input Features.* The relative position and edge feature perform similarly on ModelNet40 and PartNet, and combining them has no additional gains. However, on S3DIS datasets, combining both significantly outperforms the variants using each alone.
- *Reduction function.* MAX pooling performs the best, which is in accord with that in literature.

An Efficient Implementation When 1 FC Layer Is Used. Denote the weight matrix of this only FC layer as $W = [W^1, W^2] \in \mathbb{R}^{d \times (d+3)}$ where $W^1 \in \mathbb{R}^{d \times 3}$ and $W^2 \in \mathbb{R}^{d \times d}$. We have $G = W \cdot \text{concat}(\Delta\mathbf{p}_{ij}, \mathbf{f}_j) = W^1 \Delta\mathbf{p}_{ij} + W^2 \mathbf{f}_j$. Noting the computation of the second term $W^2 \mathbf{f}_j$ can be shared when point j appears in different neighborhoods, the computation complexity of this operator is significantly reduced from $(d+3)ndK$ to $nd^2 + 3ndK$.

Sweet Spots for Point-Wise MLP Methods. Regarding both the efficacy and efficiency, the sweet spot settings are applying 1 FC layer to an input combination of relative position and edge features. Table 2 also shows that using $\gamma = 2$ for this method can approach or surpass the state-of-the-art on all three datasets.

4.3 Performance Study on Adaptive Weight Based Method

Table 3 shows the ablations over several designing components within this method type, including the number of fully connected (FC) layers, choice of input features, the reduction function and whether to do weight normalization.

Table 3. Evaluating different settings of the adaptive weight based methods. $dp*$ denotes the 9-dimensional position vector as in [16]. "Sweet spot" denotes balanced settings regarding both efficacy and efficiency. The accuracy on PartNet test set is not tested for ablations to avoid tuning the test set.

Method	γ	Input				#FC	$R(\cdot)$	S.M.	ModelNet	S3DIS	PartNet (val)
		$\{dp\}$	$\{df\}$	$\{dp, df\}$	$\{dp^*\}$						
PConv [34]	–	✓				2	SUM		–	58.3	–
FlexConv [5]	–	✓				1	SUM		90.2	56.6	–
Sweet spot	8	✓				1	AVG		92.7	62.6	50.0
Sweet spot*	2	✓				1	AVG		93.0	66.5	50.1
FC num	8	✓				2	AVG		92.6	61.3	49.9
	8	✓				3	AVG		92.5	58.5	49.6
Input	8		✓			1	AVG		85.3	46.6	46.9
	8			✓		1	AVG		82.2	55.7	46.4
	8				✓	1	AVG		92.1	57.0	49.1
Reduction	8	✓				1	SUM		92.6	61.7	49.1
	8	✓				1	MAX		92.4	62.3	49.7
SoftMax	8	✓				1	AVG	✓	91.7	55.9	45.8

We adopt architecture hyper-parameters as: block repeat factor $N_r = 1$, base width $C = 144$, and bottleneck ratio $\gamma = 8$.

We can draw the following conclusions:

- *Number of FC layers.* Using 1 FC layer performs noticeably better than that using 2 or 3 layers on S3DIS, and is comparable on ModelNet40 and PartNet.
- *Input features.* Using relative positions alone performs best on all datasets. The accuracy slightly drops with additional position features [16]. The edge features harm the performance, probably because it hinders the effective learning of adaptive weights from relative positions.
- *Reduction function.* MAX and AVG functions perform slightly better than SUM function, probably because the MAX and AVG functions are more insensitive to varying neighbor size. We use AVG function by default.
- *SoftMax normalization.* The accuracy significantly drops by SoftMax normalization, probably because the positive weights after normalization let kernels act as low-pass filters and may cause the over-smoothing problem [13].

Sweet Spots for Adaptive Weight Based Methods. The best performance is achieved by applying 1 FC layer without SoftMax normalization on relative positions alone to compute the adaptive weights. This method also approaches or surpasses the state-of-the-art on all three datasets using a deep residual network.

Discussions. Table 1 indicates that the three local aggregation operator types with appropriate settings all achieve the state-of-the-art performance on three representative datasets using the same deep residual architectures. With 16×

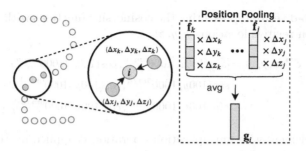

Fig. 2. Illustration of the proposed position pooling (PosPool) operator.

less parameters and computations (marked by "S"), they also perform competitive compared with the previous state-of-the-art. The sweet spots of different operators also favor a simplicity principle, that the relative position alone and 1 FC layer perform well in most scenarios.

While recent study in point cloud analysis mainly lies in inventing new local aggregation operators, the above results indicate that some of them may worth re-investigation under deeper and residual architectures. These results also stimulate a question: could a much simpler local aggregation operator achieve similar accuracy as the sophisticated ones? In the following section, we will try to answer this question by presenting an extremely simple local aggregation operator.

5   PosPool: An Extremely Simple Local Aggregation Operator

In this section, we present a new local aggregation operator, which is extremely simple with no learnable weights.

The new operator is illustrated in Fig. 2. For each neighboring point j, it combines the relative position $\Delta \mathbf{p}_{ij}$ and point feature $\mathbf{f}_j$ by element-wise multiplication. Considering the dimensional difference between the 3-dimensional $\Delta \mathbf{p}_{ij}$ and d-dimensional $\mathbf{f}_j$, the multiplication is applied group-wise that $\Delta \mathbf{p}_{ij}$'s scalars $[\Delta x_{ij}, \Delta y_{ij}, \Delta z_{ij}]$ are multiplied to 1/3 channels of $\mathbf{f}_j$, respectively, as

$$G(\Delta \mathbf{p}_{ij}, \mathbf{f}_j) = \text{Concat}\left[\Delta x_{ij}\mathbf{f}_j^0; \Delta y_{ij}\mathbf{f}_j^1; \Delta z_{ij}\mathbf{f}_j^2\right], \tag{8}$$

where $\mathbf{f}_j^{0,1,2}$ are the 3 sub-vectors equally split from $\mathbf{f}_j$, as $\mathbf{f}_j = \left[\mathbf{f}_j^0; \mathbf{f}_j^1; \mathbf{f}_j^2\right]$.

The operator is named position pooling (PosPool), featured by its property of no learnable weight. It also reserves the *permutation/translation invariance* property which is favorable for point cloud analysis.

A Variant. We also consider a variant of position pooling operator which is slightly more complex, but maintains the no learnable weight property. Instead of using 3-d relative coordinates, we embed the coordinates into a vector with the same dimension as point feature $\mathbf{f}_{ij}$ using cosine/sine functions, similar as in [31]. The embedding is concatenated from $d/6$ group of 6-dimensional vectors, with the mth 6-d vector representing the cosine/sine functions with a wave length of $1000^{6m/d}$ on relative locations x, y, z:

$$\mathcal{E}^m(x, y, z) = [sin(100x/1000^{6m/d}), cos(100x/1000^{6m/d}),$$
$$sin(100y/1000^{6m/d}), cos(100y/1000^{6m/d}),$$
$$sin(100z/1000^{6m/d}), cos(100z/1000^{6m/d})]. \qquad (9)$$

Then an element-wise multiplication operation is applied on the embedding $\mathcal{E}$ and the point feature $\mathbf{f}_{ij}$:

$$G(\Delta \mathbf{p}_{ij}, \mathbf{f}_j) = \mathcal{E} \odot \mathbf{f}_{ij}. \qquad (10)$$

The resulting operator also does not have any learnable weights, and is set as a variant of position pooling layer. We find this variant performs slightly better than the direct multiplication in Eq. (8) in some scenarios. We will show more variants in Appendix A3.

Complexity Analysis. The space complexity $\mathcal{O}(\text{space}) = 0$, as there are no learnable weights. The time complexity is also small $\mathcal{O}(\text{time}) = ndK$. Due to the no learnable weight nature, it may also potentially ease the hardware implementation, which does not require an adaption to different learnt weights.

6   Experiments

6.1   Benchmark Settings

In this section, we detailed the three benchmark datasets with varying scales of training data, task outputs (classification and semantic segmentation) and scenarios (CAD models and real scenes).

- *ModelNet40* [37] is a 3D classification benchmark. This dataset consists of 12,311 meshed CAD models from 40 classes. We follow the official data splitting scheme in [37] for training/testing. We adopt an input resolution of 5,000 and a base grid size of 2 cm.
- *S3DIS* [1] is a real indoor scene segmentation dataset with 6 large scale indoor areas captured from 3 different buildings. 273 million points are annotated and classified into 13 classes. We follow [28] and use Area-5 as the test scene and all others for training. In both training and test, we segment small subclouds in spheres with radius of 2 m. In training, the spheres are randomly selected in scenes. In test, we select spheres regularly in the point clouds. We adopt a base grid size of 4 cm.
- *PartNet* [19] is a more recent challenging benchmark for large-scale fine-grained part segmentation. This dataset consists of pre-sampled point clouds of 26,671 3D object models in 24 object categories, with each object containing 18 parts on average. This dataset is officially split into three parts: 70% training, 10% validation, and 20% test sets. We train our model with official

training dataset and then conduct the comparison study on the validation set on 17 categories with fine-grained annotation. We also report the best accuracies of different methods on the test set. We use the 10,000 points provided with the datasets as input, and the base grid size is set as 2 cm.

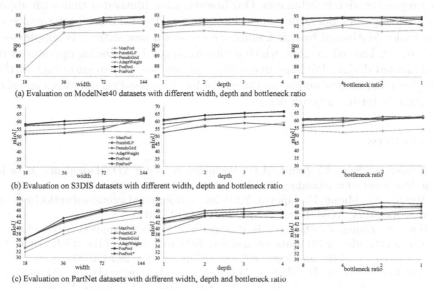

(a) Evaluation on ModelNet40 datasets with different width, depth and bottleneck ratio

(b) Evaluation on S3DIS datasets with different width, depth and bottleneck ratio

(c) Evaluation on PartNet datasets with different width, depth and bottleneck ratio

Fig. 3. Accuracy of different methods with varying width (C), depth $(N_r + 1)$ and bottleneck ratio (γ) on three benchmark datasets.

The training/inference settings are detailed in Appendix A1. Note for PartNet datasets, while in [19] independent networks are trained for 17 different shapes, we adopt a shared backbone and independent 3 fully connected layers for part segmentation of different categories and train all the categories together, which significantly facilitate the evaluation on this dataset. We note using the shared backbone network achieves similar accuracy than the methods training different shapes independently.

6.2   Comparing Operators with Varying Architecture Capacity

Figure 3 shows comparison of different local aggregation operators using architectures with different model capacity on three benchmarks, by varying the network width, depth and bottleneck ratio. Detailed experimental settings are presented in Appendix A2. It can be seen: the PosPool operators achieve top or close-to-top performances using varying network hyper-parameters on all datasets, showing its strong stability and adaptability. While the other more sophisticated operators may achieve similar accuracy with the PosPool layers on some datasets or settings, their performance are less stable across scenarios and model capacity. For example, the accuracy of the "AdaptWeight" method will drop significantly on S3DIS when the model capacity is reduced by either the width, depth or bottleneck ratio.

7   Conclusion

This paper studies existing local aggregation operators in depth via a carefully designed common testbed that consists of a deep residual architecture and three representative benchmarks. Our investigation illustrates that with appropriate settings, all operators can achieve the state-of-the-art performance on three tasks. Motivated by this finding, we present a new extremely simple operator without learned weights, which performs as good as existing operators with sophisticated design. We hope our study and new design can encourage further rethinking and understanding on the role of local aggregation operators and shed new light to future network design.

References

1. Armeni, I., Sax, A., Zamir, A.R., Savarese, S.: Joint 2D–3D-semantic data for indoor scene understanding. ArXiv e-prints, February 2017
2. Atzmon, M., Maron, H., Lipman, Y.: Point convolutional neural networks by extension operators. ACM Trans. Graph. **37**(4), 1–12 (2018)
3. Feng, Y., Zhang, Z., Zhao, X., Ji, R., Gao, Y.: GVCNN: group-view convolutional neural networks for 3D shape recognition. In: Proceedings of the IEEE Conference on Computer Vision and Pattern Recognition, pp. 264–272 (2018)
4. Gadelha, M., Wang, R., Maji, S.: Multiresolution tree networks for 3D point cloud processing. In: Ferrari, V., Hebert, M., Sminchisescu, C., Weiss, Y. (eds.) ECCV 2018. LNCS, vol. 11211, pp. 105–122. Springer, Cham (2018). https://doi.org/10.1007/978-3-030-01234-2_7
5. Groh, F., Wieschollek, P., Lensch, H.P.A.: Flex-Convolution. In: Jawahar, C.V., Li, H., Mori, G., Schindler, K. (eds.) ACCV 2018. LNCS, vol. 11361, pp. 105–122. Springer, Cham (2019). https://doi.org/10.1007/978-3-030-20887-5_7
6. Guo, H., Wang, J., Gao, Y., Li, J., Lu, H.: Multi-view 3D object retrieval with deep embedding network. IEEE Trans. Image Process. **25**(12), 5526–5537 (2016)
7. He, K., Zhang, X., Ren, S., Sun, J.: Deep residual learning for image recognition (2015)
8. Hermosilla, P., Ritschel, T., Váquez, P.P., Vinacua, A., Ropinski, T.: Monte Carlo convolution for learning on non-uniformly sampled point clouds. ACM Trans. Graph. **37**(6), 1–12 (2018)
9. Hua, B.S., Tran, M.K., Yeung, S.K.: Pointwise convolutional neural networks. In: Proceedings of the IEEE Conference on Computer Vision and Pattern Recognition, pp. 984–993 (2018)
10. Klokov, R., Lempitsky, V.: Escape from cells: deep KD-networks for the recognition of 3D point cloud models. In: Proceedings of the IEEE International Conference on Computer Vision, pp. 863–872 (2017)
11. Komarichev, A., Zhong, Z., Hua, J.: A-CNN: annularly convolutional neural networks on point clouds. In: Proceedings of the IEEE Conference on Computer Vision and Pattern Recognition, pp. 7421–7430 (2019)
12. Lan, S., Yu, R., Yu, G., Davis, L.S.: Modeling local geometric structure of 3D point clouds using geo-CNN. In: Proceedings of the IEEE Conference on Computer Vision and Pattern Recognition, pp. 998–1008 (2019)
13. Li, G., Müller, M., Thabet, A., Ghanem, B.: Can GCNs go as deep as CNNs? arXiv preprint arXiv:1904.03751 (2019)

14. Li, Y., Bu, R., Sun, M., Wu, W., Di, X., Chen, B.: PointCNN: convolution on X-transformed points. In: Advances in Neural Information Processing Systems, pp. 820–830 (2018)
15. Liu, Y., Fan, B., Meng, G., Lu, J., Xiang, S., Pan, C.: DensePoint: learning densely contextual representation for efficient point cloud processing. In: Proceedings of the IEEE International Conference on Computer Vision, pp. 5239–5248 (2019)
16. Liu, Y., Fan, B., Xiang, S., Pan, C.: Relation-shape convolutional neural network for point cloud analysis. In: Proceedings of the IEEE Conference on Computer Vision and Pattern Recognition, pp. 8895–8904 (2019)
17. Mao, J., Wang, X., Li, H.: Interpolated convolutional networks for 3D point cloud understanding. In: Proceedings of the IEEE International Conference on Computer Vision, pp. 1578–1587 (2019)
18. Maturana, D., Scherer, S.: VoxNet: a 3D convolutional neural network for real-time object recognition. In: 2015 IEEE/RSJ International Conference on Intelligent Robots and Systems (IROS), pp. 922–928. IEEE (2015)
19. Mo, K., et al.: PartNet: a large-scale benchmark for fine-grained and hierarchical part-level 3D object understanding. In: CVPR (2019)
20. Qi, C.R., Su, H., Mo, K., Guibas, L.J.: PointNet: deep learning on point sets for 3D classification and segmentation. In: Proceedings of the IEEE Conference on Computer Vision and Pattern Recognition, pp. 652–660 (2017)
21. Qi, C.R., Su, H., Nießner, M., Dai, A., Yan, M., Guibas, L.J.: Volumetric and multi-view CNNs for object classification on 3D data. In: Proceedings of the IEEE Conference on Computer Vision and Pattern Recognition, pp. 5648–5656 (2016)
22. Qi, C.R., Yi, L., Su, H., Guibas, L.J.: PointNet++: deep hierarchical feature learning on point sets in a metric space. In: NIPS (2017)
23. Riegler, G., Osman Ulusoy, A., Geiger, A.: OctNet: learning deep 3D representations at high resolutions. In: Proceedings of the IEEE Conference on Computer Vision and Pattern Recognition, pp. 3577–3586 (2017)
24. Ronneberger, O., Fischer, P., Brox, T.: U-Net: convolutional networks for biomedical image segmentation. In: Navab, N., Hornegger, J., Wells, W.M., Frangi, A.F. (eds.) MICCAI 2015. LNCS, vol. 9351, pp. 234–241. Springer, Cham (2015). https://doi.org/10.1007/978-3-319-24574-4_28
25. Su, H., Maji, S., Kalogerakis, E., Learned-Miller, E.: Multi-view convolutional neural networks for 3D shape recognition. In: Proceedings of the IEEE International Conference on Computer Vision, pp. 945–953 (2015)
26. Tatarchenko, M., Dosovitskiy, A., Brox, T.: Octree generating networks: efficient convolutional architectures for high-resolution 3D outputs. In: Proceedings of the IEEE International Conference on Computer Vision, pp. 2088–2096 (2017)
27. Tatarchenko, M., Park, J., Koltun, V., Zhou, Q.Y.: Tangent convolutions for dense prediction in 3D. In: Proceedings of the IEEE Conference on Computer Vision and Pattern Recognition, pp. 3887–3896 (2018)
28. Tchapmi, L., Choy, C., Armeni, I., Gwak, J., Savarese, S.: SEGCloud: semantic segmentation of 3D point clouds. In: 2017 International Conference on 3D Vision (3DV), pp. 537–547. IEEE (2017)
29. Thomas, H., Goulette, F., Deschaud, J.E., Marcotegui, B., LeGall, Y.: Semantic classification of 3D point clouds with multiscale spherical neighborhoods. In: 2018 International Conference on 3D Vision (3DV), September 2018
30. Thomas, H., Qi, C.R., Deschaud, J.E., Marcotegui, B., Goulette, F., Guibas, L.J.: KPConv: flexible and deformable convolution for point clouds. In: Proceedings of the IEEE International Conference on Computer Vision, pp. 6411–6420 (2019)

31. Vaswani, A., et al.: Attention is all you need (2017)
32. Wang, L., Huang, Y., Hou, Y., Zhang, S., Shan, J.: Graph attention convolution for point cloud semantic segmentation. In: Proceedings of the IEEE Conference on Computer Vision and Pattern Recognition, pp. 10296–10305 (2019)
33. Wang, P.S., Liu, Y., Guo, Y.X., Sun, C.Y., Tong, X.: O-CNN: Octree-based convolutional neural networks for 3D shape analysis. ACM Trans. Graph. (TOG) 36(4), 72 (2017)
34. Wang, S., Suo, S., Ma, W.C., Pokrovsky, A., Urtasun, R.: Deep parametric continuous convolutional neural networks. In: Proceedings of the IEEE Conference on Computer Vision and Pattern Recognition, pp. 2589–2597 (2018)
35. Wang, Y., Sun, Y., Liu, Z., Sarma, S.E., Bronstein, M.M., Solomon, J.M.: Dynamic graph CNN for learning on point clouds. ACM Trans. Graph. (TOG) 38(5), 1–12 (2019)
36. Wu, W., Qi, Z., Fuxin, L.: PointConv: deep convolutional networks on 3D point clouds. In: Proceedings of the IEEE Conference on Computer Vision and Pattern Recognition, pp. 9621–9630 (2019)
37. Wu, Z., et al: 3D ShapeNets: a deep representation for volumetric shapes. In: CVPR (2015)
38. Zhang, Z., Hua, B.S., Yeung, S.K.: ShellNet: efficient point cloud convolutional neural networks using concentric shells statistics. In: Proceedings of the IEEE International Conference on Computer Vision, pp. 1607–1616 (2019)

Look Here! A Parametric Learning Based Approach to Redirect Visual Attention

Youssef A. Mejjati[1]([✉]), Celso F. Gomez[2], Kwang In Kim[3], Eli Shechtman[2], and Zoya Bylinskii[2]

[1] University of Bath, Bath, UK
yam28@bath.ac.uk
[2] Adobe Research, Seattle, USA
[3] UNIST, Ulsan, South Korea

Abstract. Across photography, marketing, and website design, being able to direct the viewer's attention is a powerful tool. Motivated by professional workflows, we introduce an automatic method to make an image region more attention-capturing via subtle image edits that maintain realism and fidelity to the original. From an input image and a user-provided mask, our GazeShiftNet model predicts a distinct set of global parametric transformations to be applied to the foreground and background image regions separately. We present the results of quantitative and qualitative experiments that demonstrate improvements over prior state-of-the-art. In contrast to existing attention shifting algorithms, our global parametric approach better preserves image semantics and avoids typical generative artifacts. Our edits enable inference at interactive rates on any image size, and easily generalize to videos. Extensions of our model allow for multi-style edits and the ability to both increase and attenuate attention in an image region. Furthermore, users can customize the edited images by dialing the edits up or down via interpolations in parameter space. This paper presents a practical tool that can simplify future image editing pipelines.

Keywords: Automatic image editing · Visual attention · Adversarial networks

1 Introduction

Photographers, advertisers, and educators seek to control the attention of their audiences, redirecting it to the content that matters most. Professionals working with images accomplish this via subtle adjustments to the contrast, tone,

---

Work done while Youssef was interning at Adobe Research.

Electronic supplementary material The online version of this chapter (https://doi.org/10.1007/978-3-030-58592-1_21) contains supplementary material, which is available to authorized users.

© Springer Nature Switzerland AG 2020
A. Vedaldi et al. (Eds.): ECCV 2020, LNCS 12368, pp. 343–361, 2020.
https://doi.org/10.1007/978-3-030-58592-1_21

color, etc., of the relevant image regions to make them "pop-out". Motivated by professional workflows, we propose an automated learning based approach that predicts a set of global parametric edits to apply to an image to redirect a viewer's attention towards (or away from) a specific image region. Importantly, our approach is constrained, via an adversarial module, to produce realistic edits that remain faithful to the photographer's intentions and original image semantics. To ensure that the edited image successfully redirects attention, we use a state of the art saliency model during training.

Fig. 1. GazeShiftNet takes an image and binary mask as input and predicts a set of parameters (sharpening, exposure, contrast, tone, and color curves) that are sequentially applied to the image to produce the output. The transformed image *subtly* redirects visual attention towards the mask region, seen from the saliency maps. A user can then tune the edits up or down (as shown on the right) at interactive rates, using the saliency slider. (Color figure online)

While a glowing red arrow in an image would surely attract attention, this would not be practical for many use cases. Our GazeShiftNet model is specifically designed to be a practical solution according to the following criteria: (1) the model predictably redirects attention towards (or away from) image pixels denoted by a user-provided binary mask, (2) the proposed edits conserve the image semantics to maintain realism, (3) the model performs consistently and robustly across a variety of image content (e.g., objects and people), and (4) image edits are predicted at interactive rates, for use within applications.

Our solution involves applying global parametric transformations - sharpening, exposure, contrast, tone, and color - to the image, and employing an adversarial training strategy [15], rather than modifying each pixel separately [9,14]. Our network predicts two sets of parameters to apply to the foreground and background of the image, demarcated by the input mask. The goal is to create a "pop-out" effect, which cannot be achieved by a single global transformation to the entire image. In sum, our model takes the form of a parametric generator followed by a cascade of differentiable layers that apply common photo editing transformations, mimicking a professional workflow. The choice and implementation of these transformations is motivated by related work on image enhancement [3,18]. We train GazeShiftNet on a subset of the MS-COCO dataset [26], and present quantitative and qualitative results, including from user studies, on three different datasets. Compared to existing attention shifting methods [9,14,16,28,30], our approach successfully shifts viewers' attention, while achieving more realistic results that conserve the original image semantics and do not suffer from local color and texture artifacts.

We demonstrate the advantages of our global parametric approach for future applications. In particular, inference and editing take place at interactive rates, independent of image resolution, as the global transformations are predicted on a low resolution image using a feed-forward pass (rather than an optimization procedure [16,28,30]) and can be seamlessly applied to a high resolution version. Users can tweak the final result using photo editing sliders that control the parameters applied, since the interpolation takes place in parameter space. Our method also works for video data by applying the parameters predicted for the first frame to the rest of the video, thereby ensuring temporal continuity. Finally, we present extensions of our model to produce image edits of different styles, and the option to both increase and attenuate visual attention in an image region.

2 Related Work

While image editing has been a popular research topic for many years [2,8,17,37], recent breakthroughs have been made possible by generative adversarial networks [15], including in image generation [21,35], image-to-image translation [31,49], and style transfer [13,19]. However, very few image editing methods exist for redirecting the viewer's attention towards a specific image region.

Among such methods, Gatys et al. [14] use an encoder-decoder model that takes an image and target saliency map as inputs, and generates a new image satisfying the target saliency map. This approach has a number of weaknesses: (1) operating directly in pixel space often produces artifacts in the final image, (2) a target saliency map is required as input, which is not straightforward for a user to provide, and (3) the edits from the generator are limited to a fixed image resolution. Chen et al. [9] similarly use an encoder-decoder model, but with an additional cyclic loss to stabilize the training procedure and reduce artifacts. However, this approach still suffers from the last two issues above, making it inconvenient for practical editing scenarios.

Su et al. [39] use smoothened power maps with steerable pyramids to equalize texture. Mechrez et al. [30] use patches from the same image to increase the saliency of a given region. The space of possible edits is naturally limited to appearances of other pixels in the image. Furthermore, this approach is very time consuming and requires an optimization per image, with compute time scaling with image resolution.

Related to our approach Hagiwara et al. [16] predict color and intensity changes in RGB space and add them pixel-wise to the original image.

Mateescu et al. [28] use LAB space and adjust the hue within the object mask to maximize the KL divergence between the color distributions inside and outside the mask. Both methods require computationally intensive optimization at test time. Additionally, while these algorithms are based on heuristics that successfully redirect human attention, they do not preserve the semantics of the image, and can generate color anomalies affecting image realism [28].

Wong et al. [43] also use an optimization strategy to predict parameters such as saturation, sharpness and brightness to apply on different segments of an image, which require significant human supervision to generate. Compared to other parametric methods [16,28], this method struggles to effectively redirect the viewer's attention [30]. In contrast, our method uses a deep neural network to predict global parametric transformations like exposure and contrast to apply to the image, all while preserving the semantics of the image, avoiding artifacts, and significantly reducing computation time.

Similar parametric transformations as ours are used for other image processing tasks, including for photo enhancement [7] and for recovering an image from its raw counterpart by emulating the image processing pipeline [18]. Tsai et al. [40] tackle a related problem, to make the composition of image foreground and background as natural as possible, by enhancing the foreground with respect to its background using an encoder-decoder network. We rather use an encoder-decoder network to enhance the foreground with the goal of redirecting the viewer's attention.

Attention modeling has also seen significant progress in the past few years thanks to deep neural networks [4,6]. Attention models are becoming increasingly more accurate and practical for applications including object detection [41,42], object recognition [25,33], content aware image re-targeting [1,50], graphic design [5,38,47], image captioning [10,48], and action recognition [24,29]. While the application of such models has proven prolific, relatively little effort has been dedicated towards automatically manipulating attention, i.e. creating new image content that satisfies a given attention objective.

3   Method

Motivation: Given an image and region of interest, our goal is to automatically edit the image to make the selected region more attention-capturing. To learn how professionals complete this task, we ran an exploratory study on http://www.usertesting.com, providing participants with image-mask pairs, and asking them to edit the images in Adobe's Photoshop to make the masked regions more attention-capturing. We collected edits from 5 different participants on 30 high-resolution stock photographs (see Supplemental Material). Lessons learned: (1) image semantics are preserved, (2) edits are mostly restricted to parametric transformations, and (3) different operations are applied to image pixels inside and outside the mask. We designed our computational approach accordingly. Our approach (1) strives to maintain fidelity/faithfulness to the input image, while (2) applying global parametric transformations to the image, such that, (3) one set of transformations is applied to the mask (foreground), and a separate set is applied to the background.

Computational Approach: Given an input image I and a binary mask $\mathbf{m}$, our goal is to generate a new image I' redirecting viewer attention towards the image region specified by $\mathbf{m}$. We generate I' by sequentially applying a set of parametric transformations commonly used in photo editing software and by computational photography algorithms [7,18,22,44]. We apply the following ordered sequence of parameters: *sharpening, exposure, contrast, tone adjustment,* and *color adjustment* [18]. We discuss the order of operations in Sect. 4.4, and provide mathematical definitions in the Supplemental Material.

One possible approach is to train an encoder to predict a set of parameters to apply to the foreground region within a mask $\mathbf{m}$. However, we found that predicting a set of parameters both for the foreground $\mathbf{p}_f$ and for the background $\mathbf{p}_b$, is especially effective in cluttered scenes. Where multiple regions in the input image may have high saliency, attenuating the saliency of the background can be easier than increasing the saliency of the foreground. Formally, our encoder G_E is a neural network with a shared feature extractor and two specialized heads, predicting two sets of parameters: $G_E(I, \mathbf{m}) = (\mathbf{p}_f, \mathbf{p}_b)$.

Model Architecture: Our generator G is composed of an encoder G_E and a decoder G_D, where G_E predicts global parametric transformations conditioned on I and $\mathbf{m}$, and G_D applies these transformations to the image to produce the final edited image. The pipeline is as follows (Fig. 2): we concatenate I with the mask $\mathbf{m}$ and feed this to a down-sampling convolutional neural network. Providing the mask as input allows the network to focus on the region to be enhanced. Furthermore, we reinforce the mask conditioning by applying the concatenation layer-wise throughout the convolutional part of the network. We bilinearly down-sample the mask before concatenating on each hidden layer to fit the corresponding input dimensions. The resulting representation is then fed to a series of fully connected layers via global average pooling. Global average pooling provides global information about high level features in the image, which is useful for predicting global parametric transformations. The foreground and background parameters are then predicted by two fully connected network heads.

The convolutional part of G_E encodes the semantics of the image and the region specified by the mask, and is the shared part of our network. The subsequent fully connected networks are the specialized heads that leverage this shared knowledge to predict separate parameters for the foreground and background image regions.

The decoder, G_D, applies the predicted parameters to the input image. G_D consists of a sequence of fixed (non-trainable) differentiable functions applied separately to the foreground and background image regions, demarcated by the mask $\mathbf{m}$. Specifically, we sequentially generate a pair of intermediate images $I'_f(i)$ and $I'_b(i)$ from $\mathbf{p}_f$ and $\mathbf{p}_b$, respectively, based on a fixed order of operations. At iteration i:

$$I'_f(i) = G_D(I'_f(i-1), \mathbf{p}_f(i)) \circ \mathbf{m} + G_D(I'_b(i-1), \mathbf{p}_b(i)) \circ (1-\mathbf{m}), \quad (1)$$

$$I'_b(i) = G_D(I'_b(i-1), \mathbf{p}_b(i)) \circ \mathbf{m} + G_D(I'_f(i-1), \mathbf{p}_f(i)) \circ (1-\mathbf{m}), \quad (2)$$

where ∘ refers to element-wise multiplication, and $I'_f(0) = I'_b(0) = I$. This decoding process is illustrated on the right side of Fig. 2. The final products of this sequential editing process process, I'_f and I'_b, are blended to synthesize the final image I', using $\mathbf{m}$:

$$I' = I'_f \circ \mathbf{m} + I'_b \circ (1 - \mathbf{m}). \tag{3}$$

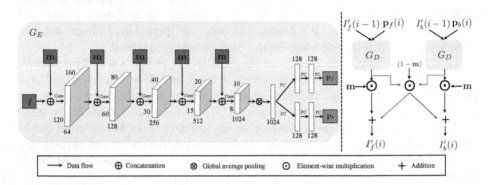

Fig. 2. GazeShiftNet architecture. Left: G_E takes the image I and the mask $\mathbf{m}$ as inputs and encodes them through a series of convolutional layers, then predicts the foreground and background parameters using fully connected network heads. Right: G_D applies a series of differentiable functions sequentially using the predicted parameters from G_E and creates the intermediate images $I'_f(i)$ and $I'_b(i)$ at iteration i.

Losses: We constrain our image transformations to respect the following properties: (1) I' should be a realistic image, without deviating significantly from the input I, (2) viewer attention should be redirected towards the image region specified by $\mathbf{m}$. To ensure the first property, we use a critic D, which differentiates between real and generated images. We train D adversarially with G_E using a hinge GAN loss known for its stability [20,32,45]:

$$\mathcal{L}_D(\Theta_D) = -\mathbb{E}_{I,\mathbf{m}}\big[\min\{0, -D(I') - 1\}\big] - \mathbb{E}_I\big[\min\{0, D(I) - 1\}\big],$$
$$\mathcal{L}_G(\Theta_G) = -\mathbb{E}_{I,\mathbf{m}}\big[D(I')\big], \tag{4}$$

where Θ_D and Θ_G are the learnable weights of D and G_E, respectively.

To ensure the second property, we use a state-of-the-art deep saliency model [11], termed $\mathbf{S}$ throughout the paper, as a proxy of viewer attention. We use $\mathbf{S}$ to compute the saliency map of the output image, $S_{I'} = \mathbf{S}(I', \Theta_S)$, where Θ_S are the parameters of $\mathbf{S}$, and calculate the attention loss in the mask as:

$$\mathcal{L}_{att}(S_{I'}, \mathbf{m}) = -\frac{1}{\sum_{i,j} \mathbf{m}_{[ij]}} \sum_{(i,j)\in \mathbf{m}} S_{I'[ij]} \mathbf{m}_{[ij]}, \tag{5}$$

by iterating over all mask pixels (i, j). Normalization by the area of the masked region gives equal importance to regions of any size. The saliency maps obtained from **S** are normalized via softmax such that they sum to one. This ensures that increasing the saliency of the foreground necessarily decreases the saliency of the background.

Minimizing Eq. 5 alone would lead to unrealistic results, but in combination with the adversarial loss in Eq. 4, the algorithm tries to redirect viewer attention as much as possible while maintaining the realism of the output I'. The overall loss becomes:

$$\mathcal{L}(\Theta_G, \Theta_D) = \mathcal{L}_G(\Theta_G) + \mathcal{L}_D(\Theta_D) + \lambda_s \mathcal{L}_{att}(S_{I'}, \mathbf{m}), \qquad (6)$$

where λ_s controls the weight of the attention loss, which we empirically set to 2.5×10^4. Attention loss values are very small due to softmax normalization of the saliency maps.

Training Dataset: We selected the MS-COCO dataset [26] because of the semantic diversity it contains, and the object segmentations that could be used as input masks to our algorithm. We curated the dataset to produce a set of image-mask pairs appropriate for our task. Specifically, we used the saliency model **S** to select images where the masked region is not already too salient, so that we can train our algorithm to shift attention to these regions. In addition, we discarded images containing only one mask instance or where the mask was too small or too large. Shifting attention to very small regions would likely require unrealistic image edits. Instances that are too large are often already quite salient. For the same reason, images with only one object segment (e.g., a single plane in the sky) may have no other regions to shift attention towards/away from. If a training image contained multiple masks satisfying all these conditions, we randomly picked one and discarded the others to ensure a diverse training set without over-representing any images. Following this process, we ended up with 49,949 image-mask pairs for the training set and 6,519 pairs for the validation set. We call this curated dataset 'CoCoClutter' to emphasize that the images contain multiple objects, without being dominated by any one object. We trained our model with a batch size of 4 for 37,500 iterations, sufficient for the model to converge. We used a learning rate of 1e−5, linearly decayed towards 0 starting halfway through training.

4   Evaluation

In this section, we compare GazeShiftNet to competing approaches on two datasets, to measure the ability of each method to shift visual attention to the target image region, and to produce image edits that are realistic and faithful to the original image. We first consider representative examples from all the methods and discuss common behaviors. Next, we evaluate the methods using computational metrics that measure saliency shift and fidelity, and finally, we present the results of three user studies to measure attention shift, image realism, and fidelity. We conclude with runtime comparisons of all the methods, and ablation experiments run on our model.

Input   Our   MEC [30]   HAG [16]   HOR [28]   GAT [14]

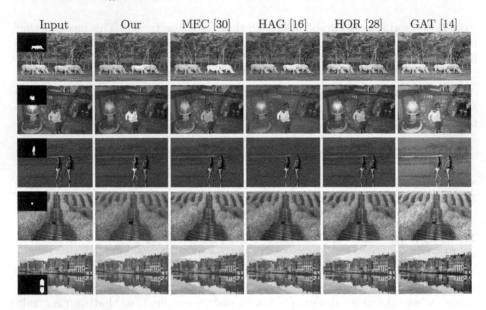

Fig. 3. Model comparisons on the Mechrez dataset. More in the Supplemental Material.

4.1   Qualitative Comparisons

We present visual results obtained from our approach and competing methods on the Mechrez and CoCoClutter datasets in Figs. 3 and 4, respectively. Here we discuss the key properties of each method.

MEC (Mechrez et al. [30][1]): Being patch-based, this approach is limited to reusing colors and textures from the same image. This approach does not always preserve the authenticity of the original photograph (note the change in shirt colors in Fig. 3, rows 2 and 3). Changing image semantics may be unwanted behavior for some applications. MEC is not included in Fig. 4 because the code they provided did not reproduce their results, and led to significantly subpar quality images.

HAG (Hagiwara et al. [16]): This approach alters the intensity and color values in an image, most often relying on changing the background, with only slight modifications to the foreground. This can result in big changes to the image content and low fidelity to the original (note the change of background image hue in Fig. 3, rows 1–4, which has removed background texture in row 2).

HOR (Mateescu et al. [28]): This approach modifies the masked region only to maximize the distribution separation between colors of the foreground and

---

[1] Because the provided code could not reproduce the high quality results presented in their paper, for favorable comparison, we directly used images from their project page: https://webee.technion.ac.il/labs/cgm/Computer-Graphics-Multimedia/ Software/saliencyManipulation/.

background, which often results in substantial color anomalies and loss of realism (striking examples can be found in Fig. 3, rows 3 and 5, and Fig. 4, all rows).

GAT (Gatys et al. [14]**):** This approach is trained using a probabilistic attention model, and requires a full target saliency map as input, which is not practical for a user to produce. To make this approach comparable with ours, we used their attention model [14] to compute a map for our generated images, which we then used as the input target map for their method. From the last column of Figs. 3 and 4, we see that GAT both struggles with making objects more salient and produces many artifacts.

In comparison to these approaches, our method does not change the hue of objects or semantics of the image, remaining faithful to the input image while succeeding to emphasize the desired image region. In the next section, we will quantify these observations.

| Input | Mask | Our | HAG [16] | HOR [28] | GAT [14] |

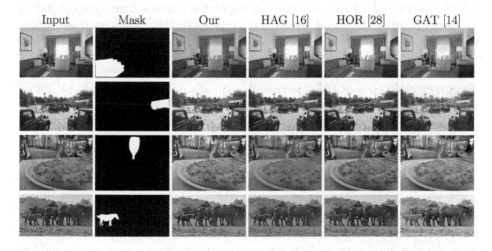

Fig. 4. Results on CoCoClutter dataset. More in the Supplemental Material.

4.2 Quantitative Comparisons

We compare our approach to competing methods based on two criteria: (1) whether a given method successfully shifts visual attention towards desired image regions, and (2) whether the generated images remain realistic and maintain fidelity/faithfulness to the original photograph. We first performed a set of analyses using computational measures - i.e., with a computational model of saliency as a proxy for visual attention, and the LPIPS similarity metric as a measure of fidelity. Next, we ran a set of user studies on the top-performing models to (1) validate whether human attention indeed shifts towards the desired region, and (2) whether humans rate the generated images as both being realistic and having high fidelity to the original photographs.

We performed all quantitative comparisons on two evaluation datasets: a subset of images from the Mechrez dataset [30] and images from our CoCoClutter validation set. We sampled 64 images from the Mechrez dataset, corresponding to 46 images from their *object enhancement* collection and 18 images from their *saliency shift* collection. We selected images containing multiple objects, in order to evaluate the ability of a given algorithm to shift attention to different image regions. For the computational measures, we used the entire CoCoClutter validation set, while for the user studies, we randomly sampled a set of 50 images with clean masks (well-segmented objects).

Computational Evaluation: To evaluate the ability of algorithms to successfully shift attention, we first use the initial and final saliency maps, corresponding to the original and edited images, to measure the mean increase of attention inside the mask area (Tables 1a, 1b, *Saliency increase - Absolute*). We also measure the relative increase of attention by normalizing by the initial saliency map (*Saliency increase - Relative[2]*). Second, we use measures previously used to evaluate attention shift [30], the Pearson correlation and weighed F-beta [27] between the final saliency map and the binary mask (*Similarity to mask*). Finally, as a measure of fidelity of the edited image to the original image, we use the LPIPS metric [46]. To have more insight about the behaviour of each algorithm, we compute LPIPS on the entire image (*Full*), on the background only (*BG*) and on the foreground only (*FG*). Note that Mechrez et al. [30] does not appear in Table 1b as their code did not produce usable results on this dataset.

The three sets of computational measures used are complementary. *Saliency increase* evaluates the increase in saliency values for the foreground, independently of changes to the background. *Similarity to mask* considers the extreme case where the ground truth saliency map would be defined by the mask, hence taking into account changes in saliency both for the foreground and background. Finally, LPIPS measures how different the edited images are compared to the originals. An ideal model has high *Saliency increase* and *Similarity to mask*, and low LPIPS.

Our approach performs the best on all computational measures of saliency across Mechrez (Table 1a) and CoCoClutter (Table 1b) datasets. MEC [30] comes in second due to its ability to replace foreground patches (from the mask) with salient patches from the same image. However, this often comes at the cost of sacrificing the colors and semantics of the original image. HAG [16] is third overall, but suffers from high LPIPS scores (LPIPS *Full* and *BG*) as it often heavily modifies the original colors and details of the background. LPIPS scores for the foreground are generally very small, suggesting that HAG relies mostly on background modifications in its attention shifting pipeline. In contrast, HOR [28] achieves very high LPIPS scores on the foreground due to the severe color artifacts it creates. At the same time, this method achieves the lowest LPIPS scores when considering the entire image (LPIPS *Full*) because it leaves the background untouched, which

---

[2] Relative saliency increases can grow large when the corresponding instance has an average initial saliency value near zero.

usually covers the largest area of the image. Finally GAT [14] performs the worst in terms of shifting attention, while simultaneously heavily modifying the image, resulting in the highest LPIPS scores and lowest *Saliency increase* and *Similarity to mask* scores. Overall, our algorithm is most successful at shifting computational attention, all while conserving original image properties, including, colors, textures, and semantics. It is in the top two models when considering LPIPS scores.

User Studies: We reinforce the findings from the computational measures by collecting human data using Amazon's Mechanical Turk for three tasks: visual attention, image fidelity, and image realism. Our first set of studies measured shifts in human attention, when images are modified by the various methods compared. We used the same crowdsourced gaze tracking method, *CodeCharts* [34], that was used to collect training data for the saliency model [11] we adapted in GazeShiftNet. In CodeCharts, participants are asked to look at an image for a few seconds, then a jittered grid of alphanumeric triplets ("codes") is flashed for a brief interval, and the participant is subsequently prompted to enter the code seen last. This task design captures the area where a participant was looking at the moment when the image disappeared. CodeCharts has been shown to approximate human eye movements collected using an eye tracker [34], which we use to evaluate the ability of algorithms to shift human attention by modifying images.

Table 1. (a) Computational evaluation on Mechrez dataset. (b) Computational evaluation on CoCoClutter dataset. The top two performing models according to each metric are highlighted in green (darker is better). (c) Left: Run time averaged over 30 high resolution images; Right: As GAT is unable to run directly on high resolution images, we use low resolution versions of the same images. (d) Ablation studies on Mechrez dataset. Our chosen method involves applying sharpening, exposure, contrast, tone adjustment, and color adjustment (in that order) to both foreground and background. Ablations consider a subset of parameters and different orderings, as well as application to either one of foreground/background. Darker green colors indicate better scores, darker red colors indicate worse scores. All metrics except run time are multiplied by 100 for legibility.

(a)

Model	LPIPS ↓			Saliency increase ↑		Similarity to mask ↑	
	Full	BG	FG	Absolute	Relative	WFB	CC
Our	5.96	5.10	0.81	3.80	35.77	11.30	30.43
MEC [30]	8.95	7.65	1.29	3.42	28.39	10.42	26.25
HOR [28]	1.60	0	1.41	1.74	16.29	9.85	24.43
HAG [16]	11.08	10.68	0.37	2.24	21.75	10.33	26.12
GAT [14]	25.64	24.72	1.11	0.34	3.05	9.59	22.98

(b)

Model	LPIPS ↓			Saliency increase ↑		Similarity to mask ↑	
	Full	BG	FG	Absolute	Relative	WFB	CC
Our	4.87	2.84	1.95	1.99	25957.92	11.81	20.65
HAG [16]	7.37	6.58	0.69	1.30	24419.15	11.22	18.89
HOR [28]	4.00	0	3.61	1.27	11065.47	11.27	18.98
GAT [30]	30.07	27.08	3.61	-0.05	2920.75	10.23	15.85

(c)

Model	Avg. run time
Our	8.87s
HOR [28]	31.82s
HAG [16]	4743.54s
MEC [30]	>1 day

Model	Avg. run time
Our	1.54s
GAT [14]	4.34s

(d)

Model	LPIPS ↓	Saliency increase ↑	
	Full	Absolute	Relative
Our	5.96	3.80	35.77
Parameter ablations			
tone + color	10.97	5.33	56.19
sharp + exp + cont	1.76	1.47	13.00
color	9.95	3.50	37.25
our + saturation	9.70	4.84	48.76
Fixed parameters	2.28	1.95	17.23
fg/bg ablations			
bg-only	2.31	0.54	3.50
fg-only	1.19	2.53	29.40
order of operations ablations			
col,ton,con,exp,sha	6.74	4.01	35.36
ton,col,sha,exp,con	8.04	3.73	32.47
con,exp,sha,ton,col	6.55	3.65	36.98
ton,col,sha,con,exp	7.87	3.72	30.49

We used CodeCharts to collect human attention data on 64 images from the Mechrez dataset and 50 from the CoCoClutter dataset. We collected attention data on the original images and on the edited images produced by each method. We obtained gaze points from an average of 50 participants per image, that we converted into an attention heatmap for the image. We then compared the attention heatmaps of the original images to those of the edited images, to evaluate the relative attention increase achievable by each method. These values are plotted on the x-axes of Figs. 5a and b (left) for the CoCoClutter and Mechrez datasets, respectively. The table of values is also available in the Supplemental Material. Based on these study results, GazeShiftNet achieves a greater attention shift than HAG, a smaller one than HOR, and a comparable one to MEC. These results are unsurprising given the algorithm behaviors: HAG mostly modifies the background which leads to less noticeable change to the masked objects, whereas HOR often drastically changes the color of the foreground object making it significantly stand out from the rest of the image. From Fig. 5, we can see that our method achieves a balance between shifting attention and maintaining image fidelity, as described below.

The next set of user studies measured image fidelity. We asked an average of 25 participants to evaluate each edited image compared to its original using the following options: not edited (3), slightly edited (2), moderately edited (1), definitely edited (0). In parentheses are the fidelity scores we assigned to each answer (higher is better). Fidelity score distributions across images, averaged over study participants, are visualized as box plots in Fig. 5a and b (middle) for the CoCoClutter and Mechrez datasets, respectively. Both GazeShiftNet and HAG achieve fidelity scores that are statistically significantly higher than HOR on the CoCoClutter dataset, and higher than MEC on the Mechrez dataset ($p < 0.001$). No other pairwise comparisons were significant. The human fidelity judgements are intended to evaluate the same aspect of models as the LPIPS computational metric. Because the fidelity scores of the models correspond most to their LPIPS FG scores from Tables 1a and 1b, we suspect that study participants focused more on the foreground regions of the edited images when judging fidelity.

While fidelity measured how similar an edited image is compared to its original, we also ran a user study to evaluate the realism of the edited images in isolation. We asked an average of 25 participants to evaluate whether each image is: definitely not edited (3), probably not edited (2), probably edited (1), definitely edited (0). In parentheses are the realism scores we assigned to each answer (higher is better). Realism score distributions across images, averaged over study participants, can be found in Fig. 5a and b (right) for the CoCoClutter and Mechrez image datasets, respectively. On Mechrez, all methods perform comparably in terms of realism. This supports the findings from Mechrez et al. [30], who similarly found that when evaluated using a realism user study, the algorithms performed comparably. However, on the CoCoClutter dataset, differences across methods are more pronounced. GazeShiftNet and HAG obtained statistically significantly higher realism scores than HOR ($p < 0.001$). The reason for these differences is that the Mechrez dataset masks often cover only part of an object, which conserves realism when edits are applied (e.g., an edited blue shirt is no less realistic than the original gray one; though fidelity suffers in this case), but the CoCoClutter dataset masks include full objects, which exposes significant failure methods of other approaches (such as when the HOR method turns entire people blue).

Across fidelity and realism, our approach achieves the smallest standard deviations in scores, and the lower range of scores starts higher (Fig. 5). This shows that GazeShiftNet behaves more consistently across a variety of image types, with fewer catastrophic failures than some of the other methods, making it a more practical method overall.

As additional validation of our approach, we compared its performance to that of professional users on the 30 high-resolution images from our exploratory studies (Sect. 3). The results show that while users produce edits that are judged to have more realism and fidelity (and correspondingly lower LPIPS *FG*) scores, our model is able to achieve both higher saliency and attention shift scores than users on average (although some users produce quite effective attention shifts in photos). We note that our model achieves the highest increase in saliency/attention than HAG and HOR on this image dataset. Detailed study results can be found in the Supplemental Material.

4.3   Run Time Comparisons

We timed each algorithm on 30 high-resolution images ($[648, 1332] \times 1500$ pixels) obtained from Adobe Stock and manually annotated to include a masked object per image. Average run times over the 30 images are listed in Table 1c. Our algorithm is significantly faster than HAG and MEC, which are iterative optimization-based algorithms. In fact, MEC took more than one day to process the 30 images. HOR is fairly quick but still takes more than a second per image. GAT cannot run on high resolution images, so for this comparison we used the same 30 images in resolutions corresponding to the required neural network input sizes. Because their network is deeper and their method additionally requires computing a target saliency map as an input, their computation time is nearly 3 times ours.

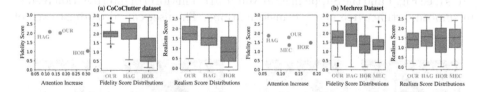

Fig. 5. Results of user studies on three separate crowdsourcing tasks - measuring image fidelity, realism, and human attention - on two datasets: (a) CoCoClutter and (b) Mechrez. Different methods trade-off average image fidelity for average attention increase, and vice versa, whereas our approach achieves a good balance of both (left). Box plots of the fidelity score distributions (middle) and realism score distributions (right) demonstrate the variability in the performance of each method. Ours performs most consistently, with the smallest range of scores, i.e., fewer failure modes. All methods perform similarly in realism on the Mechrez dataset because of the nature of the objects selected for emphasis (see text for details). (Color figure online)

4.4 Ablation Studies

Table 1d includes a summary of the ablation tests evaluating our design choices: (1) the set of parametric transformations, (2) whether to apply parametric transformations to foreground, background, or both, (3) the ordering of transformations applied to images. As it is not immediately obvious how to balance the trade-off between realism and attention shift across our different ablations, we identify the five best models according to three metrics (LPIPS, absolute and relative saliency increase), and choose a high-performing model across all criteria. In Table 1d the best performing models are highlighted in green (darker is better), and the worst performing models are highlighted in red (darker is worst). Our final model selection achieves good performance across the metrics. A detailed discussion of the ablation experiments can be found in the Supplemental Material.

5 Extensions

Our method could be extended further, making it even more flexible. We discuss additional use cases and extensions below.

Application to Videos: GazeShiftNet can be seamlessly generalized to videos by predicting parameters on the first frame, and applying them to subsequent video frames (provided we have a common, segmented object across frames). We find this produces good results on the rest of the frames while avoiding flickering. In contrast, most competing attention shifting algorithms [14,16,30] would need to be run on each frame separately, as their transformations cannot be transferred across images. We show in Fig. 6 snapshots from a video from the DAVIS dataset [36].

Interactive Image Editing: GazeShiftNet, being parametric, makes it easy to hand control of the edited image back to the user. By interpolating between the predicted parameters and the set of parameters where $I' = I$, we can let a user dial the edits up or down. We built a prototype of an interface where a user can interact with a single saliency slider that interpolates the parameters at interactive rates (see Fig. 1 and the Supplemental Material). This form of interpolation results in smooth, artifact-free image transformations. The user can also adjust each of

Fig. 6. Our method can be seamlessly applied to video frames. We predict parameters for the first frame and transfer them to all subsequent frames containing the same segmented object (in this case, the snowboard).

the parameter sliders separately. Importantly, while parameters can be predicted on smaller image sizes, final transformations can be applied to professional high-resolution photography at interactive rates. These properties do not hold for non-parameteric methods [14, 30]. Optimization-based methods [16, 28] do not allow for interactive and artifact-free post-processing edits.

Stochastic Parameter Generation: There isn't a single way to enhance an object in an image, and different users may choose to apply different sets of transformations when editing an image (two far right columns in Fig. 7). Motivated by such natural variability, GazeShiftNet can be extended to produce results in 'multiple styles' by predicting different parameter values for the transformations. We introduce stochasticity using a latent vector $z \in \mathbb{R}^{10}$, randomly sampled from a Gaussian distribution. This vector z is first tiled to match the mask dimensions, and then concatenated as an input, similarly to how the mask is handled. In fact, the architecture is the same as in Fig. 2, but replacing $\mathbf{m}$ with $\mathbf{mz}$ (where $\mathbf{mz}$ is the concatenation of $\mathbf{m}$ and tiled z: $\mathbf{z} \in \mathbf{R}^{h \times w \times 10}$). To force the network to actually use $\mathbf{z}$ in producing the output parameters $\mathbf{p}_f$ and $\mathbf{p}_b$, we add an additional loss encouraging z to be reconstructed from $\mathbf{p}$ (the concatenation of $\mathbf{p}_f$ and $\mathbf{p}_b$): $\mathcal{L}_r(z, \mathbf{p}_f, \mathbf{p}_b) = \frac{1}{10}\|z - \mathrm{ENC}(\mathbf{p})\|_1$, where ENC is an encoder formed by a series of fully connected layers. We can then randomly sample the latent vector z to produce diverse edits (Fig. 7).

| Input | z_1 | z_2 | z_3 | user 1 | user 2 |

Fig. 7. Sampling different latent z_i in our model results in stochastic variations, all of which achieve the same saliency objective, but with different 'styles' of edits. We include two sample edits done by professional users from our exploratory studies.

Decreasing Human Attention: While GazeShiftNet has been trained to shift viewer attention *towards* a specific region in the image, it can also be trained to achieve the opposite: shifting attention *away* from an image region. This can be useful for distractor attenuation [12, 23], e.g., reducing the prominence of passer-bys in personal photo collections. Towards this goal, we extend our network to have two additional heads to output the parameters $\mathbf{p}_f^{dec}$ and $\mathbf{p}_b^{dec}$. We then generate two images, I'_{dec} using $\mathbf{p}_f^{dec}$ and $\mathbf{p}_b^{dec}$ and I'_{inc} using $\mathbf{p}_f^{inc}$ and $\mathbf{p}_b^{inc}$. The new attention loss (Eq. 5) becomes: $\mathcal{L}'_{att}(S_{I'_{inc}}, S_{I'_{dec}}, \mathbf{m}) = \mathcal{L}_{att}(S_{I'_{inc}}, \mathbf{m}) - \mathcal{L}_{att}(S_{I'_{dec}}, \mathbf{m})$. To successfully train this model, we had to discard from the training set all masks where the average computational saliency was too small. Such masks were not suitable for the objective of decreasing saliency. In addition, we trained this model for double the amount of time, with hyper-parameter $\lambda_s = 2 \times 10^4$. Results from this network are visualized in Fig. 8.

| Input | ↑ Attention | Saliency Map | ↓ Attention | Saliency Map |

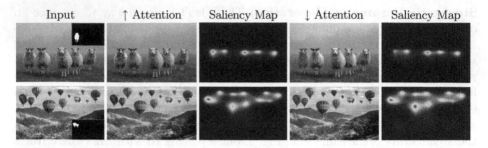

Fig. 8. Using the input image and corresponding mask, we generate two images, one to shift visual attention towards the mask (col. 2) as seen from the saliency map in col. 3, and one to shift attention away from the mask (col. 4, saliency map in col. 5).

6   Conclusion

We presented a practical method for automatically editing images (with extensions to videos) in a subtle way, while effectively redirecting viewer attention. We demonstrated that our method achieves a good balance between shifting the attention - of saliency models and human participants alike - while maintaining the realism and fidelity of the original image (i.e., the photographer's intent, semantics/colors of the original image). Having a practical method depends on balancing these objectives, and achieving consistent and robust results across a variety of images. We also showed significant improvements in computation time over past approaches. Most importantly, our global parametric approach allows running our method on high-resolution images and videos at interactive rates, as well as interpolating in parameter space to hand control of the edits back to a user within an editing interface. Such an effective and practical approach to image editing can benefit many applications, including website design, effective marketing campaigns, and image enhancement. These tasks are typically completed by professionals using advanced image editing software. Our automated approach can simplify, and has the potential to replace, some professional image editing workflows.

Acknowledgements. Y. A. Mejjati thanks the Marie Sklodowska-Curie grant No 665992, and the Centre for Doctoral Training in Digital Entertainment (CDE), EP/L016540/1. K. I. Kim thanks Institute of Information & communications Technology Planning Evaluation (IITP) grant (No. 20200013360011001, Artificial Intelligence Graduate School support (UNIST)) funded by the Korea government (MSIT).

References

1. Achanta, R., Süsstrunk, S.: Saliency detection for content-aware image resizing. In: ICIP (2009)
2. Barnes, C., Shechtman, E., Finkelstein, A., Goldman, D.B.: Patchmatch: A randomized correspondence algorithm for structural image editing. In: TOG (2009)
3. Bianco, S., Cusano, C., Piccoli, F., Schettini, R.: Learning parametric functions for color image enhancement. In: International Workshop on Computational Color Imaging (2019)
4. Borji, A.: Saliency prediction in the deep learning era: successes and limitations. TPAMI (2019)
5. Bylinskii, Z., et al.: Learning visual importance for graphic designs and data visualizations. In: UIST (2017)
6. Bylinskii, Z., Recasens, A., Borji, A., Oliva, A., Torralba, A., Durand, F.: Where should saliency models look next? In: ECCV (2016)
7. Chandakkar, P.S., Li, B.: A structured approach to predicting image enhancement parameters. In: WACV (2016)
8. Chen, S.E., Williams, L.: View interpolation for image synthesis. In: SIGGRAPH (1993)
9. Chen, Y.C., Chang, K.J., Tsai, Y.H., Wang, Y.C.F., Chiu, W.C.: Guide your eyes: learning image manipulation under saliency guidance. In: BMVC (2019)
10. Cornia, M., Baraldi, L., Cucchiara, R.: Show, control and tell: a framework for generating controllable and grounded captions. In: CVPR (2019)
11. Fosco, C., et al.: How much time do you have? modeling multi-duration saliency. In: CVPR (2020)
12. Fried, O., Shechtman, E., Goldman, D.B., Finkelstein, A.: Finding distractors in images. In: CVPR (2015)
13. Gatys, L.A., Ecker, A.S., Bethge, M.: Image style transfer using convolutional neural networks. In: CVPR (2016)
14. Gatys, L.A., Kümmerer, M., Wallis, T.S., Bethge, M.: Guiding human gaze with convolutional neural networks. arXiv preprint arXiv:1712.06492 (2017)
15. Goodfellow, I., et al.: Generative adversarial nets. In: NeurIPS (2014)
16. Hagiwara, A., Sugimoto, A., Kawamoto, K.: Saliency-based image editing for guiding visual attention. In: Proceedings of the 1st International Workshop on Pervasive Eye Tracking & Mobile Eye-based Interaction (2011)
17. Hertzmann, A., Jacobs, C.E., Oliver, N., Curless, B., Salesin, D.H.: Image analogies. In: SIGGRAPH (2001)
18. Hu, Y., He, H., Xu, C., Wang, B., Lin, S.: Exposure: a white-box photo post-processing framework. In: SIGGRAPH (2018)
19. Huang, X., Belongie, S.: Arbitrary style transfer in real-time with adaptive instance normalization. In: ICCV (2017)
20. Jolicoeur-Martineau, A.: The relativistic discriminator: a key element missing from standard GAN. In: ICLR (2019)
21. Karras, T., Laine, S., Aila, T.: A style-based generator architecture for generative adversarial networks. In: CVPR (2019)
22. Kaufman, L., Lischinski, D., Werman, M.: Content-aware automatic photo enhancement. In: Computer Graphics Forum (2012)
23. Kolkin, N.I., Shakhnarovich, G., Shechtman, E.: Training deep networks to be spatially sensitive. In: ICCV (2017)

24. Koutras, P., Maragos, P.: SUSiNEt: see, understand and summarize it. In: CVPR Workshops (2019)
25. Li, N., Zhao, X., Yang, Y., Zou, X.: Objects classification by learning-based visual saliency model and convolutional neural network. Comput. Intell. Neurosci. **2016**, 1–12 (2016)
26. Lin, T.Y., et al.: Microsoft COCO: common objects in context. In: ECCV (2014)
27. Margolin, R., Zelnik-Manor, L., Tal, A.: How to evaluate foreground maps? In: CVPR (2014)
28. Mateescu, V.A., Bajić, I.V.: Attention retargeting by color manipulation in images. In: Proceedings of the 1st International Workshop on Perception Inspired Video Processing (2014)
29. Mathe, S., Sminchisescu, C.: Dynamic eye movement datasets and learnt saliency models for visual action recognition. In: ECCV (2012)
30. Mechrez, R., Shechtman, E., Zelnik-Manor, L.: Saliency driven image manipulation. Mach. Vis. Appl. **30**(2), 189–202 (2019). https://doi.org/10.1007/s00138-018-01000-w
31. Mejjati, Y.A., Richardt, C., Tompkin, J., Cosker, D., Kim, K.I.: Unsupervised attention-guided image-to-image translation. In: NeurIPS (2018)
32. Miyato, T., Kataoka, T., Koyama, M., Yoshida, Y.: Spectral normalization for generative adversarial networks. In: ICLR (2018)
33. Moosmann, F., Larlus, D., Jurie, F.: Learning saliency maps for object categorization. In: ECCV (2006)
34. Newman, A., et al.: TurkEyes: a web-based toolbox for crowdsourcing attention data. In: ACM CHI Conference on Human Factors in Computing Systems (2020)
35. Park, T., Liu, M.Y., Wang, T.C., Zhu, J.Y.: GauGAN: semantic image synthesis with spatially adaptive normalization. In: SIGGRAPH (2019)
36. Perazzi, F., Pont-Tuset, J., McWilliams, B., Van Gool, L., Gross, M., Sorkine-Hornung, A.: A benchmark dataset and evaluation methodology for video object segmentation. In: CVPR (2016)
37. Pérez, P., Gangnet, M., Blake, A.: Poisson image editing. TOG (2003)
38. Shen, C., Zhao, Q.: Webpage saliency. In: ECCV (2014)
39. Su, S.L., Durand, F., Agrawala, M.: De-emphasis of distracting image regions using texture power maps. In: ICCV Workshops (2005)
40. Tsai, Y.H., Shen, X., Lin, Z., Sunkavalli, K., Lu, X., Yang, M.H.: Deep image harmonization. In: CVPR (2017)
41. Wang, W., Shen, J., Cheng, M.M., Shao, L.: An iterative and cooperative top-down and bottom-up inference network for salient object detection. In: CVPR (2019)
42. Wang, W., Zhao, S., Shen, J., Hoi, S.C.H., Borji, A.: Salient object detection with pyramid attention and salient edges. In: CVPR (2019)
43. Wong, L.K., Low, K.L.: Saliency retargeting: an approach to enhance image aesthetics. In: WACV-Workshop (2011)
44. Yan, Z., Zhang, H., Wang, B., Paris, S., Yu, Y.: Automatic photo adjustment using deep neural networks. TOG **35**, 1–15 (2016)
45. Zhang, H., Goodfellow, I., Metaxas, D., Odena, A.: Self-attention generative adversarial networks. In: ICML (2019)
46. Zhang, R., Isola, P., Efros, A.A., Shechtman, E., Wang, O.: The unreasonable effectiveness of deep features as a perceptual metric. In: CVPR (2018)
47. Zheng, Q., Jiao, J., Cao, Y., Lau, R.W.: Task-driven webpage saliency. In: ECCV (2018)
48. Zhou, L., Zhang, Y., Jiang, Y., Zhang, T., Fan, W.: Re-caption: saliency-enhanced image captioning through two-phase learning. IEEE Trans. Image Process. (2020)

49. Zhu, J.Y., Park, T., Isola, P., Efros, A.A.: Unpaired image-to-image translation using cycle-consistent adversarial networks. In: ICCV (2017)
50. Zünd, F., Pritch, Y., Sorkine-Hornung, A., Mangold, S., Gross, T.: Content-aware compression using saliency-driven image retargeting. In: ICIP (2013)

Variational Diffusion Autoencoders with Random Walk Sampling

Henry Li[1(✉)], Ofir Lindenbaum[1], Xiuyuan Cheng[2], and Alexander Cloninger[3]

[1] Applied Mathematics, Yale University, New Haven, CT 06520, USA
{henry.li,ofir.lindenbaum}@yale.edu
[2] Department of Mathematics, Duke University, Durham, NC 27708, USA
xiuyuan.cheng@duke.edu
[3] Department of Mathematics and Halicioğlu Data Science Institute,
University of California San Diego, La Jolla, CA 92093, USA
acloninger@ucsd.edu

Abstract. Variational autoencoders (VAEs) and generative adversarial networks (GANs) enjoy an intuitive connection to manifold learning: in training the decoder/generator is optimized to approximate a homeomorphism between the data distribution and the sampling space. This is a construction that strives to define the data manifold. A major obstacle to VAEs and GANs, however, is choosing a suitable prior that matches the data topology. Well-known consequences of poorly picked priors are posterior and mode collapse. To our knowledge, no existing method sidesteps this user choice. Conversely, *diffusion maps* automatically infer the data topology and enjoy a rigorous connection to manifold learning, but do not scale easily or provide the inverse homeomorphism (i.e. decoder/generator). We propose a method (https://github.com/lihenryhfl/vdae) that combines these approaches into a generative model that inherits the asymptotic guarantees of *diffusion maps* while preserving the scalability of deep models. We prove approximation theoretic results for the dimension dependence of our proposed method. Finally, we demonstrate the effectiveness of our method with various real and synthetic datasets.

Keywords: Deep learning · Variational inference · Manifold learning · Image and video synthesis · Generative models · Unsupervised learning

1 Introduction

Generative models such as variational autoencoders (VAEs, [19]) and generative adversarial networks (GANs, [10]) have made it possible to sample

H. Li and O. Lindenbaum—Equal contribution.

Electronic supplementary material The online version of this chapter (https://doi.org/10.1007/978-3-030-58592-1_22) contains supplementary material, which is available to authorized users.

© Springer Nature Switzerland AG 2020
A. Vedaldi et al. (Eds.): ECCV 2020, LNCS 12368, pp. 362–378, 2020.
https://doi.org/10.1007/978-3-030-58592-1_22

remarkably realistic points from complex high dimensional distributions at low computational cost. While the theoretical framework behind the two methods are different—one is derived from variational inference and the other from game theory—they both involve learning smooth mappings from a user-defined prior $p(z)$ to the data $p(x)$.

When $p(z)$ is supported on a Euclidean space (e.g. $p(z)$ is Gaussian or uniform) and the $p(x)$ is supported on a manifold (i.e. the Manifold Hypothesis, see [8, 30]), VAEs and GANs become manifold learning methods, as manifolds themselves are defined as sets that are locally homeomorphic to Euclidean space. Thus the learning of such homeomorphisms may shed light on the success of VAEs and GANs in modeling complex distributions.

This connection to manifold learning also offers a reason why these generative models fail—when they do fail. Known as *posterior collapse* in VAEs [1, 14, 33, 48] and *mode collapse* in GANs [11], both describe cases where the learned mapping collapses large parts of the input to a single point in the output. This violates the bijective requirement of a homeomorphism. It also results in degenerate latent spaces and poor generative performance.

A major cause of such failings is when the geometries of the prior and target data do not agree. We explore this issue of *prior mismatch* and previous treatments of it in Sect. 3. Given their connection to manifolds, it is natural to draw from classical approaches in manifold learning to improve deep generative models. One of the most principled methods is kernel-based manifold learning [4, 36, 38]. This involves embedding data drawn from a manifold $X \subset \mathcal{M}_X$ into a space spanned by the leading eigenfunctions of a kernel on $\mathcal{M}_X$. We focus specifically on *diffusion maps*, where [6] show that normalizations of the kernel define a diffusion process that has a uniform stationary distribution over the data manifold. Therefore, drawing from this stationary distribution samples uniformly from the data manifold. This property was used in [24] to smoothly interpolate between missing parts of the manifold. However, despite its strong theoretical guarantees, *diffusion maps* are poorly equipped for large scale generative modeling as they do not scale well with dataset size. Moreover, acquiring the inverse mapping from the embedding space—a crucial component of a generative model—is traditionally a very expensive procedure [5, 21, 28].

In this paper we address issues in variational inference and manifold learning by combining ideas from both. The theory in manifold learning allows us to recognize and correct *prior mismatch*, whereas variational inference provides a method to construct a generative model, which also offers an efficient approximation to the inverse *diffusion map*.

Our Contributions: 1) We introduce the locally bi-Lipschitz property, a necessary condition of a homeomorphism, for measuring the stability of a mapping between latent and data distributions. **2)** We introduce variational diffusion autoencoders (VDAEs), a class of variational autoencoders that, instead of directly reconstructing the input, have an encoder-decoder that approximates one discretized time-step of the diffusion process on the data manifold (with respect to a user defined kernel k). **3)** We prove approximation theoretic bounds for deep neural networks learning such diffusion processes, and show that these networks define random walks with certain desirable properties, including well-defined transition and stationary distributions. **4)** Finally, we demonstrate the utility of the VDAE framework on a set of real and synthetic datasets, and show that they have superior performance and satisfy the locally bi-Lipschitz property (Fig. 1).

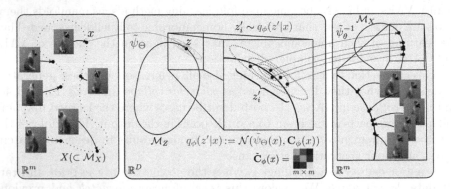

Fig. 1. A diagram depicting one step of the diffusion process modeled by the variational diffusion autoencoder (VDAE). The *diffusion* and *inverse diffusion maps* ψ, ψ^{-1}, as well as the covariance $\mathbf{C}$ of the random walk on $\mathcal{M}_Z$, are all approximated by neural networks. Images on the leftmost panel are actually generated by our method.

2   Background

Variational inference (VI, [18,46]) combines Bayesian statistics and latent variable models to approximate some probability density $p(x)$. VI exploits a latent variable structure in the assumed data generation process, that the observations $x \sim p(x)$ are conditionally distributed given unobserved latent variables z. By modeling the conditional distribution, then marginalizing over z, as in

$$p_\theta(x) = \int_z p_\theta(x|z)p(z)dz, \tag{1}$$

we obtain the model evidence, or likelihood that x could have been drawn from $p_\theta(x)$. Maximizing the model evidence (Eq. 1) leads to an algorithm for finding likely approximations of $p(x)$. The cost of computing this integral scales exponentially with the dimension of z and thus becomes intractable with high latent dimensions. Therefore we replace the model evidence (Eq. 1) with the evidence lower bound (ELBO):

$$\log p_\theta(x) \geq -D_{KL}(q(z|x)||p(z)) + \mathbb{E}_{z \sim q(z|x)}[\log p_\theta(x|z)], \tag{2}$$

where $q(z|x)$ is usually an approximation of $p_\theta(z|x)$. Maximizing the ELBO is sped up by taking stochastic gradients [16], and further accelerated by learning a global function approximator q_ϕ in an autoencoding structure [19].

Diffusion maps [6] refer to a class of kernel methods that perform non-linear dimensionality reduction on a set of observations $X \subseteq \mathcal{M}_X$, where $\mathcal{M}_X$ is the assumed data manifold equipped with measure μ. Let $x, y \in X$; given a symmetric and non-negative kernel k, *diffusion maps* involve analyzing the induced random walk on the graph of X, where the transition probabilities $P(y|x)$ are captured by the probability kernel $p(x,y) = k(x,y)/d(x)$, where $d(x) = \int_X k(x,y)d\mu(y)$ is the weighted degree of x. The diffusion map itself is defined as $\psi_D(x) := [\lambda_1 f_1(x), \lambda_2 f_2(x), ..., \lambda_D f_D(x)]$, where $\{f_i\}_{1 \leq i \leq D}$ and $\{\lambda_i\}_{1 \leq i \leq D}$ are the first D eigenfunctions and eigenvalues of p. An important construction in *diffusion maps* is the *diffusion distance*:

$$D(x,y)^2 = \int (p(x,u) - p(y,u))^2 \frac{d\mu(u)}{\pi(u)}, \tag{3}$$

where $\pi(u) = d(u)/\sum_{z \in X} d(z)$ is the stationary distribution of u. Intuitively, $D(x,y)$ measures the difference between the diffusion processes emanating from x and y. A key property of ψ_D is that it embeds the data $X \in \mathbb{R}^m$ into the Euclidean space $\mathbb{R}^D$ so that the diffusion distance is approximated by Euclidean distance (up to relative accuracy $\frac{\lambda_D}{\lambda_1}$). Therefore, the arbitrarily complex random walk induced by k on $\mathcal{M}_X$ becomes an isotropic Gaussian random walk on $\psi(\mathcal{M}_X)$.

SpectralNet [40] is a neural network approximation of the *diffusion map* ψ_D that enjoys a major computational speedup. The eigenfunctions $f_1, f_2, \ldots, f_D$ that compose ψ_D are learned by optimizing a custom loss function that stochastically maximizes the Rayleigh quotient for each f_i while enforcing the orthogonality of all $f_i \in \{f_n\}_{n=1}^D$ via a custom orthogonalization layer. As a result, the training and computation of ψ is linear in dataset and model size (as opposed to $O(n^3)$). We will use this algorithm to obtain our diffusion embedding prior.

Locally Bi-Lipschitz Coordinates by Kernel Eigenfunctions. The construction of local coordinates of Riemannian manifolds $\mathcal{M}_X$ by eigenfunctions of the diffusion kernel is analyzed in [17]. They establish, for all $x \in \mathcal{M}_X$, the existence of some neighborhood $U(x)$ and d spectral coordinates given $U(x)$ that define a bi-Lipschitz mapping from $U(x)$ to $\mathbb{R}^d$. With a smooth compact Riemannian manifold, we can let $U(x) = B(x, \delta r_{\text{in}})$, where δ is some constant and the *inradius* r_{in} is the radius of the largest ball around x still contained in $\mathcal{M}_X$. Note that δ is uniform for all x, whereas the indices of the d spectral coordinates as well as the local bi-Lipschitz constants may depend on x and are order $O(r_{\text{in}}^{-1})$. For completeness we give a simplified statement of the [17] result in the Appendix.

Using the compactness of the manifold, one can always cover the manifold with m many neighborhoods (geodesic balls) on which the bi-Lipschitz property in [17] holds. As a result, there are a total of D spectral coordinates, $D \leq md$ (in practice D is much smaller than md, since the selected spectral coordinates in the proof of [17] tend to be low-frequency ones, and thus the selection on different neighborhoods tend to overlap), such that on each of the m neighborhoods, there exists a subset of d spectral coordinates out of the D ones which are bi-Lipschitz on the neighborhood. We observe empirically that the bi-Lipschitz constants can be bounded uniformly from below and above (see Sect. 6.4) (Fig. 2).

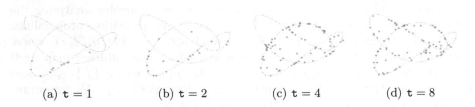

(a) t = 1 (b) t = 2 (c) t = 4 (d) t = 8

Fig. 2. An example of the diffusion random walk simulated by our method on a 3D loop dataset. t is the number of steps taken in the random walk.

3   Motivation and Related Work

In this section we justify the key idea of our method: diagnosing and correcting *prior mismatch*, a failure case of VAE and GAN training when $p(z)$ and $p(x)$ are not topologically isomorphic. Intuitively, we would like the latent distribution to have three nice properties: (1) *realizability*, that every point in the data distribution can be realized as a point in the latent distribution; (2) *validity*, that every point in the latent distribution maps to a unique valid point in the data distribution (even if it is not in the training set); and (3) *smoothness*, that points in the latent distribution vary in the intrinsic coordinate system in some smooth and coherent way.

These properties are precisely those enjoyed by a latent distribution that is homeomorphic to the data distribution. *Validity* implies injectivity, *realizability* implies surjectivity, *smoothness* implies continuity; and a mapping between topological spaces that is injective, surjective, and continuous is a homeomorphism. Therefore, studying algorithms that encourage approximations of homeomorphisms is of fundamental interest.

Though the Gaussian distribution for $p(z)$ is mathematically elegant and computationally expedient, there are many datasets for which it is ill-suited. Spherical distributions are known to be superior for modeling directional data [9,26], which can be found in fields as diverse as bioinformatics [13], geology [32], materials science [20], natural image processing [3], and many preprocessed datasets[1]. For data supported on more complex manifolds, the literature is sparse, even though it is well-known that data often lie on such manifolds [8,30]. In general, any manifold-supported distribution that is not globally homeomorphic to Euclidean space will not satisfy conditions (1–3) above.

Previous research on alleviating *prior mismatch* exists in various forms, and has focused on increasing the family of tractable latent distributions for generative models. [7,47] consider VAEs with the von-Mises Fisher distribution,

---

[1] Any dataset where the data points have been normalized to be unit length becomes a subset of a hypersphere.

a geometrically hyperspherical prior, and [43] consider mixtures of priors. [35] propose a method that, like our method, also samples from the prior via a diffusion process over a manifold. However, their method requires very explicit knowledge of the manifold (including its projection map, scalar curvature constant, and volume), and give up an exact estimation of the KL divergence. [34] avoids mode collapse by lower bounding the KL-divergence term away from zero to avoid overfitting. Similarly, [25] focuses on avoiding mode collapse by using class-conditional generative models, however it requires label supervision and does not provide any guarantees that the latent space generated is homeomorphic to the data space. Finally, [15] propose the re-scaling of various terms in the ELBO to augment the latent space—often to surprisingly great effect on latent feature discovery—but are restricted to the case where the latent features are independent.

While these methods expand the repertoire of feasible priors, they all require explicit user knowledge of the data topology. On the other hand, our method allows the user to be agnostic to this choice of topology; they only need to specify an affinity kernel k for local pairwise similarities. We achieve this by employing ideas from both *diffusion maps* and variational inference, resulting in a fully data-driven approach to latent distribution selection in deep generative models.

4   Method

In this section we propose the variational diffusion autoencoder (VDAE), a class of generative models built from ideas in *variational inference* and *diffusion maps*. Given the data manifold $\mathcal{M}_X$, observations $X \subset \mathcal{M}_X$, and a kernel k, VDAEs model the geometry of X by approximating a random walk over the latent diffusion manifold $\mathcal{M}_Z := \psi(\mathcal{M}_X)$. The model is trained by maximizing the *local evidence*: the evidence (i.e. log-likelihood) of each point given its random walk neighborhood. Points are generated from the trained model by sampling from π, the stationary distribution of the resulting random walk.

Starting from some point $x \in X$, we can think of one step of the walk as the composition of three functions: **1)** the approximate diffusion map $\widetilde{\psi}_\omega :$ $\mathcal{M}_X \to \mathcal{M}_Z$ parameterized by ω, **2)** the stochastic function that samples from the diffusion random walk $z' \sim q_\phi(z'|x) = \mathcal{N}(\widetilde{\psi}_\omega(x), \widetilde{\mathbf{C}}_\phi(x))$ on $\mathcal{M}_Z$, and **3)** the approximate inverse diffusion map $\widetilde{\psi}_\theta^{-1} : \mathcal{M}_Z \to \mathcal{M}_X$ that generates $x' \sim$ $p(x'|z') = \mathcal{N}(\widetilde{\psi}_\theta^{-1}(z'), cI)$ where c is a fixed, user-defined hyperparameter usually set to 1.

Note that Euclidean distances in $\mathcal{M}_Z$ approximate single-step random walk distances on $\mathcal{M}_X$ due to properties of the diffusion map embedding (see Sect. 2 and [6]). These properties are inherited by our method via the SpectralNet algorithm, since $\widetilde{\psi}_\omega|_{\mathcal{M}_X} : \mathcal{M}_X \to \mathcal{M}_Z$ satisfies the *locally bi-Lipschitz property*. This bi-Lipschitz property also reduces the need for regularization, and leads to guarantees of the ability of the VDAE to avoid posterior and mode collapse (see Sect. 5).

In short, to model a diffusion random walk over $\mathcal{M}_Z$, we must learn the functions $\widetilde{\psi}_\omega, \widetilde{\psi}_\theta^{-1}$, and $\widetilde{\mathbf{C}}_\phi$ that approximate the diffusion map, the inverse diffusion map, and the covariance of the random walk on $\mathcal{M}_Z$, at all points $z \in \mathcal{M}_Z$. SpectralNet gives us $\widetilde{\psi}_\omega$. To learn $\widetilde{\psi}_\theta^{-1}$ and $\widetilde{\mathbf{C}}_\phi$, we use variational inference.

4.1   The Lower Bound

Formally, let us define $U_x := B_d(x, \delta) \cap \mathcal{M}_X$, where $B_d(x, \delta)$ is the δ-ball around x with respect to $d(\cdot, \cdot)$, the diffusion distance on $\mathcal{M}_Z$. For each $x \in X$ we define the *local evidence* of x as

$$\mathbb{E}_{x' \sim p(x'|x)|_{U_x}} \log p_\theta(x'|x), \tag{4}$$

where $p(x'|x)|_{U_x}$ restricts $p(x'|x)$ to U_x. This gives the *local evidence lower bound*

$$\log p_\theta(x'|x) \geq \underbrace{-D_{KL}(q_\phi(z'|x)||p_\theta(z'|x))}_{\text{divergence from true diffusion probabilities}} + \underbrace{\mathbb{E}_{z' \sim q_\phi(z'|x)} \log p_\theta(x'|z')}_{\text{neighborhood reconstruction error}}, \tag{5}$$

which produces the empirical loss function $\tilde{\mathcal{L}}_{\text{VDAE}} = -D_{KL}(q_\phi(z'|x)||p_\theta(z'|x)) + \log p_\theta(x'|z_i')$, where $z_i' = g_{\phi,\Theta}(x, \epsilon_i)$, $\epsilon_i \sim \mathcal{N}(0, I)$. The function $g_{\phi,\Theta}$ is deterministic and differentiable, depending on $\widetilde{\psi}_\omega$ and $\widetilde{\mathbf{C}}_\phi$, that generates q_ϕ by the reparameterization trick[2].

4.2   The Sampling Procedure

Composing $q_\phi(z'|x)(\approx p_\theta(z'|x))$ with $p_\theta(x'|z')$ gives us an approximation of $p_\theta(x'|x)$. Then the simple, parallelizable, and fast random walk based sampling procedure naturally arises: initialize with an arbitrary point on the manifold $x_0 \in \mathcal{M}_X$ (e.g. from the dataset X), pick suitably large N, and for $n = 1, \ldots, N$ draw $x_n \sim p(x|x_{n-1})$. See Sect. 6.2 for examples of points drawn from this procedure.

4.3   A Practical Implementation

We now introduce a practical implementation, considering the case where $\widetilde{\psi}_\omega(x)$, $q_\phi(z'|x)$ and $p_\theta(x'|z')$ are neural network functions.

The **neighborhood reconstruction error** $\mathbb{E}_{z' \sim q_\phi(z'|x)} \log p_\theta(x'|z')$ should be differentiated from the *self* reconstruction error in VAEs, i.e. reconstructing x' vs x. Since $q_\phi(z'|x)$ models the neighborhood of $\widetilde{\psi}_\omega(x)$, we may sample q_ϕ to obtain z' (the neighbor of x in the latent space). Assuming ψ^{-1} exists, we have $x' \sim p_\theta(x'|x)(\approx \widetilde{\psi}_\theta^{-1}(q_\phi(z'|x)))$. To make this practical, we can approximate x' by finding the closest data point to x' in random walk distance (due to the

---

[2] Though q depends on ϕ and ω, we will use $q_\phi := q_{\phi,\omega}$ to be consistent with existing VAE notation and to indicate that ω is not learned by variational inference.

aforementioned advantages of the latent space). In other words, we approximate empirically by

$$x' \approx \arg\min_{y \in A} |\widetilde{\psi}_\omega(y) - z'|_d^2, \quad z' \sim q_\phi(z'|x), \tag{6}$$

where $A \subseteq X$ is the training batch.

On the other hand, the **divergence of random walk distributions**, $-D_{KL}(q_\phi(z'|x)\|p_\theta(z'|x))$, can be modeled simply as the divergence of two Gaussian kernels defined on $\mathcal{M}_Z$. Though $p_\theta(z'|x)$ is intractable, the diffusion map ψ gives us the diffusion embedding Z, which is an approximation of the true distribution of $p_\theta(z'|x)$ in a neighborhood around $z = \psi(x)$. We estimate the first and second moments of this distribution in $\mathbb{R}^D$ by computing the local Mahalanobis distance of points in the neighborhood. Then, by minimizing the KL divergence between $q_\phi(z'|x)$ and the one implied by this Mahalanobis distance, we obtain the loss:

$$- D_{KL}(q_\phi(z'|x)\|p_\theta(z'|x)) = -\log\frac{|\alpha\Sigma_*|}{|\widetilde{\mathbf{C}}_\phi|} + d - \text{tr}\{(\alpha\Sigma_*)^{-1}\widetilde{\mathbf{C}}_\phi\}, \tag{7}$$

where $\widetilde{\mathbf{C}}_\phi(x)$ is a neural network function, $\Sigma_*(x) = \text{Cov}(B_d(\psi(x), \delta) \cap Z)$ is the covariance of the points in a neighborhood of $z = \psi(x) \in Z$, and α is a scaling parameter controlling the random walk step size. Note that the covariance $\widetilde{\mathbf{C}}_\phi(x)$ does not have to be diagonal, and in fact is most likely not. Combining Eqs. 6 and 7 we obtain Algorithm 1.

Since we use neural networks to approximate the random walk induced by the composition of $q_\phi(z'|x)$ and $p_\theta(x'|z')$, the generation procedure is highly parallelizable. This leads naturally to a sampling procedure for this random walk (Algorithm 2). We observe that the random walk enjoys rapid mixing properties—it only takes several iterations of the random walk to sample from all of $\mathcal{M}_Z{}^3$.

Finally, we describe a practical method for computing the local bi-Lipschitz property. (In Sect. 6.4 we then perform comparisons with this method.) Let Z and X be the latent and generated data distributions of our model f (i.e. $f : Z \rightarrow X$). We define, for each $z \in Z$ and $k \in \mathbb{N}$, the function $\texttt{bilip}_k(z)$:

$$\texttt{bilip}_k(z) = \min\{K : \frac{1}{K} \le \frac{d_x(f(z), f(z'))}{d_Z(z, z')} \le K\},$$

for all $z' \in U_{z,k} \cap Z$, where d_X and d_Z are metrics on X and Z, and $U_{z,k}$ is the k-nearest neighborhood of z. Intuitively, increasing values of K characterize an increasing tendency to *stretc.h* or *compress* regions of space. By analyzing statistics of the local bi-Lipschitz measure at all points in a latent space Z, we gain insight into how well-behaved a mapping f is.

---

[3] For all experiments in Sect. 6, the number of steps required to draw from π is less than 10.

$\omega, \phi, \theta \leftarrow$ Initialize parameters
Obtain parameters ω for the approximate diffusion map $\widetilde{\psi}_\omega$ via SpectralNet [40]
while not converged **do**
    $A \leftarrow$ Random batch from X
    for $x \in A$ **do**
        $z \leftarrow p_\phi(z'|\widetilde{\psi}_\omega(x))$ {Random walk step}
        $x' \leftarrow \arg\min_{y \in A \setminus \{x\}} |\widetilde{\psi}_\omega(y) - z'|_d^2$ {Find batch neighbors}
        $g \leftarrow g + \frac{1}{|A|}\nabla_{\phi,\theta}\log p_\theta(x'|x)$ {Compute Eq. (5)}
    end for
    Update ϕ, θ using g
end while

Algorithm 1: VDAE training

$X_0 \leftarrow$ Initialize with points $X_0 \subset X$; $t \leftarrow 0$
while $p(X_0) \not\approx \pi$ **do**
    for $x_t \in X$ **do**
        $z_{t+1} \sim p_\phi(z'|\widetilde{\psi}_\omega(x_t))$ {Random walk step}
        $x_{t+1} \sim p_\theta(x|z_{t+1})$ {Map back to input space}
    end for
    $t \leftarrow t + 1$
end while

Algorithm 2: VDAE sampling

4.4 Comparison to Variational Inference (VI)

Traditional VI involves maximizing the joint log-evidence of each data point x_i in a given dataset via the ELBO (see Sect. 2). Our method differs in both the training and evaluation steps.

In training, our setup is the same as above, except our likelihood is a conditional likelihood $p(x'|x)|_{U_x}$, where x' is in the diffusion neighborhood of x. Thus we maximize the local log-evidence of each data point $\mathbb{E}_{x' \sim p(x'|x_i)}\log p_\theta(x'|x_i)$, which can be lower bounded by Eq. (5). Thus our prior is $p(z'|x)$ and our posterior is $p(z'|x', x) = p(x', z'|x)/p(x'|x)$, and we train an approximate posterior $q_\phi(z'|x)$ and a recognition model $p_\theta(x'|z')$.

In evaluation, we draw from the stationary distribution $p(z')$ of the diffusion random walk on the latent manifold $\mathcal{M}_z = \psi(\mathcal{M}_x)$. We then leverage the latent variable structure of our model to draw a sample $x = p_\theta(x'|z')p(z')$, where $p_\theta(x'|x_i)$ is the recognition model.

5 Theory

In this section, we show that the desired diffusion and inverse diffusion maps $\psi : \mathcal{M}_X \to \mathcal{M}_Z$ and $\psi^{-1} : \mathcal{M}_Z \to \mathcal{M}_X$ can be approximated by neural networks, where the network complexity is bounded by quantities related to the intrinsic geometry of the manifold.

The capacity of the encoder $\tilde{\psi}$ has already been considered in [39] and [29]. Thus we focus on the capacity of the decoder $\tilde{\psi}^{-1}$. The following theorem is proved in Appendix A.3, based on the result in [17].

Theorem 1. *Let $\mathcal{M}_X \subset \mathbb{R}^m$ be a smooth d-dimensional manifold, $\psi(\mathcal{M}_X) \subset \mathbb{R}^D$ be the diffusion map for $D \geq d$ large enough to have a subset of coordinates that are locally bi-Lipschitz. Let $\mathbf{X} = [X_1, ..., X_m]$ be the set of all m extrinsic coordinates of the manifold. Then there exists a sparsely-connected ReLU network f_N, with $4DC_{\mathcal{M}_X}$ nodes in the first layer, $8dmN$ nodes in the second layer, and $2mN$ nodes in the third layer, and m nodes in the output layer, such that*

$$\|\mathbf{X}(\psi(x)) - f_N(\psi(x))\|_{L^2(\psi(\mathcal{M}_X))} \leq \sqrt{m}C_\psi/\sqrt{N}, \tag{8}$$

where the norm is interpreted as $\|F\|^2_{L^2(\psi(\mathcal{M}))} := \int \|F(\psi(x))\|^2_2 d\psi(x)$. Here C_ψ depends on how sparsely $X(\psi(x))|_{U_i}$ can be represented in terms of the ReLU wavelet frame on each neighborhood U_i, and $C_{\mathcal{M}_X}$ on the curvature and dimension of the manifold $\mathcal{M}_X$.

Theorem 1 guarantees the existence and size of a decoder network for learning a manifold. Together with the main theorem in [39], we obtain guarantees for both the encoder and decoder on manifold-valued data. The proof is built on two properties of ReLU neural networks: 1) their ability to split curved domains into small, almost Euclidean patches, 2) their ability to build differences of bump functions on each patch, which allows one to borrow approximation results from the theory of wavelets on spaces of homogeneous type. The proof also crucially uses the bi-Lipschitz property of the diffusion embedding [17]. The key insight of Theorem 1 is that, because of the bi-Lipschitz property, the coordinates of the manifold in the ambient space $\mathbb{R}^m$ can be thought of as functions of the diffusion coordinates. We show that because each coordinate function X_i is Lipschitz, the ReLU wavelet coefficients of X_i are necessarily ℓ^1. This allows us to use the existing guarantees of [39] to complete the desired bound.

We also discuss the connections between the distribution at each point in diffusion map space, $q_\phi(z|x)$, and the result of this distribution after being decoded through the decoder network $f_N(z)$ for $z \sim q_\phi(z|X)$. Similar to [41], we characterize the covariance matrix $Cov(f_N(z)) := \mathbb{E}_{z \in q_\phi(z|x)}[f_N(z)f_N(z)^T]$. The following theorem is proved in Appendix A.3.

Theorem 2. *Let f_N be a neural network approximation to $\mathbf{X}$ as in Theorem 1, such that it approximates the extrinsic manifold coordinates. Let $C \in \mathbb{R}^{m \times m}$ be the covariance matrix $C = \mathbb{E}_{z \in q_\phi(z|x)}[f_N(z)f_N(z)^T]$. Let $q_\phi(z|x) \sim N(\psi(x), \Sigma)$ with small enough Σ that there exists a patch $U_{z_0} \subset \mathcal{M}$ around z_0 satisfying the bi-Lipschitz property of [17], and such that $Pr(z \sim q_\phi(z|x) \notin \psi(U_{z_0})) < \epsilon$. Then the number of eigenvalues of C greater than ϵ is at most d, and $C = J_{z_0}\Sigma J_{z_0}^T + O(\epsilon)$ where J_{z_0} is the $m \times D$ Jacobian matrix at z_0.*

Theorem 2 establishes the relationship between the covariance matrices used in the sampling procedure and their image under the decoder f_N to approximate

Fig. 3. We consider the rotating bulldog example. Images are drawn from the latent distribution and plotted in terms of the 2D latent space of each model. From left to right: VDAE, SVAE, β-VAE, WGAN.

ψ^{-1}. Similar to [41], we are able to sample according to a multivariate normal distribution in the latent space. Thus, the resulting cloud in the data space is distorted (to first order) by the local Jacobian of the map f_N. The key insight of Theorem 2 is from combining this idea with the observation of [17]: that ψ^{-1} depends locally only on d of the coordinates in the D dimensional latent space.

6   Experimental Results

In this section we explore various properties of the VDAE and compare it against several deep generative methods on a selection of real and synthetic datasets. Unless otherwise noted, all comparisons are against the Wasserstein GAN (WGAN), β-VAE, and hyperspherical VAE (SVAE). Each model is trained with the same architecture across all experiments (see Sect. A.6).

6.1   Video Generation with Rigid-Body Motion

We first consider the task of generating new frames from videos of rigid-body motion, and examine the latent spaces of videos with known topological structure to demonstrate the homeomorphic properties of the VDAE. We consider two examples, the rotating bulldog example [23] and the COIL-20 dataset. [31].

The rotating bulldog example consists of 200 frames of a color video (each frame is $100 \times 80 \times 3$) of a spinning figurine. The rotation of the bulldog and the fixed background create a data manifold that is topologically circular, corresponding to the single degree of variation (the rotation angle parameter) in the dataset. For all methods we consider a 2 dimensional latent space. In Fig. 3 we present 300 generated samples by displaying them on a scatter plot with coordinates corresponding to their latent dimensions z_1 and z_2. In the Appendix Table A.1, we evaluate the quality of the generated images using the Frechet inception distance (FID).

The COIL-20 data set consists of 360 images of five different rotating objects displayed against on a black background (each frame is $448 \times 416 \times 1$). This yields several low dimensional manifolds, one for each object, and results in a difficult data set for traditional generative models given its small size and the complex geometric structure. For all comparisons, we use 10 dimensional latent space. The resulting images are embedded with tSNE and plotted in

Fig. 4. From left to right, the first three scatterplots show examples of distributions reconstructed from a random walk on $\mathcal{M}_Z$ (via Algorithm 2) given a single seed point drawn from X. The next three are examples of a single burst drawn from $p_\theta(x|z)$. The distributions are a loop (a, d), sphere (b, e), and the Stanford bunny (c, f).

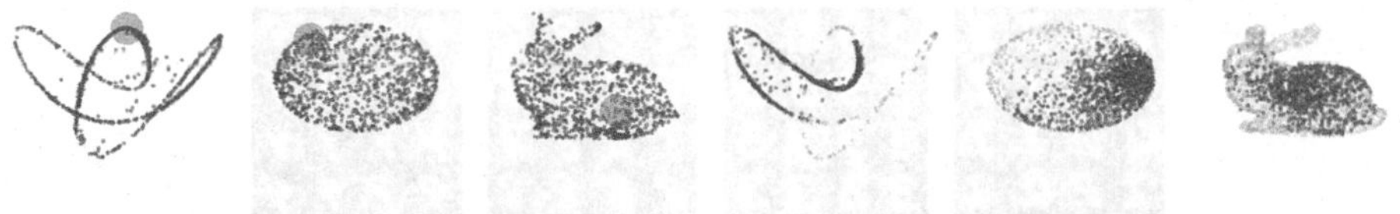

Fig. A.3. Note that, while other methods generate images that topologically mimic the fixed latent distribution of the model (e.g. $\mathcal{N}(0, I_d)$, $\text{Uniform}(0, 1)^d$), our method generates images that remain true to the actual topological structure of the dataset.

6.2   Data Generation from Uniformly Sampled Manifolds

In the next experiment, we visualize the results of the sampling procedure in Algorithm 2 on three synthetic manifolds. As discussed in Sect. 4.2, we randomly select an initial seed point, then recursively sample from $p_\theta(x'|x)$ to simulate a random walk on the manifold.

In Fig. 4(a–c) for three different manifolds, the location of the initial seed point is highlighted, then 20 steps of the random walk are taken, and the resulting generated points are displayed. The generated points remain on the manifold even after this large number of resampling iterations, and the distribution of sampled points converges to a uniform stationary distribution on the manifold. Moreover, we observe that this stationary distribution is reached quickly, within 5–10 iterations. In (d–f) of the same Fig. 4, we show $p_\theta(x'|x)$ by drawing a large number of points from a single-step random walk starting from the same seed point. As can be seen, a single step of $p_\theta(x'|x)$ covers a large part of the latent space.

6.3   Cluster Conditional Data Generation

In this section, we deal with the problem of generating samples from data with multiple clusters in an unsupervised fashion (i.e. no a priori knowledge of the cluster structure). Clustered data creates a problem for many generative models, as the topology of the latent space (i.e. normal distribution) differs from the topology of the data space with multiple clusters.

First we show that our method is capable of generating new points from a particular cluster given an input point from that cluster. This generation is done in an unsupervised fashion, which is a different setting from the approach of conditional VAEs [42] that require training labels. We demonstrate this property on MNIST [22] in Fig. 5, and show that the newly generated points after a short diffusion time remain in the same class as the seeded image.

Fig. 5. An example of cluster conditional sampling with our method, given a seed point (top left of each image grid). The VDAE is able to produce examples via the random walk that stay approximately within the cluster of the seed point, without any supervised knowledge of the cluster.

The problem of addressing differing topologies between the data and the latent space of a generative model has been acknowledged in recent works on rejection sampling [2,45]. Rejection sampling of neural networks consists of generating a large collection of samples using a standard GAN, and then designing a probabilistic algorithm to decide in a *post-hoc* fashion whether the points were truly in the support of the data distribution $p(x)$.

In the following experiment, we compare to a standard example in the literature for rejection sampling in generative models (see [2]). The data consists of nine bounded spherical densities with significant minimal separation, lying on a 5×5 grid. A GAN struggles to avoid generating points in the gaps between these densities, and thus requires the post-sampling rejection analysis described in [2]. Conversely, our model creates a latent space that separates each of these clusters into their own coordinates and generates only points that in the neighborhood of the support of $p(x)$. Figure 6 shows that this results in significantly fewer points generated in the gaps between clusters. Our VDAE architecture is described in Sect. A.6, GAN and DRS-GAN architectures are as described in [2].

6.4   Quantitative Comparisons of Generative Models

For this comparison, we consider seven datasets: three synthetic (circle, torus, Stanford bunny [44]) four involving natural images (rotating bulldog, Frey faces, MNIST, COIL-20). The β parameter in the β-VAE is optimized via a cross validation procedure. see Appendix for a complete description of the datasets. We report the mean and standard deviation of the Gromov-Wasserstein distance [27] and median bi-Lipschitz over 5 runs in Table 1. We further evaluate the results using kernel Maximum Mean Discrepancy [12], see Table A.2 in the Appendix.

By constraining our latent space to be the diffusion embedding of the data, our method finds a mapping that automatically enjoys the homeomorphic properties of an ideal mapping, and this is reflected in the low values of the local bi-Lipschitz constant. Conversely, other methods do not consider the topology of the data in the prior distribution. This is especially apparent in the β-VAE and SVAE, which must generate from the entirety of the input distribution X because they minimize a reconstruction loss. Interestingly, the mode collapse tendency of GANs alleviate the pathology of the bi-Lipschitz constant by allowing the GAN to focus on a subset of the distribution—but this comes at the cost

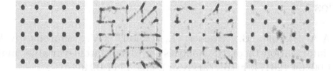

Fig. 6. Comparison of samples from our method against several others on a 5×5 Gaussian grid. Left-right are original data, GAN, DRS-GAN, and VDAE (our method). GAN and DRS-GAN samples taken from [2].

Table 1. Left: means and standard deviations of the Gromov-Wasserstein (G-W) distance between original and generated samples. Right: medians of the bi-Lipschitz measure.

G-W	WGAN	β-VAE	SVAE	VDAE	biLip	WGAN	β-VAE	SVAE	VDAE
Circle	14.9 (6.8)	46.1 (9.7)	7.9 (2.2)	**2.6 (1.3)**	Circle	4.6	3.7	3.6	**3.1**
Torus	6.4 (1.9)	11.7 (1.6)	23.4 (2.8)	**4.9 (0.5)**	Torus	**3.3**	7.9	9.5	4.8
Bunny	11.4 (3.9)	32.8 (5.9)	14.3 (5.5)	**2.9 (1.1)**	Bunny	5.6	34.4	35.6	**5.5**
Bulldog	117.3 (8.4)	61.3 (9.7)	53.9 (7.6)	**15.3 (1.7)**	Bulldog	17.4	7.6	12.9	**6.8**
Frey	18.1 (2.9)	19.8 (4.6)	13.4 (3.6)	**9.7 (3.3)**	Frey	37	33.3	39.4	**29.7**
MNIST	**3.6 (0.9)**	10.2 (3.3)	15.2 (4.9)	14.4 (3.5)	MNIST	1.9	**1.6**	6.7	8.4
COIL-20	16.5 (2.4)	23.8 (5.9)	32.1 (4.9)	**11.8 (2.1)**	COIL-20	4.7	3.8	8.4	**3.1**

of collapse to a few modes of the dataset. Our method is able to reconstruct the entirety of X while simultaneously maintaining a low local bi-Lipschitz constant.

7 Discussion

In this work, we have shown that VDAEs provide an intuitive, effective, and mathematically rigorous solution to *prior mismatch*, which is a common cause for posterior collapse in latent variable models. Unlike prior works, we do not require user specification of the prior—our method infers the prior geometry directly from the data, and we observe that it achieves state-of-the-art results on several real and synthetic datasets. Finally, our work points to several directions for future research: (1) can we leverage recent architectural advances to VAEs to further improve VDAE performance, and (2) can we leverage manifold learning techniques to improve latent representations in other methods?

Acknowledgements. This work was supported by NSF DMS grants 1819222 and 1818945. AC is also partially supported by NSF (DMS-2012266), and Russell Sage Foundation (grant 2196). XC is also partially supported by NSF (DMS-1820827), NIH (Grant R01GM131642), and the Alfred P. Sloan Foundation.

References

1. Alemi, A.A., Poole, B., Fischer, I., Dillon, J.V., Saurous, R.A., Murphy, K.: An information-theoretic analysis of deep latent-variable models. CoRR abs/1711.00464 (2017). http://arxiv.org/abs/1711.00464
2. Azadi, S., Olsson, C., Darrell, T., Goodfellow, I., Odena, A.: Discriminator rejection sampling. arXiv preprint arXiv:1810.06758 (2018)
3. Bahlmann, C.: Directional features in online handwriting recognition. Pattern Recogn. **39**(1), 115–125 (2006)
4. Belkin, M., Niyogi, P.: Laplacian eigenmaps and spectral techniques for embedding and clustering. In: Advances in Neural Information Processing Systems, pp. 585–591 (2002)
5. Cloninger, A., Czaja, W., Doster, T.: The pre-image problem for Laplacian eigenmaps utilizing l 1 regularization with applications to data fusion. Inverse Prob. **33**(7), 074006 (2017)
6. Coifman, R.R., Lafon, S.: Diffusion maps. Appl. Comput. Harmonic Anal. **21**(1), 5–30 (2006)
7. Davidson, T.R., Falorsi, L., De Cao, N., Kipf, T., Tomczak, J.M.: Hyperspherical variational auto-encoders. In: Uncertainty in Artificial Intelligence (UAI) (2018)
8. Fefferman, C., Mitter, S., Narayanan, H.: Testing the manifold hypothesis. J. Am. Math. Soc. **29**(4), 983–1049 (2016)
9. Fisher, N.I., Lewis, T., Embleton, B.J.: Statistical Analysis of Spherical Data. Cambridge University Press, Cambridge (1993)
10. Goodfellow, I., et al.: Generative adversarial nets. In: Advances in Neural Information Processing Systems (NIPS), vol. 27, pp. 2672–2680 (2014)
11. Goodfellow, I.J.: NIPS 2016 tutorial: generative adversarial networks. CoRR abs/1701.00160 (2017). http://arxiv.org/abs/1701.00160
12. Gretton, A., Borgwardt, K.M., Rasch, M.J., Schölkopf, B., Smola, A.: A Kernel two-sample test. J. Mach. Learn. Res. **13**, 723–773 (2012)
13. Hamelryck, T., Kent, J.T., Krogh, A.: Sampling realistic protein conformations using local structural bias. PLoS Comput. Biol. **2**(9), e131 (2006)
14. He, J., Spokoyny, D., Neubig, G., Berg-Kirkpatrick, T.: Lagging inference networks and posterior collapse in variational autoencoders. CoRR abs/1901.05534 (2019). http://arxiv.org/abs/1901.05534
15. Higgins, I., et al.: beta-VAE: learning basic visual concepts with a constrained variational framework. In: ICLR, vol. 2, no. 5, p. 6 (2017)
16. Hoffman, M.D., Blei, D.M., Wang, C., Paisley, J.: Stochastic variational inference. J. Mach. Learn. Res. **14**(1), 1303–1347 (2013)
17. Jones, P.W., Maggioni, M., Schul, R.: Manifold parametrizations by eigenfunctions of the Laplacian and heat kernels. Proc. Nat. Acad. Sci. **105**(6), 1803–1808 (2008)
18. Jordan, M.I., Ghahramani, Z., Jaakkola, T.S., Saul, L.K.: An introduction to variational methods for graphical models. Mach. Learn. **37**(2), 183–233 (1999)
19. Kingma, D.P., Welling, M.: Auto-encoding variational bayes. arXiv preprint arXiv:1312.6114 (2013)
20. Krieger Lassen, N., Juul Jensen, D., Conradsen, K.: On the statistical analysis of orientation data. Acta Crystallogr. A **50**(6), 741–748 (1994)
21. Kwok, J.Y., Tsang, I.H.: The pre-image problem in kernel methods. IEEE Trans. Neural Networks **15**(6), 1517–1525 (2004)
22. LeCun, Y., Bottou, L., Bengio, Y., Haffner, P.: Gradient-based learning applied to document recognition. Proc. IEEE **86**(11), 2278–2324 (1998)

23. Lederman, R.R., Talmon, R.: Learning the geometry of common latent variables using alternating-diffusion. Appl. Comput. Harmonic Anal. **44**(3), 509–536 (2018)
24. Lindenbaum, O., Stanley, J., Wolf, G., Krishnaswamy, S.: Geometry based data generation. In: Advances in Neural Information Processing Systems, pp. 1400–1411 (2018)
25. Mao, Q., Lee, H.Y., Tseng, H.Y., Ma, S., Yang, M.H.: Mode seeking generative adversarial networks for diverse image synthesis. In: Proceedings of the IEEE Conference on Computer Vision and Pattern Recognition, pp. 1429–1437 (2019)
26. Mardia, K.V.: Statistics of Directional Data. Academic Press, Cambridge (2014)
27. Mémoli, F.: Gromov-Wasserstein distances and the metric approach to object matching. Found. Comput. Math. **11**(4), 417–487 (2011). https://doi.org/10.1007/s10208-011-9093-5
28. Mika, S., Schölkopf, B., Smola, A.J., Müller, K.R., Scholz, M., Rätsch, G.: Kernel PCA and de-noising in feature spaces. In: Advances in Neural Information Processing Systems, pp. 536–542 (1999)
29. Mishne, G., Shaham, U., Cloninger, A., Cohen, I.: Diffusion nets. Appl. Comput. Harmonic Anal. **47**, 259–285 (2019)
30. Narayanan, H., Mitter, S.: Sample complexity of testing the manifold hypothesis. In: Advances in Neural Information Processing Systems, pp. 1786–1794 (2010)
31. Nene, S.A., Nayar, S.K., Murase, H., et al.: Columbia object image library (COIL-20) (1996)
32. Peel, D., Whiten, W.J., McLachlan, G.J.: Fitting mixtures of Kent distributions to aid in joint set identification. J. Am. Stat. Assoc. **96**(453), 56–63 (2001)
33. Razavi, A., van den Oord, A., Poole, B., Vinyals, O.: Preventing posterior collapse with delta-VAEs. CoRR abs/1901.03416 (2019). http://arxiv.org/abs/1901.03416
34. Razavi, A., Oord, A.v.d., Poole, B., Vinyals, O.: Preventing posterior collapse with delta-VAEs. arXiv preprint arXiv:1901.03416 (2019)
35. Rey, L.A.P., Menkovski, V., Portegies, J.W.: Diffusion variational autoencoders. CoRR abs/1901.08991 (2019). http://arxiv.org/abs/1901.08991
36. Roweis, S.T., Saul, L.K.: Nonlinear dimensionality reduction by locally linear embedding. Science **290**(5500), 2323–2326 (2000)
37. Sajjadi, M.S., Bachem, O., Lucic, M., Bousquet, O., Gelly, S.: Assessing generative models via precision and recall. In: Advances in Neural Information Processing Systems, pp. 5228–5237 (2018)
38. Schölkopf, B., Smola, A., Müller, K.R.: Nonlinear component analysis as a Kernel eigenvalue problem. Neural Comput. **10**(5), 1299–1319 (1998)
39. Shaham, U., Cloninger, A., Coifman, R.R.: Provable approximation properties for deep neural networks. Appl. Comput. Harmonic Anal. **44**(3), 537–557 (2018)
40. Shaham, U., Stanton, K., Li, H., Nadler, B., Basri, R., Kluger, Y.: SpectralNet: spectral clustering using deep neural networks. arXiv preprint arXiv:1801.01587 (2018)
41. Singer, A., Coifman, R.R.: Non-linear independent component analysis with diffusion maps. Appl. Comput. Harmonic Anal. **25**(2), 226–239 (2008)
42. Sohn, K., Lee, H., Yan, X.: Learning structured output representation using deep conditional generative models. In: Advances in Neural Information Processing Systems, pp. 3483–3491 (2015)
43. Tomczak, J.M., Welling, M.: VAE with a VampPrior. arXiv preprint arXiv:1705.07120 (2017)
44. Turk, G., Levoy, M.: Zippered polygon meshes from range images. In: Proceedings of the 21st Annual Conference on Computer Graphics and Interactive Techniques, pp. 311–318 (1994)

45. Turner, R., Hung, J., Saatci, Y., Yosinski, J.: Metropolis-Hastings generative adversarial networks. arXiv preprint arXiv:1811.11357 (2018)
46. Wainwright, M.J., Jordan, M.I., et al.: Graphical models, exponential families, and variational inference. Found. Trends® Mach. Learn. **1**(1–2), 1–305 (2008)
47. Xu, J., Durrett, G.: Spherical latent spaces for stable variational autoencoders. arXiv preprint arXiv:1808.10805 (2018)
48. Zhao, S., Song, J., Ermon, S.: InfoVAE: information maximizing variational autoencoders. CoRR abs/1706.02262 (2017). http://arxiv.org/abs/1706.02262

Adaptive Variance Based Label Distribution Learning for Facial Age Estimation

Xin Wen[1,2(✉)], Biying Li[1,2], Haiyun Guo[1,3], Zhiwei Liu[1,2], Guosheng Hu[5], Ming Tang[1], and Jinqiao Wang[1,2,4]

[1] National Laboratory of Pattern Recognition, Institute of Automation, Chinese Academy of Sciences, Beijing, China
{xin.wen,biying.li,haiyun.guo,zhiwei.liu,tangm,jqwang}@nlpr.ia.ac.cn
[2] School of Artificial Intelligence, University of Chinese Academy of Sciences, Beijing, China
[3] ObjectEye Inc., Beijing, China
[4] NEXWISE Co., Ltd., Guangzhou, China
[5] Anyvision, Belfast, UK
huguosheng100@gmail.com

Abstract. Estimating age from a single facial image is a classic and challenging topic in computer vision. One of its most intractable issues is label ambiguity, i.e., face images from adjacent age of the same person are often indistinguishable. Some existing methods adopt distribution learning to tackle this issue by exploiting the semantic correlation between age labels. Actually, most of them set a fixed value to the variance of Gaussian label distribution for all the images. However, the variance is closely related to the correlation between adjacent ages and should vary across ages and identities. To model a sample-specific variance, in this paper, we propose an adaptive variance based distribution learning (AVDL) method for facial age estimation. AVDL introduces the data-driven optimization framework, meta-learning, to achieve this. Specifically, AVDL performs a meta gradient descent step on the variable (i.e. variance) to minimize the loss on a clean unbiased validation set. By adaptively learning proper variance for each sample, our method can approximate the true age probability distribution more effectively. Extensive experiments on FG-NET and MORPH II datasets show the superiority of our proposed approach to the existing state-of-the-art methods.

Keywords: Age estimation · Distribution learning · Meta-learning

1 Introduction

Age estimation is a challenging and hot research topic, which is to predict the person's age from his/her facial image. It has a lot of potential applications, including demographic statistics collection, commercial user management, video

© Springer Nature Switzerland AG 2020
A. Vedaldi et al. (Eds.): ECCV 2020, LNCS 12368, pp. 379–395, 2020.
https://doi.org/10.1007/978-3-030-58592-1_23

security surveillance, etc. However, there are numerous internal or external factors that affect the estimation results, including the race, illumination, image quality and so on. Besides, facial images from adjacent ages of the same person, especially for adults, usually look similar, resulting in the label ambiguity.

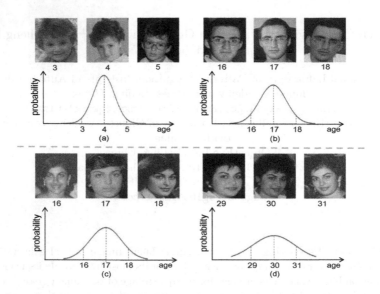

Fig. 1. The motivation of the proposed method. In each subfigure, the age probability distribution in the lower part corresponds to the middle image in the upper. The images above the dotted line belong to the same person and so do the images below the dotted line. On the one hand, by comparing (a) with (b) or (c) with (d), we can see that the facial appearance variation between adjacent ages of the same person varies at different ages. Correspondingly, the variance of the age probability distribution should differ across ages. On the other hand, by comparing (b) with (c), we can see that even at the same age, the aging process between different persons differs, thus the variance also varies across different persons

Recently, several deep learning methods have been proposed to improve the performance of facial age estimation. The most common methods model the face age prediction as a classification or a regression problem. The classification based methods treat each age as an independent class, which ignores the adjacent relationship between classes. Considering the continuity of age, regression methods predict age according to the extracted features. However, as presented by previous work [31,33], the regression methods face the overfitting problem, which is caused by the randomness of the human aging process and the ambiguous mapping between facial appearance and the actual age. In addition, some ranking based methods are proposed to achieve more accurate age estimation. Those approaches make use of individuals' ordinal information and employ multiple binary classifiers to determine the final age of the input image. Furthermore, Geng et al. [8,13] propose the label distribution learning (LDL) method which assumes that the real age can be represented by a discrete distribution. As their experiments show, it can help improve age estimation using Kullback-Leibler (K-L) divergence to measure the similarity between the predicted and ground truth distribution.

For the label distribution learning methods, the mean of the distribution is the ground truth age. However, the variance of the distribution is usually unknown for a face image. The previous methods often treat variance as a hyper-parameter and simply set it to a fixed value for all images. We think these methods are suboptimal because the variance is highly related to the correlation between adjacent ages and should vary across different ages and different persons, as illustrated in Fig. 1. The assumption that all the images sharing the same variance potentially degrades the model performance.

To tackle the above issues, in this paper, we propose a novel adaptive variance based distribution learning method (AVDL) for age estimation. Specifically, we introduce meta-learning which utilizes validation set as meta-objective and is applicable to online hyper-parameter adaptation work [28], to model *sample-specific* variance and thus better approximate true age probability distribution. As Fig. 2 shows, we firstly select a small validation set. For each iteration, with a disturbing variable added to variance, we use K-L loss as the training loss to update the training model parameter. Then we share the updated parameter with validation model and use predicted expectation age and ground truth on validation set to get L1 loss as the meta-objective. With this meta-guider, the disturbing variable is updated by gradient descent and adaptively find the proper variance with which model could perform better on validation set. The main contributions of this work can be summarized as follows:

- We propose a novel adaptive variance based distribution learning (AVDL) method for facial age estimation. AVDL can effectively model the correlation between adjacent ages and better approximate the age label distribution.
- Unlike the existing deep models which assume the variance across ages and identities is the same, we introduce a data-driven method, meta-learning, to learn the *sample-specific* variance. To our knowledge, we are the first deep model using meta-learning method to adaptively learn different variances for different samples.
- Extensive experiments on FG-NET and MORPH II datasets show the superiority of our proposed approach to the existing state-of-the-art methods.

2   Related Work

2.1   Facial Age Estimation

In recent years, with rapid development of convolution neural network (CNN) in computer vision tasks, such as facial landmark detection [23], face recognition [3,38], pedestrian attribute [35], semantic segmentation [45,46], deep learning methods were also improved the performance of age estimation. Here we briefly review some representative works in the facial age estimation field. Dex et al. [30]

regarded the facial age estimation as a classification problem and predicted ages with the expectation of ages weighted by classification probability. Tan et al. [33] proposed an age group classification method called age group-n-encoding method. However, these classification methods ignored the adjacent relationship between classes or groups. To overcome this, Niu et al. [24] proposed a multiple output CNN learning algorithm which took account of the ordinal information of ages for estimation. Shen et al. [32] proposed Deep Regression Forests by extending differentiable decision trees to deal with regression. Furthermore, Li et al. [22] proposed BridgeNet, which consists of local regressors and gating networks, to effectively explore the continuous relationship between age labels. Tan et al. [34] proposed a complex Deep Hybrid-Aligned Architecture (DHAA) that consists of global, local and global-local branches and jointly optimized the architecture with complementary information. Besides, Xie et al. [39] proposed two ensemble learning methods both utilized ordinal regression modeling for age estimation.

2.2   Distribution Learning

Distribution learning is a learning method proposed to solve the problem of label ambiguity [10], which has been utilized in a number of recognition tasks, such as head pose estimation [8,12], and age estimation [20,41]. Geng et al. [11,13] proposed two adaptive label distribution learning (ALDL) algorithms, i.e. IIS-ALDL and BFGS-ALDL, to iteratively learn the estimation function parameters and the label distributions variance. Though ALDL used an adaptive variance learning, our proposed method is different in three ways. Firstly, ALDL utilized traditional optimization method like BFGS while ours uses deep learning and CNN. Secondly, ALDL chose better samples in current training iteration to estimate new variance while our method uses meta-learning to get adaptive variance. The third point is ALDL updated variance only by estimating the training sample, which may cause overfitting. Our adaptive variance is supervised by validation set to be more general. Distribution learning of label was also used to remedy the shortage of training data with exact ages. Hou et al. [20] proposed a semi-supervised adaptive label distribution learning method. It used unlabeled data to enhance the label distribution adaptation to find a proper variance for each age. However, aging tendencies varies and variances of people at the same age could be different. Gao et al. [9] jointly used LDL and expectation regression to alleviate the inconsistency between training and testing. Moreover, Pan et al. [25] proposed a mean-variance loss for robust age estimation. Li et al. [21] proposed label distribution refinery to adaptively estimate the age distributions without assumptions about the form of label distribution, barely took into account the correlation of adjacent ages. While our method used Gaussian label distribution with adaptively meta-learned variance, which pays more attention to neighboring ages and ordinal information.

2.3 Meta-learning

Our proposed AVDL is an instantiation of meta-learning [1,36], i.e., learning to learn. According to the type of leveraged meta data, this concept can be classified to several types [37] including transferring knowledge from empirically similar tasks, transferring trained model parameters between tasks, building meta-models to learn data characteristics and learn purely from model evaluations. Model Agnostic Meta-Learning (MAML) [7] learned a model parameter initialization to perform better on target tasks. With the guidance of meta information, MAML took one gradient descent step on meta-objective to update model parameters [16]. The idea of using validation loss as meta-objective was applied in few-shot learning [27]. With reference to few-shot learning, Ren et al. [28] proposed a reweighting method (L2RW) guided by validation set. This method tried to solve the problem that data imbalance and label noise are both in the training set. The crucial criteria of L2RW is a small unbiased clean validation set which was taken as the supervisor of learning sample weight. As validation set performance measures the quality of hyper-parameters, taking it as meta-objective could not only be applied to sample reweighting but also to any other online hyper-parameter adaptation tasks. Inspired by this, we propose AVDL to incorporate validation set based meta-learning and label distribution learning to adaptively learn the label variance.

3 Methodology

In this section, we firstly give a description of the label distribution learning (LDL) method in age estimation. Then we introduce our adaptive variance based distribution learning(AVDL) method based on meta-learning framework.

3.1 The Label Distribution Learning Problem Revisit

Let X denote an input image with ground truth label y, $y \in \{0, 1, ..., 100\}$. The model is trained to predict a value as close to the ground truth label as possible. For traditional age estimation method, the ground truth is an integer. While in LDL method, to express the ambiguity of labels, Gao et al. [8] transform the real value y to a normal distribution $\mathbf{p}(y, \sigma)$ to denote the new ground truth. Mean value is set to the ground truth label y and σ is the normal distribution variance. Here we adopt the boldface lowercase letters like $\mathbf{p}(y, \sigma)$ to denote vectors, and use $p_k(y, \sigma)$ ($k \in [0, 100]$) to represent the k-th element of $\mathbf{p}(y, \sigma)$:

$$p_k(y, \sigma) = \frac{1}{\sqrt{2\pi}\sigma} \exp(-\frac{(k - y)^2}{2\sigma^2}) \tag{1}$$

where p_k is the probability that the true age is k years old. It represents the connection between the class k and y in a normal distribution view.

In the training process, assuming $G(*, \theta)$ as the classification function of the trained estimation model, θ represents the model parameters, $\mathbf{z}(X, \theta) = G(X, \theta)$

transforms the input X to the classification vector $\mathbf{z}(X, \theta)$. A softmax function is utilized to transfer $\mathbf{z}(X, \theta)$ into a probability distribution $\hat{\mathbf{p}}(X, \theta)$, the k-th element of which can be denoted by:

$$\hat{p}_k(X, \theta) = \frac{\exp(z_k(X, \theta))}{\sum_n \exp(z_n(X, \theta))} \tag{2}$$

where $z_k(X, \theta)$ is the k-th element of $\mathbf{z}(X, \theta)$.

LDL tries to generate the predicted softmax probability distribution as similar to the ground truth distribution as possible. So the Kullback-Leibler (K-L) divergence is employed to measure the difference between the predicted distribution and ground-truth distribution [8]:

$$L_{KL}(X, y, \theta, \sigma) = \sum_k p_k(y, \sigma) ln \frac{p_k(y, \sigma)}{\hat{p}_k(X, \theta)} \tag{3}$$

Then the K-L loss is used to update model parameters with SGD optimizer.

LDL method aims to construct a normal distribution of ground truth to approximate the real distribution, the key of which is the variance σ. For most LDL methods, this hyper-parameter is simply set to a fixed value, 2.0 in most cases. However, in fact, the variances for different people, or people of different ages could not be absolutely the same. So we propose a method to search proper variance for each image.

3.2   Adaptive Distribution Learning Based on Meta-learning

In machine learning, the loss on validation set is one of the guiders to adjust hyper-parameters for generalization. Therefore, using a clean unbiased validation set can help train a more general model. However, traditional training mode usually tunes the hyper-parameter manually. Inspired by the meta-learning work [28], we propose the adaptive variance based distribution learning (AVDL) algorithm guided by validation set, which offers an effective strategy to learn the sample-specific variance.

As we mentioned in Sect. 3.1, the most important hyper-parameter of LDL is the variance σ. Because our goal is to search for proper σ of each image while training, in this section we use σ to represent the variance vector for a batch of training data. The optimal σ in each iteration depends on the optimal model parameter θ:

$$\theta^*(\sigma) = \arg\min_\theta L_{KL}(X_{tr}, y_{tr}, \theta, \sigma) \tag{4}$$

$$\sigma^* = \arg\min_{\sigma, \sigma \geq 0} L_1(X_{val}, y_{val}, \theta^*, \sigma) \tag{5}$$

where $L_1(X_{val}, y_{val}, \theta^*, \sigma)$ denotes the validation loss. X_{tr} is the training input image while y_{tr} is its label. X_{val} is the validation input image while its label

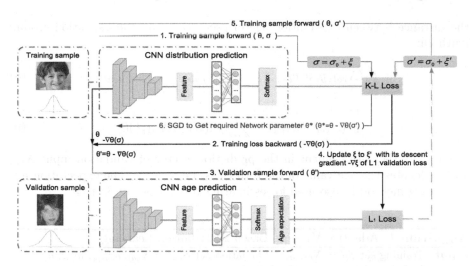

Fig. 2. Computation graph of AVDL in one iteration. The ground truth of each input image is transformed to a normal distribution. The model on top is for training and the other is for validation. The train model and validation model share the network architecture and parameters. The training loss is K-L loss while the validation loss is L1 loss. Process 1, 2, 3 belongs to traditional training steps. Perturbing variable ξ is added to initial distribution variance to get variance σ. By adding the training gradient descent step $- \bigtriangledown \theta$, the training model parameter θ is updated to θ' and is assigned to the validation model. Process 4 uses the descent gradient of ξ in validation loss to get the modified ξ' and σ'. Process 5, 6 shows the improved forward and backward computation with a proper variance σ'.

is y_{val}. To solve the optimization problem, we divided the training process into several process. Figure 2 shows the computation graph of our proposed method.

We choose a fixed number of images with correct labels from each class in the training set n to make a small unbiased validation set with m images, $m \ll n$. We utilize σ_i to denote variance for i-th image while we set the initial value of variances of all images to a fixed value σ_{i0}. To search a proper variance, we perturb each σ_i by ξ_i:

$$\sigma_i = \sigma_{i0} + \xi_i \tag{6}$$

where ξ_i is the i-th component of perturbing vector ξ which is set to 0 for initialization. Clearly, searching a proper σ is equal to searching a proper ξ.

Firstly, as Fig. 2 process 1, 2 and 3 show, in the t-th iteration, the input training batch calculates K-L loss as described in Sect. 3.1 with a perturbed σ. Update the model parameter θ_t with SGD to get $\hat{\theta}_{t+1}$:

$$\hat{\theta}_{t+1} = \theta_t - \alpha \bigtriangledown_\theta L_{KL}(X_{tr}, y_{tr}, \theta_t, \sigma) \tag{7}$$

α is the descent step.

The training loss is related to distribution. To compensate the lack of constrain in the final predicted age value, we adopt L1 loss on validation to measure the distance between expectation age of prediction and the validation ground truth [9]:

$$L_1(X_{val}, y_{val}, \hat{\theta}_{t+1}, \xi) = \left| \hat{y}^*(X_{val}, \hat{\theta}_{t+1}, \xi) - y_{val} \right| \tag{8}$$

$$\hat{y}^*(X_{val}, \hat{\theta}_{t+1}, \xi) = \sum_k \hat{p}_k(X_{val}, \hat{\theta}_{t+1}, \xi) l_k \tag{9}$$

where $\hat{p}_k$ is the k-th element in the prediction vector of validation input X_{val} and l_k denotes the age value of the k-th class, i.e. $l_k \in \mathcal{Y}$. The expectation age computing method is also used for estimating test images in Sect. 4.

Algorithm 1. Adaptive Variance Based Distribution Learning

Input: Training set $\mathcal{S}_{tr} = X_{tr}, y_{tr_n}$; Validation set $\mathcal{S}_{val} = X_{val}, y_{val_m}$, $m \ll n$;
        Initial model parameter θ_0; Initial variance σ_0
Output: Final model parameter θ_T
1: **for** t = 0,1,2...T-1 **do**
2:     Sample training batch $\mathcal{S}_{tr,t}$ from $\mathcal{S}_{tr}$
3:     Sample validation batch $\mathcal{S}_{val,t}$ from $\mathcal{S}_{val}$
4:     $\xi \leftarrow 0$
5:     $\sigma \leftarrow \sigma_0 + \xi$
6:     $L_{KL}(X_{tr}, y_{tr}, \theta_t, \sigma) \leftarrow \text{NetForward}(X_{tr}, y_{tr}, \theta_t, \sigma)$
7:     $\hat{\theta}_{t+1}(\sigma) \leftarrow \theta_t - \alpha \nabla_{\theta_t} L_{KL}(X_{tr}, y_{tr}, \theta_t, \sigma)$ % $\hat{\theta}_{t+1}$ is a function of σ
8:     $L_1(X_{val}, y_{val}, \hat{\theta}_{t+1}(\sigma), \sigma) \leftarrow \text{NetForward}(X_{val}, y_{val}, \hat{\theta}_{t+1}(\sigma), \sigma)$
9:     $\hat{\xi} \leftarrow \xi - \beta \nabla_{\xi} L_1(X_{val}, y_{val}, \hat{\theta}_{t+1}(\sigma), \sigma)$ % the gradient of ξ on $L1$ loss equals
    to the gradient of σ on $L1$ loss
10:    $\hat{\sigma} \leftarrow \sigma_0 + \hat{\xi}$ % modify σ adaptively
11:    $\hat{L}_{KL}(X_{tr}, y_{tr}, \theta_t, \hat{\sigma}) \leftarrow \text{Forward}(X_{tr}, y_{tr}, \theta_t, \hat{\sigma})$
12:    $\theta_{t+1} \leftarrow SGD(\hat{L}_{KL}(X_{tr}, y_{tr}, \theta_t, \hat{\sigma}), \theta_t)$ % update with SGD optimizer
13: **end for**

The better hyper-parameter means better validation performance. In that, we update the perturbation ξ with gradient descent step:

$$\hat{\xi} = \xi - \beta \nabla_{\xi} L_1(X_{val}, y_{val}, \hat{\theta}_{t+1}, \xi) \tag{10}$$

where β is the descent step size. This step is corresponding to the process 4 in Fig. 2. Due to the non-negativity restriction of σ, we normalize the ξ into the range $[-1, 1]$, using the mapping $\xi_i \rightarrow \frac{2\xi_i - \max(\xi) - \min(\xi)}{\max(\xi) - \min(\xi)}$. Then update the variance σ according to Eq. (6). In the third step of training, with the modified variance, we calculate forward K-L loss of the training input, then update model parameter with SGD optimizer, as the process 5, 6 in Fig. 2 shows.

We listed step-by-step pseudo code in Algorithm 1. According to step 9 in Algorithm 1, there is a two-stage deviation computation of variable ξ. PyTorch autograd mechanism can achieve this operation handily.

4 Experiments

In this section, we first introduce the datasets used in the experiments, i.e., MORPH II [29], FG-NET [26] and IMDB-WIKI [31]. Then we detail the experiment settings. Next, we validate the superiority of our approach with comparisons to the state-of-the-art facial age estimation methods. Finally, we conduct some ablation studies on our method.

4.1 Datasets

Morph II is the most popular dataset for age estimation. The dataset contains 55,134 color facial images of 13,000 individuals whose ages range from 16 to 77. On this dataset, we employ three typical protocols for evaluation: **Setting I: 80-20 protocol.** We randomly divide this dataset into two non-overlapped parts, i.e., 80% for training and 20% for testing. **Setting II: Partial 80-20 protocol.** Following the experimental setting in [33], we extract a subset of 5,493 facial images from Caucasian descent, these images are splitted into two parts: 80% of facial images for training and 20% for testing. **Setting III: S1-S2-S3 protocol.** Similar to [22,33], Morph II dataset is splitted into three non-overlapped subsets S1, S2 and S3, and all experiments are repeated twice. Firstly, train on S1 and test on S2+S3. Then, train on S2 and test on S1+S3. The performance of the two experiments and their average MAE are shown respectively.

FG-NET contains 1,002 color or gray facial images of 82 individuals whose ages are ranging from 0 to 69. We follow a widely used leave-one-person-out (LOPO) protocol [4,25] in our experiments, and the average performance over 82 splits is reported.

IMDB-WIKI is the largest facial image dataset with age labels, which consists of 523,051 images in total. This dataset is constituted of two parts: IMDB (460,723 images) and WIKI (62,328 images). We follow the practice in [22] and use this dataset to pretrain our model. Specifically, We remove non-face images and partial multi-face images. Finally, about 270,000 images are reserved.

4.2 Implementation Details

We use the detection algorithm in [44] to obtain the face detection box and five facial landmark coordinates, which are then used to align the input facial image of the network. We resize the input image to 224×224.

Following the settings in [9], we augment the face images with random horizontal flipping, scaling, rotating and translating during training time. For testing, we input both the image and its flipped version to the network, and then average their predictions as the final results.

We adopt ResNet-18 [19] as our backbone network and pretrain the network on IMDB-WIKI dataset for better initialization. We use the SGD optimizer with batch size 32 to optimize the network. The weight decay and the momentum are set to 0.0005 and 0.9. The initial learning rate is set to 0.01 and decays by 0.1 for every 20 epochs. We set the initial value of variances of all images to 1, and train the deep convolution neural network with PyTorch on 4 GTX TITAN X GPUs.

Table 1. The comparisons between the proposed method and other state-of-the-art methods on MORPH II under Setting I. Bold indicates the best (*indicates the model was pre-trained on the IMDB-WIKI dataset; †indicates the model was pre-trained on the MS-Celeb-1M dataset [17])

Method	MAE	Parameters	Year
ORCNN [24]	3.27	479.7K	2016
RGAN* [6]	2.61	–	2017
VGG-16 CNN + LDAE* [2]	2.35	138M	2017
SSR-Net* [40]	3.16	40.9K	2018
DRFs [32]	2.17	138M	2018
M-V Loss* [25]	2.16	138M	2018
DLDL-V2† [9]	1.97	3.7M	2018
C3AE* [43]	2.75	39.7K	2019
DHAA [34]	**1.91**	100M	2019
AVDL*	1.94	11M	–

4.3   Evaluation Criteria

According to previous works [31,33], we measure the performance of age estimation by the Mean Absolute Error (MAE) which is calculated using the average of the absolute errors between estimated age and chronological age.

4.4   Comparisons with State-of-the-Arts

On Morph II. We first compare the proposed method with other state-of-the-art methods on MORPH II dataset in Setting I, as illustrated in Table 1. We achieve the second best performance, which is slightly lower than DHAA [34] by 0.03. It is worth to note that DHAA is large and complex, their parameters are around 10 times larger than ours, though without additional face dataset for pre-training. Moreover, using the same pre-training dataset, we surpass the M-V Loss by a significant margin of 0.22.

Table 2 shows the test result under Setting II. We achieve the best performance, which is slightly higher than BridgeNet [22] by 0.01. Nevertheless, we have fewer parameters than BridgeNet. That is to say, we achieve the performance nearly to theirs with a significantly lower model complexity at the same time. As Table 3 shows, we achieve MAE of 2.53 under Setting III. Our method performs much better than the current state-of-the-art. All of the above comparisons consistently demonstrate the effectiveness of the proposed method.

Table 2. The comparisons between the proposed method and other state-of-the-art methods on MORPH II dataset (Setting II) and FG-NET dataset. Bold indicates the best (*indicates the model was pre-trained on the IMDB-WIKI dataset)

Method	MORPH II	FG-NET	Parameters	Year
OHRANK [4]	6.07	4.48	–	2011
CA-SVR [5]	5.88	4.67	–	2013
Human [18]	6.30	4.70	–	2015
DEX* [31]	2.68	3.09	138M	2018
DRFs [32]	2.91	3.85	138M	2018
M-V Loss* [25]	–	2.68	138M	2018
AGEn* [33]	2.52	2.96	138M	2018
C3AE* [43]	–	2.95	39.7K	2019
BridgeNet* [22]	2.38	2.56	138M	2019
DHAA* [34]	–	2.59	100M	2019
AVDL*	**2.37**	**2.32**	11M	–

On FG-NET. As shown in Table 2, we compare our model with state-of-the-art models on FG-Net. Our method achieves the lowest MAE of 2.32, which improves the state-of-the-art performance by a large margin of 0.24. Experimental results show that our method is effective even when there are only a few training images.

4.5 Ablation Study

In this subsection, we conduct ablation study on MORPH II dataset under Setting I to conduct ablation study.

The Superiority of Adaptive Variance to Fixed Variance Value. We train a set of baseline models, which all adopt ResNet-18 and K-L divergence loss but with different fixed variance values. Theoretically, the larger variance indicates the smoother distribution which refers to the stronger correlation in that age group. In comparison, the smaller variance represents the sharper distribution and the weaker correlation. If the variance is set too high, i.e., the label distribution is too smooth, the age estimation may not perform well. As Fig. 3 shows, the MAE increases along with the growth of variance when it is higher than 3, which indicates the worse performance. When the variance reduces to 0, it assumes there is no correlation between ages which is similar to the assumption

Table 3. The comparisons between the proposed method and other state-of-the-art methods on MORPH II under Setting III. Bold indicates the best (*indicates the model was pre-trained on the IMDB-WIKI dataset)

Method	Train	Test	MAE	Avg
KPLS [14]	S1	S2+S3	4.21	4.18
	S2	S1+S3	4.15	
BIF+KCCA [15]	S1	S2+S3	4.00	3.98
	S2	S1+S3	3.95	
CPLF [42]	S1	S2+S3	3.72	3.63
	S2	S1+S3	3.54	
DRFs [32]	S1	S2+S3	–	2.98
	S2	S1+S3	–	
DOEL [39]	S1	S2+S3	–	2.75
	S2	S1+S3	–	
AGEn* [33]	S1	S2+S3	2.82	2.70
	S2	S1+S3	2.58	
BridegNet* [22]	S1	S2+S3	2.74	2.63
	S2	S1+S3	2.51	
AVDL*	S1	S2+S3	**2.64**	**2.53**
	S2	S1+S3	**2.41**	

when regarding age estimation as classification problem. However, considering the gradual changing of face in aging, taking a proper use of age correlation can help age estimation. As illustrated in Fig. 3, when the fixed variance is less than 3, the MAE fluctuates. It validates that setting a fixed variance is suboptimal because the age correlation can not be the same for different people in different ages. The best performance of baseline is achieved with a variance of 3. However, it is still much worse than our proposed method, AVDL. In Fig. 3, we also show the performance achieved by training the ResNet-18 with cross-entropy loss, which is our baseline method by treating age estimation as classification task. In summary, Fig. 3 demonstrates the superiority of the adaptive variance. Actually, for each dataset and experiment setting, our approach is compared to the baseline method with fixed variance. We observed from Fig. 3 that the variation in fixed variance value within a certain range has little impact on performance, due to limited time, we only search the best variance for MORPH II(Setting I) and apply it to other experiments. In addition, the baseline with the fixed variance of MORPH II(Setting II), MORPH II(Setting III), and FGNET are 2.66, 2.79, 2.64, respectively.

The Influence of Different Sample Number in Validation Set. As [28] shows, a balanced meta dataset could provide balanced class knowledge. For the

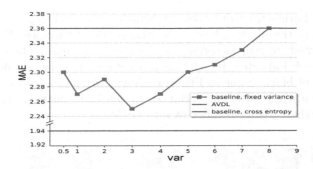

Fig. 3. The MAE results on MORPH II under Setting I. The blue line denotes the results of the baseline model trained with different fixed variance. The red line is the result of the baseline model trained with cross-entropy loss and the green one is the result of AVDL

Table 4. The performance comparison on selecting different number of facial images of each age to form validation set

Number of images	1	2	3
MAE of AVDL	1.98	1.96	1.94

same purpose, we choose an unbiased validation set as meta dataset. As for the composition of the clean validation set, we try different sizes of validation set. We respectively random select 1, 2 and 3 images from each class in the training set to form the validation set for experiment. From Table 4, we can find that the larger the validation set is, the better the model performs. However, since all validation set is used in each iteration, it needs more time and memory as the size of validation set increases. Considering the time and space cost, for each dataset setting, we randomly chose three image from each class to form the validation set.

4.6 Visualization and Discussion

Considering the affordance and credibility, here we display some visual results of AVDL in age estimation and variance adaptation.

We use the learned variance of samples to show the effectiveness of AVDL and to justify our motivation. Under the Setting I on MORPH II, each age, ranging from 16 to 60, possesses a group of face images belonging to different person identities. While there is no person whose images covering the full age range. We select images of several persons at different ages with their adapted variances in Fig. 4(a). As [11] mentioned, the age variances of younger or older people tend to be smaller than those of middle age. And the variances vary between people in the same age group. Besides, the variance in Fig. 4(b) shows the visualization of the adjusted variances in a mini-batch. The initial variance for each sample, as indicated in Sect. 4.2, is set to 1. The learned variances are shown in the horizontal band above in which each block represents a sample and the color of the block indicates the magnitude of variance. The blocks are arranged from left to right according to the ages of samples. The band below is the legend which indicates the relationship between the magnitude of variance and the color of block. Same as Fig. 4(a) shows, the variance in young age and old age is smaller. Besides, the variances in the band fluctuate slightly which demonstrates the variance is different for people.

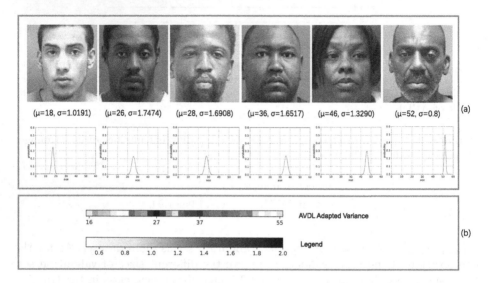

Fig. 4. Examples of age estimation results by AVDL. (a) shows some samples at different ages on Morph II with adapted variances. According to the Gaussian curves, it can be proved that for younger and older people, the variances tend to be smaller while for middle age, larger. (b) uses heat map to visualize the adaptively learned variances σ corresponding to different ages.

5   Conclusions

In this paper, we propose a novel method for age estimation, named adaptive variance based distribution learning (AVDL). AVDL introduces meta-learning to adaptively adjust the variance for each image in single iteration. It achieves better performances than others on multiple age estimation datasets. Our experiments also show that AVDL can guide variance to get close to real facial aging law. The idea that using meta-learning to guide key hyper-parameters is inspirational and we will explore more possibilities of it.

Acknowledgments. This work was supported by Key-Area Research and Development Program of Guangdong Province (No. 2019B010153001), National Natural Science Foundation of China (No. 61772527,61806200,61976210), China Postdoctoral science Foundation (No. 2019M660859), Open Project of Key Laboratory of Ministry of Public Security for Road Traffic Safety (No. 2020ZDSYSKFKT04).

References

1. Andrychowicz, M., et al.: Learning to learn by gradient descent by gradient descent. In: Advances in Neural Information Processing Systems, pp. 3981–3989 (2016)
2. Antipov, G., Baccouche, M., Berrani, S.A., Dugelay, J.L.: Effective training of convolutional neural networks for face-based gender and age prediction. Pattern Recogn. **72**, 15–26 (2017)
3. Cao, D., Zhu, X., Huang, X., Guo, J., Lei, Z.: Domain balancing: face recognition on long-tailed domains. In: Proceedings of the IEEE/CVF Conference on Computer Vision and Pattern Recognition, pp. 5671–5679 (2020)
4. Chang, K.Y., Chen, C.S., Hung, Y.P.: Ordinal hyperplanes ranker with cost sensitivities for age estimation. In: 2011 IEEE Conference on Computer Vision and Pattern Recognition (CVPR), pp. 585–592. IEEE (2011)
5. Chen, K., Gong, S., Xiang, T., Change Loy, C.: Cumulative attribute space for age and crowd density estimation. In: Proceedings of the IEEE Conference on Computer Vision and Pattern Recognition, pp. 2467–2474 (2013)
6. Duan, M., Li, K., Li, K.: An ensemble CNN2ELM for age estimation. IEEE Trans. Inf. Forensics Secur. **13**(3), 758–772 (2017)
7. Finn, C., Abbeel, P., Levine, S.: Model-agnostic meta-learning for fast adaptation of deep networks. In: Proceedings of the 34th International Conference on Machine Learning, vol. 70, pp. 1126–1135 (2017). JMLR.org
8. Gao, B.B., Xing, C., Xie, C.W., Wu, J., Geng, X.: Deep label distribution learning with label ambiguity. IEEE Trans. Image Proc. **26**(6), 2825–2838 (2017)
9. Gao, B.B., Zhou, H.Y., Wu, J., Geng, X.: Age estimation using expectation of label distribution learning. In: IJCAI, pp. 712–718 (2018)
10. Geng, X.: Label distribution learning. IEEE Trans. Knowl. Data Eng. **28**(7), 1734–1748 (2016)
11. Geng, X., Wang, Q., Xia, Y.: Facial age estimation by adaptive label distribution learning. In: 2014 22nd International Conference on Pattern Recognition, pp. 4465–4470. IEEE (2014)
12. Geng, X., Xia, Y.: Head pose estimation based on multivariate label distribution. In: Proceedings of the IEEE Conference on Computer Vision and Pattern Recognition, pp. 1837–1842 (2014)
13. Geng, X., Yin, C., Zhou, Z.H.: Facial age estimation by learning from label distributions. IEEE Trans. Pattern Anal. Mach. Intell. **35**(10), 2401–2412 (2013)
14. Guo, G., Mu, G.: Simultaneous dimensionality reduction and human age estimation via kernel partial least squares regression. In: CVPR 2011, pp. 657–664. IEEE (2011)
15. Guo, G., Mu, G.: Joint estimation of age, gender and ethnicity: CCA vs. PLS. In: 2013 10th IEEE International Conference and Workshops on Automatic Face and Gesture Recognition (FG), pp. 1–6. IEEE (2013)
16. Guo, J., Zhu, X., Zhao, C., Cao, D., Lei, Z., Li, S.Z.: Learning meta face recognition in unseen domains. In: Proceedings of the IEEE/CVF Conference on Computer Vision and Pattern Recognition, pp. 6163–6172 (2020)
17. Guo, Yandong., Zhang, Lei., Hu, Yuxiao., He, Xiaodong, Gao, Jianfeng: MS-Celeb-1M: a dataset and benchmark for large-scale face recognition. In: Leibe, Bastian, Matas, Jiri, Sebe, Nicu, Welling, Max (eds.) ECCV 2016. LNCS, vol. 9907, pp. 87–102. Springer, Cham (2016). https://doi.org/10.1007/978-3-319-46487-9_6
18. Han, H., Otto, C., Liu, X., Jain, A.K.: Demographic estimation from face images: human vs. machine performance. IEEE Trans. Pattern Anal. Mach. Intell. **37**(6), 1148–1161 (2014)

19. He, K., Zhang, X., Ren, S., Sun, J.: Deep residual learning for image recognition. In: Proceedings of the IEEE Conference on Computer Vision and Pattern Recognition, pp. 770–778 (2016)
20. Hou, P., Geng, X., Huo, Z.W., Lv, J.Q.: Semi-supervised adaptive label distribution learning for facial age estimation. In: Thirty-First AAAI Conference on Artificial Intelligence (2017)
21. Li, P., Hu, Y., Wu, X., He, R., Sun, Z.: Deep label refinement for age estimation. Pattern Recogn. **100**, 107178 (2020)
22. Li, W., Lu, J., Feng, J., Xu, C., Zhou, J., Tian, Q.: BridgeNet: a continuity-aware probabilistic network for age estimation. In: Proceedings of the IEEE Conference on Computer Vision and Pattern Recognition, pp. 1145–1154 (2019)
23. Liu, Z., et al.: Semantic alignment: finding semantically consistent ground-truth for facial landmark detection. In: 2019 IEEE/CVF Conference on Computer Vision and Pattern Recognition (CVPR), pp. 3462–3471 (2019)
24. Niu, Z., Zhou, M., Wang, L., Gao, X., Hua, G.: Ordinal regression with multiple output CNN for age estimation. In: Proceedings of the IEEE Conference on Computer Vision and Pattern Recognition, pp. 4920–4928 (2016)
25. Pan, H., Han, H., Shan, S., Chen, X.: Mean-variance loss for deep age estimation from a face. In: Proceedings of the IEEE Conference on Computer Vision and Pattern Recognition, pp. 5285–5294 (2018)
26. Panis, G., Lanitis, A., Tsapatsoulis, N., Cootes, T.F.: Overview of research on facial ageing using the FG-NET ageing database. IET Biometrics **5**(2), 37–46 (2016)
27. Ren, M., et al.: Meta-learning for semi-supervised few-shot classification. arXiv preprint arXiv:1803.00676 (2018)
28. Ren, M., Zeng, W., Yang, B., Urtasun, R.: Learning to reweight examples for robust deep learning. arXiv preprint arXiv:1803.09050 (2018)
29. Ricanek, K., Tesafaye, T.: Morph: a longitudinal image database of normal adult age-progression. In: 7th International Conference on Automatic Face and Gesture Recognition (FGR 2006), pp. 341–345. IEEE (2006)
30. Rothe, R., Timofte, R., Van Gool, L.: DEX: deep expectation of apparent age from a single image. In: The IEEE International Conference on Computer Vision (ICCV) Workshops, December 2015
31. Rothe, R., Timofte, R., Van Gool, L.: Deep expectation of real and apparent age from a single image without facial landmarks. Int. J. Comput. Vision **126**(2–4), 144–157 (2018)
32. Shen, W., Guo, Y., Wang, Y., Zhao, K., Wang, B., Yuille, A.L.: Deep regression forests for age estimation. In: Proceedings of the IEEE Conference on Computer Vision and Pattern Recognition, pp. 2304–2313 (2018)
33. Tan, Z., Wan, J., Lei, Z., Zhi, R., Guo, G., Li, S.Z.: Efficient group-n encoding and decoding for facial age estimation. IEEE Trans. Pattern Anal. Mach. Intell. **40**(11), 2610–2623 (2017)
34. Tan, Z., Yang, Y., Wan, J., Guo, G., Li, S.Z.: Deeply-learned hybrid representations for facial age estimation. In: IJCAI, pp. 3548–3554 (2019)
35. Tan, Z., Yang, Y., Wan, J., Wan, H., Guo, G., Li, S.: Attention-based pedestrian attribute analysis. IEEE Trans. Image Process. **PP**, 1–1 (2019). https://doi.org/10.1109/TIP.2019.2919199
36. Thrun, S., Pratt, L.: Learning to Learn. Springer Science & Business Media (2012)
37. Vanschoren, J.: Meta-learning: a survey. arXiv preprint arXiv:1810.03548 (2018)
38. Wang, G., Han, H., Shan, S., Chen, X.: Cross-domain face presentation attack detection via multi-domain disentangled representation learning. In: IEEE/CVF Conference on Computer Vision and Pattern Recognition (CVPR), June 2020

39. Xie, J.C., Pun, C.M.: Deep and ordinal ensemble learning for human age estimation from facial images. IEEE Trans. Inf. Forensics Secur. **15**, 2361–2374 (2020)
40. Yang, T.Y., Huang, Y.H., Lin, Y.Y., Hsiu, P.C., Chuang, Y.Y.: SSR-Net: a compact soft stagewise regression network for age estimation. In: IJCAI, vol. 5, p. 7 (2018)
41. Yang, X., et al.: Deep label distribution learning for apparent age estimation. In: Proceedings of the IEEE International Conference on Computer Vision Workshops, pp. 102–108 (2015)
42. Yi, Dong., Lei, Zhen, Li, Stan Z.: Age estimation by multi-scale convolutional network. In: Cremers, Daniel, Reid, Ian, Saito, Hideo, Yang, Ming-Hsuan (eds.) ACCV 2014. LNCS, vol. 9005, pp. 144–158. Springer, Cham (2015). https://doi.org/10.1007/978-3-319-16811-1_10
43. Zhang, C., Liu, S., Xu, X., Zhu, C.: C3AE: exploring the limits of compact model for age estimation. In: Proceedings of the IEEE Conference on Computer Vision and Pattern Recognition, pp. 12587–12596 (2019)
44. Zhao, X., Liang, X., Zhao, C., Tang, M., Wang, J.: Real-time multi-scale face detector on embedded devices. Sensors **19**(9), 2158 (2019)
45. Zhu, B., Chen, Y., Tang, M., Wang, J.: Progressive cognitive human parsing. In: Thirty-Second AAAI Conference on Artificial Intelligence (2018)
46. Zhu, B., Chen, Y., Wang, J., Liu, S., Zhang, B., Tang, M.: Fast deep matting for portrait animation on mobile phone. In: Proceedings of the 25th ACM International Conference on Multimedia, pp. 297–305 (2017)

Connecting the Dots: Detecting Adversarial Perturbations Using Context Inconsistency

Shasha Li[1(✉)], Shitong Zhu[1], Sudipta Paul[1], Amit Roy-Chowdhury[1],
Chengyu Song[1], Srikanth Krishnamurthy[1], Ananthram Swami[2],
and Kevin S. Chan[2]

[1] University of California, Riverside, USA
{sli057,szhu014,spaul007}@ucr.edu, amitrc@ece.ucr.edu,
{csong,krish}@cs.ucr.edu
[2] US Army CCDC Army Research Laboratory, Adelphi, USA
ananthram.swami.civ@mail.mil, kevin.s.chan.civ@mail.mil

Abstract. There has been a recent surge in research on adversarial perturbations that defeat Deep Neural Networks (DNNs) in machine vision; most of these perturbation-based attacks target object classifiers. Inspired by the observation that humans are able to recognize objects that appear out of place in a scene or along with other unlikely objects, we augment the DNN with a system that learns context consistency rules during training and checks for the violations of the same during testing. Our approach builds a set of auto-encoders, one for each object class, appropriately trained so as to output a discrepancy between the input and output if an added adversarial perturbation violates context consistency rules. Experiments on PASCAL VOC and MS COCO show that our method effectively detects various adversarial attacks and achieves high ROC-AUC (over 0.95 in most cases); this corresponds to over 20% improvement over a state-of-the-art context-agnostic method.

Keywords: Object detection · Adversarial perturbation · Context

1 Introduction

Recent studies have shown that Deep Neural Networks (DNNs), which are the state-of-the-art tools for a wide range of tasks [10,18,25,37,44], are vulnerable to adversarial perturbation attacks [28,49]. In the visual domain, such adversarial perturbations can be digital or physical. The former refers to adding (quasi-) imperceptible digital noises to an image to cause a DNN to misclassify an object in the image; the latter refers to physically altering an object so that the captured

Electronic supplementary material The online version of this chapter (https://doi.org/10.1007/978-3-030-58592-1_24) contains supplementary material, which is available to authorized users.

image of that object is misclassified. In general, adversarial perturbations are not readily noticeable by humans, but cause the machine to fail at its task.

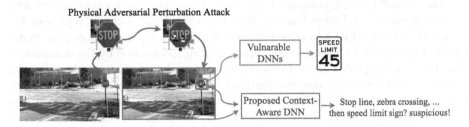

Fig. 1. An example of how our proposed context-aware defense mechanism works. Previous studies [13,43] have shown how small alterations (graffiti, patches etc.) to a stop sign make a vulnerable DNN classify it as a speed limit. We posit that a stop sign exists within the wider context of a scene (e.g., zebra crossing which is usually not seen with a speed limit sign). Thus, the scene context can be used to make the DNN more robust against such attacks.

To defend against such attacks, our observation is that the misclassification caused by adversarial perturbations is often *out-of-context*. To illustrate, consider the traffic crossing scene in Fig. 1; a stop sign often co-exists with a stop line, zebra crossing, street nameplate and other characteristics of a road intersection. Such co-existence relationships, together with the background, create a *context* that can be captured by human vision systems. Specifically, if one (physically) replaces the stop sign with a speed limit sign, humans can recognize the anomaly that the speed limit sign does not fit in the scene. If a DNN module can also learn such relationships (i.e., the context), it should also be able to deduce if the (mis)classification result (i.e., the speed limit sign) is out of context.

Inspired by these observations and the fact that context has been used very successfully in recognition problems, we propose to use *context inconsistency* to detect adversarial perturbation attacks. This defense strategy complements existing defense methods [17,24,33], and can cope with both digital and physical perturbations. To the best of our knowledge, it is the first strategy to defend object detection systems by considering objects "within the context of a scene."

We realize a system that checks for context inconsistencies caused by adversarial perturbations, and apply this approach for the defense of object detection systems; our work is motivated by a rich literature on context-aware object recognition systems [4,11,22,35]. We assume a framework for object detection similar to [42], where the system first proposes many regions that potentially contain objects, which are then classified. In brief, our approach accounts for four types of relationships among the regions, all of which together form the context for each proposed region: a) regions corresponding to the same object (*spatial context*); b) regions corresponding to other objects likely to co-exist within a scene (*object-object context*; c) the regions likely to co-exist with the background (*object-background context*); and d) the consistency of the regions within the holistic scene (*object-scene context*). Our approach constructs a fully connected graph with the proposed regions and a super-region node which represents the scene. In this graph, each node has, what we call an associated context profile.

The *context profile* is composed of node features (i.e., the original feature used for classification) and edge features (i.e., context). Node features represent the region of interest (RoI) and edge features encode how the current region relates to other regions in its feature space representation. Motivated by the observation that the context profile of each object category is almost always unique, we use an auto-encoder to learn the distribution of the context profile of each category. In testing, the auto-encoder checks whether the classification result is consistent with the testing context profile. In particular, if a proposed region (say of class A) contains adversarial perturbations that cause the DNN of the object detector to misclassify it as class B, using the auto-encoder of class B to reconstruct the testing context profile of class A will result in a high reconstruction error. Based on this, we can conclude that the classification result is suspicious.

The main contributions of our work are the following.

- To the best of our knowledge we are the first to propose using context inconsistency to detect adversarial perturbations in object classification tasks.
- We design and realize a DNN-based adversarial detection system that automatically extracts context for each region, and checks its consistency with a learned context distribution of the corresponding category.
- We conduct extensive experiments on both digital and physical perturbation attacks with three different adversarial targets on two large-scale datasets - PASCAL VOC [12] and Microsoft COCO [32]. Our method yields high detection performance in all the test cases; the ROC-AUC is over 0.95 in most cases, which is 20–35% higher than a state-of-the-art method [47] that does not use context in detecting adversarial perturbations.

2   Related Work

We review closely-related work and its relationship to our approach.

Object Detection, which seeks to locate and classify object instances in images/videos, has been extensively studied [31,34,41,42]. Faster R-CNN [42] is a state-of-the-art DNN-based object detector that we build upon. It initially proposes class-agnostic bounding boxes called region proposals (first stage), and then outputs the classification result for each of them in the second stage.

Adversarial Perturbations on Object Detection, and in particular physical perturbations targeting DNN-based object detectors, have been studied recently [6,43,48] (in addition to those targeting image classifiers [1,13,26]). Besides mis-categorization attacks, two new types of attacks have emerged against object detectors: the *hiding attack* and the *appearing attack* [6,43] (see Sect. 3.1 for more details). While defenses have been proposed against digital adversarial perturbations in image classification, our work focuses on both digital and physical adversarial attacks on object detection systems, which is an open and challenging problem.

Table 1. Comparison of existing detection-based defenses; since FeatureSqueeze [47] meets all the basic requirements of our approach, it is used as a baseline in the experimental analysis.

Detection	Beyond MNIST CIFAR	Do not need perturbed samples for training	Extensibility to object detection
PCAWhiten [19]	✗	✓	✗, PCA is not feasible on large regions
GaussianMix [14]	✗	✗	✗, Fixed-sized inputs are required
Steganalysis [33]	✓	✗	✗, Unsatisfactory performance on small regions
ConvStat [38]	✗	✗	✓
SafeNet [36]	✓	✗	✓
PCAConv [29]	✓	✗	✗, Fixed-sized inputs are required
SimpleNet [16]	✗	✗	✓
AdapDenoise [30]	✓	✗	✓
FeatureSqueeze [47]	✓	✓	✓

Adversarial Defense has been proposed for coping with digital perturbation attacks in the image domain. Detection-based defenses aim to distinguish perturbed images from normal ones. *Statistics based detection methods* rely on extracted features that have different distributions across clean images and perturbed ones [14, 19, 33]. *Prediction inconsistency based detection methods* process the images and check for consistency between predictions on the original images and processed versions [30, 47]. *Other methods train a second binary classifier* to distinguish perturbed inputs from clean ones [29, 36, 38]. However many of these are effective only on small and simple datasets like MNIST and CIFAR-10 [5]. Most of them need large amounts of perturbed samples for training, and very few can be easily extended to region-level perturbation detection, which is the goal of our method. Table 1 summarizes the differences between our method and the other defense methods; we extend FeatureSqueeze [47], considered a state-of-the-art detection method, which squeezes the input features by both reducing the color bit depth of each pixel and spatially smoothening the input images, to work at the region-level and use this as a baseline (with this extension its performance is directly comparable to that of our approach).

Context Learning for Object Detection has been studied widely [4, 11, 21, 40, 45]. Earlier works that incorporate context information into DNN-based object detectors [9, 15, 39] use object relations in post-processing, where the detected objects are re-scored by considering object relations. Some recent works [7, 27] perform sequential reasoning, i.e., objects detected earlier are used to help find objects later. The state-of-the-art approaches based on recurrent units [35] or neural attention models [22] process a set of objects using interactions between their appearance features and geometry. Our proposed context learning framework falls into this type, and among these, [35] is the one most related to our work. We go beyond the context learning method to define the context profile and use context inconsistency checks to detect attacks.

3   Methodology

3.1   Problem Definition and Framework Overview

We propose to detect adversarial perturbation attacks by recognizing the context inconsistencies they cause, i.e., by connecting the dots with respect to whether the object fits within the scene and in association with other entities in the scene.

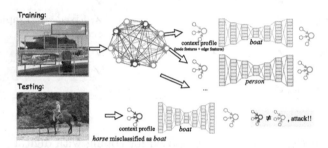

Fig. 2. Training phase: a fully connected graph is built to connect the regions of the scene image – details in Fig. 3; context information relating to each object category is collected and used to train auto-encoders. Testing phase: the context profile is extracted for each region and input to the corresponding auto-encoder to check if it matches with benign distribution.

Threat Model. We assume a strong white-box attack against the two-stage Faster R-CNN model where both the training data and the parameters of the model are known to the attacker. Since there are no existing attacks against the first stage (i.e., region proposals), we do not consider such attacks. The attacker's goal is to cause the second stage of the object detector to malfunction by adding digital or physical perturbations to *one* object instance/background region. There are three types of attacks [6,43,48]:

- *Miscategorization attacks* make the object detector miscategorize the perturbed object as belonging to a different category.
- *Hiding attacks* make the object detector fail in recognizing the presence of the perturbed object, which happens when the confidence score is low or the object is recognized as background.
- *Appearing attacks* make the object detector wrongly conclude that the perturbed background region contains an object of a desired category.

Framework Overview. We assume that we can get the region proposal results from the first stage of the Faster R-CNN model and the prediction results for each region from its second stage. We denote the input scene image as I and the region proposals as $R_I = [r_1, r_2, ..., r_N]$, where N is the total number of proposals of I. During the training phase, we have the ground truth category label and bounding box for each r_i, denoted as $S_I = [s_1, s_2, ..., s_N]$. The Faster R-CNN's predictions on proposed regions are denoted as $\tilde{S}_I$. Our goal as an attack detector is to identify perturbed regions from all the proposed regions.

Figure 2 shows the workflow of our framework. We use a structured DNN model to build a fully connected graph on the proposed regions to model the context of a scene image. We name this as Structure ContExt ModEl, or SCEME in short. In SCEME, we combine the node features and edge features of each node r_i, to form its context profile. We use auto-encoders to detect context inconsistencies as outliers. Specifically, during the training phase, for each category, we train a separate auto-encoder to capture the distribution of the benign context profile of that category. We also have an auto-encoder for the background category to detect hiding attacks. During testing, we extract the context profile for each proposed region. We then select the corresponding auto-encoder based on the prediction result of the Faster R-CNN model and check if the testing context profile belongs to the benign distribution. If the reconstruction error rate is higher than a threshold, we posit that the corresponding region contains adversarial perturbations. In what follows, we describe each step of SCEME in detail.

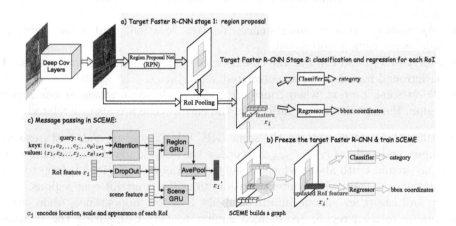

Fig. 3. (a) The attack target model, the Faster R-CNN, is a two-stage detector. (b) SCEME is built upon the proposed regions from the first stage of the Faster R-CNN, and updates the RoI features by message passing across regions. (c) Zooming in on SCEME shows how it fuses context information into each RoI, by updating RoI features via Region and Scene GRUs.

3.2   Constructing SCEME

In this subsection, we describe the design of the fully connected graph and the associated message passing mechanism in SCEME. Conceptually, SCEME builds a fully connected graph on each scene image. Each node is a region proposal generated by the first stage of the target object detector, plus the scene node. The initial node features, r_i, are the RoI pooling features of the corresponding region. The node features are then updated ($r_i \rightarrow r_i'$) using message passing from other nodes. After convergence, the updated node features r_i' are used as inputs to a regressor towards refining the bounding box coordinates and a classifier to predict the category, as shown in Fig. 3(b). Driven by the object detection objective, we train SCEME and the following regressor and classifier together. We freeze the weights of the target Faster R-CNN during the training. To force SCEME to rely more on context information instead of the appearance information (i.e., node features) when performing object detection, we apply a dropout function [20] on the node features before inputing into SCEME, during the training phase. At the end of training, SCEME should be able to have better object detection performance than the target Faster R-CNN since it explicitly uses the context information from other regions to update the appearance features of each region via message passing. This is observed in our implementation.

We use Gated Recurrent Units (GRU) [8] with attention [2] as the message passing mechanism in SCEME. For each proposed region, relationships with other regions and the whole scene form four kinds of context:

- *Same-object context*: for regions over the same object, the classification results should be consistent;
- *Object-object context*: co-existence, relative location, and scale between objects are usually correlated;
- *Object-background context*: the co-existence of the objects and the associated background regions are also correlated;
- *Object-scene context*: when considering the whole scene image as one super region, the co-existence of objects in the entire scene are also correlated.

To utilize object-scene context, the scene GRU takes the scene node features s as the input, and updates $r_i \rightarrow r_{scene}$. To utilize the other kinds of context, since we have no ground truth about which object/background the regions belong to, we use attention to learn what context category to utilize from different regions. The query and key (they encode information like location, appearance, scale, etc.) pertaining to each region are defined similar to [35]. Comparing the relative location, scale and co-existence between the query of the current region and the keys of all the other regions, the attention system assigns different attention scores to each region, i.e., it updates r_i, utilizing different amount of information from $\{r_j\}_{j \neq i}$. Thus, r_j is first weighted by the attention scores and then all r_j are summed up as the input to the Region GRU to update $r_i \rightarrow r_{regions}$ as shown in Fig. 3(c). The corresponding output, $r_{regions}$ and r_{scene}, are then combined via the average pooling function to get the final updated RoI feature vector r'.

3.3   Context Profile

In this subsection, we describe how we extract a context profile in SCEME. Recall that a context profile consists of node features r and edge features, where the edge features describe how r is updated. Before introducing the edge features that we use, we describe in detail how message passing is done with GRU [8].

Algorithm 1: SCEME: Training phase

    Input   : $\{R_I, S_I, \tilde{S}_I\}_{I \in TrainSet}$
    Output: SCEME, $AutoEncoder_c$ for each object category c, and $thresh_{err}$
1 SCEME $\leftarrow$ TrainSCEME($\{R_I, S_I\}_{I \in TrainSet}$)
2 ContextProfiles[c] = [] for each object category c
3 **for** *each* $R_I = [r_1, r_2, ...]$ **do**
4 | $X_I = [x_1, x_2, ...] \leftarrow$ ExtractContextProfiles($SCEME, R_I$)
5 | **for** *each region, its prediction, and its context profile* $\{r_j, \tilde{s}_j, x_j\}$ **do**
6 | | $\tilde{c} \leftarrow$ GetPredictedCategory($\tilde{s}_j$)
7 | | ContextProfiles[$\tilde{c}$] $\leftarrow$ ContextProfiles[$\tilde{c}$] $+ x_j$
8 | **end**
9 **end**
10 **for** *each category c* **do**
11 | $AutoEncoder_c \leftarrow$ TrainAutoEncoder($ContextProfiles[c]$)
12 **end**
13 $thresh_{err} =$ GetErrThreshold($\{AutoEncoder_c\}$)
14 **return** SCEME, $\{AutoEncoder_c\}$, $thresh_{err}$

A GRU is a memory cell that can remember the initial node features r and then fuse incoming messages from other nodes into a meaningful representation. Let us consider the GRU that takes the feature vector v (from other nodes) as the input, and updates the current node features r. Note that r and v have the same dimensions since both are from RoI pooling. GRU computes two gates given v and r, for message fusion. The reset gate γ_r drops or enhances information in the initial memory based on its relevance to the incoming message v. The update gate γ_u controls how much of the initial memory needs to be carried over to the next memory state, thus allowing a more effective representation. In other words, γ_r and γ_u are two vectors of the same dimension as r and v, which are learned by the model to decide what information should be passed to the next memory state given the current memory state and the incoming message. Therefore, we use the gate vectors as the edge features in the context profile. There are, in total, four gate feature vectors from both the Scene GRU and the Region GRU. Therefore, we define the context profile of a proposed region as $x = [r, \gamma_{u1}, \gamma_{u2}, \gamma_{r1}, \gamma_{r2}]$.

3.4   AutoEncoder for Learning Context Profile Distribution

In benign settings, all context profiles of a given category must be similar to each other. For example, stop sign features exist with features of road signs and zebra crossings. Therefore, the context profile of a stop sign corresponds to a unique distribution that accounts for these characteristics. When a stop sign is misclassified as a speed limit sign, its context profile should not fit with the distribution corresponding to that of the speed limit sign category.

For each category, we use a separate auto-encoder (architecture shown in the supplementary material) to learn the distribution of its context profile. The input to the auto-encoder is the context profile $x = [r, \gamma_{u1}, \gamma_{u2}, \gamma_{r1}, \gamma_{r2}]$. A fully connected layer is first used to compress the node features (r) and edge features $([\gamma_{u1}, \gamma_{u2}, \gamma_{r1}, \gamma_{r2}])$ separately. This is followed by two convolution layers, wherein the node and edge features are combined to learn the joint compression. Two fully connected layers are then used to further compress the joint features. These layers form a bottleneck that drives the encoder to learn the true relationships between the features and get rid of redundant information. SmoothL1Loss, as defined in [23, 46], between the input and the output is used to train the auto-encoder, which is a common practice.

Once trained, we can detect adversarial perturbation attacks by appropriately thresholding the reconstruction error. Giving a new context profile during testing, if a) the node features are not aligned with the corresponding distribution of benign node features, or b) the edge features are not aligned with the corresponding distribution of benign edge features, or c) the joint distribution between the node features and the edge features is violated, the auto-encoder will not be able to reconstruct the features using its learned distribution/relation. In other words, a reconstruction error that is larger than the chosen threshold would indicate either an appearance discrepancy or a context discrepancy between the input and output of the auto-encoder.

Algorithm 2: SCEME: Testing phase

    Input : R_I, $\tilde{S}_I$, SCEME, $\{AutoEncoder_c\}$, $thresh_{err}$
    Output: perturbed regions $PerturbedSet$
1 $PerturbedSet = []$
2 $X_I = $ ExtractContextProfiles($SCEME$, R_I)
3 **for** *each region, its prediction, and its context profile* $\{r_j, \tilde{s}_j, x_j\}$ **do**
4     $\tilde{c} \leftarrow$ GetPredictedCategory($\tilde{s}_j$)
5   err $= $ GetAutoEncoderReconErr($AutoEncoder_{\tilde{c}}, x_j$)
6   **if** *err* $> thresh_{err}$ **then**
7   region $\leftarrow$ GetRegion($\tilde{s}_j$)
8         $PerturbedSet \leftarrow PerturbedSet+$ region
9 **end**
10 **return** $PerturbedSet$

An overview of the approach (training and testing phases) is captured in Algorithms 1 and 2.

4 Experimental Analysis

We conduct comprehensive experiments on two large-scale object detection datasets to evaluate the proposed method, SCEME, against six different adversarial attacks, viz., digital miscategorization attack, digital hiding attack, digital appearing attack, physical miscategorization attack, physical hiding attack, and physical appearing attack, on Faster R-CNN (the general idea can be applied more broadly). We analyze how different kinds of context contribute to the detection performance. We also provide a case study for detecting physical perturbations on stop signs, which has been used widely as a motivating example.

4.1 Implementation Details

Datasets. We use both PASCAL VOC [12] and MS COCO [32]. PASCAL VOC contains 20 object categories. Each image, on average, has 1.4 categories and 2.3 instances [32]. We use *voc07trainval* and *voc12trainval* as training datasets and the evaluations are carried out on *voc07test*. MS COCO contains 80 categories. Each image, on average, has 3.5 categories and 7.7 instances. *coco14train* and *coco14valminusminival* are used for training, and the evaluations are carried out on *coco14minival*. Note that COCO has few examples for certain categories. To make sure we have enough number of context profiles to learn the distribution, we train 11 auto-encoders for the 11 categories that have the largest numbers of extracted context profiles. Details are provided in the supplementary material.

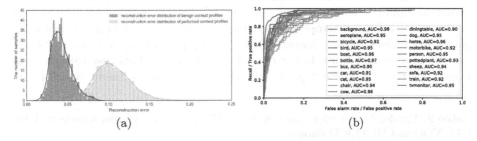

(a)   (b)

Fig. 4. (a) Reconstruction errors of benign aeroplane context profiles are generally smaller than those of the context profiles of digitally perturbed objects that are misclassified as an aeroplane. (b) Thresholding the reconstruction error, we get the detection ROC curves for all the categories on PASCAL VOC dataset.

Attack Implementations. For digital attacks, we use the standard iterative fast gradient sign method (IFGSM) [26] and constrain the perturbation location within the ground truth bounding box of the object instance. Because our defense depends on contextual information, it is not sensitive to how the perturbation is generated. We compare the performance against perturbations generated by a different method (FGSM) in the supplementary material. We use the physical attacks proposed in [13, 43], where perturbation stickers are constrained to be on the object surface; the color of the stickers should be printable, and the pattern of the stickers should be smooth. For evaluations on a large scale, we do not print or add stickers physically; we add them digitally onto the scene image. This favors attackers since they can control how their physical perturbations are captured.

Defense Implementation. Momentum optimizer with momentum 0.9 is used to train SCEME. The learning rate is 5e−4 and decays every 80k iterations at a decay rate of 0.1. The training finishes after 250k iterations. Adam optimizer is used to train auto-encoders. The learning rate is 1e−4 and reduced by 0.1 when the training loss stops decreasing for 2 epochs. Training finishes after 10 epochs.

4.2   Evaluation of Detection Performance

Evaluation Metric. We extract the context profile for each proposed region, feed it to its corresponding auto-encoder and threshold the reconstruction error to detect adversarial perturbations. Therefore, we evaluate the detection performance at the region level. Benign/negative regions are the regions proposed from clean objects; perturbed/positive regions are the regions relating to perturbed objects. We report Area Under Curve (AUC) of Receiver Operating Characteristic Curve (ROC) to evaluate the detection performance. Note that there can be multiple regions of a perturbed object. If any of these regions is detected, it is a successful perturbation detection. For hiding attacks, there is a possibility of no proposed region; however, it occurs rarely (less than 1%).

Visualizing the Reconstruction Error. We plot the reconstruction error of benign aeroplane context profiles and that of digitally perturbed objects that are misclassified as an aeroplane. As shown in Fig. 4(a), the context profiles of perturbed regions do not conform with the benign distribution of aeroplanes' context profiles and cause larger reconstruction errors. This test validates our hypothesis that the context profile of each category has a unique distribution.

Table 2. The detection performance (ROC-AUC) against six different attacks on PASCAL VOC and MS COCO dataset

Method	Digital perturbation			Physical perturbation		
	Miscateg	Hiding	Appearing	Miscateg	Hiding	Appearing
Results on PASCAL VOC						
FeatureSqueeze [47]	0.724	0.620	0.597	0.779	0.661	0.653
Co-occurGraph [3]	0.675	–	–	0.810	–	–
SCEME (node features only)	0.866	0.976	0.828	0.947	0.964	0.927
SCEME	0.938	0.981	0.869	0.973	0.976	0.970
Results on MS COCO						
FeatureSqueeze [47]	0.681	0.682	0.578	0.699	0.687	0.540
Co-occurGraph [3]	0.605	–	–	0.546	–	–
SCEME (node features only)	0.901	0.976	0.810	0.972	0.954	0.971
SCEME	0.959	0.984	0.886	0.989	0.968	0.989

The auto-encoder that learns from the context profile of class A will not reconstruct class B well.

Detection Performance. Thresholding the reconstruction error, we plot the ROC curve for "aeroplane" and other object categories tested on PASCAL VOC dataset, in Fig. 4(b). The AUCs for all 21 categories (including background) are all over 90%. This means that all the categories have their unique context profile distributions, and the reconstruction error of their auto-encoders effectively detect perturbations. The detection performance results, against six attacks on PASCAL VOC and MS COCO, are shown in Table 2. Three baselines are considered.

- *FeatureSqueeze* [47]. As discussed in Table 1, many existing adversarial perturbation detection methods are not effective beyond simple datasets. Most require perturbed samples while training, and only few can be extended to region-level perturbation detection. We extend FeatureSqueeze, one of the state-of-the-art methods, that is not limited by these, for the object detection task. Implementation details are provided in the supplementary material.
- *Co-occurGraph* [3]. We also consider a non-deep graph model where co-occurrence context is represented, as a baseline. We check the inconsistency between the relational information in the training data and testing images to detect attacks. Details are in the supplementary material. Note that the co-occurrence statistics of background class cannot be modeled, and so this approach is inapplicable for detecting hiding and appearing attacks.
- *SCEME (node features only)*. Only node features are used to train the auto-encoders (instead of using context profiles with both node features for region representation and edge features for contextual relation representation). Note that the node features already implicitly contain context information since, with Faster R-CNN, the receptive field of neurons grows with depth and eventually covers the entire image. We use this baseline to quantify the improvement we achieve by explicitly modeling context information with SCEME.

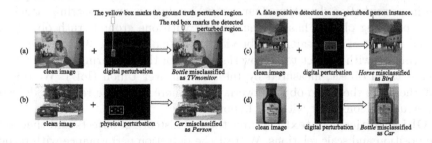

Fig. 5. A few interesting examples. SCEME successfully detects both digital and physical perturbations as shown in (a) and (b). (c) shows that the horse misclassification affects the context profile of person and leads to false positive detection on the person instance. (d) Appearance information and spatial context are used to successfully detect perturbations.

Our method SCEME, yields high AUC on both datasets and for all six attacks; many of them are over 0.95. The detection performance of SCEME is consistently better than that of FeatureSqueeze, by over 20%. Compared to Co-occurGraph, the performance of our method in detecting miscategorization attacks, is better by over 15%. Importantly, SCEME is able to detect hiding and appearing attacks and detect perturbations in images with one object, which is not feasible with Co-occurGraph. Using node features yields good detection performance and further using edge features, improves performance by up to 8% for some attacks.

Examples of Detection Results. We visualize the detected perturbed regions for both digital and physical miscategorization attack in Fig. 5. The reconstruction error threshold is chosen to make the false positive rate 0.2%. SCEME successfully detects both digital and physical perturbations as shown in Fig. 5(a)and(b). The misclassification of the perturbed object could affect the context information of another coexisting benign object and lead to a false perturbation detection on the benign object as shown in Fig. 5(c). We observe that this rarely happens. In most cases, although some part of the object-object context gets violated, the appearance representation and other context would help in making the right detection. When there are not many object-object context relationships as shown in Fig. 5(d), appearance information and spatial context are mainly used to detect a perturbation.

4.3 Analysis of Different Contextual Relations

In this subsection, we analyze what roles different kinds of context features play.

Spatial context consistency means that nearby regions of the same object should yield consistent prediction. We do two kinds of analysis. The first one is to observe the correlations between the adversarial detection performance and the number of regions proposed by the target Faster R-CNN for the perturbed object. Figure 6(a) shows that the detection performance improves when more regions are proposed for the object and this correlation is not observed for the baseline method (for both datasets). This indicates that spatial context plays a role in perturbation detection. Our second analysis is on appearing attacks. If the "appearing object" has a large overlap with one ground truth object, the spatial context of that region will be violated. We plot in Fig. 6(b) the detection performance with respect to the overlap between the appearing object and the ground truth object, measured by Intersection over Union (IoU). We observe that the more these two objects overlap, the more likely the region is detected as perturbed, consistent with our hypothesis.

Object-object context captures the co-existence of objects and their relative position and scale relations. We test the detection performance with respect to the number of objects in the scene images. As shown in Fig. 6(c), in most cases, the detection performance of SCEME first drops or stays stable, and then improves. We believe that the reason is as follows: initially, as the number of objects increases, the object-object context is weak and so is the spatial context

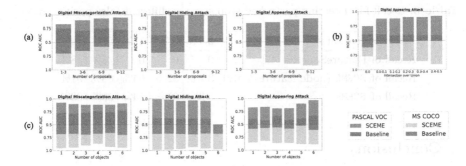

Fig. 6. Subfigures are diverging bar charts. They start with ROC-AUC = 0.5 and diverge in both upper and lower directions: upper parts are results on PASCAL VOC and lower parts are on MS COCO. For each dataset, we show both the results from the FeatureSqueeze baseline and SCEME, using overlay bars. (a) The more the regions proposed, the better our detection performs, as there is more utilizable spatial context; (b) the larger the overlapped region between the "appearing object" and another object, the better our detection performs, as the spatial context violation becomes larger and detectable (we only analyze the appearing attack here); (c) the more the objects, the better our detection performs generally, as there is more utilizable object-object context (performance slightly saturates at first due to inadequate spatial context).

as the size of the objects gets smaller with more of them; however, as the number of objects increases, the object-object context dominates and performance improves.

4.4   Case Study on Stop Sign

We revisit the stop sign example and provide quantitative results to validate that context information helps defend against perturbations. We get 1000 perturbed stop sign examples, all of which are misclassified by the Faster RCNN, from the COCO dataset. The baselines and SCEME, are tested for detecting the perturbations. If we set a lower reconstruction error threshold, we will have a better chance of detecting the perturbed stop signs. However, there will be higher false positives, which means wrong categorization of clean regions as perturbed. Thus, to compare the methods, we constrain the threshold of each method so as to meet a certain *False Positive Rate* (FPR), and compute the *recall* achieved, i.e., out of the 1000 samples, how many are detected as perturbed? The results are shown in Table 3. FeatureSqueeze [47] cannot detect any perturbation until a FPR 5% is chosen. SCEME detects 54% of the perturbed stop signs with a FPR of 0.1%. Further, compared to its ablated version (that only uses node features), our method detects almost twice as many perturbed samples when the FPR required is very low (which is the case in many real-world applications).

Table 3. Recall for detecting perturbed stop signs at different false positive rate.

False positive rate	0.1%	0.5%	1%	5%	10%
Recall of FeatureSqueeze [47]	0	0	0	3%	8%
Recall of SCEME (node features only)	33%	52%	64%	83%	91%
Recall of SCEME	54%	67%	74%	89%	93%

5   Conclusions

Inspired by how humans can associate objects with where and how they appear within a scene, we propose to detect adversarial perturbations by recognizing context inconsistencies they cause in the input to a machine learning system. We propose SCEME, which automatically learns four kinds of context, encompassing relationships within the scene and to the scene holistically. Subsequently, we check for inconsistencies within these context types, and flag those inputs as adversarial. Our experiments show that our method is extremely effective in detecting a variety of attacks on two large scale datasets and improves the detection performance by over 20% compared to a state-of-the-art, context agnostic method.

Acknowledgments. This research was partially sponsored by ONR grant N00014-19-1-2264 through the Science of AI program, and by the U.S. Army Combat Capabilities Development Command Army Research Laboratory under Cooperative Agreement Number W911NF-13-2-0045 (ARL Cyber Security CRA). The views and conclusions contained in this document are those of the authors and should not be interpreted as representing the official policies, either expressed or implied, of the Combat Capabilities Development Command Army Research Laboratory or the U.S. Government. The U.S. Government is authorized to reproduce and distribute reprints for Government purposes notwithstanding any copyright notation hereon.

References

1. Athalye, A., Engstrom, L., Ilyas, A., Kwok, K.: Synthesizing robust adversarial examples. arXiv preprint arXiv:1707.07397 (2017)
2. Bahdanau, D., Cho, K., Bengio, Y.: Neural machine translation by jointly learning to align and translate. arXiv preprint arXiv:1409.0473 (2014)
3. Bappy, J.H., Paul, S., Roy-Chowdhury, A.K.: Online adaptation for joint scene and object classification. In: Leibe, B., Matas, J., Sebe, N., Welling, M. (eds.) ECCV 2016. LNCS, vol. 9912, pp. 227–243. Springer, Cham (2016). https://doi.org/10.1007/978-3-319-46484-8_14
4. Barnea, E., Ben-Shahar, O.: Exploring the bounds of the utility of context for object detection. In: Proceedings of the IEEE Conference on Computer Vision and Pattern Recognition, pp. 7412–7420 (2019)
5. Carlini, N., Wagner, D.: Adversarial examples are not easily detected: bypassing ten detection methods. In: Proceedings of the 10th ACM Workshop on Artificial Intelligence and Security, pp. 3–14 (2017)

6. Chen, S.-T., Cornelius, C., Martin, J., Chau, D.H.P.: ShapeShifter: robust physical adversarial attack on faster R-CNN object detector. In: Berlingerio, M., Bonchi, F., Gärtner, T., Hurley, N., Ifrim, G. (eds.) ECML PKDD 2018. LNCS (LNAI), vol. 11051, pp. 52–68. Springer, Cham (2019). https://doi.org/10.1007/978-3-030-10925-7_4

7. Chen, X., Gupta, A.: Spatial memory for context reasoning in object detection. In: Proceedings of the IEEE International Conference on Computer Vision, pp. 4086–4096 (2017)

8. Cho, K., Van Merriënboer, B., Bahdanau, D., Bengio, Y.: On the properties of neural machine translation: encoder-decoder approaches. arXiv preprint arXiv:1409.1259 (2014)

9. Choi, M.J., Torralba, A., Willsky, A.S.: A tree-based context model for object recognition. IEEE Trans. Pattern Anal. Mach. Intell. **34**(2), 240–252 (2011)

10. Devlin, J., Chang, M.W., Lee, K., Toutanova, K.: BERT: pre-training of deep bidirectional transformers for language understanding. arXiv preprint arXiv:1810.04805 (2018)

11. Dvornik, N., Mairal, J., Schmid, C.: Modeling visual context is key to augmenting object detection datasets. In: Proceedings of the European Conference on Computer Vision (ECCV), pp. 364–380 (2018)

12. Everingham, M., Van Gool, L., Williams, C.K., Winn, J., Zisserman, A.: The pascal visual object classes (VOC) challenge. Int. J. Comput. Vision **88**(2), 303–338 (2010)

13. Eykholt, K., et al.: Robust physical-world attacks on deep learning visual classification. In: Proceedings of the IEEE Conference on Computer Vision and Pattern Recognition, pp. 1625–1634 (2018)

14. Feinman, R., Curtin, R.R., Shintre, S., Gardner, A.B.: Detecting adversarial samples from artifacts. arXiv preprint arXiv:1703.00410 (2017)

15. Felzenszwalb, P.F., Girshick, R.B., McAllester, D., Ramanan, D.: Object detection with discriminatively trained part-based models. IEEE Trans. Pattern Anal. Mach. Intell. **32**(9), 1627–1645 (2009)

16. Gong, Z., Wang, W., Ku, W.S.: Adversarial and clean data are not twins. arXiv preprint arXiv:1704.04960 (2017)

17. Goodfellow, I.J., Shlens, J., Szegedy, C.: Explaining and harnessing adversarial examples. arXiv preprint arXiv:1412.6572 (2014)

18. He, K., Zhang, X., Ren, S., Sun, J.: Deep residual learning for image recognition. In: Proceedings of the IEEE Conference on Computer Vision and Pattern Recognition, pp. 770–778 (2016)

19. Hendrycks, D., Gimpel, K.: Early methods for detecting adversarial images. arXiv preprint arXiv:1608.00530 (2016)

20. Hinton, G.E., Srivastava, N., Krizhevsky, A., Sutskever, I., Salakhutdinov, R.R.: Improving neural networks by preventing co-adaptation of feature detectors. arXiv preprint arXiv:1207.0580 (2012)

21. Hollingworth, A.: Does consistent scene context facilitate object perception? J. Exp. Psychol. Gen **127**(4), 398 (1998)

22. Hu, H., Gu, J., Zhang, Z., Dai, J., Wei, Y.: Relation networks for object detection. In: Proceedings of the IEEE Conference on Computer Vision and Pattern Recognition, pp. 3588–3597 (2018)

23. Huber, P.J.: Robust estimation of a location parameter. In: Kotz, S., Johnson, N.L. (eds.) Breakthroughs in Statistics, pp. 492–518. Springer, New York (1992). https://doi.org/10.1007/978-1-4612-4380-9_35

24. Jia, X., Wei, X., Cao, X., Foroosh, H.: ComDefend: an efficient image compression model to defend adversarial examples. In: Proceedings of the IEEE Conference on Computer Vision and Pattern Recognition, pp. 6084–6092 (2019)
25. Jin, D., Gao, S., Kao, J.Y., Chung, T., Hakkani-tur, D.: MMM: multi-stage multi-task learning for multi-choice reading comprehension. arXiv preprint arXiv:1910.00458 (2019)
26. Kurakin, A., Goodfellow, I., Bengio, S.: Adversarial examples in the physical world. arXiv preprint arXiv:1607.02533 (2016)
27. Li, J., et al.: Attentive contexts for object detection. IEEE Trans. Multimedia 19(5), 944–954 (2016)
28. Li, S., et al.: Stealthy adversarial perturbations against real-time video classification systems. In: NDSS (2019)
29. Li, X., Li, F.: Adversarial examples detection in deep networks with convolutional filter statistics. In: Proceedings of the IEEE International Conference on Computer Vision, pp. 5764–5772 (2017)
30. Liang, B., Li, H., Su, M., Li, X., Shi, W., Wang, X.: Detecting adversarial image examples in deep neural networks with adaptive noise reduction. IEEE Trans. Dependable Secure Comput. (2018)
31. Lin, T.Y., Goyal, P., Girshick, R., He, K., Dollár, P.: Focal loss for dense object detection. In: Proceedings of the IEEE International Conference on Computer Vision, pp. 2980–2988 (2017)
32. Lin, T.-Y., et al.: Microsoft COCO: common objects in context. In: Fleet, D., Pajdla, T., Schiele, B., Tuytelaars, T. (eds.) ECCV 2014. LNCS, vol. 8693, pp. 740–755. Springer, Cham (2014). https://doi.org/10.1007/978-3-319-10602-1_48
33. Liu, J., et al.: Detection based defense against adversarial examples from the steganalysis point of view. In: Proceedings of the IEEE Conference on Computer Vision and Pattern Recognition, pp. 4825–4834 (2019)
34. Liu, W., et al.: SSD: single shot MultiBox detector. In: Leibe, B., Matas, J., Sebe, N., Welling, M. (eds.) ECCV 2016. LNCS, vol. 9905, pp. 21–37. Springer, Cham (2016). https://doi.org/10.1007/978-3-319-46448-0_2
35. Liu, Y., Wang, R., Shan, S., Chen, X.: Structure inference net: object detection using scene-level context and instance-level relationships. In: Proceedings of the IEEE Conference on Computer Vision and Pattern Recognition, pp. 6985–6994 (2018)
36. Lu, J., Issaranon, T., Forsyth, D.: SafetyNet: detecting and rejecting adversarial examples robustly. In: Proceedings of the IEEE International Conference on Computer Vision, pp. 446–454 (2017)
37. McCool, C., Perez, T., Upcroft, B.: Mixtures of lightweight deep convolutional neural networks: applied to agricultural robotics. IEEE Robot. Autom. Lett. 2(3), 1344–1351 (2017)
38. Metzen, J.H., Genewein, T., Fischer, V., Bischoff, B.: On detecting adversarial perturbations. arXiv preprint arXiv:1702.04267 (2017)
39. Mottaghi, R., et al.: The role of context for object detection and semantic segmentation in the wild. In: Proceedings of the IEEE Conference on Computer Vision and Pattern Recognition, pp. 891–898 (2014)
40. Oliva, A., Torralba, A., Castelhano, M.S., Henderson, J.M.: Top-down control of visual attention in object detection. In: Proceedings 2003 International Conference on Image Processing (Cat. No. 03CH37429), vol. 1, pp. I-253. IEEE (2003)
41. Redmon, J., Divvala, S., Girshick, R., Farhadi, A.: You only look once: unified, real-time object detection. In: Proceedings of the IEEE Conference on Computer Vision and Pattern Recognition, pp. 779–788 (2016)

42. Ren, S., He, K., Girshick, R., Sun, J.: Faster R-CNN: towards real-time object detection with region proposal networks. In: Advances in Neural Information Processing Systems, pp. 91–99 (2015)
43. Song, D., et al.: Physical adversarial examples for object detectors. In: 12th USENIX Workshop on Offensive Technologies (WOOT 2018) (2018)
44. Szegedy, C., Vanhoucke, V., Ioffe, S., Shlens, J., Wojna, Z.: Rethinking the inception architecture for computer vision. In: Proceedings of the IEEE Conference on Computer Vision and Pattern Recognition, pp. 2818–2826 (2016)
45. Torralba, A.: Contextual priming for object detection. Int. J. Comput. Vision 53(2), 169–191 (2003)
46. Xie, J., Yang, J., Ding, C., Li, W.: High accuracy individual identification model of crested ibis (Nipponia Nippon) based on autoencoder with self-attention. IEEE Access 8, 41062–41070 (2020)
47. Xu, W., Evans, D., Qi, Y.: Feature squeezing: detecting adversarial examples in deep neural networks. arXiv preprint arXiv:1704.01155 (2017)
48. Zhao, Y., Zhu, H., Liang, R., Shen, Q., Zhang, S., Chen, K.: Seeing isn't believing: towards more robust adversarial attack against real world object detectors. In: Proceedings of the 2019 ACM SIGSAC Conference on Computer and Communications Security, pp. 1989–2004 (2019)
49. Zhu, S., et al.: A4: evading learning-based adblockers. arXiv preprint arXiv:2001.10999 (2020)

Perceive, Predict, and Plan: Safe Motion Planning Through Interpretable Semantic Representations

Abbas Sadat[1]([✉]), Sergio Casas[1,2], Mengye Ren[1,2], Xinyu Wu[1],
Pranaab Dhawan[1], and Raquel Urtasun[1,2]

[1] Uber ATG, Toronto, Canada
{asadat,sergio.casas,mren3,xinyuw,pdhawan,urtasun}@uber.com
[2] University of Toronto, Toronto, Canada

Abstract. In this paper we propose a novel end-to-end learnable network that performs joint perception, prediction and motion planning for self-driving vehicles and produces interpretable intermediate representations. Unlike existing neural motion planners, our motion planning costs are consistent with our perception and prediction estimates. This is achieved by a novel differentiable semantic occupancy representation that is explicitly used as cost by the motion planning process. Our network is learned end-to-end from human demonstrations. The experiments in a large-scale manual-driving dataset and closed-loop simulation show that the proposed model significantly outperforms state-of-the-art planners in imitating the human behaviors while producing much safer trajectories.

1 Introduction

The goal of an autonomy system is to take the output of the sensors, a map, and a high-level route, and produce a safe and comfortable ride. Meanwhile, producing interpretable intermediate representations that can explain why the vehicle performed a certain maneuver is very important in safety critical applications such as self-driving, particularly if a bad event was to happen. Traditional autonomy stacks produce interpretable representations through the perception and prediction modules in the form of bounding boxes as well as distributions over their future motion [1–6]. However, the perception module involves thresholding detection confidence scores and running Non-Maximum Supression (NMS) to trade off the precision and recall of the object detector, which cause information loss that could result in unsafe situations, e.g., if a solid object is below the threshold. To handle this, software stacks in industry rely on a secondary

---

A. Sadat and S. Casas—Equal contribution.

Electronic supplementary material The online version of this chapter (https://doi.org/10.1007/978-3-030-58592-1_25) contains supplementary material, which is available to authorized users.

© Springer Nature Switzerland AG 2020
A. Vedaldi et al. (Eds.): ECCV 2020, LNCS 12368, pp. 414–430, 2020.
https://doi.org/10.1007/978-3-030-58592-1_25

fail safe system that tries to catch all mistakes from perception. This system is however trained separately and it is not easy to decide which system to trust.

First attempts to perform end-to-end neural motion planning did not produce interpretable representations [7], and instead focused on producing accurate control outputs that mimic how humans drive [8]. Recent approaches [9–11], have tried to incorporate interpretability. The neural motion planner of [10] shared feature representations between perception, prediction and motion planning. However it can produce inconsistent estimates between the modules, as it is framed as a multi-task learning problem with separate headers between the tasks. As a consequence, the motion planner might ignore detections or motion forecasts, resulting in unsafe behaviors.

In this paper we take a different approach, and exploit a novel semantic layer as our intermediate interpretable representation. Our approach is designed with safety in mind, and thus does not rely on detection and/or thresholded activations. Instead, we propose a flexible yet efficient representation that can capture different shapes (not just rectangular objects) and can handle low-confidence objects. In particular, we generate a set of probabilistic semantic occupancy layers over space and time, capturing locations of objects of different classes (i.e., vehicles, bicyclists, and pedestrians) as well as potentially occluded ones. Our motion planner can then use this intermediate representation to penalize maneuvers that intersect regions with higher occupancy probability. Importantly, our interpretable representation is differentiable, enabling end-to-end learning of the full autonomy system (i.e., from raw sensor data to planned trajectory). Additionally, as opposed to other neural motion planners [10], our approach can utilize the intended high-level route not only to plan a trajectory that achieves the goal, but also to further differentiate semantically between on-coming or conflicting traffic. This allows the motion planner to potentially learn the risk with respect to a particular semantic class (e.g., moving close to an oncoming vehicle compared to a parked vehicle).

We demonstrate the effectiveness of our approach on a large-scale dataset that consists of smooth manual-driving in challenging urban scenarios. Furthermore, we use a state-of-the-art sensor simulation to perform closed-loop evaluations of driving behavior produced by our proposed model. We show that our method is capable of imitating human trajectories more closely than existing approaches while yielding much lower collision rate.

2   Related Work

End-to-End Self-Driving: There is a vast literature on end-to-end approaches to tackle self-driving. [7] pioneered this field using a single neural network to directly output driving control command. More recently, with the success of deep learning, direct control based methods have advanced with deeper network architectures, more complex sensor inputs, and scalable learning methods [12–15]. Although directly outputting driving command is a general solution, it may have stability and robustness issues [16]. Another line of work first outputs the cost map of future trajectories, and then a trajectory is recovered by looking for local minima on the cost map. The cost map may be parameterized as a simple linear combination of hand crafted costs [17,18], or can be defined in a general non-parametric form [10]. More recently, cost map based approaches have been shown to adapt better to more challenging environments. [19] proposes to output a navigation cost map without localization under a weakly supervised learning environment. [20] has exploited CNNs to facilitate better sampling in complex driving environments. [21–23] explore ways to perceive and map the environment in an end-to-end framework with planning, but do not predict how the world might unroll in the future. In contrast, our planner relies on interpretable cost terms that use the predicted semantic occupancy maps and hence maintains interpretability and differentiability.

Imitation Learning and Inverse Reinforcement Learning: Our proposed learning algorithm is an instantiation of max-margin planning [24], which is closely related to imitation learning and inverse reinforcement learning. *Imitation learning* attempts to directly regress the control commands from human demonstrations [7,8,11,12,14,25–27]. As this can be a very difficult regression problem needsing large amounts of training data, [15] investigates the possibility of transferring knowledge from a simulated environment.

Instead of regressing driving control commands, *max-margin planning* reasons about the cost associated with each output trajectory [10,17,18,24]. It tries to make the human driving trajectories the least costly among all possible trajectories, and penalizes for any violations. It also considers the task loss as in any behavioral differences in the trajectory representation. *Inverse reinforcement learning* (IRL) is similar to max-margin planning, where the best trajectory is replaced by a distribution over trajectories that is characterized by their energy [28,29]. Generative adversarial models have also been explored in the field of IRL and imitation learning [30], so that the model learns to generate trajectories that look similar to human demonstrations judged by a classifier network.

Multi-task Learning: Our end-to-end framework adopts multi-task learning, where we train the model on a joint objective of object detection, occupancy forecasting, and motion planning. Multi-task learning has been shown to help extract more useful information from training data by exploiting task relatedness. [31,32] showed that detection and tracking can be trained together, and [2] applies a joint detector and trajectory predictor into a single model in the context of self-driving. This was further extended by [3] to also predict the high-level intention of actors. More recently, [10] further included a cost map based motion planner in the joint model. These works show that joint learning on a multi-task objective helps individual tasks due to better data utilization and shared features, while saving computation.

Perception and Motion Prediction: The majority of previous works have adopted bounding-box detection and trajectory prediction to reason about the future state of a driving scene [2–6,33–38]. As there are multiple possible futures, these methods either generate a fixed number of modes with probabilities and/or draw samples to characterize the trajectory distribution. In robotics, occupancy grids have been a popular representation of free space. In [39], a framework is proposed to estimate occupancy probability of each grid-cell independently using range sensor data. This approach is later extended in [40] to model dependencies among neighboring cells. [41] performs dynamic occupancy grid prediction at the scene-level, but it does not predict how the scene might evolve in the future. [42] proposes a discrete residual flow to predict the distribution over a pedestrian's future position in the form of occupancy maps. Similarly, in [43] agent-centric occupancy grids are predicted from past trajectories using ConvLSTMs, and multiple trajectories are then sampled to form possible futures. [44] further improves this output parameterization by predicting a continuous offset to mitigate discretization errors, and proposes a procedure to extract trajectory samples. Different from these methods, our proposed semantic perception and future prediction is instance-free and directly produces occupancy layers for the entire scene, rather than for each actor instance, which makes it efficient. Moreover, since no thresholding is employed on the detection scores, our model allows passing low probability objects to the motion planner and hence improving safety of self-driving.

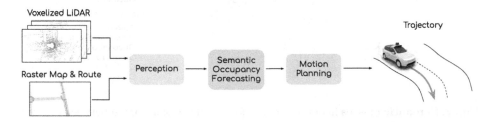

Fig. 1. Overview of our proposed end-to-end learnable autonomy system that takes raw sensor data, an HD map and a high level route as input and produces safe maneuvers for the self-driving vehicle via our novel semantic interpretable intermediate representations.

3   End-to-End Interpretable Neural Motion Planner

In this paper we propose an end-to-end approach to self-driving. Importantly, our model produces intermediate representations that are designed for safe planning and decision-making, together with interpretability. Towards this goal, we exploit the map, the intended route (high level plan to go from point A to point B), and the raw LiDAR point-cloud to generate an intermediate semantic occupancy representation over space and time (i.e., present and future time steps). These interpretable occupancy layers inform the motion planner about potential objects, including those with low probability, allowing perception of objects of arbitrary shape, rather than just bounding boxes. This is in contrast to existing approaches [3, 10, 25] that rely on object detectors that threshold activations and produce objects with only bounding box shapes. Note that thresholding activations is very problematic for safety, as if an object is below the threshold it will not be detected, potentially resulting in a collision.

Our semantic activations are very interpretable. In particular, we generate occupancy layers for each class of vehicles, bicyclists, and pedestrians, as well as occlusion layers which predict occluded objects. Furthermore, using the planned route of the self-driving vehicle (SDV), we can semantically differentiate vehicles by their interaction with our intended route (e.g., oncoming traffic vs. crossing). This not only adds to the interpretability of the perception outputs, but can potentially help the planner learn different subcosts for each category (e.g., different safety buffers for parked vehicles vs oncoming traffic). We refer the reader to Fig. 2 for an exhaustive list of the classes that we predict in the different layers of our occupancy maps.

Semantic hierarchy

Pedestrians
    Visible/Partially visible
    Occluded

Bikes
    Visible/Partially visible
    Occluded

Vehicles

    Parked
    On conflicting lane
    Oncoming traffic
    On the route
    Occluded
    Others

Fig. 2. Semantic classes in our occupancy forecasting. Colors match between drawing and hierarchy. Shadowed area corresponds to the SDV route. Black vehicle, pedestrian and bike icons represent the agents' true current location. (Color figure online)

Our sample-based learnable motion planner then takes these occupancy predictions and evaluates the associated risk of different maneuvers to find a safe and comfortable trajectory for the SDV. This is accomplished through an interpretable cost function used to cost motion-plan samples, which can efficiently exploit the occupancy information. Importantly, our proposed autonomy model is trained end-to-end to imitate human driving while avoiding collisions and traffic infractions. Figure 1 shows an overview of our proposed approach.

3.1   Perceiving and Forecasting Semantic Occupancies

Our model exploits LiDAR point clouds and HD maps to predict marginal distributions of semantic occupancy over time, as shown in Fig. 3. These are spatio-temporal, probabilistic, and instance-free representations of the present and future that capture whether a spatial region is occupied by any dynamic agent belonging to a semantic group at discrete time steps. Note that this representation naturally captures multi-modality in the future behavior of actors by placing probability mass on different spatial regions at future time steps, which is important as the future might unroll in very different ways (e.g., vehicle in front of the SDV brakes/accelerates, a pedestrian jaywalks/stays in the sidewalk).

Input Representation: We use several consecutive LiDAR sweeps as well as HD maps (including lane graphs) as input to our model as they bring complementary information. Following [10], we voxelize $T_p = 10$ past LiDAR point clouds in bird's eye view (BEV) with a resolution of $a = 0.2$ m/voxel. Our region of interest is $W = 140$ m long (70 m front and behind of the SDV), $H = 80$ m wide (40 to each side of the SDV), and $Z = 5$ m tall; obtaining a 3D tensor of size $(\frac{H}{a}, \frac{W}{a}, \frac{Z}{a}, \cdot T_p)$. As proposed in [3], we concatenate height and time along the channel dimension to avoid using 3D convolutions or a recurrent model, thus saving memory and computation. Leveraging map information is very important to have a safe motion planner as we need to drive according to traffic rules such as stop signs, traffic lights and lane markers. Maps are also very relevant for perception and motion forecasting since they provide a strong prior on the presence as well as the future motion of traffic participants (e.g., vehicles and bikes normally follow lanes, pedestrians usually use sidewalks/crosswalks). To exploit HD maps, we adopt the representation proposed in [3] and rasterize different semantic elements (e.g., roads, lanes, intersections, crossings) into different binary channels to enable separate reasoning about the distinct elements. For instance, the state of a traffic light (green, yellow, red) is rasterized in 3 different channels, facilitating traffic flow reasoning at intersections. All in all, we obtain a 3D tensor of size $(\frac{H}{a}, \frac{W}{a}, C)$, with $C = 17$ binary channels for the map.

Backbone Network: We combine ideas in [45] and [3] to build a multi-resolution, two-stream backbone network that extracts features from the LiDAR voxelization and map raster. One stream processes LiDAR while the other one processes the map. Each stream is composed of 4 residual blocks with number of layers (2, 2, 3, 6) and stride (1, 2, 2, 2) respectively. Thus, the features after each residual block are $\mathcal{F}_{1x}, \mathcal{F}_{2x}, \mathcal{F}_{4x}, \mathcal{F}_{8x}$, where the subscript indicates the downsampling factor from the input in BEV. The features from the different blocks are then concatenated at 4x downsampling by max pooling higher resolution ones $\mathcal{F}_{1x}, \mathcal{F}_{2x}$ and interpolating $\mathcal{F}_{8x}$, as proposed by [45]. The only difference between the two streams is that the LiDAR one uses more features (32, 64, 128, 256) versus (16, 32, 64, 128) on the map stream. We give more capacity to the LiDAR branch as the input is much higher dimensional than the raster map, and the backbone is responsible for aggregating geometric information from different past LiDAR sweeps to extract good appearance and motion cues. Finally, the LiDAR and map features are fused by concatenation along the feature dimension followed by a final residual block of 4 convolutional layers with no downsampling, which outputs a tensor $\mathcal{F}$ with 256 features.

Semantic Occupancy Forecasting: Predicting the future motion of traffic participants is very challenging as actors can perform complex motions and there is a lot of uncertainty due to both partial observability (because of sensor occlusion or noise) as well as the multi-modal nature of the possible outcomes.

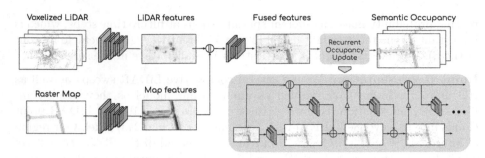

Fig. 3. Inference diagram of our proposed perception and recurrent occupancy forecasting model. ‖ symbolizes concatenation along the feature dimension, ⊕ element-wise sum and △ bilinear interpolation used to downscale the occupancy.

Many existing approaches have modeled the underlying distribution in a parametric way (e.g., Gaussian, mixture of Gaussians) [4,6]. While efficient, this incorporates strong assumptions, lacks expressivity and is prone to instabilities during optimization (see [42]). Jain et al. [42] use non-parametric occupancy distributions for each instance (i.e., actor) naturally capturing complex multi-modal distributions. However, this is a computationally and memory inefficient representation that scales poorly with the number of actors, which can be hundreds in crowded scenes. In contrast, in this paper we propose a novel representation, where groups of actors are modeled with a single non-parametric distribution of future semantic occupancy. This removes the need for both detection and tracking and is both efficient and effective as shown in our experiments.

In particular, these actors are grouped semantically in a hierarchy. We consider vehicles, pedestrians and bikes as the root semantic classes $\mathcal{C}$, as shown in Fig. 2. For each root category, we consider mutually-exclusive subclasses which include a negative (not occupied) subclass. Note that the root categories are not mutually exclusive as actors that belong to different classes can share the same occupied space (e.g., pedestrian getting in or out of a car). We create these subdivisions because we wish to learn different planning costs for each of these subclass occupancies, given that such subcategories have very different semantics for driving. For instance, parked vehicles require a smaller safety buffer than a fast moving vehicle in a lane that conflicts with the SDV route, since they are not likely to move and therefore the uncertainty around them is smaller. We do not subdivide the pedestrians and bikes (with riders) by their semantic location (road/sidewalk) or behavior (stationary/moving), as they are vulnerable road users and thus we want to make sure we plan a safe maneuver around them, no matter their actions. We model fully occluded traffic participants (i.e., vehicles, pedestrians, bikes) through additional occupancy maps, just by adding one more subcategory. This can then be used for motion planning to exert caution.

More precisely, we represent the occupancy of each class $c \in \mathcal{C}$ as a collection of categorical random variables $o_{i,j}^{t,c}$ over space and time. Space is discretized into a BEV spatial grid on the ground plane with a resolution of $0.4\,\mathrm{m/pixel}$, where i, j denotes the spatial location. Time is discretized into 11 evenly spaced horizons into the future, ranging from 0 to 5 s, every 0.5 s.

To obtain the output logits $l^{t,c}$ of these spatio-temporal discrete distributions we employ a multi-scale context fusion by performing two parallel fully convolutional networks with different dilation rates. One stream performs regular 2D convolutions over $\mathcal{F}_{2x}$, providing very local, fine-grained features needed to make accurate predictions in the recent future. The other stream takes the coarser features $\mathcal{F}$ and performs dilated 2D convolutions to obtain a bigger receptive field that is able to place occupancy mass far away from the initial actor location for those that move fast. We then concatenate the two feature maps into $\mathcal{F}_{\mathrm{occ}}$. Finally, we design an efficient recurrent occupancy update for each root class to output the logits for all its subclasses

$$l^{t,c} = l^{t-1,c} + \mathcal{U}_\theta^t(\mathcal{F}_{\mathrm{occ}} \parallel \mathcal{I}(l^{0:t-1,c}))$$

where $\mathcal{U}_\theta^t$ is a neural network that contains a transposed convolution to upsample the resolution by 2, $l^{0:t-1,c}$ are the predicted logits up to timestep $t-1$, $\mathcal{I}$ is a 2x bilinear interpolation, and $\parallel$ represents feature-wise concatenation. We perform the recurrence at a lower resolution to reduce the memory impact. We refer the reader to Fig. 3 for a detailed illustration of the recurrency. Recurrent convolutions provide the right inductive bias to express the intuition that further future horizons need a bigger receptive field, given that actors could have moved away from their starting location. Finally, to output the categorical distribution $o_{i,j}^{t,c}$ we use a softmax across the mutually-exclusive subclasses of the root class c, for each space grid cell i, j and time horizon t.

3.2 Motion Planning

Given the occupancy predictions $\mathbf{o}$ and the input data $\mathbf{x}$ in the form of the HD-map, the high level route, traffic-lights states, and the kinematic state of the SDV, we perform motion planning by sampling a diverse set of trajectories for the ego-car and pick the one that minimizes a learned cost function as follows:

$$\tau^* = \underset{\tau}{\mathrm{argmin}} \, f(\tau, \mathbf{x}, o; w) \qquad (1)$$

Here w represents the learnable parameters of the planner. The objective function f is composed of subcosts, f_o, that make sure the trajectory is safe with regards to the semantic occupancy forecasts, as well as other subcosts, f_r related to comfort, traffic rules and progress in the route (see Fig. 4). Thus

$$f(\tau, \mathbf{x}, o; w) = f_o(\tau, \mathbf{x}, o; w_o) + f_r(\tau, \mathbf{x}, o; w_r) \qquad (2)$$

with $w = (w_r, w_o)$ the vector of all learnable parameters for the motion planner. We now describe the safety costs in details, as it is one of our major contributions. We include a very brief explanation of f_r and refer the reader to the supplementary material for more details.

Safety Cost: The SDV should not collide with other objects on the road and needs to navigate cautiously when it is uncertain. For this purpose, we use the predicted semantic-occupancy o to penalize trajectories that intersect occupied regions. In particular, at each time step t of trajectory τ, we find all the cells in the occupancy layer that have intersection with the SDV polygon (with a safety margin indicated by parameter λ), and conservatively use the value of the cell with maximum probability as occupancy subcost, denoted by $o_c(\tau, t, \lambda)$. Then the safety cost is computed by

$$f_o(\tau, o) = \sum_t \sum_c w_c o_c(\tau, t, 0) + w_{cv} o_c(\tau, t, \lambda) v(\tau, t) \qquad (3)$$

with w_c and w_{cv} the weighting parameters. Note that the first term penalizes trajectories that intersect regions with high occupancy probability whereas the second term penalizes high-velocity motion in areas with uncertain occupancy.

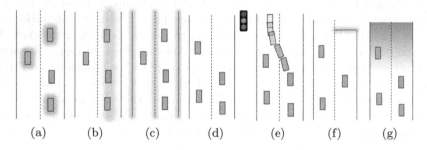

(a)   (b)   (c)   (d)   (e)   (f)   (g)

Fig. 4. Examples of the motion planner cost functions: (a) collision, (b) driving-path, (c) lane boundary, (d) traffic-light, (e) comfort, (f) route, (g) progress.

Traffic Rules, Comfort and Route Progress Costs: The trajectory of the SDV must obey traffic rules. We use the information available in the map to penalize trajectories that violate the lane boundaries, road boundaries, stop signs, red traffic-lights, speed-limit, and do not stay close to the lane center. As it is common in self-driving systems, the mission route is given to our planner as a sequence of lanes that the SDV needs to follow to reach the destination. We penalize the number of lane-changes required to switch to these lanes. This encourages behaviors that are consistent with the input route. Additionally, in order to promote comfortable driving, we penalize trajectories for acceleration and violation thereof, lateral acceleration and violation thereof, jerk and violation thereof, curvature and its first and second derivatives. Note that the violations are computed with regards to a predefined threshold that is considered comfortable. Figure 4 shows some of the described cost functions.

Trajectory Parametrization and Sampling: The output of the motion planner is a sequence of vehicle states that describes how the SDV should move within the planning horizon. At each planning iteration, a set of sampled trajectories are evaluated using the cost function in Eq. 2, and the one with minimum cost is selected for execution. It is important that the sampled set, while being small enough to allow real-time computation, cover various maneuvers such as lane-following, lane-changes, and nudging encroaching objects. Hence, to achieve this efficiently, we choose a sampling approach that is aware of the lane structures. In particular, we follow the trajectory parameterization and sampling procedure proposed in [17,46], where trajectories are sampled by combining longitudinal motion and lateral deviations relative to a particular lane (e.g., current SDV lane, right lane). Consequently, the sampled trajectories correspond to appropriate lane-based driving with variations in lateral motions which can be applied to many traffic scenarios. The details of the sampling algorithm are presented in the supplementary material.

3.3   Learning

We trained our full model of perception, prediction and planning end-to-end. The final goal is to be able to drive safely and comfortably similar to human demonstrations. Additionally, the model should forecast the semantic occupancy distributions that are similar to what happened in the real scene. We thus learn the model parameters by exploiting these two loss functions:

$$\mathcal{L} = \lambda_S \mathcal{L}_S + \lambda_M \mathcal{L}_M \tag{4}$$

Semantic Occupancy Loss: This loss is defined as the cross entropy between the ground truth distribution p and the predicted distribution q_ϕ of the semantic occupancy random variables $o_{i,j}^{t,c}$.

$$\mathcal{L}_S = H(p, q_\theta) = -\sum_t \sum_{c \in \mathcal{C}} \sum_{i,j \in \mathcal{S}^{t,c}} \sum_{o_{i,j}^{t,c} \in \mathcal{O}^c} p(o_{i,j}^{t,c}) \log q_\theta(o_{i,j}^{t,c}) \tag{5}$$

Due to the highly imbalanced data in terms of spatial occupancy since the majority of the space is free, we obtain the subset of spatial locations $\mathcal{S}^{t,c}$ at time t for class c by performing hard negative mining.

Planning Loss: Since selecting the minimum-cost trajectory within a discrete set is not differentiable, we use the max-margin loss to penalize trajectories that have small cost and are different from the human driving trajectory or are unsafe. Let $\mathbf{x}$ and τ_h be the input and human trajectory respectively for a given example. We utilize the max margin loss to encourage the human driving trajectory to have smaller cost f than other trajectories. In particular,

$$\mathcal{L}_M = \max_\tau \left[f_r(\mathbf{x}, \tau_h) - f_r(\mathbf{x}, \tau) + l_{\text{im}} + \sum_t \left[f_o^t(\mathbf{x}, \tau_h) - f_o^t(\mathbf{x}, \tau) + l_o^t \right]_+ \right]_+ \tag{6}$$

where f_o^t is the occupancy cost function at time step t, f_r is the rest of the planning subcosts as defined in Sect. 3.2 (note that we omitted o and w from

f for brevity), and $[]_+$ represents the ReLU function. The imitation task-loss l_{im} measures the ℓ_1 distance between trajectory τ and the ground-truth for the entire horizon, and the safety task-loss l_o^t accounts for collisions and their severity at each trajectory step.

4   Experimental Evaluation

Dataset and Training: We train our models using our large-scale dataset that includes challenging scenarios where the operators are instructed to drive smoothly and in a safe manner. It contains 6100 scenarios for the training set, while validation and test sets contain 500 and 1500 scenarios. Each scenario is 25 s. Compared to KITTI [47], our dataset has 33x more hours of driving and 42x more objects. We use exponentiated gradient decent to update the planner parameters and Adam optimizer for the occupancy forecasting. We scale the gradient that is passed to perception and prediction from the planner to avoid instability in P&P training.

Baselines: We compare against the following baselines: **ACC** which performs a simple car-following behavior using the measured state of the lead vehicle. **Imitation Learning (IL)** where the future positions of the SDV are predicted directly from the fused LiDAR and map features (Fig. 3), and is trained using L2 loss. **NMP** [10] where a planning cost-map is predicted from the fused features directly and detection and predictions are treated only as an axillary task. **PLT**: which is the joint behavior-trajectory planning method of [17], where planning is accomplished using a combination of interpretable subcosts, including collision costs with regards to predicted trajectories of actors. However, the detection and prediction modules are trained separately from the planner.

Metrics: Planning metrics include **cumulative collision rate** indicating the percentage of collisions with ground-truth bounding-boxes of the actors at each trajectory time step, **L2 distance** to human trajectory which indicates how well the model imitated the human driving, **jerk** and **lateral acceleration** which show how comfortable the produced trajectories are. We also measure the **progress** of the SDV along the route.

Results: The first set of experiments are performed in an open-loop setting in which the LiDAR data up to the current timestamp is passed to the model and the generated trajectory is assumed to be executed by the ego vehicle for the 5sec planning horizon (as opposed to closed-loop execution where the trajectory is constantly replanned as new sensor data becomes available). Table 1 shows the planing metrics for our proposed method and the baselines. It shows that our proposed model (P3) outperforms all the baselines in (almost) all planning metrics. In particular, our motion planner generates much safer trajectories, with 40% less collisions at 5 s compared to PLT. It also outperforms NMP by a very significant margin, which could be due to our consistent use of perception and prediction outputs in motion planning, as opposed to the free-form cost volume from sensor data in [10]. Another aspect that we observed to improve safety was the temporally smoother occupancies output by our recurrent occupancy update as opposed to a convolutional one. A more nuanced detail that could also contribute to our increased safety is the pooling of the cost on the space occupied by the SDV, as opposed to the simple indexing on its centroid previously proposed by [10]. Our model also produces less jerk which indicates the effectiveness of including multiple interpretable subcosts in the planning objective. Besides, our model exhibits much closer behavior to human compared to IL that has been optimized to match human trajectories. The progress metric also shows that our model is less aggressive compared to the other baselines and similar to IL.

Table 1. Comparison against other methods

Model	Collisions rate (%)			L2 human			Jerk	Lat. acc.	Progress
	1 s	3 s	5 s	@1 s	@3 s	@5 s			
ACC	0.31	2.00	8.73	0.20	1.75	5.16	1.74	2.87	29.3
IL	1.47	4.33	12.29	0.33	1.46	3.37	–	–	27.5
NMP [10]	0.17	0.72	5.22	0.23	2.19	5.61	4.36	**2.86**	31.7
PLT [17]	0.07	0.40	2.94	**0.18**	1.35	3.80	1.52	3.03	28.0
P3 (ours)	**0.05**	**0.17**	**1.78**	**0.18**	**1.18**	**3.34**	**1.27**	2.89	27.6

Table 2. Ablation study

ID	e2e	Prediction		Collision rate (%)			L2 human			Jerk	Lat. acc.	Progress
		Traj.	Occup.	1 s	3 s	5 s	@1 s	@3 s	@5 s			
$\mathcal{M}_1$		✓		0.07	0.40	2.94	**0.18**	1.35	3.80	1.52	3.03	28.0
$\mathcal{M}_2$	✓	✓		**0.05**	0.32	2.21	**0.18**	1.35	3.65	1.50	2.85	27.8
$\mathcal{M}_3$	✓	✓	✓	**0.05**	0.22	2.36	**0.18**	1.27	3.64	1.38	2.93	28.0
$\mathcal{M}_4$			✓	**0.05**	0.20	1.96	**0.18**	1.21	3.49	**1.23**	**2.78**	27.3
$\mathcal{M}_5$	✓		✓	**0.05**	**0.17**	**1.78**	**0.18**	**1.18**	**3.34**	1.27	2.89	27.6

Ablation Study: We report the result of the ablation study in Table 2. Our best model is $\mathcal{M}_5$ (P3 in Table 1) corresponds to the semantic occupancy and the motion planner being jointly trained. $\mathcal{M}_1$ and $\mathcal{M}_2$ perform detection and multi-modal trajectory prediction which is used in motion planner to form collision costs. Overall, end-to-end training of perception and planning modules improve safety as indicated by the collision metrics. Furthermore, using occupancy representation yields much better performance in driving metrics. The progress metric also indicates that the occupancy model is not overly cautious and the advancement in the route is similar to other models. Note that we also include $\mathcal{M}_3$ which is similar to $\mathcal{M}_1$, but the predicted trajectories are rasterized to form an occupancy representation for motion planning.

Qualitative results: Figure 5 shows examples of generated semantic occupancy layers at different time horizons for two traffic scenarios (refer to the caption for corresponding color of each semantic class). In Fig. 5(c), for example, we can see multiple modes in the prediction of a vehicle with corresponding semantics of conflicting and oncoming. In the bottom scene, our model is able to recognize the occluded region on the right end of the intersection. Furthermore, the oncoming vehicle (red color) which has a low initial velocity is predicted with large uncertainty which is visible in Fig. 5(f).

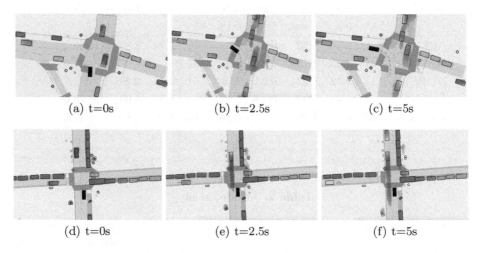

(a) t=0s   (b) t=2.5s   (c) t=5s

(d) t=0s   (e) t=2.5s   (f) t=5s

Fig. 5. Qualitative results: The colors red, orange, dark-green, dark-blue, and purple respectively shows vehicle with sub-categories of oncoming, conflicting, on-route, stationary and others. Pedestrians and bicyclists are shown with light green and brown colors. Cyan color is used to show occlusion for all classes. We also show the ground-truth bounding boxes of all actors, and planned trajectory of the ego vehicle (solid black rectangle) (Color figure online)

Closed-Loop Evaluation: We also perform experiments in a closed-loop simulated environment leveraging realistic LiDAR simulation [48]. At each simulation time-step, the simulated LiDAR point-cloud is passed to our model and a trajectory is planned for the ego vehicle and is executed by the simulation for 100 ms. This process continues iteratively for 15 s of simulation. We tested our models in a scenario with one or two initially-occluded non-compliant actors with trajectories that are in conflict with the route of the ego-vehicle (see Fig 6(a)). By varying the initial velocity and along-the-lane location of each actor, we

Table 3. Closed-loop Evaluation Results See Table 2 for definition of $\mathcal{M}_2$ and $\mathcal{M}_5$. The collision rate shows the percentage of the simulation runs where the SDV had collision with other actors. The rest of the metrics show the mean value over all the simulation steps.

ID	Collision rate (%)	Jerk	Lateral acceleration	Acceleration	Deceleration	Speed
$\mathcal{M}_2$	18.5	4.08	0.24	0.94	-0.79	9.1
$\mathcal{M}_5$	**9.8**	**1.85**	0.17	0.50	-0.50	8.6

created 80 highly challenging traffic scenes (12k frames) for our tests. We compared the performance of our proposed end-to-end autonomy system that uses semantic occupancy with the alternative trajectory-based method ($\mathcal{M}_5$ and $\mathcal{M}_2$ respectively). As shown in Table 3, our full approach can react safely to the non-compliant vehicles, resulting in less collisions than $\mathcal{M}_2$. This cautious behavior is also reflected in the rest of the metrics such as jerk, acceleration, and velocity where $\mathcal{M}_2$ exhibits more aggressive behavior. Figure 6 demonstrates an example run of the simulation. As the SDV approaches the intersection, the non-reactive vehicle, which is turning right, becomes visible (Fig. 6(c)). The planner generates a lane-change trajectory to avoid the slow-moving vehicle (Fig. 6(d)).

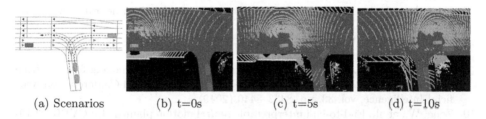

(a) Scenarios   (b) t=0s   (c) t=5s   (d) t=10s

Fig. 6. Closed-loop scenario: The general setup of the closed-loop evaluation is shown in (a) where the SDV (gray vehicle) is approaching an intersection with 3 potential non-reactive vehicles (orange colored) with conflicting routes. The variations of the scenario are generated by including 1 or 2 of the indicated vehicles with various initial speed and location. (b, c, d) show and example of the simulation run at different time horizons. (Color figure online)

5   Conclusion

In this paper, we have proposed an end-to-end perception, prediction and motion planning model that generates safe trajectories for the SDV from raw sensor data. Importantly, our model not only produces interpretable intermediate representations, but also the generated ego-vehicle trajectories are consistent with the perception and prediction outputs. Furthermore, unlike most previous approaches that employ thresholded activations in detection and trajectory prediction of objects, we use semantic occupancy layers that are able to carry information about low probability objects to the motion planning module. Our experiments on a large dataset of challenging scenarios and closed-loop simulations showed that the proposed method, while exhibiting human-like driving behavior, is significantly safer than the state-of-the-art learnable planners.

References

1. Liang, M., Yang, B., Chen, Y., Hu, R., Urtasun, R.: Multi-task multi-sensor fusion for 3d object detection. In: CVPR (2019)
2. Luo, W., Yang, B., Urtasun, R.: Fast and furious: real time end-to-end 3d detection, tracking and motion forecasting with a single convolutional net. In: CVPR (2018)
3. Casas, S., Luo, W., Urtasun, R.: IntentNet: learning to predict intention from raw sensor data. In: CoRL (2018)
4. Chai, Y., Sapp, B., Bansal, M., Anguelov, D.: MultiPath: multiple probabilistic anchor trajectory hypotheses for behavior prediction. arXiv preprint arXiv:1910.05449 (2019)
5. Tang, C., Salakhutdinov, R.R.: Multiple futures prediction. In: Advances in Neural Information Processing Systems, pp. 15398–15408 (2019)
6. Casas, S., Gulino, C., Liao, R., Urtasun, R.: Spatially-aware graph neural networks for relational behavior forecasting from sensor data. arXiv preprint arXiv:1910.08233 (2019)
7. Pomerleau, D.A.: ALVINN: an autonomous land vehicle in a neural network. In: NIPS (1989)
8. Chen, D., Zhou, B., Koltun, V., Krähenbühl, P.: Learning by cheating. arXiv preprint arXiv:1912.12294 (2019)
9. Hou, Y., Ma, Z., Liu, C., Loy, C.C.: Learning to steer by mimicking features from heterogeneous auxiliary networks. In: Proceedings of the AAAI Conference on Artificial Intelligence, vol. 33, pp. 8433–8440 (2019)
10. Zeng, W., et al.: End-to-end interpretable neural motion planner. In: CVPR (2019)
11. Rhinehart, N., McAllister, R., Levine, S.: Deep imitative models for flexible inference, planning, and control. arXiv preprint arXiv:1810.06544 (2018)
12. Bojarski, M., et al.: End to end learning for self-driving cars. arXiv preprint arXiv:1604.07316 (2016)
13. Kendall, A., et al..: Learning to drive in a day. arXiv preprint arXiv:1807.00412 (2018)
14. Codevilla, F., Miiller, M., López, A., Koltun, V., Dosovitskiy, A.: End-to-end driving via conditional imitation learning. In: ICRA (2018)
15. Müller, M., Dosovitskiy, A., Ghanem, B., Koltun, V.: Driving policy transfer via modularity and abstraction. arXiv preprint arXiv:1804.09364 (2018)
16. Codevilla, F., Santana, E., López, A.M., Gaidon, A.: Exploring the limitations of behavior cloning for autonomous driving. In: Proceedings of the IEEE International Conference on Computer Vision, pp. 9329–9338 (2019)
17. Sadat, A., Ren, M., Pokrovsky, A., Lin, Y.C., Yumer, E., Urtasun, R.: Jointly learnable behavior and trajectory planning for self-driving vehicles. In: IROS (2018)
18. Fan, H., Xia, Z., Liu, C., Chen, Y., Kong, Q.: An auto-tuning framework for autonomous vehicles. arXiv preprint arXiv:1808.04913 (2018)

19. Ma, H., Wang, Y., Tang, L., Kodagoda, S., Xiong, R.: Towards navigation without precise localization: weakly supervised learning of goal-directed navigation cost map. CoRR abs/1906.02468 (2019)
20. Banzhaf, H., Sanzenbacher, P., Baumann, U., Zöllner, J.M.: Learning to predict ego-vehicle poses for sampling-based nonholonomic motion planning. IEEE Robot. Autom. Lett. 4(2), 1053–1060 (2019)
21. Gupta, S., Davidson, J., Levine, S., Sukthankar, R., Malik, J.: Cognitive mapping and planning for visual navigation. In: Proceedings of the IEEE Conference on Computer Vision and Pattern Recognition, pp. 2616–2625 (2017)
22. Parisotto, E., Salakhutdinov, R.: Neural map: structured memory for deep reinforcement learning. arXiv preprint arXiv:1702.08360 (2017)
23. Khan, A., Zhang, C., Atanasov, N., Karydis, K., Kumar, V., Lee, D.D.: Memory augmented control networks. arXiv preprint arXiv:1709.05706 (2017)
24. Ratliff, N.D., Bagnell, J.A., Zinkevich, M.A.: Maximum margin planning. In: ICML (2006)
25. Bansal, M., Krizhevsky, A., Ogale, A.: ChauffeurNet: learning to drive by imitating the best and synthesizing the worst. arXiv preprint arXiv:1812.03079 (2018)
26. Zhao, A., He, T., Liang, Y., Huang, H., Broeck, G.V.d., Soatto, S.: LaTeS: latent space distillation for teacher-student driving policy learning. arXiv preprint arXiv:1912.02973 (2019)
27. Hawke, J., et al.: Urban driving with conditional imitation learning. arXiv preprint arXiv:1912.00177 (2019)
28. Ziebart, B.D., Maas, A.L., Bagnell, J.A., Dey, A.K.: Maximum entropy inverse reinforcement learning. In: AAAI (2008)
29. Wulfmeier, M., Ondruska, P., Posner, I.: Deep inverse reinforcement learning. CoRR abs/1507.04888 (2015)
30. Ho, J., Ermon, S.: Generative adversarial imitation learning. In: NIPS (2016)
31. Feichtenhofer, C., Pinz, A., Zisserman, A.: Detect to track and track to detect. In: 2017 IEEE International Conference on Computer Vision (ICCV), pp. 3057–3065. IEEE (2017)
32. Frossard, D., Kee, E., Urtasun, R.: DeepSignals: predicting intent of drivers through visual signals. In: 2019 International Conference on Robotics and Automation (ICRA), pp. 9697–9703. IEEE (2019)
33. Phan-Minh, T., Grigore, E.C., Boulton, F.A., Beijbom, O., Wolff, E.M.: CoverNet: multimodal behavior prediction using trajectory sets. arXiv preprint arXiv:1911.10298 (2019)
34. Liang, M., et al.: PnPNet: end-to-end perception and prediction with tracking in the loop. In: Proceedings of the IEEE/CVF Conference on Computer Vision and Pattern Recognition (CVPR), June 2020
35. Zhao, T., et al.: Multi-agent tensor fusion for contextual trajectory prediction. In: Proceedings of the IEEE Conference on Computer Vision and Pattern Recognition, pp. 12126–12134 (2019)
36. Casas, S., Gulino, C., Suo, S., Urtasun, R.: The importance of prior knowledge in precise multimodal prediction. arXiv preprint arXiv:2006.02636 (2020)
37. Li, L., et al.: End-to-end contextual perception and prediction with interaction transformer. In: 2020 IEEE/RSJ International Conference on Intelligent Robots and Systems (IROS) (2020)
38. Casas, S., Gulino, C., Suo, S., Luo, K., Liao, R., Urtasun, R.: Implicit latent variable model for scene-consistent motion forecasting (2020)
39. Elfes, A.: Using occupancy grids for mobile robot perception and navigation. Computer 22(6), 46–57 (1989)

40. Thrun, S.: Learning occupancy grid maps with forward sensor models. Auton. Robots **15**(2), 111–127 (2003)
41. Hoermann, S., Bach, M., Dietmayer, K.: Dynamic occupancy grid prediction for urban autonomous driving: a deep learning approach with fully automatic labeling. In: 2018 IEEE International Conference on Robotics and Automation (ICRA), pp. 2056–2063. IEEE (2018)
42. Jain, A., et al.: Discrete residual flow for probabilistic pedestrian behavior prediction. arXiv preprint arXiv:1910.08041 (2019)
43. Ridel, D., Deo, N., Wolf, D., Trivedi, M.: Scene compliant trajectory forecast with agent-centric spatio-temporal grids. IEEE Robot. Autom. Lett. **5**(2), 2816–2823 (2020)
44. Liang, J., Jiang, L., Murphy, K., Yu, T., Hauptmann, A.: The garden of forking paths: towards multi-future trajectory prediction. In: Proceedings of the IEEE/CVF Conference on Computer Vision and Pattern Recognition, pp. 10508–10518 (2020)
45. Yang, B., Luo, W., Urtasun, R.: PIXOR: real-time 3d object detection from point clouds. In: CVPR (2018)
46. Werling, M., Ziegler, J., Kammel, S., Thrun, S.: Optimal trajectory generation for dynamic street scenarios in a frenet frame. In: ICRA (2010)
47. Geiger, A., Lenz, P., Urtasun, R.: Are we ready for autonomous driving? The KITTI vision benchmark suite. In: 2012 IEEE Conference on Computer Vision and Pattern Recognition, pp. 3354–3361. IEEE (2012)
48. Manivasagam, S., et al.: LiDARsim: realistic lidar simulation by leveraging the real world. arXiv (2020)

VarSR: Variational Super-Resolution Network for Very Low Resolution Images

Sangeek Hyun[1] and Jae-Pil Heo[1,2(✉)]

[1] Department of Artificial Intelligence, Sungkyunkwan University,
Suwon, South Korea
jaepilheo@skku.edu
[2] Department of Computer Science and Engineering, Sungkyunkwan University,
Suwon, South Korea

Abstract. As is well known, single image super-resolution (SR) is an ill-posed problem where multiple high resolution (HR) images can be matched to one low resolution (LR) image due to the difference in their representation capabilities. Such many-to-one nature is particularly magnified when super-resolving with large upscaling factors from very low dimensional domains such as 8×8 resolution where detailed information of HR is hardly discovered. Most existing methods are optimized for deterministic generation of SR images under pre-defined objectives such as pixel-level reconstruction and thus limited to the one-to-one correspondence between LR and SR images against the nature. In this paper, we propose VarSR, Variational Super Resolution Network, that matches latent distributions of LR and HR images to recover the missing details. Specifically, we draw samples from the learned common latent distribution of LR and HR to generate diverse SR images as the many-to-one relationship. Experimental results validate that our method can produce more accurate and perceptually plausible SR images from very low resolutions compared to the deterministic techniques.

Keywords: Single image super resolution · Variational super resolution · Very low resolution image

1 Introduction

Single image super-resolution (SISR) is a fundamental computer vision problem and has a broad range of real-world applications. The recent advances of deep convolutional neural networks (CNNs) have produced great progress in SISR. There has been a lot of CNN architectures to improve the performance of SR in terms of accuracy and time. On the other hand, developing new objectives and quality measures for SR algorithms other than traditional reconstruction

Electronic supplementary material The online version of this chapter (https://doi.org/10.1007/978-3-030-58592-1_26) contains supplementary material, which is available to authorized users.

errors also has been an active research topic of late years. Notable examples include perceptual similarity, GAN-based loss, and even learned metrics. Thanks to sustained research efforts and recent breakthroughs, the current SR techniques can produce super-resolved images comparable with the original high resolution (HR) ones.

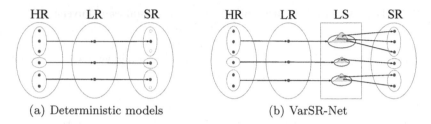

(a) Deterministic models   (b) VarSR-Net

Fig. 1. This figure illustrates the main difference between the deterministic super-resolution models and our proposed VarSR-Net. Although the ill-posed nature of the problem, the deterministic models produce an unique solution so thus they are not able to generate diverse super-resolved results as the high resolution images. However, our proposed VarSR-Net matches the latent distributions of low and high resolution images to produce diverse super-resolved outputs by sampling multiple latent variables from the shared distribution. Note that, 'LS' indicates the matched latent space.

Despite the success of recent SISR techniques, in this paper, we point out the problem of the deterministic mechanism of those methods. SISR is widely known as an ill-posed problem where multiple HR images can share a single matched low resolution (LR) image and thus the super-resolved image is not necessary to be unique. This is mainly caused by the difference in the representation capabilities between LR and HR image spaces. This many-to-one or one-to-many nature of SISR problem is especially magnified when dealing with large upscaling factors from very low resolution images such as 8×8 pixels, since extremely small LR images hardly preserve the detailed information of HR images. However, existing SISR solutions produce a deterministic one-to-one correspondence between LR and SR images against the nature of the problem, since they are mostly optimized for point-to-point error minimization based on strictly paired training samples of LR and its HR ground truth images.

In this paper, we address the aforementioned problem by proposing a novel VarSR-Net that enables us to generate diverse SR images that better reflect the ill-posed nature of the problem. Specifically, we first introduce two different latent variables for LR and HR images which encode their contents, respectively. Two latent variables are learned to have the shared representation so that the HR and LR images can be mapped to a common feature space by minimizing the KL-divergence to bridge the gap between the description capabilities of LR and HR images. Our SR module receives the latent variables as input and it is trained to produce pixel-level accurate and perceptually plausible upscaled images as the ordinary SR techniques. Since the HR images have a higher degree of diversity than LR images, the latent variables of HR images are much denser than LR ones. Thus, we draw multiple samples from the learned common latent distribution to generate diverse SR images in the inference stage. Figure 1 illustrates the main difference between the deterministic techniques and ours. To our knowledge, it is the first attempt to model and match the latent distributions of LR and HR images and generate diverse SR images from very low resolution LR images.

One might ask why we argue the importance of diverse SR image generation. In theoretical aspect, it is natural that a single LR image is not necessary to match with an unique HR image as we mentioned. On the other hand, we present the following real-world scenario. Suppose that we have a very low resolution face image of the criminal and we need to super-resolve the LR image before searching in the face image database. In this circumstance, we could miss the chance to identify the criminal if the single result of the deterministic SR model is poor. However, the possibility increases if the SR model produces multiple outputs close to the HR image so that we can perform search many times to make up a sufficient sized shortlist. Similar scenarios can be written such as investigation of the numbers on the very low resolution vehicle license plate images.

Our main contributions are summarized as follows:

- We highlight the problem of deterministic super-resolution approaches which paid less attention to the ill-posed nature of the SR task. Those methods assuming one-to-one correspondence between LR and HR images can fail to recover the detailed information of HR images which is hardly discovered in low dimensional LR images.
- We introduce VarSR, Variational Super-Resolution Network, capable of generating diverse SR results from a single LR image by sampling multiple latent variables from the learned common latent distribution of LR and HR domains.
- Our extensive evaluation with various quality measures validates that our method can produce more accurate and perceptually plausible SR images compared to the deterministic SR techniques.

2   Related Work

2.1   Image Super-Resolution

Recent deep-learning based single image super-resolution (SISR) techniques can be broadly categorized into two directions, advances in neural network architectures and objective functions.

In aspect of network architectures, Dong et al. [8] successfully developed a three-layered CNN model to SISR and showed superior performance over the handcrafted algorithms. After that, many advanced deep architectures for SR have been proposed. Ledig et al. [18] and Zhang et al. [38] adopt residual blocks and dense blocks toward more accurate SR image reconstruction, respectively. The laplacian pyramid network [16] is proposed to generate multi-scale SR images progressively in one feed-forward pass. Tai et al. [29] reduced the number of parameters of SR networks without loss of depths by using convolutional layers recursively.

In the early stages of deep-learning based SISR, most of the networks [8,14] are trained under L_p distances as their loss function for pixel-level reconstruction of the HR images. However, models learned for the reconstruction objective alone tend to produce blurry results. To resolve this problem, Johnson et al. [13] proposed the perceptual loss defined by the L_2 distance of activation maps extracted from CNN models pre-trained on large-scale datasets. In addition to the perceptual loss, SRGAN [18] incorporated the adversarial loss [9] to recover realistic image textures. Furthermore, there have been numbers of SR objectives including texture matching [25], semantic prior [31], and rank loss [37].

To our knowledge, the diversity of SR images is not discussed in the aforementioned previous work. We believe our study can provide a good starting point for the community to have more attention toward the new research direction.

2.2   Super-Resolving Very Low Resolution Images

The SR methods specialized for very low resolution images have mostly focused on human face images with 8×8 or 16×16 pixels [5,12,34,35,40]. Yu et al. [33] exploited facial attributes to train SR networks based on conditional GAN [22]. The structural prior of face such as facial landmarks is actively utilized for super-resolving face images [3,4,32]. There also have been several attempts to utilize person identities as a constraint when training SR models [7,10]. Note that, the aforementioned recent techniques tailored for the face image SR exploit additional supervision while our VarSR network is trained totally unsupervised manner.

On the other hand, Dahl et al. [5] proposed recursive learning that generates SR images pixel-by-pixel based on the autoregressive model [23]. Their fully probabilistic model generates diverse results, however, it suffers from the expensive sampling costs proportional to the image size and the number of inferences. On the contrary, our VarSR network can generate diverse and plausible images in one feed-forward path. Moreover, their objective solely concentrates to generate realistic images while our method focuses on the accurate reconstruction of original images.

2.3   Multimodal Generative Models

Deep-learning models produce a deterministic output unless there are no components for stochasticity. Therefore, there has been a series of work injecting stochasticity into conditional generative models (e.g. adding noise to input) in various tasks. For instance, VAE-based techniques to model the uncertainty were proposed and achieved the state-of-the-art performance in the video prediction task [2,6]. The BicycleGAN [39] encodes the styles of images as low dimensional latent variables and utilizes randomly sampled latent vectors in the image-to-image translation task. Lee et al. [19] proposed the DRIT by extending BicycleGAN for unpaired image-to-image translation based on a disentangled representation. Unlike the aforementioned methods for improving multimodality in the video prediction or image-to-image translation tasks, our method is the first attempt to introduce latent variables toward diversity in the image super-resolution task.

3   Our Approach

3.1   Motivation

In this paper, we mainly focus on generating multiple SR images with diversity for a single LR image. Before introducing our proposed solution, we clarify our motivation:

Single image super resolution is an ill-posed problem, since a single LR image and multiple HR images can correspond. The one-to-many relationship is basically coming from the difference in the representation capabilities of two domains. The information of HR images cannot be retained by the LR representation without any loss, thus perfect estimation of HR images is inherently impossible. However, most existing SR techniques are optimized to produce deterministic outputs regardless of the nature of the problem.

The problem is especially magnified when super-resolving with high upscaling factors from very low resolution images such as 8×8 pixels where it is extremely hard to discover the high-frequency details of HR images. In this circumstance, it is highly probable that several HR images can be matched to a single LR image. Therefore, a super-resolved image is not necessary to be unique but should be diverse.

There are critical real-world applications including surveillance systems that inferring diverse SR results is beneficial. One practical scenario is described in Sect. 1. In those applications, it is preferred to have a set of candidate SR images highly likely to contain an element very close to the target HR image, rather than a single deterministic result which moderately recovers the HR image.

3.2   Variational Super-Resolution Network

According to our motivation, we propose Variational Super-Resolution Network (VarSR-Net), specialized for the SR tasks from very low resolution images such as 8×8 dimensions where the details of HR images are hardly noticeable, by matching the latent distributions of LR and HR images.

A super-resolution network g_{SR} is a generator to produce a single output image $\hat{I}_H$ for a given low resolution image I_L:

$$\hat{I}_H = g_{SR}(I_L). \tag{1}$$

The network is trained to minimize the distortion (i.e. reconstruction error) between a super-resolved image $\hat{I}_H$ and a corresponding ground-truth high resolution image I_H. Additional objectives such as perceptual similarities or discrimination scores based on GAN can be assigned to produce better imitations. However, the missing detailed information of I_H hardly encoded in I_L makes the super-resolution models fail to infer reliable outputs, especially for very low resolution input.

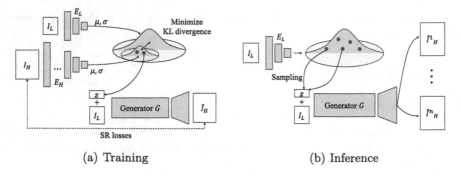

(a) Training   (b) Inference

Fig. 2. This figure illustrates the training and inference procedures of VarSR-Net. The HR and LR encoders are stochastically modeled to output parametric distributions. Specifically, encoders estimate a multivariate Gaussian distribution $\mathcal{N}(\mu, \sigma)$ where the their input highly likely to belong to. (a) The E_H and G are trained as an encoder-decoder structure. Specifically, given a pair of training sample (I_L, I_H), the E_H is trained to extract features of I_H while the generator $G(I_L, z)$ is learned to reconstruct I_H from I_L and z sampled from $\mathcal{N}(\mu, \sigma)$ estimated by E_H. Meanwhile, the E_L is learned to minimize the KL divergence with E_H to match the latent distributions of LR and HR images. (b) In inference stage, E_L estimates the latent distribution $\mathcal{N}(\mu, \sigma)$ from the input LR image. Diverse SR images are then produced by $G(I_L, z_i)$ where z_i are randomly drawn latent variables from $\mathcal{N}(\mu, \sigma)$ predicted by E_L.

In order to address this problem, a simple choice is to give an additional latent variable to our SR sub-module g_{SR} that describes the information of I_H to resolve the ambiguity that I_L has:

$$\hat{I}_H = g_{SR}\Big(I_L, E_H(I_H)\Big), \tag{2}$$

where $E_H(\cdot)$ is an encoder to extract features from the high resolution images. The SR model g_{SR} then learns to super-resolve a low resolution image I_L much more accurately by exploiting the feature extracted from I_H, since the features $E_H(I_H)$ provide the missing but strong cues for recovering HR images. Furthermore, joint learning of g_{SR} and $E_H(I_H)$ encourages the feature extractor to focus more on the information complementary to I_L.

However, the aforementioned approach (Eq. 2) is totally contradictory since the generator requires I_H or its features $E_H(I_H)$ to produce I_H. In order to resolve such a circular logic, we need to estimate $E_H(\cdot)$ only based on a low resolution image I_L.

We introduce another encoder $E_L(\cdot)$ for low resolution images. As our motivation and the ill-posed nature of super-resolution problem, the feature extracted from a single LR image needs to be matched to several features of HR images. However, it cannot be fulfilled by deterministic encoders. Therefore, we model the latent representation of both encoders $E_H(\cdot)$ and $E_L(\cdot)$ as multivariate Gaussian distributions where we can sample multiple latent variables to provide one-to-many mappings as follows:

$$E_L(x) = [\mu_x, \sigma_x] \text{ and } E_H(x) = [\mu_x, \sigma_x], \tag{3}$$

where x is either of LR and HR image, and both μ_x and σ_x are D-dimensional vector. Note that, D is the dimensionality of the latent representation. We also denote $E^\mu(x) = \mu_x$ and $E^\sigma(x) = \sigma_x$ for the sake of simplicity.

Our key idea is to match the two latent distributions of $E_H(\cdot)$ and $E_L(\cdot)$. In other words, we aim to realize that the sampled latent variables from E_L are highly likely to be ones sampled from E_H. To this end, we train both encoders to minimize the KL divergence between two distributions of $E_H(I_H)$ and $E_L(I_L)$. We further explain more details about the KL divergence term later.

In training phase, the input to our generator $G(\cdot, \cdot)$ is pairs of (I_L, z) where z is a sampled variable from $\mathcal{N}(E_H^\mu(I_H), E_H^\sigma(I_H))$, since the HR images are available. On the other hand, in testing phase, the generator receives a pair of (I_L, z) where z is drawn from $\mathcal{N}(E_L^\mu(I_L), E_L^\sigma(I_L))$. Specifically, the n diverse SR images $\{\hat{I}_H^1, ..., \hat{I}_H^n\}$ super-resolved from a LR image I_L are obtained as the follows:

$$\hat{I}_H^i = G(I_L, z_i), \quad z_i \sim \mathcal{N}\Big(E_L^\mu(I_L), E_L^\sigma(I_L)\Big), \tag{4}$$

for $i \in \{1, ..., n\}$. Although we utilize the latent distribution predicted by the LR encoder E_L in the inference stage, our generator $G(\cdot, \cdot)$ is capable to recover the information of HR images since E_L is trained to share the common latent distribution with the HR encoder E_H (Fig. 2).

Relation with CVAE. Conditional Variational AutoEncoder (CVAE) [27] approximates the conditional distribution $p_\theta(x|y)$ where x is data and y is a condition. The conditional generative process of the model is as follows; for a given condition y, latent variable z is drawn from the prior distribution $p_\theta(z|y)$, and the output x is generated from the distribution $p_\theta(x|y, z)$. This process allows to generate diverse outputs $\{x_i\}$ through the sampling of multiple latent variables $\{z_i\}$. Variation lower bound for CVAE is defined as:

$$L_{CVAE}(x, y; \theta, \phi) = \mathbb{E}_{q_\phi(z|x,y)} \log p_\theta(x|y, z)$$
$$- D_{KL}(q_\phi(z|x, y) \| p_\theta(z|y)) \leq \log p_\theta(x|y), \tag{5}$$

where $q_\phi(z|x, y)$ is the approximated distribution of the true posterior and D_{KL} is the KL divergence.

If we assume that a high resolution image contains all the information of its low resolution counterpart, we can translate our VarSR-Net network to CVAE architecture; x as high resolution image I_H, y as low resolution image I_L, $p_\theta(z|y)$ as a LR encoder $E_L(I_L)$, $q_\phi(z|x, y)$ as a HR encoder $E_H(I_H)$, and $p_\theta(x|y, z)$ as a generator network $G(I_L, z)$. Also, the term $\log p_\theta(x|y, z)$ can be replaced with losses used in previous SR works such as pixel-level reconstruction or perceptual loss. This interpretation gives a theoretical support for our model that maximizes conditional log-likelihood of observed data.

3.3   Objective Functions

We train the entire model in an end-to-end fashion based on a weighted combination of KL divergence and pixel-level losses. The pixel-level reconstruction loss which guides to reduce the distortion of a super-resolved image against its high resolution counterpart encourages the HR encoder $E_H(\cdot)$ to extract the informative features of a high resolution image, while the KL divergence loss minimizes the divergence between latent feature distributions of the high and low resolution images.

Specifically, a pixel loss minimizes the pixel-wise L_2 distances between a ground-truth high resolution image I_H and a super-resolved image. Note that, the super-resolved output is generated with latent variables sampled from a high resolution encoder E_H in training time. Therefore, the pixel-level loss is formulated as follows:

$$\mathcal{L}_{\text{pixel}} = \frac{1}{r^2 HW} \sum_{x=1}^{rH} \sum_{y=1}^{rW} \left(I_H^{x,y} - G(I_L, z)^{x,y} \right)^2, \ z \sim \mathcal{N}\left(E_H^\mu(I_H), E_H^\sigma(I_H) \right), \ (6)$$

where r is an upscaling factor, and H, W are height and width of I_L, respectively.

KL divergence loss has the most important role in our framework. It enables the low resolution encoder E_L to infer the latent variables which pretend ones from E_H. The KL divergence loss is formulated as follows:

$$\mathcal{L}_{\text{KL}} = D_{KL}\left(q(z|I_H) \ \| \ p(z|I_L) \right), \tag{7}$$

where $q(z|I_H)$ and $p(z|I_L)$ are the latent feature distributions of $E_H(\cdot)$ and $E_L(\cdot)$, respectively.

Furthermore, we also apply the adversarial loss to recover the realistic texture of high resolution images. We especially adopt the Improved Wasserstein GAN (WGAN-GP) [1,11] as follows:

$$\mathcal{L}_{\text{adv}} = \mathop{\mathbb{E}}_{\hat{I} \sim \mathbb{P}_g} [D(\hat{I})] - \mathop{\mathbb{E}}_{I \sim \mathbb{P}_r} [D(I)] + \delta \mathop{\mathbb{E}}_{\hat{I} \sim \mathbb{P}_{\hat{I}}} [(\|\nabla_{\hat{I}} D(\hat{I})\|_2 - 1)^2], \tag{8}$$

where D is a critic network. $\mathbb{P}_r$ and $\mathbb{P}_g$ are HR and SR data distributions, respectively. Plus, $\mathbb{P}_{\hat{I}}$ is the distribution of images sampled uniformly along straight lines connecting pairs of points from $\mathbb{P}_r$ and $\mathbb{P}_g$. We use $\delta = 10$ for our experiments.

Finally, our final loss function is given as follows:

$$\mathcal{L} = \lambda_{\text{pixel}} \mathcal{L}_{\text{pixel}} + \lambda_{\text{KL}} \mathcal{L}_{\text{KL}} + \lambda_{\text{adv}} \mathcal{L}_{\text{adv}}. \tag{9}$$

where λ_{pixel}, λ_{KL}, and λ_{adv} are hyper-parameters that balance three different loss terms.

4   Experiments

We evaluate our model quantitatively and qualitatively in the human face and digit datasets. Details of each dataset are introduced in the following section. We compare our model against the pixel recursive super resolution (PRSR) [5] which is an auto-regressive model for super-resolution, and MR-GAN [20] that reduces the mode-collapse problem in conditional GAN by replacing the mean squared error (MSE) loss with the momentum reconstruction loss. Unfortunately, PRSR is not tested in the face super-resolution task in our experiments, since it requires very expensive sampling cost for the images of 64×64 pixels. In addition, we utilize SRGAN [18] as the baseline deterministic SR technique for a face super-resolution task. For digit datasets, we use an autoencoder with skip-connections for entire methods since the input low resolution images of digit datasets have extremely low dimensionality such as 2×4 pixels. Therefore, we denote the deterministic baseline as "Det." for digit datasets instead of using SRGAN.

4.1   Datasets

Human Face Dataset. We adopt Celebrity Face Attributes (CelebA) [21] dataset for face super-resolution task. This dataset contains about 200K celebrity facial images. Among them, 100K images are used for training, and other 1K images without any overlap with the training set are utilized as a testing set.

We used a cropped version of CelebA to focus on learning various facial attributes unsupervisingly. We also set a spatial resolution of 64×64 for high resolution image, and 8×8 for low resolution image. For a fair comparison, we allocate 8 residual blocks for all models as did in MR-GAN [20].

Digit Datasets. For super-resolving digits, we use two datasets: MNIST [17] and license plate (LP) [28] datasets. The MNIST dataset [17] contains handwritten digit images with a resolution of 28×28. There are 60K and 10K images as training and testing sets, respectively. To make the digits unrecognizable in low resolution, all images are downsampled to the resolution of 6×6. The LP dataset [28] is originally designed for the vehicle re-identification task based on low quality LP images labeled in character-level. We collect about 110K and 7K character images cropped from the LP images for training and testing, respectively. Each character image is downsampled to a resolution of 2×4. We set the upscaling scale factor to 4 for digit datasets.

4.2   Implementation Details

We implement our VarSR-Net based on the architectures of U-net [24] and SRGAN [18]. Further architectural details are available in supplementary materials. We set the dimension for latent representation to 256 and 64 for face and digits datasets, respectively. Adam optimizer [15] is utilized with $\beta_1 = 0.9$,

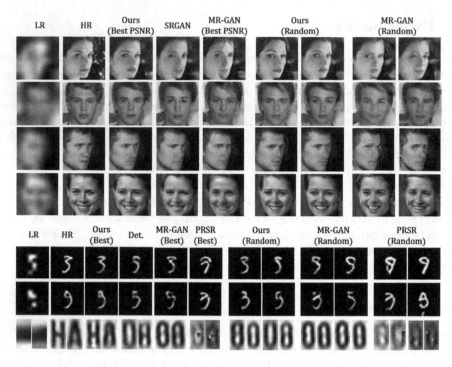

Fig. 3. Qualitative comparison of different super-resolution methods. Images with the highest PSNR scores among stochastically generated diverse results are reported with '(Best PSNR)' or '(Best)'. Randomly sampled images from stochastic models are reported with '(Random)'.

$\beta_2 = 0.99$ and set the learning rate $1e-4$ which is decayed once throughout entire training. As SRGAN [18] did, we first train the model without the adversarial loss to avoid to falling into undesirable local minima. We borrow the architecture of SRGAN's discriminator. However, we remove 3 convolution layers from the discriminator for digit datasets. For normalization layer, we adopt Instance Normalization [30]. Our hyper-parameters to balance the magnitudes of each terms in loss function (Eq. 9) are set as follows; $\lambda_{pixel} = 1$, and $\lambda_{adv} = 0.001$ for all datasets, and $\lambda_{KL} = 0.05$ for LP dataset, $\lambda_{KL} = 0.01$ for MNIST dataset, and $\lambda_{KL} = 0.01, 0.02$ for CelebA dataset. In addition, we adopt perceptual loss [13] to realistic texture of images like SRGAN. We add one subpixel convolution layer before the output layer for SRGAN and MR-GAN to deal with the upscaling scale factor of 8 in the face dataset.

4.3   Evaluation Metrics

Since VarSR-Net is not developed to generate a deterministic result, we perform the evaluations based on the mean and best scores among diverse super-resolved images. Traditional image quality measures including PSNR, SSIM, and MSE are

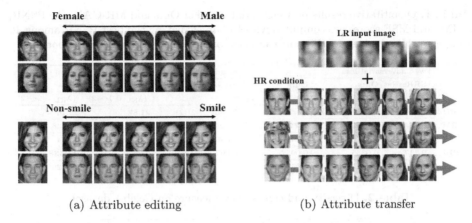

(a) Attribute editing   (b) Attribute transfer

Fig. 4. Qualitative results in face attribute editing and transfer by latent vector manipulations. (a) Given a latent vector z_{LR} of a LR image I_L downsampled from the leftmost HR image, we generate super-resolved images by $G(I_L, z_k)$ with $z_k = z_{LR} + ks\bar{Z}^{att}$, where $\bar{Z}^{att}$ is the mean of latent vectors $E_H(\cdot)$ of HR images having a particular attribute such as "male" or "smile", and s is a scaling constant. The 5 super-resolved images are corresponding to different $k \in \{-2, -1, 0, 1, 2\}$. (b) We denote a pair of LR and HR images, I_L^j and I_H^k, at the jth column on the uppermost row and k^{th} row on the leftmost column, respectively. The images $G(I_L^j, E_H(I_H^k))$ super-resolved from I_L^j with the latent vector $E_H(I_H^k)$ of I_H^k are presented at j^{th} column of kth row of the central 3×5 image array to validate that the attributes of HR images can be transferred to super-resolved images by its latent variable.

used. In addition, we perform the image classification in digit datasets to quantify how well the models produce semantically reliable outputs. For the face dataset, we use the perceptual image quality metrics to quantitatively measure the capabilities of tested methods to generate perceptually plausible images. Specifically, we utilize LPIPS score [36] and the distance between features extracted by a face verification network FaceNet [26]. In addition, we also measure the diversity of super-resolved face images as the average LPIPS distance among multiple resulting images according to Zhu et al. [39].

4.4 Qualitative Results

The qualitative results on the human face and digit datasets are shown in Fig. 3. The deterministic baseline suffers from the blurriness of outputs due to the inherited uncertainty of one-to-many mapping and generates inappropriate attributes. Besides, MR-GAN and PRSR succeed to generate diverse outputs, however, the most resulting images are perceptually unsatisfied while our VarSR-Net produces diverse and visually realistic images.

Furthermore, we validate that our common latent distribution shared by LR and HR domains reflects high-level semantics. We specifically perform attribute editing and transfer via manipulation of latent vectors, and the results are shown

Table 1. Quantitative results on CelebA dataset. For Ours and MR-GAN, the PSNR, SSIM, and MSE scores are computed with the image having the best PSNR among 10 samples, while the best LPIPS and Facenet scores within 10 samples are reported.

	w/o adversarial loss			w/adversarial loss			
	SRResNet	Ours	Ours	SRGAN	MR-GAN	Ours	Ours
		($\lambda_{KL} = 1e-2$)	($\lambda_{KL} = 2e-2$)			($\lambda_{KL} = 1e-2$)	($\lambda_{KL} = 2e-2$)
PSNR	22.57	22.46	**22.74**	22.28	21.21	21.87	22.14
SSIM	0.7242	0.7162	**0.7278**	0.7042	0.6476	0.6855	0.6948
MSE	73.93	72.87	**72.30**	74.49	77.77	74.64	73.74
LPIPS	0.1172	0.0885	0.0927	0.0679	0.0591	0.0539	**0.0538**
Facenet	0.0463	0.0425	0.0430	0.0463	0.0434	**0.0422**	0.0426

Table 2. Diversity and consistency measure on CelebA dataset.

	SRGAN	MR-GAN	Ours	Ours
			($\lambda_{KL} = 1e-2$)	($\lambda_{KL} = 2e-2$)
Diversity	0.0000	**0.0665**	0.0353	0.0238
Consistency	3.595	4.139	**3.402**	3.411

in Fig. 4. We observe that the edited results successfully reflect the attributes encoded by the conditional latent vectors. More importantly, edited images maintain the original characteristics of the given input LR images. This confirms that the latent vectors estimated by encoders describe the high-level semantics that is complementary to the information LR images have.

4.5   Quantitative Results

We perform the quantitative experiments on the human face dataset. We generate 10 SR images for each LR image input. The tested methods are categorized into two groups depending on; training with adversarial loss or not. The results of traditional metrics are shown in Table 1. Without adversarial loss, our model with $\lambda_{KL} = 2e-2$ achieves the highest scores in PSNR/SSIM/MSE, and the model with $\lambda_{KL} = 1e-2$ also shows comparable results to the baselines. Unlike the aforementioned cases, SRGAN shows better performance than ours when trained with adversarial loss. However, our models are still significantly better than another stochastic model MR-GAN and accomplished higher scores in perceptual metrics.

In order to measure diversity, we compute the average LPIPS scores among generated samples. Furthermore, we also evaluate the consistency of generated images by measuring the pixel-level L_1 distance between an input low resolution image and the downsampled version of generated SR images. A higher consistency score indicates that resulting SRs are less relevant to the corresponding HR image. Note that, the same downsampling scheme to construct the training set is utilized to measure the consistency. As reported in Table 2, we observe that MR-GAN generates more diverse images but shows much lower consistency

Table 3. Quantitative results on digit datasets. The PSNR, SSIM, and MSE scores of PRSR, MR-GAN, and Ours are computed with the image having the best PSNR among 5 samples. 'Det.' denotes a deterministic model.

	MNIST			LP		
	PSNR	SSIM	MSE	PSNR	SSIM	MSE
Det.	20.64	0.8250	15.81	21.16	0.9354	88.50
PRSR	18.04	0.7637	15.62	19.22	0.8793	94.45
MR-GAN	21.05	0.8512	15.19	21.54	0.9422	87.93
Ours	**22.00**	**0.8642**	**14.97**	**22.00**	**0.9494**	**85.95**

Table 4. Classification results of super-resolved images in MNIST and LP datasets. "Best" is measured by the criteria that there is at least one correctly classified image among 5 sampled images, and "Mean" is the average softmax values of 5 samples. Note that, the classification accuracies for ground truth HR images are 98.04% and 99.24% for MNIST and LP datasets, respectively.

	MNIST		LP	
	Best	Mean	Best	Mean
Det.	84.74	84.74	93.14	**93.14**
PRSR	84.42	70.22	96.10	92.19
MR-GAN	92.73	84.44	95.53	91.87
Ours	**92.79**	**85.58**	**96.33**	93.01

compared to our method. It is mainly because the MR-GAN generates many samples less relevant to the HR ground truth. On the other hand, our method shows even higher consistency score than the deterministic model.

We now perform the quantitative experiments on digit datasets. We sample 5 images for each low resolution image in digit datasets. In Table 3, our model shows superior performance in traditional metrics in both datasets. Also, we observe that the proposed model achieves higher classification accuracy compared to baseline models as reported in Table 4. Those results support that the deterministic model generates semantically or visually incorrect images in both MNIST and LP datasets. On the other hand, MR-GAN generates diverse outputs that contain correctly classified images, but shows low scores in distortion measures as shown in Table 3. PRSR records the lowest score except for classification results in the LP dataset. Note that, a low score of "Mean" in the tables do not degrade our method, because the proposed model can suggest the correct outputs which indicated by the "Best" in Table 4.

One may argue that the "Best" accuracy in Table 4 can be unfair because higher score can be achieved if the model produce just diverse digits regardless of the input. For instance, if a model always produces 5 different digits, the accuracy of "Best" must be largely improved. Therefore, we measure the number

Table 5. Semantic consistency. The number of distinct classes within 5 generated samples in digit datasets is measured to evaluate semantic consistency. Note that, the class labels are predicted by a pre-trained digit classification model. In more than 90% of cases, the 5 samples produced by our method belong to less than 2 different semantic classes. The scores of "Best" are measured by a criteria that there is at least one correctly classified image among 5 sampled images.

		The number of distinct classes among 5 samples				
		1	2	3	4	5
MNIST	Ratio	73.52%	20.18%	4.86%	0.83%	0.06%
	Best	94.83	88.02	82.95	82.18	87.96
LP	Ratio	89.29%	9.67%	0.94%	0.08%	0.00%
	Best	96.85	92.24	89.92	96.66	–

of distinct classes within 5 predictions and report in Table 5. In most cases, 5 sampled images belong to one or two classes. This validates that our model does not just generate diverse images but produce semantically reasonable results.

5   Conclusion

In this paper, we have highlighted the ill-posed nature of the single image super-resolution (SR) problem where multiple high resolution (HR) images can have a common matched low resolution (LR) image due to the difference in their representation capabilities. Despite of the many-to-one nature of the problem, most previous super-resolution techniques deterministically generate outputs. To establish the diversity of super-resolved images, we have modeled stochastic latent distributions for both LR and HR domains and train our VarSR-Net to match two distributions by adopting the KL divergence. The LR encoder learns to imitate the latent distribution of the HR encoder while the HR encoder is trained to extract the informative features of HR images. To produce diverse super-resolved images, we sample multiple latent variables from the parametric distribution estimated by the LR encoder. We intensively evaluate our proposed VarSR-Net against deterministic SR models, and the experimental results validate that our method is capable to produce more accurate and perceptually plausible SR images from very low resolution images. To our knowledge, our VarSR-Net is the first stochastic attempt to overcome the underdetermined characteristic of the SR problem. We believe that our work will stimulate the researcher to pay more attention to resolve the ill-posed nature of the SR problem.

Acknowledgements. This work was supported in part by Samsung Research Funding & Incubation Center for Future Technology (SRFC-IT1901-01), Police Lab (NRF-2018M3E2A1081572), and AI Graduate School Support Program (MSIT/IITP 2019-0-00421).

References

1. Arjovsky, M., Chintala, S., Bottou, L.: Wasserstein GAN. arXiv preprint arXiv:1701.07875 (2017)
2. Babaeizadeh, M., Finn, C., Erhan, D., Campbell, R.H., Levine, S.: Stochastic variational video prediction. arXiv preprint arXiv:1710.11252 (2017)
3. Bulat, A., Tzimiropoulos, G.: Super-FAN: integrated facial landmark localization and super-resolution of real-world low resolution faces in arbitrary poses with GANs. In: Proceedings of the IEEE Conference on Computer Vision and Pattern Recognition, pp. 109–117 (2018)
4. Chen, Y., Tai, Y., Liu, X., Shen, C., Yang, J.: FSRNet: end-to-end learning face super-resolution with facial priors. In: Proceedings of the IEEE Conference on Computer Vision and Pattern Recognition, pp. 2492–2501 (2018)
5. Dahl, R., Norouzi, M., Shlens, J.: Pixel recursive super resolution. In: Proceedings of the IEEE International Conference on Computer Vision, pp. 5439–5448 (2017)
6. Denton, E., Fergus, R.: Stochastic video generation with a learned prior. arXiv preprint arXiv:1802.07687 (2018)
7. Dogan, B., Gu, S., Timofte, R.: Exemplar guided face image super-resolution without facial landmarks. In: Proceedings of the IEEE Conference on Computer Vision and Pattern Recognition Workshops (2019)
8. Dong, C., Loy, C.C., He, K., Tang, X.: Image super-resolution using deep convolutional networks. IEEE Trans. Pattern Anal. Mach. Intell. **38**(2), 295–307 (2015)
9. Goodfellow, I., et al.: Generative adversarial nets. In: Advances in Neural Information Processing Systems, pp. 2672–2680 (2014)
10. Grm, K., Scheirer, W.J., Štruc, V.: Face hallucination using cascaded super-resolution and identity priors. IEEE Trans. Image Process. **29**(1), 2150–2165 (2019)
11. Gulrajani, I., Ahmed, F., Arjovsky, M., Dumoulin, V., Courville, A.C.: Improved training of Wasserstein GANs. In: Advances in Neural Information Processing Systems, pp. 5767–5777 (2017)
12. Huang, H., He, R., Sun, Z., Tan, T.: Wavelet-SRNet: a wavelet-based CNN for multi-scale face super resolution. In: Proceedings of the IEEE International Conference on Computer Vision, pp. 1689–1697 (2017)
13. Vogel, J., Auinger, A., Riedl, R.: Cardiovascular, neurophysiological, and biochemical stress indicators: a short review for information systems researchers. In: Davis, F.D., Riedl, R., vom Brocke, J., Léger, P.-M., Randolph, A.B. (eds.) Information Systems and Neuroscience. LNISO, vol. 29, pp. 259–273. Springer, Cham (2019). https://doi.org/10.1007/978-3-030-01087-4_31
14. Kim, J., Kwon Lee, J., Mu Lee, K.: Accurate image super-resolution using very deep convolutional networks. In: Proceedings of the IEEE Conference on Computer Vision and Pattern Recognition, pp. 1646–1654 (2016)
15. Kingma, D.P., Ba, J.: Adam: a method for stochastic optimization. arXiv preprint arXiv:1412.6980 (2014)
16. Lai, W.S., Huang, J.B., Ahuja, N., Yang, M.H.: Deep Laplacian pyramid networks for fast and accurate super-resolution. In: Proceedings of the IEEE Conference on Computer Vision and Pattern Recognition, pp. 624–632 (2017)
17. LeCun, Y., Cortes, C.: MNIST handwritten digit database (2010). http://yann.lecun.com/exdb/mnist/
18. Ledig, C., et al.: Photo-realistic single image super-resolution using a generative adversarial network. In: Proceedings of the IEEE Conference on Computer Vision and Pattern Recognition, pp. 4681–4690 (2017)

19. Lee, H.Y., Tseng, H.Y., Huang, J.B., Singh, M., Yang, M.H.: Diverse image-to-image translation via disentangled representations. In: Proceedings of the European Conference on Computer Vision (ECCV), pp. 35–51 (2018)
20. Lee, S., Ha, J., Kim, G.: Harmonizing maximum likelihood with GANs for multi-modal conditional generation. arXiv preprint arXiv:1902.09225 (2019)
21. Liu, Z., Luo, P., Wang, X., Tang, X.: Deep learning face attributes in the wild. In: Proceedings of International Conference on Computer Vision (ICCV), December 2015
22. Mirza, M., Osindero, S.: Conditional generative adversarial nets. arXiv preprint arXiv:1411.1784 (2014)
23. Van den Oord, A., Kalchbrenner, N., Espeholt, L., Vinyals, O., Graves, A., et al.: Conditional image generation with PixelCNN decoders. In: Advances in Neural Information Processing Systems, pp. 4790–4798 (2016)
24. Ronneberger, O., Fischer, P., Brox, T.: U-Net: convolutional networks for biomedical image segmentation. In: Navab, N., Hornegger, J., Wells, W.M., Frangi, A.F. (eds.) MICCAI 2015. LNCS, vol. 9351, pp. 234–241. Springer, Cham (2015). https://doi.org/10.1007/978-3-319-24574-4_28
25. Sajjadi, M.S., Scholkopf, B., Hirsch, M.: EnhanceNet: single image super-resolution through automated texture synthesis. In: Proceedings of the IEEE International Conference on Computer Vision, pp. 4491–4500 (2017)
26. Schroff, F., Kalenichenko, D., Philbin, J.: FaceNet: a unified embedding for face recognition and clustering. In: Proceedings of the IEEE Conference on Computer Vision and Pattern Recognition, pp. 815–823 (2015)
27. Sohn, K., Lee, H., Yan, X.: Learning structured output representation using deep conditional generative models. In: Advances in Neural Information Processing Systems, pp. 3483–3491 (2015)
28. Španhel, J., Sochor, J., Juránek, R., Herout, A., Maršík, L., Zemčík, P.: Holistic recognition of low quality license plates by CNN using track annotated data. In: 2017 14th IEEE International Conference on Advanced Video and Signal Based Surveillance (AVSS), pp. 1–6. IEEE (2017). https://doi.org/10.1109/AVSS.2017.8078501
29. Tai, Y., Yang, J., Liu, X.: Image super-resolution via deep recursive residual network. In: Proceedings of the IEEE Conference on Computer Vision and Pattern Recognition, pp. 3147–3155 (2017)
30. Ulyanov, D., Vedaldi, A., Lempitsky, V.: Instance normalization: the missing ingredient for fast stylization. arXiv preprint arXiv:1607.08022 (2016)
31. Wang, X., Yu, K., Dong, C., Change Loy, C.: Recovering realistic texture in image super-resolution by deep spatial feature transform. In: Proceedings of the IEEE Conference on Computer Vision and Pattern Recognition, pp. 606–615 (2018)
32. Yu, X., Fernando, B., Ghanem, B., Porikli, F., Hartley, R.: Face super-resolution guided by facial component heatmaps. In: Proceedings of the European Conference on Computer Vision (ECCV), pp. 217–233 (2018)
33. Yu, X., Fernando, B., Hartley, R., Porikli, F.: Super-resolving very low-resolution face images with supplementary attributes. In: Proceedings of the IEEE Conference on Computer Vision and Pattern Recognition, pp. 908–917 (2018)
34. Yu, X., Porikli, F.: Ultra-resolving face images by discriminative generative networks. In: Leibe, B., Matas, J., Sebe, N., Welling, M. (eds.) ECCV 2016. LNCS, vol. 9909, pp. 318–333. Springer, Cham (2016). https://doi.org/10.1007/978-3-319-46454-1_20

35. Yu, X., Porikli, F.: Hallucinating very low-resolution unaligned and noisy face images by transformative discriminative autoencoders. In: Proceedings of the IEEE Conference on Computer Vision and Pattern Recognition, pp. 3760–3768 (2017)
36. Zhang, R., Isola, P., Efros, A.A., Shechtman, E., Wang, O.: The unreasonable effectiveness of deep features as a perceptual metric. In: Proceedings of the IEEE Conference on Computer Vision and Pattern Recognition, pp. 586–595 (2018)
37. Zhang, W., Liu, Y., Dong, C., Qiao, Y.: RankSRGAN: generative adversarial networks with ranker for image super-resolution. In: Proceedings of the IEEE International Conference on Computer Vision, pp. 3096–3105 (2019)
38. Zhang, Y., Tian, Y., Kong, Y., Zhong, B., Fu, Y.: Residual dense network for image super-resolution. In: Proceedings of the IEEE Conference on Computer Vision and Pattern Recognition, pp. 2472–2481 (2018)
39. Zhu, J.Y., et al.: Toward multimodal image-to-image translation. In: Advances in Neural Information Processing Systems, pp. 465–476 (2017)
40. Zhu, S., Liu, S., Loy, C.C., Tang, X.: Deep cascaded bi-network for face hallucination. In: Leibe, B., Matas, J., Sebe, N., Welling, M. (eds.) ECCV 2016. LNCS, vol. 9909, pp. 614–630. Springer, Cham (2016). https://doi.org/10.1007/978-3-319-46454-1_37

Co-heterogeneous and Adaptive Segmentation from Multi-source and Multi-phase CT Imaging Data: A Study on Pathological Liver and Lesion Segmentation

Ashwin Raju[1,2], Chi-Tung Cheng[3], Yuankai Huo[1], Jinzheng Cai[1],
Junzhou Huang[2], Jing Xiao[4], Le Lu[1], ChienHung Liao[3],
and Adam P. Harrison[1(✉)]

[1] PAII Inc., Bethesda, MD, USA
adampharrison070@paii-labs.com
[2] The University of Texas at Arlington, Arlington, TX, USA
[3] Chang Gung Memorial Hospital, Linkou, Taiwan, ROC
[4] PingAn Technology, Shenzhen, China

Abstract. Within medical imaging, organ/pathology segmentation models trained on current publicly available and fully-annotated datasets usually do not well-represent the heterogeneous modalities, phases, pathologies, and clinical scenarios encountered in real environments. On the other hand, there are tremendous amounts of unlabelled patient imaging scans stored by many modern clinical centers. In this work, we present a novel segmentation strategy, co-heterogenous and adaptive segmentation (CHASe), which only requires a small labeled cohort of *single* phase data to adapt to any unlabeled cohort of heterogenous *multi-phase* data with possibly new clinical scenarios and pathologies. To do this, we develop a versatile framework that fuses appearance-based semi-supervision, mask-based adversarial domain adaptation, and pseudo-labeling. We also introduce co-heterogeneous training, which is a novel integration of co-training and hetero-modality learning. We evaluate CHASe using a clinically comprehensive and challenging dataset of multi-phase computed tomography (CT) imaging studies (1147 patients and 4577 3D volumes), with a test set of 100 patients. Compared to previous state-of-the-art baselines, CHASe can further improve pathological liver mask Dice-Sørensen coefficients by ranges of 4.2% to 9.4%, depending on the phase combinations, *e.g.*, from 84.6% to 94.0% on non-contrast CTs.

Keywords: Multi-phase segmentation · Semi-supervised learning · Co-training · Domain adaptation · Liver and lesion segmentation

Electronic supplementary material The online version of this chapter (https://doi.org/10.1007/978-3-030-58592-1_27) contains supplementary material, which is available to authorized users.

A. Vedaldi et al. (Eds.): ECCV 2020, LNCS 12368, pp. 448–465, 2020.
https://doi.org/10.1007/978-3-030-58592-1_27

1   Introduction

Delineating anatomical structures is an important task within medical imaging, *e.g.*, to generate biomarkers, quantify or track disease progression, or to plan radiation therapy. Manual delineation is prohibitively expensive, which has led to a considerable body of work on automatic segmentation. However, a perennial problem is that models trained on available image/mask pairs, *e.g.*, publicly available data, do not always reflect clinical conditions upon deployment, *e.g.*, the present pathologies, patient characteristics, scanners, and imaging protocols. This leads to potentially drastic performance gaps. When multi-modal or multi-phase imagery is present these challenges are further compounded, as datasets may differ in their composition of available modalities or consist of heterogeneous combinations of modalities. The challenges then are in both managing new patient/disease variations and in handling heterogeneous multi-modal data. Ideally segmentation models can be deployed without first annotating large swathes of additional data matching deployment scenarios. This is our goal, where we introduce co-heterogenous and adaptive segmentation (CHASe), which can adapt models trained on single-modal and public data to produce state-of-the-art results on *multi-phase and multi-source* clinical data *with no extra annotation cost.*

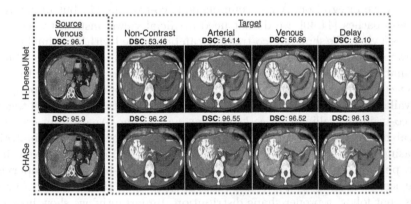

Fig. 1. Ground truth and predictions are rendered in green and red, respectively. Despite performing excellently on labeled source data, state-of-the-art fully-supervised models [25] can struggle on cohorts of multi-phase data with novel conditions, *e.g.*, the patient shown here with splenomegaly and a TACE-treated tumor. CHASe can adapt such models to perform on new data. (Color figure online)

Our motivation stems from challenges in handling the wildly variable data found in picture archiving and communication systems (PACSs), which closely follows deployment scenarios. In this study, we focus on segmenting pathological livers and lesions from dynamic computed tomography (CT). Liver disease and cancer are major morbidities, driving efforts toward better detection and characterization methods. The dynamic contrast CT protocol images a patient under multiple time-points after a contrast agent is injected, which is critical for characterizing liver lesions [30]. Because accurate segmentation produces important volumetric biomarkers [3,12], there is a rich body of work on automatic segmentation [10,20,23,25,28,36,41,46,48], particularly for CT. Despite this, all publicly available data [3,7,11,18,38] is limited to venous-phase (single-channel) CTs. Moreover, when lesions are present, they are typically limited to hepatocellular carcinoma (HCC) or metastasized tumors, lacking representation of intrahepatic cholangiocellular carcinoma (ICC) or the large bevy of benign lesion types. Additionally, public data may not represent other important scenarios, *e.g.*,the transarterial chemoembolization (TACE) of lesions or splenomegaly, which produce highly distinct imaging patterns. As Fig. 1 illustrates, even impressive leading entries [25] within the public LiTS challenge [3], can struggle on clinical PACS data, particularly when applied to non-venous contrast phases.

To meet this challenge, we integrate together powerful, but complementary, strategies: hetero-modality learning, appearance-based consistency constraints, mask-based adversarial domain adaptation (ADA), and pseudo-labeling. The result is a semi-supervised model trained on smaller-scale supervised public *venous-phase* data [3,7,11,18,38] and large-scale unsupervised *multi-phase* data. Crucially, we articulate non-obvious innovations that avoid serious problems from a naive integration. A key component is co-training [4], but unlike recent deep approaches [44,49], we do not need artificial views, instead treating each phase as a view. We show how co-training can be adopted with a minimal increase of parameters. Second, since CT studies from clinical datasets may exhibit any combination of phases, ideally liver segmentation should also be able to accept whatever combination is available, with performance topping out as more phases are available. To accomplish this, we fuse hetero-modality learning [15] together with co-training, calling this *co-heterogeneous training*. Apart from creating a natural hetero-phase model, this has the added advantage of combinatorially increasing the number of views for co-training from 4 to 15, boosting even single-phase performance. To complement these appearance-based semi-supervision strategies, we also apply pixel-wise ADA [40], guiding the network to predict masks that follow a proper shape distribution. Importantly, we show how ADA can be applied to co-heterogeneous training with no extra computational cost over adapting a single phase. Finally, we address edge cases using a principled pseudo-labelling technique specific to pathological organ segmentation. These innovations combine to produce a powerful approach we call CHASe.

We apply CHASe to a challenging unlabelled dataset of 1147 dynamic-contrast CT studies of patients, all with liver lesions. The dataset, extracted directly from a hospital PACS, exhibits many features not seen in public single-phase data and consists of a heterogeneous mixture of non-contrast (NC), arterial (A), venous (V), and delay (D) phase CTs. With a test set of 100 studies, *this is the largest, and arguably the most challenging, evaluation of multi-phase pathological liver segmentation to date.* Compared to strong fully-supervised baselines [14,25] only trained on public data, CHASe can dramatically boost segmentation performance on non-venous phases, *e.g.*, moving the mean Dice-Sørensen coefficient (DSC) from 84.5 to 94.0 on NC CTs. Importantly, performance is also boosted on V phases, *i.e.*, from 90.7 mean DSC to 94.9. Inputting all available phases to CHASe maximizes performance, matching desired behavior. Importantly, CHASe also significantly improves robustness, operating with much greater reliability and without deleterious outliers. Since CHASe is general-purpose, it can be applied to other datasets or even other organs with ease.

2   Related Work

Liver and Liver Lesion Segmentation. In the past decade, several works addressed liver and lesion segmentation using traditional texture and statistical features [10,23,41]. With the advent of deep learning, fully convolutional networks (FCNs) have quickly become dominant. These include 2D [2,36], 2.5D [13,43], 3D [20,28,45,48], and hybrid [25,46] FCN-like architectures. Some reported results show that 3D models can improve over 2D ones, but these improvements are sometimes marginal [20].

Like related works, we also use FCNs. However, all prior works are trained and evaluated on venous-phase CTs in a fully-supervised manner. In contrast, we aim to robustly segment a large cohort of multi-phase CT PACS data in a semi-supervised manner. As such, our work is orthogonal to much of the state-of-the-art, and can, in principle, incorporate any future fully-supervised solution as a starting point.

Semi-supervised Learning. Annotating medical volumetric data is time consuming, spurring research on semi-supervised solutions [39]. In *co-training*, predictions of different "views" of the same unlabelled sample are enforced to be consistent [4]. Recent works integrate co-training to deep-learning [31,44,49]. While CHASe is related to these works, it uses different contrast phases as views and therefore has no need for artificial view creation [31,44,49]. More importantly, CHASe effects a stronger appearance-based semi-supervision by fusing co-training with hetero-modality learning (co-heterogenous learning). In addition, CHASe complements this appearance-based strategy via prediction-based ADA, resulting in significantly increased performance.

Adversarial domain adaptation (ADA) for semantic segmentation has also received recent attention in medical imaging [39]. The main idea is to align the distribution of predictions or features between the source and target domains, with many successful approaches [6,9,40,42]. Like Tsai *et al.*, we align the distribution of *mask shapes* of source and target predictions [40]. Unlike Tsai *et al.*, we use ADA in conjunction with appearance-based semi-supervision. In doing so, we show how ADA can effectively adapt 15 different hetero-phase predictions at the same computational cost as a single-view variant. Moreover, we demonstrate the need for complementary semi-supervised strategies in order to create a robust and practical medical segmentation system.

In *self-learning*, a network is first trained with labelled data. The trained model then produces pseudo labels for the unlabelled data, which is then added to the labelled dataset via some scheme [24]. This approach has seen success within medical imaging [1,29,47], but it is important to guard against "confirmation bias" [26], which can compound initial misclassifications [39]. While we thus avoid this approach, we do show later that co-training can also be cast as self-learning with consensus-based pseudo-labels. Finally, like more standard self-learning, we also generate pseudo-labels to finetune our model. But, these are designed to deduce and correct likely mistakes, so they do not follow the regular "confirmation"-based self-learning framework.

Hetero-Modality Learning. In medical image acquisition, multiple phases or modalities are common, *e.g.*, dynamic CT. It is also common to encounter hetero-modality data, *i.e.*, data with possible missing modalities [15]. Ideally a segmentation model can use whatever phases/modalities are present in the study, with performance improving the more phases are available. CHASe uses hetero-modal fusion for fully-supervised FCNs [15]; however it fuses it with co-training, thereby using multi-modal learning to perform appearance-based learning from *unlabelled* data. Additionally, this *co-heterogeneous training* combinatorially increases the number of views for co-training, significantly boosting even single-phase performance by augmenting the training data. To the best of our knowledge, we are the first to propose co-heterogenous training.

3   Method

Although CHASe is not specific to any organ, we will assume the liver for this work. We assume we are given a curated and labelled dataset of CTs and masks, *e.g.*, from public data sources. We denote this dataset $\mathcal{D}_\ell = \{\mathcal{X}_i, Y_i\}_{i=1}^{N_\ell}$, with $\mathcal{X}_i$ denoting the set of available phases and $Y_i(k) \in \{0, 1, 2\}$ indicating background, liver, and lesion for all pixel/voxel indices k. Here, without loss of generality, we assume the CTs are all V-phase, *i.e.*, $\mathcal{X}_i = V_i \; \forall \mathcal{X}_i \in \mathcal{D}_\ell$. We also assume we are given a large cohort of unlabelled multi-phase CTs from a challenging and uncurated clinical source, *e.g.*, a PACS. We denote this dataset $\mathcal{D}_u = \{\mathcal{X}_i\}_{i=1}^{N_u}$, where $\mathcal{X}_i = \{NC_i, A_i, V_i, D_i\}$ for instances with all contrast phases. When appropriate, we drop the i for simplicity. Our goal is to learn a segmentation model which can accept any combination of phases from the target domain and robustly delineate liver or liver lesions, despite possible differences in morbidities, patient characteristics, and contrast phases between the two datasets.

Figure 2 CHASe, which integrates supervised learning, co-heterogenous training, ADA, and specialized pseudo-labelling. We first start by training a standard fully-supervised segmentation model using the labelled data. Then under CHASe, we finetune the model using consistency and ADA losses:

$$\mathcal{L} = \mathcal{L}_{seg} + \mathcal{L}_{cons} + \lambda_{adv}\mathcal{L}_{adv}, \tag{1}$$

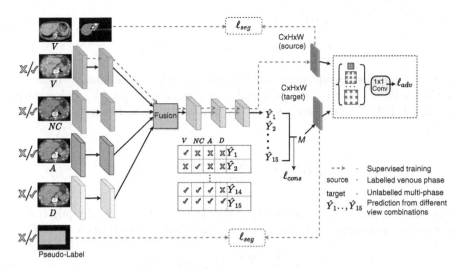

Fig. 2. Overview of CHASe. As shown at the top, labelled data, V-phase in our experiments, are trained using standard segmentation losses. Multi-phase unlabelled data are inputted into an efficient co-heterogenous pipeline, that minimizes divergence between mask predictions across all 15 phase combinations. ADA and specialized pseudo-labelling are also applied. Not shown are the deeply-supervised intermediate outputs of the PHNN backbone.

where $\mathcal{L}$, $\mathcal{L}_{seg}$, $\mathcal{L}_{cons}$, $\mathcal{L}_{adv}$ are the overall, supervised, co-heterogenous, and ADA losses, respectively. For adversarial optimization, a discriminator loss, $\mathcal{L}_d$, is also deployed in competition with (1). We elaborate on each of the above losses below. Throughout, we minimize hyper-parameters by using standard architectures and little loss-balancing.

3.1 Backbone

CHASe relies on an FCN backbone, $f(.)$, for its functionality and can accommodate any leading choice, including encoder/decoder backbones [28,32]. Instead, we adopt the deeply-supervised progressive holistically nested network (PHNN) framework [14], which has demonstrated leading segmentation performance for many anatomical structures [5,14,21,22,34], sometimes even outperforming U-Net [21,22]. Importantly, PHNN has roughly half the parameters and activation maps of an equivalent encoder/decoder. Since we aim to include additional components for semi-supervised learning, this lightweightedness is important.

In brief, PHNN relies on deep supervision in lieu of a decoder and assumes the FCN can be broken into stages based on pooling layers. Here, we assume there are five FCN stages, which matches popular FCN configurations. PHNN produces a sequence of logits, $a^{(m)}$, using 1×1 convolutions and upsamplings operating on the terminal backbone activations of each stage. Sharing similarities to residual connections [16], predictions are generated for each stage using a progressive scheme that adds to the previous stage's activations:

$$\hat{Y}^{(1)} = \sigma(a^{(1)}), \tag{2}$$

$$\hat{Y}^{(m)} = \sigma(a^{(m)} + a^{(m-1)}) \ \forall m > 1, \tag{3}$$

where $\sigma(.)$ denotes the softmax and $\hat{Y}^{(\cdot)}$ represents the predictions, with the final stage's predictions acting as the actual segmentation output, $\hat{Y}$. Being deeply supervised, PHNN optimizes a loss at each stage, with higher weights for later stages:

$$\ell_{seg}(f(V), Y) = \sum_{j=1}^{5} \frac{m}{5} \ell_{ce}\left(\hat{Y}^m, Y\right), \tag{4}$$

where we use pixel-wise cross-entropy loss (weighted by prevalence), $\ell_{ce}(.,.)$. Prior to any semi-supervised learning, this backbone is pre-trained using $\mathcal{D}_\ell$:

$$\mathcal{L}_{seg} = \frac{1}{N_\ell} \sum_{V,Y \in \mathcal{D}_\ell} \ell_{seg}(f(V), Y). \tag{5}$$

Our supplementary visually depicts the PHNN structure.

3.2   Co-training

With a pretrained fully-supervised backbone, the task is now to leverage the unlabelled cohort of dynamic CT data, $\mathcal{D}_u$. We employ the ubiquitous strategy of enforcing consistency. Because dynamic CT consists of the four NC, A, V, and D phases, each of which is matched to same mask, Y, they can be regarded as different views of the same data. This provides for a natural co-training setup [4] to penalize inconsistencies of the mask predictions across different phases.

To do this, we must create predictions for each phase. As Fig. 2 illustrates, we accomplish this by using phase-specific FCN stages, $i.e.$, the first two low-level stages, and then use a shared set of weights for the later semantic stages. Because convolutional weights are more numerous at later stages, this allows for an efficient multi-phase setup. All layer weights are initialized using the fully-supervised V-phase weights from Sect. 3.1, including the phase-specific layers. Note that activations across phases remain distinct. Despite the distinct activations, for convenience we abuse notation and use $\hat{y} = f(\mathcal{X})$ to denote the generation of all phase predictions for one data instance. When all four phases are available in $\mathcal{X}$, then $\hat{y}$ corresponds to $\{\hat{Y}^{NC}, \hat{Y}^A, \hat{Y}^V, \hat{Y}^D\}$.

Like Qiao $et\ al.$ [31] we use the Jensen-Shannon divergence (JSD) [27] to penalize inconsistencies. However, because it will be useful later we devise the JSD by first deriving a consensus prediction:

$$M = \frac{1}{|\hat{y}|} \sum_{\hat{Y} \in \hat{y}} \hat{Y}. \tag{6}$$

The JSD is then the divergence between the consensus and each phase prediction:

$$\ell_{cons}(f(\mathcal{X})) = \frac{1}{|\hat{\mathcal{Y}}|} \sum_{\hat{Y} \in \hat{y}} \sum_{k \in \Omega} KL(\hat{Y}(k) \| M(k)), \tag{7}$$

$$\mathcal{L}_{cons} = \frac{1}{N_u} \sum_{\mathcal{X} \in \mathcal{D}^u} \ell_{cons}(f(\mathcal{X})), \tag{8}$$

where Ω denotes the spatial domain and $KL(. \| .)$ is the Kullback-Leibler divergence across label classes. (8) thus casts co-training as a form of self-learning, where pseudo-labels correspond to the consensus prediction in (6). For the deeply-supervised PHNN, we only calculate the JSD across the final prediction.

3.3 Co-heterogeneous Training

While the loss in (8) effectively incorporates multiple phases of the unlabelled data, it is not completely satisfactory. Namely, each phase must still be inputted separately into the network, and there is no guarantee of a consistent output. Despite only having single-phase *labelled* data, ideally, the network should be adapted for multi-phase operation on $\mathcal{D}_u$, meaning it should be able to consume whatever contrast phases are available and output a unified prediction that is stronger as more phases are available.

To do this, we use hetero-modality image segmentation (HeMIS)-style feature fusion [15], which can predict masks given any arbitrary combination of contrast phases. To do this, a set of phase-specific layers produce a set of phase-specific activations, $\mathcal{A}$, whose cardinality depends on the number of inputs. The activations are then fused together using first- and second-order statistics, which are flexible enough to handle any number of inputs:

$$\mathbf{a}_{fuse} = \mathrm{concat}(\mu(\mathcal{A}), \mathrm{var}(\mathcal{A})), \tag{9}$$

where $\mathbf{a}_{fuse}$ denotes the fused feature, and the mean and variance are taken across the available phases. When only one phase is available, the variance features are set to 0. To fuse intermediate predictions, an additional necessity for deeply-supervised networks, we simply take their mean.

For choosing a fusion point, the co-training setup of Sect. 3.2, with its phase-specific layers, already offers a natural fusion location. We can then readily combine hetero-phase learning with co-training, re-defining a "view" to mean any possible combination of the four contrast phases. This has the added benefit of combinatorially increasing the number of co-training views. More formally, we use $\mathcal{X}^* = \mathcal{P}(\mathcal{X}) \setminus \{\varnothing\}$ to denote all possible contrast-phase combinations, where $\mathcal{P}(.)$ is the powerset operator. The corresponding predictions we denote as $\hat{\mathcal{Y}}^*$. When a data instance has all four phases, then the cardinality of $\mathcal{X}^*$ and $\hat{\mathcal{Y}}^*$ is 15, which is a drastic increase in views. With hetero-modality fusion in place, the consensus prediction and co-training loss of (6) and (7), respectively, can be supplanted by ones that use $\hat{\mathcal{Y}}^*$:

$$M = \frac{1}{|\hat{\mathcal{Y}}*|} \sum_{\hat{Y} \in \hat{\mathcal{Y}}*} \hat{Y}, \tag{10}$$

$$\ell_{cons}(f(\mathcal{X})) = \frac{1}{|\hat{\mathcal{Y}}*|} \sum_{\hat{Y} \in \hat{\mathcal{Y}}*} \sum_{k \in \Omega} KL(\hat{Y}(k) \parallel M(k)). \tag{11}$$

When only single-phase combinations are used, (10) and (11) reduce to standard co-training. To the best of our knowledge we are the first to combine co-training with hetero-modal learning. This combined workflow is graphically depicted in Fig. 2.

3.4   Adversarial Domain Adaptation

The co-heterogeneous training of Sect. 3.3 is highly effective. Yet, it relies on accurate consensus predictions, which may struggle to handle significant appearance variations in $\mathcal{D}_u$ that are not represented in $\mathcal{D}_\ell$. Mask-based ADA offers an a complementary approach that trains a network to output masks that follow a *prediction-based* distribution learned from labelled data [40]. Since liver shapes between $\mathcal{D}_u$ and $\mathcal{D}_\ell$ should follow similar distributions, this provides an effective learning strategy that is not as confounded by differences in appearance. Following Tsai *et al.* [40], we can train a discriminator to classify whether a softmax output originates from a labelled- or unlabelled-dataset prediction. However, because we have a combinatorial number (15) of possible input phase combinations, *i.e.*, $\hat{\mathcal{X}}*$, naively domain-adapting all corresponding predictions is prohibitively expensive. Fortunately, the formulations of (7) and (11) offer an effective and efficient solution. Namely, we can train the discriminator on the consensus prediction, M. This adapts the combinatorial number of possible predictions *at the same computational cost as performing ADA on only a single prediction.*

More formally, let $d(.)$ be defined as an FCN discriminator, then the discriminator loss can be expressed as

$$\mathcal{L}_d = \frac{1}{N_\ell} \sum_{\mathcal{D}_\ell} \ell_{ce}(d(\hat{Y}^V), \mathbf{1}) + \frac{1}{N_u} \sum_{\mathcal{D}_u} \ell_{ce}(d(M, \mathbf{0})), \tag{12}$$

where ℓ_{ce} represents a pixel-wise cross-entropy loss. The opposing labels pushes the discriminator to differentiate semi-supervised consensus predictions from fully-supervised variants. Unlike natural image ADA [40], we do not wish to naively train the discriminator on all output classes, as it not reasonable to expect similar distributions of liver *lesion* shapes across datasets. Instead we train the discriminator on the *liver region*, *i.e.*, the union of healthy liver and lesion tissue predictions. Finally, when minimizing (12), we only optimize the discriminator weights. The segmentation network can now be tasked with fooling the discriminator, through the addition of an adversarial loss:

$$\mathcal{L}_{adv} = \frac{1}{N_u} \sum_{\mathcal{D}_u} \ell_{ce}(d(M, \mathbf{1})),$$ (13)

where the ground-truth labels for ℓ_{ce} have been flipped from (12). Note that here we use single-level ADA as we found the multi-level variant [40] failed to offer significant enough improvements to offset the added complexity. When minimizing (13), or (1) for that matter, the discriminator weights are frozen. We empirically set λ_{adv} to 0.001.

3.5 Pseudo-Labelling

By integrating co-heterogeneous training and ADA, CHASe can robustly segment challenging multi-phase unlabelled data. However some scenarios still present challenging edge cases, *e.g.*, lesions treated with TACE. See the supplementary for some visualizations. To manage these cases, we use a simple, but effective, domain-specific pseudo-labelling.

First, after convergence of (1), we produce predictions on $\mathcal{D}_u$ using all available phases and extract any resulting 3D holes in the liver region (healthy tissue plus lesion) greater than 100 voxels. Since there should never be 3D holes, these are mistakes. Under the assumption that healthy tissue in both datasets should equally represented, we treat these holes as missing "lesion" predictions. We can then create a pseudo-label, Y_h, that indicates lesion at the hole, *with all others regions being ignored*. This produces a new "holes" dataset, $\mathcal{D}_h = \{\mathcal{X}, Y_h\}_{i=1}^{N_h}$, using image sets extracted from $\mathcal{D}_u$. We then finetune the model using (1), but replace the segmentation loss of (5) by

$$\mathcal{L}_{seg} = \frac{1}{N_\ell} \sum_{V,Y \in \mathcal{D}_\ell} \ell_{seg}(f(V), Y)$$

$$+ \frac{\lambda_h}{N_h} \sum_{\mathcal{X}, Y_h \in \mathcal{D}_h} \sum_{X \in \mathcal{X}^*} \ell_{seg}(f(X), Y_h),$$ (14)

where we empirically set λ_h to 0.01 for all experiments. We found results were not sensitive to this value. While the hole-based pseudo-labels do not capture all errors, they only have to capture enough of missing appearances to guide CHASe's training to better handle recalcitrant edge cases.

4   Results

Datasets. To execute CHASe, we require datasets of single-phase labelled and multi-phase unlabelled studies, $\mathcal{D}_u$ and $\mathcal{D}_\ell$, respectively. The goal is to robustly segment patient studies from $\mathcal{D}_u$ while only having training mask labels from the less representative $\mathcal{D}_\ell$ dataset. 1) For $\mathcal{D}_u$, we collected 1147 multi-phase dynamic CT studies (4577 volumes in total) directly from the PACS of Chang Gung Memorial Hospital (CGMH). The only selection criteria were patients with biopsied or resected liver lesions, with dynamic contrast CTs taken within one month before the procedure. Patients may have ICC, HCC, benign or metastasized tumors, along with co-occuring maladies, such as liver fibrosis, splenomegaly, or TACE-treated tumors. Thus, $\mathcal{D}_u$ directly reflects the variability found in clinical scenarios. We used the DEEDS algorithm [19] to correct any misalignments. 2) For $\mathcal{D}_\ell$, we collected 235 V-phase CT studies collected from as many public sources as we could locate [3,7,11,18]. This is a superset of the LiTS training data [3], and consists of a mixture of healthy and pathological livers, with only HCC and metastasis represented.

Evaluation Protocols. 1) To evaluate performance on $\mathcal{D}_u$, 47 and 100 studies were randomly selected for validation and testing, respectively, with 90 test studies having all four phases. Given the extreme care required for lesion annotation, *e.g.*, the four readers used in the LiTS dataset [3], only the liver region, *i.e.*, union of healthy liver and lesion tissue, of the $\mathcal{D}_u$ evaluation sets was annotated by a clinician. For each patient study, this was performed independently for each phase, with a final *study-wise* mask generated via majority voting. 2) We also evaluate whether the CHASe strategy of learning from unlabelled data can also improve performance on $\mathcal{D}_\ell$. To do this, we split $\mathcal{D}_\ell$, with 70%/10%/20% for training, validation, and testing, respectively, resulting in 47 test volumes. To measure CHASe's impact on *lesion* segmentation, we use the $\mathcal{D}_\ell$ test set.

Backbone Network. We used a 2D segmentation backbone, an effective choice for many organs [14,25,33,35], due to its simplicity and efficiency. We opt for the popular ResNet-50 [17]-based DeepLabv2 [8] network with PHNN-style deep supervision. We also tried VGG-16 [37], which also performed well and its results can be found in the supplementary. To create 3D masks we simply stack 2D predictions. **CHASe training.** We randomly sample multi-phase slices and, from them, randomly sample four out of the 15 phase combinations from $\mathcal{X}^*$ to stochastically minimize (11) and (13). For standard co-training baselines, we sample all available phases to minimize (7). **Discriminator Network.** We use an atrous spatial pyramid pooling (ASPP) layer, employing dilation rates of 1,2,3,4 with a kernel size of 3 and a leaky ReLU with negative slope 0.2 as our activation function. After a 1×1 convolution, a sigmoid layer classifies whether a pixel belongs to the labelled or unlabelled dataset. *Specific details on data pre-processing, learning rates and schedules can be found in the supplementary material.*

4.1   Pathological Liver Segmentation

We first measure the performance of CHASe on segmenting pathological livers from the unlabeled CGMH PACS dataset, *i.e.*, $\mathcal{D}_u$. We use PHNN trained only on $\mathcal{D}_\ell$ as a baseline, testing against different unlabeled learning baselines, *i.e.*, co-training [4], co-heterogeneous training, ADA [40], and hole-based pseudo-labelling. We measure the mean DSC and average symmetric surface distance (ASSD). For non hetero-modality variants, we use majority voting across each single-phase prediction to produce a multi-phase output. We also test against

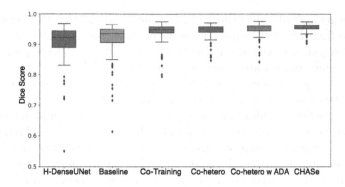

Fig. 3. Box and whisker plot. Shown is the distribution of DSCs of pathological liver segmentation on the CGMH PACS when using *all available* phases for inference.

the publicly available hybrid H-DenseUNet model [25], one of the best published models. It uses a cascade of 2D and 3D networks.

Table 1. Pathological liver segmentation. Mean DSC and ASSD scores on the CGMH PACS dataset are tabulated across different contrast phase inputs. "All" means all available phases are used as input. Number of samples are in parentheses.

Models	NC (96)		A (98)		V (97)		D (98)		All (100)	
	DSC	ASSD	DSC	ASSD	DSC	ASSD	DSC	ASSD	DSC	ASSD
HDenseUNet [25]	85.2	3.25	90.1	2.19	90.7	2.61	85.2	2.91	89.9	2.59
Baseline [14]	84.6	2.97	90.3	1.23	90.7	1.18	86.7	2.12	91.4	1.21
Baseline w pseudo	89.4	1.97	90.5	1.34	90.9	1.29	90.6	2.03	91.9	1.27
Baseline w ADA [40]	90.9	1.34	91.9	1.13	91.5	1.14	90.9	1.65	92.6	1.03
Co-training [31]	92.8	0.95	93.4	0.84	93.4	0.83	92.4	0.90	94.0	0.92
Co-hetero	93.4	0.81	93.7	0.77	94.5	0.79	93.6	0.86	94.7	0.89
Co-hetero w ADA	93.8	0.81	93.9	0.79	94.8	**0.66**	93.9	0.81	95.0	0.68
CHASe	**94.0**	**0.79**	**94.2**	**0.74**	**94.9**	**0.66**	**94.1**	**0.80**	**95.4**	**0.63**

As Table 1 indicates, despite being only a single 2D network, our PHNN baseline is strong, comparing similarly to the cascaded 2D/3D H-DenseUNet on our dataset[1]. However, both H-DenseUNet and our PHNN baseline still struggle to perform well on the CGMH dataset, particularly on non V-phases, indicating that training on public data alone is not sufficient. In contrast, through its principled semi-supervised approach, CHASe is able to dramatically increase performance, producing boosts of 9.4%, 3.9%, 4.2%, 7.4%, and 4.0% in mean DSCs for inputs of NC, A, V, D, and all phases, respectively. As can also be seen,

---

[1] A caveat is that the public H-DenseUNet model was only trained on the LiTS subset of $\mathcal{D}_\ell$.

all components contribute to these improvements, indicating the importance of each to the final result. Compared to established baselines of co-training and ADA, CHASe garners marked improvements. In addition, CHASe performs more strongly as more phases are available, something the baseline models are not always able to do. Results across all 15 possible combinations, found in our supplementary material, also demonstrate this trend.

More compelling results can be found in Fig. 3's box and whisker plots. As can be seen, each component is not only able to reduce variability, but more importantly significantly improves worst-case results. These same trends are seen across all possible phase combinations. Compared to improvements in mean DSCs, these worst-case reductions, with commensurate boosts in reliability, can often be more impactful for clinical applications. Unlike CHASe, most prior work on pathological liver segmentation is fully-supervised. Wang *et al.* report 96.4% DSC on 26 LiTS volumes and Yang *et al.* [45] report 95% DSC on 50 test volumes with unclear healthy vs pathological status. We achieve comparable, or better, DSCs on 100 pathological multi-phase test studies. As such, we articulate a versatile strategy to use and learn from the vast amounts of uncurated multi-phase clinical data housed within hospitals.

These quantitative results are supported by qualitative examples in Fig. 4. As the first two rows demonstrate, H-DenseUNet [25] and our baseline can perform inconsistently across contrast phases, with both being confused by the splenomegaly (overly large spleen) of the patient. The CHASe components are able to correct these issues. The third row in Fig. 4 depicts an example of a TACE-treated lesion, not seen in the public dataset and demonstrates how CHASe's components can progressively correct the under-segmentation. Finally, the last row depicts the *worst-case* performance of CHASe. Despite this unfavorable selection, CHASe is still able to predict better masks than the alternatives. Of note, CHASe is able to provide tangible improvements in consistency and reliability, robustly predicting even when presented with image features not seen in $\mathcal{D}_\ell$. More qualitative examples can be found in our supplementary material.

4.2 Liver and Lesion Segmentation

We also investigate whether using CHASe on unlabelled data can boost performance on the labelled data, *i.e.*, the $\mathcal{D}_\ell$ test set of 47 single-phase V volumes. Note, we include results from the public H-DenseUNet [25] implementation, even though it was only trained on LiTS and included some of our test instances originating from LiTS in its training set.

As Table 2 indicates, each CHASe component progressively boosts performance, with lesion scores being the most dramatic. The one exception is that the holes-based pseudo-labelling produces a small decrease in mean lesion scores. Yet, box and whisker plots, included in our supplementary material, indicate that holes-based pseudo-labelling boosts *median* values while reducing variability. Direct comparisons against other works, all typically using the LiTS challenge, are not possible, given the differences in evaluation data. nnUNet [20], the

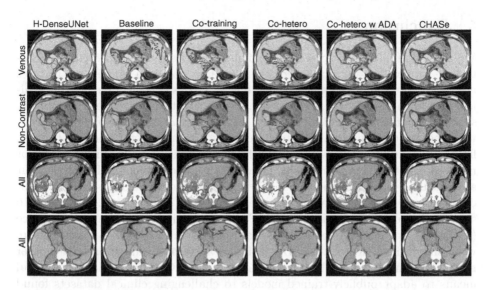

Fig. 4. Qualitative results. The first two rows are captured from the *same patient* across different contrast phases. The third and fourth row shows the performance when all available phases are included. **Green** and **red** curves depict the ground truth and segmentation predictions, respectively. (Color figure online)

Table 2. Ablation study on public data. Presented are test set DSC scores with their standard deviation of healthy liver, lesion, and liver region.

Model	Liver	Lesion	Liver region
HDenseUNet [25]	96.5 ± 2.0	51.7 ± 19.4	96.8 ± 1.8
Baseline [14]	96.3 ± 2.2	47.5 ± 24.1	96.6 ± 2.1
Co-training	96.3 ± 1.8	51.9 ± 20.5	96.7 ± 1.7
Co-hetero	96.4 ± 1.5	53.2 ± 19.1	96.7 ± 1.4
Co-hetero w ADA	96.5 ± 1.5	$\mathbf{61.0 \pm 17.2}$	97.0 ± 1.3
CHASe	$\mathbf{96.8 \pm 1.3}$	60.3 ± 18.0	$\mathbf{97.1 \pm 1.1}$

winner of the Medical Decathlon, reported 61% and 74% DSCs for their own validation and challenge test set, respectively. However, 57% of the patients in our test set are healthy, compared to the 3% in LiTS. More healthy cases will tend to make it a harder lesion evaluation set, as any amount of false positives will produce DSC scores of zero. For unhealthy cases, CHASe's lesion mean DSC is 61.9% compared to 53.2% for PHNN. CHASe allows a standard backbone, with no bells or whistles, to achieve dramatic boosts in lesion segmentation performance. As such, these results broaden the applicability of CHASe, suggesting it can even improve the *source*-domain performance of fully-supervised models.

5   Conclusion

We presented CHASe, a powerful semi-supervised approach to organ segmentation. Clinical datasets often comprise multi-phase data and image features not represented in single-phase public datasets. Designed to manage this challenging domain shift, CHASe can adapt publicly trained models to robustly segment multi-phase clinical datasets *with no extra annotation*. To do this, we integrate co-training and hetero-modality into a co-heterogeneous training framework. Additionally, we propose a highly computationally efficient ADA for multi-view setups and a principled holes-based pseudo-labeling. To validate our approach, we apply CHASe to a highly challenging dataset of 1147 multi-phase dynamic contrast CT volumes of patients, all with liver lesions. Compared to strong fully-supervised baselines, CHASe dramatically boosts mean performance (>9% in NC DSCs), while also drastically improving worse-case scores. Future work should investigate 2.5D/3D backbones and apply this approach to other medical organs. Even so, these results indicate that CHASe provides a powerful means to adapt publicly-trained models to challenging clinical datasets found "in-the-wild".

References

1. Bai, W., et al.: Semi-supervised learning for network-based cardiac MR image segmentation. In: Descoteaux, M., Maier-Hein, L., Franz, A., Jannin, P., Collins, D.L., Duchesne, S. (eds.) MICCAI 2017. LNCS, vol. 10434, pp. 253–260. Springer, Cham (2017). https://doi.org/10.1007/978-3-319-66185-8_29
2. Ben-Cohen, A., Diamant, I., Klang, E., Amitai, M., Greenspan, H.: Fully convolutional network for liver segmentation and lesions detection. In: Carneiro, G., et al. (eds.) LABELS/DLMIA -2016. LNCS, vol. 10008, pp. 77–85. Springer, Cham (2016). https://doi.org/10.1007/978-3-319-46976-8_9
3. Bilic, P., et al.: The liver tumor segmentation benchmark (LiTS). arXiv:1901.04056 (2019). http://arxiv.org/abs/1901.04056, arXiv: 1901.04056
4. Blum, A., Mitchell, T.: Combining labeled and unlabeled data with co-training. In: Proceedings of the eleventh annual conference on Computational learning theory, pp. 92–100. Citeseer (1998)
5. Cai, J., et al.: Accurate weakly-supervised deep lesion segmentation using large-scale clinical annotations: slice-propagated 3D mask generation from 2D RECIST. In: Frangi, A.F., Schnabel, J.A., Davatzikos, C., Alberola-López, C., Fichtinger, G. (eds.) MICCAI 2018. LNCS, vol. 11073, pp. 396–404. Springer, Cham (2018). https://doi.org/10.1007/978-3-030-00937-3_46
6. Chang, W.L., Wang, H.P., Peng, W.H., Chiu, W.C.: All about structure: adapting structural information across domains for boosting semantic segmentation. In: Proceedings of the IEEE Conference on Computer Vision and Pattern Recognition, pp. 1900–1909 (2019)
7. Chaos: Chaos - combined (CT-MR) healthy abdominal organ segmentation (2019). https://chaos.grand-challenge.org/Combined_Healthy_Abdominal_Organ_Segmentation

8. Chen, L.C., Papandreou, G., Kokkinos, I., Murphy, K., Yuille, A.L.: DeepLab: semantic image segmentation with deep convolutional nets, atrous convolution, and fully connected CRFs. arXiv:1606.00915 (2016)

9. Chen, Y.C., Lin, Y.Y., Yang, M.H., Huang, J.B.: CrDoCo: pixel-level domain transfer with cross-domain consistency. In: Proceedings of the IEEE Conference on Computer Vision and Pattern Recognition, pp. 1791–1800 (2019)

10. Conze, P.H., et al.: Scale-adaptive supervoxel-based random forests for liver tumor segmentation in dynamic contrast-enhanced CT scans. Int. J. Comput. Assist. Radiol. Surg. 12(2), 223–233 (2017). https://doi.org/10.1007/s11548-016-1493-1

11. Gibson, E., et al.: Multi-organ abdominal CT reference standard segmentations (2018). https://doi.org/10.5281/zenodo.1169361. This data set was developed as part of independent research supported by Cancer Research UK (Multidisciplinary C28070/A19985) and the National Institute for Health Research UCL/UCL Hospitals Biomedical Research Centre

12. Gotra, A., et al.: Liver segmentation: indications, techniques and future directions. Insights Imaging 8(4), 377–392 (2017). https://doi.org/10.1007/s13244-017-0558-1

13. Han, X.: Automatic liver lesion segmentation using a deep convolutional neural network method. arXiv preprint arXiv:1704.07239 (2017)

14. Harrison, A.P., Xu, Z., George, K., Lu, L., Summers, R.M., Mollura, D.J.: Progressive and multi-path holistically nested neural networks for pathological lung segmentation from CT images. In: Descoteaux, M., Maier-Hein, L., Franz, A., Jannin, P., Collins, D.L., Duchesne, S. (eds.) MICCAI 2017. LNCS, vol. 10435, pp. 621–629. Springer, Cham (2017). https://doi.org/10.1007/978-3-319-66179-7_71

15. Havaei, M., Guizard, N., Chapados, N., Bengio, Y.: HeMIS: hetero-modal image segmentation. In: Ourselin, S., Joskowicz, L., Sabuncu, M.R., Unal, G., Wells, W. (eds.) MICCAI 2016. LNCS, vol. 9901, pp. 469–477. Springer, Cham (2016). https://doi.org/10.1007/978-3-319-46723-8_54

16. He, K., Zhang, X., Ren, S., Sun, J.: Deep residual learning for image recognition. In: 2016 IEEE Conference on Computer Vision and Pattern Recognition (CVPR), pp. 770–778 (2015)

17. He, K., Zhang, X., Ren, S., Sun, J.: Deep residual learning for image recognition. In: Proceedings of the IEEE Conference on Computer Vision and Pattern Recognition, pp. 770 778 (2016)

18. Heimann, T., et al.: Comparison and evaluation of methods for liver segmentation from CT datasets. IEEE Trans. Med. Imaging 28(8), 1251–1265 (2009). https://doi.org/10.1109/TMI.2009.2013851

19. Heinrich, M.P., Jenkinson, M., Brady, M., Schnabel, J.A.: MRF-based deformable registration and ventilation estimation of lung CT. IEEE Trans. Med. Imaging 32, 1239–1248 (2013)

20. Isensee, F., et al.: nnU-Net: self-adapting framework for u-net-based medical image segmentation. arXiv preprint arXiv:1809.10486 (2018)

21. Jin, D., et al.: Accurate esophageal gross tumor volume segmentation in PET/CT using two-stream chained 3D deep network fusion. In: Shen, D., et al. (eds.) MICCAI 2019. LNCS, vol. 11765, pp. 182–191. Springer, Cham (2019). https://doi.org/10.1007/978-3-030-32245-8_21

22. Jin, D., et al.: Deep esophageal clinical target volume delineation using encoded 3D spatial context of tumors, lymph nodes, and organs at risk. In: Shen, D., et al. (eds.) MICCAI 2019. LNCS, vol. 11769, pp. 603–612. Springer, Cham (2019). https://doi.org/10.1007/978-3-030-32226-7_67

23. Kuo, C., Cheng, S., Lin, C., Hsiao, K., Lee, S.: Texture-based treatment prediction by automatic liver tumor segmentation on computed tomography. In: 2017 International Conference on Computer, Information and Telecommunication Systems (CITS), pp. 128–132 (2017). https://doi.org/10.1109/CITS.2017.8035318

24. Lee, D.H.: Pseudo-label : the simple and efficient semi-supervised learning method for deep neural networks. In: ICML 2013 Workshop : Challenges in Representation Learning (WREPL) (2013)

25. Li, X., Chen, H., Qi, X., Dou, Q., Fu, C.W., Heng, P.A.: H-DenseUNet: hybrid densely connected UNet for liver and tumor segmentation from CT volumes. IEEE Trans. Med. Imaging 37(12), 2663–2674 (2018)

26. Li, Y., Liu, L., Tan, R.T.: Decoupled certainty-driven consistency loss for semi-supervised learning (2019)

27. Lin, J.: Divergence measures based on the Shannon entropy. IEEE Trans. Inf. Theory 37(1), 145–151 (1991)

28. Milletari, F., Navab, N., Ahmadi, S.A.: V-Net: fully convolutional neural networks for volumetric medical image segmentation. In: 2016 Fourth International Conference on 3D Vision (3DV), pp. 565–571. IEEE (2016)

29. Min, S., Chen, X., Zha, Z.J., Wu, F., Zhang, Y.: A two-stream mutual attention network for semi-supervised biomedical segmentation with noisy labels (2019)

30. Oliva, M., Saini, S.: Liver cancer imaging: role of CT, MRI, US and PET. Cancer Imaging Official Publ. Int. Cancer Imaging Soc. 4(Spec No A), S42–6 (2004). https://doi.org/10.1102/1470-7330.2004.0011

31. Qiao, S., Shen, W., Zhang, Z., Wang, B., Yuille, A.: Deep co-training for semi-supervised image recognition. In: Proceedings of the European Conference on Computer Vision (ECCV), pp. 135–152 (2018)

32. Ronneberger, O., Fischer, P., Brox, T.: U-Net: convolutional networks for biomedical image segmentation. In: Navab, N., Hornegger, J., Wells, W.M., Frangi, A.F. (eds.) MICCAI 2015. LNCS, vol. 9351, pp. 234–241. Springer, Cham (2015). https://doi.org/10.1007/978-3-319-24574-4_28

33. Roth, H.R., et al.: DeepOrgan: multi-level deep convolutional networks for automated pancreas segmentation. In: Navab, N., Hornegger, J., Wells, W.M., Frangi, A.F. (eds.) MICCAI 2015. LNCS, vol. 9349, pp. 556–564. Springer, Cham (2015). https://doi.org/10.1007/978-3-319-24553-9_68

34. Roth, H.R., et al.: Spatial aggregation of holistically-nested convolutional neural networks for automated pancreas localization and segmentation. Med. Image Anal. 45, 94–107 (2018)

35. Roth, H.R., et al.: Spatial aggregation of holistically-nested convolutional neural networks for automated pancreas localization and segmentation. Med. Image Anal. 45, 94–107 (2018). https://doi.org/10.1016/j.media.2018.01.006, http://www.sciencedirect.com/science/article/pii/S1361841518300215

36. Roth, K., Konopczyński, T., Hesser, J.: Liver lesion segmentation with slice-wise 2D Tiramisu and Tversky loss function. arXiv preprint arXiv:1905.03639 (2019)

37. Simonyan, K., Zisserman, A.: Very deep convolutional networks for large-scale image recognition. arXiv preprint arXiv:1409.1556 (2014)

38. Soler, L., et al.: 3D image reconstruction for comparison of algorithm database: a patient specific anatomical and medical image database. Technical report, IRCAD, Strasbourg, France (2010)

39. Tajbakhsh, N., Jeyaseelan, L., Li, Q., Chiang, J., Wu, Z., Ding, X.: Embracing imperfect datasets: a review of deep learning solutions for medical image segmentation. Med. Image Anal. 63, 101693 (2019)

40. Tsai, Y.H., Hung, W.C., Schulter, S., Sohn, K., Yang, M.H., Chandraker, M.: Learning to adapt structured output space for semantic segmentation. In: Proceedings of the IEEE Conference on Computer Vision and Pattern Recognition, pp. 7472–7481 (2018)

41. Vorontsov, E., Abi-Jaoudeh, N., Kadoury, S.: Metastatic liver tumor segmentation using texture-based omni-directional deformable surface models. In: Yoshida, H., Nappi, J., Saini, S. (eds.) Abdominal Imaging. Computational and Clinical Applications. ABD-MICCAI 2014. Lecture Notes in Computer Science, vol. 8676, pp. 74–83. Springer, Cham (2014). https://doi.org/10.1007/978-3-319-13692-9_7

42. Vu, T.H., Jain, H., Bucher, M., Cord, M., Pérez, P.: Advent: adversarial entropy minimization for domain adaptation in semantic segmentation. In: Proceedings of the IEEE Conference on Computer Vision and Pattern Recognition, pp. 2517–2526 (2019)

43. Wang, R., Cao, S., Ma, K., Meng, D., Zheng, Y.: Pairwise semantic segmentation via conjugate fully convolutional network. In: Shen, D., et al. (eds.) MICCAI 2019. LNCS, vol. 11769, pp. 157–165. Springer, Cham (2019). https://doi.org/10.1007/978-3-030-32226-7_18

44. Xia, Y., et al.: 3D semi-supervised learning with uncertainty-aware multi-view co-training. arXiv preprint arXiv:1811.12506 (2018)

45. Yang, D., et al.: Automatic liver segmentation using an adversarial image-to-image network. In: Descoteaux, M., Maier-Hein, L., Franz, A., Jannin, P., Collins, D.L., Duchesne, S. (eds.) MICCAI 2017. LNCS, vol. 10435, pp. 507–515. Springer, Cham (2017). https://doi.org/10.1007/978-3-319-66179-7_58

46. Zhang, J., Xie, Y., Zhang, P., Chen, H., Xia, Y., Shen, C.: Light-weight hybrid convolutional network for liver tumor segmentation. In: Proceedings of the Twenty-Eighth International Joint Conference on Artificial Intelligence, IJCAI-19, pp. 4271–4277. International Joint Conferences on Artificial Intelligence Organization (2019). https://doi.org/10.24963/ijcai.2019/593

47. Zhang, L., Gopalakrishnan, V., Lu, L., Summers, R.M., Moss, J., Yao, J.: Self-learning to detect and segment cysts in lung CT images without manual annotation. In: 2018 IEEE 15th International Symposium on Biomedical Imaging (ISBI 2018), pp. 1100–1103. IEEE (2018)

48. Zhang, Q., Fan, Y., Wan, J., Liu, Y.: An efficient and clinical-oriented 3D liver segmentation method. IEEE Access 5, 18737–18744 (2017). https://doi.org/10.1109/ACCESS.2017.2754298

49. Zhou, Y., et al.: Semi-supervised 3D abdominal multi-organ segmentation via deep multi-planar co-training. In: 2019 IEEE Winter Conference on Applications of Computer Vision (WACV), pp. 121–140. IEEE (2019)

Towards Recognizing Unseen Categories in Unseen Domains

Massimiliano Mancini[1,2](✉) [iD], Zeynep Akata[2] [iD], Elisa Ricci[3,4] [iD], and Barbara Caputo[5,6] [iD]

[1] Sapienza University of Rome, Rome, Italy
mancini@diag.uniroma1.it
[2] University of Tübingen, Tübingen, Germany
[3] University of Trento, Trento, Italy
[4] Fondazione Bruno Kessler, Trento, Italy
[5] Politecnico di Torino, Turin, Italy
[6] Italian Institute of Technology, Genova, Italy

Abstract. Current deep visual recognition systems suffer from severe performance degradation when they encounter new images from classes and scenarios unseen during training. Hence, the core challenge of Zero-Shot Learning (ZSL) is to cope with the *semantic-shift* whereas the main challenge of Domain Adaptation and Domain Generalization (DG) is the *domain-shift*. While historically ZSL and DG tasks are tackled in isolation, this work develops with the ambitious goal of solving them jointly, i.e. by recognizing *unseen visual concepts in unseen domains*. We present CuMix (**Cu**rriculum **Mix**up for recognizing unseen categories in unseen domains), a holistic algorithm to tackle ZSL, DG and ZSL+DG. The key idea of CuMix is to simulate the test-time domain and semantic shift using images and features from unseen domains and categories generated by *mixing up* the multiple source domains and categories available during training. Moreover, a curriculum-based mixing policy is devised to generate increasingly complex training samples. Results on standard ZSL and DG datasets and on ZSL+DG using the DomainNet benchmark demonstrate the effectiveness of our approach.

Keywords: Zero-Shot Learning · Domain Generalization

1 Introduction

Despite their astonishing success in several applications [12,34], deep visual models perform poorly for the classes and scenarios that are unseen during training. Most existing approaches are based on the assumptions that (a) training and test data come from the same underlying distribution, i.e. domain shift, and (b)

Electronic supplementary material The online version of this chapter (https:// doi.org/10.1007/978-3-030-58592-1_28) contains supplementary material, which is available to authorized users.

A. Vedaldi et al. (Eds.): ECCV 2020, LNCS 12368, pp. 466–483, 2020.
https://doi.org/10.1007/978-3-030-58592-1_28

the set of classes seen during training constitute the only classes that will be seen at test time, i.e. semantic shift. These assumptions rarely hold in practice and, in addition to depicting different semantic categories, training and test images may differ significantly in terms of visual appearance in the real world.

To address these limitations, research efforts have been devoted to designing deep architectures able to cope with varying visual appearance [7] and with novel semantic concepts [47]. In particular, the domain-shift problem [15] has been addressed by proposing domain adaptation (DA) models [7] that assume the availability target domain data during training. To circumvent this assumption, a recent trend has been to move to more complex scenarios where the adaptation problem must be either tackled online [14,24], with the help of target domain descriptions [23], auxiliary data [32] or multiple source domains [25,26,36]. For instance, domain generalization (DG) methods [5,19,21] aim to learn domain-agnostic prediction models and to generalize to any unseen target domain.

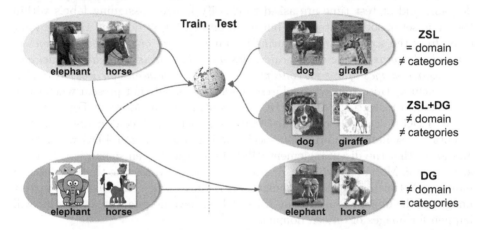

Fig. 1. Our ZSL+DG problem. During training we have images of multiple categories (*e.g. elephant,horse*) and domains (*e.g. photo, cartoon*). At test time, we want to recognize unseen categories (*e.g. dog, giraffe*), as in ZSL, in unseen domains (*e.g. paintings*), as in DG, exploiting side information describing seen and unseen categories.

Regarding semantic knowledge, multiple works have designed approaches for extending deep architectures to handle new categories and new tasks. For instance, continual learning methods [18] attempt to sequentially learn new tasks while retaining previous knowledge, tackling the catastrophic forgetting issue. Similarly, in open-world recognition [4] the goal is to detect unseen categories and successfully incorporate them into the model. Another research thread is Zero-Shot Learning (ZSL) [47], where the goal is to recognize objects unseen during training given external information about the novel classes provided in forms of semantic attributes [17], visual descriptions [2] or word embeddings [27].

Despite these significant efforts, an open research question is whether we can tackle the two problems jointly. Indeed, due to the large variability of visual concepts in the real world, in terms of both semantics and acquisition conditions, it is impossible to construct a training set capturing such variability. This calls for a holistic approach addressing them together. Consider for instance the case depicted in Fig. 1. A system trained to recognize elephants and horses from realistic images and cartoons might be able to recognize the same categories in another visual domain, like art paintings (Fig. 1, bottom) or it might be able to describe other quadrupeds in the same training visual domains (Fig. 1, top). On the other hand, how to deal with the case where new animals are shown in a new visual domain is not clear.

To our knowledge, our work is the first attempt to answer this question, proposing a method that is able to *recognize unseen semantic categories in unseen domains*. In particular, our goal is to jointly tackle ZSL and DG (see Fig. 1). ZSL algorithms usually receive as input a set of images with their associated semantic descriptions, and learn the relationship between an image and its semantic attributes. Likewise, DG approaches are trained on multiple source domains and at test time are asked to classify images, assigning labels within the same set of source categories but in an unseen target domain. Here we want to address the scenario where, during training, *we are given a set of images of multiple domains and semantic categories and our goal is to build a model able to recognize images of unseen concepts, as in ZSL, in unseen domains, as in DG.*

To achieve this, we need to address challenges usually not present when these two classical tasks, i.e. ZSL and DG, are considered in isolation. For instance, while in DG we can rely on the fact that the multiple source domains permit to disentangle semantic and domain-specific information, in ZSL+DG we have no guarantee that the disentanglement will hold for the unseen semantic categories at test time. Moreover, while in ZSL it is reasonable to assume that the learned mapping between images and semantic attributes will generalize also to test images of the unseen concepts, in ZSL+DG we have no guarantee that this will happen for images of unseen domains.

To overcome these issues, during training we simulate both the semantic and the domain shift we will encounter at test time. Since explicitly generating images of unseen domains and concepts is an ill-posed problem, we sidestep this issue and we synthesize unseen domains and concepts by interpolating existing ones. To do so, we revisit the *mixup* [53] algorithm as a tool to obtain partially unseen categories and domains. Indeed, by randomly mixing samples of different categories we obtain new samples which do not belong to a single one of the available categories during training. Similarly, by mixing samples of different domains, we obtain new samples which do not belong to a single source domain available during training.

Under this perspective, mixing samples of both different domains and classes allows to obtain samples that cannot be categorized in a single class and domain of the one available during training, thus they are *novel* both for the semantic and their visual representation. Since higher levels of abstraction contain more task-related information, we perform *mixup* at both image and feature level, showing experimentally the need for this choice. Moreover, we introduce a curriculum-based mixing strategy to generate increasingly complex training samples.

We show that our CuMix (**Curriculum Mix**up for recognizing unseen categories in unseen domains) model obtains state-of-the-art performances in both ZSL and DG in standard benchmarks and it can be effectively applied to the combination of the two tasks, recognizing unseen categories in unseen domains.[1]

To summarize, our contributions are as follows. (i) We introduce the ZSL+DG scenario, a first step towards recognizing unseen categories in unseen domains. (ii) Being the first holistic method able to address ZSL, DG, and the two tasks together, our method is based on simulating new domains and categories during training by mixing the available training domains and classes both at image and feature level. The mixing strategy becomes increasingly more challenging during training, in a curriculum fashion. (iii) Through our extensive evaluations and analysis, we show the effectiveness of our approach in all three settings: namely ZSL, DG and ZSL+DG.

2   Related Works

Domain Generalization (DG). Over the past years the research community has put considerable efforts into developing methods to contrast the domain shift. Opposite to domain adaptation [7], where it is assumed that target data are available in the training phase, the key idea behind DG is to learn a domain agnostic model to be applied to any unseen target domain.

Previous DG methods can be broadly grouped into four main categories. The first category comprises methods which attempt to learn domain-invariant feature representations [28] by considering specific alignment losses, such as maximum mean discrepancy (MMD), adversarial loss [22] or self-supervised losses [5]. The second category of methods [15,19] develop from the idea of creating deep architectures where both domain-agnostic and domain-specific parameters are learned on source domains. After training, only the domain-agnostic part is retained and used for processing target data. The third category devises specific optimization strategies or training procedures in order to enhance the generalization ability of the source model to unseen target data. For instance, in [20] a meta-learning approach is proposed for DG. Differently, in [21] an episodic training procedure is presented to learn models robust to the domain shift. The latter category comprises methods which introduce data and feature augmentation strategies to synthesize novel samples and improve the generalization capability of the learned model [39,42,43]. These strategies are mostly based on adversarial training [39,43].

Our work is related to the latter category since we also generate synthetic samples with the purpose of learning more robust target models. However, differently from previous methods, we specifically employ mixup to perturb feature representations. Recently, works have considered mixup in the context of domain adaptation [51] to *e.g.* reinforce the judgments of a domain discrimination. However, we employ mixup from a different perspective *i.e.* simulating semantic and

---

[1] The code is available at https://github.com/mancinimassimiliano/CuMix.

domain shift we will encounter at test time. To this extent, we are not aware of previous methods using mixup for DG and ZSL.

Zero-Shot Learning (ZSL). Traditional ZSL approaches attempt to learn a projection function mapping images/visual features to a semantic embedding space where classification is performed. This idea is achieved by directly predicting image attributes e.g. [17] or by learning a linear mapping through margin-based objective functions [1,2]. Other approaches explored the use of non-linear multi-modal embeddings [45], intermediate projection spaces [54,55] or similarity-based interpolation of base classifiers [6]. Recently, various methods tackled ZSL from a generative point of view considering Generative Adversarial Networks [48], Variational Autoencoders (VAE) [38] or both of them [50]. While none of these approaches explicitly tackled the domain shift, i.e. visual appearance changes among different domains/datasets, various methods proposed to use domain adaptation technique, e.g. to refine the semantic embedding space, aligning semantic and projected visual features [38] or, in transductive scenarios, to cope with the inherent domain shift existing among the appearance of attributes in different categories [9,10,16]. For instance, in [38] a distance among visual and semantic embedding projected in the VAE latent space is minimized. In [16] the problem is addressed through a regularised sparse coding framework, while in [9] a multi-view hypergraph label propagation framework is introduced.

Recently, works have considered also coupling ZSL and DA in a transductive setting. For instance, in [56] a semantic guided discrepancy measure is employed to cope with the asymmetric label space among source and target domains. In the context of image retrieval, multiple works addressed the sketch-based image retrieval problem [8,52], even across multiple domains. In [40] the authors proposed a method to perform cross-domain image retrieval by training domain-specific experts. While these approaches integrated DA and ZSL, none of them considered the more complex scenario of DG, where no target data are available.

3   Method

In this section, we first formalize the Zero-Shot Learning under Domain Generalization (ZSL+DG). We then describe our approach, CuMix , which, by performing curriculum learning through mixup, simulates the domain- and semantic-shift the network will encounter at test time, and can be holistically applied to ZSL, DG and ZSL+DG.

3.1   Problem Formulation

In the ZSL+DG problem, the goal is to recognize unseen categories (as in ZSL) in unseen domains (as in DG). Formally, let $\mathcal{X}$ denote the input space (e.g. the image space), $\mathcal{Y}$ the set of possible classes and $\mathcal{D}$ the set of possible domains. During training, we are given a set $\mathcal{S} = \{(x_i, y_i, d_i)\}_{i=1}^n$ where $x_i \in \mathcal{X}$, $y_i \in \mathcal{Y}^s$ and $d_i \in \mathcal{D}^s$. Note that $\mathcal{Y}^s \subset \mathcal{Y}$ and $\mathcal{D}^s \subset \mathcal{D}$ and, as in standard DG, we have multiple source domains (*i.e.* $\mathcal{D}^s = \cup_{j=1}^m d_j$, with $m > 1$) with different distributions *i.e.* $p_{\mathcal{X}}(x|d_i) \neq p_{\mathcal{X}}(x|d_j)$, $\forall i \neq j$.

Given $\mathcal{S}$ our goal is to learn a function h mapping an image x of domains $\mathcal{D}^u \subset \mathcal{D}$ to its corresponding label in a set of classes $\mathcal{Y}^u \subset \mathcal{Y}$. Note that in standard ZSL, while the set of train and test domains are shared, *i.e.* $\mathcal{D}^s \equiv \mathcal{D}^u$, the label sets are disjoint *i.e.* $\mathcal{Y}^s \cap \mathcal{Y}^u \equiv \emptyset$, thus $\mathcal{Y}^u$ is a set of *unseen* classes. On the other hand, in DG we have a shared output space, *i.e.* $\mathcal{Y}^s \equiv \mathcal{Y}^u$, but a disjoint set of domains between training and test *i.e.* $\mathcal{D}^s \cap \mathcal{D}^u \equiv \emptyset$, thus $\mathcal{D}^u$ is a set of *unseen* domains. Since the goal of our work is to recognize unseen classes in unseen domains, we unify the settings of DG and ZSL, considering both semantic- and domain-shift at test time *i.e.* $\mathcal{Y}^s \cap \mathcal{Y}^u \equiv \emptyset$ and $\mathcal{D}^s \cap \mathcal{D}^u \equiv \emptyset$.

In the following we divide the function h into three parts: f, mapping images into a feature space $\mathcal{Z}$, *i.e.* $f : \mathcal{X} \to \mathcal{Z}$, g going from $\mathcal{Z}$ to a semantic embedding space $\mathcal{E}$, *i.e.* $g : \mathcal{Z} \to \mathcal{E}$, and an embedding function $\omega : \mathcal{Y}^t \to \mathcal{E}$ where $\mathcal{Y}^t \equiv \mathcal{Y}^s$ during training and $\mathcal{Y}^t \equiv \mathcal{Y}^u$ at test time. Note that ω is a learned classifier for DG while it is a fixed semantic embedding function in ZSL, mapping classes into their vectorized representation extracted from external sources. Given an image x, the final class prediction is obtained as follows:

$$y^* = \text{argmax}_y \omega(y)^{\mathsf{T}} g(f(x)). \tag{1}$$

In this formulation, f can be any learnable feature extractor (*e.g.* a deep neural network), while g any ZSL predictor (*e.g.* a semantic projection layer, as in [46] or a compatibility function among visual features and labels, as in [1,2]). The first solution to address the ZSL+DG problem could be training a classifier using the aggregation of data from all source domains. In particular, for each sample we could minimize a loss function of the form:

$$\mathcal{L}_{\text{AGG}}(x_i, y_i) = \sum_{y \in \mathcal{Y}^s} \ell(\omega(y)^{\mathsf{T}} g(f(x_i)), y_i) \tag{2}$$

with ℓ an arbitrary loss function, *e.g.* the cross-entropy loss. In the following, we show how we can use the input to Eq. (2) to effectively recognize unseen categories in unseen domains.

3.2   Simulating Unseen Domains and Concepts Through *Mixup*

The fundamental problem of ZSL+DG is that, during training, we have neither access to visual data associated to categories in $\mathcal{Y}^u$ nor to data of the unseen domains $\mathcal{D}^u$. One way to overcome this issue in ZSL is to generate samples of unseen classes by learning a generative function conditioned on the semantic embeddings in $\mathcal{W} = \{\omega(y)|y \in \mathcal{Y}^s\}$ [48,50]. However, since no description is available for the unseen target domain(s) in $\mathcal{D}^u$, this strategy is not feasible in ZSL+DG. On the other hand, previous works on DG proposed to synthesize images of unseen domains through adversarial strategies of data augmentation [39,43]. However, these strategies are not applied to ZSL since they cannot easily be extended to generate data for unseen semantic categories $\mathcal{Y}^u$.

To circumvent this issue, we introduce a strategy to simulate, during training, novel domains and semantic concepts by interpolating from the ones available in $\mathcal{D}^s$ and $\mathcal{Y}^s$. Simulating novel domains and classes allows to train the network to cope with both semantic- and domain-shift, the same situation our model will face at test time. Since explicitly generating inputs of novel domains and categories is a complex task, in this work we propose to achieve this goal, by *mixing* images and features of different classes and domains, revisiting the popular *mixup* [53] strategy.

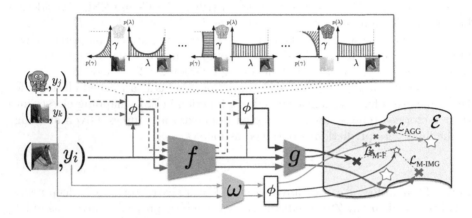

Fig. 2. Our CuMix Framework. Given an image (bottom, *horse*, *photo*), we randomly sample one image from the same (middle, *photo*) and one from another (top, *cartoon*) domain. The samples are mixed through ϕ (white blocks) both at image and feature level, with their features and labels projected into the embedding space $\mathcal{E}$ (by g and ω respectively) and there compared to compute our final objective. Note that ϕ varies during training (top part), changing the mixing ratios in and across domains.

In practice, given two elements a_i and a_j of the same space (*e.g.* $a_i, a_j \in \mathcal{X}$), *mixup* [53] defines a mixing function φ as follows:

$$\varphi(a_i, a_j) = \lambda \cdot a_i + (1 - \lambda) \cdot a_j \qquad (3)$$

with λ sampled from a beta distribution, *i.e.* $\lambda \sim \text{Beta}(\beta, \beta)$, with β an hyperparameter. Given two samples (x_i, y_i) and (x_j, y_j) randomly drawn from a training set $\mathcal{T}$, a new loss term is defined as:

$$\mathcal{L}_{\text{MIXUP}}((x_i, y_i), (x_j, y_j)) = \mathcal{L}_{\text{AGG}}(\varphi(x_i, x_j), \varphi(\bar{y}_i, \bar{y}_j)) \qquad (4)$$

where $\bar{y}_i \in \Re^{|\mathcal{Y}^s|}$ is the one-hot vectorized representation of label y_i. Note that, when mixing two samples and label vectors with φ, a single λ is drawn and applied within φ in both image and label spaces. The loss defined in Eq. (4) forces the network to disentangle the various semantic components (*i.e.* y_i and y_j) contained in the mixed inputs (*i.e.* x_i and x_j) plus the ratio λ used to mix them.

This auxiliar task acts as a strong regularizer that helps the network to *e.g.* being more robust against adversarial examples [53]. Note however that the function φ creates input and targets which do not represent a single semantic concept in $\mathcal{T}$ but contains characteristics taken from multiple samples and categories, synthesising a *new* semantic concept from the interpolation of existing ones.

For recognizing unseen concepts in unseen domains at test time, we revisit φ to obtain both cross-domain and cross-semantic mixes during training, simulating both semantic- and domain-shift. While simulating the semantic-shift is a by-product of the original *mixup* formulation, here we explicitly revisit φ in order to perform cross-domain mixups. In particular, instead of considering a pair of samples from our training set, we consider a triplet (x_i, y_i, d_i), (x_j, y_j, d_j) and (x_k, y_k, d_k). Given (x_i, y_i, d_i), the other two elements of the triplet are randomly sampled from $\mathcal{S}$, with the only constraint that $d_i = d_k, i \neq k$ and $d_j \neq d_i$. In this way, the triplet contains two samples of the same domain (*i.e.* d_i) and a third of a different one (*i.e.* d_j). Then, our mixing function ϕ is defined as follows:

$$\phi(a_i, a_j, a_k) = \lambda a_i + (1 - \lambda)(\gamma a_j + (1 - \gamma)a_k) \tag{5}$$

with γ sampled from a Bernoulli distribution $\gamma \sim \mathcal{B}(\alpha)$ and a representing either the input x or the vectorized version of the label y, *i.e.* $\bar{y}$. Note that we introduced a term γ which allows to perform either intra-domain (with $\gamma = 0$) or cross-domain (with $\gamma = 1$) mixes.

To learn a feature extractor f and a semantic projection layer g robust to domain- and semantic-shift, we propose to use ϕ to simulate both samples and features of novel domains and classes during training. Namely, we simulate the semantic- and domain-shift at two levels, i.e. image and class levels. Given a sample $(x_i, y_i, d_i) \in \mathcal{S}$ we define the following loss:

$$\mathcal{L}_{\text{M-IMG}}(x_i, y_i, d_i) = \mathcal{L}_{\text{AGG}}(\phi(x_i, x_j, x_k), \phi(\bar{y}_i, \bar{y}_j, \bar{y}_k)). \tag{6}$$

where (x_i, y_i, d_i), (x_j, y_j, d_j), (x_k, y_k, d_k) are randomly sampled from $\mathcal{S}$, with $d_i = d_k$ and $d_j \neq d_k$. The loss term in Eq. (6) enforces the feature extractor to effectively process inputs of mixed domains/semantics obtained through ϕ. Additionally, to also act at classification level, we design another loss which forces the semantic consistency of mixed features in $\mathcal{E}$. This loss term is defined as:

$$\mathcal{L}_{\text{M-F}}(x_i, y_i, d_i) = \sum_{y \in \mathcal{Y}^s} \ell\Big(\omega(y)^\top g\big(\phi(f(x_i), f(x_j), f(x_k))\big), \phi(\bar{y}_i, \bar{y}_j, \bar{y}_k)\Big) \tag{7}$$

where, as before, (x_j, y_j, d_j), $(x_k, y_k, d_k) \sim \mathcal{S}$, with $d_i = d_k, i \neq k$ and $d_j \neq d_k$ and ℓ is a generic loss function *e.g.* the cross-entropy loss. This second loss term forces the classifier ω and the semantic projection layer g to be robust to features with mixed domains and semantics.

While we can simply use a fixed mixing function ϕ, as defined in Eq. (5), for Eq. (6) and Eq. (7), we found that it is more beneficial to devise a dynamic ϕ which changes its behaviour during training, in a curriculum fashion. Intuitively, minimizing the two objectives defined in Eq. (6) and Eq. (7) requires our model to correctly disentangle the various semantic components used to form the mixed samples. While this is a complex task even for intra-domain mixes (*i.e.* when only the semantic is mixed), mixing samples across domains makes the task even harder, requiring to isolate also domain specific factors. To effectively tackle this task, we choose to act on the mixing function ϕ. In particular, we want our ϕ to create mixed samples with progressively increased degree of mixing both with respect to content and domain, in a curriculum-based fashion.

During training we regulate both α (weighting the probability of cross-domain mixes) and β (modifying the probability distribution of the mix ratio λ), changing the probability distribution of the mixing ratio λ and of the cross-domain mix γ. In particular, given a warm-up step of N epochs and being s the current epoch we set $\beta = \min(\frac{s}{N}\beta_{max}, \beta_{max}))$, with β_{max} as hyperparameter, while $\alpha = \max(0, \min(\frac{s-N}{N}, 1))$. As a consequence, the learning process is made of three phases, with a smooth transition among them. We start by solving the plain classification task on a single domain (*i.e.* $s < N$, $\alpha = 0$, $\beta = \frac{s}{N}\beta_{max}$,). In the subsequent step ($N \leq s < 2N$) samples of the same domains are mixed randomly, with possibly different semantics (*i.e.* $\alpha = \frac{s-N}{N}$, $\beta = \beta_{max}$). In the third phase ($s \geq 2N$), we mix up samples of different domains (*i.e.* $\alpha = 1$), simulating the domain shift the predictor will face at test time. Figure 2, shows a representation of how ϕ varies during training (top, white block).

Final Objective. The full training procedure, is represented in Fig. 2. Given a training sample (x_i, y_i, d_i), we randomly draw other two samples, (x_j, y_j, d_j) and (x_k, y_k, d_k), with $d_i = d_k, i \neq k$ and $d_j \neq d_i$, feed them to ϕ and obtain the first mixed input. We then feed x_i, x_j, x_k and the mixed sample through f, to extract their respective features. At this point we use features extracted from other two randomly drawn samples (in the figure, and just for simplicity, x_j and x_k with same mixing ratios λ and γ), to obtain the feature level mixed features needed to build the objective in Eq. (7). Finally, the features of x_i and the two mixed variants at image and feature level, are fed to the semantic projection layer g, which maps them to the embedding space $\mathcal{E}$. At the same time, the labels in $\mathcal{Y}^s$ are projected in $\mathcal{E}$ through ω. Finally, the objectives defined in Eq. (2), Eq. (6) and Eq.(7) functions are then computed in the semantic embedding space. Our final objective is:

$$\mathcal{L}_{\text{CuMIX}}(\mathcal{S}) = |\mathcal{S}|^{-1} \sum_{(x_i, y_i, d_i) \in \mathcal{S}} \mathcal{L}_{\text{AGG}}(x_i, y_i) + \eta_I \mathcal{L}_{\text{M-IMG}}(x_i, y_i, d_i) + \eta_F \mathcal{L}_{\text{M-F}}(x_i, y_i, d_i)$$

$$(8)$$

with η_I and η_F hyperparameters weighting the importance of the two terms. As $\ell(x, y)$ in both $\mathcal{L}_{\text{AGG}}$, $\mathcal{L}_{\text{M-IMG}}$ and $\mathcal{L}_{\text{M-F}}$, we use the standard cross-entropy loss, even if any ZSL objective can be applied. Finally, we highlight that the optimization is performed batch-wise, thus also the sampling of the triplet considers the current batch and not the full training set $\mathcal{S}$. Moreover, while in Fig. 2 we show for simplicity that the same samples are drawn for $\mathcal{L}_{\text{M-IMG}}$ and $\mathcal{L}_{\text{M-F}}$, in practice, given a sample, the random sampling procedure of the other two members of the triplet is held-out twice, one at the image level and one at the feature level. Similarly, the sampling of the mixing ratios λ and cross domain factor γ of ϕ is held-out sample-wise and twice, one at image level and one at feature level. As in Eq. (3), λ and γ are kept fixed across mixed inputs/features and their respective targets in the label space.

Discussion. We present similarities between our framework with DG and ZSL methods. In particular, presenting the classifier with noisy features extracted by a non-domain specialist network, has a similar goal as the episodic strategy for DG described in [21]. On the other hand, here we sidestep the need to train domain experts by directly presenting as input to our classifier features of novel domains that we obtain by interpolating the available sources samples. Our method is also linked to *mixup* approaches developed in DA [51]. Differently from them, we use *mixup* to simulate unseen domains rather then to progressively align the source to the given target data.

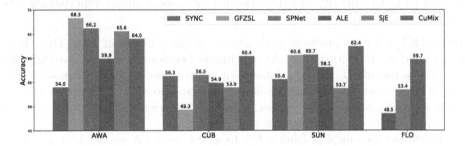

Fig. 3. ZSL results on CUB, SUN, AWA and FLO datasets with ResNet-101 features.

Our method is also related to ZSL frameworks based on feature generation [48,50]. While the quality of our synthesized samples is lower since we do not exploit attributes for conditional generation, we have a lower computational cost. In fact, during training we simulate the test-time semantic shift without generating samples of unseen classes. Moreover, we do not require additional training phases on the generated samples or the availability of unseen class attributes to be available beforehand.

4   Experimental Results

4.1   Datasets and Implementation Details

We assess CuMix in three scenarios: ZSL, DG and the proposed ZSL+DG setting.

ZSL. We conduct experiments on four standard benchmarks: Caltech-UCSD-Birds 200-2011 (CUB) [44], SUN attribute (SUN) [31], Animals with Attributes (AWA) [17] and Oxford Flowers (FLO) [29]. CUB contains 11,788 images of 200 bird species, with 312 attributes, SUN 14,430 images of 717 scenes annotated with 102 attributes, and AWA 30,475 images of 50 animal categories with 85 attributes. Finally, FLO is a fine-grained dataset of flowers, containing 8,189 images of 102 categories. As semantic representation, we use the visual descriptions of [35], following [46,48]. For each dataset, we use the train, validation and test split provided by [47]. In all the settings we employ features extracted from the second-last layer of a ResNet-101 [13] pretrained on ImageNet as image representation. For CuMix , we consider f as the identity function and as g a simple fully connected layer, perform our version of mixup directly at the feature-level while applying our alignment loss in the embedding space. All hyperparameters have been set following [47].

DG. We perform experiments on the PACS dataset [19] with 9,991 images of 7 semantic classes in 4 different visual domains, *art paintings*, *cartoons*, *photos* and *sketches*. For this experiment we use the standard train and test split defined in [19], with the same validation protocol. We use as base architecture a ResNet-18 [13] pretrained on ImageNet. For our model, we consider f to be the ResNet-18 while g to be the identity function. We use the same training hyperparameters and protocol of [21].

ZSL+DG. Since no previous work addressed the problem of ZSL+DG, there is no benchmark on this task. As a valuable benchmark, we choose Domain-Net [33], a recently introduced dataset for multi-source domain adaptation [33] with a large variety of domains, visual concepts and possible descriptions. It contains approximately 600'000 images from 345 categories and 6 domains, *clipart*, *infograph*, *painting*, *quickdraw*, *real* and *sketch*.

To convert this dataset from a DA to a ZSL scenario, we need to define an unseen set of classes. Since our method uses a network pretrained on ImageNet [37], the set of unseen classes can not contain any of the classes present in ImageNet following the good practices in [49]. We build our validation + test set with 100 classes that contain at least 40 images per domain and that has no overlap with ImageNet. We reserve 45 of these classes for the unseen test set, matching the number used in [40], and the remaining 55 classes for the unseen validation set. The remaining 245 classes are used as seen classes during training.

We set the hyperparameters of each method by training on all the images of the seen classes on a *subset* of the source domains and validating on all the images of the validation set from the held-out source domain. After the hyperparameters are set, we retrain the model on the training set, i.e. 245 classes, and validation set, i.e. 55 classes, of a total number of 300 classes. Finally, we report the final results on the 45 unseen classes. As semantic representation we use word2vec embeddings [27] extracted from the Google News corpus and $L2$-normalized, following [40]. For all the baselines and our method, we employ as base architecture a ResNet-50 [13] pretrained on ImageNet, using the same number of epochs and SGD with momentum as optimizer, with the same hyperparameters of [40].

4.2 Results

ZSL. In the ZSL scenario, we choose as baselines standard inductive methods plus more recent approaches. In particular we report the results of ALE [1], SJE [2], SYNC [6], GFZSL [41] and SPNet [46]. ALE [1] and SJE [2] are linear compatibility methods using a ranking loss and the structural SVM loss respectively. SYNC [6] learns a mapping from the feature space and the semantic embedding space by means of phantom classes and a weighted graph. GFZSL [41] employs a generative framework where each class-conditional distribution is modeled as a multivariate Gaussian. Finally, SPNet [46] learns a semantic projection function from the feature space through the image embedding space by minimizing the standard cross-entropy loss.

Our results grouped by datasets are reported in Fig. 3. Our model achieves performance either superior or comparable to the state-of-the-art in all benchmarks but AWA. We believe that in AWA learning a better alignment between visual features and attributes may not be as effective as improving the quality of the visual features. Especially, although the names of the test classes do not appear in the training set of ImageNet, for AWA being a non-fine-grained dataset, the information content of the test classes is likely represented by the ImageNet training classes. Moreover, for non-fine-grained datasets, finding labeled training data may not be as challenging as it is in fine-grained datasets. Hence, we argue that zero-shot learning is of higher practical interest in fine-grained settings. Indeed our proposed model is effective in fine-grained scenarios (*i.e.* CUB, SUN, FLO) where it consistently outperforms the state-of-the-art approaches.

Table 1. Domain Generalization accuracies on PACS with ResNet-18.

Target	AGG	DANN [11]	MLDG [20]	CrossGrad [39]	MetaReg [3]	JiGen [5]	Epi-FCR [21]	CuMix
Photo	94.9	94.0	94.3	94.0	94.3	**96.0**	93.9	95.1
Art	76.1	81.3	79.5	78.7	79.5	79.4	82.1	**82.3**
Cartoon	73.8	73.8	**77.3**	73.3	75.4	75.3	77.0	76.5
Sketch	69.4	**74.3**	71.5	65.1	72.2	71.4	73.0	72.6
Average	78.5	80.8	80.7	80.7	77.8	80.4	81.5	**81.6**

These results show that our model based on *mixup* achieves competitive performances on ZSL by simulating the semantic shift the classifier will experience at test time. To this extent, our approach is the first to show that mixup can be a powerful regularization strategy for the challenging ZSL setting.

DG. The second series of experiments consider the standard DG scenario. Here we test our model on the PACS dataset using a ResNet-18 architecture. As baselines for DG we consider the standard model trained on all source domains together (AGG), the adversarial strategies in [11] (DANN) and [39], the meta learning-based strategy MLDG [20] and MetaReg [3]. Moreover we consider the episodic strategy presented in [21] (Epi-FCR).

As shown in Table 1, our model achieves competitive results comparable to the state-of-the-art episodic strategy Epi-FCR [21]. Remarkable is the gain obtained with respect to the adversarial augmentation strategy CrossGrad [39]. Indeed, synthesizing novel domains for domain generalization is an ill-posed problem, since the concept of unseen domain is hard to capture. However, with CuMix we are able to simulate inputs/features of novel domains by simply interpolating the information available in the samples of our sources. Despite containing information available in the original sources, our approach allows to produce a model more robust to domain shift.

Another interesting comparison is against the self-supervised approach JiGen [5]. Similarly to [5] we employ an additional task to achieve higher generalization abilities to unseen domains. While in [5] the JigSaw puzzles [30] are used as a secondary self-supervised task, here we employ the mixed samples/features in the same manner. The improvement in the performances of our method highlights that recognizing the semantic of mixed samples acts as a more powerful secondary task to improve robustness to unseen domains.

Finally, it is worth noting that CuMix performs a form of episodic training, similar to Epi-FCR [21]. However, while Epi-FCR considers multiple domain-specific architectures (to simulate the domain experts needed to build the episodes), we require a single domain agnostic architecture. We build our episodes by making the *mixup* among images/features of different domains increasingly more drastic. Despite not requiring any domain experts, CuMix achieves comparable performances to Epi-FCR, showing the efficacy of our strategy to simulate unseen domain shifts.

Table 2. Ablation on PACS dataset with ResNet-18 as backbone.

$\mathcal{L}_{AGG}$	$\mathcal{L}_{M\text{-}IMG}$	$\mathcal{L}_{M\text{-}F}$	Curriculum	Art	Cartoon	Photo	Sketch	Avg.
✓				76.1	73.8	94.9	69.4	78.5
✓	✓			78.4	72.7	94.7	59.5	76.3
✓		✓		81.8	**76.5**	94.9	71.2	81.1
✓	✓	✓		**82.7**	75.4	**95.4**	71.5	81.2
✓	✓	✓	✓	82.3	**76.5**	95.1	**72.6**	**81.6**

Ablation Study. In this section, we ablate the various components of our method. We performed the ablation on the PACS benchmark for DG, since this allows us to show how different choices act on the generalization to unseen domains. In particular, we ablate the following implementation choices: 1) mixing samples at the image level, feature level or both 2) impact of our curriculum-based strategy for mixing features and samples.

As shown in Table 2, mixing samples at feature level produces a clear gain on the results with respect to the baseline, while mixing samples only at image level can even harm the performance. This happens particularly in the *sketch* domain, where mixing samples at feature level produces a gain of 2% while at image level we observe a drop of 10% with respect to the baseline. This could be explained by mixing samples at image level producing inputs that are too noisy for the network and not representative of the actual shift experienced at test time. Mixing samples at feature level instead, after multiple layers of abstractions, allows to better synthesize the information contained in the different samples, leading to more reliable features for the classifier. Using both of them allows to obtain higher results in almost all domains.

Finally, we analyze the impact of the curriculum-based strategy for mixing samples and features. As the table shows, adding the curriculum strategy allows to boost the performances for the most difficult cases (i.e. sketches) producing a further accuracy boost. Moreover, applying this strategy allows to stabilize the training procedure, as demonstrated experimentally.

ZSL+DG. On the proposed ZSL+DG setting we use the DomainNet dataset, training on five out of six domains and reporting the average per-class accuracy on the held-out one. We report the results for all possible target domains but one, *i.e.* real photos, since our backbone has been pretrained on ImageNet, thus the photo domain is not an unseen one. Since no previous methods addressed the ZSL+DG problem, in this work we consider simple baselines derived from the literature of both ZSL and DG. The first baseline is a standard ZSL model without any DG algorithm (i.e. the standard AGG): as ZSL method we consider SPNet [46]. The second baseline is a DG approach coupled with a ZSL algorithm. To this extent we select the state-of-the-art Epi-FCR as the DG approach, coupling it with SPNet. As a reference, we also evaluate the performance of standard *mixup* coupled with SPNet.

Table 3. ZSL+DG scenario on the DomainNet dataset with ResNet-50 as backbone.

Method	Clipart	Infograph	Painting	Quickdraw	Sketch	Avg.
SPNet	26.0	16.9	23.8	8.2	21.8	19.4
mixup+SPNet	27.2	16.9	24.7	8.5	21.3	19.7
Epi-FCR+SPNet	26.4	16.7	24.6	9.2	**23.2**	20.0
CuMix	**27.6**	**17.8**	**25.5**	**9.9**	22.6	**20.7**

As shown in Table 3, our method achieves competitive performances in ZSL+DG setting when compared to a state-of-the-art approach for DG (Epi-FCR) coupled with a state-of-the-art one for ZSL (SPNet), outperforming this baseline in almost all settings but *sketch* and, in average by almost 1%. Particularly interesting are the results on the *infograph* and *quickdraw* domains. These two domains are the ones where the shift is more evident as highlighted by the lower results of the baseline. In these scenarios, our model consistently outperforms the competitors, with a remarkable gain of more than 1.5% in average accuracy per class with respect to the ZSL only baseline. We want to highlight also that DomainNet is a challenging dataset, where almost all standard DA approaches are ineffective or can even lead to negative transfer [33]. Our method however is able to overcome the unseen domain shift at test time, improving the performance of the baselines in all scenarios. Our model consistently outperforms SPNet coupled with the standard *mixup* strategy in every scenario. This demonstrates the efficacy of the choices in CuMix for revisiting *mixup* in order to recognize unseen categories in unseen domains.

5 Conclusions

In this work, we propose the novel ZSL+DG scenario. In this setting, during training, we are given a set of images of multiple domains and semantic categories and our goal is to build a model able to recognize unseen concepts, as in ZSL, in unseen domains, as in DG. To solve this problem we design CuMix, the first algorithm which can be holistically and effectively applied to DG, ZSL and ZSL+DG. CuMix is based on simulating inputs and features of new domains and categories during training by mixing the available source domains and classes, both at image and feature level. Experiments on public benchmarks show the effectiveness of CuMix, achieving state-of-the-art performances in almost all settings in all tasks. Future works will investigate the use of alternative data-augmentation schemes in the ZSL+DG setting.

Acknowledgment. We thank the ELLIS Ph.D. student program and the ERC grants 637076 - RoboExNovo (B.C.) and 853489 - DEXIM (Z.A.). This work has been partially funded by the DFG under Germany's Excellence Strategy – EXC number 2064/1 – Project number 390727645.

References

1. Akata, Z., Perronnin, F., Harchaoui, Z., Schmid, C.: Label-embedding for attribute-based classification. In: Proceedings of the IEEE Conference on Computer Vision and Pattern Recognition, pp. 819–826 (2013)
2. Akata, Z., Reed, S., Walter, D., Lee, H., Schiele, B.: Evaluation of output embeddings for fine-grained image classification. In: Proceedings of the IEEE Conference on Computer Vision and Pattern Recognition, pp. 2927–2936 (2015)
3. Balaji, Y., Sankaranarayanan, S., Chellappa, R.: MetaReg: towards domain generalization using meta-regularization. In: Advances in Neural Information Processing Systems, pp. 998–1008 (2018)
4. Bendale, A., Boult, T.: Towards open world recognition. In: Proceedings of the IEEE Conference on Computer Vision and Pattern Recognition, pp. 1893–1902 (2015)
5. Carlucci, F.M., D'Innocente, A., Bucci, S., Caputo, B., Tommasi, T.: Domain generalization by solving jigsaw puzzles. In: Proceedings of the IEEE Conference on Computer Vision and Pattern Recognition, pp. 2229–2238 (2019)

6. Changpinyo, S., Chao, W.L., Gong, B., Sha, F.: Synthesized classifiers for zero-shot learning. In: Proceedings of the IEEE Conference on Computer Vision and Pattern Recognition, pp. 5327–5336 (2016)
7. Csurka, G.: A comprehensive survey on domain adaptation for visual applications. In: Csurka, G. (ed.) Domain Adaptation in Computer Vision Applications. ACVPR, pp. 1–35. Springer, Cham (2017). https://doi.org/10.1007/978-3-319-58347-1_1
8. Dutta, A., Akata, Z.: Semantically tied paired cycle consistency for zero-shot sketch-based image retrieval. In: Proceedings of the IEEE Conference on Computer Vision and Pattern Recognition, pp. 5089–5098 (2019)
9. Fu, Y., Hospedales, T.M., Xiang, T., Gong, S.: Transductive multi-view zero-shot learning. IEEE Trans. Pattern Anal. Mach. Intell. **37**(11), 2332–2345 (2015)
10. Gan, C., Yang, T., Gong, B.: Learning attributes equals multi-source domain generalization. In: Proceedings of the IEEE Conference on Computer Vision and Pattern Recognition, pp. 87–97 (2016)
11. Ganin, Y., et al.: Domain-adversarial training of neural networks. J. Mach. Learn. Res. **17**(1), 2096–2030 (2016)
12. Girshick, R.: Fast R-CNN. In: Proceedings of the IEEE International Conference on Computer Vision, pp. 1440–1448 (2015)
13. He, K., Zhang, X., Ren, S., Sun, J.: Deep residual learning for image recognition. In: Proceedings of the IEEE Conference on Computer Vision and Pattern Recognition, pp. 770–778 (2016)
14. Hoffman, J., Darrell, T., Saenko, K.: Continuous manifold based adaptation for evolving visual domains. In: Proceedings of the IEEE Conference on Computer Vision and Pattern Recognition, pp. 867–874 (2014)
15. Khosla, A., Zhou, T., Malisiewicz, T., Efros, A.A., Torralba, A.: Undoing the damage of dataset bias. In: European Conference on Computer Vision, pp. 158–171 (2012)
16. Kodirov, E., Xiang, T., Fu, Z., Gong, S.: Unsupervised domain adaptation for zero-shot learning. In: Proceedings of the IEEE International Conference on Computer Vision, pp. 2452–2460 (2015)
17. Lampert, C.H., Nickisch, H., Harmeling, S.: Attribute-based classification for zero-shot visual object categorization. IEEE Trans. Pattern Anal. Mach. Intell. **36**(3), 453–465 (2013)
18. Lange, M.D., et al.: Continual learning: a comparative study on how to defy forgetting in classification tasks. arXiv:1909.08383 (2019)
19. Li, D., Yang, Y., Song, Y.Z., Hospedales, T.M.: Deeper, broader and artier domain generalization. In: Proceedings of the IEEE International Conference on Computer Vision, pp. 5542–5550 (2017)
20. Li, D., Yang, Y., Song, Y.Z., Hospedales, T.M.: Learning to generalize: meta-learning for domain generalization. In: Thirty-Second AAAI Conference on Artificial Intelligence (2018)
21. Li, D., Zhang, J., Yang, Y., Liu, C., Song, Y.Z., Hospedales, T.M.: Episodic training for domain generalization. In: Proceedings of the IEEE International Conference on Computer Vision, pp. 1446–1455 (2019)
22. Li, H., Jialin Pan, S., Wang, S., Kot, A.C.: Domain generalization with adversarial feature learning. In: Proceedings of the IEEE Conference on Computer Vision and Pattern Recognition, pp. 5400–5409 (2018)
23. Mancini, M., Bulo, S.R., Caputo, B., Ricci, E.: AdaGraph: unifying predictive and continuous domain adaptation through graphs. In: Proceedings of the IEEE Conference on Computer Vision and Pattern Recognition, pp. 6568–6577 (2019)

24. Mancini, M., Karaoguz, H., Ricci, E., Jensfelt, P., Caputo, B.: Kitting in the wild through online domain adaptation. In: IEEE/RSJ International Conference on Intelligent Robots and Systems (IROS), pp. 1103–1109 (2018)
25. Mancini, M., Porzi, L., Bulo, S.R., Caputo, B., Ricci, E.: Inferring latent domains for unsupervised deep domain adaptation. IEEE Trans. Pattern Anal. Mach. Intell. (2019)
26. Mancini, M., Porzi, L., Rota Bulò, S., Caputo, B., Ricci, E.: Boosting domain adaptation by discovering latent domains. In: Proceedings of the IEEE Conference on Computer Vision and Pattern Recognition, pp. 3771–3780 (2018)
27. Mikolov, T., Chen, K., Corrado, G., Dean, J.: Efficient estimation of word representations in vector space. In: 1st International Conference on Learning Representations Workshop Track Proceedings (2013)
28. Muandet, K., Balduzzi, D., Schölkopf, B.: Domain generalization via invariant feature representation. In: International Conference on Machine Learning, pp. 10–18 (2013)
29. Nilsback, M.E., Zisserman, A.: Automated flower classification over a large number of classes. In: 2008 Sixth Indian Conference on Computer Vision, Graphics & Image Processing, pp. 722–729. IEEE (2008)
30. Noroozi, M., Favaro, P.: Unsupervised learning of visual representations by solving jigsaw puzzles. In: Leibe, B., Matas, J., Sebe, N., Welling, M. (eds.) ECCV 2016. LNCS, vol. 9910, pp. 69–84. Springer, Cham (2016). https://doi.org/10.1007/978-3-319-46466-4_5
31. Patterson, G., Hays, J.: Sun attribute database: discovering, annotating, and recognizing scene attributes. In: 2012 IEEE Conference on Computer Vision and Pattern Recognition, pp. 2751–2758. IEEE (2012)
32. Peng, K.C., Wu, Z., Ernst, J.: Zero-shot deep domain adaptation. In: Proceedings of the European Conference on Computer Vision (ECCV), pp. 764–781 (2018)
33. Peng, X., Bai, Q., Xia, X., Huang, Z., Saenko, K., Wang, B.: Moment matching for multi-source domain adaptation. In: Proceedings of the IEEE International Conference on Computer Vision, pp. 1406–1415 (2019)
34. Redmon, J., Divvala, S., Girshick, R., Farhadi, A.: You only look once: unified, real-time object detection. In: Proceedings of the IEEE Conference on Computer Vision and Pattern Recognition, pp. 779–788 (2016)
35. Reed, S., Akata, Z., Lee, H., Schiele, B.: Learning deep representations of fine-grained visual descriptions. In: Proceedings of the IEEE Conference on Computer Vision and Pattern Recognition, pp. 49–58 (2016)
36. Roy, S., Siarohin, A., Sangineto, E., Bulo, S.R., Sebe, N., Ricci, E.: Unsupervised domain adaptation using feature-whitening and consensus loss. In: Proceedings of the IEEE Conference on Computer Vision and Pattern Recognition, pp. 9471–9480 (2019)
37. Russakovsky, O., et al.: ImageNet large scale visual recognition challenge. Int. J. Comput. Vision 115(3), 211–252 (2015)
38. Schonfeld, E., Ebrahimi, S., Sinha, S., Darrell, T., Akata, Z.: Generalized zero-and few-shot learning via aligned variational autoencoders. In: Proceedings of the IEEE Conference on Computer Vision and Pattern Recognition, pp. 8247–8255 (2019)
39. Shankar, S., Piratla, V., Chakrabarti, S., Chaudhuri, S., Jyothi, P., Sarawagi, S.: Generalizing across domains via cross-gradient training. In: International Conference on Learning Representations (2018)
40. Thong, W., Mettes, P., Snoek, C.G.: Open cross-domain visual search. arXiv preprint arXiv:1911.08621 (2019)

41. Verma, V.K., Rai, P.: A simple exponential family framework for zero-shot learning. In: Ceci, M., Hollmén, J., Todorovski, L., Vens, C., Džeroski, S. (eds.) ECML PKDD 2017. LNCS (LNAI), vol. 10535, pp. 792–808. Springer, Cham (2017). https://doi.org/10.1007/978-3-319-71246-8_48
42. Volpi, R., Murino, V.: Addressing model vulnerability to distributional shifts over image transformation sets. In: Proceedings of the IEEE International Conference on Computer Vision, pp. 7980–7989 (2019)
43. Volpi, R., Namkoong, H., Sener, O., Duchi, J.C., Murino, V., Savarese, S.: Generalizing to unseen domains via adversarial data augmentation. In: Advances in Neural Information Processing Systems, pp. 5334–5344 (2018)
44. Welinder, P., et al.: Caltech-UCSD birds 200 (2010)
45. Xian, Y., Akata, Z., Sharma, G., Nguyen, Q., Hein, M., Schiele, B.: Latent embeddings for zero-shot classification. In: Proceedings of the IEEE Conference on Computer Vision and Pattern Recognition, pp. 69–77 (2016)
46. Xian, Y., Choudhury, S., He, Y., Schiele, B., Akata, Z.: Semantic projection network for zero-and few-label semantic segmentation. In: Proceedings of the IEEE Conference on Computer Vision and Pattern Recognition, pp. 8256–8265 (2019)
47. Xian, Y., Lampert, C.H., Schiele, B., Akata, Z.: Zero-shot learning-a comprehensive evaluation of the good, the bad and the ugly. IEEE Trans. Pattern Anal. Mach. Intell. **41**(9), 2251–2265 (2018)
48. Xian, Y., Lorenz, T., Schiele, B., Akata, Z.: Feature generating networks for zero-shot learning. In: Proceedings of the IEEE Conference on Computer Vision and Pattern Recognition, pp. 5542–5551 (2018)
49. Xian, Y., Schiele, B., Akata, Z.: Zero-shot learning-the good, the bad and the ugly. In: Proceedings of the IEEE Conference on Computer Vision and Pattern Recognition, pp. 4582–4591 (2017)
50. Xian, Y., Sharma, S., Schiele, B., Akata, Z.: f-VAEGAN-D2: a feature generating framework for any-shot learning. In: Proceedings of the IEEE Conference on Computer Vision and Pattern Recognition, pp. 10275–10284 (2019)
51. Xu, M., Zhang, J., Ni, B., Li, T., Wang, C., Tian, Q., Zhang, W.: Adversarial domain adaptation with domain mixup. In: The Thirty-Fourth AAAI Conference on Artificial Intelligence, pp. 6502–6509. AAAI Press (2020)
52. Yelamarthi, S.K., Reddy, S.K., Mishra, A., Mittal, A.: A zero-shot framework for sketch based image retrieval. In: Ferrari, V., Hebert, M., Sminchisescu, C., Weiss, Y. (eds.) ECCV 2018. LNCS, vol. 11208, pp. 316–333. Springer, Cham (2018). https://doi.org/10.1007/978-3-030-01225-0_19
53. Zhang, H., Cisse, M., Dauphin, Y.N., Lopez-Paz, D.: mixup: beyond empirical risk minimization. In: International Conference on Learning Representations (2018)
54. Zhang, Z., Saligrama, V.: Zero-shot learning via semantic similarity embedding. In: Proceedings of the IEEE International Conference on Computer Vision, pp. 4166–4174 (2015)
55. Zhang, Z., Saligrama, V.: Zero-shot learning via joint latent similarity embedding. In: Proceedings of the IEEE Conference on Computer Vision and Pattern Recognition, pp. 6034–6042 (2016)
56. Zhuo, J., Wang, S., Cui, S., Huang, Q.: Unsupervised open domain recognition by semantic discrepancy minimization. In: Proceedings of the IEEE Conference on Computer Vision and Pattern Recognition, pp. 750–759 (2019)

Square Attack: A Query-Efficient Black-Box Adversarial Attack via Random Search

Maksym Andriushchenko[1(✉)], Francesco Croce[2], Nicolas Flammarion[1], and Matthias Hein[2]

[1] EPFL, Lausanne, Switzerland
maksym.andriushchenko@epfl.ch
[2] University of Tübingen, Tübingen, Germany

Abstract. We propose the *Square Attack*, a score-based black-box l_2- and l_∞-adversarial attack that does not rely on local gradient information and thus is not affected by gradient masking. Square Attack is based on a randomized search scheme which selects localized square-shaped updates at random positions so that at each iteration the perturbation is situated approximately at the boundary of the feasible set. Our method is significantly more query efficient and achieves a higher success rate compared to the state-of-the-art methods, especially in the untargeted setting. In particular, on ImageNet we improve the average query efficiency in the untargeted setting for various deep networks by a factor of at least 1.8 and up to 3 compared to the recent state-of-the-art l_∞-attack of Al-Dujaili & O'Reilly (2020). Moreover, although our attack is *black-box*, it can also outperform gradient-based *white-box* attacks on the standard benchmarks achieving a new state-of-the-art in terms of the success rate. The code of our attack is available at https://github.com/max-andr/square-attack.

1 Introduction

Adversarial examples are of particular concern when it comes to applications of machine learning which are safety-critical. Many defenses against adversarial examples have been proposed [1,5,7,23,32,40,56] but with limited success, as new more powerful attacks could break many of them [4,10,12,35,57]. In particular, gradient obfuscation or masking [4,35] is often the reason why seemingly robust models turn out to be non-robust in the end. Gradient-based attacks are most often affected by this phenomenon (white-box attacks but also black-box attacks based on finite difference approximations [35]). Thus it is important to

---

M. Andriushchenko and F. Croce—Equal contribution.

Electronic supplementary material The online version of this chapter (https://doi.org/10.1007/978-3-030-58592-1_29) contains supplementary material, which is available to authorized users.

© Springer Nature Switzerland AG 2020
A. Vedaldi et al. (Eds.): ECCV 2020, LNCS 12368, pp. 484–501, 2020.
https://doi.org/10.1007/978-3-030-58592-1_29

have attacks which are based on different principles. Black-box attacks have recently become more popular [8,36,46] as their attack strategies are quite different from the ones employed for adversarial training, where often PGD-type attacks [32] are used. However, a big weakness currently is that these black-box attacks need to query the classifier too many times before they find adversarial examples, and their success rate is sometimes significantly lower than that of white-box attacks.

In this paper we propose Square Attack, a score-based adversarial attack, i.e. it can query the probability distribution over the classes predicted by a classifier but has no access to the underlying model. The Square Attack exploits random search[1]   [41,43] which is one of the simplest approaches for black-box optimization. Due to a particular sampling distribution, it requires significantly fewer queries compared to the state-of-the-art black-box methods (see Fig. 1) in the score-based threat model while outperforming them in terms of *success rate*, i.e. the percentage of successful adversarial examples. This is achieved by a combination of a particular initialization strategy and our square-shaped updates. We motivate why these updates are particularly suited to attack

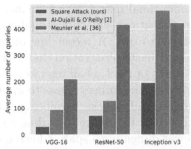

Fig. 1. Avg. number of queries of successful untargeted l_∞-attacks on three ImageNet models for three score-based black-box attacks. Square Attack outperforms all other attacks by large margin

neural networks and provide convergence guarantees for a variant of our method. In an extensive evaluation with untargeted and targeted attacks, three datasets (MNIST, CIFAR-10, ImageNet), normal and robust models, we show that Square Attack outperforms state-of-the-art methods in the l_2- and l_∞-threat model.

2   Related Work

We discuss black-box attacks with l_2- and l_∞-perturbations since our attack focuses on this setting. Although attacks for other norms, e.g. l_0, exist [16,36], they are often algorithmically different due to the geometry of the perturbations.

l_2- and l_∞-**Score-Based Attacks.** Score-based black-box attacks have only access to the score predicted by a classifier for each class for a given input. Most of such attacks in the literature are based on gradient estimation through finite differences. The first papers in this direction [6,27,51] propose attacks which approximate the gradient by sampling from some noise distribution around the point. While this approach can be successful, it requires many queries of the classifier, particularly in high-dimensional input spaces as in image classification. Thus, improved techniques reduce the dimension of the search space via using the principal components of the data [6], searching for perturbations in

---

[1] It is an iterative procedure different from random sampling inside the feasible region.

the latent space of an auto-encoder [50] or using a low-dimensional noise distribution [28]. Other attacks exploit evolutionary strategies or random search, e.g. [3] use a genetic algorithm to generate adversarial examples and alleviate gradient masking as they can reduce the robust accuracy on randomization- and discretization-based defenses. The l_2-attack of [25] can be seen as a variant of random search which chooses the search directions in an orthonormal basis and tests up to two candidate updates at each step. However, their algorithm can have suboptimal query efficiency since it adds at every step only small (in l_2-norm) modifications, and suboptimal updates cannot be undone as they are orthogonal to each other. A recent line of work has pursued black-box attacks which are based on the observation that successful adversarial perturbations are attained at corners of the l_∞-ball intersected with the image space $[0, 1]^d$ [2,34,44]. Searching over the corners allows to apply discrete optimization techniques to generate adversarial attacks, significantly improving the query efficiency. Both [44] and [2] divide the image according to some coarse grid, perform local search in this lower dimensional space allowing componentwise changes only of $-\epsilon$ and ϵ, then refine the grid and repeat the scheme. In [2] such a procedure is motivated as an estimation of the gradient signs. Recently, [34] proposed several attacks based on evolutionary algorithms, using discrete and continuous optimization, achieving nearly state-of-the-art query efficiency for the l_∞-norm. In order to reduce the dimensionality of the search space, they use the "tiling trick" of [28] where they divide the perturbation into a set of squares and modify the values in these squares with evolutionary algorithms. A related idea also appeared earlier in [22] where they introduced black rectangle-shaped perturbations for generating adversarial occlusions. In [34], as in [28], both size and position of the squares are fixed at the beginning and not optimized. Despite their effectiveness for the l_∞-norm, these discrete optimization based attacks are not straightforward to adapt to the l_2-norm. Finally, approaches based on Bayesian optimization exist, e.g. [45], but show competitive performance only in a low-query regime.

Different Threat and Knowledge Models. We focus on l_p-*norm-bounded* adversarial perturbations (for other perturbations such as rotations, translations, occlusions in the black-box setting see, e.g., [22]). Perturbations with *minimal* l_p-norm are considered in [13,50] but require significantly more queries than norm-bounded ones. Thus we do not compare to them, except for [25] which has competitive query efficiency while aiming at small perturbations.

In other cases the attacker has a different knowledge of the classifier. A more restrictive scenario, considered by *decision-based* attacks [8,9,11,14,24], is when the attacker can query only the decision of the classifier, but not the predicted scores. Other works use more permissive threat models, e.g., when the attacker already has a substitute model similar to the target one [15,20,39,47,52] and thus can generate adversarial examples for the substitute model and then transfer them to the target model. Related to this, [52] suggest to refine this approach by running a black-box gradient estimation attack in a subspace spanned by the gradients of substitute models. However, the gain in query efficiency given by such extra knowledge does not account for the computational cost required to train the substitute models, particularly high on ImageNet-scale. Finally, [31] use extra information on the target data distribution to train a model that predicts adversarial images that are then refined by gradient estimation attacks.

3   Square Attack

In the following we recall the definitions of the adversarial examples in the threat model we consider and present our black-box attacks for the l_∞- and l_2-norms.

Algorithm 1: The Square Attack via random search

Input: classifier f, point $x \in \mathbb{R}^d$, image size w, number of color channels c,
      l_p-radius ϵ, label $y \in \{1, \ldots, K\}$, number of iterations N
Output: approximate minimizer $\hat{x} \in \mathbb{R}^d$ of the problem stated in Eq. (1)

1   $\hat{x} \leftarrow init(x)$, $l^* \leftarrow L(f(x), y)$, $i \leftarrow 1$
2   **while** $i < N$ **and** $\hat{x}$ *is not adversarial* **do**
3      $h^{(i)} \leftarrow$ side length of the square to modify (according to some schedule)
4      $\delta \sim P(\epsilon, h^{(i)}, w, c, \hat{x}, x)$ (see Alg. 2 and 3 for the sampling distributions)
5      $\hat{x}_{\text{new}} \leftarrow$ Project $\hat{x} + \delta$ onto $\{z \in \mathbb{R}^d : \|z - x\|_p \le \epsilon\} \cap [0,1]^d$
6      $l_{\text{new}} \leftarrow L(f(\hat{x}_{\text{new}}), y)$
7   **if** $l_{new} < l^*$ **then** $\hat{x} \leftarrow \hat{x}_{\text{new}}$, $l^* \leftarrow l_{\text{new}}$;
8      $i \leftarrow i + 1$
9 **end**

3.1   Adversarial Examples in the l_p-threat Model

Let $f : [0,1]^d \rightarrow \mathbb{R}^K$ be a classifier, where d is the input dimension, K the number of classes and $f_k(x)$ is the predicted score that x belongs to class k. The classifier assigns class $\arg\max_{k=1,\ldots,K} f_k(x)$ to the input x. The goal of an *untargeted* attack is to change the correctly predicted class y for the point x. A point $\hat{x}$ is called an *adversarial example* with an l_p-norm bound of ϵ for x if

$$\arg\max_{k=1,\ldots,K} f_k(\hat{x}) \ne y, \quad \|\hat{x} - x\|_p \le \epsilon \quad \text{and} \quad \hat{x} \in [0,1]^d,$$

where we have added the additional constraint that $\hat{x}$ is an image. The task of finding $\hat{x}$ can be rephrased as solving the constrained optimization problem

$$\min_{\hat{x} \in [0,1]^d} L(f(\hat{x}), y), \quad \text{s.t.} \quad \|\hat{x} - x\|_p \le \epsilon, \tag{1}$$

for a loss L. In our experiments, we use $L(f(\hat{x}), y) = f_y(\hat{x}) - \max_{k \ne y} f_k(\hat{x})$.

The goal of *targeted* attacks is instead to change the decision of the classifier to a particular class t, i.e., to find $\hat{x}$ so that $\arg\max_k f_k(\hat{x}) = t$ under the same constraints on $\hat{x}$. We further discuss the targeted attacks in Sup. E.1.

3.2   General Algorithmic Scheme of the Square Attack

Square Attack is based on *random search* which is a well known iterative technique in optimization introduced by Rastrigin in 1963 [41]. The main idea of the algorithm is to sample a random update δ at each iteration, and to add this update to the current iterate $\hat{x}$ if it improves the objective function. Despite its simplicity, random search performs well in many situations [54] and does not depend on gradient information from the objective function g.

Many variants of random search have been introduced [33, 42, 43], which differ mainly in how the random perturbation is chosen at each iteration (the original scheme samples uniformly on a hypersphere of fixed radius). For our goal of crafting adversarial examples we come up with two sampling distributions specific to the l_∞- and the l_2-attack (Sect. 3.3 and Sect. 3.4), which we integrate in the classic random search procedure. These sampling distributions are motivated by both how images are processed by neural networks with convolutional filters and the shape of the l_p-balls for different p. Additionally, since the considered objective is non-convex when using neural networks, a good initialization is particularly important. We then introduce a specific one for better query efficiency.

Our proposed scheme differs from classical random search by the fact that the perturbations $\hat{x} - x$ are constructed such that for every iteration they lie on the boundary of the l_∞- or l_2-ball before projection onto the image domain $[0,1]^d$. Thus we are using the perturbation budget almost maximally at each step. Moreover, the changes are localized in the image in the sense that at each step we modify just a small fraction of contiguous pixels shaped into **squares**. Our overall scheme is presented in Algorithm 1. First, the algorithm picks the side length $h^{(i)}$ of the square to be modified (step 3), which is decreasing according to an a priori fixed schedule. This is in analogy to the step-size reduction in gradient-based optimization. Then in step 4 we sample a new update δ and add it to the current iterate (step 5). If the resulting loss (obtained in step 6) is smaller than the best loss so far, the change is accepted otherwise discarded. Since we are interested in query efficiency, the algorithm stops as soon as an adversarial example is found. The time complexity of the algorithm is dominated by the evaluation of $f(\hat{x}_{\text{new}})$, which is performed at most N times, with N total number of iterations. We plot the resulting adversarial perturbations in Fig. 3 and additionally in Sup. E where we also show imperceptible perturbations.

We note that previous works [28, 34, 44] generate perturbations containing squares. However, while those use a fixed grid on which the squares are constrained, we optimize the position of the squares as well as the color, making our attack more flexible and effective. Moreover, unlike previous works, we motivate squared perturbations with the structure of the convolutional filters (see Sect. 4).

Size of the Squares. Given images of size $w \times w$, let $p \in [0,1]$ be the percentage of elements of x to be modified. The length h of the side of the squares used is given by the closest positive integer to $\sqrt{p \cdot w^2}$ (and $h \geq 3$ for the l_2-attack). Then, the initial p is the only free parameter of our scheme. With $N = 10000$

iterations available, we halve the value of p at $i \in \{10, 50, 200, 1000, 2000, 4000, 6000, 8000\}$ iterations. For different N we rescale the schedule accordingly.

3.3 The l_∞-Square Attack

Initialization. As initialization we use vertical stripes of width one where the color of each stripe is sampled uniformly at random from $\{-\epsilon, \epsilon\}^c$ (c number of color channels). We found that convolutional networks are particularly sensitive to such perturbations, see also [53] for a detailed discussion on the sensitivity of neural networks to various types of high frequency perturbations.

Sampling Distribution. Similar to [44] we observe that successful l_∞-perturbations usually have values $\pm\epsilon$ in all the components (note that this does not hold perfectly due to the image constraints $\hat{x} \in [0,1]^d$). In particular, it holds

$$\hat{x}_i \in \{\max\{0, x_i - \epsilon\}, \min\{1, x_i + \epsilon\}\}.$$

Our sampling distribution P for the l_∞-norm described in Algorithm 2 selects sparse updates of $\hat{x}$ with $\|\delta\|_0 = h \cdot h \cdot c$ where $\delta \in \{-2\epsilon, 0, 2\epsilon\}^d$ and the non-zero elements are grouped to form a square. In this way, after the projection onto the l_∞-ball of radius ϵ (Step 5 of Algorithm 1) all components i for which $\epsilon \leq x_i \leq 1 - \epsilon$ satisfy $\hat{x}_i \in \{x_i - \epsilon, x_i + \epsilon\}$, i.e. differ from the original point x in each element either by ϵ or $-\epsilon$. Thus $\hat{x} - x$ is situated at

---
Algorithm 2: Sampling distribution P for l_∞-norm

Input: maximal norm ϵ, window size h, image size w, color channels c
Output: New update δ
1 $\delta \leftarrow$ array of zeros of size $w \times w \times c$
2 sample uniformly
   $\quad r, s \in \{0, \dots, w - h\} \subset \mathbb{N}$
3 **for** $i = 1, \dots, c$ **do**
4 $\quad \rho \leftarrow Uniform(\{-2\epsilon, 2\epsilon\})$
5 $\quad \delta_{r+1:r+h,\ s+1:s+h,\ i} \leftarrow \rho \cdot \mathbb{1}_{h \times h}$
6 **end**

---

one of the corners of the l_∞-ball (modulo the components which are close to the boundary). Note that all projections are done by clipping. Moreover, we fix the elements of δ belonging to the same color channel to have the same sign, since we observed that neural networks are particularly sensitive to such perturbations (see Sect. 4.3).

3.4 The l_2-Square Attack

Initialization. The l_2-perturbation is initialized by generating a 5×5 grid-like tiling by squares of the image, where the perturbation on each tile has the shape described next in the sampling distribution. The resulting perturbation $\hat{x} - x$ is rescaled to have l_2-norm ϵ and the resulting $\hat{x}$ is projected onto $[0,1]^d$ by clipping.

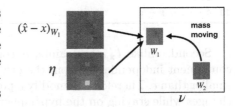

Fig. 2. Perturbation of the l_2-attack

Sampling Distribution. First, let us notice that the adversarial perturbations typically found for the l_2-norm tend to be much more localized than those for the l_∞-norm [49], in the sense that large changes are applied on some pixels of the original image, while many others are minimally modified. To mimic this feature we introduce a new update η which has two "centers" with large absolute value and opposite signs, while the other components have lower absolute values as one gets farther away from the centers, but never reaching zero (see Fig. 2 for one example with $h = 8$ of the resulting update η). In this way the modifications are localized and with high contrast between the different halves, which we found to improve the query efficiency. Concretely, we define $\eta^{(h_1,h_2)} \in \mathbb{R}^{h_1 \times h_2}$ (for some $h_1, h_2 \in \mathbb{N}_+$ such that $h_1 \geq h_2$) for every $1 \leq r \leq h_1, 1 \leq s \leq h_2$ as

$$\eta_{r,s}^{(h_1,h_2)} = \sum_{k=0}^{M(r,s)} \frac{1}{(n+1-k)^2}, \quad \text{with } n = \left\lfloor \frac{h_1}{2} \right\rfloor,$$

and $M(r,s) = n - \max\{|r - \lfloor \frac{h_1}{2} \rfloor - 1|, |s - \lfloor \frac{h_2}{2} \rfloor - 1|\}$. The intermediate square update $\eta \in \mathbb{R}^{h \times h}$ is then selected uniformly at random from either

$$\eta = \left(\eta^{(h,k)}, -\eta^{(h,h-k)} \right), \quad \text{with } k = \lfloor h/2 \rfloor, \tag{2}$$

or its transpose (corresponding to a rotation of 90°).

Algorithm 3: Sampling distribution P for l_2-norm

Input: maximal norm ϵ, window size h, image size w, number of color channels c, current image $\hat{x}$, original image x

Output: New update δ

1 $\nu \leftarrow \hat{x} - x$
2 sample uniformly $r_1, s_1, r_2, s_2 \in \{0, \ldots, w - h\}$
3 $W_1 := r_1 + 1 : r_1 + h, s_1 + 1 : s_1 + h, \ W_2 := r_2 + 1 : r_2 + h, s_2 + 1 : s_2 + h$
4 $\epsilon_{unused}^2 \leftarrow \epsilon^2 - \|\nu\|_2^2, \quad \eta^* \leftarrow \eta/\|\eta\|_2$ with η as in (2)
5 **for** $i = 1, \ldots, c$ **do**
6 $\quad \rho \leftarrow Uniform(\{-1, 1\})$
7 $\quad \nu_{temp} \leftarrow \rho\eta^* + \nu_{W_1,i}/\|\nu_{W_1,i}\|_2$
8 $\quad \epsilon_{avail}^i \leftarrow \sqrt{\|\nu_{W_1 \cup W_2,i}\|_2^2 + \epsilon_{unused}^2/c}$
9 $\quad \nu_{W_2,i} \leftarrow 0, \quad \nu_{W_1,i} \leftarrow (\nu_{temp}/\|\nu_{temp}\|_2)\epsilon_{avail}^i$
10 **end**
11 $\delta \leftarrow x + \nu - \hat{x}$

Second, unlike l_∞-constraints, l_2-constraints do not allow to perturb each component independently from the others as the overall l_2-norm must be kept smaller than ϵ. Therefore, to modify a perturbation $\hat{x} - x$ of norm ϵ with localized changes while staying on the hypersphere, we have to "move the mass" of $\hat{x} - x$ from one location to another. Thus, our scheme consists in randomly selecting two squared windows in the current perturbation $\nu = \hat{x} - x$, namely ν_{W_1} and

Fig. 3. Visualization of the adversarial perturbations and examples found by the l_∞- and l_2-versions of the Square Attack on ResNet-50

ν_{W_2}, setting $\nu_{W_2} = 0$ and using the budget of $\|\nu_{W_2}\|_2$ to increase the total perturbation of ν_{W_1}. Note that the perturbation of W_1 is then a combination of the existing perturbation plus the new generated η. We report the details of this scheme in Algorithm 3 where step 4 allows to utilize the budget of l_2-norm lost after the projection onto $[0,1]^d$. The update δ output by the algorithm is such that the next iterate $\hat{x}_{new} = \hat{x} + \delta$ (before projection onto $[0,1]^d$ by clipping) belongs to the hypersphere $B_2(x, \epsilon)$ as stated in the following proposition.

Proposition 1. *Let δ be the output of Algorithm 3. Then $\|\hat{x} + \delta - x\|_2 = \epsilon$.*

4   Theoretical and Empirical Justification of the Method

We provide high-level theoretical justifications and empirical evidence regarding the algorithmic choices in Square Attack, with focus on the l_∞-version (the l_2-version is significantly harder to analyze).

4.1   Convergence Analysis of Random Search

First, we want to study the convergence of the random search algorithm when considering an L-smooth objective function g (such as neural networks with activation functions like softplus, swish, ELU, etc.) on the whole space $\mathbb{R}^d$ (without projection[2]) under the following assumptions on the update δ_t drawn from the sampling distribution P_t:

$$\mathbb{E}\|\delta_t\|_2^2 \leq \gamma_t^2 C \quad \text{and} \quad \mathbb{E}|\langle \delta_t, v \rangle| \geq \tilde{C}\gamma_t\|v\|_2, \ \forall v \in \mathbb{R}^d, \tag{3}$$

where γ_t is the step size at iteration t, $C, \tilde{C} > 0$ some constants and $\langle \cdot, \cdot \rangle$ denotes the inner product. We obtain the following result, similar to existing convergence rates for zeroth-order methods [21,37,38]:

Proposition 2. *Suppose that $\mathbb{E}[\delta_t] = 0$ and the assumptions in Eq. (3) hold. Then for step-sizes $\gamma_t = \gamma/\sqrt{T}$, we have*

$$\min_{t=0,\dots,T} \mathbb{E}\|\nabla g(x_t)\|_2 \leq \frac{2}{\gamma\tilde{C}\sqrt{T}}\left(g(x_0) - \mathbb{E}g(x_{T+1}) + \frac{\gamma^2 CL}{2}\right).$$

---

[2] Nonconvex constrained optimization under noisy oracles is notoriously harder [19].

This basically shows for T large enough one can make the gradient arbitrary small, meaning that the random search algorithm converges to a critical point of g (one cannot hope for much stronger results in non-convex optimization without stronger conditions).

Unfortunately, the second assumption in Eq. (3) does not directly hold for our sampling distribution P for the l_∞-norm (see Sup. A.3), but holds for a similar one, P^{multiple}, where each component of the update δ is drawn uniformly at random from $\{-2\epsilon, 2\epsilon\}$. In fact we show in Sup. A.4, using the Khintchine inequality [26], that

$$\mathbb{E}\|\delta_t\|_2^2 \leq 4c\varepsilon^2 h^2 \text{ and } \mathbb{E}|\langle \delta_t, v \rangle| \geq \frac{\sqrt{2}c\varepsilon h^2}{d}\|v\|_2, \ \forall v \in \mathbb{R}^d.$$

Moreover, while P^{multiple} performs worse than the distribution used in Algorithm 2, we show in Sect. 4.3 that it already reaches state-of-the-art results.

4.2   Why Squares?

Previous works [34,44] build their l_∞-attacks by iteratively adding square modifications. Likewise we change square-shaped regions of the image for both our l_∞- and l_2-attacks—with the difference that we can sample any square subset of the input, while the grid of the possible squares is fixed in [34,44]. This leads naturally to wonder why squares are superior to other shapes, e.g. rectangles.

Let us consider the l_∞-threat model, with bound ϵ, input space $\mathbb{R}^{d\times d}$ and a convolutional filter $w \in \mathbb{R}^{s\times s}$ with entries unknown to the attacker. Let $\delta \in \mathbb{R}^{d\times d}$ be the sparse update with $\|\delta\|_0 = k \geq s^2$ and $\|\delta\|_\infty \leq \epsilon$. We denote by $S(a,b)$ the index set of the rectangular support of δ with $|S(a,b)| = k$ and shape $a \times b$. We want to provide intuition why sparse square-shaped updates are superior to rectangular ones in the sense of reaching a maximal change in the activations of the first convolutional layer.

Let $z = \delta * w \in \mathbb{R}^{d\times d}$ denote the output of the convolutional layer for the update δ. The l_∞-norm of z is the maximal componentwise change of the convolutional layer:

$$\|z\|_\infty = \max_{u,v} |z_{u,v}| = \max_{u,v} \left| \sum_{i,j=1}^{s} \delta_{u-\lfloor\frac{s}{2}\rfloor+i, v-\lfloor\frac{s}{2}\rfloor+j} \cdot w_{i,j} \right|$$

$$\leq \max_{u,v} \epsilon \sum_{i,j} |w_{i,j}| \mathbb{1}_{(u-\lfloor\frac{s}{2}\rfloor+i, v-\lfloor\frac{s}{2}\rfloor+j)\in S(a,b)},$$

where elements with indices exceeding the size of the matrix are set to zero. Note that the indicator function attains 1 only for the non-zero elements of δ involved in the convolution to get $z_{u,v}$. Thus, to have the largest upper bound possible on $|z_{u,v}|$, for some (u,v), we need the largest possible amount of components of δ with indices in

$$C(u,v) = \left\{ (u - \lfloor\frac{s}{2}\rfloor + i, v - \lfloor\frac{s}{2}\rfloor + j) : i,j = 1,\ldots,s \right\}$$

to be non-zero (that is in $S(a, b)$).

Therefore, it is desirable to have the shape $S(a, b)$ of the perturbation δ selected so to maximize the number N of convolutional filters $w \in \mathbb{R}^{s \times s}$ which fit into the rectangle $a \times b$. Let $\mathcal{F}$ be the family of the objects that can be defined as the union of axis-aligned rectangles with vertices on $\mathbb{N}^2$, and $\mathcal{G} \subset \mathcal{F}$ be the squares of $\mathcal{F}$ of shape $s \times s$ with $s \geq 2$. We have the following proposition:

Proposition 3. *Among the elements of $\mathcal{F}$ with area $k \geq s^2$, those which contain the largest number of elements of $\mathcal{G}$ have*

$$N^* = (a - s + 1)(b - s + 1) + (r - s + 1)^+ \tag{4}$$

of them, with $a = \left\lfloor \sqrt{k} \right\rfloor$, $b = \left\lfloor \frac{k}{a} \right\rfloor$, $r = k - ab$ and $z^+ = \max\{z, 0\}$.

This proposition states that, if we can modify only k elements of δ, then shaping them to form (approximately) a square allows to maximize the number of pairs (u, v) for which $|S(a, b) \cap C(u, v)| = s^2$. If $k = l^2$ then $a = b = l$ are the optimal values for the shape of the perturbation update, i.e. the shape is exactly a square.

Table 1. Ablation study of the l_∞-Square Attack which shows how the individual design decisions improve the performance. The fourth row corresponds to the method for which we have shown convergence guarantees in Sect. 4.1. The last row corresponds to our final l_∞-attack. c indicates the number of color channels, h the length of the side of the squares, so that "# random sign" c represents updates with constant sign for each color, while $c \cdot h^2$ updates with signs sampled independently of each other

Update shape	# random signs	Initialization	Failure rate	Avg. queries	Median queries
random	$c \cdot h^2$	vert. stripes	0.0%	401	48
random	$c \cdot h^2$	uniform rand.	0.0%	393	132
random	c	vert. stripes	0.0%	339	53
square	$c \cdot h^2$	vert. stripes	0.0%	153	15
rectangle	c	vert. stripes	0.0%	93	16
square	c	uniform rand.	0.0%	91	26
square	c	vert. stripes	0.0%	**73**	**11**

4.3   Ablation Study

We perform an ablation study to show how the individual design decisions for the sampling distribution of the random search improve the performance of l_∞-Square Attack, confirming the theoretical arguments above. The comparison is done for an l_∞-threat model of radius $\epsilon = 0.05$ on $1,000$ test points for a ResNet-50 model trained normally on ImageNet (see Sect. 5 for details) with a query limit of $10,000$ and results are shown in Table 1. Our sampling distribution is special in two aspects: i) we use localized update shapes in form of squares and ii) the update is constant in each color channel. First, one can observe that our update shape "square" performs better than "rectangle" as we discussed in the previous section, and it is significantly better than "random" (the same amount of pixels is perturbed, but selected randomly in the image). This holds both for c (constant sign per color channel) and $c \cdot h^2$ (every pixel and color channel is changed independently of each other), with an improvement in terms of average queries of 339 to 73 and 401 to 153 respectively. Moreover, with updates of the same shape, the constant sign over color channels is better than selecting it uniformly at random (improvement in average queries: 401 to 339 and 153 to 73). In total the algorithm with "square-c" needs more than 5× less average queries than "random-$c \cdot h^2$", showing that our sampling distribution is the key to the high query efficiency of Square Attack.

The last innovation of our random search scheme is the initialization, crucial element of every non-convex optimization algorithm. Our method ("square-c") with the vertical stripes initialization improves over a uniform initialization on average by $\approx 25\%$ and, especially, median number of queries (more than halved).

We want to also highlight that the sampling distribution "square-$c \cdot h^2$" for which we shown convergence guarantees in Sect. 4.1 performs already better in terms of the success rate and the median number of queries than the state of the art (see Sect. 5). For a more detailed ablation, also for our l_2-attack, see Sup. C.

Table 2. Results of **untargeted** attacks on ImageNet with a limit of 10,000 queries. For the l_∞-attack we set the norm bound $\epsilon = 0.05$ and for the l_2-attack $\epsilon = 5$. Models: normally trained **I**: Inception v3, **R**: ResNet-50, **V**: VGG-16-BN. The Square Attack outperforms for both threat models all other methods in terms of success rate and query efficiency. The missing entries correspond to the results taken from the original paper where some models were not reported

Norm	Attack	Failure rate			Avg. queries			Med. queries		
		I	R	V	I	R	V	I	R	V
l_∞	Bandits [28]	3.4%	1.4%	2.0%	957	727	394	218	136	36
	Parsimonious [44]	1.5%	–	–	722	–	–	237	–	–
	DFO$_c$–CMA [34]	0.8%	**0.0%**	0.1%	630	270	219	259	143	107
	DFO$_d$–Diag. CMA [34]	2.3%	1.2%	0.5%	424	417	211	**20**	20	2
	SignHunter [2]	1.0%	0.1%	0.3%	471	129	95	95	39	43
	Square attack	**0.3%**	**0.0%**	**0.0%**	**197**	**73**	**31**	24	**11**	1
l_2	Bandits [28]	9.8%	6.8%	10.2%	1486	939	511	660	392	196
	SimBA-DCT [25]	35.5%	12.7%	7.9%	**651**	**582**	452	564	467	360
	Square attack	**7.1%**	**0.7%**	**0.8%**	1100	616	**377**	385	170	109

5 Experiments

In this section we show the effectiveness of the Square Attack. Here we concentrate on **untargeted** attacks since our primary goal is query efficient robustness evaluation, while the **targeted** attacks are postponed to the supplement. First,

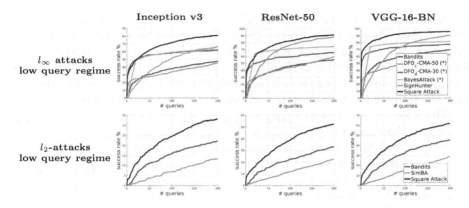

Fig. 4. Success rate in the low-query regime (up to 200 queries). * denotes the results obtained via personal communication with the authors and evaluated on 500 and 10,000 randomly sampled points for BayesAttack [45] and DFO [34] methods, respectively

we follow the standard setup [28, 34] of comparing black-box attacks on three ImageNet models in terms of success rate and query efficiency for the l_∞- and l_2-untargeted attacks (Sect. 5.1). Second, we show that our *black-box* attack can even outperform *white-box* PGD attacks on several models (Sect. 5.2). Finally, in the supplement we provide more experimental details (Sup. B), a stability study of our attack for different parameters (Sup. C) and random seeds (Sup. D), and additional results including the experiments for targeted attacks (Sup. E).

5.1 Evaluation on ImageNet

We compare the Square Attack to state-of-the-art score-based black-box attacks (without any extra information such as surrogate models) on three pretrained models in PyTorch (Inception v3, ResNet-50, VGG-16-BN) using 1,000 images from the ImageNet validation set. Unless mentioned otherwise, we use the code from the other papers with their suggested parameters. As it is standard in the literature, we give a budget of 10,000 queries per point to find an adversarial perturbation of l_p-norm at most ϵ. We report the *average* and *median* number of queries each attack requires to craft an adversarial example, together with the *failure rate*. All query statistics are computed only for successful attacks on the points which were originally correctly classified.

Tables 2 and 3 show that the Square Attack, despite its simplicity, achieves in all the cases (models and norms) the **lowest failure rate**, (<1% everywhere except for the l_2-attack on Inception v3), and almost always requires **fewer queries** than the competitors to succeed. Figure 4 shows the progression of the success rate of the attacks over the first 200 queries. Even in the low query regime the Square Attack outperforms the competitors for both norms. Finally, we highlight that the only hyperparameter of our attack, p, regulating the size of the squares, is set for all the models to 0.05 for l_∞ and 0.1 for l_2-perturbations.

l_∞-**Attacks.** We compare our attack to Bandits [29], Parsimonious [44], DFO_c/DFO_d [34], and SignHunter [2]. In Table 2 we report the results of the l_∞-attacks with norm bound of $\epsilon = 0.05$. The Square Attack always has the lowest failure rate, notably 0.0% in 2 out of 3 cases, and the lowest query consumption. Interestingly, our attack has median equal 1 on VGG-16-BN, meaning that the proposed initialization is particularly effective for this model.

The closest competitor in terms of the *average* number of queries is SignHunter [2], which still needs on average between 1.8 and 3 times more queries to find adversarial examples and has a higher failure rate than our attack. Moreover, the median number of queries of SignHunter is much worse

Table 3. Query statistics for untargeted l_2-attacks computed for the points for which all three attacks are successful for fair comparison

Attack	Avg. queries			Med. queries		
	I	R	V	I	R	V
Bandits [28]	536	635	398	368	314	177
SimBA-DCT [25]	647	563	421	552	446	332
Square attack	**352**	**287**	**217**	**181**	**116**	**80**

than for our method (e.g. 43 vs 1 on VGG). We note that although DFO_c–CMA [34] is competitive to our attack in terms of *median* queries, it has a significantly higher failure rate and between 2 and 7 times worse average number of queries. Additionally, our method is also more effective in the low-query regime (Fig. 4) than other methods (including [45]) on all the models.

l_2-**Attacks.** We compare our attack to Bandits [28] and SimBA [25] for $\epsilon = 5$, while we do not consider SignHunter [2] since it is not as competitive as for the l_∞-norm, and in particular worse than Bandits on ImageNet (see Fig. 2 in [2]).

As Table 2 and Fig. 4 show, the Square Attack outperforms by a large margin the other methods in terms of failure rate, and achieves the lowest median number of queries for all the models and the lowest average one for VGG-16-BN. However, since it has a significantly lower failure rate, the statistics of the Square Attack are biased by the "hard" cases where the competitors fail. Then, we recompute the same statistics considering only the points where all the attacks are successful (Table 3). In this case, our method improves by at least 1.5× the average and by at least 2× the median number of queries.

5.2   Square Attack Can Be More Accurate Than White-Box Attacks

Here we test our attack on problems which are challenging for both white-box PGD and other black-box attacks. We use for evaluation *robust accuracy*, defined as the worst-case accuracy of a classifier when an attack perturbs each input in some l_p-ball. We show that our algorithm outperforms the competitors both on

Table 4. On the robust models of [32] and [55] on MNIST l_∞-Square Attack with $\epsilon = 0.3$ achieves state-of-the-art (SOTA) results outperforming white-box attacks

Model	Robust accuracy	
	SOTA	**Square**
Madry et al. [32]	**88.13%**	88.25%
TRADES [55]	93.33%	**92.58%**

state-of-the-art robust models and defenses that induce different types of gradient masking. Thus, our attack is useful to evaluate robustness without introducing adaptive attacks designed for each model separately.

Outperforming White-Box Attacks on Robust Models. The models obtained with the adversarial training of [32] and TRADES [55] are standard benchmarks to test adversarial attacks, which means that many papers have tried to reduce their robust accuracy, without limit on the computational budget and primarily via white-box attacks. We test our l_∞-Square Attack on these robust models on MNIST at $\epsilon = 0.3$, using $p = 0.8$, 20k queries and 50 random restarts, i.e., we run our attack 50 times and consider it successful if any of the runs finds an adversarial example (Table 4). On the model of Madry et al. [32] Square Attack is only 0.12% far from the *white-box* state-of-the-art, achieving the second best result (also outperforming the 91.47% of SignHunter [2] by a large margin). On the TRADES benchmark [57], our method obtains a new SOTA of 92.58% robust accuracy outperforming the white-box attack of [17]. Additionally, the subsequent work of [18] uses the Square Attack as part of their *AutoAttack* where they show that the Square Attack outperforms other white-box attacks on 9 out of 9 MNIST models they evaluated. Thus, our black-box attack can be also useful for robustness evaluation of new defenses in the setting where gradient-based attacks require many restarts and iterations.

Resistance to Gradient Masking. In Table 5 we report the robust accuracy at different thresholds ϵ of the l_∞-adversarially trained models on MNIST of [32] for the l_2-threat model. It is known that the PGD is ineffective since it suffers from gradient masking [48]. Unlike PGD and other black-box attacks, our Square Attack does not suffer from gradient masking and yields robust accuracy close to zero for $\epsilon = 2.5$, with only a single run. Moreover, the l_2-version of SignHunter [2] fails to accurately assess the robustness because the method optimizes only over the extreme points of the l_∞-ball of radius $\epsilon/\sqrt{d}$ embedded in the target l_2-ball.

Table 5. l_2-robustness of the l_∞-adversarially trained models of [32] at different thresholds ϵ. PGD is shown with 1, 10, 100 random restarts. The black-box attacks are given a 10 k queries budget (see the supplement for details)

ϵ_2	Robust accuracy						
	White-box			Black-box			
	PGD_1	PGD_{10}	PGD_{100}	SignHunter	Bandits	SimBA	**Square**
2.0	79.6%	67.4%	**59.8%**	95.9%	80.1%	87.6%	**16.7%**
2.5	69.2%	51.3%	**36.0%**	94.9%	32.4%	75.8%	**2.4%**
3.0	57.6%	29.8%	**12.7%**	93.8%	12.5%	58.1%	**0.6%**

Table 6. l_∞-robustness of Clean Logit Pairing (CLP), Logit Squeezing (LSQ) [30]. The Square Attack is competitive to white-box PGD with many restarts (R = 10,000, R = 100 on MNIST, CIFAR-10 resp.) and more effective than black-box attacks [2, 28]

ϵ_∞	Model	Robust accuracy				
		White-box		Black-box		
		PGD_1	PGD_R	Bandits	SignHunter	**Square**
0.3	CLP_{MNIST}	62.4%	**4.1%**	33.3%	62.1%	**6.1%**
	LSQ_{MNIST}	70.6%	**5.0%**	37.3%	65.7%	**2.6%**
16/255	CLP_{CIFAR}	2.8%	**0.0%**	14.3%	**0.1%**	0.2%
	LSQ_{CIFAR}	27.0%	**1.7%**	27.7%	13.2%	**7.2%**

Attacking Clean Logit Pairing and Logit Squeezing. These two l_∞ defenses proposed in [30] were broken in [35]. However, [35] needed up to 10k restarts of PGD which is computationally prohibitive. Using the publicly available models from [35], we run the Square Attack with $p = 0.3$ and 20k query limit (results in Table 6). We obtain robust accuracy similar to PGD_R in most cases, but with a *single run*, i.e. without additional restarts. At the same time, although on some models Bandits and SignHunter outperform PGD_1, they on average achieve significantly worse results than the Square Attack. This again shows the utility of the Square Attack to accurately assess robustness.

6   Conclusion

We have presented a simple black-box attack which outperforms by a large margin the state-of-the-art both in terms of query efficiency and success rate. Our results suggest that our attack is useful *even in comparison to white-box attacks* to better estimate the robustness of models that exhibit gradient masking.

Acknowledgements. We thank L. Meunier and S. N. Shukla for providing the data for Fig. 6. M.A. thanks A. Modas for fruitful discussions. M.H and F.C. acknowledge support by the Tue.AI Center (FKZ: 01IS18039A), DFG TRR 248, project number 389792660 and DFG EXC 2064/1, project number 390727645.

References

1. Akhtar, N., Mian, A.: Threat of adversarial attacks on deep learning in computer vision: a survey. IEEE Access **6**, 14410–14430 (2018)
2. Al-Dujaili, A., O'Reilly, U.M.: There are no bit parts for sign bits in black-box attacks. In: ICLR (2020)
3. Alzantot, M., Sharma, Y., Chakraborty, S., Srivastava, M.: GenAttack: practical black-box attacks with gradient-free optimization. In: Genetic and Evolutionary Computation Conference (GECCO) (2019)

4. Athalye, A., Carlini, N., Wagner, D.A.: Obfuscated gradients give a false sense of security: circumventing defenses to adversarial examples. In: ICML (2018)
5. Bastani, O., Ioannou, Y., Lampropoulos, L., Vytiniotis, D., Nori, A., Criminisi, A.: Measuring neural net robustness with constraints. In: NeurIPS (2016)
6. Bhagoji, A.N., He, W., Li, B., Song, D.: Practical black-box attacks on deep neural networks using efficient query mechanisms. In: Ferrari, V., Hebert, M., Sminchisescu, C., Weiss, Y. (eds.) ECCV 2018. LNCS, vol. 11216, pp. 158–174. Springer, Cham (2018). https://doi.org/10.1007/978-3-030-01258-8_10
7. Biggio, B., Roli, F.: Wild patterns: ten years after the rise of adversarial machine learning. Pattern Recogn. **84**, 317–331 (2018)
8. Brendel, W., Rauber, J., Bethge, M.: Decision-based adversarial attacks: reliable attacks against black-box machine learning models. In: ICLR (2018)
9. Brunner, T., Diehl, F., Le, M.T., Knoll, A.: Guessing smart: biased sampling for efficient black-box adversarial attacks. In: ICCV (2019)
10. Carlini, N., Wagner, D.: Adversarial examples are not easily detected: bypassing ten detection methods. In: ACM Workshop on Artificial Intelligence and Security (2017)
11. Chen, J., Jordan, M.I., J., W.M.: HopSkipJumpAttack: a query-efficient decision-based attack (2019). arXiv preprint arXiv:1904.02144
12. Chen, P., Sharma, Y., Zhang, H., Yi, J., Hsieh, C.: EAD: elastic-net attacks to deep neural networks via adversarial examples. In: AAAI (2018)
13. Chen, P.Y., Zhang, H., Sharma, Y., Yi, J., Hsieh, C.J.: ZOO: zeroth order optimization based black-box attacks to deep neural networks without training substitute models. In: 10th ACM Workshop on Artificial Intelligence and Security - AISec 2017. ACM Press (2017)
14. Cheng, M., Le, T., Chen, P.Y., Yi, J., Zhang, H., Hsieh, C.J.: Query-efficient hard-label black-box attack: an optimization-based approach. In: ICLR (2019)
15. Cheng, S., Dong, Y., Pang, T., Su, H., Zhu, J.: Improving black-box adversarial attacks with a transfer-based prior. In: NeurIPS (2019)
16. Croce, F., Hein, M.: Sparse and imperceivable adversarial attacks. In: ICCV (2019)
17. Croce, F., Hein, M.: Minimally distorted adversarial examples with a fast adaptive boundary attack. In: ICML (2020)
18. Croce, F., Hein, M.: Reliable evaluation of adversarial robustness with an ensemble of diverse parameter-free attacks. In: ICML (2020)
19. Davis, D., Drusvyatskiy, D.: Stochastic model-based minimization of weakly convex functions. SIAM J. Optim. **29**(1), 207–239 (2019)
20. Du, J., Zhang, H., Zhou, J.T., Yang, Y., Feng, J.: Query-efficient meta attack to deep neural networks. In: ICLR (2020)
21. Duchi, J., Jordan, M., Wainwright, M., Wibisono, A.: Optimal rates for zero-order convex optimization: the power of two function evaluations. IEEE Trans. Inf. Theory **61**(5), 2788–2806 (2015)
22. Fawzi, A., Frossard, P.: Measuring the effect of nuisance variables on classifiers. In: British Machine Vision Conference (BMVC) (2016)
23. Gu, S., Rigazio, L.: Towards deep neural network architectures robust to adversarial examples. In: ICLR Workshop (2015)
24. Guo, C., Frank, J.S., Weinberger, K.Q.: Low frequency adversarial perturbation. In: UAI (2019)
25. Guo, C., Gardner, J.R., You, Y., Wilson, A.G., Weinberger, K.Q.: Simple black-box adversarial attacks. In: ICML (2019)
26. Haagerup, U.: The best constants in the Khintchine inequality. Studia Math. **70**(3), 231–283 (1981)

27. Ilyas, A., Engstrom, L., Athalye, A., Lin, J.: Black-box adversarial attacks with limited queries and information. In: ICML (2018)
28. Ilyas, A., Engstrom, L., Madry, A.: Prior convictions: black-box adversarial attacks with bandits and priors. In: ICLR (2019)
29. Ilyas, A., Santurkar, S., Tsipras, D., Engstrom, L., Tran, B., Madry, A.: Adversarial examples are not bugs, they are features. In: NeurIPS (2019)
30. Kannan, H., Kurakin, A., Goodfellow, I.: Adversarial logit pairing (2018). arXiv preprint arXiv:1803.06373
31. Li, Y., Li, L., Wang, L., Zhang, T., Gong, B.: NATTACK: learning the distributions of adversarial examples for an improved black-box attack on deep neural networks. In: ICML (2019)
32. Madry, A., Makelov, A., Schmidt, L., Tsipras, D., Vladu, A.: Towards deep learning models resistant to adversarial attacks. In: ICLR (2018)
33. Matyas, J.: Random optimization. Autom. Remote Control **26**(2), 246–253 (1965)
34. Meunier, L., Atif, J., Teytaud, O.: Yet another but more efficient black-box adversarial attack: tiling and evolution strategies (2019). arXiv preprint, arXiv:1910.02244
35. Mosbach, M., Andriushchenko, M., Trost, T., Hein, M., Klakow, D.: Logit pairing methods can fool gradient-based attacks. In: NeurIPS 2018 Workshop on Security in Machine Learning (2018)
36. Narodytska, N., Kasiviswanathan, S.: Simple black-box adversarial attacks on deep neural networks. In: CVPR Workshops (2017)
37. Nemirovsky, A.S., Yudin, D.B.: Problem Complexity and Method Efficiency in Optimization. Wiley-Interscience Series in Discrete Mathematics. Wiley, Hoboken (1983)
38. Nesterov, Y., Spokoiny, V.: Random gradient-free minimization of convex functions. Found. Comput. Math. **17**(2), 527–566 (2017). https://doi.org/10.1007/s10208-015-9296-2
39. Papernot, N., McDaniel, P., Goodfellow, I.: Transferability in machine learning: from phenomena to black-box attacks using adversarial samples (2016). arXiv preprint arXiv:1605.07277
40. Papernot, N., McDaniel, P., Wu, X., Jha, S., Swami, A.: Distillation as a defense to adversarial perturbations against deep networks. In: IEEE Symposium on Security & Privacy (2016)
41. Rastrigin, L.: The convergence of the random search method in the extremal control of a many parameter system. Autom. Remote Control **24**, 1337–1342 (1963)
42. Schrack, G., Choit, M.: Optimized relative step size random searches. Math. Program. **10**, 230–244 (1976). https://doi.org/10.1007/BF01580669
43. Schumer, M., Steiglitz, K.: Adaptive step size random search. IEEE Trans. Automat. Control **13**(3), 270–276 (1968)
44. Seungyong, M., Gaon, A., Hyun, O.S.: Parsimonious black-box adversarial attacks via efficient combinatorial optimization. In: ICML (2019)
45. Shukla, S.N., Sahu, A.K., Willmott, D., Kolter, Z.: Black-box adversarial attacks with Bayesian optimization (2019). arXiv preprint arXiv:1909.13857
46. Su, J., Vargas, D., Sakurai, K.: One pixel attack for fooling deep neural networks. IEEE Trans. Evol. Comput. **23**, 828–841 (2019)
47. Suya, F., Chi, J., Evans, D., Tian, Y.: Hybrid batch attacks: finding black-box adversarial examples with limited queries (2019). arXiv preprint, arXiv:1908.07000
48. Tramèr, F., Boneh, D.: Adversarial training and robustness for multiple perturbations. In: NeurIPS (2019)

49. Tsipras, D., Santurkar, S., Engstrom, L., Turner, A., Madry, A.: Robustness may be at odds with accuracy. In: ICLR (2019)
50. Tu, C.C., et al.: Autozoom: autoencoder-based zeroth order optimization method for attacking black-box neural networks. In: AAAI Conference on Artificial Intelligence (2019)
51. Uesato, J., O'Donoghue, B., Van den Oord, A., Kohli, P.: Adversarial risk and the dangers of evaluating against weak attacks. In: ICML (2018)
52. Yan, Z., Guo, Y., Zhang, C.: Subspace attack: exploiting promising subspaces for query-efficient black-box attacks. In: NeurIPS (2019)
53. Yin, D., Lopes, R.G., Shlens, J., Cubuk, E.D., Gilmer, J.: A Fourier perspective on model robustness in computer vision. In: NeurIPS (2019)
54. Zabinsky, Z.B.: Random search algorithms. In: Wiley Encyclopedia of Operations Research and Management Science (2010)
55. Zhang, H., Yu, Y., Jiao, J., Xing, E.P., Ghaoui, L.E., Jordan, M.I.: Theoretically principled trade-off between robustness and accuracy. In: ICML (2019)
56. Zheng, S., Song, Y., Leung, T., Goodfellow, I.J.: Improving the robustness of deep neural networks via stability training. In: CVPR (2016)
57. Zheng, T., Chen, C., Ren, K.: Distributionally adversarial attack. In: AAAI (2019)

You Are Here: Geolocation by Embedding Maps and Images

Noe Samano[✉], Mengjie Zhou, and Andrew Calway

University of Bristol, Bristol, UK
obed.samanoabonce@bristol.ac.uk

Abstract. We present a novel approach to geolocalising panoramic images on a 2-D cartographic map based on learning a low dimensional embedded space, which allows a comparison between an image captured at a location and local neighbourhoods of the map. The representation is not sufficiently discriminatory to allow localisation from a single image, but when concatenated along a route, localisation converges quickly, with over 90% accuracy being achieved for routes of around 200 m in length when using Google Street View and Open Street Map data. The method generalises a previous fixed semantic feature based approach and achieves significantly higher localisation accuracy and faster convergence.

Keywords: Geolocalisation · Image-map embeddings · Cross domain localisation · Representation learning

1 Introduction

We consider the problem of geolocalising ground panoramic images on a 2-D cartographic map without using GPS, i.e. determining the geographic coordinates at which the images where captured. As illustrated in Fig. 1, we seek to do this by linking the semantic information on the map to the content in the image, hence localising the latter. This is akin to the human skill of interpreting maps for way-finding using detailed survey maps or the ubiquitous You Are Here schematic maps found in cities and tourist attractions. It contrasts with previous works on image geolocalisation and the related problem of visual place recognition, which are based on matching the location image with a large database of georeferenced images, either ground or aerial (see Sect. 2).

The motivation for using maps as the reference source, in preference to georeferenced images, is multifold. Comprehensive map data covering large areas is readily available, contrasting with the difficulty in obtaining sufficient ground images. Although large amounts of aerial imagery is available (in fact, it is often used to generate maps), using it or ground images for localisation means having to overcome the invariance challenges faced when matching images taken at the same location at different times. In contrast, maps provide a semantic representation which is to a large extent independent of capture time, and hence offers the

© Springer Nature Switzerland AG 2020
A. Vedaldi et al. (Eds.): ECCV 2020, LNCS 12368, pp. 502–518, 2020.
https://doi.org/10.1007/978-3-030-58592-1_30

potential for more robust matching. Techniques for relating image data to spatial semantics are also likely to be needed as human-robot interaction becomes increasingly sophisticated, such as endowing robots with the ability to produce visual descriptions to aid human way-finding.

Fig. 1. Images (centre) in different viewing directions within a panorama (left) captured at the location circled on the map (right), where the arrow indicates the heading direction (top left centre image). The junction, buildings and park area on the map are visible in the images, illustrating the features we seek to leverage for geolocalisation.

We adopt a learning approach to the problem. It builds upon and generalises the work of Panphattarasap and Calway [28], who represent locations by binary descriptors indicating the presence or not of pre-selected semantic features (junctions and building gaps). This limits applicability to areas rich in those features and for which they provide sufficient discrimination. In contrast, we seek to learn descriptions that optimally link images to local map areas, without pre-assumptions as to which features are important, with the dual aim of generalising and increasing discrimination.

To do so, we learn a low dimensional embedded vector space (16-D) within which corresponding image and map tile pairs are close (a map tile is an image of a small region of the map), as illustrated in Fig. 2a. Images from Google Street View (GSV) [24,32] and map tiles from Open Street Map (OSM) [27] are used for training, and we use a Siamese-like network with triplet loss function to derive the embedded space. The network gives embedded space vectors for images and map tiles, which we treat as descriptors, and provide a means of assessing the similarity between an image and potential map locations. However, not surprisingly, the descriptors are not sufficiently discriminatory to localise a single image, since many places share similar map tiles.

Instead, analogous to that found in [28], it is the pattern of descriptors along routes that allows localisation, with the required route length dependent on local characteristics. Localisation therefore proceeds by comparing descriptors captured along a route with those derived from sequences of map tiles along all routes in the map, yielding a ranked list of likely locations and routes (see Fig. 2b). As in [28], we consider the specific problem of geolocalising images known to lie at regular intervals along roads, mimicking images taken from a vehicle or by a pedestrian, for example, which allows us to carry out large scale experiments using GSV/OSM data. It also means that we can use a naive route matching algorithm to investigate the discriminatory power of the representation. In summary, we demonstrate the following key findings:

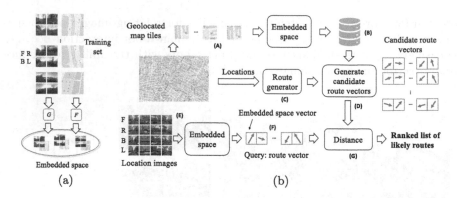

Fig. 2. (a) We learn transformations G and F which embed location images (in four orthogonal directions within a panorama - F, R, B and L) and map tiles into a low dimensional vector space (16-D), within which corresponding image/map tile pairs are close; (b) Geolocalisation: embedded vectors are computed for all map tiles (A), giving descriptors that are stored with their locations (B); all possible routes in the map are computed (C) and route descriptors generated from the map tiles (D); descriptors are computed for images along test routes (E) giving a query route descriptor (F), which is compared with all route descriptors (G), yielding a ranked list of likely routes.

1. It is possible to learn an embedded space which links location images to map tiles, achieving top-1% recall of between 72%–82%.
2. The space can be used to geolocalise image sequences along a route, achieving over 90% accuracy for routes of approximately 200 m within all testing areas of size 2–3 km^2. This compares with under 60% when using the approach in [28] for areas not rich in the chosen semantic features.
3. The approach gives increased accuracy for shorter routes and similar performance whether using turn information or not, demonstrating its ability to generalise and increase discrimination.

We discuss related work next, followed in Sect. 3 by details of the network architecture and learning. Section 4 gives details of the geolocalisation and Sect. 5 presents results of experiments using GSV and OSM data, including a comparison of performance with the method used in [28].

2   Related Work

The key insight identified in [28] is that map semantics along a route uniquely identify location. 4-bit descriptors indicating the presence or not of semantic features (junctions and building gaps) are used to represent locations and convolutional neural network (CNN) classifiers trained on GSV/OSM data detect feature presence in orthogonal viewing directions. As in our approach, descriptors are concatenated along routes to give localisation by comparison with a database of route descriptors derived from the map. 85% accuracy is reported for test routes up to 200 m in length using GSV/OSM data when descriptors are

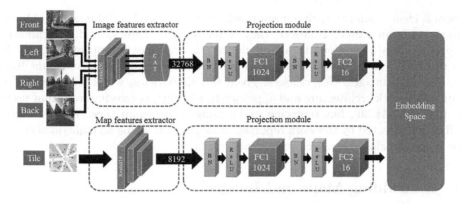

Fig. 3. Network architecture for embedded space learning within which corresponding location 4-images (front, left, right and back viewing directions) and map tiles are close. Each is processed independently via a sub-network consisting of feature extraction and projection layers, resulting in a 16-D embedded space. See text for further details.

combined with turn information along routes, although this was for a small test set (150 routes) within a densely built environment, which aligned well with the chosen semantic features. Notably, accuracy dropped to 45% when turn information was not used. The same 4-bit descriptors are also used in a particle filtering implementation in [37], although results suggest that localisation is slower to converge, possibly due to the limited route memory within the filter. A key motivation of our work is to generalise these approaches to avoid the reliance on specific semantic features.

Semantic features are also used in [5] for geolocalising images by matching with GIS data. Coarse spatial layouts of known object classes are extracted from images and matched with GIS data at candidate locations using probabilistic descriptors. Top-20% recall values of 70–80% are reported, but this drops to 30% for top-10% recall, significantly lower than that achieved here and in [28], albeit for a much larger test area but smaller test set (50 locations).

Map data has been used for self-driving vehicle navigation [1, 4, 9, 12, 21, 22, 30]. In [30], learning from GSV/OSM data is used to recognise semantic features such as junctions, bike lanes, etc., to verify GPS, whilst in [4,9] visual odometry is used to track map location by matching with road topology, extended in [22] to incorporate semantic cues such as junctions and sun position. End-to-end learning is used in [1,12,21], combining images with map road topology to predict steering control commands. Map semantics have also been used with GPS for 6-D pose estimation from images [2,3,6,13,25], although the focus is on metric pose estimation as opposed to geolocalisation.

As noted above, our use of maps as a reference for geolocalisation contrasts with previous approaches, which are based on georeferenced ground or aerial images, see for example [10,15,16,19,33,34,36] and the large body of work on visual place recognition [20]. Many of these approaches learn embedded spaces to achieve ground-to-aerial cross-view matching, see e.g. [15,16,19,34,36], and there is clear similarity with our approach since maps and aerial images are both overhead 'plan views'. However, they differ in that maps provide a semantic representation as opposed to a time dependent snapshot of appearance, which as previously noted is an advantage when seeking to provide an invariant link with ground images. Nevertheless, given the similarities, we chose to adopt a similar network architecture and approach to learning, borrowing from that used in [15,16,19,34,36], but replacing aerial images with map tiles. We were also initially motivated by the work reported in [31], which learns an embedded space for map tiles alone to reflect semantic similarity.

3   Embedding Maps and Images

We now describe the model, data sets and methodology used to learn the embedded space. Our data inputs are panoramic images from GSV and RGB map tiles corresponding to local neighbourhoods of a 2-D map rendered from OSM.

3.1   Network Model

Our network architecture is shown in Fig. 3. It has a Siamese-like form, with two independent sub-networks, one for location images and one for map tiles, each having a feature extractor and a projection module. We keep the weights in corresponding layers independent since each sub-network is processing information from very different domains.

In the location image sub-network, the feature extractor module is based on the Resnet50 architecture [11], from which we have removed the average pooling and the last convolutional layers to produce a 4×4 feature map of 512-D local descriptors. The input to the module is four images corresponding to orthogonal front, left, right and back views (w.r.t the vehicle heading direction) cropped from a GSV panorama. Feature maps for each view are flattened and concatenated and then input to the projection module.

The map tile feature extractor also uses a residual network to produce a 512-D feature map, although we use Resnet18 [11] in place of Resnet50, since map tiles contain considerably fewer details than location images. The module input is a map tile corresponding to the same location from which the GSV panorama was taken and the output is a descriptor vector.

In both sub-networks, the projection module consists of two fully connected layers, both preceded by batch normalization [17] and ReLu activation [26], which reduce descriptor dimension down to the embedding size (we used 16-D) and help to project semantically similar inputs near to each other.

3.2   Data Sets

For training and testing, we used the StreetLearn data set [24,32], which contains 113,767 panoramic images extracted from GSV in the cities of New York (Manhattan) and Pittsburgh, U.S. Metadata is provided for all images, including geographic coordinates, neighbours and heading direction (yaw). Corresponding map tiles for each GSV image location were rendered from OSM using Mapnik [23], with all text and building numbering removed.

To generate the training and testing datasets, we adopted the following procedure. First, using Breadth-First Search (BFS) and the same central panoramas as described in [24], we generated three testing subsets from areas in Hudson River (HR), Union Square (US) and Wall Street (WS), each containing 5,000 locations and covering $3.25\,\mathrm{km}^2$, $2.77\,\mathrm{km}^2$ and $2.33\,\mathrm{km}^2$, respectively. The remaining locations in Manhattan, together with all locations from Pittsburgh, were used for training. For each location, we rendered two 256×256 pixel map tiles at different scales centred on the location coordinates and with the map projection aligned with the GSV heading direction to ensure a geographic match between the data domains. We used two scales to account for the fact that the relevance of semantic features in the map is dependent on visibility, i.e. a map tile at a larger scale may not include features visible in a location image, whereas a map tile at a smaller scale may include non-visible features. In this work we used map tiles with scales corresponding to geographic areas of approximately $152 \times 152\,\mathrm{m}^2$ (small scale, S1) and $76 \times 76\,\mathrm{m}^2$ (large scale, S2). In total, the training set consisted of 98,767 panorama images and 197,534 map tiles, two tiles for each location, and three testing data sets each with 5,000 panoramas and 10,000 map tiles.

3.3 Training

We trained our model in an end-to-end way as all network parameters, including feature extraction, and projection layers, are updated at the same time. Since we do not have categories, or equivalently every location is a category, we train the model using an unsupervised method based on triplet loss metric learning [29], similar to that used in [15]. Note that we have to take care when considering data augmentation, since we need to maintain position and point of view relationships between the two domains, e.g. warping a map tile would require a suitable transformation of the corresponding location images, which is non-trivial to compute. Hence we limit augmentation to small changes in the scale of the map tiles and the viewing directions when cropping the panoramic images (see below). To form triplets, we take examples of matched and unmatched image/map tile pairs inside every batch.

We generate matched pairs as follows. Let X and Y denote the set of training tiles and panoramic images, respectively, and let L be the set of all locations. We start the training process by taking n random locations to form a subset L_B of locations in the training batch. Then, for each location $l_i \in L_B$, we take the associated panoramic image $y_i \in Y$ and pick a map tile $x_i \in \{x_i^1, x_i^2\}$ at random, where x_i^1 and x_i^2 are the two map tiles at different scales for location i. We then apply data augmentation to generate $1 \leq k \leq K$ different image/map tile pairs for each location l_i. Specifically, in the case of map tiles, we apply $T_x : x_i \rightarrow x_{ik}$, where T_x applies a further random small scale change to x_i and resizes it to 128×128 pixels, i.e. we randomly generate map tiles with random scales around that of the two 'reference' tiles. In the case of the panoramic images, we apply $T_y : y_i \rightarrow y_{ik} = (y_{ik}^F, y_{ik}^L, y_{ik}^R, y_{ik}^B)$ that takes y_i and crops four 128×128 pixel images in the front, left, right and back viewing directions, respectively, relative to the vehicle heading direction, with a random component in the cropping parameters to provide a degree of visual variation. We denote y_{ik} as a '4-image'. In addition, we incorporate a standard colour normalisation and a random horizontal flip in both T_x and T_y.

In summary, at each training step, there are N_B locations in the batch, each with K 4-image/map tile pairs. This gives a total of $N_B^2 K^2$ pairs, from which we can generate $N_B K^2$ matched pairs and $N_B(N_B - 1)K^2$ unmatched pairs. Hence, the ratio of matched versus unmatched pairs depends on the number of locations in the batch.

3.4   Loss Function

For the loss function, we used the weighted soft-margin ranking loss proposed in [15], a variation of that in [14,34] to address the problem of having to select the margin parameter in a traditional triplet loss [29]. It is defined as $\mathcal{L}_{wgt}(d) = ln(1 + e^{\alpha d})$, where d is the difference between a matched pair descriptor distance (Euclidean) and an unmatched pair descriptor distance, and α is a weighting factor that helps to improve convergence [15]. We also adopted a similar strategy to [35] and included bidirectional cross-domain and intra-domain ranking constraints to force the network to preserve embedding structure in both data representations. Our loss function is therefore given by

$$\mathcal{L}(X,Y) = \sum_{i,j,k,l,m} \lambda_1 \mathcal{J}_{wgt}(\mathbf{x}_{ik}, \mathbf{y}_{il}, \mathbf{y}_{jm}) + \lambda_2 \mathcal{J}_{wgt}(\mathbf{y}_{ik}, \mathbf{x}_{il}, \mathbf{x}_{jm})$$
$$+ \sum_{i,j,k,l,m,k \neq l} \lambda_3 \mathcal{J}_{wgt}(\mathbf{x}_{ik}, \mathbf{x}_{il}, \mathbf{x}_{jm}) + \lambda_4 \mathcal{J}_{wgt}(\mathbf{y}_{ik}, \mathbf{y}_{il}, \mathbf{y}_{jm}) \tag{1}$$

where $\mathcal{J}_{wgt}(\mathbf{x}, \mathbf{y}, \mathbf{z}) = \mathcal{L}_{wgt}(d(\mathbf{x}, \mathbf{y}) - d(\mathbf{x}, \mathbf{z}))$, $\mathbf{x}_{ik}$ and $\mathbf{y}_{ik}$ denote the embedded vectors corresponding to the kth augmentation of the map tile and panoramic image at location i, respectively, i.e. the outputs from the sub-networks in Fig. 3, and $d(.,.)$ denotes Euclidean distance. Note that i and j refer to different locations, i.e. $1 \leq i, j \leq N$ and $i \neq j$, and k, l and m refer to the K augmentations per location, i.e. $1 \leq k, l, m \leq K$. The values of $\lambda_1, \lambda_2, \lambda_3$ and λ_4 are weighting factors to control the influence of each constraint in the loss function.

3.5   Implementation

To train our model, we initialized Resnet50 and Resnet18 with Places365 [38] and ImageNet [7] weights, respectively. In the loss function, values of $\lambda_1 = \lambda_2 = \lambda_3 = \lambda_4 = 1.0$ achieved the best results. The embedding size was set to 16-D following ablation studies (see below) and we forced the embedded space to reside in an hyper-sphere manifold by performing L2-normalization on the network outputs and then scaling by a factor of 32. Although scaling does not affect nearest neighbor searching, it has an influence in the training process since it changes the steepness of the logistic curve [34]. We set the value of α in $\mathcal{L}_{wgt}(d)$ to be 0.2, as we found that larger values in combination with a large scaling factor of the manifold, sometimes led to training collapse. The number of locations N_B in a batch was set to 10, and the number of augmentations at each location K was set to 5. Hence, in each learning step, there were 250 matched pairs and 2250 unmatched pairs. To form our triplets, we followed the batch all mining strategy discussed in [8,14] and averaged the loss value over all possible triplets in the batch. We trained the model for 10 epochs using the Adam optimizer [18] with a 4×10^{-5} learning rate.

4   Geolocalisation Using Embedded Descriptors

We now describe using the learned embedded space for geolocalising images. The key principle is motivated by the approach used in [28] and illustrated in Fig. 2b: along a route consisting of adjacent locations, the pattern of embedded descriptors obtained from the sequence of captured images enables the route to be uniquely distinguished from all other routes. The route length required for successful localisation is dependent on the semantic characteristics, which are reflected in both the images and the map. We emphasise again that we are considering the case of discrete locations at regular intervals along roads, enabling us to adopt the naive search approach to localisation described below.

Localisation proceeds as follows. Given a 4-image from each panorama along a route, we derive embedded descriptors via the learned model described above and compare the concatenation of the descriptors with those derived from sequences of map tiles corresponding to possible routes on the map. The closest gives the most likely route, as illustrated in Fig. 2b. The methodology follows that in [28] and so we provide a summary below and adopt similar notation for clarity.

For an area with N locations, we define a route of length m as a list of m adjacent locations, $r_m = (l_{\gamma(1)}, .., l_{\gamma(m)})$, where $l_{\gamma(i)} \in L$ for $1 \leq \gamma(i) \leq N$ and the $\gamma(i)$, $1 \leq i \leq m$, define the route trajectory. We limit the number of possible routes by considering only routes without loops or direction reversal. Given a panoramic image $y_{\gamma(i)}$ at location $l_{\gamma(i)}$, we extract a 4-image such that the front image (F) is aligned with the heading direction and use the learned model to obtain an embedded descriptor $\mathbf{y}_{\gamma(i)}$ and hence a route descriptor $s_m^y = (\mathbf{y}_{\gamma(1)}, .., \mathbf{y}_{\gamma(m)})$. Similarly, map routes with locations $\zeta(i)$ have route descriptors $s_m^x = (\mathbf{x}_{\zeta(1)}, .., \mathbf{x}_{\zeta(m)})$ derived from the map tiles along the route. Note from earlier that these are also aligned with the heading direction. Let S_m^x denote the set of route descriptors corresponding to all possible routes. Localisation then follows by finding the route descriptor $\hat{s}_m^x$ that is closest to s_m^y, i.e.

$$\hat{s}_m^x = \underset{s_m^x \in S_m^x}{\arg\min} \, DIST(s_m^x, s_m^y) \tag{2}$$

where $DIST(s_m^x, s_m^y)$ denotes the sum of the Euclidean distances between corresponding descriptors

$$DIST(s_m^x, s_m^y) = \sum_{i=1}^{m} d(\mathbf{x}_{\zeta(i)}, \mathbf{y}_{\gamma(i)}) \tag{3}$$

In practice, we apply the above minimisation each time a new panoramic image is obtained along the route being traversed, giving an estimated route at each time step. This raises the question as to when successful localisation has been achieved, i.e. at what route length can we have sufficient confidence in the location estimate, which we investigate in the experiments. The results demonstrate that the required length of route depends on a number of factors, including area size, local appearance and environmental characteristics.

We also investigated the use of turn information along a route as a means of improving localisation accuracy, as adopted in [28]. Specifically, again using similar notation to that in [28], we define a binary turn pattern along a route r_m as $t_{m-1} = (t_{\gamma(1)}, .., t_{\gamma(m-1)})$, where the ith component of t indicates whether there is a turn between locations $l_{\gamma(i)}$ and $l_{\gamma(i+1)}$. Turn patterns can be estimated from the heading directions along a route and computed from OSM map data. This provides an additional constraint to that in Eq. 2, in that we require the best route estimate to have a matching turn pattern to that being traversed.

Finally, note that although we adopt a brute force search approach to the above, the recursive link between routes as the length increases means that it can be efficiently implemented by careful storage and searching of route indices $\zeta(i)$. There is also significant potential for culling routes that have low matching scores at each time step. We did a preliminary test of this idea in Fig. 9b and intend to investigate it further in future work.

5 Experiments

We evaluated the approach both in terms of the quality of the learned embedded space, i.e. its ability to link location images with appropriate map tiles, and the accuracy of localisation that can be achieved using embedded descriptors. We compared the latter with that achieved using binary semantic descriptors in [28].

5.1 Embedding Maps and Images

We investigated the recall performance when using the embedded space to retrieve corresponding map tiles given location images, i.e. how likely is the corresponding map tile to be the closest within the space. Top-$k\%$ recall plots are shown in Fig. 4a, with top 1% recall values in brackets, where top-$k\%$ recall is the fraction of cases in which the ground truth tile is within the top $k\%$ of best estimates. These results indicate that the network is doing well, with over 70% top-1% recall. This is confirmed in Fig. 4b, which shows precision-recall curves generated from balanced datasets of matched and unmatched pairs and by varying the distance cut-off threshold used when searching for close map tiles.

Note that performance is slightly worse for Wall Street, which reflects the fact that the area is noticeably different from those in the training set, with irregular intersections, motorways, and tunnels, making the area more challenging. We also noted significant mismatch between information in the map tiles and that in the location images, possibly resulting from redevelopment in the area yet to be reflected in OSM. Nevertheless, even under these conditions, our model is able to assign high rank to corresponding map tiles.

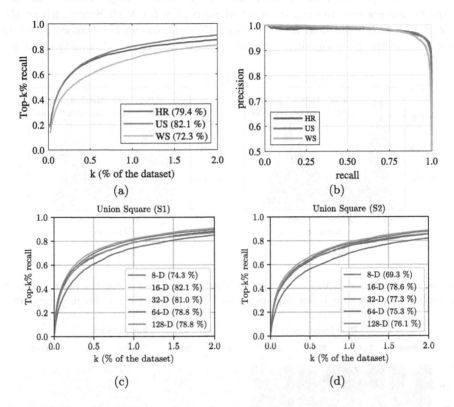

Fig. 4. (a) Top $k\%$ recall when learned embedded space is used to retrieve map tiles given a location image, where $k\%$ is the fraction of the dataset size. Top 1% recall values are shown in brackets; (b) Precision-recall curves generated by varying the distance cut-off threshold; (c–d) Top-k% recall for Union Square using different space dimension sizes and map tile scales, S1 (c) and S2 (d).

The above results were obtained using an embedded space dimension of 16 and using small scale (S1) map tiles when generating descriptors for localisation. We found that these values gave the best performance as illustrated in Figs. 4c–d, which show the top-k recall when using different dimension sizes and when using the two scales for Union Square. Similar results were obtained for the other datasets and these settings were used in all of the experiments.

We also examined how well the model separated matched and unmatched image/map tile pairs. Figure 5 shows histograms of the distance between matched and unmatched pairs inside the embedded space for Union Square and Wall Street. As expected, distances between unmatched pairs are larger, confirming that training has been effective and able to generalise outside the training set. Note also the larger overlapping area for Wall street, confirming the above comments regarding discrimination in the area.

To further illustrate the performance of the model, Fig. 6 shows examples of query images and the top-5 retrieved map tiles for Wall Street, where the correct map tile is outlined in green. Note that the model has learned to relate semantics of the two domains, e.g. buildings (a–b), parks (b), junctions (a–b),

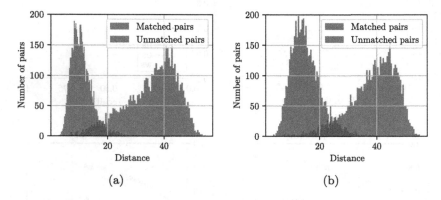

Fig. 5. Histograms of distance between matched (green) and unmatched pairs (red) in the embedding space for Union Square (a) and Wall Street (b). (Color figure online)

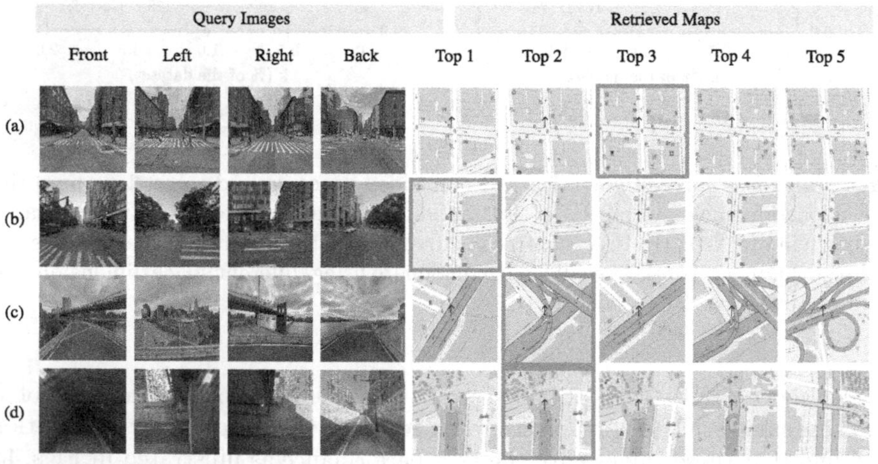

	Query Images				Retrieved Maps				
	Front	Left	Right	Back	Top 1	Top 2	Top 3	Top 4	Top 5

Fig. 6. Top-5 retrieved examples (right) given a query 4-image (left) for Wall Street. The green frame encloses the true map location and the blue arrow in the center of the map tile represent the position and heading. (Color figure online)

rivers (c), and tunnels (d). In future we intend to investigate performance further, especially w.r.t optimising the architecture using appropriate ablation studies.

5.2 Localisation

To evaluate geolocalisation performance, we randomly simulated 500 routes in each of the three test areas and applied the algorithm described in Sect. 4. In the route generation process, we excluded locations that were labelled as tunnels or motorways in OSM since we found that otherwise it generated many routes with very little appearance variation, consisting of long stretches tunnels or overpasses, which were near-impossible to distinguish from image data alone. To calculate localisation accuracy, we recorded the number of routes successfully localised at every step as a function of route length. We deemed a route to have been successfully localised if and only if the last 5 locations corresponds with those of the ground truth, i.e. the final 5 locations along the route.

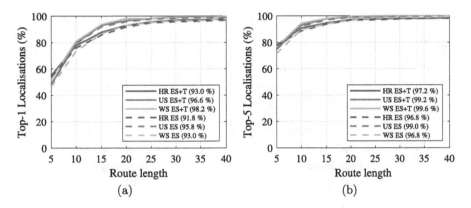

Fig. 7. (a) Top-1 and (b) top-5 localisation accuracy versus route length for all three datsets. Dashed lines indicate using embedded space descriptors only (ES) and solid lines indicate including turn information (ES+T).

Figure 7 shows top-1 and top-5 localisation accuracy versus route length when using embedded descriptors only and when combined with turn patterns. Top-k accuracy for route length m indicates the number of cases when a correct route, i.e. one meeting the above success criterion w.r.t the ground truth, was within the top k most likely routes. Accuracy values for route lengths of 20 locations (approximately 200 m) are shown in brackets. These results demonstrate that the method is performing exceptionally well, achieving over 90% top-1 accuracy across all datasets for route lengths of 20, even when not using turn information.

For shorter routes, although the top-1 accuracy drops, note from Fig. 7b, that the method still ranks correct routes highly, with over around 90% within the top-5 for route lengths of 10. Note also that the overall performance of the method on the three test sets is comparable, indicating that the model is generalising well to unseen environments. It is also interesting to note that for Wall Street, the inclusion of turn information has a greater impact, reflecting the fact that the area contains significantly more variation in possible route configurations.

We also compared the method with the BSD approach in [28]. We trained binary classifiers to detect junctions and building gaps, using the training set described in Sect. 3.2 and labels obtained from OSM data in the same manner as described in [28]. Experiments with different classifier architectures showed that Resnet18 [11] gave the best performance, achieving for example precision/recall values of 0.6/0.75 and 0.74/0.8 for junctions and gaps, respectively, when tested on Hudson River, which are similar values to that given in [28]. We used the classifiers to detect the presence or not of the features in each 4-image, i.e. junctions in the F and R images and gaps in the L and R images (cf. Sect. 3.3), and hence derived the 4-bit image-based BSD for each location. Map BSDs were generated in the same way as labels when generating the training set.

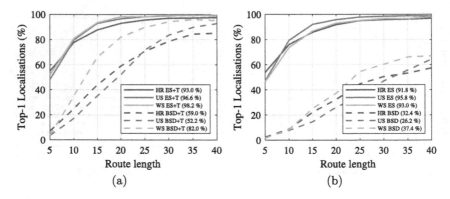

Fig. 8. Comparison of top-1 localisation accuracy between using our method and the BSD method in [28] when (a) using descriptors combined with turn information and (b) descriptors only.

Figure 8 shows top-1 accuracy versus route length using our method and the BSD method, both with and without turn information. Further comparison is shown in Fig. 9a, which gives more detailed analysis of successful localisation rates for 4 route lengths. Also shown is the accuracy achieved when using only turn information, i.e. matching routes based only on the road pattern, which as also noted in [28], performs poorly, even for long routes.

These results demonstrate that our method gives superior performance when compared with using BSDs. Although the latter performs reasonably well for Wall Street with turns, comparable to that reported in [28] for a similar dense urban area with plenty of junctions and building gaps, its performance falls significantly in the other areas within which those features are less prevalent. In contrast, our method performs equally well in all areas, demonstrating that it is generalising well. Also notable is that the difference in performance is greater when not using turn information. We believe that this is due to the greater generality provided by the embedded space descriptors in contrast to the fixed features used in the BSD approach. Note also that our method is able to localise shorter routes with and without turns, again suggesting better generalisation.

We also investigated the robustness of our method by considering the impact of culling a significant percentage of route candidates in the searching process. Figure 9b shows a comparison of top-1 accuracy when using 100% of route candidates and when discarding 50% of them at each motion step until only 100 are left. Note the similarity of the curves, indicating that route discrimination occurs early and is maintained as routes grow.

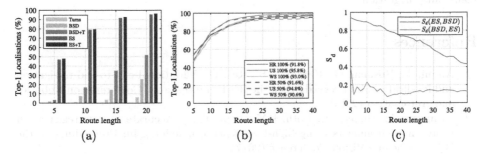

Fig. 9. (a) Top-1 localisation accuracy for 4 route lengths for the ES and BSD methods (with and without turn information) and when using only turn information for Union Square; (b) Accuracy when keeping 100% (solid) and 50% (dashed) of route candidates between updates; (c) Relative difference score between localisation sets for ES and BSD plotted against route length for Union Square. High scores for short routes suggests that our method is learning different semantic cues compared to BSD. (Color figure online)

Finally, to investigate the extent to which our model is using different information than BSD, we defined a score to measure the percentage of routes that our model is correctly localising but the BSD method is not and vice-versa. Formally, let S_a and S_b denote the set of routes successfully localised using methods a and b, respectively, where $a, b \in \{BSD, ES\}$. We define the difference score as $S_d(a, b) = |S_a \backslash S_b| / |S_a|$. A value of one would mean that none of the routes localised using method a were localised by method b, whereas a value near zero would mean that all routes localised in S_a were also in S_b. Results for both $S_d(ES, BSD)$ (blue) and $S_d(BSD, ES)$ (red) are shown in Fig. 9c for Union Square. These results clearly show that our method is using different information, enabling it to generalise and increase discrimination.

6   Conclusions

We have presented a novel method for correlating panorama location images and 2D cartographic map tiles into a common low dimensional space using a deep learning approach. This allows us to compare both domains directly using Euclidean distance. Furthermore, we have shown how this space can be used for geolocalisation using image to map tile matching defined along routes. Results indicate that the method can achieve over 90 % accuracy when detecting routes of approximately 200 m length in urban areas of over 2 km^2. Moreover, the approach significantly outperforms a previous method based on hand-crafted semantic features, demonstrating greater discrimination and generalisation.

Acknowledgments. We gratefully acknowledge the support of the Mexican Council of Science and Technology (CONACyT) and the Chinese Scholarship Council (CSC). We are also grateful to Google and the DeepMind StreetLearn project for use of the StreetLearn GSV dataset, and also to the Reviewers and Area Chairs for their comments and suggestions.

References

1. Amini, A., Rosman, G., Karaman, S., Rus, D.: Variational end-to-end navigation and localization. In: Proceedings of the IEEE International Conference on Robotics and Automation (2019)
2. Armagan, A., Hirzer, M., Roth, P.M., Lepetit, V.: Accurate camera registration in urban environments using high-level feature matching. In: Proceedings of the British Machine Vision Conference (2017)
3. Arth, C., Pirchheim, C., Ventura, J., Schmalstieg, D., Lepetit, V.: Instant outdoor localization and SLAM initialization from 2.5D maps. IEEE Trans. Vis. Comput. Graph. **21**(11), 1309–1318 (2015)
4. Brubaker, M.A., Geiger, A., Urtasun, R.: Map-based probabilistic visual self-localization. IEEE Trans. Pattern Anal. Mach. Intell. **38**(4), 652–665 (2016)
5. Castaldo, F., Zamir, A., Angst, R., Palmieri, F.A.N., Savarese, S.: Semantic cross-view matching. In: Proceedings of the ICCV Workshop: Vision from Satellite to Street (2015)
6. Cham, T.J., Ciptadi, A., Tan, W.C., Pham, M.T., Chia, L.T.: Estimating camera pose from a single urban ground-view omnidirectional image and a 2D building outline map. In: Proceedings of the IEEE Conference on Computer Vision and Pattern Recognition (2010)
7. Deng, J., Dong, W., Socher, R., Li, L.J., Li, K., Fei-Fei, L.: ImageNet: a large-scale hierarchical image database. In: Proceedings of the IEEE Conference on Computer Vision and Pattern Recognition (2009)
8. Ding, S., Lin, L., Wang, G., Chao, H.: Deep feature learning with relative distance comparison for person re-identification. Pattern Recogn. **48**(10), 2993–3003 (2015)
9. Floros, G., van der Zander, B., Leibe, B.: OpenStreetSLAM: global vehicle localization using OpenStreetMaps. In: Proceedings of the IEEE International Conference on Robotics and Automation (2013)
10. Hays, J., Efros, A.: IM2GPS: estimating geographic information from a single image. In: Proceedings of the IEEE Conference on Computer Vision and Pattern Recognition, pp. 1–8, July 2008
11. He, K., Zhang, X., Ren, S., Sun, J.: Deep residual learning for image recognition. In: Proceedings of the IEEE Conference on Computer Vision and Pattern Recognition (2016)

12. Hecker, S., Dai, D., Gool, L.V.: End-to-end learning of driving models with surround-view cameras and route planners. In: Proceedings of the European Conference on Computer Vision (2018)
13. Hentschel, M., Wagner, B.: Autonomous robot navigation based on OpenStreetMap geodata. In: Proceedings of the IEEE International Conference on Intelligent Transportation Systems (2010)
14. Hermans, A., Beyer, L., Leibe, B.: In defense of the triplet loss for person re-identification. arXiv:1703.07737 (2017)
15. Hu, S., Feng, M., Nguyen, R.M., Hee Lee, G.: CVM-Net: cross-view matching network for image-based ground-to-aerial geo-localization. In: Proceedings of the IEEE Conference on Computer Vision and Pattern Recognition (2018)
16. Hu, S., Lee, G.H.: Image-based geo-localization using satellite imagery. Int. J. Comput. Vis. **128**(5), 1205–1219 (2019). https://doi.org/10.1007/s11263-019-01186-0
17. Ioffe, S., Szegedy, C.: Batch normalization: accelerating deep network training by reducing internal covariate shift. arXiv:1502.03167 (2015)
18. Kingma, D.P., Ba, J.: Adam: a method for stochastic optimization. arXiv:1412.6980 (2014)
19. Lin, T., Yin Cui, Belongie, S., Hays, J.: Learning deep representations for ground-to-aerial geolocalization. In: Proceedings of the IEEE Conference on Computer Vision and Pattern Recognition (2015)
20. Lowry, S., et al.: Visual place recognition: a survey. IEEE Trans. Robot. **32**, 1–19 (2016)
21. Ma, H., Wang, Y., Tang, L., Kodagoda, S., Xiong, R.: Towards navigation without precise localization: weakly supervised learning of goal-directed navigation cost map. arXiv:1906.02468 (2019)
22. Ma, W.C., Wang, S., Brubaker, M.A., Fidler, S., Urtasun, R.: Find your way by observing the sun and other semantic cues. In: Proceedings of the IEEE International Conference on Robotics and Automation (2017)
23. Mapnik. https://mapnik.org
24. Mirowski, P., et al.: The StreetLearn environment and dataset. arXiv:1903.01292 (2019)
25. Mousavian, A., Kosecka, J.: Semantic image based geolocation given a map. arXiv:1609.00278 (2016)
26. Nair, V., Hinton, G.E.: Rectified linear units improve restricted Boltzmann machines. In: Proceedings of the International Conference on Machine Learning (2010)
27. Open Street Map. https://www.openstreetmap.org
28. Panphattarasap, P., Calway, A.: Automated map reading: image based localisation in 2-D maps using binary semantic descriptors. In: Proceedings of the IEEE/RSJ International Conference on Intelligent Robots and Systems (2018)
29. Schroff, F., Kalenichenko, D., Philbin, J.: FaceNet: a unified embedding for face recognition and clustering. In: Proceedings of the IEEE Conference on Computer Vision and Pattern Recognition (2015)
30. Seff, A., Xiao, J.: Learning from maps: visual common sense for autonomous driving. arXiv:1611.08583 (2016)
31. Spruyt, V.: Loc2vec: Learning location embeddings with triplet-loss networks. Sentiance web article. https://www.sentiance.com/2018/05/03/venue-mapping/ (2018)
32. StreetLearn. https://sites.google.com/view/streetlearn/

33. Tian, Y., Chen, C., Shah, M.: Cross-view image matching for geo-localization in urban environments. In: Proceedings of the IEEE Conference on Computer Vision and Pattern Recognition (2017)
34. Vo, N.N., Hays, J.: Localizing and orienting street views using overhead imagery. In: Leibe, B., Matas, J., Sebe, N., Welling, M. (eds.) ECCV 2016. LNCS, vol. 9905, pp. 494–509. Springer, Cham (2016). https://doi.org/10.1007/978-3-319-46448-0_30
35. Wang, L., Li, Y., Lazebnik, S.: Learning deep structure-preserving image-text embeddings. In: Proceedings of the IEEE Conference on Computer Vision and Pattern Recognition (2016)
36. Workman, S., Souvenir, R., Jacobs, N.: Wide-area image geolocalization with aerial reference imagery. In: Proceedings of the IEEE International Conference on Computer Vision (2015)
37. Yan, F., Vysotska, O., Stachniss, C.: Global localization on OpenStreetMap using 4-bit semantic descriptors. In: Proceedings of the European Conference on Mobile Robots (2019)
38. Zhou, B., Lapedriza, A., Khosla, A., Oliva, A., Torralba, A.: Places: a 10 million image database for scene recognition. IEEE Trans. Pattern Anal. Mach. Intell. **40**(6), 1452–1464 (2017)

Segmentations-Leak: Membership Inference Attacks and Defenses in Semantic Image Segmentation

Yang He[1,2](✉), Shadi Rahimian[1], Bernt Schiele[2], and Mario Fritz[1]

[1] CISPA Helmholtz Center for Information Security, Saarbrücken, Germany
{yang.he,shadi.rahimian,fritz}@cispa.saarland
[2] Max Planck Institute for Informatics, Saarland Informatics Campus, Saarbrücken, Germany
schiele@mpi-int.mpg.de

Abstract. Today's success of state of the art methods for semantic segmentation is driven by large datasets. Data is considered an important asset that needs to be protected, as the collection and annotation of such datasets comes at significant efforts and associated costs. In addition, visual data might contain private or sensitive information, that makes it equally unsuited for public release. Unfortunately, recent work on membership inference in the broader area of adversarial machine learning and inference attacks on machine learning models has shown that even black box classifiers leak information on the dataset that they were trained on. We show that such membership inference attacks can be successfully carried out on complex, state of the art models for semantic segmentation. In order to mitigate the associated risks, we also study a series of defenses against such membership inference attacks and find effective counter measures against the existing risks with little effect on the utility of the segmentation method. Finally, we extensively evaluate our attacks and defenses on a range of relevant real-world datasets: Cityscapes, BDD100K, and Mapillary Vistas. Our source code and demos are available at https://github.com/SSAW14/segmentation_membership_inference.

Keywords: Membership inference · Data privacy and security · Forensics · Semantic segmentation

1 Introduction

The availability of large datasets is playing a key role in today's state of the art computer vision methods ranging from image classification (e.g. ImageNet [7]),

Electronic supplementary material The online version of this chapter (https://doi.org/10.1007/978-3-030-58592-1_31) contains supplementary material, which is available to authorized users.

A. Vedaldi et al. (Eds.): ECCV 2020, LNCS 12368, pp. 519–535, 2020.
https://doi.org/10.1007/978-3-030-58592-1_31

over semantic segmentation [6,21,35], to visual question answering [2]. Therefore, research and industry alike have recognized the importance of large-scale datasets [7,15,31,39] to push performance of computer vision algorithms. However, data collection and in particular annotation and curation of large datasets comes at a substantial cost. There are sizable efforts from the research community [6,11,35], and also industry has picked up the task of collection (e.g. [21]) as well as providing annotation services such as Amazon MTurk, which in turn can be monetized and constitutes important assets to companies.

Consequently, such assets need protection e.g. as part of intellectual property and it should be controlled which parts are made public (e.g. for research purposes) and which part remain private. Based on these datasets, high performing models are trained and then made public (e.g. as black box models) via an API or as part of a product. One might assume that the information of the training set remains contained within the trained parameters of the model and therefore remains private. Beyond the aspect of intellectual property, data might also include private information that were captured as part of the data collection process, which are sensitive and important for safe and clean services.

Unfortunately, recent work on membership inference attacks [26,27,29] has shown that even a black box model leaks information of the training data, *aiming to infer if a particular sample was used as part of the training data or not.* Such approaches have shown high success rates on a range of **classification** tasks and have equally proven to be hard to fully prevent (= defend). While this constitutes a potential threat to the machine learning model, it can also potentially be used as a forensics technique to detect a potentially unauthorized use of data.

However, we are still missing even a basic understanding on if and how these membership attack vectors extend to semantic segmentation, which is a basic computer vision task and has broad applications [4,13,16,17,37]. Hence, we propose and study first membership inference attacks and defenses for semantic segmentation. To reach this goal, we design an attack pipeline based on per-patch analysis, and discover (1) not all the areas of an input are helpful to membership inference, (2) structural information itself leaks membership privacy and (3) effective defense mechanisms exists that can reduce the effectiveness of these attacks substantially. Accordingly, we highlight our contributions to **segmentation** task and review relevant work.

1.1 Contributions

Our main contributions are as follows. (1) We present the first work on membership inference attacks against semantic segmentation models under different data/model assumptions. (2) We show structural outputs of segmentation have severe risks of leaking membership. Our proposed structured loss maps achieve the best attack results. (3) We present a range of defense methods to reduce membership leakage. In the end, we show feasible solutions to protect against membership attacks. (4) Extensive comparisons and ablation studies are provided in order to shed light on the core challenges of membership inference attacks for semantic segmentation.

1.2   Related Work

Recent attacks against machine learning models have drawn much attention to communities focusing on attacking model functionality (e.g., adversarial attacks [10,18,19,23,30,34]), or stealing functionality [24] or configurations [22] of a model. In this paper, we detail the topics of data privacy and security in the following.

Membership Inference Attack. Membership inference attacks have been successfully achieved in many problems and domains, varying from biomedical data [3], locations [25], purchasing records [27], and images [29].

It has been shown that machine learning models can be attacked to infer the membership status of their training data. Shokri et al. [29] proposed membership inference attacks against classification models utilizing multiple shadow models to mimic behaviors of the victim model. Shadow models were trained by querying the victim model using examples with higher confidences from the victim model. Hence, a binary classifier was trained with information from shadow models, and applied to attack the victim. Further, Salem et al. [27] demonstrated only one shadow model is enough to reach similar results rather than multiple shadow models. They also show that underlying distributions of data used to train shadow models and the victim can be different, which allows for attacks under relaxed assumptions. In addition, learning free attacks were proposed, which constitutes a low-skill attack without knowledge about the model and data distribution priors. Salem et al. [27] proposed to directly set a threshold on the confidence scores of predictions to recognize memberships. Sablayrolles et al. [26] set a threshold on loss values and achieved quite successful results. While prior work has only studied classification models so far, our contribution is to show the differences between segmentation and classification models and present the first study of attacks and defenses on semantic segmentation models based on new methods. Although the segmentation problem can be understood as pixel-wise classification, it turns out the derived information is weak and needs to be aggregated over a patch or even the full image for a successful attack. Beyond this, we propose the first dedicated attacks that fully leverage the information of the full segmentation output and hence lead to even stronger attack vectors.

Privacy-Preserving Machine Learning. The goal of these techniques is to reduce information leakage with limited access to training data, which have been applied to deep learning [1,28]. Differential privacy [9] allows learning the statistical properties of a dataset while preserving the privacy of the individual data points in it. Jayaraman et al. [14] discussed the connection between the effectiveness of differential privacy and membership inference in practice. Besides, Nasr et al. [20] provided membership protection for a classifier by training a coupled attacker in an adversary manner. Zhang et al. [36] obfuscated training data before feeding them to the model training task, which hides the statistical properties of an original dataset by adding random noises or providing new samples. In our work, we compare a series of defense approaches to mitigate membership leakage in semantic segmentation.

2 Attacks Against Black-Box Semantic Segmentation Models

Membership inference is to attack a **victim**, aiming to determine whether a particular data point was part of the training data of the victim. Such attacks exploit overfitting artifacts on training data [26,27,29]. Typical machine learning models tend to be overconfident on data points that were seen during the training. Such overfitting issues lead to characteristic patterns and distributions of confidence scores [27,29] or loss values [26] which has facilitated membership inference attacks. As a result, successful attacks against classification models can be achieved according to a **shadow** model trained by a malicious attacker, mimicking the overfitting patterns and distribution gaps.

We show how such attacks can equally be constructed against models for semantic segmentation with a specially designed pipeline and representations. While such models can be understood as pixel-wise classification, it turns out that the information that can be derived from a single pixel is rather weak. Hence, we develop a method that aggregates such information over patches and full images to arrive at stronger attacks. We first describe our pipeline for attacking segmentation models, and then present two attack settings exploited in our study, which have different constraints during attacks. Furthermore, we discuss our evaluation methodology, and then show evaluation results.

Algorithm 1. Training an attacker

Input: $\mathcal{D}^S = \{(X_i, Y_i)\}_i$, $\mathbf{V}$, *Epoch*
Output: Per-patch attacker $\mathbf{A_P}$
1: Query $\mathcal{D}^S$ with $\mathbf{V}$;
2: Partition $\mathcal{D}^S$ into $\mathcal{D}^S_{in}$, $\mathcal{D}^S_{out}$;
3: Train a shadow model $\mathbf{S}$ with $\mathcal{D}^S_{in}$;
4: Initialize $\mathbf{A_P}$;
5: **for** $i = 1; i \le Epoch; i + +$ **do**
6:    **for** $j = 1; j \le |\mathcal{D}^S|; j + +$ **do**
7:        Crop a patch $(\hat{X}_j, \hat{Y}_j)$ from (X_j, Y_j);
8:        **if** $((X_j, Y_j) \in \mathcal{D}^S_{in})$ **then**
9:            $\mathbf{A_P}(\mathbf{S}(\hat{X}_j), \hat{Y}_j) \xrightarrow{\text{learn}} 1$
10:       **else**
11:           $\mathbf{A_P}(\mathbf{S}(\hat{X}_j), \hat{Y}_j) \xrightarrow{\text{learn}} 0$
12:       **end if**
13:   **end for**
14: **end for**
15: **return** $\mathbf{A_P}$;

Algorithm 2. Testing (Membership Inference)

Input: Testing pair (X, Y), $\mathbf{V}$, $\mathbf{A_P}$, N, τ
Output: Image-level inference result $\mathbf{A}$
1: $\mathbf{A} = 0$; $i = 0$;
2: **while** $i < N$ **do**
3:    Crop $(\hat{X}, \hat{Y})$ from (X, Y); // patch selection
4:    **if** $\text{Mean}(\mathbf{V}(\hat{X}) \otimes \hat{Y}) > \tau)$ **then**
5:        **continue**; // reject too confident patches
6:    **end if**
7:    $\mathbf{A} = \mathbf{A} + \mathbf{A_P}(\mathbf{V}(\hat{X}), \hat{Y})/N$;
      $i++$;
8: **end while**
9: **return** $\mathbf{A}$;

2.1 Methods

Our approach infers image-level membership information based on observing predictions of segmentation models and correct labels. In this section, we describe our membership inference pipeline based on per-patch analysis, as summarized in Algorithm 1 and 2. Further, several design choices are discussed that significantly contribute to the success of the attack and help to understand the essence in attacking semantic segmentation models with structured outputs.

Notation. We define the notation used through the paper. Let $\mathcal{D}^{\{V,S\}} = \{(X_i, Y_i)\}_i$ be two datasets including images $X \in \mathcal{R}^{H \times W \times 3}$ and densely annotated GTs $Y \in \mathcal{R}^{H \times W \times C}$ with one-hot vectors, where C is the number of predefined labels. For each dataset, we partition it into two parts for proving different membership status, i.e., $\mathcal{D}^{\{V,S\}} = \mathcal{D}_{in}^{\{V,S\}} \cup \mathcal{D}_{out}^{\{V,S\}}$. The victim model, which is trained on $\mathcal{D}_{in}^V$ and we aim for attacking, is denoted as **V**. To achieve attacks, we build a shadow semantic segmentation model **S** with $\mathcal{D}_{in}^S$, for training an attacker. Let P be the posterior of a segmentation output, i.e., P = **S**(X) or **V**(X), depending on the stages of membership inference. Our per-patch attacker is denoted as $\mathbf{A_P}(P, Y)$, taking P and Y as the inputs and outputs a binary classification score for membership status. Finally, the image-level attacker is denoted as **A**.

Training. Our method is built upon a per-patch attacker $\mathbf{A_P}$, as described in Algorithm 1. In line with previous work on membership inference [27,29], we construct a shadow model **S** that is to some extent similar to **V** and therefore is expected to exhibit similar behaviour and artifacts w.r.t. membership. In addition, **S** aims to capture semantic relations and dependencies between different classes in structured outputs and provide training data to the patch classifier with known membership labels. We prepare a dataset $\mathcal{D}^S$ with the same label space to $\mathcal{D}^V$, and then **S** is trained on $\mathcal{D}_{in}^V$. The exact assumptions of our knowledge on **V** that inform the construction of **S** are detailed in Sect. 2.2.

Construction of Per-patch Attacker. **S** provides training data for the per-patch attacker $\mathbf{A_P}$, as we have complete membership information of **S**. This allows us to train $\mathbf{A_P}$ by achieving the binary In/Out classification on the data pairs from $\mathcal{D}_{in}^S$ and $\mathcal{D}_{out}^S$. $\mathbf{A_P}$ can be any architecture taking image-like data as inputs, and we discuss different data representations to train it as follows.

Data Representation. We apply two representations of a data pair (X, Y) over segmentation models, as the inputs of a per-patch attacker. In other words, we train a classifier to compare the differences between P=**S**(X) and Y to determine the membership status of (X, Y) in training the classifiers.

1. *Concatenation.* We concatenate P and Y over the channel dimension, leading to a representation with size $H \times W \times 2C$.
2. *Structured loss map.* The structured loss map (SLM $\in \mathbf{R}^{H \times W \times 1}$) computes the cross-entropy loss values at all the locations (i, j), where $\mathrm{SLM}(i, j) = -\sum_{c=1}^{C} Y(i, j, c) \cdot \log(P(i, j, c))$. Previous work [26] shows the success of applying a threshold on the loss value of an image pair for image classification, and this method can be easily applied to semantic segmentation. Despite this, we show keeping structures of loss maps is still crucial to the success of attacking semantic segmentation.

Testing. Given a data pair (X, Y) to determine if it was used to train **V** (i.e., $(X, Y) \in \mathcal{D}_{in}^V$), we are able to crop a patch $(\hat{X}, \hat{Y})$ from the pair, and thus the inference result is $\mathbf{A_P}(\mathbf{V}(\hat{X}), \hat{Y})$ according to the representation used in the training. In order to further amplify the attack, we aggregate the information of the per-patch attack on an image-level, therefore, the final inference result is calculated by

$$\mathbf{A} = \frac{1}{N} \sum_{i=1}^{N} \mathbf{A_P}(\mathbf{V}(\hat{X}^i), \hat{Y}^i), \qquad (1)$$

where $(\hat{X}^i, \hat{Y}^i)$ is the i-th cropped patch from $(\hat{X}, \hat{Y})$.

Selection of Patches. As our method is based on scoring each patch, the selection of patches plays an important role in obtaining stronger attacks. Besides, it also helps us to understand which patches are particularly important for determine the membership status of an example. Therefore, we study the influence of different patch selection schemes with the following choices:

1. *Sliding windows.* We crop patches on a regular grid with a fixed step size.
2. *Random locations.* We sample patches uniformly across the image.
3. *Random locations with rejection.* We emphasize the importance of different patches for recognizing membership is not alike, therefore, this scheme aims to reject patches, which do not contribute to final results or even provide misleading information. We observe the patches with very strong confidences or very small loss should be omitted. For example, road area counts for most pixels of an image and are segmented very well, therefore, this scheme tends to select the bordering areas between a road and other classes, instead of the center of a road.

   As summarized in Algorithm 2, we construct image-level membership inference attacks according to per-patch attacks, allowing us to leverage distinct patches for successful attacks. Our pipeline is flexible to image sizes and aspect ratios if different image sizes exist in a dataset and even cross multiple datasets.

2.2   Attack Settings

In our method, we train a shadow segmentation model $\mathbf{S}$ and an attacker $\mathbf{A}$ for attacking a victim segmentation model $\mathbf{V}$. Our two attack settings differ in the knowledge on data distribution and model selection for training $\mathbf{V}$ and $\mathbf{S}$.

Data and Model Dependent Attacks: This attack assumes that the victims model can be queried at training time of an attacker. Besides, this setting allows to train a shadow model with the same architecture to the victim. Specifically, $\mathbf{S}$ and $\mathbf{V}$ have the same learning protocol and post-processing techniques during inference. Further, this attack assumes the data distributions of $\mathcal{D}^V$ and $\mathcal{D}^S$ are also identical, which comes from the same database. Last, query with a victim model is allowed to split $\mathcal{D}^S$ into $\mathcal{D}_{in}^S$ and $\mathcal{D}_{out}^S$, as listed in the 1-st line of Algorithm 1, that we use the examples with stronger confidences to build $\mathcal{D}_{in}^S$.

Data and Model Independent Attacks: For this attack, we only know the victim model's functionality and a defined label space. There is no query process for constructing training set for $\mathbf{S}$, instead, $\mathbf{S}$ is able to be trained with a dataset of the different distribution, which leads to a cheaper and more practical attack.

Furthermore, the model configuration and training protocol of the victim are unknown. The goal of the shadow model is to capture the membership status for each example, and provide training data for attack model **A**. Particularly, model and data distribution are completely different to victims, even there is no query process, which might be detected on the server. Therefore, we highlight the severity of information leakage in this simplified attack.

2.3 Evaluation Methodology

We evaluate the performance of membership inference attacks with **precision-recall** curves and receiver operating characteristic (**ROC**) curves. We regard the images used during training as positive examples, and negatives if not. Therefore, given a testing set with M image pairs used to train a model and N pairs not used, random guess with probability 0.5/0.5 for both classes is able to achieve precision $\frac{M}{M+N}$ and recall 0.5. We set different thresholds in a classifier and compare its precision-recall curve to the random guess performance, to observe if attacks are successful. Similarly, we draw the random guess behavior in a ROC curve, which is the diagonal of a plot. Furthermore, to compare different attacks quantitatively, we apply maximum **F-score** ($\frac{2 \cdot \text{precison} \cdot \text{recall}}{\text{precison} + \text{recall}}$) in precision-recall curves and **AUC-score** in ROC curves to evaluate attack performance. Last, our method is based on per-patch attacks, therefore, we employ the same metrics for per-patch evaluation, to help us understand and compare different attacks, as well as defense methods in Sect. 3, exhaustively.

2.4 Evaluation Results

Data and Architectures. We conduct the experiments on street scene semantic segmentation between various datasets, including *Cityscapes* [6], *BDD100K* [35] and *Mapillary Vistas* [21], which are captured in different countries under diverse weathers and image qualities, providing multiple domains. Besides, we apply PSPNet [37], UperNet [33], Deeplab-v3+ [5] and DPC [4] as our segmentation models. For per-patch attacker $\mathbf{A_P}$, we train a ResNet-50 [12] from scratch, allowing us to visualize the regions contributing to the recognition of membership for an example by class activation mapping [38]. In detail, the size of inputs for ResNet-50 is 90×90 in spatial, corresponding to 713×713 image patches.

Comparison Methods. To demonstrate the effectiveness of specific considerations for segmentation models, we compare our pipeline to previous attackers for classification models [26,27]. For [27], we adapt their shadow model based attacker, by regarding each location as a classification problem. We also test their learning-free attacker by only considering the mean of confidence scores of a prediction. Besides, we compare the proposed method with [26], which employs a threshold on the loss value.

Setup for Data and Model Dependent Attacks. For dependent attacks, we conduct experiments with *Cityscapes* and PSPNet (a.k.a. PSP→PSP). We split

Cityscapes into four parts, i.e., $\mathcal{D}_{in}^V$, $\mathcal{D}_{out}^V$, $\mathcal{D}_{in}^S$ and $\mathcal{D}_{out}^V$, where the sizes of those sets are as follows: $|\mathcal{D}_{in}^V| = 1488$, $|\mathcal{D}_{out}^V| = 912$, $|\mathcal{D}_{in}^S| = 555$ and $|\mathcal{D}_{out}^S| = 520$. We train a victim model from ImageNet [7] pretrained models and lead to 59.88 mean IoU (mIoU) for segmentation. For evaluation of per-patch attacks, we sample 29760 patches from $\mathcal{D}_{in}^V$ and 30096 patches from $\mathcal{D}_{out}^V$. Therefore, this setting leads to 62% and 49.7% precision for image-level and per-patch attacks in random guess. The resulting F-scores for image-level and per-patch attacks are 55.36% and 49.85% respectively, which are drawn in Fig. 1.

Table 1. Data and model descriptions of victim and shadow models for independent attacks.

Dataset	Model	Backbone	In/Out
Cityscapes (Victim)	PSPNet [37] UperNet [33]	ResNet-101 [12]	2975/500
BDD100K (Shadow) *Mapillary* (Shadow)	Deeplab-v3+ [5] DPC [4]	Xception-71 [32]	4k/(3k + 1k) 10k/(8k + 2k)

Table 2. Comparison of different attackers (in %). We compare our attackers to previous methods, including the learning-based attacker [27]* and learning-free attackers by applying a threshold on a confidence score [27]⁺ or a loss value [26]. "→" means the attacks with a shadow model of the left, and the victims are the right, which can be PSPNet [37], UperNet [33], DPC [4] or Deeplab-v3+ [5].

Methods	Dependent		Independent								
	[37]→[37]		[5]→[37]		[5]→[33]		[4]→[37]		[4]→[33]	
	F	AUC	F	AUC	F	AUC	F	AUC	F	AUC
Adapted [27]*	77.2	67.2	92.4	63.5	92.3	62.6	–	–	–	–
Adapted [27]⁺	77.4	62.0	92.3	63.4	92.3	59.2	92.3	63.4	92.3	59.2
Adapted [26]	82.2	74.9	94.4	81.4	93.0	72.4	94.4	81.4	93.0	72.4
Ours (C+GT, Full)	80.6	81.2	94.5	85.0	92.8	71.8	93.2	73.5	92.6	68.8
Ours (Loss, Full)	84.2	82.6	**95.7**	89.1	93.2	76.3	93.1	73.5	92.4	68.3
Ours (C+GT, Random)	83.4	82.7	95.0	86.1	95.4	88.5	92.9	74.9	94.4	**85.5**
Ours (Loss, Random)	84.8	84.6	**95.7**	90.8	95.8	94.3	94.0	77.7	93.3	79.4
Ours (C+GT, Rejection)	83.3	83.0	94.9	86.3	95.3	91.2	93.5	76.3	94.4	86.1
Ours (Loss, Rejection)	86.7	87.1	95.9	91.1	96.2	94.9	94.1	77.8	93.5	82.0

Setup for Data and Model Independent Attacks. For independent attacks, we employ different segmentation models for shadow models and victims, as summarized in Table 1. Particularly, *BDD100K* has completely compatible label space to *Cityscapes* of 19 classes, but *Mapillary Vistas* has 65 labels. We merge the some classes from *Mapillary Vistas* into *Cityscapes*, and ignore the others. For victim models, we train a PSPNet and an UperNet using the official split of

Cityscapes, leading to 79.7 and 76.6 mIoU for segmentation. For shadow models, we apply our splits with balanced In/Out distribution to train a binary classifier. In the end, the F-score of random guess for image-level independent attack is 63.13%. Comparing it to the numbers in Table 2, we observe all the attackers obtain much higher F-score than 63.13%, which shows the severe information leakage of semantic segmentation models.

Results. Results of the different versions of our model as well as comparision to previous work in presented in Table 2. While previous work on membership inference targets classification models [26,27], we facilitate a comparison to these approaches by extending them to the segmentation scenario. [27] proposes a learning-based attacker and a learning-free attacker. We train their learning-based attacker with 1×1 vector inputs, and test on all pixel locations. Final image-level attacks are obtained by averaging the binary classification scores of all locations. Similar to our method, we test different settings, and it fails to achieve attacks with the shadow model DPC [4] in Table 1. Besides, we test their learning-free attacker by averaging the confidence scores of all locations. Equally, we facilitate a comparison to [26] where we use the loss map for the segmentation output. For our methods, we report the numbers for last two patch selection strategies with sampling 10 patches. Besides, we also perform attacks with full image inputs using our binary classifiers, which have a global average pooling in the end and are able to handle different sizes of inputs. We emphasize that the ratio of In/Out testing examples are different for dependent and independent attacks, therefore, the numbers between them cannot be compared. We conclude

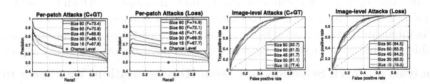

Fig. 1. Evaluation of the **importance of spatial structures** for PSP→PSP, starting from our final model (Size 90).

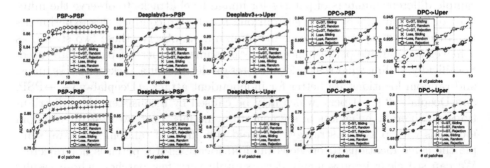

Fig. 2. Image-level comparison results w.r.t. **patch selection** and **data representations**, under varying patch numbers.

that recent models for semantic segmentation are susceptible to membership inference attacks with AUC scores of the attacker up to 87.1% in the dependent and 94.9% in the independent setting. Overall, we observe that our loss-based method with rejection scheme performs best in most settings and measures.

Importance of Spatial Structures. Key to strong attacks is exploiting the structural information of an output from a segmentation model. Hence, we conduct attacks with gradually reduced structural inputs in our dependent attacks in order to analyze the importance of this structural information for our goal. Our final model takes 90×90 blocks as inputs for per-patch attackers. Therefore, we crop sub-blocks from our final model with input sizes of 60, 45, 30, 15 for providing different level of structures. We compare the precision-recall curves for per-patch attacks and ROC curves for image-level attacks in Fig. 1. We note that all the feature vectors in the blocks of different sizes have the same scale of receptive fields. We apply the same architecture of per-patch attacker for sizes 90-30, but modify the ResNet-50 with fewer pooling operations for size 15, because its spatial size is too small. First, we compare the per-patch attack performance, and are able to observe that attacks become harder with decreasing patch sizes, where smaller patches provide less structures. Second, we compare image-level attacks for them, where random selection strategy is applied to integrate all patches. We sample 5, 20, 20, 30, 30 random patches for size 90, 60, 45, 30, and 15 to integrate image-level results. Consequently, size 90 achieves the best performance, even though other attackers obtain very close image-level results. Last, we highlight that our concatenation-based attacker degenerates to previous work [27] with 1×1 vector inputs. We observe that 1×1 inputs keep this decreasing trend and achieve worse results than size 15, which can be found in Table 2. From this results, we conclude that structures are of great importance in membership inference attacks for semantic segmentation, so that an attacker is able to mine some In/Out confidence or loss patterns over an array input.

Analysis of Patch Selection and Data Representation. We test our three sampling strategies and two representations as depicted in Sect. 2.1. Figure 2 plots the image-level comparison results. For sliding windows, we sample at least 6 patches to guarantee an entire image can be covered. For random locations, we sample different numbers of patches for image-level attacks to observe the influence of patch numbers, starting from one patch. We conduct this experiments for 3 times and report the mean. In summary, we observe these two strategies achieves comparable performance when the same numbers of patches are used. Specifically, sliding windows perform better on dependent attacks with loss maps, and random locations are better for independent attacks (Deeplab-v3+ $\rightarrow$PSP, and Deeplab-v3+ $\rightarrow$Uper), which may be caused by inconsistent data distributions or different behaviors of segmentation models. Last, we test our random locations with rejection strategy. To avoid the affect of random seeds, we sample the same locations to previous random locations if a patch is not rejected. We can see clear improvements if we sample very few patches, whose results are sensitive to sampled locations. In street scenes, road has a large portion of pixels, therefore, it tend to sample a road patch, which has the highest accuracy over all the classes and less discrimination for In/Out classification. After ignoring those patches, performance is improved because the rejection helps us avoid those less informative patches. To conclude, not all the regions contribute to successful attacks for segmentation, that we need a regime to determine membership status of an image, instead of processing the whole like previous work for classification.

Comparing our patch-based attacks to the full image attacks, we realize using full images as inputs makes performance significantly decreased, even though the same classifier is applied. The classification for full images may be affected by misleading areas. Hence, partitioning an image into many patches helps focus on local patterns and makes a better decision. Besides, we observe that our rejection scheme achieves better performance than random scheme, which further supports our argument on the difference between segmentation and classification. In addition, our concatenation-based attacker outperforms [27], which demonstrates the importance of spatial structures, similar to Fig. 1. From our results, [26] is able to obtain acceptable performance but worse than our structured loss map-based attackers, which hold the structural information. Finally, our novel structured loss maps achieve better results than concatenation and other methods [26,27] in most cases.

3   Defenses

To mitigate the membership leakage and protect the authority of a model, we study several defenses for semantic segmentation models, while maintaining their utility with little performance degeneration. When a model is deployed, a service provider has all rights to access the model and data. Our work shows for the first time a feasible solution for protecting very large semantic segmentation model. As a consequence, we manipulate the model in training or testing stages by reducing the distribution gaps between training data and others w.r.t. confidence scores of predictions or loss values, including Argmax, Gauss, Dropout and DPSGD. The first two methods can be applied in any segmentation models and last two can be applied in deep neural networks.

Settings. We analyze the performance of image-level attacks according to random locations in this section, which are easily compared to the results without defenses in Table 2. Because Gaussian noises, or dropout will change output distributions, rejection scheme may sample different patches, we only test the random location schemes and sample patches at the same locations for different defenses, and keep consistent to previous attacks.

For dependent attacks, our shadow and victim models have the same post-processing and learning protocol, as claimed in Sect. 2.2. In other words, we employ the same defenses and strength factors for them in this setting. For independent attacks, we report the settings of Deeplab-v3+ →PSP and Deeplab-v3+ →Uper, which are the most successful. We only employ defenses on victim models as protections for released black-box semantic segmentation models.

Evaluation Methodology. Due to the different ratios for In/Out examples in various settings, we do not report their F-scores in this section. Instead, we only employ AUC-score to compare different defense methods, expecting to reduce the original attacks' AUC-scores to 0.5, that random guess in all the settings hold this number. Furthermore, an ideal defense is supposed to make attacks hard and preserve segmentation utility at the same time. Therefore, we apply mIoU [17] to evaluate the segmentation performance and jointly compare different defense methods w.r.t. capability of membership protection and utility of segmentation.

3.1   Methods and Results

Argmax. It only returns predicted labels instead of posteriors for an image. We use one-hot vectors to complete attacks for our methods and others [26,27]. Obviously, previous learning-free attacker [27] based on confidence scores fails to recognize membership states, because every example has confidence 1. In Fig. 3, we show the comparison results for all the other methods. Because argmax is very easy to be noticed, we train binary classifiers for independent attacks with argmax operation as well. In general, argmax only reduces membership leakage in segmentation models a little for all the attackers. A model already leaks information when it only returns predicted labels. To conclude, we highlight the difference to protecting classification, that argmax cannot successfully protect the membership privacy for segmentation.

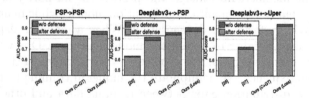

Fig. 3. Performance comparison for Argmax defense.

Gauss. To hide overfitting artifacts or patterns, we add Gaussian noises on the posteriors with different variances, varying from 0.01 to 0.1 with step 0.01 for independent attacks. To further test the defense for dependent attacks, we add very strong noises to variance 0.4. After noising, we set the values into 0 in case they are smaller than 0, and then normalize each location individually. Segmentation performance is decreased with stronger noises, therefore, we show the joint privacy-segmentation plots in Fig. 4(a) to observe the defense behaviors as well as the maintained utility of the segmentation method. First, we observe Gauss protects PSPNet and UperNet in independent attacks successfully, which reduces AUC-scores from 0.9 to less than 0.6, while only losing 0.2 mIoU. Second, we observe our loss-based attackers are more sensitive to Gaussian noises. Despite stronger attacks of structured loss maps, they are easier to protect with Gaussian noises. Finally, we realize this defense is hard to mitigate leakage for dependent attacks. Even though we employ very strong noises for this, losing mIoU from 59.88 to 23.17, it still has more than 0.75 AUC-scores for both attacks. To conclude, Gauss is hard to protect a model when the noises of the same distribution are added to victim and shadow models, and binary classifiers can pick useful information from noisy inputs.

Dropout. It is used to avoid overfitting in training a deep neural networks, that we applied in training our victim model with dropout ratio 0.1. However, it does not hide membership from our studies in Sect. 2. Therefore, we enable dropout operation during testing to blur a prediction. We realize a network still produces decent results when we use a different dropout ratio. Hence, we apply dropout ratio 0.1, 0.5 and 0.9 to obfuscate a prediction at different degrees. We show the joint plots in Fig. 4(b). From our study, we observe enabling dropout during test is able to slightly mitigate membership leakage, but segmentation performance decreases a lot when a large ratio is applied.

DPSGD. Differential Privacy SGD (DPSGD) [1] adds Gaussian noises on the clipped gradients for individual examples of a training batch, in a way that the learnt parameters and hence all derived results such as predictions are differentially private. We apply DPSGD in our study to protect a model. Before training, we collect gradient statistics over entire training data for different layers of a network, and set individual clipping factors for all the layers. Next, we train PSPNet with Gaussian variances 10^{-3}, 4×10^{-3} for dependent settings, and variances 10^{-3}, 4×10^{-3}, 8×10^{-3} for independent settings. For UperNet, we train with 10^{-6}, 10^{-3}, 3×10^{-3}, 6×10^{-3}. Theoretically, the Gaussian noises used in our model is not enough to guarantee a tight differential privacy bound [8]. However, there is a gap between theoretical garuantees and emperical defenses. Prior work has shown practical defenses from small gaussian noise and hence loose bounds [14]. In our work, we demonstrate this in semantic segmentation models and show the utility-privacy plots in Fig. 4(c). We observe that DPSGD successfully protect memberships in all the settings, in particular, it will not hamper the utility of segmentation models which only reduces 1.12, 1.36 and 0.75 mIoU when noises with 1e−6 are applied in three rows. Therefore, we recommend DPSGD to train a segmentation model for protecting membership privacy in practise.

Summary of Defenses. In spite of the success of membership inference under various settings, we point out feasible solutions which can significantly reduce the risk of information leakage. (1) Adding Gaussian noises helps prevent leakage in independent settings from unknown attackers. Tradeoff between model degeneration and information leakage is able to be considered to choose a suitable noise level. Hence, we recommend this method as a basic protection without further costs to prevent potential independent attacks, which are very cheap to implement. (2) For neural networks, we suggest applying DPSGD to train a model, which mitigates the leakage in all the settings with limited model degeneration, even though it adds noises on the gradients during training and hence requires increased training time.

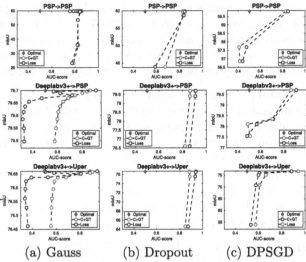

(a) Gauss   (b) Dropout   (c) DPSGD

Fig. 4. Joint plots for different defenses. x-axis is the AUC-score for membership protection and y-axis is the mIoU for segmentation utility. optimal defenses achieve 0.5 AUC-score while preserving segmentation utility, as drawn with the red diamond.

3.2 Interpretability

One of difference between attacking segmentation and classification is on the input form of the binary classifier, where the input of segmentation can be regarded as an image. Hence, our method can provide interpretations for different examples, indicating important regions for recognizing membership status. Besides, interpretations also help us to understand and compare different defenses. We apply class activation maps (CAMs) [38] to highlight the areas that help to detect examples/patches from training set in Fig. 5. Besides, we also compare the activation areas before and after defenses with structured loss maps.

First, we observe our attacker is able to mine some regions with specific objects or intersections between two classes, even our attacker has no interaction with a victim. Second, we compare the attacker's different behaviors for those defenses. We can see argmax can simply change the CAM to different intensities, but still hold the major layout of the original CAM. For Gaussian noises, we employ variance 0.1 here, and can apparently observe noises on the structural

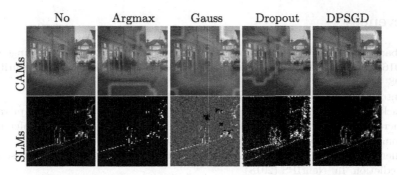

Fig. 5. Class activation maps (CAMs) and structured loss maps (SLMs) for independent attack Deeplab-v3+ →Uper.

loss map for all the pixel locations, therefore, it makes all the examples have a similar CAM. For dropout, it will change structured loss maps in many places and then change the CAM. In particular, it changes the locations with strong loss values more than others. For DPSGD, we can see it has very similar loss maps to the original model. The only differences are on some regions hard to segment. Even DPSGD changes the loss maps a little, the final CAMs are able to change a lot for some examples, therefore, it helps defend stealing memberships while preserving segmentation performance very well.

4   Conclusion

We have provided the first membership inference attacks and defenses for semantic segmentation models by extending previous membership attacker for classification and proposing a new specific representation (i.e., structured loss maps). Our study is conducted under two different settings with various model/data assumptions. We show that spatial structures are important to achieve successful attacks in segmentation, and our structured loss maps achieve the best results among all. Besides, we study defense methods to reduce membership leakage and provide safe segmentation. As a result, we suggest to add Gaussian noises on the posteriors in inference, or apply differential privacy in training. We hope that our work contributes to the awareness of novel threats that modern deep learning models pose – such as leakage of information on the training data. Our contributions shows that such threats can be mitigated with little impact on the utility of the overall model.

Acknowledgement. The authors thank Apratim Bhattacharyya, Hui-Po Wang and Zhongjie Yu for their valuable feedback to improve the quality of this manuscript.

References

1. Abadi, M., et al.: Deep learning with differential privacy. In: Proceedings of the 2016 ACM SIGSAC Conference on Computer and Communications Security, pp. 308–318 (2016)
2. Antol, S., et al.: VQA: visual question answering. In: ICCV (2015)
3. Backes, M., Berrang, P., Humbert, M., Manoharan, P.: Membership privacy in microrna-based studies. In: Proceedings of the 2016 ACM SIGSAC Conference on Computer and Communications Security (2016)
4. Chen, L.C., et al.: Searching for efficient multi-scale architectures for dense image prediction. In: NeurIPS (2018)
5. Chen, L.-C., Zhu, Y., Papandreou, G., Schroff, F., Adam, H.: Encoder-decoder with Atrous separable convolution for semantic image segmentation. In: Ferrari, V., Hebert, M., Sminchisescu, C., Weiss, Y. (eds.) ECCV 2018. LNCS, vol. 11211, pp. 833–851. Springer, Cham (2018). https://doi.org/10.1007/978-3-030-01234-2_49
6. Cordts, M., et al.: The cityscapes dataset for semantic urban scene understanding. In: CVPR (2016)
7. Deng, J., Dong, W., Socher, R., Li, L.J., Li, K., Fei-Fei, L.: ImageNet: a large-scale hierarchical image database. In: CVPR (2009)
8. Dwork, C.: Differential privacy: a survey of results. In: Agrawal, M., Du, D., Duan, Z., Li, A. (eds.) TAMC 2008. LNCS, vol. 4978, pp. 1–19. Springer, Heidelberg (2008). https://doi.org/10.1007/978-3-540-79228-4_1
9. van Tilborg, H.C.A., Jajodia, S. (eds.): Encyclopedia of Cryptography and Security. Springer, Boston (2011). https://doi.org/10.1007/978-1-4419-5906-5
10. Fischer, V., Kumar, M.C., Metzen, J.H., Brox, T.: Adversarial examples for semantic image segmentation. arXiv preprint arXiv:1703.01101 (2017)
11. Geiger, A., Lenz, P., Urtasun, R.: Are we ready for autonomous driving? The KITTI vision benchmark suite. In: CVPR (2012)
12. He, K., Zhang, X., Ren, S., Sun, J.: Deep residual learning for image recognition. In: CVPR (2016)
13. He, Y., Chiu, W.C., Keuper, M., Fritz, M.: STD2P: RGBD semantic segmentation using spatio-temporal data-driven pooling. In: CVPR (2017)
14. Jayaraman, B., Evans, D.: Evaluating differentially private machine learning in practice. In: 28th {USENIX} Security Symposium ({USENIX} Security 2019), pp. 1895–1912 (2019)
15. Kay, W., et al.: The kinetics human action video dataset. arXiv preprint arXiv:1705.06950 (2017)
16. Lin, G., Milan, A., Shen, C., Reid, I.: RefineNet: multi-path refinement networks for high-resolution semantic segmentation. In: CVPR (2017)
17. Long, J., Shelhamer, E., Darrell, T.: Fully convolutional networks for semantic segmentation. In: CVPR (2015)
18. Moosavi-Dezfooli, S.M., Fawzi, A., Fawzi, O., Frossard, P.: Universal adversarial perturbations. In: CVPR (2017)
19. Moosavi-Dezfooli, S.M., Fawzi, A., Frossard, P.: DeepFool: a simple and accurate method to fool deep neural networks. In: CVPR (2016)
20. Nasr, M., Shokri, R., Houmansadr, A.: Machine learning with membership privacy using adversarial regularization. In: Proceedings of the 2018 ACM SIGSAC Conference on Computer and Communications Security (2018)
21. Neuhold, G., Ollmann, T., Rota Bulo, S., Kontschieder, P.: The mapillary vistas dataset for semantic understanding of street scenes. In: ICCV (2017)

22. Oh, S.J., Schiele, B., Fritz, M.: Towards reverse-engineering black-box neural networks. In: Samek, W., Montavon, G., Vedaldi, A., Hansen, L.K., Müller, K.-R. (eds.) Explainable AI: Interpreting, Explaining and Visualizing Deep Learning. LNCS (LNAI), vol. 11700, pp. 121–144. Springer, Cham (2019). https://doi.org/10.1007/978-3-030-28954-6_7
23. Oh, S.J., Fritz, M., Schiele, B.: Adversarial image perturbation for privacy protection a game theory perspective. In: ICCV (2017)
24. Orekondy, T., Schiele, B., Fritz, M.: Knockoff Nets: stealing functionality of black-box models. In: CVPR (2019)
25. Pyrgelis, A., Troncoso, C., De Cristofaro, E.: Knock knock, who's there? membership inference on aggregate location data. NDSS (2018)
26. Sablayrolles, A., Douze, M., Schmid, C., Ollivier, Y., Jegou, H.: White-box vs black-box: Bayes optimal strategies for membership inference. In: ICML (2019)
27. Salem, A., Zhang, Y., Humbert, M., Fritz, M., Backes, M.: Ml-leaks: Model and data independent membership inference attacks and defenses on machine learning models. In: NDSS (2019)
28. Shokri, R., Shmatikov, V.: Privacy-preserving deep learning. In: Proceedings of the 22nd ACM SIGSAC Conference on Computer and Communications Security, pp. 1310–1321 (2015)
29. Shokri, R., Stronati, M., Song, C., Shmatikov, V.: Membership inference attacks against machine learning models. In: IEEE Symposium on Security and Privacy (SP) (2017)
30. Stutz, D., Hein, M., Schiele, B.: Disentangling adversarial robustness and generalization. In: CVPR (2019)
31. Sun, C., Shrivastava, A., Singh, S., Gupta, A.: Revisiting unreasonable effectiveness of data in deep learning era. In: ICCV (2017)
32. Szegedy, C., Vanhoucke, V., Ioffe, S., Shlens, J., Wojna, Z.: Rethinking the inception architecture for computer vision. In: CVPR (2016)
33. Xiao, T., Liu, Y., Zhou, B., Jiang, Y., Sun, J.: Unified perceptual parsing for scene understanding. In: ECCV (2018)
34. Xie, C., Wang, J., Zhang, Z., Zhou, Y., Xie, L., Yuille, A.: Adversarial examples for semantic segmentation and object detection. In: ICCV (2017)
35. Yu, F., et al.: BDD100K: a diverse driving video database with scalable annotation tooling. arXiv preprint arXiv:1805.04687 (2018)
36. Zhang, T.: Privacy-preserving machine learning through data obfuscation. arXiv preprint arXiv:1807.01860 (2018)
37. Zhao, H., Shi, J., Qi, X., Wang, X., Jia, J.: Pyramid scene parsing network. In: CVPR (2017)
38. Zhou, B., Khosla, A., Lapedriza, A., Oliva, A., Torralba, A.: Learning deep features for discriminative localization. In: CVPR (2016)
39. Zhou, B., Lapedriza, A., Khosla, A., Oliva, A., Torralba, A.: Places: a 10 million image database for scene recognition. IEEE T-PAMI **40**, 1452–1464 (2017)

From Image to Stability: Learning Dynamics from Human Pose

Jesse Scott[1]([⊠]), Bharadwaj Ravichandran[1], Christopher Funk[2],
Robert T. Collins[1], and Yanxi Liu[1]

[1] School of EECS, Pennsylvania State University, State College, USA
{jus121,bzr49}@psu.edu,
{rcollins,yanxi}@cse.psu.edu
[2] Kitware, Inc., New York, USA
christopher.funk@kitware.com

Abstract. We propose and validate two end-to-end deep learning architectures to learn foot pressure distribution maps (dynamics) from 2D or 3D human pose (kinematics). The networks are trained using 1.36 million synchronized pose+pressure data pairs from 10 subjects performing multiple takes of a 5-min long choreographed Taiji sequence. Using leave-one-subject-out cross validation, we demonstrate reliable and repeatable foot pressure prediction, setting the first baseline for solving a non-obvious pose to pressure cross-modality mapping problem in computer vision. Furthermore, we compute and quantitatively validate Center of Pressure (CoP) and Base of Support (BoS), two key components for stability analysis, from the predicted foot pressure distributions.

Keywords: Stability · Center of Pressure · Base of support · Foot pressure estimation · 3D human pose · Deep regression models

1 Introduction

Current computer vision research on human pose focuses on extracting skeletal kinematics from images or video [9,12,15,16,20,24,47,68]. A more effective analysis of human movement should take into account the dynamics of the human body [59]. Understanding body dynamics such as foot pressure is essential to study the effects of perturbations caused by external forces and torques on the human postural system, which change body equilibrium in static posture and during locomotion [73]. Computing stability from visual data could unlock a wide range of applications in the fields of healthcare, kinesiology, and robotics.

J. Scott and B. Ravichandran—Co-First Authors.

Electronic supplementary material The online version of this chapter (https:// doi.org/10.1007/978-3-030-58592-1_32) contains supplementary material, which is available to authorized users.

© Springer Nature Switzerland AG 2020
A. Vedaldi et al. (Eds.): ECCV 2020, LNCS 12368, pp. 536–554, 2020.
https://doi.org/10.1007/978-3-030-58592-1_32

For understanding the relation between a body pose and the corresponding foot pressure of a human subject (Fig. 1), we explore two deep convolutional residual architectures, PressNet and PressNet-Simple (Fig. 5), and train them on a dataset containing 1,363,400 data pairs of body pose with corresponding foot pressure measurements. Body pose is input to a network as either 2D or 3D human joint locations extracted from the Openpose [12] Body25 model (3D joints are derived by 2-view stereo triangulation of 2D joints). Each network predicts a foot pressure heatmap as output, providing an estimated distribution of pressure applied at different foot locations (Fig. 1, stage 6).

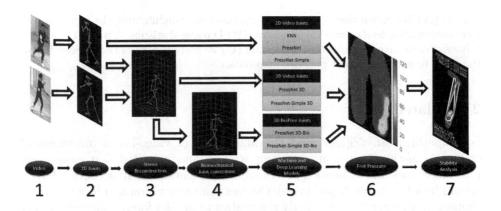

Fig. 1. Our PressNet and PressNet-Simple networks learn to predict a foot pressure heatmap from 2D or 3D human body joints. We also compute Center of Pressure (CoP) and Base of Support (BoS), two key components for stability estimation, from predicted foot pressure distributions (rightmost).

The main contributions of this work include **1) Novelty**: Our PressNet and PressNet-Simple networks are the first vision-based networks to regress foot pressure (dynamics) from 3D or 2D body pose (kinematics). Furthermore, we introduce a 3D Pose Estimation Network (BioPose) that enhances body joint positions biomechanically. **2) Dataset**: We have collected the largest synchronized Video, motion capture (MoCap), and foot pressure dataset of a 5-minute long, complex human movement sequence (Fig. 2). **3) Application**: For validation, Center of Pressure (CoP) and Base of Support (BoS), two key components in the analysis of stability, are bench-marked for potential future applications.

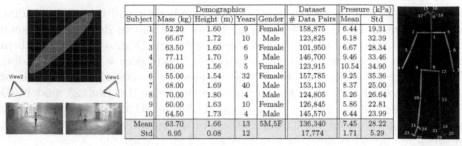

	Demographics				Dataset	Pressure (kPa)	
Subject	Mass (kg)	Height (m)	Years	Gender	# Data Pairs	Mean	Std
1	52.20	1.60	9	Female	158,875	6.44	19.31
2	66.67	1.72	10	Male	123,825	6.18	32.39
3	63.50	1.60	6	Female	101,950	6.67	28.34
4	77.11	1.70	9	Male	146,700	9.46	33.46
5	60.00	1.56	5	Female	123,915	10.54	34.90
6	55.00	1.54	32	Female	157,785	9.25	35.36
7	68.00	1.69	40	Male	153,130	8.37	25.00
8	70.00	1.80	4	Male	124,805	5.26	26.64
9	60.00	1.63	10	Female	126,845	5.86	22.81
10	64.50	1.73	4	Male	145,570	6.44	23.99
Mean	63.70	1.66	13	5M,5F	136,340	7.45	28.22
Std	6.95	0.08	12		17,774	1.71	5.29

(A) Two Views   (B) Subject Demographics   (C) Joints

Fig. 2. (A) Top-down view of the motion capture space highlighting the region of performance relative to the two video cameras. **(B)** Dataset statistics. A total of 1,363,400 frames of data have been collected, providing **(C)** 25 body joints from each video frame, time-synchronized with foot pressure map data.

2   Related Work

Seethapathi et al. [59] reviewed the limitations of video-based measurement of human motion for use in movement science, and indicated that more accurate kinematics and estimation of dynamics information, such as contact forces, should be a key research goal in order to use computer vision as a tool in biomechanics. In this paper, we use body kinematics to predict foot pressure dynamics and to develop a quantitative method to analyze human stability from video. Earlier work in computer vision and graphics has incorporated dynamics equations into models of human motion and person tracking, and has even estimated contact forces from video and kinematics [7,8,39,42,71], but their estimates of contact dynamics tend to be simple force vectors rather than full foot pressure maps, as in our work.

Studying human stability during standing and locomotion [3,19,38] is typically addressed by direct measurement of foot pressure using force plates or insole foot pressure sensors. Previous studies have shown that foot pressure patterns can be used to discriminate between walking subjects [50,70]. Instability of the CoP of a standing person is an indication of postural sway and thus a measure of a person's ability to maintain balance [27,28,37,49]. Grimm et al. [23] predicts the pose of a patient using foot pressure mats. The authors of [45] and [55] evaluate foot pressure patterns of 1,000 subjects ages 3 to 101 and determine there is a significant difference in the contact area by gender but not in magnitude of foot pressure for adults. As a result, the force applied by females is lower but is accounted for by female mass also being significantly lower. In [11], a depth regularization model is trained to estimate dynamics of hand movement from 2D joints obtained from RGB video cameras. Stability analysis of 3D printed models is presented in [4,53,54]. Although these are some insightful ways to analyze stability, there has been no vision-based or deep learning approach to tackle this problem.

Table 1. Comparison of our dataset with other available human pose datasets.

Name	# Data samples	# Subjects	Scenario	MoCap joints	Image joints	Foot pressure	Humans per image
Human3.6M [32,33]	3,600,000	11	Indoor	✓	✓	–	Single
HumanEva [61]	80,000	4	Indoor	✓	✓	–	Single
MPII Human Pose [2]	25,000	N/A	Indoor/ Outdoor	–	✓	–	Single/ Multiple
MS-COCO [40]	200,000	N/A	Indoor/ Outdoor	–	✓	–	Multiple
PoseTrack [34]	66,000	N/A	Indoor/ Outdoor	–	✓	–	Multiple
Taiji stability (Ours)	1,363,400	10	Indoor	✓	✓	✓	Single

Estimation of 2D body pose in images is a well-studied problem in computer vision, with state-of-the-art methods being based on deep networks [9,12,15,16, 20,22,24,30,47,66,68]. We adopt one of the more popular approaches, CMU's OpenPose [12], to compute the 2D pose input to our networks. Success in 2D human pose estimation also has encouraged researchers to detect 3D skeletons by extending existing 2D human pose detectors [6,14,44,46,48,62,75] or by directly using image features [1,51,57,64,74]. Martinez et al. [44] showed that given high-quality 2D joint information, the process of lifting 2D pose to 3D pose can be done efficiently using a relatively simple deep feed-forward network. All these papers concentrate on pose estimation by learning to infer joint angles or joint locations, which can be broadly classified as learning basic kinematics of a body skeleton. These methods do not predict external torques/forces exerted by the environment, balance, or physical interaction of the body within the scene.

Table 1 shows a summary of available pose datasets. Human3.6M dataset [32,33] is one of the top public datasets that is used for pose estimation tasks. It has 3.6 million frames of synchronized 3D MoCap-based human pose and video data. This data was collected from 6 male and 5 female subjects for 17 different scenarios. HumanEva [61] is a similar dataset with MoCap and video data, but it is smaller in size than the Human3.6M dataset. Another dataset that is widely popular is the MPII Human Pose dataset [2]. This dataset consists of 25,000 images containing over 40,000 individuals with Ground Truth (GT) human body joints, covering over 410 human activities. The Posetrack [34] dataset consists of about 1,400 video sequences with over 66,000 annotated video frames and 276,000 body pose annotations. The MS-COCO [40] dataset has more than 200,000 images and 250,000 individuals with labeled keypoints of human pose.

A major motivation for computing foot pressure maps from pose is to estimate body stability from video in the wild, accurately and economically, rather than in a biomechanics lab. Fundamental elements used in stability analysis (Fig. 3C) include Center of Mass (CoM), Base of Support (BoS), and Center of Pressure (CoP). The relative locations of CoP, BoS, and CoM have been identified as a determinant of stability in a variety of tasks [27,28,49].

3   Our Approach and Motivation

Mapping from human pose to foot pressure (Fig. 1) is an ill-posed problem. On the one hand, similar poses of different subjects can yield different foot pressure maps (Fig. 3B) due to differences in movement, mass, height, gender, and foot shape. On the other hand, PCA analysis (Fig. 3A) suggests the top principal components capture statistically similar "modes" of variation across subjects. Thus, we formulate our problem as learning foot pressure distribution conditioned on human pose rather than trying to directly regress precise foot pressure magnitude. For simplicity, we assume the conditional distribution of pressure given pose is Gaussian, with a mean that can be learned through deep learning regression using MSE and KL Divergence loss. Our networks are trained to map from pose, encoded as 25 joint locations (2D or 3D), to the mean of a corresponding foot pressure map intensity distribution (Fig. 5).

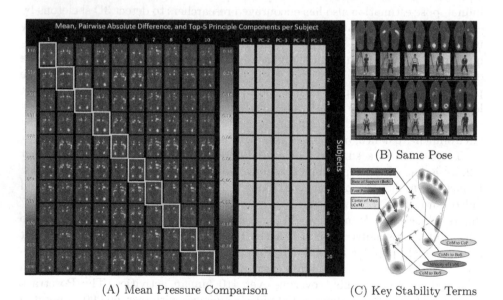

(A) Mean Pressure Comparison   (C) Key Stability Terms

Fig. 3. (A) Left: Pairwise absolute differences between mean foot pressure across all subjects with inter-subject comparison of differences in pressure magnitude and spatial distribution. Mean pressure is provided on the diagonal (yellow boxes). Right: Top-5 Principal Components of foot pressure data per subject. **(B)** Same "starting pose" yields different foot pressure for different subjects. **(C)** Basic concepts in stability analysis, including Center of Pressure, Center of Mass, and Base of Support. (Color figure online)

Fig. 4. Example Taiji poses similar to ordinary movements: 1 - standing with hand behind, 2 - standing with two arms down, 3 - step to left, 4 - bump (arm) to left, 5 - bump (arm) to right, 6 - push to left, 7 - push to right, 8 - left kick, and 9 - right kick.

3.1 Data Collection and Pre-processing

We have collected a tri-modal dataset containing synchronized video, motion capture, and foot pressure data (Fig. 2) of 24-form simplified *Taiji Quan* (Taiji or Tai Chi) [72]. Justifications for this choice include 1) that Taiji is a low-cost, low-impact, slow, and hands-free movement sequence, aiming at enhanced balance; meanwhile, it contains ordinary body poses and movements such as stand, turn, pull, push, bump, and kick (Fig. 4) [72]; 2) Simplified 24-form Taiji is practiced worldwide by people of every gender, race and ages; 3) the Taiji routine (5 min) is significantly longer than existing publicly available motion capture (MoCap) sequences in the computer vision community (Sect. 2).

Pose Extraction: Synchronized video is collected at 50 fps from two Vicon Vue cameras. Locations for 2D body joints are first estimated in each video frame using the OpenPose Body25 model, which uses non-parametric representations called Part Affinity Fields to regress joint positions and body segment connections between the joints [12]. The output from OpenPose has X, Y pixel coordinates and confidence of prediction for each of the 25 modeled joints. To generate 3D joint locations, a confidence-weighted stereo triangulation is performed on 2D Openpose joints across the two synchronized and calibrated camera views. Finally, the 3D joints are corrected spatially using a deep regression network, named BioPose, trained separately to predict offsets between triangulated Open-Pose joints and biomechanical joints computed from motion capture data using the Vicon Plug-in-Gait module (Tables 2A and 2B). Pose detectors and BioPose corrections were tested and evaluated in detail by [56]; showing that OpenPose is more biomechanically accurate than HRNet [63] and Biopose correction of OpenPose creates the most biomechanical accuracy joint locations.

Foot Pressure: Foot pressure is collected at 100 fps using a Tekscan F-Scan insole pressure measurement system. Each subject wears a pair of canvas shoes with cut-to-fit capacitive pressure measurement insoles. Maximum recorded pressure values are clipped at an upper bound of 862 kPa based on the technical limits of the pressure measurement sensors. The foot pressure heatmaps are 2-channel images of size 60×21 (Fig. 1) and have been evaluated as accurate measurement sensors by [29].

Dataset Statistics: Figure 2B presents demographic information of the 10 subjects. Each subject performs two to three sessions of 24-form Taiji at an

Table 2. (A) Subject-wise and **(B)** Joint-wise L2 distance error of 3D pose data. The mean, std, min, median, and max are provided (in mm) for both OpenPose and BioPose joint locations as compared to motion capture joint data. Difference shows percentage improvement of BioPose over OpenPose.

Subject-wise L2 error relative to Motion Capture								
Sub	OpenPose(mm)			BioPose(mm)			Difference(%)		
#	mean	std	med	mean	std	med	mean	std	med
1	52.6	54.1	43.3	33.7	27.5	28.4	35.9	49.2	34.4
2	54.3	62.5	42.3	35.9	33.7	29.5	33.9	46.2	30.3
3	54.2	49.9	43.5	34.4	27.4	29.5	36.5	45.0	32.2
4	57.6	55.1	46.4	33.8	22.9	30.0	41.4	**58.5**	35.3
5	54.7	78.6	42.3	37.9	68.4	28.4	30.8	13.0	33.0
6	51.9	47.4	43.0	33.0	24.5	28.5	36.5	48.3	33.8
7	53.4	46.7	45.3	30.8	21.8	26.4	42.2	53.4	41.8
8	51.3	46.7	43.2	30.4	21.2	26.3	40.7	54.6	39.2
9	55.5	51.6	45.8	31.4	30.5	25.5	**43.3**	41.0	**44.3**
10	50.9	51.4	42.2	31.0	24.8	26.0	39.1	51.7	38.5
Mean	53.6	54.4	43.7	33.2	30.3	27.8	38.0	44.4	36.3
Std	2.1	9.8	1.5	2.4	14.0	1.7	4.0	12.7	4.5

(A) Subject-wise

Joint-wise L2 error relative to Motion Capture								
Joint	OpenPose(mm)			BioPose(mm)			Difference(%)		
Location	mean	std	med	mean	std	med	mean	std	med
Rshoulder	34.7	18.7	33.0	26.4	16.6	24.7	24.0	11.3	25.3
Relbow	52.0	49.7	41.1	34.1	28.3	27.8	34.5	43.0	32.2
Rwrist	67.9	85.8	44.9	42.2	40.5	31.9	37.8	52.8	29.1
Lshoulder	40.9	21.6	39.3	28.2	17.0	26.4	31.2	21.2	32.7
Lelbow	65.5	60.5	46.7	37.9	29.9	31.0	42.1	50.5	33.5
Lwrist	87.9	100.5	51.1	49.5	43.8	36.7	43.7	**56.4**	28.3
Rhip	55.3	22.7	53.3	34.9	18.5	33.9	36.9	18.5	36.3
Rknee	51.3	34.5	48.9	28.2	24.3	25.0	45.2	29.6	48.8
Rankle	49.7	47.8	44.2	26.9	31.5	22.3	46.0	34.0	**49.6**
Lhip	59.9	24.9	59.3	32.0	18.0	30.8	**46.5**	27.5	48.1
Lknee	39.7	28.6	37.1	32.4	25.6	29.7	18.3	10.6	20.0
Lankle	42.1	41.2	37.6	27.8	32.6	23.2	33.9	21.0	38.2
Mean	53.6	54.4	43.7	33.2	30.3	27.8	38.0	44.4	36.3

(B) Joint-wise

average of four repeated performances per session. The dataset contains a total of 1,363,400 frames of synchronized video body pose and foot pressure maps. This new dataset captures significant statistical variations: 1) Diversity in the subjects in terms of gender, age, mass, height, and years of experience in Taiji practice for amateurs and professionals. 2) Kernel density plots (on project page) of the distributions of body joint locations show the subject performances are statistically similar to one another spatially. 3) PCA analysis (Fig. 3) of foot pressure highlights that each subject has a unique pressure distribution relative to other subjects, but the top principal components encode similar modes of variation (e.g., variability in left/right foot pressure, toe/heel pressure, etc.).

Preprocessing: Body joints are centered by subtracting off the hip center joint location (making it the origin) to remove camera-specific offsets during video recording. Other joint locations are normalized per body joint by subtracting each dimension (2D or 3D) by the mean and dividing by its standard deviation, leading to a zero-mean, unit-variance distribution.

Foot pressure data is recorded in kilopascals (kPa) at discretized sensor locations (prexels) on the shoe insoles. Prexel values are clipped between 0 to 862 kPa based on pressure sensor technology limitations. The clipped data is further normalized by dividing each prexel by its max intensity value over the entire training set. The left and right normalized foot pressure maps are concatenated as two channels to form a ground truth heatmap of size ($60 \times 21 \times 2$), with prexel intensities in the range $[0, 1]$.

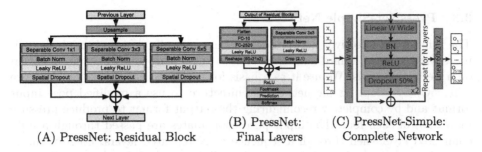

(A) PressNet: Residual Block   (B) PressNet: Final Layers   (C) PressNet-Simple: Complete Network

Fig. 5. Our foot pressure regression architectures have a 96-coordinate input representing 24 3D joint locations and confidences ($24 \times 4 = 96$) and a 2520-prexel output representing 60×21 pressure maps for both feet ($60 \times 21 \times 2 = 2520$). (**A**) A residual block, one of the building blocks of PressNet network, upsamples the data and computes features. (**B**) Final set of layers of PressNet include a fully connected layer and a concurrent branch to preserve spatial consistency. (**C**) The PressNet-Simple network architecture is defined by two hyperparameters: the depth (# of layers, N) of the network and the width (# of fully connected nodes, W) of those layers.

3.2   PressNet Network

The design of our PressNet network (Figs. 5A and 5B) is initially motivated by the residual generator of the Improved Wasserstein GAN [25]. We use a generator-inspired architecture because our input is 1D body joints and the output is a 2D foot pressure heatmap. This design aids in capturing information at different resolutions, acting like a decoder network for feature extraction. The primary aim of this network is to extract features without loss of spatial information across different scales.

PressNet is a feed forward Convolutional Neural Network with an input layer that is a flattened vector of joint coordinates of size 96×1 (24 joints $\times$ 4, consisting of x,y,z coordinates and joint detection confidences). The input is processed through a fully connected layer with an output dimension of 6144×1. This output is reshaped into an image of size 4×3 with 512 channels. The network contains four residual convolution blocks that perform nearest neighbor upsampling. Each residual block of PressNet (Fig. 5A) has three parallel convolution layers with kernel sizes 5×5, 3×3 and 1×1. The number of channels of each residual block is progressively halved as the resolution doubles, starting at 512 channels and decreasing to 64. The output of the final residual block is split and sent to a convolutional branch and a fully connected branch (Fig. 5B). The convolutional branch serves to preserve spatial coherence while the fully connected branch has a field of view over the entire prediction. PressNet contains separable convolutions [17], batch normalization (BN) [31], spatial dropouts [67] and leaky ReLU [43] layers.

3.3   PressNet-Simple Network

The "simple yet effective" network of [44] was originally designed to jointly estimate the unobserved third dimension of a set of 2D body joint coordinates (pose)
on a per frame basis. We use it as a basis for our PressNet-Simple architecture
(Fig. 5C) by adapting the network architecture to use a modified pose input
format and by completely reconfiguring the output format to produce pressure
map data of each foot. The input pose coordinates are passed through a fully
connected layer then through a sequence of N repeated layers. Each of the N
layers has two iterations of the sequence: fully connected, batch normalization,
ReLU, and 50% dropout layers. The result of each of the N layer sequences
is then added to the results from the previous layer sequence (N-1) and finally
passed through a 2520 fully connected layer to produce the output foot pressure.
The PressNet-Simple architecture is configured via two hyper-parameters: the
depth (# of layers, N) of the network and the width (# of fully connected nodes,
W) of those layers. For this study, through empirical testing, it was determined
that the optimal hyper-parameters are N = 4 and W = 2560. Because of the
sequential nature of this network with fully connected layers, this network architecture does not maintain the spatial coherence that PressNet has established
with upsampling and convolutional layers.

3.4   Training Details

We use a Leave-One-Subject-Out (LOSO) cross-validation to determine how the
network generalizes to an unseen individual. Furthermore, the training data is
split sequentially in a 9:1 ratio where the smaller split is used as validation data.

PressNet is trained for 35 epochs with a piecewise learning rate starting at
10^{-4} and a batch size of 32 and takes 7.5 h to train each LOSO data split on a
NVIDIA Tesla P100 GPU. A binary footmask (produced by the foot pressure
capturing system) is element-wise multiplied with the predictions of the network.
This enables the network to not have to learn the approximate shape of the foot
in the course of training, only the pressure distribution. The learning rate is
reduced by 50% after every 13 epochs to ensure a decrease in validation loss
with training. KL Divergence (KL) is used as the loss function along with Adam
Optimizer for supervision, as we are learning the distribution of prexels [5].

PressNet-Simple is trained with an initial learning rate of 10^{-4} for 40 epochs
at a batch size of 128. PressNet-Simple takes 3 to 3.5 h to train each LOSO data
split on an NVIDIA TitanX GPU with 12GB of memory. The learning rate is
reduced by 75% every 7 epochs, and MSE loss is used with the Adam Optimizer.

Table 3. Analysis for each network architecture by subject using error metrics: Mean Absolute Error (MAE), Similarity (SIM), KL Divergence (KLD), and Information Gain (IG). Results from inference on input poses with 25 valid joints are included in the evaluation. Best values are shown in bold. Arrow indicates direction of better value. Networks Evaluated are KNN 2D with K = 5 (K5), PressNet-Simple 2D (PNS2), PressNet-Simple 3D (PNS3), PressNet-Simple 3D using BioPose(PNS3B), and PressNet 3D using BioPose and KL loss (PN3BK).

Sub #	Mean Absolute Error (MAE) ↓					Similarity (SIM) ↑					KL Divergence (KLD) ↓					Information Gain (IG) ↑				
						Mean Error of Estimated Foot Pressure relative to Measured Ground Truth														
1	9.56	7.55	7.21	**7.06**	N/A	0.30	0.44	0.49	**0.50**	0.48	17.02	2.79	2.04	2.65	**1.16**	-4.35	-0.74	-0.50	-0.77	**-0.27**
2	9.98	**9.44**	10.10	9.90	N/A	0.22	0.23	0.25	**0.28**	0.27	14.34	5.21	4.39	3.50	**1.98**	-1.23	-0.41	-0.34	-0.27	**-0.14**
3	9.73	8.82	**7.77**	7.87	N/A	0.31	0.39	0.44	**0.44**	0.43	14.80	2.95	1.61	2.02	**0.43**	-2.76	-0.54	-0.29	-0.39	**-0.18**
4	10.91	9.38	**9.12**	10.07	N/A	0.32	0.45	**0.48**	0.42	0.43	14.42	1.88	1.59	3.53	**0.43**	-3.70	-0.52	-0.42	-0.92	**-0.23**
5	11.21	10.14	9.28	**9.22**	N/A	0.40	0.47	**0.55**	0.53	0.50	12.19	1.94	1.51	1.96	**0.50**	-2.65	-0.44	-0.36	-0.46	**-0.21**
6	11.06	9.98	10.04	**9.13**	N/A	0.38	0.45	0.45	**0.51**	0.44	10.48	1.54	1.07	1.60	**0.44**	-1.56	-0.22	-0.14	-0.25	**-0.12**
7	11.08	10.18	**8.97**	9.14	N/A	0.26	0.34	0.42	**0.42**	0.39	18.49	4.44	2.70	3.19	**1.87**	-4.39	-0.98	-0.65	-0.74	**-0.37**
8	8.94	8.31	**7.32**	7.75	N/A	0.30	0.33	0.34	**0.38**	0.38	13.32	3.32	3.02	2.48	**1.62**	-1.51	-0.34	-0.32	-0.29	**-0.17**
9	9.26	8.24	**7.43**	7.63	N/A	0.32	0.37	**0.47**	0.45	0.43	13.09	2.82	1.45	2.04	**1.14**	-2.16	-0.46	-0.25	-0.35	**-0.16**
10	8.82	7.44	7.53	**7.26**	N/A	0.33	**0.44**	0.42	0.41	0.41	15.30	3.18	3.14	3.81	**1.50**	-2.84	-0.00	-0.65	-0.73	**-0.24**
Mean	10.06	8.95	**8.48**	8.50	N/A	0.31	0.39	0.43	**0.43**	0.42	14.35	3.01	2.25	2.68	**1.11**	-2.71	-0.53	-0.39	-0.52	**-0.21**
Std	**0.89**	0.98	1.09	1.05	N/A	**0.05**	0.07	0.08	0.07	0.06	2.18	1.08	0.98	0.74	**0.59**	1.09	0.20	0.16	0.24	**0.07**
Model	K5	PNS2	PNS3	PNS3B	PN3BK	K5	PNS2	PNS3	PNS3B	PN3BK	K5	PNS2	PNS3	PNS3B	PN3BK	K5	PNS2	PNS3	PNS3B	PN3BK

4 Evaluation and Visualization of Results

4.1 Quantitative Evaluation

KNN Baseline: KNN provides a convenient data-driven (memory-based, non-linear) way to directly map between two different modalities; in this case, pose to pressure. As the number of data samples becomes large, even simple NN (aka 1-NN) retrieval can perform surprisingly well, both theoretically [18] and empirically [65], due to the "Unreasonable Effectiveness of Data" [52]. The main drawback of KNN is the high cost of computing distances between a pose query and all samples in a large dataset; thus, we use it only as a baseline in our work.

The distance metric for KNN is the sum of Euclidean distances between corresponding normalized body joint locations. The foot pressure maps corresponding to these nearest neighbors are combined as a weighted average to generate the output pressure map prediction, using inverse distance weighting [60]. Empirical tests with K ranging from 1 to 50 showed that error reduces as K increases with diminishing improvements. Results from KNN-based pressure estimation, where K = 5 and K = 50, are included in Fig. 7 for comparison with the deep learning based methods.

Evaluation Measures: We use six quantitative measures to evaluate the performance of the trained networks and KNN baseline:

1. Mean Absolute Error (MAE in kPa) between estimated foot pressure maps and measured ground truth pressure.
2. Three metrics for spatial distribution of the learned foot pressure map: Similarity, KL Divergence, and Information Gain [10].
3. Two measures on accuracy for estimated *Center of Pressure* (CoP) and *Base of Support* (BoS), which are directly related to the computation of stability

Table 4. Cross-view validation using PNS trained on camera View 1 to predict pressure maps from View 2 (Fig. 2A). The reported metrics are Mean Absolute Error (MAE), Similarity (SIM), KL Divergence (KLD), and Information Gain (IG). Only outputs generated from input poses with 25 valid joints are included in the evaluation. Best values are shown in bold. Best is defined as closer to 1 for SIM and closer to 0 for MAE, KLD, and IG. Arrow indicates direction of better value.

Sub	MAE ↓		SIM ↑		KLD ↓		IG ↑	
#	View 1	View 2	View 1	View 2	View 1	View 2	View 1	View 2
				Foot Pressure Error Metrics for Viewpoint Comparison (Mean / Std)				
1	**7.55 / 1.54**	9.15 / 1.94	**0.44 / 0.10**	0.36 / 0.12	**2.79 / 1.99**	4.27 / 3.23	**-0.74 / 0.50**	-1.00 / 0.74
2	**9.44 / 2.93**	9.94 / 2.76	0.23 / 0.15	**0.25 / 0.16**	5.21 / 5.38	**4.65 / 5.72**	-0.41 / 0.41	**-0.36 / 0.41**
3	8.82 / 2.37	8.82 / 2.51	0.39 / 0.13	**0.41 / 0.12**	2.95 / 2.70	**2.54 / 1.89**	-0.54 / 0.47	**-0.51 / 0.46**
4	**9.38 / 2.42**	9.76 / 2.55	**0.45 / 0.11**	0.45 / 0.12	**1.88 / 1.39**	2.16 / 1.53	**-0.52 / 0.45**	-0.57 / 0.51
5	**10.14 / 2.53**	10.15 / 2.61	0.47 / 0.13	**0.49 / 0.13**	1.94 / 1.52	**1.77 / 1.52**	-0.44 / 0.37	**-0.39 / 0.35**
6	**9.98 / 2.35**	10.67 / 2.45	**0.45 / 0.12**	0.44 / 0.12	1.54 / 1.35	**1.41 / 1.44**	-0.22 / 0.23	**-0.18 / 0.20**
7	10.18 / 2.23	**10.08 / 2.16**	0.34 / 0.09	**0.36 / 0.11**	4.44 / 2.61	**3.65 / 2.22**	-0.98 / 0.63	**-0.81 / 0.47**
8	8.31 / 3.58	**8.30 / 3.55**	0.33 / 0.13	**0.36 / 0.14**	3.32 / 2.63	**2.62 / 2.36**	-0.34 / 0.27	**-0.29 / 0.24**
9	8.24 / 2.88	**7.77 / 2.41**	0.37 / 0.13	**0.40 / 0.13**	2.82 / 2.58	**2.28 / 2.00**	-0.46 / 0.40	**-0.37 / 0.34**
10	**7.44 / 2.51**	7.68 / 2.53	0.44 / 0.14	**0.46 / 0.14**	3.18 / 3.22	**2.62 / 2.23**	-0.60 / 0.48	**-0.52 / 0.44**
Mean	**8.95 / 2.53**	9.23 / 2.55	0.39 / 0.12	**0.40 / 0.13**	3.01 / 2.54	**2.80 / 2.41**	-0.53 / 0.42	**-0.50 / 0.42**
Std	**0.98 / 0.50**	1.00 / 0.40	0.07 / 0.02	**0.07 / 0.01**	1.08 / 1.12	**1.01 / 1.21**	**0.20 / 0.11**	0.23 / 0.14

(Fig. 3C). We use ℓ_2 distance (in mm)/IoU (in %) between the estimated CoP from learned foot pressure maps and the CoP calculated directly from ground truth foot pressure to quantify CoP and BoS quality, respectively.

Evaluation of Predicted Pressure Maps: Table 3 shows our evaluation results using the first four metrics above on the KNN baseline and variations of PNS and PN. For each pressure prediction method, only frames that have 25 detected joints are included in the analysis to minimize confounding factors on the method's effectiveness. KL Divergence and Information Gain both show the advantages of networks on learning statistical distributions of the input data. The key takeaway is that both networks excel at predicting the spatial distribution of ground truth pressure more so than the overall magnitude.

Cross-View Validation. Our networks using 2D joint data were trained on View 1 (Fig. 2A) of the two video cameras. Table 4 presents results of running a 2D network to predict foot pressure on images from the other camera, View 2. Results are similar to those in Table 3, indicating that both networks are robust to viewpoint and subject position/orientation relative to the camera view. PressNet-Simple appears less affected by viewpoint than PressNet, based on Similarity, KL Divergence, and Information Gain measures.

Distance to CoP: As a step towards estimating stability from pose, Center of Pressure (CoP) is computed from predicted foot pressure maps and compared to ground truth CoP locations computed from insole foot pressure readings. CoP is calculated as the weighted mean of the pressure elements (prexels) in the XY ground plane. A systematic threshold starting from 0 kPa is applied to both the

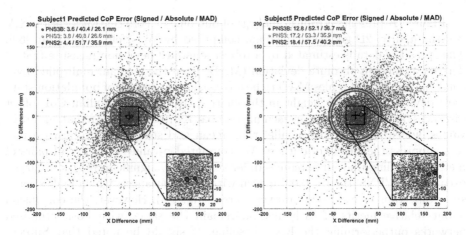

Fig. 6. 2D offsets between ground truth CoP (black cross) and CoP predicted by PNS2 (blue), PNS3 (red), and PNS3B (green). Large dots plot the mean of each scatter plot distribution; that they appear close to the Ground Truth (GT) indicates relatively symmetrical distributions of spatial error around the GT CoP. The concentric circles represent mean (solid) and median (dashed) offset distances as an error radius, and they indicate PNS3 and PNS3B CoP estimates cluster more tightly about the GT than PNS2. (Color figure online)

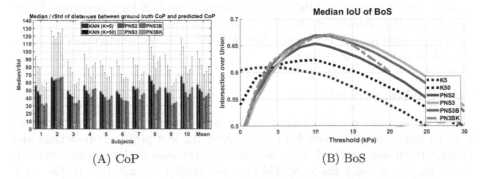

(A) CoP   (B) BoS

Fig. 7. (A) Comparison of CoP offset distance errors across methods and subjects, characterized by robust estimation Median/rStd (robust STD). **(B)** Comparison of BoS using Intersection over Union (IoU) relative to ground truth over a range of pressure thresholds (Note: PN3BK has a normalized threshold scaled relative to kPa and therefore a shorter line). All results differ statistically significantly from one another. See more details in text.

ground truth and the predicted pressure, similar to the procedure used in [36]. Figure 6 provides scatter plots of CoP offset errors for each method. Also shown on the plot are the mean of the offset error distributions, and an error radius for each method derived from the mean and median of the offset distances. This figure highlights the similarities across subjects as well as the clear improvement that both deep learning networks make over KNN in more accurately predicting CoP.

Figure 7A presents a robust analysis of outcomes where central location X,Y is estimated by 2D geometric median, computed by Weiszfeld's algorithm [69]. The spread of data is estimated by a robust standard deviation measure (rStd), derived as median absolute deviation (MAD) from the median, multiplied by a constant 1.4826 that scales MAD to be a *consistent estimator* of population standard deviation [58]. Bar height in the chart corresponds to median CoP offset distances, and the whiskers on each median bar represent rStd. Median and rStd values, which are generally smaller than mean and std because robust estimators suppress the effects of outliers. We computed p-values between all pairs of methods, finding that all outcomes differ statistically significantly ($p << 0.001$), except PNS3/PNS3B ($p = 0.009$), even with large variances, which makes sense since our train/test set sizes are very large (Fig. 2B). However, the conclusion about relative merits of each method remain the same, with both proposed networks outperforming the KNN baseline. It should be noted that Subject 2 consistently under-performs for all methods, which may be an indication of inaccuracy in the input pose or ground truth pressure data, requiring further investigation. Fig. 7B presents an analysis of the Base of Support (BoS) resulting from the predicted foot pressures relative to the ground truth pressure using the Intersection over Union (IoU). With identical overlap (IoU = 1) being the goal, the results indicate that all networks outperform KNN (K5 and K50) with the PNS2 under-performing all 3D methods with 65–68% overlap. The X-axis presents the threshold (in kPa) used to calculate the BoS for comparison for all but the PN3BK model, which is on a normalized and unitless scale as part of data processing for KL Divergence loss. PN3BK is scaled relative in Fig. 7B to provide easy visual comparison.

4.2 Qualitative Evaluation

Figure 8 visualizes ground truth, foot pressure predictions, and their BoS and CoP for some sample frames. For each subject, the foot pressure predictions and ground truth are rendered with independent pressure scales (weight related) based on the pressure range needed for each subject. In addition to the qualitative comparison by visualization, the respective mean absolute errors with respect to ground truth frames have been calculated and included in Table 3.

Finally, we show preliminary results on exploring the potential use of estimated foot-pressure distributions to obtain classic stability measures defined in kinesiology [13,26–28,35,41]. Motivated by findings in medicine via randomized trials that Taiji intervention may improve stability of certain populations [21], we observe convincing correlations between years of Taiji practice and two stability measures (Fig. 3C) for two different Taiji poses (Fig. 9). The trend of correlation is consistent with previous work; as with [21], the more Taiji practice, the more stable a subject is. Figure 9A shows negative correlation, meaning CoM and CoP align better for more experienced Taiji subjects. Figure 9B shows positive correlation, meaning: the most experienced subjects can maintain better stability when CoM's Time To Collision (TTC) with BoS is longer.

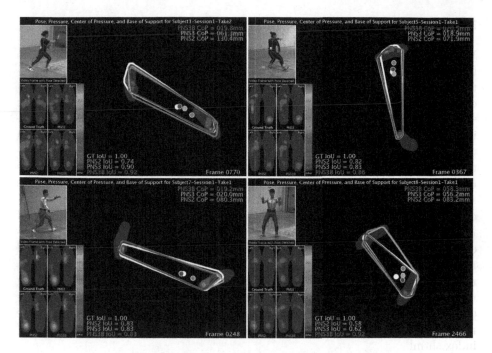

Fig. 8. Sample output frames showing the ground truth and estimated Center of Pressure (CoP) and Base of Support (BoS). Foot pressure is scaled for each subject based on their range of pressure. BoS and CoP of Ground Truth (white), PNS (yellow), PNS3 (red), and PNS3B (green) plotted as an overlaid on the floor plane. Intersection over Union (IoU) and distance to Ground Truth CoP (mm) are used to quantify the quality of BoS and CoP estimation, respectively. (Color figure online)

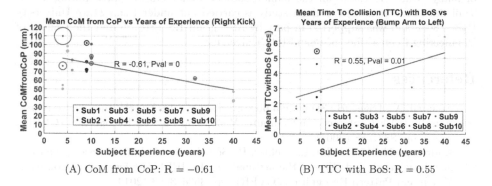

(A) CoM from CoP: R = −0.61   (B) TTC with BoS: R = 0.55

Fig. 9. Correlation between the number of years of Taiji experience/training N and two different stability metrics can be seen here on two different Taiji poses. They seem to confirm the general observation that the more experienced Taiji practitioners are more stable, where N is larger: **(A)** distance between CoM and CoP is smaller; and **(B)** the time to reach the boundary of BoS is longer.

5 Conclusions

We present a fully validated approach to estimate foot pressure distributions from 2D/3D human body pose. Given the multi-faceted complexity of this pose-to-foot pressure mapping problem, we have gained several insights from this exercise: (1) KNN is a reasonable baseline predictor, while deep learning networks surpass KNN statistically significantly; (2) 3D pose input has a high positive impact over 2D, albeit with an upfront higher computational cost; (3) system performance is surprisingly stable across subject weight, gender, height variations, and the number of subjects in the training/testing data; (4) networks trained on one camera view produce comparable results when tested on images from a different view, confirming that the Taiji dataset provides adequate orientation generalization information; (5) quantitative evaluations on a subset of *ordinary poses* indicates that networks trained on Taiji movements can be generalized to non-Taiji-specific poses; and (6) correlation between the quantified results of deep learning networks support our initial hypothesis that learning a mapping from kinematics to dynamics from static images is feasible, opening up a door for precision computer vision devoted to human body centered sciences. Access to implementation and dataset details are available through the project website: http://vision.cse.psu.edu/research/dynamicsFromKinematics/index.shtml.

Acknowledgments. We thank our collaborator Professor John Challis for offering his Kinesiology perspectives and helpful discussions. We thank Master Sitan Chen of Sitan Tai Chi School; Dr. Hesheng Bao of Win-Win Kung Fu Culture Center, Inc.; Professor Zuofeng Sun of School of Physical Education, Hebei Normal University; and Professor Pingping Xie of Tianjin University of Sport for their expertise and support of many years. We thank all the human subject volunteers for participating in our Taiji data collection. This human study is carried out under Penn State University IRB #8085 and was supported in part by NSF grant IIS-1218729 and the College of Engineering Dean's office of Penn State University.

References

1. Agarwal, A., Triggs, B.: 3D human pose from silhouettes by relevance vector regression. In: IEEE Conference on Computer Vision and Pattern Recognition (CVPR), vol. 2, pp. 882–888 (2004)
2. Andriluka, M., Pishchulin, L., Gehler, P., Schiele, B.: 2D human pose estimation: new benchmark and state of the art analysis. In: IEEE Conference on Computer Vision and Pattern Recognition (CVPR), pp. 3686–3693 (2014)
3. Arvin, M., Hoozemans, M., Pijnappels, M., Duysens, J., Verschueren, S., Van Dieen, J.: Where to step? Contributions of stance leg muscle spindle afference to planning of mediolateral foot placement for balance control in young and older adults. Front. Physiol. **9**, 1134 (2018)
4. Bächer, M., Whiting, E., Bickel, B., Sorkine-Hornung, O.: Spin-it: optimizing moment of inertia for spinnable objects. ACM Trans. Graph. **33**(4), 1–10 (2014)
5. Bishop, C.M.: Pattern Recognition and Machine Learning (Information Science and Statistics). Springer, Heidelberg (2006)

6. Bogo, F., Kanazawa, A., Lassner, C., Gehler, P., Romero, J., Black, M.J.: Keep it SMPL: automatic estimation of 3D human pose and shape from a single image. In: Leibe, B., Matas, J., Sebe, N., Welling, M. (eds.) ECCV 2016. LNCS, vol. 9909, pp. 561–578. Springer, Cham (2016). https://doi.org/10.1007/978-3-319-46454-1_34

7. Brubaker, M.A., Sigal, L., Fleet, D.J.: Estimating contact dynamics. In: IEEE International Conference on Computer Vision (ICCV), pp. 2389–2396 (2009)

8. Brubaker, M., Fleet, D., Hertzmann, A.: Physics-based person tracking using the anthropomorphic walker. Int. J. Comput. Vis. (IJCV) **87**(1), 140–155 (2010). https://doi.org/10.1007/s11263-009-0274-5

9. Bulat, A., Tzimiropoulos, G.: Human pose estimation via convolutional part heatmap regression. In: Leibe, B., Matas, J., Sebe, N., Welling, M. (eds.) ECCV 2016. LNCS, vol. 9911, pp. 717–732. Springer, Cham (2016). https://doi.org/10.1007/978-3-319-46478-7_44

10. Bylinskii, Z., Judd, T., Oliva, A., Torralba, A., Durand, F.: What do different evaluation metrics tell us about saliency models? IEEE Trans. Pattern Anal. Mach. Intell. (PAMI) **41**(3), 740–757 (2019)

11. Cai, Y., Ge, L., Cai, J., Yuan, J.: Weakly-supervised 3D hand pose estimation from monocular RGB images. In: Ferrari, V., Hebert, M., Sminchisescu, C., Weiss, Y. (eds.) ECCV 2018. LNCS, vol. 11210, pp. 678–694. Springer, Cham (2018). https://doi.org/10.1007/978-3-030-01231-1_41

12. Cao, Z., Simon, T., Wei, S.E., Sheikh, Y.: Realtime multi-person 2D pose estimation using part affinity fields. In: IEEE Conference on Computer Vision and Pattern Recognition (CVPR), pp. 1302–1310 (2017)

13. Chaudhry, H., Bukiet, B., Ji, Z., Findley, T.: Measurement of balance in computer posturography: comparison of methods - a brief review. J. Bodywork Mov. Ther. **15**(1), 82–91 (2011)

14. Chen, C.H., Ramanan, D.: 3D human pose estimation = 2D pose estimation + matching. In: IEEE Conference on Computer Vision and Pattern Recognition (CVPR), pp. 5759–5767 (2017)

15. Chen, W., et al.: Synthesizing training images for boosting human 3D pose estimation. In: IEEE International Conference on 3D Vision (3DV), pp. 479–488 (2016)

16. Chen, X., Yuille, A.L.: Articulated pose estimation by a graphical model with image dependent pairwise relations. In: Advances in Neural Information Processing Systems (NIPS), pp. 1736–1744 (2014)

17. Chollet, F.: Xception: Deep learning with depthwise separable convolutions. In: IEEE Conference on Computer Vision and Pattern Recognition (CVPR) (2017)

18. Cover, T., Hart, P.: Nearest neighbor pattern classification. IEEE Trans. Inf. Theory **13**(1), 21–27 (1967)

19. Eckardt, N., Rosenblatt, N.J.: Healthy aging does not impair lower extremity motor flexibility while walking across an uneven surface. Hum. Mov. Sci. **62**, 67–80 (2018)

20. Fan, X., Zheng, K., Lin, Y., Wang, S.: Combining local appearance and holistic view: dual-source deep neural networks for human pose estimation. In: IEEE Conference on Computer Vision and Pattern Recognition (CVPR), pp. 1347–1355 (2015)

21. Fuzhong, L., et al.: Tai Chi and postural stability in patients with Parkinson's disease. N. Engl. J. Med. **366**(6), 511–519 (2012)

22. Gilbert, A., Trumble, M., Malleson, C., Hilton, A., Collomosse, J.: Fusing visual and inertial sensors with semantics for 3D human pose estimation. Int. J. Comput. Vis. (IJCV) **127**, 381–397 (2019)

23. Grimm, R., Sukkau, J., Hornegger, J., Greiner, G.: Automatic patient pose estimation using pressure sensing mattresses. In: Handels, H., Ehrhardt, J., Deserno, T., Meinzer, H., Tolxdorff, T. (eds.) Bildverarbeitung für die Medizin, pp. 409–413. Springer, Heidelberg (2011). https://doi.org/10.1007/978-3-642-19335-4_84

24. Güler, R.A., Neverova, N., Kokkinos, I.: DensePose: dense human pose estimation in the wild. In: IEEE Conference on Computer Vision and Pattern Recognition (CVPR). pp. 7297–7306 (2018)

25. Gulrajani, I., Ahmed, F., Arjovsky, M., Dumoulin, V., Courville, A.C.: Improved training of Wasserstein GANs. In: Advances in Neural Information Processing Systems (NIPS), pp. 5767–5777 (2017)

26. Hof, A., Gazendam, M., Sinke, W.: The condition for dynamic stability. J. Biomech. **38**(1), 1–8 (2005)

27. Hof, A.L.: The equations of motion for a standing human reveal three mechanisms for balance. J. Biomech. **40**(2), 451–457 (2007)

28. Hof, A.L.: The "extrapolated center of mass" concept suggests a simple control of balance in walking. Hum. Mov. Sci. **27**(1), 112–125 (2008)

29. Hsiao, H., Guan, J., Weatherly, M.: Accuracy and precision of two in-shoe pressure measurement systems. Ergonomics **45**(8), 537–555 (2002)

30. Huang, G., Liu, Z., Van Der Maaten, L., Weinberger, K.Q.: Densely connected convolutional networks. In: IEEE Conference on Computer Vision and Pattern Recognition (CVPR), pp. 2261–2269 (2017)

31. Ioffe, S., Szegedy, C.: Batch normalization: accelerating deep network training by reducing internal covariate shift. In: Proceedings of the International Conference on Machine Learning (ICML), vol. 37, pp. 448–456 (2015)

32. Ionescu, C., Li, F., Sminchisescu, C.: Latent structured models for human pose estimation. In: IEEE International Conference on Computer Vision (ICCV), pp. 2220–2227 (2011)

33. Ionescu, C., Papava, D., Olaru, V., Sminchisescu, C.: Human3.6m: large scale datasets and predictive methods for 3D human sensing in natural environments. IEEE Trans. Pattern Anal. Mach. Intell. (PAMI) **36**(7), 1325–1339 (2014)

34. Iqbal, U., Milan, A., Gall, J.: PoseTrack: joint multi-person pose estimation and tracking. In: IEEE Conference on Computer Vision and Pattern Recognition (CVPR), pp. 2011–2020 (2017)

35. Jian, Y., Winter, D.A., Ishac, M.G., Gilchrist, L.: Trajectory of the body COG and COP during initiation and termination of gait. Gait Posture **1**(1), 9–22 (1993)

36. Keijsers, N., Stolwijk, N., Nienhuis, B., Duysens, J.: A new method to normalize plantar pressure measurements for foot size and foot progression angle. J. Biomech. **42**(1), 87–90 (2009)

37. Ko, J.H., Wang, Z., Challis, J.H., Newell, K.M.: Compensatory mechanisms of balance to the scaling of arm-swing frequency. J. Biomech. **48**(14), 3825–3829 (2015)

38. Lemaire, E.D., Biswas, A., Kofman, J.: Plantar pressure parameters for dynamic gait stability analysis. In: IEEE Engineering in Medicine and Biology Society (EMBS), pp. 4465–4468 (2006)

39. Li, Z., Sedlar, J., Carpentier, J., Laptev, I., Mansard, N., Sivic, J.: Estimating 3D motion and forces of person-object interactions from monocular video. In: IEEE Conference Computer Vision and Pattern Recognition (CVPR), pp. 8632–8641 (2019)

40. Lin, T.-Y., et al.: Microsoft COCO: common objects in context. In: Fleet, D., Pajdla, T., Schiele, B., Tuytelaars, T. (eds.) ECCV 2014. LNCS, vol. 8693, pp. 740–755. Springer, Cham (2014). https://doi.org/10.1007/978-3-319-10602-1_48

41. Lugade, V., Lin, V., Chou, L.S.: Center of mass and base of support interaction during gait. Gait Posture **33**(3), 406–411 (2011)
42. Lv, X., Chai, J., Xia, S.: Data driven inverse dynamics for human motion. ACM Trans. Graph. **35**(6), 1–12 (2016)
43. Maas, A.L., Hannun, A.Y., Ng, A.Y.: Rectifier nonlinearities improve neural network acoustic models. In: ICML Workshop on Deep Learning for Audio, Speech and Language Processing (2013)
44. Martinez, J., Hossain, R., Romero, J., Little, J.J.: A simple yet effective baseline for 3D human pose estimation. In: IEEE International Conference on Computer Vision (ICCV), pp. 2659–2668 (2017)
45. McKay, M.J., et al.: Spatiotemporal and plantar pressure patterns of 1000 healthy individuals aged 3–101 years. Gait Posture **58**, 78–87 (2017)
46. Moreno-Noguer, F.: 3D human pose estimation from a single image via distance matrix regression. In: IEEE Conference on Computer Vision and Pattern Recognition (CVPR), pp. 1561–1570 (2017)
47. Newell, A., Yang, K., Deng, J.: Stacked hourglass networks for human pose estimation. In: Leibe, B., Matas, J., Sebe, N., Welling, M. (eds.) ECCV 2016. LNCS, vol. 9912, pp. 483–499. Springer, Cham (2016). https://doi.org/10.1007/978-3-319-46484-8_29
48. Nie, B.X., Wei, P., Zhu, S.C.: Monocular 3D human pose estimation by predicting depth on joints. In: IEEE International Conference on Computer Vision (ICCV), pp. 3467–3475 (2017)
49. Pai, Y.C.: Movement termination and stability in standing. Exerc. Sport Sci. Rev. **31**(1), 19–25 (2003)
50. Pataky, T., Mu, T., Bosch, K., Rosenbaum, D., Goulermas, J.: Gait recognition: highly unique dynamic plantar pressure patterns among 104 individuals. J. Royal Soc. Interface **9**, 790–800 (2012)
51. Pavlakos, G., Zhou, X., Derpanis, K.G., Daniilidis, K.: Coarse-to-fine volumetric prediction for single-image 3D human pose. In: IEEE Conference on Computer Vision and Pattern Recognition (CVPR), pp. 1263–1272. IEEE (2017)
52. Pereira, F., Norvig, P., Halevy, A.: The unreasonable effectiveness of data. IEEE Intell. Syst. **24**(02), 8–12 (2009)
53. Prévost, R., Bacher, M., Jarosz, W., Sorkine-Hornung, O.: Balancing 3D models with movable masses. In: Conference on Vision, Modeling and Visualization (VMV 2016), pp. 9–16. Eurographics Association (2016)
54. Prévost, R., Whiting, E., Lefebvre, S., Sorkine-Hornung, O.: Make it stand: balancing shapes for 3D fabrication. ACM Trans. Graph. **32**(4), 1–10 (2013)
55. Putti, A., Arnold, G., Abboud, R.: Foot pressure differences in men and women. Foot Ankle Surg. **16**(1), 21–24 (2010)
56. Ravichandran, B.: BioPose-3D and PressNet-KL: A Path to Understanding Human Pose Stability from Video. Master's thesis, Computer Science and Engineering, The Pennsylvania State University (2020)
57. Rogez, G., Weinzaepfel, P., Schmid, C.: LCR-Net++: multi-person 2D and 3D pose detection in natural images. IEEE Trans. Pattern Anal. Mach. Intell. (PAMI) **42**(5), 1146–1161 (2020)
58. Rousseeuw, P.J., Croux, C.: Alternatives to the median absolute deviation. J. Am. Stat. Assoc. **88**(424), 1273–1283 (1993)
59. Seethapathi, N., Wang, S., Saluja, R., Blohm, G., Körding, K.P.: Movement science needs different pose tracking algorithms. CoRR abs/1907.10226 (2019)
60. Shepard, D.: A two-dimensional interpolation function for irregularly-spaced data. In: Proceedings of the ACM National Conference (ACM 1968), pp. 517–524 (1968)

61. Sigal, L., Balan, A., Black, M.J.: HumanEva: synchronized video and motion capture dataset and baseline algorithm for evaluation of articulated human motion. Int. J. Comput. Vis. **87**(1), 4–27 (2010)
62. Simo-Serra, E., Ramisa, A., Alenyà, G., Torras, C., Moreno-Noguer, F.: Single image 3D human pose estimation from noisy observations. In: IEEE Conference on Computer Vision and Pattern Recognition (CVPR), pp. 2673–2680 (2012)
63. Sun, K., Xiao, B., Liu, D., Wang, J.: Deep high-resolution representation learning for human pose estimation. In: IEEE Conference on Computer Vision and Pattern Recognition (CVPR), pp. 5693–5703 (2019)
64. Sun, X., Shang, J., Liang, S., Wei, Y.: Compositional human pose regression. In: IEEE International Conference on Computer Vision (ICCV), pp. 2621–2630 (2017)
65. Tatarchenko, M., Richter, S.R., Ranftl, R., Li, Z., Koltun, V., Brox, T.: What do single-view 3D reconstruction networks learn? In: IEEE Conference on Computer Vision and Pattern Recognition (CVPR), pp. 3405–3414 (2019)
66. Tompson, J., Jain, A., LeCun, Y., Bregler, C.: Joint training of a convolutional network and a graphical model for human pose estimation. In: Advances in Neural Information Processing Systems (NIPS), pp. 1799–1807 (2014)
67. Tompson, J., Goroshin, R., Jain, A., LeCun, Y., Bregler, C.: Efficient object localization using convolutional networks. In: IEEE Conference on Computer Vision and Pattern Recognition (CVPR), pp. 648–656 (2015)
68. Toshev, A., Szegedy, C.: DeepPose: human pose estimation via deep neural networks. In: IEEE Conference on Computer Vision and Pattern Recognition (CVPR), pp. 1653–1660 (2014)
69. Vardi, Y., Zhang, C.H.: The multivariate L_1-median and associated data depth. Proc. Natl. Acad. Sci. **97**(4), 1423–1426 (2000)
70. Vera-Rodriguez, R., Mason, J.S.D., Fierrez, J., Ortega-Garcia, J.: Comparative analysis and fusion of spatiotemporal information for footstep recognition. IEEE Trans. Pattern Anal. Mach. Intell. (PAMI) **35**, 823–34 (2013)
71. Vondrak, M., Sigal, L., Jenkins, O.C.: Physical simulation for probabilistic motion tracking. In: IEEE Conference on Computer Vision and Pattern Recognition (CVPR), pp. 1–8 (2008)
72. Wang, C., Bannuru, R., Ramel, J., Kupelnick, B., Scott, T., Schmid, C.: Tai Chi on psychological well-being: systematic review and meta-analysis. BMC Complement. Altern. Med. **10**, 23 (2010)
73. Winter, D.A.: Human balance and posture control during standing and walking. Gait Posture **3**, 193–214 (1995)
74. Zhou, X., Zhu, M., Leonardos, S., Derpanis, K.G., Daniilidis, K.: Sparseness meets deepness: 3D human pose estimation from monocular video. In: IEEE Conference on Computer Vision and Pattern Recognition (CVPR), pp. 4966–4975 (2016)
75. Zhou, X., Huang, Q., Sun, X., Xue, X., Wei, Y.: Towards 3D human pose estimation in the wild: a weakly-supervised approach. In: IEEE International Conference on Computer Vision (ICCV), pp. 398–407 (2017)

LevelSet R-CNN: A Deep Variational Method for Instance Segmentation

Namdar Homayounfar[1,2(✉)], Yuwen Xiong[1,2(✉)], Justin Liang[1(✉)],
Wei-Chiu Ma[1,3], and Raquel Urtasun[1,2]

[1] Uber Advanced Technologies Group, Pittsburgh, USA
{namdar,yuwen,justin.liang,weichiu,urtasun}@uber.com
[2] University of Toronto, Toronto, Canada
[3] MIT, Cambridge, USA

Abstract. Obtaining precise instance segmentation masks is of high importance in many modern applications such as robotic manipulation and autonomous driving. Currently, many state of the art models are based on the Mask R-CNN framework which, while very powerful, outputs masks at low resolutions which could result in imprecise boundaries. On the other hand, classic variational methods for segmentation impose desirable global and local data and geometry constraints on the masks by optimizing an energy functional. While mathematically elegant, their direct dependence on good initialization, non-robust image cues and manual setting of hyperparameters renders them unsuitable for modern applications. We propose LevelSet R-CNN, which combines the best of both worlds by obtaining powerful feature representations that are combined in an end-to-end manner with a variational segmentation framework. We demonstrate the effectiveness of our approach on COCO and Cityscapes datasets.

1 Introduction

Instance segmentation, the task of detecting and categorizing the pixels of unique countable objects in an image, is of paramount interest in many computer vision applications such as medical imaging [67], photo editing [68], pose estimation [50], robotic manipulation [21] and autonomous driving [69]. With the advent of deep learning [38] and its tremendous success in object classification and detection tasks [24,58,59], the computer vision community has made great strides in instance segmentation [2,3,10,30,47,64,66].

Currently, the prevailing instance segmentation approaches are based on the Mask R-CNN [27] framework which detects and classifies objects in the image and further processes each instance to produce a binary segmentation mask. While achieving impressive results in many benchmarks, the predicted masks

---

Electronic supplementary material The online version of this chapter (https://doi.org/10.1007/978-3-030-58592-1_33) contains supplementary material, which is available to authorized users.

A. Vedaldi et al. (Eds.): ECCV 2020, LNCS 12368, pp. 555–571, 2020.
https://doi.org/10.1007/978-3-030-58592-1_33

are produced at a low resolution and label predictions are independent per pixel, which could result in imprecise boundaries and irregular object discontinuities.

In contrast, traditional variational segmentation methods [6,7,31] are explicitly designed to delineate the boundaries of objects and handle complicated topologies. They first encode desired geometric properties into an energy functional and then evolve an initial contour according to the minimization landscape of the energy functional. One seminal work in this direction is the *Chan-Vese* [7] level set method, which formulates the segmentation problem as a partitioning task where the goal is to divide the image into two regions, each of which has similar intensity values. Through an energy formulation, Chan-Vese can produce good results even from a coarse initialization. However, in the real world, the photometric values may not be consistent, for example due to illumination changes and varying textures, rendering this method impractical for modern challenging applications.

With these problems in mind, we propose *LevelSet R-CNN*, a novel deep structured model that combines the strengths of modern deep learning with the energy based Chan-Vese segmentation framework. Specifically, we build our model in a multi-task setting following the Mask R-CNN framework: four different heads are utilized based on Feature Pyramid Network (FPN) [43] to output object localization and classification, a truncated signed distance function (TSDF) as the mask initialization, a set of instance-aware energy hyperparameters, and a deep object feature embedding, as shown in Fig. 1. These intermediate outputs are then passed into a differentiable unrolled optimization module to refine the predicted TSDF mask of each detected object by minimizing the Chan-Vese energy functional. This results in more precise object masks at higher resolutions.

We evaluate the effectiveness of our method on the challenging Cityscapes [15] instance segmentation task, where we achieve state-of-the-art results. We show also improvements over the baseline on the COCO [44] and the higher quality LVIS [25] datasets. Finally, we evaluate our model choices through extensive ablation studies.

2   Related Work

Instance Segmentation: Current modern instance segmentation methods can be classified as being either a top down or a bottom up approach. In a top down approach [5,8,9,22,33], region proposals for each instance are generated and a voting process is used to determine which ones to keep. Masks are predicted from these proposals to obtain the final instance segmentation output. For example, [17] uses a cascade of networks to predict boxes, estimate masks and categorize objects in a sequential manner so that the convolutional features are shared between the tasks. In [40], the authors use position sensitive inside/outside score maps to jointly perform detection and object segmentation. Recently, Mask R-CNN [27] augments Faster R-CNN [59] to achieve very strong instance segmentation performance across benchmarks. Following this paper, the authors in [30] optimize the scores of the bounding boxes to match the mask IoU, [47] adds a bottom to top aggregation path to allow for better information flow to improve the performance and [35, 66] extend it to panoptic segmentation. In [39] the authors improve an initial segmentation by fine-tuning it using a recurrent unit [14] that mimics level set evolution. Our approach is also top down. Here we add structure to the output space of Mask R-CNN by optimizing an explicit energy functional that incorporates geometrical constraints.

The bottom up approaches [4, 20, 36, 54, 64] typically perform segmentation by grouping the feature embeddings of individual instances without any early stage object proposals. In [42], the authors develop a model that predicts the category confidence, instance number and instance location and use a normalized spectral clustering algorithm [55] to group the instances together. In [69, 70], a CNN outputs instance labels followed by a Markov Random Field to achieve a coherent and consistent labeling of the global image. In [3], the authors exploit a CNN to output a deep watershed energy which can be thresholded to obtain the instance components. [46] use a sequence of neural networks to solve a subgrouping problem that gradually increase in complexity to group pixels of the same instance. In [32], the authors propose a multi task framework that as a subtask groups pixels by regressing a vector pointing towards the object's center. [53] are able to achieve real time instance segmentation by introducing and using a new clustering loss that encourages pixels to point towards an optimal region around the instance center. While bottom up approaches have a much simpler design than top down methods, they usually underperform in standard metrics such as average precision and recall. In our work, we cluster feature embeddings of an instance by differentiable optimization of an energy functional embedded within a state of the art top down approach.

Variational Methods: The classic pioneering active contour models (ACM) of [31] formulate the segmentation task as the minimization of an energy functional w.r.t. an explicit contour parametrization of the boundaries. This energy functional is comprised of a data term that moves the contours to areas of high gradient in the image. Furthermore, it regularizes the contour in terms of its smoothness and curvature. The shortcomings of ACM is that it is sensitive to initialization and requires heuristics such as re-sampling of points to handle changes of topology of the contour. The Level Set frameworks of [19, 56] overcome these challenges by formulating the segmentation task as finding the zero-level crossing of a higher dimensional function. In this framework, the contours of an object are implicitly defined as the zero crossing of an embedding function such as the TSDF. This eliminates the need for heuristics to handle complicated object topologies [16]. In this work, we build upon the level set framework put forward by Chan and Vese [7] where we exploit neural networks to learn robust features and optimization schedules from data.

In recent years, several works have explored combining these classical variational methods with neural networks. In the context of building segmentation from aerial images, CNNs have been deployed to output the energy terms used to evolve an active contour and develop a deep structured model that can be learned end-to-end [13,51]. In [26,45], the authors predict the offset to an initial circle to obtain object polygons and use a differentiable renderer to compare with the ground truth mask in the presence of a ground truth bounding box. However, they are not minimizing an explicit energy functional. These works focus on a simpler setting than us where detection is eschewed in favor of using ground truth boxes and a dedicated neural network for segmentation. Moreover, they parameterize the output space with explicit polygons which are not able to handle multi component objects without heuristics. In our work, we tackle the full instance segmentation setting with a single backbone and also use implicit level sets that can naturally handle complicated topologies without heuristics.

In the context of implicit contours, certain works have explored leveraging level sets in neural networks either as a post processing step to obtain ground truth data, or as a loss function for deep neural networks. In a semi-supervised setting, initial masks have been predicted for unlabeled data then further refined with level set evolution to create a quasi ground truth label [63]. The authors in [11,29,34] employ level set energies as a loss function for saliency estimation and semantic segmentation. In contrast, we employ level set optimization as a differentiable module within a deep neural network. In the experimental section, we evaluate the efficacy of using a level set loss function for the task of instance segmentation. The closest work to ours is [65], where the authors embed a different level set optimization framework within a neural network for the task of annotator in the loop foreground segmentation. There are several key differences: (i) The energy formulation is different, whereas their work is built upon the edge based method of [6] to push the contour to the boundaries, we exploit the region based approach of [7] which imposes uniformity of object masks. (ii) their setting requires ground truth object bounding boxes to output the features used in the level set optimization, while we embed the optimization within Mask R-CNN to build on top of shared features. Note that our setting is much more challenging. In the experimental section, we extend their method to the setting of instance segmentation and compare to our proposed model.

3   Overview of Chan-Vese Segmentation

In this section we provide a brief overview of the classic Chan-Vese level set segmentation method [7], which we later combine in a differentiable manner with Mask R-CNN. Chan-Vese is a region based segmentation approach which is capable of segmenting objects with complex topologies, e.g., holes and multiple components. This method operates globally on image intensities and is not dependent on local well-defined edge information. At a high level, Chan-Vese partitions an image to foreground and background segments by minimizing an energy functional that encourages regions to have uniform intensity values.

Let I be an image defined on the image plane $\Omega \subset \mathbb{R}^2$. Suppose I contains only one object that we wish to segment. Let $\phi : \Omega \to \mathbb{R}$ be the truncated signed distance function (TSDF) to the boundaries of this object taking positive values inside the object and negative outside. Let the curve C correspond to, possibly multi-component, boundaries of this object. The curve C can implicitly be defined as the zero crossing of ϕ, i.e., $C = \{x \in \mathbb{R}^2 \mid \phi(x) = 0\}$.

The core idea is to evolve an initial TSDF ϕ_0 by minimizing an energy functional E such that the zero crossing C of the minimizer coincides with the object boundaries. In the Chan-Vese [7] framework, the energy functional is defined as:

$$E(\phi, c_1, c_2) = \lambda_1 \int_\Omega \|I(x) - c_1\|^2 \, H(\phi(x)) dx$$
$$+ \lambda_2 \int_\Omega \|I(x) - c_2\|^2 \, (1 - H(\phi(x))) dx + \mu \int_\Omega \delta(\phi(x)) \, \|\nabla \phi(x)\| \, dx \quad (1)$$

where H and δ are Heaviside and Dirac delta functions respectively. The first two terms encourage the image intensity values inside and outside of the object to be close to constants c_1 and c_2 respectively. These terms impose a partitioning of the image to two regions of similar intensity values. The last term regularizes the length of the zero level set C. The parameters μ, λ_1 and λ_2 are positive global hyperparameters that regulate the contribution of each energy term.

The minimization of Eq. (1) is achieved by alternatively optimizing the function ϕ and the constants c_1 and c_2. In particular, by holding ϕ fixed, the minimizer of Eq. (1) w.r.t. c_1 and c_2 is given by:

$$c_1(\phi) = \frac{\int_\Omega I(x) H(\phi(x)) dx}{\int_\Omega H(\phi(x)) dx} \quad , \quad c_2(\phi) = \frac{\int_\Omega I(x)(1 - H(\phi(x))) dx}{\int_\Omega (1 - H(\phi(x))) dx} \quad (2)$$

We thus observe that c_1 and c_2 correspond to the average of the intensity values inside and outside of the object respectively.

Next, by holding c_1 and c_2 fixed and introducing an artificial time constant $t \geq 0$, we compute the functional derivative of E w.r.t. ϕ:

$$\frac{\partial \phi(\varepsilon)}{\partial t} = \delta_\varepsilon(\phi)\left(\mu \operatorname{div}(\frac{\nabla \phi}{\|\nabla \phi\|}) - \lambda_1 \|I - c_1\|^2 + \lambda_2 \|I - c_2\|^2\right) \quad (3)$$

where div is the divergence operator, ∇ is the spatial derivative and we have used a soft version of H and δ defined as:

$$H_\varepsilon(z) = \frac{1}{2}\left(1 + \frac{2}{\pi}\arctan(\frac{z}{\varepsilon})\right) \quad , \quad \delta_\varepsilon(z) = \frac{1}{\pi} \cdot \frac{\varepsilon^2}{\varepsilon^2 + z^2} \quad (4)$$

Finally, the update step of ϕ is given by:

$$\phi_n = \phi_{n-1} + \Delta t \frac{\partial \phi(\varepsilon)}{\partial t} \quad (5)$$

The alternating optimization is repeated for N iterations. This procedure draws similarities to clustering techniques such as K-Means, where the optimization involves alternating assignments and cluster center computations.

While Chan-Vese segmentation is mathematically elegant and powerful, working directly on image intensities is not robust due to factors such as lighting, different textures, motion blur or backgrounds that have similar intensities to the foreground. Moreover, the energy and optimization hyperparameters such as μ, λ_1 and λ_2, that balance the energy terms, and ε and Δt that regulate the gradient descent have to be manually adjusted depending on the image and domain. Furthermore, different objects have different optimal hyperparameters as their appearance and resolution might be very different. As a consequence, this method is not used in modern segmentation algorithms. In this paper, we leverage the power of deep learning to learn high dimensional object representations where the representations of pixels of the same object instance cluster together. We also learn complex inference schedules via data dependent adaptive hyperparameters for the energy terms and the optimization.

4   LevelSet R-CNN

In this section, we develop a deep structured model for the task of instance segmentation by combining the strengths of modern deep neural networks with the classical continuous energy based Chan-Vese [7] segmentation framework. In particular, we build on top of Mask R-CNN [27], which has been widely adopted for object localization and segmentation. However, the masks it produces suffer from low resolution resulting in segmentations that roughly have the right shape but are not precise. Moreover, pixel predictions are independent and there is no explicit mechanism encouraging neighboring pixels to have the same label. On the other hand, the Chan-Vese segmentation framework provides an elegant mathematical approach for global region based segmentation which encourages the pixels within the object to have the same label. However, it suffers in the presence of objects with different appearances within the instance, as it relies on non-robust intensity cues. In this paper we take the best of both worlds by combining these two paradigms.

We build on top of Mask R-CNN to first locate the objects in the image from the detection branch. Next, for each detected RoI corresponding to that object, we predict an initial TSDF ϕ_0, the set of hyperparameters $\{\mu, \lambda_1, \lambda_2\}$ for the energy terms and $\{\varepsilon, \Delta t\}$ for the optimization, and finally a deep feature embedding F that will replace the image intensities in (1). These predictions in turn will be fed into the Chan-Vese module where the costs are created and the optimization is unrolled for N steps as layers of a feedforward neural network. This module will output an evolved TSDF ϕ_N for each object such that its zero crossing corresponds to the boundaries of this object.

In what follows, we first describe how we build on top of Mask R-CNN, and then discuss how inference is performed in the deep Chan-Vese module. Finally, we will describe how learning is done in an end-to-end manner.

4.1   LevelSet R-CNN Architecture

Here we describe the specifics of the backbone and the additional heads of our model that provide the necessary components for the Chan-Vese optimization. The model architecture is presented in Fig. 1.

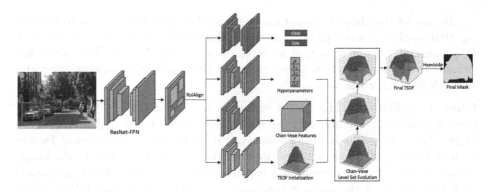

Fig. 1. LevelSet R-CNN for Instance Segmentation: We build on top of Mask R-CNN to first detect and classify all objects in the image. Then for each detection, the corresponding RoI is fed to a series of convolutions to obtain a truncated signed distance function (TSDF) initialization, a deep feature tensor, and a set of instance aware adaptive hyperparameters. These in turn are inputted into an unrolled Chan-Vese level set optimization procedure which outputs a final TSDF. We obtain a mask by applying the Heaviside function to the TSDF.

Backbone, Object Localization and Classification: As our shared backbone we employ a Residual Network [28] augmented with an FPN [43] and an RPN [59] that provides object region proposals. For object localization and classification, we maintain the original head structure of Mask R-CNN where RoIs are passed through a series of fully connected layers to output bounding box coordinates and object classification scores. Next, using *RoIAlign* [27] we extract features from the backbone that are further processed by the *initial TSDF head*, *hyperparameter head*, and the *Chan-Vese features head*. We denote the features corresponding to a RoI by r_m for $m \in \{1, \ldots, M\}$. We refer the reader to the supplementary material for the exact architectural details.

Initial TSDF Head: We replace the binary output of the mask head of Mask R-CNN to produce a TSDF output instead. Specifically, each pixel of the 28×28 output provides the signed ℓ_2 distance to the closest point on the object boundary. Furthermore, we threshold the values to a fixed symmetric range and normalize to $[-1, 1]$. This output is upsampled to 112×112 and used as the initial TSDF $\phi_0(r_m)$ in Eq. (1).

Hyperparameter Head: Each object instance could benefit from an adaptive set of hyperparameters for the energy terms and optimization steps. To achieve this, we output $\lambda_1(r_m)$ and $\lambda_2(r_m)$ to adaptively balance the influence of the foreground and background pixels in Eq. 1 for the object. We also predict $\mu(r_m)$ to regulate the length of its boundary. For the optimization hyperparameters, we output a separate $\varepsilon_n(r_m)$ for each of the N iterations. As shown in Eqs. 3 and 5, larger values of $\varepsilon_n(r_m)$ update the TSDF ϕ_n more globally and smaller values focus the evolution on the boundaries. Similarly, we output N step sizes $\Delta t_n(r_m)$ for each gradient descent step.

To predict the above hyperparameters, we add an additional head to the RoI r_m that applies a series of convolutions followed by average pooling and two fully connected layers to output a vector of dimension $2N + 3$. To ensure that these hyperparameters are positive, we found that applying a *sigmoid* layer and multiplying by 2 works well.

Chan-Vese Features Head: The energy in Eq. (1) encourages partitioning of the image based on the uniformity of image intensities I inside and outside of the object. However, image intensity values can be non-regular due to many factors such as lighting, different textures, motion blur, etc. Hence, we map the image intensities to a higher dimensional feature embedding space which is learned such that pixels of the same instance are close together in embedding space. We achieve this by passing the RoI r_m through a sequence of convolutions and upsampling layers to output a feature embedding $F(r_m)$ of dimension $C \times H \times W$. In our experiments we found $C = 64$ and $H = W = 112$ to be the most efficient in terms of memory for training and inference. The feature embedding $F(r_m)$ will replace the image intensities I in Eq. 1.

Chan-Vese Optimization as a Recurrent Net: After obtaining the initial TSDF, the set of hyperparameters, and the Chan-Vese feature map, we optimize the following deep energy functional E_m for each RoI r_m:

$$E_m(\phi, c_1, c_2) = \lambda_1(r_m) \int_{\Omega_m} \|F(r_m)(x) - c_1\|^2 H(\phi(x))dx$$
$$+ \lambda_2(r_m) \int_{\Omega_m} \|F(r_m)(x) - c_2\|^2 (1 - H(\phi(x)))dx$$
$$+ \mu(r_m) \int_{\Omega_m} \delta_\varepsilon(\phi(x)) \|\nabla\phi(x)\| \, dx \tag{6}$$

Note that the integration is over the image subset $\Omega_m \subset \Omega$ corresponding to r_m. We perform alternating optimization of ϕ and c_1, c_2. We implement the ϕ update step:

$$\phi_n = \phi_{n-1} + \Delta t_n(r_m) \frac{\partial\phi(\varepsilon_n(r_m))}{\partial t} \tag{7}$$

for $n = 1, \ldots, N$ as a set of feedforward layers with

$$\frac{\partial\phi(\varepsilon_n(r_m))}{\partial t} = \delta_{\varepsilon_n(r_m)}(\phi)\left(\mu(r_m)\text{div}(\frac{\nabla\phi}{\|\nabla\phi\|}) - \lambda_1(r_m)\|F(r_m) - c_1\|^2\right.$$
$$\left. + \lambda_2(r_m)\|F(r_m) - c_2\|^2\right) \tag{8}$$

In practice, we implement the gradient and the divergence term by using the Sobel operator [61] and the integration as a sum on the discrete image grid. At each update step, the constants c_1 and c_2 have closed-form updates as:

$$c_1(\phi) = \frac{\int_{\Omega_m} F(r_m)(x)H(\phi(x))dx}{\int_{\Omega_m} H(\phi(x))dx}, c_2(\phi) = \frac{\int_{\Omega_m} F(r_m)(x)(1 - H(\phi(x)))dx}{\int_{\Omega_m}(1 - H(\phi(x)))dx} \tag{9}$$

Here c_1 and c_2 are vectors where each element is the average of the corresponding feature embedding channel inside or outside of the object in the ROI respectively.

4.2 Learning

We train our model jointly in an end-to-end manner, as the Mask R-CNN backbone, the three extra heads, and the deep Chan-Vese recurrent network are all fully differentiable. We employ the standard regression and cross-entropy losses for the bounding box and classification components of both the RPN and the detection/classification heads of the backbone. For training the weights of the initial TSDF head, the hyperparameter head and the Chan-Vese features head, we apply the following loss, which is a mix of l_1 and binary cross-entropy BCE, to the initial and final TSDFs ϕ_0 and ϕ_N:

$$\ell_{TSDF}(\phi_{\{0,N\}}, \phi_{GT}, M_{GT}) = \left\| \phi_{\{0,N\}} - \phi_{GT} \right\|_1 + BCE(H_\varepsilon(\phi_{\{0,N\}}), M_{GT})$$

Here M_{GT} and ϕ_{GT} are the ground truth mask and TSDF targets. In order to apply BCE on ϕ_0 and ϕ_N, similar to [65] we map them to $[0,1]$ with the soft Heaviside function and $\varepsilon = 0.1$. During backpropagation, the loss gradient from ϕ_N flows through the unrolled level set optimization and then through the Chan-Vese features head, the hyperparameter head, and the initial TSDF head.

5 Experimental Evaluation

In this section, we describe the datasets, implementation details and the metrics and compare our approach with the state-of-the-art. Next, we study the various aspects of our proposed approach through ablations.

Datasets: We evaluate our model on Cityscapes [15] and COCO [44] datasets. Cityscapes contains very precise annotations for 8 categories split into 2975 train, 500 validation and 1525 test images of resolution 1024×2048. The COCO dataset has 80 categories with 118k images in the `train2017` set for training and 5k images in the `val2017` set for evaluation. However, as demonstrated quantitatively by [25], COCO does not consistently provide accurate object annotations rendering mask quality evaluation of a method not indicative. As such, we follow the approach of [37] and also evaluate our model on the COCO sub-categories of the validation set of the LVIS dataset [25] with our model trained only on COCO. Note that LVIS re-annotates all the COCO validation images with high quality masks which makes it suitable for evaluating mask improvements. We follow this protocol since the LVIS dataset has more than 1000 categories and is designed for large vocabulary instance segmentation which is still in its infancy and an exciting topic for future research.

Table 1. Instance segmentation on Cityscapes val and test sets: This table shows our instance segmentation results on Cityscape on val and test. We report models trained on Cityscapes with and without COCO/Mapillary pre-training as well the methods that use horizontal flipping (F) or multiscale (MS) inference at test time.

	Training data	AP_{val}	AP_{test}	AP_{test}^{50}	person	rider	car	truck	bus	train	mcycle	bcycle
DWT [3]	fine	21.2	19.4	35.3	15.5	14.1	31.5	22.5	27.0	22.9	13.9	8.0
Kendall et al. [32]	fine	–	21.6	39.0	19.2	21.4	36.6	18.8	26.8	15.9	19.4	14.5
Arnab et al. [2]	fine	–	23.4	45.2	21.0	18.4	31.7	22.8	31.1	**31.0**	19.6	11.7
SGN [46]	fine+coarse	29.2	25.0	44.9	21.8	20.1	39.4	24.8	33.2	30.8	17.7	12.4
PolyRNN++ [1]	fine	–	25.5	45.5	29.4	21.8	48.3	21.2	32.3	23.7	13.6	13.6
Mask R-CNN [27]	fine	31.5	26.2	49.9	30.5	23.7	46.9	22.8	32.2	18.6	19.1	16.0
BShapeNet+ [33]	fine	–	27.3	50.4	29.7	23.4	46.7	26.1	33.3	24.8	20.3	14.1
GMIS [48]	fine+coarse	–	27.3	45.6	31.5	25.2	42.3	21.8	37.2	28.9	18.8	12.8
Neven et al. [53]	fine	–	27.6	50.9	34.5	26.1	52.4	21.7	31.2	16.4	20.1	18.9
PANet [47]	fine	36.5	31.8	57.1	36.8	**30.4**	**54.8**	27.0	36.3	25.5	22.6	**20.8**
Ours	fine	**37.9**	**33.3**	**58.2**	**37.0**	29.2	54.6	**30.4**	**39.4**	30.2	**25.5**	20.3
AdaptIS [62] (F)	fine	36.3	32.5	52.5	31.4	29.1	50.0	31.6	41.7	**39.4**	24.7	12.1
SSAP [23] (MS+F)	fine	37.3	32.7	51.8	35.4	25.5	55.9	**33.2**	**43.9**	31.9	19.5	16.2
Pan-DL [12] (MS+F)	fine	38.5	34.6	57.3	34.3	28.9	55.1	32.8	41.5	36.6	26.3	21.6
Ours (MS+F)	fine	**40.0**	**35.8**	**61.2**	**40.5**	**31.7**	**56.9**	31.4	42.4	32.5	**28.6**	**22.2**
Mask R-CNN [27]	fine+COCO	36.4	32.0	58.1	34.8	27.0	49.1	30.1	40.9	30.9	24.1	18.7
BShapeNet+ [33]	fine+COCO	–	32.9	58.8	36.6	24.8	50.4	33.7	41.0	33.7	25.4	17.8
UPSNet [66]	fine+COCO	37.8	33.0	59.7	35.9	27.4	51.9	31.8	43.1	31.4	23.8	19.1
PANet [47]	fine+COCO	41.4	36.4	63.1	41.5	33.6	58.2	31.8	45.3	28.7	28.2	24.1
Pan-DL [12]	fine+MV	42.5	39.0	64.0	36.0	30.2	56.7	**41.5**	**50.8**	**42.5**	30.4	23.7
Polytransform [41]	fine+COCO	**44.6**	**40.1**	**65.9**	42.4	**34.8**	58.5	39.8	50.0	41.3	30.9	23.4
Ours (COCO)	fine+COCO	43.3	40.0	65.7	**43.4**	33.9	**59.0**	37.6	49.4	39.4	**32.5**	**24.9**

Implementation Details: For Cityscapes, we follow [27] and adopt multi-scale training where we resize the input image in a way that the length of the shorter edge is randomly sampled from [800, 1024]. We train the model on 8 GPUs for 24 K iterations with a learning rate of 0.01, decayed to 0.001 at 18 K iterations. We set the loss weights for the initial and final TSDF output to 1 and 5 in the multitask objective. For COCO, following [27], we train the model without multi-scaling on 16 GPUs for 90 K iterations with a learning rate of 0.02 decayed by a factor of 10 at 60 K and 80 K iterations. We set the loss weights for the initial and final TSDF output to 0.2 and 1 in the multitask objective. For both datasets, we set the weight decay as 0.0001, with mini-batch size of 8. We employ WideResNet-38 [60] on Cityscapes test set and ResNet-50 [28] in all the other experiments. For the level set optimization, we unroll the optimization for 3 steps. Finally, we simply apply the Heaviside function to the TSDF output to obtain a mask. Note that if we apply marching squares [49] to the final TSDF instead, we could obtain sub-pixel accuracy for the boundaries. However for simplicity and since the AP metric of COCO and Cityscapes requires binary masks for evaluation, we simply threshold our TSDFs using the heaviside function.

Evaluation Metrics: We report the standard AP metric of [44] on both Cityscapes and COCO. For LVIS, we report the federated average precision metric denoted by AP* [25] on the COCO subcategories.

Table 2. LevelSet R-CNN vs. Mask R-CNN: We report mask AP for both COCO and Cityscapes on the val set with `Resnet-50` backbone. We also report the federated AP, i.e. AP*, of the LVIS dataset with COCO subcategories trained only COCO. For our model we report both the initial and final mask results after optimization.

	Cityscapes AP	COCO AP	LVIS AP*
Mask R-CNN	32.3	33.8	35.6
Ours (Initial Mask)	35.4	33.7	35.8
Ours	**36.2**	**34.3**	**36.4**

Table 3. LevelSet R-CNN vs. deep level set methods on the val set of Cityscapes: Using level sets as a loss function (LS Loss) or using the geodesic level sets (DELSE).

	LS Loss [29]	DELSE [65]	DELSE [65] + HP head	Ours
Initial Mask AP	–	34.6	34.9	**35.4**
Final Mask AP	34.3	34.6	35	**36.2**

Cityscapes Test: We compare LevelSet R-CNN against published state-of-the-art (SOTA) methods on Cityscapes in Table 1. LevelSet R-CNN outperforms all previous methods that are trained on Cityscapes data without test time augmentations achieving a new state-of-the-art performance by 1.5 AP over PANet [47]. We also compare against models that adopt multiscale (MS) and horizontal flipping (F) at test time. We improve upon the state-of-the-art, Panoptic-Deeplab [12], by 1.2 AP. Next, we evaluate against models that pretrain on external datasets such as COCO [44] or Mapillary Vistas [52]. For a fair comparison, we follow the exact setting of Polytransform [41] which was the state-of-the-art at the time of submission. In particular, we use a `WideResNet-38` backbone with deformable convolutions [18] and PANet modifications [47]. We train on COCO for 270000 iterations with a learning rate of 0.02 decayed by a factor of 10 at 210000 and 250000 iterations on 16 GPUs. On Cityscapes, we finetuned for 6000 iterations on 8 GPUs with a learning rate of 0.01 decayed to 0.001 at 4000 iterations. As shown in Table 1, our performance is comparable with Polytransform.

AP Improvements Across Datasets: In Table 2, we compare Mask R-CNN with our initial and final mask outputs on the validation sets of Cityscapes, COCO and LVIS. All models employ the `Resnet-50` backbone. LevelSet R-CNN outperforms Mask R-CNN on all datasets. Note that while Levelset R-CNN was only trained on COCO and not with the precise boundaries of LVIS, it improves upon the baseline by 0.8 AP*. We also see an improvement of about 4 AP on Cityscapes.

Different Deep Level Set Formulations: To further justify our deep region based level set formulation, we compare with two different variations of level sets applied to the task of instance segmentation. [29] use the Chan-Vese energy as a

Table 4. Backproping through the initial mask from the final TSDF loss ℓ_{TSDF} on the val set of Cityscapes

	Detach ϕ_0	Backprop Thru ϕ_0
Initial Mask AP	33	**35.4**
Final Mask AP	34.2	**36.2**

Table 5. Boundary metric: on val set of Cityscapes: We evaluate the AF at thresholds of 1 and 2 pixels against Mask R-CNN across two backbones.

	Backbone	AF_1	AF_2
Mask R-CNN [27]	Resnet-50	40.2	57.6
Ours	Resnet-50	**45.8**	**63.1**
Mask R-CNN [27]	WideResnet-38	42.7	59.9
Ours	WideResnet-38	**46.8**	**64.6**

loss function for salient object detection. Here, we employ their loss for instance segmentation. In particular, we shift the mask output of Mask R-CNN by -0.5, apply the soft Heaviside and pass to the Chan-Vese energy loss function. In Table 3 we observe that LevelSet R-CNN improves the level set loss by about 2 AP. Next, we combine the deep edge based level set of [65], referred to as DELSE, with Mask R-CNN by changing the mask head to a TSDF head and adding two more heads: the velocity head that predicts the direction to the object boundaries and the modulation head which regulates the effect of the curvature term on the object boundaries. We evaluate in two settings: 1) Similar to their work, we use hand-tuned hyperpameters and use their exact loss functions, i.e., L_2 for the initial TSDF, class balanced cross entropy for the final TSDF and L_2 on angular domain. 2) To remove the effect of loss functions and hyperparameters choices and provide the most fair comparison, we add our hyperparameter head to their model and use our loss functions with the exception of having an extra loss function for the velocity head. In Table 3, we observe that LevelSet R-CNN has a 1.2 AP improvement over the DELSE formulation. Moreover, we obtain 0.8 AP improvement over the initial mask whereas DELSE gain is 0.1 AP.

Passing Gradients Through the Initial TSDF: Table 4 shows that by passing the gradient from ϕ_N through the initial TSDF head, we improve the AP of both the initial TSDF ϕ_0 and the final TSDF ϕ_N. As an alternative we could have detached ϕ_0 from the computation graph so that it does not take supervision from the final TSDF ϕ_N; Passing the gradient improves ϕ_0 by 2.4 AP and ϕ_N by 2 AP. Finally, we see that the AP of the initial TSDF when passing gradients is higher than the AP of the final TSDF when not passing the gradient by 1.2 AP. This suggests that the hyperparameter head, the Chan-Vese features head and the unrolled optimization, can also be used during training for improving the performance of the mask head and discarded during inference.

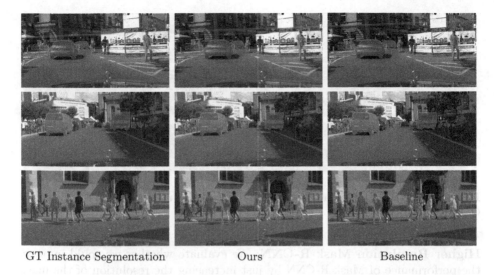

GT Instance Segmentation   Ours   Baseline

Fig. 2. We showcase qualitative instance segmentation results of our model on the Cityscapes validation set. We can see that our method produces masks with higher quality when bounding box results are similar.

Boundary Metric: In addition, to evaluate the capacity of our model in improving the boundaries of objects, we adapt the boundary metric of DAVIS [57] to our task. In particular, for a True Positive detection, we compute F1 between the prediction and ground truth boundary pixels at thresholds of 1 and 2 pixels. Similar to AP, we obtain the True Positives at IoUs in range [0.5, 0.95] at 0.05 increments. The F1s are averaged over all the classes and thresholds and are denoted by (AF_1) and (AF_2) for thresholds of 1 and 2 pixels away. In Table 5, we observe that our method is able to improve the boundaries of the objects by at least 4 AF at each threshold compared to the baseline across the two backbones `Resnet-50` and `WideResnet-38`.

Mask R-CNN with Different Training Targets: We modify the mask head of Mask R-CNN to output a TSDF instead of a binary mask and we train with ℓ_{TSDF} rather than BCE as loss function. To understand the dependence of the Mask R-CNN performance on this TSDF target ℓ_{TSDF}, we trained a model with only the mask head modification and without the other Chan-Vese components (i.e., the adaptive hyperparameter head and the deep Chan-Vese module and unrolled optimization). We obtain the same AP of 32.3 for model with ℓ_{TSDF} as the original Mask R-CNN. This indicates that the model improvements do not simply come from changing the loss of the mask head.

Effect of the Hyperparameter Head: To verify the importance of learning adaptive hyperparameters per object instance, we perform an ablation where we remove the hyperparameter head and just learn a set of global hyperparameters for the whole dataset. The adaptive hyper parameter head achieves 36.2 AP vs. the 35.4 AP of a global set giving a boost of 0.8 AP.

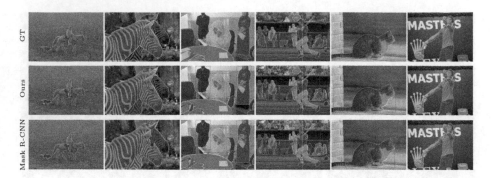

Fig. 3. We showcase qualitative instance segmentation results of our model on the COCO validation set.

Higher Resolution Mask R-CNN: We evaluate whether we could improve the performance of Mask R-CNN by just increasing the resolution of the mask head. We train Mask R-CNN on Cityscapes with `ResNet-50` backbone at 112×112 resolution which is the same resolution of the Chan-Vese features and our final TSDF ϕ_N. Interestingly the performance drops by 2.2 AP from 32.3. We hypothesize that by increasing the resolution, the ratio of non-boundary pixels vs. boundary pixels will become higher and they dominate the loss function gradients leading to worse masks. In our proposed method however, there is a global competition between the foreground/background regions to minimize the energy and hence we are able to increase the resolution.

Inference Time: LevelSet R-CNN with a `ResNet-50` runs on average at 182 ms vs. Mask R-CNN at 145 ms on GTX 1080 ti on images of dimension 1024×2048.

Qualitative Results: As shown in Figs. 2 and 3 we observe mask boundary and region improvements compared to the baseline.

6 Conclusion

In this paper, we proposed *LevelSet R-CNN* which combines the strengths of modern deep learning based Mask R-CNN and classical energy based Chan-Vese level set segmentation framework in an end-to-end manner. In particular, we utilize four heads based on FPN to obtain each detected object, an initial level set, deep robust feature representation for the Chan-Vese energy data terms and a set of instance dependent hyperparameters that balance the energy terms and schedule the optimization procedure. We demonstrated the effectiveness of our method on COCO and Cityscapes showing improvements on both datasets.

References

1. Acuna, D., Ling, H., Kar, A., Fidler, S.: Efficient interactive annotation of segmentation datasets with polygon-RNN++. In: CVPR (2018)

2. Arnab, A., Torr, P.H.S.: Pixelwise instance segmentation with a dynamically instantiated network. In: CVPR (2017)
3. Bai, M., Urtasun, R.: Deep watershed transform for instance segmentation. In: CVPR (2017)
4. Brabandere, B.D., Neven, D., Gool, L.V.: Semantic instance segmentation with a discriminative loss function. In: CVPR (2017)
5. Cai, Z., Vasconcelos, N.: Cascade R-CNN: delving into high quality object detection. In: CVPR (2018)
6. Caselles, V., Kimmel, R., Sapiro, G.: Geodesic active contours. Int. J. Comput. Vis. **22**(1), 61–79 (1997)
7. Chan, T.F., Vese, L.A.: Active contours without edges. IEEE Trans. Image Process. **10**(2), 266–277 (2001)
8. Chen, K., et al.: Hybrid task cascade for instance segmentation. In: CVPR (2019)
9. Chen, L.C., Hermans, A., Papandreou, G., Schroff, F., Wang, P., Adam, H.: Masklab: instance segmentation by refining object detection with semantic and direction features. In: CVPR (2018)
10. Chen, X., Girshick, R.B., He, K., Dollár, P.: TensorMask: a foundation for dense object segmentation. In: ICCV (2019)
11. Chen, X., Williams, B.M., Vallabhaneni, S.R., Czanner, G., Williams, R.S., Zheng, Y.: Learning active contour models for medical image segmentation. In: CVPR (2019)
12. Cheng, B., et al.: Panoptic-DeepLab: a simple, strong, and fast baseline for bottom-up panoptic segmentation. In: CVPR (2020)
13. Cheng, D., Liao, R., Fidler, S., Urtasun, R.: DARNet: deep active ray network for building segmentation. In: CVPR (2019)
14. Cho, K., van Merrienboer, B., Gulcehre, C., Bougares, F., Schwenk, H., Bengio, Y.: Learning phrase representations using RNN encoder-decoder for statistical machine translation. In: EMNLP 2014 (2014)
15. Cordts, M., et al.: The cityscapes dataset for semantic urban scene understanding. In: CVPR (2016)
16. Cremers, D., Rousson, M., Deriche, R.: A review of statistical approaches to level set segmentation: integrating color, texture, motion and shape. Int. J. Comput. Vis. **72**(2), 195–215 (2007)
17. Dai, J., He, K., Sun, J.: Instance-aware semantic segmentation via multi-task network cascades. In: CVPR (2015)
18. Dai, J., et al.: Deformable convolutional networks. In: ICCV (2017)
19. Dervieux, A., Thomasset, F.: A finite element method for the simulation of a Rayleigh-Taylor instability. In: Rautmann, R. (ed.) Approximation Methods for Navier-Stokes Problems. LNM, vol. 771, pp. 145–158. Springer, Heidelberg (1980). https://doi.org/10.1007/BFb0086904
20. Fathi, A., et al.: Semantic instance segmentation via deep metric learning. ArXiv (2017)
21. Fazeli, N., Oller, M., Wu, J., Wu, Z., Tenenbaum, J.B., Rodriguez, A.: See, feel, act: hierarchical learning for complex manipulation skills with multisensory fusion. Sci. Robot. (2019)
22. Fu, C.Y., Shvets, M., Berg, A.C.: RetinaMask: learning to predict masks improves state-of-the-art single-shot detection for free. ArXiv (2019)
23. Gao, N., et al.: SSAP: single-shot instance segmentation with affinity pyramid. In: ICCV (2019)
24. Girshick, R.B.: Fast R-CNN. In: ICCV (2015)

25. Gupta, A., Dollar, P., Girshick, R.: LVIS: a dataset for large vocabulary instance segmentation. In: CVPR (2019)
26. Gur, S., Shaharabany, T., Wolf, L.: End to end trainable active contours via differentiable rendering. In: ICLR (2020)
27. He, K., Gkioxari, G., Dollár, P., Girshick, R.: Mask R-CNN. In: CVPR (2017)
28. He, K., Zhang, X., Ren, S., Sun, J.: Deep residual learning for image recognition. In: CVPR (2015)
29. Hu, P., Shuai, B., Liu, J., Wang, G.: Deep level sets for salient object detection. In: CVPR (2017)
30. Huang, Z., Huang, L., Gong, Y., Huang, C., Wang, X.: Mask scoring R-CNN. In: CVPR (2019)
31. Kass, M., Witkin, A., Terzopoulos, D.: Snakes: active contour models. Int. J. Comput. Vis. **1**(4), 321–331 (1988)
32. Kendall, A., Gal, Y., Cipolla, R.: Multi-task learning using uncertainty to weigh losses for scene geometry and semantics. In: CVPR (2018)
33. Kim, H.Y., Kang, B.R.: BshapeNet: object detection and instance segmentation with bounding shape masks. Pattern Recogn. Lett. **131**, 449–455 (2020)
34. Kim, Y., Kim, S., Kim, T., Kim, C.: CNN-based semantic segmentation using level set loss. In: WACV (2019)
35. Kirillov, A., He, K., Girshick, R., Rother, C., Dollar, P.: Panoptic segmentation. In: CVPR (2019)
36. Kirillov, A., Levinkov, E., Andres, B., Savchynskyy, B., Rother, C.: InstanceCut: from edges to instances with multicut. In: CVPR (2017)
37. Kirillov, A., Wu, Y., He, K., Girshick, R.: Pointrend: image segmentation as rendering. In: ECCV (2020)
38. Krizhevsky, A., Sutskever, I., Hinton, G.E.: ImageNet classification with deep convolutional neural networks. In: NeurIPS (2012)
39. Le, T.H.N., Quach, K.G., Luu, K., Duong, C.N., Savvides, M.: Reformulating level sets as deep recurrent neural network approach to semantic segmentation. IEEE Trans. Image Process. **27**(5), 2393–2407 (2018)
40. Li, Y., Qi, H., Dai, J., Ji, X., Wei, Y.: Fully convolutional instance-aware semantic segmentation. In: CVPR (2017)
41. Liang, J., Homayounfar, N., Ma, W.C., Xiong, Y., Hu, R., Urtasun, R.: PolyTransform: deep polygon transformer for instance segmentation. In: CVPR (2020)
42. Liang, X., Lin, L., Wei, Y., Shen, X., Yang, J., Yan, S.: Proposal-free network for instance-level object segmentation. IEEE Trans. Pattern Anal. Mach. Intell. **40**(12), 2978–2991 (2018)
43. Lin, T.Y., Dollár, P., Girshick, R.B., He, K., Hariharan, B., Belongie, S.J.: Feature pyramid networks for object detection. In: CVPR (2016)
44. Lin, T.-Y., et al.: Microsoft COCO: common objects in context. In: Fleet, D., Pajdla, T., Schiele, B., Tuytelaars, T. (eds.) ECCV 2014. LNCS, vol. 8693, pp. 740–755. Springer, Cham (2014). https://doi.org/10.1007/978-3-319-10602-1_48
45. Ling, H., Gao, J., Kar, A., Chen, W., Fidler, S.: Fast interactive object annotation with curve-GCN. In: CVPR (2019)
46. Liu, S., Jia, J., Fidler, S., Urtasun, R.: SGN: sequential grouping networks for instance segmentation. In: ICCV (2017)
47. Liu, S., Qi, L., Qin, H., Shi, J., Jia, J.: Path aggregation network for instance segmentation. In: CVPR (2018)
48. Liu, Y., et al.: Affinity derivation and graph merge for instance segmentation. In: ECCV (2018)

49. Lorensen, W.E., Cline, H.E.: Marching cubes: a high resolution 3D surface construction algorithm. In: SIGGRAPH (1987)
50. Ma, W.C., Wang, S., Hu, R., Xiong, Y., Urtasun, R.: Deep rigid instance scene flow. In: CVPR (2019)
51. Marcos, D., et al.: Learning deep structured active contours end-to-end. In: CVPR (2018)
52. Neuhold, G., Ollmann, T., Rota Bulò, S., Kontschieder, P.: The mapillary vistas dataset for semantic understanding of street scenes. In: ICCV (2017)
53. Neven, D., Brabandere, B.D., Proesmans, M., Gool, L.V.: Instance segmentation by jointly optimizing spatial embeddings and clustering bandwidth. In: CVPR, June 2019
54. Newell, A., Huang, Z., Deng, J.: Associative embedding: end-to-end learning for joint detection and grouping. In: NeurIPS (2017)
55. Ng, A.Y., Jordan, M.I., Weiss, Y.: On spectral clustering: analysis and an algorithm. In: NeurIPS (2001)
56. Osher, S., Sethian, J.A.: Fronts propagating with curvature-dependent speed: algorithms based on Hamilton-Jacobi formulations. J. Comput. Phys. **79**(1), 12–49 (1988)
57. Perazzi, F., Pont-Tuset, J., McWilliams, B., Van Gool, L., Gross, M., Sorkine-Hornung, A.: A benchmark dataset and evaluation methodology for video object segmentation. In: CVPR (2016)
58. Redmon, J., Divvala, S.K., Girshick, R.B., Farhadi, A.: You only look once: unified, real-time object detection. In: CVPR (2015)
59. Ren, S., He, K., Girshick, R., Sun, J.: Faster R-CNN: towards real-time object detection with region proposal networks. In: NeurIPS (2015)
60. Rota Bulò, S., Porzi, L., Kontschieder, P.: In-place activated batchnorm for memory-optimized training of DNNs. In: CVPR (2018)
61. Sobel, I.: An isotropic 3×3 image gradient operator. Presentation at Stanford A.I. Project 1968, February 2014
62. Sofiiuk, K., Barinova, O., Konushin, A.: Adaptis: adaptive instance selection network. In: ICCV (2019)
63. Tang, M., Valipour, S., Zhang, Z., Cobzas, D., Jagersand, M.: A deep level set method for image segmentation. In: Cardoso, M.J., et al. (eds.) DLMIA/ML-CDS -2017. LNCS, vol. 10553, pp. 126–134. Springer, Cham (2017). https://doi.org/10.1007/978-3-319-67558-9_15
64. Uhrig, J., Cordts, M., Franke, U., Brox, T.: Pixel-level encoding and depth layering for instance-level semantic labeling. In: GCPR (2016)
65. Wang, Z., Acuna, D., Ling, H., Kar, A., Fidler, S.: Object instance annotation with deep extreme level set evolution. In: CVPR (2019)
66. Xiong, Y., et al.: UPSNet: a unified panoptic segmentation network. In: CVPR (2019)
67. Xu, Y., et al.: Gland instance segmentation by deep multichannel neural networks. In: MICCAI (2016)
68. Yao, S., et al.: 3D-aware scene manipulation via inverse graphics. In: NeurIPS (2018)
69. Zhang, Z., Fidler, S., Urtasun, R.: Instance-level segmentation for autonomous driving with deep densely connected MRFs. In: CVPR (2016)
70. Zhang, Z., Schwing, A.G., Fidler, S., Urtasun, R.: Monocular object instance segmentation and depth ordering with CNNs. In: ICCV (2015)

Efficient Scale-Permuted Backbone with Learned Resource Distribution

Xianzhi Du[✉], Tsung-Yi Lin, Pengchong Jin, Yin Cui, Mingxing Tan,
Quoc Le, and Xiaodan Song

Google Research, Brain Team, Mountain View, USA
{xianzhi,tsungyi,pengchong,yincui,tanmingxing,qvl,xiaodansong}@google.com

Abstract. Recently, SpineNet has demonstrated promising results on object detection and image classification over ResNet model. However, it is unclear if the improvement adds up when combining scale-permuted backbone with advanced efficient operations and compound scaling. Furthermore, SpineNet is built with a uniform resource distribution over operations. While this strategy seems to be prevalent for scale-decreased models, it may not be an optimal design for scale-permuted models. In this work, we propose a simple technique to combine efficient operations and compound scaling with a previously learned scale-permuted architecture. We demonstrate the efficiency of scale-permuted model can be further improved by learning a resource distribution over the entire network. The resulting efficient scale-permuted models outperform state-of-the-art EfficientNet-based models on object detection and achieve competitive performance on image classification and semantic segmentation.

Keywords: Scale-permuted model · Object detection

1 Introduction

The scale-permuted network proposed by Du *et al.* [4] opens up the design of a new family of meta-architecture that allows wiring features with a scale-permuted ordering in convolutional neural network. The scale-permuted architecture achieves promising results on visual recognition and localization by significantly outperforming its scale-decreased counterpart when using the same residual operations but different architecture topology. Concurrently, EfficientNet-based models [23,24] demonstrate state-of-the-art performance using an advanced MBconv operation and the compound model scaling rule, while still adopting a scale-decreased backbone architecture design. A natural question is: *can we obtain new state-of-the-art performance by combining scale-permuted architecture and efficient operations?*

Electronic supplementary material The online version of this chapter (https://doi.org/10.1007/978-3-030-58592-1_34) contains supplementary material, which is available to authorized users.

A. Vedaldi et al. (Eds.): ECCV 2020, LNCS 12368, pp. 572–586, 2020.
https://doi.org/10.1007/978-3-030-58592-1_34

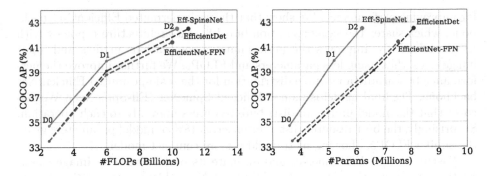

Fig. 1. Eff-SpineNet achieves better FLOPs *vs.* AP and Params *vs.* AP trade-off curves for regular-size object detection comparing to state-of-the-art scale-decreased models EfficientNet-FPN and EfficientDet. All models adopt the RetinaNet framework [15]

In this paper, we decompose the model design into three parts: (1) architecture topology; (2) operation; (3) resource distribution. The architecture topology describes the wiring and the resolution of features. The operation defines the transformation (*e.g.*, convolution and ReLU) applied to the features. The resource distribution indicates the computation allocated for each operation. Our study begins with directly combining the scale-permuted architecture topology from [4] and efficient operations from [23]. Unlike the previous works, we purposely *do not* perform any neural architecture search because the architecture topology and operation have been extensively studied and learned by sophisticated neural architecture search algorithms respectively. Instead of designing a joint search space for learning an even more tailored model, we are curious if the scale-permuted architecture and efficient operations are *generic* in the status quo and can directly be used to build the state-of-the-art model.

Despite having the learned advanced architecture topology and operation, the resource distribution has not been well studied in isolation in existing works. In [4], the resource distribution is nearly uniform for all operations, regardless of the resolution and location of a feature in the architecture. In [23], the search space contains only a few hand-selected feature dimensions for each operation and the neural architecture search algorithm is learned to select the best one. This greatly limits the possible resource distribution over the entire network. In this work, we propose a search algorithm that learns the resource distribution with the fixed architecture topology and operation. Given the target resource budget, we propose to learn the percentage of total computation allocated to each operation. In contrast to learning the absolute feature dimension, our resource targeted algorithm has the advantage of exploring a wider range of resource distribution in a manageable search space size.

We mainly conduct experiments on object detection using COCO dataset [16]. We carefully study the improvements brought by the architecture topology and operation and discover that simply combining scale-permuted architecture and MBConv operation outperforms EfficientDet [24]. The experiment results show that the architecture topology and operation are complementary for improving performance. We show that the scale-permuted EfficientNet backbone, which shares the same operation but different architecture topology with EfficientNet-FPN, improves the performance across various models and input image sizes while using less parameters and FLOPs. We further improve the performance by learning a resource distribution for the scale-permuted EfficientNet backbone. The final model is named Efficient SpineNet (Eff-SpineNet). We discover that the model prefers to distribute resources unevenly to each operation. Surprisingly, the best resource distribution saves 18% of model parameters given the similar FLOPs, allowing us to build a more compact model.

Lastly, we take Eff-SpineNet and evaluate its performance on image classification and semantic segmentation. Eff-SpineNet achieves competitive results on both tasks. Interestingly, we find that Eff-SpineNet is able to retain the performance with less parameters. Compared with EfficientNet that is specifically designed for image classification, Eff-SpineNet has around 35% less parameters under the same FLOPs, while the Top-1 ImageNet accuracy drops by less than 1–2%. For semantic segmentation, Eff-SpineNet models achieve comparable mIOU on PASCAL VOC val 2012 to popular semantic segmentation networks, such as the DeepLab family [1,2], while using 95% less FLOPs. To summarize, these observations show that Eff-SpineNet is versatile and is able to transfer to other visual tasks including image classification and semantic segmentation.

2   Related Work

Scale-Permuted Network: Multi-scale feature representations have been the core research topic for object detection and segmentation. The dominant paradigm is to have a strong backbone model with a lightweight decoder such as feature pyramid networks [14]. Recently, many work has discovered performance improved with a stronger decoder [6,18,26]. Inspired by NAS-FPN [6], SpineNet [4] proposes a scale-permuted backbone architecture that removes the distinction of encoder and decoder and allows the scales of intermediate feature maps to increase or decrease anytime, and demonstrates promising performance on object detection and image classification. Auto-DeepLab [17] is another example that builds scale-permuted models for semantic segmentation.

Efficient Operation: The efficiency is the utmost important problem for mobile-size convolution model. The efficient operations have been extensively studied in the MobileNet family [10,10,21,22]. Spare depthwise convolution and the inverted bottleneck block are the core ideas for efficient mobile size network. MnasNet [22] and EfficientNet [23] takes a step further to develop MBConv operation based on the mobile inverted bottleneck in [21]. EfficientNet shows that the models with MBConv operations not only achieving the state-of-the-art in ImageNet challenge but also very efficient. Recently, EfficientDet [24] builds object detection models based on the EfficientNet backbone model and achieves impressive detection accuracy and computation efficiency.

Resource-Aware Neural Architecture Search: In neural architecture search, adding resource constraints is critical to avoid the bias to choose a model with

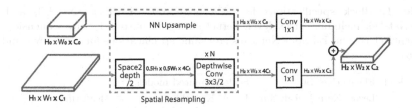

Fig. 2. Resampling operation

higher computation. MnasNet [22] introduces multi-objective rewards that optimize the model accuracy while penalizes models that violate the constraints. CR-NAS [13] searches for the best resource allocation by learning the number of blocks allocated in each resolution stage and the dilated convolution kernel.

3 Method

In this section, we first describe how to combine the scale-permuted architecture topology [4] and efficient operation MBConv [23]. Then, we introduce feature resampling and fusion operations in the efficient scale-permuted model. Lastly, we propose a search method to learn resource distribution for building Eff-SpineNet.

3.1 Scale-Permuted Architecture with Efficient Operations

We first combine SpinetNet-49 architecture topology with MBconv blocks. We start with permuting the EfficientNet-B0 model. The goal here is to build an efficient scale-permuted model, SP-EfficientNet-B0, that has the similar computation and parameters as the EfficientNet-B0 baseline. We follow the idea of the compound scaling rule in EfficientNet to create 5 higher capacity models.

SP-EfficientNet-B0: We attempt to replace all the residual and bottleneck blocks in SpineNet-49 with MBconv blocks. One design decision is how to assign the convolution kernel size and feature dimension when applying MBConv to scale-permuted architecture. Given SpineNet-49 has already had a large receptive field, we decide to fix the kernel size to 3 for all MBConv operations. To obtain a model with similar size as EfficientNet-B0, we obtain the feature dimension for each level by averaging the feature dimensions over all blocks at the corresponding levels in Efficient-B0. Since the L_6 and L_7 blocks does not have a corresponding feature in EfficientNet, we follow [4] to set them to have the same feature dimension as the L_5 block. The detailed network specifications of the SP-EfficientNet-B0 model is presented in Table 1.

Compound Scaling for Scale-Permuted Network: We follow the compound scaling rule proposed in [23] to scale up the SP-EfficientNet-B0 model. We use the rule to compute the number of blocks for each feature level, feature

Table 1. Block specifications for EfficentNet-B0, SP-EfficientNet-B0, and Eff-SpineNet-D0, including block level, kernel size, and output feature dimension. SP-EfficientNet-B0 and Eff-SpineNet-D0 share same specifications for block level and kernel size

Block id	EfficientNet-B0			Scale-permuted models			
	Level	Kernel	feat. dim	Level	Kernel	feat. dim	
						SP-EfficientNet-B0	Eff-SpineNet-D0
1	L_1	3×3	16	L_1	3×3	16	16
2	L_2	3×3	24	L_2	3×3	24	24
3	L_2	3×3	24	L_2	3×3	24	16
4	L_3	5×5	40	L_2	3×3	24	16
5	L_3	5×5	40	L_4	3×3	96	104
6	L_4	3×3	80	L_3	3×3	40	48
7	L_4	3×3	80	L_4	3×3	96	120
8	L_4	3×3	80	L_6	3×3	152	40
9	L_4	5×5	112	L_4	3×3	96	120
10	L_4	5×5	112	L_5	3×3	152	168
11	L_4	5×5	112	L_7	3×3	152	96
12	L_5	5×5	192	L_5	3×3	152	192
13	L_5	5×5	192	L_5	3×3	152	136
14	L_5	5×5	192	L_4	3×3	96	104
15	L_5	5×5	192	L_3	3×3	40	40
16	L_5	3×3	320	L_5	3×3	152	136
17	–	–	–	L_7	3×3	152	136
18	–	–	–	L_6	3×3	152	40

dimension, and input image size. Since the number of blocks for a level after scaling may be more than the blocks at the corresponding level in SP-EfficientNet-B0 model, we uniformly repeat the blocks in SP-EfficientNet-B0 model. If the scaled number of blocks is not the multiple of those in SP-EfficientNet-B0, we add the remainder blocks one-by-one in the bottom up ordering. The detailed model scaling specifications are given in Table 2.

3.2   Feature Resampling and Fusion

Given the MBConv output feature dimension is much lower compared to residual and bottleneck blocks, we redesign the feature resampling method. And we adopt the fusion method from EfficientDet [24].

Resampling Method: Since MBConv has a small output feature dimension, it removes the need of the scaling factor α in SpineNet to reduce the computation. Compared to SpineNet, the 1×1 convolution to reduce input feature dimension is removed. Besides, we find using the space-to-depth operation followed by stride 2 convolutions preserves more information than the original design, with a small increase of computation. The new resampling strategy is shown in Fig. 2.

Table 2. Model scaling method for Eff-SpineNet models. **input size:** Input resolution. **feat. mult.:** Feature dimension multiplier for convolutional layers in backbone. **block repeat:** Number of repeats for each block in backbone. The 18 blocks are ordered from left to right. **feat. dim.:** Feature dimension for separable convolutional layers in subnets. **#layers:** Number of separable convolutional layer in subnets

Model id	Scale-permuted backbone			Subnets	
	Input size	feat. mult	Block repeat	feat. dim	#layers
M0	256	0.4	{1,1,1,1,1,1,1,1,1,1,1,1,1,1,1,1,1,1}	24	3
M1	384	0.5	{1,1,1,1,1,1,1,1,1,1,1,1,1,1,1,1,1,1}	40	3
M2	384	0.8	{1,1,1,1,1,1,1,1,1,1,1,1,1,1,1,1,1,1}	64	3
D0	512	1.0	{1,1,1,1,1,1,1,1,1,1,1,1,1,1,1,1,1,1}	64	3
D1	640	1.0	{2,2,1,1,2,2,2,1,1,2,1,2,1,1,1,1,1,1}	88	3
D2	768	1.1	{2,2,1,1,2,2,2,1,1,2,1,2,1,1,1,1,1,1}	112	3

Weighted Block Fusion: As shown in [24], input features at different resolutions or network building stages may contribute unequally during feature fusion. We apply the fast normalized fusion strategy introduced in [24] to block fusion in SpineNet. The method is shown in Eq. 1:

$$B^{out} = \frac{\sum_i w_i \times B_i^{in}}{\sum_j w_j + 0.001},$$ (1)

where B^{in} and B^{out} represent the input blocks and the output block respectively. w is the weight to be learned for each input block.

3.3 Learning Resource Distribution

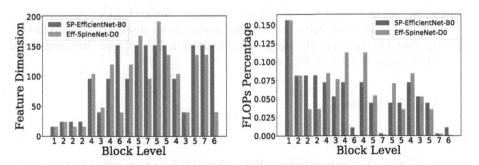

Fig. 3. Comparisons of SP-EfficientNet-B0 and Eff-SpineNet-D0 in feature dimension distribution (left) and resource distribution (right). The 18 blocks are plotted in order from left to right with block level shown in the x-axis

Typically, the conventional architecture gradually increases feature dimension with decreasing spatial resolution of feature [9,10,21,23,25]. However, this design may be sub-optimal for a scale-permuted network. In this section, we propose a simple yet effective search method for resource reallocation. We learn the resource distribution through adjusting the feature dimension of MBConv blocks. In the search space, we fix the total FLOPs of MBConv blocks in the entire model, and learn a scale multiplier of feature dimension for each block in SP-EfficientNet-B0.

Consider c_i to be the feature dimension of MBConv block i, the FLOPs can be computed as $\mathcal{F}_i \simeq C_i \times c_i^2$, where C_i is a constant that depends on height, width, and expansion ratio of a given block.

$$
\begin{aligned}
\mathcal{F}_i &= H_i \times W_i \times (2 \times c_i^2 \times r + k^2 \times c_i \times r) \\
&\simeq H_i \times W_i \times c_i^2 \times 2 \times r \\
&\simeq C_i \times c_i^2,
\end{aligned}
\tag{2}
$$

where H_i, W_i is the height and width of the feature map, r is the expansion ratio and k is the kernel size in a MBConv block.

We propose to learn a multiplier α_i that adjusts the resource distribution over the entire model with a target total desired computation $\mathcal{F}_t$

$$
\mathcal{F}_t = \sum_i \alpha_i \mathcal{F}_i
\tag{3}
$$

In our experiment, we simply set $\mathcal{F}_t = \sum_i \mathcal{F}_i$.

Learning α_i can be challenging because α_i can be any positive real number. Here, we propose to learn a multiplier β_i which is selected from a set of N positive numbers $\{\beta^1, \beta^2, ..., \beta^N\}$. Then, we can represent α_i as a function of β_i which satisfies the Eq. 3.

$$
\alpha_i = \frac{\mathcal{F}_t}{\sum_k \beta_k \mathcal{F}_k} \beta_i
\tag{4}
$$

Finally, we use α_i to modify the feature dimension for each block $\hat{c}_i = \sqrt{\alpha_i} c_i$.

Using this resource distribution learning strategy, we discover our final model, Eff-SpineNet-D0. We show the model specification in Table 1 and the comparison with SP-EfficientNet-B0 in Fig. 3.

4   Applications

4.1   Object Detection

We use Eff-SpineNet as backbone in RetinaNet [15] for one-stage object detection and in Mask R-CNN [8] for two-stage object detection and instance segmentation. The feature map of the 5 output blocks $\{P_3, P_4, P_5, P_6, P_7\}$ are used as the multi-scale feature pyramid. Similar to [24], we design a heuristic scaling rule to maintain a balance in computation between backbone and subnets during model scaling and use separable convolutions in all subnets. In RetinaNet, we gradually use more convolutional layers and a larger feature dimension for each layer in the box and class subnets for a larger Eff-SpineNet model. In Mask R-CNN, the same scaling rule is applied to convolutional layers in the RPN, Fast R-CNN and mask subnets. In addition, a fully connected layer is added after convolutional layers in the Fast R-CNN subnet and we apply the scaling to adjust its dimension to 256 for D0 and D1, and 512 for D2. Details are shown in Table 2.

4.2 Image Classification

We directly utilize all feature maps from P_3 to P_7 to build the classification model. Different from the object detection models shown in Table 2, we set the feature dimension to 256 for all models. The final feature vector is generated by nearest-neighbor upsampling and averaging all feature maps to the same size as P_3 followed by the global average pooling. We apply a linear classifier on the 256-dimensional feature vector and train the classification model with softmax cross-entropy loss.

4.3 Semantic Image Segmentation

In this subsection, we explore Eff-SpineNet for the task of semantic image segmentation. We apply nearest-neighbor upsampling to match the sizes of all feature maps in $\{P_3, P_4, P_5, P_6, P_7\}$ to $P3$ then take the average. The averaged feature map P at output stride 8 is used as the final feature map from Eff-SpineNet. We further apply separable convolutional layers before the pixel-level prediction layer. The number of layers and feature dimension for each layer are fixed to be the same as the subnets in object detection, shown in Table 2.

5 Experimental Results

We present experimental results on object detection, image classification, and semantic segmentation to demonstrate the effectiveness and generality of the proposed Eff-SpineNet models. For object detection, we evaluate Eff-SpineNet on COCO bounding box detection [16]. We train all models on the COCO train2017 split and report results on the COCO val2017 split. For image classification, we train Eff-SpineNet models on ImageNet ILSVRC-2012 and report Top-1 and Top-5 validation accuracy. For semantic segmentation, we follow the common practice to train Eff-SpineNet on PASCAL VOC 2012 with augmented 10,582 training images and report mIOU on 1,449 val set images (Table 3).

5.1   Object Detection

5.1.1   Experimental Settings

Table 3. One-stage object detection results on the COCO benchmark. We compare using different backbones with RetinaNet on single model without test-time augmentation. FLOPs is represented by Multi-Adds

Model	#FLOPs	#Params	AP	AP_{50}	AP_{75}	AP_S	AP_M	AP_L
Eff-SpineNet-D0	**2.5B**	**3.6M**	**34.7**	**53.1**	**37.0**	**15.2**	**38.7**	**52.8**
EfficientNet-B0-FPN	2.5B	3.7M	33.5	52.8	35.4	14.5	37.5	50.7
EfficientDet-D0 [24]	2.5B	3.9M	33.5	–	–	–	–	–
Eff-SpineNet-D1	**6.0B**	**5.2M**	**39.9**	**59.6**	**42.5**	**43.5**	**21.1**	**57.5**
EfficientNet-B1-FPN	5.8B	6.3M	38.8	59.1	41.4	20.2	43.0	55.7
EfficientDet-D1 [24]	6.0B	6.6M	39.1	–	–	–	–	–
Eff-SpineNet-D2	**10.3B**	**6.2M**	**42.5**	**62.0**	**46.0**	**24.5**	**46.4**	**57.6**
EfficientNet-B2-FPN	10.0B	7.5M	41.4	62.3	44.1	24.4	45.4	56.8
EfficientDet-D2 [24]	11.0B	8.1M	42.5	–	–	–	–	–
ResNet-50-FPN [4]	96.8B	34.0M	42.3	61.9	45.9	23.9	46.1	58.5
SpineNet-49 [4]	85.4B	28.5M	44.3	63.8	47.6	25.9	47.7	61.1

Table 4. Ablation studies on advanced training strategies for Eff-SpineNet-D2 and ResNet50-FPN. We begins with 72 epochs training steps and multi-scale training [0.8, 1.2] as the baseline. **SE:** squeeze and excitation; **ms train:** large-scale multi-scale [0.1, 2.0] and extend training steps that attain the best performance (650 epochs for Eff-SpineNet-D2 and 250 epochs for ResNet-50-FPN); **Swish:** Swish activation that replaces ReLU; **SD:** stochastic depth

Model	Baseline	+SE	+ms train	+Swish	+SD
Eff-SpineNet-D2	32.2	32.6(+0.4)	40.1(+7.4)	42.1 (+2.0)	42.5(+0.4)
ResNet-50-FPN	37.0	N/A	40.4 (+3.4)	40.7 (+0.3)	42.3(+1.6)

Training Details: We generally follow the training protocol in [4,24] to train all models for the proposed method, EfficientNet-FPN, and SpineNet on COCO train2017 from scratch. We train on Cloud TPU v3 devices using standard stochastic gradient descent (SGD) with 4e-5 weight decay and 0.9 momentum. We apply batch size 256 and stepwise learning rate with 0.28 initial learning rate that decays to $0.1\times$ and $0.01\times$ at the last 30 and 10 epochs. All models are trained for 650 epochs, which we observe the model starts to overfit and hurt performance after 650 epochs. We apply synchronized batch normalization with 0.99 momentum, swish activation [19], and stochastic depth [12]. To pre-process training data, we resize the long side of an image to the target

Table 5. Impact of longer training schedule using advanced training strategies when training a model from scratch

Model	72 epoch	200 epoch	350 epoch	500 epoch	650 epoch
Eff-SpineNet-D2	34.8	40.0 (+5.2)	41.4 (+1.4)	42.1 (+0.7)	42.5 (+0.4)

Table 6. Two-stage object detection results on COCO. We compare using different backbones with Mask R-CNN on single model

Model	#FLOPs	#Params	AP	AP_{50}	AP_{75}	AP^{mask}	AP^{mask}_{50}	AP^{mask}_{75}
Eff-SpineNet-D0	4.7B	4.6M	35.0	54.0	37.3	30.5	50.2	32.2
Eff-SpineNet-D1	9.2B	6.4M	40.7	60.9	44.1	35.0	56.9	36.8
Eff-SpineNet-D2	16.0B	9.2M	42.9	63.5	46.5	37.3	60.2	39.1

image size described in Table 2 then pad the short side with zeros to make it a square image. Horizontal flipping and multiscale augmentation [0.1, 2.0] are implemented during training.

Search Details: We design our search space $\{\beta_1, \beta_2, ..., \beta_N\}$ as $\{1, 5, 10, 15, 20\}$ in this work to cover a wide range of possible resource distributions with a manageable search space size. We follow [4,23] to implement the reinforcement learning based search method [27]. In brief, we reserve 7392 images from COCO `train2017` as the validation set for searching and use other images for training. Sampled models at the D0 scale are used for proxy task training with the same training settings described above. AP on the reserved set of proxy tasks trained for 4.5k iterations is collected as rewards. The best architecture is collected after 5k architectures have been sampled.

5.1.2 Object Detection Results

RetinaNet: Our main results are presented on the COCO bounding box detection task with RetinaNet. Compared to our architecture-wise baseline EfficientNet-FPN models, our models consistently achieve 1–2% AP gain from scale D0 to D2 while using less computations. The FLOPs *vs.* AP curve and the Params *vs.* AP curve among Eff-SpineNet and other state-of-the-art one-stage object detectors are shown in Fig. 1 and Fig. 4 respectively.

Mask R-CNN: We evaluate Eff-SpineNet models with Mask R-CNN on the COCO bounding box detection and instance-level segmentation task. The results of Eff-SpineNet D0, D1, and D2 models are shown in Table 6.

5.1.3 Mobile-Size Object Detection Results

The results of Eff-SpineNet-M0/1/2 models are presented in Table 7 and the FLOPs *vs.* AP curve is plotted in Fig. 1. Eff-SpineNet models are able to consistent use less resources while surpassing all other state-of-the-art mobile-size

Table 7. Mobile-size object detection results on COCO. Eff-SpineNet models achieve the new state-of-the-art FLOPs *vs.* AP trade-off curve

Backbone model	#FLOPs	#Params	AP	AP$_S$	AP$_M$	AP$_L$
Eff-SpineNet-M0	**0.15B**	**0.67M**	**17.3**	**2.2**	**16.4**	**33.0**
MobileNetV3-Small-SSDLite [10]	0.16B	1.77M	16.0	–	–	–
Eff-SpineNet-M1	**0.51B**	**0.99M**	**25.0**	**7.4**	**27.3**	42.0
MobileNetV3-SSD [10]	0.51B	3.22M	22.0	–	–	–
MobileNetV2 + MnasFPN	0.53B	1.29M	23.8	–	–	–
MnasNet-A1-SSD [22]	0.8B	4.9M	23.0	3.8	21.7	42.0
Eff-SpineNet-M2	**0.97B**	**2.36M**	**29.2**	**9.7**	**32.7**	**48.0**
MobileNetV2-NAS-FPN [6]	0.98B	2.62M	25.7	–	–	–
MobileNetV2-FPN [21]	1.01B	2.20M	24.3	–	–	–

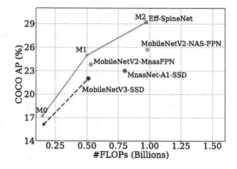

Fig. 4. A comparison of Eff-SpineNet and other state-of-the-art detectors on mobile-size object detection. Eff-SpineNet models outperform the other detectors at various scales

object detectors by large margin. In particular, our Eff-SpineNet-M2 achieves 29.2% AP with 0.97B FLOPs, attaining the new state-of-the-art for mobile-size object detection.

5.1.4   Ablation Studies

Ablation Studies on Advanced Training Strategies: We conduct detailed ablation studies on the advanced training features used in this paper and [4,23]. Starting from the final Eff-SpineNet-D2 model, we gradually remove one feature at a time: 1) removing stochastic depth in model training leads to 0.4 AP drop; 2) replacing swish activation with ReLU leads to 2.0 AP drop; 3) using less aggressive multi-scale training strategy with 72 training epochs leads to 7.5 AP drop; 4) removing squeeze and excitation [11] layers from all MBConv blocks leads to 0.4 AP drop. We further perform the ablation studies to ResNet-50-FPN and the results are shown in Table 4.

Impact of Longer Training Schedule: We conduct experiments by adopting different training epochs for Eff-SpineNet-D2. We train all models from scratch on COCO 2017train and report AP on COCO 2017val. The results are presented in Table 5. We show that prolonging the training epochs from 72 to 650 gradually improves the performance of Eff-SpineNet-D2 by 7.7% AP. Except training schedule, the other training strategies are the same as Sect. 5.1.1.

Table 8. Improvement from learning a better resource distribution. All models are evaluated on COCO 2017val

Model	#FLOPs	#Params	AP	AP_{50}	AP_{75}	AP_S	AP_M	AP_L
SP-EfficientNet-B0	2.4B	4.4M	33.0	50.3	34.7	13.0	38.4	51.7
Eff-SpineNet-D0	2.5B	3.6M	33.8	51.3	35.8	13.6	39.3	52.4

Table 9. An ablation study of the two architecture improvements in Eff-SpineNet

Model	Weighted fusion	Space-to-depth	#FLOPs	#Params	AP
Eff-SpineNet-D0	✓	✓	2.5B	3.6M	33.8
Model 1	✓	–	2.4B	3.3M	33.1
Model 2	–	–	2.4B	3.3M	32.8

Learning Resource Distribution: From the final architecture discovered by NAS shown in Table 1, we observe that parameters in low-level L_2 blocks and high-level $\{L_6, L_7\}$ block, are reallocated to middle-level $\{L_3, L_4, L_5\}$ blocks. Since the high-level blocks are low in resolution, by doing so, the number of parameters in the network is significantly reduced from 4.4M to 3.6M while the total FLOPs remains roughly the same. Learning resource distribution also brings a 0.8% AP gain. The performance improvements from SP-EfficientNet-B0 to Eff-SpineNet-D0 is shown in Table 8.

Architecture Improvements: We conduct ablation studies for the two techniques, resampling method based on the space-to-depth operation and weighted block fusion, introduced to SpineNet's scale-permuted architecture with Eff-SpineNet-D0. As shown in Table 9, the performance drops by 0.7% AP if we remove the new resampling method. The performance further drops by 0.3% AP if we remove weighted block fusion.

5.1.5   A Study of the Proposed Search Algorithm

We visualize some of the randomly sampled architectures in the search phase. The FLOPs *vs.* AP plot and the Params *vs.* AP plot are presented in Fig. 5. From the FLOPs *vs.* AP plot, we can observe that the FLOPs of the sampled architectures fall into a range of ±10% of our target FLOPs because of the proposed search algorithm. We can also observe from the Params *vs.* AP plot that the number of parameters in sampled architectures are reduced in most cases.

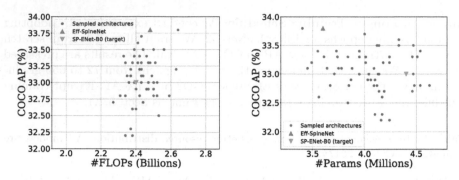

Fig. 5. FLOPs *vs.* AP (left) and Params *vs.* AP (right) plots for architectures sampled throughout the searching phase. The *x*-axes are plotted within a ±20% range to the centers

Table 10. Comparison between Eff-SpineNet and EfficientNet on ImageNet classification

Model	Input resolution	Feature dim	#FLOPs	#Params	Top-1	Top-5
Eff-SpineNet-D0	224 × 224	**256**	**0.38B**	**3.57M**	**75.3**	**92.4**
EfficientNet-B0	224 × 224	1280	0.39B	5.30M	77.3	93.5
Eff-SpineNet-D1	240 × 240	**256**	**0.70B**	**4.97M**	**77.7**	**93.6**
EfficientNet-B1	240 × 240	1280	0.70B	7.80M	79.2	94.5
Eff-SpineNet-D2	256 × 256	**256**	**0.89B**	**5.83M**	**78.5**	**94.2**
EfficientNet-B2	260 × 260	1280	1.00B	9.20M	80.3	95.0

5.2   Image Classification

We conduct image classification experiments on ImageNet ILSVRC 2012 [3,20] with Eff-SpineNet, following the same training strategy used in EfficientNet [23]. We scale the input size with respect to different model size by roughly following the compound scaling [23] and adjusting it to be the closest multiples of 16.

We compare Eff-SpineNet with EfficientNet in all aspects in Table 10. At the same FLOPs, Eff-SpineNet is able to save around 35% parameters at the cost of 1.5–2% drop in top-1 accuracy. We hypothesize this is likely due to the fact that higher level features (P_6 and P_7) do not contain enough spatial resolution for small input size. For 224 × 224 input size, the spatial resolution of P_6 and P_7 is only 4 × 4 and 2 × 2 respectively. We will explore how to construct better scale-permuted models for image classification in the future.

5.3   Semantic Segmentation

We present experimental results of employing Eff-SpineNet as backbones for semantic segmentation. We conduct the experiments with evaluation metric mIOU on PASCAL VOC 2012 [5] with extra annotated images from [7]. For training implementations, we generally follow the settings in Sect. 5.1.1. In brief,

Table 11. Semantic segmentation result comparisons of Eff-SpineNet and other popular semantic segmentation networks on the PASCAL VOC 2012 `val` set

Model	ImageNet pre-train	COCO pre-train	output stride	#FLOPs	mIOU
MobileNetv2 + DeepLabv3	–	✓	16	2.8B	75.3
ResNet-101 + DeepLabv3	✓	✓	8	81.0B	80.5
Eff-SpineNet-D0	–	✓	8	**2.1B**	**76.0**
Eff-SpineNet-D2	–	✓	8	**3.8B**	**78.0**

we fine-tune all models from the COCO bounding box detection pre-trained models for 10k iterations with batch size 256 with cosine learning rate. We set the initial learning to 0.05 and a linear learning rate warmup is applied for the first 500 iterations. We fix the input crop size to 512×512 for all Eff-SpineNet models without strictly following the scaling rule described in Table 2.

Our results on PASCAL VOC 2012 `val` set are presented in Table 11. Eff-SpineNet-D0 slightly outperforms MobileNetv2 with DeepLabv3 [1,21] by 0.7 mIOU while using 25% less FLOPs. Our D2 model is able to attain comparable mIOU with other popular semantic segmentation networks, such as ResNet101 with DeepLabv3 [1], at the same output stride while using 95% less FLOPs.

6   Conclusion

In this paper, we propose to decompose model design into architecture topology, operation, and resource distribution. We show that simply combining scale-permuted architecture topology and efficient operations achieves new state-of-the-art in object detection, showing the benefits of efficient operation and scale-permuted architecture are complementary. The model can be further improved by learning the resource distribution over the entire network. The resulting Eff-SpineNet is a versatile backbone model that can be also applied to image classification and semantic segmentation tasks, attaining competitive performance, proving Eff-SpineNet is a versatile backbone model that can be easily applied to many tasks without extra architecture design.

References

1. Chen, L.C., Papandreou, G., Schroff, F., Adam, H.: Rethinking Atrous convolution for semantic image segmentation. ArXiv abs/1706.05587 (2017)
2. Chen, L.-C., Zhu, Y., Papandreou, G., Schroff, F., Adam, H.: Encoder-decoder with Atrous separable convolution for semantic image segmentation. In: Ferrari, V., Hebert, M., Sminchisescu, C., Weiss, Y. (eds.) ECCV 2018. LNCS, vol. 11211, pp. 833–851. Springer, Cham (2018). https://doi.org/10.1007/978-3-030-01234-2_49
3. Deng, J., Dong, W., Socher, R., Li, L.J., Li, K., Fei-Fei, L.: ImageNet: a large-scale hierarchical image database. In: CVPR (2009)
4. Du, X., et al.: SpineNet: learning scale-permuted backbone for recognition and localization (2019)

5. Everingham, M., Eslami, S.M.A., Van Gool, L., Williams, C.K.I., Winn, J., Zisserman, A.: The PASCAL visual object classes challenge: a retrospective. Int. J. Comput. Vis. **111**(1), 98–136 (2014). https://doi.org/10.1007/s11263-014-0733-5
6. Ghiasi, G., Lin, T.Y., Le, Q.V.: NAS-FPN: learning scalable feature pyramid architecture for object detection. In: CVPR (2019)
7. Hariharan, B., Arbelaez, P., Bourdev, L., Maji, S., Malik, J.: Semantic contours from inverse detectors. In: International Conference on Computer Vision (ICCV) (2011)
8. He, K., Gkioxari, G., Dollár, P., Girshick, R.: Mask R-CNN. In: ICCV (2017)
9. He, K., Zhang, X., Ren, S., Sun, J.: Deep residual learning for image recognition. In: CVPR (2016)
10. Howard, A., et al.: Searching for mobilenetv3. In: ICCV (2019)
11. Hu, J., Shen, L., Sun, G.: Squeeze-and-excitation networks. In: 2018 IEEE/CVF Conference on Computer Vision and Pattern Recognitionm, June 2018. https://doi.org/10.1109/CVPR.2018.00745
12. Huang, G., Sun, Yu., Liu, Z., Sedra, D., Weinberger, K.Q.: Deep networks with stochastic depth. In: Leibe, B., Matas, J., Sebe, N., Welling, M. (eds.) ECCV 2016. LNCS, vol. 9908, pp. 646–661. Springer, Cham (2016). https://doi.org/10.1007/978-3-319-46493-0_39
13. Liang, F., et al.: Computation reallocation for object detection. In: International Conference on Learning Representations (2020). https://openreview.net/forum?id=SkxLFaNKwB
14. Lin, T.Y., Dollár, P., Girshick, R., He, K., Hariharan, B., Belongie, S.: Feature pyramid networks for object detection. In: CVPR (2017)
15. Lin, T.Y., Goyal, P., Girshick, R., He, K., Dollár, P.: Focal loss for dense object detection. In: ICCV (2017)
16. Lin, T.-Y., et al.: Microsoft COCO: common objects in context. In: Fleet, D., Pajdla, T., Schiele, B., Tuytelaars, T. (eds.) ECCV 2014. LNCS, vol. 8693, pp. 740–755. Springer, Cham (2014). https://doi.org/10.1007/978-3-319-10602-1_48
17. Liu, C., et al.: Auto-DeepLab : hierarchical neural architecture search for semantic image segmentation. In: CVPR (2019)
18. Liu, S., Qi, L., Qin, H., Shi, J., Jia, J.: Path aggregation network for instance segmentation. In: CVPR (2018)
19. Ramachandran, P., Zoph, B., Le, Q.V.: Searching for activation functions (2017)
20. Russakovsky, O., et al.: ImageNet large scale visual recognition challenge. Int. J. Comput. Vis. **115**(3), 211–252 (2015). https://doi.org/10.1007/s11263-015-0816-y
21. Sandler, M., Howard, A., Zhu, M., Zhmoginov, A., Chen, L.C.: MobileNetv 2: inverted residuals and linear bottlenecks. In: CVPR (2018)
22. Tan, M., et al.: MnasNet: platform-aware neural architecture search for mobile. In: CVPR (2019)
23. Tan, M., Le, Q.: EfficientNet: Rethinking model scaling for convolutional neural networks. In: Proceedings of the 36th International Conference on Machine Learning. Proceedings of Machine Learning Research
24. Tan, M., Pang, R., Le, Q.V.: EfficientDet: scalable and efficient object detection (2019)
25. Xie, S., Girshick, R., Dollár, P., Tu, Z., He, K.: Aggregated residual transformations for deep neural networks. In: CVPR (2017)
26. Zhao, Q., et al.: M2Det: a single-shot object detector based on multi-level feature pyramid network. In: AAAI (2019)
27. Zoph, B., Le, Q.V.: Neural architecture search with reinforcement learning. In: ICLR (2017)

Reducing Distributional Uncertainty by Mutual Information Maximisation and Transferable Feature Learning

Jian Gao[1,2], Yang Hua[1(✉)], Guosheng Hu[1,2], Chi Wang[1,2],
and Neil M. Robertson[1]

[1] EEECS/ECIT, Queen's University Belfast, Belfast, UK
{jgao05,y.hua,cwang38,n.robertson}@qub.ac.uk
[2] Anyvision, Belfast, UK
huguosheng100@gmail.com

Abstract. Distributional uncertainty exists broadly in many real-world applications, one of which in the form of domain discrepancy. Yet in the existing literature, the mathematical definition of it is missing. In this paper, we propose to formulate the distributional uncertainty both between the source(s) and target domain(s) and within each domain using mutual information. Further, to reduce distributional uncertainty (e.g. domain discrepancy), we (1) maximise the mutual information between source and target domains and (2) propose a transferable feature learning scheme, balancing two complementary and discriminative feature learning processes (general texture learning and self-supervised transferable shape learning) according to the uncertainty. We conduct extensive experiments on both domain adaption and domain generalisation using challenging common benchmarks: Office-Home and Domain-Net. Results show the great effectiveness of the proposed method and its superiority over the state-of-the-art methods.

Keywords: Distributional uncertainty · Domain discrepancy · Mutual information · Object shape · Self-supervised learning

1 Introduction

A fundamental assumption in machine learning is the similarity of training and test distribution. Various algorithms have been proposed based on this assumption, Convolutional Neural Networks (CNNs) among which achieved huge success together with large scale training data. However, opposed to this ideal setting, distributional uncertainty exists broadly in almost every real-world problem.

Domain discrepancy is a type of distributional uncertainty in which the dissimilarity of two domains (distributions) is considered. Unsupervised Domain

Electronic supplementary material The online version of this chapter (https://doi.org/10.1007/978-3-030-58592-1_35) contains supplementary material, which is available to authorized users.

Fig. 1. Examples of the Effiel Tower presented in different styles from the six domains in DomainNet dataset [38]. Left to right: quickdraw, sketch, clipart, infograph, painting and real. *Best viewed in colour.*

Adaptation (UDA) aims at resolving the domain discrepancy problem and enhancing model transferability, while not requiring any labels for target domain. Different approaches have been proposed to tackle it, such as direct minimisation of domain discrepancy [32] and domain adversarial learning [15]. Though promising progress has been made, some critical issues still remain.

First of all, there lacks a unified and quantified explanation for the concept of domain discrepancy. Existing methods either minimise the distributional difference between classifier outputs [44], or try to align intermediate features [49] from different distributions. Yet a rigorous definition of domain discrepancy and a precise measurement of it are missing.

Secondly, most methods are restricted to aligning a single source and a single target domain at a time, as they assume that training and test data each follows a single distribution. However, this is often not true in practice. In some tasks, the presence of multiple distinct distributions in training is almost unavoidable, for example, different camera-angle sub-domains in person re-identification tasks [40]. Directly applying a single source to single target adaptation method to such tasks is problematic. Since the adaptation performance across multiple domains is bounded by the worst model obtained from a source domain that is least similar to the target [56].

Further, contemporary methods try to align the output distributions by one or more classifiers [35], while leaving the feature learning entirely handled by CNNs. However one main drawback of CNNs is its lack of regularisation in learning generalised and transferable features [39]. For example, CNNs are heavily biased towards learning textures which may change dramatically across domains, while neglecting object structural features such as shape that is often more consistent [17]. For example in Fig. 1, the Effiel Tower appears in visually diametrically different image styles but its shape remains consistent. The situation becomes even worse when a distributional dissimilarity lies between the training and the test data [38].

From our observations, the above issues are inherently caused by the same fact which is the distributional uncertainty. The domain discrepancy in UDA describes the distributional uncertainty *between source and target*. Single source adaptation methods fail to consider the distributional uncertainty *within the source samples*. The lack of regularisation in CNNs can be compensated by a reduction in the distributional uncertainty of training data, such as providing certain prior knowledge about the distribution.

In this paper, we propose to resolve the above issues by reducing distributional uncertainty, combing Mutual Information Maximisation and Transferable Feature Learning (abbreviated as **MIMTFL**). We formulate the estimation of distributional uncertainty using Mutual Information (MI). During training, we calculate MI over each batch of samples to exam its uncertainty. We learn to reduce the uncertainty by maximising MI between source and target, while considering uncertainty within the source samples. We further leverage a self-supervised transferable feature learning scheme by enforcing a balance between texture and object shape features. According to the estimated uncertainty level, the network learns to automatically balance texture and shape features for better transferability.

In summary, we propose the following contributions:

- A formulation to measure distributional uncertainty as a unified definition of domain discrepancy. The proposed distributional uncertainty measurement using mutual information is mathematically grounded, and generalise to not only discrepancy between different domains, but also disagreement within source.
- A self-supervised transferable feature learning strategy that utilises MI to automatically balance the learning of texture and shape features for better generalisation.
- Extensive results under various settings on two large-scale multi-domain adaptation benchmarks with state-of-the-art performance to prove the effectiveness of the proposed method.

2 Related Work

2.1 Unsupervised Domain Adaptation

The main challenges in UDA come from two aspects. First, how to make the source model transferable to the target in view of the distributional gap between the two. Secondly, how to make use of the unlabelled target samples. For the first, learning a transferable representation for both the source and target has attracted much attention. Methods including DDC [50], DAN [30] and JAN [33] align the domains by minimising a domain discrepancy measurement between them. Domain adversarial learning [15,49] seems to be effective where the gradients from a domain discriminator network trying to distinguish source and target samples are reversed. It is also found that aligning the first, second and even higher order moments of source and target distributions is helpful [43,47].

To use the unlabelled target samples, many semi-supervised learning techniques are introduced into UDA algorithms. Pseudo-Labelling [28] and Label-Propagation [58] are found to be useful to estimate the true labels of target samples [5]. Another effective solution is the Mean-Teacher [48] model in which an unsupervised consistency loss is enforced between a student model prediction and a teacher model prediction [14].

2.2 Distributional Uncertainty

The study of distributional uncertainty spreads through various fields of science and engineering. In control theory, researchers apply distributional uncertainty analysis to enhance robustness of controllers [34]. Evidence imprecision and uncertainty modelling using fuzzy sets is shown critical in medical diagnosis [46]. Probabilistic machine learning values the representation and manipulation of uncertainty in both data distributions and models, as it plays a central role in scientific data analysis [18].

The connection between distributional uncertainty estimation and generalisation has recently been uplifted significantly too. In modern deep learning, explicit modelling of distributional uncertainty within the Bayesian framework makes the network more robust to noisy data as well as achieving better generalisation results on difficult computer vision tasks, such as semantic segmentation and depth regression [24]. Kendall et al. [25] found that using uncertainty weighting in multi-task learning allows effective simultaneous learning of various tasks, which even outperforms individually learned models for each task. Uncertainty-based reliability analysis in 3D vehicle detection from point cloud data brings steady improvement in the model's performance in adverse environment, such as heavy occlusions, which is of great importance to promoting safe autonomous driving [13]. Examples of benefits in generalisation can be found throughout a variety of vision applications including optical flow estimation [22], people tracking in traffic scenes [2], so on and so forth.

2.3 Self-supervised Visual Feature Learning

As a promising solution towards eliminating the need of costly human annotations, self-supervised learning methods learn visual features from unlabelled images on auxiliary tasks. The supervision signals for the auxiliary tasks are usually automatically obtained without requiring any human labelling effort. In other words, it absorbs advantages from supervised learning methods with accurate labels, while saving the tedious labour for labelling.

So far, outstanding progress has been made in a wide range of vision tasks with self-supervised visual feature learning. Zhang et al. [55] found learning to colourise of de-coloured images eminently improve generalisation in the downstream recognition task. With an image inpainting auxiliary task, the network captures not just appearance but also the semantics of visual structures [37]. Another popular method to learn structural features in an image is extracting patches from it, and learning to predict their relative spatial locations, so-called solving a jigsaw puzzle. The learned visual representation is seen to perform incredibly well in object detection tasks [10], as well as effectively increasing generalisation in standard Domain Generalisation tasks [4].

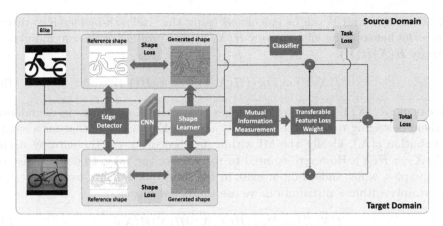

Fig. 2. System pipeline: A reference shape is obtained for all the samples using an edge detector in advance. During training, both the source and the target sample batch are fed to the backbone feature extractor CNN. A shape loss is calculated via comparing the generated shape by the shape learner and reference shape for each sample. The CNN extracted features are also fed into a mutual information measurement module, to estimate both inter-domain and intra-domain distributional uncertainty for source and target domain. The estimated uncertainty is then used to automatically weight the shape loss. The overall objective consists of the classifier task loss on the labelled source samples, and the weighted shape loss on all the samples. *Best viewed in colour.*

3 Methodology

We hereby introduce the details of **MIMTFL**, as illustrated in Fig. 2. We first elucidate the measurement of distributional uncertainty formulation using MI in Sect. 3.1. Then move on to the MI-guided self-supervised learning of transferable features including our proposed shape learning method, detailed in Sect. 3.2.

3.1 Distributional Uncertainty Measurement Using MI

Preliminaries. For distributional uncertainty estimation, we are interested in the problem: given two batches of samples, how can we know whether they follow the same distribution? Specifically, how can we measure the similarity between them? In information theory, the measure of uncertainty on a distribution $p(X)$ is the entropy H of X [45] on the sample space $\mathcal{X}$, given as:

$$H(X) = -\sum_{x \in \mathcal{X}} p(x) \log p(x) = -\mathbb{E}[\log p(X)]. \tag{1}$$

Uncertainty Measurement. Mutual Information measures the reduction in uncertainty for one variable X given a known value of another variable Y, which is defined by:

$$I(X; Y) = H(X) - H(X|Y). \tag{2}$$

MI between X and Y can be calculated using the Kullback-Leibler(KL) divergence [26] between the joint entropy $H(X, Y)$ and product of the two individual entropy $H(X)H(Y)$.

$$I(X; Y) = D_{KL}(H(X, Y) \| H(X)H(Y)). \tag{3}$$

We calculate $I(X; Y)$ between two batches of samples X and Y as the confidence of them belonging to the same distribution using Eq. (3). Note that for a single distribution $p(X)$, ideally, the MI within its observations is the entropy itself: $I(X; X) = H(X)$. However, we need to re-evaluate the value to detect noise in the samples, which indeed often exist in real-world. Therefore, to measure the uncertainty within a distribution, we use the below formula:

$$I(X; X) = D_{KL}(H(X, X) \| H(X)H(X)). \tag{4}$$

In practice, the higher the measured MI, the lower the distributional uncertainty. During learning, we maximise the MI between source and target to reduce the uncertainty and learn a more generalised model. Note that although there is also possibility that the MI between distributions [1] can be learnt instead of calculated, it cannot be directly applied here. Since this would require the distributions to be known, while we have one of the distributions (the target) unknown.

3.2 Transferable Feature Learning

MI-Guided Transferable Feature Learning. Geirhos et al. [17] found that CNNs recognise objects mainly according to their texture, while overlooking other structural features such as shape and edges. This texture bias is observed repeatedly with different network architectures across different tasks [41,53]. To increase model transferability, we need to reduce such bias by introducing appropriately balanced learning of texture and other non-texture features. In tasks where texture varies drastically across observations, learning of other transferable features should be attenuated. Whilst if texture may serve as a generic feature across different distributions, the learning of texture should not be diminished.

We propose to further utilise the distributional uncertainty measured by mutual information in Sect. 3.1 as the controlling factor for transferable feature learning. To be specific, we adopt the below learning object:

$$\mathcal{L}_{total} = \mathcal{L}_{cls}(X) + \lambda_X \cdot \mathcal{L}_{trans}(X) + \lambda_Y \cdot \mathcal{L}_{trans}(Y), \tag{5}$$

$$\lambda_X = \frac{I(X; X)}{I(X; Y)}, \lambda_Y = \frac{I(Y; Y)}{I(Y; X)}, \tag{6}$$

where $\mathcal{L}_{cls}$ is the supervised classification loss on the labelled source samples X and $\mathcal{L}_{trans}$ the auxiliary loss for explicit learning of non-texture transferable features on both X and unlabelled target samples Y.

The weighting term λ_X and λ_Y, decided by MI, dedicates to balance between texture and non-texture feature learning. The intuition is that, when MI is small which indicates high distributional uncertainty, learning of more transferable features can be beneficial. Whilst larger MI manifests a successful transfer within the network and thus alleviating the need for complementary features other than texture.

The selection of a proper $\mathcal{L}_{trans}$ can be exhaustive. One option is increasing the texture diversity in the training data. For example, some involve comprehensive pre-defined image augmentations [7,9,54], some apply learned style transfer [17,29]. However, the improvement in their generalisation comes at the cost of huge computations in retrieving the enormous potential sample space. Not to mention that, diverse training samples are often required in the first place to learn appropriate augmentation/transformation methods.

Self-Supervised Visual Feature Learning. Since the requirement for exhaustive annotations has been identified as a major bottleneck for deep learning, the concept of self-supervised learning is proposed as a promising solution [10]. In self-supervised learning, rich information in the input that may be ignored by the designated task learner is further mined by an auxiliary task. And the auxiliary task can be trained without requiring any human labelling effort. Examples of such tasks are colourisation of de-coloured images [27,55], jigsaw puzzle [4,10] and inpainting [37]. Empirical results using these auxiliary tasks are found helpful in regularising the learning procedure and notably improving generalisation [20]. Hence, we propose to employ self-supervised learning methods to explicitly enforce non-texture feature learning.

One thing to note for our selection of the auxiliary loss $\mathcal{L}_{trans}$ is that, no labels should be required for the unlabelled target samples Y. And this fits seamlessly into the setting of self-supervised learning. In theory, the aforementioned tasks should be compatible within our framework, where their corresponding learning objectives can be the choice of $\mathcal{L}_{trans}$ in our MI-guided transferable feature learning.

While most of these methods are proven effective to regularise the network learning to be less biased towards easy-to-fit texture, none of them exploits object shape explicitly. This is, however, contradictory to the way by which human recognise visual objects. Neuroscience study [19] found that, training human participants on recognising objects composed of certain object shapes significantly improves their performance in recognising other different objects containing the same shapes. Motivated by the fact that shape plays a vital role in human's visual perception and recognition paradigm, we propose a new self-supervised shape learning task to better mimic human vision using CNNs. Note that our new shape learning method serves as one new potential candidate for the choice of $\mathcal{L}_{trans}$, and is in parallel with the aforementioned ones.

Self-Supervised Object Shape Learning. The target of object shape learning is to embed object structural information into the un-constrained features that the network learns. Learning object shape as a complementary feature is advantageous not only because it is more interpretable. In the existence of sharp change in texture and colour, recognition using object shape is more reliable as it is often more consistent and independent of the frequent appearance variations.

Based on the principle that we would like to avoid requiring any extra labelling effort in creating a reference shape for each training sample, we design a self-supervised shape learning scheme by creating reference shapes using edge detectors. Edge detection as a traditional low-level vision task has been studied thoroughly with mature tools formulated. The overall object shape can be obtained by running an edge detector on an image. Here, the target of our

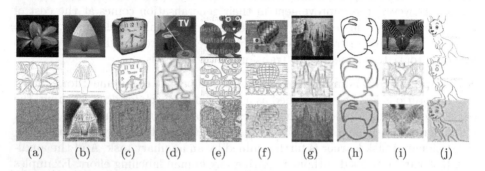

(a)   (b)   (c)   (d)   (e)   (f)   (g)   (h)   (i)   (j)

Fig. 3. Examples in Office-Home (a-d) and DomainNet dataset (e-h): 1st row - original, 2nd row - reference shape and 3rd row - generated shape image trio.

learning is a set of shape-embedded generic features that can contribute to the visual recognition end. Motivated by the Perceptual Loss [23], we define a feature descriptor f that captures the low-level feature of an image. A shape learner learns to generate a shape image given an input object image. Examples of shapes extracted by traditional edge detector and generated shape by our framework are illustrated in Fig. 3. Then we measure the perceptual difference between low-level features of the generated shape f_{gen} and the reference shape f_{ref} as:

$$\mathcal{L}_{shape} = \mathcal{D}(f_{gen}, f_{ref}), \tag{7}$$

where $\mathcal{D}$ is a deviation measure for the low-level features.

Compared to pixel-level loss, our shape loss gives more freedom to the shape learner to keep certain degrees of texture-related details that can benefit the recognition process in some cases (such as the shade in Fig. 3(b)). On the contrary, strictly constraining the shape learner to reference shape at the pixel-level obviously would lose this discriminative information.

4   Experiments

In this section, we provide a detailed empirical analysis of MIMTFL. We conduct extensive experiments on two large scale datasets with varying distributional uncertainties. To analyse the effectiveness of MIMTFL, we consider these three aspects: i) the level of distributional uncertainty exists in these existing evaluation benchmarks; ii) the benefit of uncertainty reduction and iii) the benefit of explicit shape learning.

4.1 Settings

Unsupervised Domain Adaptation (UDA). In UDA experiments, the training set consists of two types of data: labelled samples from one or more distributions (source), and unlabelled samples from one or more distributions (target). We further divide these tasks into three groups, which correspond to most common real-world scenarios, with the difficulty level in each group magnified compared to the previous: i) **Single-to-single** adaptation in which we train using a single labelled source distribution and a single unlabelled target distribution. We evaluate the effect of our self-supervised shape feature learning on model transferability with limited samples, and the generalisation capacity of our model across these different tasks. ii) **Multi-to-single** adaptation where the labelled source samples come from multiple distributions, whilst the unlabelled target samples follow a single distribution. We validate the benefits of our framework to simultaneous learning from multiple disjoint distributions. iii) **Multi-to-multi** adaptation with both the labelled source and unlabelled target are drawn from multiple distributions. We verify the effect of our proposed method with up-scaled unlabelled target data and even fewer labelled source data.

Domain Generalisation (DG). DG is a more challenging task since the target is completely absent during training. Under this setting, we can only estimate the distributional shift within the labelled source domain, yet still transfer the learned model to the target at test time. To achieve a reasonable result on target, a model has to learn as much as possible the most transferable and generalised features from the given source distributions.

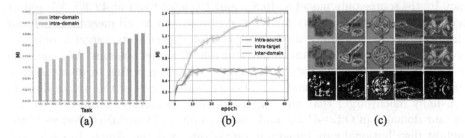

Fig. 4. Measured MI during training in Office-Home experiments, the higher the MI, the lower the distributional uncertainty. (a) Initial MI both intra and inter-domain in Office-Home Dataset, (b) the change of MI during training in R→P experiment, (c) visualised features without (1st row) and with (2nd row) shape learning, and the difference between them (3rd row). *Best viewed in colour.*

4.2   Datasets

Office-Home. The dataset [51] parents four distinct domains in 65 object categories: Art (A), Clipart (C), Product (P) and RealWorld (R). The number of images in each domain is: 2427, 4365, 4439 and 4357, respectively. For all the experiments, we use all the samples in each domain as either source or target. Note that, although we treat each domain as a single distribution following common practice, there still exist varying levels of out-of-distribution examples in some domains. For instance, the Clipart domain which contains artistic images in various forms and patterns exhibits a rather high distributional uncertainty even within the domain itself. This is clearly indicated in its low measured MI value shown in Fig. 4(a).

DomainNet. As the largest and the most challenging benchmarking dataset for DA till today, DomainNet dataset [38] includes six domains: clipart (c), infograph (i), painting (p), quickdraw (q), real (r) and sketch (s). The samples spread through 345 diverse categories and each domain is divided into a train split for training and a test split for test only. The difficulty of this dataset comes from several aspects: i) highly abstract and artistic images, ii) diverse domains, and iii) severely unbalanced domains.

4.3   Implementation Details

Domain Adaptation. We apply MIMTFL on both Office-Home and Domain-Net for all the three types of tasks. In single-to-single adaptation experiments, we use each domain as source and another domain as the target in turn for all the domains, ending up with 12 single-to-single tasks for Office-Home and 30 single-to-single tasks for DomainNet. Specifically, for better evaluation of different levels of distributional uncertainty, we divide 30 single-to-single adaptation tasks in DomainNet into two groups: 15 difficult tasks with average target accuracy by the source only model at 10.8% and 15 easier tasks at 42.3%. We report the mean accuracy separately for the two groups and an all mean accuracy for all of the 30 tasks. In multi-to-single adaptation experiments, target is only one domain while the source comes from multiple different domains. We use each domain as the target and the rest as source in turn, resulting in four tasks for Office-Home and six for DomainNet. The combined scale of the source domain is usually much larger than the target. In multi-to-multi experiments, we use all four domains in Office-Home and four domains in DomainNet that yield the highest distributional gap: infograph, quickdraw, real and sketch. In each task, two domains are chosen as source and the other two as target.

Domain Generalisation. To compare with existing domain generalisation methods, we conduct single-to-single and multi-to-single experiments on Office-Home dataset. Under this setting the shape loss is only applied to the source samples, where no target samples are used in training.

Reference Shape Extraction. While many well-developed edge detectors including Canny [3] are generally applicable, we use a structure-aware edge detector [11] here as an example. Each image is input into the algorithm, and the output is a single channel image under the same resolution of the input. Object shapes are described by the edges, resulting in a sketch-lized representation of the original image. For most images, the extracted edges well depict the object shapes. For all our experiments, we use the Structured Forests edge detector implemented in OpenCV[1].

Network and Training Setups. We use the original ResNet [21] architecture with a single FC layer as our backbone network for all experiments. For the relatively smaller Office-Home dataset, we use ResNet-50 for UDA experiments and ResNet-18 for DG experiments. For DomainNet experiments, we use ResNet-101 backbone. Following common practice, ImageNet [8] pre-trained model is used as backbone network initialisation. The shape loss compares shape features extracted from the first convolutional block output in ResNet. The weight of the shape loss is calculated using the discrepancy between output features of the first residual block output in ResNet. The shape learner is composed of two layers of deconvolution, with output channels 128 and 3, respectively. The generated shape image is exactly the same dimension with the input image, fixed as $224 \times 224 \times 3$. We use SGD [42] optimiser with learning rate 0.0001 for the ImageNet pre-trained layers and 0.001 for the FC layer in all the experiments. We apply basic data augmentation routine including random cropping and horizontal flipping only. All implementations and experiments are based on PyTorch [36].

Benchmarking. We compare our method with representative and latest state-of-the-art algorithms that study the problem of distributional discrepancy. Following the UDA setting, SE [14], CDAN [31] and DWT [43] are the latest top-performing DA methods on the Office-Home dataset. DANN [16] and ADDA [49] are representative domain adversarial learning methods, while JAN [33] and MCD [44] focus on discrepancy minimisation. For a fair evaluation in Office-Home experiments, we also report result using our method combined with the Min-Entropy Consensus loss (MEC) [43] to compare with aggregated methods: DWT+MEC [43] and BSP+CDAN [6]. For DG setting, JiGen[4] uses self-supervised task of solving jigsaw puzzles to increase model transferability and is the state-of-art on Office-Home dataset. D-SAMs [12] is a representative method that targets at bridging multiple distinct distributions in training for DG. All the results are reported in top 1 accuracy and cited from the original papers.

4.4 Ablation Study

Uncertainty Analysis in Existing Datasets. We compare the average uncertainty, indicated by MI both between different domains and within the same domain illustrated in Fig. 4(a). We observe clearly that, in Office-Home dataset, the visually most diverse domain - Clipart yields the lowest MI even within itself.

---

[1] https://opencv.org/.

This is consistent with the adaptation results, where we see that tasks involving the Clipart domain are generally more difficult than others.

Component Analysis of MIMTFL. To further understand the effectiveness of MIMTFL, we conduct ablation studies on its two main components, namely **MIM** for MI maximisation and **TFL** for transferable feature learning. Figure 4 (c) clearly shows that with the shape learning objective, the network focuses more on object shape rather than textures. From the results in Table 1, we observe that without either module, the adaptation results on almost all the tasks drop significantly. Especially on those difficult tasks where the source only model accuracy is lower, such as P → C and A → C, adding in the MI module increases model transferability by a large margin. The change of measured MI during training is plotted in Fig. 4(b), which shows a steady increase of the mutual information between domains, indicating the reduction in the uncertainty. Specifically, we observe that the value of the MI yields a higher impact on those more difficult adaptation tasks, where the distributional uncertainty is higher. Our MI maximisation gives the best performance on these challenging tasks. As in reality, it is difficult to manually tune the hyperparameter for transferable feature learning, which is a strong demonstration of the benefit of our adaptive weighting according to calculated distributional uncertainty.

4.5   Results

Unsupervised Domain Adaptation. The results are shown in Table 1 and 2. We list the task column in the ascending order of their source only model performance. Although the Office-Home benchmark is highly competitive, it is observed that MIMTFL is able to achieve top performance. Specifically, our method achieves significant improvement on those tasks with lower source only accuracy. Such as A→C task, it outperforms previous SOTA with a 3.3% absolute gain. When combined with the MEC loss, it further boosts the average target accuracy and produces the best result among all methods in 6 out of all the 12 tasks. It is worth noting that, DWT and DWT+MEC engage affine transformation and Gaussian blurring as additional data augmentation during training, while our reported results are without such sophisticated augmentation. Additionally, we report a source only result where no target data is used in training, namely MIMTFL$_{src}$. Compared with latest method such as DDAIG [57], our method proves strong generalisation capacity in this experiment.

On the more challenging DomainNet dataset, firstly, we observe that in many tasks the adaptation actually harms the target performance (i.e., negative transfer). For instance, the images in "infograph" domain are largely occupied by texts, while the object to be recognised are highly abstracted or in various artistic styles. This could lead to failure in CNN learned features to capture the true commonalities in objects. The high accuracy of adaptation to the "real" domain can be mainly due to the use of ImageNet pre-trained model in the backbone networks. While for "quickdraw" domain, the perquisite from ImageNet model

Table 1. Results on Office-Home single-to-single UDA (all with ResNet-50 backbone).

Method	P→C	A→C	C→A	P→A	R→C	C→P	C→R	A→P	R→A	A→R	P→R	R→P	mean
Source only	31.2	34.9	37.4	38.5	41.2	41.9	46.2	50.0	53.9	58.0	60.4	59.9	46.1
DDAIG [57]	36.8	40.8	43.7	40.0	44.5	55.6	56.9	53.6	56.3	65.4	63.5	73.8	52.6
MIMTFL*$_{src}$	**48.5**	**51.5**	**57.1**	**53.2**	**52.2**	**65.6**	**67.6**	**67.9**	**66.2**	**74.8**	**78.8**	**74.1**	**63.1**
DANN [16]	43.7	45.6	47.0	46.1	51.8	58.5	60.9	59.3	63.2	70.1	68.5	76.8	57.6
JAN [33]	43.4	45.9	50.4	45.8	52.4	59.7	61.0	61.2	63.9	68.9	70.3	76.8	58.3
SE [14]	41.5	43.2	55.0	50.4	49.5	59.0	64.5	60.2	64.9	70.4	68.9	75.2	58.6
CDAN [31]	49.1	50.6	55.7	51.8	**56.9**	62.7	64.2	65.9	68.2	73.4	74.5	80.7	62.8
DWT [43]	49.5	50.8	58.9	**57.2**	55.3	65.6	60.2	**72.0**	**70.1**	**75.8**	**78.3**	78.2	64.3
TFL	48.6	52.5	57.2	54.8	51.4	66.2	68.6	68.4	66.5	75.0	74.0	78.2	63.4
MIM	48.8	52.1	57.1	52.8	52.6	66.0	58.5	67.5	65.9	74.2	73.5	78.1	63.1
MIMTFL	**51.1**	**54.1**	**59.1**	55.8	55.4	**66.9**	**69.4**	68.4	67.8	75.2	74.6	**79.1**	**64.7**
BSP+DANN [6]	49.6	51.4	56.0	57.0	57.1	67.8	58.8	68.3	70.4	75.9	75.8	80.6	64.9
DWT+MEC [43]	47.9	54.7	56.9	54.8	54.9	68.5	59.8	**72.3**	68.6	**77.2**	**78.1**	81.2	65.4
BSP+CDAN [6]	50.2	52.0	58.0	**58.6**	59.3	70.3	70.2	68.6	**72.2**	76.1	77.6	**81.9**	66.3
MIMTFL+MEC	**54.9**	**56.9**	**61.2**	**58.6**	**59.4**	70.0	**71.6**	70.3	69.8	75.6	77.5	80.4	**67.2**

Table 2. Results on DomainNet single-to-single UDA (all with ResNet-101 backbone).

Method	q→i	q→p	i→q	q→r	p→q	r→q	q→c	q→s	s→q	c→q	s→i	p→i	c→i	r→i	i→s	mean
Source only	**0.9**	1.4	**3.6**	4.1	**4.9**	**6.4**	**7.0**	**8.3**	**10.9**	**11.1**	**15.4**	**18.7**	**19.3**	**22.2**	**27.9**	**10.8**
MIMTFL$_{src}$	0.8	**4.8**	3.2	**4.3**	4.8	6.0	6.7	7.6	10.8	10.2	13.9	16.5	15.4	19.1	27.2	10.1
ADDA [49]	2.6	5.4	3.2	9.9	**8.4**	**12.1**	15.7	11.9	**14.9**	3.2	8.9	9.5	11.2	14.5	14.6	9.7
MCD [44]	3.0	**7.0**	1.5	11.5	1.9	2.2	15.0	10.2	3.8	1.6	13.7	14.8	14.2	**19.6**	18.0	9.2
DANN [16]	2.0	4.4	**3.8**	9.8	5.5	6.3	11.8	8.4	10.4	9.5	13.9	15.1	15.5	17.9	25.7	10.7
TFL	1.5	3.6	2.3	9.0	3.1	5.1	14.5	8.0	11.1	9.9	15.4	**16.2**	**15.9**	18.9	27.8	10.8
MIMTFL	**3.1**	5.0	2.9	**16.0**	4.2	5.8	**18.8**	**13.8**	12.3	**10.7**	**16.5**	14.7	15.1	19.0	**31.0**	**12.6**

Method	i→c	i→p	p→s	s→p	c→p	r→s	p→c	c→s	i→r	s→c	s→r	r→p	c→r	p→r	r→c	mean	all mean
Source only	30.2	**31.2**	36.3	**37.0**	**37.5**	**38.8**	39.6	41.0	44.0	46.9	47.0	48.4	**49.4**	**52.2**	54.5	42.3	**26.6**
MIMTFL$_{src}$	**33.1**	30.8	**36.8**	35.1	36.0	37.9	**41.8**	**42.3**	**46.4**	**49.4**	**47.2**	**50.0**	47.4	51.7	**56.4**	**42.8**	26.3
ADDA [49]	19.1	16.4	25.4	25.2	24.1	25.7	31.2	30.7	26.9	35.3	37.6	39.5	29.1	41.9	39.1	29.8	19.8
MCD [44]	23.6	21.2	28.4	27.6	26.1	29.3	34.4	33.8	36.7	41.2	34.8	42.6	42.6	45.0	50.5	34.5	21.9
DANN [16]	31.8	30.2	35.1	34.5	34.8	37.3	39.6	41.4	44.8	47.9	46.8	47.5	47.0	50.8	54.6	41.6	26.1
TFL	**33.7**	**32.0**	**36.8**	39.9	**36.0**	37.1	**43.2**	41.9	46.9	**52.2**	50.9	**50.0**	47.5	**52.4**	**56.5**	**43.8**	27.3
MIMTFL	32.1	31.0	**36.8**	**40.3**	35.6	**39.4**	40.1	**43.1**	**48.5**	51.7	**53.5**	48.5	**47.6**	51.5	55.4	43.7	**28.1**

Table 3. Results on Office-Home multi-to-single UDA (all with ResNet-50 backbone).

Method	ACP→R	ACR→P	APR→C	CPR→A	mean
Source only	81.7	80.1	58.5	69.4	72.4
SE [14]	79.2	76.3	54.3	68.8	69.7
DWT+MEC [43]	**83.8**	**83.9**	59.1	**73.0**	74.9
MIMTFL+MEC	83.1	81.9	**64.3**	72.6	**75.5**

Table 4. Results on DomainNet multi-to-single UDA (all with ResNet-101 backbone).
* indicates multi-source adaptation methods

Method	ipqrs→c	cpqrs→i	ciqrs→p	ciprs→q	cipqs→r	cipqr→s	mean
Source only	39.6	8.2	33.9	11.8	41.6	23.1	26.4
SE [14]	24.7	3.9	12.7	7.1	22.8	9.1	13.4
ADDA [49]	47.5	11.4	36.7	14.7	49.1	33.5	32.2
MCD [44]	54.3	22.1	45.7	7.6	58.4	43.5	38.6
DANN [16]	58.1	21.0	51.1	10.3	66.2	49.3	42.7
DCTN* [52]	48.6	23.5	48.8	7.2	53.5	47.3	38.2
M^3SDA* [38]	57.2	24.2	51.6	5.2	61.6	49.6	41.6
MIMTFL	**67.2**	**25.0**	**54.4**	**13.4**	**67.0**	**54.1**	**46.8**

Table 5. Results on Office-Home multi-to-multi UDA (all with ResNet-50 backbone).

Method	PR→AC	AP→CR	AR→CP	CP→AR	AC→PR	CR→AP	mean
Source only	59.7	68.9	69.0	72.8	74.9	76.0	70.2
SE [14]	54.6	62.4	63.4	70.1	70.0	73.4	65.6
DWT+MEC [43]	59.1	69.9	68.8	**76.2**	**78.8**	**77.6**	71.7
MIMTFL+MEC	**62.9**	**70.7**	**70.4**	75.7	76.1	77.3	**72.2**

Table 6. Results on DomainNet multi-to-multi UDA (all with ResNet-101 backbone).

Method	rs→iq	ir→qs	qr→is	is→qr	qs→ir	iq→rs	mean
Source only	14.0	17.6	31.0	35.4	36.2	42.0	29.4
ADDA [49]	4.2	2.9	15.9	20.6	27.7	17.2	14.7
MCD [44]	12.2	15.4	27.4	29.3	36.6	33.8	25.8
MIMTFL	**14.3**	**17.9**	**31.9**	**36.3**	**43.1**	**43.6**	**31.2**

is of little help, and the network is forced to learn almost from scratch. Under such challenges, MIMTFL improves performance in 8 out of the 15 difficult tasks. While ADDA fails to bring any improvement on 7 tasks and MCD on 10, MIMTFL only fails on two, comparing to the source only model. Here since all the other methods focus on aligning the classifier output distributions using general CNN features, the effectiveness of our proposed MI-guided transferable feature learning is clearly proven.

Results of multiple-to-single tasks are in Table 3 and 4. While MIMTFL easily outperforms other methods on both datasets, we observe that it is especially outstanding for the difficult tasks. In the most difficult task APR→C in Office-Home experiments, MIMTFL + MEC improves the absolute target accuracy from source only model by 5.8%, bringing a 9.9% gain. In DomainNet experiments, our method creates new SOTA results in all of the six tasks, astoundingly boosting the source only model by 77%. Note that DCTN [52] and M^3SDA [38] are specifically designed multi-source adaptation algorithms that consider collaborative learning among multiple source domains. This proves the superiority of our formulation of distributional uncertainty for multiple source domains.

Results of multi-to-multi tasks are shown in Table 5 and 6. In Office-Home experiments, our model is further proven to work well especially on the more difficult tasks, while other distribution alignment methods, such as SE, fail to even beat the source only model. In the more challenging DomainNet experiments, we observe negative transfer here again in ADDA and MCD. Since the scale of the target domain is multiplied, the adaptation becomes even harder. Results show that MIMTFL is able to improve the source model and performs the best in all the six tasks.

Domain Generalisation. We observe in results presented in Table 7 that our method easily outperforms existing state-of-the-art specialised DG methods. Specifically, we find that our shape learning is more effective than JiGen using jigsaw puzzle as transferable feature learning.

Table 7. Results on Office-Home multi-to-single DG (all with ResNet-18 backbone).

Method	ACP→R	ACR→P	APR→C	CPR→A	mean
D-SAMs [12]	71.5	69.2	44.4	**58.0**	60.8
JiGen [4]	72.8	71.5	47.5	53.0	61.2
MIMTFL	**74.4**	**73.1**	**51.1**	53.3	**63.0**

5   Conclusions

In this paper, we propose a theory grounded formulation for the definition of domain discrepancy using distributional uncertainty. We maximise the mutual information between source and target. In addition, we propose to enhance transferable feature learning in CNNs by balancing texture and non-texture feature learning with the measured uncertainty. To explicitly learn non-texture features, we propose a novel self-supervised object shape learning method, which can be used in parallel with many existing self-supervised visual feature learning methods. Our idea is thoroughly experimented and validated through extensive experiments.

References

1. Belghazi, M.I., et al.: Mutual information neural estimation. In: ICML (2018)
2. Bhattacharyya, A., Fritz, M., Schiele, B.: Long-term on-board prediction of people in traffic scenes under uncertainty. In: CVPR (2018)
3. Canny, J.: A computational approach to edge detection. IEEE Trans. Pattern Anal. Mach. Intell. **6**, 679–698 (1986)
4. Carlucci, F.M., D'Innocente, A., Bucci, S., Caputo, B., Tommasi, T.: Domain generalization by solving jigsaw puzzles. In: CVPR (2019)
5. Chen, C., et al: Progressive feature alignment for unsupervised domain adaptation. In: CVPR (2019)
6. Chen, X., Wang, S., Long, M., Wang, J.: Transferability vs. discriminability: Batch spectral penalization for adversarial domain adaptation. In: ICML (2019)
7. Cubuk, E.D., Zoph, B., Mane, D., Vasudevan, V., Le, Q.V.: Autoaugment: learning augmentation strategies from data. In: CVPR (2019)
8. Deng, J., Dong, W., Socher, R., Li, L.J., Li, K., Li, F.F.: Imagenet: a large-scale hierarchical image database. In: CVPR (2009)
9. DeVries, T., Taylor, G.W.: Improved regularization of convolutional neural networks with cutout. arXiv preprint arXiv:1708.04552 (2017)
10. Doersch, C., Gupta, A., Efros, A.A.: Unsupervised visual representation learning by context prediction. In: ICCV (2015)
11. Dollár, P., Zitnick, C.L.: Fast edge detection using structured forests. IEEE Trans. Pattern Anal. Mach. Intell. **37**(8), 1558–1570 (2014)
12. D'Innocente, A., Caputo, B.: Domain generalization with domain-specific aggregation modules. In: Brox, T., Bruhn, A., Fritz, M. (eds.) GCPR 2018. LNCS, vol. 11269, pp. 187–198. Springer, Cham (2019). https://doi.org/10.1007/978-3-030-12939-2_14
13. Feng, D., Rosenbaum, L., Dietmayer, K.: Towards safe autonomous driving: capture uncertainty in the deep neural network for lidar 3D vehicle detection. In: IEEE Intelligent Transportation Systems Conference (ITSC) (2018)
14. French, G., Mackiewicz, M., Fisher, M.: Self-ensembling for visual domain adaptation. In: ICLR (2018)
15. Ganin, Y., Lempitsky, V.: Unsupervised domain adaptation by backpropagation. In: ICML (2015)
16. Ganin, Y., et al.: Domain-adversarial training of neural networks. J. Mach. Learn. Res. **17**(1), 2030–2096 (2016)
17. Geirhos, R., Rubisch, P., Michaelis, C., Bethge, M., Wichmann, F.A., Brendel, W.: ImageNet-trained CNNs are biased towards texture; increasing shape bias improves accuracy and robustness. In: ICLR (2019)
18. Ghahramani, Z.: Probabilistic machine learning and artificial intelligence. Nature **521**(7553), 452–459 (2015)
19. Gölcü, D., Gilbert, C.D.: Perceptual learning of object shape. J. Neurosci. **29**(43), 13621–13629 (2009)
20. Goyal, P., Mahajan, D., Gupta, A., Misra, I.: Scaling and benchmarking self-supervised visual representation learning. In: ICCV (2019)
21. He, K., Zhang, X., Ren, S., Sun, J.: Deep residual learning for image recognition. In: CVPR (2016)
22. Ilg, E., et al.: Uncertainty estimates and multi-hypotheses networks for optical flow. In: Ferrari, V., Hebert, M., Sminchisescu, C., Weiss, Y. (eds.) ECCV 2018. LNCS, vol. 11211, pp. 677–693. Springer, Cham (2018). https://doi.org/10.1007/978-3-030-01234-2_40

23. Johnson, J., Alahi, A., Fei-Fei, L.: Perceptual losses for real-time style transfer and super-resolution. In: Leibe, B., Matas, J., Sebe, N., Welling, M. (eds.) ECCV 2016. LNCS, vol. 9906, pp. 694–711. Springer, Cham (2016). https://doi.org/10.1007/978-3-319-46475-6_43

24. Kendall, A., Gal, Y.: What uncertainties do we need in Bayesian deep learning for computer vision? In: NIPS (2017)

25. Kendall, A., Gal, Y., Cipolla, R.: Multi-task learning using uncertainty to weigh losses for scene geometry and semantics. In: CVPR (2018)

26. Kullback, S., Leibler, R.A.: On information and sufficiency. Ann. Math. Stat. **22**(1), 79–86 (1951)

27. Larsson, G., Maire, M., Shakhnarovich, G.: Colorization as a proxy task for visual understanding. In: CVPR (2017)

28. Lee, D.H.: Pseudo-label: the simple and efficient semi-supervised learning method for deep neural networks. In: ICML Workshop (2013)

29. Lee, H., Kim, H.E., Nam, H.: SRM: a style-based recalibration module for convolutional neural networks. In: ICCV (2019)

30. Long, M., Cao, Y., Wang, J., Jordan, M.: Learning transferable features with deep adaptation networks. In: ICML (2015)

31. Long, M., Cao, Z., Wang, J., Jordan, M.I.: Conditional adversarial domain adaptation. In: NeurIPS (2018)

32. Long, M., Wang, J., Ding, G., Sun, J., Yu, P.S.: Transfer feature learning with joint distribution adaptation. In: ICCV (2013)

33. Long, M., Zhu, H., Wang, J., Jordan, M.I.: Deep transfer learning with joint adaptation networks. In: ICML (2017)

34. Nagy, Z., Braatz, R.D.: Distributional uncertainty analysis using power series and polynomial chaos expansions. J. Process Control **17**(3), 229–240 (2007)

35. Pan, Y., Yao, T., Li, Y., Wang, Y., Ngo, C.W., Mei, T.: Transferrable prototypical networks for unsupervised domain adaptation. In: CVPR (2019)

36. Paszke, A., et al.: Automatic differentiation in Pytorch. In: NIPS Workshop (2017)

37. Pathak, D., Krahenbuhl, P., Donahue, J., Darrell, T., Efros, A.A.: Context encoders: feature learning by inpainting. In: CVPR (2016)

38. Peng, X., Bai, Q., Xia, X., Huang, Z., Saenko, K., Wang, B.: Moment matching for multi-source domain adaptation. In: ICCV (2019)

39. Pereyra, G., Tucker, G., Chorowski, J., Kaiser, L., Hinton, G.: Regularizing neural networks by penalizing confident output distributions. In: ICLR Workshop (2017)

40. Qi, L., Wang, L., Huo, J., Zhou, L., Shi, Y., Gao, Y.: A novel unsupervised camera-aware domain adaptation framework for person re-identification. In: ICCV (2019)

41. Ringer, S., Williams, W., Ash, T., Francis, R., MacLeod, D.: Texture bias of CNNs limits few-shot classification performance. arXiv preprint arXiv:1910.08519 (2019)

42. Robbins, H., Monro, S.: A stochastic approximation method. Ann. Math. Stat. **22**(3), 400–407 (1951)

43. Roy, S., Siarohin, A., Sangineto, E., Bulo, S.R., Sebe, N., Ricci, E.: Unsupervised domain adaptation using feature-whitening and consensus loss. In: CVPR (2019)

44. Saito, K., Watanabe, K., Ushiku, Y., Harada, T.: Maximum classifier discrepancy for unsupervised domain adaptation. In: CVPR (2018)

45. Shannon, C.E.: A mathematical theory of communication. Bell Syst. Tech. J. **27**(3), 379–423 (1948)

46. Straszecka, E.: Combining uncertainty and imprecision in models of medical diagnosis. Inf. Sci. **176**(20), 3026–3059 (2006)

47. Sun, B., Saenko, K.: Deep CORAL: correlation alignment for deep domain adaptation. In: Hua, G., Jégou, H. (eds.) ECCV 2016. LNCS, vol. 9915, pp. 443–450. Springer, Cham (2016). https://doi.org/10.1007/978-3-319-49409-8_35
48. Tarvainen, A., Valpola, H.: Mean teachers are better role models: weight-averaged consistency targets improve semi-supervised deep learning results. In: NIPS (2017)
49. Tzeng, E., Hoffman, J., Saenko, K., Darrell, T.: Adversarial discriminative domain adaptation. In: CVPR (2017)
50. Tzeng, E., Hoffman, J., Zhang, N., Saenko, K., Darrell, T.: Deep domain confusion: Maximizing for domain invariance. arXiv preprint arXiv:1412.3474 (2014)
51. Ramakrishnan, R., Nagabandi, B., Eusebio, J., Chakraborty, S., Venkateswara, H., Panchanathan, S.: Deep hashing network for unsupervised domain adaptation. In: Venkateswara, H., Panchanathan, S. (eds.) Domain Adaptation in Computer Vision with Deep Learning, pp. 57–74. Springer, Cham (2020). https://doi.org/10.1007/978-3-030-45529-3_4
52. Xu, R., Chen, Z., Zuo, W., Yan, J., Lin, L.: Deep cocktail network: multi-source unsupervised domain adaptation with category shift. In: CVPR (2018)
53. Zaech, J.N., Dai, D., Hahner, M., Van Gool, L.: Texture underfitting for domain adaptation. In: IEEE Intelligent Transportation Systems Conference (ITSC) (2019)
54. Zhang, H., Cisse, M., Dauphin, Y.N., Lopez-Paz, D.: mixup: beyond empirical risk minimization. In: ICLR (2018)
55. Zhang, R., Isola, P., Efros, A.A.: Colorful image colorization. In: Leibe, B., Matas, J., Sebe, N., Welling, M. (eds.) ECCV 2016. LNCS, vol. 9907, pp. 649–666. Springer, Cham (2016). https://doi.org/10.1007/978-3-319-46487-9_40
56. Zhao, H., Zhang, S., Wu, G., Moura, J.M., Costeira, J.P., Gordon, G.J.: Adversarial multiple source domain adaptation. In: NeurIPS (2018)
57. Zhou, K., Yang, Y., Hospedales, T.M., Xiang, T.: Deep domain-adversarial image generation for domain generalisation. In: AAAI (2020)
58. Zhu, X., Ghahramani, Z.: Learning from labeled and unlabeled data with label propagation. Technical report CMU-CALD-02-107 (2002)

Bridging Knowledge Graphs to Generate Scene Graphs

Alireza Zareian[✉], Svebor Karaman, and Shih-Fu Chang

Columbia University, New York, NY 10027, USA
{az2407,sk4089,sc250}@columbia.edu

Abstract. Scene graphs are powerful representations that parse images into their abstract semantic elements, *i.e.*, objects and their interactions, which facilitates visual comprehension and explainable reasoning. On the other hand, commonsense knowledge graphs are rich repositories that encode how the world is structured, and how general concepts interact. In this paper, we present a unified formulation of these two constructs, where a scene graph is seen as an image-conditioned instantiation of a commonsense knowledge graph. Based on this new perspective, we reformulate scene graph generation as the inference of a bridge between the scene and commonsense graphs, where each entity or predicate instance in the scene graph has to be linked to its corresponding entity or predicate class in the commonsense graph. To this end, we propose a novel graph-based neural network that iteratively propagates information between the two graphs, as well as within each of them, while gradually refining their bridge in each iteration. Our Graph Bridging Network, GB-NET, successively infers edges and nodes, allowing to simultaneously exploit and refine the rich, heterogeneous structure of the interconnected scene and commonsense graphs. Through extensive experimentation, we showcase the superior accuracy of GB-NET compared to the most recent methods, resulting in a new state of the art. We publicly release the source code of our method (https://github.com/alirezazareian/gbnet).

1 Introduction

Extracting structured, symbolic, semantic representations from data has a long history in Natural Language Processing (NLP), under the umbrella terms *semantic parsing* at the sentence level [8,9] and *information extraction* at the document level [22,41]. The resulting *semantic graphs* or *knowledge graphs* have many applications such as question answering [7,17] and information retrieval [6,50]. In computer vision, Xu *et al.* have recently called attention to the task of Scene Graph Generation (SGG) [44], which aims at extracting a symbolic, graphical representation from a given image, where every node corresponds to a

---

Electronic supplementary material The online version of this chapter (https://doi.org/10.1007/978-3-030-58592-1_36) contains supplementary material, which is available to authorized users.

A. Vedaldi et al. (Eds.): ECCV 2020, LNCS 12368, pp. 606–623, 2020.
https://doi.org/10.1007/978-3-030-58592-1_36

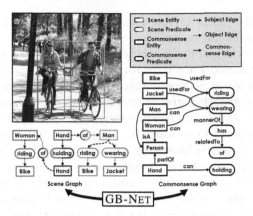

Fig. 1. Left: An example of a Visual Genome image and its ground truth scene graph. Right: A relevant portion of the commonsense graph. In this paper we formulate the task of Scene Graph Generation as the problem of creating a bridge between these two graphs. Such bridge not only classifies each scene entity and predicate, but also creates an inter-connected heterogeneous graph whose rich structure is exploited by our method (GB-NET).

localized and categorized object (entity), and every edge encodes a pairwise interaction (predicate). This has inspired two lines of follow-up work, some improving the performance on SGG [2,10,11,23,24,31,43,47,52], and others exploiting such rich structures for down-stream tasks such as Visual Question Answering (VQA) [12,38,39,54], image captioning [48,49], image retrieval [15,37], and image synthesis [14]. In VQA for instance, SGG not only improves performance, but also promotes interpretability and enables explainable reasoning [38].

Although several methods have been proposed, the state-of-the-art performance for SGG is still far from acceptable. Most recently, [2] achieves only 16% mean recall, for matching the top 100 predicted subject-predicate-object triples against ground truth triples. This suggests the current SGG methods are insufficient to address the complexity of this task. Recently, a few papers have attempted to use external *commonsense* knowledge to advance SGG [2,10,52], as well as other domains [3,16]. This commonsense can range from curated knowledge bases such as ConceptNet [27], ontologies such as WordNet [30], or automatically extracted facts such as co-occurrence frequencies [52]. The key message of those works is that a prior knowledge about the world can be very helpful when perceiving a complex scene. If we know the relationship of a Person and a Bike is most likely riding, we can more easily disambiguate between riding, on, and attachedTo, and classify their relationship more accurately. Similarly, if we know a Man and a Woman are both sub-types of Person, even if we only see Man-riding-Bike in training data, we can generalize and recognize a Woman-riding-Bike triplet at test time. Although this idea is intuitively promising, existing methods that implement it have major limitations, as detailed in Sect. 2, and we address those in the proposed method.

More specifically, recent methods either use ad-hoc heuristics to integrate limited types of commonsense into the scene graph generation process [2,52], or fail to exploit the rich, graphical structure of commonsense knowledge [10]. To devise a general framework for incorporating any type of graphical knowledge into the process of scene understanding, we take inspiration from early works on knowledge representation and applying structured grammars to computer vision problems [32,40,55], and redefine those concepts in the light of the recent advances in graph-based deep learning. Simply put, we formulate both scene and commonsense graphs as knowledge graphs with entity and predicate nodes, and various types of edges. A scene graph node represents an entity or predicate *instance* in a specific image, while a commonsense graph node represents an entity or predicate *class*, which is a general concept independent of the image. Similarly, a scene graph edge indicates the participation of an entity instance (*e.g.* as a subject or object) in a predicate instance in a scene, while a commonsense edge states a general fact about the interaction of two concepts in the world. Figure 1 shows an example scene graph and commonsense graph side by side.

Based on this unified perspective, we reformulate the problem of scene graph generation from entity and predicate classification into the novel problem of bridging those two graphs. More specifically, we propose a method that given an image, initializes potential entity and predicate nodes, and then classifies each node by connecting it to its corresponding class node in the commonsense graph, through an edge we call a *bridge*. This establishes a connectivity between instance-level, visual knowledge and generic, commonsense knowledge. To incorporate the rich combination of visual and commonsense information in the SGG process, we propose a novel graphical neural network, that iteratively propagates messages between the scene and commonsense graphs, as well as within each of them, while gradually refining the bridge in each iteration. Our Graph Bridging Network, GB-NET, successively infers edges and nodes, allowing to simultaneously exploit and refine the rich, heterogeneous structure of the interconnected scene and commonsense graphs.

To evaluate the effectiveness of our method, we conduct extensive experiments on the Visual Genome [20] dataset. The proposed GB-NET outperforms the state of the art consistently in various performance metrics. Through ablative studies, we show how each of the proposed ideas contribute to the results. We also publicly release a comprehensive software package based on [52] and [2], to reproduce all the numbers reported in this paper. We provide further quantitative, qualitative, and speed analysis in our Supplementary Material, as well as additional implementation details.

2   Related Work

2.1   Scene Graph Generation

Most SGG methods are based on an object detection backbone that extracts region proposals from the input image. They utilize some kind of information propagation module to incorporate context, and then classify each region to an object class, as well as each pair of regions to a relation class [2,44,47,52]. Our method has two key differences with this conventional process: firstly, our information propagation network operates on a larger graph which consists of not only object nodes, but also predicate nodes and commonsense graph nodes, and has a more complex structure. Secondly, we do not classify each object and relation using classifiers, but instead use a pairwise matching mechanism to connect them to corresponding class nodes in the commonsense graph.

More recently, a few methods [2,10,52] have used external knowledge to enhance scene graph generation. This external knowledge is sometimes referred to as "commonsense", because it encodes ontological knowledge about classes, rather than specific instances. Despite encouraging results, these methods have major limitations. Specifically, [52] used triplet frequency to bias the logits of their predicate classifier, and [2] used such frequencies to initialize edge weights on their graphs. Such external priors have been also shown beneficial for recognizing objects [45,46] and relationships [26,53], that are building blocks for SGG. Nevertheless, neither of those methods can incorporate other types or knowledge, such as the semantic hierarchy concepts, or object affordances. Gu *et al.* [10] propose a more general way to incorporate knowledge in SGG, by retrieving a set of relevant facts for each object from a pool of commonsense facts. However, their method does not utilize the structure of the commonsense graph, and treats knowledge as a set of triplets. Our method considers commonsense as a general graph with several types of edges, explicitly integrates that graph with the scene graph by connecting corresponding nodes, and incorporates the rich structure of commonsense by graphical message passing.

2.2 Graph-Based Neural Networks

By Graph-based Neural Networks (GNN), we refer to the family of models that take a graph as input, and iteratively update the representation of each node by applying a learnable function (a.k.a., message) on the node's neighbors. Graph Convolutional Networks (GCN) [19], Gated Graph Neural Networks (GGNN) [25], and others are all specific implementations of this general model. Most SGG methods use some variant of GNNs to propagate information between region proposals [2,24,44,47]. Our message passing method, detailed in Sect. 4, resembles GGNN but instead of propagating messages through a static graph, we update (some) edges as well. Few methods exist that dynamically update edges during message passing [35,51], but we are the first to refine edges between a scene graph and an external knowledge graph.

Apart from SGG, GNNs have been used in several other computer vision tasks, often in order to propagate context information across different objects in a scene. For instance, [28] injects a GNN into a Faster R-CNN [36] framework to contextualize the features of region proposals before classifying them. This improves the results since the presence of a `table` can affect the detection of a `chair`. On the other hand, some methods utilize GNNs on graphs that represent the ontology of concepts, rather than objects in a scene [16,21,29,42].

This often enables generalization to unseen or infrequent concepts by incorporating their relationship with frequently seen concepts. More similarly to our work, Chen *et al.* [3] were the first to bring those two ideas together, and form a graph by objects in an image as well as object classes in the ontology. Nevertheless, the class nodes in that work were merely an auxiliary means to improve object features before classification. In contrast, we classify the nodes by explicitly inferring their connection to their corresponding class nodes. Moreover, we iteratively refine the bridge between scene and commonsense graphs to enhance our prediction. Furthermore, their task only involves objects and object classes, while we explore a more complex structure where predicates play an important role as well.

3   Problem Formulation

In this section, we first formalize the concepts of knowledge graph in general, and commonsense graph and scene graph in particular. Leveraging their similarities, we then reformulate the problem of scene graph generation as bridging these two graphs.

3.1   Knowledge Graphs

We define a knowledge graph as a set of entity and predicate nodes $(\mathcal{N}_{\mathrm{E}}, \mathcal{N}_{\mathrm{P}})$, each with a semantic label, and a set of directed, weighted edges $\mathcal{E}$ from a predefined set of types. Denoting by Δ a node type (here, either entity E or predicate P), the set of edges encoding the relation r between nodes of type Δ and Δ' is defined as

$$\mathcal{E}_r^{\Delta \to \Delta'} \subseteq \mathcal{N}_\Delta \times \mathcal{N}_{\Delta'} \to \mathbb{R}. \tag{1}$$

A *commonsense graph* is a type of knowledge graph in which each node represents the general concept of its semantic label, and hence each semantic label (entity or predicate class) appears in exactly one node. In such a graph, each edge encodes a relational fact involving a pair of concepts, such as Hand-partOf-Person and Cup-usedFor-Drinking. Formally, we define the set of commonsense entity (CE) nodes $\mathcal{N}_{\mathrm{CE}}$ and commonsense predicate (CP) nodes $\mathcal{N}_{\mathrm{CP}}$ as all entity and predicate classes in our task. Commonsense edges $\mathcal{E}_{\mathrm{C}}$ consist of 4 distinct subsets, depending on the source and destination node type:

$$\mathcal{E}_{\mathrm{C}} = \{\mathcal{E}_r^{\mathrm{CE} \to \mathrm{CP}}\} \cup \{\mathcal{E}_r^{\mathrm{CP} \to \mathrm{CE}}\} \cup$$
$$\{\mathcal{E}_r^{\mathrm{CE} \to \mathrm{CE}}\} \cup \{\mathcal{E}_r^{\mathrm{CP} \to \mathrm{CP}}\}. \tag{2}$$

A *scene graph* is a different type of knowledge graph where: (a) each scene entity (SE) node is associated with a bounding box, referring to an image region, (b) each scene predicate (SP) node is associated with an ordered pair of SE nodes, namely a subject and an object, and (c) there are two types of undirected edges which connect each SP to its corresponding subject and object respectively.

Here because we define knowledge edges to be directed, we model each undirected subject or object edge as two directed edges in the opposite directions, each with a distinct type. More specifically,

$$
\begin{aligned}
\mathcal{N}_{\mathrm{SE}} &\subseteq [0,1]^4 \times \mathcal{N}_{\mathrm{CE}}, \\
\mathcal{N}_{\mathrm{SP}} &\subseteq \mathcal{N}_{\mathrm{SE}} \times \mathcal{N}_{\mathrm{SE}} \times \mathcal{N}_{\mathrm{CP}}, \\
\mathcal{E}_{\mathrm{S}} &= \{\mathcal{E}^{\mathrm{SE}\to\mathrm{SP}}_{\mathtt{subjectOf}}, \mathcal{E}^{\mathrm{SE}\to\mathrm{SP}}_{\mathtt{objectOf}}, \\
&\qquad \mathcal{E}^{\mathrm{SP}\to\mathrm{SE}}_{\mathtt{hasSubject}}, \mathcal{E}^{\mathrm{SP}\to\mathrm{SE}}_{\mathtt{hasObject}}\},
\end{aligned}
\tag{3}
$$

where $[0,1]^4$ is the set of possible bounding boxes, and $\mathcal{N}_{\mathrm{SE}} \times \mathcal{N}_{\mathrm{SE}} \times \mathcal{N}_{\mathrm{CP}}$ is the set of all possible triples that consist of two scene entity nodes and a scene predicate node. Figure 1 shows an example of scene graph and commonsense graph side by side, to make their similarities clearer. Here we assume every scene graph node has a label that exists in the commonsense graph, since in reality some objects and predicates might belong to background classes, we consider a special commonsense node as background entity and another for background predicate.

3.2   Bridging Knowledge Graphs

Considering the similarity between the commonsense and scene graph formulations, we make a subtle refinement in the formulation to bridge these two graphs. Specifically, we remove the class from SE and SP nodes and instead encode it into a set of *bridge* edges $\mathcal{E}_{\mathrm{B}}$ that connect each SE or SP node to its corresponding class, *i.e.*, a CE or CP node respectively:

$$
\begin{aligned}
\mathcal{N}^{?}_{\mathrm{SE}} &\subseteq [0,1]^4, \\
\mathcal{N}^{?}_{\mathrm{SP}} &\subseteq \mathcal{N}_{\mathrm{SE}} \times \mathcal{N}_{\mathrm{SE}}, \\
\mathcal{E}_{\mathrm{B}} &= \{\mathcal{E}^{\mathrm{SE}\to\mathrm{CE}}_{\mathtt{classifiedTo}}, \mathcal{E}^{\mathrm{SP}\to\mathrm{CP}}_{\mathtt{classifiedTo}}, \\
&\qquad \mathcal{E}^{\mathrm{CE}\to\mathrm{SE}}_{\mathtt{hasInstance}}, \mathcal{E}^{\mathrm{CP}\to\mathrm{SP}}_{\mathtt{hasInstance}}\},
\end{aligned}
\tag{4}
$$

where $.^{?}$ means the nodes are implicit, *i.e.*, their classes are unknown. Each edge of type `classifiedTo`, connects an entity or predicate to its corresponding label in the commonsense graph, and has a reverse edge of type `hasInstance` which connects the commonsense node back to the instance. Based on this reformulation, we can define the problem of SGG as the extraction of implicit entity and predicate nodes from the image (hereafter called *scene graph proposal*), and then classifying them by connecting each entity or predicate to the corresponding node in the commonsense graph. Accordingly, Given an input image I and a provided and fixed commonsense graph, the goal of SGG with commonsense knowledge is to maximize

$$
\begin{aligned}
&p(\mathcal{N}_{\mathrm{SE}}, \mathcal{N}_{\mathrm{SP}}, \mathcal{E}_{\mathrm{S}} | I, \mathcal{N}_{\mathrm{CE}}, \mathcal{N}_{\mathrm{CP}}, \mathcal{E}_{\mathrm{C}}) \\
&= p(\mathcal{N}^{?}_{\mathrm{SE}}, \mathcal{N}^{?}_{\mathrm{SP}}, \mathcal{E}_{\mathrm{S}} | I) \times \\
&\quad p(\mathcal{E}_{\mathrm{B}} | I, \mathcal{N}_{\mathrm{CE}}, \mathcal{N}_{\mathrm{CP}}, \mathcal{E}_{\mathrm{C}}, \mathcal{N}^{?}_{\mathrm{SE}}, \mathcal{N}^{?}_{\mathrm{SP}}, \mathcal{E}_{\mathrm{S}}).
\end{aligned}
\tag{5}
$$

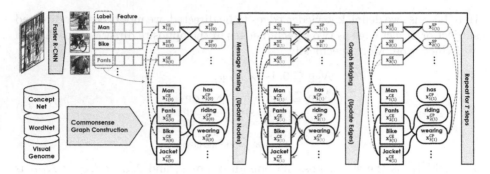

Fig. 2. An illustrative example of the GB-NET process. First, we initialize the scene graph and entity bridges using a Faster R-CNN. Then we propagate messages to update node representations, and use them to update the entity and predicate bridges. This is repeated T times and the final bridge determines the output label of each node.

In this paper, the first term is implemented as a region proposal network that infers $\mathcal{N}_{SE}^{?}$ given the image, followed by a simple predicate proposal algorithm that considers all possible entity pairs as $\mathcal{N}_{SP}^{?}$. The second term is fulfilled by the proposed GB-NET which infers bridge edges by incorporating the rich structure of the scene and commonsense graphs. Note that unlike most existing methods [2,52], we do not factorize this into predicting entity classes given the image, and then predicate classes given entities. Therefore, our formulation is more general and allows the proposed method to classify entities and predicates jointly.

4   Method

The proposed method is illustrated in Fig. 2. Given an image, our model first applies a Faster R-CNN [36] to detect objects, and represents them as scene entity (SE) nodes. It also creates a scene predicate (SP) node for each pair of entities, which forms a scene graph proposal, yet to be classified. Given this graph and a background commonsense graph, each with fixed internal connectivity, our goal is to create *bridge* edges between the two graphs that connect each instance (SE and SP node) to its corresponding class (CE and CP node). To this end, our model initializes entity bridges by connecting each SE to the CE that matches the label predicted by Faster R-CNN, and propagates messages among all nodes, through every edge type with dedicated message passing parameters. Given the updated node representations, it computes a pairwise similarity between every SP node and every CP node, and finds maximal similarity pairs to connect scene predicates to their corresponding classes, via predicate bridges. It also does the same for entity nodes to potentially refine their bridges too. Given the new bridges, it propagates messages again, and repeats this process for a predefined number of steps. The final state of the bridge determines which class each node belongs to, resulting in the output scene graph.

4.1 Graph Initialization

The object detector outputs a set of n detected objects, each with a bounding box b_j, a label distribution p_j and an RoI-aligned [36] feature vector $\mathbf{v}_j$. Then we allocate a *scene entity node* (SE) for each object, and a *scene predicate node* (SP) for each pair of objects, representing the potential predicate with the two entities as its subject and object. Each entity is initialized using its RoI features $\mathbf{v}_j$, and each predicate is initialized using the RoI features $\mathbf{u}_j$ of a bounding box enclosing the union of its subject and object. Formally, we can write, *i.e.*,

$$\mathbf{x}_j^{\mathrm{SE}} = \phi_{\mathrm{init}}^{\mathrm{SE}}(\mathbf{v}_j) , \quad \text{and} \quad \mathbf{x}_j^{\mathrm{SP}} = \phi_{\mathrm{init}}^{\mathrm{SP}}(\mathbf{u}_j), \tag{6}$$

where $\phi_{\mathrm{init}}^{\mathrm{SE}}$ and $\phi_{\mathrm{init}}^{\mathrm{SP}}$ are two fully connected networks that are branched from the backbone after ROI-align. To form a scene graph proposal, we connect each predicate node to its subject and object via labeled edges. Specifically, we define the following 4 edge types: for a triplet $s - p - o$, we connect p to s using a hasSubject edge, p to o using a hasObject edge, s to p using a subjectOf edge, and o to p using an objectOf edge. The reason we have two directions as separate types is that in the message passing phase, the way we use predicate information to update entities should be different from the way we use entities to update predicates.

On the other hand, we initialize the commonsense graph with *commonsense entity nodes* (CE) and *commonsense predicate nodes* (CP) using a linear projection of their word embeddings:

$$\mathbf{x}_i^{\mathrm{CE}} = \phi_{\mathrm{init}}^{\mathrm{CE}}(\mathbf{e}_i^n) , \quad \text{and} \quad \mathbf{x}_i^{\mathrm{CP}} = \phi_{\mathrm{init}}^{\mathrm{CP}}(\mathbf{e}_i^p). \tag{7}$$

The commonsense graph also has various types of edges, such as UsedFor and PartOf, as detailed in Sect. 5.2. Our method is independent of the types of commonsense edges, and can utilize any provided graph from any source.

So far, we have two isolated graphs, scene and commonsense. An SE node representing a detected Person intuitively refers to the Person concept in the ontology, and hence the Person node in the commonsense graph. Therefore, we connect each SE node to the CE node that corresponds the semantic label predicted by Faster R-CNN, via a classifiedTo edge type. Instead of a hard classification, we connect each entity to top K_{bridge} classes using p_j (class distribution predicted by Faster R-CNN) as weights. We also create a reverse connection from each CE node to corresponding SE nodes, using an hasInstance edge, but with the same weights p_j. As mentioned earlier, this is to make sure information flows from commonsense to scene as well as scene to commonsense, but not in the same way. We similarly define two other edge types, classifiedTo and hasInstance for predicates, which are initially an empty set, and will be updated to bridge SP nodes to CP nodes as we explain in the following. These 4 edge types can be seen as flexible *bridges* that connect the two fixed graphs, which are considered latent variables to be determined by the model.

This forms a heterogeneous graph with four types of nodes (SE, SP, CE, and CP) and various types of edges: scene graph edges $\mathcal{E}_{\mathrm{S}}$ such as subjectOf, commonsense edges $\mathcal{E}_{\mathrm{C}}$ such as usedFor, and bridge edges $\mathcal{E}_{\mathrm{B}}$ such as classifiedTo.

Next, we explain how our proposed method updates node representations and bridge edges, while keeps commonsense and scene edges constant.

4.2 Successive Message Passing and Bridging

Given a heterogeneous graph as described above, we employ a variant of GGNN [25] to propagate information among nodes. First, each node representation is fed into a fully connected network to compute *outgoing* messages, that is

$$\mathbf{m}_i^{\Delta\rightarrow} = \phi_{\text{send}}^{\Delta}(\mathbf{x}_i^{\Delta}), \tag{8}$$

for each i and node type Δ, where ϕ_{send} is a trainable *send head* which has shared weights across nodes of each type. After computing outgoing messages, we send them through all outgoing edges, multiplying by the edge weight. Then for each node, we aggregate incoming messages, by first adding across edges of the same type, and then concatenating across edge types. We compute the *incoming* message for each node by applying another fully connected network on the aggregated messages:

$$\mathbf{m}_j^{\Delta\leftarrow} = \phi_{\text{receive}}^{\Delta}\left(\bigcup_{\Delta'} \overset{\mathcal{E}_k \in \mathcal{E}^{\Delta'\rightarrow\Delta}}{\bigcup} \sum_{(i,j,a_{ij}^k)\in\mathcal{E}_k} a_{ij}^k \mathbf{m}_i^{\Delta'\rightarrow}\right), \tag{9}$$

where ϕ_{receive} is a trainable *receive head* and $\cup$ denotes concatenation. Note that the first concatenation is over all 4 node types, the second concatenation is over all edge types from Δ' to Δ, and the sum is over all edges of that type, where i and j are the head and tail nodes, and a_{ij}^k is the edge weight. Given the incoming message for each node, we update the representation of the node using a Gated Recurrent Unit (GRU) update rule, following [4]:

$$\begin{aligned}
\mathbf{z}_j^{\Delta} &= \sigma\big(W_z^{\Delta}\mathbf{m}_j^{\Delta\leftarrow} + U_z^{\Delta}\mathbf{x}_j^{\Delta}\big),\\
\mathbf{r}_j^{\Delta} &= \sigma\big(W_r^{\Delta}\mathbf{m}_j^{\Delta\leftarrow} + U_r^{\Delta}\mathbf{x}_j^{\Delta}\big),\\
\mathbf{h}_j^{\Delta} &= \tanh\big(W_h^{\Delta}\mathbf{m}_j^{\Delta\leftarrow} + U_h^{\Delta}(\mathbf{r}_j^{\Delta}\odot\mathbf{x}_j^{\Delta})\big),\\
\mathbf{x}_j^{\Delta} &\Leftarrow (1-\mathbf{z}_j^{\Delta})\odot\mathbf{x}_j^{\Delta} + \mathbf{z}_j^{\Delta}\odot\mathbf{h}_j^{\Delta},
\end{aligned} \tag{10}$$

where σ is the sigmoid function, and W^{Δ} and U^{Δ} are trainable matrices that are shared across nodes of the same type, but distinct for each node type Δ. This update rule can be seen as an extension of GGNN [25] to heterogeneous graphs, with a more complex message aggregation strategy. Note that $\Leftarrow$ means we update the node representation. Mathematically, this means $\mathbf{x}_{j(t+1)}^{\Delta} = U(\mathbf{x}_{j(t)}^{\Delta})$, where U is the aforementioned update rule and (t) denotes iteration number. For simplicity, we drop this subscript throughout this paper.

So far, we have explained how to update node representations using graph edges. Now using the new node representations, we should update the bridge edges $\mathcal{E}_B$ that connect scene nodes to commonsense nodes. To this end, we compute a pairwise similarity from each SE to all CE nodes, and from each SP to all CP nodes.

$$\mathbf{a}_{ij}^{\mathrm{EB}} = \frac{\exp\langle \mathbf{x}_i^{\mathrm{SE}}, \mathbf{x}_j^{\mathrm{CE}} \rangle_{\mathrm{EB}}}{\sum_{j'} \exp\langle \mathbf{x}_i^{\mathrm{SE}}, \mathbf{x}_{j'}^{\mathrm{CE}} \rangle_{\mathrm{EB}}} \ , \quad \text{where} \quad \langle \mathbf{x}, \mathbf{y} \rangle_{\mathrm{EB}} = \phi_{\mathrm{att}}^{\mathrm{SE}}(\mathbf{x})^T \phi_{\mathrm{att}}^{\mathrm{CE}}(\mathbf{y}), \quad (11)$$

and similarly for predicates,

$$\mathbf{a}_{ij}^{\mathrm{PB}} = \frac{\exp\langle \mathbf{x}_i^{\mathrm{SP}}, \mathbf{x}_j^{\mathrm{CP}} \rangle_{\mathrm{PB}}}{\sum_{j'} \exp\langle \mathbf{x}_i^{\mathrm{SP}}, \mathbf{x}_{j'}^{\mathrm{CP}} \rangle_{\mathrm{PB}}} \ , \quad \text{where} \quad \langle \mathbf{x}, \mathbf{y} \rangle_{\mathrm{PB}} = \phi_{\mathrm{att}}^{\mathrm{SP}}(\mathbf{x})^T \phi_{\mathrm{att}}^{\mathrm{CP}}(\mathbf{y}). \quad (12)$$

Here $\phi_{\mathrm{att}}^{\Delta}$ is a fully connected network that resembles *attention head* in transformers. Note that since $\phi_{\mathrm{att}}^{\Delta}$ is not shared across node types, our similarity metric is asymmetric. We use each $\mathbf{a}_{ij}^{\mathrm{EB}}$ to set the edge weight of the `classifiedTo` edge from $\mathbf{x}_i^{\mathrm{SE}}$ to $\mathbf{x}_j^{\mathrm{CE}}$, as well as the `hasInstance` edge from $\mathbf{x}_j^{\mathrm{CE}}$ to $\mathbf{x}_i^{\mathrm{SE}}$. Similarly we use each $\mathbf{a}_{ij}^{\mathrm{PB}}$ to set the weight of edges between $\mathbf{x}_i^{\mathrm{SP}}$ and $\mathbf{x}_j^{\mathrm{CP}}$. In preliminary experiments we realised that such fully connected bridges hurt performance in large graphs. Hence, we only keep the top K_{bridge} values of $\mathbf{a}_{ij}^{\mathrm{EB}}$ for each i, and set the rest to zero. We do the same thing for predicates, keeping the top K_{bridge} values of $\mathbf{a}_{ij}^{\mathrm{PB}}$ for each i. Given the updated bridges, we propagate messages again to update node representations, and iterate for a fixed number of steps, T. The final values of $\mathbf{a}_{ij}^{\mathrm{EB}}$ and $\mathbf{a}_{ij}^{\mathrm{PB}}$ are the outputs of our model, which can be used to classify each entity and predicate in the scene graph.

4.3 Training

We closely follow [2] which itself follows [52] for training procedure. Specifically, given the output and ground truth graphs, we align output entities and predicates to ground truth counterparts. To align entities we use IoU and predicates will be aligned naturally since they correspond to aligned pairs of entities. Then we use the output probability scores of each node to define a cross-entropy loss. The sum of all node-level loss values will be the objective function to be minimized using Adam [18].

Due to the highly imbalanced predicate statistics in Visual Genome, we observed that best-performing models usually concentrate their performance merely on the most frequent classes such as `on` and `wearing`. To alleviate this, we modify the basic cross-entropy objective that is commonly used by assigning an importance weight to each class. We follow the recently proposed class-balanced loss [5] where the weight of each class is inversely proportional to its frequency. More specifically, we use the following loss function for each predicate node:

$$\mathcal{L}_i^P = -\frac{1-\beta}{1-\beta^{n_j}} \log \mathbf{a}_{ij}^{\mathrm{PB}}, \quad (13)$$

where j is the class index of the ground truth predicate aligned with i, n_j is the frequency of class j in training data, and β is a hyperparameter. Note that

$\beta = 0$ leads to a regular cross-entropy loss, and the more it approaches 1, the more strictly it suppresses frequent classes. To be fair in comparison with other methods, we include a variant of our method without reweighting, which still outperforms all other methods.

5   Experiments

Following the literature, we use the large-scale Visual Genome benchmark [20] to evaluate our method. We first show our GB-NET outperforms the state of the art, by extensively evaluating it on 24 performance metrics. Then we present an ablation study to illustrate how each innovation contributes to the performance. In the Supplementary Material, we also provide a per-class performance breakdown to show the consistency and robustness of our performance across frequent and rare classes. That is accompanied by a computational speed analysis, and several qualitative examples of our generated graphs compared to the state of the art, side by side.

5.1   Task Description

Visual Genome [20] consists of 108,077 images with annotated objects (entities) and pairwise relationships (predicates), which is then post-processed by [44] to create scene graphs. They use the most frequent 150 entity classes and 50 predicate classes to filter the annotations. Figure 1 shows an example of their post-processed scene graphs which we use as ground truth. We closely follow their evaluation settings such as train and test splits.

The task of scene graph generation, as described in Sect. 4, is equivalent to the SGGEN scenario proposed by [44] and followed ever since. Given an image, the task of SGGEN is to jointly infer entities and predicates from scratch. Since this task is limited by the quality of the object proposals, [44] also introduced two other tasks that more clearly evaluate entity and predicate recognition. In SGCLS, we take localization (here region proposal network) out of the picture, by providing the model with ground truth bounding boxes during test, simulating a *perfect* proposal model. In PREDCLS, we take object detection for granted, and provide the model with not only ground truth bounding boxes, but also their true entity class. In each task, the main evaluation metric is average per-image recall of the top K subject-predicate-object triplets. The confidence of a triplet that is used for ranking is computed by multiplying the classification confidence of all three elements. Given the ground truth scene graph, each predicate forms a triplet, which we match against the top K triplets in the output scene graph. A triplet is matched if all three elements are classified correctly, and the bounding boxes of subject and object match with an IoU of at least 0.5. Besides the choice of K, there are two other choices to be made: (1) Whether or not to enforce the so-called *Graph Constraint* (GC), which limits the top K triplets to only one predicate for each ordered entity pair, and (2) Whether to compute the recall for each predicate class separately and take the mean (mR), or compute a single recall for all triplets (R) [2]. We comprehensively report both mean and overall recall, both with and without GC, and conventionally use both 50 and 100 for K, resulting in 8 metrics for each task, 24 in total.

5.2 Implementation Details

We use three-layer fully connected networks with ReLU activation for all trainable networks ϕ_{init}, ϕ_{send}, ϕ_{receive} and ϕ_{att}. We set the dimension of node representations to 1024, and perform 3 message passing steps, except in ablation experiments where we try 1, 2 and 3. We tried various values for β. Generally the higher it is, mean recall improves and recall falls. We found 0.999 is a good trade-off, and chose $K_{\text{bridge}} = 5$ empirically. All hyperparameters are tuned using a validation set randomly selected from training data. We borrow the Faster R-CNN trained by [52] and shared among all our baselines, which has a VGG-16 backbone and predicts 128 proposals.

In our commonsense graph, the nodes are the 151 entity classes and 51 predicate classes that are fixed by [44], including background. We use the GloVE [33] embedding of category titles to initialize their node representation (via ϕ_{init}), and fix GloVE during training. We compile our commonsense edges from three sources, WordNet [30], ConceptNet [27], and Visual Genome. To summarize, there are three groups of edge types in our commonsense graph. We have SimilarTo from WordNet hierarchy, we have PartOf, RelatedTo, IsA, MannerOf, and UsedFor from ConceptNet, and finally from VG training data we have conditional probabilities of subject given predicate, predicate given subject, subject given object, *etc.* We explain the details in the supplementary material. The process of compiling and pruning the knowledge graph is semi-automatic and takes less than a day from a single person. We make it publicly available as a part of our code. We have also tried using each individual source (e.g. only ConceptNet) independently, which requires less effort, and does not significantly impact the performance. There are also recent approaches to automate the process of commonsense knowledge graph construction [1,13], which can be utilized to further reduce the manual labor.

5.3 Main Results

Table 1 summarizes our results in comparison to the state of the art. IMP+ refers to the re-implementation of [44] by [52] using their new Faster R-CNN backbone. That method does not use any external knowledge and only uses message passing among the entities and predicates and then classifies each. Hence, it can be seen as a strong, but knowledge-free baseline. FREQ is a simple baseline proposed by [52], which predicts the most frequent predicate for any given pair of entity classes, solely based on statistics from the training data. FREQ surprisingly outperforms IMP+, confirming the efficacy of commonsense in SGG.

SMN [52] applies bi-directional LSTMs on top of the entity features, then classifies each entity and each pair. They bias their classifier logits using statistics from FREQ, which improves their total recall significantly, at the expense

Table 1. Evaluation in terms of mean and overall triplet recall, at top 50 and top 100, with and without Graph Constraint (GC), for the three tasks of SGGEN, SGCLS and PREDCLS. Numbers are in percentage. All baseline numbers were borrowed from [2]. Top two methods for each metric is shown in **bold** and *italic* respectively.

Task	Metric	GC	Method					
			IMP+	FREQ	SMN	KERN	GB-NET	GB-NET-β
SGGEN	mR@50	Y	3.8	4.3	5.3	*6.4*	6.1	**7.1**
		N	5.4	5.9	9.3	*11.7*	9.8	**11.7**
	mR@100	Y	4.8	5.6	6.1	7.3	*7.3*	**8.5**
		N	8.0	8.9	12.9	*16.0*	14.0	**16.6**
	R@50	Y	20.7	23.5	**27.2**	*27.1*	26.4	26.3
		N	22.0	25.3	*30.5*	**30.9**	29.4	29.3
	R@100	Y	24.5	27.6	**30.3**	29.8	*30.0*	29.9
		N	27.4	30.9	*35.8*	**35.8**	35.1	35.0
SGCLS	mR@50	Y	5.8	6.8	7.1	9.4	*9.6*	**12.7**
		N	12.1	13.5	15.4	19.8	*21.4*	**25.6**
	mR@100	Y	6.0	7.8	7.6	10.0	*10.2*	**13.4**
		N	16.9	19.6	20.6	26.2	*29.1*	**32.1**
	R@50	Y	34.6	32.4	35.8	36.7	**38.0**	*37.3*
		N	43.4	40.5	44.5	45.9	**47.7**	*46.9*
	R@100	Y	35.4	34.0	36.5	37.4	**38.8**	*38.0*
		N	47.2	43.7	47.7	49.0	**51.1**	*50.3*
PREDCLS	mR@50	Y	9.8	13.3	13.3	17.7	*19.3*	**22.1**
		N	20.3	24.8	27.5	36.3	*41.1*	**44.5**
	mR@100	Y	10.5	15.8	14.4	19.2	*20.9*	**24.0**
		N	28.9	37.3	37.9	49.0	*55.4*	**58.7**
	R@50	Y	59.3	59.9	65.2	65.8	**66.6**	*66.6*
		N	75.2	71.3	81.1	81.9	**83.6**	*83.5*
	R@100	Y	61.3	64.1	67.1	67.6	**68.2**	*68.2*
		N	83.6	81.2	88.3	88.9	**90.5**	*90.3*

of higher bias against less frequent classes, as revealed by [2]. More recently, KERN [2] encodes VG statistics into the edge weights of the graph, which is then incorporated by propagating messages. Since it encodes statistics more implicitly, KERN is less biased compared to SMN, which improves mR. Our method improves both R and mR significantly, and our class-balanced model, GB-NET-β, further enhances mR (+2.7% in average) without hurting R by much (−0.2%).

We observed that the state of the art performance has been saturated in the SGGEN setting, especially for overall recall. This is partly because object detection performance is a bottleneck that limits the performance. It is worth noting

Table 2. Ablation study on Visual Genome. All numbers are in percentage, and graph constraint is enforced.

Method	SGGen				PredCls			
	mR@50	mR@100	R@50	R@100	mR@50	mR@100	R@50	R@100
No Knowledge	5.5	6.6	25.3	28.8	15.4	16.8	62.5	64.5
$T = 1$	5.6	6.7	24.9	28.5	15.6	17.1	62.1	64.2
$T = 2$	5.7	6.9	26.1	29.7	18.2	19.7	66.7	68.4
GB-Net	**6.1**	**7.3**	**26.4**	**30.0**	**18.2**	**19.7**	**67.0**	**68.6**

that mean recall is a more important metric than overall recall, since most SGG methods tend to score a high overall recall by investing on few most frequent classes, and ignoring the rest [2]. As shown in Table 1, our method achieves significant improvements in mean recall. We provide in-depth performance analysis by comparing our recall per predicate class with that of the state of the art, as well as qualitative analysis in the Supplementary Material.

There are other recent SGG methods that are not used for comparison here, because their evaluation settings are not identical to ours, and their code is not publicly available to the best of our knowledge [10,34]. For instance, [34] reports only 8 out of our 24 evaluation metrics, and although our method is superior in 6 metrics out of those 8, that is not sufficient to fairly compare the two methods.

5.4 Ablation Study

To further explain our performance improvement, Table 2 compares our full method with its weaker variants. Specifically, to investigate the effectiveness of commonsense knowledge, we remove the commonsense graph and instead classify each node in our graph using a 2-layer fully connected classifier after message passing. This negatively impacts performance in all metrics, proving our method is able to exploit commonsense knowledge through the proposed bridging technique. Moreover, to highlight the importance of our proposed message passing and bridge refinement process, we repeated the experiments with fewer steps. We observe the performance drops significantly with fewer steps, proving the effectiveness of our model, but it saturates as we go beyond 3 steps.

6 Conclusion

We proposed a new method for Scene Graph Generation that incorporates external commonsense knowledge in a novel, graphical neural framework. We unified the formulation of scene graph and commonsense graph as two types of knowledge graph, which are fused into a single graph through a dynamic message passing and bridging algorithm. Our method iteratively propagates messages to update nodes, then compares nodes to update bridge edges, and repeats until the two graphs are carefully connected. Through extensive experiments, we showed our method outperforms the state of the art in various metrics.

Acknowledgement. This work was supported in part by Contract N6600119C4032 (NIWC and DARPA). The views expressed are those of the authors and do not reflect the official policy of the Department of Defense or the U.S. Government.

References

1. Bosselut, A., Rashkin, H., Sap, M., Malaviya, C., Celikyilmaz, A., Choi, Y.: COMET: commonsense transformers for automatic knowledge graph construction. arXiv preprint arXiv:1906.05317 (2019)
2. Chen, T., Yu, W., Chen, R., Lin, L.: Knowledge-embedded routing network for scene graph generation. In: Proceedings of the IEEE Conference on Computer Vision and Pattern Recognition, pp. 6163–6171 (2019)
3. Chen, X., Li, L.J., Fei-Fei, L., Gupta, A.: Iterative visual reasoning beyond convolutions. In: Proceedings of the IEEE Conference on Computer Vision and Pattern Recognition, pp. 7239–7248 (2018)
4. Cho, K., et al.: Learning phrase representations using RNN encoder-decoder for statistical machine translation. arXiv preprint arXiv:1406.1078 (2014)
5. Cui, Y., Jia, M., Lin, T.Y., Song, Y., Belongie, S.: Class-balanced loss based on effective number of samples. In: Proceedings of the IEEE Conference on Computer Vision and Pattern Recognition, pp. 9268–9277 (2019)
6. Dietz, L., Kotov, A., Meij, E.: Utilizing knowledge graphs for text-centric information retrieval. In: The 41st International ACM SIGIR Conference on Research & Development in Information Retrieval, pp. 1387–1390. ACM (2018)
7. Fader, A., Zettlemoyer, L., Etzioni, O.: Open question answering over curated and extracted knowledge bases. In: Proceedings of the 20th ACM SIGKDD International Conference on Knowledge Discovery and Data Mining, pp. 1156–1165. ACM (2014)
8. Flanigan, J., Thomson, S., Carbonell, J., Dyer, C., Smith, N.A.: A discriminative graph-based parser for the abstract meaning representation. In: Proceedings of the 52nd Annual Meeting of the Association for Computational Linguistics (Volume 1: Long Papers), pp. 1426–1436 (2014)
9. Gardner, M., Dasigi, P., Iyer, S., Suhr, A., Zettlemoyer, L.: Neural semantic parsing. In: Proceedings of the 56th Annual Meeting of the Association for Computational Linguistics: Tutorial Abstracts, pp. 17–18 (2018)
10. Gu, J., Zhao, H., Lin, Z., Li, S., Cai, J., Ling, M.: Scene graph generation with external knowledge and image reconstruction. In: Proceedings of the IEEE Conference on Computer Vision and Pattern Recognition, pp. 1969–1978 (2019)
11. Herzig, R., Raboh, M., Chechik, G., Berant, J., Globerson, A.: Mapping images to scene graphs with permutation-invariant structured prediction. In: Advances in Neural Information Processing Systems, pp. 7211–7221 (2018)
12. Hudson, D.A., Manning, C.D.: Learning by abstraction: the neural state machine. arXiv preprint arXiv:1907.03950 (2019)
13. Ilievski, F., Szekely, P., Cheng, J., Zhang, F., Qasemi, E.: Consolidating commonsense knowledge. arXiv preprint arXiv:2006.06114 (2020)
14. Johnson, J., Gupta, A., Fei-Fei, L.: Image generation from scene graphs. In: Proceedings of the IEEE Conference on Computer Vision and Pattern Recognition, pp. 1219–1228 (2018)
15. Johnson, J., et al.: Image retrieval using scene graphs. In: Proceedings of the IEEE Conference on Computer Vision and Pattern Recognition, pp. 3668–3678 (2015)

16. Kato, K., Li, Y., Gupta, A.: Compositional learning for human object interaction. In: Ferrari, V., Hebert, M., Sminchisescu, C., Weiss, Y. (eds.) Computer Vision – ECCV 2018. LNCS, vol. 11218, pp. 247–264. Springer, Cham (2018). https://doi.org/10.1007/978-3-030-01264-9_15

17. Khashabi, D., Khot, T., Sabharwal, A., Roth, D.: Question answering as global reasoning over semantic abstractions. In: Thirty-Second AAAI Conference on Artificial Intelligence (2018)

18. Kingma, D.P., Ba, J.: Adam: a method for stochastic optimization. arXiv preprint arXiv:1412.6980 (2014)

19. Kipf, T.N., Welling, M.: Semi-supervised classification with graph convolutional networks. arXiv preprint arXiv:1609.02907 (2016)

20. Krishna, R., et al.: Visual genome: connecting language and vision using crowd-sourced dense image annotations. Int. J. Comput. Vis. **123**(1), 32–73 (2017)

21. Lee, C.W., Fang, W., Yeh, C.K., Frank Wang, Y.C.: Multi-label zero-shot learning with structured knowledge graphs. In: Proceedings of the IEEE Conference on Computer Vision and Pattern Recognition, pp. 1576–1585 (2018)

22. Li, M., et al.: Multilingual entity, relation, event and human value extraction. In: Proceedings of the 2019 Conference of the North American Chapter of the Association for Computational Linguistics (Demonstrations), pp. 110–115 (2019)

23. Li, Y., Ouyang, W., Zhou, B., Shi, J., Zhang, C., Wang, X.: Factorizable net: an efficient subgraph-based framework for scene graph generation. In: Ferrari, V., Hebert, M., Sminchisescu, C., Weiss, Y. (eds.) ECCV 2018. LNCS, vol. 11205, pp. 346–363. Springer, Cham (2018). https://doi.org/10.1007/978-3-030-01246-5_21

24. Li, Y., Ouyang, W., Zhou, B., Wang, K., Wang, X.: Scene graph generation from objects, phrases and region captions. In: Proceedings of the IEEE International Conference on Computer Vision, pp. 1261–1270 (2017)

25. Li, Y., Tarlow, D., Brockschmidt, M., Zemel, R.: Gated graph sequence neural networks. arXiv preprint arXiv:1511.05493 (2015)

26. Liang, K., Guo, Y., Chang, H., Chen, X.: Visual relationship detection with deep structural ranking. In: Thirty-Second AAAI Conference on Artificial Intelligence (2018)

27. Liu, H., Singh, P.: ConceptNet—a practical commonsense reasoning tool-kit. BT Technol. J. **22**(4), 211–226 (2004)

28. Liu, Y., Wang, R., Shan, S., Chen, X.: Structure inference net: object detection using scene-level context and instance-level relationships. In: Proceedings of the IEEE Conference on Computer Vision and Pattern Recognition, pp. 6985–6994 (2018)

29. Marino, K., Salakhutdinov, R., Gupta, A.: The more you know: using knowledge graphs for image classification. arXiv preprint arXiv:1612.04844 (2016)

30. Miller, G.A.: WordNet: a lexical database for English. Commun. ACM **38**(11), 39–41 (1995)

31. Newell, A., Deng, J.: Pixels to graphs by associative embedding. In: Advances in Neural Information Processing Systems, pp. 2171–2180 (2017)

32. Pei, M., Jia, Y., Zhu, S.C.: Parsing video events with goal inference and intent prediction. In: 2011 International Conference on Computer Vision, pp. 487–494. IEEE (2011)

33. Pennington, J., Socher, R., Manning, C.: Glove: global vectors for word representation. In: Proceedings of the 2014 Conference on Empirical Methods in Natural Language Processing (EMNLP), pp. 1532–1543 (2014)

34. Qi, M., Li, W., Yang, Z., Wang, Y., Luo, J.: Attentive relational networks for mapping images to scene graphs. In: Proceedings of the IEEE Conference on Computer Vision and Pattern Recognition, pp. 3957–3966 (2019)
35. Qi, S., Wang, W., Jia, B., Shen, J., Zhu, S.-C.: Learning human-object interactions by graph parsing neural networks. In: Ferrari, V., Hebert, M., Sminchisescu, C., Weiss, Y. (eds.) ECCV 2018. LNCS, vol. 11213, pp. 407–423. Springer, Cham (2018). https://doi.org/10.1007/978-3-030-01240-3_25
36. Ren, S., He, K., Girshick, R., Sun, J.: Faster R-CNN: towards real-time object detection with region proposal networks. In: Advances in Neural Information Processing Systems, pp. 91–99 (2015)
37. Schuster, S., Krishna, R., Chang, A., Fei-Fei, L., Manning, C.D.: Generating semantically precise scene graphs from textual descriptions for improved image retrieval. In: Proceedings of the Fourth Workshop on Vision and Language, pp. 70–80 (2015)
38. Shi, J., Zhang, H., Li, J.: Explainable and explicit visual reasoning over scene graphs. In: Proceedings of the IEEE Conference on Computer Vision and Pattern Recognition, pp. 8376–8384 (2019)
39. Teney, D., Liu, L., van den Hengel, A.: Graph-structured representations for visual question answering. In: Proceedings of the IEEE Conference on Computer Vision and Pattern Recognition, pp. 1–9 (2017)
40. Tu, K., Meng, M., Lee, M.W., Choe, T.E., Zhu, S.C.: Joint video and text parsing for understanding events and answering queries. IEEE MultiMedia 21(2), 42–70 (2014)
41. Wadden, D., Wennberg, U., Luan, Y., Hajishirzi, H.: Entity, relation, and event extraction with contextualized span representations. arXiv preprint arXiv:1909.03546 (2019)
42. Wang, X., Ye, Y., Gupta, A.: Zero-shot recognition via semantic embeddings and knowledge graphs. In: Proceedings of the IEEE Conference on Computer Vision and Pattern Recognition, pp. 6857–6866 (2018)
43. Woo, S., Kim, D., Cho, D., Kweon, I.S.: Linknet: relational embedding for scene graph. In: Advances in Neural Information Processing Systems, pp. 560–570 (2018)
44. Xu, D., Zhu, Y., Choy, C.B., Fei-Fei, L.: Scene graph generation by iterative message passing. In: Proceedings of the IEEE Conference on Computer Vision and Pattern Recognition, pp. 5410–5419 (2017)
45. Xu, H., Jiang, C., Liang, X., Li, Z.: Spatial-aware graph relation network for large-scale object detection. In: Proceedings of the IEEE Conference on Computer Vision and Pattern Recognition, pp. 9298–9307 (2019)
46. Xu, H., Jiang, C., Liang, X., Lin, L., Li, Z.: Reasoning-RCNN: unifying adaptive global reasoning into large-scale object detection. In: Proceedings of the IEEE Conference on Computer Vision and Pattern Recognition, pp. 6419–6428 (2019)
47. Yang, J., Lu, J., Lee, S., Batra, D., Parikh, D.: Graph R-CNN for scene graph generation. In: Ferrari, V., Hebert, M., Sminchisescu, C., Weiss, Y. (eds.) ECCV 2018. LNCS, vol. 11205, pp. 690–706. Springer, Cham (2018). https://doi.org/10.1007/978-3-030-01246-5_41
48. Yang, X., Tang, K., Zhang, H., Cai, J.: Auto-encoding scene graphs for image captioning. In: Proceedings of the IEEE Conference on Computer Vision and Pattern Recognition, pp. 10685–10694 (2019)
49. Yao, T., Pan, Y., Li, Y., Mei, T.: Exploring visual relationship for image captioning. In: Ferrari, V., Hebert, M., Sminchisescu, C., Weiss, Y. (eds.) Computer Vision – ECCV 2018. LNCS, vol. 11218, pp. 711–727. Springer, Cham (2018). https://doi.org/10.1007/978-3-030-01264-9_42

50. Yu, J., Lu, Y., Qin, Z., Zhang, W., Liu, Y., Tan, J., Guo, L.: Modeling text with graph convolutional network for cross-modal information retrieval. In: Hong, R., Cheng, W.-H., Yamasaki, T., Wang, M., Ngo, C.-W. (eds.) PCM 2018. LNCS, vol. 11164, pp. 223–234. Springer, Cham (2018). https://doi.org/10.1007/978-3-030-00776-8_21
51. Zareian, A., Karaman, S., Chang, S.F.: Weakly supervised visual semantic parsing. In: Proceedings of the IEEE/CVF Conference on Computer Vision and Pattern Recognition, pp. 3736–3745 (2020)
52. Zellers, R., Yatskar, M., Thomson, S., Choi, Y.: Neural motifs: scene graph parsing with global context. In: Proceedings of the IEEE Conference on Computer Vision and Pattern Recognition, pp. 5831–5840 (2018)
53. Zhan, Y., Yu, J., Yu, T., Tao, D.: On exploring undetermined relationships for visual relationship detection. In: Proceedings of the IEEE Conference on Computer Vision and Pattern Recognition, pp. 5128–5137 (2019)
54. Zhang, C., Chao, W.L., Xuan, D.: An empirical study on leveraging scene graphs for visual question answering. arXiv preprint arXiv:1907.12133 (2019)
55. Zhao, Y., Zhu, S.C.: Image parsing with stochastic scene grammar. In: Advances in Neural Information Processing Systems, pp. 73–81 (2011)

Implicit Latent Variable Model for Scene-Consistent Motion Forecasting

Sergio Casas[1,2], Cole Gulino[1], Simon Suo[1,2(✉)], Katie Luo[1], Renjie Liao[1,2], and Raquel Urtasun[1,2]

[1] Uber ATG, Toronto, Canada
{sergio.casas,cgulino,suo,katie.luo,katie.luo,rjliao,urtasun}@uber.com
[2] University of Toronto, Toronto, Canada

Abstract. In order to plan a safe maneuver an autonomous vehicle must accurately perceive its environment, and understand the interactions among traffic participants. In this paper, we aim to learn scene-consistent motion forecasts of complex urban traffic directly from sensor data. In particular, we propose to characterize the joint distribution over future trajectories via an implicit latent variable model. We model the scene as an interaction graph and employ powerful graph neural networks to learn a distributed latent representation of the scene. Coupled with a deterministic decoder, we obtain trajectory samples that are consistent across traffic participants, achieving state-of-the-art results in motion forecasting and interaction understanding. Last but not least, we demonstrate that our motion forecasts result in safer and more comfortable motion planning.

1 Introduction

Self driving vehicles (SDV) have the potential to make a broad impact in our society, providing a safer and more efficient solution to transportation. A critical component for autonomous driving is the ability to perceive the world and forecast all possible future instantiations of the scene. 3D perception algorithms have improved incredibly fast in recent years [29, 33, 42, 52, 53, 56], yielding very accurate object detections surrounding the SDV. However, producing multi-modal motion forecasts that precisely capture multiple plausible futures consistently for all actors in the scene remains a very open problem.

The complexity is immense: the future is inherently uncertain as actor behaviors are influenced not only by their own individual goals and intentions but also by the other actors' actions. For instance, an actor at an intersection may choose to turn right or go straight due to its own destination, and yield or go if the

---

S. Casas, C. Gulino and S. Suo—Denotes equal contribution.

Electronic supplementary material The online version of this chapter (https://doi.org/10.1007/978-3-030-58592-1_37) contains supplementary material, which is available to authorized users.

A. Vedaldi et al. (Eds.): ECCV 2020, LNCS 12368, pp. 624–641, 2020.
https://doi.org/10.1007/978-3-030-58592-1_37

behavior of a nearby traffic participant is aggressive or conservative. Moreover, unobserved traffic rules such as the future traffic light states heavily affect the traffic (see Fig. 1). It is clear that all these aspects cannot be directly observed and require complex reasoning about the scene as a whole, including its geometry, topology and the interaction between multiple agents.

(a) Sample 1: protected left turn   (b) Sample 2: horizontal traffic flow

Fig. 1. Two scene-consistent future trajectory samples from our model. Ground truth trajectories are shown as white polylines.

In an autonomy system, detections and motion forecasts for other actors in the scene are typically passed as obstacles to a motion-planner [43,48] in order to plan a safe maneuver. Importantly, the distribution over future trajectories needs to cover the ground-truth for the plan to be safe, but also must exhibit low enough entropy such that a comfortable ride with reasonable progress is achieved. Thus in complex urban environments the SDV should reason about multiple futures separately [15,20,27], and plan proactively by understanding how its own actions might influence other actors' behaviors [16,39]. Furthermore, as self-driving vehicles get closer to full autonomy, closed-loop simulation is becoming increasingly critical not only for testing but also for training. In a self-driving simulator [6,14,38], smart-actor models [3,4,7,51] are responsible for generating stochastic joint behaviors that are realistic at a scene-level, with actors obeying to underlying scene dynamics with complex interactions.

These applications require learning a joint distribution over actors' future trajectories that characterizes how the scene might unroll as a whole. Since this is generally intractable, many motion forecasting approaches [9,11,12,18] assume marginal independence across actors' future trajectories, failing to get scene-consistent futures. Alternatively, auto-regressive formulations [45,50] model interactions at the output level, but require sequential sampling which results in slow inference and compounding errors [47].

To overcome these challenges, we propose a novel way to characterize the joint distribution over motion forecasts via an implicit latent variable model (ILVM). We aim to recover a latent space that can summarize all the unobserved scene dynamics given input sensor data. This is challenging given that (i) modern roads present very complex geometries and topologies that make every inter-section unique, (ii) the dynamic environment is only partially observed through sensor returns, and (iii) the number of actors in a scene is variable. To address these, we model the scene as an interaction graph [9,22,26,32], where nodes are traffic participants. We then partition the scene latent space into a distributed representation among actors. We leverage graph neural networks (GNN) [2] both to encode the full scene into the latent space as well as to decode latent samples into socially consistent future trajectories. We frame the decoding of all actors' trajectories as a deterministic mapping from the inputs and scene latent sam-ples, making the latent variables capture all the stochasticity in our generative process. Furthermore, this allows us to perform efficient inference via parallel sampling.

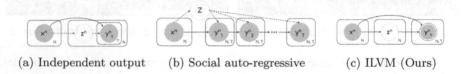

(a) Independent output   (b) Social auto-regressive   (c) ILVM (Ours)

Fig. 2. Graphical models of trajectory distribution. Dashed arrows/circles denote that only some approaches within the group use those components. Double circle in (c) denotes that it is a deterministic mapping of its inputs.

We show that our ILVM significantly outperforms the motion forecasting state-of-the-art in ATG4D [53] and NUSCENES [8]. We observe that our ILVM is able to generate scene-consistent samples (see Fig. 1) while producing less entropic joint distributions that also better cover the ground-truth. Moreover, when using our scene-consistent motion forecasts, a state-of-the-art motion plan-ner [48] can plan safer and more comfortable trajectories.

2   Related Work

In this section, we review recent advances in motion forecasting, with a focus on realistic approaches that predict from sensor data, explicitly reason about the multi-modality of the output distribution, or model multi-agent interactions.

In traditional self-driving stacks, an object detection module is responsible for recognizing other traffic participants in the scene, followed by a motion fore-casting module that predicts how the scene might unroll given the current state of each actor. However, the actor state is typically a very compact representa-tion that includes pose, velocity, and acceleration. As a consequence, it is hard to incorporate uncertainty due to sensor noise or occlusion.

We follow the works of [10,35,54], which unified these two tasks by hav-ing a single fully convolutional backbone network predict both the current and future states for each pixel in a bird's eye view grid directly from a voxelized

LiDAR point-cloud and semantic raster of an HD map. This approach naturally propagates uncertainty between the two tasks in the feature space, without the need of explicit intermediate representations. While these models reason about uncertainty in sensor observations, they neglect inherent uncertainty in the actors' future behavior. [9,32] add agent-agent interaction reasoning to this framework. [9] introduces spatially-aware graph neural networks that aggregate features from neighboring actors in the scene to predict a single trajectory per actor with gaussian waypoints, assuming marginal independence across actors. This approach is still limited in expressivity since (i) a uni-modal characterization of the future is insufficient for downstream motion planning to make safe decisions, and (ii) modeling the marginal distribution per actor cannot provide trajectory samples that are consistent across actors.

Another research stream [1,13,19,26,30,37,44,45,50] has focused on the problem of multi-agent trajectory prediction from perfect perception, i.e., assuming that the ground-truth past trajectory of all actors' is given. Unfortunately, this is not realistic in self-driving vehicles, which rely on imperfect perception with noise that translates into failures such as false positive and false negative detections and id switches in tracking. Nonetheless, these methods have proposed output parameterizations that can predict multi-modal distributions over future trajectories, which are applicable to our end-to-end perception and prediction setting.

Various factorizations of the joint distribution over N actors' trajectories $p(Y|X) = p(y_1, \cdots, y_N | x_1, \cdots, x_N)$ with different levels of independence assumptions have been proposed to sidestep the intractability of $p(Y|X)$. The simplest approximation is to assume *independent futures* across actors and time steps $p(Y|X) = \prod_n \prod_t p(y_n^t | X)$, as shown in Fig. 2a. Some approaches directly regress the parameters of a mixture of Gaussians over time [11,12,34], which provides efficient sampling but can suffer from low expressivity and unstable optimization. Non-parametric approaches [23,24,41,46] have also been proposed to characterize the multi-modality of one actor's individual behavior. These approaches either score trajectory samples from a finite set [41,55] with limited coverage or predict an occupancy grid at different future horizons [23,24,46], which is very memory consuming. [44] proposed to learn a one-step policy that predicts the next waypoint based on the previous history, avoiding the time independence assumption. Variational methods [18,31] inspired by [25,49] have also been proposed to learn an actor independent latent space to capture unobserved actor dynamics such as goals. Unfortunately, none of these methods can accurately characterize the joint distribution in interactive situations, since the generative process is independent per actor.

An alternative approach to better characterize the behavior of multiple actors jointly is *autoregressive generation* with social mechanisms [1,45], which predict the distribution over the next trajectory waypoint of each actor conditioned on the previous states of all actors $p(Y|X) = \prod_n \prod_t p\left(y_n^t | Y^{0:t-1}, X\right)$. This approach has been enhanced by introducing latent variables [22,26,50], as in Fig. 2b. In particular, [26] introduces discrete latent variables to model pairwise relationships in an interaction graph, while in [22,50] they capture per-actor high-level actions. Autoregressive approaches, however, suffer from compounding errors [28,40,47]. During training, the model is fed the ground-truth $Y^{0:t-1}$, while during inference, the model must rely on approximate samples from the learned distribution. While scheduled sampling [5] has been proposed to mitigate this issue, the objective function underlying this method is improper [21] and pushes the conditional distributions $p(y_n^t|Y^{0:t-1})$ to model the marginal distributions $p(y_n^t)$ instead. Moreover, these methods require sequential sampling, which is not amenable to real-time applications such as self-driving.

In contrast to previous works, we propose to model interaction in a scene latent space that captures all sources of uncertainty, and use a deterministic decoder to characterize an implicit joint distribution over all actors' future trajectories without any independence assumptions at the output level, as shown in Fig. 2c. This design features efficient parallel sampling, high expressivity and yields trajectory samples that are substantially more consistent across actors.

3   Scene Level Reasoning for Motion Forecasting

In this section we introduce our approach to model **the joint distribution** $P(Y|X)$ **over** N **actors' future trajectories** $Y = \{y_1, y_2, \cdots, y_N\}$ given each actor's local context $X = \{x_1, x_2, \cdots, x_N\}$ extracted from sensor data and HD maps. An actor's trajectory y_n is composed of 2D waypoints over time y_n^t in the coordinate frame defined by the actor's current position and heading. In the following, we first explain our implicit latent variable model, then introduce our concrete architecture including the actor feature extraction from sensor data, and finally explain how to train our model in an end-to-end manner.

3.1   Implicit Latent Variable Model with Deterministic Decoder

We formulate the generative process of future trajectories over actors with a latent variable model:

$$P(Y|X) = \int_Z P(Y|X, Z)P(Z|X)dZ$$

where Z is a latent variable that captures unobserved scene dynamics such as actor goals and style, multi-agent interactions, or future traffic light states.

We propose to use a **deterministic mapping** $Y = f(X, Z)$ to implicitly characterize $P(Y|X, Z)$, instead of explicitly representing it in a parametric form. This approach allows us to avoid factorizing $P(Y|X, Z)$ (as in Fig. 2a or Fig. 2b) and sidestep the associated shortcomings discussed in Sect. 2. In this framework, generating scene-consistent future trajectories Y across actors is simple and highly efficient, since it only requires one stage of parallel sampling:

1. Draw latent scene samples from prior $Z \sim P(Z|X)$
2. Decode with the deterministic decoder $Y = f(X, Z)$

We emphasize that this modeling choice encourages the latent Z to capture *all* stochasticity in our generative process. To this end, we leverage a *continuous latent* Z for high expressivity. This stands in contrast to previous methods [22,26,50], where discrete latent Z are employed to model discrete high-level actions or pairwise interactions, and an explicit $P(Y|X,Z)$ to represent continuous uncertainty.

Producing a latent space that can capture all the uncertainties in any scenario is challenging: scenarios vary drastically in the number of actors N, the road topology as well as traffic rules. To mitigate this challenge, we propose to partition the scene latent as $Z = \{z_1, z_2, \cdots, z_N\}$, obtaining a distributed representation where z_n is anchored to actor n in an interaction graph with traffic participants as nodes. The distributed representation has the benefit of naturally scaling the capacity of the latent space as the number of actors grow. Furthermore, the anchoring gives the model an inductive bias that eases the learning of a scene latent space. Intuitively, each latent z_n encodes unobserved dynamics most relevant to actor n, including interactions with neighboring actors and traffic rules that apply in its locality. We represent each z_n as a diagonal multivariate gaussian $z_n \sim \mathcal{N}\left(\left[\mu_n^1(X), \cdots, \mu_n^D(X)\right], \mathrm{diag}\left(\left[\sigma_n^1(X), \cdots, \sigma_n^D(X)\right]\right)\right)$, as is common with variational models [25,49]. We emphasize that although factorized, the latent space is not marginally independent across actors since each z_n is conditioned on all $x_1, \cdots, x_N$ as shown in the graphical model in Fig. 2c.

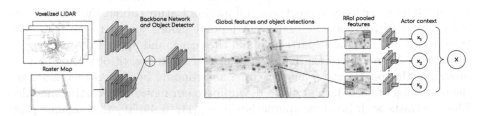

Fig. 3. Actor feature extraction. Given LiDAR and maps, our backbone CNN detects the actors in the scene, and individual feature vectors per actor are extracted via RRoI Align [36], followed by a CNN with spatial pooling.

Since integration over Z is intractable, we exploit amortized variational inference [25,49]. By introducing an encoder distribution $Q(Z|X,Y)$ to approximate the true posterior $P(Z|X,Y)$, the learning problem can be reformulated as a maximization of the Evidence Lower BOund (ELBO). Please visit the supplementary for a more thorough description of variational inference.

3.2 Joint Perception and Motion Forecasting Architecture

Our architecture consists of an actor feature extractor that detects objects in the scene and provides rich representations of each actor (Fig. 3), encoder/prior modules that infer a scene latent space at training/inference respectively, and a decoder that predicts the actors' future trajectories (Fig. 4). To implement the prior, encoder and decoder modules, we leverage a flexible scene interaction module (SIM) as our building block for relational reasoning (Algorithm 1).

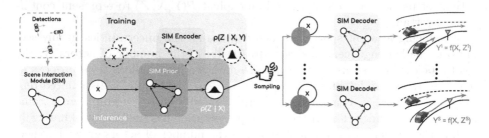

Fig. 4. Our implicit latent variable model encodes the scene into a latent space, from which it can efficiently sample multiple future realizations in parallel, each with socially consistent trajectories.

Actor Feature Extractor: Figure 3 shows how we extract per actor features $X = \{x_1, x_2, \cdots, x_N\}$ from raw sensor data and HD maps in a differentiable manner, such that perception and motion forecasting can be trained jointly end-to-end. We use a CNN-based perception backbone network architecture inspired by [10,53] to extract rich geometrical and motion features about the whole scene from a past history of voxelized LiDAR point clouds and a raster map. We then detect [53] the traffic participants in the scene, and apply Rotated Region of Interest Align [36] to the backbone features around each object detection, providing the local context for all actors, as proposed by [9]. As mentioned at the beginning of Sect. 3, this will be the input to our motion forecasting module. This contrasts with previous approaches (e.g., [11,13,45,50]) that assume past trajectories for each actor are given. We refer the reader to the supplementary material for more details about our perception module, including the backbone architecture and detection parameterization.

Scene Interaction Module (SIM): This is a core building block of our encoder, prior, and decoder networks, as shown in Fig. 4. Once we have extracted individual actor features, we can frame the scene as a fully-connected interaction graph where each traffic participant is a node. Inspired by [9], we use a spatially-aware graph neural network to model multi-agent dynamics, as described in Algorithm 1. Our SIM performs a single round of message passing to update the nodes' representation, taking into account spatiotemporal relationships.

Encoder: To approximate the true posterior latent distribution $P(Z|X,Y)$, we introduce an approximate posterior $q_\phi(Z|X,Y)$, implemented by our SIM and parameterized by ϕ. This network is also commonly known as recognition network, since it receives the target output variable Y as an input, and thus it can *recognize* the scene dynamics that are unobserved by the prior $p_\gamma(Z|X)$.

Note that the encoder can only be used during training, since it requires access to the ground-truth future trajectories. We initialize the node representations as $h_n = \text{MLP}(x_n \oplus \text{GRU}(y_n))$, where $\oplus$ denotes concatenation along the feature dimension. After running one round of message passing, the scene interaction module predicts the distribution over scene latent variables $Z = \{z_1, z_2, \cdots, z_N\}$. We stress that despite anchoring each partition of the scene latent to an actor, each individual z_n contains information about the full scene, since each final node representation is dependent on the whole input X because of the message propagation in the fully-connected interaction graph.

Algorithm 1. Scene Interaction Module (SIM)

Input: Initial hidden state for all of the actors in the scene $\{h_0, h_1, \cdots, h_N\}$. BEV coordinates (centroid and heading) of the detected bounding boxes $\{c_0, c_1, ..., c_N\}$.
Output: Feature vector per node $\{o_0, o_1, \cdots, o_N\}$.

1: Construct fully-connected interaction graph $G = (V, E)$ from detections
2: Compute pairwise coordinate transformations $T(c_u, c_v)$, $\forall (u, v) \in E$
3: **for** $(u, v) \in E$ **do**   ▷ Compute message for every edge in the graph in parallel
4:     $m_{u \to v} = \text{MLP}(h_u, h_v, T(c_u, c_v))$
5: **for** $v \in V$ **do**   ▷ Update node states in parallel
6:     $a_v = \text{MaxPooling}(\{m_{u \to v} : u \in \mathbf{N}(v)\})$ ▷ Aggregate messages from neighbors
7:     $h_v' = \text{GRU}(h_v, a_v)$ ▷ Update the hidden state
8:     $o_v = \text{MLP}(h_v')$ ▷ Compute outputs
9: **return** $\{o_0, o_1, \cdots, o_N\}$

Prior: The prior network $p_\gamma(Z|X)$ is responsible for approximating the prior distribution of the scene latent variable Z at inference time. Similar to the encoder, we model the scene-level latent space with our SIM, where the only difference is that the initial node representations in the graph propagation are $h_n = \text{MLP}(x_n)$, since y_n is not available at inference time.

Deterministic Decoder: Recall that our scene latent has been partitioned into a distributed representation $Z = \{z_1, z_2, \cdots, z_N\}$. To leverage actor features and distributed latents from the whole scene, we parameterize the decoder with another SIM. We can then predict the s-th realization of the future at a scene level via message passing, where each actor trajectory y_n^s takes into account a sample from all the partitions of the scene latent $Z^s = \{z_1^s, \cdots, z_n^s\}$ as well as all actors' features X, enabling reasoning about multi-agent interactions such as car following, yielding, etc. More precisely, given each actor context x_n, we initialize its node representation for the decoder graph propagation as $h_n^s = \text{MLP}(x_n \oplus z_n^s)$. After a round of message passing in our SIM, $h_n'^s$ contains an updated representation of actor n that takes into account the underlying dynamics of the whole scene summarized in Z^s. Finally, the s-th trajectory sample for actor n is deterministically decoded $y_n^s = \text{MLP}(h_n'^s)$ by the SIM output function, without additional sampling steps. The trajectory-level scene sample is simply the collection of all actor trajectories $Y^s = \{y_1^s, \ldots, y_N^s\}$. We can generate S possible futures for all actors in the scene in parallel by batching S scene latent samples.

In this fashion, our model implicitly characterizes the joint distribution over actors' trajectories, achieving superior scene-level consistency. In the experiments section we ablate the design choices in the encoder, prior and decoder, and show that although all of them are important, the deterministic decoder is the key contribution towards socially-consistent trajectories.

3.3 Learning

Our perception and prediction model can be trained end-to-end using stochastic gradient descent. In particular, we minimize a multi-task loss for detection and motion forecasting: $\mathcal{L} = \mathcal{L}_{\text{det}} + \lambda \cdot \mathcal{L}_{\text{forecast}}$

Detection: For the detection classification branch we employ a binary cross entropy loss with hard negative mining $\mathcal{L}_{\text{cla}}$. We select all positive examples from the ground-truth and 3 times as many negative examples. For box fitting, we apply a smooth ℓ_1 loss $\mathcal{L}_{\text{reg}}$ to each of the 5 parameters $(x_i, y_i, w_i, h_i, \phi_i)$ of the bounding boxes anchored to a positive example i. The overall detection loss is a linear combination $\mathcal{L}_{\text{det}} = \mathcal{L}_{\text{cla}} + \alpha \cdot \mathcal{L}_{\text{reg}}$.

Motion Forecasting: We adapt the variational learning objective of the CVAE framework [49] and optimize the evidence-based lower bound (ELBO) of the log-likelihood $\log P(Y|X)$. In our case, due to the deterministic decoder leading to an implicit distribution over Y, we use Huber loss ℓ_δ as the reconstruction loss, and reweight the KL term with β as proposed by [17]:

$$\mathcal{L}_{\text{forecast}} = \sum_n^N \sum_t^T \ell_\delta(y_n^t - y_{n,GT}^t) + \beta \cdot \text{KL}\left(q_\phi\left(Z|X, Y_{GT}\right) || p_\gamma\left(Z|X\right)\right)$$

where the first term minimizes the reconstruction error between all the trajectories in the scene $Y = \{y_n^t | \forall n, t\} = f_\theta(Z)$, $Z \sim q_\phi(Z|X, Y_{GT})$ and their corresponding ground-truth Y_{GT}, and the second term brings the privileged *posterior* $q_\phi(Z|X, Y_{GT})$ and the approximate *prior* $p_\gamma(Z|X)$ distributions closer.

4   Experimental Evaluation

In this section, we first explain the metrics and baselines we use for evaluation. Next, we compare our model against state-of-the-art motion forecasting algorithms on predicting the future 5 s trajectories on two real-world datasets: ATG4D [53] and NUSCENES [8] (see supplementary for details). Then, we measure the impact on motion planning. Finally, we carry out an ablation study to understand which part of our model contributes the most.

4.1 Scene Level Motion Forecasting Metrics

Previous methods use sample quality metrics at the actor level such as the popular minimum/mean average displacement error (minADE/meanADE). However, these metrics only evaluate the quality of the underlying marginal distribution per actor. For instance, minADE takes the trajectory sample that best fits the ground-truth of each actor independently, which does not measure the consistency between different actors' sample trajectories and can be easily cheated by predicting high entropy distributions that cover all the space but are not precise.

Table 1. [ATG4D] Scene-level motion forecasting ($S = 15$ samples)

Type	Model	SCR$_{5s}$ (%)	min SFDE(m)	min SADE(m)	mean SFDE(m)	mean SADE(m)
Indep. output	SpAGNN [9]	8.19	2.83	1.34	4.37	1.92
	RulesRoad [18]	6.66	2.71	1.32	4.21	1.84
	MTP [12]	3.98	1.91	0.95	3.11	1.37
	MultiPath [11]	4.41	1.97	0.95	3.14	1.36
	R2P2-MA [44]	4.63	2.13	1.09	3.27	1.49
Social	SocialLSTM [1]	6.13	2.75	1.38	4.05	1.83
auto-regressive	NRI [26]	7.00	2.68	1.43	3.81	1.74
	ESP [45]	2.67	1.91	0.97	2.84	1.29
	MFP [50]	5.15	2.35	1.13	3.35	1.45
	ILVM	**0.70**	**1.53**	**0.76**	**2.27**	**1.02**

We propose scene-level sample quality metrics to evaluate how well the models capture the joint distribution over future outcomes. To this end, we define a scene-level counterpart of the popular minimum/mean average displacement error. We emphasize that in this context, each scene sample $s \in 1, ..., S$ is a collection of N future trajectories, one for each actor in the scene.

$$\text{minSADE} = \min_{s \in 1...S} \frac{1}{NT} \sum_{n=1}^{N} \sum_{t=1}^{T} ||y_{n,GT}^t - y_{n,s}^t||^2$$

$$\text{meanSADE} = \frac{1}{NTS} \sum_{s=1}^{S} \sum_{n=1}^{N} \sum_{t=1}^{T} ||y_{n,GT}^t - y_{n,s}^t||^2$$

We also compute their final counterparts minSFDE and meanSFDE, which evaluate only the motion forecasts at the final timestep (i.e. at 5 s).

Furthermore, to evaluate the consistency of the motion forecasts we propose to measure the scene collision rate (SCR). It measures the percentage of trajectory samples that collide with any other trajectory in the same scene sample s. Two trajectory samples are considered in collision if the overlap between their future bounding boxes at any time step is higher than a small IOU threshold ε_{IOU}. To compute this, we first obtain the bounding boxes for future time steps $\{b_{i,s}^t\}$. The size of the bounding boxes are the same as their object detections and the future headings are extracted by finite differences on the trajectory samples.

$$\mathrm{SCR}_T = \frac{1}{NS} \sum_{s=1}^{S} \sum_{i=1}^{N} \min\left(1, \sum_{j>i}^{N} \sum_{t=1}^{T} \mathbb{1}\left[IoU(b_{i,s}^t, b_{j,s}^t) > \varepsilon_{IOU}\right]\right)$$

Finally, to perform a fair comparison in motion forecasting metrics, which are evaluated on true positive detections, we follow [9] and operate the object detector at 90% recall point for all models in ATG4D and 80% in NuSCENES.

Table 2. [nuScenes] **Scene-level motion forecasting** ($S = 15$ samples)

Type	Model	SCR$_{5s}$ (%)	min SFDE(m)	min SADE(m)	mean SFDE(m)	mean SADE(m)
Indep. output	SpAGNN [9]	7.54	2.07	1.00	3.85	1.82
	RulesRoad [18]	5.67	2.10	1.01	3.55	1.67
	MTP [12]	8.68	1.86	0.91	3.86	1.85
	MultiPath [11]	7.31	2.01	0.95	3.50	1.65
	R2P2-MA [44]	4.56	2.25	1.08	3.47	1.67
Social auto-regressive	SocialLSTM [1]	6.45	2.71	1.33	4.20	2.05
	NRI [26]	5.98	2.54	1.28	3.91	1.88
	ESP [45]	5.09	2.16	1.07	3.46	1.67
	MFP [50]	4.94	2.74	1.30	4.11	1.95
	ILVM	**1.91**	**1.84**	**0.86**	**2.99**	**1.43**

4.2 Baselines

In this section, we discuss the state of the art motion forecasting models that we use as baselines. It is important to note that most baselines are designed for motion forecasting given perfect perception, i.e., ground-truth past trajectories. However, this is not realistic in self-driving vehicles, which rely on imperfect noisy perception. Thus, we adapt them to the realistic setting by replacing their past trajectory encoders with our extracted actor features (see Fig. 3) and training end-to-end with our perception backbone (see supplementary for details).

Independent Output: We benchmark against SpAGNN [9], MTP [12], MULTIPATH [11], RULESROAD [18], and R2P2-MA [44]. Since the trajectory sampling process from these models is independent per actor, we define a scene sample s by drawing one sample for each actor in the scene.

Social Auto-Regressive: We compare against SocialLSTM [1], ESP [45], MFP [50], and NRI [26]. It is worth sharing that for these baselines to achieve competitive results we had to perturb the ground-truth trajectories with white noise during training. This is because these models suffer from a distributional shift between training and inference, as explained in Sect. 2. We note that white noise was more effective than teacher forcing [28] or scheduled sampling [5].

4.3   Motion Forecasting Results

Experimental results for motion forecasting in the ATG4D dataset (with $S = 15$ samples) are shown in Table 1. Our ILVM outperforms the baselines across all metrics. Very notably, it **achieves a 75% reduction in collision rate** with respect to the strongest baseline in this metric (ESP [45]), thus highlighting the better characterization of the joint distribution across actors (which also translates into scene-consistent samples). Our model is also much more precise (20% reduction in meanSFDE) while exhibiting better coverage of the ground-truth data (19% reduction in minSFDE). We include an analysis of how the minSADE and minSFDE vary across different number of samples S in the supplementary.

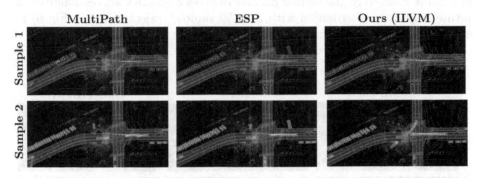

Fig. 5. Scene-level samples. Our latent variable model captures underlying scene dynamics at the intersection level (i.e. yield vs. go)

Figure 5 shows individual samples. We heuristically select the two most distinct samples for visualization to show diverse realizations of the future. The baseline models capture variations in individual actors' future, but do not capture the yielding interaction at the intersection, which our model does. In addition, Fig. 6 showcases the full distribution learned by the models. More concretely, this plot shows a Monte Carlo estimation of the marginal distribution per actor, where 50 samples are drawn from each model. Transparency in the plots illustrates the probability density at a given location. These examples support the same conclusion taken from the quantitative results and highlight the ability of our model to understand complex road geometries and the multi-modal behaviors they induce. This is particularly interesting since all models share the same representation of the environment and backbone architecture.

To show that our improvements generalize to a dataset with a different distribution of motions and road topologies, we validate our method on NUSCENES. Table 2, shows that ILVM brings improvements over the baselines across all metrics. In particular, we observe significant gains in scene-consistency (SCR) and precision metrics (meanSADE and meanSFDE).

4.4   Motion Planning Results

To validate the system-level impact of different perception and prediction models, we use the state-of-the-art learnable motion planner of [48] to plan a trajectory for the SDV (τ_{SDV}):

$$\tau_{SDV} = \arg\min_{\tau \in T} \mathbb{E}_{p(Y|X)} \left[c(\tau, Y \setminus y_{SDV}) \right] \approx \arg\min_{\tau \in T} c(\tau, \{Y^s \setminus y_{SDV}^s : \forall s \in 1 \ldots S\})$$

where $p(Y|X)$ is the distribution over future trajectories output by the perception and prediction model, T is a predefined set of SDV trajectories given the map and high-level route, and c is a costing function that measures safety and comfort taking into account the motion forecasts for the rest of the vehicles. More concretely, the motion planner receives a Monte Carlo estimate of the future trajectory distribution with $S = 50$ sample trajectories (see Fig. 6) for every detected vehicle (excluding the SDV), which are considered obstacles in order to approximate the expected cost of plans $\tau \in T$.

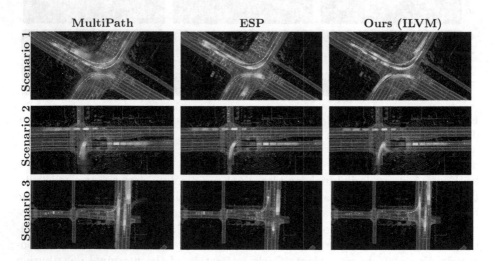

Fig. 6. Motion forecasting visualizations of 50 samples. Time is encoded in the rainbow color map ranging from red (0 s) to pink (5 s). (Color figure online)

The experiments in Table 3 measure how different motion forecasts translate into the safety and comfort of the SDV trajectory (τ_{SDV}), an impact often overlooked by previous works. Our motion forecasts (ILVM) enable the motion planner to execute significantly safer and more comfortable trajectories. We notice that the ego-motion plans make similar progress across models, but our approach produces the closest trajectories to the expert demonstrations (lowest ℓ_2 distance at 5 s into the future), while yielding much fewer collisions. We include planning qualitative results in our supplementary material.

4.5 Ablation Study

Implicit vs. Explicit Decoder: We ablate ILVM ($\mathcal{M}_0$ in Table 4) by replacing the proposed implicit decoder with an explicit decoder that produces a full covariance bi-variate Gaussian per waypoint, and the reconstruction loss with Negative Log Likelihood. This gives us $\mathcal{M}_1$, where ancestral sampling is used for inference: first sample latent, then sample output. Here, we can see that assuming conditional independence across actor at the output level significantly degrades all aspects of the motion forecasting performance. Most notably, the high scene collision rate shows that the samples are no longer socially consistent.

Table 3. [ATG4D] System level performance (ego-motion planning)

Type	Model	Collision (% up to 5 s)	L2 human (m @ 5 s)	Lat. acc. (m/s^2)	Jerk (m/s^3)	Progress (m @ 5 s)
Indep. output	SpAGNN [9]	4.19	5.98	2.94	2.90	32.37
	RulesRoad [18]	4.04	5.83	2.84	2.76	32.50
	MTP [12]	3.10	5.67	2.83	2.66	33.14
	MultiPath [11]	3.30	5.58	2.73	2.57	32.99
	R2P2-MA [45]	3.71	5.65	2.84	2.53	33.90
Social auto-regressive	SocialLSTM [1]	4.22	5.92	2.76	2.66	32.60
	NRI [26]	4.94	5.73	2.78	2.55	33.43
	ESP [45]	3.13	5.48	2.76	2.44	33.74
	MFP [50]	4.14	5.57	2.61	2.43	32.94
	ILVM	**2.64**	**5.33**	**2.59**	**2.30**	33.72

Table 4. [ATG4D] Motion forecasting ablation study ($S = 15$ samples)

ID	Learned prior	Implicit output	SIM Encoder	SIM Decoder	SCR_{5s}	min SFDE	min SADE	mean SFDE	mean SADE
$\mathcal{M}_0$	✓	✓	✓	✓	**0.70**	**1.53**	**0.76**	**2.27**	**1.02**
$\mathcal{M}_1$	✓		✓	✓	8.46	2.66	1.31	4.17	1.80
$\mathcal{M}_2$		✓	✓	✓	1.10	1.53	0.76	2.43	1.08
$\mathcal{M}_3$	✓	✓		✓	1.03	1.57	0.78	2.42	1.08
$\mathcal{M}_4$	✓	✓	✓		1.52	1.67	0.81	2.44	1.09
$\mathcal{M}_5$	✓	✓			1.74	1.66	0.81	2.43	1.08

Learned vs. Fixed Prior: A comparison between $\mathcal{M}_0$ and $\mathcal{M}_2$ in Table 4 shows that using a learned prior network $P(Z|X)$ achieves a better precision diversity trade-off compared to using a fixed prior distribution of isotropic Gaussians.

ILVM Architecture: In Table 4, $\mathcal{M}_3$ ablates the SIM encoder and prior networks by replacing them with MLPs that model $p(z_n|x_n)$ and $p(z_n|x_n, y_n)$ at the actor-level, respectively. $\mathcal{M}_4$ replaces the SIM decoder by an MLP per actor $y_n^s = \text{MLP}(X, z_n^s)$. Finally, $\mathcal{M}_5$ applies the changes in $\mathcal{M}_3$ and $\mathcal{M}_4$. These experiments show that both the graph based prior/encoder and decoder are important for our latent variable model. In particular, the large gap in scene level collision demonstrates that our proposed SIM encoder and decoder capture scene-level understanding that is not present in the ablations with independent assumptions at the latent or output level.

5   Conclusion and Future Work

We have proposed a latent variable model to obtain an implicit joint distribution over actor trajectories that characterizes the dependencies over their future behaviors. Our model achieves fast parallel sampling of the joint trajectory space and produces scene-consistent motion forecasts. We have demonstrated the effectiveness of our method on two challenging datasets by significantly improving over state-of-the-art motion forecasting models on scene-level sample quality metrics. Our method achieves much more precise predictions that are more socially consistent. We also show that our method produces significant improvements in motion planning, even though the planner does not make explicit use of the strong consistency of our scenes. We leave it to future work to design a motion planner to better utilize joint distributions over trajectories.

References

1. Alahi, A., Goel, K., Ramanathan, V., Robicquet, A., Fei-Fei, L., Savarese, S.: Social LSTM: human trajectory prediction in crowded spaces. In: Proceedings of the IEEE CVPR (2016)
2. Battaglia, P.W., et al.: Relational inductive biases, deep learning, and graph networks (2018)
3. Behbahani, F., et al.: Learning from demonstration in the wild. In: 2019 International Conference on Robotics and Automation (ICRA), May 2019. https://doi.org/10.1109/icra.2019.8794412
4. Behrisch, M., Bieker, L., Erdmann, J., Krajzewicz, D.: Sumo-simulation of urban mobility: an overview. In: Proceedings of SIMUL 2011, The Third International Conference on Advances in System Simulation. ThinkMind (2011)
5. Bengio, S., Vinyals, O., Jaitly, N., Shazeer, N.: Scheduled sampling for sequence prediction with recurrent neural networks. In: Advances in Neural Information Processing Systems, pp. 1171–1179 (2015)
6. Best, A., Narang, S., Pasqualin, L., Barber, D., Manocha, D.: AutonoVi-Sim: autonomous vehicle simulation platform with weather, sensing, and traffic control. In: 2018 IEEE/CVF Conference on Computer Vision and Pattern Recognition Workshops (CVPRW), pp. 1161–11618 (2018)
7. Bhattacharyya, R.P., Phillips, D.J., Wulfe, B., Morton, J., Kuefler, A., Kochenderfer, M.J.: Multi-agent imitation learning for driving simulation. In: 2018 IEEE/RSJ International Conference on Intelligent Robots and Systems (IROS), October 2018. https://doi.org/10.1109/iros.2018.8593758

8. Caesar, H., et al.: nuScenes: a multimodal dataset for autonomous driving. arXiv preprint arXiv:1903.11027 (2019)
9. Casas, S., Gulino, C., Liao, R., Urtasun, R.: Spatially-aware graph neural networks for relational behavior forecasting from sensor data. arXiv preprint arXiv:1910.08233 (2019)
10. Casas, S., Luo, W., Urtasun, R.: IntentNet: learning to predict intention from raw sensor data. In: Conference on Robot Learning (2018)
11. Chai, Y., Sapp, B., Bansal, M., Anguelov, D.: Multipath: multiple probabilistic anchor trajectory hypotheses for behavior prediction. arXiv preprint arXiv:1910.05449 (2019)
12. Cui, H., et al.: Multimodal trajectory predictions for autonomous driving using deep convolutional networks. arXiv preprint arXiv:1809.10732 (2018)
13. Djuric, N., et al.: Motion prediction of traffic actors for autonomous driving using deep convolutional networks. arXiv preprint arXiv:1808.05819 (2018)
14. Dosovitskiy, A., Ros, G., Codevilla, F., Lopez, A., Koltun, V.: CARLA: an open urban driving simulator. In: Proceedings of the 1st Annual Conference on Robot Learning, pp. 1–16 (2017)
15. Hardy, J., Campbell, M.: Contingency planning over probabilistic obstacle predictions for autonomous road vehicles. IEEE Trans. Robot. **29**, 913–929 (2013)
16. Henaff, M., Canziani, A., LeCun, Y.: Model-predictive policy learning with uncertainty regularization for driving in dense traffic. arXiv preprint arXiv:1901.02705 (2019)
17. Higgins, I., et al.: beta-VAE: learning basic visual concepts with a constrained variational framework
18. Hong, J., Sapp, B., Philbin, J.: Rules of the road: predicting driving behavior with a convolutional model of semantic interactions. In: The IEEE Conference on Computer Vision and Pattern Recognition (CVPR), June 2019
19. Hoshen, Y.: VAIN: attentional multi-agent predictive modeling. In: Advances in Neural Information Processing Systems, pp. 2701–2711 (2017)
20. Hubmann, C., Schulz, J., Becker, M., Althoff, D., Stiller, C.: Automated driving in uncertain environments: planning with interaction and uncertain maneuver prediction. IEEE Trans. Intell. Veh. **3**(1), 5–17 (2018)
21. Huszár, F.: How (not) to train your generative model: Scheduled sampling, likelihood, adversary? arXiv preprint arXiv:1511.05101 (2015)
22. Ivanovic, B., Pavone, M.: The trajectron: probabilistic multi-agent trajectory modeling with dynamic spatiotemporal graphs. In: Proceedings of the IEEE International Conference on Computer Vision, pp. 2375–2384 (2019)
23. Jain, A., Casas, S., Liao, R., Xiong, Y., Feng, S., Segal, S., Urtasun, R.: Discrete residual flow for probabilistic pedestrian behavior prediction. arXiv preprint arXiv:1910.08041 (2019)
24. Kim, B., Kang, C.M., Kim, J., Lee, S.H., Chung, C.C., Choi, J.W.: Probabilistic vehicle trajectory prediction over occupancy grid map via recurrent neural network. In: 2017 IEEE 20th International Conference on Intelligent Transportation Systems (ITSC), pp. 399–404. IEEE (2017)
25. Kingma, D.P., Welling, M.: Auto-encoding variational Bayes. In: ICLR (2013)
26. Kipf, T., Fetaya, E., Wang, K.C., Welling, M., Zemel, R.: Neural relational inference for interacting systems. arXiv preprint arXiv:1802.04687 (2018)
27. Klingelschmitt, S., Damerow, F., Eggert, J.: Managing the complexity of inner-city scenes: an efficient situation hypotheses selection scheme. In: 2015 IEEE intelligent vehicles symposium (IV), pp. 1232–1239. IEEE (2015)

28. Lamb, A.M., Alias Parth Goyal, A.G., Zhang, Y., Zhang, S., Courville, A.C., Bengio, Y.: Professor forcing: a new algorithm for training recurrent networks. In: Lee, D.D., Sugiyama, M., Luxburg, U.V., Guyon, I., Garnett, R. (eds.) Advances in Neural Information Processing Systems 29. Curran Associates, Inc. (2016). http://papers.nips.cc/paper/6099-professor-forcing-a-new-algorithm-for-training-recurrent-networks.pdf
29. Lang, A.H., Vora, S., Caesar, H., Zhou, L., Yang, J., Beijbom, O.: PointPillars: fast encoders for object detection from point clouds. In: Proceedings of the IEEE Conference on Computer Vision and Pattern Recognition, pp. 12697–12705 (2019)
30. Le, H.M., Yue, Y., Carr, P., Lucey, P.: Coordinated multi-agent imitation learning (2017)
31. Lee, N., Choi, W., Vernaza, P., Choy, C.B., Torr, P.H., Chandraker, M.: Desire: distant future prediction in dynamic scenes with interacting agents. In: Proceedings of the IEEE CVPR (2017)
32. Li, L., Yang, B., Liang, M., Zeng, W., Ren, M., Segal, S., Urtasun, R.: End-to-end contextual perception and prediction with interaction transformer. 2020 IEEE/RSJ International Conference on Intelligent Robots and Systems (IROS) (2020)
33. Liang, M., Yang, B., Chen, Y., Hu, R., Urtasun, R.: Multi-task multi-sensor fusion for 3D object detection. In: Proceedings of the IEEE CVPR (2019)
34. Liang, M., Yang, B., Hu, R., Chen, Y., Liao, R., Feng, S., Urtasun, R.: Learning lane graph representations for motion forecasting. In: ECCV (2020)
35. Luo, W., Yang, B., Urtasun, R.: Fast and furious: real time end-to-end 3D detection, tracking and motion forecasting with a single convolutional net. In: Proceedings of the IEEE CVPR (2018)
36. Ma, J., et al.: Arbitrary-oriented scene text detection via rotation proposals. IEEE Trans. Multimed. **20**, 3111–3122 (2018)
37. Ma, W.C., Huang, D.A., Lee, N., Kitani, K.M.: Forecasting interactive dynamics of pedestrians with fictitious play. In: Proceedings of the IEEE CVPR (2017)
38. Martinez, M., Sitawarin, C., Finch, K., Meincke, L., Yablonski, A., Kornhauser, A.: Beyond grand theft auto V for training, testing and enhancing deep learning in self driving cars (2017)
39. Okamoto, M., Perona, P., Khiat, A.: DDT: deep driving tree for proactive planning in interactive scenarios. In: 2018 21st International Conference on Intelligent Transportation Systems (ITSC), pp. 656–661. IEEE (2018)
40. Osa, T., Pajarinen, J., Neumann, G., Bagnell, J.A., Abbeel, P., Peters, J., et al.: An algorithmic perspective on imitation learning. Found. Trends® Rob. (2018)
41. Phan-Minh, T., Grigore, E.C., Boulton, F.A., Beijbom, O., Wolff, E.M.: CoverNet: multimodal behavior prediction using trajectory sets. arXiv preprint arXiv:1911.10298 (2019)
42. Qi, C.R., Su, H., Mo, K., Guibas, L.J.: PointNet: deep learning on point sets for 3D classification and segmentation. In: Proceedings of the IEEE CVPR (2017)
43. Ratliff, N.D., Bagnell, J.A., Zinkevich, M.A.: Maximum margin planning. In: Proceedings of the 23rd International Conference on Machine Learning, pp. 729–736 (2006)
44. Rhinehart, N., Kitani, K.M., Vernaza, P.: R2P2: a ReparameteRized pushforward policy for diverse, precise generative path forecasting. In: Ferrari, V., Hebert, M., Sminchisescu, C., Weiss, Y. (eds.) ECCV 2018. LNCS, vol. 11217, pp. 794–811. Springer, Cham (2018). https://doi.org/10.1007/978-3-030-01261-8_47
45. Rhinehart, N., McAllister, R., Kitani, K., Levine, S.: PRECOG: PREdiction conditioned on goals in visual multi-agent settings. arXiv e-prints arXiv:1905.01296, May 2019

46. Ridel, D., Deo, N., Wolf, D., Trivedi, M.: Scene compliant trajectory forecast with agent-centric spatio-temporal grids. IEEE Robot. Autom. Lett. **5**, 2816–2823 (2020)
47. Ross, S., Gordon, G., Bagnell, D.: A reduction of imitation learning and structured prediction to no-regret online learning. In: Proceedings of the Fourteenth International Conference on Artificial Intelligence And Statistics, pp. 627–635 (2011)
48. Sadat, A., Ren, M., Pokrovsky, A., Lin, Y.C., Yumer, E., Urtasun, R.: Jointly learnable behavior and trajectory planning for self-driving vehicles. arXiv preprint arXiv:1910.04586 (2019)
49. Sohn, K., Lee, H., Yan, X.: Learning structured output representation using deep conditional generative models. In: Advances in Neural Information Processing Systems, pp. 3483–3491 (2015)
50. Tang, C., Salakhutdinov, R.R.: Multiple futures prediction. In: Advances in Neural Information Processing Systems, pp. 15398–15408 (2019)
51. Treiber, M., Hennecke, A., Helbing, D.: Congested traffic states in empirical observations and microscopic simulations. Phys. Rev. E **62**, 1805 (2000)
52. Yang, B., Guo, R., Liang, M., Sergio, C., Urtasun, R.: Exploiting radar for robust perception of dynamic objects. In: ECCV (2020)
53. Yang, B., Luo, W., Urtasun, R.: Pixor: Real-time 3D object detection from point clouds. In: Proceedings of the IEEE CVPR (2018)
54. Zeng, W., Luo, W., Suo, S., Sadat, A., Yang, B., Casas, S., Urtasun, R.: End-to-end interpretable neural motion planner. In: Proceedings of the IEEE CVPR (2019)
55. Zeng, W., Wang, S., Liao, R., Chen, Y., Yang, B., Urtasun, R.: DSDNet: deep structured self-driving network. In: ECCV (2020)
56. Zhou, Y., et al.: End-to-end multi-view fusion for 3D object detection in lidar point clouds. arXiv preprint arXiv:1910.06528 (2019)

Learning Visual Commonsense for Robust Scene Graph Generation

Alireza Zareian[(⊠)], Zhecan Wang, Haoxuan You, and Shih-Fu Chang

Columbia University, New York, NY 10027, USA
{az2407,zw2627,hy2612,sc250}@columbia.edu

Abstract. Scene graph generation models understand the scene through object and predicate recognition, but are prone to mistakes due to the challenges of perception in the wild. Perception errors often lead to nonsensical compositions in the output scene graph, which do not follow real-world rules and patterns, and can be corrected using commonsense knowledge. We propose the first method to acquire visual commonsense such as affordance and intuitive physics automatically from data, and use that to improve the robustness of scene understanding. To this end, we extend Transformer models to incorporate the structure of scene graphs, and train our Global-Local Attention Transformer on a scene graph corpus. Once trained, our model can be applied on any scene graph generation model and correct its obvious mistakes, resulting in more semantically plausible scene graphs. Through extensive experiments, we show our model learns commonsense better than any alternative, and improves the accuracy of state-of-the-art scene graph generation methods.

1   Introduction

In recent computer vision literature, there is a growing interest in incorporating commonsense reasoning and background knowledge into the process of visual recognition and scene understanding [8,9,13,31,33]. In Scene Graph Generation (SGG), for instance, external knowledge bases [7] and dataset statistics [2,34] have been utilized to improve the accuracy of entity (object) and predicate (relation) recognition. The effect of these techniques is usually to correct obvious perception errors, and replace with more plausible alternatives. For instance, Fig. 1 (top) shows an SGG model mistakenly classifies a bird as a bear, possibly due to the dim lighting and small object size. However, a commonsense model can correctly predict bird, because bear on branch is a less common situation, less aligned with intuitive physics, or contrary to animal behavior.

Nevertheless, existing methods to incorporate commonsense into the process of visual recognition have two major limitations. Firstly, they rely on an

---

A. Zareian, Z. Wang and H. You—Equal contribution.

Electronic supplementary material The online version of this chapter (https://doi.org/10.1007/978-3-030-58592-1_38) contains supplementary material, which is available to authorized users.

A. Vedaldi et al. (Eds.): ECCV 2020, LNCS 12368, pp. 642–657, 2020.
https://doi.org/10.1007/978-3-030-58592-1_38

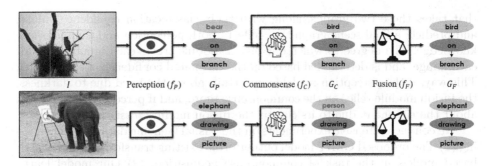

Fig. 1. Overview of the proposed method: We propose a commonsense model that takes a scene graph generated by a perception model and refines that to make it more plausible. Then a fusion module compares the perception and commonsense outputs and generates a final graph, incorporating both signals.

external source of commonsense, such as crowd-sourced or automatically mined commonsense rules, which tend to be incomplete and inaccurate [7], or statistics directly gathered from training data, which are limited to simple heuristics such as co-occurrence frequency [2]. In this paper, we propose the first method to learn graphical commonsense automatically from a scene graph corpus, which does not require external knowledge, and **acquires** commonsense by learning complex, structured patterns beyond simple heuristics.

Secondly, most existing methods are strongly vulnerable to data bias as they integrate data-driven commonsense knowledge into data-driven neural networks. For instance, the commonsense model in Fig. 1 mistakes the `elephant` for a `person`, in order to avoid the bizarre triplet `elephant drawing picture`, while the `elephant` is quite clear visually, and the perception model already recognizes it correctly. None of the existing efforts to equip scene understanding with commonsense have studied the fundamental question of whether to trust perception or commonsense, *i.e.*, what you see versus what you expect. In this paper, we propose a way to disentangle perception and commonsense into two separately trained models, and introduce a method to exploit the disagreement between those two models to achieve the best of both worlds.

To this end, we first propose a mathematical formalization of visual commonsense, as a problem of auto-encoding perturbed scene graphs. Based on the new formalism, we propose a novel method to learn visual commonsense from annotated scene graphs. We extend recently successful transformers [23] by adding local attention heads to enable them to encode the structure of a scene graph, and we train them on a corpus of annotated scene graphs to predict missing elements of a scene via a masking framework similar to BERT [5]. As illustrated in Fig. 2, our commonsense model learns to use its experience to imagine which entity or predicate could replace the mask, considering the structure and context of a given scene graph. Once trained, it can be stacked on top of any perception (*i.e.*, SGG) model to correct nonsensical mistakes in the generated scene graphs.

The output of the perception and commonsense models can be seen as two generated scene graphs with potential disagreements. We devise a fusion module that takes those two graphs, along with their classification confidence values, and predicts a final scene graph that reflects both perception and commonsense knowledge. The degree to which our fusion module trusts each input varies for each image, and is determined based on the estimated confidence of each model. This way, if the perception model is uncertain about the `bird` due to darkness, the fusion module relies on the commonsense more, and if perception is confident about the `elephant` due to its clarity, the fusion module *trusts its eyes*.

We conduct extensive experiments on the Visual Genome datasets [12], showing (1) The proposed GLAT model outperforms existing transformers and graph-based models in the task of commonsense acquisition; (2) Our model learns various types of commonsense that are absent in SGG models, such as object affordance and intuitive physics; (3) The proposed model is robust to dataset bias, and shows commonsensical behavior even in rare and zero-shot scenarios; (4) The proposed GLAT and Fusion mechanism can be applied on any SGG method to correct their mistakes and improve their accuracy. The main contributions of this paper are the following:

- We propose the first method for learning structured visual commonsense, Global-Local Attention Transformer (GLAT), which does not require any external knowledge, and outperforms conventional transformers and graph-based networks.
- We propose a cascaded fusion architecture for Scene Graph Generation, which disentangles commonsense reasoning from visual perception, and integrates them in a way that is robust to the failure of each component.
- We report experiments that showcase our model's unique ability of learning commonsense without picking up dataset bias, and its utility in downstream scene understanding.

2   Related Work

2.1   Commonsense in Computer Vision

Incorporating commonsense knowledge has been explored in various computer vision tasks such as object recognition [3,14,28], object detection [13], semantic segmentation [19], action recognition [9], visual relation detection [31], scene graph generation [2,7,34], and visual question answering [18,22]. There are two aspects to study about these methods: where their commonsense comes from, and how they use it.

Most methods either adopt an external curated knowledge base such as ConceptNet [7,14,18,19,21,28], or acquire commonsense automatically by collecting statistics over an often annotated corpus [2,3,13,22,31,34]. Nevertheless, the former group are limited to incomplete external knowledge, and the latter are based on ad-hoc, hard-coded heuristics such as the co-occurrence frequency of categories. Our method is the first to formulate visual commonsense as a machine learning task, and train a graph-based neural network to solve it. There are a third group of works that focus on a particular type of commonsense by designing a specialized model, such as intuitive physics [6], or object affordance [4]. We put forth a more general framework that includes but is not limited to physics and affordance, by exploiting scene graphs as a versatile semantic representation. The most similar to our work is [26], which only models object co-occurrence patterns, while we also incorporate object relationships and scene graph structure.

When it comes to utilizing commonsense, existing methods integrate it within the inference pipeline, either by retrieving a set of relevant facts from a knowledge base and feeding as additional features to the model [7,18,22], or by employing a graph-based message propagation process to embed the structure of the knowledge graph within the intermediate representations of the model [2,3,9,14,28]. Some other methods distill the knowledge during training through auxiliary objectives, making the inference simple and free of external knowledge [19,31]. Nevertheless, in all those approaches, commonsense is seamlessly infused into the model and cannot be disentangled. This makes it hard to study and evaluate commonsense and perception separately, or control their influence. Few methods have modeled commonsense as a standalone module which is late-fused into the prediction of the perception model [13,34]. Yet, we are the first to devise separate perception and commonsense models, and adaptively weigh their importance based on their confidence, before fusing their predictions.

2.2 Commonsense in Scene Graph Generation

Zellers *et al.* [34] were the first to explicitly incorporate commonsense into the process of scene graph generation. They biased predicate classification logits using a pre-computed frequency prior that is a static distribution, given each entity class pair. Although this significantly improved their overall accuracy, the improvement is mainly due to the fact that they favor frequent triplets over others, which is statistically rewarding. Even if their model classifies the relation between a `person` and a `hat` as `holding`, their frequency bias would most likely change that to `wearing`, which is more frequent.

More recently, Chen *et al.* [2] employed a less explicit way to incorporate the frequency prior within the process of entity and predicate classification. They embed the frequencies into the edge weights of their inference graph, and utilize those weights within their message propagation process. This improves the results especially on less frequent predicates, since it less strictly enforces the statistics on the final decision. However, this way commonsense is integrated implicitly into the SGG model and cannot be probed or studied in isolation. We remove the adverse effect of statistical bias while keeping the commonsense model disentangled from perception.

Gu *et al.* [7] exploits ConceptNet [21] rather than dataset statistics, which is a large-scale knowledge graph comprising relational facts about concepts, *e.g.* `dog is-a animal` or `fork is-used-for eating`. Given each detected object, they retrieve ConceptNet facts involving that object class, and employ a recurrent neural net and an attention mechanism to encode those facts into the object features, before classifying objects and predicates. Nevertheless, ConceptNet is not exhaustive, since it is extremely hard to compile all commonsense facts. Our method does not depend on a limited source of external knowledge, and acquires commonsense automatically, via a generalizable neural network.

2.3 Transformers and Graph-Based Neural Networks

Transformers were originally proposed to replace recurrent neural networks for machine translation, by stacking several layers of multi-head attention [23]. Ever since, transformers have been successful in various vision and language tasks [5,16,27]. Particularly, BERT [5] randomly replaces some words from a given sentence with a special MASK token and tries to reconstruct those words. Through this self-supervised game, BERT acquires natural language, and can transfer its language knowledge to perform well in other NLP tasks. We use a similar self-supervised strategy to learn to complete missing pieces of a scene graph. Rather than language, our model acquires the ability to imagine a scene in a structured, semantic way, which is a hallmark of human commonsense.

Transformers treat their input as a set of tokens, and discard any form of structure among them. To preserve the order of tokens in a sentence, BERT augments the initial embedding of each token with a position embedding before feeding into transformers. Scene graphs, on the other hand, have a more complex structure that cannot be embedded in such a trivial way. Recently, Graph-based Neural Networks (GNN) have been successful to encode graph structures into node representations, by applying several layers of neighborhood aggregation. More specifically, each layer of a GNN represents each node by a trainable function that takes the node as well as its neighbors as input. Graph convolutional nets [11], gated graph neural nets [15], and graph attention nets [24] all implement this idea with different computational models for neighborhood aggregation. GNNs have been widely utilized for scene graph generation by incorporating context [29,30,32], but we are the first to exploit GNNs to learn visual commonsense.

We adopt graph attention nets due to their similarity to transformers in using attention. The main difference of graph attention nets to transformers is that instead of representing each node by an attention over all other nodes, they only compute an attention over immediate neighbors. Inspired by that, we use a BERT-like transformer network, but replace half of its attention heads by local attention, simply by enforcing the attention between non-neighbor nodes to zero. Through ablation experiments in Sect. 4, we show the proposed Global-Local Attention Transformers (GLAT) outperforms conventional transformers, as well as widely used graph-based models such as graph convolution nets and graph attention nets.

3 Method

In this section, we first formalize the task, and propose a novel formulation of visual commonsense in connection with visual perception. We then provide an overview of the proposed architecture (Fig. 1), followed by an in-depth description of each proposed module.

We define a scene graph as $G = (\mathcal{N}_e, \mathcal{N}_p, \mathcal{E}_s, \mathcal{E}_o)$, where $\mathcal{N}_e$ is a set of entity nodes, $\mathcal{N}_p$ is a set of predicate nodes, $\mathcal{E}_s$ is a set of edges from each predicate to its subject (which is an entity node), and $\mathcal{E}_o$ is a set of edges from each predicate to its object (that also is an entity node). Each entity node is represented with an entity class $c_e \in \mathcal{C}_e$ and a bounding box $b \in [0,1]^4$, while each predicate node is represented with a predicate class $c_p \in \mathcal{C}_p$ and is connected to exactly one subject and one object. Note that this formulation of scene graph is slightly different from the conventional one [29], as we formulate predicates as nodes rather than edges. This tweak does not cause any limitation since every scene graph can be converted from the conventional representation to our representation. However, this formulation allows multiple predicates between the same pair of entities, and it also enables us to define a unified attention over all nodes no matter entity or predicate.

Given a training dataset with many images $I \in [0,1]^{h \times w \times c}$ paired with ground truth scene graphs G_T, our goal is to train a model that takes a new image and predicts a scene graph that maximizes $p(G|I)$. This is equivalent of maximizing $p(I|G)p(G)$, which breaks the problem into what we call *perception* and *commonsense*. In our proposed intuition, commonsense is the mankind's ability to predict which situations are possible and which are not, or in other words, what makes *sense* and what does not. This can be seen as a prior distribution $p(G)$ over all possible situations in the world, represented as scene graphs. Perception, on the other hand, is the ability to form symbolic belief from raw sensory data, which are respectively G and I in our case. Although the goal of computer vision is to solve the Maximum a Posteriori (MAP) problem (maximizing $p(G|I)$), neural nets often fail to estimate the posterior, unless the prior is explicitly enforced in the model definition [17]. This is while in computer vision, the prior is often overlooked, or inaccurately considered to be a uniform distribution, making MAP equivalent to Maximum Likelihood (ML), *i.e.*, finding G that maximizes $p(I|G)$ [20].

We propose the first method to explicitly approximate the MAP inference by devising an explicit prior model (commonsense). Since posterior inference is intractable, we propose a two-stage framework as an approximation: We first adopt any off-the-shelve SGG model as the *perception model*, which takes an input image and produces a perception-driven scene graph, G_P, that approximately maximizes the likelihood. Then we propose a *commonsense model*, which takes G_P as input, and produces a commonsense-driven scene graph, G_C, to approximately maximize the posterior, i.e.,

$$G_P = f_P(I) \approx \operatorname*{argmax}_G p(I|G), \tag{1}$$

$$G_C = f_C(G_P) \approx \operatorname*{argmax}_G p(I|G)p(G), \tag{2}$$

where f_P and f_C are the perception and commonsense models. The commonsense model can be seen as a graph-based extension of denoising autoencoders [25],

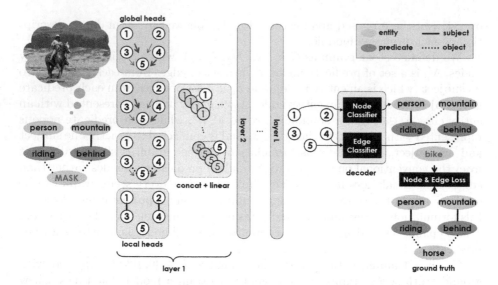

Fig. 2. The proposed Global-Local Attention Transformer (GLAT), and its training framework: We augment transformers with local attention heads to help them encode the structure of scene graphs within node embeddings. The decoder takes the embeddings of a perturbed scene graph and reconstructs the correct scene graph without having access to the image. Note this figure only shows the commonsense block of our overall pipeline shown in Fig. 1.

which evidently can learn the generative distribution of data [1,10], that is $p(G)$ in our case. Accordingly, f_C can take any scene graph as input and produce a more plausible graph by only slightly changing the input. A key design choice here is the fact that f_C does not take the image as input. Otherwise, it would be hard to ensure it is purely learning commonsense and not perception.

Ideally, G_C is the best decision to make, since it maximizes the posterior distribution. However, in practice autoencoders tend to under-represent long-tailed distributions and only capture the modes. This means the commonsense model may fail to predict less common structures, in favor of more statistically rewarding alternatives. To alleviate this problem, we propose a *fusion module* that takes G_P and G_C as input, and outputs a fused scene graph, G_F, which is the final output of our system. This can be seen as a decision-making agent that has to decide how much to trust each model, based on how confident they are.

Figure 1 illustrates an overview of the proposed architecture. In the rest of this section, we elaborate each module in detail.

3.1   Global-Local Attention Transformers

We propose the first graph-based visual commonsense model, which learns a generative distribution over the semantic structure of real-world scenes, through a denoising autoencoder framework. Inspired by BERT [5], which reconstructs masked tokens in a sentence through stacked layers of multi-head attention, we propose Global-Local Attention Transformers (GLAT) that take a graph with masked nodes as input, and reconstructs the missing nodes. Figure 2 illustrates how GLAT works. Given an input scene graph G_P, we represent node i as a one-hot vector $x_i^{(0)}$, that includes entity and predicate categories, as well as a special MASK class. We stack node representations as rows of a matrix $X^{(0)}$ for notation purposes.

GLAT takes $X^{(0)}$ as input and represents each node by encoding the structure and context. To this end, it applies L layers of multi-head attention on the input nodes. Each layer l creates new node representations $X^{(l)}$, by applying a linear layer on the concatenated output of that layer's attention heads. More specifically,

$$X^{(l)} = \underset{h \in \mathcal{H}_l}{\text{concatenate}} \left[h(X^{(l-1)}) \right] \times W_l + b_l, \tag{3}$$

where $\mathcal{H}_l$ is the set of attention heads for layer l, W_l and b_l are trainable fusion weights and bias for that layer, and the concatenation operates along columns. We use two types of attention head, namely global and local. Each node can attend to all other nodes through global attention, while only its neighbors through local attention. We further divide local heads based on the type of edge they use, in order to differentiate the way subjects and objects interact with predicates, and vice versa. Therefore, we can write:

$$\mathcal{H}_l = \mathcal{H}_l^G \cup \mathcal{H}_l^{LS} \cup \mathcal{H}_l^{LO}. \tag{4}$$

All heads within each subset are identical, except they have distinct parameters that are initialized and trained independently. Each global head h^G operates as a typical self-attention would:

$$h^G(X) = \left[q(X)^T k(X) \right] v(X), \tag{5}$$

where q, k, v are query, key, and value heads, each a fully connected network, typically (but not necessarily) with a single linear layer. A local attention is the same, except queries can only interact with keys of their immediate neighbor nodes. For instance in subject heads,

$$h^{LS}(X) = \left[q(X)^T k(X) \odot A_s \right] v(X), \tag{6}$$

where A_s is the adjacency matrix of subject edges, which is 1 between from each predicate to its subject and vice versa, and 0 elsewhere. We similarly define A_o and h^{LO} for object edges.

Once we get contextualized, structure-aware representations $x_i^{(L)}$ for each node i, we devise a simple decoder to generate the output scene graph G_C, using a fully connected network that classifies each node to an entity or predicate class, and another fully connected network that classifies each pair of nodes into an edge type (subject, object or no edge). We train the encoder and decoder end-to-end, by randomly adding noise to annotated scene graphs from Visual Genome, feeding the noisy graph to GLAT, reconstructing nodes and edges, and comparing each with the original scene graph before perturbation. We train the network using two cross-entropy loss terms on the node and edge classifiers. The details of training including the perturbation process are explained in Sect. 4.1.

3.2   Fusing Perception and Commonsense

The perception and commonsense models each predict the output node categories using a classifier that computes a probability distribution over all classes by applying a softmax on its logits. The class with highest probability is chosen and assigned a confidence score equal to its softmax probability. More specifically, node i from G_P has a logit vector L_i^P that has $|\mathcal{C}_e|$ or $|\mathcal{C}_p|$ dimensions depending of whether it is an entity node or predicate node. Similarly node i from G_C has a logit vector L_i^C. Note that these two nodes correspond to the same entity or predicate in the image, since GLAT does not change the order of nodes. Then the confidence of each node can be written as

$$q_i^P = \max_j \frac{\exp(L_i^P[j])}{\sum_k \exp(L_i^P[k])}, \tag{7}$$

and similarly q_i^C is defined given L_i^C.

The fusion module takes each node of G_P and the corresponding node of G_C, and computes a new logit vector for that node, as a weighted average of L_i^P and L_i^C. The weights determine the contribution of each model in the final prediction, and thus have to be proportional to the confidence of each model. Therefore, we compute the fused logits as:

$$L_i^F = \frac{q_i^P L_i^P + q_i^C L_i^C}{q_i^P + q_i^C}. \tag{8}$$

Finally, a softmax is applied on L_i^F to compute the final classification distribution for node i.

4   Experiments

In this section, we describe our experiments on the Visual Genome (VG) dataset in detail. We first evaluate how well our GLAT model learns visual commonsense, by comparing it to other models on the task of masked scene graph reconstruction. Then we provide a statistical analysis of our model prediction to show the kinds of commonsense knowledge it acquires, and distinguish it from bias. Next, we evaluate how effective GLAT and our fusion mechanism are for the downstream task of SGG, when applied on various perception models. We also provide several examples of how the commonsense model corrects the perceived output, and how the fusion model combines the two.

4.1   Implementation Details

We train the perception and commonsense models separately using the ground truth scene graphs G_T from VG [12], particularly the version most widely used for SGG [29], which has 150 entity and 50 predicate classes. We then stack commonsense on top of perception and fine-tune it on VG, this time with actual scene graphs generated by perception, to adapt to the downstream task. The fusion module does not have trainable parameters and is thus only used during inference. We use the 75k VG scene graphs for training all models, and use the other 25k for test. We hold a small portion of the train set for validation. Our GLAT model (and other baselines when applicable) have 6 layers, each with 8 attention heads, and has a 300-dimensional representation for each node. While training GLAT, we randomly mask 30% of the nodes, which is the average number of nodes mistaken by a typical SGG model. We average the classification loss over all nodes and edges classified by the decoder, no matter masked or not. For fine-tuning and inference, we prune the output of the perception model before feeding to GLAT, by keeping the top 100 most confident predicates and all entities connected to those.

4.2   Evaluating Commonsense

Once GLAT is trained, we evaluate it on the same task of reconstructing ground truth VG graphs that are perturbed by randomly masking 30% of their nodes. We evaluate the accuracy of our model in classifying the masked nodes, and report the accuracy (Table 1) separately for entity nodes and predicate nodes, as well as overall. This is a good measure of how well the model has learned commonsense, because it mimics mankind's ability to imagine what would a real-world scene look like, given some context. In Fig. 2, for instance, given the fact that there is a person riding something that is masked, we can immediately tell it is probably a bike, a motorcycle, or a horse. If we also know there is a mountain behind the masked object, and the masked object has a face and legs (not shown in the figure for brevity), then we can more certainly imagine it is a horse. By incorporating the global context of the scene, as well as the local structure of the graph, GLAT is able to effectively imagine the scene and predict the class of the entity or predicate that was masked, at a significantly higher accuracy compared to all baselines.

More specifically, we compare GLAT to: (1) A transformer [5] that is the same as our model, except it only has global heads; (2) A Graph Attention Net [24] which is also the same as our model, but only with local heads; and (3) A Graph Convolutional Network [11], which has only one local head at each layer, and the attention is fixed to be equal for all neighbors of each node. We also compare our method with the frequency prior used by Zellers et al. [34], which can only be applied for masked predicates, and simply predicts the most frequent predicate given its subject and object. As Table 1 shows, our method significantly outperforms all aforementioned baselines, which are a good representative of any existing method to learn semantic graph reconstruction.

Table 1. Ablation study on Visual Genome. All numbers are in percentage, and graph constraint is enforced

Method	Entity	Predicate	Both
Triplet frequency [34]	–	44.4	–
Graph convolutional nets [11] (local-only, fixed attention)	8.7	43.4	19.7
Graph attention nets [24] (local-only)	12.0	45.0	22.3
Transformers [5] (global-only)	14.0	42.3	22.9
Global-local attention transformers (ours)	**22.3**	**60.7**	**34.4**

To provide a better sense of the commonsense knowledge our model learns, we apply GLAT on the entire VG test set, using the procedure detailed below (Sect. 4.3), and collect its prediction statistics in a diverse set of situations. We elaborate using an example, shown in the top left cell of Table 2. Out of all triplets from all scene graphs produced by our model, we collect those triplets that match the certain template of person [X] horse, and show our sorted top 5 predictions in terms of frequency. The 5 predicates most often predicted by our method between a person and a horse are on, riding, near, watching, and behind. These are all possible interactions between a person and a horse, and all follow the affordance properties of both person and horse. Nevertheless, when we get the same statistics from the output of a state-of-the-art scene graph generation model (IMP [29]), we observe that it frequently predicts person wearing horse, which does not follow the affordance of horse. This can be attributed to the high frequency of wearing in VG annotation, which biases the IMP model, while our commonsense model is prone to such bias, and has learned affordances through the self-supervised training framework.

Table 2 provides several more scenarios like this, demonstrating our proficiency in three types of commonsense: object affordance, intuitive physics, and object composition. As an example of physics, we choose the triplet template [X] under bed, and show that our model predicts plausible objects such as pot, shoe, drawer, book, and sneaker. This is while IMP predicts bed under bed, counter under bed, and sink under bed, which are all physically counter-intuitive. More interestingly, one of our frequent predictions, book under bed, is a composition that does not exist in training data, suggesting the knowledge acquired by GLAT is not merely a biased memory of frequent compositions in training data.

The last type of commonsense in our illustration is object composition, *i.e.*, the fact that certain objects are physical parts of other objects. For [X] has ear, we predict head, cat, elephant, zebra, and person, out of which head has ear and person has ear are not within the 10 most frequent triplets in training data that match the template. Yet our model frequently predicts them, demonstrating its unbiased knowledge. Not to mention, 4 out of 5 top predictions made by IMP are nonsensical.

Table 2. Prediction statistics of our method compared to IMP [29] in various situations, showcasing our model's commonsense knowledge, and its robustness to dataset bias. Each row is designated for a certain type of commonsense, and has three examples in three pairs of columns. Each pair of columns show the top 5 most frequent triplets matching a certain template from our model's prediction, compared to IMP. **Black** triplets are commonsensically correct, **red** triplets are wrong, **blue** are commonsensically correct but statistically rare in training data, and **green** are correct but never seen in training data.

	Template 1		Template 2		Template 3	
	IMP + GLAT	IMP	IMP + GLAT	IMP	IMP + GLAT	IMP
Object Affordance	person on horse person riding horse person near horse person watching horse person behind horse	person on horse person riding horse person near horse person wearing horse person behind horse	person has flower person with flower person near flower person watching flower person holding flower	person on flower person has flower person near flower person wearing flower person holding flower	person sitting on chair person sitting on bench person sitting on seat person sitting on stand person sitting on rock	person sitting on chair person sitting on bench person sitting on table person sitting on seat person sitting on person
Intuitive Physics	orange in basket fruit in basket banana in basket food in basket paper in basket	basket in basket man in basket woman in basket orange in basket fruit in basket	airplane above plane airplane above car airplane above bird airplane above beach airplane above vehicle	airplane above table	pot under bed shoe under bed drawer under bed book under bed sneaker under bed	bed under bed drawer under bed counter under bed sink under bed shoe under bed
Object Composition	head has ear cat has ear elephant has ear zebra has ear person has ear	leg has ear head has ear ear has ear tree has ear nose has ear	skier has head skier has hand skier has leg skier has arm skier has face	skier has leg skier has arm skier has pole skier has person skier has tree	leg of person leg of man leg of giraffe leg of zebra leg of elephant	leg of leg leg of man leg of person leg of tree leg of head

4.3   Evaluating Scene Graph Generation

Now that we showed the efficacy of GLAT in learning visual commonsense and correcting perturbed scene graphs, we apply and evaluate it on the downstream task of scene graph generation. We adopt existing SGG models as our perception model, and compare their output G_P, to the ones corrected by our commonsense model G_C, as well as the final output of our system after fusion G_F. We compare those 3 outputs for 3 different choices of perception model, all of which have competitive state-of-the-art performance. More specifically, we use Iterative Message Passing (IMP [29]) as a strong baseline that is not augmented by commonsense. We also use Stacked Neural Motifs (SNM [34]) that late-fuse a frequency prior with their output, and Knowledge-Embedded Routing Networks (KERN [2]) that encode frequency prior within their internal message passing.

To evaluate, we conventionally compute the mean recall of the top 50 (mR@50) and top 100 (mR@100) triplets predicted by each model. Each triplet is considered correct if the subject, predicate, and object are all classified correctly, and the bounding box of the subject and object have more than 50% overlap (intersection over union) with the ground truth. We compute the recall for the triplets of each predicate class separately, and average over classes. The aforementioned metrics are measured in 2 sub-tasks: (1) SGCLS is the main scenario where we classify entities and predicates given annotated bounding boxes. This way the performance is not limited by proposal quality. (2) PREDCLS provides the model with ground truth object labels, which helps evaluation focus on predicate recognition accuracy. Table 3 shows the full comparison of all methods on all metrics. We observe that GLAT improves the performance of IMP which does not have commonsense, but does not significantly change the performance of

Table 3. The mean recall of our method compared to the state of the art on the task of scene graph generation, evaluated on the Visual Genome dataset [29], following the experiment settings of [34]. All baseline numbers were borrowed from [2], and all numbers are in percentage

Method	PREDCLS		SGCLS	
	mR@50	mR@100	mR@50	mR@100
IMP [29]	9.8	10.5	5.8	6.0
IMP + GLAT	11.1	11.9	6.2	6.5
IMP + GLAT + Fusion	**12.1**	**12.9**	**6.6**	**7.0**
SNM [34]	13.3	14.4	7.1	7.5
SNM + GLAT	13.6	14.6	7.3	7.8
SNM + GLAT + Fusion	**14.1**	**15.3**	**7.5**	**7.9**
KERN [2]	17.7	19.2	9.4	10.0
KERN + GLAT	17.6	19.1	9.3	10.0
KERN + GLAT + Fusion	**17.8**	**19.3**	**9.9**	**10.4**

SNM and KERN which already use dataset statistics. However, our full model which uses both the output of the perception model as well as commonsense model consistently improves SGG performance. In the supplementary material, we provide a more detailed analysis by breaking the results down into subgroups based on triplet frequency, and showing our performance boost is consistent in frequent and rare situations.

Finally, we provide several examples in Fig. 3 to illustrate how our commonsense model fixes perception errors in difficult scenarios, and improves the robustness of our model. To save space, we merge the three scene graphs predicted by the perception, commonsense, and fusion models into a single graph, and emphasize any node or edge where these three models disagree. In example (a), the chair is not fully visible, and the visible part does not visually show the action of sitting, thus the perception model incorrectly predicts wearing, which is likely to be also affected by the bias due to the prevalence of wearing annotations in Visual Genome. However, it is trivial for the commonsense model that the affordance of chair is sitting. The fusion module correctly prefers the output of the commonsense model, due to its higher confidence. In (b), the perception model mistakes the head of the bird for a hat, due to the complexity of the lighting and the similarity of foreground and background colors. This might be also affected by the bias of head instances in VG, which are usually human heads, and the fact that hat instances typically co-occur with a head. Nevertheless, our commonsense model has the knowledge of object composition and knows brids typically have heads but not hats. Example (c) is an unusual case of holding, in terms of visual attributes such as arm pose. Hence, the perception model fails to predict holding correctly, while our commonsense model corrects that mistake by incorporating the affordance of bottle. Finally, in (d), the person is perceived under the tower due to the camera angle, but for the

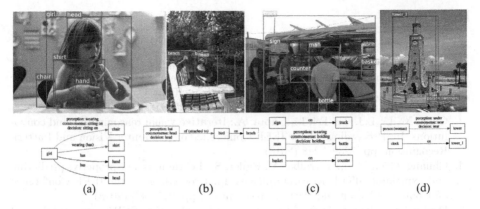

Fig. 3. Example scene graphs generated by the perception, commonsense, and fusion modules, merged into one graph. Entities are shown as rectangular nodes and predicates are shown as directed edges from subject to object. For entities and predicates that are identically classified by the perception and commonsense model, we simply show the predicted label. But in cases where the perception and commonsense models disagree, we show both of their predictions as well as the final output chosen by the fusion module. We show mistakes in red, with the ground truth in parentheses. (Color figure online)

commonsense model that is unlikely due to intuitive physics. Hence, it corrects the mistake and the fusion module accepts that fix. More examples are provided in the supplementary material.

5   Conclusion

We presented the first method to learn visual commonsense automatically from a scene graph corpus. Our method learns structured commonsense patterns, rather than simple co-occurrence statistics, through a novel self-supervised training strategy. Our unique way of augmenting transformers with local attention heads significantly outperforms transformers, as well as widely used graph-based models such as graph convolutional nets. Furthermore, we proposed a novel architecture for scene graph generation, which consists of two individual models, perception and commonsense, which are trained differently, and can complement each other under uncertainty, improving the overall robustness. To this end, we proposed a fusion mechanism to combine the output of those two models based on their confidences, and showed our model correctly determines when to trust its perception and when to fall back on its commonsense. Experiments show the effectiveness of our method for scene graph generation, and encourage future work to apply the same methodology on other computer vision tasks.

Acknowledgement. This work was supported in part by Contract N6600119C4032 (NIWC and DARPA). The views expressed are those of the authors and do not reflect the official policy of the Department of Defense or the U.S. Government.

References

1. Alain, G., Bengio, Y.: What regularized auto-encoders learn from the data-generating distribution. J. Mach. Learn. Res. **15**(1), 3563–3593 (2014)
2. Chen, T., Yu, W., Chen, R., Lin, L.: Knowledge-embedded routing network for scene graph generation. In: Proceedings of the IEEE Conference on Computer Vision and Pattern Recognition, pp. 6163–6171 (2019)
3. Chen, X., Li, L.J., Fei-Fei, L., Gupta, A.: Iterative visual reasoning beyond convolutions. In: Proceedings of the IEEE Conference on Computer Vision and Pattern Recognition, pp. 7239–7248 (2018)
4. Chuang, C.Y., Li, J., Torralba, A., Fidler, S.: Learning to act properly: predicting and explaining affordances from images. In: Proceedings of the IEEE Conference on Computer Vision and Pattern Recognition, pp. 975–983 (2018)
5. Devlin, J., Chang, M.W., Lee, K., Toutanova, K.: BERT: pre-training of deep bidirectional transformers for language understanding. arXiv preprint arXiv:1810.04805 (2018)
6. Groth, O., Fuchs, F.B., Posner, I., Vedaldi, A.: ShapeStacks: learning vision-based physical intuition for generalised object stacking. In: Ferrari, V., Hebert, M., Sminchisescu, C., Weiss, Y. (eds.) ECCV 2018. LNCS, vol. 11205, pp. 724–739. Springer, Cham (2018). https://doi.org/10.1007/978-3-030-01246-5_43
7. Gu, J., Zhao, H., Lin, Z., Li, S., Cai, J., Ling, M.: Scene graph generation with external knowledge and image reconstruction. In: Proceedings of the IEEE Conference on Computer Vision and Pattern Recognition, pp. 1969–1978 (2019)
8. Jiang, C., Xu, H., Liang, X., Lin, L.: Hybrid knowledge routed modules for large-scale object detection. In: Advances in Neural Information Processing Systems, pp. 1552–1563 (2018)
9. Kato, K., Li, Y., Gupta, A.: Compositional learning for human object interaction. In: Ferrari, V., Hebert, M., Sminchisescu, C., Weiss, Y. (eds.) Computer Vision – ECCV 2018. LNCS, vol. 11218, pp. 247–264. Springer, Cham (2018). https://doi.org/10.1007/978-3-030-01264-9_15
10. Kingma, D.P., Welling, M.: Auto-encoding variational bayes. arXiv preprint arXiv:1312.6114 (2013)
11. Kipf, T.N., Welling, M.: Semi-supervised classification with graph convolutional networks. arXiv preprint arXiv:1609.02907 (2016)
12. Krishna, R., et al.: Visual genome: connecting language and vision using crowd-sourced dense image annotations. Int. J. Comput. Vis. **123**(1), 32–73 (2017)
13. Singh, K.K., Divvala, S., Farhadi, A., Lee, Y.J.: DOCK: detecting objects by transferring common-sense knowledge. In: Ferrari, V., Hebert, M., Sminchisescu, C., Weiss, Y. (eds.) ECCV 2018. LNCS, vol. 11217, pp. 506–522. Springer, Cham (2018). https://doi.org/10.1007/978-3-030-01261-8_30
14. Lee, C.W., Fang, W., Yeh, C.K., Frank Wang, Y.C.: Multi-label zero-shot learning with structured knowledge graphs. In: Proceedings of the IEEE Conference on Computer Vision and Pattern Recognition, pp. 1576–1585 (2018)
15. Li, Y., Tarlow, D., Brockschmidt, M., Zemel, R.: Gated graph sequence neural networks. arXiv preprint arXiv:1511.05493 (2015)
16. Lu, J., Batra, D., Parikh, D., Lee, S.: ViLBERT: pretraining task-agnostic visiolinguistic representations for vision-and-language tasks. In: Advances in Neural Information Processing Systems, pp. 13–23 (2019)
17. Malinin, A., Gales, M.: Predictive uncertainty estimation via prior networks. In: Advances in Neural Information Processing Systems, pp. 7047–7058 (2018)

18. Narasimhan, M., Schwing, A.G.: Straight to the facts: learning knowledge base retrieval for factual visual question answering. In: Ferrari, V., Hebert, M., Sminchisescu, C., Weiss, Y. (eds.) ECCV 2018. LNCS, vol. 11212, pp. 460–477. Springer, Cham (2018). https://doi.org/10.1007/978-3-030-01237-3_28

19. Qi, M., Wang, Y., Qin, J., Li, A.: KE-GAN: knowledge embedded generative adversarial networks for semi-supervised scene parsing. In: Proceedings of the IEEE Conference on Computer Vision and Pattern Recognition, pp. 5237–5246 (2019)

20. Romaszko, L., Williams, C.K., Moreno, P., Kohli, P.: Vision-as-inverse-graphics: Obtaining a rich 3D explanation of a scene from a single image. In: Proceedings of the IEEE International Conference on Computer Vision Workshops, pp. 851–859 (2017)

21. Speer, R., Chin, J., Havasi, C.: Conceptnet 5.5: an open multilingual graph of general knowledge. In: Thirty-First AAAI Conference on Artificial Intelligence (2017)

22. Su, Z., Zhu, C., Dong, Y., Cai, D., Chen, Y., Li, J.: Learning visual knowledge memory networks for visual question answering. In: Proceedings of the IEEE Conference on Computer Vision and Pattern Recognition, pp. 7736–7745 (2018)

23. Vaswani, A., et al.: Attention is all you need. In: Advances in Neural Information Processing Systems, pp. 5998–6008 (2017)

24. Veličković, P., Cucurull, G., Casanova, A., Romero, A., Lio, P., Bengio, Y.: Graph attention networks. arXiv preprint arXiv:1710.10903 (2017)

25. Vincent, P., Larochelle, H., Bengio, Y., Manzagol, P.A.: Extracting and composing robust features with denoising autoencoders. In: Proceedings of the 25th International Conference on Machine Learning, pp. 1096–1103 (2008)

26. Wang, T., Huang, J., Zhang, H., Sun, Q.: Visual commonsense R-CNN. In: Proceedings of the IEEE/CVF Conference on Computer Vision and Pattern Recognition, pp. 10760–10770 (2020)

27. Wang, X., Girshick, R., Gupta, A., He, K.: Non-local neural networks. In: Proceedings of the IEEE Conference on Computer Vision and Pattern Recognition, pp. 7794–7803 (2018)

28. Wang, X., Ye, Y., Gupta, A.: Zero-shot recognition via semantic embeddings and knowledge graphs. In: Proceedings of the IEEE Conference on Computer Vision and Pattern Recognition, pp. 6857–6866 (2018)

29. Xu, D., Zhu, Y., Choy, C.B., Fei-Fei, L.: Scene graph generation by iterative message passing. In: Proceedings of the IEEE Conference on Computer Vision and Pattern Recognition, pp. 5410–5419 (2017)

30. Yang, J., Lu, J., Lee, S., Batra, D., Parikh, D.: Graph R-CNN for scene graph generation. In: Ferrari, V., Hebert, M., Sminchisescu, C., Weiss, Y. (eds.) ECCV 2018. LNCS, vol. 11205, pp. 690–706. Springer, Cham (2018). https://doi.org/10.1007/978-3-030-01246-5_41

31. Yu, R., Li, A., Morariu, V.I., Davis, L.S.: Visual relationship detection with internal and external linguistic knowledge distillation. In: Proceedings of the IEEE International Conference on Computer Vision, pp. 1974–1982 (2017)

32. Zareian, A., Karaman, S., Chang, S.F.: Weakly supervised visual semantic parsing. In: Proceedings of the IEEE/CVF Conference on Computer Vision and Pattern Recognition, pp. 3736–3745 (2020)

33. Zellers, R., Bisk, Y., Farhadi, A., Choi, Y.: From recognition to cognition: visual commonsense reasoning. In: Proceedings of the IEEE Conference on Computer Vision and Pattern Recognition, pp. 6720–6731 (2019)

34. Zellers, R., Yatskar, M., Thomson, S., Choi, Y.: Neural motifs: scene graph parsing with global context. In: Proceedings of the IEEE Conference on Computer Vision and Pattern Recognition, pp. 5831–5840 (2018)

MPCC: Matching Priors and Conditionals for Clustering

Nicolás Astorga[1,4]([envelope]) [iD], Pablo Huijse[2,4]([envelope]) [iD], Pavlos Protopapas[3],
and Pablo Estévez[1,4]([envelope]) [iD]

[1] Department of Electrical Engineering, Universidad de Chile, Santiago, Chile
`nicolas.astorga.r@ug.uchile.cl, pestevez@ing.uchile.cl`
[2] Informatics Institute, Faculty of Engineering Sciences,
Universidad Austral de Chile, Valdivia, Chile
`phuijse@inf.uach.cl`
[3] Institute for Applied Computational Science, Harvard University, Cambridge, USA
`pavlos@seas.harvard.edu`
[4] Millennium Institute of Astrophysics, Santiago, Chile

Abstract. Clustering is a fundamental task in unsupervised learning
that depends heavily on the data representation that is used. Deep
generative models have appeared as a promising tool to learn informative low-dimensional data representations. We propose Matching Priors
and Conditionals for Clustering (MPCC), a GAN-based model with an
encoder to infer latent variables and cluster categories from data, and
a flexible decoder to generate samples from a conditional latent space.
With MPCC we demonstrate that a deep generative model can be com-
petitive/superior against discriminative methods in clustering tasks surpassing the state of the art over a diverse set of benchmark datasets.
Our experiments show that adding a learnable prior and augmenting the
number of encoder updates improve the quality of the generated samples,
obtaining an inception score of 9.49 ± 0.15 and improving the Fréchet
inception distance over the state of the art by a 46.9% in CIFAR10.

1 Introduction

Clustering is a fundamental unsupervised learning problem that aims to group
the input data based on a similarity criterion. Traditionally, clustering models are trained on a transformed low-dimensional version of the original data
obtained via feature-engineering or dimensionality reduction *e.g.* PCA. Hence
the performance of clustering relies heavily on the quality of the feature space
representation. In recent years deep generative models have been successful in
learning low-dimensional representations from complex data distributions, and
two particular models have gained wide attention: The Variational Autoencoder
(VAE) [27,47] and the Generative Adversarial Network (GAN) [14].

Electronic supplementary material The online version of this chapter (https://
doi.org/10.1007/978-3-030-58592-1_39) contains supplementary material, which is
available to authorized users.

© Springer Nature Switzerland AG 2020
A. Vedaldi et al. (Eds.): ECCV 2020, LNCS 12368, pp. 658–677, 2020.
https://doi.org/10.1007/978-3-030-58592-1_39

In VAE an encoder and a decoder network pair is trained to map the data to a low-dimensional latent space, and to reconstruct it back from the latent space, respectively. The encoder is used for inference while the decoder is used for generation. The main limitations of the standard VAE are the restrictive assumptions associated with the explicit distributions of the encoder and decoder outputs. For the latter this translates empirically as loss of detail in the generator output. In GAN a generator network that samples from latent space is trained to mimic the underlying data distribution while a discriminator is trained to detect whether the generated samples are true or synthetic (fake). This adversarial training strategy avoids explicit assumptions on the distribution of the generator, allowing GANs to produce the most realistic synthetic outputs up to date [3,25,26]. The weaknesses of the standard GAN are the lack of inference capabilities and the difficulties associated with training (*e.g.* mode collapse).

One would like to combine the strengths of these two models, *i.e.* to be able to infer the latent variables directly from data and to have a flexible decoder that learns faithful data distributions. Additionally, we would like to train simultaneously for feature extraction and clustering as this performs better according to [59,61,62]. Extensions of the standard VAE that modify the prior distribution to make it suitable for clustering have been proposed in [8,23,24], although they still suffer from too restrictive generator models. On the other hand, the standard GAN has been extended to infer categories [6,33,52]. Other works have extended GAN to infer the posterior distribution of the latent variables reporting good results in both reconstruction and generation [5,9,12,34,35,53]. These models do inference and have flexible generators but were not designed for clustering.

In this paper we propose a model able to learn good representations for clustering in latent space. The model is called Matching Priors and Conditionals for Clustering (MPCC). This is a GAN-based model with (a) a learnable mixture of distributions as prior for the generator, (b) an encoder to infer the latent variables from the data and (c) a clustering network to infer the cluster membership from the latent variables. Code are available at github.com/jumpynitro/MPCC.

2   Background

MPCC is based on a matching joint distribution optimization framework. Let us denote $q(x)$ as the true distribution and $p(z)$ the prior, where $x \in \mathcal{X}$ is the observed variable and $z \in \mathcal{Z}$ is the latent variable, respectively. $q(x)$ and $p(z)$ stand for the marginalization of the inference model $q(x,z)$ and generative model $p(x,z)$, respectively. If the joint distributions $q(x,z)$ and $p(x,z)$ match then it is guaranteed that all the conditionals and marginals also match. Intuitively this means that we can reach one domain starting from the other, *i.e.*, we have an encoder that allows us to reach the latent variables $p(z) \approx q(z) = \mathbb{E}_{q(x)}[q(z|x)]$ and a generator that approximates the real distribution $q(x) \approx p(x) = \mathbb{E}_{p(z)}[p(x|z)]$. Notice that the latter approximation corresponds to a GAN optimization problem. In the case of vanilla GAN the Jensen-Shannon divergence $D_{JS}(q(x)\|p(x))$ is minimized, but other distances can be used [1,16,41,45].

Although other classifications can be done [65], we recognize that the joint distribution matching problem can be divided in three general categories: i) matching the joints directly, ii) matching conditionals in $\mathcal{Z}$ and marginals in $\mathcal{X}$, and iii) matching conditionals in $\mathcal{X}$ and marginals in $\mathcal{Z}$. The straight forward approach is to minimize the distance between the joint distributions using a fully adversarial optimization such as [9,10,12], which yields competitive results but still shows difficulties in reconstruction tasks likely affecting unsupervised representation learning. According to [34] these issues are related to the lack of an explicit optimization of the conditional distributions.

Recent works [35,49,65] have shown that the VAE [27] loss function (ELBO) is related to matching the inference and generative joint distributions. This can be demonstrated for the Kullback–Leibler (KL) divergence of p from q, which we refer as forward KL, as follows:

$$
\begin{aligned}
& D_{KL}(q(z,x)\|p(z,x)) \\
=\ & \mathbb{E}_{q(x)}[D_{KL}(q(z|x)\|p(z|x))] + D_{KL}(q(x)\|p(x)) \\
=\ & \mathbb{E}_{q(x)}\mathbb{E}_{q(z|x)}[-\log p(x|z)] + \mathbb{E}_{q(x)}[D_{KL}(q(z|x)\|p(z))] + \mathbb{E}_{q(x)}[\log q(x)] \\
=\ & \mathbb{E}_{q(x)}[-\text{ELBO}] + \mathbb{E}_{q(x)}[\log q(x)], \quad\quad\quad\quad\quad\quad\quad\quad (1)
\end{aligned}
$$

hence maximizing the ELBO can be seen as matching the conditionals in latent space $\mathcal{Z}$ and the marginals in data space $\mathcal{X}$ (see the second line in Eq. 1). The proof for the first equivalence in Eq. 1 can be found in the Appendix A.

In order to avoid latent collapse and the parametric assumptions of VAE, AIM [35] proposed the opposite, *i.e.* to match the conditionals in data space and the marginals in latent space. Starting from the KL divergence of q from p, which we refer as reverse KL, they obtained the following:

$$
\begin{aligned}
& D_{KL}(p(z,x)\|q(z,x)) \\
=\ & \mathbb{E}_{p(z)}[D_{KL}(p(x|z)\|q(x|z))] + D_{KL}(p(z)\|q(z)) \\
=\ & \mathbb{E}_{p(z)}\mathbb{E}_{p(x|z)}[-\log q(z|x)] + \mathbb{E}_{p(z)}[D_{KL}(p(x|z)\|q(x))] + \mathbb{E}_{p(z)}[\log p(z)], \quad (2)
\end{aligned}
$$

where $p(z)$ is a fixed parametric distribution hence $\mathbb{E}_{p(z)}[\log p(z)]$ is constant. Therefore [35] achieves the matching of joint distributions by minimizing $D_{KL}(p(x|z)\|q(x))$ to learn the real domain, and maximizing the likelihood of the encoder $\mathbb{E}_{p(x|z)}[\log q(z|x)]$. This allows obtaining an overall better performance than [9,12,34] in terms of reconstruction and generation scores. This method matches the conditional distribution explicitly, uses a flexible generator [14] and avoids latent collapse problems [38].

Lot of research has been done in unsupervised and semi supervised learning using straight forward joint distribution optimization [9,10,12,21], and even more for conditional in latent space decomposition [27,28,39,40]. In this work we explore the representation capabilities of the decomposition proposed in [35]. Our main contributions are:

- A mathematical derivation that allows us to have a varied mixture of distributions in latent space enforcing its clustering capabilities. Based on this derivation we developed a new generative model for clustering called MPCC, trained by matching prior and conditional distributions jointly.
- A comparison with the state-of-the-art showing that MPCC outperforms generative and discriminative models in terms of clustering accuracy and generation quality.
- An ablation study of the most relevant parameters of MPCC and a comparison with the AIM baseline [35] using state of the art architectures [10].

3   Method

3.1   Model Definition

MPCC extends the usual joint distribution of variables $x \in \mathcal{X}$ and $z \in \mathcal{Z}$ incorporating an additional latent variable, $y \in \mathcal{Y}$, which represents a given cluster. We specify the graphical models for *generation* and *inference* as

$$\bullet\, p(x, z, y) = p(y)p(z|y)p(x|z, y),$$
$$\bullet\, q(x, z, y) = q(y|z)q(z|x)q(x),$$

respectively. The only assumption in the graphical model is $q(y|z) = q(y|z, x)$, *i.e.* z contains all the information from x that is necessary to estimate y.

For generation, we seek to match the decoder $p(x|z, y)$ to the real data distribution $q(x)$. The latent variable is defined by the conditional distributions $p(z|y)$ which in general can be any distribution under certain restrictions (Sect. 3.3). The marginal distribution $p(y)$ is defined as multinomial with weight probabilities ϕ. Note that under this graphical model the latent space becomes multimodal defined by a mixture of distributions $p(z) = \sum_y p(y)p(z|y)$.

In the inference procedure the latent variables are obtained by the conditional posterior $q(z|x)$ using the empirical data distribution $q(x)$. The distribution $q(y|z)$ is a posterior approximation of the cluster membership of the data.

We call our model Matching Priors and Conditionals for Clustering (MPCC) and we optimize it by minimizing the reverse Kullback-Leibler divergence of the conditionals and priors between the inference and generative networks as follows:

$$
\begin{aligned}
& D_{KL}\left(p(x, z, y)||q(x, z, y)\right) \\
= \;& \mathbb{E}_{p(z,y)}[D_{KL}(p(x|z, y)||q(x|z, y))] \\
& + \mathbb{E}_{p(y)}[D_{KL}(p(z|y)||q(z|y))] + D_{KL}(p(y)||q(y)).
\end{aligned}
\tag{3}
$$

The proof for Eq. (3) can be found in Appendix A. In the following sections we derive a tractable expression for Eq. (3) and present the MPCC algorithm.

3.2   Loss Function

Because $q(y)$, $q(z|y)$ and $q(x|z,y)$ are impossible to sample from, we derive a closed-form solution for Eq. (3). In particular for any fixed y and z we can decompose $D_{KL}(p(x|z,y)||q(x|z,y))$ as follows:

$$
\begin{aligned}
& D_{KL}(p(x|z,y)||q(x|z,y)) \\
= \; & \mathbb{E}_{p(x|z,y)}\left[\log\frac{p(x|z,y)}{q(x)}\frac{q(z,y)}{q(z,y|x)}\right] \\
= \; & \mathbb{E}_{p(x|z,y)}\left[\log\frac{p(x|z,y)}{q(x)} - \log q(y|z) - \log q(z|x) + \log q(z|y) + \log q(y)\right].
\end{aligned}
$$
$$(4)$$

Adding $\log p(z|y) + \log p(y) - \log q(z|y) - \log q(y)$ to both sides of Eq. (4) and taking the expectation with respect to $p(z,y)$ the Eq. (3) is recovered. After adding these terms and taking the expectation we can collect the resulting right hand side of Eq. (4) as follows:

$$
\begin{aligned}
& \mathbb{E}_{p(z,y)}[D_{KL}(p(x|z,y)||q(x|z,y)) + D_{KL}(p(z|y)||q(z|y)) + D_{KL}(p(y)||q(y))] \\
= \; & \underbrace{\mathbb{E}_{p(y)p(z|y)}[D_{KL}(p(x|z,y)||q(x))]}_{\textbf{Loss I}} + \underbrace{\mathbb{E}_{p(y)p(z|y)p(x|z,y)}[-\log q(z|x) - \log q(y|z)]}_{\textbf{Loss II}} \\
& + \underbrace{\mathbb{E}_{p(z|y)p(y)}[\log p(y) + \log p(z|y)]}_{\textbf{Loss III}},
\end{aligned}
$$
$$(5)$$

where **Loss I** seeks to match the true distribution $q(x)$, **Loss II** is related to the variational approximation of the latent variables and **Loss III** is associated with the distribution of the cluster parameters. The right hand term of Eq. (5) is a loss function, composed of three terms with distributions that we can sample from. In the next section we explain the strategy to optimize each of the terms of the proposed loss function.

MPCC follows the idea that the data space $\mathcal{X}$ is compressed in the latent space $\mathcal{Z}$ and a separation in this space will likely partition the data in the most representatives clusters $p(z|y)$. The separability of these conditional distributions will be enforced by $q(y|z)$ which also backpropagates through the parameters of $p(z|y)$. The connection with the data space is through the decoder $p(x|z,y)$ for generation and the encoder $q(z|x)$ for inference.

3.3   Optimizing MPCC

In what follows we describe the assumptions made in the distributions of the graphical model and how to optimize Eq. 5. For simplicity we assume the conditional $p(z|y)$ to be a Gaussian distribution, but other distributions could be used with the only restriction being that their entropy should have a closed-form or at least a bound (second term in **Loss III**). In our experiments the latent variable $z|y \sim \mathcal{N}(\mu_y, \sigma_y^2)$ is sampled using the reparameterization trick

[27], *i.e.* $z = \mu_y + \sigma_y \odot \epsilon$ where $\epsilon \sim \mathcal{N}(0, I)$ and $\odot$ is the Hadamard product. The parameters μ_y, σ_y^2 are learnable and they are conditioned on y. Under Gaussian conditional distribution the latent space becomes a GMM, as we can observe mathematically $p(z) = \sum_y p(y)p(z|y) = \sum_y p(y)\mathcal{N}(\mu_y, \sigma_y^2)$.

The distribution $p(x|z, y)$ is modeled by a neural network and trained via adversarial learning, *i.e.* it does not require parametric assumptions. The inferential distribution $q(z|x)$ is also modeled by a neural network and its distribution is assumed Gaussian for simplicity. The categorical distribution $q(y|z)$ may also be modeled by a neural network but we propose a simpler approach based on the membership from the latent variable z to the Gaussian components. A diagram of the proposed model considering these assumptions is shown in Fig. 1. We now expand on this for each of the losses in Eq. 5.

Loss I: Instead of minimizing the Kullback-Leibler divergence shown in the first term on the right hand of Eq. (5) we choose to match the conditional decoder $p(x|z, y)$ with the empirical data distribution $q(x)$ using a generative adversarial approach. The GAN loss function can be formulated as [11]

$$\max_D \mathbb{E}_{x \sim q(x)}[f(D(x))] + \mathbb{E}_{\tilde{x} \sim p(x,z,y)}[g(D(\tilde{x}))],$$
$$\min_G \mathbb{E}_{\tilde{x} \sim p(x,z,y)}[h(D(\tilde{x}))], \tag{6}$$

where D and G are the discriminator and generator networks, respectively, and tilde is used to denote sampled variables. For all our experiments we use the hinge loss function [36,55], *i.e.* $f = -\min(0, o - 1)$, $g = \min(0, -o - 1)$ and $h = -o$, being o the output of the discriminator. The parameters and distribution associated with **Loss I** are colored in blue in Fig. 1.

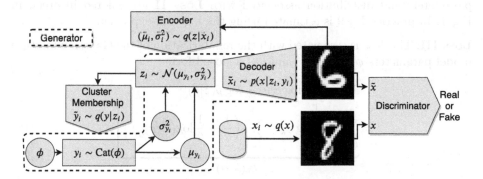

Fig. 1. Diagram of the MPCC model. The blue colored elements are associated with **Loss I** (Eq. 6). The green colored elements are associated with **Loss II** (Eq. 7 and 8). The red colored elements are associated with **Loss III** (Eq. 10). The dashed line corresponds to the generator (GMM plus decoder). (Color figure online)

Loss II: The first term of this loss is estimated through Monte Carlo sampling as

$$\mathbb{E}_{p(y)p(z|y)p(x|z,y)}[-\log q(z|x)]$$

$$= \mathbb{E}_{y_i \sim p(y), z_i \sim p(z|y=y_i), \tilde{x}_i \sim p(x|z=z_i,y=y_i)} \underbrace{\left[\sum_{j=1}^{J} \frac{1}{2} \log(2\pi\tilde{\sigma}_{ij}^2) + \frac{(z_{ij} - \tilde{\mu}_{ij})^2}{2\tilde{\sigma}_{ij}^2} \right]}_{L_q(\tilde{\mu}_i, \tilde{\sigma}_i^2, z_i)}, \quad (7)$$

where J is the dimensionality of the latent variable z. By minimizing Eq. (7) we are maximizing the log-likelihood of the encoder $q(z|x)$ with respect to the Gaussian prior $p(z|y)$. This reconstruction error is estimated by matching the samples $z_i \sim p(z|y = y_i)$ with the Gaussian distribution $(\tilde{\mu}_i, \tilde{\sigma}_i^2) \sim q(z|x = \tilde{x}_i)$, where $\tilde{x}_i$ is the decoded representation of z_i.

The second term of **Loss II** is equivalent to the cross-entropy between the sampled label $y_i \sim p(y)$ and the estimated cluster membership $\tilde{y}_i$

$$L_c(y_i, \tilde{y}_i) = -\sum_{k=1}^{K} y_{ik} \log \tilde{y}_{ik}, \quad (8)$$

where K is the number of clusters and

$$\tilde{y}_{im} = q(y = m|z = z_i) = \frac{\mathcal{N}(z_i|\mu_m, \sigma_m^2)}{\sum_{k=1}^{K} \mathcal{N}(z_i|\mu_k, \sigma_k^2)}, \quad (9)$$

is the membership of z_i to the m-th cluster. The parameters μ_m and σ_m^2 are learnable, and $m \in [1, \dots, K]$ is the index corresponding to each cluster. The parameters and distribution associated with **Loss II** are colored in green in Fig. 1. In practice Eq. 9 is estimated using the log-sum-exp trick.

Loss III: This loss is associated with the regularization of the Gaussian mixture model parameters ϕ, μ and σ^2 and has a closed form

$$\mathbb{E}_{p(y)p(z|y)}[\log p(y) + \log p(z|y)]$$

$$= \underbrace{\sum_{k=1}^{K} \phi_k \left[\log \phi_k - \sum_{j=1}^{J} \left(\frac{1}{2} + \frac{1}{2} \log(2\pi\sigma_{kj}^2) \right) \right]}_{L_p(\phi, \sigma^2)}, \quad (10)$$

where the first term corresponds to the entropy maximization of the mixture weights, *i.e.* in general every Gaussian will not collapse to less than K modes of the data distribution which is a solution with lower entropy. In our experiments we fix $\phi_k = 1/K$, *i.e.* ϕ is not learnable. The second term is a regularization for the variance (entropy) of each Gaussian which avoids the collapse of $p(z|y)$. The parameters associated with **Loss III** are shown in red in Fig. 1.

Loss I scale differs from that of the terms associated with the latent variables. To balance all terms we multiply Eq. (7) by one over the dimensionality of x^1 and the second term of Eq. (10) by one over the dimensionality of the latent variables. During training **Loss III** is weighted by a constant factor λ_p. We explain how this constant is set in Sect. 5.3. The full procedure to train the MPCC model is summarized in Algorithm 1. Note that MPCC is scalable in the number of clusters since Eq. 7 is a Monte Carlo approximation in y and the cost of Eq. 10 is low since J is small in comparison to the data dimensionality.

Algorithm 1. MPCC algorithm

1: K, $J \leftarrow$ Set number of clusters and latent dimensionality
2: η, $\eta_p \leftarrow$ Set learning rates
3: $\theta_g, \theta_d, \theta_e \leftarrow$ Initialize network parameters
4: $\phi, \mu, \sigma^2 \leftarrow$ Initialize GMM parameters
5: $\theta_c \leftarrow [\phi, \mu, \sigma^2]$
6: **repeat**
7:      **for** D_{steps} **do**
8:          $x_1, \ldots, x_n \sim q(x)$ ▷ Draw n samples from empirical distribution
9:          $y_1, \ldots, y_n \sim p(y)$ ▷ Draw n samples from categorical prior
10:         $z_i \sim p(z|y = y_i)$, $i = 1, \ldots, n$ ▷ Draw n samples from Gaussian conditional prior
11:         $\tilde{x}_i \sim p(x|z = z_i, y = y_i)$, $i = 1, \ldots, n$ ▷ Generate samples using generator network
12:         $\theta_d \leftarrow \theta_d + \eta \nabla_{\theta_d} \left[\frac{1}{n} \sum_{j=1}^{n} f(D(x_j)) + \frac{1}{n} \sum_{i=1}^{n} g(D(\tilde{x}_i)) \right]$ ▷ Gradient update on
    discriminator network
13:     **end for**
14:     $y_1, \ldots, y_n \sim p(y)$ ▷ Draw n samples from categorical prior
15:     $z_i \sim p(z|y = y_i)$, $i = 1, \ldots, n$ ▷ Draw n samples from Gaussian conditional prior
16:     $\tilde{x}_i \sim p(x|z = z_i, y = y_i)$, $i = 1, \ldots, n$ ▷ Generate samples using generator network
17:     $(\theta_g, \theta_c) \leftarrow (\theta_g, \theta_c) - \eta \nabla_{(\theta_g, \theta_c)} \frac{1}{n} \sum_{i=1}^{n} h(D(\tilde{x}_i))$ ▷ Gradient update on generator network
18:     **for** E_{steps} **do**
19:         $y_1, \ldots, y_n \sim p(y)$ ▷ Draw n samples from categorical prior
20:         $z_i \sim p(z|y = y_i)$, $i = 1, \ldots, n$ ▷ Draw n samples from Gaussian conditional prior
21:         $\tilde{x}_i \sim p(x|z = z_i, y = y_i)$, $i = 1, \ldots, n$ ▷ Generate samples using generator network
22:         $(\tilde{\mu}_i, \tilde{\sigma}_i^2) \sim q(z|x = \tilde{x}_i)$, $i = 1, \ldots, n$ ▷ Encode $\tilde{x}$ to obtain mean and variance
23:         $\theta_e \leftarrow \theta_e - \eta \nabla_{\theta_e} \frac{1}{n} \sum_{i=1}^{n} L_q(\tilde{\mu}_i, \tilde{\sigma}_i^2, z_i)$ ▷ Gradient update on encoder network
24:         **if** first E_{step} **then**
25:             $\tilde{y}_i \sim q(y|z = z_i)$, $i = 1, \ldots, n$
26:             $\theta_c \leftarrow \theta_c - \eta_p \nabla_{\theta_c} \left[\frac{1}{n} \sum_{i=1}^{n} L_c(y_i, \tilde{y}_i) + \lambda_p \cdot L_p(\phi, \sigma^2)) \right]$ ▷ Gradient update on Prior
    parameters
27:         **end if**
28:     **end for**
29: **until** convergence

---

4   Related Methods

In Sect. 3 we showed that the latent space of MPCC it is reduced to a GMM under Gaussian conditional distribution. Because all the experiments are performed based on this assumption in this section we summarize the literature of generative and autoencoding models that consider GMMs.

---

[1] If x is an image then its dimensionality would be *channels* × *height* × *width*.

The combination of generative models and GMMs is not new. Several methods have applied GMM in autoencoding [56,66] or GAN [17,48] applications without clustering purposes. Other approaches have performed clustering but are not directly comparable since they use mixtures of various generators and discriminators [64] or fixed priors with ad-hoc set parameters [2].

Among the related works on generative models for clustering the closest approaches to our proposal are ClusterGAN [43] and Variational Deep Embedding [23]. ClusterGAN differs from our model in that it sets the dimensions of the latent space as either continuous or categorical while MPCC uses a continuous latent space which is conditioned on the categorical variable y. On the other hand, Variational Deep Embedding (VADE) differs greatly in the training procedure, despite its similar theoretical basis. VADE, as a variational autoencoder model, matches the joint distributions in the forward KL sense $D_{KL}(q(x,z,y)\|p(x,z,y))$ by matching the posteriors and the marginals in data space as demonstrated in Appendix B. MPCC optimizes the reverse KL, *i.e.* matching the priors in latent space and conditionals in data space. Optimizing different KLs yield notably different decompositions and thus training procedures. For the forward KL [23] in addition to the challenges in scaling to larger dimension (Sect. 2) it is more difficult to generalize the latent space to any multimodal distribution, we briefly discuss the reasons for this in Appendix B.

5   Experiments

5.1   Quantitative Comparison

Following [59], the performance of MPCC is measured using the clustering accuracy metric in which each cluster is assigned to the most frequent class in the cluster. Formally this is defined as

$$\text{ACC} = \max_{m\in\mathcal{M}} \frac{\sum_{i=1}^{N} \mathbb{1}\{y_i = m(c_i)\}}{N},\tag{11}$$

where N is the total number of samples, y_i is the ground truth, $c_i = \arg\max_k q(y = k|z = z_i)$ is the predicted cluster and $\mathcal{M}$ is the space of all possible mappings between clusters and labels.

To measure the quality of the samples generated by MPCC we use the inception score (IS) [50] and the Fréchet inception distance (FID) [19].

5.2   Datasets

In order to evaluate MPCC we performed clustering in five benchmark datasets: a handwritten digit dataset (MNIST, [32]), a handwritten character dataset (Omniglot, [31]), two color image dataset (CIFAR-10 and CIFAR-100 [30]) and a fashion products image dataset (Fashion-MNIST, [58]). For CIFAR-100 we consider the 20 superclasses. Omniglot was created using the procedure described in [20]. Because the task is fully unsupervised we concatenate the training and test sets as frequently done in the area [4,20,59]. All datasets have 10 classes except for Omniglot and CIFAR-20 with 100 and 20 respectively. All images were resized to 32×32 and rescaled to $[-1,1]$ in order to use similar architectures. The CIFAR-10 experiments where IS and FID are reported (Tables 1, 2 and 3) were trained using only the training set (50,000 examples) for a fair comparison with the literature. For all clustering experiments we use the same number of cluster as the datasets classes.

5.3   Empirical Details

Our architecture is based on optimization techniques used in the BigGAN [3][2], we found that simpler architectures such as DCGAN [46] were not able to learn complex distributions like CIFAR-10 while optimizing the parameters of the prior. Architecture details are given in Appendix C. We consider parameter sharing between the encoder and discriminator, and we test the importance of this in Sect. 5.5. We set $D_{steps} = 4$ (see Algorithm 1), we found that using smaller values of D_{steps} causes mode collapse problems when training on CIFAR-10 (see Appendix D). A similar effect can be observed when choosing a low number of latent dimensions, therefore we set $J = 128$ in all CIFAR-10 experiments. We made small changes in the architecture and optimization parameters depending on the dataset (see Appendix C).

We observed the same relation between batch size and (IS, FID) reported in [3]. However we found artifacts that hurt accuracy performance when using batch size larger than 50. For simplicity we used this value in all experiments.

We consider a weighting factor λ_p for **Loss III** (Eq. 10). We observed that if $\lambda_p = 1$, the standard deviation of the prior σ would increase monotonically, hindering training. On the other hand if λ_p is too small, σ decreases, collapsing at some point. We found empirically that a value of $\lambda_p = 0.01$ combined with a minimum threshold for σ of 0.5 allow the algorithm to converge to good solutions.

The parameter settings indicated above were fixed for all experiments and didn't show a big effect in accuracy performance. In Sect. 5.4 we explore the parameters that most affect the training. We trained all experiments for 75,000 iterations, except for MNIST and Omniglot which iterate for 125,000. For unconditional and conditional training we kept the model of the last iteration.

5.4   Ablation Study

We found that E_{steps}, the number of encoder updates per epoch, and η_p, the learning rate of the prior parameters, are the most relevant hyperparameters to obtain high accuracy and generation quality. Increasing E_{steps} improves the estimation of $q(z|x)$ since the prior and generator parameters are changing constantly. Rows 1–3 of Table 1 show that the reconstruction error (MSE) decreases

---

[2] https://github.com/ajbrock/BigGAN-PyTorch.

Table 1. E_{steps} correspond to the encoder updates and η_p to the learning rate of the prior parameters. The scale of MSE is in 10^{-3}. The statistics were obtained for at least three runs

E_{steps}	η_p	Acc %	IS	FID	MSE
1	2e−4	41.31 ±5.74	8.82 ±0.07	11.38 ±0.23	1.34 ±0.96
2	2e−4	38.67 ±3.52	9.02 ±0.05	9.66 ±3.98	1.01 ±1.11
4	2e−4	38.27 ±2.46	9.25 ±0.09	7.50 ±0.43	0.331 ±0.09
4	4e−4	52.58 ±5.30	9.44 ±0.06	6.55 ±0.33	0.48 ±0.22
4	6e−4	61.99 ±4.96	9.49 ±0.15	6.59 ±0.45	1.04 ±1.03

Table 2. Comparison of MPCC and AIM-MPCC methods with sharing parameters (S) and without sharing (NS) on the CIFAR-10 dataset. The scale of MSE is in 10^{-3}. The statistics were obtained for five runs

Model	Acc %	IS	FID	MSE
AIM-MPCC (NS)	–	8.24 ±0.07	21.55 ±1.47	1.52 ±0.84
AIM-MPCC (S)	–	9.09 ±0.04	10.42 ±0.36	1.64 ±1.42
MPCC (S)	61.99 ±4.96	9.49 ±0.15	6.59 ±0.45	1.04 ±1.03

with E_{steps}. Generation quality metrics (IS, FID) also improve with larger values of E_{steps} due to the shared parameters between encoder and discriminator.

At initialization the GMM components might not be separated. We observed that the clustering accuracy drops when the generators learns a good approximation of the real distribution before the clusters are separated. To avoid this we use a larger learning rate for the parameters of the GMM prior with respect to the parameters of the generator, encoder and discriminator. Rows 4–6 of Table 1 show that the clustering accuracy increases for larger values of η_p.

5.5   Comparison Between GMM Prior and Normal Prior

Using the best configuration found in the ablation study, we performed a comparison with AIM [35], whose results are shown in Table 2. We can consider AIM as a particular case of MPCC where a standard Normal prior is used instead of the GMM prior. AIM does not perform clustering therefore we compare it with MPCC in terms of reconstruction and generation quality. We use the same architecture and parameter settings of MPCC for AIM and we denote this model as AIM-MPCC. To extend our analysis further, Table 2 includes the results of using parameter sharing between the encoder and the discriminator (Appendix C), an idea that was considered but not fully explored in [35].

Note that AIM-MPCC (NS) is considered a baseline because the prior is Gaussian and the encoder doesn't share parameters with the discriminator thus the existence of the encoder doesn't affect the generation quality. In Table 2 we can observe the relevance of parameter sharing, with this configuration (E_{steps} = 4) the baseline improves by 0.85 (IS) and 11.13 (FID) points. Adding the GMM in the prior improves an additional 0.4 (IS) and 3.93 (FID) points. In total when using the GMM Prior and parameter sharing with additional encoder updates we improve the baseline from 21.55 to 6.59 (69.4% improvement) in terms of FID score and 1.25 points (15.2% improvement) in terms of IS. It is important to notice that these techniques are general and easily applied to any GAN scheme.

5.6   Generation Quality of MPCC

Using the configuration of row five from Table 1 we compare MPCC with nine state of the art methods, surpassing them in terms of IS and FID scores in both the unsupervised and supervised setting, as shown in Table 3. The unconditional generation is the most significant with an improvement of 46.9% (FID) over state-of-the-art (SOTA), AutoGAN [13]. Most notably its performance is better than the current best conditional method (BigGAN).

Table 3. Inception and FID scores for CIFAR-10, in unconditional and conditional training. Higher IS is better. Lower FID is better. [†]: Average of 10 runs. [‡]: Best of many runs. [††]: Average of 5 runs. Results without symbols are not specified

Model	IS	FID	Model	IS	FID
DCGAN [46]	6.64 ± 0.14	–	WGAN-GP [16]	8.42 ± 0.10	–
SN-GAN[†] [42]	8.22 ± 0.05	21.7 ± 2.1	SN-GAN [†] [42]	8.60 ± 0.08	17.5
AutoGAN [13]	8.55 ± 0.10	12.42	Splitting GAN [‡] [15]	8.87 ± 0.09	–
PG-GAN [‡] [25]	8.80 ± 0.05	–	CA-GAN [†] [44]	9.17 ± 0.13	–
NCSN [51]	8.91	25.32	BigGAN [3]	9.22	–
MPCC[††]	9.49 ± 0.15	6.59 ± 0.45	**MPCC**[††]	9.55 ± 0.08	5.69 ± 0.17

(a) Unconditional (unsupervised) generation	(b) Conditional (supervised) generation

5.7   Clustering Experiments

Table 4 shows the clustering results for the selected benchmarks. We observe that in all the available benchmarks MPCC outperform the related methods, VADE [23] and ClusterGAN [43]. In more complex datasets such as CIFAR10, MPCC notably surpass discriminative based models (*e.g.* [20,22]) which are the most competitive methods in the current literature. For benchmarks with more classes the margin is even larger obtaining improvements over the SOTA of $\sim$42% and $\sim$9.7% points in Omniglot and CIFAR-20 respectively, demonstrating empirically the scalability of MPCC when using a high number of clusters.

It can be observed that for all datasets our proposed method either achieves or surpasses the SOTA in terms of clustering. Figures 2 and 3 show examples of generated and reconstructed images, respectively, using the MPCC model with the highest accuracy in the MNIST and CIFAR-10 datasets.

Table 4. Clustering accuracies for several methods and datasets. All the results of CIFAR-20 dataset were extracted from [22], the results of IMSAT and DEC from [20], the results of InfoGAN and ClusterGAN from [43] and the remaining from their respective papers. †: average of 5 or more runs. ‡: best of 5 runs. §: best of 10 or more runs. $^\parallel$: best of 3 runs. Results without symbols are not specified

Methods	Datasets				
	MNIST	Onmiglot	FMNIST	CIFAR-10	CIFAR-20
DEC [59]	$84.3^§$	$5.3 \pm 0.3^\dagger$	–	$46.9 \pm 0.9^\dagger$	18.5
VADE [23]	$94.46^§$	–	–	–	–
InfoGAN [5]	$89.0^\ddagger$	–	$61.0^\ddagger$	–	–
ClusterGAN [43]	$95.0^\ddagger$	–	$63.0^\ddagger$	–	–
DAC [4]	$97.75^\parallel$	–	–	$52.18^\parallel$	23.8
IMSAT (VAT) [20]	$98.4 \pm 0.4^\dagger$	$24.0 \pm 0.9^\dagger$	–	$45.6 \pm 0.8^\dagger$	–
ADC [18]	$98.7 \pm 0.6^\dagger$	–	–	$29.3 \pm 1.5^\dagger$	16.0
SCAE [29]	$98.5 \pm 0.10^\dagger$	–	–	$33.48 \pm 0.3^\dagger$	–
IIC [22]	$98.4 \pm 0.65^\dagger$	–	–	$57.6 \pm 0.3^\dagger$	$25.5 \pm 0.46^\dagger$
MPCC (Five runs)	98.48 ± 0.52	65.87 ± 1.46	62.56 ± 4.16	64.25 ± 5.31	35.21 ± 1.69
MPCC (Best three runs)	98.76 ± 0.03	66.95 ± 0.62	64.99 ± 2.22	67.73 ± 2.50	36.51 ± 0.71

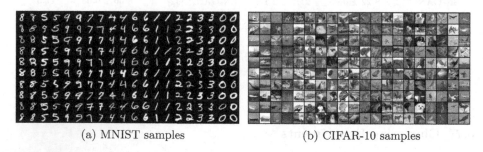

(a) MNIST samples   (b) CIFAR-10 samples

Fig. 2. Generated images for a) MNIST and b) CIFAR-10 datasets, respectively. Every two columns we set a different value for the categorical latent variable y. *i.e.* the samples shown correspond to a different conditional latent space $z \sim p(z|y)$.

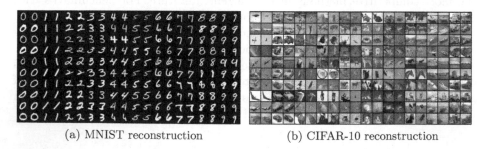

(a) MNIST reconstruction   (b) CIFAR-10 reconstruction

Fig. 3. Reconstructions for a) MNIST, and b) CIFAR-10 datasets, respectively. Odd columns represent real data and even columns correspond to their reconstructions.

6   Discussion

Our results show that MPCC achieves a superior performance with respect to the SOTA on both clustering and generation quality. We note that the current SOTA on unsupervised and semisupervised learning relies on consistency training [60] and/or data augmentation [22], $i.e.$ techniques that are complementary to MPCC and could be used to further improve our results.

To the best of our knowledge MPCC is the first deep generative clustering model capable of dealing with more complex distributions such as CIFAR-10/20 and the first to report clustering accuracy on these datasets. Additionally we empirically prove the scalability of MPCC showing significant improvements in datasets with a larger number of classes, 20 in case of CIFAR-20 and 100 in case of Omniglot, such scalability has not been proven for the current literature on generative models [23,43,52,63].

Our experiments show that MPCC's key innovations: GMM prior, loss function and optimization scheme ($e.g.$ extra encoder updates with parameter sharing) are not only relevant to achieve a good clustering accuracy but also allows us to obtain unprecedented results in terms of generation quality (Table 3). Which translates in improvements of 69.4% over the baseline (Table 2) and 46.9% over the SOTA (Table 3) in terms of FID score. We think that the exceptional generation capabilities of MPCC are related to the support that each cluster covers of the real domain. Since each cluster learns a subset of the real distribution the interpolation between two points within a cluster is smoother compared to the case where no latent separation exists. The latter is explained by the learnable shared features which exploit the similarities existing in a cluster and are not present in a fixed global prior ($i.e.$ ALI, AIM).

The high generation quality can be appreciated in Fig. 2 (more samples in Appendix E), where many clusters sample consistently different classes. However we can still see some classes mixed in some clusters, for example in columns 7–8 with cats and dogs. MPCC also presents a competitive performance in terms of conditional distribution matching (Fig. 3). The errors observed in reconstruction are semantic and similar to those observed in [10].

MPCC opens the possibility of future research in many relevant topics which are out of the scope of this paper. Based on our experiments the most important extensions are: 1) Experiment with other conditional distributions $p(z|y)$, $e.g.$ other exponential-family distributions or other flexible distributions by bounding their entropy (Sect. 3). This can be suitable for more expressive priors as it's shown in recent work [57]. 2) Experiment with imbalanced distribution of classes by changing ϕ accordingly, we consider this to be a relevant problem in the unsupervised setting which only a few works have addressed [54]. 3) Experiment with higher resolution datasets such as ImageNet [7] or CelebA [37]. Current works on clustering have not focus their attention to higher-resolution due to its complexity, MPCC is a promising approach to tackle this task from a semantic perspective [10].

7   Conclusions

We developed a new clustering algorithm called MPCC, derived from a joint distribution matching perspective with a latent space modeled by a mixture distribution. As a deep generative model this algorithm is suitable for interpretable representations, having both an inference and a generative network. The inference network allows us to infer the cluster membership and latent variables from data, while the generator performs sampling conditioned on the cluster category.

An important contribution of this work lays in the solid mathematical and optimization framework on which MPCC is based. This framework is general and we recognize several opportunities to further enhance our model. The results obtained with MPCC improve over the state of the art in clustering methods. Most notably, our model is able to generate samples with an unprecedented high quality, surpassing the state of the art performances in both conditional and unconditional training in the CIFAR-10 dataset.

Acknowledgement. Pablo Huijse and Pablo Estevez acknowledge financial support from the Chilean National Agency for Research and Development (ANID) through grants FONDECYT 1170305 and 1171678, respectively. Also, the authors acknowledge support from the ANID's Millennium Science Initiative through grant IC12009, awarded to the Millennium Institute of Astrophysics, MAS. We acknowledge the Innov-ING 2030 project of the Faculty of Engineering Sciences UACh for the support and funding provided for the development of this work. We thank Pavlos Protopapas and the IACS for hosting Nicolás Astorga through the development of this research.

References

1. Arjovsky, M., Chintala, S., Bottou, L.: Wasserstein generative adversarial networks. In: Precup, D., Teh, Y.W. (eds.) Proceedings of the 34th International Conference on Machine Learning. Proceedings of Machine Learning Research, vol. 70, pp. 214–223. PMLR, International Convention Centre, Sydney, Australia, 06–11 August 2017. http://proceedings.mlr.press/v70/arjovsky17a.html
2. Ben-Yosef, M., Weinshall, D.: Gaussian mixture generative adversarial networks for diverse datasets, and the unsupervised clustering of images. CoRR abs/1808.10356 (2018). http://arxiv.org/abs/1808.10356
3. Brock, A., Donahue, J., Simonyan, K.: Large scale GAN training for high fidelity natural image synthesis. In: International Conference on Learning Representations (2019). https://openreview.net/forum?id=B1xsqj09Fm
4. Chang, J., Wang, L., Meng, G., Xiang, S., Pan, C.: Deep adaptive image clustering. In: 2017 IEEE International Conference on Computer Vision (ICCV), pp. 5880–5888, October 2017. https://doi.org/10.1109/ICCV.2017.626
5. Chen, X., Duan, Y., Houthooft, R., Schulman, J., Sutskever, I., Abbeel, P.: Info-GAN: interpretable representation learning by information maximizing generative adversarial nets. In: Advances in Neural Information Processing Systems, pp. 2172–2180 (2016)

6. Dai, Z., Yang, Z., Yang, F., Cohen, W.W., Salakhutdinov, R.R.: Good semi-supervised learning that requires a bad GAN. In: Guyon, I., et al. (eds.) Advances in Neural Information Processing Systems 30, pp. 6510–6520. Curran Associates, Inc. (2017). http://papers.nips.cc/paper/7229-good-semi-supervised-learning-that-requires-a-bad-gan.pdf

7. Deng, J., Dong, W., Socher, R., Li, L.J., Li, K., Fei-Fei, L.: ImageNet: a large-scale hierarchical image database. In: CVPR09 (2009)

8. Dilokthanakul, N., Mediano, P.A., Garnelo, M., Lee, M.C., Salimbeni, H., Arulkumaran, K., Shanahan, M.: Deep unsupervised clustering with Gaussian mixture variational autoencoders. arXiv preprint arXiv:1611.02648 (2016)

9. Donahue, J., Krähenbühl, P., Darrell, T.: Adversarial feature learning. In: 5th International Conference on Learning Representations, ICLR 2017, Toulon, France, 24–26 April 2017, Conference Track Proceedings (2017). https://openreview.net/forum?id=BJtNZAFgg

10. Donahue, J., Simonyan, K.: Large scale adversarial representation learning. ArXiv abs/1907.02544 (2019)

11. Dong, H.W., Yang, Y.H.: Towards a deeper understanding of adversarial losses. ArXiv abs/1901.08753 (2019)

12. Dumoulin, V., et al.: Adversarially learned inference. In: 5th International Conference on Learning Representations, ICLR 2017, Toulon, France, April 24–26, 2017, Conference Track Proceedings (2017). https://openreview.net/forum?id=B1ElR4cgg

13. Gong, X., Chang, S., Jiang, Y., Wang, Z.: AutoGAN: neural architecture search for generative adversarial networks. In: The IEEE International Conference on Computer Vision (ICCV), October 2019

14. Goodfellow, I., et al.: Generative adversarial nets. In: Advances in Neural Information Processing Systems, pp. 2672–2680 (2014)

15. Grinblat, G.L., Uzal, L.C., Granitto, P.M.: Class-splitting generative adversarial networks (2017)

16. Gulrajani, I., Ahmed, F., Arjovsky, M., Dumoulin, V., Courville, A.: Improved training of Wasserstein GANs. In: Proceedings of the 31st International Conference on Neural Information Processing Systems, NIPS 2017, pp. 5769–5779., Curran Associates Inc., USA (2017). http://dl.acm.org/citation.cfm?id=3295222.3295327

17. Gurumurthy, S., Kiran Sarvadevabhatla, R., Venkatesh Babu, R.: DeLiGAN: generative adversarial networks for diverse and limited data. In: The IEEE Conference on Computer Vision and Pattern Recognition (CVPR), July 2017

18. Haeusser, P., Plapp, J., Golkov, V., Aljalbout, E., Cremers, D.: Associative deep clustering: training a classification network with no labels. In: Brox, T., Bruhn, A., Fritz, M. (eds.) GCPR 2018. LNCS, vol. 11269, pp. 18–32. Springer, Cham (2019). https://doi.org/10.1007/978-3-030-12939-2_2

19. Heusel, M., Ramsauer, H., Unterthiner, T., Nessler, B., Hochreiter, S.: GANs trained by a two time-scale update rule converge to a local Nash equilibrium. In: Guyon, I., rt al. (eds.) Advances in Neural Information Processing Systems 30, pp. 6626–6637. Curran Associates, Inc. (2017)

20. Hu, W., Miyato, T., Tokui, S., Matsumoto, E., Sugiyama, M.: Learning discrete representations via information maximizing self-augmented training. In: Proceedings of the 34th International Conference on Machine Learning - Volume 70, ICML 2017, pp. 1558–1567. JMLR.org (2017). http://dl.acm.org/citation.cfm?id=3305381.3305542

21. Jesson, A., Low-Kam, C., Nair, T., Soudan, F., Chandelier, F., Chapados, N.: Adversarially learned mixture model. CoRR abs/1807.05344 (2018). http://arxiv.org/abs/1807.05344
22. Ji, X., Henriques, J.F., Vedaldi, A.: Invariant information clustering for unsupervised image classification and segmentation. In: Proceedings of the IEEE International Conference on Computer Vision, pp. 9865–9874 (2019)
23. Jiang, Z., Zheng, Y., Tan, H., Tang, B., Zhou, H.: Variational deep embedding: an unsupervised and generative approach to clustering. In: Proceedings of the 26th International Joint Conference on Artificial Intelligence, pp. 1965–1972. AAAI Press (2017)
24. Johnson, M.J., Duvenaud, D.K., Wiltschko, A., Adams, R.P., Datta, S.R.: Composing graphical models with neural networks for structured representations and fast inference. In: Lee, D.D., Sugiyama, M., Luxburg, U.V., Guyon, I., Garnett, R. (eds.) Advances in Neural Information Processing Systems 29, pp. 2946–2954. Curran Associates, Inc. (2016)
25. Karras, T., Aila, T., Laine, S., Lehtinen, J.: Progressive growing of GANs for improved quality, stability, and variation. In: International Conference on Learning Representations (2018). https://openreview.net/forum?id=Hk99zCeAb
26. Karras, T., Laine, S., Aila, T.: A style-based generator architecture for generative adversarial networks. CoRR abs/1812.04948 (2018). http://arxiv.org/abs/1812.04948
27. Kingma, D.P., Welling, M.: Auto-encoding variational bayes. In: 2nd International Conference on Learning Representations, ICLR 2014, Banff, AB, Canada, 14–16 April 2014, Conference Track Proceedings (2014). http://arxiv.org/abs/1312.6114
28. Kingma, D.P., Mohamed, S., Jimenez Rezende, D., Welling, M.: Semi-supervised learning with deep generative models. In: Ghahramani, Z., Welling, M., Cortes, C., Lawrence, N.D., Weinberger, K.Q. (eds.) Advances in Neural Information Processing Systems 27, pp. 3581–3589. Curran Associates, Inc. (2014). http://papers.nips.cc/paper/5352-semi-supervised-learning-with-deep-generative-models.pdf
29. Kosiorek, A., Sabour, S., Teh, Y.W., Hinton, G.E.: Stacked capsule autoencoders. In: Wallach, H., Larochelle, H., Beygelzimer, A., d'Alché-Buc, F., Fox, E., Garnett, R. (eds.) Advances in Neural Information Processing Systems 32, pp. 15512–15522. Curran Associates, Inc. (2019). http://papers.nips.cc/paper/9684-stacked-capsule-autoencoders.pdf
30. Krizhevsky, A.: Learning multiple layers of features from tiny images. Technical report, Citeseer (2009)
31. Lake, B.M., Salakhutdinov, R., Tenenbaum, J.B.: Human-level concept learning through probabilistic program induction. Science **350**(6266), 1332–1338 (2015)
32. LeCun, Y., Bottou, L., Bengio, Y., Haffner, P., et al.: Gradient-based learning applied to document recognition. Proc. IEEE **86**(11), 2278–2324 (1998)
33. LI, C., Xu, T., Zhu, J., Zhang, B.: Triple generative adversarial nets. In: Guyon, I., et al. (eds.) Advances in Neural Information Processing Systems 30, pp. 4088–4098. Curran Associates, Inc. (2017). http://papers.nips.cc/paper/6997-triple-generative-adversarial-nets.pdf
34. Li, C., Liu, H., Chen, C., Pu, Y., Chen, L., Henao, R., Carin, L.: Alice: towards understanding adversarial learning for joint distribution matching. In: Advances in Neural Information Processing Systems, pp. 5495–5503 (2017)
35. Li, H., Wang, Y., Chen, C., Gao, J.: AIM: adversarial inference by matching priors and conditionals (2019). https://openreview.net/forum?id=rJx_b3RqY7
36. Lim, J.H., Ye, J.C.: Geometric GAN (2017)

37. Liu, Z., Luo, P., Wang, X., Tang, X.: Deep learning face attributes in the wild. In: Proceedings of International Conference on Computer Vision (ICCV), December 2015
38. Lucas, J., Tucker, G., Grosse, R.B., Norouzi, M.: Understanding posterior collapse in generative latent variable models. In: DGS@ICLR (2019)
39. Maaløe, L., Fraccaro, M., Liévin, V., Winther, O.: BIVA: a very deep hierarchy of latent variables for generative modeling (2019)
40. Maaløe, L., Sønderby, C.K., Sønderby, S.K., Winther, O.: Auxiliary deep generative models. In: Balcan, M.F., Weinberger, K.Q. (eds.) Proceedings of The 33rd International Conference on Machine Learning. Proceedings of Machine Learning Research, vol. 48, New York, New York, USA, 20–22 June 2016, pp. 1445–1453. PMLR (2016). http://proceedings.mlr.press/v48/maaloe16.html
41. Mao, X., Li, Q., Xie, H., Lau, R., Wang, Z., Smolley, S.: Least squares generative adversarial networks. In: Proceedings of the IEEE International Conference on Computer Vision, pp. 2813–2821, October 2017. https://doi.org/10.1109/ICCV. 2017.304
42. Miyato, T., Kataoka, T., Koyama, M., Yoshida, Y.: Spectral normalization for generative adversarial networks. In: International Conference on Learning Representations (2018). https://openreview.net/forum?id=B1QRgziT-
43. Mukherjee, S., Asnani, H., Lin, E., Kannan, S.: Clustergan: latent space clustering in generative adversarial networks. In: The Thirty-Third AAAI Conference on Artificial Intelligence, AAAI 2019, The Thirty-First Innovative Applications of Artificial Intelligence Conference, IAAI 2019, The Ninth AAAI Symposium on Educational Advances in Artificial Intelligence, EAAI 2019, Honolulu, Hawaii, USA, 27 January–1 February 2019, pp. 4610–4617. AAAI Press (2019). https://doi.org/ 10.1609/aaai.v33i01.33014610
44. Ni, Y., Song, D., Zhang, X., Wu, H., Liao, L.: CAGAN: consistent adversarial training enhanced GANs. In: Proceedings of the Twenty-Seventh International Joint Conference on Artificial Intelligence, IJCAI-18, pp. 2588–2594. International Joint Conferences on Artificial Intelligence Organization, July 2018. https://doi. org/10.24963/ijcai.2018/359
45. Nowozin, S., Cseke, B., Tomioka, R.: f-GAN: training generative neural samplers using variational divergence minimization. In: Lee, D.D., Sugiyama, M., Luxburg, U.V., Guyon, I., Garnett, R. (eds.) Advances in Neural Information Processing Systems 29, pp. 271–279. Curran Associates, Inc. (2016)
46. Radford, A., Metz, L., Chintala, S.: Unsupervised representation learning with deep convolutional generative adversarial networks. In: 4th International Conference on Learning Representations, ICLR 2016, San Juan, Puerto Rico, 2–4 May 2016, Conference Track Proceedings (2016). http://arxiv.org/abs/1511.06434
47. Rezende, D.J., Mohamed, S., Wierstra, D.: Stochastic backpropagation and approximate inference in deep generative models. In: Proceedings of the 31th International Conference on Machine Learning, ICML 2014, Beijing, China, 21–26 June 2014, pp. 1278–1286 (2014). http://jmlr.org/proceedings/papers/v32/rezende14. html
48. Richardson, E., et al. (eds.): Advances in Neural Information Processing Systems 31, pp. 5847–5858. Curran Associates, Inc. (2018). http://papers.nips.cc/paper/ 7826-on-gans-and-gmms.pdf
49. Rosca, M., Lakshminarayanan, B., Mohamed, S.: Distribution matching in variational inference. CoRR abs/1802.06847 (2018)

50. Salimans, T., et al.: Improved techniques for training gans. In: Lee, D.D., Sugiyama, M., Luxburg, U.V., Guyon, I., Garnett, R. (eds.) Advances in Neural Information Processing Systems 29, pp. 2234–2242. Curran Associates, Inc. (2016). http://papers.nips.cc/paper/6125-improved-techniques-for-training-gans.pdf
51. Song, Y., Ermon, S.: Generative modeling by estimating gradients of the data distribution (2019)
52. Springenberg, J.T.: Unsupervised and semi-supervised learning with categorical generative adversarial networks. In: 4th International Conference on Learning Representations, ICLR 2016, San Juan, Puerto Rico, 2–4 May 2016, Conference Track Proceedings (2016). http://arxiv.org/abs/1511.06390
53. Srivastava, A., Valkov, L., Russell, C., Gutmann, M.U., Sutton, C.: VEEGAN: reducing mode collapse in GANs using implicit variational learning. In: Advances in Neural Information Processing Systems, pp. 3308–3318 (2017)
54. Tao, Y., Takagi, K., Nakata, K.: RDEC: integrating regularization into deep embedded clustering for imbalanced datasets. In: Zhu, J., Takeuchi, I. (eds.) Proceedings of The 10th Asian Conference on Machine Learning. Proceedings of Machine Learning Research, vol. 95, 14–16 November 2018, pp. 49–64. PMLR. http://proceedings.mlr.press/v95/tao18a.html
55. Tran, D., Ranganath, R., Blei, D.M.: Hierarchical implicit models and likelihood-free variational inference (2017)
56. Wang, L., Schwing, A., Lazebnik, S.: Diverse and accurate image description using a variational auto-encoder with an additive gaussian encoding space. In: Guyon, I., et al. (eds.) Advances in Neural Information Processing Systems 30, pp. 5756–5766. Curran Associates, Inc. (2017)
57. Wu, Y., Donahue, J., Balduzzi, D., Simonyan, K., Lillicrap, T.P.: LOGAN: latent optimisation for generative adversarial networks. CoRR abs/1912.00953 (2019). http://arxiv.org/abs/1912.00953
58. Xiao, H., Rasul, K., Vollgraf, R.: Fashion-MNIST: a novel image dataset for benchmarking machine learning algorithms. arXiv preprint arXiv:1708.07747 (2017)
59. Xie, J., Girshick, R.B., Farhadi, A.: Unsupervised deep embedding for clustering analysis. In: Proceedings of the 33nd International Conference on Machine Learning, ICML 2016, New York City, NY, USA, June 19–24, 2016, pp. 478–487 (2016). http://jmlr.org/proceedings/papers/v48/xieb16.html
60. Xie, Q., Dai, Z., Hovy, E., Luong, M.T., Le, Q.V.: Unsupervised data augmentation for consistency training. arXiv preprint arXiv:1904.12848 (2019)
61. Yang, B., Fu, X., Sidiropoulos, N.D., Hong, M.: Towards k-means-friendly spaces: Simultaneous deep learning and clustering. In: Proceedings of the 34th International Conference on Machine Learning-Volume 70, pp. 3861–3870. JMLR. org (2017)
62. Yang, J., Parikh, D., Batra, D.: Joint unsupervised learning of deep representations and image clusters. In: Proceedings of the IEEE Conference on Computer Vision and Pattern Recognition, pp. 5147–5156 (2016)
63. Yang, L., Cheung, N.M., Li, J., Fang, J.: Deep clustering by Gaussian mixture variational autoencoders with graph embedding. In: The IEEE International Conference on Computer Vision (ICCV), October 2019
64. Yu, Y., Zhou, W.J.: Mixture of GANs for clustering. In: Proceedings of the Twenty-Seventh International Joint Conference on Artificial Intelligence, IJCAI-18, pp. 3047–3053. International Joint Conferences on Artificial Intelligence Organization, July 2018. https://doi.org/10.24963/ijcai.2018/423

65. Zhao, S., Song, J., Ermon, S.: The information autoencoding family: a lagrangian perspective on latent variable generative models. arXiv preprint arXiv:1806.06514 (2018)
66. Zong, B., Song, Q., Min, M.R., Cheng, W., Lumezanu, C., Cho, D., Chen, H.: Deep autoencoding gaussian mixture model for unsupervised anomaly detection. In: International Conference on Learning Representations (2018). https://openreview.net/forum?id=BJJLHbb0-

POINTAR: Efficient Lighting Estimation for Mobile Augmented Reality

Yiqin Zhao$^{(\boxtimes)}$ and Tian Guo$^{(\boxtimes)}$

Worcester Polytechnic Institute, Worcester, USA
{yzhao11,tian}@wpi.edu

Abstract. We propose an efficient lighting estimation pipeline that is suitable to run on modern mobile devices, with comparable resource complexities to state-of-the-art mobile deep learning models. Our pipeline, POINTAR, takes a single RGB-D image captured from the mobile camera and a 2D location in that image, and estimates 2nd degree spherical harmonics coefficients. This estimated spherical harmonics coefficients can be directly utilized by rendering engines for supporting spatially variant indoor lighting, in the context of augmented reality. Our key insight is to formulate the lighting estimation as a point cloud-based learning problem directly from point clouds, which is in part inspired by the Monte Carlo integration leveraged by real-time spherical harmonics lighting. While existing approaches estimate lighting information with complex deep learning pipelines, our method focuses on reducing the computational complexity. Through both quantitative and qualitative experiments, we demonstrate that POINTAR achieves lower lighting estimation errors compared to state-of-the-art methods. Further, our method requires an order of magnitude lower resource, comparable to that of mobile-specific DNNs.

Keywords: Lighting estimation · Deep learning · Mobile AR

1 Introduction

In this paper, we describe the problem of lighting estimation in the context of mobile augmented reality (AR) applications for indoor scene. We focus on recovering scene lighting, from a partial view, to a representation within the widely used image-based lighting model [10]. Accurate lighting estimation positively impacts realistic rendering, making it an important task in real-world mobile AR scenarios, e.g., furniture shopping apps that allow user to place a chosen piece in a physical environment.

In the image-based lighting model, to *obtain* the lighting information at a given position in the physical environment, one would use a 360° panoramic camera that can capture incoming lighting from every direction. However, commodity mobile phones often lack such panoramic camera, making it challenging to directly obtain accurate lighting information and necessitating the task of.

© Springer Nature Switzerland AG 2020
A. Vedaldi et al. (Eds.): ECCV 2020, LNCS 12368, pp. 678–693, 2020.
https://doi.org/10.1007/978-3-030-58592-1_40

Spatially Variant Lighting

Ground Truth

Fig. 1. Rendering examples of PointAR in spatially variant lighting conditions. Row 1 shows the Stanford bunnies that were lit using the spherical harmonics coefficients predicted by PointAR. Row 2 shows the ground truth rendering results.

There are three key challenges when estimating lighting information for mobile AR applications. First, the AR application needs to estimate the lighting at the rendering location, i.e., where the 3D object will be placed, from the camera view captured by the mobile device. Second, as the mobile camera often only has a limited field of view (FoV), i.e., less than 360°, the AR application needs to derive or estimate the lighting information outside the FoV. Lastly, as lighting information is used for rendering, the estimation should be fast enough and ideally to match the frame rate of 3D object rendering.

Recently proposed learning-based lighting estimation approaches [12,13,25] did not consider the aforementioned unique challenges of supporting lighting estimation for Mobile AR. Gardner et al. proposed a simple transformation to tackle the spatial difference between observation and rendering positions [12]. However, the proposed transformation did not use the depth information and therefore can lead to image distortion. Garon et al. improved the spatial lighting estimations with a two-branches neural network that was reported to perform well on a laptop GPU but not on mobile devices [13]. Song et al. further improved the estimation accuracy by decomposing the pipeline into differentiable sub-tasks [25]. However, the overall network is large in size and has high computational complexity, which makes it ill-suited for running on mobile phones.

Our key insight is to break down the lighting estimation problem into two sub-problems: *(i)* geometry-aware view transformation and *(ii)* point-cloud based learning from limited scene. At a high level, geometry-aware view transformation handles the task of applying spatial transformation to a camera view with a mathematical model. In other words, we skip the use of neural networks for considering scene geometry, unlike previous methods that approached the lighting estimation with a monolithic network [12,13]. Stripping down the complexity of lighting estimation is crucial as it makes designing mobile-oriented learning models possible. Our key idea for learning lighting information directly from point clouds, instead of images, is in part inspired by the use of Monte Carlo Integration in the real-time spherical harmonics lighting calculation.

Concretely, we propose a mobile-friendly lighting estimation pipeline PointAR that combines both physical knowledge and neural network. We rethink and

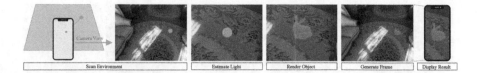

Fig. 2. Lighting estimation workflow in AR applications. Lighting estimation starts from a camera view captured by a user's mobile phone camera. The captured photo, together with a screen coordinate (e.g., provided by the user through touchscreen), is then passed to a lighting estimation algorithm. The estimated lighting information is then used to render 3D objects, which are then combined with the original camera view into a 2D image frame.

redefine the lighting estimation pipeline by leveraging an efficient mathematical model to tackle the view transformation and a compact deep learning model for point cloud-based lighting estimation. Our two-stage lighting estimation for mobile AR has the promise of realistic rendering effects and fast estimation speed.

PointAR takes the input of an RGB-D image and a 2D pixel coordinate (i.e., observation position) and outputs the 2nd degree spherical harmonics (SH) coefficients (i.e., a compact lighting representation of diffuse irradiance map) at a world position. The estimated SH coefficients can be directly used for rendering 3D objects, even under spatially variant lighting conditions. Figure 1 shows the visually satisfying rendering effects of Stanford bunnies at three locations with different lighting conditions. In summary, PointAR circumvents the hardware limitation (i.e., 360° cameras) and enables fast lighting estimation on commodity mobile phones.

We evaluated our method by training on a point cloud dataset generated from large-scale real-world datasets called Matterport3D and the one from Neural Illumination [7,25]. Compared to recently proposed lighting estimation approaches, our method PointAR achieved up to 31.3% better irradiance map $l2$ loss with one order of magnitude smaller and faster model. Further, PointAR produces comparable rendering effects to ground truth and has the promise to run efficiently on commodity mobile devices.

2   Mobile AR Lighting Estimation and Its Challenges

We describe how lighting estimation, i.e., recovering scene lighting based on limited scene information, fits into the mobile augmented workflow from an end-user's perspective. The description here focuses on how lighting estimation component will be used while a human user is interacting with a mobile AR application such as a furniture shopping app. Understanding the role played by lighting estimation module can underscore the challenges and inform the design principles of mobile lighting estimation.

Figure 2 shows how a mobile user with a multi-cameras mobile phone, such as iPhone 11 or Samsung S10, interacts with the mobile AR application.

Such mobile devices are increasingly popular and can capture image in RGB-D format, i.e., with depth information. The user can tap the mobile screen to place the 3D virtual object, such as a couch, on the detected surface. To achieve a realistic rendering of the 3D object, i.e., seamlessly blending to the physical environment, the mobile device leverages lighting estimation methods such as those provided by AR frameworks [1,5]. The estimated lighting information will then be used by the rendering engine to relit the 3D object.

In this work, we target estimating indoor lighting which can change both spatially, e.g., due to user movement, and temporally, e.g., due to additional light sources. To provide good end-user experiences, the mobile AR application often needs to re-estimate lighting information in a rate that matches desired fresh rate measured in frame per second (fps). This calls for fast lighting estimation method that can finish execute in less than 33 ms (assuming 30 fps).

However, lighting estimation for mobile AR comes with three inherent challenges that might benefit from deep learning approaches [11–13,25]. First, obtaining accurate lighting information for mobile devices is challenging as it requires access to the 360° panorama of the rendering position; mobile devices at best can obtain the lighting information at the device location, also referred to as *observation position*, through the use of ambient light sensor or both front- and rear-facing cameras to expand limited field-of-view (FoV). Second, as indoor lighting often varies spatially, directly using lighting information at the observation location to render a 3D object can lead to undesirable visual effect. Third, battery-powered commodity mobile devices, targeted by our work, have limited hardware supports when comparing to specialized devices such as Microsoft HoloLens. This further complicates the network designs and emphasizes the importance of mobile resource efficiency.

3   Problem Formulation

We formulate the *lighting estimation in mobile augmented reality* as a SH coefficients regression problem as $h : h(g(f(C, D, I), r)) = S_r$ where $g : g(P_o, r) = P_r$ and $f: f(C, D, I) = P$. Specifically, we decompose the problem into two stages to achieve the goal of fast lighting estimation on commodity mobile phones.

The first stage starts with an operation $f(C, D, I)$ that generates a point cloud P_o at observation position o. This operation takes three inputs: *(i)* an RGB image, represented as C, *(ii)* the corresponding depth image, represented as D; and *(iii)* the mobile camera intrinsic I. Then $g(P_o, r)$ takes both P_o and the rendering position r, and leverages a linear translation T to generate a point cloud P_r centered at r. In essence, this transformation simulates the process of re-centering the camera from user's current position o to the rendering position r. We describe the geometric-aware transformation design in Sect. 4.1.

For the second stage, we formulate the lighting estimation as a point cloud based learning problem h that takes an incomplete point cloud P_r and outputs 2nd degree SH coefficients S_r. We describe the end-to-end point cloud learning in Sects. 4.2 to Sect. 4.4.

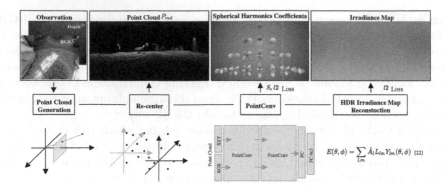

Fig. 3. POINTAR pipeline composition and components. We first transform the camera view into a point cloud centered at the observation position, as described in Sect. 4.1. Then we use a compact neural network described in Sect. 4.2 to estimate SH coefficients at the rendering position. We use $l2$ loss on both estimated SH coefficients and HDR irradiance map reconstructed from spherical harmonics coefficients for evaluation.

4   POINTAR Pipeline Architecture

In this section, we describe our two-stage lighting estimation pipeline POINTAR, as shown in Fig. 3, that is used to model h. The first stage includes a point cloud generation and a geometry transformation modules (Sect. 4.1). The second stage corresponds to a deep learning model for estimating lighting information, represented as SH coefficients (Sect. 4.2). Compared to traditional end-to-end neural network designs [12,13,25], POINTAR is more resource efficient (as illustrated in Sect. 5) and exhibits better interpretability. We describe our dataset generation in Sect. 4.3 and our design rationale in Sect. 4.4.

4.1   Point Cloud Generation

The transformation module takes an RGB-D image, represented as (C, D), and the rendering location r and outputs a point cloud of the rendering location. Our key insights for using point cloud data format are two-folds: *(i)* point cloud can effectively represent the environment geometry and support view transformation; *(ii)* point cloud resembles the Monte Carlo Integration optimization used for real-time spherical harmonics calculation.

Figure 4 compares the warped results between traditional sphere warping [12] and our point cloud based approach. Our approach achieved better warping effect by circumventing the distortion problem associated with the use of RGB images. More importantly, our point cloud transformation only requires a simple linear matrix operation, and therefore has the promise to achieve fast lighting estimation for heterogeneous mobile hardware, during inference.

Figure 5 shows example video frames when transforming an RGB-D image at observation position by recentering and rotating the generated point cloud,

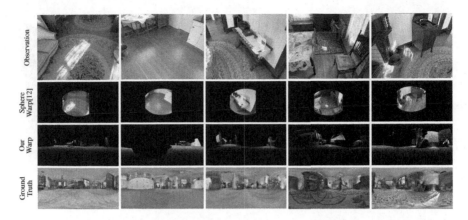

Fig. 4. Comparison of different warping methods. To transform mobile camera view spatially, i.e., from observation to rendering location, we leverage the point cloud which was generated from the RGB-D image. Our approach is not impacted by distortion and therefore is more accurate than spherical warping that only considers RGB image [12].

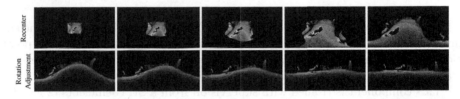

Fig. 5. Example video frames showing point cloud transformation process. Row 1 shows the recenter process that simulates a linear camera movement in 3D space. Row 2 shows the rotation adjustment on the recentered point cloud.

to the rendering position. We detail the process of generating point cloud and its corresponding SH coefficients from a large-scale real-world indoor dataset in Sect. 4.3.

4.2 From Point Cloud to SH Coefficients Estimation

Our second component takes the point cloud P_r at rendering position r and estimates the SH coefficients S_r which is a compact representation of lighting information at location r. Our choice of learning SH coefficients S_r *directly*, instead of other representations such as image style irradiance map (Fig. 6), is largely driven by our design goal, i.e,. efficient rendering in commodity mobile phones. Formulating the illumination as a pixel-wise regression problem [25] often requires complex neural network designs. As mobile augmented reality applications often have a tight time budget, e.g., 33 ms for 30 fps UI update, it can be challenging to support the use of such neural networks directly on mobile devices. Additionally, popular rendering engines such as Unreal Engine support rendering 3D objects directly with SH coefficients.

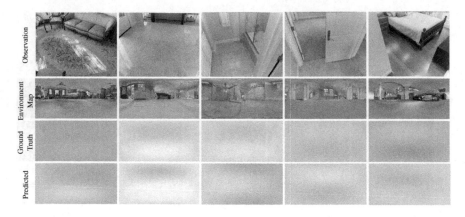

Fig. 6. Irradiance map comparison between ground truth and predicted ones. Row 1 shows the observation captured by mobile camera. Row 2 shows environment map generated from the dataset. Row 3 shows the irradiance map generated from the environment map using spherical harmonics convolution. Row 4 shows the reconstructed irradiance map from SH coefficients predicted by POINTAR.

To train $h : h(P_r, r) = S_r$, we chose the PointConv [28] model, an efficient implementation for building deep convolutional networks directly on 3D point clouds. It uses multi-layer perceptrons to approximate convolutional filters by training on local point coordinates.

This component is trained with supervision from a SH coefficients $l2$ loss L_S as defined in Eq. (1), similar to Garon et al. [13].

$$L_S = \frac{1}{27} \sum_{c=1}^{3} \sum_{l=0}^{2} \sum_{m=-l}^{l} (i_{l,c}^{m*} - i_{l,c}^{m}), \tag{1}$$

where c is the color channel (RGB), l and m are the degree and order of SH coefficients. We chose to target 2nd degree SH coefficients as it is sufficient for our focused application scenarios, i.e., diffuse irradiance learning [24]. We envision that POINTAR will run in tandem with existing techniques such as environment probe in modern mobile AR frameworks to support both specular and diffuse material rendering.

4.3 Dataset Generation of Point Clouds and SH Coefficients

Next, we describe how we generated a large-scale real-world training dataset by leveraging two existing datasets, i.e., Matterport3D [7] and Neural Illumination [25] datasets. Briefly, Matterport3D contains 194,400 RGB-D images forming 10,800 panoramic views for indoor scenes. Each RGB-D panorama contains aligned color and *depth* images (of size 1028×1024) for 18 viewpoints. The Neural Illumination dataset was derived from Matterport3D and contains additional relationship between images at observation and rendering locations.

The first step of our dataset creation process is to transform the observation RGB-D images, i.e., captured at user locations, in Matterport3D dataset into point cloud format. To do so, we leveraged the pinhole camera model [9] and camera intrinsics of each photo in the dataset. For each RGB-D photo, we first recovered a small portion of missing depth data by preprocessing the corresponding depth image with the cross bilateral filter. Then we calculated the 3D point cloud coordinates (x, y, z) as:

$$x = \frac{(u - cx) * z}{fx}, \quad y = \frac{(v - cy) * z}{fy},$$

where z is the depth value in the RGB-D photo, u and v are the photo pixel coordinates, fx and fy are the vertical and horizontal camera focal length, cx and cy are the photo optical center.

With the above transformation, we generated the point cloud P_o for each observation position. Then, we applied a linear translation T to P_o to transform the view at observation position to the rendering position. Specifically, T is determined by using the pixel coordinates of each rendering position on observation image from the Neural Illumination dataset in order to calculate a vector to the locale point. To represent the rendering position for a 3D object, we used a scale factor. This allows us *(i)* to compensate for the position difference between the placement and the ground truth locations; *(ii)* to account for the potentially inaccurate depth information that was introduced by IR sensors. Currently, we used 0.95 based on empirical observation of point cloud projection.

We also used a rotation operation that aligns the recentered point cloud P_o with ground truth environment maps in our dataset. Both the recenter and rotation operations are needed during the inference to achieve spatially-variant lighting estimation and to account for the geometry surface and camera pose.

Finally, for each panorama at the rendering position, we extracted 2nd degree SH coefficients as 27 float numbers to represent the irradiance ground truth. On generated point clouds, we also performed projection and consequently generated respective 2D panorama images which will be used in the ablation study.

4.4   Design and Training Discussions

Learning from Point Cloud. Our choices to learn lighting information *directly* from point cloud and estimating diffuse environment map are driven by mobile-specific challenges and inference time requirement. For example, it is challenging to construct detailed environment map from a limited scene view captured by the mobile phone camera. This in turns can lead to more complex neural networks that might not be suitable to run on mobile devices. Furthermore, neural network generated environment maps may be subject to distortion and unexpected shape. This might lead to reflective textures during rendering and can significantly affect the AR end-user experience.

One intuitive approach is to formulate the learning process as an image learning task by projecting the transformed point cloud into a panorama. This is because image-based lighting models [10] commonly use 360° panoramic view to calculate lighting from every direction of the rendering position. However, learning from projected point cloud can be challenging due to potential image distortion and missing pixel values. We compare this formulation to POINTAR in Sect. 5.

Our idea to learn diffuse lighting from the point cloud representation is in part inspired by how real-time diffuse lighting calculation has been optimized in modern 3D rendering engines. Ramamoorthi et al. proposed the use of spherical harmonics convolution to speedup the irradiance calculation [24]. Compared to diffuse convolution, spherical harmonics convolution is approximately $O(\frac{T}{9})$ times faster where T is the number of texels in the irradiance environment map [24]. However, as spherical harmonics convolution still includes integral operation, performing it directly on large environment maps might hurt real-time performance. Consequently, Monte Carlo Integration was proposed as an optimization to speed up lighting calculation through spherical harmonics convolution by uniformly sampling pixels from the environment map.

In short, Monte Carlo Integration demonstrates the feasibility to calculate the incoming irradiance with enough uniformly sampled points of the environment map. In our problem setting of mobile AR, we have limited samples of the environment map which makes it nature to formulate as a data-driven learning problem with a neural network.

Directly learning from point cloud representation can be difficult due to the sparsity of point cloud data [21]. We chose a recently proposed PointConv [28] architecture as an example point cloud-based neural network for lighting estimation (the second stage of POINTAR). Other point cloud learning approaches might also be used [19,23,30].

Training Dataset. Training a neural network to accurately estimate lighting requires a large amount of real-world indoor 3D scenes that represent complicated indoor geometries and lighting variances. Furthermore, in the context of mobile AR, each training data item needs to be organized as a tuple of *(i)* a RGB-D photo (C, D) captured by mobile camera at the observation position to represent the user's observation; *(ii)* a 360° panorama E at the rendering position for extracting lighting information ground truth; *(iii)* a relation R between (C, D) and E to map the pixel coordinates at the observation location to the ones at the rendering position, as well as the distance between these pixel coordinates.

Existing datasets all fall short to satisfy our learning requirements [6,7,25]. For example, Matterport3D provides a large amount of real-world panorama images which can be used to extract lighting information to serve as ground truth. However, this dataset does not include either observation nor relation data. The Neural Illumination dataset has the format that is closest to our requirements but is still missing information such as point clouds.

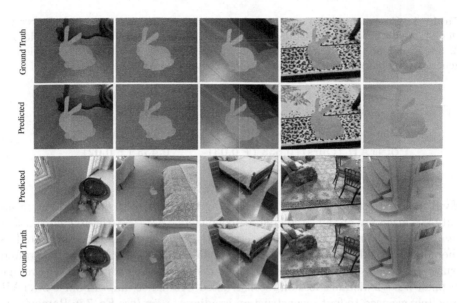

Fig. 7. Rendering example from PoinтAR. Comparison between rendering a 3D object with ground truth irradiance map and irradiance map generated by PoinтAR. Row 1 and 2 show the comparison between a closeup look of the rendered objects. Row 3 and 4 show the comparison between rendered objects in original observations. Note we did not cast shadow to highlight lighting-related rendering effects to avoid misleading visual effects.

In this work, we leveraged both Matterport3D and the Neural Illumination datasets to generate a dataset that consists of point cloud data P and SH coefficients S. Each data entry is a five-item tuple represented as $((C, D), E, R, P, S)$. However, training directly on the point clouds generated from observation images are very large, e.g., 1310720 points per observation image, which complicates model training with can be inefficient as each observation image contains 1310720 points, requiring large amount of GPU memory. In our current implementation, we uniformly down-sampled each point cloud to 1280 points to reduce resource consumption during training and inference. Our uniform down-sampling method is consistent with the one used in the PointConv paper [28]. Similar to what was demonstrated by Wu et al. [28], we also observed that reducing the point cloud size, i.e., the number of points, does not necessarily lead to worse prediction but can reduce GPU memory consumption linearly during training.

5   Evaluation

We trained and tested PoinтAR on our generated dataset (Sect. 4.3) by following the train/test split method described in previous work [25]. Our evaluations include end-to-end comparisons (Fig. 7) to recent works on lighting estimation, ablation studies of our pipeline design, and resource complexities compared to commonly used mobile DNNs.

Table 1. Comparison to state-of-the-art networks. Our approach PointAR (highlighted in green) achieved the lowest loss for both spherical harmonics coefficients $l2$ and irradiance map $l2$. Note Song et al. used traditional diffuse convolution to generate irradiance map and did not use SH coefficients.

Method	SH coefficients $l2$ loss	Irradiance map $l2$ loss
Song et al. [25]	N/A	0.619
Garon et al. [13]	1.10 ($\pm$ 0.1)	0.63 ($\pm$ 0.03)
PointAR (Ours)	**0.633 ($\pm$ 0.03)**	**0.433 ($\pm$ 0.02)**

Evaluation Metrics. We use the following two metrics to quantitatively evaluate the accuracy on our predicted SH coefficients S_r: *(i) SH coefficients $l2$ distance loss* is the average of all SH coefficients $l2$ distance, which is calculated as the numerical difference between the S_r predicted by PointAR and the ground truth. *(ii) Reconstructed irradiance map $l2$ distance loss* is defined as the average of pixel-wise $l2$ distance loss on reconstructed irradiance map. In our PointAR pipeline, we need to first reconstruct an irradiance map (see Eq. 7 in Ramamoorthi et al. [24]) and then compare to the ground truth extracted directly from our dataset, to calculate the $l2$ loss. We compare this metric to the diffuse loss, defined by Song et al. [25], as both representing the reconstruction quality of irradiance map.

Hyperparameter Configurations. We adopted a similar model architecture used by Wu et al. [28] for ModelNet40 classification [29]. We used a set of hyperparameters, i.e., 2 PointConv blocks, each with multilayer perceptron setup of (64, 128) and (128, 256), based on accuracy and memory requirement.

Comparisons to State-of-the-Art. We preformed quantitative comparison experiments with two state-of-the-art end-to-end deep learning model architectures: *(i)* Song et al. [25]; and *(ii)* Garon et al. [13]. Table 1 shows the comparison on two loss metrics.

Song et al. estimates the irradiance by using a neural network pipeline that decomposes the lighting estimation task into four sub-tasks: *(i)* estimate geometry, *(ii)* observation warp, *(iii)* LDR completion, and *(iv)* HDR illumination. As we used the same dataset as Song et al., we obtained the corresponding irradiance map $l2$ loss from the paper. However, since Song et al. used the traditional diffuse convolution to obtain irradiance map, the paper did not include SH coefficients $l2$ loss. Garon et al. estimates the SH coefficients represented lighting and a locale pixel coordinate of a given input image by training a two-branch convolutional neural network with end-to-end supervision. We reproduced the network architecture and trained on our dataset, excluding point clouds and relation E.

Table 2. Comparison to variants. Row 1 and 2 compare the lighting estimation accuracy with two input formats: point cloud and projected point cloud panorama image. Row 3 to 6 compare the lighting estimation accuracy with different downsampled input point clouds.

Method	SH coefficients $l2$ loss	Irradiance map $l2$ loss
Projected Point Cloud + ResNet50	0.781 ($\pm$ 0.015)	0.535 ($\pm$ 0.02)
PointAR (Point Cloud + PointConv)	**0.633 ($\pm$ 0.03)**	**0.433 ($\pm$ 0.02)**
512 points	0.668 ($\pm$ 0.02)	0.479 ($\pm$ 0.02)
768 points	0.660 ($\pm$ 0.02)	0.465 ($\pm$ 0.02)
1024 points	0.658 ($\pm$ 0.03)	0.441 ($\pm$ 0.02)
1280 points (PointAR)	**0.633 ($\pm$ 0.03)**	**0.433 ($\pm$ 0.02)**

Table 1 shows that our PointAR achieved 31.3% and 30% lower irradiance map $l2$ loss compared to Garon et al. and Song et al., respectively. We attribute such improvement to PointAR's ability in handling spatially variant lighting with effective point cloud transformation. Further, the slight improvement (1.7%) on irradiance map $l2$ loss achieved by Song et al. over Garon et al. is likely due to the use of depth and geometry information.

Comparisons to Variants. To understand the impact of neural network architecture on the lighting estimation, we further conducted two experiments: *(i)* learning from projected point cloud and *(ii)* learning from point clouds with different number of points. Table 2 and Table 3 compare the model accuracy and complexity, respectively.

In the first experiment, we study the learning accuracy with two different data representations, i.e., projected point cloud and point cloud, of 3D environment. We compare learning accuracy between learning from point cloud directly with PointConv and learning projected point cloud panorama with ResNet50, which was used in Song et al. [25]. In this experiment, we observed that learning from projected point cloud resulted in lower accuracy (i.e., higher $l2$ losses) despite that ResNet50 requires an order of magnitude more parameters (i.e., memory) and MACs (i.e., computational requirements) than PointConv. The accuracy difference is most likely due to the need to handle image distortion, caused by point cloud projection. Even though there might be other more suitable convolution kernels than the one used in ResNet50 for handing image distortion, the high computational complexity still makes them infeasible to directly run on mobile phones. In summary, our experiment shows that learning lighting estimation from point cloud achieved better performance than traditional image-based learning.

In the second experiment, we evaluate the performance difference on a serial of down-sampled point clouds. From Table 3, we observe that the multiply accumulates (MACs) decreases proportionally to the number of sampled points, while parameters remain the same. This is because the total parameters of

Table 3. Comparison of model complexities. Row 1 to 4 compare resource complexity of ResNet50 (which is used as one component in Song et al. [25]) and mobile-oriented DNNs to that of POINTAR. Row 5 to 8 compare the complexity with different downsampled input point clouds.

Model	Parameters (M)	MACs (M)
ResNet50 [16]	25.56	4120
MobileNet v1 1.0_224 [17]	4.24	569
SqueezeNet 1_0 [18]	1.25	830
POINTAR (Ours)	**1.42**	**790**
512 points	1.42	320
768 points	1.42	470
1024 points	1.42	630
1280 points (POINTAR)	**1.42**	**790**

convolution layers in PointConv block do not change based on input data size, while the number of MACs depends on input size. Furthermore, we observe comparable prediction accuracy using down-sampled point cloud sizes to POINTAR, as shown in Table 2. An additional benefit of downsampled point cloud is the training speed, as less GPU memory is needed for the model and larger batch sizes can be used. In summary, our results suggest the potential benefit for carefully choosing the number of sampled points to trade-off goals such as training time, inference time, and inference accuracy.

Complexity Comparisons. To demonstrate the efficiency of our point cloud-based lighting estimation approach, we compare the resource requirements of POINTAR to state-of-the-art mobile neural network architectures [17,18]. We chose the number of parameters and the computational complexity as proxies to the inference time [26]. Compared to the popular mobile-oriented models MobileNet [17], our POINTAR only needs about 33.5% memory and 1.39X of multiple accumulates (MACs) operations, as shown in Table 3. Further, as MobileNet was shown to produce inference results in less than 10 ms [2], it indicates POINTAR's potential to deliver real-time performance. Similar observations can be made when comparing to another mobile-oriented model SqueezeNet [18].

6   Related Work

Lighting estimation has been a long-standing challenge in both computer vision and computer graphics. A large body of work [11,12,15,25,31] has been proposed to address various challenges and more recently for enabling real-time AR on commodity mobile devices [13].

Learning-Based Approaches. Recent works all formulated the indoor lighting estimation problem by learning directly from a single image, using end-to-end neural networks [12,13,25]. Gardner et al. trained on an image dataset that does not contain depth information and their model only outputs one estimate for one image [12]. Consequently, their model lacks the ability to handle spatially varying lighting information [12]. Similarly, Cheng et al. proposed to learn a single lighting estimate in the form of SH coefficients for an image by leveraging both the front and rear cameras of a mobile device [8]. In contrast, Garon et al. proposed a network with a global and a local branches for estimating spatially-varying lighting information by training on indoor RGB images [13]. However, the model still took about 20 ms to run on a mobile GPU card and might not work for older mobile devices. Song et al. [25] proposed a fully differential network that consists of four components for learning respective subtasks, e.g., 3D geometric structure of the scene. Although it was shown to work well for spatially-varying lighting, this network is too complex to run on most of the mobile devices. In contrast, our PointAR not only can estimate lighting for a given locale (spatially-variance), but can do so quickly with a compact point cloud-based network. Further, our work generated a dataset of which each scene contains a dense set of observations in the form of (point cloud, SH coefficients).

Mobile AR. Companies are providing basic support for developing AR applications for commodity mobile devices in the form of development toolkits [1,5]. However, to achieve seamless AR experience, there are still a number of mobile-specific challenges. For example, it is important to detect and track the positions of physical objects so as to better overlay the virtual 3D objects [3,20,27]. Apicharttrisorn et al. [3] proposed a framework that achieves energy-efficiency object detection by only using DNNs as needed, and leverages lightweight tracking method opportunistically. Due to its impact on visual coherence, lighting estimation for AR has received increasing attention in recent years [4,14,22]. GLEAM proposed a non-deep learning based mobile illumination estimation framework that relies on physical light probes [22]. Our work shares similar performance goals, i.e., being able to run AR tasks on mobile devices, and design philosophy, i.e., by reducing the reliance on complex DNNs. Unlike prior studies, our work also focuses on rethinking and redesigning lighting estimation, an important AR task for realistic rendering, by being mobile-aware from the outset.

7 Conclusion and Future Work

In this work, we described a two-stage lighting estimation pipeline PointAR that consists of an efficient mathematical model and a compact deep learning model. PointAR provides spatially-variant lighting estimation in the form of SH coefficients at any given 2D locations of an indoor scene.

Our current focus is to improve the lighting estimation accuracy for each camera view captured by mobile devices, given the real-time budgets.

However, mobile AR applications need to run on heterogeneous resources, e.g., lack of mobile GPU support, and have different use cases, e.g., 60 fps instead of 30 fps, which might require further performance optimizations. As part of the future work, we will explore the temporal and spatial correlation of image captures, as well as built-in mobile sensors for energy-aware performance optimization.

Acknowledgement. This work was supported in part by NSF Grants #1755659 and #1815619.

References

1. ARCore. https://developers.google.com/ar. Accessed 3 Mar 2020
2. TensorFlow Mobile and IoT Hosted Models. https://www.tensorflow.org/lite/guide/hosted_models
3. Apicharttrisorn, K., Ran, X., Chen, J., Krishnamurthy, S.V., Roy-Chowdhury, A.K.: Frugal following: power thrifty object detection and tracking for mobile augmented reality. In: Proceedings of the 17th Conference on Embedded Networked Sensor Systems, SenSys 2019, pp. 96–109. ACM, New York (2019)
4. Apple: adding realistic reflections to an AR experience. https://developer.apple.com/documentation/arkit/adding_realistic_reflections_to_an_ar_experience. Accessed 10 July 2020
5. Apple Inc: Augmented reality - apple developer. https://developer.apple.com/augmented-reality/. Accessed 3 Mar 2020
6. Armeni, I., Sax, A., Zamir, A.R., Savarese, S.: Joint 2D–3D-semantic data for indoor scene understanding. ArXiv e-prints, February 2017
7. Chang, A., et al.: Matterport3D: learning from RGB-D data in indoor environments. arXiv preprint arXiv:1709.06158 (2017)
8. Cheng, D., Shi, J., Chen, Y., Deng, X., Zhang, X.: Learning scene illumination by pairwise photos from rear and front mobile cameras. Comput. Graph. Forum **37**(7), 213–221 (2018). http://dblp.uni-trier.de/db/journals/cgf/cgf37.html#ChengSCDZ18
9. Chuang, Y.Y.: Camera calibration (2005)
10. Debevec, P.: Image-based lighting. In: ACM SIGGRAPH 2006 Courses, pp. 4-es (2006)
11. Gardner, M.A., Hold-Geoffroy, Y., Sunkavalli, K., Gagne, C., Lalonde, J.F.: Deep parametric indoor lighting estimation. In: The IEEE International Conference on Computer Vision (ICCV), October 2019
12. Gardner, M., et al.: Learning to predict indoor illumination from a single image. ACM Trans. Graph. **36**(6), 14 (2017). https://doi.org/10.1145/3130800.3130891. Article No. 176
13. Garon, M., Sunkavalli, K., Hadap, S., Carr, N., Lalonde, J.: Fast spatially-varying indoor lighting estimation. In: CVPR (2019)
14. Google.   https://developers.google.com/ar/develop/unity/light-estimation/developer-guide-unity
15. Gruber, L., Richter-Trummer, T., Schmalstieg, D.: Real-time photometric registration from arbitrary geometry. In: 2012 IEEE International Symposium on Mixed and Augmented Reality (ISMAR), pp. 119–128. IEEE (2012)
16. He, K., Zhang, X., Ren, S., Sun, J.: Deep residual learning for image recognition. arXiv preprint arXiv:1512.03385 (2015)

17. Howard, A.G., et al.: MobileNets: efficient convolutional neural networks for mobile vision applications (2017). http://arxiv.org/abs/1704.04861, cite arxiv:1704.04861
18. Iandola, F.N., Han, S., Moskewicz, M.W., Ashraf, K., Dally, W.J., Keutzer, K.: SqueezeNet: AlexNet-level accuracy with 50x fewer parameters and <0.5mb model size (2016). http://arxiv.org/abs/1602.07360, cite arxiv:1602.07360Comment. In ICLR Format
19. Li, Y., Bu, R., Sun, M., Wu, W., Di, X., Chen, B.: PointCNN: convolution on x-transformed points. In: Advances in Neural Information Processing Systems, pp. 820–830 (2018)
20. Liu, L., Li, H., Gruteser, M.: Edge assisted real-time object detection for mobile augmented reality. In: The 25th Annual International Conference on Mobile Computing and Networking (MobiCom 2019) (2019)
21. Liu, W., Sun, J., Li, W., Hu, T., Wang, P.: Deep learning on point clouds and its application: a survey. Sensors 19(19), 4188 (2019)
22. Prakash, S., Bahremand, A., Nguyen, L.D., LiKamWa, R.: GLEAM: an illumination estimation framework for real-time photorealistic augmented reality on mobile devices. In: Proceedings of the 17th Annual International Conference on Mobile Systems, Applications, and Services, MobiSys 2019, pp. 142–154. Association for Computing Machinery, New York, June 2019
23. Qi, C.R., Su, H., Mo, K., Guibas, L.J.: PointNet: deep learning on point sets for 3D classification and segmentation. arXiv preprint arXiv:1612.00593 (2016)
24. Ramamoorthi, R., Hanrahan, P.: An efficient representation for irradiance environment maps. In: Proceedings of the 28th Annual Conference on Computer Graphics and Interactive Techniques - SIGGRAPH 2001, pp. 497–500. ACM Press, Not Known (2001). https://doi.org/10.1145/383259.383317. http://portal.acm.org/citation.cfm?doid=383259.383317
25. Song, S., Funkhouser, T.: Neural illumination: lighting prediction for indoor environments. In: CVPR (2019)
26. Sze, V., Chen, Y., Yang, T., Emer, J.S.: Efficient processing of deep neural networks: a tutorial and survey. CoRR (2017). http://arxiv.org/abs/1703.09039
27. Tulloch, A., et al.: Enabling full body AR with mask R-CNN2Go - facebook research, January 2018. https://research.fb.com/blog/2018/01/enabling-full-body-ar-with-mask-r-cnn2go/. Accessed 3 Mar 2020
28. Wu, W., Qi, Z., Fuxin, L.: PointConv: deep convolutional networks on 3D point clouds. In: The IEEE Conference on Computer Vision and Pattern Recognition (CVPR), June 2019
29. Wu, Z., et al.: 3D ShapeNets: a deep representation for volumetric shapes. In: Proceedings of the IEEE Conference on Computer Vision and Pattern Recognition, pp. 1912–1920 (2015)
30. Xu, Y., Fan, T., Xu, M., Zeng, L., Qiao, Y.: SpiderCNN: deep learning on point sets with parameterized convolutional filters. arXiv preprint arXiv:1803.11527 (2018)
31. Zhang, E., Cohen, M.F., Curless, B.: Discovering point lights with intensity distance fields. In: Proceedings of the IEEE Conference on Computer Vision and Pattern Recognition, pp. 6635–6643 (2018)

Discrete Point Flow Networks for Efficient Point Cloud Generation

Roman Klokov[1]([✉])(iD), Edmond Boyer[1](iD), and Jakob Verbeek[2](iD)

[1] Univ. Grenoble Alpes, Inria, CNRS, Grenoble INP, LJK, 38000 Grenoble, France
{roman.klokov,edmond.boyer}@inria.fr
[2] Facebook AI Research, Paris, France
jjverbeek@fb.com

Abstract. Generative models have proven effective at modeling 3D shapes and their statistical variations. In this paper we investigate their application to point clouds, a 3D shape representation widely used in computer vision for which, however, only few generative models have yet been proposed. We introduce a latent variable model that builds on normalizing flows with affine coupling layers to generate 3D point clouds of an arbitrary size given a latent shape representation. To evaluate its benefits for shape modeling we apply this model for generation, autoencoding, and single-view shape reconstruction tasks. We improve over recent GAN-based models in terms of most metrics that assess generation and autoencoding. Compared to recent work based on continuous flows, our model offers a significant speedup in both training and inference times for similar or better performance. For single-view shape reconstruction we also obtain results on par with state-of-the-art voxel, point cloud, and mesh-based methods.

Keywords: Generative modeling · Normalizing flows · 3D shape modeling · Point cloud generation · Single view reconstruction

1 Introduction

Generative shape models are used in numerous computer vision applications where they allow to encode 3D shape variations with respect to different attributes, such as shape classes or shape deformations, as well as to infer shapes from partial observations, for instance from a single or a few images. Central to shape models is the representation chosen for shapes that can be extrinsic, for example the ubiquitous voxels and octrees, or intrinsic as with meshes and point clouds. While extrinsic representations enable relatively straightforward extensions of 2D deep learning techniques to 3D, they suffer from their inherent trade-off between precision and complexity. This is why 3D shapes are often represented using intrinsic models, among which point clouds are a natural and versatile solution, serving as a basis for many 3D capturing methods, including most multi-view stereo and range sensing methods, *e.g.* kinect.

Electronic supplementary material The online version of this chapter (https://doi.org/10.1007/978-3-030-58592-1_41) contains supplementary material, which is available to authorized users.

A. Vedaldi et al. (Eds.): ECCV 2020, LNCS 12368, pp. 694–710, 2020.
https://doi.org/10.1007/978-3-030-58592-1_41

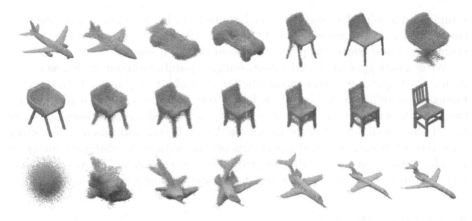

Fig. 1. Top: Point clouds sampled from DPF-Net for the ShapeNet classes *airplane*, *car*, and *chair*. Middle: Latent space interpolation between two point clouds from the test set. Bottom: Deformation of points across the flow steps.

Following the success of CNNs for 2D computer vision problems, many deep learning models have been proposed that can handle 3D data. This includes works on voxel grids [5,10,14,16,28,34,51,52], octrees [45,47], meshes [21,36,48, 50], point clouds [13,22,27,33,41,42,49], and implicit functions [35,38]. While they provide effective tools to build predictive models of 3D shapes, *e.g.* from a single image, we investigate in this paper the less extensively explored and more generic problem of probabilistic generative 3D shape modeling.

Significant advances have been made in generative modeling of natural images using deep neural networks with convolutional architectures. Consequently, they can easily be adapted to generative shape models which are based on extrinsic representations, using regular 3D convolutional layers [5,28,51]. On the other hand, their extensions to intrinsic representations, such as point clouds and meshes, are less obvious and, to the best of our knowledge, so far, only Yang *et al.* [53] have studied generative models from which arbitrary size point clouds can be sampled without any conditioning information.

We explore a hierarchical latent variable model that treats the points as exchangeable variables, which allows us to model and sample point clouds of an arbitrary size. Within this framework, each point cloud is considered as a sample from a shape-specific distribution over the 3D surface of the object, and these distributions are embedded in a latent space. To sample a point cloud, first, a vector is sampled in the latent shape space, and then, any desired number of 3D points can be sampled i.i.d. conditioned on the latent shape representation. Our model shares the high-level structure with PointFlow [53], but differs in the underlying network architectures, reducing the training and sampling time by more than an order of magnitude. In particular, our model builds on discrete normalizing flows with affine coupling layers [12] rather than continuous flows, and FiLM conditioning layers [39] to construct a flexible density on 3D points given the latent shape representation. In Fig. 1 we illustrate diverse point clouds sampled from our class-specific models, interpolation between point clouds in the learned latent shape space, and the sequential process by which the discrete normalizing flow warps the points to obtain the final shape.

We evaluate generative and autoencoding capabilities of our model, as well as its use for single-view shape reconstruction. We obtain similar or better performance compared to GAN-based models in terms of metrics that assess generation and autoencoding. Compared to recent work based on continuous flows, our model offers a significant speedup, for similar or better performance. For single view reconstruction our model performs on par with state-of-the-art methods, yet allows to reconstruct with arbitrarily large point clouds. Moreover, we analyze various design choices regarding data splitting and normalization.

2   Related Work

Generative Models. Deep neural networks have sparked significant progress in generative modeling. The most widely adopted models are variational autoencoders (VAEs) [26,44], generative adversarial networks (GANs) [15,23], and normalizing flows [11,12]. All three approaches share the basic principle of defining a latent variable z with a simple prior, *e.g.* unit Gaussian, and construct a complex conditional $p(x|z)$ on data x by means of deep neural networks. Maximum likelihood training of the resulting marginal $p(x)$ is generally intractable due to the non-linearities. To train the model, VAEs rely on an amortized inference network that produces a variational posterior $q(z|x)$. GANs, on the other hand, use a discriminator network to distinguish training examples and model samples, and use it as a signal to train the generative model. Alternatively to previous approaches, normalizing flow models rely on invertible neural network architectures to avoid the intractability of the marginalization altogether. In this case, the likelihood can be computed exactly by the means of the change of variable formula, and latent variables can be inferred deterministically. A variety of different normalizing flows has recently been proposed, see *e.g.* [2,8,12,17,24,25,43]. See [29,37] for recent comprehensive reviews on normalizing flows.

Affine coupling layers [12] allow for a computation of the inverse in a closed form that is as efficient as the function itself. Within this approach, the activations A^ℓ in layer ℓ are partitioned in two groups, A_1^ℓ and A_2^ℓ. The first group is unchanged, and used to update the other group by scaling and translation, *i.e.* $A_2^{\ell+1} = A_2^\ell \odot s(A_1^\ell) + t(A_1^\ell)$, where $\odot$ denotes element-wise multiplication, and $s(\cdot)$ and $t(\cdot)$ can be arbitrary (non-invertible) neural networks. The inverse is trivially obtained by subtraction and division, since $A_1^{\ell+1} = A_1^\ell$. Many coupling layers with changing variable partitioning can be stacked to construct a complex invertible flow. Training and sampling the model require to compute the flow reverse directions, and since affine coupling layers are equally efficient in both directions, it means that both processes are fast. This is in contrast to some other normalizing flows, such as invertible ResNets [2,8], or planar and radial flows [43], for which the inverse flow does not have a closed form.

Neural ordinary differential equations were recently proposed as a generalization of deep residual networks (ResNets, [19,20]) in the limit of infinite depth [8]. Chen *et al.* [8] demonstrated that Neural ODEs can be used to define normalizing flow, which are referred to as "continuous normalizing flows".

Several conditional flow-based models have recently been proposed for vision tasks. In [31] flows conditioned on an input image are used for image segmentation, inpainting, denoising, and depth refinement. Their model is trained directly via maximum likelihood estimation, as their model does not include a global latent variable. Similarly, C-Flow [40] does not involve a global latent variable, and rather than treating point clouds as sets, it sorts the points to a regular pixel grid, and applies 2D normalizing flows for single image point cloud reconstruction. A conditional VAE model, where flow is used to define a flexible distribution on the latent variable given the conditioning data was introduced by [3]. This is similar in structure to our model for single view reconstruction. Their experiments, however, concern the prediction of point trajectories in 2D for hand-written digits, and traffic participants such as pedestrians and cars. The generative image model of [32] is related to ours, as a VAE model with a flow-based decoder. The application contexts, RGB images *vs.* point clouds, and resulting architectures are, however, quite different.

Point Cloud Generating Networks. Deep learning models for point cloud processing have received significant attention in recent years, see *e.g.* [1,18,27,30, 41,42,46,53,54]. The PointNet architecture of Qi *et al.* [41] was the first to propose a deep network for recognition of point clouds. The points are first processed in an identical and independent manner by an MLP, and global max-pooling is used to aggregate the per-point information. KD-Net [27] and PointNet++ [42] add a notion of spatial proximity to the architecture, replacing global max-pooling with local aggregation. While these models can interpret point clouds, they cannot generate them.

Early point cloud generating networks [1,13] produce point clouds with a fixed number of points n, by using a network with $n \times 3$ outputs. AtlasNet [18] mitigates this limitation by using a set of k square 2D patches, and deforming each of these non-linearly by using k patch-specific MLPs that takes as input 2D patch coordinates as well as a global shape representation. The shape vector is obtained from a point cloud encoder network (for autoencoding), or from a CNN trained for single-view image reconstruction. The point cloud GAN (PC-GAN, [30]) is related, but uses a single generator that takes a global shape vector as input together with (arbitrarily many) samples from a unit Gaussian. Similarly to [13], AtlasNet is a conditional model, that generates point clouds *given* another point cloud or an image. In contrast, PC-GAN includes a second generator that models a distribution on the latent shape space, so it can generate point clouds in an unconditional manner.

The high-level hierarchical latent variable structure of PointFlow [53] is similar to PC-GAN. Rather than using adversarial training, however, they train the model using a VAE-like approach in which an inference network produces an approximate posterior on the latent shape representation. Moreover, they use continuous normalizing flows [7] to define a prior on the shape space, and a conditional distribution on 3D points given a latent shape representation. Our work is based on the same high-level VAE-like structure as PointFlow, but differs in the design of the network components. Most importantly, we make use of efficient "discrete" affine coupling layers, avoiding the use of computationally expensive ODE solvers for training and generation needed for the "continuous" flows, resulting in a significant speed-up to train and sample from the model. We describe our "Discrete Point Flow Networks" in the next section.

3   Discrete Point Flow Networks

In this section we first present the high-level hierarchical latent variable model, followed by a more detailed description of the model components in Sect. 3.2.

3.1   Hierarchical Latent Variable Model for Point Cloud Generation

Our goal is to define a generative model over point clouds of variable size that represent 3D shapes. The defining characteristics of point clouds are that the number of points may vary from one cloud to another, and that there is no inherent ordering among the points.

Let $X = \{x_1, \ldots, x_n\}$ be a point cloud with $x_i \in \mathbb{R}^d$, where $d = 3$ for point clouds for 3D shapes. The dimension d may be larger in some cases, e.g. $d = 6$ when modeling 3D points equipped with surface normals. An exchangeable distribution is one that is invariant to permutations of the data, i.e.

$$p(x_1, \ldots, x_n) = p(x_{\pi_1}, \ldots, x_{\pi_n}), \tag{1}$$

where π is a permutation of the integers $1, \ldots, n$. Note that independence implies exchangeability, but the reverse does not hold.

De Finetti's representation theorem states that any exchangeable distribution can be written as a factored distribution, conditioned on a latent variable:

$$p(X) = \int_z p_\psi(z) \prod_{x \in X} p_\theta(x|z) \mathrm{d}z. \tag{2}$$

In the case of 3D point cloud modeling, the latent variable z can be thought of as an element in an abstract shape space, sampled from a prior $p_\psi(z)$. This construction allows for point clouds of different cardinality, since conditioned on the shape representation z, the elements of the point cloud are sampled i.i.d. Given this general framework, also adopted in [30,53], the challenge is to:

1. Design a flexible model so that the conditional distribution $p_\theta(x|z)$ concentrates around the surface of the object represented by z.
2. Mitigate the intractability of the integral in Eq. 2 during training when using, e.g., deep neural networks to construct $p_\theta(x|z)$.

Before we consider the design of $p_\theta(x|z)$ and $p_\psi(z)$ in Sect. 3.2, we describe how to deal with the integral in Eq. 2 using the VAE framework [26].

We efficiently approximate the intractable posterior $p(z|X)$ with an amortized inference network $q_\phi(z|X)$. The approximate posterior allows us to define a variational bound on the likelihood in Eq. 2 as using Jensen's inequality [4]:

$$\ln p(X) \geq \sum_{x \in X} \mathbb{E}_{q_\phi(z|X)}[\ln p_\theta(x|z)] - \mathcal{D}_{\mathrm{KL}}(q_\phi(z|X)||p_\psi(z)) \equiv -\mathcal{F}. \quad (3)$$

The first term aims to reconstruct the points $x \in X$ using shape representations sampled from $q_\phi(z|X)$, whereas the second term ensures that the approximate posterior cannot arbitrarily deviate from the prior. Following [26,44] we use Monte Carlo sampling and the reparametrization trick to jointly minimize the loss $\mathcal{F}$ over θ, ψ and ϕ using stochastic gradient descent. The distributions $q_\phi(z|X)$, $p_\theta(x|z)$, and $p_\psi(z)$ and the underlying network architectures that make up the loss are detailed in the following section.

3.2   Design of Model Components

Shape-Conditional Point Distribution. The density on points for a given latent shape, $p_\theta(x|z)$, needs to be flexible enough to concentrate its support around the surface of the 3D shape. To this end we construct a conditional form of normalizing flows based on affine coupling layers [12].

Let $y \in \mathbb{R}^3$ denote a latent variable for each 3D point x, with a Gaussian conditional distribution given by $p_\theta(y|z) = \mathcal{N}(y; \nu_\theta(z), \mathrm{diag}(\omega_\theta(z)))$, where $\nu_\theta(z)$ and $\omega_\theta(z)$ are non-linear functions of z. In the affine coupling layer, we partition the coordinates of y in two groups, y^c and y^u, and update y^u by affine transformation conditioned on y^c and the latent shape representation z, i.e. $x^u = y^u \odot s_\theta(y^c, z) + t_\theta(y^c, z)$, while leaving the conditioning coordinates unchanged, i.e. $x^c = y^c$. To achieve the desired expressivity, we stack many affine coupling layers, cycling through the six possible partitionings of the three coordinates. Each coupling layer in the resulting flow $f_\theta(x; z)$ is conditioned on the latent shape representation z by the means of the FiLM conditioning mechanism [39] in the scaling and translation functions.

In practice, the scaling and translation functions are implemented by MLPs, which inflate the dimensionality of y^c to D_{inf}, and then deflate it to the dimensionality of y^u. Simultaneously, a separate MLP takes the latent variable z and outputs conditioning coefficients of size D_{inf}, with which we multiply and shift the inflated hidden units in the scaling and translation functions. In Fig. 2 we provide an overview of the architecture of our conditional coupling layers.

Using $f_\theta(x; z)$ to denote the invertible flow network that maps x to y, the change of variable formula allows to write the density of 3D points x given z as:

$$p_\theta(x|z) = \mathcal{N}\left(f_\theta(x; z); \nu_\theta(z), \mathrm{diag}(\omega_\theta(z))\right) \left|\det\left(\frac{\partial f_\theta(x; z)}{\partial x^\top}\right)\right|, \quad (4)$$

which enters into the loss defined in Eq. 3.

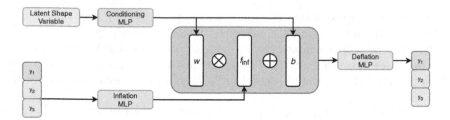

Fig. 2. Architecture of our conditional affine coupling layer applied to a single 3D point, with red dimension of the point being updated given the blue ones. (Color figure online)

Amortized Inference Network. The amortized inference network $q_\phi(z|X)$ takes a point cloud and produces a distribution on the latent shape representation. We use a permutation invariant design based on the PointNet architecture for shape classification [41]. As an output, the model produces the mean and diagonal covariance matrix of a Gaussian on $z \in \mathbb{R}^D$, *i.e.* $q_\phi(z|X) = \mathcal{N}(z; \mu_\phi(X), \text{diag}(\sigma_\phi(X)))$.

Latent Shape Prior. Rather than using a unit Gaussian prior in the latent space, as is common in deep generative models, we learn a more expressive prior $p_\psi(z)$ by means of another normalizing flow $g_\psi(z)$ based on affine coupling layers, similar to [9,53]. In our experiments it reduces the KL divergence in Eq. 3 by adapting the prior to fit the marginal posterior $\sum_X q_\phi(z|X)$, rather than forcing the inference network to induce a unit Gaussian marginal posterior, resulting in improved generative performance. Using this construction, we obtain the KL divergence as:

$$\mathcal{D}_{\text{KL}}(q_\phi(z|X)||p_\psi(z)) = \mathbb{E}q_\phi(z|X)\ln p_\psi(z) - \mathcal{H}(q_\phi(z|X)) \quad (5)$$

$$= \mathbb{E}q_\phi(z|X)\ln \mathcal{N}(g_\psi(z); \eta, \text{diag}(\kappa)) + \ln\left|\det\left(\frac{\partial g_\psi(z)}{\partial z^\top}\right)\right| - \mathbf{1}^\top \ln \sigma_\phi(X), \quad (6)$$

where we use Monte Carlo sampling to approximate the expectation.

Single-View Reconstruction Architecture. For single-view reconstruction, we follow [28] and define the model as:

$$p(X|v) = \int_z p_\psi(z|v) \prod_{x \in X} p_\theta(x|z)\text{d}z, \quad (7)$$

where we replaced the latent shape prior $p_\psi(z)$ with an image conditioned one, $p_\psi(z|v)$. In this case, the latent shape flow g_ψ does not deform a parametric Gaussian, but rather a Gaussian whose mean and variance are computed from an image v by a CNN encoder. We train the model by optimizing a variational bound similar to Eq. 3, and using the PointNet inference network to obtain an approximate posterior. Figure 3 provides an overview of the data flow between the model components for training and point cloud generation.

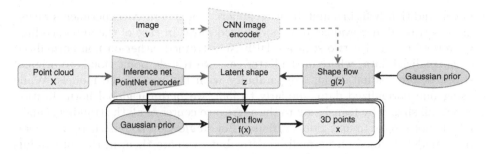

Fig. 3. Overview of DPF-Net: arrows indicate data flow to sample new point clouds (blue) and point cloud autoencoding (red), black arrows are used in both processes. During training flow modules are traversed in the reverse direction. For single-view reconstruction the shape prior is conditioned on the image (dashed). (Color figure online)

4   Experiments

Datasets. In order to provide a comparison with prior academic studies on shape generation, we perform experiments on subsets of ShapeNet [6] dataset. For autoencoding we use the ShapeNetCore.v2, containing roughly 55k meshes from 55 classes. In the generative setting, following [53], we use single class subsets (*airplanes*, *cars*, and *chairs*) from the same dataset. For the single-view reconstruction task we used a subset of 13 major classes of ShapeNetCore.v1 from Choy *et al.* [10], which comes with rendered 137×137 images from 24 randomized viewpoints per shape. We substitute the voxel grids provided by Choy *et al.* with the original meshes to sample point clouds for training and evaluation.

Data Split. For autoencoding we use a random split of the data, distributed across train, validation, and test sets in a 70/10/20 proportion per class. In the generative setting, we use single class subsets from the same random split. By using a random data split per class we intentionally deviate from the official ShapeNet data split, used in [53], which splits into significantly different data subsets per class. For example, in case of airplanes, training and validation sets mostly contain regular passenger aircraft, while the test set is populated with fighter jets and spaceships. While such a split could be useful in the context of autoencoding to assess out-of-distribution generalization, a significant mismatch between training and test set is undesirable for evaluation of generative models which are supposed to fit the training distribution. For single-view reconstruction we use the train/test split from [10].

Normalization. The original meshes in the ShapeNet are not normalized for position and scale which negatively affects the reconstruction quality. We therefore, additionally use a normalized version of the dataset, where we preprocess each mesh separately so that the sampled point clouds are (approximately) zero mean, and tightly fit in a unit diameter sphere. For generative experiments we use models trained and evaluated on normalized data. In case of the autoencoding, we report results for two separate DPF-Nets trained either on non-normalized or normalized data, where in the latter case we rescale point clouds to original scales before evaluation for comparability with the rest of the models. While using non-normalized data, similarly to [53], we perform global normalization across all shapes by translation to the aggregate center of all the training shapes, but do not rescale to unit global variance. For single-view reconstruction we compare models trained on normalized data, but evaluate them in the unit radius sphere scale for comparability with related work.

Point clouds are uniformly sampled from the meshes by sampling polygons with a probability proportional to their area, and then uniformly sampling a point per each selected polygon. Unlike previous works using precomputed point clouds, we perform this procedure on the fly, thus obtaining a different random point cloud each time we process a 3D shape. During both training and evaluation we sample two point clouds for each shape: one is used as input to the inference network, the other for optimization or evaluation of the decoder. We use $2,048$, $2,048$, and $2,500$ points for training and quantitative evaluation for generative, autoencoding, and single-view reconstruction tasks accordingly.

Evaluation Metrics. We follow the standard protocol [1,13,53] and use Chamfer distance (CD) and earth mover's distance (EMD) to assess point cloud reconstructions. To measure the generative properties we follow [1,53], and use metrics to compare equally sized *sets* of generated and reference point clouds:

- The Jensen-Shannon divergence (JSD) compares the marginal distributions obtained by taking the union of all generated (or reference) point clouds, and quantizing the distributions to a voxel grid.
- The Minimum matching distance (MMD) computes the average distance of reference point clouds to their nearest (in CD/EMD) generated point cloud.
- Coverage (COV) is the fraction of reference point clouds matched by minimum CD/EMD distance by at least one generated point cloud.
- 1-nearest neighbour accuracy (1-NNA) classifies generated and reference point clouds as belonging to either of these two sets using a leave-one-out 1-nearest neighbor classifier (using CD/EMD). Ideal accuracy is 50%.

For single-view reconstruction we additionally report F1-score. For more detailed descriptions of these metrics we refer to [1,53].

4.1   Experimental Setup

Inference Network. We use a PointNet encoder [41] with the number of features progressing over layers as $3-64-128-256-512$, followed by a max-pooling across points, and two fully-connected layers of sizes 512 and D, where D is the size of the latent space. We use $D = 512$ in the autoencoding and single-view reconstruction experiments, and $D = 128$ for generative modeling.

Latent Shape Prior. We use 14 affine coupling layers to construct the latent space prior. In the coupling layers, we alternate between two orthogonal partitioning schemes: odd and even dimensions are split in different groups; the first $D/2$ dimension go in one group, and the remaining in the other. For single-view reconstruction we use ResNet18 image encoder, similarly to [18,49,50].

Point Decoder. The point decoder $p(x|z)$ starts with a three-dimensional Gaussian with mean and variances computed from z by an MLP with two hidden layers. This Gaussian is transformed by 63 of our conditioned affine coupling layers, each consisting of (i) two fully-connected layers that map z to the FiLM conditioning coefficients, each of size 64, (ii) two fully-connected layers that inflate input dimensionality to 64 hidden units (at which point the FiLM conditioning is performed), (iii) a final fully-connected layer that deflate the dimensionality to compute the scaling and translation functions.

Baselines. We retrained AtlasNet [18], l-GAN-CD/EMD [1], PointFlow [53], and DCG [49] with our split of the ShapeNet dataset, using the implementation provided by the authors and our data processing pipeline. For improved comparability we also modified the point cloud encoders in all models to match each other (except for l-GANs, since it significantly worsened their results). To match other approaches, we used $2,048$ points per cloud for AtlasNet in the autoencoding task and consequently set the number of learned primitives to 16.

Oracles. For all the tasks we also provide an "oracle" to assess the best possible performance values. For autoencoding and single-view reconstruction, the oracle samples a second point cloud from the ground truth mesh, rather than generating a point cloud. For generative modeling, the oracle uses the point clouds from the training set, instead of sampling point clouds from the model.

Our code is publicly available at https://github.com/Regenerator/dpf-nets.

4.2   Generative Modeling Evaluation

Efficiency Comparison with PointFlow. We compare our DPF-Networks in terms of computational efficiency and memory footprint to PointFlow [53]. We train both models for point cloud generation, and report the number of parameters and total training time. To compute the training memory footprint, training time, and generation time per sample (point cloud), we divided total GPU memory occupied during training, batch iteration time, and batch generation time, respectively, by the batch size.

Both models were run on a single TITAN RTX GPU. We estimate the total training time for PointFlow after the initial 100 epochs of training, which took 2 days, and assume that the full training procedure requires $4,000$ epochs, as reported by the authors of PointFlow. We observed that the training procedure slowed down over the course of training, because ODE-solver gradually increases the number of iterations to meet the required tolerance. Thus, all timings in Table 1 for PointFlow should be understood as lower bounds.

Table 1. Efficiency comparison for DPF-Nets and PointFlow generative models.

Model	Nr. params., 10^6	Mem. footprint, Mb/sample	Tr. time, ms/sample	Total tr. time, days	Gen. time, ms/sample
PointFlow [53]	1.63	470	500	80	150
DPF-Net (Ours)	3.76	370	16	1.1	4

From the results in Table 1 we see that even though DPF-Networks have more parameters, the associated training memory footprint is lower and, our model is approximately 30 times faster both in training and inference iterations, and can be trained in a single day.

Quantitative Results. We compare to l-GANs and PointFlow models and report oracle performance as a reference. Given the prohibitive computation cost of complete PointFlow training, we provide results obtained after training for four days, which is four times the full training time of DPF-Net in the same setting. In order to account for random sampling every model is evaluated using ten different sets of generated objects, each of the size of the test set. Thus, for each metric we report mean values over ten runs. In addition to the best values, in Table 2 we also write in bold results that are within two standard deviations of the best result.

Table 2. Generative modeling results. Oracle results are underlined when they are not the best. JSD and MMD-EMD are multiplied by 10^2, MMD-CD by 10^4.

Category	Model	JSD↓	MMD↓		COV↑, %		1-NNA↓, %	
			CD	EMD	CD	EMD	CD	EMD
Airplane	l-GAN-CD [1]	2.76	**5.69**	5.16	39.5	17.1	72.9	92.1
	l-GAN-EMD [1]	1.77	6.05	**4.15**	39.7	40.4	75.7	73.0
	PointFlow [53]	1.42	6.05	4.32	**44.7**	**48.4**	**70.9**	68.4
	DPF-Nets (Ours)	**0.94**	6.07	4.26	**46.8**	**48.4**	**70.6**	**67.0**
	Oracle	0.50	5.97	3.98	51.4	52.7	49.8	48.2
Car	l-GAN-CD [1]	2.65	**8.83**	5.36	41.3	15.9	**62.6**	92.7
	l-GAN-EMD [1]	1.31	9.00	**4.40**	38.3	32.9	65.2	**63.2**
	PointFlow [53]	0.59	9.53	4.71	**42.3**	35.8	70.1	74.2
	DPF-Nets (Ours)	**0.45**	9.59	4.61	**43.4**	**45.7**	70.3	**64.3**
	Oracle	0.37	9.24	4.56	52.8	52.7	50.9	50.5
Chair	l-GAN-CD [1]	3.65	**16.66**	7.91	42.3	17.1	68.5	96.5
	l-GAN-EMD [1]	1.27	**16.78**	**5.75**	44.3	43.8	66.6	67.8
	PointFlow [53]	1.51	17.15	6.20	43.3	46.5	67.0	70.4
	DPF-Nets (Ours)	**1.01**	17.08	6.14	**46.9**	**48.5**	**63.5**	**64.8**
	Oracle	0.49	16.39	5.71	52.8	53.4	49.7	49.6

Overall, DPF-Networks yield the best results in terms of JSD, COV-CD/EMD and 1-NNA-CD/EMD, except for the 1-NNA-CD for *car*. This confirms that our DPF-Network is capable of generating more realistic and diverse sets of point clouds, for random samples from our model see Fig. 1.

L-GAN-CD experiences mode collapses, and generates objects with good CD values, but with very poor coverage and 1-NNA in terms of the EMD metric. PointFlow shows performance similar to ours, except for JSD, while being significantly slower in both training and sampling. In contrast to the evaluations performed in [53], based on the official split, in our experiments the oracle obtains the best performances for all metrics, except for MMD (see underlined results). We believe that this highlights the fact the MMD metric does not favor diversity in the generated point clouds, but instead favors point clouds with low CD/EMD distances to all the reference shapes. If the generated point clouds contain a subset of high quality modes from the test subset, the metric can yield good results, even better than the oracle. DPF-Nets and PointFlow yield qualitatively similar point cloud samples, we provide a comparison of samples in the supplementary material.

Table 3. Autoencoding results. † - results from [53] on the official split, ∗ - results for equal training time as DPF-Net on the random split.

Metric	CD $\times 10^4$	EMD $\times 10^2$
l-GAN-CD [1]	7.07	7.70
l-GAN-EMD [1]	9.18	5.30
AtlasNet [18]	**5.66**	5.81
PointFlow† [53]	7.54	5.18
PointFlow∗	10.22	6.58
DPF-Net, orig.	6.85	5.06
DPF-Net, norm.	6.17	**4.37**
Oracle	3.10	3.13

4.3   Autoencoding Evaluation

We compare DPF-networks with other models in terms of autoencoding performance in Table 3. Similarly to generative experiments, we restricted the training time of PointFlow, this time, to match the training time of our approach which was approximately a week. Among models trained on non-normalized data, DPF-Net (orig.) achieve the best results in the EMD metric and second best in the CD metric, outperformed only by the non-generative AtlasNet which is trained by optimization of the CD metric. The DPF-Net outperforms both l-GANs which were specifically optimized for the CD/EMD metrics, while being trained by

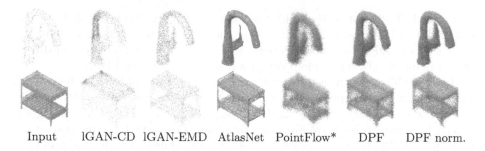

Input   lGAN-CD lGAN-EMD AtlasNet PointFlow*   DPF   DPF norm.

Fig. 4. Qualitative comparison of the models from Table 3 for the autoencoding task with sparse (top) and dense (bottom) inputs.

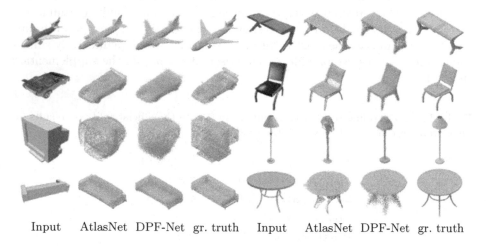

Input   AtlasNet DPF-Net gr. truth   Input   AtlasNet DPF-Net gr. truth

Fig. 5. Qualitative comparison for single-view reconstruction task.

optimization of the likelihood lower bound. Importantly DPF-Nets outperform PointFlow in both metrics under the same and extended computational budget.

When our model is trained on normalized data, results significantly improve, achieving state-of-the-art among generative models for both metrics. This underlines the importance of proper data normalization for shape modeling.

In Fig. 4 we qualitatively compare our autoencoding results to l-GANs, Atlas-Net, and PointFlow. All approaches can work with arbitrary size inputs, but only AtlasNet, PointFlow, and DPF-Nets can reconstruct with arbitrary density. In this comparison we use 512 and 32, 768 points as sparse and dense inputs, while reconstructing fixed 2, 048 points for l-GANs and 32, 768 point for AtlasNet, PointFlow and DPF-Nets. Models with better CD values (l-GAN-CD, AtlasNet) tend to concentrate points in some regions of reconstructed shapes, while models with better EMD values (l-GAN-EMD, DPF-Nets) distribute points more evenly. While AtlasNet achieves best CD, its reconstructions contain sharp plane-like artifacts. Our DPF-Nets produce overall smoother reconstructions, but, on the other hand, suffer from more noise.

4.4 Single-View Reconstruction

In this section we test DPF-Nets on the inference of 3D point clouds from single images. The architecture used for this specific task is depicted in Fig. 3 and detailed in Sect. 3.2. We compare our results to recent state-of-the-art methods in the field. This includes: the voxel-based PRN [28], point cloud-based approaches of AtlasNet [18] and DCG [49], and the mesh-based Pixel2Mesh [50].

Although convenient, in general, comparison to voxel-based approaches should be taken with a grain of salt, since it is biased. To compute the proposed metrics either ground truth or reconstructed voxelized shapes are fed to the marching cubes algorithm to obtain final meshes which are used to sample point clouds. Resulting ground truth meshes in that case are crude approximations of the original meshes, used in the evaluation of the point cloud and mesh-based approaches. Moreover, there are cases both in voxelized data and reconstructions, when the marching cubes algorithm fails to output meshes.

The results in Table 4 show that DPF-Net clearly outperforms earlier works in terms of the EMD metric. It also achieves best results in terms of the F1-score among point cloud and mesh-based models. In terms of CD, similarly to autoencoding it is outperformed only by AtlasNet with a small margin. This validates the ability of normalizing flows to capture complex distributions in 3D and to model shape surfaces.

Qualitative single-view reconstruction results can be found in Fig. 5. Note that a single reconstruction model has been trained across all 13 classes for both AtlasNet and DPF-Net. Similarly to the autoencoding task, compared to AtlasNet our approach produces more evenly distributed point clouds without sharp dense clusters, but introduces more noise.

Table 4. Single-view reconstruction results. [†]: results taken from [50].

Model	CD$\downarrow$, $\times 10^3$	EMD$\downarrow$, $\times 10^2$	F1$\uparrow$,$\tau = 0.001$,%
PRN [28]	7.56	11.00	**53.1**
AtlasNet [18]	**5.34**	12.54	52.2
DCG [49]	6.35	18.94	45.7
Pixel2Mesh[†] [50]	5.91	13.80	–
DPF-Nets	5.51	**10.95**	52.4
Oracle	1.10	5.70	84.0

5   Conclusion

We presented DPF-Networks, a generative model for point clouds of arbitrary size. DPF-Nets are based on a latent variable model and use normalizing flows with affine coupling layers to construct a flexible, yet tractable, shape conditional density on 3D points, and an expressive latent shape space prior. They are trained akin to VAEs, using a permutation invariant point cloud encoder as approximate posterior distribution over the latent shape space.

The evaluation on the ShapeNet dataset demonstrates that DPF-nets improve generative performance metrics over previous work in most metrics and classes. Compared to a recent related work based on continuous normalizing flows, our model is between one and two orders of magnitude faster to train and sample from. Applied to single view reconstruction, DPF-Nets outperform state-of-the-art methods, hence showing promising capabilities in 3D shape modeling.

Acknowledgements. Work done while Jakob Verbeek was at INRIA.

References

1. Achlioptas, P., Diamanti, O., Mitliagkas, I., Guibas, L.: Learning representations and generative models for 3D point clouds. In: ICML (2018)
2. Behrmann, J., Grathwohl, W., Chen, R., Duvenaud, D., Jacobsen, J.H.: Invertible residual networks. In: ICML (2019)
3. Bhattacharyya, A., Hanselmann, M., Fritz, M., Schiele, B., Straehle, C.N.: Conditional flow variational autoencoders for structured sequence prediction. In: NeurIPS Workshop on Machine Learning for Autonomous Driving (2019)
4. Bishop, C.: Pattern Recognition and Machine Learning. Springer, New York (2006)
5. Brock, A., Lim, T., Ritchie, J., Weston, N.: Generative and discriminative voxel modeling with convolutional neural networks. In: NeurIPS 3D deep learning workshop (2016)
6. Chang, J.R., Chen, Y.S.: Batch-normalized maxout network in network. In: ICML (2016)
7. Chen, R., Behrmann, J., Duvenaud, D., Jacobsen, J.H.: Residual flows for invertible generative modeling. In: NeurIPS (2019)
8. Chen, T., Rubanova, Y., Bettencourt, J., Duvenaud, D.: Neural ordinary differential equations. In: NeurIPS (2018)
9. Chen, X., et al.: Variational lossy autoencoder. In: ICLR (2017)
10. Choy, C.B., Xu, D., Gwak, J.Y., Chen, K., Savarese, S.: 3D-R2N2: a unified approach for single and multi-view 3D object reconstruction. In: Leibe, B., Matas, J., Sebe, N., Welling, M. (eds.) ECCV 2016. LNCS, vol. 9912, pp. 628–644. Springer, Cham (2016). https://doi.org/10.1007/978-3-319-46484-8_38
11. Deco, G., Brauer, W.: Higher order statistical decorrelation without information loss. In: NeurIPS (1995)
12. Dinh, L., Sohl-Dickstein, J., Bengio, S.: Density estimation using Real NVP. In: ICLR (2017)
13. Fan, H., Su, H., Guibas, L.: A point set generation network for 3D object reconstruction from a single image. In: CVPR (2017)

14. Girdhar, R., Fouhey, D.F., Rodriguez, M., Gupta, A.: Learning a predictable and generative vector representation for objects. In: Leibe, B., Matas, J., Sebe, N., Welling, M. (eds.) ECCV 2016. LNCS, vol. 9910, pp. 484–499. Springer, Cham (2016). https://doi.org/10.1007/978-3-319-46466-4_29
15. Goodfellow, I., et al.: Generative adversarial nets. In: NeurIPS (2014)
16. Graham, B., Engelcke, M., van der Maaten, L.: 3D semantic segmentation with submanifold sparse convolutional networks. In: CVPR (2018)
17. Grathwohl, W., Chen, R., Bettencourt, J., Sutskever, I., Duvenaud, D.: FFJORD: free-form continuous dynamics for scalable reversible generative models. In: ICLR (2019)
18. Groueix, T., Fisher, M., Kim, V., Russell, B., Aubry, M.: A papier-mâché approach to learning 3D surface generation. In: CVPR (2018)
19. He, K., Zhang, X., Ren, S., Sun, J.: Deep residual learning for image recognition. In: CVPR (2016)
20. He, K., Zhang, X., Ren, S., Sun, J.: Identity mappings in deep residual networks. In: Leibe, B., Matas, J., Sebe, N., Welling, M. (eds.) ECCV 2016. LNCS, vol. 9908, pp. 630–645. Springer, Cham (2016). https://doi.org/10.1007/978-3-319-46493-0_38
21. Henderson, P., Ferrari, V.: Learning to generate and reconstruct 3D meshes with only 2D supervision. In: BMVC (2018)
22. Insafutdinov, E., Dosovitskiy, A.: Unsupervised learning of shape and pose with differentiable point clouds. In: NeurIPS (2018)
23. Karras, T., Aila, T., Laine, S., Lehtinen, J.: Progressive growing of GANSs for improved quality, stability, and variation. In: ICLR (2018)
24. Kingma, D., Dhariwal, P.: Glow: generative flow with invertible 1×1 convolutions. In: NeurIPS (2018)
25. Kingma, D., Salimans, T., Jozefowicz, R., Chen, X., Sutskever, I., Welling, M.: Improved variational inference with inverse autoregressive flow. In: NeurIPS (2016)
26. Kingma, D., Welling, M.: Auto-encoding variational Bayes. In: ICLR (2014)
27. Klokov, R., Lempitsky, V.: Escape from cells: deep Kd-networks for the recognition of 3D point cloud models. In: ICCV (2017)
28. Klokov, R., Verbeek, J., Boyer, E.: Probabilistic reconstruction networks for 3D shape inference from a single image. In: BMVC (2019)
29. Kobyzev, I., Prince, S., Brubaker, M.: Normalizing flows: an introduction and review of current methods. arXiv preprint (2019)
30. Li, C.L., Zaheer, M., Zhang, Y., Poczos, B., Salakhutdinov, R.: Point cloud GAN. arXiv preprint (2018)
31. Lu, Y., Huang, B.: Structured output learning with conditional generative flows. In: AAAI (2020)
32. Lucas, T., Shmelkov, K., Alahari, K., Schmid, C., Verbeek, J.: Adaptive density estimation for generative models. In: NeurIPS (2019)
33. Mandikal, P., Navaneet, K., Agarwal, M., Babu, R.: 3D-LMNet: latent embedding matching for accurate and diverse 3D point cloud reconstruction from a single image. In: BMVC (2018)
34. Maturana, D., Scherer, S.: VoxNet: A 3D convolutional neural network for real-time object recognition (2015)
35. Michalkiewicz, M., Pontes, J., Jack, D., Baktashmotlagh, M., Eriksson, A.: Implicit surface representations as layers in neural networks. In: ICCV (2019)
36. Monti, F., Boscaini, D., Masci, J., Rodolà, E., Svoboda, J., Bronstein, M.: Geometric deep learning on graphs and manifolds using mixture model CNNs. In: CVPR (2017)

37. Papamakarios, G., Nalisnick, E., Rezende, D., Mohamed, S., Lakshminarayanan, B.: Normalizing flows for probabilistic modeling and inference. arXiv preprint (2019)
38. Park, J., Florence, P., Straub, J., Newcombe, R., Lovegrove, S.: DeepSDF: learning continuous signed distance functions for shape representation. In: CVPR (2019)
39. Perez, E., Strub, F., Vries, H.D., Dumoulin, V., Courville, A.: FiLM: visual reasoning with a general conditioning layer. In: AAAI (2018)
40. Pumarola, A., Popov, S., Moreno-Noguer, F., Ferrari, V.: C-Flow: conditional generative flow models for images and 3D point clouds. In: CVPR (2020)
41. Qi, C., Su, H., Mo, K., Guibas, L.: PointNet: deep learning on point sets for 3D classification and segmentation. In: CVPR (2017)
42. Qi, C., Yi, L., Su, H., Guibas, L.: PointNet++: deep hierarchical feature learning on point sets in a metric space. In: NeurIPS (2017)
43. Rezende, D., Mohamed, S.: Variational inference with normalizing flows. In: ICML (2015)
44. Rezende, D., Mohamed, S., Wierstra, D.: Stochastic backpropagation and approximate inference in deep generative models. In: ICML (2014)
45. Riegler, G., Ulusoy, A., Geiger, A.: OctNet: learning deep 3D representations at high resolutions. In: CVPR (2017)
46. Su, H., Jampani, V., Sun, D., Maji, S., Kalogerakis, E., Yang, M.H., Kautz, J.: SPLATNet: sparse lattice networks for point cloud processing. In: CVPR (2018)
47. Tatarchenko, M., Dosovitskiy, A., Brox, T.: Octree generating networks: efficient convolutional architectures for high-resolution 3D outputs. In: ICCV (2017)
48. Verma, N., Boyer, E., Verbeek, J.: FeaStNet: feature-steered graph convolutions for 3D shape analysis. In: CVPR (2018)
49. Wang, K., Chen, K., Jia, K.: Deep cascade generation on point sets. In: IJCAI (2019)
50. Wang, N., Zhang, Y., Li, Z., Fu, Y., Liu, W., Jiang, Y.-G.: Pixel2Mesh: generating 3D mesh models from single RGB images. In: Ferrari, V., Hebert, M., Sminchisescu, C., Weiss, Y. (eds.) ECCV 2018. LNCS, vol. 11215, pp. 55–71. Springer, Cham (2018). https://doi.org/10.1007/978-3-030-01252-6_4
51. Wu, J., Zhang, C., Xue, T., Freeman, W., Tenenbaum, J.: Learning a probabilistic latent space of object shapes via 3D generative-adversarial modeling. In: NeurIPS (2016)
52. Wu, Z., Song, S., Khosla, A., Yu, F., Zhang, L., Tang, X., Xiao, J.: 3D ShapeNets: a deep representation for volumetric shapes. In: CVPR (2015)
53. Yang, G., Huang, X., Hao, Z., Liu, M.Y., Belongie, S., Hariharan, B.: PointFlow: 3D point cloud generation with continuous normalizing flows. In: ICCV (2019)
54. Zaheer, M., Kottur, S., Ravanbakhsh, S., Poczos, B., Salakhutdinov, R., Smola, A.: Deep sets. In: NeurIPS (2017)

Accelerating Deep Learning with Millions of Classes

Zhuoning Yuan[1][✉], Zhishuai Guo[1], Xiaotian Yu[3], Xiaoyu Wang[2],
and Tianbao Yang[1]

[1] The University of Iowa, Iowa City, IA 52242, USA
{zhuoning-yuan,zhishuai-guo,tianbao-yang}@uiowa.edu
[2] The Chinese University of Hong Kong (Shenzhen), Shenzhen, China
[3] Shenzhen, China

Abstract. Deep learning has achieved remarkable success in many classification tasks because of its great power of representation learning for complex data. However, it remains challenging when extending to classification tasks with millions of classes. Previous studies are focused on solving this problem in a distributed fashion or using a sampling-based approach to reduce the computational cost caused by the softmax layer. However, these approaches still need high GPU memory in order to work with large models and it is non-trivial to extend them to parallel settings. To address these issues, we propose an efficient training framework to handle extreme classification tasks based on *Random Projection*. The key idea is that we first train a slimmed model with a random projected softmax classifier and then we recover it to the original classifier. We also show a theoretical guarantee that this recovered classifier can approximate the original classifier with a small error. Later, we extend our framework to parallel settings by adopting a communication reduction technique. In our experiments, we demonstrate that the proposed framework is able to train deep learning models with millions of classes and achieve above 10× speedup compared to existing approaches.

Keywords: Large-scale learning · Deep learning · Random projection · Acceleration in deep learning

1 Introduction

One of the biggest advantages of deep learning is to improve the capability of modeling complex data over conventional approaches. In practice, deep learning has achieved the state-of-the-art performance on various applications in computer vision and natural language processing [8,12]. With the dramatic increase of scale in data, extreme classification emerges as a challenging task. This leads to some new challenges: (1) the extremely large model size, i.e., the model size

Electronic supplementary material The online version of this chapter (https://doi.org/10.1007/978-3-030-58592-1_42) contains supplementary material, which is available to authorized users.

A. Vedaldi et al. (Eds.): ECCV 2020, LNCS 12368, pp. 711–726, 2020.
https://doi.org/10.1007/978-3-030-58592-1_42

is dominated by softmax layer, which is proportional to the number of classes; (2) the expensive cost to train such large models. A widely used benchmark dataset, ImageNet [19], contains only 1000 classes. However, for applications such as facial recognition and language modeling, the number of classes can easily reach a million-scale, e.g., Megaface [18] and Google Billion Words [5]. To reduce the cost of training, we utilize a well-known dimensionality reduction technique, *Random Projection*, e.g., a high-dimensional feature **x** can be reduced to a low-dimensional one through a random matrix R. To understand the idea behind the random projection, the key result is built on Johnson-Lindenstrauss lemma [2], which reveals that any high dimensional data can be mapped onto a lower-dimensional space while the distance between data points is approximately persevered. Compared to other reduction methods. e.g., SVD, random projection has a lower computational cost and thus it is being widely used in image and text tasks [3]. We summarize the existing works of accelerating training for extreme classification in the following sections.

1.1 Naive Methods

Typically, a single machine is not enough to handle a large-scale task efficiently and therefore the most straightforward solution is to directly increase the number of workers and distribute the task over all workers. As a result, it leads to an immediate improvement of training efficiency. For example, a distributed system of 256 GPUs can train ResNet50 on ImageNet with batch size of 8192 in one hour with a top1 validation accuracy of 75.7% [9]. Recently, it has been extended to a larger system of 2176 Tesla V100 GPUs, which takes only 224 s for training to achieve a similar performance [22]. Considering the trade-off between model precision and training efficiency, mixed precision training [16] is another solution to handle the large-scale tasks with lower memory demands. The key idea is to convert the model weights, activations and gradients to lower precision, e.g., float16, during forward and backward pass, but a copy of weights in float32 is maintained during parameter updates. In addition, it is also possible to distribute the computation process of softmax layer based on matrix partition to multiple parallel workers in order to overcome the difficulty of training large models on large-scale datasets [7].

1.2 Sampling-Based Methods

In the applications of Natural Language Processing, there are two common ways to tackle extreme classification: sampling-based and softmax-based approaches. These works are generally inspired by an important fact, Zipf's law, which tells that there are a small number of vocabularies (classes) with high occurrence and a very large number of vocabularies (classes) with low occurrence in the dataset. This rule also applies to many facial datasets, e.g., Celeb and MegaFace as shown in Fig. 1. For sampling-based methods, Zhang et al. [27] identified that only 1%–2% classes are active during training on facial recognition tasks. Accordingly, they proposed an algorithm dynamically selecting those active classes to speed

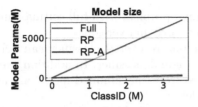 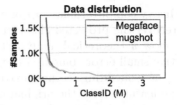

Fig. 1. Illustrations of model size by varying the number of classes and the class distribution of two facial datasets. Figure 1 shows the size of model with full softmax increases dramatically as the increase of number of classes compared to slimmed model with projected softmax. Figure 2 reveals that facial datasets follow long-tailed distribution.

up the training. Their experiments show a significant reduction on GPU memory cost and an improvement of training efficiency. The key idea of softmax-based approaches is to convert a multi-class task to a binary task by sampling a small set of negative samples along with the positive sample according to Zipf's law, e.g., noise-contrastive estimation and sampled softmax [1,11]. In addition, some works [10,18] indicate that pretraining on a small subset sampled from the original dataset by class frequency is also a useful sampling-based strategy to speed up model converge compared to training from scratch for large models.

1.3 Challenges

We summarize the existing challenges to tackle the extreme classification from two perspectives as follows:

1. **Memory limitations**. As the increase of model size, the GPU memory demand to train such model grows rapidly as illustrated in Fig. 1. Sampling-based approaches can reduce this cost by computing sampled softmax scores, but the model size remains unchanged. Although we can apply additional compression techniques to reduce model size, it may also suffer a performance drop.
2. **Communication limitations**. To extend the training to parallel settings, if we naively increase the number of workers in the system, the communication cost among these workers can be a new bottleneck for the training. Unfortunately, the most existing approaches fail to consider these scenarios and are unlikely to be deployed in practice and obtain a linear speedup for training efficiency.

1.4 Our Contributions

The main contributions of our work to tackle the above issues are summarized as follows:

1. We propose a two-stage algorithm to tackle extreme classification. We first solve a slimmed problem with random projected softmax classifier to avoid

the heavy computational cost of original problem with full softmax and then we recover the projected classifier to its original version.

2. We show a theoretical analysis that our recovered classifier can achieve a relative small error compared to the original classifier. In addition, we verify that the embedding features extracted from the slimmed model are as good as the features from the original model.

3. We present a feasible solution with communication reduction technique to extend the proposed framework to parallel settings and we show a nearly linear speed-up of our algorithm with respect to the number of workers as well as much less communication cost.

4. We conduct a comprehensive empirical study to verify our algorithm on different scales of datasets, and we also investigate the important factors that may affect the performance by various ablation studies.

2   Proposed Approach

In this section, we formally introduce our two-stage proposed framework as described in Algorithm 1 and illustrated in Fig. 2. At the first stage, we aim to quickly solve the slimmed problem with random projected softmax layer to avoid the heavy computational cost when the number of classes is large. At the second stage, we propose a dual recovery approach to recover the projected softmax classifier to its original version. In addition, we adopt a communication reduction technique to further accelerate the training in parallel settings. We present the details of these approaches in the following sections.

2.1   Solving Problem with Projected Softmax Layer

Firstly, we formulate the slimmed problem as

$$\min_{\mathbf{z}} \mathcal{L}(\mathbf{z}) = \sum_{i=1}^{n} \mathcal{L}(\mathbf{z}; \mathbf{x}_i, y_i), \tag{1}$$

where $\mathcal{L}(\mathbf{z}; \mathbf{x}_i, y_i)$ is a loss function with a regularization term, $(\mathbf{x}_i, y_i)$ is a pair of feature and label for a random sample i out of n samples, and $\mathbf{z}^*$ is the solution of the problem (1), including the feature extractor $f(\mathbf{x})$ and the corresponding projected softmax classifier $\widehat{W}$. Using a Gaussian random matrix $R^{d \times m}$, where d is the embedding size and m is the random projection size, we generate a low-dimensional representation for each input feature $f(\mathbf{x}_i)$ by

$$\widehat{f}(\mathbf{x}_i) = \frac{1}{\sqrt{m}} R^{\top} f(\mathbf{x}_i) \tag{2}$$

We specify $\widehat{W} = (\hat{\mathbf{w}}_1, \ldots, \hat{\mathbf{w}}_C) \in \mathbb{R}^{m \times C}$, where C is the number of classes. For a feature extractor $f(\cdot)$ and random projection matrix R, to learn the projected softmax classifier can be formulated as follows,

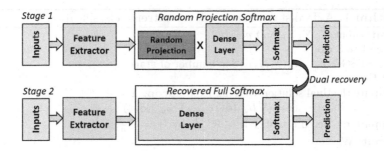

Fig. 2. The training framework of random projected softmax. In stage one, we reduce full softmax to projected softmax to speed up the training. In the second stage, we finetune the recovered softmax layer to further boost the performance.

$$\min_{\widehat{W}} \frac{\lambda}{2}\|\widehat{W}\|_F^2 + \sum_{i=1}^n \ell_{y_i}(\hat{\mathbf{o}}_i) \tag{3}$$

where $\hat{\mathbf{o}}_i = (\hat{o}_{i,1}, \dots, \hat{o}_{i,C}) = \left(\hat{\mathbf{w}}_1^\top \hat{f}(\mathbf{x}_i), \hat{\mathbf{w}}_2^\top \hat{f}(\mathbf{x}_i), \dots, \hat{\mathbf{w}}_C^\top \hat{f}(\mathbf{x}_i)\right)$. Since the feature extractor and the projected classifier are trained together, we could suppose that the learned projected classifier is optimal to the learned features.

Remarks: When C is sufficiently large, e.g., $C = 10^6$, solving the original model can be computational expensive and time consuming. Instead, solving slimmed model with projected softmax layer is more efficient. In addition, we argue that the learned embedding features from the slimmed model are as good as the embedding features from original model and we verify this in the experiments.

2.2 Solving Recovery Problem

Denote $W \in \mathbb{R}^{d \times C}$ as the original softmax classifier without projection corresponding to feature extractor $f(\mathbf{x})$. Noting that the size of $\widehat{W}$ is much less than the size of W, it is expected that the original classifier can produce better predictions than the projected classifier. Therefore, in the second stage, we recover W from $\widehat{W}$ by fixing feature extractor $f(\mathbf{x})$. In this setting, the problem can be formulated as follows,

$$\min_{W} \frac{\lambda}{2}\|W\|_F^2 + \sum_{i=1}^n \ell_{y_i}(\mathbf{o}_i), \tag{4}$$

where $\mathbf{o}_i = (o_{i,1}, \dots, o_{i,C}) = \left(\mathbf{w}_1^\top f(\mathbf{x}_i), \dots, \mathbf{w}_C^\top f(\mathbf{x}_i)\right)$ and λ is a parameter for the regularized term. Note that the problem (3) is a projected version of the problem (4). Next, we show that we can recover the projected softmax classifier to the original softmax classifier with a small error for any feature extractor $f(\cdot)$.

Recall that $f(\mathbf{x})$ is the embedding features of the data $\mathbf{x}_i$ generated by the network. $\mathbf{o}_i$ is the scores of $\mathbf{x}_i$ over C classes, and $\mathbf{o}_i = (o_{i,1}, \dots, o_{i,C}) = \left(\mathbf{w}_1^\top f(\mathbf{x}_i), \dots, \mathbf{w}_C^\top f(\mathbf{x}_i)\right)^T = \mathbf{W}^\top f(\mathbf{x}_i)$. For any i, we write $\ell_{y_i}(\mathbf{o}_i)$ in its conjugate form as

---

Algorithm 1. A Two-stage framework for extreme classification

---

1: **input:** input X, Y, $\mathbf{z}_0$, and m
2: **for** s=1 to S **do**
3:    $\mathbf{z}_s$ =PSGD$(\mathbf{z}_{s-1}, \eta_s, T_s, I_s)$ {S is number of stages }
4: **end for**
5: Compute the dual solution $\widehat{\Theta}$ using $\mathbf{z}_S$ by **prop. 1**,
6:             $\widehat{\Theta} = (\nabla L_{y_1}(\hat{\mathbf{o}}_1), \ldots, \nabla L_{y_n}(\hat{\mathbf{o}}_n))$
7: Compute recovered classifier $\widetilde{W} \in \mathbb{R}^{d \times K}$ by $\widetilde{W} = -\frac{1}{n\lambda} X\widehat{\Theta}$
8: **output:** $\widetilde{W}$

$$\ell_{y_i}(\mathbf{o}_i) = \max_{\boldsymbol{\theta}_i}[\mathbf{o}_i^T \boldsymbol{\theta}_i - \ell_{y_i}^*(\boldsymbol{\theta}_i)] \tag{5}$$

Plugging (5) to (4), we get the dual optimization problem of (4)

$$\max_{\Theta} -\sum_{i=1}^{n} \ell_{y_i}^*(\boldsymbol{\theta}_i) - \frac{1}{2\lambda}\|X\Theta^{*T}\|_F^2, \tag{6}$$

where $\Theta = (\boldsymbol{\theta}_1, \ldots, \boldsymbol{\theta}_n)$. We denote $W^* \in \mathbb{R}^{d \times C}$ as the optimal primal solution to (4), and denote $\Theta^* \in \mathbb{R}^{C \times n}$ as the optimal dual solution to (6) and the following proposition connects W^* and Θ^*.

Proposition 1. *Let $W^* \in \mathbb{R}^{d \times C}$ be the optimal primal solution to (4) and $\Theta^* \in \mathbb{R}^n$ be the optimal dual solution to (6). We have*

$$W^* = -\frac{1}{\lambda}X\Theta^{*T}, \text{ and } \Theta^* = (\nabla L_{y_1}(\mathbf{o}_1^*), \ldots, \nabla L_{y_n}(\mathbf{o}_n^*)), \tag{7}$$

where $\mathbf{o}_i^* = W^{*T}\mathbf{x}_i$, $i = 1, \ldots, n$.

Algorithm 2. PSGD: Proximal Stochastic Gradient Descent $(\mathbf{z}_0, \eta, T, I)$

1: **for** $t = 1$ to T **do**
2:    Each machine k updates its local solution in parallel:
3:    $\mathbf{z}_{t+1}^k = \mathbf{z}_t^k - \eta(\nabla\psi(\mathbf{z}_t^k; \xi_t^k) + \frac{1}{\gamma}(\mathbf{z}_t^k - \mathbf{z}_0))$
4:    **if** $t + 1 \mod I = 0$ **then**
5:       $\mathbf{z}_{t+1}^k = \frac{1}{K}\sum_{k=1}^{K}\mathbf{z}_{t+1}^k$ {K is the total number of workers}
6:    **end if**
7: **end for**

Similarly, we have the following proposition for primal and dual solution of the projected softmax classifier.

Proposition 2. *Let $\widehat{W}^* \in \mathbb{R}^{m \times C}$ be the optimal primal solution of the projected problem and $\widehat{\Theta}^* \in \mathbb{R}^{n \times C}$ be the optimal dual solution for the projected problem. We have*

$$\widehat{W}^* = -\frac{1}{\lambda}\frac{1}{m}R^T X\widehat{\Theta}^{*T}, \text{ and } \widehat{\Theta}^* = (\nabla L_{y_1}(\hat{\mathbf{o}}_1^*), \ldots, \nabla L_{y_n}(\hat{\mathbf{o}}_n^*)), \tag{8}$$

where $\hat{\mathbf{o}}_i^* = \frac{1}{m}(\mathbf{R}\mathbf{z}_i^*)^T \mathbf{x}_i, \ldots, \frac{1}{m}(\mathbf{R}\mathbf{z}_C^*)^T \mathbf{x}_i, \ i = 1, \ldots, n.$

In the following theorem, we show that we can recover $\widehat{W}^*$ to W^* with a small error.

Theorem 1. *We recover the optimal solution from dual variables* $\widetilde{W} = -\frac{1}{\lambda}\mathbf{X}\widehat{\Theta}^{*T}$. *For any* $0 < \epsilon \leq 1/2$, *with a probability at least* $1 - \delta$, *we have*

$$\|\widetilde{W} - W^*\|_F \leq \frac{\epsilon}{1 - \epsilon}\|W^*\|_F, \tag{9}$$

provided

$$m \geq \frac{(r + 1)\log(2r/\delta)}{c_0 \epsilon^2},$$

where constant c_0 *is at least* $1/4$, *and* r *is the rank of* $f(X) = f(\mathbf{x}_1, \ldots, \mathbf{x}_n)$.

Remarks: The above theorem shows that the recovery error between recovered softmax classifier and original softmax classifier is bounded with a relative small error, which implies that we can use the recovered softmax classifier $\widetilde{W}$ to approximate the original W. The proof is skipped and included in the supplement.

2.3 Parallel Training with Communication Reduction

For deep learning in practice, the scales of model and dataset can easily reach a million or billion level, which makes it very inefficient to train on a single GPU. To further accelerate the training, a parallel multi-worker system is a common solution. In this setting, each worker performs an individual task concurrently and updates the local solution by the average of all individuals (e.g., gradient or weight averaging) at each iteration. For the moderate scale classification task, the averaging cost is usually ignored compared to other costs. However, this is not the case for extreme classification. The communication cost of averaging can be a real bottleneck for overall training process. To address this issue, we adopt a strategy inspired by *Parallel Restarted SGD* [23] to reduce the number of communication rounds between local workers and the gradient algorithm with the communication reduction is presented in Algorithm 2.

Here, we use the same formulation as in the first stage of our approach:

$$\min_{\mathbf{z}} \mathcal{L}(\mathbf{z}) = \sum_{i=1}^{n} \mathcal{L}(\mathbf{z}; \mathbf{x}_i, y_i), \tag{10}$$

where $\mathcal{L}(\mathbf{z}; \mathbf{x}_i, y_i)$ is a loss function. Next, we analyze the convergence rate of $\mathcal{L}(\mathbf{z})$ for Algorithm 1 with Algorithm 2 and we show how distributed machines can speed up the training and how we reduce the communication cost.

We make the following assumptions for analysis:

Assumption 1
(1) Loss function $\mathcal{L}(\mathbf{z}; \xi)$ is L-smooth on $\mathbf{z}$ for any i. Thus it is also L-weakly convex.
(2) $\|\nabla\mathcal{L}(\mathbf{z}) - \nabla\mathcal{L}(\mathbf{z}; \xi)\|_2^2 \le \sigma^2$ for any i.
(3) $\mathcal{L}(\mathbf{z})$ satisfies μ-PL condition, i.e., $\mu(\mathcal{L}(\mathbf{z}) - \mathcal{L}(\mathbf{z}_\mathcal{L}^*)) \le \frac{1}{2}\|\nabla\mathcal{L}(\mathbf{z})\|^2$.

The analysis of one call of Algorithm 2 is shown in the following lemma.

Lemma 1 (Analysis of Algorithm 2). *Suppose Assumption 1 holds and $\|\nabla\mathcal{L}(\mathbf{z}; \mathbf{x}_i, y_i) + \frac{1}{\gamma}(\mathbf{z} - \mathbf{z}_0)\|_2 \le B$. Let $\psi(\mathbf{z}) = \sum_{i=1}^{n}\mathcal{L}(\mathbf{z}; \mathbf{x}_i, y_i) + \frac{1}{2\gamma}\|\mathbf{z} - \mathbf{z}_0\|^2$. By running Algorithm 2, taking $\gamma = \frac{1}{2L}$, and $\eta \le \frac{1}{3L}$, we have*

$$E[\mathcal{L}(\mathbf{z}) - \mathcal{L}(\mathbf{z}_*)] \le \frac{\|\mathbf{z}_0 - \mathbf{z}^*\|^2}{2\eta T} + 2L\eta^2 I^2 B^2 \mathbb{I}_{I>1} + \frac{\eta\sigma^2}{K}.$$

With this lemma, we have the following theorem to show the convergence of the Algorithm 1.

Theorem 2. *Suppose the same condition in Lemma 2 holds. Set $\gamma = \frac{1}{2L}$, $\eta_0 \le \frac{1}{3LK}$ and $c_1 = \frac{\mu/L}{5+\mu/L}$. Take $\eta_s = \eta_0 K \exp(-(s-1)c_1)$, $T_s = \frac{2}{L\eta_0 K}\exp((s-1)c_1)$ and $I_s = \max(1, \frac{1}{\sqrt{K\eta_s}})$. To return $\bar{\mathbf{z}}_S$ such that $E[\mathcal{L}(\bar{\mathbf{z}}_S) - \mathcal{L}(\mathbf{z}_*)] \le \epsilon$, the number of iterations is at most $\tilde{O}\left(\max(\frac{1}{\mu\epsilon\eta_0 K}, \frac{1}{\mu^2 K\epsilon})\right)$ and the number of communications is at most $\tilde{O}\left(\max\left(\frac{K}{\mu} + \frac{1}{\mu(\eta_0\epsilon)^{1/2}}, \frac{K}{\mu} + \frac{1}{\mu^{3/2}\epsilon^{1/2}}\right)\right)$, where $\tilde{O}$ suppresses logarithmic factors and appropriate constants.*

Remark. Take $\eta_0 = \frac{1}{3LK}$. If $K \le O(\frac{1}{\mu})$, then number of iteration is dominated by $O(\frac{L}{\mu^2 K\epsilon})$ since μ is very small in deep learning [24]. Then we can see that there is a linear speedup over the convergence speed with respect to the number of machines K. In addition, the number of communications is much less than the number of total iterations. When $I = 1$, Algorithm 2 reduces to standard parallel SGD.

3   Experiments

3.1   Implementation Details

Tasks. We conduct a comprehensive study to verify the effectiveness of our proposed methods on two tasks. **Representation Learning**: we aim to justify the quality of the feature extractor trained on slimmed model with projected softmax classifier and the quality of the feature extractor from the same model with different parallel settings. The validation process is conducted on a facial verification task using only embedding features and the recovery producers are ignored. For training, we implement our algorithms on three facial datasets,

Celeb [10], Megaface [18] and Mugshot [6]. For validation, we test our models on three benchmark sets, Labelled Faces in the Wild (LFW), AgeDatabase (AgeDB) and Celebritiesin Frontal Profile (CFP). **Classification**: we aim to justify the quality of the recovered softmax classifier. Training and evaluation are both done on classification datasets, such as, ImageNet [22], Penn Treebank (PTB) [20] and Google Billion Words (GBW) [5] with a varying number of classes from one thousand to one million. The detailed descriptions of the above datasets are summarized in Table 1.

Table 1. Datasets summary. "M" and "K" indicate millions and thousands.

Datasets	#samples/words	#classes	Task
ImageNet	1.2 M	1 K	Classification
Penn Treebank	929 K	10 K	Classification
Google Billion Words	800 M	0.79 M	Classification
MS-Celeb-1M	5.3 M	93 K	Representation
MegaFace2	4.7 M	0.67 M	Representation
Mugshot-1M	3.5 M	1 M	Representation

Experiment Setup. For the preprocessing steps, we follow the similar producers of their original works for all datasets. For ImageNet, we apply random cropping and left-right flipping on raw images, resulting in 224 × 224 random images at each iteration. For facial datasets, we augment the images by left-right flipping and resize them to 112 × 112. For language modeling datasets, we reshape the data by time steps in order to fit the time-series model (LSTM). The facial and language datasets are sorted by the class occurrence in order to implement sampling-based methods. We adopt the ResNet50 and two variants of LSTM as backbones and we use multiple 11 GB GTX2080-Ti or a 32 GB V100 for training.

Table 2. Model performance varying the size of random projection m.

RP(m)	Celeb	Megaface	Mugshot	Imagenet	RP(m)	PTB	GBW
10	0.934	0.815	0.631	0.696	10	137.59	122.16
100	0.965	0.845	0.671	0.740	100	86.15	52.28
1000	0.960	0.831	0.664	0.739	**500**	83.98	46.34

Parameters. We employ the original version of ResNet50 [13] with the embedding size of 2048 for all images classification tasks. The baselines are trained using SGD and L2 regularization with a weight decay of 0.0001. The learning rate is scheduled in a stage-wise fashion: starting with a initial learning rate 0.1, dividing it by 10 at {30, 60, 90} and {10, 16} epoch for ImageNet and Facial datasets [7]. For language modeling tasks, we use the configurations of LSTM [14] from [17,25]. For GBW, Adagrad with initial learning rate of 0.2 is used for optimization. The size of word embedding and feature embedding are {1500, 1500} and {1024, 1024} for PTB and GBW, respectively. The number of time steps for LSTM model is set to 35 and 20 respectively. The initial random seeds are fixed for all experiments.

Evaluation Metrics. We use *GPU Hrs* to measure the training speed for all experiments. The speedup is defined as T_2/T_1 in order to measure the relative improvement of the first algorithm with respect to the second algorithm, where T_1 is the runtime for the first algorithm and T_2 is the runtime for the second algorithm. For representation learning, we report *verification accuracy* for face verification task, identifying whether two given face images are the same person. For language modeling, we evaluate *perplexity*, i.e., $2^{-\sum_x p(x)\log_2 p(x)}$, where $p(x)$ is the output probability of a sample x.

3.2 Ablation Study on the Choice of Random Projection Size

We first explore the effect of the model performance by varying the random projection size m. Considering both model performance and training efficiency, we set the upper bound of projection size m to be the half of embedding size d. The embedding sizes of ResNet50/LSTM are {2048, 1500, 1024}, thus we set $m = \{10, 100, 500/1000\}$. For facial datasets, we report the average score of three validation sets. Table 2 shows the model behaviors of different choice of m for all datasets. The results tell that the best choice of m is 100 for Celeb, MegaFace and Mugshot. The best choice for language datasets is $m = 500$ with the lowest test perplexity. We fix $m = 100/500$ for the following experiments.

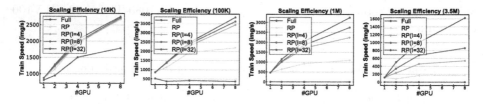

Fig. 3. Multi-GPU scaling efficiency on different settings. The RP softmax has a clear advantage over Full softmax when number of classes is over 100K.

3.3 Ablation Study on Multi-GPU Scaling Efficiency

We further explore the runtime by varying number of classes in a 8-GPU GTX2080-Ti workstation. We choose ResNet50 as the main network which has around 23.5 million parameters. We compare the full softmax (Full) with the random projected (RP) softmax, e.g., $m = 100$ under different number of communication reduction rounds $I = \{4, 8, 32\}$. We fit the synthetic images ($112 \times 112 \times 3$) to GPU memory before training to avoid the penitential issues caused by data I/O operations. We choose number of classes C to be $\{10K, 100K, 1000K, 3500K\}$, leading to model size: Full = $\{43.98M, 228.30M, 2071M, 7801M\}$, RP =$\{24.5M, 33.5M, 123.5M, 345M\}$. Figure 3 shows the scaling efficiency over the number of GPUs used. We can observe that the cost of communication (averaging operations) can be ignored when the number of classes C is small and we obtain a significant performance boost by directly increasing the number of GPUs. When C reaches million level, vanilla Full and RP softmax have terrible scaling performance but RP softmax with communication reduction I has a clear advantage. By choosing a proper I, RP softmax can achieve nearly linear speedup when $C = 3.5$ millions.

3.4 Ablation Study on Communication Rounds

From the results in previous section, we have found that both Full softmax and RP softmax ($I = 1$) have limitations when C is large. Next, we investigate the proper choice of communication round I in order to achieve a good balance in both training speed and model performance. We first vary I by fixing K (#gpus). We compare $I = \{1, 4, 8, 32\}$ on $K = \{2, 4, 8\}$. We conduct the experiments on MegaFace, which contains 0.67M classes and all models are trained with the same number of epochs. In Fig. 4, $I = 1$ ($K = 1$) is the baseline trained on a single GPU with batch size of 256. As the increase of number K, we also use the increasing batch size $K \times 256$ with initial learning rate $K \times \eta_0$. From the results, the algorithms with $I >= 1$ generally coverage much faster than baselines. In particular, $I = 8$ seems to be the best choice considering both speed and performance. However, we also noticed that there is a significant performance drop when $I = 32$, which may indicate the upper limits of the choice of I. In the last figure, we show the result by varying K and fixing I.

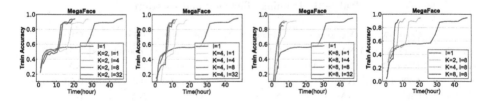

Fig. 4. Performance of different choice of I and K. For the first three figures, we fix K and vary I. For the last figure, we fix I and vary K.

4   Evaluation Results

We compare our proposed approach with several baselines. The full description of all methods is listed below:

1. **Full-D**: full softmax with data parallel
2. **Full-D-C**: full softmax with data and center parallel
3. **S-S**: sampled softmax [15]
4. **Pretrained**: pretrained model on a random sampled subset of full dataset
5. **RP**: random projected softmax classifier
6. **RP-A**: an adaptive variant of random projected softmax classifier

4.1   Results on Speedup

We first examine the results of the training speed (GPU hour) of the listed methods. We set a fixed batch size for all experiments on the same dataset, leading to a different number of utilized GPUs due to the varying model size. The detailed evaluations are presented in Table 3. For the pair of two numbers in the table, the first number denotes GPU hour per epoch and the second number denotes the speedup. We set the speedup for the baseline to **1.00** for the convenience. NA denotes the method is not applicable due to memory issues or other reasons. From the results, we can see that as the increase of the number of classes, RP softmax approaches achieve significant improvement for training efficiency on facial recognition tasks. This is reasonable since we directly reduce the model parameters on softmax layer. Unlike ResNet50, RP softmax on LSTM achieves a lower speedup on a similar scale dataset, e.g., MegaFace v.s. GBW. This is due to the difference between two network architectures: ResNet has 23.5M while LSTM has 450M parameters for the model excluding the softmax layer. Thus, we utilize the RP softmax with sampled softmax, which achieves a speedup of 12.5× v.s. 2.97× without sampled softmax.

Table 3. Training efficiency and speedup

	Celeb-1M	MegaFace	Mugshot-1M	ImageNet	PTB	GBW
Full-D	14.18 (1.00×)	NA	NA	2.38 (1.00×)	0.061(1.00×)	3.51 (1.00×)
Full-D-C	4.96 (2.86×)	27.36 (1.00×)	28.96 (1.00×)	2.36 (1.00×)	NA	NA
S-S	4.69 (3.02×)	NA	NA	2.36 (1.00×)	0.061(1.00×)	0.53 (6.62×)
Pretrained	NA	NA	NA	NA	NA	NA
RP	**2.39 (5.93×)**	**2.53 (10.83×)**	**2.94 (11.84×)**	2.36 (1.00×)	0.055(1.11×)	**0.26 (12.5×)**
RP-A	2.60 (5.46×)	2.72 (10.03×)	3.32 (10.48×)	NA	0.055(1.11×)	1.32 (2.66×)

Table 4. Model performance on validation sets

	Celeb-1M			MegaFace			Mugshot-1M			ImageNet	PTB	GBW
	LFW	CFP	AgeDB	LFW	CFP	AgeDB	LFW	CFP	AgeDB	Val acc	Perplexity	Perplexity
Full-D	99.55	97.03	94.97	NA	NA	NA	NA	NA	NA	**75.69**	78.26	59.07
Full-D-C	99.57	**97.34**	95.03	95.58	79.63	71.25	84.33	66.60	50.82	75.12	NA	NA
S-S	99.03	94.10	88.05	NA	NA	NA	NA	NA	NA	70.11	79.99	47.48
Pretrained	98.91	87.80	92.53	NA	NA	NA	NA	NA	NA	NA	NA	NA
RP	**99.62**	96.80	**95.22**	**96.60**	**82.46**	**74.40**	**90.70**	**69.71**	**57.27**	73.98	83.98	**46.38**
Recovery	NA	NA	NA	NA	NA	NA	NA	NA	NA	74.56	**74.78**	46.56
RP-A	99.48	97.14	95.42	96.95	83.99	74.85	90.83	68.97	59.23	NA	81.76	74.3

4.2   Results on Model Performance

Representation Learning. Our focus is to verify the quality of the feature extractor trained by slimmed model with projected softmax classifier. We report the verification accuracy based on embedding features on three sets in Table 4. From the table, we observed some counter-intuitive results that RP-based approaches outperform full softmax approaches in general. Recall our observations in the Fig. 1 that there is a dramatic increase in the model size for full softmax approaches, which potentially links to another fact that the extremely large models are more difficult to train than slimmed models with the limited choice of step size, batch size, number of epochs. This explains why RP-based methods achieve better performance. Meanwhile, the results also verify our previous argument that the slimmed model can also generate high-quality embedding features as the original model. In terms of training speed, RP-based methods spend much less time on training than original baselines in overall.

Classification. We compute the classification scores for evaluation on validation sets. For the results in Table 4, we summarize several key observations: 1) RP softmax achieves better score than S-S softmax in GBW while there is no performance boost on PTB. 2) Training with RP softmax on ImageNet has no improvement on both performance and speed. The first observation can be explained by a similar reason that the extremely large model is more difficult to train but we can actually improve the score by recovery approach. The second observation on ImageNet is due to the parameters in softmax classifier only takes a very small portion of entire model and thus projected softmax will not lead to a significant parameter reduction and improvement on training speed. In addition, ImageNet doesn't follow the long-tailed distribution, and thus random projected softmax may carry some negative effect on the model performance.

4.3   Results on Recovered Softmax Classifier

From previous section, we have noticed that training with random projected softmax suffers a performance drop on the datasets with a small number of classes. Thus, we investigate whether our recovery approach can recover the projected softmax classifier $\hat{\mathbf{W}}$ to full softmax classifier $\mathbf{W}$ to improve the performance. We use pretrained model from stage 1 to recover the full classifier $\mathbf{W}$ from $m \times C$ to $d \times C$. In this case, $L_{y_i}(\mathbf{o}_i) = -\log \frac{\exp([\mathbf{o}_i]_{y_i})}{\sum_{k=1}^{C} \exp([\mathbf{o}_i]_k)}$, the recovery formula is shown as follow

$$\widetilde{W} = -\frac{1}{\lambda}\mathbf{X}(P(\widehat{W}^*) - \mathbf{Y})^T \tag{11}$$

where $P(\widehat{W}^*) \in \mathbb{R}^{n \times C}$ indicates the predicted probability by $\widehat{W}^*$, i.e., $[P(\widehat{W}^*)]_{i,j} = \frac{\exp \mathbf{w}_i^{*T}\mathbf{x}_j}{\sum_{c=1}^{C} \exp \mathbf{w}_c^{*T}\mathbf{x}_j}$, and $\mathbf{Y} \in \{0,1\}^{C \times n}$ denotes the labels in one-hot encoding. Then, we finetine the recovered softmax classifier for a few epochs. In Table 4, the recovered classifier improves 1% accuracy after 4 epochs training on ImageNet, which is slightly worse than the baseline. For PTB, we achieve a test perplexity of 78.29 after training for 18 epochs and achieve a perplexity of 74.78 after training for 33 epochs. For GBW, we train 10 h to recover the projected classifier and achieve a competitive test perplexity of 46.56. The above results indicate that the recovered classifier has a similar or even better decision-making capability compared to original classifier.

4.4   Results on Parallel Training with Communication Reduction

We explore the optimal choice of I and K to accelerate the parallel training on Mugshot and Celeb datasets. We denote the proposed Algorithm 2 as RP with model averaging (**RP-model**) and compare it with three baselines, Full-D, Full-D-C, RP with gradient averaging (**RP-grad**). We compare $K = \{2, 4, 8\}$ and $I = \{4, 8\}$ and use the increasing batch size and learning rate of all experiments for each dataset. The results are reported in Table 5 in terms of verification accuracy and training time. It shows that RP-model outperforms the baselines when $K = 2$ on Mugshot and Celeb. Further, as we increase K to 8, training time per epoch can be reduced by a large margin of 90%, e.g., from 3.62 to 0.35 hrs when $I = 8$ on Mugshot and from 3.55 to 0.39 h when $I = 8$ on Celeb. In other words, we can train one million classification task in just few hours instead of days for Full-D. However, we also observe some performance drops when I, K are relatively large, but the results are still competitive to Full-D/Full-D-C.

4.5   An Adaptive Variant of Random Projected Softmax

Inspired by [1,27], we propose an adaptive variant of RP softmax, called adaptive RP softmax (RP-A). The intuition is simple: we assign a larger projection size to "head" classes and a smaller projection size to "tail" classes since "head" classes contain more information than "tail" classes. For implementation, we group all classes into clusters according to their occurrence, and assign each cluster with a varying projection size m from large to small. For example, we group all classes into four clusters with $m = \{100, 70, 40, 10\}$ in our experiments. For the results in Table 4, RP-A outperforms the baselines in some settings.

Table 5. Model performance on parallel training.

Methods	K	I	Mugshot			Train time		K	I	Celeb			Train time	
			LFW	CFP	AgeDB	GPU	Hrs			LFW	CFP	AgeDB	GPU	Hrs
Full-D	8	NA	NA	NA	NA	NA	NA	4	1	99.57	97.34	95.03	14.18	3.55
Full-D-C	8	1	84.33	66.60	50.82	28.96	3.62	4	1	99.55	97.03	94.97	4.96	1.24
RP-grad	2	1	90.70	69.71	57.27	2.94	1.47	2	1	99.62	96.68	95.22	2.39	1.20
RP-model	2	4	90.75	68.77	59.23	1.95	0.98	2	4	99.70	97.45	95.91	2.04	1.02
RP-model	2	8	91.03	70.57	61.05	1.90	0.95	2	8	99.60	97.07	95.46	2.08	1.04
RP-model	4	4	90.72	69.14	60.96	2.42	0.61	4	4	99.68	97.25	96.00	2.47	0.62
RP-model	4	8	89.80	67.21	59.93	2.09	0.52	4	8	99.65	97.34	95.78	2.35	0.59
RP-model	8	4	88.20	67.65	59.61	3.47	0.43	8	4	99.66	97.25	95.76	3.29	0.41
RP-model	8	8	85.45	67.20	56.23	2.83	0.35	8	8	99.60	97.11	95.05	3.12	0.39

5 Conclusion

In this paper, we proposed an effective framework to tackle extreme classification problem. Our framework is able to train deep learning models with millions of classes on various tasks. In our numerical experiments, we have verified that the slimmed model with projected softmax classifier can also generate high-quality embedding features as good as embedding features trained from the original model and meanwhile our methods spend less time on training. In addition, we also demonstrate that the recovered softmax classifier can achieve a competitive or even better classification performance. Finally, we successfully extend our framework to large-scale parallel settings and it achieves good results in terms of training efficiency and model performance.

References

1. Blanc, G., Rendle, S.: Adaptive sampled softmax with kernel based sampling. arXiv preprint arXiv:1712.00527 (2017)
2. Borwein, J., Lewis, A.S.: Convex Analysis and Nonlinear Optimization: Theory and Examples. Springer, New York (2010)
3. Boutsidis, C., Zouzias, A., Drineas, P.: Random projections for k-means clustering. In: Advances in Neural Information Processing Systems, pp. 298–306 (2010)
4. Cesa-Bianchi, N., Lugosi, G.: Prediction, Learning, and Games. Cambridge University Press, Cambridge (2006)
5. Chelba, C., et al.: One billion word benchmark for measuring progress in statistical language modeling. arXiv preprint arXiv:1312.3005 (2013)
6. Data.gov: Nist mugshot identification database mid. NIST Mugshot Identification Database MID - NIST Special Database 18 (2016)
7. Deng, J., Guo, J., Xue, N., Zafeiriou, S.: ArcFace: additive angular margin loss for deep face recognition. In: Proceedings of the IEEE Conference on Computer Vision and Pattern Recognition, pp. 4690–4699 (2019)
8. Devlin, J., Chang, M.W., Lee, K., Toutanova, K.: Bert: pre-training of deep bidirectional transformers for language understanding. arXiv preprint arXiv:1810.04805 (2018)

9. Goyal, P., et al.: Accurate, large minibatch SGD: training ImageNet in 1 hour. arXiv preprint arXiv:1706.02677 (2017)
10. Guo, Y., Zhang, L., Hu, Y., He, X., Gao, J.: MS-Celeb-1M: a dataset and benchmark for large-scale face recognition. In: Leibe, B., Matas, J., Sebe, N., Welling, M. (eds.) ECCV 2016. LNCS, vol. 9907, pp. 87–102. Springer, Cham (2016). https://doi.org/10.1007/978-3-319-46487-9_6
11. Gutmann, M., Hyvärinen, A.: Noise-contrastive estimation: a new estimation principle for unnormalized statistical models. In: Proceedings of the Thirteenth International Conference on Artificial Intelligence and Statistics, pp. 297–304 (2010)
12. He, K., Gkioxari, G., Dollár, P., Girshick, R.: Mask R-CNN. In: Proceedings of the IEEE International Conference on Computer Vision, pp. 2961–2969 (2017)
13. He, K., Zhang, X., Ren, S., Sun, J.: Deep residual learning for image recognition. In: Proceedings of the IEEE Conference on Computer Vision and Pattern Recognition, pp. 770–778 (2016)
14. Hochreiter, S., Schmidhuber, J.: Long short-term memory. Neural Comput. 9(8), 1735–1780 (1997)
15. Jean, S., Cho, K., Memisevic, R., Bengio, Y.: On using very large target vocabulary for neural machine translation. arXiv preprint arXiv:1412.2007 (2014)
16. Jia, X., et al.: Highly scalable deep learning training system with mixed-precision: training ImageNet in four minutes. arXiv preprint arXiv:1807.11205 (2018)
17. Jozefowicz, R., Vinyals, O., Schuster, M., Shazeer, N., Wu, Y.: Exploring the limits of language modeling. arXiv preprint arXiv:1602.02410 (2016)
18. Kemelmacher-Shlizerman, I., Seitz, S.M., Miller, D., Brossard, E.: The MegaFace benchmark: 1 million faces for recognition at scale. In: Proceedings of the IEEE Conference on Computer Vision and Pattern Recognition, pp. 4873–4882 (2016)
19. Krizhevsky, A., Sutskever, I., Hinton, G.E.: ImageNet classification with deep convolutional neural networks. In: Advances in Neural Information Processing Systems, pp. 1097–1105 (2012)
20. Miltsakaki, E., Prasad, R., Joshi, A.K., Webber, B.L.: The PENN discourse treebank. In: LREC (2004)
21. Nesterov, Y.E.: Introductory Lectures on Convex Optimization - A Basic Course, Applied Optimization, vol. 87. Springer, New York (2004). https://doi.org/10.1007/978-1-4419-8853-9
22. You, Y., Zhang, Z., Hsieh, C.J., Demmel, J., Keutzer, K.: ImageNet training in minutes. In: Proceedings of the 47th International Conference on Parallel Processing, p. 1. ACM (2018)
23. Yu, H., Yang, S., Zhu, S.: Parallel restarted SGD with faster convergence and less communication: demystifying why model averaging works for deep learning. In: Proceedings of the AAAI Conference on Artificial Intelligence, vol. 33, pp. 5693–5700 (2019)
24. Yuan, Z., Yan, Y., Jin, R., Yang, T.: Stagewise training accelerates convergence of testing error over SGD. In: Advances in Neural Information Processing Systems 32: Annual Conference on Neural Information Processing Systems 2019, NeurIPS 2019, Vancouver, BC, Canada, 8–14 December 2019, pp. 2604–2614 (2019)
25. Zaremba, W., Sutskever, I., Vinyals, O.: Recurrent neural network regularization. arXiv preprint arXiv:1409.2329 (2014)
26. Zhang, L., Mahdavi, M., Jin, R., Yang, T., Zhu, S.: Recovering the optimal solution by dual random projection. In: Conference on Learning Theory, pp. 135–157 (2013)
27. Zhang, X., Yang, L., Yan, J., Lin, D.: Accelerated training for massive classification via dynamic class selection. In: Thirty-Second AAAI Conference on Artificial Intelligence (2018)

Password-Conditioned Anonymization and Deanonymization with Face Identity Transformers

Xiuye Gu[1,2](✉) [iD], Weixin Luo[2,3] [iD], Michael S. Ryoo[4] [iD], and Yong Jae Lee[2] [iD]

[1] Stanford University, Stanford, USA
laoreja0922@gmail.com
[2] UC Davis, Davis, USA
[3] ShanghaiTech, Shanghai, China
[4] Stony Brook University, Stony Brook, USA

Abstract. Cameras are prevalent in our daily lives, and enable many useful systems built upon computer vision technologies such as smart cameras and home robots for service applications. However, there is also an increasing societal concern as the captured images/videos may contain privacy-sensitive information (*e.g.*, face identity). We propose a novel *face identity transformer* which enables automated photo-realistic password-based anonymization and deanonymization of human faces appearing in visual data. Our face identity transformer is trained to (1) remove face identity information after anonymization, (2) recover the original face when given the correct password, and (3) return a wrong—but photo-realistic—face given a wrong password. With our carefully designed password scheme and multi-task learning objective, we achieve both anonymization and deanonymization using the same single network. Extensive experiments show that our method enables multimodal password conditioned anonymizations and deanonymizations, without sacrificing privacy compared to existing anonymization methods.

1 Introduction

As computer vision technology is becoming more integrated into our daily lives, addressing privacy and security questions is becoming more important than ever. For example, smart cameras and robots in homes are widely being used, but their recorded videos often contain sensitive information of their users. In the worst case, a hacker could intrude these devices and gain access to private information.

Recent anonymization techniques aim to alleviate such privacy concerns by redacting privacy-sensitive data like face identity information. Some methods [3,27] perform low-level image processing such as extreme downsampling, image masking, etc. A recent paper proposes to *learn* a face anonymizer that

Electronic supplementary material The online version of this chapter (https://doi.org/10.1007/978-3-030-58592-1_43) contains supplementary material, which is available to authorized users.

© Springer Nature Switzerland AG 2020
A. Vedaldi et al. (Eds.): ECCV 2020, LNCS 12368, pp. 727–743, 2020.
https://doi.org/10.1007/978-3-030-58592-1_43

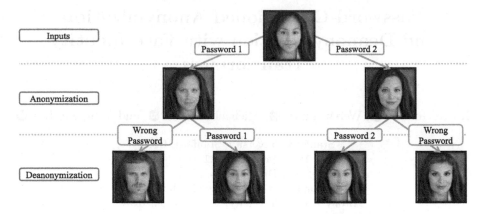

Fig. 1. Our system never stores users' faces on disk, and instead only stores the anonymized faces. When a user provides a correct recovery password, s/he will get the deanonymized face back. If a hacker invading their privacy inputs a wrong password, s/he will get a face whose identity is different from the original as well as the anonymized face. The photo-realism of the modified faces is meant to fool the hacker by providing no clues as to whether the real face was recovered.

modifies the identity of a face while preserving activity relevant information [24]. However, none of these techniques consider the fact that the video/image owner (and his/her friends, family, law enforcement, etc.) may want to see the *original* identities and not the anonymized ones. For example, people may not want their real faces to be saved directly on home security cameras due to privacy concerns; however, remote family members may want to see the real faces from time to time. Or when crimes arise, to catch criminals, police need to see their real faces.

This problem poses an interesting tradeoff between privacy and accessibility. On the one hand, we would like a system that can anonymize sensitive regions (face identity) so that even if a hacker were to gain access to such data, they would not be able to know who the person is (without additional identity revealing meta-data). On the other hand, the owner of the visual data inherently wants to see the original data, not the anonymized one.

To address this issue, we introduce a novel *face identity transformer* that can both *anonymize and deanonymize (recover)* the original image, while maintaining privacy. We design a discrete password space, in which the password conditions the identity change. Specifically, given an original face, our face identity transformer outputs different anonymized face images with different passwords (Fig. 1 Anonymization). Then, given an anonymized face, the original face is recovered only if the correct password is provided (Fig. 1 Deanonymization, 'Password 1/2'). We further increase security as follows: Given an anonymized face, if a wrong password is provided, then it changes to a new identity, which is still different from the original identity (Fig. 1 Deanonymization, 'Wrong Password'). Moreover, each wrong password maps to a unique identity. In this way, we provide security via ambiguity: even if a hacker guesses the correct password, it is extremely difficult to know that without having access to any other identity revealing meta-data, since each password—regardless of whether it is correct or not—always leads to a different realistic identity.

To enforce the face identity transformer to output different anonymized face identities with different passwords, we optimize a multi-task learning objective, which includes maximizing the feature-level dissimilarity between pairs of anonymized faces that have different passwords and fooling a face classifier. To enforce it to recover the original face with the correct password, we train it to anonymize and then recover the correct identity only when given the correct password, and to produce a new identity otherwise. Lastly, we maximize the feature dissimilarity between an anonymized face and its deanonymized face with a wrong password so that the identity always changes. Moreover, considering the limited memory space on devices, we propose to use the same single transformer to serve both anonymization and deanonymization purposes.

We note that our approach is related to cryptosystems like RSA [25]. The key difference is that cryptosystems do not produce encryptions that are visually recognizable to human eyes. However, in various scenarios, users may want to understand what is happening in anonymized visual data. For example, people may share photos/videos over public social media with anonymized faces, but only their real-life friends have the passwords and can see their real faces to protect identity information. Moreover, with photorealistic anonymizations, one can *easily apply existing computer vision based recognition algorithms on the anonymized images* as we demonstrate in Sect. 5.5. In this way, it could work with e.g., smart cameras that use CV algorithms to analyze content but in a privacy-preserving way, unlike other schemes (*e.g.*, homomorphic encryption) that require developing new ad-hoc recognition methods specific to nonphotorealistic modifications, in which accuracy may suffer.

In our approach, only the anonymized data is saved to disk (*i.e.*, conceptually, the anonymization would happen at the hardware-level via an embedded chipset – the actual implementation of which is outside the scope of this work). The advantage of this concept is that the hacker could never have direct access to the original data. Finally, although there may be other identity-revealing information such as gait, clothing, background, etc., our work entirely focuses on improving privacy of face identity information, but would be complementary to systems that focus on those other aspects.

Our experiments on CASIA [32], LFW [11], and FFHQ [13] show that the proposed method enables multimodal face anonymization as well as recovery of original face images, without sacrificing privacy compared to existing advanced anonymization [24] and classical image processing techniques including masking, noising, and blurring, etc. *Please see* https://youtu.be/FrYmf-CL4yk *and Fig. 6 in the supp for image/video in the wild results.*

2   Related Work

Privacy-Preserving Visual Recognition. This is the problem of detecting humans, their actions, and objects without accessing user-sensitive information in images/videos. Some methods employ extreme low-resolution downsampling to

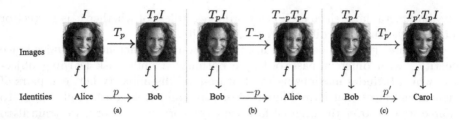

Fig. 2. Privacy-preserving properties that our face identity transformer T learns. (a) Anonymization stage. (b) Deanonymization stage with correct recovery password. (c) Deanonymization stage with incorrect recovery password.

hide sensitive details [6,26,27,30] but suffer from lower recognition performance in downstream tasks. More recent work propose a head inpainting obfuscation technique [29], a four-stage pipeline that first obfuscates facial attributes and then synthesizes faces [15], and a video anonymizer that performs pixel-level modifications to remove people's identity while preserving motion and object information for activity detection [24]. Unlike our approach, none of these existing work employ a password scheme to condition the anonymization, and also do not perform deanonymization to recover the original face. Moreover, even if one could brute-forcely train a deanonymizer for these methods, there is no way to provide wrong recoveries upon wrong passwords, as our method does.

Security/cryptography research on privacy-preserving recognition is also related *e.g.*, [8,9]. The key difference is that these methods encrypt data in a secure but visually-uninterpretable way, whereas our goal is to anonymize the data in a way that is still interpretable to humans and existing computer vision techniques can still be applied. Differential privacy [1,33] is also related but its focus is on protecting privacy in the training data whereas ours is on anonymizing visual data during the inference stage.

Face Image Manipulation and Conditional GANs. Our work builds upon advances in pixel-level synthesis and editing of realistic human faces [2,13,14,20, 21,28] and conditional GANs [5,7,12,19,22,23,34,39], but we differ significantly in our goal, which is to completely change the identity of a face (and also recover the original) for privacy-preserving visual recognition.

3   Desiderata

Our face identity transformer T takes as input a face image $I \in \Phi$ and a user-defined password $p \in P$, where Φ and P denote the face image domain and password domain. We use the notation $T_p I$ to denote the transformed image with input image I and password p. Before diving into the details, we first outline desired properties of a privacy-preserving face identity transformer.

Minimal Memory Consumption. Considering the limited memory space on most camera systems, a *single* face identity transformer that can both anonymize and deanonymize faces is desirable.

Photo-Realism. We would like the transformer to maintain photo-realism for any transformed face image:

$$T_p I \in \Phi, \quad \forall p \in P, \forall I \in \Phi. \tag{1}$$

Photo-realism has three benefits: 1) a human who views the transformed images will still be able to interpret them; 2) one can easily apply existing computer vision algorithms on the transformed images; and 3) it's possible to confuse a hacker, since photo-realism can no longer be used as a cue to differentiate the original face from an anonymized one.

Compatibility with Background. The background $B(\cdot)$ of the transformed face should be the same as the original:

$$B(T_p I) = B(I), \quad \forall p \in P, \forall I \in \Phi. \tag{2}$$

This will ensure that there are no artifacts between the face region and the rest of the image (*i.e.*, it will not be obvious that the image has been altered).

Anonymization with Passwords. Let $f : \Phi \to \Gamma$ denote the function mapping face images to people's identities. We would like to condition anonymization via a password p:

$$f(T_p I) \neq f(I), \quad \forall p \in P, \forall I \in \Phi. \tag{3}$$

Deanonymization with Inverse Passwords. We should recover the original identity only when the correct password is provided. To achieve our goal of minimal memory consumption, we can model the additive inverse of the password used for anonymization as the correct password for deanonymization. In this way, we can use the same transformer for deanonymization, *i.e.* we model $T_{-p} = T_p^{-1}$:

$$f(T_{-p} T_p I) = f(T_p^{-1} T_p I) = f(I), \ \forall p \in P, \forall I \in \Phi. \tag{4}$$

Wrong Deanonymization with Wrong Inverse Passwords. We would like the transformer to change the anonymized identity into a *different* identity that is different from both the original as well as the anonymized image when given a wrong inverse password:

$$f(T_{p'} T_p I) \neq f(I), \quad \forall p, p' \in P, p' \neq -p, \forall I \in \Phi, \tag{5}$$

$$f(T_{p'} T_p I) \neq f(T_p I), \ \forall p, p' \in P, p' \neq -p, \forall I \in \Phi. \tag{6}$$

In this way, whether the password is correct or not, the identity is always changed so as to confuse the hacker.

Diversity. The image I should be transformed to different identities with different passwords, to increase security in both anonymization and deanonymization. Otherwise, if multiple passwords produce the same identity, a hacker could realize that the photo is anonymized or his attempts have failed in deanonymization:

$$f(T_{p_1} I) \neq f(T_{p_2} I), \ \forall p_1, p_2 \in P, p_1 \neq p_2, \forall I \in \Phi. \tag{7}$$

Figure 2 summarizes our desiderata for anonymization and deanonymization.

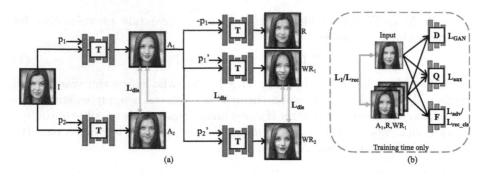

(a)   (b)

Fig. 3. (a) Face identity transformer network architecture. (b) Objectives we apply to synthesized images during training (not included in (a) for clarity). I: Input image, $A_{1,2}$: Anonymized faces, R: Recovered face, $WR_{1,2}$: Wrongly Recovered faces. $\mathcal{L}_{feat}$ is the sum of the three $\mathcal{L}_{dis}$'s.

4   Approach: *Face Identity Transformer*

Our face identity transformer T is a conditional GAN trained with a multi-task learning objective. It is conditioned on both the input image I and an input password p. Importantly, the function of p is different from the usual random noise vector z in GANs: z simply tries to model the distribution of the input data, while p in our case makes the transformer hold the desired privacy-preserving properties (Eq. 3–7). We next explain our password scheme, multimodal identity generation, and multi-task learning objective.

4.1   Password Scheme

We use an N-bit string $p \in \{0,1\}^N$ as our password format, which consists of 2^N unique passwords. Given image $I \in \mathbb{R}^{H \times W \times 3}$, we form the input to the transformer as a depthwise concatenation $(I, p) \in \mathbb{R}^{H \times W \times (3+N)}$, where p is replicated in every pixel location. To make the transformer condition its identity change on the input password, we design an auxiliary network $Q(I, T_p I) = \hat{p}$. It learns to predict the embedded password from the input and transformed image pair, and thus maximizes the mutual information between the injected password and the identity change in the image domain, similar to InfoGAN [7]. We use cross entropy loss for the classifier Q, and denote it as $\mathcal{L}_{aux}(T, Q)$. See supp Sec. 1 for the detailed formula.

4.2   Multimodal Identity Change

Conditional GANs with random noise do not produce highly stochastic outputs [12,18]. To overcome this, BicycleGAN [38] uses an explicitly-encoded multimodality strategy similar to our auxiliary network Q. However, even with Q, we only observe multimodality on colors and textures as in [38], but not on high-level face identity.

Thus, to induce diverse high-level identity changes, we propose an explicit feature dissimilarity loss. Specifically, we use a face recognition model F to extract deep embeddings of the faces, and minimize their cosine similarity when they are associated with different passwords:

$$\mathcal{L}_{dis}(M_1, M_2) = \max\left(0, \cos\left(F_{embed}(M_1), F_{embed}(M_2)\right)\right), \qquad (8)$$

where cos is cosine similarity, and M_1 and M_2 are two transformed face images with two different passwords. We do not penalize pairs whose cosine similarity is less than 0; $i.e.$, it is enough for the faces to be different up to a certain point.

We apply the dissimilarly loss between (1) two anonymized faces with different passwords, (2) two incorrectly deanonymized faces given different wrong passwords, and (3) the anonymized face and wrongly recovered face:

$$\begin{aligned}
\mathcal{L}_{feat}(T) = {} & \mathbb{E}_{(I, p_1 \neq p_2)} \mathcal{L}_{dis}(T_{p_1}I, T_{p_2}I) \\
& + \mathbb{E}_{(I, p_1' \neq p_2', p_1' \neq -p, p_2' \neq -p)} \mathcal{L}_{dis}(T_{p_1'}T_pI, T_{p_2'}T_pI) \\
& + \mathbb{E}_{(I, p' \neq -p)} \mathcal{L}_{dis}(T_pI, T_{p'}T_pI).
\end{aligned} \qquad (9)$$

This loss can be easily satisfied when the model outputs extremely different content that do not necessarily look like a face, and thus can adversely affect other desideratum ($e.g.$, photo-realism) of a privacy-preserving face identity transformer. We next introduce a multi-task learning objective to restrict the outputs to lie on the face manifold, as a form of regularization.

4.3   Multi-task Learning Objective

We describe our multi-task objective that further aids identity change, identity recovery, and photo-realism.

Face Classification Adversarial Loss. We apply the face classification adversarial loss from [24], which helps change the input face's identity. We apply it on both the transformed face T_pI as well as the reconstructed face with wrong recovery password $T_{p'}T_pI$:

$$\begin{aligned}
\mathcal{L}_{adv}(T, F) = {} & - \mathbb{E}_I \mathcal{L}_{CE}\left(F(I), y_I\right) - \mathbb{E}_{(I,p)} \mathcal{L}_{CE}\left(F(T_pI), y_I\right) \\
& - \mathbb{E}_{(I, p' \neq -p)} \mathcal{L}_{CE}\left(F(T_{p'}T_pI), y_I\right),
\end{aligned} \qquad (10)$$

where F is the face classifier, y_I is face identity label, and $\mathcal{L}_{CE}$ denotes cross entropy loss.

Similar to the dissimilarity loss ($\mathcal{L}_{dis}$), this loss pushes the transformed face to have a different identity. The key difference is that this loss requires face identity labels so cannot be used to push $T_{p_1}I$ and $T_{p_2}I$ to have different identities, but has the advantage of utilizing supervised learning so that it can change the identity more directly.

Reconstruction Losses. We use L_1 reconstruction loss for deanonymization:

$$\mathcal{L}_{rec}(T) = \|T_{-p}T_pI - I\|_1. \tag{11}$$

With the L_1 loss alone, we find the reconstruction to be often blurry. Hence, we also introduce a face classification loss $\mathcal{L}_{rec\_cls}$ on the reconstructed face to enforce the transformer to recover the high-frequency identity information:

$$\mathcal{L}_{rec\_cls}(T, F) = \mathbb{E}_{(I,p)}\mathcal{L}_{CE}\big(F(T_{-p}T_pI), y_I\big). \tag{12}$$

This loss enforces the reconstructed face $T_{-p}T_pI$ to be predicted as having the same identity as I by face classifier F.

Background Preservation Loss. For any transformed face, we try to preserve its original background. To this end, we apply another L_1 loss (with lower weight):

$$\mathcal{L}_1(T) = \|T_pI - I\|_1 + \|T_{p'}T_pI - I\|_1. \tag{13}$$

Although employing a face segmentation algorithm is an option, we find that applying $\mathcal{L}_1$ on the whole image works well to preserve the background.

Photo-Realism Loss. We use a photo-realism adversarial loss $\mathcal{L}_{GAN}$ [10] on generated images to help model the distribution of real faces. Specifically, we use PatchGAN [12] to restrict the discriminator D's attention to the structure in local image patches. To stabilize training, we use LSGAN [17]:

$$\max_D \mathcal{L}_{GAN}(D) = -\frac{1}{2}\mathbb{E}_I[(D(I) - 1)^2] - \frac{1}{2}\mathbb{E}_{(I,p)}[D(T_pI)^2] \tag{14}$$

$$\min_T \mathcal{L}_{GAN}(T) = \mathbb{E}_{(I,p)}[(D(T_pI) - 1)^2] \tag{15}$$

4.4   Full Objective

Overall, our full objective is:

$$\begin{aligned}\mathcal{L} = \ &\lambda_{aux}\mathcal{L}_{aux}(T, Q) + \lambda_{feat}\mathcal{L}_{feat}(T) \\ &+ \lambda_{adv}\mathcal{L}_{adv}(T, F) + \lambda_{rec\_cls}\mathcal{L}_{rec\_cls}(T, F) \\ &+ \lambda_{rec}\mathcal{L}_{rec}(T) + \lambda_{L_1}\mathcal{L}_1(T) + \mathcal{L}_{GAN}(T, D).\end{aligned} \tag{16}$$

We optimize the following minimax problem to obtain our face identity transformer:

$$T^* = \arg\min_{T,Q} \max_{D,F} \mathcal{L} \tag{17}$$

Training. Figure 3 shows our network for training. For each input I, we randomly sample two different passwords for anonymization and two incorrect passwords for wrong recoveries, and then impose $\mathcal{L}_{dis}$ on the generated pairs and enforce $\mathcal{L}_{dis}$ between the anonymization and wrong reconstruction. We observe that during training, the auxiliary networks and backprop can consume a lot of GPU memory, which limits batch size. We propose a strategy based on symmetry: except for the feature dissimilarity loss, we apply all other losses only to the first anonymization and first wrong recovery, which empirically works well.

We adopt a two-stage training strategy for the minimax problem [10]. In the discriminator's stage, we fix the parameters of T, Q, and update D, F; in the generator's stage, we fix D, F, and update T, Q.

Inference. During testing, the transformer T takes as input a user-defined password and a face image, anonymizes the face, and saves it to disk. When the user/hacker wants to see the original image, the transformer takes the recovery password and the anonymized image, and either outputs the identity-recovered image or a hacker-fooling image depending on password correctness. Throughout the whole process, the original images and passwords are never saved on disk for privacy reasons.

5   Experiments

In this section, we demonstrate that our face identity transformer achieves password conditioned anonymization and deanonymization with photo-realism and multimodality. We also conduct ablation studies to analyze each module/loss.

Implementation Details. Our identity transformer T is built upon the network from [37]. We use size 128×128 for both inputs and outputs. We subtract 0.5 from p before inputting it to the transformer to make the password channels have zero mean. We set $N = 16$. We use the pretrained SphereFace [16] as our face recognition network F for both deep embedding extraction in the feature dissimilarity loss and face classification adversarial training. For each stage, we use two PatchGAN discriminators [12] that have identical structure but operate at different image scales to improve photo-realism. The coarser discriminator is shared among all stages, while three separate finer discriminators are used for anonymization, reconstruction, and wrong recovery. To improve stability, we use a buffer of 500 generated images when updating D. We set $\lambda_{aux} = 1$, $\lambda_{feat} = 2$, $\lambda_{adv} = 2$, $\lambda_{rec\_cls} = 1$, $\lambda_{L_1} = 10$ and $\lambda_{rec} = 100$, based on qualitative observations.

Datasets. 1) CASIA [32] has 454,590 face images belonging to 10,574 identities. We split the dataset into training/validation/testing subsets made up of 80%/10%/10% identities. We use the validation set to select our model. All reported results are on the test set. 2) LFW [11] has 13,233 face images belonging to 5,749 identities. As our network is never trained on LFW, we evaluate on the entire LFW to test generalization ability. 3) FFHQ [13] is a high-quality face

Table 1. Privacy-preserving ability comparison. Our method is the only one that supports password-conditioned face (de)anonymization without sacrificing privacy.

Method	Anonymize?	Deanonymize?	Password-conditioned?
Ren *et al.* [24]	✔	✔	✘
Super-pixel	✔	✔	✘
Edge	✔	✔	✘
Blur	✔	✘	✘
Noise	✔	✔	✘
Masked	✔	✘	✘
Ours	✔	✔	✔

dataset for benchmarking GANs. It is not a face recognition dataset, so we use it to only test generalization. We directly test our model on its validation set at 128×128 resolution, which contains 10,000 images.

Evaluation Metrics. **Face Verification Accuracy:** We measure our transformer's identity changing ability with a standard binary face verification test, which scores whether a pair of images have the same identity or not. Since different face recognition models may have different biases, we use two popular pretrained face recognition models: SphereFace [16] and VGGFace2 [4].

Face Recovery Quality: We measure face recovery quality using **LPIPS distance** [36], which measures perceptual similarity between two images based on deep features, and **DSSIM** [31], which is a commonly-used low-level perceptual metric. We also use pixel-level L_1 and L_2 distance.

AMT Perceptual Studies: We use Amazon Mechanical Turk (AMT) to test how well our method 1) changes and recovers identities, 2) achieves photorealism, and 3) attains multimodal anonymizations, as judged by human raters.

Runtime: On a single Titan V, averaged over CASIA testset, runtime is 0.0266 sec/batch with 12 images per batch. Though we use multiple auxiliary networks to help achieve our desiderata, they are all discarded during inference time.

5.1   Anonymization and Deanonymization

To our knowledge, *no prior work achieves password-conditioned anonymization and deanonymization on visual data like ours*, see Table 1. Hence, we cannot directly compare with any existing method on generating *multimodal* anonymizations and deanonymizations.

Despite this, we want to ensure that our method does no worse than existing methods in terms of anonymization and deanonymization (setting aside the password conditioning capability). To demonstrate this, following [24], we compare to the following baselines: **Ren *et al.* [24]:** a learned face anonymizer that

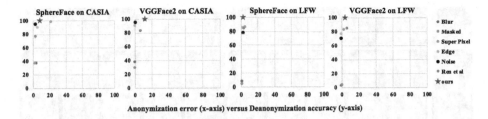

Fig. 4. Anonymization vs. deanonymization quality, measured by face verification error/accuracy on CASIA and LFW. Top-left corner is ideal. This result shows that we don't sacrifice (de)anonymization ability by introducing password conditioning.

Table 2. CASIA and LFW reconstruction error. Ours produces best deanonymizations.

Method	CASIA				LFW			
	LPIPS	DSSIM	L_1	L_2	LPIPS	DSSIM	L_1	L_2
Ren et al.	0.08	0.07	0.06	0.009	0.08	0.07	0.06	0.010
Superpixel	0.09	0.10	0.06	0.01	0.10	0.11	0.07	0.02
Edge	0.25	0.24	0.25	0.24	0.28	0.26	0.29	0.18
Blur	0.30	0.21	0.12	0.04	0.34	0.24	0.14	0.05
Noise	0.12	0.12	0.07	0.01	0.13	0.12	0.08	0.01
Masked	0.10	0.09	0.07	0.02	0.16	0.13	0.10	0.05
Ours	**0.03**	**0.03**	**0.04**	**0.004**	**0.04**	**0.03**	**0.04**	**0.004**

maintains action detection accuracy; **Superpixel** [3]: each pixel's RGB value is replaced with its superpixel's mean RGB value; **Edge** [3]: face regions are replaced with corresponding edge maps; **Blur** [27]: images are downsampled to extreme low-resolution (8×8) and then upsampled back; **Noise**: strong Gaussian noise ($\sigma^2 = 0.5$) is added to the image; **Masked**: face areas ($0.6\times$ of the face image) are masked out.

We also train deanonymizers for each baseline (*i.e.*, to recover the original face), by using the same generator architecture with our reconstruction and photo-realism losses. Please refer to supp Fig. 1 for a qualitative example of the baselines and their anonymizations/deanonymizations.

Figure 4 shows anonymization vs. deanonymization (recovery) quality on CASIA and LFW using SphereFace and VGGFace2 as our face recognizers. Our approach performs competitively to Ren *et al.* [24], "Superpixel", "Edge", "Blur", "Noise", and "Masked" when considering both anonymization and deanonymization quality together. This result confirms that we do not sacrifice the ability to anonymize/deanonymize by introducing password-conditioning. In fact, in terms of reconstruction (deanonymization) quality (Table 2), our method outperforms the baselines by a large margin because we train our identity transformer to do anonymization and deanonymization in conjunction in an end-to-end way.

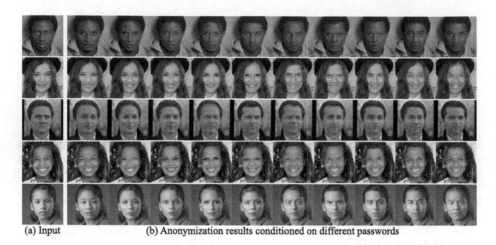

(a) Input   (b) Anonymization results conditioned on different passwords

Fig. 5. Multimodality results on CASIA. We observe a wide range of identity changes with different passwords.

Lastly, we perform AMT perceptual studies to rate our anonymizations and deanonymizations. Specifically, we randomly sample 150 testing images (I), and generate for each image: an anonymized face with a random password (A), a recovered face with correct inverse password (R), and a recovered face with wrong password (WR). We then distribute 600 I vs A, I vs R, I vs WR, and A vs WR pairs to turkers and ask "Are they the same person?". For each pair, we collect responses from 3 different turkers and take the majority as the answer to reduce noise.

The turkers reported **4.7%/100%/0.7%/1.3%** on $IvsA/IvsR/IvsWR$ $/AvsWR$. (low, high, low, low is ideal.) This further shows our method obtains the desired password-conditioned anonymization/deanonymization goals. We show all failure pairs for $IvsA$ in supp Sec. 5 and analyze the error there.

5.2   Photo-Realism

To evaluate whether our (de)anonymization affects photo-realism, we conduct AMT user studies. We follow the same perceptual study protocol from [37] and test on both anonymizations and wrong recoveries. For each test, we randomly generate 100 "real vs. fake" pairs. For each pair, we average responses from 10 unique turkers. Turkers label our anonymizations as being more real than a real face **28.9%** of the time, and label our wrong reconstructions as more real than a real face **15.4%** of the time. (Chance performance is 50%.) This shows that our generated images are quite photo-realistic.

5.3   Multimodality

We next evaluate our model's ability to create different faces given different passwords. Figure 5 shows qualitative results. Our transformer successfully changes

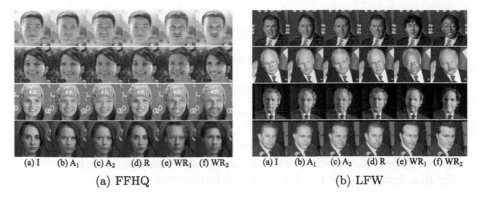

| (a) I | (b) A$_1$ | (c) A$_2$ | (d) R | (e) WR$_1$ | (f) WR$_2$ | | (a) I | (b) A$_1$ | (c) A$_2$ | (d) R | (e) WR$_1$ | (f) WR$_2$ |

(a) FFHQ   (b) LFW

Fig. 6. FFHQ and LFW generalization results. I: original image, $A_{1,2}$: anonymized faces using different passwords, $R/WR_{1,2}$: recovered faces with correct/wrong passwords.

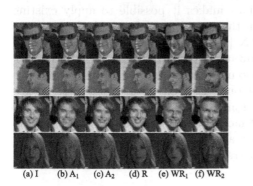

| (a) I | (b) A$_1$ | (c) A$_2$ | (d) R | (e) WR$_1$ | (f) WR$_2$ |

| (a) I | (b) A$_1$ | (c) A$_2$ | (d) R | (e) WR$_1$ | (f) WR$_2$ |

Fig. 7. Hard cases on CASIA. See Fig. 6 caption for key.

Fig. 8. Typical failures of each ablation. See Fig. 6 caption for key.

the identity into a broad spectrum of different identities, from women to men, from young to old, *etc*.

We quantitatively evaluate multimodality through an AMT perceptual study. We ask AMT workers to compare 150 A_1 *vs* A_2 and 150 WR_1 *vs* WR_2 pairs (pairs of anonymized/wrong-recovered faces with different passwords generated from the same input image) and ask "are they the same person?". The turkers reported "yes" only **12.2%** and **2.7%** of the time, respectively (lower is better). The results show that our transformer does quite well in generating different identities given different passwords.

5.4   Generalization and Difficult Cases

Figure 6 shows generalization results on FFHQ and LFW using our model trained on CASIA. Without any fine-tuning, our model achieves good generalization performance on both the high quality FFHQ dataset and the LFW dataset where resolution is usually lower.

Table 3. Average pixel difference in detected coordinates of face bounding boxes and 5 keypoints between transformed faces (A, R, WR) and input face (I).

Avg spatial coordinate difference	CASIA	LFW	FFHQ
Bounding boxes	1.81	1.62	1.91
Keypoints	0.94	0.76	0.89

Figure 7 shows hard-case qualitative results on CASIA. Our method works well even if the faces are with occlusions (sunglasses), with extreme poses, vague, under dim light, etc. We provide more qualitative results in supp.

5.5 Applying CV Algorithms on Transformed Faces

Unlike most traditional anonymization algorithms [3, 27], our choice of achieving photo-realism on the (de)anonymizations makes it possible to apply existing computer vision algorithms directly on the transformed faces. To demonstrate this, we apply an off-the-shelf MTCNN [35] face bounding box and keypoint detector on the transformed faces. Qualitative detection results (see supp Fig. 5) are good. Quantitatively, although we do not have the ground truth annotations for transformed faces, we observe that our (de)anonymizations mostly do not change the head/keypoints' positions from the input faces so we can compare the detection results between the input faces and the transformed faces. Results are shown in Table 3, which shows that a face detection algorithm trained on real images performs accurately on our transformed faces.

5.6 Ablation Studies

Finally, we evaluate the contribution of each component and loss in our model. Here, original image (I), anonymized face with two different passwords $(A_{1,2})$, recovered face with correct inverse password (R), and recovered faces with wrong passwords $(WR_{1,2})$:

w/o $\mathcal{L}_{\mathbf{dis}}$: We remove feature dissimilarity loss on (A_1, A_2) and (WR_1, WR_2).
w/o WR: We do not explicitly train to produce wrong reconstructions.
w/o $\mathcal{L}_{\mathbf{aux}}$: We remove the password-predicting auxiliary network Q, but still embed the passwords.
w/o $\mathcal{L}_{\mathbf{rec\_cls}}$: We remove the face classification loss on the reconstruction.

Figure 8 shows the typical drawbacks of each ablation model. w/o $\mathcal{L}_{dis}$ shows that $\mathcal{L}_{dis}$ is necessary to achieve semantic-level multimodality on both anonymization and wrong reconstruction. w/o WR shows that without training for wrong reconstructions, the transformer fails to conceal identities when given incorrect passwords. w/o $\mathcal{L}_{aux}$ verifies the importance of the auxiliary network, which helps improve photo-realism and we also observe it helps with multimodality. Without $\mathcal{L}_{rec\_cls}$, the reconstruction quality suffers because of unbalanced losses.

6   Discussion

We presented a novel privacy-preserving face identity transformer with a password embedding scheme, multimodal identity change, and a multi-task learning objective. We feel that this paper has shown the promise of password-conditioned face anonymization and deanonymization to address the privacy versus accessibility tradeoff. Although relatively rare, we sometimes notice artifacts that look similar to general GAN artifacts. They tend to arise due to the difficulty of image generation itself – we believe they can be greatly reduced with more advances in image synthesis research, which can be (orthogonally) plugged into our system.

Acknowledgements. This work was supported in part by NSF IIS-1812850, NSF IIS-1812943, NSF CNS-1814985, NSF CAREER IIS-1751206, AWS ML Research Award, and Google Cloud Platform research credits. We thank Jason Ren, UC Davis labmates, and the reviewers for constructive discussions.

References

1. Abadi, M., et al.: Deep learning with differential privacy. In: CCS (2016)
2. Bao, J., Chen, D., Wen, F., Li, H., Hua, G.: Towards open-set identity preserving face synthesis. In: CVPR (2018)
3. Butler, D.J., Huang, J., Roesner, F., Cakmak, M.: The privacy-utility tradeoff for remotely teleoperated robots. In: ICHRI (2015)
4. Cao, Q., Shen, L., Xie, W., Parkhi, O.M., Zisserman, A.: VGGFace2: a dataset for recognising faces across pose and age. In: FG (2018)
5. Cao, Y., Liu, B., Long, M., Wang, J.: HashGAN: deep learning to hash with pair conditional Wasserstein GAN. In: CVPR (2018)
6. Chen, J., Wu, J., Konrad, J., Ishwar, P.: Semi-coupled two-stream fusion convnets for action recognition at extremely low resolutions. In: WACV (2017)
7. Chen, X., Duan, Y., Houthooft, R., Schulman, J., Sutskever, I., Abbeel, P.: Info-GAN: interpretable representation learning by information maximizing generative adversarial nets. In: NeurIPS (2016)
8. Erkin, Z., Franz, M., Guajardo, J., Katzenbeisser, S., Lagendijk, I., Toft, T.: Privacy-preserving face recognition. In: Goldberg, I., Atallah, M.J. (eds.) PETS 2009. LNCS, vol. 5672, pp. 235–253. Springer, Heidelberg (2009). https://doi.org/10.1007/978-3-642-03168-7_14
9. Gilad-Bachrach, R., Dowlin, N., Laine, K., Lauter, K., Naehrig, M., Wernsing, J.: CryptoNets: applying neural networks to encrypted data with high throughput and accuracy. In: ICML (2016)
10. Goodfellow, I., et al.: Generative adversarial nets. In: NeurIPS (2014)
11. Huang, G.B., Mattar, M., Berg, T., Learned-Miller, E.: Labeled faces in the wild: a database for studying face recognition in unconstrained environments. In: Workshop on Faces in 'Real-Life' Images (2008)
12. Isola, P., Zhu, J.Y., Zhou, T., Efros, A.A.: Image-to-image translation with conditional adversarial networks. In: CVPR (2017)
13. Karras, T., Laine, S., Aila, T.: A style-based generator architecture for generative adversarial networks. arXiv:1812.04948 (2018)
14. Larsen, A.B.L., Sønderby, S.K., Larochelle, H., Winther, O.: Autoencoding beyond pixels using a learned similarity metric. arXiv:1512.09300 (2015)

15. Li, T., Lin, L.: AnonymousNet: natural face de-identification with measurable privacy. In: CVPR Workshops (2019)
16. Liu, W., Wen, Y., Yu, Z., Li, M., Raj, B., Song, L.: SphereFace: deep hypersphere embedding for face recognition. In: CVPR (2017)
17. Mao, X., Li, Q., Xie, H., Lau, R.Y., Wang, Z., Paul Smolley, S.: Least squares generative adversarial networks. In: ICCV (2017)
18. Mathieu, M., Couprie, C., LeCun, Y.: Deep multi-scale video prediction beyond mean square error. arXiv:1511.05440 (2015)
19. Mirza, M., Osindero, S.: Conditional generative adversarial nets. arXiv:1411.1784 (2014)
20. Perarnau, G., Van De Weijer, J., Raducanu, B., Álvarez, J.M.: Invertible conditional GANs for image editing. arXiv:1611.06355 (2016)
21. Pumarola, A., Agudo, A., Martinez, A.M., Sanfeliu, A., Moreno-Noguer, F.: GANimation: anatomically-aware facial animation from a single image. In: Ferrari, V., Hebert, M., Sminchisescu, C., Weiss, Y. (eds.) ECCV 2018. LNCS, vol. 11214, pp. 835–851. Springer, Cham (2018). https://doi.org/10.1007/978-3-030-01249-6_50
22. Reed, S., Akata, Z., Yan, X., Logeswaran, L., Schiele, B., Lee, H.: Generative adversarial text to image synthesis. arXiv:1605.05396 (2016)
23. Regmi, K., Borji, A.: Cross-view image synthesis using conditional GANs. In: CVPR (2018)
24. Ren, Z., Lee, Y.J., Ryoo, M.S.: Learning to anonymize faces for privacy preserving action detection. In: Ferrari, V., Hebert, M., Sminchisescu, C., Weiss, Y. (eds.) ECCV 2018. LNCS, vol. 11205, pp. 639–655. Springer, Cham (2018). https://doi.org/10.1007/978-3-030-01246-5_38
25. Rivest, R.L., Shamir, A., Adleman, L.: A method for obtaining digital signatures and public-key cryptosystems. Commun. ACM **21**, 120–126 (1978)
26. Ryoo, M.S., Kim, K., Yang, H.J.: Extreme low resolution activity recognition with multi-siamese embedding learning. In: AAAI (2018)
27. Ryoo, M.S., Rothrock, B., Fleming, C., Yang, H.J.: Privacy-preserving human activity recognition from extreme low resolution. In: AAAI (2017)
28. Shen, W., Liu, R.: Learning residual images for face attribute manipulation. In: CVPR (2017)
29. Sun, Q., Ma, L., Joon Oh, S., Van Gool, L., Schiele, B., Fritz, M.: Natural and effective obfuscation by head inpainting. In: CVPR (2018)
30. Wang, Z., Chang, S., Yang, Y., Liu, D., Huang, T.S.: Studying very low resolution recognition using deep networks. In: CVPR (2016)
31. Wang, Z., Bovik, A.C., Sheikh, H.R., Simoncelli, E.P., et al.: Image quality assessment: from error visibility to structural similarity. IEEE TIP **13**(4), 600–612 (2004)
32. Yi, D., Lei, Z., Liao, S., Li, S.Z.: Learning face representation from scratch. arXiv:1411.7923 (2014)
33. Yonetani, R., Naresh Boddeti, V., Kitani, K.M., Sato, Y.: Privacy-preserving visual learning using doubly permuted homomorphic encryption. In: ICCV (2017)
34. Zhang, H., et al.: StackGAN: text to photo-realistic image synthesis with stacked generative adversarial networks. In: ICCV (2017)
35. Zhang, K., Zhang, Z., Li, Z., Qiao, Y.: Joint face detection and alignment using multitask cascaded convolutional networks. IEEE Signal Process. Lett. **23**(10), 1499–1503 (2016)
36. Zhang, R., Isola, P., Efros, A.A., Shechtman, E., Wang, O.: The unreasonable effectiveness of deep features as a perceptual metric. In: CVPR (2018)

37. Zhu, J.Y., Park, T., Isola, P., Efros, A.A.: Unpaired image-to-image translation using cycle-consistent adversarial networks. In: ICCV (2017)
38. Zhu, J.Y., et al.: Toward multimodal image-to-image translation. In: NeurIPS (2017)
39. Zhu, S., Urtasun, R., Fidler, S., Lin, D., Change Loy, C.: Be your own Prada: fashion synthesis with structural coherence. In: ICCV (2017)

Inertial Safety from Structured Light

Sizhuo Ma and Mohit Gupta[✉]

University of Wisconsin-Madison, Madison, WI 53706, USA
{sizhuoma,mohitg}@cs.wisc.edu

Abstract. We present inertial safety maps (ISM), a novel scene representation designed for fast detection of obstacles in scenarios involving camera or scene motion, such as robot navigation and human-robot interaction. ISM is a motion-centric representation that encodes both scene geometry and motion; different camera motion results in different ISMs for the same scene. We show that ISM can be estimated with a two-camera stereo setup without explicitly recovering scene depths, by measuring differential changes in disparity over time. We develop an active, single-shot structured light-based approach for robustly measuring ISM in challenging scenarios with textureless objects and complex geometries. The proposed approach is computationally light-weight, and can detect intricate obstacles (e.g., thin wire fences) by processing high-resolution images at high-speeds with limited computational resources. ISM can be readily integrated with depth and range maps as a complementary scene representation, potentially enabling high-speed navigation and robotic manipulation in extreme environments, with minimal device complexity.

1 Introduction

Imagine a drone flying through a forest or a robot arm repairing a complex machine part. In order to determine if they are on a collision course with an obstacle, they require knowledge of the 3D structure of the surroundings, as well as their own motion. Although classical approaches such as SLAM [5, 29] can recover 3D geometry and motion, doing so at high-speeds needed for fast collision avoidance is often prohibitively expensive. While a full 3D map and precise motion may be needed for long-term navigation policies (and for other applications such as augmented reality), it may not be critical for making short-term but time-critical decisions like detection and avoidance of obstacles.

In this paper, we propose *weak 3D cameras* which recover scene representations that are less informative than 3D maps, but can be captured considerably faster, with limited power and computation budgets. These weak 3D cameras are based on inertial safety maps (ISM), a novel scene representation tailored for time-critical and resource-constrained applications such as fast collision avoidance. ISM, for each pixel, is defined as the product of scene depth and time-to-contact (TTC) – time it will take the camera to collide with the scene if it keeps

Electronic supplementary material The online version of this chapter (https://doi.org/10.1007/978-3-030-58592-1_44) contains supplementary material, which is available to authorized users.

A. Vedaldi et al. (Eds.): ECCV 2020, LNCS 12368, pp. 744–761, 2020.
https://doi.org/10.1007/978-3-030-58592-1_44

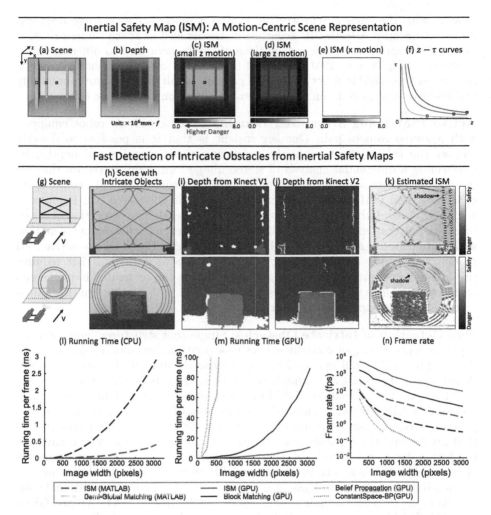

Fig. 1. Inertial safety map (ISM) is a motion-centric scene representation tailored for fast collision avoidance. **(a–b)** An example scene – a room with several pillars. **(c–e)** For the same scene, different camera motion results in different ISMs. A low value of ISM indicates a higher likelihood of collision, whereas higher values convey safety in the immediate future. **(f)** For a given value of ISM, the possible (z, τ) pairs lie on a hyperbolic curve called the $z - \tau$ curve, which can be used for navigation policy design. **(g–h)** Scenes with intricate objects that are a few millimeters thick being observed from a distance of 1.5 m. **(i–j)** Conventional depth cameras based on structured-light or time-of-flight have low spatial resolution, and cannot resolve these thin objects. **(k)** The proposed ISM estimation algorithm, due to its low computational complexity, can use high-resolution images for detecting intricate obstacles, while maintaining high speeds. With conventional 3D imaging techniques, increasing the resolution comes with increased device complexity or high computational cost, often precluding real-time performance. **(l–n)** Quantitative timing comparisons show that for the same image resolution, the unoptimized CPU, and the GPU implementation of the proposed method are up to one order of magnitude faster than existing matching methods.

moving with the current velocity [19]. ISM is a *motion-centric scene represen-tation*; it encodes *both* the 3D scene geometry as well as scene-camera relative motion.[1] Given a scene, different motions of the camera lead to different ISMs, as shown in Fig. 1(a–e). ISM lends itself to intuitive interpretations that can be readily incorporated in robot navigation policies; small ISM indicates potentially imminent danger of collision, whereas large values convey relative safety.

Active ISM Using Structured Illumination: Consider a robot equipped with a stereo camera pair. Our key insight is that it is possible to directly recover the ISM without explicitly computing the disparities and depths, by performing a differential analysis of stereo image formation. Based on this, we develop a theoretical model of active ISM, the structured light (SL) counterpart of stereo-based ISM, where one of the cameras from the stereo pair is replaced with a projector (which is treated as an inverse camera). The projection of coded intensity patterns enables robust estimation of ISM even in challenging scenarios, such as textureless and geometrically complex objects. Based on the active ISM model, we develop a practical *single-shot* ISM recovery method that requires projecting and capturing only a single image[2]. This can be readily implemented with low complexity optical devices (e.g., an LED with a static mask), and is amenable to high-speed motion scenarios.

Fast Detection of Intricate Obstacles: Single-shot structured light methods typically require computationally expensive algorithms for computing correspon-dences. Since ISM requires only *differential disparity* and not absolute correspon-dences, we design a fast algorithm based on Fourier analysis of the images, which enables real-time estimation of ISM even for high-resolution images using only commodity hardware with a limited computational budget. As a result, the pro-posed approaches can detect intricate obstacles (e.g., tree twigs, thin wire fences) that are beyond the capabilities of commodity 3D cameras [43], which have low resolution due to constraints on device complexity and computational resources (Fig. 1(g–k)). While it is theoretically possible to increase the resolution of con-ventional 3D imaging techniques, it often comes with a high computational cost. For example, we perform timing comparisons of CPU-based MATLAB and GPU-based CUDA C++ implementations of our approach, both of which are up to one order of magnitude faster than current methods *at the same spatial resolution* (Fig. 1(l–n)). With such computational benefits across computing architectures, ISMs can help robots with limited computation budgets navigate challenging environments with intricate obstacles.

Scope and Limitations: ISM should not be seen as a replacement for conven-tional scene representations such as depth maps, which are needed for long-term path planning. Instead, ISM should be considered a complementary representa-tion that can be recovered at lower time and power budgets, but only pro-vides conservative collision estimation. In general, it is not possible to recover depth maps from ISMs. However, it is possible to recover depth map from the

---

[1] In contrast, a 3D map is a motion-invariant scene representation.

[2] The method is "single-shot" in that we compute N ISMs from $N + 1$ frames (single-shot except one initial frame).

same captured data used for estimating single-shot active ISM, albeit with a higher computational cost. In future, we envision navigation policies with parallel threads utilizing the same data – a *fast* thread to estimate ISM that makes fast navigation decisions such as braking and collision avoidance, and a *slow* thread to create a full 3D map to aid high-level navigation. Developing such policies, although beyond the scope of this paper, is an important next step.

2   Related Work

Single-Shot Structured Light 3D Imaging. Single-shot structured light (SL) methods project only one pattern to recover depths and thus are suitable for dynamic scenes. Common patterns include sinusoids [36,37], de Bruijn stripes [23,33], grids [21,34], or random dots [43]. These methods often rely on computationally expensive search algorithms, and cannot operate at high frame rates. Recently, Fanello *et al.* [6] treated SL matching as a classification problem and showed depth recovery at 1 kHz for 1 MP images. Our goal is different: We define a motion-centric safety measure that is fundamentally easier to compute than depths. An interesting next step is to apply learning techniques to further increase the efficiency of ISM computation. Furukawa *et al.* [9] has a similar idea of utilizing the disparity change due to object motion, but require one or more color projectors, which increases the hardware complexity and reduces robustness for scenes with non-uniform color distributions [38].

Collision Detection Based on Other Modalities. Proximity sensors based on various modalities (LiDAR [12], ultrasound [35], RADAR [1], programmable light curtain [2,40]) either measure a single global proximity value or require mechanical scanning for generating a 2D map. Time-of-flight (ToF) [15] and Doppler ToF-based methods [16] require correlation sensors. Because of their hardware complexities, all these methods are usually limited in resolution (see Fig. 1 for examples). Navigation based on optical flow [4,11] and time-to-contact [19,28,41] uses passive sensors and is not suitable for textureless scenes and low-light environment. Our method is active, single-shot, has low hardware and computational complexity, and recovers a high-resolution 2D safety map with fine details that can be used for avoiding obstacles with thin structures.

Metrics Used for Collision Avoidance. The level of safety for a robot to navigate without collision depends on not only distance to obstacles (depth map), but also speed, mass, physical size, *etc.* [27]. In robotics, several works [17,20,22,44] have proposed safety metrics for collision avoidance. In this paper, we propose a novel safety metric that captures both the geometry and the motion aspects of safety, and can be estimated from visual data with minimal computation requirements. Deploying this metric in real-world robotic applications is an exciting direction for future work.

3   Inertial Safety Map

In this section, we present *inertial safety map* (ISM), a novel representation of the scene for collision avoidance in scenarios involving scene or camera motion (e.g., robot navigation). Consider a rectified binocular stereo setup observing a scene. Suppose a scene point $\mathbf{R} = (x, y, z)$ projects to pixel (u, v) in the right view, and pixel $(u + \Upsilon, v)$ in the left view. Υ is called the disparity of $\mathbf{R}$. The disparity Υ and the depth z of $\mathbf{R}$ are related by the triangulation equation:

$$z = \frac{fb}{\Upsilon}, \tag{1}$$

where f is the focal length of the cameras (assumed same for both cameras). b is the baseline of the stereo setup. Stereo algorithms compute depths z by estimating corresponding pixels and disparity Υ between the stereo image pair, which often requires computationally intensive search and optimization algorithms [6].

3.1   Differential Analysis of Triangulation Equation

Suppose the scene point $\mathbf{R}$ moves with respect to the camera pair due to scene/camera motion. Due to this relative motion, the disparity Υ may change over time. Our key observation is that, although computing absolute disparities Υ may be expensive, it is possible to efficiently recover *differential changes in disparity* $\Delta\Upsilon$ due to small motion. For instance, a differential disparity change may be estimated by searching in a small local window instead of the entire epipolar line [3]. Later in Sect. 5, we will discuss an approach for fast computation of differential disparity change in an active stereo (coded structured light) system.

What Information is Recoverable from Disparity Change? We address this question by performing a differential analysis of the triangulation equation (Eq. 1). By re-writing $\Upsilon = \frac{fb}{z}$ as a function of depth z from Eq. 1, and taking the derivative of Υ with respect to time t, we get:

$$\frac{d\Upsilon}{dt} = -\frac{fb}{z^2}\frac{dz}{dt}. \tag{2}$$

Assuming the time difference Δt between two successive frames is small, we can multiply both sides by Δt and get

$$\Delta\Upsilon = -\frac{fb}{z^2}\Delta z, \tag{3}$$

where $\Delta\Upsilon$ and Δz are the changes in the disparity and depth of point $\mathbf{R}$, respectively, due to relative scene-camera motion. Next, we define *time-to-contact* $\tau = -\frac{z}{\Delta z}$, as the time it will take for the camera (the image plane of the right camera) to collide with point $\mathbf{R}$ if the relative velocity between the camera and the point remains the same [19]. By substituting in the above equation, and rearranging the terms, we get the following key relationship:

$$\boxed{z \cdot \tau = \frac{fb}{\Delta\Upsilon}.} \tag{4}$$

Assuming a calibrated stereo system (known focal length f and baseline b), the right-hand side involves only one unknown $\Delta\Upsilon$, which we assume can be computed efficiently. The left-hand side is the product of two quantities that are indicators of the chances of keeping moving safely. Intuitively, a large value of the product of z and τ indicates low chances of collision. Based on this intuition, we define the *inertial safety measure* for collision avoidance as follows:

Definition. The *inertial safety measure* S of a scene point with respect to its relative motion to the stereo camera is defined as the product of its depth z and time to contact τ,

$$S = z \cdot \tau = \frac{fb}{\Delta \Upsilon}. \tag{5}$$

which can be computed from camera parameters f and b, and disparity change $\Delta \Upsilon$. The *inertial safety map (ISM)* is a per-pixel map of inertial safety measure.

The inertial safety measure S encodes the level of safety *if the camera keeps its current motion*. By estimating $\Delta \Upsilon$, we can compute S for collision avoidance. For example, a robot can detect obstacles and get around them by identifying image regions with low values of S, without explicitly computing depths.

3.2 Inertial Safety Map: Interpretations

Imagine a fast-moving drone navigating around pillars in a room (Fig. 1). For this simulated scene, we plot the ground truth ISMs for three different motions. The unit of ISM is $mm \times f$, where depth is in mm and the time-to-contact (TTC) is expressed in terms of the number of frames before collision. Darker colors represent low values of the ISM, and therefore a higher level of danger of collision. All values higher than a threshold, or less than zero (due to camera moving away from the scene) are mapped to white.

A Motion-Centric Scene Representation: From Figs. 1(c–e), we observe that for the same scene, different motions results in different ISMs. This is because the TTC depends on the z-velocity. The amount of z-motion between frames is doubled in (d) compared to (c), so the TTC is halved for every pixel. In (e), since there is no z-motion, the ISM is $+\infty$ everywhere. Thus, the inertial safety map can be considered a *motion-centric scene representation* as it depends both on the scene's geometry, as well as the relative scene-camera motion; it encodes the degree of safety (from collision) if the camera/scene keeps its *current motion*.

$z - \tau$ **Curve:** A given value of the ISM corresponds to an infinite number of possible $z - \tau$ pairs, which trace out a hyperbolic curve called the $z - \tau$ curve in the 2D $z - \tau$ space (Fig. 1(c,f)). The $z - \tau$ curves corresponding to three highlighted scene points are plotted in (f), with the exact (z, τ) values indicated by the colored rectangles. Although we cannot determine the true z and τ values from an estimate of the ISM, the ISM can be used as a fast and conservative safety check in robot navigation policies. This is because when ISM is high, both z and τ have to be high, so the robot is safe. When ISM is low, it can be due to either high z and low τ, or low z and high τ. This ambiguity can be resolved by designing a more sophisticated navigation policy, or by triggering a full depth recovery algorithm for collision avoidance. See the supplementary technical report for a detailed discussion.

4   Active Inertial Safety Map

So far we have defined the inertial safety map for a passive two-camera stereo system. However, the ISM can be generalized to any two-view imaging system, including active methods such as structured light (SL), where the second camera is replaced by a projector (an inverse camera). In SL, a coded light pattern is projected to enable robust scene recovery even in challenging scenarios including lack of scene texture and insufficient lighting. In this section, we develop mathematical model and approaches for *active ISM*, i.e., recovering ISM using SL. Specifically, we consider active ISM recovery from *single-shot SL* where a single image is captured with a single projected pattern.

Consider a projector-camera system with a horizontal baseline For ease of analysis, we assume that the projector projects a pattern with 1D translational symmetry, i.e., all the projector pixels in a column have the same intensity. Such patterns are used in several SL 3D imaging systems [23,37], and can be expressed as a 1D function $P(c)$, where c is the projector column index.

Suppose a scene point $\mathbf{R}$ is illuminated by projector column index c, and imaged at camera pixel (u, v) at time t. The intensity of pixel (u, v) at t is:

$$i(u, v, t) = \alpha(u, v, t)\, P(c) + \beta(u, v, t)\,, \tag{6}$$

where $\alpha(u, v, t)$ encapsulates the reflectance properties of point $\mathbf{R}$, and $\beta(u, v, t)$ is the intensity component due to ambient light. The projector column index c, camera pixel index u and the disparity Υ are related as:

$$\Upsilon(u, v, t) = c - u. \tag{7}$$

Suppose point $\mathbf{R}$ moves with respect to the camera (due to camera or scene motion) from time t to $t + \Delta t$. After motion, let the point be illuminated by projector column index $c + \Delta c$, and imaged at camera pixel $(u + \Delta u, v + \Delta v)$. Similar to Eq. 6, the observed intensity of point $\mathbf{R}$ at $t + \Delta t$ is given as:

$$i(u + \Delta u, v + \Delta v, t + \Delta t) = \alpha' P(c + \Delta c) + \beta'\,, \tag{8}$$

where $\alpha' = \alpha(u + \Delta u, v + \Delta v, t + \Delta t), \beta' = \beta(u + \Delta u, v + \Delta v, t + \Delta t)$. The disparity of point $\mathbf{R}$ after motion is then:

$$\Upsilon(u + \Delta u, v + \Delta v, t + \Delta t) = (c + \Delta c) - (u + \Delta u)\,. \tag{9}$$

Recovering the ISM requires measuring the disparity change $\Delta \Upsilon$, which is the difference between the new and old disparity, i.e., $\Delta \Upsilon = \Upsilon(u + \Delta u, v + \Delta v, t + \Delta t) - \Upsilon(u, v, t)$. From Eqs. 7 and 9, we get:

$$\Delta \Upsilon = \Delta c - \Delta u\,. \tag{10}$$

Computational Considerations: To compute $\Delta \Upsilon$, we need to estimate both the "texture flow" (Δu) and the "illumination flow" Δc, which are the projected motion of the scene point on the camera's and projector's image planes [39]. This problem is challenging due to several non-linearly coupled unknowns for each pixel $(\alpha, \beta, \Upsilon, \Delta u, \Delta c)$. Solutions require expensive nonlinear optimization and therefore are not suitable for applications with limited computational budget and extreme timing requirements.

To make the computation tractable, instead of considering the disparity change $\Delta \Upsilon$ of a *fixed scene point*, we consider $\Delta \Upsilon$ of a *fixed pixel* in the camera image. At time $t + \Delta t$, we analyze the image intensity at the *same* pixel (u, v):

$$i(u, v, t + \Delta t) = \alpha(u, v, t + \Delta t) \, P(c + \Delta \Upsilon_R) + \beta(u, v, t + \Delta t),$$

where $\alpha(u, v, t + \Delta t)$ and $\beta(u, v, t + \Delta t)$ are the reflectance and ambient terms for the scene point imaged at pixel (u, v) after motion, and

$$\Delta \Upsilon_R = \Upsilon(u, v, t + \Delta t) - \Upsilon(u, v, t) \tag{11}$$

is the disparity change along the camera ray at pixel (u, v). This definition of *active ISM* does not compute correspondences between frames, instead relying on a differential analysis that estimates the differential depth change between two frames at each pixel. The resulting active ISM provides a *conservative* measure of danger that detects all potential collisions: When a collision is about to happen, the depth at the corresponding pixel will decrease to zero. Therefore, the ISM will be small at the pixel at some point before collision. As we show in Sect. 5, it is possible to estimate $\Delta \Upsilon_R$ with simple (linear) analytic expressions that can be computed extremely fast with limited computational resources.

ISM Estimation Under Sharp Depth Variations: Due to relative scene-camera motion, pixel (u, v) may image different scene points at times t and $t + \Delta t$. As a result, the computed ISM value may result in overly conservative collision warnings, especially at depth edges where the depth changes significantly across frames. This issue can be mitigated by spatio-temporal filtering the estimated ISM. See the supplementary report for a detailed discussion.

5   ISM from Single-Shot Structured Light

In this section, we present practical approaches for computing active ISM from single-shot structured light. One way to estimate ISM is to directly estimate scene disparities by using globally-unique patterns such as de Bruijn [23,33] and random patterns [6]. Once disparities Υ are computed before and after motion, ISM can be trivially computed by taking their difference (Eq. 11). However, single-shot SL methods typically requires computationally intensive algorithms which are not suitable for scenarios with limited computational budget.

5.1   Fast Fourier Domain Computation of ISM

We propose a fast method for computing ISM, based on projecting a 1D high-frequency sinusoid pattern. A pictorial summary of the method is shown in Fig. 2. Let the projected sinusoid pattern be given as:

$$P(c) = 0.5 + 0.5 \cos(\omega c), \tag{12}$$

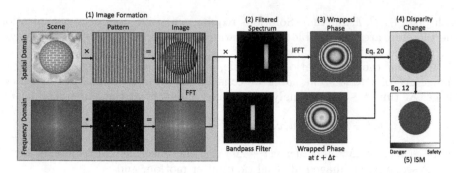

Fig. 2. Overview of active ISM recovery method. (1) The scene is illuminated by a high-frequency sinusoidal pattern, which can mathematically be represented as a multiplication in the spatial domain and a convolution in the frequency domain (ignoring ambient light). The frequency domain images are plotted in log scale. (2) A bandpass filter is applied to extract the term G. (3) Inverse FFT is used to get wrapped phase maps at time t and $t + \Delta t$. (4) Combine both maps to get the disparity change (Eq. 19). (5) Compute ISM (Eq. 5).

where ω is the angular frequency of the sinusoid. Substituting in Eq. 6, the intensity at pixel (u, v) is:

$$i = \alpha + \alpha \cos(\omega u + \omega \Upsilon) + \beta, \tag{13}$$

where abusing the notation, the constant 0.5 is absorbed with α. For brevity, we drop the indices (u, v, t). Equation 13 is an underconstrained nonlinear equation in three unknowns (α, β and Υ), and thus, challenging to solve directly.

Solving Eq. 13 by Linearizing and Regularizing: The cos term on the right hand side can be expanded into a sum of complex functions:

$$
\begin{aligned}
i &= \alpha + \alpha \cdot \left(e^{j\omega(u+\Upsilon)} + e^{-j\omega(u+\Upsilon)}\right)/2 + \beta \\
&= \underbrace{\alpha + \beta}_{f} + \underbrace{0.5\,\alpha\,e^{j\omega\Upsilon}}_{g}\,e^{j\omega u} + \underbrace{0.5\,\alpha\,e^{-j\omega\Upsilon}}_{g^*}\,e^{-j\omega u},
\end{aligned}
\tag{14}
$$

where $j = \sqrt{-1}$. The above is now a linear equation in three unknowns f, g and g^* (conjugate of g):

$$i(u, v, t) = f(u, v, t) + e^{j\omega u}\, g(u, v, t) + e^{-j\omega u}\, g^*(u, v, t). \tag{15}$$

Regularizing by Assuming Global Smoothness: One way to solve Eq. 15 is to make the restrictive assumption that the variables f, g and g^* are locally constant and stack the equations in a local neighborhood into a linear system (similar to Lucas-Kanade image alignment [26]). However, inspired by Fourier transform profilometry [36,37], we make a less restrictive assumption that scene reflectance and depths (and thus, α, β and Υ) vary smoothly horizontally (in each row) compared to the pattern frequency ω. In this case, f and g are band-limited signals. Taking the 2D Fourier Transform with respect to u and v:

$$I(\omega_u, \omega_v, t) = F(\omega_u, \omega_v, t) + G(\omega_u - \omega, \omega_v, t) \\ + G^*(\omega_u + \omega, \omega_v, t). \tag{16}$$

The spectra F, G and G^* can be separated from each other by ω, as shown in Fig. 2. We extract the spectrum $G(\omega_u - \omega, \omega_v, t)$ by applying a bandpass filter (a 2D Hanning window as in [24]) and transform it back to the primal domain. The complex signal g is recovered as:

$$g(u, v, t) = 0.5\, \alpha(u, v, t)\, e^{j\omega \Upsilon(u, v, t)}. \tag{17}$$

The disparity can be estimated from the complex argument:

$$\hat{\Upsilon} = \arg(g(u, v, t))/\omega. \tag{18}$$

Oriented 2D Filter: Consider a tilted thin thread in front of a wall, as shown in Fig. 3(a). The disparity no longer changes smoothly in each row, which makes the spectra inseparable by using simple vertical bandpass filters (Fig. 3(b)). However, if we filter the spectrum using an oriented 2D filter, as shown in (c), it is still possible to separate the signal G from F, as shown in Fig. 3(d). As a result, the estimated ISM (Fig. 3(e)) is more accurate than using a vertical filter (Fig. 3(f)). In practice, it is possible to divide the image into small patches, and filter each patch using different oriented filters to detect thin structures with different orientations. See the supplementary report for details.

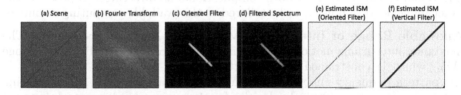

Fig. 3. Oriented 2D filter for resolving thin structures with different orientations. (a) A tilted thin thread in front of a wall. Disparity changes abruptly in each row. **(b–d)** The spectra are no longer separable by using simple vertical bandpass filters. Instead, they can be separated using oriented filters. **(e–f)** An oriented filter recovers the ISM with higher accuracy as compared to a vertical filter.

Need for Phase Unwrapping? It may appear from first glance that absolute disparity Υ can be recovered from Eq. 18. To recover the absolute disparity, one needs to recover the absolute phase $\phi_u = \omega \Upsilon$. However, we can only recover the wrapped phase ϕ_w, related to ϕ_u by $\phi_u = \phi_w + 2k\pi$ for some unknown $k \in \mathbb{N}$.[3]

---

[3] It is possible to recover absolute phase using a unit-frequency sinusoid, however at a considerably lower phase-recovery precision than high-frequency sinusoids.

Absolute phase ϕ_u can be recovered by spatial phase unwrapping methods [10], which require global reasoning and are highly computationally intensive.

Fortunately, to compute the ISM S_E, we only need to compute the disparity change $\Delta\Upsilon_R$, instead of the absolute disparity Υ. Assuming a small change in Υ across consecutive frames, *i.e.*, $\Delta\Upsilon_R \in (-\pi/\omega, \pi/\omega]$, it is possible to compute $\Delta\Upsilon_R$ by taking the difference of wrapped phases:

$$\Delta\Upsilon_R(u, v, t) = \mathrm{wrap}(\arg(g(u,v,t+\Delta t)) - \arg(g(u,v,t)))/\omega, \tag{19}$$

where $\mathrm{wrap}(\phi)$ is a function that wraps a phase to the principal values. As a result, a dense S_E map can be computed efficiently *without phase unwrapping*.

5.2   Practical Considerations

Computational Efficiency: ISMs can be computed at high speeds even for very high resolution images, as demonstrated in Fig. 1(l–n). A direct comparison between ISM and other single-shot SL methods is difficult since their code is usually not publicly available. Instead, we compare the computational speeds of the proposed ISM algorithm and a few widely-used stereo matching algorithms. The CPU (MATLAB) implementation of the method is up to one order of magnitude faster than MATLAB's semi-global matching algorithm [18]. We also develop a GPU implementation of the proposed method, which is able to reach 1 kfps at 1 megapixel resolution, and achieves real-time performance even for very high resolution (90 fps at 9 megapixel), which is 9x faster than OpenCV's CUDA implementation of block matching and considerably faster than belief propagation (BP) [7] and constant-space BP [42]. See the supplementary report for details on the experiment setting. These comparisons demonstrate the computational benefit of ISM for both computational architectures. In practice, the exact implementation needs to be tailored to the available computing resources.

Allowable Range of Inter-Frame Motion vs. Pattern Frequency: The maximum inter-frame motion is determined by the recoverable disparity change $\Delta\Upsilon_R$, which is constrained by the pattern period $\Lambda = \frac{2\pi}{\omega}$ as $\Delta\Upsilon_R < \frac{\Lambda}{2}$. A low pattern frequency enables recovering a wider range of $\Delta\Upsilon_R$, thus allowing faster camera motion. Typical velocities for the current hardware prototype are 1–10 cm/s (captured at 30 fps). On the other hand, using a higher frequency pattern can separate f and g with wider frequency bands, which means higher robustness for scenes with high-frequency textures or cluttered geometry. Finally, although our image model (Eq. 6) assumes only direct lighting, in practice, inter-reflections may result in erroneous estimates of S_E map. Using a high-frequency pattern also mitigates this effect [13, 30].

6   Experimental Results

Simulations. Figure 4 shows three simulated scenes that emulate different robot navigation scenarios. Our method is able to estimate the ISM of the thin, complex geometry of bamboos, tree branches and warehouse racks.

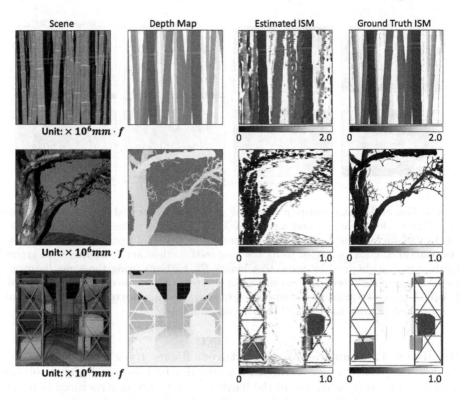

| Scene | Depth Map | Estimated ISM | Ground Truth ISM |

Fig. 4. Simulation results for example robot navigation scenarios through intricate obstacles. The proposed approach can recover the ISM for scenes with complex, overlapping thin structures (bamboos, tree branches, warehouse racks).

Experiment Setup. We build a prototype structured light system using a Canon DSLR camera and an Epson 3LCD projector. The projector projects a 1920×1080 high frequency sinusoidal pattern with a period of 8 pixels. After rectification, a 3714×2182 captured image is used for ISM computation. The camera-projector baseline is 353 mm. Details on calibration and rectification can be found in the supplementary technical report.

Ground Truth Comparison. Figure 5 shows the comparison between the ISM estimated by our method and the ground truth, which is obtained by projecting a sequence of binary-coded SL patterns. The camera translates for roughly 3 mm along the z-axis. Our method correctly estimates the ISM for scenes with planar and curved surfaces with strong depth edges.

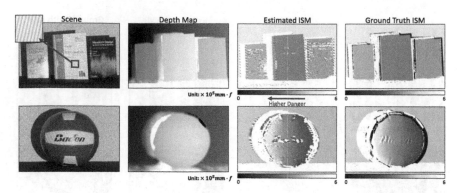

Fig. 5. Ground truth comparison. Our prototype structured light system consists of a Canon DSLR camera and an Epson 3LCD projector. The projector projects a 1920×1080 high-frequency sinusoidal pattern with a period of 8 pixels. **Zoom in to see the pattern.** ISMs estimated using the proposed method are compared with ground truth, which is obtained by projecting binary SL patterns using the same hardware. Depth maps of the scenes are also shown for comparison (not used in computing ISM). **(Top)** A piecewise planar scene consisting of three books. **(Bottom)** A spherical ball. Our method recovers the ISMs of both scenes accurately.

Resolving Extremely Thin Structures. Figure 1(g–k) demonstrates our method's capability of recovering thin structures by processing high-resolution images, which is possible due to the hardware simplicity of structured light and the computational efficiency of the proposed algorithm. The thinnest part of the scenes are 4 mm and 1.5 mm, which could be challenging to resolve from 1.5 m away. We also show results for two commodity depth cameras: Kinect V1 and V2, whose spatial resolutions are 640×480 and 512×424 respectively. From the same distance, the depth cameras are only able to partially recover the thicker parts of the fence in the top scene and completely miss the rings in the bottom scene. This is not meant to be a direct comparison of the three approaches, because the data is acquired from different cameras. With a higher resolution, depth cameras may also be able to recover the scene details, albeit at a higher computational cost, as shown in the timing comparisons in Fig. 1.

Navigation Sequences. Figure 6 shows a simulated sequence (top) where a drone flies through thin threads, and a real captured sequence which emulates a robot navigating around a pillar (bottom). Trajectories are manually planned in both examples. As the robot approaches the threads and the pillar, danger is detected from the estimated ISM, and the robot reacts accordingly to avoid collision.

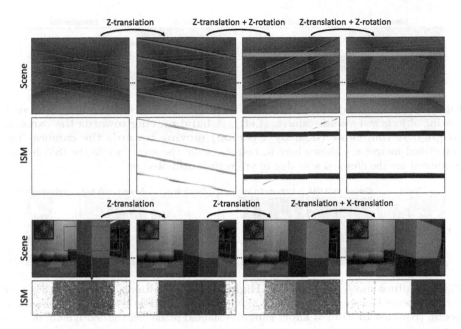

Fig. 6. Navigation sequences with manually planned trajectories. (Top) A simulated sequence where a drone flies through thin threads. As the drone detects the threads, it aligns its pose to be parallel with the threads to avoid collision. **(Bottom)** A real video sequence where a robot navigates around a pillar. The unrectified images are shown here to better convey the scene, while the ISM is only computed for the cropped area due to projector's field-of-view. The robot moves forward (first three frames), detects the pillar and moves to the left to circumvent it (last frame).

Detecting Object Motion. ISMs can also be used to detect collision due to moving objects. For example, in a co-robot scenario where robots and human workers collaborate in the same environment [8,25], it is important to prevent collisions between human and robot arms for safety. Figure 7 shows two examples where a moving human hand and a thin cable are correctly detected in the ISMs. The thin cable is an intricate obstacle, which is challenging to detect with current depth cameras. The estimated ISMs can be used to avoid collision with human co-workers and intricate dynamic objects in the working environment. **All complete video sequences can be found in the supplementary video. See the technical report for additional results.**

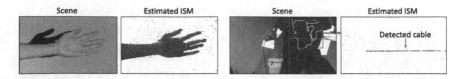

Fig. 7. Detecting object motion. ISMs can also be used to detect collisions between moving objects and a static camera. **(Left): A hand moving towards the camera. (Right): A thin cable (held by a person) moving towards the camera.** The unrectified images are shown here to better convey the scene, while the ISM is only computed for the cropped area due to projector's field-of-view.

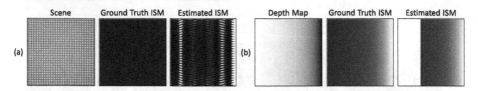

Fig. 8. Failure modes. Scene (a): High-frequency albedo. A plane with a very high-frequency bi-sinusoidal pattern, which violates the albedo smoothness assumption. **Scene (b): Fast motion at short range.** A slanted plane (see the depth map) moving fast in z-direction towards the camera. Disparity change in the closest part of the scene wraps around the period and the estimated ISM becomes negative (shown as white).

7 Limitations and Future Outlook

Failure Modes. The proposed method may fail to estimate the ISM correctly when the assumptions made in Sect. 5.1 are not satisfied. This happens when the albedo varies too quickly, or when the objects are moving too fast at a short distance, causing disparity changes too abruptly (Fig. 8). See the supplementary technical report for a quantitative analysis.

Resolving Vertical Depth Edges. Our method cannot accurately recover thin structures that are nearly vertical because the depth varies abruptly along the epipolar lines, which violates the smoothness assumption. A potential solution is to have one camera and two projectors in an L-configuration such that both horizontal and vertical epipolar lines are available.

Performance in Outdoor Settings: Outdoor deployment under sunlight is challenging for all power-limited active imaging systems due to photon noise from sunlight. It is possible to mitigate this issue by spatio-temporal illumination coding [14], as well as joint illumination and image coding [31,32] to enable ISM recovery under strong sunlight.

Acknowledgement. This research is supported in part by the DARPA REVEAL program and a Wisconsin Alumni Research Foundation (WARF) Fall Competition award.

References

1. Azevedo, S., McEwan, T.E.: Micropower impulse radar. IEEE Potentials **16**(2), 15–20 (1997)
2. Bartels, J.R., Wang, J.: Agile depth sensing using triangulation light curtains. In: International Conference on Computer Vision (ICCV), pp. 7899–7907. IEEE (2019)
3. Čech, J., Sanchez-Riera, J., Horaud, R.: Scene flow estimation by growing correspondence seeds. In: IEEE Conference on Computer Vision and Pattern Recognition (CVPR), pp. 3129–3136. IEEE (2011)
4. Coombs, D., Herman, M., Hong, T.H., Nashman, M.: Real-time obstacle avoidance using central flow divergence, and peripheral flow. IEEE Trans. Robot. Autom. **14**(1), 49–59 (1998)
5. Engel, J., Schöps, T., Cremers, D.: LSD-SLAM: large-scale direct monocular SLAM. In: Fleet, D., Pajdla, T., Schiele, B., Tuytelaars, T. (eds.) ECCV 2014. LNCS, vol. 8690, pp. 834–849. Springer, Cham (2014). https://doi.org/10.1007/978-3-319-10605-2_54
6. Fanello, S.R., et al.: HyperDepth: learning depth from structured light without matching. In: IEEE Conference on Computer Vision and Pattern Recognition (CVPR), Las Vegas, NV, USA, pp. 5441–5450. IEEE, June 2016
7. Felzenszwalb, P.F., Huttenlocher, D.P.: Efficient belief propagation for early vision. Int. J. Comput. Vis. (IJCV) **70**(1), 41–54 (2006)
8. Flacco, F., Kroger, T., De Luca, A., Khatib, O.: A depth space approach to human-robot collision avoidance. In: IEEE International Conference on Robotics and Automation (ICRA), Saint Paul, MN, pp. 338–345. IEEE, May 2012
9. Furukawa, R., Sagawa, R., Kawasaki, H.: Depth estimation using structured light flow — analysis of projected pattern flow on an object's surface. In: IEEE International Conference on Computer Vision (ICCV), pp. 4650–4658. IEEE, October 2017
10. Gorthi, S.S., Rastogi, P.: Fringe projection techniques: whither we are? Opt. Lasers Eng. **48**(2), 133–140 (2010)
11. Green, W.E., Oh, P.Y.: Optic-flow-based collision avoidance. IEEE Robot. Autom. Mag. **15**(1), 96–103 (2008)
12. Grewal, H., Matthews, A., Tea, R., George, K.: LIDAR-based autonomous wheelchair. In: IEEE Sensors Applications Symposium (SAS), Glassboro, NJ, USA, pp. 1–6. IEEE (2017)
13. Gupta, M., Nayar, S.K.: Micro phase shifting. In: IEEE Conference on Computer Vision and Pattern Recognition (CVPR), Providence, RI, pp. 813–820. IEEE, June 2012
14. Gupta, M., Yin, Q., Nayar, S.K.: Structured light in sunlight. In: IEEE International Conference on Computer Vision (ICCV), Sydney, Australia, pp. 545–552. IEEE, December 2013
15. Hansard, M., Lee, S., Choi, O., Horaud, R.: Time of Flight Cameras: Principles, Methods, and Applications. Springer, London (2012). https://doi.org/10.1007/978-1-4471-4658-2
16. Heide, F., Heidrich, W., Hullin, M., Wetzstein, G.: Doppler time-of-flight imaging. ACM Trans. Graph. **34**(4), 1–11 (2015)
17. Heinzmann, J., Zelinsky, A.: Quantitative safety guarantees for physical human-robot interaction. Int. J. Robot. Res. **22**(7–8), 479–504 (2003)
18. Hirschmuller, H.: Accurate and efficient stereo processing by semi-global matching and mutual information. In: IEEE Computer Society Conference on Computer Vision and Pattern Recognition (CVPR), vol. 2, pp. 807–814, San Diego, CA, USA. IEEE (2005)

19. Horn, B., Fang, Y.F.Y., Masaki, I.: Time to contact relative to a planar surface. In: IEEE Intelligent Vehicles Symposium, pp. 68–74. IEEE (2007)
20. Ikuta, K., Ishii, H., Nokata, M.: Safety evaluation method of design and control for human-care robots. In: International Symposium on Micromechatronics and Human Science (MHS), pp. 119–127. IEEE (2000)
21. Kawasaki, H., Furukawa, R., Sagawa, R., Yagi, Y.: Dynamic scene shape reconstruction using a single structured light pattern. In: IEEE Conference on Computer Vision and Pattern Recognition (CVPR), pp. 1–8, Anchorage, AK, USA. IEEE, June 2008
22. Lacevic, B., Rocco, P.: Kinetostatic danger field - a novel safety assessment for human-robot interaction. In: IEEE/RSJ International Conference on Intelligent Robots and Systems (IROS), Taipei, pp. 2169–2174. IEEE, October 2010
23. Zhang, L., Curless, B., Seitz, S.: Rapid shape acquisition using color structured light and multi-pass dynamic programming. In: 3D Data Processing Visualization and Transmission, pp. 24–36. IEEE Computur Society (2002)
24. Lin, J.F., Su, X.Y.: Two-dimensional Fourier transform profilometry for the automatic measurement of three-dimensional object shapes. Opt. Eng. **34**(11), 3297 (1995)
25. Liu, C., Tomizuka, M.: Algorithmic safety measures for intelligent industrial co-robots. In: IEEE International Conference on Robotics and Automation (ICRA), Stockholm, Sweden, pp. 3095–3102. IEEE, May 2016
26. Lucas, B.D., Kanade, T.: An iterative image registration technique with an application to stereo vision. In: International Joint Conference on Artificial Intelligence (IJCAI), Vancouver, British Columbia, Canada, pp. 674–679 (1981)
27. Marvel, J.A.: Performance metrics of speed and separation monitoring in shared workspaces. IEEE Trans. Autom. Sci. Eng. **10**(2), 405–414 (2013)
28. Muller, D., Pauli, J., Nunn, C., Gormer, S., Muller-Schneiders, S.: Time to contact estimation using interest points. In: International IEEE Conference on Intelligent Transportation Systems (ITSC), St. Louis, pp. 1–6. IEEE, October 2009
29. Mur-Artal, R., Tardos, J.D.: ORB-SLAM2: an open-source SLAM system for monocular, stereo, and RGB-D cameras. IEEE Trans. Robot. **33**(5), 1255–1262 (2017)
30. Nayar, S.K., Krishnan, G., Grossberg, M.D., Raskar, R.: Fast separation of direct and global components of a scene using high frequency illumination. ACM Trans. Graph. **25**(3), 935–944 (2006)
31. O'Toole, M., Achar, S., Narasimhan, S.G., Kutulakos, K.N.: Homogeneous codes for energy-efficient illumination and imaging. ACM Trans. Graph. **34**(4), 1–13 (2015)
32. O'Toole, M., Mather, J., Kutulakos, K.N.: 3D shape and indirect appearance by structured light transport. In: IEEE Conference on Computer Vision and Pattern Recognition (CVPR), pp. 3246–3253. IEEE (2014)
33. Pagès, J., Salvi, J., Collewet, C., Forest, J.: Optimised De Bruijn patterns for one-shot shape acquisition. Image Vis. Comput. **23**(8), 707–720 (2005)
34. Sagawa, R., Ota, Y., Yagi, Y., Furukawa, R., Asada, N., Kawasaki, H.: Dense 3D reconstruction method using a single pattern for fast moving object. In: IEEE International Conference on Computer Vision (ICCV), pp. 1779–1786. IEEE, September 2009
35. Min, S.D., Kim, J.K., Shin, H.S., Yun, Y.H., Lee, C.K., Lee, M.: Noncontact respiration rate measurement system using an ultrasonic proximity sensor. IEEE Sens. J. **10**(11), 1732–1739 (2010)
36. Takeda, M., Ina, H., Kobayashi, S.: Fourier-transform method of fringe-pattern analysis for computer-based topography and interferometry. J. Opt. Soc. Am. **72**(1), 156 (1982)

37. Takeda, M., Mutoh, K.: Fourier transform profilometry for the automatic measurement of 3-D object shapes. Appl. Opt. **22**(24), 3977 (1983)
38. Van der Jeught, S., Dirckx, J.J.: Real-time structured light profilometry: a review. Opt. Lasers Engineering **87**, 18–31 (2016)
39. Vo, M., Narasimhan, S.G., Sheikh, Y.: Separating texture and illumination for single-shot structured light reconstruction. In: IEEE Conference on Computer Vision and Pattern Recognition Workshops, pp. 433–440. IEEE, June 2014
40. Wang, J., Bartels, J., Whittaker, W., Sankaranarayanan, A.C., Narasimhan, S.G.: Programmable triangulation light curtains. In: Ferrari, V., Hebert, M., Sminchisescu, C., Weiss, Y. (eds.) ECCV 2018. LNCS, vol. 11207, pp. 20–35. Springer, Cham (2018). https://doi.org/10.1007/978-3-030-01219-9_2
41. Watanabe, Y., Sakaue, F., Sato, J.: Time-to-contact from image intensity. In: IEEE Conference on Computer Vision and Pattern Recognition (CVPR), Boston, MA, USA, pp. 4176–4183. IEEE, June 2015
42. Yang, Q., Wang, L., Ahuja, N.: A constant-space belief propagation algorithm for stereo matching. In: IEEE Computer Society Conference on Computer Vision and Pattern Recognition (CVPR). San Francisco, CA, USA, pp. 1458–1465. IEEE, June 2010
43. Zhang, Z.: Microsoft kinect sensor and its effect. IEEE Multimedia **19**(2), 4–10 (2012)
44. Zinn, M., Khatib, O., Roth, B., Salisbury, J.: Playing it safe. IEEE Robot. Autom. Mag. **11**(2), 12–21 (2004)

PointTriNet: Learned Triangulation of 3D Point Sets

Nicholas Sharp[1]([✉]) [iD] and Maks Ovsjanikov[2] [iD]

[1] Carnegie Mellon University, Pittsburgh, USA
nsharp@cs.cmu.edu
[2] LIX, École Polytechnique, IP Paris, Palaiseau, France

Abstract. This work considers a new task in geometric deep learning: generating a triangulation among a set of points in 3D space. We present PointTriNet, a differentiable and scalable approach enabling point set triangulation as a layer in 3D learning pipelines. The method iteratively applies two neural networks: a classification network predicts whether a candidate triangle should appear in the triangulation, while a proposal network suggests additional candidates. Both networks are structured as PointNets over nearby points and triangles, using a novel triangle-relative input encoding. Since these learning problems operate on local geometric data, our method is efficient and scalable, and generalizes to unseen shape categories. Our networks are trained in an unsupervised manner from a collection of shapes represented as point clouds. We demonstrate the effectiveness of this approach for classical meshing tasks, robustness to outliers, and as a component in end-to-end learning systems.

Keywords: Geometric learning · Triangulation · Geometry processing

1 Introduction

Generating surface meshes is a fundamental problem in visual computing; meshes are essential for tasks ranging from visualization to simulation and shape analysis. In this work, we specifically consider generating surfaces via *point set triangulation*, where a fixed vertex set is given as input and a collection of triangles among those points is returned as output.

Point set triangulation has been widely studied as a classical problem in computational geometry, typically in the context of surface reconstruction from sampled point clouds such as 3D range scan data [6,29,44]. The key property of point set triangulation is that rather than allowing *any* mesh as output, the output is constrained to only those meshes which have the input points as

Code is available at github.com/nmwsharp/learned-triangulation.

Electronic supplementary material The online version of this chapter (https:// doi.org/10.1007/978-3-030-58592-1_45) contains supplementary material, which is available to authorized users.

A. Vedaldi et al. (Eds.): ECCV 2020, LNCS 12368, pp. 762–778, 2020.
https://doi.org/10.1007/978-3-030-58592-1_45

their vertices. When reconstructing surfaces from sensor data this constraint is artificial, and makes the problem needlessly difficult: there simply may not exist any desirable mesh which has the input points as its vertex set. For this reason most practical methods for surface reconstruction have pivoted to volumetric techniques like Poisson reconstruction [27,28] or learned implicit functions, and extract a mesh only as a post-process [13,40].

However, recent advances in point-based geometric learning motivate a new setting for point set triangulation, where the points are generated not as the output of a sensor, but rather as the output of a learned procedure [1,42]. If such point sets could be easily triangulated, then any learning-based method capable of generating points could be immediately augmented to generate a mesh. Using point-based representations avoids the implicit smoothing and computational cost that come from working in a volumetric grid, and a differentiable meshing block enables end-to-end training to optimize for the final surface mesh.

Unfortunately, classical techniques from computational geometry are not differentiable, and recent learning-based 3D reconstruction methods are typically either restricted to particular shape classes [13] or too computationally expensive to use as a component of a larger procedure [16]. The focus of our work is to overcome these challenges, and design the first point set triangulation scheme suitable for geometric learning. Several aspects make this problem difficult: first the method must be differentiable, yet the choice of triangles is discrete—we will need to generalize the output to a distribution over possible triangles. The second major challenge is scalability; exhaustive exploration is prohibitively expensive even for moderately-sized point sets. Finally, the method should be generalizable, not tied to specific shape categories.

Our main observation is that while triangulation is a global problem, it is driven largely by local considerations. For example, the local circumscribing-circle test in 2D is sufficient to identify faces in a Delaunay triangulation. Based on this intuition, we propose a novel neural network that predicts the probability that a single triangle should appear in a triangulation, and a second, similarly-structured network to suggest neighboring triangles as new candidates. Iteratively applying these two networks generates a coherent triangle mesh.

Our approach yields differentiable triangle probabilities; gradients can be evaluated via ordinary backpropagation. The use of a proposal network allows our method to efficiently identify candidate triangles, and thus scale to very large inputs. Moreover, since our problems are local and geometric in nature, our method can easily be applied to surface patches across arbitrary shape classes. We validate these properties through extensive experiments, demonstrating that our method is competitive with classical approaches. Finally, we highlight new applications in geometric learning made possible by our differentiable approach.

2 Related Work

Surface reconstruction and meshing are among the oldest and most researched areas of computational geometry, computer vision and related fields—their full overview is beyond the scope of this paper. Below we review methods most closely related to ours and refer the interested readers to recent surveys and monographs [6,29,44] for a more in-depth discussion.

A classical approach for surface reconstruction is to estimate a volumetric representation, e.g. a signed distance function, and then extract a mesh [15,25]. Both of these steps have been extended to improve robustness and accuracy, including [27,36,46] among many others (see also related techniques in [38]). These approaches work well on densely sampled surfaces, but both require normal orientation, which is notoriously difficult to estimate in practice, and do not preserve the input point samples, often leading to loss of detail.

Other approaches, based on Delaunay triangulations [9,10,30], alpha shapes [7,19] or Voronoi diagrams [3,4] often come with strong theoretical guarantees and perform well for densely sampled surfaces that satisfy certain conditions (see [18] for an overview of classical techniques). Several of these approaches, e.g. [2,7] also preserve the input point set, but, as we show below can struggle in the presence of poorly sampled data. Interestingly, there are several NP hardness results for surface reconstruction [5,8], which both explain the prevalence of heuristics, and point towards the use of *data-driven* priors. More fundamentally, classical approaches are not differentiable by nature and thus do not allow end-to-end training or backpropagation of the mesh with respect to input positions.

More closely related to ours are recent learning-based methods that aim to exploit the growing collections of geometric objects, which often come with known mesh structure. Similarly to the classical methods, most approaches in this area are based on volumetric representation. This includes, for example, a fully differentiable variant of the marching cubes [31] for computing a mesh via voxel grid occupancy prediction [22,35], and recent generative models for implicit surface prediction [13,20]. Unlike these works, our focus is on directly meshing an input point set. This allows our method to scale better with input complexity, avoid over-smoothing inherent in voxel-based techniques, and differentiate with respect to point positions.

Other surface generation methods include template-based techniques which fit a fixed mesh to the input [26,32,34], or deform a simple template while updating its connectivity [39,47]. More recent representations include fitting parameterized surface patches to points [23,48], or decomposing space in to convex sets [12,17]. Most importantly, none of the methods directly triangulate an arbitrary input point set. More generally, many of these schemes are limited to specific shape categories or topologies, or are too prohibitively expensive to use as a building-block in a larger system.

Perhaps most closely related to ours is the recent Scan2Mesh technique [16], which uses a graph-based formulation to generate triangles in a mesh. Though both generate meshes, there are many differences between this approach and ours, including that Scan2Mesh is applied to a volumetric representation, as opposed to the point-based representation used here, and Scan2Mesh does not attempt to generate watertight meshes. Several of the ideas introduced in this work could perhaps be combined with Scan2Mesh for further benefit, for instance using our proposal network (Sect. 3.3) to avoid quadratic complexity.

Finally, we note that our *proposal* and *classifier* architecture is inspired by classical approaches for object detection in images [21,43].

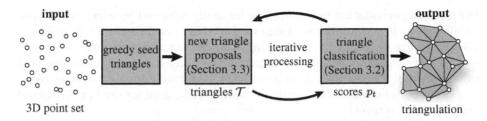

Fig. 1. An overview of our PointTriNet pipeline, which generates a triangulation by alternating between proposing new triangles and classifying them, each with a neural network. The classifier network identifies triangles which should appear in the triangulation, while the proposal network generates new candidates.

3   Method

Our method, called *PointTriNet*, has two essential components, both of which are learned. The first is a *classification* network (Sect. 3.2), which takes a candidate triangle as input, and outputs a score, which we interpret as the probability of the triangle appearing to the triangulation. The second is a *proposal* network (Sect. 3.3), which suggests likely neighbor triangles for an existing triangle. PointTriNet alternates between these networks, classifying candidate triangles and proposing new candidates, iteratively generating the output mesh (Fig. 1).

3.1   Triangle-Relative Coordinates

Rather than treating mesh generation as a global learning problem, we pose it as a local problem of predicting a single triangle, and make many such predictions. Both our classification and proposal subproblems are defined with respect to a particular query triangle: either classifying the triangle as a member of the mesh, or proposing candidates adjacent to the triangle. These problems depend on both the triangle's geometry and nearby neighborhood, but it is not immediately obvious how to encode both of these quantities for a neural network while preserving the expected invariants. Our solution is to design an encoding of points in the neighborhood of a triangle, *relative* to the triangle's geometry.

For any point p in the neighborhood of a triangle t, our encoding is given by Cartesian coordinates x', y', z' in a frame aligned with the first edge and normal of t, as well as the point's barycentric coordinates u, v, w after projecting in to the plane of the triangle. This results in the following encoding function, which outputs a 6-dimensional vector for p relative to t

$$\texttt{encode(t, p)} \rightarrow [x', y', z', u, v, w]. \tag{1}$$

Using both Cartesian and barycentric coordinates is essential to simultaneously encode *both* the shape of the triangle and the location of a neighbor point; either individually would not suffice. This encoding is invariant to scale and rigid transformations, but not yet invariant to the permutation of the triangle's vertices.

Per-triangle permutation invariance can be easily achieved by averaging results for all 6 possible permutations, though we find this to be unnecessary in practice. Finally, to encode a nearby neighboring triangle using this scheme, we encode each of the neighbor's vertices, then take a max and min along these encoded values, yielding 12 coordinates. This encoding is easy to evaluate, provides rigid- and (if desired) permutation- invariance, and is well-suited for use in point-based learning architectures.

3.2   Classifier Network

The primary tool in our method is a network which classifies whether a single query triangle belongs in the output triangulation. This classification is a function of the nearby points, as well as nearby triangles and their previous classification scores, as shown in Fig. 2.

More precisely, for a given query triangle we gather the n nearest points and m nearest triangles, as measured from triangle barycenters (we use $n = m = 64$ throughout). These neighboring points and triangles are then encoded relative to the query triangle as described in Sect. 3.1, and for the nearby triangles we additionally concatenate previous classification scores for a total $12 + 1 = 13$ coordinates per triangle. The result is a set of nearby encoded points $\mathcal{N}_{\text{point}} = \{\mathbb{R}^6\}_n$ and the set of nearby encoded triangles $\mathcal{N}_{\text{tri}} = \{\mathbb{R}^{13}\}_m$. We then learn a function

$$f : \mathcal{N}_{\text{point}}, \mathcal{N}_{\text{tri}} \rightarrow [0, 1] \tag{2}$$

which we interpret as the probability that the query triangle appears in the triangulation. We model this function as a PointNet [41], using the multi-layer perceptron (MLP) architecture shown in Fig. 2. Here, our careful problem formulation and input encoding enables the use of a small, ordinary PointNet without spatial hierarchies or learned rotations—the inputs are already localized to a small neighborhood and encoded in a rigid-invariant manner.

3.3   Proposal Network

The classifier network is effective at identifying good triangles, but requires a set of candidates as input. Naively enumerating all $O(n^3)$ possibilities does not scale, and simple heuristics miss important triangles.

Our approach identifies candidates via second *proposal* network. For each edge of a candidate triangle, the proposal network suggests new neighboring triangles across that edge. These proposals are learned by predicting a scalar function $\mathcal{V} \rightarrow [0, 1]$ on the nearby vertices. Formally, for an edge ij in triangle ijk, we learn the probability that a nearby vertex l forms a neighboring triangle ijl. These probabilities are then sampled to generate new candidates, which are assigned an initial probability as the product of the probability for the triangle that generated them and their learned sampling probability.

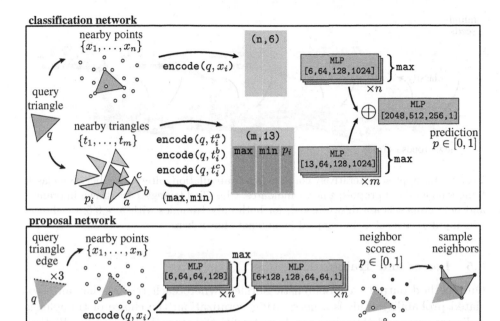

Fig. 2. The core of our technique is a network which classifies whether a given triangle belongs in a triangulation, as a function of nearby points and already-classified candidates. All coordinates are encoded relative to the query triangle. A second, similarly-structured network proposes new triangle candidates for subsequent classification.

The proposal network is again a PointNet over the local neighborhood, though it takes only points as input, and outputs a per-point result. We use the same triangle-relative point encoding (Sect. 3.1), but here the ordering of query triangle vertices ijk serves as a feature—the network always predicts neighbors across edge ij. Applying the network to the three cyclic permutations of ijk yields three sets of predictions, potential neighbors across each edge.

3.4 Iterative Prediction

To triangulate a point set, we alternate between proposing new candidates and classifying them. At each iteration, each candidate triangle t has probability $p_t \in [0,1]$, the probabilities from each round are features in the next, akin to a recurrent neural network. We initialize with greedy seed triangles among the nearest neighbors of each vertex, then for each iteration we first classify candidates, then generate new candidates, retaining only the highest-probability candidates (Fig. 3). If multiple instances of the same triangle arise, we discard lower probability instances. In our experiments, we iterate 5 times, sample 4 neighbors across each edge per iteration, and retain $12|V|$ candidates. If desired, the output can be post-processed to fill small holes (Fig. 4 and 7, *right*).

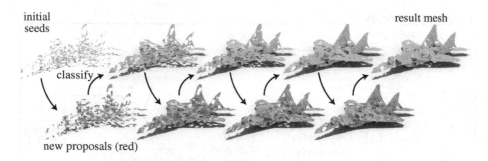

initial
seeds

result mesh

classify

new proposals (red)

Fig. 3. Our approach iteratively constructs a triangulation, alternating between classifying triangles and proposing new candidates. Classified triangles are drawn in orange, and new proposals are drawn in red, for both only triangles with $p > 0.5$ are shown. The input vertex set is 1k sampled points, not shown here.

3.5 Losses and Training

Our method outputs a probabilistic surface, where each triangle has an associated probability. This is a necessary generalization: the choice of triangles is a discrete set, and thus does not have traditional smooth derivatives. We thus make use of probabilistic losses detailed below, which are essentially the expected value of common geometric losses, like expected Chamfer distance.

For training, we assume a known, ground-truth surface $\mathcal{S}$, and the probabilistic surface $\mathcal{T}$ with probabilities $p : \mathcal{T} \to [0, 1]$ predicted by our algorithm. Note that we do not assume to be given a ground truth triangulation of $\mathcal{S}$, only some representation of the underlying surface sufficient to measure distance. In the losses below we assume $\mathcal{S}$ is represented by a sampled point cloud, but other representations like an implicit function could easily be substituted as training data. Notice that our method is thus unsupervised, in the sense that it learns to generate triangulations given un-annotated point clouds as input—it is not trained to match existing triangulations.

Expected Chamfer Distance (Forward). The distance from the ground truth surface to our predicted surface is measured as

$$\mathcal{L}_{\vec{C}} := E[\int_{x \in \mathcal{S}} \min_{y \in \mathcal{T}} d(x, y)] = \int_{x \in \mathcal{S}} E[\min_{y \in \mathcal{T}} d(x, y)]$$

$$= \int_{x \in \mathcal{S}} \int_{y \in \mathcal{T}} \gamma(x, y) d(x, y) \tag{3}$$

where $\gamma(x, y)$ denotes the probability that y is the closest point in $\mathcal{T}$ to x. Discretely, we evaluate this by sampling points x on $\mathcal{S}$, then for each point sorting all triangles in $\mathcal{T}$ by distance from x. A cumulative product of probabilities for these sorted triangles gives the probability that each triangle is the closest to x, and taking an expectation of distance under these probabilities yields the desired expected distance from the sample to our surface. In practice we truncate to the $k = 32$ nearest triangles.

Expected Chamfer Distance (Reverse). The distance from our predicted surface to the ground truth surface is measured as

$$\mathcal{L}_{\overline{C}} := E[\int_{y \in \mathcal{T}} \min_{x \in \mathcal{S}} d(x, y)] = \int_{y \in \mathcal{T}} p(y) \min_{x \in \mathcal{S}} d(x, y) \tag{4}$$

Discretely, we evaluate this by sampling points y on $\mathcal{T}$, then for each point we measure the distance to $\mathcal{S}$ and sum, weighting by the probability $p(y)$. Similar losses have appeared in 3D reconstruction, e.g. [31].

Overlap Kernel Loss. To discourage triangles from overlapping in space, we define a spatial kernel around each triangle

$$g_t(x) = p_t \max\left(0, 1 - \frac{d_n(x)}{d_e(x)}\right) \tag{5}$$

where x is an arbitrary point in space, p_t is the probability of triangle t, d_n is the distance in the normal direction from the triangle, and d_e is the signed perpendicular distance from the triangle's edge (see supplement for details). In a good triangulation, for any point x on the surface there will be exactly one triangle t for which $g_t(x) \approx 1$ while $g_{t'}(x) \approx 0$ for all other triangles t'. This is modeled by the loss

$$\mathcal{L}_O := \int_{x \in \mathcal{T}} \left(-1 + \sum_{t \in \mathcal{T}} g_t(x)\right)^2 + \left(-1 + \max_{t \in \mathcal{T}} g_t(x)\right)^2 \tag{6}$$

which is minimized when exactly one kernel contributes a value of 1 at x. Discretely, we sample points x on the surface of the generated triangulation $\mathcal{T}$ to evaluate the loss. One could instead sample x on the ground truth surface, but using the generated triangulation $\mathcal{T}$ makes the loss a regularizer, applicable in generative settings.

Watertight Loss. To encourage a watertight mesh, we explicitly maximize the probability that each edge in the triangulation is watertight, via

$$\mathcal{L}_W := \sum_{ij \in \mathcal{E}} p_{ij}(1 - p_{ij}^{\text{water}}) \tag{7}$$

where $\mathcal{E}$ is the set of edges, p_{ij} denotes the probability that ij appears in the triangulation, and p_{ij}^{water} denotes the probability that ij is watertight. An expression for evaluating this loss from triangle probabilities is given in the supplementary material. Notice that this loss does not directly penalize vertex-manifoldness—we observe that in practice almost all watertight configurations are also vertex-manifold, and thus simply encourage watertightness.

Network Training. The full loss for training the classification network is given by

$$\mathcal{L} := \mathcal{L}_{\vec{C}} + \mathcal{L}_{\overline{C}} + \lambda_O \mathcal{L}_O + \lambda_W \mathcal{L}_W, \tag{8}$$

where we use $\lambda_O = 0.01$ and $\lambda_W = 1$ for all experiments. All losses are normalized by the surface area or number of elements as appropriate, for scale invariance. During training, we backpropagate gradients only through the last iteration of classification. This strategy proves effective here because the network must be able to classify the most useful triangles regardless of the quality of the current candidate set.

We train the proposal network simultaneously with the classifier by encouraging its suggestions to match the classification scores. Intuitively, this means that the network attempts to propose triangles which will receive a high classification score on the next iteration. During training, we perform one last round of proposal to generate a suggestion set P with probabilities u_t. Rather than merging these triangles into the candidate set T as usual, we instead evaluate the loss

$$\mathcal{L}_M := \frac{1}{|P \cap T|} \sum_{t \in P \cap T} (u_t - p_t)^2 \tag{9}$$

and backpropagate through this loss only with respect to the proposal scores u_t. During training we also augment the proposal network to predict a random neighbor with 25% probability, to encourage diversity in the proposal set.

4   Experiments

Architecture and Training. The layer sizes of our networks are given in Fig. 2; we use ReLU activation functions throughout [37], except for a final single sigmoid activation to generate predictions. We use dropout with $p = 0.5$ on only the hidden layers of the prediction MLPs [45]. Neither batch nor layer normalization are used, as they were observed to negatively impact both result quality and the rate of convergence.

We train using the ADAM optimizer with a constant learning rate of 0.0001, and batch size of 8. Our networks are trained for 3 epochs over 20k training samples, which amounts to ≈5 h on a single RTX 2070 GPU.

Datasets. We validate our approach on the ShapeNetCore V2 dataset [11], which consists of about 50,000 3D models. We use the dataset-recommended 80%/20% train/test split, and further reserve 20% of the training models as a validation set. Our method is geometric in nature, so unlike semantic networks which operate per-category, we train and test simultaneously on all of ShapeNet, and do not need to subdivide by category.

We uniformly sample 1k points on the surface of each model, which will serve as the vertex set to be triangulated, and then separately sample another 10k points from each model, which serve as the representation of the surface when evaluating loss functions. Note that we *do not* use the mesh structure of the dataset during our training procedure.

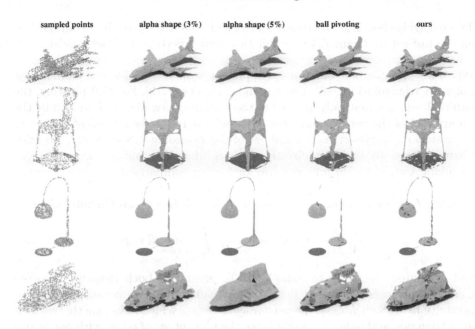

Fig. 4. A selection of outputs from PointTriNet and baselines on ShapeNet. The input is 1k points sampled on the surface of the shape, and the output is a mesh using those points as vertices. Constructing point set triangulations which are both accurate and watertight is still an open problem, but our method shows that a differentiable, learning-based approach can yield results competitive with classical computational geometry schemes.

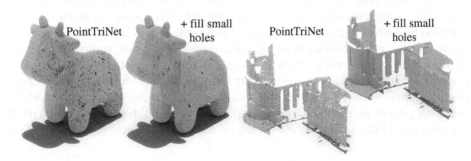

Fig. 5. Our method is data-driven, but its geometric nature leads to excellent generalization. *Left:* reconstructing a synthetic model with vertices of varying regularity and density. *Right:* reconstructing a 3D range scan of a cathedral (from [24]). Both use our networks trained on uniformly-sampled ShapeNet in Sect. 4.

To generate training examples in small neighborhoods, we choose a random center point on some shape, and gather the 256 nearest vertex samples (among the 1k), along with all surface samples (among the 10k) in the same radius. We use 20k such neighborhoods as a training set, with another 5k for validation.

These neighborhoods are used only for efficient training—to evaluate our method on the test set we triangulate the full 1k sampled vertices for each model.

Baselines. We compare our learned approach against two classical methods for point set triangulation: ball pivoting and α-shapes [7,19]. For ball-pivoting, the ball radius is automatically guessed as the bounding box diagonal divided by the square root of the vertex count. For α-shapes, we report two different choices of the radius parameter α, as 3% and 5% of the bounding box diagonal. No prior learning-based methods can directly serve as a baseline, but in the supplement we adapt a variant of Scan2Mesh [16] for further comparison.

Metrics. Geometric accuracy is assessed with a bi-directional Chamfer distance

$$\frac{1}{|\mathcal{A}|} \int_{x \in \mathcal{A}} \min_{y \in \mathcal{B}} d(x, y) + \frac{1}{|\mathcal{B}|} \int_{y \in \mathcal{B}} \min_{x \in \mathcal{A}} d(x, y) \qquad (10)$$

and evaluated discretely by sampling 10k points on both meshes. Reported Chamfer distances are scaled by 100× for readability. We measure mesh connectivity by *watertightness*, the percentage of edges which have exactly two incident triangles, and (edge-) *manifoldness*, the percentage of edges with one or two incident triangles—with this terminology mesh boundaries are manifold but not watertight. For our method, we take the triangles with $p > 0.9$ as the output.

4.1 Results

The most basic task for PointTriNet is to generate a mesh for a point set sampled from some underlying shape. Table 1 shows the effectiveness of our method on the sampled ShapeNet dataset, compared to the baseline approaches. Our method is geometrically more accurate than all of the classical schemes, though it achieves somewhat less regular connectivity than ball pivoting. We note that it may not be possible for any algorithm to significantly improve on these metrics: an imperfect (e.g., random) vertex sampling likely does not admit any triangulation which has both perfect geometric accuracy and perfect manifoldness. The supplement contains an ablation study, a breakdown by class, and triangle quality statistics.

Table 1. When evaluating on sampled ShapeNet, PointTriNet outperforms classical methods in geometric accuracy, though it has moderately less watertight connectivity.

	Chamfer	Watertight	Manifold
Ours	**0.7417**	77.0%	97.4%
Ballpivot	1.3440	**84.1%**	**100.0%**
Alpha-3	1.1099	49.8%	61.3%
Alpha-5	0.9195	47.6%	53.0%

Performance. Triangulating a point set with PointTriNet amounts to evaluating standard MLPs, in contrast to methods which e.g. solve an optimization problem for each input (Sect. 2). In our unoptimized implementation, triangulating 1000 points takes about 1 s on an RTX 2070 GPU. To reduce the memory footprint for large point sets (Fig. 5), one can evaluate the MLPs independently over many small patches of the input.

Outliers. An immediate benefit of a learned approach is that it can adapt to deficiencies in the input data. We add synthetic noise to our sampled ShapeNet dataset by perturbing 25% of the points according to a Gaussian with 2%-bounding-box deviation, and train our method from scratch on this data (Table 2). The classical baselines are not designed to handle noise, so our approach yields an even larger improvement in geometric accuracy in this setting.

Table 2. Evaluation on sampled ShapeNet, with noise added to a subset of the samples. Our method can adapt to the noise, further improving geometric accuracy.

	Chamfer	Watertight	Manifold
Ours	**1.0257**	74.3%	98.0%
Ballpivot	1.5418	**83.9%**	**100%**
Alpha-3	1.3260	50.8%	61.5%
Alpha-5	1.2309	47.5%	52.7%

Geometry Processing. A key benefit of our strategy is that PointTriNet directly generates a standard triangle mesh. Although these meshes are not necessarily manifold (a property shared with classical methods like α-shapes), they are still suitable for many standard algorithms in geometry processing (Fig. 6).

4.2 PointTriNet in Geometric Learning

Our method is designed to be a new building block in geometric learning.

Learned Vertex Improvement. We design a network which additionally learns a position update at each of the input vertices to further improve the quality of the output mesh, inspired by methods like [42]. More precisely, we introduce an additional MLP per-triangle, structured identically to the prediction MLP except the last layer has dimension $\mathbb{R}^{3\times2}$: an offset in the triangle's tangent basis for each of its three vertices, accumulated to yield a per-vertex update in the output mesh. This network, coupled with the base PointTriNet, is trained end to end as described above, including an additional loss term which encourages mesh quality by penalizing the deviation of the edge lengths. Figure 7 shows how these learned offsets further improve mesh quality with a more natural distribution of vertices.

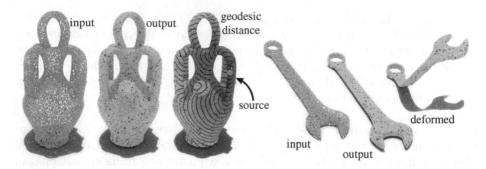

Fig. 6. PointTriNet directly generates a triangle mesh, opening the door to many standard geometric algorithms. *Left*: geodesic distance is computed from a source point [14]. *Right*: a reconstructed object is deformed via an anchor at one endpoint [33].

Generative Triangulation. PointTriNet enables triangulation as a component in end-to-end systems—algorithms designed to output points can be augmented to generate meshes useful for downstream applications. As a proof-of-concept, we construct a shape autoencoder which directly outputs a mesh using our technique. The encoder is a PointNet mapping input points to a 128 dimensional latent space, then decoding to 512 vertices with the architecture of [1]; these vertices are then triangulated with PointTriNet. We evaluate this scheme on the planes class of ShapeNet, as shown in Fig. 8. The point encoder/decoder and PointTriNet are pretrained as in [1] and Table 1, respectively, and the full system is then trained end-to-end for 3 epochs with the losses in Sect. 3.5.

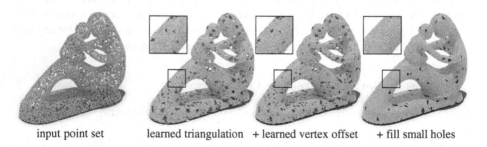

Fig. 7. Our method can be incorporated with other learning-based techniques. Here, we additionally learn a position offset at points to further improve output mesh quality.

This autoencoder is intentionally simplistic, and we leave a full investigation of its effectiveness to future work. However, it serves to demonstrate that PointTriNet can indeed directly generate mesh outputs from point-based geometric learning, while nearly all past approaches instead relied on volumetric representations or template deformations.

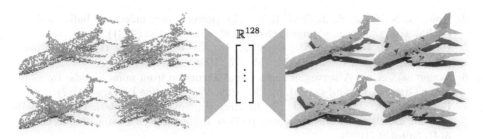

Fig. 8. A preliminary example of a an autoencoder, which uses our method to directly output a mesh without any intermediate volumetric representations. Input point clouds (*left*) are decoded to meshes with 512 vertices (*right*).

5 Limitations and Future Work

We introduce PointTriNet, a data-driven approach to address the classical problem of triangulating a given set of points. Our scheme uses point-based neural network architectures, and is shown to be accurate, generalizable, and scalable.

The meshes resulting from our method are competitive in accuracy with state-of-the-art point set triangulation methods from computational geometry. Crucially, however, our approach is fully differentiable, enabling novel applications such as building an autoencoder network with a triangle mesh output.

Our meshes are still noticeably less smooth than those extracted from volumetric representations—this is an inherent challenge of the strictly-harder point set triangulation problem, as opposed to general surface reconstruction. By incorporating topological priors, hole-filling, and building upon the differentiability of the meshing network, future work could further improve mesh quality.

We believe that this new approach to a classical problem will spur further research in geometric learning, and prove to be a useful component in a wide variety of applications.

Acknowledgements. The authors are grateful to Marie-Julie Rakotosaona and Keenan Crane for fruitful initial discussions, and to Angela Dai for assistance comparing with Scan2Mesh. Parts of this work were supported by an NSF Graduate Research Fellowship, the KAUST OSR Award No. CRG-2017-3426 and the ERC Starting Grant No. 758800 (EXPROTEA).

References

1. Achlioptas, P., Diamanti, O., Mitliagkas, I., Guibas, L.: Learning representations and generative models for 3D point clouds. arXiv preprint arXiv:1707.02392 (2017)
2. Amenta, N., Bern, M.: Surface reconstruction by Voronoi filtering. Discrete Comput. Geom. **22**(4), 481–504 (1999)
3. Amenta, N., Bern, M., Kamvysselis, M.: A new Voronoi-based surface reconstruction algorithm. In: Proceedings of the SIGGRAPH, pp. 415–421 (1998)

4. Amenta, N., Choi, S., Kolluri, R.K.: The power crust, unions of balls, and the medial axis transform. Comput. Geom. **19**(2–3), 127–153 (2001)
5. Barequet, G., Dickerson, M., Eppstein, D.: On triangulating three-dimensional polygons. Comput. Geom. **10**(3), 155–170 (1998)
6. Berger, M., et al.: A survey of surface reconstruction from point clouds. In: Computer Graphics Forum, vol. 36, pp. 301–329. Wiley Online Library (2017)
7. Bernardini, F., Mittleman, J., Rushmeier, H., Silva, C., Taubin, G.: The ball-pivoting algorithm for surface reconstruction. IEEE Trans. Visual. Comput. Graph. **5**(4), 349–359 (1999)
8. Biedl, T., Durocher, S., Snoeyink, J.: Reconstructing polygons from scanner data. Theoret. Comput. Sci. **412**(32), 4161–4172 (2011)
9. Boissonnat, J.D.: Geometric structures for three-dimensional shape representation. ACM Trans. Graph. (TOG) **3**(4), 266–286 (1984)
10. Boissonnat, J.D., Oudot, S.: Provably good sampling and meshing of surfaces. Graph. Models **67**(5), 405–451 (2005)
11. Chang, A.X., et al.: ShapeNet: an information-rich 3D model repository. arXiv preprint arXiv:1512.03012 (2015)
12. Chen, Z., Tagliasacchi, A., Zhang, H.: BSP-Net: generating compact meshes via binary space partitioning. In: Proceedings of the IEEE/CVF Conference on Computer Vision and Pattern Recognition, pp. 45–54 (2020)
13. Chen, Z., Zhang, H.: Learning implicit fields for generative shape modeling. In: Proceedings of the IEEE Conference on Computer Vision and Pattern Recognition, pp. 5939–5948 (2019)
14. Crane, K., Weischedel, C., Wardetzky, M.: The heat method for distance computation. Commun. ACM **60**(11), 90–99 (2017)
15. Curless, B., Levoy, M.: A volumetric method for building complex models from range images. In: Proceedings of the 23rd Annual Conference on Computer Graphics and Interactive Techniques, pp. 303–312. In: Proceedings of SIGGRAPH (1996)
16. Dai, A., Nießner, M.: Scan2Mesh: from unstructured range scans to 3D meshes. In: Proceedings of the IEEE Conference on Computer Vision and Pattern Recognition, pp. 5574–5583 (2019)
17. Deng, B., Genova, K., Yazdani, S., Bouaziz, S., Hinton, G., Tagliasacchi, A.: CvxNet: Learnable convex decomposition. In: Proceedings of the IEEE/CVF Conference on Computer Vision and Pattern Recognition, pp. 31–44 (2020)
18. Dey, T.K.: Curve and Surface Reconstruction: Algorithms with Mathematical Analysis, vol. 23. Cambridge University Press, Cambridge (2006)
19. Edelsbrunner, H., Mücke, E.P.: Three-dimensional alpha shapes. ACM Trans. Graph. (TOG) **13**(1), 43–72 (1994)
20. Genova, K., Cole, F., Vlasic, D., Sarna, A., Freeman, W.T., Funkhouser, T.: Learning shape templates with structured implicit functions. arXiv preprint arXiv:1904.06447 (2019)
21. Girshick, R., Donahue, J., Darrell, T., Malik, J.: Rich feature hierarchies for accurate object detection and semantic segmentation. In: Proceedings of the IEEE Conference on Computer Vision and Pattern Recognition, pp. 580–587 (2014)
22. Gkioxari, G., Malik, J., Johnson, J.: Mesh R-CNN. arXiv preprint arXiv:1906.02739 (2019)
23. Groueix, T., Fisher, M., Kim, V.G., Russell, B.C., Aubry, M.: A papier-mâché approach to learning 3D surface generation. In: Proceedings of the IEEE Conference on Computer Vision and Pattern Recognition, pp. 216–224 (2018)

24. Hackel, T., Savinov, N., Ladicky, L., Wegner, J.D., Schindler, K., Pollefeys, M.: Semantic3d. net: a new large-scale point cloud classification benchmark. arXiv preprint arXiv:1704.03847 (2017)

25. Hoppe, H., DeRose, T., Duchamp, T., McDonald, J., Stuetzle, W.: Surface reconstruction from unorganized points, vol. 26. ACM (1992)

26. Kanazawa, A., Tulsiani, S., Efros, A.A., Malik, J.: Learning category-specific mesh reconstruction from image collections. In: Ferrari, V., Hebert, M., Sminchisescu, C., Weiss, Y. (eds.) ECCV 2018. LNCS, vol. 11219, pp. 386–402. Springer, Cham (2018). https://doi.org/10.1007/978-3-030-01267-0_23

27. Kazhdan, M., Bolitho, M., Hoppe, H.: Poisson surface reconstruction. In: Proceedings of the Fourth Eurographics Symposium on Geometry Processing, vol. 7 (2006)

28. Kazhdan, M., Hoppe, H.: Screened poisson surface reconstruction. ACM Transactions on Graphics (ToG) $32(3)$, 1–13 (2013)

29. Khatamian, A., Arabnia, H.R.: Survey on 3D surface reconstruction. J. Inf. Process. Syst. $12(3)$, 338–357 (2016)

30. Kolluri, R., Shewchuk, J.R., O'Brien, J.F.: Spectral surface reconstruction from noisy point clouds. In: Proceedings of the 2004 Eurographics/ACM SIGGRAPH Symposium on Geometry Processing, pp. 11–21. ACM (2004)

31. Liao, Y., Donne, S., Geiger, A.: Deep marching cubes: learning explicit surface representations. In: Proceedings of the IEEE Conference on Computer Vision and Pattern Recognition, pp. 2916–2925 (2018)

32. Lin, C.H., et al.: Photometric mesh optimization for video-aligned 3D object reconstruction. In: Proceedings of the IEEE Conference on Computer Vision and Pattern Recognition, pp. 969–978 (2019)

33. Lipman, Y., Sorkine, O., Cohen-Or, D., Levin, D., Rossi, C., Seidel, H.P.: Differential coordinates for interactive mesh editing. In: Proceedings Shape Modeling Applications, 2004, pp. 181–190. IEEE (2004)

34. Litany, O., Bronstein, A., Bronstein, M., Makadia, A.: Deformable shape completion with graph convolutional autoencoders. In: Proceedings of the IEEE Conference on Computer Vision and Pattern Recognition, pp. 1886–1895 (2018)

35. Mescheder, L., Oechsle, M., Niemeyer, M., Nowozin, S., Geiger, A.: Occupancy networks: Learning 3D reconstruction in function space. In: Proceedings of the IEEE Conference on Computer Vision and Pattern Recognition, pp. 4460–4470 (2019)

36. Mullen, P., De Goes, F., Desbrun, M., Cohen-Steiner, D., Alliez, P.: Signing the unsigned: Robust surface reconstruction from raw pointsets. In: Computer Graphics Forum, vol. 29, pp. 1733–1741. Wiley Online Library (2010)

37. Nair, V., Hinton, G.E.: Rectified linear units improve restricted Boltzmann machines. In: Proceedings of the 27th International Conference on Machine Learning (ICML 2010), pp. 807–814 (2010)

38. Newman, T.S., Yi, H.: A survey of the marching cubes algorithm. Comput. Graph. $30(5)$, 854–879 (2006)

39. Pan, J., Han, X., Chen, W., Tang, J., Jia, K.: Deep mesh reconstruction from single RGB images via topology modification networks. In: Proceedings of the IEEE International Conference on Computer Vision, pp. 9964–9973 (2019)

40. Park, J.J., Florence, P., Straub, J., Newcombe, R., Lovegrove, S.: DeepSDF: learning continuous signed distance functions for shape representation. In: Proceedings of the IEEE Conference on Computer Vision and Pattern Recognition, pp. 165–174 (2019)

41. Qi, C.R., Su, H., Mo, K., Guibas, L.J.: Pointnet: deep learning on point sets for 3D classification and segmentation. In: Proceedings of the IEEE Conference on Computer Vision and Pattern Recognition, pp. 652–660 (2017)
42. Rakotosaona, M.J., La Barbera, V., Guerrero, P., Mitra, N.J., Ovsjanikov, M.: Pointcleannet: learning to denoise and remove outliers from dense point clouds. In: Computer Graphics Forum. Wiley Online Library (2019)
43. Ren, S., He, K., Girshick, R., Sun, J.: Faster R-CNN: towards real-time object detection with region proposal networks. In: Advances in Neural Information Processing Systems, pp. 91–99 (2015)
44. Shewchuk, J., Dey, T.K., Cheng, S.W.: Delaunay Mesh Generation. Chapman and Hall/CRC, Boca Raton (2016)
45. Srivastava, N., Hinton, G., Krizhevsky, A., Sutskever, I., Salakhutdinov, R.: Dropout: a simple way to prevent neural networks from overfitting. J. Mach. Learn. Res. **15**(1), 1929–1958 (2014)
46. Treece, G.M., Prager, R.W., Gee, A.H.: Regularised marching tetrahedra: improved ISO-surface extraction. Comput. Graph. **23**(4), 583–598 (1999)
47. Wang, N., Zhang, Y., Li, Z., Fu, Y., Liu, W., Jiang, Y.G.: Pixel2Mesh: Generating 3D mesh models from single RGB images. In: Proceedings of the European Conference on Computer Vision (ECCV), pp. 52–67 (2018)
48. Williams, F., Schneider, T., Silva, C., Zorin, D., Bruna, J., Panozzo, D.: Deep geometric prior for surface reconstruction. In: Proceedings of the IEEE Conference on Computer Vision and Pattern Recognition, pp. 10130–10139 (2019)

Toward Unsupervised, Multi-object Discovery in Large-Scale Image Collections

Huy V. Vo[1,2,3(✉)], Patrick Pérez[3], and Jean Ponce[1,2]

[1] Inria, Paris, France
van-huy.vo@inria.fr
[2] Département d'informatique de l'ENS, ENS, CNRS, PSL University, Paris, France
[3] Valeo.ai, Paris, France

Abstract. This paper addresses the problem of discovering the objects present in a collection of images without any supervision. We build on the optimization approach of Vo *et al.* [34] with several key novelties: (1) We propose a novel saliency-based region proposal algorithm that achieves significantly higher overlap with ground-truth objects than other competitive methods. This procedure leverages off-the-shelf CNN features trained on classification tasks without any bounding box information, but is otherwise unsupervised. (2) We exploit the inherent hierarchical structure of proposals as an effective regularizer for the approach to object discovery of [34], boosting its performance to significantly improve over the state of the art on several standard benchmarks. (3) We adopt a two-stage strategy to select promising proposals using small random sets of images before using the whole image collection to discover the objects it depicts, allowing us to tackle, for the first time (to the best of our knowledge), the discovery of multiple objects in each one of the pictures making up datasets with up to 20,000 images, an over five-fold increase compared to existing methods, and a first step toward true large-scale unsupervised image interpretation.

Keywords: Object discovery · Large-scale · Optimization · Region proposals · Unsupervised learning

1 Introduction

Object discovery, that is finding the location of salient objects in images without using any source of supervision, is a fundamental scientific problem in computer vision. It is also potentially an important practical one, since any effective solution would serve as a reliable free source of supervision for other tasks such as

---

Electronic supplementary material The online version of this chapter (https://doi.org/10.1007/978-3-030-58592-1_46) contains supplementary material, which is available to authorized users.

© Springer Nature Switzerland AG 2020
A. Vedaldi et al. (Eds.): ECCV 2020, LNCS 12368, pp. 779–795, 2020.
https://doi.org/10.1007/978-3-030-58592-1_46

object categorization, object detection and the like. While many of these tasks can be tackled using massive amounts of annotated data, the manual annotation process is complex and expensive at large scales. Combining the discovery results with a limited amount of annotated data in a semi-supervised setting is a promising alternative to current data-hungry supervised approaches [35].

Vo *et al.* [34] posit that image collections possess an implicit graph structure. The pictures themselves are the nodes, and an edge links two images when they share similar visual content. They propose the object and structure discovery framework (OSD) to localize objects and find the graph structure simultaneously by solving an optimization problem. Though demonstrating promising results, [34] has several shortcomings, e.g., the use of supervised region proposals, the limitation in addressing large image collections (See Sect. 2). Our work is built on OSD, aims to alleviate its limitations and improves it to effectively discover multiple objects in large image collections. Our contributions are:

- We propose a simple but effective method for generating region proposals directly from CNN features (themselves trained beforehand on some auxiliary task [29] *without* bounding boxes) in an unsupervised way (Sect. 3.1). Our algorithm gives on average half the number of region proposals per image compared to selective search [33], edgeboxes [40] or randomized Prim [23], yet significantly outperforms these off-the-shelf region proposals in object discovery (Table 3).

- Leveraging the intrinsic structure of region proposals generated by our method allows us to add an additional constraint into the OSD formulation that acts as a regularizer on its behavior (Sect. 3.2). This new formulation (rOSD) significantly outperforms the original algorithm and allows us to effectively perform multi-object discovery, a setting never studied before (to the best of our knowledge) in the literature.

- We propose a two-stage algorithm to make rOSD applicable to large image collections (Sect. 3.3). In the first stage, rOSD is used to choose a small set of good region proposals for each image. In the second stage, these proposals and the full image collection are fed to rOSD to find the objects and the image graph structure.

- We demonstrate that our approach yields significant improvements over the state of the art in object discovery (Tables 4 and 5). We also run our two-stage algorithm on a new and much larger dataset with 20,000 images and show that it significantly outperforms plain OSD in this setting (Table 7).

The only supervisory signal used in our setting are the image labels used to train CNN features in an auxiliary classification task (see [21,35] for similar approaches in the related colocalization domain). We use CNN features trained on ImageNet classification [29], *without* any bounding box information. Our region proposal and object discovery algorithms are otherwise fully unsupervised.

2   Related Work

Region proposals have been used in object detection/discovery to serve as object priors and reduce the search space. In most cases, they are found either by a bottom-up approach in which low-level cues are aggregated to rank a large set of boxes obtained with sliding window approaches [1,33,40] and return the top windows as proposals, or by training a model to classify them (as in randomized Prim [23], see also [26]), with *bounding box supervision*. Edgeboxes [40] and selective search [33] are popular off-the-shelf algorithms that are used to generate region proposals in object detection [13,14], weakly supervised object detection [7,31] or image colocalization [21]. Note, however, that the features used to generate proposals in these algorithms and those representing them in the downstream tasks are generally different in nature: Typically, region proposals are generated from low-level features such as color and texture [33] or edge density [40], but CNN features are used to represent them in downstream tasks. However, the Region Proposal Network in Faster-RCNN [26] shows that proposals generated directly from the features used in the object detection task itself give a great boost in performance. In the object discovery setting, we therefore propose a novel approach for generating region proposals in an unsupervised way from CNN features trained on an auxiliary classification task without bounding box information. Features from CNNs trained on large-scale image classification have also been used to localize object in the weakly supervised setting. Zhou *et al.* [39] and Selvaraju *et al.* [28] fine-tune a pre-trained CNN to classify images and construct class activation maps, as weighted sums of convolutional feature maps or their gradient with respect to the classification loss, for localizing objects in these images. Tang *et al.* [32] generate region proposals to perform weakly supervised object detection on a set of labelled images by training a proposal network using the images' labels as supervision. Contrary to these works, we generate region proposals using only pre-trained CNN features without fine-tuning the feature extractor. Moreover, our region proposals come with a nice intrinsic structure which can be exploited to improve object discovery performance.

Early work on object discovery [12,15,20,27,30] focused on a restricted setting where images are from only a few distinctive object classes. Cho *et al.* [6] propose an approach for object and structure discovery by combining a part-based matching technique and an iterative *match-then-localize* algorithm, using off-the-shelf region proposals as primitives for matching. Vo *et al.* [34] reformulate [6] in an optimization framework and obtain significantly better performance. Image colocalization can be seen as a narrow setting of object discovery where all images in the collection contain objects from the same class. Observing that supervised object detectors often assign high scores to only a small number of region proposals, Li *et al.* [21] propose to mimic this behavior by training a classifier to minimize the entropy of the scores it gives to region proposals. Wei *et al.* [35] localize objects by clustering pixels with high activations in feature maps from CNNs pre-trained in ImageNet. All of the above works, however, focus on discovering only the main object in the images and target small-to-medium-scale datasets. Our approach is based on a modified version of the OSD formulation of Vo *et al.* [34] and pre-trained CNN features for object discovery, offers an effective and efficient solution to discover multiple objects in images in large-scale datasets. The recent work of Hsu *et al.* [18] for instance co-segmentation can also be adapted for localizing multiple objects in images. However, it requires input images to contain an object of a single dominant class while images may instead contain several objects from different categories in our setting.

Object and Structure Discovery (OSD) [34]. Since our work is built on [34], we give a short recap of this work in this section. Given a collection of n images, possibly containing objects from different categories, each equipped with p region proposals (which can be obtained using selective search [33], edgeboxes [40], randomized Prim [23], etc.) and a set of potential neighbors, the unsupervised object and structure discovery problem (OSD) is formalized in [34] as follows: Let us define the variable e as an element of $\{0,1\}^{n \times n}$ with a zero diagonal, such that $e_{ij} = 1$ when images i and j are linked by a (directional) edge, and $e_{ij} = 0$ otherwise, and the variable x as an element of $\{0,1\}^{n \times p}$, with $x_i^k = 1$ when region proposal number k corresponds to visual content shared with neighbors of image i in the graph. This leads to the following optimization problem:

$$\max_{x,e} S(x,e) = \sum_{i=1}^{n} \sum_{j \in N(i)} e_{ij} x_i^T S_{ij} x_j, \text{ s.t. } \sum_{k=1}^{p} x_i^k \leq \nu \text{ and } \sum_{j \neq i} e_{ij} \leq \tau \; \forall i, \quad (1)$$

where $N(i)$ is the set of potential neighbors of image i, S_{ij} is a $p \times p$ matrix whose entry S_{ij}^{kl} measures the similarity between regions k and l of images i and j, and ν and τ are predefined constants corresponding respectively to the maximum number of objects present in an image and to the maximum number of neighbors an image may have. This is however a hard combinatorial optimization problem. As shown in [34], an approximate solution can be found by (a) a dual gradient ascent algorithm for a continuous relaxation of Eq. (1) with exact updates obtained by maximizing a supermodular cubic pseudo-Boolean function [4,24], (b) a simple greedy scheme, or (c) a combination thereof. Since solving the continuous relaxation of Eq. (1) is computationally expensive and may be less effective for large datasets [34], we only consider the version (b) of OSD in our analysis.

OSD has some limitations: (1) Although the algorithm itself is fully unsupervised, it gives by far its best results with region proposals from randomized Prim [23], a region proposal algorithm trained with bounding box supervision. (2) Vo *et al.* use whitened HOG (WHO) [16] to represent region proposals in their implementation although CNN features work better on the similar image colocalization problem [21,35]. In our experiments, naively switching to CNN features does not give consistent improvement on common benchmarks. OSD with CNN features gives a CorLoc of 82.9, 71.5 and 42.8 compared to 87.1, 71.2 and 39.5 given by OSD with WHO, respectively on OD, VOC_6x2 and VOC_all data sets respectively. (3) Finally, due to its high memory cost, the algorithm cannot be applied to large datasets without compromising its final performance.

In the next section, we describe our approach to addressing these limitations, as well as extending OSD to solve multi-object discovery.

3   Proposed Approach

3.1   Region Proposals from CNN Features

We address the limitation of using off-the-shelf region proposals of [34] with insights gained from the remarkably effective method for image colocalization proposed by Wei *et al.* [35]: CNN features pre-trained for an auxiliary task, such as ImageNet classification, give a strong, *category-independent* signal for unsupervised tasks. In retrospect, this insight is not particularly surprising, and it is implicit in several successful approaches to image retrieval [38] or co-saliency detection [2,3,19,36]. Wei *et al.* [35] use it to great effect in the image colocalization task. Feeding an image to a pre-trained convolutional neural network yields a set of feature maps represented as a 3D tensor (e.g., a convolutional layer of VGG16 [29] or ResNet [17]). Wei *et al.* [35] observe that the "image" obtained by simply adding the feature maps gives hints to the locations of the objects it contains, and identify objects by clustering pixels with high activation. Similar but different from them, we observe that local maxima in the above "images" correspond to salient parts of objects in the original image and propose to exploit this observation for generating region proposals directly from CNN features. As we do not make use of any annotated bounding boxes, our region proposal itself is indeed unsupervised. Our method consists of the following steps. First, we feed the image to a pre-trained convolutional neural network to obtain a 3D tensor of size $(H \times W \times D)$, noted F. Adding elements of the tensor along its depth dimension yields a $(H \times W)$ 2D saliency map, noted as s_g (*global* saliency map), showing salient locations in the image with each location in s_g being represented by the corresponding D-dimensional feature vector from F. Next, we find robust local maxima in the previous saliency map using *persistence*, a measure used in topological data analysis [5,8,9,25,41] to find critical points of a function (see Sect. 4.2 for details). We find regions around each local maximum y using a *local* saliency map s_y of the same size as the global one. The value at any location in s_y is the dot product between normalized feature vectors at that location and the local maximum. By construction, the local saliency map highlights locations that are likely to belong to the same object as the corresponding local maximum. Finally, for each local saliency map, we discard all locations with scores below some threshold and the bounding box around the connected component containing the corresponding local maximum is returned as a region proposal. By varying the threshold, we can obtain tens of region proposals per local saliency map. An example illustrating the whole process is shown in Fig. 1.

3.2   Regularized OSD

Due to the greedy nature of OSD [34], its block-coordinate ascent iterations are prone to bad local maxima. Vo *et al.* [34] attempt to resolve this problem by

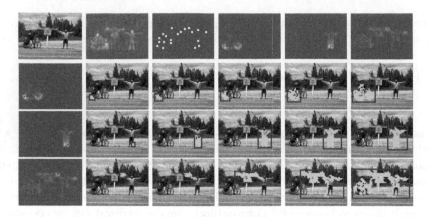

Fig. 1. Illustration of the unsupervised region proposal generation process. The top row shows the original image, the global saliency map s_g, local maxima of s_g and three local saliency maps s_y from three local maxima (marked by red stars). The next three rows illustrate the proposal generation process on the local saliency maps: From left to right, we show in green the connected component formed by pixels with saliency above decreasing thresholds and, in red, the corresponding region proposals. (Color figure online)

using a larger value of ν in the optimization than the actual number of objects they intend to retrieve (which is one in their case) to diversify the set of retained regions in each iteration. The final region in each image is then chosen amongst its retained regions in a post processing step by ranking these using a new score solely based on their similarity to the retained regions in the image's neighbors. Increasing ν in fact gives limited help in diversifying the set of retained regions. Since there is redundancy in object proposals with many highly overlapping regions, the ν retained regions are often nearly identical (see supplementary document for a visual illustration). This phenomenon also prevents OSD from retrieving multiple objects in images. One can use the ranking in OSD's post processing step with non-maximum suppression to return more than one region from ν retained regions but since ν regions are often highly overlapping, this fails to localize multiple objects.

By construction, proposals produced by our approach also contain many highly overlapping regions, especially those generated from the same local maximum in the saliency map. However, they come with a nice intrinsic structure: Proposals in an image can be partitioned into groups labelled by the local maximum from which they are generated. Naturally, it makes sense to impose that at most one region in a group is retained in OSD since they are supposed to correspond to the same object. This additional constraint also conveniently helps to diversify the set of proposals returned by the block-coordinate ascent procedure by avoiding to retain highly overlapping regions. Concretely, let G_{ig} be the set of region proposals in image i generated from the g-th local maximum in its global saliency map s_g, with $1 \leq g \leq L_i$ where L_i is the number of local maxima in s_g, we propose to add the constraints $\sum_{k \in G_{ig}} x_i^k \leq 1 \; \forall i, g$ to Eq. (1). We coin the new formulation regularized OSD (rOSD). Similar to OSD, a solution to rOSD can be obtained by a greedy block-coordinate ascent algorithm whose iterations are illustrated in the supplementary document. We will demonstrate the effectiveness of rOSD compared to OSD and the state of the art in Sect. 4.

3.3 Large-Scale Object Discovery

The optimization algorithm of Vo *et al.* [34] requires loading all score matrices S_{ij} into the memory (they can also be computed on-the-fly but at an unacceptable computational cost). The corresponding memory cost is $M = (\sum_{i=1}^n |N(i)|) \times K$, decided by two main factors: The number of image pairs considered $\sum_{i=1}^n |N(i)|$ and the number of positive entries K in matrices S_{ij}. To reduce the cost on larger datasets, Vo *et al.* [34] pre-filter the neighborhood of each image ($|N(i)| \leq 100$ for classes with more than 1000 images) and limit K to 1000. This value of K is approximately the average number of proposals in each image, and it is intentionally chosen to make sure that S_{ij} is not too sparse in the sense that approximately every proposal in image i should have a positive match with some proposal in image j. Further reducing the number of positive entries in score matrices is likely to hurt the performance (Table 7) while a number of 100 potential neighbors is already small and can not be significantly lowered. Effectively scaling up OSD[1] therefore requires lowering considerably the number of proposals it uses. To this end, we propose two different interpretations of the image graph and exploit both to scale up OSD.

Two Different Interpretations of the Image Graph. The *image graph* $G = (x, e)$ obtained by solving Eq. (1) can be interpreted as capturing the "true" structure of the input image collection. In this case, ν is typically small (say, 1 to 5) and the discovered "objects" correspond to maximal cliques of G, with instances given by active regions ($x_i^k = 1$) associated with nodes in the clique. But it can also be interpreted as a *proxy* for that structure. In this case, we typically take ν larger (say, 50). The active regions found for each node x_i of G are interpreted as the most promising regions in the corresponding image and the active edges e_{ij} link it to other images supporting that choice. We dub this variant *proxy* OSD.

For small image collections, it makes sense to run OSD only. For large ones, we propose instead to split the data into random groups with fixed size, run proxy OSD on each group to select the most promising region proposals in the corresponding images, then run OSD using these proposals. Using this two-stage algorithm, we reduce significantly the number of image pairs in each run of the first stage, thus permitting the use of denser score matrices in these runs. In the second stage, since only a very small number of region proposals are considered in each image, we need to keep only a few positive entries in each score matrix and are able to run OSD on the entire image collection. Our approach for large-scale object discovery is summarized the supplementary material.

---

[1] Since the analysis in this section applies to both OSD and rOSD, we refer to both as OSD for ease of notation.

4   Experiments

4.1   Datasets and Metrics

Similar to previous works on object discovery [6,34] and image colocalization [21,35], we evaluate object discovery performance with our proposals on four datasets: Object Discovery (OD), VOC_6x2, VOC_all and VOC12. OD is a small dataset with three classes *airplane, car* and *horse*, and 100 images per class, among which 18, 11 and 7 images are outliers (images not including an object of the corresponding class) respectively. VOC_all is a subset of the PASCAL VOC 2007 dataset [11] obtained by eliminating all images containing only *difficult* or *truncated* objects as well as *difficult* or *truncated* objects in retained images. It has 3550 images and 6661 objects. VOC_6x2 is a subset of VOC_all which contains images of 6 classes *aeroplane, bicycle, boat, bus, horse* and *motorbike* divided into 2 views *left* and *right*. In total, VOC_6x2 contains 463 images of 12 classes. VOC12 is a subset of the PASCAL VOC 2012 dataset [10] and obtained in the same way as VOC_all. It contains 7838 images and figures 13957 objects. For large-scale experiments, we randomly choose 20000 images from the training set of COCO [22] and eliminate those containing only *crowd* bounding boxes as well as bounding boxes marked as *crowd* in retained images. The resulting dataset, which we call COCO_20k, has 19817 images and 143951 objects.

As single-object discovery and colocalization performance measure, we use *correct localization* (CorLoc) defined as the percentage of images correctly localized. In our context, this means the intersection over union (IoU) between one of the ground-truth regions and one of the predicted regions in the image is greater than 0.5. Since CorLoc does not take into account multiple detections per image, for multi-object discovery, we use instead *detection rate* at the IoU threshold of 0.5 as measure of performance. Given some threshold ζ, detection rate at $IoU = \zeta$ is the percentage of ground-truth bounding boxes that have an IoU with one of the retained proposals greater than ζ. We run the experiments in both the colocalization setting, where the algorithm is run separately on each class of the dataset, and the average CorLoc/detection rate over all classes is computed as the overall performance measure on the dataset, and the true *discovery* setting where the whole dataset is considered as a single class.

4.2   Implementation Details

Features. We test our methods with the pre-trained CNN features from VGG16 and VGG19 [29]. For generating region proposals, we apply the algorithm described in Sect. 3.1 separately to the layers right before the last two max pooling layers of the networks (*relu4_3* and *relu5_3* in VGG16, *relu4_4* and *relu5_4* in VGG19), then fuse proposals generated from the two layers as our final set of proposals. Note that using CNN features at multiple layers is important as different layers capture different visual patterns in images [37]. One could also use more layers from VGG16 (e.g., layers *relu3_3, relu4_2* or *relu5_2*) but we only use two for the sake of efficiency. In experiments with OSD, we extract features for the region proposals by applying the RoI pooling operator introduced in Fast-RCNN [13] to layer $relu5\_3$ of VGG16.

Region Proposal Generation Process. For finding robust local maxima of the global saliency maps s_g, we rank its locations using persistence [5, 8, 9, 25, 41]. Concretely, we consider s_g as a 2D image and each location in it as a pixel. We associate with each pixel a cluster (the 4-neighborhood connected component of pixels that contains it), together with a "birth" (its own saliency) and "death" time (the highest value for which one of the pixels in its cluster also belongs to the cluster of a pixel with higher saliency, or, if no such location exists, the lowest saliency value in the map). The persistence of a pixel is defined as the difference between its birth and death times. A sorted list of pixels in decreasing persistence order is computed, and the local maxima are chosen as the top pixels in the list. For additional robustness, we also apply non maximum suppression on the list over a 3×3 neighborhood. Since the saliency map created from CNN feature maps can be very noisy, we eliminate locations with score in s_g below $\alpha \max s_g$ before computing the persistence to obtain only good local maxima. We also eliminate locations with score smaller than the average score in s_y and whose score in s_g is smaller than β times the average score in s_g. We choose the value of the pair (α, β) in $\{0.3, 0.5\} \times \{0.5, 1\}$ by conducting small-scale object discovery on VOC_6x2. We find that $(\alpha, \beta) = (0.3, 0.5)$ yields the best performance and gives local saliency maps that are not fragmented while eliminating well irrelevant locations across settings and datasets. We take up to 20 local maxima (after non-maximum suppression) and use 50 linearly spaced thresholds between the lowest and the highest scores in each local saliency map to generate proposals.

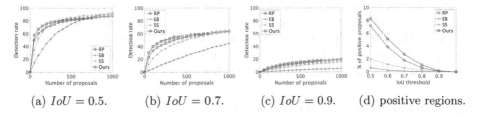

(a) $IoU = 0.5$. (b) $IoU = 0.7$. (c) $IoU = 0.9$. (d) positive regions.

Fig. 2. Quality of proposals by different methods. (a–c): Detection rate by number of proposals at different IoU thresholds of randomized Prim (RP) [23], edgeboxes (EB) [40], selective search (SS) [33] and ours; (d): Percentage of positive proposals for the four methods.

Object Discovery Experiments. For single-object colocalization and discovery, following [34], we use $\nu = 5, \tau = 10$ and apply the OSD's post processing to obtain the final localization result. For multi-object setting, we use $\nu = 50$, $\tau = 10$ and apply the post processing with non-maximum suppression at $IoU = 0.7$ to retain at most 5 regions in the final result. On large classes/datasets, we pre-filter the set of neighbors that are considered in the optimization for each image, using the cosine similarity between features from the fully connected layer *fc6* of the pre-trained network, following [2]. The number of potential neighbors of each image is fixed to 50 in all experiments where the pre-filtering is necessary.

4.3   Region Proposal Evaluation

Following other works on region proposals [23,33,40], we evaluate the quality of our proposals on PASCAL VOC 2007 using the *detection rate* at various IoU thresholds. But since we intend to later use our proposals for object discovery, unlike other works, we evaluate directly our proposals on VOC_all instead of the test set of VOC 2007 to reveal the link between the quality of proposals and the object discovery performance. Figure 2(a–c) shows the performance of different proposals on VOC_all. It can be seen that our method performs better than others at a very high overlap threshold (0.9) regardless of the number of proposals allowed. At medium threshold (0.7), our proposals are on par (or better for fewer than 500 proposals) with those from selective search [33] and randomized Prim [23] and much better than those from edgeboxes [40]. At a small threshold (0.5), our method is still on par with randomized Prim and edgeboxes, but does not fare as well as selective search. It should be noted that randomized Prim is supervised whereas the others are unsupervised.

In OSD, localizing an object in an image means singling out a *positive* proposal, that is, a proposal having an IoU greater than some threshold with object bounding boxes. It is therefore easier to localize the object if the percentage of positive region proposals is larger. As shown by Fig. 2(d), our method performs very well according to this criterion: Over 8% of our proposals are positive at an IoU threshold of 0.5, and over 3% are still positive for an IoU of 0.7. Also, randomized Prim and our method are by far better than selective search and edgeboxes, which explains the superior object discovery performance of the former over the latter (*cf.* [34] and Table 3). Note that region proposals with a high percentage of positive ones could also be used in other tasks, i.e., weakly supervised object detection, but this is left for future work.

4.4   Object Discovery Performance

Single-Object Colocalization and Discovery. An important component of OSD is the similarity model used to compute score matrices S_{ij}, which, in [34], is the Probabilistic Hough Matching (PHM) algorithm [6]. Vo *et al.* [34] introduce two scores, *confidence* score and *standout* score, but use only the latter for it gives better performance. Since our new proposals come with different statistics, we test both scores in our experiments. Table 1 compares colocalization performance on OD, VOC_6x2 and VOC_all of OSD using the confidence and standout scores as well as our proposals. It can be seen that on VOC_6x2 and VOC_all, the confidence score does better than the standout score, while on OD, the latter does better. This is in fact not particularly surprising since images in

OD generally contain bigger objects (relative to image size) than those in the other datasets. In fact, although the standout score is used on all datasets in [6] and [34], the authors adjust the parameter γ (see [6]) used in computing their standout score to favor larger regions when running their models on OD. In all of our experiments from now on, we use the standout score on OD and the confidence score on other datasets (VOC_6x2, VOC_all, VOC12 and COCO_20k).

Our proposal generation process introduces a few hyper-parameters. Apart from α and β, two other important hyper-parameters are the number of local maxima u and the number of thresholds v which together control the number of proposals p per image returned by the process. We study their influence on the colocalization performance by conducting experiments on VOC_6x2 and report the results in Table 2. It shows that the colocalization performance does not depend much on the values of these parameters. Using $(u = 50, v = 100)$ actually gives the best performance but with twice as many proposals as $(u = 20, v = 50)$. For efficiency, we use $u = 20$ and $v = 50$ in all of our experiments.

Table 1. Colocalization performance with our proposals in different configurations of OSD

Config	Confidence	Standout
OD	83.7 ± 0.4	**89.0 ± 0.6**
VOC_6x2	**73.6 ± 0.6**	64.1 ± 0.3
VOC_all	**44.7 ± 0.3**	41.4 ± 0.1

Table 2. Colocalization performance for different values of hyper-parameters

(u, v)	(20,50)	(20,100)	(50,50)	(50,100)
CorLoc	73.6 ± 0.8	73.4 ± 0.7	73.3 ± 1.1	74.2 ± 0.8
p	760	882	1294	1507

Table 3. Single-object colocalization and discovery performance of OSD with different types of proposals. We use VGG16 features to represent regions in these experiments

Region proposals	Colocalization			Discovery		
	OD	VOC_6x2	VOC_all	OD	VOC_6x2	VOC_all
Edgeboxes [40]	81.6 ± 0.3	54.2 ± 0.3	29.7 ± 0.1	81.4 ± 0.3	55.2 ± 0.3	32.6 ± 0.1
Selective search [33]	82.2 ± 0.2	54.5 ± 0.3	30.9 ± 0.1	81.3 ± 0.3	57.8 ± 0.2	33.0 ± 0.1
Randomized Prim [23]	82.9 ± 0.3	71.5 ± 0.3	42.8 ± 0.1	82.5 ± 0.1	70.6 ± 0.4	44.5 ± 0.1
Ours (OSD)	**89.0 ± 0.6**	**73.6 ± 0.6**	44.7 ± 0.3	**87.8 ± 0.4**	69.2 ± 0.5	48.7 ± 0.3
Ours (rOSD)	**89.0 ± 0.5**	73.3 ± 0.5	**45.8 ± 0.3**	87.6 ± 0.3	71.1 ± 0.8	49.2 ± 0.2

We report in Table 3 the performance of OSD and rOSD on OD, VOC_6x2 and VOC_all with different types of proposals. It can be seen that our proposals give the best results on all datasets among all types of proposals with significant margins: 6.1%, 2.1% and 3.0% in colocalization and 5.3%, 0.5% and 4.7% in discovery, respectively. It is also noticeable that our proposals not only fare much better than the unsupervised ones (selective search and edgeboxes) but outperform those generated by randomized Prim, an algorithm trained with bounding box annotation.

We compare OSD and rOSD using our region proposals to the state of the art in Table 4 (colocalization) and Table 5 (discovery). In their experiments, Wei *et al.* [35] only use features from VGG19. We have conducted experiments with features from both VGG16 and VGG19 but only present experiment results with VGG19 features in comparisons with [35] due to the space limit. A more comprehensive comparison with features from VGG16 is included in the supplementary material. It can be seen that our use of CNN features (for both creating proposals and representing them in OSD) consistently improves the performance compared to the original OSD [34]. It is also noticeable that rOSD performs significantly better than OSD on the two large datasets (VOC_all and VOC12) while on the two smaller ones (OD and VOC_6x2), their performances are comparable. It is due to the fact that images in OD and VOC_6x2 mostly contain only one well-positioned object thus bad local maxima are not a big problem in the optimization while images in VOC_all and VOC12 contain much more complex scenes and the optimization works better with more regularization. In overall, we obtain the best results on the two smaller datasets, fare better than [21] but are behind [35] on VOC_all and VOC12 in the colocalization setting. It should be noticed that while methods for image colocalization [21,35] suppose that images in the collection come from the same category and explicitly exploit this assumption, rOSD is intended to deal with the much more difficult and general object discovery task. Indeed, in discovery setting, rOSD outperforms [35] by a large margin, 5.9% and 4.9% respectively on VOC_all and VOC12.

Table 4. Single-object colocalization performance of our approach compared to the state of the art. Note that Wei *et al.* [35] outperform our method on VOC_all and VOC12 in this case, but the situation is clearly reversed in the much more difficult discovery setting, as demonstrated in Table 5

Method	Features	OD	VOc_6x2	VOC_all	VOC12
Cho *et al.* [6]	WHO	84.2	67.6	37.6	–
Vo *et al.* [34]	WHO	87.1 ± 0.5	71.2 ± 0.6	39.5 ± 0.1	–
Li *et al.* [21]	VGG19	–	–	41.9	45.6
Wei *et al.* [35]	VGG19	87.9	67.7	**48.7**	**51.1**
Ours (OSD)	VGG19	**90.3 ± 0.3**	75.3 ± 0.7	45.6 ± 0.3	47.8 ± 0.2
Ours (rOSD)	VGG19	90.2 ± 0.3	**76.1 ± 0.7**	46.7 ± 0.2	49.2 ± 0.1

Table 5. Single-object discovery performance on the datasets with our proposals compared to the state of the art

Method	Features	OD	VOC_6x2	VOC_all	VOC12
Cho *et al.* [6]	WHO	82.2	55.9	37.6	–
Vo *et al.* [34]	WHO	82.3 ± 0.3	62.5 ± 0.6	40.7 ± 0.2	–
Wei *et al.* [35]	VGG19	75.0	54.0	43.4	46.3
Ours (OSD)	VGG19	89.1 ± 0.4	71.9 ± 0.7	47.9 ± 0.3	49.2 ± 0.2
Ours (rOSD)	VGG19	**89.2 ± 0.4**	**72.5 ± 0.5**	**49.3 ± 0.2**	**51.2 ± 0.2**

Multi-object Colocalization and Discovery. We demonstrate the effectiveness of rOSD in multi-object colocalization and discovery on VOC_all and VOC12 datasets, which contain images with multiple objects. We compare the performance of OSD and rOSD to Wei *et al.* [35] in Table 6. Although [35] tackles only the single-object colocalization problem, we modify their method to have a reasonable baseline for the multi-object colocalization and discovery problem. Concretely, we take the bounding boxes around the 5 largest connected components of positive locations in the image's *indicator matrix* [35] as the localization results. It can be seen that our method obtains the best performance with significant margins to the closest competitor across all datasets and settings. It is also noticeable that rOSD, again, significantly outperforms OSD in this task. An illustration of multi-object discovery is shown in Fig. 3. For a fair comparison, we use high values of ν (50) and IoU (0.7) in the multi-object experiments to make sure that both OSD and rOSD return approximately 5 regions per image. Images may of course contain fewer than 5 objects. In such cases, OSD and rOSD usually return overlapping boxes around the actual objects. We can often eliminate these overlapping boxes and obtain better qualitative results by using smaller ν and IoU threshold values. It can be seen in Fig. 3 that with $\nu = 25$ and $IoU = 0.3$, rOSD is able to return bounding boxes around objects

Table 6. Multi-object colocalization and discovery performance of rOSD compared to competitors on VOC_all and VOC12 datasets

Method	Features	Colocalization		Discovery	
		VOC_all	VOC12	VOC_all	VOC12
Vo *et al.* [34]	WHO	40.7 ± 0.1	–	30.7 ± 0.1	–
Wei *et al.* [35]	VGG19	43.3	45.5	28.1	30.3
Ours (OSD)	VGG19	46.8 ± 0.1	47.9 ± 0.0	34.8 ± 0.0	36.8 ± 0.0
Ours (rOSD)	VGG19	**49.4 ± 0.1**	**51.5 ± 0.1**	**37.6 ± 0.1**	**40.4 ± 0.1**

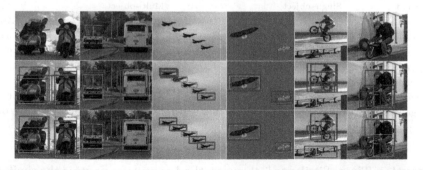

Fig. 3. Qualitative multi-object discovery results obtained with rOSD. White boxes are ground truth objects and red ones are our predictions. Original images are in the first row. Results with $\nu = 50$ and $IoU = 0.7$ are in the second row. Results with $\nu = 25$ and $IoU = 0.3$ are in the third row. (Color figure online)

without many overlapping regions. Note however that the quantitative results may worsen due to the reduced number of regions returned and the fact that many images contain objects that highly overlap, e.g., the last two columns of Fig. 3. In such cases, a small IoU threshold prevents discovering all of these objects. See supplementary document for more visualizations and details.

Large-Scale Object Discovery. We apply our large-scale algorithm in the discovery setting on VOC_all, VOC12 and COCO_20k which are randomly partitioned respectively into 5, 10 and 20 parts of roughly equal sizes. In the first stage of all experiments, we prefilter the initial neighborhood of images and keep only 50 potential neighbors. We choose $\nu = 50$ and keep K_1 (which are 250, 500 and 1000 respectively on VOC_all, VOC12 and COCO_20k) positive entries in each score matrix. In the second stage, we run rOSD (OSD) on the entire datasets with $\nu = 5$, limit the number of potential neighbors to 50 and use score matrices with only 50 positive entries. We choose K_1 such that each run in the first stage and the OSD run in the second stage have the same memory cost, hence the values of K chosen above. As baselines, we have applied rOSD (OSD) directly to the datasets, keeping 50 positive entries (baseline 1) and 1000 positive entries (baseline 2) in score matrices. Table 7 shows the object discovery performance on VOC_all, VOC12 and COCO_20k for our large-scale algorithm compared to the baselines. It can be seen that our large-scale two-stage rOSD algorithm yields significant performance gains over the baseline 1, obtains an improvement of 6.6%, 9.3% and 4.0% in single-object discovery and 2.9%, 4.0% and 0.4% in multi-object discovery, respectively on VOC_all, VOC12 and COCO_20k. Interestingly, large-scale rOSD also outperforms the baseline 2, which has a much higher memory cost, on VOC_all and VOC12.

Table 7. Performance of our large-scale algorithm compared to the baselines. Our method and baseline 1 have the same memory cost, which is much smaller than the cost of baseline 2. Also, due to memory limits, we cannot run baseline 2 on COCO_20k

Method	Single-object			Multi-object		
	VOC_all	VOC12	COCO_20k	VOC_all	VOC12	COCO_20k
Baseline 1 (OSD)	41.1 ± 0.3	40.5 ± 0.2	43.6 ± 0.2	31.4 ± 0.1	32.4 ± 0.0	10.5 ± 0.0
Baseline 1 (rOSD)	42.8 ± 0.3	42.6 ± 0.2	44.5 ± 0.1	35.4 ± 0.2	37.2 ± 0.1	11.6 ± 0.0
Baseline 2 (OSD)	47.9 ± 0.3	49.2 ± 0.2	–	34.8 ± 0.0	36.8 ± 0.0	–
Baseline 2 (rOSD)	49.3 ± 0.2	51.2 ± 0.2	–	37.6 ± 0.1	40.4 ± 0.1	–
Large-scale OSD	45.5 ± 0.3	46.3 ± 0.2	46.9 ± 0.1	34.6 ± 0.0	36.9 ± 0.0	11.1 ± 0.0
Large-scale rOSD	**49.4 ± 0.1**	**51.9 ± 0.1**	**48.5 ± 0.1**	**38.3 ± 0.0**	**41.2 ± 0.1**	**12.0 ± 0.0**

Execution Time. Similar to [34], our method requires computing the similarity scores for a large number of image pairs which makes it computationally costly. It takes in total 478 paralellizable CPU hours, 300 unparalellizable CPU seconds and 1 GPU hour to run single-object discovery on VOC_all with 3550 images. This is more costly compared to only 812 GPU seconds needed by DDT+ [35]

but is less costly than [34] using CNN features. The latter requires 546 paralelliz-able CPU hours, 250 unparalellizable CPU seconds and 4 GPU hours. Note that the unparallelizable computational cost, which comes from the main OSD algorithm, grows very fast (at least linearly in theory, it takes 2.3 h on COCO_20k in practice) with the data set's size and is the time bottleneck in large scale.

5   Conclusion

We have presented an unsupervised algorithm for generating region proposals from CNN features trained on an auxiliary and unrelated task. Our proposals come with an intrinsic structure which can be leveraged as an additional regularization in the OSD framework of Vo *et al.* [34]. The combination of our proposals and regularized OSD gives comparable results to the current state of the art in image colocalization, sets a new state-of-the-art single-object discovery and has proven effective in the multi-object discovery. We have also successfully extended OSD to the large-scale case and show that our method yields significantly better performance than plain OSD. Future work will be dedicated to investigating other applications of our region proposals.

Acknowledgments. This work was supported in part by the Inria/NYU collaboration, the Louis Vuitton/ENS chair on artificial intelligence and the French government under management of Agence Nationale de la Recherche as part of the "Investissements d'avenir" program, reference ANR19-P3IA-0001 (PRAIRIE 3IA Institute). Huy V. Vo was supported in part by a Valeo/Prairie CIFRE PhD Fellowship.

References

1. Alexe, B., Deselaers, T., Ferrari, V.: Measuring the objectness of image windows. IEEE Trans. Pattern Anal. Mach. Intell. (TPAMI) **34**, 2189–2202 (2012)
2. Babenko, A., Slesarev, A., Chigorin, A., Lempitsky, V.: Neural codes for image retrieval. In: Fleet, D., Pajdla, T., Schiele, B., Tuytelaars, T. (eds.) ECCV 2014. LNCS, vol. 8689, pp. 584–599. Springer, Cham (2014). https://doi.org/10.1007/978-3-319-10590-1_38
3. Babenko, A., Lempitsky, V.: Aggregating deep convolutional features for image retrieval. In: Proceedings of the International Conference on Computer Vision (ICCV) (2015)
4. Bach, F.: Learning with submodular functions: a convex optimization perspective. Found. Trends Mach. Learn. **6**(2–3), 145–373 (2013)
5. Chazal, F., Guibas, L.J., Oudot, S.Y., Skraba, P.: Persistence-based clustering in Riemannian manifolds. J. ACM **60**(6), 41:1–41:38 (2013)
6. Cho, M., Kwak, S., Schmid, C., Ponce, J.: Unsupervised object discovery and localization in the wild: part-based matching with bottom-up region proposals. In: Proceedings of the Conference on Computer Vision and Pattern Recognition (CVPR) (2015)
7. Cinbis, R., Verbeek, J., Schmid, C.: Weakly supervised object localization with multi-fold multiple instance learning. IEEE Trans. Pattern Anal. Mach. Intell. (TPAMI) **39**, 189–203 (2017)

8. Edelsbrunner, H., Harer, J.: Computational Topology: An Introduction. AMS Press, Providence (2009)
9. Edelsbrunner, H., Letscher, D., Zomorodian, A.: Topological persistence and simplification. Discrete Comput. Geometry **28**, 511–533 (2002). https://doi.org/10.1007/s00454-002-2885-2
10. Everingham, M., Van Gool, L., Williams, C.K.I., Winn, J., Zisserman, A.: The PASCAL Visual Object Classes Challenge 2012 (VOC2012) Results. http://www.pascal-network.org/challenges/VOC/voc2012/workshop/index.html
11. Everingham, M., Van Gool, L., Williams, C.K.I., Winn, J., Zisserman, A.: The PASCAL Visual Object Classes Challenge 2007 (VOC2007) Results
12. Faktor, A., Irani, M.: "Clustering by composition" – unsupervised discovery of image categories. In: Fitzgibbon, A., Lazebnik, S., Perona, P., Sato, Y., Schmid, C. (eds.) ECCV 2012. LNCS, vol. 7578, pp. 474–487. Springer, Heidelberg (2012). https://doi.org/10.1007/978-3-642-33786-4_35
13. Girshick, R.: Fast R-CNN. In: Proceedings of the International Conference on Computer Vision (ICCV) (2015)
14. Girshick, R., Donahue, J., Darrell, T., Malik, J.: Rich feature hierarchies for accurate object detection and semantic segmentation. In: Proceedings of the Conference on Computer Vision and Pattern Recognition (CVPR) (2014)
15. Grauman, K., Darrell, T.: Unsupervised learning of categories from sets of partially matching image features. In: Proceedings of the Conference on Computer Vision and Pattern Recognition (CVPR) (2006)
16. Hariharan, B., Malik, J., Ramanan, D.: Discriminative decorrelation for clustering and classification. In: Proceedings of the European Conference on Computer Vision (ECCV) (2012)
17. He, K., Zhang, X., Ren, S., Sun, J.: Deep residual learning for image recognition. In: Proceedings of the Conference on Computer Vision and Pattern Recognition (CVPR) (2016)
18. Hsu, K.J., Lin, Y.Y., Chuang, Y.Y.: Deepco[3]: deep instance co-segmentation by co-peak search and co-saliency detection. In: IEEE Computer Society Conference on Computer Vision and Pattern Recognition (CVPR) (2019)
19. Hsu, K.J., Tsai, C.C., Lin, Y.Y., Qian, X., Chuang, Y.Y.: Unsupervised CNN-based co-saliency detection with graphical optimization. In: Proceedings of the European Conference on Computer Vision (ECCV) (2018)
20. Kim, G., Torralba, A.: Unsupervised detection of regions of interest using iterative link analysis. In: Proceedings of the Conference on Neural Information Processing Systems (NeurIPS) (2009)
21. Li, Y., Liu, L., Shen, C., Hengel, A.: Image co-localization by mimicking a good detector's confidence score distribution. In: Proceedings of the European Conference on Computer Vision (ECCV) (2016)
22. Lin, T.-Y., et al.: Microsoft COCO: common objects in context. In: Fleet, D., Pajdla, T., Schiele, B., Tuytelaars, T. (eds.) ECCV 2014. LNCS, vol. 8693, pp. 740–755. Springer, Cham (2014). https://doi.org/10.1007/978-3-319-10602-1_48
23. Manen, S., Guillaumin, M., Van Gool, L.: Prime object proposals with randomized Prim's algorithm. In: Proceedings of the International Conference on Computer Vision (ICCV) (2013)
24. Nedić, A., Ozdaglar, A.: Approximate primal solutions and rate analysis for dual subgradient methods. SIAM J. Optim. **19**(4), 1757–1780 (2009)
25. Oudot, S.: Persistence Theory: From Quiver Representations to Data Analysis. AMS Surveys and Monographs (2015)

26. Ren, S., He, K., Girshick, R., Sun, J.: Faster R-CNN: towards real-time object detection with region proposal networks. In: Proceedings of the Conference on Neural Information Processing Systems (NeurIPS) (2015)
27. Russell, B., Freeman, W., Efros, A., Sivic, J., Zisserman, A.: Using multiplesegmentations to discover objects and their extent in image collections. In: Proceedings of the Conference on Computer Vision and Pattern Recognition(CVPR) (2006)
28. Selvaraju, R.R., Cogswell, M., Das, A., Vedantam, R., Parikh, D., Batra, D.: Grad-CAM: visual explanations from deep networks via gradient-based localization. In: Proceedings of the International Conference on Computer Vision (ICCV) (2017)
29. Simonyan, K., Zisserman, A.: Very deep convolutional networks for large-scale image recognition. In: Proceedings of the International Conference on Learning Representations (ICLR) (2015)
30. Sivic, J., Russell, B.C., Efros, A.A., Zisserman, A., Freeman, W.T.: Discovering object categories in image collections. In: Proceedings of the International Conference on Computer Vision (ICCV) (2005)
31. Tang, P., et al.: PCL: proposal cluster learning for weakly supervised object detection. IEEE Trans. Pattern Anal. Mach. Intell. (TPAMI) $42(1)$, 176–191 (2020)
32. Tang, P., et al.: Weakly supervised region proposal network and object detection. In: Ferrari, V., Hebert, M., Sminchisescu, C., Weiss, Y. (eds.) ECCV 2018. LNCS, vol. 11215, pp. 370–386. Springer, Cham (2018). https://doi.org/10.1007/978-3-030-01252-6_22
33. Uijlings, J.R.R., van de Sande, K.E.A., Gevers, T., Smeulders, A.W.M.: Selective search for object recognition. Int. J. Comput. Vis. $104(2)$, 154–171 (2013)
34. Vo, H.V., et al.: Unsupervised image matching and object discovery as optimization. In: Proceedings of the Conference on Computer Vision and Pattern Recognition (CVPR) (2019)
35. Wei, X.S., Zhang, C.L., Wu, J., Shen, C., Zhou, Z.H.: Unsupervised object discovery and co-localization by deep descriptor transforming. Pattern Recogn. (PR) 88, 113–126 (2019)
36. Wei, X.S., Luo, J.H., Wu, J., Zhou, Z.H.: Selective convolutional descriptor aggregation for fine-grained image retrieval. IEEE Trans. Image Process. $26(6)$, 2868–2881 (2017)
37. Zeiler, M.D., Fergus, R.: Visualizing and understanding convolutional networks. In: Fleet, D., Pajdla, T., Schiele, B., Tuytelaars, T. (eds.) ECCV 2014. LNCS, vol. 8689, pp. 818–833. Springer, Cham (2014). https://doi.org/10.1007/978-3-319-10590-1_53
38. Zhang, D., Meng, D., Li, C., Jiang, L., Zhao, Q., Han, J.: A self-paced multiple-instance learning framework for co-saliency detection. In: Proceedings of the International Conference on Computer Vision (ICCV) (2015)
39. Zhou, B., Khosla, A., Lapedriza, A., Oliva, A., Torralba, A.: Learning deep features for discriminative localization. In: Proceedings of the Conference on Computer Vision and Pattern Recognition (CVPR) (2016)
40. Zitnick, C.L., Dollár, P.: Edge boxes: locating object proposals from edges. In: Fleet, D., Pajdla, T., Schiele, B., Tuytelaars, T. (eds.) ECCV 2014. LNCS, vol. 8693, pp. 391–405. Springer, Cham (2014). https://doi.org/10.1007/978-3-319-10602-1_26
41. Zomorodian, A., Carlsson, G.: Computing persistent homology. Discrete Comput. Geom. 33, 249–274 (2005)

Author Index

Printed in the United States
By Bookmasters